# Mathematical Statistics for Economics and Business

## Second Edition

Ron C. Mittelhammer

# Mathematical Statistics for Economics and Business

## Second Edition

With 93 Illustrations

Ron C. Mittelhammer
School of Economic Sciences
Washington State University
Pullman, Washington
USA

ISBN 978-1-4899-8950-5 ISBN 978-1-4614-5022-1 (eBook)
DOI 10.1007/978-1-4614-5022-1
Springer New York Heidelberg Dordrecht London



Printed on acid-free paper

Springer is part of Springer Science+Business Media (www.springer.com)

*To my wife Linda,*
*and to the memory of Carl, Edith, Dolly,*
*and Ralph.*

# Preface to the Second Edition of Mathematical Statistics for Economics and Business

The general objectives of the second edition of *Mathematical Statistics for Economics and Business* remain the same as the first, namely, to provide a rigorous and accessible foundation in the principles of probability and in statistical estimation and inference concepts for beginning graduate students and advanced undergraduate students studying primarily in the fields of economics and business. Since its publication, the first edition of the book has found use by those from other disciplines as well, including the social sciences (e.g., psychology and sociology), applied mathematics, and statistics, even though many of the applied examples in later chapters have a decidedly "economics and business" feel (although the examples are chosen in such a way that they are fairly well "self-contained" and understandable for those who have not studied either discipline in substantial detail).

The general philosophy regarding how and why the book was originally written was presented in the preface to the first edition and in large measure could be inserted at this point for motivating the fundamental rationale for the second edition. This philosophy includes the necessity of having a conceptual base of probability and statistical theory to be able to fully understand the application and interpretation of applied econometric and business statistics methods, coupled with the need to have a treatment of the subject that, while rigorous, also assumes an accessible level of prerequisites that can be expected to have been met by a large majority of graduate students entering the fields. The choice of topic coverage is also deliberate and decidedly chosen to form the fundamental foundation on which econometric and business statistics methodology is built. With the ongoing expansion, in both scope and depth, of econometric and statistical methodology for quantitative analyses in both economics and business, it has never been more important, and many are now thinking absolutely essential, that a base of formal probability and statistics understanding become part of student training to enable effective reading of the literature and success in the fields.

Regarding the nature of the updates and revisions that have been made in producing the second edition, many of the basic probability and statistical concepts remain in common with the first edition. The fundamental base of probability and statistics principles needed for later study of econometrics, business statistics, and a myriad of stochastic applications of economic and business theory largely intersects the topics covered in the first edition. While a few topics were deleted in the second edition as being less central to that foundation, many more have been added. These include the following: greater detail on the issue of parametric, semiparametric, and nonparametric models; an introduction to nonlinear least squares methods; Stieltjes integration has been added strategically in some contexts where continuous and discrete random variable properties could be clearly and efficiently motivated in parallel; additional testing methodology for the ubiquitous normality assumption; clearer differentiation of parametric and semiparametric testing of hypotheses; as well as many other refinements in topic coverage appropriate for applications in economics and business.

Perhaps the most important revision of the text has been in terms of the organization, exposition, and overall usability of the material. Reacting to the feedback of a host of professors, instructors, and individual readers of the first edition, the presentation of both the previous and new material has been notably reorganized and rewritten to make the text easier to study and teach from. At the highest level, the compartmentalization of topics is now better and easier to navigate through. All theorems and examples are now titled to provide a better foreshadowing of the content of the results and/or the nature of what is being illustrated. Some topics have been reordered to improve the flow of reading and understanding (i.e., the relatively more esoteric concept of events that cannot be assigned probability consistently has been moved to the end of a chapter and the review of elements of real analysis has been moved from the beginning of the asymptotic theory chapter to the appendix of the book), and in some cases, selected proofs of theorems that were essentially pure mathematics and that did little to bolster the understanding of statistical concepts were moved to chapter appendices to improve readability of the chapter text. A large number of new and expanded exercises/problems have been added to the chapters.

While a number of texts focused on statistical foundations of estimation and inference are available, *Mathematical Statistics for Economics and Business* is a text whose level of presentation, assumed prerequisites, examples and problems, and topic coverage will continue to provide a solid foundation for future study of econometrics, business statistics, and general stochastic economic and business theory and application. With its redesigned topic organization, additional topic coverage, revision of exposition, expanded set of problems, and continued focus on accessibility and motivation, the book will provide a conceptual foundation on which students can base their future study and understanding of rigorous econometric and statistical applications, and it can also serve as an accessible refresher for practicing professionals who wish to reestablish their understanding of the foundations on which all of econometrics, business statistics, and stochastic economic and business theory are based.

## Acknowledgments

In addition to all of the acknowledgments presented in the first edition, which certainly remain deserving of inclusion here, I would like to thank Ms. Danielle Engelhardt, whose enormous skills in typing, formatting, and proof-checking of the text material and whose always cheerful and positive "can-do" personality made the revision experience a much more enjoyable and efficient process. I am also indebted to Dr. Miguel Henry-Osorio for proofreading every character of every page of material and pointing out corrections, in addition to making some expositional suggestions that were very helpful to the revision process. Mr. Sherzod Akhundjanov also provided expert proof-checking, for which I am very grateful. I also thank Haylee and Hanna Gecas for their constant monitoring of my progress on the book revision and for making sure that I did not stray too far from the targeted timeline for the effort. I also wish to thank my colleague Dr. Tom Marsh, who utilized the first edition of this book for many years in the teaching of his econometrics classes and who provided me with helpful feedback on student learning from and topic coverage in the book. Finally, a deep thank you for the many comments and helpful suggestions I continued to receive over the years from my many doctoral students, the students who attended my statistics and econometrics classes here at the university; the many additional questions and comments I received from students elsewhere; and the input received from a host of individuals all over the world – the revision of the book has benefitted substantially from your input. Thank you all.

# Preface (First Edition)

This book is designed to provide beginning graduate students and advanced undergraduates with a rigorous and accessible foundation in the principles of probability and mathematical statistics underlying statistical inference in the fields of business and economics. The book assumes no prior knowledge of probability or statistics and effectively builds the subject "from the ground up." Students who complete their studies of the topics in this text will have acquired the necessary background to achieve a mature and enduring understanding of statistical and econometric methods of inference and will be well equipped to read and comprehend graduate-level econometrics texts. Additionally, this text serves as an effective bridge to a more advanced study of both mathematical statistics and econometric theory and methods. The book will also be of interest to researchers who desire a decidedly business and economic-based treatment of the subject in terms of its topics, depth, breadth, examples, and problems.

Without the unifying foundations that come with training in probability and mathematical statistics, students in statistics and econometrics classes too often perceive the subject matter as a potpourri of formulae and techniques applied to a collection of special cases. The details of the cases and their solutions quickly fade for those who do not understand the reasons for using the procedures they attempt to apply. Many institutions now recognize the need for a more rigorous study of probability and mathematical statistics principles in order to prepare students for a higher-level, longer-lasting understanding of the statistical techniques employed in the fields of business and economics. Furthermore, quantitative analysis in these fields has progressed to the point where a deeper understanding of the principles of probability and statistics is now virtually *necessary* for one to read and contribute successfully to quantitative research in economics and business. Contemporary students themselves know this and need little convincing from advisors that substantial statistical training must be acquired in order to compete successfully with their peers and to become effective researchers. Despite these observations, there are very few rigorous books on probability and mathematical statistics foundations that are also written with the needs of business and economics students in mind.

This book is the culmination of 15 years of teaching graduate level statistics and econometrics classes for students who are beginning graduate programs in business (primarily finance, marketing, accounting, and decision sciences), economics, and agricultural economics. When I originally took on the teaching assignment in this area, I cycled through a number of very good texts in mathematical statistics searching for an appropriate exposition for beginning graduate students. With the help of my students, I ultimately realized that the available textbook presentations were optimizing the wrong objective functions for our purposes! Some books were too elementary; other presentations did not cover multivariate topics in sufficient detail, and proofs of important results were omitted occasionally because they were "obvious" or "clear" or "beyond the scope of the text." In most cases, they were neither obvious nor clear to students, and in many cases, useful and accessible proofs of the most important results can and should be provided at this level of instruction. Sufficient asymptotic theory was often lacking and/or tersely developed. At the extreme, material was presented in a sterile mathematical context at a level that was inaccessible to most beginning graduate students while nonetheless leaving notable gaps in topic coverage of particular interest to business and economics students. Noting these problems, gaps, and excesses, I began to teach the course from lecture notes that I had created and iteratively refined them as I interacted with scores of students who provided me with feedback regarding what was working—and what wasn't—with regard to topics, proofs, problems, and exposition. I am deeply indebted to the hundreds of students who persevered through, and contributed to, the many revisions and continual sophistication of my notes. Their influence has had a substantial impact on the text: It is a time-tested and class-tested product. Other students at a similar stage of development should find it honest, accessible, and informative.

Instructors attempting to teach a rigorous course in mathematical statistics soon learn that the typical new graduate student in economics and business is thoroughly intelligent, but often lacks the sophisticated mathematical training that facilitates understanding and assimilation of the mathematical concepts involved in *mathematical* statistics. My experience has been that these students can understand and become functional with sophisticated concepts in mathematical statistics if their backgrounds are respected and the material is presented carefully and thoroughly, using a realistic level of mathematics. Furthermore, it has been my experience that most students are actually eager to see proofs of propositions, as opposed to merely accepting statements on faith, so long as the proofs do not insult the integrity of the nonmathematician.

Additionally, students almost always remark that the understanding and the long-term memory of a stated result are enhanced by first having worked through a formal proof of a proposition and then working through examples and problems that require the result to be applied.

With the preceding observations in mind, the prerequisites for the book include only the usual introductory college-level courses in basic calculus (including univariate integration and differentiation, partial differentiation, and multivariate integration of the iterated integral type) and basic matrix algebra. The text is largely self-contained for students with this preparation. A significant effort has been made to present proofs in ways that are accessible. Care has been taken to choose methods and types of proofs that exercise and extend the learning process regarding statistical results and concepts learned prior to the introduction of the proof. A generous number of examples are presented with a substantial amount of detail to illustrate the application of major theories, concepts, and methods. The problems at the end of the chapters are chosen to provide an additional perspective to the learning process. The majority of the problems are word problems designed to challenge the reader to become adept at what is generally the most difficult hurdle—translating descriptions of statistical problems arising in business and economic settings into a form that lends itself to solutions based on mathematical statistics principles. I have also warned students through the use of asterisks (*) when a proof, concept, example, or problem may be stretching the bounds of the prerequisites so as not to frustrate the otherwise diligent reader, and to indicate when the help of an instructor or additional readings may be useful.

The book is designed to be versatile. The course that inspired this book is a semester-long four-credit intensive mathematical statistics foundation course. I do not lecture on all of the topics contained in the book in the 50 contact hours available in the semester. The topics that I do not cover are taught in the first half of a subsequent semester-long three-credit course in statistics and econometric methods. I have tended to treat Chapters 1–4 in detail, and I recommend that this material be thoroughly understood before venturing into the statistical inference portion of the book. Thereafter, the choice of topics is flexible. For example, the instructor can control the depth at which asymptotic theory is taught by her choice of whether the starred topics in Chapter 5 are discussed. While random sampling, empirical distribution functions, and sample moments should be covered in Chapter 6, the instructor has leeway in the degree of emphasis that she places on other topics in the chapter. Point estimation and hypothesis testing topics can then be mixed and matched with a minimal amount of back-referencing between the respective chapters.

Distinguishing features of this book include the care with which topics are introduced, motivated, and built upon one another; use of the appropriate level of mathematics; the generous level of detail provided in the proofs; and a familiar business and economics context for examples and problems. This text is bit longer than some of the others in the field. The additional length comes from additional explanation, and detail in examples, problems, and proofs, and not from a proliferation of topics which are merely surveyed rather than fully developed. As I see it, a survey of statistical techniques is useful only after one has the fundamental statistical background to appreciate what is being surveyed. And this book provides the necessary background.

## Acknowledgments

I am indebted to a large number of people for their encouragement and comments. Millard Hastay, now retired from the Washington State University economics faculty, is largely responsible for my unwavering curiosity and enthusiasm for the field of theoretical and applied statistics and econometrics. George Judge has been a constant source of encouragement for the book project and over the years has provided me with very valuable and selfless advice and support in all endeavors in which our paths have crossed. I thank Jim Chalfant for giving earlier drafts of chapters a trial run at Berkeley, and for providing me with valuable student and instructor feedback. Thomas Severini at Northwestern provided important and helpful critiques of content and exposition. Martin Gilchrist at Springer-Verlag provided productive and pleasurable guidance to the writing and revision of the text. I also acknowledge the steadfast support of Washington State University in the pursuit of the writing of this book. Of the multitude of past students who contributed so much to the final product and that are too numerous to name explicitly, I owe a special measure of thanks to Don Blayney, now of the Economic Research Service, and Brett Crow, currently a promising Ph.D. candidate in economics at WSU, for reviewing drafts of the text literally character by character and demanding clarification in a number of proofs and examples. I also wish to thank many past secretaries who toiled faithfully on the book project. In particular, I wish to thank Brenda Campbell, who at times literally typed morning, noon, and night to bring the manuscript to completion, without whom completing the project would have been infinitely more difficult. Finally, I thank my wife Linda, who proofread many parts of the text, provided unwavering support, sustenance, and encouragement to me throughout the project, and despite all of the trials and tribulations, remains my best friend.

# Contents

# List of Figures

# List of Tables

# 1 Elements of Probability Theory

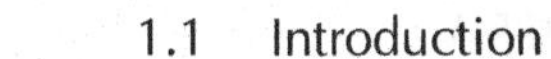



## 1.1 Introduction

The objective of this chapter is to define a quantitative measure of the propensity for an uncertain outcome to occur, or the degree of belief that a proposition or conjecture is true. This quantitative measure will be called **probability** and is relevant for quantifying such things as how likely it is that a shipment of smart phones contains less than 5 percent defectives, that a gambler will win a crap game, that next year's corn yields will exceed 80 bushels per acre, or that electricity demand in Los Angeles will exceed generating capacity on a given day. This probability concept will also be relevant for quantifying the degree of belief in such propositions as it will rain in Seattle tomorrow, Congress will raise taxes next year, and the United States will suffer another recession in the coming year.

The value of such a measure of outcome propensity or degree of belief can be substantial in the context of decision making in business, economics, government, and everyday life. In the absence of such a measure, all one can effectively say when faced with an uncertain situation is "I don't know what will happen" or "I don't know whether the proposition is true or false." A rational decision-maker will prefer to have as much information as possible about the final outcome or state of affairs associated with an uncertain situation in order to more fully consider its impacts on profits, utility, welfare, or other measures of well-being. Indeed, the problem of increasing profit, utility, or welfare through appropriate choices of production and inventory levels and scheduling, product pricing, advertising effort, trade policy, tax strategy, input or commodity purchases, technology adoption, and/or capital investment is substantially more difficult when the results of one's choice are affected by factors that are simply unknown, as opposed to occurring with varying degrees of likelihood.

Probability is a tool for distinguishing likely from unlikely outcomes or states of affairs and provides business managers, economists, legislators, consumers, and individuals with information that can be used to rank the potential results of their decisions in terms of propensity to occur or degree of validity. It then may be possible to make choices that maximize the likelihood of a desired outcome, provide a high likelihood of avoiding disastrous outcomes, or achieve a desirable expected result (where "expected" will be rigorously defined in the next chapter).

There have been many ways proposed for defining the type of quantitative measure described above, and that are useful in many respects, but they have not gained widespread acceptance and/or use. These include the concept of **belief functions** (Schaefer), **structural probability** (Fraser), and **fiducial probability** (Fisher). Four principal definitions that have found substantial degrees of acceptance and use, and that have been involved in the modern development of probability theory, include **classical probability**, **relative frequency probability**, **subjective probability**, and the **axiomatic approach**. We will briefly discuss the first three definitions, and then concentrate on the modern axiomatic approach, which will be seen to subsume the other three approaches as special cases.

Prior to our excursion into the realm of probability theory, it is helpful to examine how the terms "experiment," "sample space," "outcome," and "event" will be used in our discussion. The next section provides the necessary information.

## 1.2 Experiment, Sample Space, Outcome and Event

The term **experiment** is used very generally in the field of probability and statistics, and is not at all limited to the colloquial interpretation of the term as referring to activities that scientists perform in laboratories.

**Definition 1.1** *Experiment*

Any activity for which the outcome or final state of affairs cannot be specified in advance, but for which a set containing all potential outcomes or final states of affairs can be identified.

Thus, determining the yield per acre of a new type of wheat, observing the quantity of a commodity sold during a promotional campaign, identifying the fat percentage of a hundredweight of raw farm milk, observing tomorrow's closing Dow Jones Industrial Average on the NY Stock Exchange, or analyzing the underlying income elasticity affecting the demand for gasoline are all examples of experiments according to this definition of the term.

The final state of affairs resulting from an experiment is referred to as an **outcome**.

**Definition 1.2**
***Outcome of an Experiment***

A final result, observation, or measurement occurring from an experiment.

Thus, referring to the preceding examples of experiments, 80 bushels per acre, 2,500 units sold during a week of promotions, 3.7 percent fat per hundredweight, a DJIA of 13,500, and an income elasticity of .75 are, respectively, possible outcomes.

Prior to analyzing probabilities of outcomes of an experiment, it is necessary to identify what outcomes are *possible*. This leads to the definition of the **sample space** of an experiment.

**Definition 1.3**
***Sample Space, S***

A set that contains all possible outcomes of a given experiment.

Note that our definition of sample space, which we will henceforth denote by **S**, does not necessarily identify a unique set since we require only that the sample space *contains* all possible outcomes of an experiment. In many cases, the set of all possible outcomes will be readily identifiable and not subject to controversy, and in these cases it will be natural to refer to this set as the sample space. For example, the experiment of rolling a die and observing the number of dots facing up has a sample space that can be rather uncontroversially specified as {1, 2, 3, 4, 5, 6} (as long as one is ruling out that the die will not land on an edge!). However, defining the collection of possible outcomes of an experiment may also require some careful deliberation. For instance, in our example of measuring the fat percentage of a given hundredweight of raw farm milk, it is clear that the outcomes must reside in the set $A = \{x : 0 \leq x \leq 100\}$. However, the accuracy of our measuring device might only allow us to observe differences in fat percentages up to hundredths of a percent, and thus a smaller set containing all possible *measurable* fat percentages might be specified as $B = \{0, .01, .02, ..., 100\}$ where $B \subset A$. It might be argued further that fat percentages of greater than 20 percent and less than 1 percent will simply not occur in raw farm milk, and thus the smaller set $C = \{1,\ 1.01,\ 1.02, ...,\ 20\}$where $C \subset B \subset A$ could represent the sample space of the fat-measuring experiment. Fortunately, as the reader will come to recognize, the principal concern of practical importance is that the sample space be specified large enough to contain the set of all possible outcomes of the experiment as a subset. The sample space need not be identically equal to the set of all possible outcomes. The reader may wish to suggest appropriate sample spaces for the remaining four example experiments described above.

Consistent with set theory terminology and the fact that the sample space is indeed a set, each outcome in a sample space can also be referred to as an **element** or **member** of the sample space. In addition, the outcomes in a sample space are also sometimes referred to as **sample points**. The reader should be aware of these multiple names for the same concept, and there will be other instances ahead where concepts are referred to by multiple different names.

The sample space, as all sets, can be classified according to whether the number of elements in the set is **finite**, **countably infinite**, or **uncountably infinite.**[1] Two particular types of sample spaces will figure prominently in our study due to the fact that probabilities will ultimately be assigned using either finite mathematics, or via calculus, respectively.

**Definition 1.4**
***Discrete Sample Space***

A sample space that is finite or countably infinite.

**Definition 1.5**
***Continuous Sample Space***

An uncountably infinite sample space that consists of a continuum of points.

The fundamental entities to which probabilities will be assigned are events, which are equivalent to *subsets* in set theoretic terminology.

**Definition 1.6**
***Event***

A subset of the sample space.

Thus, events are simply collections of outcomes of an experiment. Note that a technical issue in measure theory can arise when we are dealing with uncountably infinite sample spaces, such that certain complicated subsets cannot be assigned probability in a consistent manner. For this reason, in more technical treatments of probability theory, one would define the term **event** to refer to *measureable* subsets of the sample space. We provide some background relating to this theoretical problem in the Appendix of this chapter. As a practical matter, all of the subsets to which an empirical analyst would be interested in assigning probability will be measureable, and we refrain from explicitly using this qualification henceforth.

In the special case where the event consists of a single element or outcome, we will use the special term **elementary event** to refer to the event.

**Definition 1.7**
***Elementary Event***

An event that is a singleton set, consisting of one element of the sample space.

[1] A countably infinite set is one that has an infinite number of elements that can be "counted" in the sense of being able to place the elements in a one-to-one correspondence with the positive integers. An uncountable infinite set has an infinite number of elements that cannot be counted, i.e., the elements of the set cannot be placed in a one-to-one correspondence with the positive integers.

One says that an event $A$ has occurred if the experiment results in an outcome that is a member or element of the event or subset $A$.

**Definition 1.8** *Occurrence of an Event*

An event is said to have occurred if the outcome of the experiment is an element of the event.

The real-world meaning of the statement "**the event** $\boldsymbol{A}$ **has occurred**" will be provided by the real-world definition of the set $A$. That is, verbal or mathematical statements that are utilized in a verbal or mathematical definition of set $A$, or the collection of elements or description of elements placed in brackets in an exhaustive listing of set $A$, provide the meaning of "the event A has occurred." The following examples illustrate the meaning of both **event** and the **occurrence of an event**.

**Example 1.1** *Occurrence of Dice Events*

An experiment consists of rolling a die and observing the number of dots facing up. The sample space is defined to be $S = \{1, 2, 3, 4, 5, 6\}$. Examine two events in $S$: $A_1 = \{1, 2, 3\}$and $A_2 = \{2, 4, 6\}$. Event $A_1$ has occurred if the outcome, x, of the experiment (the number of dots facing up) is such that $x \in A_1$. Then $A_1$ is an event whose occurrence means that after a roll, the number of dots facing up on the die is three or less. Event $A_2$ has occurred if the outcome, x, is such that $x \in A_2$. Then $A_2$ is an event whose occurrence means that the number of dots facing up on the die is an even number. □

**Example 1.2** *Occurrence of Survey Events*

An experiment consists of observing the percentage of a large group of consumers, representing a consumer taste panel, who prefer Schpitz beer to its closest competitor, Nickelob beer. The sample space for the experiment is specified as $S = \{x : 0 \leq x \leq 100\}$. Examine two events in $S$:$A_1 = \{x{:}x < 50\}$, and $A_2 = \{x{:}x > 75\}$. Event $A_1$ has occurred if the outcome, x, of the experiment (the actual percentage of the consumer panel preferring Schpitz beer) is such that $x \in A_1$. Then $A_1$ is an event whose occurrence means that less than 50 percent of the consumers preferred Schpitz to Nickelob or, in other words, the group of consumers preferring Schpitz were in the minority. Event $A_2$ has occurred if the outcome $x \in A_2$. Then $A_2$ is an event whose occurrence means that greater than 75 percent of the consumers preferred Schpitz to Nickelob. □

When two events have no outcomes in common, they are referred to as **disjoint events**.

**Definition 1.9** *Disjoint Events*

Events that are mutually exclusive, having no outcomes in common.

The concept is identical to the concept of mutually exclusive or disjoint sets, where $A_1$ and $A_2$are disjoint events *iff* $A_1 \cap A_2 = \emptyset$. Examples 1.1 and 1.2 can be used to illustrate the concept of **disjoint events**. In Example 1.1, it is recognized that events $A_1$ and $A_2$ are *not* mutually exclusive events, since $A_1 \cap A_2 = \{2\} \neq \emptyset$.

Events that are not mutually exclusive can occur simultaneously. Events $A_1$ and $A_2$ will occur simultaneously (which cannot be the case for mutually exclusive events) *iff* $x \in A_1 \cap A_2 = \{2\}$. In Example 1.2, events $A_1$ and $A_2$ are disjoint events since $A_1 \cap A_2 = \emptyset$. Events $A_1$ and $A_2$ cannot occur simultaneously since if the outcome is such that $x \in A_1$, then it follows that $x \notin A_2$, or if $x \in A_2$, then it follows that $x \notin A_1$.

We should emphasize that in applications it is the researcher who specifies the events in the sample space whose occurrence or lack thereof provides useful information from the researcher's viewpoint. Thus, referring to Example 1.2, if the researcher were employed by Schpitz Brewery, the identification of which beer was preferred by a majority of the participants in a taste comparison would appear to be of significant interest to the management of the brewery, and thus event $A_1$ would be of great importance. Event $A_2$ in that example might be considered important if the advertising department of Schpitz Brewery wished to utilize an advertising slogan such as "Schpitz beer is preferred to Nickelob by more than 3 to 1."

## 1.3 Nonaxiomatic Probability Definitions

There are three prominent nonaxiomatic definitions of probability that have been suggested in the course of the development of probability theory. We briefly discuss each of these alternative probability definitions. In the definition below, $N(A)$ is the size-of-set function whose value equals the number of elements that are contained in the set $A$ (see Definition A.21).

**Definition 1.10**
***Classical Probability***

Let $S$ be the sample space for an experiment having a finite number $N(S)$ of equally likely outcomes, and let $A \subset S$ be an event containing $N(A)$ elements. Then the probability of the event $A$, denoted by $P(A)$, is given by $P(A) = N(A)/N(S)$.

In the classical definition, probabilities are images of sets generated by a set function $P(\cdot)$. The domain of $P(\cdot)$ is $D(P) = \{A : A \subset S\}$, i.e., the collection of all subsets of a finite (and thus discrete) sample space, while the range is contained in the unit interval, i.e., $R(P) \subset [0, 1]$, since $N(A) \leq N(S)$. The following example illustrates the application of the classical probability concept.

**Example 1.3**
***Fair Dice and Classical Probability***

Reexamine the die-rolling experiment of Example 1.1, and now assume that the die is fair so that the outcomes in the sample space $S = \{1, 2, 3, 4, 5, 6\}$ are equally likely. The number of elements in the sample space is given by $N(S) = 6$. Let $E_i$, $i = 1,\ldots,6$, represent the elementary events in the set $S$. Then according to the classical probability definition, $P(E_i) = N(E_i)/N(S) = 1/6$ for all $i = 1,\ldots,6$, so that the probability of each elementary event is $1/6$. Referring to the events $A_1$ and $A_2$ of Example 1.1, note that

$$P(A_1) = \frac{N(A_1)}{N(S)} = \frac{3}{6} = \frac{1}{2} \text{ and } P(A_2) = \frac{N(A_2)}{N(S)} = \frac{3}{6} = \frac{1}{2}.$$

Therefore, the probability of rolling a three or less and the probability of rolling an even number are both 1/2. Note finally that $P(S) = N(S)/N(S) = 6/6 = 1$, which states that the probability of the event that the outcome of the experiment is an element of the sample space is 1, as it intuitively should be if the number 1 is to be associated with an event that will occur with certainty.□

The classical definition has two major limitations that preclude its use as the foundation on which to build a general theory of probability. First, the sample space must be finite or else $N(S) = \infty$ and, depending on the event, possibly $N(A) = \infty$ as well. Thus, probability in the classical sense is not useful for defining the probabilities of events contained in a countably infinite or uncountably infinite sample space. Another limitation of the classical definition is that outcomes of an experiment must be equally likely. Thus, for example, if in a coin-tossing experiment it cannot be assumed that the coin is fair, the classical probability definition provides no information about how probabilities should be defined. In order to relax these restrictions, we examine the relative frequency approach.

**Definition 1.11**
***Relative Frequency Probability***

Let $n$ be the number of times that an experiment is repeated under identical conditions. Let $A$ be an event in the sample space $S$, and define $n_A$ to be the number of times in $n$ repetitions of the experiment that the event $A$ occurs. Then the probability of event $A$ is equal to $P(A) = \lim_{n\to\infty} n_A/n$.

It is recognized that in the relative frequency definition, the probability of an event $A$ is the image of $A$ generated by a set function $P(\cdot)$, where the image is defined as the limiting fraction of the total number of outcomes of the $n$ experiments that are observed to be members of the set $A$. As in the classical definition of probability, the domain of $P(\cdot)$ is $D(P) = \{A : A \subset S\}$, i.e., the collection of all subsets of the sample space, while the range is contained in the unit interval, i.e., $R(P) \subset [0, 1]$, since $0 \le n_A \le n$. The following example illustrates the application of the relative frequency concept of probability.

**Example 1.4**
***Coin Tossing and Relative Frequency Probability***

Consider the following collection of coin-tossing experiments, where a coin was tossed various numbers of times and, based on the relative frequency definition of probability, the fraction of the tosses resulting in heads was recorded for each collection of experiments.[2]

[2]These experiments were actually performed by the author, except the author did not actually physically flip the coins to obtain the results listed here. Rather, the coin flips were simulated by the computer. In the coming chapters the reader will come to understand exactly how the computer might be used to simulate the coin-flipping experiment, and how to simulate other experiments as well.

| No. of tosses | No. of heads | Relative frequency |
|---|---|---|
| 100 | 48 | .4800 |
| 500 | 259 | .5180 |
| 1,000 | 489 | .4890 |
| 5,000 | 2,509 | .5018 |
| 75,000 | 37,447 | .4993 |

It would *appear* that as $n \to \infty$, the observed relative frequency of heads is approaching .5. □

The relative frequency definition enjoys some advantages over the classical definition. For one, the sample space can be an infinite set, since the ability to form the relative frequency $n_A/n$ does not depend on the underlying sample space being finite. Also, there is no need to assume that outcomes are equally likely, since the concept of $\lim_{n\to\infty} n_A/n$ does not depend on the outcomes being equally likely.

Unfortunately, there are problems with the relative frequency definition that reduce its appeal as a foundation for the development of a general theory of probability. First of all, while it is an *empirical fact* in many types of experiments, such as the coin-tossing experiment in Example 1.4, that the relative frequencies tend to stabilize as $n$ increases, how do we know that $n_A/n$ will actually converge to a limit in all cases? Indeed, how could we ever observe the limiting value if an infinite number of repetitions of the experiment are required? Furthermore, even if there is convergence to a limiting value in one sequence of experiments, how do we know that convergence to the same value will occur in another sequence of the experiments? Lacking a definitive answer to these conceptual queries, we refrain from using the relative frequency definition as the foundation for the probability concept.

A third approach to defining probability involves personal opinion, judgments, or educated guesses, and is called **subjective probability**.

**Definition 1.12** ***Subjective Probability***

A real number, $P(A)$, contained in [0,1] and chosen to express the degree of personal belief in the likelihood of occurrence or validity of event $A$, the number 1 being associated with certainty.

Like the preceding definitions of probability, subjective probabilities can be viewed as images of set functions $P(\cdot)$ having domain $D(P) = \{A : A \subset S\}$ and range $R(P) \subset [0, 1]$. Note that the subjective probability assigned to an event can obviously vary depending on who is assigning the probabilities and the personal beliefs of the individual assigning the probabilities. Even supposing that two individuals possess exactly the same information regarding the characteristics of an experiment, the way in which each individual interprets the information may result in differing probability assignments to an event $A$.

Unlike the relative frequency approach, subjective probabilities can be defined for experiments that cannot be repeated. For example, one might be assigning probability to the proposition that a recession will occur in the coming year. Defining the probability of the event "recession next year" does not conveniently fit into the relative frequency definition of probability, since one can only run the experiment of observing whether a recession occurs next year *once*. In addition, the classical definition would not apply unless a recession, or not, were equally likely, a priori. Similarly, assigning probability to the event that one or the other team will win in a Superbowl game is commonly done in various ways by many individuals, and a considerable amount of betting is based on those probability assignments. However, the particular Superbowl "experiment" cannot be repeated, nor is there usually any a priori reason to suspect that the outcomes are equally likely so that neither relative frequency nor classical probability definitions apply.

In certain problem contexts the assignment of probabilities solely on the basis of personal beliefs may be undesirable. For example, if an individual is betting on some game of chance, that individual would prefer to know the "true" likelihood of the game's various outcomes and not rely merely on his or her personal perceptions. For example, after inspecting a penny, suppose you consider the coin to be fair and (subjectively) assign a probability of ½ to each of the outcomes "heads" and "tails." However, if the penny was supplied by a ruthless gambler who altered the penny in such a way that an outcome of heads is twice as likely to occur as tails, the gambler could induce you to bet in such a way that you would lose money in the long run if you adhered to your initial subjective probability assignments and bet as if both outcomes were equally likely – the game would not be "fair."

There is another issue relating to the concept of subjective probability that needs to be considered in assessing its applicability. Why should one assume that the numbers assigned by any given individual behave in a manner that makes sense as a measure of the propensity for an uncertain outcome to occur, or the degree of belief that a proposition or conjecture is true? Indeed, they might not, and we seek criteria that individuals must follow so that the numbers they assign do make sense as probabilities.

Given that objective (classical and relative frequency approaches) and subjective probability concepts might both be useful, depending on the problem situation, we seek a probability theory that is general enough to accommodate all of the concepts of probability discussed heretofore. Such an accommodation can be achieved by defining probability in axiomatic terms.

## 1.4 Axiomatic Definition of Probability

Our objective is to devise a quantitative measure of the propensity of events to occur, or the degree of belief in various events contained in a sample space. How should one go about defining such a measure? A useful approach is to define the measure in terms of properties that are believed to be generally appropriate and/

or necessary for the measure to make sense for its intended purpose. So long as the properties are not contradictory, the properties can then be viewed collectively as a set of axioms on which to build the concept of probability.

Note, as an aside, that the approach of using a set of axioms as the foundation for a body of theory should be particularly familiar to students of business and economics. For example, the neoclassical theory of the consumer is founded on a set of behavioral assumptions, i.e., a set of axioms. The reader might recall that the axioms of comparability, transitivity, and continuity of preferences are sufficient for the existence of a utility function, the maximization of which, subject to an income constraint, depicts consumption behavior in the neoclassical theory.[3] Many other disciplines have axiomatic bases for bodies of theory.

What mathematical properties should a measure of probability possess? First of all, it seems useful for the measure to be in the form of a real-valued set function, since this would allow probabilities of events to be stated in terms of real numbers, and moreover, it is consistent with all of the prior definitions of probability reviewed in the previous section. Thus, we begin with a set function, say $P$, which has as its domain all of the events in a sample space, $S$, and has as its range a set of real numbers, i.e., we have $P : \Upsilon \rightarrow \mathbb{R}$, where $\Upsilon$ is the set of all events in $S$ and $\mathbb{R}$ denotes the real line $(-\infty, \infty)$. The set $\Upsilon$ is called the **event space.**

**Definition 1.13** ***Event space***

The set of all events in the sample space $S$.

We have in mind that the images of events under $P$ will be probabilities of the events, i.e., $P(A)$ will be the probability of the event $A \in \Upsilon$. Now, what type of properties seem appropriate to impose on the real-valued set function $P$?

Reviewing the three definitions of probability presented in Section 1.3, it is recognized that in each case, probability was defined to be a *nonnegative* number. Since each of the previous nonaxiomatic definitions of probability possess some intuitive appeal as measures of the propensity of an event to occur or the degree of belief in an event (despite our recognition of some conceptual difficulties), let us agree that the measure should be nonnegative valued. By doing so, we will have defined the first axiom to which the measure must adhere while remaining consistent with all of our previous probability definitions. Since we decided that our measure would be generated by a set function, $P$, our assumption requires that the set function be such that the image of any event $A$, $P(A)$, be a nonnegative number. Our first axiom is thus

[3] See G. Debreu (1959), *Theory of Value: An Axiomatic Analysis of Economic Equilibrium.* Cowles Monograph 17. New York: John Wiley, pp. 60–63. Note that additional axioms are generally included that are not needed for the existence of a utility function *per se,* but that lead to a simplification of the consumer maximization problem. See L. Phlips (1983), *Applied Consumption Analysis.* New York: North Holland, pp. 8–11.

**Axiom 1.1**
***Nonnegativity***

For any event $A \subset S$, $P(A) \geq 0$.

Now that we have committed to a measure that is nonnegative, what nonnegative number should the measure associate with the **certain event,** $S$?[4] There are some advantages to choosing the number 1 to denote the propensity of occurrence of, or the degree of belief in, the certain event. First, it is consistent with all of our nonaxiomatic definitions of probability discussed earlier. Second, it allows the probability of any event $A \subset S$ to be directly interpreted as a proportion of certainty. That is, if we assume that our set function is such that $P(S) = 1$, and if $P(A) = k$, say, then the measure of event $A$ relative to the measure of the certain event $S$ is $P(A)/P(S) = k/1 = k$, so that $P(A) = k\ P(S)$, and thus the event $A$ is assigned a proportion, $k$, of certainty. Our second axiom is then

**Axiom 1.2**
***Probability of the Certain Event***

$P(S) = 1$.

Regarding the value of $k$ above, it is clear that what we intuitively had in mind was a number $k \in [0,1]$. Our intuitive reasoning would be that if $S$ is the certain event, then the occurrence of $A \subset S$ surely cannot be "more than certain." That is, unless $A = S$, there are outcomes in $S$ that are not in $A$ (i.e., $S - A \neq \emptyset$). Thus, while the event $S$ will always occur, the event $A$ may or may not, and surely $A$ is no more certain to occur than $S$. This suggests that our measure must be such that $P(A) \leq 1$. However, we can proceed further and extend this argument. If $A$ and $B$ are any two events such that $A \subset B$, then following the same logic, we would require that $P(A) \leq P(B)$, since every element of $A$ is also in $B$, but $B$ may contain outcomes that are not in $A$, and thus $A$ can surely be no more likely to occur than $B$.

Investigating this line of reasoning still further, if $A \subset B$, and thus $P(B) \geq P(A)$, to what should we ascribe the remaining portion, $P(B) - P(A)$, of the probability assigned to event $B$? An intuitively obvious answer comes to mind. The set $B{-}A$ represents the additional outcomes remaining in $B$ after we remove the outcomes it has in common with $A$. Since $A$ and $B{-}A$ are disjoint and $B = A \cup (B - A)$, the event $B$ can be partitioned[5] into two disjoint events. Represented this way, $B$ can occur *iff* either $A$ or $(B - A)$ occur. If $P(B) > P(A)$, the added probability, $P(B) - P(A)$, of event $B$ occurring compared to event $A$ must be due to the probability of the occurrence of the event $B - A$. Thus, we must attribute any remaining probability measure, $P(B) - P(A)$, to the event $B - A$.

---

[4] By definition, since $S$ contains all possible outcomes of the experiment, the event $S$ is then certain to occur.

[5] A **partition** of a set $B$ is a collection of disjoint subsets of B, say $\{B_i, i \in I\}$ such that $B = \cup_{i \in I} B_i$.

Then our measure should have the property that for events $A$ and $B$ for which $A \subset B$, $P(B) = P(A) + P(B - A)$. Note that since $P(B - A) \geq 0$ by Axiom 1.1, this *implies* our previous requirement that if $A \subset B$, $P(B) \geq P(A)$. However, we have discovered much more than just another way of stating a potential third axiom. We have actually found that for *any* two disjoint events $A_1$ and $A_2$, our measure should have the property that $P(A_1 \cup A_2) = P(A_1) + P(A_2)$. To see this, define the set $B = A_1 \cup A_2$, where $A_1 \cap A_2 = \emptyset$. Then $B - A_1 = A_2$ because $A_1$ and $A_2$ are disjoint, and substituting $A_2$ for $(B - A_1)$ in $P(B) = P(A_1) + P(B - A_1)$, yields $P(A_1 \cup A_2) = P(A_1) + P(A_2)$. Thus, we have demonstrated that *probability should be additive across any two disjoint events.*

The preceding additivity argument can be extended to three *or more* disjoint events. To motivate the extension, first examine the case of three disjoint events, $A_1$, $A_2$, and $A_3$. Note that the two events $A_1 \cup A_2$ and $A_3$ are a pair of disjoint events, since

$$(A_1 \cup A_2) \cap A_3 = (A_1 \cap A_3) \cup (A_2 \cap A_3) = \emptyset \cup \emptyset = \emptyset$$

where $A_i \cap A_j = \emptyset$ for $i \neq j$ by the disjointness of $A_1$, $A_2$, and $A_3$. Then applying our probability additivity result for the case of two disjoint events results in $P(A_1 \cup A_2 \cup A_3) = P(A_1 \cup A_2) + P(A_3)$. But since $A_1$ and $A_2$ are disjoint, $P(A_1 \cup A_2) = P(A_1) + P(A_2)$, so that by substitution for $P(A_1 \cup A_2)$, we obtain $P(A_1 \cup A_2 \cup A_3) = P(A_1) + P(A_2) + P(A_3)$, which implies that probability is additive across any three disjoint events. Recognizing the sequential logic of the extension of probability additivity from two to three disjoint events, the reader can no doubt visualize the repetition of the argument ad infinitum to establish that probability should be additive across an arbitrary number of disjoint events. A concise, formal way of establishing the extension is through the use of **mathematical induction.**

**Lemma 1.1**
***Mathematical Induction Principle***

Let $P_1, P_2, P_3, \ldots$ be a sequence of propositions. Each of the propositions in the sequence is true provided

**(a)** $P_1$ is true; and

**(b)** For an arbitrary positive integer $k$, if $P_k$ were true, it would necessarily follow that $P_{k+1}$ is true.

Returning to our probability additivity argument, the first proposition in the sequence of propositions we are interested in is "$P(A_1 \cup A_2) = \sum_{i=1}^{2} P(A_i)$ for disjoint events $A_i, i = 1,2$." We have already defended the validity of this proposition. Now consider the proposition that for some $k$, "$P(\cup_{i=1}^{k} A_i) = \sum_{i=1}^{k} P(A_i)$ for disjoint events $A_i, i = 1, 2, \ldots, k$." Using the method of mathematical induction, we tentatively act as if this proposition were true, and we attempt to demonstrate that the truth of the next proposition in the sequence follows from the truth of the previous proposition, i.e., is "$P(\cup_{i=1}^{k+1} A_i) = \sum_{i=1}^{k+1} P(A_i)$ for disjoint events $A_i, i = 1, 2, \ldots, k + 1$" then true? Note that the two events $\cup_{i=1}^{k} A_i$ and $A_{k+1}$ are disjoint, since

$$\left(\bigcup_{i=1}^{k} A_i\right) \cap A_{k+1} = \bigcup_{i=1}^{k} (A_i \cap A_{k+1}) = \bigcup_{i=1}^{k} \emptyset = \emptyset$$

where $A_i \cap A_{k+1} = \emptyset \forall i \neq k + 1$ by the disjointness of the $k + 1$ events $A_1, \ldots, A_{k+1}$. But then by additivity for the two-event case,

$$P\left(\left(\bigcup_{i=1}^{k} A_i\right) \bigcup A_{k+1}\right) = P\left(\bigcup_{i=1}^{k} A_i\right) + P(A_{k+1}) = \sum_{i=1}^{k+1} P(A_i),$$

where the last equality follows from the assumed validity of probability additivity in the $k$-disjoint event case. Then by mathematical induction, we have demonstrated that $P(\cup_{i=1}^{m} A_i) = \sum_{i=1}^{m} P(A_i)$ for disjoint events $A_1, \ldots, A_m, \forall$ positive integer $m$, i.e., probability is additive across any number of disjoint events.

We finally state our probability additivity requirement as a third probability axiom, where we generalize the representation of the collection of disjoint events by utilizing an index set of subscripts rather than unnecessarily restricting ourselves to an increasing ordered integer sequence given by 1, 2, 3, . . .,$m$.

**Axiom 1.3**
***Countable Additivity***

Let $I$ be a finite or countably infinite index set of positive integers, and let $\{A_i: i \in I\}$ be a collection of disjoint events contained in $S$. Then, $P(\cup_{i \in I} A_i) = \sum_{i \in I} P(A_i)$.

The Russian mathematician A.N. Kolmogorov suggested that Axioms 1.1–1.3 provide an axiomatic foundation for probability theory.[6] As it turns out, sufficient information concerning the behavior of probability is contained in the three axioms to be able to derive from them the modern theory of probability. We begin deriving some important probability results in the next section.

In summary, we have defined the concept of probability by defining a number of properties that probabilities should possess. Specifically, probabilities will be generated by a set function that has the collection of events of a sample space, i.e., the event space, as its domain; its range will be contained in the interval [0,1]; the image of the certain event $S$ will be 1; and the probability of a countable union of disjoint events of $S$ will be equal to the sum of the probabilities of the individual events comprising the union. Any set function, $P(\cdot)$, that satisfies the

[6]See A.N. Kolmogorov (1956) *Foundations of the Theory of Probability, 2nd ed.* New York: Chelsea.

three Axioms 1.1, 1.2, and 1.3 will be called a **probability measure** or **probability set function**. The image of an event $A$ generated by a probability set function $P$ is called the **probability of event** $A$.

**Definition 1.14**
***Probability Set Function (or Probability Measure)***

A set function that adheres to the three axioms of probability.

**Definition 1.15**
***Probability***

An image of an event generated by a probability set function.

The following examples provide illustrations of probability set functions for finite, countably infinite, and uncountably infinite sample spaces.

**Example 1.5**
***Probability Set Function Verification: Discrete S***

Let $S = \{1, 2, 3, 4, 5, 6\}$ be the sample space for rolling a fair die and observing the number of dots facing up. Then $P(A) = N(A)/6$, for $A \subset S$, defines a probability set function on the events in $S$. We can verify that $P(\cdot)$ is a probability measure by noting that $P(A) \geq 0$ for all $A \subset S$, $P(S) = N(S)/6 = 6/6 = 1$ and $P(\cup_{i \in I} A_i) = \sum_{i \in I} P(A_i)$ for any collection $\{A_i, i \in I\}$ of disjoint subsets of $S$. For example, if $A_1 = \{1,2\}$ and $A_2 = \{4, 5, 6\}$, then

$$P(A_1) = \frac{N(A_1)}{6} = 2/6,$$

$$P(A_2) = \frac{N(A_2)}{6} = 3/6,$$

$$P(A_1 \cup A_2) = \frac{N(A_1 \cup A_2)}{6} = 5/6,$$

and thus $P(A_1 \cup A_2) = P(A_1) + P(A_2)$. More generally, if $\{A_i, i \in I\}$ are disjoint events, then

$$N\left(\bigcup_{i \in I} A_i\right) = \sum_{i \in I} N(A_i),$$

and thus

$$P\left(\bigcup_{i \in I} A_i\right) = \sum_{i \in I} N(A_i)/6 = \sum_{i \in I} P(A_i). \qquad \square$$

**Example 1.6**
***Probability Set Function Verification: Countable S***

Let $S = \{x\colon x \text{ is a positive integer}\}$, and examine the set function defined by $P(A) = \sum_{x \in A} (1/2)^x$ for $A \subset S$. The set function, so defined, is a probability set function since, first of all, $P(A) \geq 0$ because $P(A)$ is defined as the sum of a collection of nonnegative numbers. To verify that $P(S) = 1$, recall the following results from real analysis:

**Lemma 1.1**

$$\sum_{j=1}^{n} ar^j = \frac{a(r-r^{n+1})}{1-r} \text{ and for } |r|<1,\ \sum_{j=1}^{\infty} ar^j = \lim_{n\to\infty} \frac{a(r-r^{n+1})}{1-r} = \frac{ar}{1-r}.$$

In the case at hand, $a = 1$ and $r = 1/2$, so that $P(S) = \sum_{x=1}^{\infty} (1/2)^x = (1/2)/(1-(1/2)) = 1$. Finally, by definition of the summation operation, if $A_i$, $i \in I$, are disjoint subsets of $S$, then

$$P\left(\bigcup_{i\in I} A_i\right) = \sum_{x\in \cup_{i\in I} A_i} (1/2)^x = \sum_{i\in I}\sum_{x\in A_i} (1/2)^x = \sum_{i\in I} P(A_i). \qquad \square$$

**Example 1.7** ***Probability Set Function Verification: Uncountable S***

Let $S = \{x : 0 \leq x < \infty\}$ be the sample space corresponding to the experiment of observing the operating life, in hours, of computer memory chips produced by a chip manufacturer. Let the probability set function be given by $P(A) = \int_{x\in A} \frac{1}{2} e^{-x/2}\, dx$, with the event space, $\Upsilon$, (the domain of $P$) being the collection of all interval subsets of $S$ together with any sets that can be formed by a countable number of union, intersection, and/or complement operations applied to the interval subsets.[7]

We can verify that $P(A) \geq 0\ \forall A \in \Upsilon$, since $P(A) = \int_{x\in A} \frac{1}{2} e^{-x/2}\, dx$ has a nonnegative integrand and the integral of a nonnegative integrand is nonnegative valued.[8] It is also true that $P(S) = 1$, since $P(S) = \int_0^\infty \frac{1}{2} e^{-x/2}\, dx = -e^{-x/2} \big|_0^\infty = 1$. Finally, if $A = \cup_{i=1}^{n} A_i$, with the sets $A_1, A_2, \ldots, A_n$ being disjoint, it follows from the additivity property of the Riemann integral that

$$P\left(\bigcup_{i=1}^{n} A_i\right) = \int_{x\in\cup_{i=1}^{n} A_i} \frac{1}{2} e^{-x/2}\, dx = \sum_{i=1}^{n} \int_{x\in A_i} \frac{1}{2} e^{-x/2}\, dx = \sum_{i=1}^{n} P(A_i),$$

and so countable additivity holds. $\square$

All problems involving the assignment of probabilities to the various events in a sample space will formally share a common mathematical structure given by a three-tuple of objects, collectively referred to as the **probability space** of an experiment.

---

[7]Unlike the previous example which used a countable sample space, when the sample space is uncountable, not all subsets of $S$ can technically be considered events, i.e., there may be subsets of $S$ to which probability cannot be assigned. The collection of subsets defined here are the **Borel sets** contained in $S$, all of which can be considered events in $S$. We discuss this technical question further in the Appendix to the Chapter.

[8]We will tacitly assume, unless explicitly stated otherwise, that the orientation of integral ranges is from lowest to highest values in defining the integral over any set $A$.

**Definition 1.16**
***Probability Space***

> A probability space is the three-tuple $\{S, \Upsilon, P\}$, where $S$ is the sample space of an experiment, $\Upsilon$ is the event space, and $P$ is a probability set function having domain $\Upsilon$.

In any probabilistic analysis of an experiment, we will seek to establish

1. A universal set, $S$, that contains all of the potential outcomes or elementary events of an experiment;
2. A set of sets, $\Upsilon$, representing the collection of events or subsets of $S$ on which probability will be defined; and
3. A probability set function, $P$, that can be used to assign the appropriate probabilities to the events in $S$.

Once the probability space is defined, all of the information is available that is needed to assign probabilities to the various events of interest related to an experiment is available. As one might suspect, it is the discovery of the appropriate probability set function that represents a major challenge in the application of probability and statistics, and we examine the discovery problem in the latter half of the text when we discuss topics in inferential statistics. Our immediate goal in the remaining sections of this chapter is to establish a number of useful results in probability theory that are implied by the probability axioms.

## 1.5 Some Probability Theorems

The three axioms governing the behavior of probability set functions, together with results from set theory, can be used to prove probability theorems that contribute to the development of probability theory. In stating and proving such theorems, additional insights and truths are established about how probability assignments must behave.

**Theorem 1.1** $P(A) = 1 - P(\bar{A})$.

**Proof** By the definition of the complement of $A$, $A \cup \bar{A} = S$. Thus, by substitution, and by Axiom 1.2, $P(S) = 1 = P(A \cup \bar{A})$. However, since $A \cap \bar{A} = \emptyset$, Axiom 1.3 allows us to state that $1 = P(A) + P(\bar{A})$. Subtracting $P(\bar{A})$ from both sides obtains the result. ■

**Theorem 1.2** $P(\emptyset) = 0$.

**Proof** Let $A = \emptyset$ in Theorem 1.1. Then since $\bar{A} = S$, it follows that $P(\emptyset) = 1 - P(S) = 1 - 1 = 0$ since $P(S) = 1$ by Axiom 1.2. ■

**Theorem 1.3** *If $A \subset B$, then $P(A) \leq P(B)$ and $P(B - A) = P(B) - P(A)$.*

**Proof** Since $A \subset B$, $B = A \cup (B - A)$. The sets $A$ and $B - A$ are disjoint and thus by Axiom 1.3, $P(B) = P(A) + P(B - A)$. Since $P(B - A) \geq 0$ by Axiom 1.1, dropping $P(B - A)$ from the probability equality implies $P(B) \geq P(A)$. Subtracting $P(A)$ from both sides of the probability equality yields the second result of the theorem. ■

**Theorem 1.4** $P(A) = P(A \cap B) + P(A \cap \bar{B})$.

**Proof** $A = A \cap S = A \cap (B \cup \bar{B}) = (A \cap B) \cup (A \cap \bar{B})$ since the intersection operation is distributive and $S = B \cup \bar{B}$. Then because the events $A \cap B$ and $A \cap \bar{B}$ are disjoint, it follows by Axiom 1.3 that $P(A) = P(A \cap B) + P(A \cap \bar{B})$. ■

**Theorem 1.5** $P(A \cup B) = P(A) + P(B) - P(A \cap B)$.

**Proof** $A \cup B = (A \cup B) \cap S = (A \cup B) \cap (B \cup \bar{B}) = B \cup (A \cap \bar{B})$ since the union operation is distributive and $S = B \cup \bar{B}$. Because events $B$ and $(A \cap \bar{B})$ are disjoint, it follows by Axiom 1.3 that $P(A \cup B) = P(B) + P(A \cap \bar{B})$. However, Theorem 1.4 implies that $P(A \cap \bar{B}) = P(A) - P(A \cap B)$, and thus by substitution, $P(A \cup B) = P(A) + P(B) - P(A \cap B)$. ■

**Corrolary 1.1** $P(A \cup B) \leq P(A) + P(B)$. (*Boole's Inequality*)[9]

**Proof** Note that the corollary follows directly from Theorem 1.5 since $P(A \cap B) \geq 0$.

**Theorem 1.6** $P(A) \in [0, 1]$.

**Proof** $\emptyset \subset A$ implies $P(\emptyset) \leq P(A)$ and $A \subset S$ implies $P(A) \leq P(S)$, by Theorem 1.3. Since $P(S) = 1$ by Axiom 1.2 and $P(\emptyset) = 0$ by Theorem 1.2, we have $0 \leq P(A) \leq 1$. ■

**Theorem 1.7** $P(A \cap B) \geq 1 - P(\bar{A}) - P(\bar{B})$. (*Bonferroni's Inequality-2 event case*)[10]

**Proof** By Theorem 1.1, $P(A \cap B) = 1 - P(\overline{A \cap B})$. DeMorgan's law indicates that $\overline{A \cap B} = \bar{A} \cup \bar{B}$, and thus $P(A \cap B) = 1 - P(\bar{A} \cup \bar{B})$ by substitution. Theorem 1.5 indicates that $P(\bar{A} \cup \bar{B}) = P(\bar{A}) + P(\bar{B}) - P(\bar{A} \cap \bar{B})$, and thus $P(A \cap B) = 1 - [P(\bar{A}) + P(\bar{B}) - P(\bar{A} \cap \bar{B})] = 1 - P(\bar{A}) - P(\bar{B}) + P(\bar{A} \cap \bar{B})$, again by substitution. Finally, since $P(\bar{A} \cap \bar{B}) \geq 0$ by Axiom 1.1, we have that $P(A \cap B) \geq 1 - P(\bar{A}) - P(\bar{B})$. ■

**Theorem 1.8** $P(\cap_{i=1}^{k} A_i) \geq 1 - \sum_{i=1}^{k} P(\bar{A}_i)$. (*Bonferroni's Inequality – General*).

---

[9]Named after the English mathematician and logician George Boole.

[10]Named for the Italian mathematician, C.E. Bonferroni.

**Proof** We have already proven the validity of the proposition when $n = 2$ by Theorem 1.7. Suppose for purposes of invoking the induction principle (recall Lemma 1.1) that we assume

$$P\left(\cap_{i=1}^{k} A_i\right) \geq 1 - \sum_{i=1}^{k} P(\bar{A}_i)$$

is true. Using Theorem 1.7, we know that

$$P\left(\cap_{i=1}^{k+1} A_i\right) = P\left(\left(\cap_{i=1}^{k} A_i\right) \cap A_{k+1}\right) \geq 1 - P\left(\overline{\cap_{i=1}^{k} A_i}\right) - P(\bar{A}_{k+1}).$$

Theorem 1.1 allows us to rewrite the inequality as

$$P\left(\cap_{i=1}^{k+1} A_i\right) \geq P\left(\cap_{i=1}^{k} A_i\right) - P(\bar{A}_{k+1})$$

Then by the assumption that the Bonferroni inequality is valid for $k$ events, we may write

$$P\left(\cap_{i=1}^{k+1} A_i\right) \geq 1 - \sum_{i=1}^{k} P(\bar{A}_i) - P(\bar{A}_{k+1})$$

which implies that Bonferroni's inequality is valid for $k + 1$ events. ■

**Theorem 1.9** ***Classical Probability*** *Let $S$ be the finite sample space for an experiment having $N(S)$ equally likely outcomes, and let $A \subset S$ be an event containing $N(A)$ elements. Then $P(A) = N(A)/N(S)$.*

**Proof** Let $E_1,\ldots,E_n$ represent the $n = N(S)$ outcomes (or elementary events) in the sample space $S$. Since all outcomes are equally likely, $P(E_i) = k$, $\forall i$, and since the outcomes are disjoint and $S = \left(\cup_{i=1}^{n} E_i\right)$, we have

$$P(S) = \sum_{i=1}^{N(S)} P(E_i) = \sum_{i=1}^{N(S)} k = N(S)k = 1$$

by Axioms 1.2 and 1.3. It follows that $P(E_i) = k = 1/N(S)$ for $i = 1,\ldots,N(S)$. Let $I \subset \{1, 2,\ldots,N(S)\}$ be the index set identifying the $N(A)$ number of outcomes (or elementary events) that define the event $A$, i.e., $A = \cup_{i\in I} E_i$. Then by Axiom 1.3,

$$P(A) = \sum_{i\in I} P(E_i) = \sum_{i\in I} 1/N(S) = \frac{N(A)}{N(S)}.$$ ■

By proving Theorem 1.9, we have shown that the classical probability definition is implied by the axiomatic definition of probability. Thus, whenever the conditions of the classical probability definition apply, we are free to follow the classical prescription for assigning probabilities to events. It can also be shown that the relative frequency definition of probability is implied by the axiomatic definition. Among other things, this implies that the axiomatic foundation for probability theory provides the rationale for the existence of the limit of relative frequencies referred to in the relative frequency definition of probability. We will need to develop results relating to asymptotic theory (Chapter 5) before a proof of this proposition can be provided.

Finally, the subjective probability definition is implied by the axiomatic definition in the sense that an individual assigning subjective probabilities to events will be *required* to adhere to the axioms in making those assignments. The requirement is interpreted by subjective probabilists as a *consistency* condition for subjective probability assignments.

The example below illustrates the use of the preceding probability theorems (for an example of Theorem 1.9, recall Example 1.3).

**Example 1.8** ***Applying Probability Theorems*** Let $S = \{2, 3, ..., 12\}$ be the sample space corresponding to the experiment of rolling a pair of fair dice and observing the total number of dots facing up. We will see in Chapter 2 that the probability set function appropriate for assigning probabilities to events $A$ in $S$ is given by $P(A) = \sum_{x \in A} \frac{6-|x-7|}{36}$.

*Theorem 1.1* The event of winning a game of "craps" in Las Vegas on the first roll of the dice is the event $A = \{7, 11\}$. Using the probability set function defined above, we have that $P(A) = 8/36$. It follows immediately from the theorem that $P(\bar{A}) = 1 - P(A) = 28/36$.
*Theorem 1.2* Define the event of rolling an even number of dots, $B = \{2, 4, 6, 8, 10, 12\}$. Examine the event of rolling an even number of dots and winning the game of craps on the first roll of the dice, $A \cap B = \emptyset$. From the theorem, $P(A \cap B) = P(\emptyset) = 0$.
*Theorem 1.3* Define the elementary event $C = \{7\}$. Noting that $C \subset A$, we know from the theorem that $P(C) \leq P(A)$, and in fact, $P(C) = 6/36 < 8/36 = P(A)$.
*Theorem 1.4* Define the event of rolling a six or less, $D = \{2, 3, 4, 5, 6\}$. From the theorem, we have that $P(B) = P(B \cap D) + P(B \cap \bar{D}) = 9/36 + 9/36 = 18/36$.
*Theorem 1.5 and Corollary 1.1* Note that $A \cup D = \{2, 3, 4, 5, 6, 7, 11\}$, and a calculation using the probability set function results in $P(A \cup D) = 23/36$. The theorem indicates that the probability can be represented as $P(A \cup D) = P(A) + P(D) - P(A \cap D) = 8/36 + 15/36 - 0 = 23/36$. The corollary indicates that the probability adheres to $P(A \cup D) \leq P(A) + P(D)$, which it clearly does as a weak inequality.
*Theorem 1.6* The theorem indicates that $P(A) \in [0,1]$ for any event contained in $S$, which is how probabilities assigned via the probability set function above will behave. Note that all of the probability assignments above adhere to this restriction.
*Theorem 1.7* The theorem implies that $P(B \cap D) \geq 1 - P(\bar{B}) - P(\bar{D})$, which is true, since $9/36 \geq 1 - 18/36 - 21/36$.
*Theorem 1.8* The theorem implies that $P(\bar{A} \cap \bar{B} \cap \bar{C}) \geq 1 - P(A) - P(B) - P(C)$, which is true since $10/36 \geq 1 - 8/36 - 18/36 - 6/36$. □

## 1.6 Conditional Probability

When an experiment is conducted, the one event that is certain to occur is the event $S$ since the outcome of an experiment must be an element of the sample space. We now study the effect that additional information concerning the outcome of an experiment has on the probability of events. In particular, if it is known that the outcome of the experiment is an element of some subset, $B$, of the sample space, what is the effect of this additional information on the probabilities of events in $S$? As an example of such a situation, it appears intuitively plausible that the probability of a company earning \$10 million in annual profits would be higher if the company were randomly chosen from the list of Fortune 500 companies than if the company were chosen from among all companies in the United States. For another example, examine the experiment of tossing two fair coins in succession, and let the sample space for the experiment be defined by $S = \{(H,H), (H,T), (T,H), (T,T)\}$, where $H$ = heads and $T$ = tails. The probability (unconditional) of observing two tails is 1/4 since, by the Classical Probability theorem, $P((T,T)) = N(A)/N(S) = 1/4$, where $A = \{(T,T)\}$. However, the probability of observing two tails must be zero *given that* the outcome of the first coin toss was heads. We develop the notion of conditional probability ahead.

Suppose that we are analyzing an experiment with an associated probability space $\{S,\Upsilon,P\}$ and it is *given* that the outcome of the experiment is some element of a subset, $B$, of the sample space. How should the probability of an event, $A$, be defined given the additional information that event $B$ has occurred? By making a number of observations concerning properties that conditional probabilities should possess, we will be led to a definition of the conditional probability of event $A$ given event $B$.

First of all, since it is given that $B$ occurs, it is certain that $B$ will not occur. In effect, the sample space has been reduced to the subset $B$, i.e., the outcomes in $S$-$B$ are no longer relevant and $B$ can be interpreted as a new **conditional sample space**. Letting the symbol $P(A|B)$ represent **the conditional probability of event A, given event *B***, it follows that since the new sample space $B$ is an event that is now certain to occur, the *conditional* probability assigned to event $B$ should be 1, so that $P(B|B) = 1$. Note further that since $B$ will occur, it is clear that an event $A$ can also occur *iff* $A$ occurs concurrently with $B$, that is, *iff* $A \cap B$ occurs (see Figure 1.1). This suggests that conditional probability should be defined so that $P(A|B) = P(A \cap B|B)$ for any event $A$.

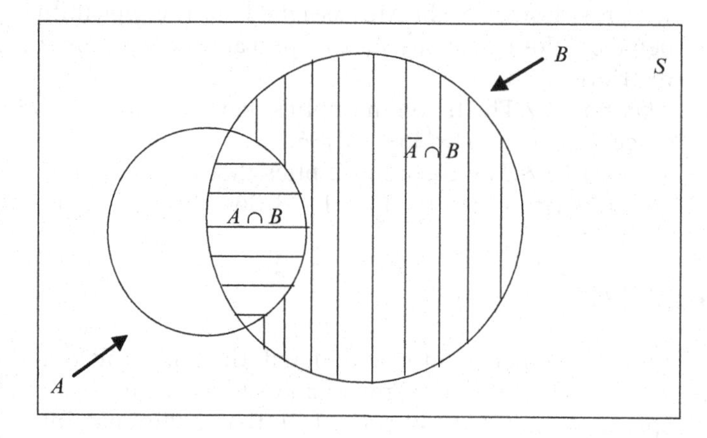


**Figure 1.1** Conditional sample space $B$.

Now note that Theorem 1.4 implies the probability equation $P(B) = P(A \cap B) + P(\bar{A} \cap B)$ since $B$ can be partitioned into the two disjoint subsets, as $B = (A \cap B) \cup (\bar{A} \cap B)$. Dividing both sides of the probability equality by $P(B)$ (assuming $P(B) \neq 0$) obtains a proportional decomposition of the probability of event $B$ as $1 = \frac{P(A \cap B)}{P(B)} + \frac{P(\bar{A} \cap B)}{P(B)}$, where the proportion $k = P(A \cap B)/P(B) \in [0,1]$ of event $B$'s probability is attributable to event $A \cap B$ with the remaining proportion, $(1 - k)$, attributable to event $\bar{A} \cap B$. Then since event $A \cap B$ accounts for a proportion, $k$, of the probability that event $B$ will occur, and since $B$ is now certain to occur and so is assigned (conditional) probability $P(B|B) = 1$, then the proportion, $k$, of this unit probability should be attributable to the event $A \cap B$. We thus assign $P(A \cap B|B) = k = P(A \cap B)/P(B)$, and since we also require that $P(A|B) = P(A \cap B|B)$, we are led to the following definition of conditional probability.

**Definition 1.17** ***Conditional Probability***

Let A and B be any two events in a sample space $S$. If $P(B) \neq 0$, then the conditional probability of event $A$, given event $B$, is given by $P(A|B) = P(A \cap B)/P(B)$.

Our intuition concerning the meaning of conditional probability can be enhanced by examining the definition in light of the classical and relative frequency definitions of probability. In an experiment for which classical probability is applicable, the probability space consists of a finite sample space, an event space that consists of all subsets of the sample space, and a probability set function that assigns probability to an event $A \subset S$ as $P(A) = N(A)/N(S)$.

Conditioning on an event $B \subset S$, the probability of an event $A$, by the definition of conditional probability, is given by

$$P(A|B) = \frac{P(A \cap B)}{P(B)} = \frac{N(A \cap B)/N(S)}{N(B)/N(S)} = \frac{N(A \cap B)}{N(B)}.$$

Since all outcomes are equally likely in this case, and since we are effectively restricting the sample space to the set $B$ by conditioning on event $B$, it stands to reason that the probability of observing $A$ is given by the number of outcomes in $B$ that result in $A$'s occurring, i.e., $N(A \cap B)$, relative to the total number of outcomes in $B$. This is, of course, consistent with the classical probability definition applied to the event $A$ in the context of the new sample space, $B$.

Regarding the relative frequency definition and conditional probability, recall that the probability set function assigns probabilities to events via $P(A) = \lim_{n \to \infty}(n_A/n)$. Conditioning on event $B \subset S$, the probability of event $A$, by the definition of conditional probability, is given by

[11]See R.C. Buck (1978) *Advanced Calculus, 3rd edition*, McGraw-Hill, p. 44. We will discuss the concept of limits in more detail in Chapter 5. For now, a more intuitive understanding of limits is sufficient.

$$P(A|B) = \frac{P(A \cap B)}{P(B)} = \frac{\lim_{n\to\infty} (n_{A\cap B}/n)}{\lim_{n\to\infty} (n_B/n)} = \lim_{n\to\infty} (n_{A\cap B}/n_B)$$

Note the last equality follows from the fact that the limit of a ratio equals the ratio of the limits if all limits exist, and if the limit in the denominator of the ratio of limits is not zero.[11] Restricting the sample space to the set $B$ by conditioning on event $B$, $P(A|B)$ is seen to equal the limiting fraction of the number of occurrences of event $B$ that also result in the occurrence of $A$. Consistent with the logic of the relative frequency definition of probability, $P(A|B)$ could then be interpreted as the limit of the frequency of observing event $A$ relative to the total number of outcomes generated from the conditional sample space $B$.

The conditional probability set function $P(\cdot|B)$ can be used to define two new probability spaces $\{S, \Upsilon, P(\cdot|B)\}$ and $\{B, \Upsilon_B, P(\cdot|B)\}$, where $\Upsilon_B$ is the collection of all events contained in the set $B$. It follows that conditional *probabilities* can be legitimately assigned by $P(\cdot|B)$ to all of the events in the original sample space $S$ as well as to all of the events in the conditional sample space $B$. One can formally demonstrate that $P(\cdot|B)$ adheres to the three probability axioms and that conditional probabilities are being generated from a legitimate probability set function. It is suggested that the reader pursue such a demonstration.

Since it can be shown that $P(A|B)$ adheres to the probability axioms, all of the theorems that were proved for unconditional probabilities apply equally well to conditional probabilities.[12] This follows because the validity of the theorems is derived from the probability axioms regardless of whether the set function is representing unconditional or conditional probabilities. Put another way, the proofs of all of the theorems would apply analogously to conditional probabilities by simply changing $P(\cdot)$ to $P(\cdot|D)$, say, in the proofs and recognizing that the probability axioms apply to the set function $P(\cdot|D)$. Note that we use the letter $D$ here to allow for the possibility that the event being conditioned on is different than the events, $A$ and $B$, that are referred to in some of the previous probability theorems. For convenience, we list the probability theorems below as they apply to conditional probabilities. It is assumed that the conditional probability is defined, i.e., $P(D) \neq 0$, and that $A$ and/or $B$ are events in either the conditional sample space $D$ or the original sample space $S$.

**Theorem 1.1^c** $P(A|D) = 1 - P(\bar{A}|D)$.

**Theorem 1.2^c** $P(\emptyset|D) = 0$.

**Theorem 1.3^c** If $A \subset B$, then $P(A|D) \leq P(B|D)$ and $P(B - A|D) = P(B|D) - P(A|D)$.

**Theorem 1.4^c** $P(A|D) = P(A \cap B|D) + P(A \cap \bar{B}|D)$.

**Theorem 1.5^c** $P(A \cup B|D) = P(A|D) + P(B|D) - P(A \cap B|D)$.

**Corollary 1.1^c** $P(A \cup B|D) \leq P(A|D) + P(B|D)$.

**Theorem 1.6^c** $P(A|D) \in [0,1]$.

**Theorem 1.7^c** $P(A \cap B|D) \geq 1 - P(\bar{A}|D) - P(\bar{B}|D)$.

[12]Note that, in a sense, all probabilities could be viewed as *conditional*, where $P(A|S)$ could be used to denote probabilities in previous sections. We will continue to use "unconditional" to refer to the case where the original sample space, $S$, has been left "unconditioned."

**Theorem 1.8**$^c$ $P(\cap_{i=1}^{k} A_i|D) \geq 1 - \sum_{i=1}^{k} P(\bar{A}_i|D)$.

The following examples illustrate the application of conditional probability.

**Example 1.9** ***Conditional Probabilities in Coin Tossing***

Consider the experiments of tossing two coins in succession, and let $S = \{(H,H), (H,T), (T,H), (T,T)\}$, where $H$ = head, $T$ = tail. Assume all outcomes are equally likely.

**(a)** What is the probability of obtaining two heads, given the first coin toss was heads?
**Answer**: $B = \{(H,H), (H,T)\}$ is the event that the first coin toss results in heads, which is our conditional sample space. $A = \{(H,H)\}$ is the event of obtaining two heads. Then $P(A|B) = P(A \cap B)/P(B) = (1/4)/(1/2) = 1/2$.

**(b)** What is the probability of obtaining two heads, given that at least one of the coins comes up heads?
**Answer**: $C = \{(H,H), (H,T), (T,H)\}$ is the event that at least one of the coins came up heads, which is our conditional sample space. Then $P(A|C) = P(A \cap C)/P(C) = (1/4)/(3/4) = 1/3$.

**(c)** What is the probability of obtaining one head and one tail given that the first coin is a tail?
**Answer**: $D = \{(T,H), (T,T)\}$ is the event that the first coin was a tail. $E = \{(T,H), (H,T)\}$ is the event of obtaining one head and one tail. Then
$P(E|D) = P(E \cap D)/P(D) = (1/4)/(1/2) = 1/2$. □

**Example 1.10** ***Conditional Probabilities in Sample Selection***

A perplexed investor must choose an investment instrument from among 15 different stocks, 10 different bonds, and 5 different mutual funds. Allowing each instrument an equal probability of being chosen, the investor randomly chooses an instrument. Given that the chosen instrument was not a bond, what is the probability that a stock was chosen?

**Answer**: We are conditioning on the event, $B$, that the instrument is either a stock or a mutual fund (i.e., not a bond), and $P(B) = 2/3$. Let $A$ represent the event that the outcome is a stock, so that $P(A \cap B) = 1/2$. Then

$$P(A|B) = P(A \cap B)/P(B) = (1/2)/(2/3) = 3/4$$

is the probability we seek. Note this makes sense from the standpoint that the conditional sample space is 20 investment instruments, of which 15 are stocks and 5 are mutual funds. The classical probability definition suggests that the probability of observing a stock in this sample space of 20 instruments is $15/20 = 3/4$. □

The definition of conditional probability can be transformed to obtain a result known as the **multiplication rule**. The multiplication rule allows one to calculate the probability of the event $A \cap B$ from knowledge of the conditional probability of event $A$, given event $B$, and the unconditional probability of $B$.

**Theorem 1.10** ***Multiplication Rule: Two Events***

*Let $A$ and $B$ be any two events in the sample space for which $P(B) \neq 0$. Then $P(A \cap B) = P(A|B)\, P(B)$.*

**Proof** Multiply both sides of $P(A|B) = P(A \cap B)/P(B)$ in Definition 1.7 by $P(B)$. ■

The multiplication rule is especially useful in cases where an experiment can be viewed as being conducted in two stages.

**Example 1.11** ***Multiplication Rule for Two Card Draw***

What is the probability of drawing two aces in succession from a well-shuffled deck of poker cards? Assume cards drawn are not replaced in the deck.

**Answer**: Let $B$ be the event that the first card drawn is an ace. Since there are four aces in a poker deck, with a total of 52 cards in the deck, $P(B) = 4/52 = 1/13$. Now let $A$ be the event that the second card drawn is an ace. Given that the first card drawn is an ace (i.e., given the event $B$), there are three aces remaining to be chosen from the remaining 51 cards, and thus the probability that the second draw is an ace, given that the first card drawn is an ace, i.e., $P(A|B)$, equals $3/51 = 1/17$. Then by the multiplication rule, the probability that both draws result in aces is given by

$$P(A \cap B) = P(A|B)P(B) = (1/17)(1/13) = 1/221.$$ □

**Example 1.12** ***Multiplication Rule in One-Stage Inspections***

As part of its quality control program, an apparel manufacturer has inspectors examine every garment the company produces. A garment is shipped to a retail outlet only if it passes inspection. The probability that a garment is defective is .02. The probability that an inspector assigns a "pass" to a defective garment is .05. What is the probability that a garment is defective and shipped to a retail outlet?

**Answer**: Let $D$ be the event that a garment is defective. Let $B$ be the event that the inspector assigns a "pass" to a garment. We know that $P(D) = .02$ and $P(B|D) = .05$. $B \cap D$ is the event that a garment is defective and is passed by the inspector. Then $P(B \cap D) = P(B|D)P(D) = (.05)(.02) = .001$. □

The multiplication rule can be extended to three or more events, in which case the probability of the intersection of all of the events can be represented as follows.

**Theorem 1.11** ***Multiplication Rule: General***

*Let $A_1, A_2, \ldots, A_n$, $n \geq 2$ be events in the sample space. Then if all of the conditional probabilities exist, $P(\cap_{i=1}^{n} A_i) = P(A_1) \prod_{i=2}^{n} P\left(A_i \mid \cap_{j=1}^{i-1} A_j\right)$.*

**Proof** We know from Theorem 1.10 that the result holds for $n = 2$ (note $\cap_{j=1}^{1} A_j = A_1$ by definition). In an attempt to invoke the mathematical induction principle assume that the result is true for $n = k$, where $k$ is some arbitrary positive integer $\geq 3$. Let $B = \cap_{i=1}^{k} A_i$. Then

$$P\left(\cap_{i=1}^{k+1} A_i\right) = P(A_{k+1} \cap B) = P(A_{k+1}\,|B)P(B)(\text{Theorem 1.11})$$

$$= \underbrace{P(A_1)\prod_{i=2}^{k} P\left(A_i \,|\, \cap_{j=1}^{i-1} A_j\right)}_{P(B)} P(A_{k+1}\,|B)(\text{assuming result holds for } n = k)$$

$$= P(A_1)\prod_{i=2}^{k+1} P\left(A_i \,|\, \cap_{j=1}^{i-1} A_j\right)(\text{substitution for } B)$$

Thus, by mathematical induction, the theorem holds. (See Definition 1.12) ■

Similar to the case of the multiplication rule for two events, the extended multiplication rule is especially useful in cases where an experiment can be viewed as being conducted in $n$ stages.

**Example 1.13**
***Multiplication Rule for Three Card Draw***

What is the probability of drawing four aces in succession from a well-shuffled deck of poker cards? Assume cards drawn are not replaced in the deck.

**Answer**: Let $A_i$ be the event that the $i$th card drawn is an ace, $i = 1, 2, 3, 4$. Then using Theorem 1.12,

$$P(\cap_{i=1}^{4} A_i) = P(A_1)P(A_2\,|\,A_1)P(A_3\,|\,A_1 \cap A_2)P(A_4\,|\,A_1 \cap A_2 \cap A_3)$$
$$= \left(\frac{4}{52}\right)\left(\frac{3}{51}\right)\left(\frac{2}{50}\right)\left(\frac{1}{49}\right) = .3693 \times 10^{-5}.$$ □

**Example 1.14**
***Multiplication Rule in Two-Stage Inspections***

Recall the garment inspection problem of Example 1.12. Suppose the retailers who market the garments of the apparel manufacturer also inspect each garment they purchase and place on sale only those for which they perceive no defects. The probability that a retailer places a defective garment on sale is .10. What is the probability that a garment is defective, shipped to the retail outlet, and placed on sale by retailers?
**Answer**: Let $A$ be the event that the retailer places a garment on sale. $A \cap B \cap D$ is the event of interest, and by Theorem 1.11,

$$P(A \cap B \cap D) = P(D)P(B|D)P(A|B \cap D) = (.02)(.05)(.10) = .0001$$ □

## 1.7 Independence

In everyday language, if one were to say that two events *are independent*, it is generally meant that the occurrence of one event does not affect the likelihood of an occurrence of the other, and vice-versa. This meaning of **independence** can be formalized within the theory of probability. We begin with the technical definition of **independent events.**

**Definition 1.18**
***Independence of Events: Two Event Case***

Let $A$ and $B$ be two events in a sample space $S$. Then $A$ and $B$ are **independent events** iff $P(A \cap B) = P(A)P(B)$. If $A$ and $B$ are not independent, $A$ and $B$ are said to be **dependent events**.

An interpretation of the independence condition in Definition 1.18 that is closely aligned with our layman's interpretation of the word *independence* is available when $P(A) > 0$ and $P(B) > 0$. In this case, $P(A \cap B) = P(A)P(B)$ implies

$$P(A|B) = P(A \cap B)/P(B) = P(A)P(B)/P(B) = P(A),$$

and

$$P(B|A) = P(B \cap A)/P(A) = P(B)P(A)/P(A) = P(B).$$

Thus the probability of event $A$ occurring is unaffected by the occurrence of event $B$, and the probability of event $B$ occurring is unaffected by the occurrence of event $A$.

If event $A$ and/or event $B$ has probability zero, then *by definition*, events $A$ and $B$ are independent. This follows immediately from the fact that if either $P(A) = 0$ or $P(B) = 0$, then $P(A \cap B) = 0 = P(A)P(B)$ (since $(A \cap B) \subset A$ and $(A \cap B) \subset B$ imply both $P(A \cap B) \leq P(A)$ and $P(A \cap B) \leq P(B)$ by Theorem 1.3, and $P(A \cap B) \geq 0$ by Axiom 1.1, so that together the inequalities imply $P(A \cap B) = 0$), and thus the independence condition is fulfilled. However, in this case one or both conditional probabilities $P(A|B)$ and $P(B|A)$ are undefined, and thus the basis no longer exists for stating that "the independence of events $A$ and $B$ implies the probability of either event is unaffected by the occurrence of the other."

If $A$ and $B$ are independent, then it follows that $A$ and $\bar{B}$, $\bar{A}$ and $B$, and $\bar{A}$ and $\bar{B}$ are also independent. This result can be demonstrated by showing that the independence definition is satisfied for each of the preceding pairs of events if independence is satisfied for $A$ and $B$. We state the result as a theorem.

**Theorem 1.12**
***Independence of Complement Event Pairs***

*If events $A$ and $B$ are independent, then events* $A$ and $\bar{B}$, $\bar{A}$ and $B$, *and* $\bar{A}$ *and* $\bar{B}$ *are also independent.*

$$\begin{aligned} P(A \cap \overline{B}) &= P(A) - P(A \cap B) \text{(Theorem 1.4)} \\ &= P(A) - P(A)P(B) \text{(Independence of } A \text{ and } B) \\ &= P(A)[1 - P(B)] \text{(Algebra)} \\ &= P(A)P(\overline{B}) \text{(Theorem 1.1)} \end{aligned}$$

$$\begin{aligned}P(\overline{A}\cap B) &= P(B) - P(A\cap B)(\text{Theorem 1.4})\\ &= P(B) - P(A)P(B)(\text{Independence of } A \text{ and } B)\\ &= P(B)[1 - P(A)](\text{Algebra})\\ &= P(\overline{A})P(B)(\text{Theorem 1.1})\end{aligned}$$

$$\begin{aligned}P(\overline{A}\cap \overline{B}) &= P(\overline{A\cup B})\ (\text{DeMorgan's laws})\\ &= 1 - P(A\cup B)\ (\text{Theorem 1.1})\\ &= 1 - (P(A) + P(B) - P(A\cap B))(\text{Theorem 1.5})\\ &= 1 - P(A) - P(B) + P(A)P(B)(\text{Independence of } A \text{ and } B)\\ &= P(\overline{A}) - P(B)[1 - P(A)](\text{Theorem 1.1 and Algebra})\\ &= P(\overline{A})[1 - P(B)](\text{Algebra and Theorem 1.1})\\ &= P(\overline{A})P(\overline{B})(\text{Theorem 1.1})\end{aligned}$$

■

The following example illustrates the concept of independence of events.

**Example 1.15** ***Independence of Gender and Occupation***

The work force of the Excelsior Corporation has the following distribution among type and gender of workers:

| | Type of worker | | | |
|---|---|---|---|---|
| **Sex** | **Sales** | **Clerical** | **Production** | **Total** |
| Male | 825 | 675 | 750 | 2,250 |
| Female | 1,675 | 825 | 250 | 2,750 |
| Total | 2,500 | 1,500 | 1,000 | 5,000 |

In order to promote loyalty to the company, the company randomly chooses a worker to receive an all-expenses paid vacation each month. Is the event of choosing a female independent of the event of choosing a clerical worker?
**Answer:** Let $F$ = event of choosing a female and $C$ = event of choosing a clerical worker. From the data in the table, we know that $P(F) = .55$ and $P(C) = .30$. Also, $P(F\cap C) = .165$. Then $P(F\cap C) = P(F)P(C)$, and the events are *independent*.
Is the event of choosing a female independent of the event of choosing a production worker?
**Answer:** Let $A$ = event of choosing a production worker. Then $P(A) = .20$ and $P(A\cap F) = .05$, and thus $P(A\cap F) = .05 \neq .11 = P(A)P(F)$, so the events are *dependent*. □

The property that $A$ and $B$ are independent events is sometimes confused with the property that sets $A$ and $B$ are disjoint. It should be noted that the two properties are distinct, but related concepts. The disjointness property is a property of events, while the independence property is a property of the

**Table 1.1** Disjointness Versus Independence

| | $A \cap B = \emptyset$ | $A \cap B \neq \emptyset$ |
|---|---|---|
| $P(A) > 0$ and $P(B) > 0$ | Dependent | Independent *iff* $P(A \cap B) = P(A)P(B)$ |
| $P(A) = 0$ and/or $P(B) = 0$ | Independent | Independent |

probability set function defined on the events. Table 1.1 presents the relationship between the two properties.

We now verify the three cases in which an immediate conclusion can be reached regarding the independence of events $A$ and $B$.

**Theorem 1.13** ***Independence and Disjointness***

1. $P(A) > 0$, $P(B) > 0$, $A \cap B = \emptyset \Longrightarrow$ *A and B are dependent.*
2. $P(A)$ and/or $P(B) = 0$, $A \cap B = \emptyset \Longrightarrow$ *A and B are independent.*
3. $P(A)$ and/or $P(B) = 0$, $A \cap B \neq \emptyset \Longrightarrow$ *A and B are independent.*

**Proof**

1. $P(A \cap B) = P(\emptyset) = 0$ by Theorem 1.2. Since $P(A \cap B) = 0 < P(A)P(B)$ because P($A$) and P($B$) are both positive, $A$ and $B$ cannot be independent events, and so they are dependent.
2. $P(A \cap B) = P(\emptyset) = 0$ by Theorem 1.2. Since $P(A)$ and/or $P(B) = 0$, then $P(A)$ $P(B) = 0$, and thus $P(A \cap B) = 0 = P(A)P(B)$. Therefore, $A$ and $B$ are independent.
3. $P(A \cap B) \leq P(A)$ and $P(A \cap B) \leq P(B)$ by Theorem 1.3, since $(A \cap B) \subset A$ and $(A \cap B) \subset B$. If $P(A)$ and/or $P(B) = 0$, then $P(A \cap B) \leq 0$. $P(A \cap B) \geq 0$ by Axiom 1.1. Then $P(A \cap B) = 0 = P(A)P(B)$, and $A$ and $B$ are independent events. ■

The concept of independent events can be generalized to more than two events.

**Definition 1.19** ***Independence of Events: n Event Case***

Let $A_1, A_2,\ldots,A_n$ be events in the sample space $S$. The events $A_1, A_2,\ldots,A_n$ are independent *iff* $P(\cap_{j \in J} A_j) = \prod_{j \in J} P(A_j)$ for all subsets $J \subset \{1, 2,\ldots,n\}$ for which $N(J) \geq 2$. If the events $A_1, A_2,\ldots,A_n$ are not independent, they are said to be dependent events.

Note that the independence concept defined in Definition 1.19 is sometimes referred to as the **joint**, **mutual**, or **complete independence** of the events $A_1$, $A_2,\ldots,A_n$ when $n \geq 3$ to emphasize that additional conditions are required beyond the condition given in Definition 1.18 applicable to pairs of events. We will refrain from using these additional adjectives and simply refer to the *independence of events*, regardless of $n$. Furthermore, note that if the condition $P(A_i \cap A_j) = P(A_i)P(A_j)$ of Definition 1.18 applies to all pairs of events in $A_1, A_2,\ldots,A_n$, the events are referred to as being **pairwise independent** whether or not they are independent in the sense of Definition 1.19.

In the case of three events, independence requires that $P(A_1 \cap A_2) = P(A_1)$ $P(A_2)$, $P(A_1 \cap A_3) = P(A_1)P(A_3)$,$P(A_2 \cap A_3) = P(A_2)P(A_3)$, and $P(A_1 \cap A_2 \cap A_3) = P(A_1)P(A_2)P(A_3)$. Note that if all three events have nonzero probability, then the reader can straightforwardly verify from the definition of conditional probability that independence implies $P(A_i|A_j) = P(A_i)\ \forall i \neq j$, $P(A_i|A_j \cap A_k) = P(A_i) \forall i \neq j \neq k$, and $P(A_i \cap A_j\ |A_k) = P(A_i \cap A_j)\ \ \forall i \neq j \neq k$. It is not as straightforward to demonstrate that independence implies $P(A_i|A_j \cup A_k) = P(A_i)$ for $i \neq j \neq k$, and so we demonstrate the result below.

$$
\begin{aligned}
P(A_i \mid A_j \cup A_k) &= \frac{P(A_i \cap (A_j \cup A_k))}{P(A_j \cup A_k)} \text{(by definition)} \\
&= \frac{P((A_i \cap A_j) \cup (A_i \cap A_k))}{P(A_j \cup A_k)} \text{(distributive law)} \\
&= \frac{P(A_i \cap A_j) + P(A_i \cap A_k) - P(A_i \cap A_j \cap A_k)}{P(A_j \cup A_k)} \text{(Theorem 1.5)} \\
&= \frac{P(A_i)P(A_j) + P(A_i)P(A_k) - P(A_i)P(A_j)P(A_k)}{P(A_j \cup A_k)} \text{(independence)} \\
&= \frac{P(A_i)\left[P(A_j) + P(A_k) - P(A_j)P(A_k)\right]}{P(A_j \cup A_k)} \text{(algebra)} \\
&= \frac{P(A_i)P(A_j \cup A_k)}{P(A_j \cup A_k)} \text{(Theorem 1.5 and independence)} \\
&= P(A_i)
\end{aligned}
$$

It is thus recognized that if events $A_1$, $A_2$, and $A_3$ are independent in the sense of Definition 1.19, and if each of the events occurs with nonzero probability, then the probability of any one of the events is unaffected by the occurrence of any of the remaining events and is also unaffected by the occurrence of the union or intersection of the remaining events. This interpretation extends in a straightforward way to cases involving four or more independent events, and the reader should attempt some of the possible extensions.

The reader may wonder whether the numerous conditions (for $n$ events, the number of conditions will be $2^n - n - 1$) cited in Definition 1.19 for independence of events are all necessary, i.e., wouldn't the one condition $P\left(\cap_{j=1}^n A_j\right) = \prod_{j=1}^n P(A_j)$ suffice and imply all of the others? Unfortunately, the answer is no – all of the conditions are required. The following example illustrates the point.

**Example 1.16** ***Need for Pairwise Independence Conditions***

The Venn diagram in Figure 1.2 summarizes the probabilities assigned to some of the events in the sample space $S$. Note that $P(A \cap B \cap C) = .15 = (.5)(.6)(.5) = P(A)P(B)P(C)$. However, $P(A \cap B) = .25 \neq .3 = P(A)P(B)$, $P(A \cap C) = .20 \neq .25 = P(A)P(C)$, $P(B \cap C) = .40 \neq .3 = P(B)P(C)$, and thus $P(A \cap B \cap C) = P(A)P(B)P(C)$ does *not* imply the pairwise independence conditions. □

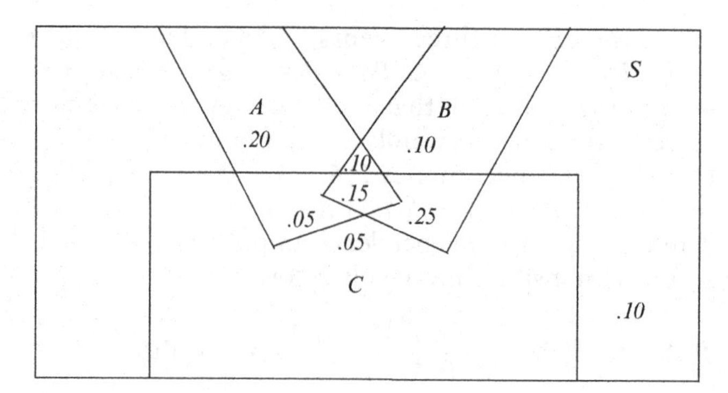


**Figure 1.2** Probability assignments in $S$.

The reader should construct an example illustrating that pairwise independence among $A$, $B$, and $C$ does *not* imply $P(A \cap B \cap C) = P(A)P(B)P(C)$.

## 1.8 Bayes' Rule

Bayes' rule, discovered by the 17th century English clergyman and mathematician, Thomas Bayes, provides an alternative representation of conditional probabilities. At first glance, the representation may appear somewhat convoluted in comparison to the representation of conditional probability given in Definition 1.17. However, the rule is well suited for providing conditional probabilities in certain experimental situations, which we will identify following the formal derivation of the rule.

Bayes' rule is actually a simple consequence of a result known as the Theorem of Total Probability, which we state and prove next.

**Theorem 1.14** ***Theorem of Total Probability*** *Let the events $B_i$, $i \in I$, be a finite or countably infinite partition of the sample space, $S$, so that $B_j \cap B_k = \emptyset$ for $j \neq k$, and $\cup_{i\in I} B_i = S$. Let $P(B_i) > 0 \forall i$. Then $P(A) = \Sigma_{i\in I} P(A \cap B_i) = \Sigma_{i\in I} P(A|B_i)P(B_i)$.*

**Proof** Since $\cup_{i\in I} B_i = S$, it follows that $A = A \cap (\cup_{i\in I} B_i) = \cup_{i\in I}(A \cap B_i)$, where we have used the fact that the intersection operation is distributive over the union operation. Now note that $(A \cap B_j) \cap (A \cap B_k) = \emptyset$ for $j \neq k$ since the $B_i$'s are disjoint. But then by Axiom (1.3), $P(A) = P(\cup_{i\in I}(A \cap B_i)) = \Sigma_{i\in I} P(A \cap B_i)$. The result of the theorem follows from applying the multiplication rule to each term, $P(A \cap B_i)$, in the sum. ■

Regarding the title of Theorem 1.14, note that the "total probability" of event $A$ is represented as the sum of the portions of $A$'s probability distributed over the events in the partition of $A$ represented by the events $A \cap B_i$, $i \in I$. We now state Bayes' rule.

**Theorem 1.15** ***Bayes' Rule*** *Let events $B_i$, $i \in I$, be a finite or countably infinite partition of the sample space, S, so that $B_j \cap B_k = \emptyset$ for $j \neq k$ and $\cup_{i \in I} B_i = S$. Let $P(B_i) > 0 \; \forall i \in I$. Then, provided $P(A) \neq 0$,*

$$P(B_j|A) = \frac{P(A|B_j)P(B_j)}{\sum_{i \in I} P(A|B_i)P(B_i)}, \forall j \in I.$$

**Proof** From the multiplication rule, $P(A \cap B_j) = P(A|B_j)P(B_j)$, and from the Theorem of Total Probability, $P(A) = \sum_{i \in I} P(A|B_i)P(B_i)$. Substituting respectively into the numerator and denominator of the representation of $P(B_j|A)$ above yields $P(A \cap B_j)/P(A)$, the conditional probability of $B_j$ given $A$. ■

**Corollory 1.2** ***Bayes' Rule: Two Event Case***

$$P(B|A) = \frac{P(A|B)P(B)}{P(A|B)P(B) + P(A|\overline{B})P(\overline{B})}$$

**Proof** This is a direct consequence of Theorem 1.15 when $I = \{1, 2\}$. ■

In the next two examples we provide illustrations of the types of experimental situations for which Bayes' rule have particularly useful application.

**Example 1.17** ***Bayes Rule in Oil Discovery*** Explorations, Inc. is in the oil well-drilling business. Let $B$ be the event that a well being drilled will produce oil and let $A$ be an event representing geological well-site characteristics that are conducive to discovering oil at a site. Suppose further that, from past experience, it is known that the unconditional probability of a successful strike when drilling for oil is .06. Also suppose it is known that when oil is discovered, the probability is .85 that the geological characteristics are given by event $A$, whereas the probability is only .4 that geological characteristics represented by $A$ are present when no oil is discovered. If event $A$ occurs at a site, what is the probability of discovering oil at the site, i.e., what is $P(B|A)$?

**Answer**: It is known that $P(B) = .06$, $P(\bar{B}) = .94$, $P(A|B) = .85$, and $P(A|\bar{B}) = .40$. Bayes's rule applies here, so that

$$P(B|A) = \frac{P(A|B)P(B)}{P(A|B)P(B) + P(A|\overline{B})P(\overline{B})} = \frac{(.85)(.06)}{(.85)(.06) + (.40)(.94)} = .12$$

Note that the occurrence of event A increases considerably the probability of discovering oil at a particular site. □

**Example 1.18** ***Bayes Rule in Assessing Blood Test Accuracy*** A blood test developed by a pharmaceutical company for detecting a certain disease is 98 percent effective in detecting the disease *given* that the disease is, in fact, present in the individual being tested. The test yields a "false positive" result (meaning a person without the disease is incorrectly indicated as having the disease) for only 1 percent of the disease-free persons tested.

If an individual is randomly chosen from the population and tested for the disease, and given that .1 percent of the population actually has the disease, what is the probability that the person tested actually has the disease if the test result is positive (i.e., the disease is indicated as being present by the test)?

**Answer**: In this case, let $A$ be the event that the test result is positive, and let $B$ be the event that the individual actually has the disease. Then, from the preceding discussion concerning the characteristics of the test, it is known that $P(A|B) = .98$, $P(B) = .001$, and $P(A|\bar{B}) = .01$. Then, an application of Bayes' rule yields

$$P(B|A) = \frac{P(A|B)P(B)}{P(A|B)P(B) + P(A|\bar{B})P(\bar{B})} = \frac{(.98)(.001)}{(.98)(.001) + (.01)(.999)} = .089$$

Thus, one has very little confidence that a positive test result implies that the disease is present. □

A common thread in the two examples, consistent with the statement of Bayes's rule itself, is that the sample space is partitioned into a collection of disjoint events $(B_i, i \in I)$, that are of interest and whose nonzero probabilities are known. Furthermore, an event occurs whose various conditional probabilities formed by conditioning on each of the events in the partition are known. Given this background information, Bayes' rule provides the means for reevaluating the probabilities of the various events in the partition of $S$, *given* the information that event $A$ occurs. The probabilities of the events in the partition are, in effect, "updated" in light of the new information provided by the occurrence of $A$. This interpretation of Bayes's rule has led to the use of the terms **prior probabilities** and **posterior probabilities** being used to refer to the $P(B_i)$'s and $P(B_i|A)$'s, respectively. That is, $P(B_i)$ is the probability of event $B_i$ in the partition of $S$ *prior* to the occurrence of event $A$, whereas $P(B_i|A)$ is the probability of $B_i$ *posterior* to, or after, event $A$ occurs.

Another prominent interpretation of Bayes' rule is that of being a tool for *inverting conditional probabilities*. In particular, note that Theorem 1.15 begins with information relating to the conditional probabilities $P(A|B_j)\ \forall j$ and then *inverts* them into the conditional probabilities $P(B_j|A)\ \forall j$. Given this interpretation, some have referred to Bayes Rule alternatively as **The Law of Inverse Probabilities.**

Returning to the oil well-drilling example, note that each elementary event in the (implied) sample space is given by a pair of observations, one being whether or not the geological characteristics of the well-site favor the discovery of oil, and the other being whether or not oil is actually discovered at the well site. The partition of the sample space that is of interest to the oil well-drilling company is the event "oil is discovered" versus the complementary event that "oil is not discovered" at the well site. The additional information used to update the prior probabilities concerning oil discovery is whether the geological characteristics of the site favor the discovery of oil. Bayes' rule can be applied to generate posterior probabilities of oil discovery because the conditional probabilities of favorable well-site characteristics being observed, *with* and *without* the condition of oil being discovered, are known. Also, Bayes' rule is inverting the conditional probabilities that geological characteristics conducive to oil production are present when oil is discovered and when it is not, to conditional probabilities that oil is discovered or not, given the presence of geological characteristics conducive to oil discovery.

The reader can provide a characterization of the sample space, partition of interest, and additional information used to update and invert probabilities in the case of the drug test example.

## 1.9 Appendix: Digression on Events that Cannot Be Assigned Probability

When we first began our discussion of the axiomatic approach to the definition of probability, we stated that probability is generated by a set function whose domain consisted of all of the events in a sample space. This collection of "all of the events in a sample space" was termed the **event space**. However, in uncountable sample spaces it can be the case that there are some very complicated events (subsets) of the sample space to which probability cannot be assigned. The issue here is whether a set function can have a domain that literally consists of *all* of the subsets of a sample space and still adhere to the three axioms of probability.

In the case of a countable sample space, the domain *can* consist of all of the subsets of the sample space and still have the set function exhibiting the properties required by the probability axioms. Henceforth, whenever we are dealing with a countable sample space, the domain will always be defined as the collection of all of the subsets of the sample space, unless explicitly stated otherwise.

The situation is more complicated in the case of an uncountably infinite sample space. In this case, the collection of all subsets of $S$ is, in a sense, so large that a set function cannot have this collection of sets for its domain and still have the probability axioms hold true for all possible applications. The problem is addressed in a field of mathematics called **measure theory** and is beyond the scope of our study. As a practical matter, essentially any subset of $S$ that will be of practical interest as an event in real-world applications will be in the domain of the probability set function. Put another way, the subsets of $S$ that are not in the domain are by definition so complicated that they will not be of interest as events in any real-world application.

While it takes a great deal of ingenuity to define a subset of an uncountably infinite sample space that is not an event, the reader may still desire a more precise and technically correct definition of the domain of the probability set function in the case of an uncountably infinite sample space so that one is certain to be referring to a collection of subsets of $S$ for which each subset can be assigned a probability. This can be done relatively straightforwardly so long as we restrict our attention to real-valued sample spaces (i.e., sample spaces whose sample points are all real numbers). Since we can always "code" the elements of a sample space with real numbers (this relates to the notion of random variables, which we address in Chapter 2), restricting attention to real-valued sample spaces does not involve any loss of generality, and so we proceed on the assumption that $S \subset \mathbb{R}^n$. Our characterization depends on the notion of Borel sets, named after the French mathematician Emile Borel, which we define next. The definition uses the concept of rectangles in $\mathbb{R}^n$, which are generalizations of intervals.

**Definition A.1**
***Rectangles in $\mathbb{R}^n$***

a. ***Closed rectangle:*** $\{(x_1,\ldots,x_n): a_i \le x_i \le b_i, i = 1,\ldots,n\}$

b. ***Open rectangle:*** $\{(x_1,\ldots,x_n): a_i < x_i < b_i, i = 1,\ldots,n\}$

c. ***Half-open/Half-closed rectangle:***

$\{(x_1, ..., x_n) : a_i < x_i \le b_i, i = 1, ..., n\}$

$\{(x_1, ..., x_n) : a_i \le x_i < b_i, i = 1, ..., n\}$

where the $a_i$'s and $b_i$'s are real numbers, with $-\infty$ or $\infty$ being admissible for strong inequalities. Clearly, rectangles are intervals when $n = 1$.

The collection of Borel sets contained in a sample space $S$ will include all of the rectangle subsets of $S$ as well as an infinite number of other sets that can be formed from them via set operations as defined below.

**Definition A.2**
***Borel Sets in S***

Let $S \subset \mathbb{R}^n$. The collection of Borel sets in $S$ consists of all closed, open, and half-open/half-closed rectangles contained in $S$, as well as any other set that can be defined by applying a countable number of union, intersection, and/or complement operations to these rectangles.

The collection of Borel sets in $S$ is an example of what is known as a **sigma-field** ($\sigma$-field), or a **sigma-algebra** ($\sigma$-algebra). A $\sigma$-field is a nonempty set of sets that is closed under countable union, intersection, and complement operations. The use of the word "closed" here means that if $A_i$, $i \in I$, all belong to the $\sigma$-field, any set formed by applying a countable number of unions, intersections, and/or complement operations to the $A_i$'s is also a set that belongs to the $\sigma$-field, where $I$ is any countable index set.

The collection of Borel sets is extremely large and will contain any subset of the real-valued sample space that will be of practical interest as events in real-world applications. In particular, *all* open and *all* closed (rectangular or nonrectangular) sets are contained in the collection of Borel sets. Most importantly, probabilities can always be assigned to Borel sets. Consequently, we will tacitly assume that the collection of Borel sets is our domain for probability set functions associated with real-valued sample spaces. However, we will continue to refer to the event space as the domain of a probability set function, and act as if the domain consisted of all subsets of the sample space, with little worry that we will ever encounter a subset of $S$ in practice that cannot be assigned probability.

## Keywords, Phrases, and Symbols

Axiomatic approach
Bayes' Rule
Bonferroni's inequality
Borel sets
$\subset$, contained in
Certain event, $S$
Classical probability
Conditional probability of event $A$, given event $B$
Conditional sample space
Continuous sample space
Discrete sample space
Element (or member)
Elementary event
$\in$, element of
Event
Event space, $\Upsilon$
Experiment
$\sigma$-field
*iff*, if and only if
$\Rightarrow$, implies that
Independence of events
$\forall$, for all
$\cap$, intersection
Inverting conditional probabilities
Joint, mutual, or complete independence of events
Mathematical induction
$\prod_{i=1}^{n}$ or $\prod_{j\in J}$, multiplication
Multiplication rule
Mutually exclusive (or disjoint) events
$\notin$, not an element of
$\emptyset$, null set
Outcome
$P(A|B)$
Pairwise independence
Partition of a set
Posterior probability
Prior probability
Probability of event $A$
Probability set function (or probability measure), $P$
Probability space
Probability theorems
Product notation
Relative frequency probability
Sample point
Sample space
Subjective probability
The Law of Inverse Probabilities
The occurrence of event $A$
$\cup$, union

## Problems

**1.** Define an appropriate sample space for each of the experiments described below:

a. At the close of business each day, the Acme Department Store's accountant counts the number of customer transactions that were made in cash. On a particular day, there were 100 customer transactions at the department store. The outcome of interest is the number of cash transactions made.

b. An Italian restaurant in the city of Spokane runs an ad in the city newspaper, *The Spokesman Review*, that contains a coupon that allows a customer to purchase two meals for the price of one for each newspaper coupon the customer has. The coupon is valid for 30 days after the ad is run. The outcome of interest is how many free meals the restaurant serves at the end of the 30-day period.

c. On a local 11 o'clock news broadcast for the town of College Station, the weather report includes the high and low temperatures, in Fahrenheit, for the preceding 24 hours. The outcome of interest is the pair of high and low temperatures on any given day.

d. A local gasoline jobber supplies a number of the area's independent gas stations with unleaded gasoline. The outcome of interest is the quantity of gasoline demanded from the jobber in any given week.

e. The mutual funds management company of Dewey, Cheatum, and Howe posts the daily closing net asset value of shares in its mutual fund on a readerboard outside of its headquarters. The outcome of interest is the posted net asset value of the shares at the end of a given day.

f. The office manager of a business specializing in copying services is counting the number of copies that a given copying machine produces before suffering a paper jam. The outcome of interest is the number of copies made before the machine suffers a paper jam.

**2.** For each of the sample spaces you have defined above, indicate whether the sample space is finite, countably infinite, or uncountably infinite. Justify your answers.

**3.** The sales team of a large car dealership in Seattle consists of the following individuals:

| Name | Sales experience | Age | Education | Married |
|---|---|---|---|---|
| Tom | 4 years | 34 | High school | Yes |
| Karen | 12 years | 31 | < High school | No |
| Frank | 21 years | 56 | College grad | Yes |

| | | | | |
|---|---|---|---|---|
| Eric | 9 years | 42 | High school | Yes |
| Wendy | 3 years | 24 | College grad | No |
| Brenda | 7 years | 29 | High school | No |
| Scott | 15 years | 44 | College grad | Yes |
| Richard | 2 years | 25 | < High school | No |

A customer visiting the dealership randomly chooses one of the salespersons to discuss the purchase of a new vehicle. Define the set and assign the probability associated with each of the following events:

a. A woman is chosen.

b. A man less than 40 years of age is chosen.

c. An individual with at least 10 years of sales experience is chosen.

d. A married College graduate is chosen.

e. A married female with a high school education and at least 5 years of sales experience is chosen.

f. An individual with at least 2 years' experience and at least 21 years of age is chosen.

**4.** Assign probabilities to the events a to f in the preceding question, but include the condition "***given that*** the individual chosen is $\geq 30$ years old."

**5.** The manager of the cost accounting department of a large computer manufacturing firm always tells three jokes during her monthly report to the board of directors in an attempt to inject a bit of levity into an otherwise sobering presentation. She has an inventory of a dozen different jokes from which she chooses three to present for any given monthly report.

a. If she chooses the three jokes randomly from the inventory of 12 each month, what is the probability that, in any given month, at least one of the three jokes will be different from the jokes she told the month before?

b. If she chooses the three jokes randomly from the inventory of 12 each month, what is the probability that, in any given month, all three jokes will be different from the three she told the month before?

**6.** Schneider's Plumbing and Heating, located in Fargo, North Dakota, has 300 accounts receivable distributed as follows:

| Current | 1–30 days past due | 31–60 days past due | 61–90 days past due | Sent for collection |
|---|---|---|---|---|
| 140 | 80 | 40 | 25 | 15 |

An auditor is coming to inspect Schneider's financial records. Included in the auditor's analysis is a randomly chosen sample of four accounts from the company's collection of accounts receivable.

a. What is the probability that all of the accounts chosen by the auditor will be current accounts?

b. What is the probability that all of the accounts chosen by the auditor will be less than or equal to 60 days past due?

c. What is the probability that all of the accounts chosen by the auditor will be more than 60 days past due?

d. What is the probability that two of the accounts will be current, and two will be 1–30 days past due?

**7.** A computer manufacturing firm produces three product lines: (1) desktop computer systems, (2) notebook computers, and (3) subnotebook computers. The sales department has convened its monthly meeting in which the four staff members of the department provide the department manager with their indications of whether sales will increase for each of the product lines in the coming month. Let $A_i$ represent the event that sales for product line $i$ $(=1, 2, \text{ or } 3)$ will increase in the coming month. The manager will consider the information of a given staff member to be usable if that information is *internally consistent*, where internally consistent in this context means consistent with the axioms and theorems of probability. Which of the staff members have provided the manager with usable information? Be sure to provide a convincing reason if you decide that a staff member's information needs to be discarded.

| Staff member | Tom | Dick | Harry | Sally |
|---|---|---|---|---|
| $P(A_1)$ | .5 | .3 | .3 | .2 |
| $P(A_2)$ | .3 | .2 | .6 | .3 |
| $P(A_3)$ | .7 | .8 | −.4 | .5 |
| $P(A_1 \cap A_2)$ | .9 | .4 | .4 | .2 |
| $P(A_1 \cap A_3)$ | .6 | .15 | .2 | .3 |
| $P(A_2 \cap A_3)$ | .15 | .1 | .1 | .4 |
| $P(A_1 \cap A_2 \cap A_3)$ | .1 | 1.5 | .05 | .1 |

**8.** A large electronics firm is attempting to hire six new electrical engineers. It has been the firm's experience that 35 percent of the college graduates who are offered positions with the firm have turned down the offer of employment. After interviewing candidates for the positions, the firm offers employment contracts to seven college graduates. What is the probability that the firm will receive acceptances of employment from one too many engineers? You may assume that the decisions of the college graduates are independent.

**9.** A computer manufacturing firm accepts a shipment of CPU chips from its suppliers only if an inspection of 5 percent of the chips, randomly chosen from the shipment, does not contain any defective chips. If a shipment contains five defective chips and there are 1,000 chips in the shipment, what is the probability that the shipment will be accepted?

**10.** The probability that a stereo shop sells at least one amplifier on a given day is .75; the probability of selling at least one CD player is .6; and the probability of selling at least one amplifier *and* at least one CD player is .5.

a. What is the probability that the stereo shop will sell at least one of the two products on a given day?

b. What is the probability that the stereo shop will sell at least one CD player, given that the shop sells at least one amplifier?

c. What is the probability that the stereo shop will sell at least one amplifier, given that the shop sells at least one CD player?

d. What is the probability that the shop sells neither of the products on a given day?

**11.** Prove that the set function defined by

$$P(A|B) = \frac{P(A \cap B)}{P(B)} \quad \textit{for } P(B) \neq 0$$

is a valid *probability* set function in the probability space $\{B, \Upsilon_B, P(\cdot|B)\}$, where $\Upsilon_B$ is the event space for the sample space $B$.

**12.** A large midwestern bank has devised a math aptitude test that it claims provides valuable input into the hiring decision for bank tellers. The bank's research indicates that 60 percent of all tellers hired by midwestern banks are classified as performing satisfactorily in the position at their initial 6-month performance review, while the rest are rated as unsatisfactory. Of the tellers whose performance is rated as satisfactory, 90 percent had passed the math aptitude test. Of the tellers who were rated unsatisfactory, only 20 percent had passed the math aptitude test.

a. What is the probability that a teller would be rated as satisfactory at her 6-month performance review, given that she passed the math aptitude test?

b. What is the probability that a teller would be rated as satisfactory at her 6-month performance review, given that she did not pass the math aptitude test?

c. Does the test seem to be an effective screening device to use in hiring tellers for the bank? Why or why not?

**13.** A large-scale firm specializing in providing temporary secretarial services to corporate clients has completed a study of the main reason why secretaries become dissatisfied with their work assignments, and how likely it is that a dissatisfied secretary will quit her job. It was found that 20 percent of all secretaries were dissatisfied with some aspect of their job assignment. Of all dissatisfied secretaries, it was found that 55 percent were dissatisfied mainly because they disliked their supervisor; 30 percent were dissatisfied mainly because they felt they were not paid enough; 10 percent were dissatisfied mainly because they disliked the type of work; and 5 percent were dissatisfied mainly because they had conflicts with other employees. The probabilities that the dissatisfied secretaries would quit their jobs were respectively .20, .30, .90, and .05.

a. Given that a dissatisfied secretary quits her job, what is the most probable main reason why she was dissatisfied with her job assignment?

b. If a secretary were chosen at random, what is the probability that she would be dissatisfied, mainly because of her pay?

c. Given that a secretary is dissatisfied with her job assignment, what is the probability that she will quit?

**14.** A clerk is maintaining three different files containing job applications submitted for three different positions currently open in the firm at which the clerk is employed. One file contains two completed applications, one file contains one complete and one incomplete application, and the third file contains two incomplete applications. The clerk wishes to examine the files and chooses one of the files at random. She then chooses at random one of the applications contained in the chosen file. If the application chosen is complete, what is the probability that the remaining application in the file is also complete?

**15.** A company manages three different mutual funds. Let $A_i$ be the event that the $i$th mutual fund increases in value on a given day. Probabilities of various events relating to the mutual funds are given as follows:

$P(A_1) = .55, P(A_2) = .60, P(A_3) = .45, P(A_1 \cup A_2) = .82,$
$P(A_1 \cup A_3) = .7525, P(A_2 \cup A_3) = .78, P(A_2 \cap A_3 \mid A_1) = .20.$

a. Are events $A_1$, $A_2$, and $A_3$ pairwise independent?

b. Are events $A_1$, $A_2$, and $A_3$ independent?

c. What is the probability that funds 1 and 2 both increase in value, given that fund 3 increases in value? Is this different from the unconditional probability that funds 1 and 2 both increase in value?

d. What is the probability that at least one mutual fund will increase in value on a given day?

**16.** Answer the following questions regarding the validity of probability assignments. If you answer false, explain why the statement is false.

a. If $P(A) = .2$, $P(B) = .3$, and $A \cap B = \emptyset$, then $P(A \cup B) = .06$. True or False?

b. If $A \cap B = \emptyset$ and $P(B) = .2$, then $P(A|B) = 0$. True or False?

c. If $P(B) = .05$, $P(A|B) = .80$, and $P(A|\bar{B}) = .5$, then $P(B|A) = .0777$ (to four digits of accuracy). True or False?

d. If $P(A) = .8$ and $P(B) = .7$, then $P(A \cap B) \geq .5$. True or False?

e. It is possible that $P(A) = .7$, $P(B) = .4$, and $A \cap B = \emptyset$. True or False?

**17.** The ZAP Electric Co. manufactures electric circuit breakers. The circuit breakers are produced on two different assembly lines in the company's Spokane plant. Assembly line I is highly automated and produces 85 percent of the plant's output. Assembly line II uses older technology that is more labor intensive, producing 15 percent of the plant's output. The probability that a circuit breaker manufactured on assembly line I is defective is .04, while the corresponding probability for assembly line II is .01.

As part of its quality control program, ZAP uses a testing device for determining whether a circuit breaker is faulty. Some important characteristics of the testing device are as follows:

$$P(A|B) = P(\bar{A}|\bar{B}) = .985,$$

where $A$ is the event that the testing device *indicates* that a circuit breaker is faulty and $B$ is the event that the circuit breaker *really is* faulty.

a. If a circuit breaker is randomly chosen from a bin containing a day's production and the circuit breaker is actually defective, what is the probability that it was produced on assembly line II?

b. What is the probability that the testing device *indicates* that a circuit breaker is *not* faulty, given that the circuit breaker really is faulty?

c. If the testing device is applied to circuit breakers produced on assembly line I, what is the probability that a circuit breaker really is faulty, given that the testing device indicates that the circuit breaker is faulty? Would you say that this is a good testing device?

**18.** The ACME Computer Co. operates three plants that manufacture notebook computers. The plants are located in Seattle, Singapore, and New York. The plants produce 20, 30, and 50 percent of the company's output, respectively. ACME attaches the labels "Seattle," "SING," or "NY" to the underside of the computer in order to identify the plant in which a notebook computer was manufactured. The computers carry a 2-year warranty, and if a customer requires repairs during the warranty period, he or she must send the computer back to the plant in which the computer was manufactured. There is also a stamp on the motherboard inside the computer which technicians at a plant can use as an additional way of identifying which plant manufactured the computer. The consumer is unable to examine this inside stamp, because if the consumer opens up the computer housing to look inside, a seal is broken which voids the warranty. Regarding quality control at the plants, the warranty-period failure rates of computers manufactured in the three plants are known to be .01, .05, and .02 for the Seattle, Singapore, and New York plants, respectively. You have bought an ACME computer, and it has failed during the warranty period. You need to send the computer back to the plant for repairs, but the label on the underside of the computer has been lost and so you don't know which plant manufactured your computer.

a. Which plant is the most probable plant to have manufactured your computer?

b. Which plant is the least probable plant to have manufactured your computer?

c. What is the probability that an ACME notebook computer will fail during the warranty period?

d. Given that an ACME computer does not fail during the warranty period, what is the probability that the computer was manufactured in New York?

**19.** The diagram below indicates how probabilities have been assigned to various subsets of the sample space $S$:

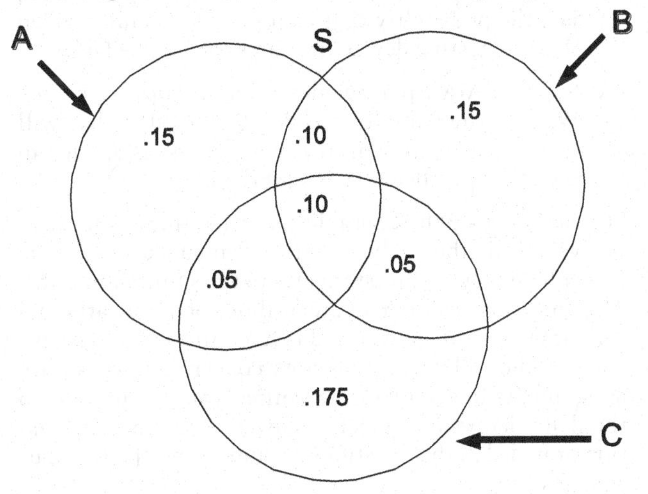


a. Are the three events $A$, $B$, and $C$ pairwise independent events?

b. Are the three events $A$, $B$, and $C$ independent events?

c. What is the value of $P(A \cap B)$? What is the value of $P(A \cap B \mid C)$?

d. Suppose event $D$ is such that $P(D) = .05$ and $D \cap (A \cup B \cup C) = \emptyset$. Are events $D$ and $A$ independent? Are events $D$ and $(A \cup B \cup C)$ independent?

e. What is the probability of event $C$, *given* $(A \cap B)$?

**20.** A large sack contains 1,000 flower seeds consisting of 300 carnations and 700 impatiens. Of the 300 carnation seeds, 200 will produce red flowers and 100 will produce white flowers. Of the 700 impatiens seeds, 400 will produce red flowers and 300 will produce white flowers.

a. If you randomly choose five seeds in succession (without replacing any seeds that have been chosen), what is the probability that these seeds will produce two impatiens with red flowers, two carnations with red flowers, and one carnation with white flowers?

b. If you randomly choose four seeds in succession (without replacing any seeds that have been chosen), what is the probability that these seeds will produce red- and white-flowered impatiens and red- and white-flowered carnations?

c. Given that four randomly chosen seeds all produce carnations, what is the probability that three are red flowered and one is white flowered?

d. Is the event of randomly choosing a carnation seed on the first draw independent of choosing an impatiens seed on the second draw?

**21.** For each case below, determine whether or not the real-valued set function $P(A)$ is in fact a *probability* set function:

a. Sample space $S = \{1, 2, 3, 4, 5, 6, 7, 8\}$, Event space $\Upsilon = \{A : A \subset S\}$, Set Function $P(A) = \Sigma_{x \in A}(x/36)$ for $A \in \Upsilon$.

b. Sample space $S = [0, 4)$, Event Space $\Upsilon = \{A : A$ is an interval subset of $S$, or any set formed by unions, intersections, or complements of these interval subsets$\}$, Set Function $P(A) = \int_{x \in A} e^{-x} dx$ for $A \in \Upsilon$.

c. Sample Space $S = \{x : x$ is a positive integer$\}$, Event Space $\Upsilon = \{A : A \subset S\}$, Set Function $P(A) = \Sigma_{x \in A}(x^2 / 10^5)$ for $A \in \Upsilon$

d. Sample Space $S = (0, 1)$, Event Space $\Upsilon = \{A : A$ is an interval subset of $S$, or any set formed by unions, intersections, or complements of these interval subsets$\}$, Set Function $P(A) = \int_{x \in A} 12x(1-x)^2 dx$ for $A \in \Upsilon$

**22.** The Smith Floor Wax Company manufactures and sells industrial-strength floor wax in the wholesale market for home care products. The factory produces 10,000 gal of floor wax daily and currently has an inventory of 5,000 gal of floor wax in its warehouse. If sales of floor wax exceed production, the company meets the excess demand by using inventory; while if sales are less than production, the company adds this excess production to inventory. The company economist provides you with the following information concerning probabilities of daily sales events, where events are measured in gallons of wax sold.
$A = [0, 5{,}000]$, $P(A) = .25$
$B = (5{,}000, 10{,}000]$, $P(B) = .65$
$C = [2{,}500, 7{,}500]$, $P(C) = .35$

$D = (5{,}000, 7{,}500]$, $P(D) = .20$

a. What is the probability that inventory will have to be used to satisfy sales on a given day?

b. What is the probability that fewer than 2,500 gal of wax will be sold on a given day?

c. What is the probability of the event $E = [0, 2{,}500) \cup (7{,}500, 10{,}000]$?

**23.** A box contains four different computer disks, labeled 1, 2, 3, and 4. Two disks are selected at random from the box "with replacement," meaning that after the first selection is made, the selected disk is returned to the box before the second selection is made. "At random" means that all disks in the box have an equal chance of being selected.

a. Define the sample space for this experiment.

b. Is the event of choosing disk 1 or 3 on the first selection independent of choosing disk 1 or 2 on the second selection? Why or why not?

c. Is the event of choosing disk 1 on the first selection independent of choosing disk 1 on the second selection? Why or why not?

d. Is the event of choosing disk 1 on both the first and second selections independent of the event that neither disk 3 nor 4 is chosen in the selection process?

**24.** The AJAX Microchip Company produces memory chips for personal computers. The company's entire production is generated from two assembly lines, labeled I and II. Assembly line I uses more rapid assembly techniques and produces 80 percent of the company's output, while assembly line II produces 20 percent of the output. The probability that a memory chip produced on assembly line I is defective is .05, while the corresponding probability for assembly line II is .01.

A memory chip is chosen at random from a bin containing a day's production. Given that the chip is found to be defective, what is the probability that the chip was made on assembly line II? (*Hint*: Can you put this problem in a form for which Bayes rule would be applicable?)

**25.** The management of the AJAX Microchip Company (mentioned in Problem 24) is interested in increasing quality control at the plant, and is considering the purchase of a testing device that can determine when a memory chip is faulty. In particular, the specifications on the device are as follows:

$$P(A|B) = P(\overline{A}|\overline{B}) = .98,$$

where $A$ is the event that the testing device *indicates* that a memory chip is faulty, and $B$ is the event that the memory chip really *is* faulty.

a. What is the probability that the testing device indicates that a memory chip is not faulty, given that the memory chip really is faulty?

b. If the testing device is applied to the memory chips produced by AJAX's assembly line I, what is $P(B|A)$, i.e., the probability that a chip really is faulty, given that the testing device indicates the chip is faulty.

c. Suppose AJAX management wants $P(B|A)$ to be .95. What is the value of $r = P(A|B) = P(\overline{A}|\overline{B})$ that will ensure this testing accuracy if the test is applied to the chips produced on assembly line I.

**26.** Let $S = [0, 5]$ be a sample space containing all possible values of the daily quantity demanded of electric power for a large midwestern city in the summer months. The units of measurement are millions of megawatts, and the capacity of the power grid is five million megawatts. Answer the following questions concerning probability assignments to events in the sample space, $S$, relating to the daily demand for electric power. Treat the information provided in the questions as cumulative. Justify your answers.

a. Given that $A = [0, 4]$, $B = [3, 5]$, $P(A) = .512$ and $P(B) = .784$, what is the probability that the power demand will be no greater than 4 million megawatts and no less than three million megawatts, i.e., what is the probability of $A \cap B$?

b. What is the probability of the event $C = [0, 3)$?

c. Can $P(D) = .6$ given that $D = [0, 2.5]$?

d. Given that $P([0, 2]) = .064$, what is the probability of the event $E = (2,4]$?

**27.** SUPERCOMP, a retail computer store, sells personal computers and printers. The number of computers and printers sold on any given day varies, with the probabilities of the various possible sales outcomes being given by the following table:

**Number of computers sold**

| | | 0 | 1 | 2 | 3 | 4 | |
|---|---|---|---|---|---|---|---|
| Number of printers sold | 0 | .03 | .03 | .02 | .02 | .01 | Probabilities of elementary events |
| | 1 | .02 | .05 | .06 | .02 | .01 | |
| | 2 | .01 | .02 | .10 | .05 | .05 | |
| | 3 | .01 | .01 | .05 | .10 | .10 | |
| | 4 | .01 | .01 | .01 | .05 | .15 | |

a. Define an appropriate sample space for the experiment of observing how many computers and printers are sold on any given day.

b. Can the information provided in the table be used to define a probability set function for assigning probabilities to all events in the sample space? Explain (briefly, but clearly).

c. What is the probability that more than two computers will be sold on any given day? What is the probability that more than two printers will be sold on any given day? In each case, define the set of outcomes in $S$ that corresponds to the stated events.

d. What is the probability of selling more than two printers, *given* that more than two computers are sold? Show your calculation.

e. What is the probability of selling more than two printers *and* more than two computers? Show your calculation.

f. What is the probability that SUPERCOMP has no sales on a given day? *Given* that SUPERCOMP sells no computers, what is the probability that it sells no printers on a given day?

**28.** Consider the experiment of tossing a fair coin (meaning heads and tails are equally likely on each toss) three times and observing the sequence of heads and tails that results. Let $H$ denote heads and $T$ denote tails.

a. Define the sample space, $S$, for this experiment.

b. Let $A$ be the event that *at least one* of the tosses results in heads. Define the appropriate subset of $S$ that defines $A$. Find $P(A)$.

c. Let B be the event that *at least two* of the tosses results in tails. Define the appropriate subset of $S$ that defines $B$. Find $P(B)$.

d. Define the probability that at least one of the tosses results in heads *and* at least two of the tosses results in tails, $P(A \cap B)$.

e. Let $C$ be the event that all three tosses result in tails. Are A and $C$ disjoint events? Are $B$ and $C$ disjoint events?

f. What is the probability of $A$ *or* $C$ occurring? What is the probability of $B$ *or* $C$ occurring? What is the probability of $A$ *or* $B$ occurring?

**29.** BuyOnLine is a large internet-based online retailer that maintains four different teams of sales representatives. The ages of unpaid invoices from each of the four sales teams is summarized in the table below.

a. If an invoice is selected randomly from the pooled set of invoices, what is the probability that it is from sales team C?

b. What is the probability that a randomly selected invoice from the pooled set of invoices is over 180 days old?

c. If an invoice is selected randomly from the pooled set of invoices, what is the probability that is over 180 days old and from sales team C?

d. If an invoice is selected randomly from the pooled set of invoices, what sales team has the lowest probability of being associated with the invoice?

| | Sales teams | | | |
|---|---|---|---|---|
| **Age of invoice** | **A** | **B** | **C** | **D** |
| Under 120 days | 34 | 103 | 45 | 97 |
| 120–180 days | 27 | 39 | 65 | 47 |
| Over 180 days | 18 | 25 | 19 | 10 |

**30.** The Port Authority of a large East Coast City is investigating the traffic flow in and out of a large train station in the middle of the city. There are six entry gates and six exit gates that travelers can use to enter or leave the train station. An experiment is to be conducted to observe the number of gates open (i.e., a person is traveling through it) at a given point in time. You can assume that each point in the sample space is *equally likely* to occur, where an outcome is characterized by the two-tuple $(x, y)$ denoting that $x$ entry gates are open and $y$ exit gates are open.

a. Define an appropriate sample space, $S$, for this experiment.

b. Define the appropriate probability set function for assigning probability to any event $A \subset S$.

c. What is the probability that one entry and one exit gate will be open?

d. What is the probability that at least half the gates will be open *in each direction*?

e. What is the probability that the same number of gates will be open in each direction?

f. What is the probability that the *total number* of gates open will be less than four?

**31.** Let $A_i$, $i = 1, \ldots, 4$, represent four events in a sample space, $S$. For each of the situations below, determine which assignment of probabilities are actually possible (i.e., do not contradict Kolmogorov's axioms), and which are not. Justify your answers.

a. $P(A_1) = .3,\ P(A_2) = .3,\ P(A_3) = .2,\ P(A_4) = .2$

b. $P(A_1) = .3,\ P(A_j) \geq P(A_1)$ for $j = 2, 3, 4$

c. $P(A_1) = .7,\ P(A_2) = .6,\ P(A_1 \cap A_2) = .1$

d. $P(A_1) = .7,\ P(A_2) = .6,\ P(A_1 \cup A_2) = .1$

e. $P(A_1) = .3, P(A_2) = .4, P(A_3) = .1, P(A_4) = .2$ where $A_i \subset A_j, i<j$

f. $P(A_1) = .4,\ P(A_2) = .3,\ P(A_1 \cup A_2) = .5$ where $A_1 \cap A_2 = \emptyset$

**32.** In each case below, determine whether the set function, *P*, is a *probability* set function.

a. $P(A) = \frac{1}{91} \sum_{x \in A} x^2$ for $A \subset S,\ S = \{1, 2, 3, 4, 5, 6\}$

b. $P(A) = \int_{x \in A} .25e^{-.25x} dx$, where $A$ is any Borel subset of $S = [0, \infty)$

c. $P(A) = \sum_{x \in A} .3^x .7^{1-x}$ for $A \subset S,\ S = \{0, 1\}$

d. $P(A) = \int_{x \in A} 4x^3 dx$ where $A$ is any Borel subset of $S = [0, 1]$

**33.** BuyOnLine is a large internet-based online retailer that maintains four different teams of sales representatives. The ages of unpaid invoices from each of the four sales teams is summarized in the table below.

a. If an invoice is selected randomly from the pooled set of invoices, what is the probability that it is from sales team C, *given* that it is from either team C or D?

b. What is the probability that two randomly selected invoices from the pooled set of invoices, selected sequentially without replacement, are both over 180 days old?

c. If an invoice is selected randomly from the pooled set of invoices, what is the probability that it is from sales team C, *given* that is it under 120 days old?

d. If an invoice is selected randomly from the pooled set of invoices, what is the probability that it is from sales team A or B, *given* that it is less than or equal to 180 days old?

| | Sales teams | | | |
|---|---|---|---|---|
| **Age of invoice** | **A** | **B** | **C** | **D** |
| Under 120 days | 34 | 103 | 45 | 97 |
| 120–180 days | 27 | 39 | 65 | 47 |
| Over 180 days | 18 | 25 | 19 | 10 |

**34.** If $P(A) = .3$, $P(B) = .4$, $P(A|B) = .3$, what is the value of

a. $P(A \cap B)$

b. $P(A \cup B)$

c. $P(\bar{A}|B)$

d. $P(A|\bar{B})$

e. $P(\bar{A}|\bar{B})$

f. $P(B|A)$

g. $P(\bar{A} \cap \bar{B})$

h. $P(\bar{A} \cup \bar{B})$

**35.** A regional airline implements a standard sales practice of "overbooking" their flights, whereby they sell more tickets for a flight then there are seats available for passengers. Their rationale for this practice is that they want to fill all of the seats on their planes for maximum profitability, and there is a positive probability that a customer who has been sold a ticket will not use their ticket on the day of the flight, so that even if there are more tickets sold than seats available, there may be sufficient seats available to accommodate the customers who actually use their tickets and take a flight on any given day. Assuming that the event that a customer actually uses their ticket is .995, the airline's planes have 100 seats, and the events that customers use their tickets on the day of the flight are jointly independent, answer the following questions relating to their overbooking practice.

a. If the airline does not overbook, and only sells 100 tickets for each of their flights, what is the probability that a given flight will fly full ?

b. Using the sales strategy in (a), what is the probability that one or more seats for a given flight will be empty?

c. For the sales strategy in (a), if the airline has 10 flights per day from the Seattle-Tacoma airport, what is the probability that all of the flights will fly full?

d. For the sales strategy in (a), what is the probability that there will be one or more empty seats among the 1,000 seats available on the airline's 10 flights from the Seattle-Tacoma airport on a given day?

**36.** An automobile manufacturer will accept a shipment of tires only if an inspection of 5 percent of the tires, randomly chosen from the shipment, does not contain any defective tires. The manufacturer receives a shipment of 500 tires, and unknown to the manufacturer, five of the tires are defective.

a. What is the probability that the shipment will be accepted?

b. What percent of the tires would need to be randomly chosen and inspected if the manufacturer wanted to reject such a shipment described above, with .90 probability?

**37.** The table below indicates the probabilities of various outcomes with regard to the size of purchases and method of payment for customers that enter to a large New York electronics store:

a. Is the event of a customer paying cash independent of the event that the customer spends $<$ \$100?

b. Given that the customer pays cash, what is the probability that the customer spends $\leq$ \$500?

c. Given that the customer pays by credit card, what is the probability that the customer spends $\leq$ \$500?

d. What is the probability that the customer pays by credit card, given that the purchase is $\leq$ \$500?

e. Given that the customer spends \$100 or more, what is the probability that the customer will not pay by cash?

| | Method of payment | | |
|---|---|---|---|
| Size of purchase | Cash | Credit card | Layaway plan |
| $<$ \$100 | .20 | .10 | .01 |
| \$100–500 | .15 | .15 | .05 |
| $>$ \$500 | .05 | .20 | .09 |

**38.** A new medical test has been developed by a major pharmaceutical manufacturer for detecting the incidence of a bacterial infection. Of the people who actually have the disease, the test will correctly indicate that the disease is present 95 percent of the time. Among people who do not have the disease, the test incorrectly indicates the disease is present 5 percent of the time. The health department of a major east coast city is contemplating making the test available to a population in which .2 percent of the individuals in the population actually have the disease.

a. For a randomly selected individual from the population, if the test indicates that the person has the disease, what is the probability that the person actually does have the disease?

b. For a randomly selected individual from the population, if the test indicates that the person does not have the disease, what is the probability that the person actually does not have the disease?

**39.** A large food processor operates three processing plants on the west coast. The plants, labeled 1, 2, and 3, differ in size, and produce 20, 35, and 45 percent of the food processor's total output of spinach, respectively. Given past history of USDA inspections for sanitation, the probability of a contaminated box of spinach emanating from each of the three plants can be assumed to be .0001, .0002, and .0005, respectively.
Contamination of the food processor's spinach product with Ecoli was identified in a box of spinach shipped to an east coast grocery store, but the bill of lading has been misplaced on the shipment so that it is not known from which plant the shipment originated from. Let $A_i, i = 1, 2, 3$, denote the events that a box of spinach came from plants 1, 2, and 3, respectively. Let the event $C$ denote

that a shipment is contaminated, and thus $\bar{C}$ denotes that a shipment is contamination free.

a. Which plant is the most probable to have produced this contaminated spinach?

b. Which plant is the least probable to have produced this contaminated spinach?

c. What is the probability that the contaminated spinach was produced at one of the two smaller plants?

**40.** This problem is the famous "**Birthday Problem**" in the statistics literature. The problem is the following: In a room of $n$ people, what is the probability that at least two people share the same birthday? You can ignore leap years, so assume there are 365 different birthday possibilities, and you can also assume that a person being born on any of the 365 days is equally likely.

a. If there are 23 people in the room, what is the probability that at least two people share the same birthday?

b. How many people need to be in the room for there to be a .99 probability that at least two people share the same birthday?

**41.** The Baseball World Series in the U.S. consists of seven games, and the first team to win four games is the winner of the series. Assume that the teams are evenly matched.

a. What is the probability that the team that wins the first game of the series will go on to win the World Series?

b. What is the probability that a team that has lost the first three games of the series will win the World Series?

**42.** The BigVision Electronic Store sells a large 73 inch diagonal big screen TV. The TV comes with a standard 1 year warranty on parts and labor so that if anything malfunctions on the TV in the first year of ownership, the company repairs or replaces the TV for free. The store also sells an "extended warranty," which a customer can purchase that extends warranty coverage on the TV for another 2 years, for a total of three years of coverage. The daily numbers of TVs and extended warranties sold, and their probabilities, are represented by the following probability space $\{S, \Upsilon, P\}$, where $x$ denotes the number of TVs sold and $y$ denotes the number of extended $\Upsilon$warranties sold:

$$S = \{(x, y) : x \text{ and } y \in \{0, 1, 2, 3, 4\}, x \geq y\}, \Upsilon = \{A : A \subset S\},$$

and

$$P(A) = \sum_{(x,y) \in A} f(x, y)$$

where the nonzero values of $f(x, y)$ are defined in the following table:

| | | Number of extended warranties | | | | |
|---|---|---|---|---|---|---|
| | | **0** | **1** | **2** | **3** | **4** |
| | 0 | 0 | 0 | 0 | 0 | 0 |
| | 1 | .02 | .03 | 0 | 0 | 0 |
| Number of TVs | 2 | .04 | .05 | .06 | 0 | 0 |
| | 3 | .06 | .07 | .08 | .09 | 0 |
| | 4 | .08 | .09 | .10 | .11 | .12 |

a. What is the probability that all of the TVs sold on a given day will be sold with extended warranties?

b. Given that $\leq 2$ TVs are sold, what is the probability that all of the TVs will be sold with extended warranties?

c. Let A be the event that no TVs are sold, and B be the event that no extended warranties are sold. Are A and B independent events?

d. What is the probability that $\geq 3$ extended warranties are sold on a given day?

e. What is the probability that more extended warranties are sold than TVs?

f. Find an algebraic representation of the function $f(x,y)$ that can be used to replace the table of values above.

# 2 Random Variables, Densities, and Cumulative Distribution Functions

## 2.1 Introduction

It is natural for the outcomes of many experiments in the real world to be measured in terms of real numbers. For example, measuring the height and weight of individuals, observing the market price and quantity demanded of a commodity, measuring the yield of a new variety of wheat, or measuring the miles per gallon achievable by a new hybrid automobile all result in real-valued outcomes. The sample spaces associated with these types of experiments are subsets of the real line or, if multiple values are needed to characterize the outcome of the experiment, subsets of $n$-dimensional real space, $\mathbb{R}^n$.

There are also experiments whose outcomes are not inherently numbers and whose sample space is not inherently a subset of a real space. For example, observing whether a tossed coin results in heads or tails, observing whether an item selected from an assembly line is defective or nondefective, observing the type of weeds growing in a garden, and observing which engine components caused an engine failure in an automobile are not experiments characterized inherently by real-valued outcomes. It will prove to be both convenient and useful to convert such sample spaces into real-valued sample spaces by associating a real number to each outcome in the original sample space. This process can be viewed as coding the outcomes of an experiment with real numbers.

Furthermore, the outcomes of an experiment may not be of direct interest in a given problem setting; instead, real-valued functions of the outcomes may be of prime importance. For example, in a game of craps, it is not the outcome of each die that is of primary importance, but rather the sum of the dots facing up determines whether a player has won or lost. As another example, if a firm is interested in calculating the profit associated with a given operation, it is the price of the product multiplied by the quantity sold, defining revenue, that will be of primary importance in the profit calculation, and not price and quantity, per se.

All of the previous situations involve the concept of a random variable, which can be used to characterize the ultimate experimental outcomes of interest as real numbers. We now develop the concept of a random variable.

## 2.2 Univariate Random Variables and Density Functions

We begin with the definition of the term **random variable** appropriate for the univariate, or one-variable, case.

**Definition 2.1** ***Univariate Random Variable***

Let $\{S, \Upsilon, P\}$ be a probability space. If $X : S \rightarrow \mathbb{R}$ is a real-valued function having as its domain the elements of $S$, then $X$ is a random variable.

A pictorial illustration of the random variable concept is given in Figure 2.1.

The reader might find it curious, and perhaps even consider it a misnomer, for the term "random variable" to be used as a label for the concept just given. The expression *random-valued function* would seem more appropriate since it is, after all, a real-valued *function* that is at the heart of the concept presented in the definition. Nonetheless, usage of "random variable" has become standard terminology, and we will use it also.

The phrase **outcome of the random variable** refers to the particular image element in the range of the random variable, $R(X)$, that occurs as a result of

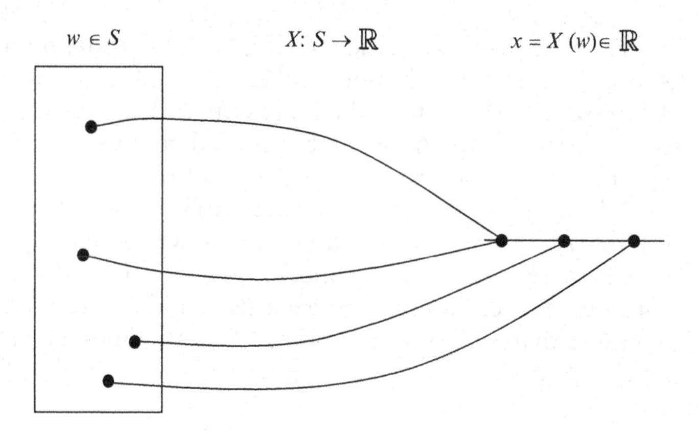


**Figure 2.1**
Random variable $X$.

observing the outcome of a given experiment, i.e., if the outcome of an experiment is $w \in S$, then the outcome of the random variable is $x = X(w)$.

**Definition 2.2**
***Random Variable Outcome***

The image $x = X(w)$ of an outcome $w \in S$ generated by a random variable $X$.

Henceforth, we will use upper case letters, such as $X$, to denote random variables and their lower case counterparts to denote an image value of the random variable, as $x = X(w)$ for $w \in S$. The letter $X$ that we use here is arbitrary, and any other symbol could be used to denote a random variable. For the most part, we will use letters in the latter part of the alphabet for representing random variables. Letters at the beginning of the alphabet will be used to denote constants, and so the expression $x = a$ will mean that the value, $x$, of the random variable, $X$, equals the constant $a$. Similarly, $x \in A$ will mean that the value of $X$ is an element of the set $A$.

If the outcomes of an experiment are real numbers to begin with, they are directly interpretable as outcomes of a random variable, since we can always represent the real-valued outcomes $w \in S$ as images of an identity function, e.g., $X(w) = w$. If the outcomes of an experiment are not initially in the form of real numbers, a random variable can always be defined that associates a real number with each outcome $w \in S$, as $X(w) = x$, and thus as we noted above, a random variable effectively codes the outcomes of a sample space with real numbers. Through the use of the random variable concept, all experiments with univariate outcomes can be ultimately interpreted as having sample spaces consisting of real-valued elements. In particular, the range of the random variable, $R(X) = \{x : x = X(w), w \in S\}$, represents a real-valued sample space for the experiment.

### 2.2.1 Probability Space Induced by a Random Variable

Given a real-valued sample space that has been defined for a given experiment via a random variable, we seek a probability space that can be used for assigning probabilities to events involving random variable outcomes. This requires that we establish how probabilities are to be assigned to subsets of the real-valued sample space $R(X)$. In so doing, we must define an appropriate *probability set function* for assigning probabilities to subsets of $R(X)$ and identify the *event space* or domain of the probability set function.

Given the probability space, $\{S, \Upsilon, P\}$, we are initially equipped to assign probabilities to events in $S$. What is the probability that an outcome of $X$ resides in the set $A \subset R(X)$? Suppose an event in $S$ can be defined, say $B$, that occurs *iff* the event $A$ in $R(X)$ occurs. Then, since the two events occur only simultaneously, they must have the same probability of occurring and we can state that $P_X(A) \equiv P(B)$ when $A \Leftrightarrow B$, where $P_X(\cdot)$ is used to denote the probability set function for assigning probability to events for outcomes of $X$. Two events that occur only simultaneously are called **equivalent events**, where the fundamental implication of the term is that the probabilities of the events are equivalent or

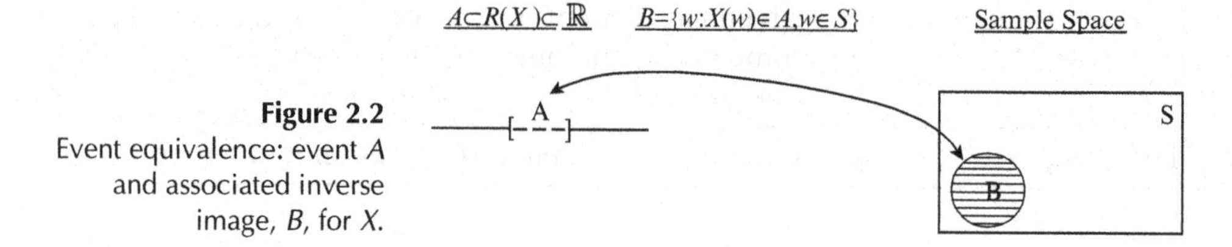


**Figure 2.2** Event equivalence: event $A$ and associated inverse image, $B$, for $X$.

the same. The event $B$ in $S$ that is equivalent to event $A$ in $R(X)$can be defined as $B = \{w{:}X(w) \in A,\ w \in S\}$, which is the set of inverse images of the elements of $A$ defined by the function $X$. By definition, $w \in B \Leftrightarrow x \in A$, and thus $A$ and $B$ are equivalent events (see Figure 2.2). It is clear that in order for two equivalent events to represent different sets of outcomes, they must reside in different probability spaces – if they resided in the same probability space, they could occur only simultaneously *iff* they were the same event.

**Definition 2.3** ***Equivalent Events***

Let $S_1$ and $S_2$ be different sample spaces. If $A \subset S_1$ occurs *iff* $B \subset S_2$ occurs, then $A$ and $B$ are said to be equivalent events.

Based on the preceding discussion, we have the following representation of probability assignments to events involving random variable outcomes:

$P_X(A) \equiv P(B)$ for $B = \{w : X(w) \in A,\ w \in S\}$.

Thus, probabilities assigned to events in $S$ are transferred to events in $R(X)$ through the functional relationship $x = X(w)$, which relates outcomes $w$ in $S$ to outcomes $x$ in $R(X)$. Note, this underscores a fundamental difference between ordinary real-valued functions and random variables, which are also real-valued functions. In particular, random variables are defined on a domain, $S$, that belongs to a probability space, $\{S, \Upsilon, P\}$, and thus the random variable function not only maps domain elements into image elements $x \in R(X)$, but it also maps probabilities of events in $S$ to events in $R(X)$. An ordinary real-valued function does only the former mapping, and its domain does not reside in a probability space, and thus there is no simultaneous probability mapping.

What is the domain of $P_X(\cdot)$, i.e., what is the event space for $X$? It is clear from the foregoing discussion that to be able to assign probabilities to a set $A \subset R(X)$ it must be the case that its associated inverse image in $S$, $B = \{w : X(w) \in A,\ w \in S\}$, can be assigned probability based on the known probability space $\{S, \Upsilon, P\}$. If not, there is no basis for assigning probability to either sets $B$ or $A$ from knowledge of the probability space $\{S, \Upsilon, P\}$. No difficulty will arise if $S$ is a finite or countably infinite sample space, since then the event space $\Upsilon$ equals the collection of *all* subsets of $S$, and *whatever* subset $B \subset S$ is associated with the subset $A \subset R(X)$, $B$ can be assigned probability. Thus, *any* real-valued function defined on a discrete sample space will generate a real-valued sample space for which all subsets can be assigned probability.

**Table 2.1** Relationship Between Original and X-Induced Probability Spaces

| Probability space | Random variable $X$: $S \to \mathbb{R}$ | Induced probability space |
|---|---|---|
| $\{S, \Upsilon, P(\cdot)\}$ | $x = X(w)$ | $\left\{\begin{array}{l} R(X) = \{x : x = X(w), w \in S\} \\ \Upsilon_X = \{A : A \text{ is an event in } R(X)\} \\ P_X(A) = P(B), B = \{w : X(w) \in A, w \in S\}, \forall A \in \Upsilon_X \end{array}\right\}$ |

Henceforth, the event space, $\Upsilon_X$, for outcomes of random variables defined on finite or countably infinite sample spaces is defined to be the set of *all* subsets of $R(X)$.

In order to avoid problems that might occur when $S$ is uncountably infinite, one can simply restrict the types of real-valued functions that are used to define random variables to those for which the problem will not occur. To this effect, a proviso is generally added, either explicitly or implicitly, to the definition of a random variable $X$ requiring the real-valued function defined on $S$ to be such that for every Borel set, $A$, contained in $R(X)$, the set $B = \{w: X(w) \in A, w \in S\}$ is an event in $S$ that can be assigned probability (which is to say, it is *measureable* in terms of probability). Then, since every Borel set $A \subset R(X)$ would be associated with an *event* $B \subset S$, every Borel set could be assigned a probability as $P_X(A) = P(B)$. Since the collection of Borel sets includes all intervals in $R(X)$ (and thus all points in $R(X)$), as well as all other sets that can be formed from the intervals by a countable number of union, intersection, and/or complement operations, the collection of Borel sets defines an event space sufficiently large for all real world applications.

In practice, it requires a great deal of ingenuity to define a random variable for which probability cannot be associated with each of the Borel sets in $R(X)$, and the types of functions that naturally arise when defining random variables in actual applications will generally satisfy the aforementioned proviso. Henceforth, we will assume that the event space, $\Upsilon_X$, for random variable outcomes consists of all Borel sets in $R(X)$ if $R(X)$ is uncountable. We add that for all practical purposes, the reader need not even unduly worry about the latter restriction to Borel sets, since any subset of an uncountable $R(X)$ that is of practical interest will be a Borel set.

In summary, a random variable induces an alternative probability space for the experiment. The **induced probability space** takes the form $\{R(X), \Upsilon_X, P_X\}$ where the range of the random variable $R(X)$ is the real-valued sample space, $\Upsilon_X$ is the event space for random variable outcomes, and $P_X$ is a probability set function defined on the events in $\Upsilon_X$. The relationship between the original and induced probability spaces associated with a random variable is summarized in Table 2.1.

**Example 2.1**
***An Induced Probability Space***

Let $S = \{1, 2, 3,\ldots,10\}$ represent the potential number of cars that a car salesperson sells in a given week, let the event space $\Upsilon$ be the set of all subsets of $S$, and let the probability set function be defined as $P(B) = (1/55)\sum_{w\in B} w$ for $B \in \Upsilon$. Suppose the salesperson's weekly pay consists of a base salary of \$100/week plus a \$100 commission for each car sold. The salesperson's weekly pay can be represented by the random variable $X(w) = 100 + 100w$, for $w \in S$. The induced probability space $\{R(X), \Upsilon_X, P_X\}$ is then characterized by $R(X) = \{200, 300, 400,\ldots,1100\}$, $\Upsilon_X = \{A\colon A \subset R(X)\}$, and $P_X(A) = (1/55)\sum_{w\in B} w$ for $B = \{w\colon (100 + 100w) \in A,\ w \in S\}$ and $A \in \Upsilon_X$. Then, for example, the event that the salesperson makes $\leq$ \$300/week, $A = \{200, 300\}$, has probability $P_X(A) = (1/55)\sum_{w\in\{1,2\}} w = (3/55)$. □

A major advantage in dealing with only real-valued sample spaces is that all of the mathematical tools developed for the real number system are available when analyzing the sample spaces. In practice, once the induced probability space has been identified, the underlying probability space $\{S, \Upsilon, P\}$ is generally ignored for purposes of defining random variable events and their probabilities. In fact, we will most often choose to deal with the induced probability space $\{R(X),\Upsilon_X,P_X\}$ directly at the outset of an experiment, paying little attention to the underlying definition of the *function* having the range $R(X)$ or to the original probability space $\{S, \Upsilon, P\}$. However, we will sometimes need to return to the formal relationship between $\{S, \Upsilon, P\}$ and $\{R(X),\Upsilon_X,P_X\}$ to facilitate the proofs of certain propositions relating to random variable properties.

Note for future reference that a real-valued function of a random variable is, itself, a random variable. This follows by definition, since a real-valued function of a random variable, say $Y$ defined by $y = Y(X(w))$ for $w \in S$, is a function of a function (i.e., a composition of functions) of the elements in a sample space $S$, which is then indirectly also a real-valued function of the elements in the sample space $S$. One might refer to such a random variable as a **composite random variable**.

### 2.2.2 Discrete Random Variables and Probability Density Functions

In practice, it is useful to have a representation of the probability set function, $P_X$ that is in the form of a well-defined algebraic formula and that does not require constant reference either to events in $S$ or to the probability set function defined on the events in $S$. A conceptually straightforward way of representing $P_X$ is available when the real-valued sample space $R(X)$ contains, at most, a countable number of elements. In this case, any subset of $R(X)$ can be represented as the union of the specific elements comprising the subset, i.e., if $A \subset R(X)$ then $A = \cup_{x\in A}\{x\}$. Since the elementary events in $A$ are clearly disjoint, we know from Axiom 1.3 that $P_X(A) = \sum_{x\in A} P_X(\{x\})$. It follows that once we know the probability of every elementary event in $R(X)$, we can assign probability to any other event in $R(X)$ by summing the probabilities of the elementary events contained in the event. This suggests that we define a *point* function $f\colon R(X)\rightarrow\mathbb{R}$ as $f(x) \equiv$ probability of $x = P_X(\{x\})\ \forall x \in R(X)$. Once $f$ is defined, then $P_X$ can be defined for all events as $P_X(A) = \sum_{x \in A} f(x)$. Furthermore, knowledge of $f(x)$eliminates the need for any further reference to the probability space $\{S, \Upsilon, P\}$ for assigning probabilities to events in $R(X)$.

In the following example we illustrate the specification of the point function, $f$.

**Example 2.2**
***Assigning Probabilities with a Point Function***

Examine the experiment of rolling a pair of dice and observing the number of dots facing up on each die. Assume the dice are fair. Letting $i$ and $j$ represent the number of dots facing up on each die, respectively, the sample space for the experiment is $S = \{(i, j) : i \text{ and } j \in \{1, 2, 3, 4, 5, 6\}\}$. Now define the random variable $x = X((i, j)) = i + j$ *for* $(i, j) \in S$. Then the following correspondence can be set up between outcomes of $X$, events in $S$, and the probability of outcomes of $X$ and events in $S$, where $w = (i, j)$:

| | $X(w) = x$ | $B_x = \{w : X(w) = x, w \in S\}$ | $f(x) = P(B_x)$ |
|---|---|---|---|
| | 2 | {(1,1)} | 1/36 |
| | 3 | {(1,2), (2,1)} | 2/36 |
| | 4 | {(1,3), (2,2), (3,1)} | 3/36 |
| | 5 | {(1,4), (2,3), (3,2), (4,1)} | 4/36 |
| | 6 | {(1,5), (2,4), (3,3), (4,2), (5,1)} | 5/36 |
| $R(X)$ | 7 | {(1,6), (2,5), (3,4), (4,3), (5,2), (6,1)} | 6/36 |
| | 8 | {(2,6), (3,5), (4,4), (5,3), (6,2)} | 5/36 |
| | 9 | {(3,6), (4,5), (5,4), (6,3)} | 4/36 |
| | 10 | {(4,6), (5,5), (6,4)} | 3/36 |
| | 11 | {(5,6), (6,5)} | 2/36 |
| | 12 | {(6,6)} | 1/36 |

The range of the random variable is $R(X) = \{2, 3, \ldots, 12\}$, which represents the collection of images of the points $(i, j) \in S$ generated by the function $x = X((i, j)) = i + j$. Probabilities of the various outcomes of $X$ are given by $f(x) = P(B_x)$, where $B_x$ is the collection of inverse images of x.

If we desired the probability of the event that $x \in A = \{7, 11\}$, then $P_X(A) = \sum_{x \in A} f(x) = f(7) + f(11) = 8/36$ (which, incidentally, is the probability of winning a game of craps on the first roll of the dice). If $A = \{2\}$, the singleton set representing "snake eyes," we find that $P_X(A) = \sum_{x \in A} f(x) = f(2) = 1/36$. □

In examining the outcomes of $X$ and their respective probabilities in Example 2.2, it is recognized that a compact algebraic specification can be suggested for $f(x)$, namely[1] $f(x) = (6 - |x - 7|)/36 I_{\{2,3,\ldots,12\}}(x)$. It is generally desirable to express the relationship between the domain and image elements of a function

[1]Notice that the algebraic specification faithfully represents the positive values of $f(x)$ in the preceding table of values, and defines $f(x)$ to equal $0 \; \forall x \notin \{2, 3, \ldots, 12\}$. Thus, the domain of $f$ is the entire real line. The reason for extending the domain of $f$ from $R(X)$ to $\mathbb{R}$ will be discussed shortly. Note that assignments of probabilities to events as $P_X(A) = \sum_{x \in A} f(x)$ are unaffected by this domain extension.

in a compact algebraic formula whenever possible, as opposed to expressing the relationship in tabular form as in Example 2.2. This is especially true if the number of elements in $R(X)$ is large. Of course, if the number of elements in the domain is infinite, the relationship cannot be represented in tabular form and must be expressed algebraically. The reader is asked to define an appropriate point function $f$ for representing probabilities of the elementary events in the sample space $R(X)$ of Example 2.1.

We emphasize that if the outcomes of the random variable $X$ are the outcomes of fundamental interest in a given experimental situation, then given that a probability set function, $P_X(A) = \sum_{x \in A} f(x)$, has been defined on the events in $R(X)$, the original probability space $\{S, \Upsilon, P\}$ is no longer needed for defining probabilities of events in $R(X)$. Note that in Example 2.2, given $f(x)$, the probability set function $P_X(A) = \sum_{x \in A} f(x)$ can be used to define probabilities for all events $A \subset R(X)$ without reference to $\{S, \Upsilon, P\}$.

The next example illustrates a case where an experiment is analyzed exclusively in terms of the probability space relating to random variable outcomes.

**Example 2.3** ***Probability Set Function Definition via Point Function***

The Bippo Lighter Co. manufactures a Piezo gas BBQ grill lighter that has a .90 probability of lighting the grill on any given attempt to use the lighter. The probability that it lights on a given trial is independent of what occurs on any other trial. Define the probability space for the experiment of observing the number of ignition trials required to obtain the first light. What is the probability that the lighter lights the grill in three or fewer trials?

**Answer:** The range of the random variable, or equivalently the real-valued sample space, can be specified as $R(X) = \{1, 2, 3, \ldots\}$. Since $R(X)$ is countable, the event space $\Upsilon_X$ will be defined as the set of all subsets of $R(X)$. The probability that the lighter lights the grill on the first attempt is clearly .90, and so $f(1) = .90$. Using independence of events, the probability it lights for the first time on the second trial is $(.10)\,(.90) = .09$, on the third trial is $(.10)^2\,(.90) = .009$, on the fourth trial is $(.10)^3\,(.90) = .0009$, and so on. In general, the probability that it takes x trials to obtain the first light is $f(x) = (.10)^{x-1}\,.90\, I_{\{1,2,3\ldots\}}(x)$. Then the probability set function is given by $P_X(A) = \sum_{x \in A} (.10)^{x-1} .90 I_{\{1,2,3,\ldots\}}(x)$. The event that the lighter lights the grill in three trials or less is represented by $A = \{1, 2, 3\}$. Then $P_X(A) = \sum_{x=1}^{3} (.10)^{x-1} .90 = .999$. □

The preceding examples illustrate the concept of a **discrete random variable** and a **discrete probability density function**, which we formalize in the following definitions.

**Definition 2.4** ***Discrete Random Variable***

> A random variable is called discrete if its range consists of a countable number of elements.

**Definition 2.5** ***Discrete Probability Density Function***

> The discrete probability density function, $f$, is defined as $f(x) \equiv$ probability of x, $\forall x \in R(X)$, and $f(x) = 0$, $\forall x \notin R(X)$.

Note that in the case of discrete random variables, some authors refer to $f(x)$ as a **probability mass function** as opposed to a **discrete probability density function**. We will continue to use the latter terminology.

It should be noted that even though there is only a countable number of elements in the range of the discrete random variable, $X$, the probability density function (PDF) defined here has the entire (uncountable) real line for its domain. The value of $f$ at a point $x$ in the range of the random variable is the probability of $x$, while the value of $f$ is zero at all other points on the real line. This definition is adopted for the sake of mathematical convenience – it standardizes the domain of all discrete density functions to be the real line while having no effect on the assignment of event probabilities made via the set function $P_X(A) = \sum_{x \in A} f(x)$. This convention will provides a considerable simplification in the definition of marginal and conditional density functions which we will examine ahead.

In our previous examples, the probability space for the experiment was a priori deducible under the stated assumptions of the problems. It is most often the case in practice that the probability space is not a priori deducible, and an important problem in statistical inference is the identification of the appropriate density function, $f(x)$, to use in defining the probability set function component of the probability space.

### 2.2.3 Continuous Random Variables and Probability Density Functions

So far, our discussion concerning the representation of $P_X$ in terms of the point function, $f(x)$, is applicable only to those random variables that have a countable number of possible outcomes. Can $P_X$ be similarly represented when the range of $X$ is uncountably infinite? Given that we can have an event $A$ defined as an uncountable subset of $R(X)$, it is clear that the summation operation over the elements of the set, (i.e., $\sum_{x \in A}$ ) is not generally defined. Thus, defining a probability set function on the events in $R(X)$ as $P(A) = \sum_{x \in A} f(x)$ will not be possible. However, integration over uncountable sets is possible, suggesting that the probability set function might be defined as $P(A) = \int_{x \in A} f(x)dx$ when $R(X)$ is uncountably infinite. In this case the point function $f(x)$ would be defined so that $\int_{x \in A} f(x)\,dx$ defines the probability of the event $A$. The following example illustrates the specification of such a point function $f(x)$ when $R(X)$ is uncountably infinite.

**Example 2.4** ***Probabilities by Integrating a Point Function***

Suppose a trucking company has observed that accidents are equally likely to occur on a certain 10-mile stretch of highway, beginning at point 0 and ending at point 10. Let $R(X) = [0, 10]$ define the real-valued sample space of potential accident points.

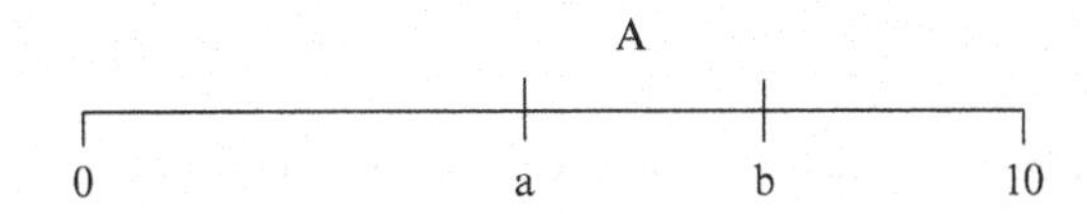

It is clear that given all points are equally likely, the probability set function should assign probabilities to intervals of highway, say $A$, in such a way that the probability of an accident is equal to the proportion of the total highway length represented by the stretch of highway, $A$, as

$$P_X(A) = \frac{\text{length of } A}{10} = \frac{b-a}{10}, \quad \text{for } A = [a, b].$$

If we wish to assign these probabilities using $P_X(A) = \int_{x \in A} f(x)dx$, we require that $\int_a^b f(x)dx \equiv \frac{b-a}{10}$ for all $0 \leq a \leq b \leq 10$. The following lemma will be useful in deriving the explicit functional form of $f(x)$:

**Lemma 2.1**
***Fundamental Theorem of Calculus***

Let $f(x)$ be a continuous function at $b$ and $a$, respectively.[2] Then $\frac{\partial \int_a^b f(x)dx}{\partial b} = f(b)$ and $\frac{\partial \int_a^b f(x)dx}{\partial a} = -f(a)$.

Applying the lemma to the preceding integral identity yields

$$\frac{\partial \int_a^b f(x)dx}{\partial b} = f(b) \equiv \frac{\partial \left(\frac{b-a}{10}\right)}{\partial b} = \frac{1}{10} \quad \forall\ b \in [0, 10],$$

which implies that the function defined by $f(x) = .1\, I_{[0,10]}(x)$ can be used to define the probability set function $P_X(A) = \int_{x \in A} .1\, dx$, for $A \in \Upsilon_X$. For an example of the use of this representation, the probability that an accident occurs in the first half of the stretch of highway, i.e., the probability of the event $A = [0, 5]$, is given by $P_X(A) = \int_0^5 .1 dx = .5$. □

The preceding example illustrates the concept of a **continuous random variable** and a **continuous probability density function,** which we formalize in the next definition.

**Definition 2.6**
***Continuous Random Variables and Continuous Probability Density Functions***

A random variable is called continuous if (1) its range is uncountably infinite, and (2) there exists a nonnegative-valued function $f(x)$, defined for all $x \in (-\infty, \infty)$, such that for any event $A \subset R(X)$, $P_X(A) = \int_{x \in A} f(x)\, dx$, and $f(x) = 0\ \forall\, x \notin R(X)$. The function $f(x)$ is called a continuous probability density function.

Clarification of a number of important characteristics of continuous random variables is warranted. First of all, note that probability in the case of a

[2] See F.S. Woods (1954) *Advanced Calculus*, Boston: Ginn and Co., p. 141. Regarding **continuity** of $f(x)$, note that $f(x)$ is continuous at a point $d \in D(f)$ if, $\forall \varepsilon > 0$, $\exists$ a number $\delta(\varepsilon) > 0$ such that if $|x - d| < \delta(\varepsilon)$, then $f(x) - f(d) < \varepsilon$. The function $f$ is continuous if it is continuous at every point in its domain. Heuristically, a function will be continuous if there are no breaks in the graph of $y = f(x)$. Put another way, if the graph of $y = f(x)$ can be completely drawn without ever lifting a pencil from the graph paper, then $f$ is a continuous function.

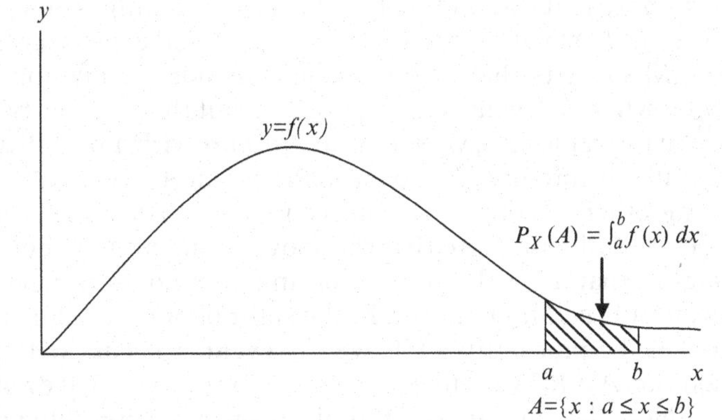


**Figure 2.3** Probability represented as area.

continuous random variable is represented by the area under the graph of the density function $f$ and above the points in the set $A$, as illustrated in Figure 2.3.

Of course, the event in question need not be an interval, but given our convention regarding the event space $\Upsilon_X$, the event will be a Borel set for which an integral can be defined. A justification for the existence of the integral for Borel sets is beyond the scope of this text, but implementation of the integration process in these cases is both natural and straightforward.[3] The next example illustrates the procedure of determining probabilities for events more complicated than a single interval.

**Example 2.5** ***Probabilities for Non-Continuous Events***

Reexamine the highway accident example (Example 2.4) where $R(X) = [0,10]$ and $f(x) = .1\, I_{[0,10]}(x)$.

**a.** What is the probability of $A = [1,2] \cup [7,9]$? The probability of $A$ is given by the area above the points in $A$ and below the graph of $f$, i.e.,

$$P_X(A) = \int_{x\in A} f(x)dx = \int_1^2 \left(\frac{1}{10}\right)dx + \int_7^9 \left(\frac{1}{10}\right)dx = .1 + .2 = .3$$

**b.** Given A defined above, what is the probability of $\bar{A} = [0,1) \cup (2,7) \cup (9,10]$? The area representing the probability in question is calculated as

$$P_X(\bar{A}) = \int_{x\in\bar{A}} f(x)dx = \int_0^1 \left(\frac{1}{10}\right)dx + \int_2^7 \left(\frac{1}{10}\right)dx + \int_9^{10} \left(\frac{1}{10}\right)dx$$
$$= .1 + .5 + .1 = .7.$$

□

A consequence of the definition of the probability set function $P_X$ in Definition 2.6 is that, for a continuous random variable, the probability of any elementary event is zero, i.e., if $A = \{a\}$, then $P_X(A) = \int_a^a f(x)dx = 0$. Note this certainly does

[3] It can be shown that Borel sets are representable as the union of a collection of disjoint intervals, some of which may be single points. The collective area in question can then be defined as the sum of the areas lying above the various intervals and below the graph of $f$.

not imply that every outcome of $X$ in $R(X)$ is impossible, since some elementary event in $R(X)$ will occur as a result of a given experiment. Instead, $P_X(\{x\}) = 0\ \forall x \in R(X)$ suggests that zero probability is not synonymous with impossibility. In cases where an event, say $A$, can occur, but the probability set function assigns the event the value zero, we say **event $A$ occurs with probability zero**. The reader might intuitively interpret this to mean that event $A$ is relatively impossible, i.e., relative to the other outcomes that can occur $(R(X) - A)$, the likelihood that $A$ would occur is essentially nil. Note that the above argument together with Theorem 1.1 then suggest that if $P_X(A) = 1$ for a continuous random variable, it does *not* follow that event $A$ is *certain* to occur. In the spirit of our preceding discussion, if event $A$ is assigned a probability of 1, we say **event $A$ occurs with probability 1**, and if in addition $A \neq R(X)$, we might interpret this to mean that event $A$ is relatively certain.

Note that an important implication of the preceding property for continuous random variables, which has already been utilized in Example 2.5b, is that the sets $[a,b]$, $(a,b]$, $[a,b)$, and $(a,b)$ are all assigned the same probability value $\int_a^b f(x)dx$ since adding or removing a finite number of elementary events to another event will be adding or removing a collection of outcomes that occur with probability zero. That is, since $[a,b] = (a,b] \cup \{a\} = [a,b) \cup \{b\} = (a,b) \cup \{a\} \cup \{b\}$, and since $P_X(\{a\}) = P_X(\{b\}) = 0$, Axiom 1.3 implies that $P_X([a,b]) = P_X((a,b]) = P_X([a,b)) = P_X((a,b))$, so that the integral $\int_a^b f(x)dx$ suffices to assign the appropriate probability to all four interval events.

There is a fundamental difference in the interpretation of the image value $f(x)$ depending on whether $f$ is a discrete or continuous PDF. In particular, while $f(x)$ is the probability of the outcome $x$ in the discrete case, $f(x)$ is *not* the probability of $x$ in the continuous case. To motivate this latter point, recognize that if $f(x)$ were the probability of outcome $x$ in the continuous case, then by our argument above, $f(x) = 0\ \ \forall x \in R(X)$ since the probability of elementary events are zero. But this would imply that for every event $A$, including the certain event $R(X)$, $P_X(A) = \int_{x \in A} f(x)dx = \int_{x \in A} 0dx = 0$, since having an integrand of 0 ensures that the integral has a zero value. The preceding property would contradict the interpretation of $P_X$ as a probability set function, and so $f(x)$ is clearly not interpretable as a probability. It is interpretable as a density function value, but nothing more – the continuous PDF must be integrated to define probabilities.

As in the discrete case, a continuous PDF has the entire real line for its domain. Again, this convention is adopted for the sake of mathematical convenience, as it standardizes the domain of all continuous density functions while leaving probabilities of events unaffected. It also simplifies the definition of marginal and conditional probability density functions, which we will examine ahead. We now provide another example of a continuous random variable together with its density function, where the latter, we will assume, has been discovered by your personnel department.

**Example 2.6**
***Probabilities of Lower and Upper Bounded Events***

Examine the experiment of observing the amount of time that passes between employee work-related injuries at a metal fabricating plant. Let $R(X) = \{x : x \geq 0\}$ represent the potential outcomes of the experiment measured in hours, and let the density of the continuous random variable be given by

$$f(x) = \frac{1}{100} e^{-x/100} I_{(0,\infty)}(x).$$

**(a)** What is the probability of the event that 100 or more hours pass between work related injuries? Letting $A = \{x: x \geq 100\}$ represent the event in question,

$$P(A) = \int_{100}^{\infty} \frac{1}{100} e^{-x/100} dx = -e^{-x/100} \Big|_{100}^{\infty} = -e^{-\infty/100} + e^{-1} = e^{-1} = .37.$$

**(b)** What is the probability that an injury occurs within 50 hours of the previous injury? Letting $B = \{x : 0 \leq x \leq 50\}$ represent the event in question,

$$P(B) = \int_{0}^{50} \frac{1}{100} e^{-x/100} dx = -e^{-x/100} \Big|_{0}^{50} = -e^{-50/100} + e^{-0} = 1 - .61 = .39.$$

□

### 2.2.4 Classes of Discrete and Continuous PDFs

In our later study of statistical inference, we will generally identify an appropriate range for a random variable based on the characteristics of a particular experiment being analyzed and have as an objective the identification of an appropriate $f(x)$ with which to complete the specification of the probability space. The fact that for all events $A \subset R(X)$ the values generated by $\sum_{x \in A} f(x)$ or $\int_{x \in A} f(x)dx$ must adhere to the probability axioms places some general restrictions on the types of functions that can be used as density functions, regardless of the specific characteristics of a given experiment. These general restrictions on the admissible choices of $f(x)$ are identified in the following definition.

**Definition 2.7**
***The Classes of Discrete and Continuous Probability Density Functions: Univariate Case***

a. **Class of Discrete Density Functions.** The function $f: \mathbb{R} \rightarrow \mathbb{R}$ is a member of the class of discrete density functions *iff* (1) the set $C = \{x: f(x) > 0, x \in \mathbb{R}\}$ (i.e., the subset of points in $\mathbb{R}$ having a positive image under $f$) is countable, (2) $f(x) = 0$ for $x \in \overline{C}$, and (3) $\sum_{x \in C} f(x) = 1$.

b. **Class of Continuous Density Functions.** The function $f: \mathbb{R} \rightarrow \mathbb{R}$ is a member of the class of continuous density functions *iff* (1) $f(x) \geq 0$ for $x \in (-\infty, \infty)$, and (2) $\int_{-\infty}^{\infty} f(x)dx = 1$.

Note for future reference that the set of outcomes for which the PDF of a random variable assigns *positive* density weightings, i.e., $\{x : f(x) > 0, x \in \mathbb{R}\}$, is called **the support of the random variable.**

**Definition 2.8**
***Support of a Random Variable***

The set $\{x : f(x) > 0, x \in \mathbb{R}\}$ is called the support of the random variable.

Thus, in Definition 2.7, the set $C$ is the support of the discrete random variable $X$ when $f(x)$ is the PDF of $X$. We will henceforth adopt the convention that the range of a random variable is synonymous with its support. This simply implies that any value of $X$ for which $f(x) = 0$(probability zero in the discrete case, and probability density equal to zero in the continuous case) is not part of the range $R(X)$, and is thus not considered a relevant outcome of the random variable.[4] We formalize this equivalence in the following definition.

**Definition 2.9**
***Support and Range Equivalence***

$$R(X) \equiv \{x : f(x) > 0 \ for \ x \in \mathbb{R}\}$$

Some clarifying remarks concerning Definition 2.7 are warranted. First, it should be noted that the definition simply identifies the respective classes of function specifications that are *candidates* for use as PDFs. The *specific* functional form of the density function appropriate for a real-world experimental situation depends on the particular characteristics of the process generating the outcomes of the experiment.

A second observation concerns the fact that the definitions focus exclusively on real-valued functions having the entire real line for their domains. As we discussed earlier, this is a convention adopted as a matter of mathematical convenience. To ensure that subsets of points outside of the range of $X$ are properly assigned zero probability, all one needs to do is to extend the domain of $f$ to the remaining points $\mathbb{R} - R(X)$ on the real line by assigning to each point a zero density weighting, i.e., $f(x) = 0$ if $x \in \overline{R(X)}$.

A final remark concerns the rationale in support of the properties that are required for a function $f$ to be considered a PDF. The properties are imposed on $f$: $\mathbb{R} \rightarrow \mathbb{R}$ to ensure that the set functions constructed from $f$, i.e., $P_X(A) = \sum_{x \in A} f(x)$ or $P_X(A) = \int_{x \in A} f(x)dx$, are in fact *probability* set functions, which of course requires that the probability assignments adhere to the axioms of probability. To motivate the *sufficiency* of these conditions, first examine the discrete case. Since $f(x) \geq 0 \ \forall x$, $P_X(A) = \sum_{x \in A} f(x) \geq 0$ for any event, $A$, and Axiom 1.1 is satisfied. Letting $R(X)$ equal the set C defined in Definition 2.7.a, it follows that $P_X(R(X)) = \sum_{x \in R(X)} f(x) = 1$, satisfying Axiom 1.2. Finally, if $\cup_{i \in I} A_i$ is the union of a collection of disjoint events indexed by the index set $I$, then summing over all of the elementary events in $A = \cup_{i \in I} A_i$ obtains $P_X(\cup_{i \in I} A_i) = \sum_{x \in A} f(x) = \sum_{i \in I} \left( \sum_{x \in A_i} f(x) \right) = \sum_{i \in I} P_X(A_i)$ . Satisfying Axiom 1.3. Thus, the three probability axioms are satisfied, and $P_X$ is a probability set function.

To motivate *sufficiency* in the continuous case, first note that Axiom 1.1 is satisfied since if $f(x) \geq 0 \ \forall x$, then $P_X(A) = \int_{x \in A} f(x)dx \geq 0$ because integrating a

[4]Note that in the discrete case, it is conceptually possible to define a random variable that has an outcome that occurs with zero probability. For example, if $f(y) = I_{[0,1]}(y)$is the density function of the continuous random variable Y, then $X = I_{[0,1)}(Y)$is a discrete random variable that takes the value 1 with probability 1 and the value 0 with probability zero. Such random variables have little use in applications, and for simplicity, we suppress this possibility in making the range of the random variable synonymous with its support.

nonnegative integrand over any interval (or Borel) set, $A$, results in a nonnegative number. Furthermore, since $\int_{-\infty}^{\infty} f(x)dx = 1$, there exists at least one event $A \subset (-\infty,\infty)$ such that $\int_{x\in A} f(x)dx = 1$ (the event can be $(-\infty,\infty)$ itself, or else there may be some other partition of $(-\infty,\infty)$ into $A \cup B$ such that $\int_{x\in A} f(x)dx = 1$ and $\int_{x\in B} f(x)dx = 0$). Letting $R(X) = A$, we have that $P_X(R(X)) = \int_{x\in R(X)} f(x)dx = 1$ and Axiom 1.2 is satisfied. Finally, if $D = \cup_{i\in I} A_i$ is the union of a collection of disjoint events indexed by the index set $I$, then by the additive property of integrals, $P_X(D) = \int_{x\in D} f(x)dx = \sum_{i\in I}\left(\int_{x\in A_i} f(x)dx\right) = \sum_{i\in I} P_X(A_i)$ satisfying Axiom 1.3. Thus, the three probability axioms are satisfied, and $P_X$ is a probability set function.

It can also be shown that the function properties presented in Definition 2.7 are actually *necessary* for the discrete case and *practically necessary* in the continuous case. For the discrete case, first recall that $f(x)$ is directly interpretable as the probability of the outcome $x$, and this requires that $f(x) \geq 0 \; \forall x \in \mathbb{R}$ (or else we would be assigning negative probabilities to some $x$'s). Second, the number of outcomes that can receive positive probability must be countable in the discrete case since $R(X)$ is countable, leading to the requirement that $C = \{x : f(x) > 0,\; x \in R\}$ is countable. Finally, $\sum_{x\in C} f(x) = 1$ is required if the probability assigned to the certain event is to equal one.

In the continuous case, it is necessary that $\int_{-\infty}^{\infty} f(x)dx = 1$. To see this, first note that $R(X)$ is the certain event, implying $P(R(X)) = 1$. Now since $R(X)$ and $\overline{R(X)}$ are disjoint, we have that $P_X(R(X) \cup \overline{R(X)}) = P_X(R(X)) + P_X(\overline{R(X)}) = 1 + P_X(\overline{R(X)})$, which implies $P_X(\overline{R(X)}) = 0$ since probabilities cannot exceed 1. But, since $R(X) \cup \overline{R(X)} = \mathbb{R}$ by definition, then $P(R(X)) \cup (\overline{R(X)}) = \int_{-\infty}^{\infty} f(x)dx = 1$. Regarding the requirement that $f(x) \geq 0$ for $x \in (-\infty,\infty)$, note that the condition is technically *not necessary*. It is known from the properties of integrals that the value of $\int_a^b f(x)dx$ is invariant to changes in the value of $f(x)$ at a finite number of isolated points, and thus $f(x)$ could technically be negative for such a finite number of $x$ values without affecting the values of the probability set function. As others do, we will ignore this technical anomaly since its practical significance in defining PDFs is nil. We thus insist, as a practical matter, on the nonnegativity of $f(x)$.

**Example 2.7** ***Verifying Probability Density Functions***

In each case below, determine whether the stated function can serve as a PDF:

**a.** $f(x) = 1/2 I_{[0,2]}(x)$.

**Answer**: The function can serve as a continuous probability density function since $f(x) \geq 0 \; \forall x \in (-\infty,\infty)$ (note $f(x) = 1/2 > 0 \; \forall x \in [0, 2]$ and $f(x) = 0$ for $x \notin [0, 2]$), and $\int_{-\infty}^{\infty} f(x)dx = \int_{-\infty}^{\infty} (1/2) I_{[0,2]}(x)dx = \int_0^2 1/2 dx = x/2|_0^2 = 1$.

**b.** $f(x) = (.3)^x(.7)^{1-x} I_{\{0,1\}}(x)$.

**Answer**: The function can serve as a discrete probability density function, since $f(x) > 0$ on the countable set $\{0, 1\}$, $\sum_{x=0}^{1} f(x) = 1$, and $f(x) = 0 \; \forall x \notin \{0, 1\}$.

**c.** $f(x) = (x^2 + 1) I_{[-1,1]}(x)$.

**Answer**: The function *cannot* serve as a PDF. While $f(x) \geq 0 \; \forall x \in (-\infty,\infty)$, the function does not integrate to 1:

$$\int_{-\infty}^{\infty} f(x)dx = \int_{-\infty}^{\infty} (x^2+1)\, I_{[-1,1]}(x)dx$$
$$= \int_{-1}^{1} (x^2+1)dx$$
$$= \frac{x^3}{3} + x\,\Big|_{-1}^{1} = \frac{8}{3} \neq 1$$

**d.** $f(x) = (3/8)(x^2+1)I_{[-1,1]}(x)$.
**Answer**: The reader should demonstrate that this function *can* serve as a continuous probability density function. Note its relationship to the function in part *c*. □

### 2.2.5 Mixed Discrete-Continuous Random Variables

The categories of *discrete* and *continuous* random variables do not exhaust the possible types of random variables. There is a category of random variable called *mixed discrete-continuous* which exhibits the characteristics of a discrete random variable for some events and the characteristics of a continuous random variable for other events. In particular, a mixed discrete-continuous random variable is such that a countable subset of the elementary events are assigned positive probabilities, as in the case of a discrete random variable, *except* the sum of the probabilities over the countable set does *not* equal 1. The remaining probability is attributable to an uncountable collection of elementary events, each elementary event being assigned zero probability, as in the case of a continuous random variable. The following example illustrates the concept of a mixed discrete-continuous random variable.

**Example 2.8** ***Operating Life as a Mixed Discrete-Continuous RV***

Let $X$ be a random variable representing the length of time, measured in units of one hundred thousand hours, that a LCD color screen for a laptop computer operates properly until failure. Assume the probability set function associated with the random variable is $P_X(A) = .25\, I_A(0) + .75 \int_{x\in A} e^{-x} I_{(0,\infty)}(x)dx$ for every event $A$ (i.e., Borel set) contained in $R(X) = [0,\infty)$.

**a.** What is the probability that the color screen is defective, i.e., it does not function properly at the outset?
**Answer**: The event in question is $A = \{0\}$. Using $P_X$, we calculate the probability to be $P_X(\{0\}) = .25\, I_{\{0\}}(0) + .75 \int_{x\in\phi} e^{-x}dx = .25$ . (Note: By definition, $\int_{x\in\phi} f(x)dx = 0$).

**b.** What is the probability that the color screen operates satisfactorily for less than 100,000 hours?
**Answer**: Here, $A = [0,1)$. Using $P_X$, we calculate $P_X([0,1)) = .25\, I_{[0,1)}(0) + .75 \int_0^1 e^{-x}\, dx = .25 + .474 = .724$.

**c.** What is the probability that the color screen operates satisfactorily for at least 50,000 hours?

**Answer**: The event in question is $A = [.5,\infty)$. The probability assigned to this event is given by $P_X([.5,\infty)) = .25 \ I_{[.5,\infty)} \ (0) + .75 \quad \int_{.5}^{\infty} e^{-x}dx = 0 + .4549 = .4549.$ □

We formalize the concept of a mixed discrete-continuous random variable in the following definition.

**Definition 2.10** ***Mixed Discrete-Continuous Random Variables***

A random variable is called mixed discrete-continuous *iff*

**a.** Its range is uncountably infinite;

**b.** There exists a countable set, $C$, of outcomes of $X$ such that $P_X(\{x\}) = f_d(x) > 0 \ \forall x \in C$, $f_d(x) = 0 \ \forall x \notin C$, and $\sum_{x \in C} f_d(x) < 1$, where the function $f_d$ is referred to as the **discrete density function component** of the probability set function of $X$;

**c.** There exists a nonnegative-valued function, $f_c$, defined for all $x \in (-\infty,\infty)$ such that for every event $B \subset R(X) - C$, $P_X(B) = \int_{x \in B} f_c(x)dx$, $f(x) = 0 \forall x \in \mathbb{R} - R(X)$, and $\int_{-\infty}^{\infty} f_c(x)dx = 1 - \sum_{x \in C} f_d(x)$, where the function $f_c$ is referred to as the **continuous density function component** of the probability set function of $X$; and

**d.** The probability set function for $X$ is given by combining or *mixing* the discrete and continuous density function components in (b) and (c) above, as $P_X(A) = \sum_{x \in A \cap C} f_d(x) + \int_{x \in A} f_c(x)dx$ for every event $A$.

To see how the definition applies to a specific experimental situation, recall Example 2.8. If we substitute $f_d(x) = .25 \ I_{\{0\}}(x)$, $C = \{0\}$, and $f_c(x) = .75e^{-x} I_{(0,\infty)}(x)$ into the definition of $P_X$ given in Definition 2.10.d, we obtain

$$P_X(A) = \sum_{x \in A \cap \{0\}} (.25I_{\{0\}}(x)) + .75 \int_{x \in A} e^{-x} \ I_{(0,\infty)} \ (x)dx = .25I_A(0) + .75 \int_{x \in A} e^{-x} I_{(0,\infty)}(x)dx,$$

which is identical to the probability set function defined in Example 2.8.

As the reader may have concluded from examining the definition, the concept of a mixed discrete-continuous random variable is more complicated than either the discrete or continuous random variable case, since there is no single PDF that can either be summed or integrated to define probabilities of events.[5] On the other hand, once the discrete and continuous random variable concepts are understood, the notion of a mixed discrete-continuous random variable is a rather straightforward conceptual extension. Note that the definition of the probability set function in Definition 2.10.d essentially implies that the probability of an event $A$ is equivalent to adding together the probabilities of the

[5] In a more advanced treatment of the subject, we could resort to more general integration methods, in which case a single *integral* could once again be used to define $P_X$. On Stieltjes integration, see R.G. Bartle (1976) *The Elements of Real Analysis*, 2nd ed., New York: John Wiley, and Section 3.2 of the next chapter.

discrete event $A \cap C$ and the continuous event $A$. Assigning probability to the event $A \cap C$ is done in a way that emulates the discrete random variable case – a real-valued function (the discrete density component) is summed over the points in the event $A \cap C$. The probability of the event $A$ is calculated in a way that emulates the continuous random variable case – a real-valued function (the continuous density component) is integrated over the points in the event $A$. Adding together the results obtained for the discrete event $A \cap C$ and the continuous event $A$ defines the probability of the "mixed" event $A$. Note that the overlap of discrete points $A \cap C$ in the event A is immaterial when the probability of A is assigned via the continuous PDF component since $\int_{x \in A-(A \cap C)} f_c(x)dx = \int_{x \in A} f_c(x)dx$, i.e., the integral over the countable points in $A \cap C$ will be zero.[6]

## 2.3 Univariate Cumulative Distribution Functions

Situations arise in practice that require finding the probability that the outcome of a random variable is less than or equal to some real number, i.e., the event in question is $\{x\colon x \leq b,\ x \in R(X)\}$ for some real number $b$. These types of probabilities are provided by the **cumulative distribution function** (CDF), which we introduce in this section.

Henceforth, we will eliminate the random variable subscript used heretofore in our probability set function notation; we will now write $P(A)$ rather than $P_X(A)$ whenever the context makes it clear to which probability space the event $A$ refers. Thus, the notation $P(A)$ will be used to represent the probability of either an event $A \subset S$ or an event $A \subset R(X)$. To economize on notation further, we introduce an **abbreviated set definition** for representing events.

**Definition 2.11**
***Abbreviated Set Definition for Events***

For an event $\{x\colon$ set defining conditions, $x \in R(X)\}$ and associated probability represented by $P(\{x\colon$ set defining conditions, $x \in R(X)\})$, the **abbreviated set definition** for the event and associated probability are respectively {set-defining conditions} and $P$(set-defining conditions), the condition $x \in R(X)$ always being tacitly assumed. Alternatively, $S$ may appear in place of $R(X)$.

For an example of an abbreviated set definition that is particularly relevant to our current discussion of CDFs, note that $\{x \leq b\}$ will be used to represent $\{x\colon x \leq b,\ x \in R(X)\}$, and $P(x \leq b)$ will be used to represent $P(\{x\colon x \leq b,\ x \in R(X)\})$.[7]

[6]There are still other types of random variables besides those we have examined, but they are rarely utilized in applied work. See T.S. Chow and H. Teicher (1978) *Probability Theory*, New York: Springer-Verlag, pp. 247–248.

[7]Alternative shorthand notation that is often used in the literature is respectively $\{X \leq b\}$ and $P(X \leq b)$. Our notation establishes a distinction between the function $X$ and a value of the function $x$.

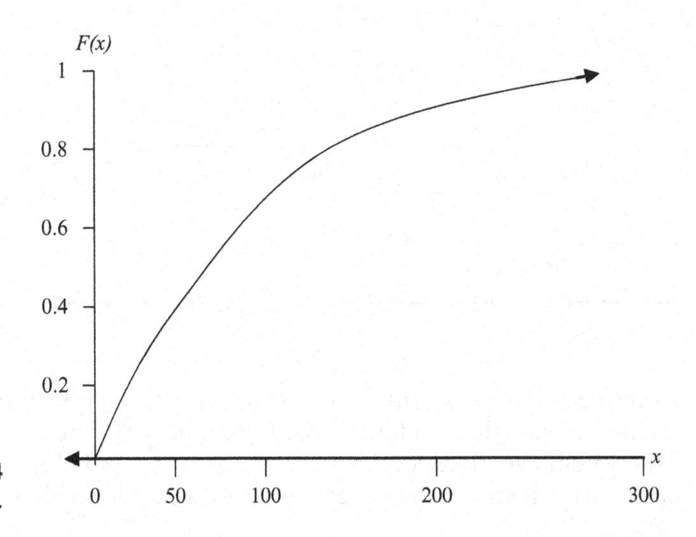


**Figure 2.4**
A CDF for a continuous $X$.

The formal definition of the cumulative distribution function, and its particular algebraic representations in the discrete, continuous, and mixed discrete-continuous cases, are given next.

**Definition 2.12**
***Univariate Cumulative Distribution Function***

The cumulative distribution function of a random variable $X$ is defined by $F(b) \equiv P(x \leq b)\ \forall b \in (-\infty,\infty)$. The functional representation of $F(b)$ in particular cases is as follows:

**a.** *Discrete*: $F(b) = \sum_{x \leq b, f(x)>0} f(x),\ b \in (-\infty,\infty)$

**b.** *Continuous*: $F(b) = \int_{-\infty}^{b} f(x)dx,\ b \in (-\infty,\infty)$

**c.** *Mixed discrete-continuous*: $F(b) = \sum_{x \leq b,\, f_d(x)>0} f_d(x) + \int_{-\infty}^{b} f_c(x)dx$, $b \in (-\infty,\infty)$.

**Example 2.9**
***CDF for Continuous RV***

Reexamine Example 2.6, where the amount of time that passes between work-related injuries is observed. We can define the cumulative distribution function for $X$ as

$$F(b) = \int_{-\infty}^{b} \frac{1}{100} e^{-x/100}\, I_{(0,\infty)}(x)dx = \left[1 - e^{-b/100}\right] I_{(0,\infty)}(b).$$

If one were interested in the event that an injury occurs within 50 hours of the previous injury, the probability would be given by

$$F(50) = [1 - e^{-50/100}]I_{(0,\infty)}(50) = 1 - .61 = .39.$$

A graph of the cumulative distribution function is given in Figure 2.4. □

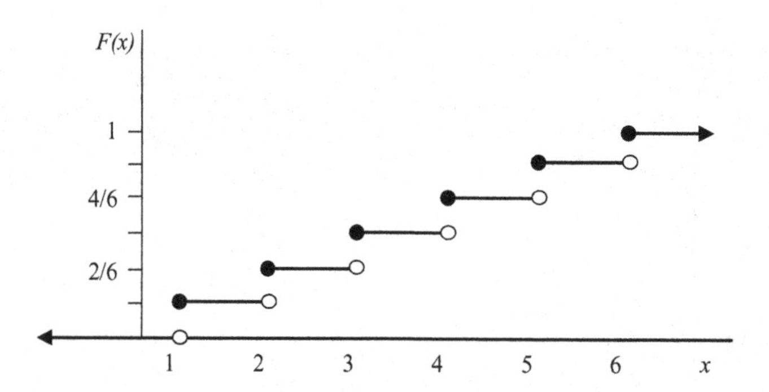


**Figure 2.5**
A CDF for a discrete $X$.

**Example 2.10**
***CDF for Discrete RV***

Examine the experiment of rolling a fair die and observing the number of dots facing up. Let the random variable $X$ represent the possible outcomes of the experiment, so that $R(X) = \{1, 2, 3, 4, 5, 6\}$ and $f(x) = 1/6\ I_{\{1,2,3,4,5,6\}}(x)$. The cumulative distribution function for $X$ can be defined as

$$F(b) = \sum_{x \leq b,\, f(x)>0} \frac{1}{6} I_{\{1,2,3,4,5,6\}}(x) = \frac{1}{6}\text{trunc}(b) I_{[0,6]}(b) + I_{(6,\infty)}(b),$$

where trunc($b$) is the **truncation function** defined by assigning to any domain element $b$ the number that results after truncating the decimal part of $b$. For example, trunc(5.97) = 5, or trunc(−2.12) = −2. If we were interested in the probability of tossing a 3 or less, the probability would be given by

$$F(3) = \frac{1}{6}\ \text{trunc}(3) I_{[0,6]}(3) + I_{(6,\infty)}(3) = \frac{1}{2} + 0 = \frac{1}{2}.$$

A graph of the cumulative distribution function is given in Figure 2.5. □

**Example 2.11**
***CDF for a Mixed Discrete Continuous RV***

Recall Example 2.8, where color screen lifetimes were represented by a mixed discrete-continuous random variable. The cumulative distribution for $X$ is given by

$$\begin{aligned} F(b) &= .25 I_{[0,\infty)}(b) + .75 \int_{-\infty}^{b} e^{-x} I_{(0,\infty)}(x) dx \\ &= .25 I_{[0,\infty)}(b) + .75 \left[1 - e^{-b}\right] I_{(0,\infty)}(b). \end{aligned}$$

If one were interested in the probability that the color screen functioned for 100,000 hours or less, the probability would be given by

$$\begin{aligned} F(1) &= .25 I_{[0,\infty)}(1) + .75\left[1 - e^{-1}\right] I_{(0,\infty)}(1) \\ &= .25 + .474 = .724. \end{aligned}$$ □

A graph of the cumulative distribution function is given in Figure 2.6.

### 2.3.1 CDF Properties

The graphs in the preceding examples illustrate some general properties of CDFs. First, CDFs have the entire real line for their domain, while their range is contained in the interval [0, 1]. Secondly, the CDF exhibits limits as

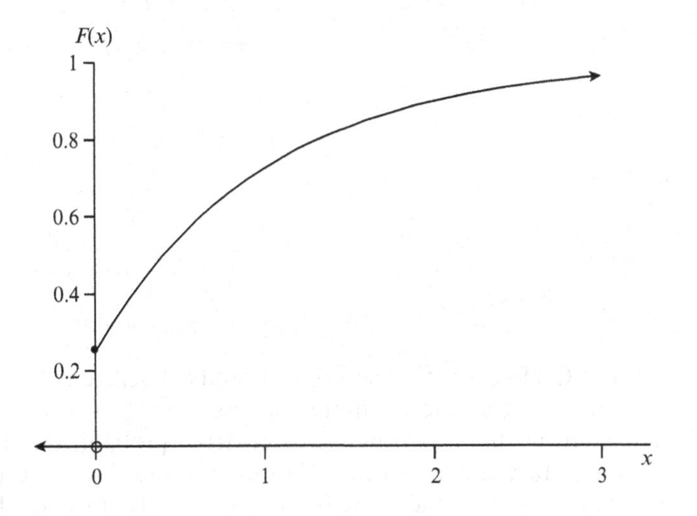


**Figure 2.6**
A CDF for a mixed discrete-continuous $X$.

$$\lim_{b\to-\infty} F(b) = \lim_{b\to-\infty} P(x \le b) = P(\emptyset) = 0$$

and

$$\lim_{b\to\infty} F(b) = \lim_{b\to\infty} P(x \le b) = P(R(X)) = 1.$$

It is also true that if $a < b$, then necessarily $F(a) = P(x \le a) \le P(x \le b) = F(b)$, which is the defining property for $F$ to be an **increasing function**, i.e., if $\forall x_i$ and $x_j$ for which $x_i < x_j$, $F(x_i) \le F(x_j)$, F is an increasing function.[8]

The CDFs of discrete, continuous, and mixed discrete-continuous random variables can be distinguished by their continuity properties and by the behavior of $F(b)$ on sets of domain elements for which $F$ is continuous. The CDF of a continuous random variable must be a continuous function on the entire real line, as illustrated in Figure 2.4, for suppose the contrary that there existed a discontinuous "jumping up" point at a point $d$. Then $P(x = d) = \lim_{b\to d^-} P(b < x \le d) = F(d) - \lim_{b\to d^-} F(b) > 0$ because of the discontinuity (see Figure 2.7), contradicting that $P(x = d) = 0 \; \forall d$ if $X$ is continuous.[9]

[8] For those readers whose recollection of the limit concept from calculus courses is not clear, it suffices here to appeal to intuition and interpret the limit of $F(b)$ as "the real number to which $F(b)$ becomes and remains infinitesimally close to as $b$ increases without bound (or as $b$ decreases without bound)." We will examine the limit concept in more detail in Chapter 5.

[9] $\lim_{b\to d^-}$ indicates that we are examining the limit as $b$ approaches $d$ from below (also called a left-hand limit). $\lim_{b\to d^+}$ would indicate the limit as $b$ approached d from above (also called a right-hand limit). For now, it will suffice for the reader to appeal to intuition and interpret $\lim_{b\to d^-} F(b)$ as "the real number to which $F(b)$ becomes and remains infinitesimally close to as $b$ increases and becomes infinitesimally close to $d$."

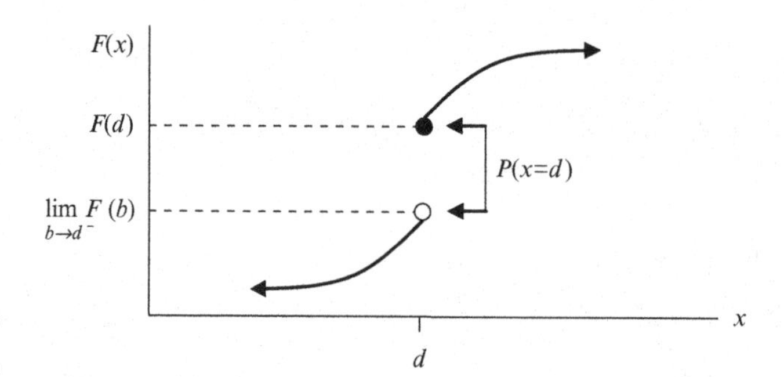


**Figure 2.7**
Discontinuity in a CDF.

The CDFs for both discrete and mixed discrete-continuous random variables exhibit a countable number of discontinuities at "jumping up" points, representing the assignments of positive probabilities to a countable number of elementary events (recall Figures 2.5 and 2.6). The discrete case is distinguished from the mixed case by the property that the CDF in the former case is a constant function on all intervals for which $F$ is continuous. The mixed case will have a CDF that is an increasing function of $x$ on one or more interval subsets of the real line.[10]

### 2.3.2 Duality Between CDFs and PDFs

A CDF can be used to derive a PDF as well as discrete and continuous density components in the mixed discrete-continuous random variable case.

**Theorem 2.1** ***Discrete PDFs from CDFs*** *Let $x_1 < x_2 < x_3 < \ldots$, be the countable collection of outcomes in the range of the discrete random variable X. Then the discrete PDF for X can be defined as*

$$f(x_1) = F(x_1),$$
$$f(x_i) = F(x_i) - F(x_{i-1}), \quad i = 2, 3, \ldots,$$
$$f(x) = 0 \text{ for } x \notin R(X).$$

**Proof** The proof follows directly from the definition of the CDF, and is left to the reader.■

Note, in a large number of empirical applications of discrete random variables, the range of the random variable exhibits an identifiable smallest value, $x_1$, as in Theorem 2.1. In cases where the range of the random variable does not have a finite smallest value, the Theorem can be restated simply as $f(x_i) = F(x_i) - F(x_{i-1})$, for $x_i > x_{i-1}$ and $f(x) = 0$ for $x \notin R(X)$.

[10] A *strictly* increasing function has $F(x_i) < F(x_j)$ when $X_i < X_j$.

**Theorem 2.2** ***Continuous PDFs from CDFs*** *Let f(x) and F(x) represent the PDF and CDF for the continuous random variable X. The density function for X can be defined as f(x) = dF(x)/dx wherever f(x) is continuous, and f(x) = 0 (or any nonnegative number) elsewhere.*

**Proof** By the fundamental theorem of the calculus (recall Lemma 2.1), it follows that

$$\frac{dF(x)}{dx} = \frac{d\int_{-\infty}^{x} f(t)dt}{dx} = f(x)$$

wherever $f(x)$ is continuous, so the first part of the theorem is demonstrated. Now, since $X$ is a continuous random variable, then $P(x \leq b) = F(b) = \int_{-\infty}^{b} f(x)dx$ exists $\forall b$ by definition. Changing the value of the nonnegative integrand at points of discontinuity will have no effect on the value of $F(b) = \int_{-\infty}^{b} f(x)dx$,[11] so that $f(x)$ can be defined arbitrarily at the points of discontinuity. ■

**Theorem 2.3** ***Density Components of a Mixed Discrete-Continuous Random Variable from CDFs*** *Let X be a mixed discrete-continuous random variable with a CDF, F. Let $x_1 < x_2 < x_3 < \ldots$ be the countable collection of outcomes of X for which F(x) is discontinuous. Then the discrete density component of X can be defined as $f_d(x_i) = F(x_i) - \lim_{b \to x_i^-} F(b)$ for $i = 1, 2, 3, \ldots$; and $f_d(x) = 0$* (for any nonnegative numbers) *elsewhere.*

*The continuous density component of X can be defined as $f_c(x) = dF(x)/dx$ wherever f(x) is continuous, and f(x) = 0 (or any nonnegative number) elsewhere.*

**Proof** The proof is a combination of the arguments used in the proofs of the preceding two theorems and is left to the reader. ■

Given Theorems 2.1–2.3, it follows that there is a complete **duality between CDFs and PDFs** whereby either function can be derived from the other. We illustrate Theorems 2.1–2.3 in the following examples.

**Example 2.12** ***Deriving Discrete PDF via Duality*** Recall Example 2.10, where the outcome of rolling a fair die is observed. We can define the discrete density function for $X$ using the CDF for $X$ as follows:

$$f(1) = F(1) = \frac{1}{6}$$

$$f(x) = \begin{cases} F(x) - F(x-1) = \frac{x}{6} - \frac{x-1}{6} = 1/6 \text{ for } x = 2,3,4,5,6, \\ 0 \text{ elsewhere} \end{cases}.$$

[11] This can be rigorously justified by the fact that under the conditions stated: (1) the (improper) Riemann integral is equivalent to a Lebesque integral; (2) the largest set of points for which f(x) can be discontinuous and still have the integral $\int_{-\infty}^{b} f(x)dx$ defined $\forall b$ has "measure zero;" and (3) the values of the integrals are unaffected by changing the values of the integrand on a set of points having "measure zero." This result applies to multivariate integrals as well. See C.W. Burill, 1972, *Measure, Integration, and Probability*, New York: McGraw-Hill, pp. 106–109, for further details.

A more compact representation of $f(x)$ can be given as $f(x) = 1/6\ I_{\{1,2,3,4,5,6\}}(x)$, which we know to be the appropriate discrete density function for the case at hand. □

**Example 2.13** ***Deriving Continuous PDF* via *Duality***

Recall Example 2.9, where the time that passes between work-related injuries is observed. We can define the continuous density function for $X$ using the stated CDF for $X$ as follows:

$$f(x) = \begin{cases} \dfrac{dF(x)}{dx} = \dfrac{d(1 - e^{-x/100})\, I_{(0,\infty)}(x)}{dx} = \dfrac{1}{100}\, e^{-x/100} & \text{for } x \in (0,\infty) \\ 0 & \text{for } x \in (-\infty, 0) \end{cases}$$

The derivative of $F(x)$ does not exist at the point $x = 0$ (recall Figure 2.4), which is a reflection of the fact that $f(x)$ is discontinuous at $x = 0$. We arbitrarily assign $f(x) = 0$ when $x = 0$ so that the density function of $x$ is ultimately defined by $f(x) = 1/100\ e^{-x/100}\ I_{(0,\infty)}(x)$, which we know to be an appropriate continuous density function for the case at hand. □

**Example 2.14** ***Deriving Mixed Discrete-Continuous PDF* via *Duality***

Recall Example 2.11, where the operating lives of notebook color screens are observed. The CDF of the mixed discrete-continuous random variable $X$ is discontinuous only at the point $x = 0$ (recall Figure 2.6). Then the discrete density component of $X$ is given by
$f_d(0) = F(0) - \lim_{b \to 0^-} F(b) = .25 - 0 = .25$ and $f_d(x) = 0$, $x \neq 0$, or alternatively,
$f_d(x) = .25 I_{\{0\}}(x)$,

which we know to be the appropriate discrete density function component in this case.

The continuous density function component can be defined as

$$f_c = \begin{cases} \dfrac{dF(x)}{dx} = .75e^{-x} & \text{for } x \in (0,\infty), \\ 0 & \text{for } x \in (-\infty, 0), \end{cases}$$

but the derivative of $F(x)$ does not exist at the point $x = 0$ (recall Figure 2.6). We arbitrarily assign $f_c(x) = 0$ when $x = 0$, so that the continuous density function component of $X$ is finally representable as $f_c(x) = .75\, e^{-x} I_{(0,\infty)}(x)$, which we know to be an appropriate continuous density function component in this case. □

## 2.4 Multivariate Random Variables, PDFs, and CDFs

In the preceding sections of this chapter, we have examined the concept of a univariate random variable, where only one real-valued function was defined on the elements of a sample space. The concept of a multivariate random variable is an extension of the univariate case, where two or more real-valued functions are concurrently defined on the elements of a given sample space. Underlying the concept of a multivariate random variable is the notion of a **real-valued vector function**, which we define now.

**Definition 2.13**
***Real-Valued Vector Function***

Let $g_i : A \rightarrow \mathbb{R}$, $i = 1,\ldots,n$, be a collection of $n$ real-valued functions, where each function is defined on the domain $A$. Then the function $\mathbf{g} : A \rightarrow \mathbb{R}^n$ defined by

$$\mathbf{y} = \begin{bmatrix} y_1 \\ \cdot \\ \cdot \\ \cdot \\ y_n \end{bmatrix} = \begin{bmatrix} g_1(w) \\ \cdot \\ \cdot \\ \cdot \\ g_n(w) \end{bmatrix} = \mathbf{g}(w), \text{ for } w \in A,$$

is called an **(*n*-dimensional) real-valued vector function**. The real-valued functions $g_1,\ldots,g_n$ are called **coordinate functions** of the vector *function* **g**.

Note that the real-valued vector function $\mathbf{g} : A \rightarrow \mathbb{R}^n$ is distinguished from the scalar function $g : A \rightarrow \mathbb{R}$ by the fact that its range elements are *n-dimensional vectors* of real numbers as opposed to scalar real numbers. The range of the real-valued vector function is given by $R(\mathbf{g}) = \{(y_1,\ldots,y_n) : y_i = g_i(w), i = 1,\ldots,n;\ w \in A\}$. We now provide a formal definition of the notion of a multivariate random variable.

**Definition 2.14**
***Multivariate (n-variate) Random Variable***

Let $\{S, \Upsilon, P\}$ be a probability space. If $\mathbf{X}: S \rightarrow \mathbb{R}^n$ is a real-valued vector function having as its domain the elements of $S$, then $\mathbf{X}$ is called a **multivariate (n-variate) random variable**.

Since the multivariate random variable is defined by

$$\underset{(n\times1)}{\mathbf{x}} = \begin{bmatrix} x_1 \\ x_2 \\ \cdot \\ \cdot \\ \cdot \\ x_n \end{bmatrix} = \begin{bmatrix} X_1(w) \\ X_2(w) \\ \cdot \\ \cdot \\ \cdot \\ X_n(w) \end{bmatrix} = \underset{(n\times1)}{\mathbf{X}(w)} \text{ for } w \in S,$$

it is admissible to interpret $\mathbf{X}$ as a collection of $n$ univariate random variables, each defined on the same probability space $\{S, \Upsilon, P\}$. The range of the $n$-variate random variable is given by $R(\mathbf{X}) = \{(x_1,\ldots,x_n) : x_i = X_i(w), i = 1,\ldots,n;\ w \in S\}$.

The multivariate random variable concept applies to any real world experiment in which more than one characteristic is observed for each outcome of the experiment. For example, upon making an observation concerning a futures trade on the Chicago Mercantile Exchange, one could record the price, quantity, delivery date, and commodity grade associated with the trade. Upon conducting a poll of registered voters, one could record various political preferences and a myriad of sociodemographic data associated with each randomly chosen interviewee. Upon making a sale, a car dealership will record the price, model, year, color, and the selections from the options list that were made by the buyer.

Definitions for the concept of *discrete* and *continuous* multivariate random variables and their associated density functions are as follows:

**Definition 2.15**
***Discrete Multivariate Random Variables and Probability Density Functions***

A multivariate random variable is called *discrete* if its range consists of a countable number of elements. The **discrete joint PDF**, $f$, for a discrete multivariate random variable $\mathbf{X} = (X_1,\ldots,X_n)$ is defined as $f(x_1,\ldots,x_n) \equiv$ {probability of $(x_1,\ldots,x_n)$} if $(x_1,\ldots,x_n) \in R(\mathbf{X})$, $f(x_1,\ldots,x_n) = 0$ otherwise.

**Definition 2.16**
***Continuous Multivariate Random Variables and Probability Density Functions***

A multivariate random variable is called *continuous* if its range is uncountably infinite and there exists a nonnegative-valued function $f(x_1,\ldots,x_n)$, defined for all $(x_1,\ldots,x_n) \in \mathbb{R}^n$, such that $P(A) = \int_{(x_1,\ldots,x_n)\in A} f(x_1,\ldots,x_n)\, dx_1 \ldots dx_n$ for any event $A \subset R(\mathbf{X})$, and $f(x_1,\ldots,x_n) = 0 \;\; \forall\, (x_1,\ldots,x_n) \notin R(\mathbf{X})$. The function $f(x_1,\ldots,x_n)$ is called a **continuous joint PDF**.

### 2.4.1 Multivariate Random Variable Properties and Classes of PDFs

A number of properties of discrete and continuous multivariate random variables, and their joint probability densities, can be identified through analogy with the univariate case. In particular, the multivariate random variable induces a new probability space, $\{R(\mathbf{X}), \Upsilon_{\mathbf{X}}, P_{\mathbf{X}}\}$, for the experiment. The rationale underlying the transition from the probability space $\{S, \Upsilon, P\}$ to the induced probability space $\{R(\mathbf{X}), \Upsilon_{\mathbf{X}}, P_{\mathbf{X}}\}$ is precisely the same as in the univariate case, except for the increased dimensionality of the elements in $R(\mathbf{X})$ in the multivariate case. The probability set function defined on the events in the event space is represented in terms of multiple summation of a PDF in the discrete case, and multiple integration of a PDF in the continuous case. In the discrete case, $f(x_1,\ldots,x_n)$ is directly interpretable as the probability of the outcome $(x_1,\ldots,x_n)$; in the continuous case the probability of each elementary event is zero and $f(x_1,\ldots,x_n)$ is not interpretable as a probability. As a matter of mathematical convenience, both density functions are defined to have the entire $n$-dimensional real space for their domains, so that $f(x_1,\ldots,x_n) = 0 \;\forall\, \mathbf{x} \notin R(\mathbf{X})$.

Regarding the classes of functions that can be used as discrete or continuous joint density functions, we provide the following generalization of Definition 2.5:

**Definition 2.17**
***The Classes of Discrete and Continuous Joint Probability Density Functions***

**a.** Class of discrete joint density functions. A function $f: \mathbb{R}^n \to \mathbb{R}$ is a member of the class of discrete joint density functions *iff*:

1. the set $C = \{(x_1,\ldots,x_n): f(x_1,\ldots,x_n) > 0,\, (x_1,\ldots,x_n) \in \mathbb{R}^n\}$ is countable;
2. $f(x_1,\ldots,x_n) = 0$ for $\mathbf{x} \in \overline{C}$; and
3. $\sum_{(x_1,\ldots,x_n)\in C} f(x_1,\ldots,x_n) = 1$.

**b.** Class of continuous joint density functions. A function $f: \mathbb{R}^n \to \mathbb{R}$ is a member of the class of continuous joint density functions *iff*:

1. $f(x_1,\ldots,x_n) \geq 0 \;\forall (x_1,\ldots,x_n) \in \mathbb{R}^n$; and
2. $\int_{\mathbf{x}\in\mathbb{R}^n} f(x_1,\ldots,x_n) dx_1 \ldots dx_n = 1$.

The reader can generalize the arguments used in the univariate case to demonstrate that the properties stated in Definition 2.17 are sufficient, as well as necessary in the discrete case and "almost necessary" in the continuous case, for set functions defined as

$$P(A) = \begin{cases} \sum_{(x_1,\dots,x_n)\in A} f(x_1,\dots,x_n) & \text{(discrete case)}, \\ \int_{(x_1,\dots,x_n)\in A} f(x_1,\dots,x_n)dx_1\dots dx_n & \text{(continuous case)} \end{cases}$$

to satisfy the probability axioms $\forall A \in \Upsilon_{\mathbf{X}}$.

Similar to the univariate case, we define the support of a multivariate random variable, and the equivalence of the range and support as follows.

**Definition 2.18** ***Support of a Multivariate Random Variable***

The set $\{\mathbf{x} : f(\mathbf{x}) > 0, \mathbf{x} \in \mathbb{R}^n\}$ is called the support of the $n \times 1$ random variable $\mathbf{X}$.

**Definition 2.19** ***Support and Range Equivalence of Multivariate Random Variables***

$R(\mathbf{X}) \equiv \{\mathbf{x} : f(\mathbf{x}) > 0 \text{ for } \mathbf{x} \in \mathbb{R}^n\}$

The following is an example of the specification of the probability space for a bivariate discrete random variable.

**Example 2.15** ***Probability Space for a Bivariate Discrete RV***

For the experiment of rolling a pair of dice in Example 2.2, distinguish the two die by letting the first die be "red" and the second "green." Thus an outcome $(i,j)$ refers to $i$ dots on the red die and $j$ dots on the green die. Define the following two random variables: $x_1 = X_1(w) = i$, and $x_2 = X_2(w) = i + j$

The range of the bivariate random variable $(X_1, X_2)$ is given by $R(\mathbf{X}) = \{(x_1, x_2): x_1 = i, x_2 = i + j, i \text{ and } j \in \{1,\dots,6\}\}$. The event space is $\Upsilon_{\mathbf{X}} = \{A: A \subset R(\mathbf{X})\}$. The correspondence between elementary events in $R(\mathbf{X})$ and elementary events in $S$ is displayed as follows:

| $x_2$ \ $x_1$ | 1 | 2 | 3 | 4 | 5 | 6 | |
|---|---|---|---|---|---|---|---|
| 2 | (1,1) | | | | | | Elementary events in $S$ |
| 3 | (1,2) | (2,1) | | | | | |
| 4 | (1,3) | (2,2) | (3,1) | | | | |
| 5 | (1,4) | (2,3) | (3,2) | (4,1) | | | |
| 6 | (1,5) | (2,4) | (3,3) | (4,2) | (5,1) | | |
| 7 | (1,6) | (2,5) | (3,4) | (4,3) | (5,2) | (6,1) | |
| 8 | | (2,6) | (3,5) | (4,4) | (5,3) | (6,2) | |
| 9 | | | (3,6) | (4,5) | (5,4) | (6,3) | |
| 10 | | | | (4,6) | (5,5) | (6,4) | |
| 11 | | | | | (5,6) | (6,5) | |
| 12 | | | | | | (6,6) | |

It follows immediately from the correspondence with the probability space $\{S, \Upsilon, P\}$ that the discrete density function for the bivariate random variable $(X_1, X_2)$ can be represented as

$$f(x_1,x_2) = \frac{1}{36} I_{\{1,\dots,6\}}(x_1)\, I_{\{1,\dots,6\}}(x_2 - x_1),$$

and the probability set function defined on the events in $R(\mathbf{X})$ is then

$$P(A) = \sum_{(x_1,x_2)\in A} f(x_1,x_2) \text{ for } A \in \Upsilon_{\mathbf{X}}.$$

Let $A = \{(x_1,x_2)\text{: } 1 \leq x_1 \leq 2,\ 2 \leq x_2 \leq 5,\ (x_1, x_2) \in R(\mathbf{X})\}$, which is the event of rolling 2 or less on the red die and a total of 5 or less on the pair of dice. Then the probability of this event is given by

$$P(A) = \sum_{(x_1,x_2)\in A} f(x_1,x_2) = \sum_{x_1=1}^{2} \sum_{x_2=x_1+1}^{5} f(x_1,x_2) = \frac{7}{36}. \qquad \square$$

The preceding example illustrates two general characteristics of the multivariate random variable concept that should be noted. First of all, even though a multivariate random variable can be viewed as a collection of univariate random variables, it is *not* necessarily the case that the range of the multivariate random variable $\mathbf{X}$ equals the Cartesian product of the ranges of the univariate random variable defining $\mathbf{X}$. Depending on the definition of the $\mathbf{X}_i$'s, either $R(\mathbf{X}) \neq \times_{i=1}^{n} R(X_i)$ and $R(\mathbf{X}) \subset \times_{i=1}^{n} R(X_i)$, or $R(\mathbf{X}) = \times_{i=1}^{n} R(X_i)$ is possible. Example 2.15 is an example of the former case, where a number of scalar outcomes that are *individually* possible for the univariate random variables $X_1$ and $X_2$ are not *simultaneously* possible as outcomes for the bivariate random variable $(X_1, X_2)$. Secondly, note that our convention of defining $f(x_1, x_2) = 0\ \forall\ (x_1, x_2) \notin R(\mathbf{X})$ allows an alternative summation expression for defining the probability of event $A$ in Example 2.15:

$$P(A) = \sum_{x_1=1}^{2} \sum_{x_2=2}^{5} f(x_1,x_2) = \frac{7}{36}.$$

We have included the point (2,2) in the summation above, which is an impossible event – we cannot roll a 2 on the red die and a total of 2 on the pair of dice, so that $(2,2) \notin R(\mathbf{X})$. Nonetheless, the probability assigned to $A$ is correct since $f(2,2) = 0$ by definition. In general, when defining the probability of an event A for an $n$-dimensional discrete random variable $\mathbf{X}$, $f(x_1,\dots,x_n)$ can be summed over the points identified in the set-defining conditions for A *without* regard for the condition that $\mathbf{x} \in R(\mathbf{X})$, since any $\mathbf{x} \notin R(\mathbf{X})$ will be such that $f(x_1,\dots,x_n) = 0$, and the value of the summation will be left unaltered. This approach can be especially convenient if set A is defined by individual, independent set-defining conditions applied to each $X_i$ in an $n$-dimensional random variable $(X_1,\dots,X_n)$, as in the preceding example. An analogous argument applies to the continuous case, with integration replacing summation.

We now present an example of the specification of the probability space for a bivariate continuous random variable.

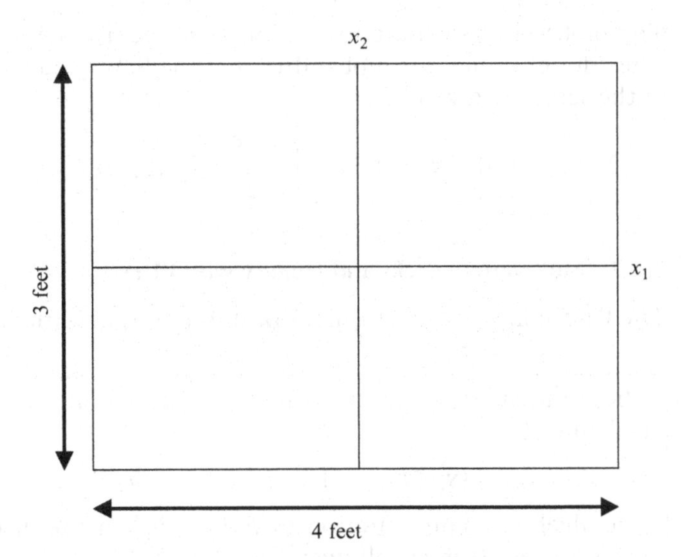


**Figure 2.8**
Television screen.

**Example 2.16**
***Probability Space for a Bivariate Continuous RV***

Your company manufactures big-screen television sets. The screens are 3 ft high by 4 ft wide rectangles that must be coated with a metallic reflective coating (see Figure 2.8). The machine that is coating the screens begins to randomly produce a coating flaw at a point on the screen surface, where all points on the screen are equally likely to be the point of the flaw. Letting (0,0) be the center of the screen, we represent the collection of potential flaw points as $R(\mathbf{X}) = \{(x_1,x_2): x_1 \in [-2,2], x_2 \in [-1.5, 1.5]\}$ □.

Clearly, the total area of the screen is $3 \cdot 4 = 12\ \text{ft}^2$, and any closed rectangle on the screen having width $W$ and height $H$ contains the proportion $WH/12$, of the total area of the screen. Since all of the points are equally likely, the probability set function defined on the events in $R(\mathbf{X})$ should assign to each closed rectangle of points a probability equal to $WH/12$ where $W$ and $H$ are, respectively, the width and height of the rectangular event. We thus seek a function $f(x_1, x_2)$ such that

$$\int_c^d \int_a^b f(x_1, x_2)\, dx_1\, dx_2 \equiv \frac{(b-a)(d-c)}{12}$$

$\forall a, b, c,$ and $d$ such that $-2 \leq a \leq b \leq 2$ and $-1.5 \leq c \leq d \leq 1.5$. Differentiating the iterated integral above, first with respect to $d$ and then with respect to $b$, yields $f(b,d) = 1/12 \ \ \forall b \in [-2,2]$ and $\forall d \in [-1.5, 1.5]$.[12] The form of the continuous PDF is then defined by the following:

$$f(x_1, x_2) = 1/12\ I_{[-2,2]}(x_1) I_{[-1.5,1.5]}(x_2).$$

[12] The differentiation is accomplished by applying Lemma 2.1 twice: once to the integral $\int_c^d \left[\int_a^b f(x_1,x_2)\, dx_1\right] dx_2$, differentiating with respect to $d$ to yield $\int_a^b f(x_1, d)\, dx_1$, and then differentiating the latter integral with respect to b to obtain $f(b,d)$. In summary, $(\partial^2/\partial b \partial d) \int_c^d \int_a^b f(x_1,x_2)\, dx_1 dx_2 = f(b,d)$.

The probability set function is thus defined as $P(A) = \int_{(x_1,x_2 \in A)} (1/12) dx_1 dx_2$. Then, for example, the probability that the flaw occurs in the upper left quarter of the screen is given by

$$P(-2 \leq x_1 \leq 0, 0 \leq x_2 \leq 1.5) = \int_0^{1.5} \int_{-2}^{0} \frac{1}{12} dx_1\, dx_2 = \int_0^{1.5} 1/6\, dx_2 = .25.$$

### 2.4.2 Multivariate CDFs and Duality with PDFs

The CDF concept can be generalized to the multivariate case as follows:

**Definition 2.20** ***Multivariate Cumulative Distribution Function***

The cumulative distribution function of an $n$-dimensional random variable $\mathbf{X}$ is defined by

$F(b_1, ..., b_n) = P(x_i \leq b_i, i = 1, ..., n)\ \forall (b_1, ..., b_n) \in \mathbb{R}^n$.

The algebraic representation of $F(b_1,...,b_n)$ in the discrete and continuous cases can be given as follows:

a. **Discrete** $\boldsymbol{X}$: $F(b_1, \ldots, b_n) = \sum\limits_{x_1 \leq b_1} \cdots \sum\limits_{\substack{x_n \leq b_n \\ f(x_1,\ldots,x_n)>0}} f(x_1, \ldots, x_n)$ for $(b_1, \ldots, b_n) \in \mathbb{R}^n$.

b. **Continuous** $\boldsymbol{X}$: $F(b_1, ..., b_n) = \int_{-\infty}^{b_n} \cdots \int_{-\infty}^{b_1} f(x_1, \ldots, x_n) dx_1, \ldots, dx_n$ for $(b_1,...,b_n) \in \mathbb{R}^n$.

Some general properties of the joint cumulative distribution function include:

1. $\lim_{b_i \to -\infty} F(b_1, \ldots, b_n) = P(\emptyset) = 0, i = 1, \ldots, n$;
2. $\lim_{b_i \to \infty \forall i} F(b_1, \ldots, b_n) = P(R(X)) = 1$;
3. $F(\mathbf{a}) \leq F(\mathbf{b})$ for $\mathbf{a} < \mathbf{b}$, where

$$\mathbf{a} = \begin{bmatrix} a_1 \\ \vdots \\ a_n \end{bmatrix} < \begin{bmatrix} b_1 \\ \vdots \\ b_n \end{bmatrix} = \mathbf{b}$$

The *vector* inequality above is taken in the usual sense to mean $a_i \leq b_i\ \forall i$, and $a_i < b_i$ for at least one $i$. The reader should convince herself that these properties follow directly from the definition of the multivariate cumulative distribution function and the probabilities of the events identified by the appropriate event-defining conditions.

Similar to the univariate case, the joint CDF can be used to derive joint discrete and continuous probability densities. For the discrete case, we discuss the result for bivariate random variables only. For multivariate random variables of three dimensions or higher, the large number of terms required in the density-defining procedure makes its use somewhat cumbersome. We state the generalization in a footnote.[13]

**Theorem 2.4** ***Discrete Bivariate PDFs from Bivariate CDFs*** *Let* $(X, Y)$ *be a discrete bivariate random variable with joint cumulative distribution function* $F(x,y)$, *and let* $x_1 < x_2 < x_3 \ldots$, *and* $y_1 < y_2 < y_3 < \ldots$, represent the possible outcomes *of X and Y. Then*

**a.** $f(x_1, y_1) = F(x_1, y_1)$;
**b.** $f(x_1, y_j) = F(x_1, y_j) - F(x_1, y_{j-1}),\ j \geq 2$;
**c.** $f(x_i, y_1) = F(x_i, y_1) - F(x_{i-1}, y_1),\ i \geq 2$; and
**d.** $f(x_i, y_j) = F(x_i, y_j) - F(x_i, y_{j-1}) - F(x_{i-1}, y_j) + F(x_{i-1}, y_{j-1}),\ i$ and $j \geq 2$.

**Proof** The proof is left to the reader. ■

As we remarked in the univariate case, if the range of the random variable is such that a lowest ordered outcome does not exist, then the definition simplifies to $f(x_i, y_j) = F(x_i, y_j) - F(x_i, y_{j-1}) - F(x_{i-1}, y_j) + F(x_{i-1}, y_{j-1}),\ \forall i$ and $j$.

**Theorem 2.5** ***Continuous Multivariate PDFs from CDFs*** *Let* $F(x_1,\ldots,x_n)$ *and* $f(x_1,\ldots,x_n)$ *represent the CDF and PDF for the continuous multivariate random variable* $\mathbf{X} = (X_1,\ldots,X_n)$. *The PDF of* $\mathbf{X}$ *can be defined as*

$$f(x_1, \ldots, x_n) = \begin{cases} \dfrac{\partial^n F(x_1, \ldots, x_n)}{\partial x_1 \ldots \partial x_n} & \text{where } f(\cdot) \text{is continuous} \\ 0 \text{ (or any nonnegative number) elsewhere.} \end{cases}$$

**Proof** The first part of the definition follows directly from an $n$-fold application of Lemma 2.1 for differentiating the iterated integral defining the joint CDF. In particular,

$$\frac{\partial^n F(x_1, \ldots, x_n)}{\partial x_1 \ldots \partial x_n} = \frac{\partial^n \int_{-\infty}^{x_n} \cdots \int_{-\infty}^{x_1} f(t_1, \ldots, t_n)\, dt_1 \ldots dt_n}{\partial x_1 \ldots \partial x_n} = f(x_1, \ldots, x_n)$$

wherever $f(\cdot)$ is continuous.

Regarding the second part of the definition, as long as the integral exists, arbitrarily changing the values of the nonnegative integrand at the points of discontinuity will not affect the value of $F(b_1, \ldots, b_n) = \int_{-\infty}^{b_n} \cdots \int_{-\infty}^{b_1} f(x_1, \ldots, x_n)$ $dx_1, \ldots, dx_n$ (recall footnote 11). ■

**Example 2.17** ***Piecewise Definition of Discrete Bivariate CDF*** Examine the experiment of tossing two fair coins independently and observing whether heads $(H)$ or tails $(T)$ occurs on each toss, so that $S = \{(H,H), (H,T), (T,H), (T,T)\}$ with all elementary events in $S$ being equally likely. Define a bivariate

[13] In the discrete $m$-dimensional case, the PDF can be defined as $f(\mathbf{x}) = F(\mathbf{x}) + \lim_{\delta \to 0_+} \left( \sum_{i=1}^{m} (-1)^i \sum_{\mathbf{v} \in S_i} F(\mathbf{x} - \delta \mathbf{v}) \right)$ where $S_i$ is the set of all of the different $(m \times 1)$ vectors that can be constructed using $i$ 1's and $m$-$i$ 0's.

random variable on the elements of $S$ by letting $x$ represent the total number of heads and $y$ represent the total number of tails resulting from the two tosses. The joint density function for the bivariate random variable $(X,Y)$ is then defined by $f(x,y) = 1/4\, I_{\{(0,2),\,(2,0)\}}(x,y) + 1/2\, I_{\{(1,1)\}}(x,y)$.

It follows from Definition 2.14 that the joint CDF for $(X,Y)$ can be represented as

$$\begin{aligned} F(b_1,b_2) &= \tfrac{1}{4} I_{[2,\infty)}(b_1)\, I_{(-\infty,1)}(b_2) + \tfrac{1}{4} I_{(-\infty,1)}(b_1)\, I_{[2,\infty)}(b_2) \\ &\quad + \tfrac{1}{2} I_{[1,2)}(b_1)\, I_{[1,2)}(b_2) + \tfrac{3}{4} I_{[2,\infty)}(b_1)\, I_{[1,2)}(b_2) \\ &\quad + \tfrac{3}{4} I_{[1,2)}(b_1)\, I_{[2,\infty)}(b_2) + I_{[2,\infty)}(b_1)\, I_{[2,\infty)}(b_2). \end{aligned}$$

The CDF no doubt appears to be somewhat "pieced together", making the definition of $F$ a rather complicated expression. Unfortunately, such piecewise functional definitions often arise when specifying joint CDFs in the discrete case, even for seemingly simple experiments such as the one at hand. To understand more clearly the underlying rationale for the preceding definition of $F$, it is useful to partition $\mathbb{R}^2$ into subsets that correspond to the events in $S$. In particular, we are interested in defining the collection of elements $w \in S$ for which $X(w) \leq b_1$ and $Y(w) \leq b_2$ is true for the various values of $(b_1,b_2) \in \mathbb{R}^2$. Examine the following table:

| $b_1$ | $b_2$ | $A = \{w: X(w) \leq b_1,\ Y(w) \leq b_2,\ w \in S\}$ | $P(A)$ |
|---|---|---|---|
| $b_1 < 1$ | $b_2 < 1$ | $\emptyset$ | 0 |
| $1 \leq b_1 < 2$ | $b_2 < 1$ | $\emptyset$ | 0 |
| $b_1 < 1$ | $1 \leq b_2 < 2$ | $\emptyset$ | 0 |
| $b_1 \geq 2$ | $b_2 < 1$ | $\{(H,H)\}$ | 1/4 |
| $b_1 < 1$ | $b_2 \geq 2$ | $\{(T,T)\}$ | 1/4 |
| $1 \leq b_1 < 2$ | $1 \leq b_2 < 2$ | $\{(H,T),\ (T,H)\}$ | 1/2 |
| $b_1 \geq 2$ | $1 \leq b_2 < 2$ | $\{(H,T),\ (T,H),\ (H,H)\}$ | 3/4 |
| $1 \leq b_1 < 2$ | $b_2 \geq 2$ | $\{(H,T),\ (T,H),\ (T,T)\}$ | 3/4 |
| $b_1 \geq 2$ | $b_2 \geq 2$ | $S$ | 1 |

The reader should convince herself using a graphical representation of $\mathbb{R}^2$ that the conditions defined on $(b_1, b_2)$ can be used to define nine disjoint subsets of $\mathbb{R}^2$ that exhaustively partition $\mathbb{R}^2$ (i.e., the union of the disjoint sets $= \mathbb{R}^2$). The reader will notice that the indicator functions used in the definition of F were based on the latter six sets of conditions on $(b_1, b_2)$ exhibited in the preceding table. If one were interested in the probability $P(x \leq 1,\ y \leq 1) = F(1,1)$, for example, the joint CDF indicates that 1/2 is the number we seek. □

**Example 2.18** ***Piecewise Definition of Continuous Bivariate CDF***

Reexamine the projection television screen example, Example 2.16. The joint CDF for the bivariate random variable $(X_1, X_2)$, whose outcome represents the location of the flaw point, is given by

$$
\begin{aligned}
F(b_1,b_2) &= \int_{-\infty}^{b_2}\int_{-\infty}^{b_1} \frac{1}{12} I_{[-2,2]}(x_1) I_{[-1.5,1.5]}(x_2)\, dx_1\, dx_2 \\
&= \frac{(b_1+2)(b_2+1.5)}{12} I_{[-2,2]}(b_1) I_{[-1.5,1.5]}(b_2) \\
&\quad + \frac{4(b_2+1.5)}{12} I_{(2,\infty]}(b_1) I_{[-1.5,1.5]}(b_2) \\
&\quad + \frac{3(b_1+2)}{12} I_{[-2,2]}(b_1) I_{(1.5,\infty)}(b_2) \\
&\quad + I_{(2,\infty)}(b_1) I_{(1.5,\infty)}(b_2).
\end{aligned}
$$

It is seen that piecewise functional definitions of joint CDFs occur in the continuous case as well. To understand the rationale for the piecewise definition, first note that if $b_1 < -2$ and/or $b_2 < -1.5$, then we are integrating over a set of $(x_1, x_2)$ points $\{(x_1, x_2): x_1 < b_1, x_2 < b_2\}$ for which the integrand has a zero value, resulting in a zero value for the definite integral. Thus, $F(b_1,b_2) = 0$ if $b_1 < -2$ and/or $b_2 < -1.5$. If $b_1 \in [-2,2]$ and $b_2 \in [-1.5, 1.5]$, then taking the effect of the indicator functions into account, the integral defining $F$ can be represented as

$$
F(b_1,b_2) = \int_{-1.5}^{b_2}\int_{-2}^{b_1} \frac{1}{12}\, dx_1\, dx_2 = \frac{(b_1+2)(b_2+1.5)}{12}
$$

which is represented by the first term in the preceding definition of $F$. If $b_1 > 2$, but $b_2 \in [-1.5, 1.5]$, then since the integrand is zero for all values of $x_1 > 2$, we can represent the integral defining $F$ as

$$
F(b_1,b_2) = \int_{-1.5}^{b_2}\int_{-2}^{2} \frac{1}{12}\, dx_1\, dx_2 = \frac{4(b_2+1.5)}{12}
$$

which is represented by the second term in our definition of $F$. If $b_2 > 1.5$, but $b_1 \in [-2,2]$, then since the integrand is zero for all values of $x_2 > 1.5$, we have that

$$
F(b_1,b_2) = \int_{-1.5}^{1.5}\int_{-2}^{b_1} \frac{1}{12}\, dx_1\, dx_2 = \frac{3(b_1+2)}{12}
$$

which is represented by the third term in our definition of $F$. Finally, if both $b_1 > 2$ and $b_2 > 1.5$, then since the integrand is zero for all values of $x_1 > 2$ and/or $x_2 > 1.5$, the integral defining $F$ can be written as

$$
F(b_1,b_2) = \int_{-1.5}^{1.5}\int_{-2}^{2} \frac{1}{12}\, dx_1\, dx_2 = 1
$$

which justifies the final term in our definition of $F$. The reader should convince herself that the preceding conditions on $(b_1, b_2)$ collectively exhaust the possible values of $(b_1, b_2) \in \mathbb{R}^2$.

If one were interested in the probability $P(x_1 \leq 1, x_2 \leq 1)$, the "relevant piece" in the definition of $F$ would be the first term, and thus $F(1,1) = \frac{(3)(2.5)}{12} = .625$. Alternatively, the probability $P(x_1 \leq 1, x_2 \leq 10)$ would be assigned using the third term in the definition of $F$, yielding $F(1,10) = .75$. □

### 2.4.3 Multivariate Mixed Discrete-Continuous and Composite Random Variables

A discussion of multivariate random variables in the mixed discrete-continuous case could be presented here. However, we choose not to do so. In fact, we will not examine the mixed case any further in this text. We are content with having introduced the mixed case in the univariate context. The problem is that in the multivariate case, representations of the relevant probability set functions – especially when dealing with the concepts of marginal and conditional densities which will be discussed subsequently – become extremely tedious and cumbersome unless one allows a more general notion of integration than that of Riemann, which would then require us to venture beyond the intended scope of this text. We thus leave further study of mixed discrete-continuous random variables to a more advanced course. Note, however, that since elements of both the discrete and continuous random variable concepts are involved in the mixed case, our continued study of the discrete and continuous cases will provide the necessary foundation on which to base further study of the mixed case.

As a final remark concerning our general discussion of multivariate random variables, note that a function (or vector function) of a multivariate random variable is also a random variable (or multivariate random variable). This follows from the same composition of functions argument that was noted in the univariate case. That is, $\mathbf{y} = \mathbf{Y}(\mathbf{X}(w))$, or $\mathbf{y} = \mathbf{Y}(X_1(w), \ldots, X_n(w))$, or

$$\underset{m\times 1}{\mathbf{y}} = \begin{bmatrix} y_1 \\ \vdots \\ y_m \end{bmatrix} = \begin{bmatrix} Y_1\left(X_1\left(w\right), \ldots, X_n\left(w\right)\right) \\ \vdots \\ Y_m\left(X_1\left(w\right), \ldots, X_n\left(w\right)\right) \end{bmatrix} = \underset{m\times 1}{\mathbf{Y}(\mathbf{X}(w))}$$

are all in the context of "functions of functions," so that ultimately $\mathbf{Y}$ is a function of the elements $w \in S$, and is therefore a random variable.[14] One might refer to these as **composite random variables**.

## 2.5 Marginal Probability Density Functions and CDFs

Suppose that we have knowledge of the probability space corresponding to an experiment involving outcomes of the $n$-dimensional random variable $\mathbf{X}_{(n)} = (X_1,\ldots,X_m, X_{m+1},\ldots,X_n)$, but our real interest lies in assigning probabilities to events involving only the $m$-dimensional random variable $\mathbf{X}_{(m)} = (X_1,\ldots,X_m)$, $m < n$. In practical terms, this relates to an experiment in which $n$ different

[14] The reader is reminded that we are suppressing the technical requirement that for every Borel set of $\mathbf{y}$ values, the associated collection of $w$ values in $S$ must constitute an *event* in $S$ for the function $\mathbf{Y}$ to be called a random variable. As we have remarked previously, this technical difficulty does not cause a problem in applied work.

characteristics were recorded for each outcome but we are specifically interested in analyzing only a subset of the characteristics. We will now examine the concept of a **marginal probability density function (MPDF)** for $\mathbf{X}_{(m)}$, which will be derived from knowledge of the joint density function for $\mathbf{X}_{(n)}$. Once defined, the MPDF can be used to identify the appropriate probability space only for the portion of the experiment characterized by the outcomes of $(X_1,\ldots,X_m)$, and we will be able to use the MPDF in the usual way (summation in the discrete case, integration in the continuous case) to assign probabilities to events concerning $(X_1,\ldots,X_m)$.

The key to understanding the definition of a marginal probability density is to establish the equivalence between events of the form $(x_1,\ldots,x_m) \in B$ in the probability space for $(X_1,\ldots,X_m)$ and events of the form $(x_1,\ldots,x_n) \in A$ in the probability space for $(X_1,\ldots,X_n)$ since it is the latter events to which we can assign probabilities knowing $f(x_1,\ldots,x_n)$.

### 2.5.1 Bivariate Case

Let $f(x_1, x_2)$ be the joint density function and $R(\mathbf{X})$ be the range of the bivariate random variable $(X_1, X_2)$. Suppose we want to assign probability to the event $x_1 \in B$. Which event for the *bivariate* random variable is equivalent to event $B$ occurring for the *univariate* random variable $X_1$? By definition, this event is given by $A = \{(x_1, x_2)\colon x_1 \in B, (x_1, x_2) \in R(\mathbf{X})\}$, i.e., the event $B$ occurs for $x_1$ *iff* the outcome of $(X_1, X_2)$ is in $A$ so that $x_1 \in B$. Then since $B$ and $A$ are *equivalent events*, the probability that we will observe $x_1 \in B$ is identically equal to the probability that we will observe $(x_1, x_2) \in A$ (recall the discussion of equivalent events in Section 2.2).

For the discrete case, the foregoing probability correspondence implies that

$$P_{X_1}(B) = P(x_1 \in B) = P(A) = \sum_{(x_1,x_2)\in A} f(x_1, x_2).$$

Our convention of defining $f(x_1, x_2) = 0 \; \forall (x_1, x_2) \notin R(\mathbf{X})$ allows the following alternative representation of $P_{X_1}(B)$:

$$P_{X_1}(B) = \sum_{x_1\in B} \sum_{x_2\in R(X_2)} f(x_1, x_2)$$

The equivalence of the two representations of $P_{X_1}(B)$ follows from the fact that the set of elementary events being summed over the latter case, $C = \{(x_1,x_2)\colon x_1 \in B, x_2 \in R(X_2)\}$ is such that $A \subset C$, and $f(x_1,x_2) = 0 \; \forall (x_1,x_2) \in C-A$. The latter representation of $P_{X_1}(B)$ leads to the following definition of the **marginal probability density** of $X_1$:

$$f_1(x_1) = \sum_{x_2\in R(X_2)} f(x_1, x_2).$$

This function, when summed over the points comprising the event $x_1 \in B$, yields the probability that $x_1 \in B$, i.e.,

$$P_{X_1}(B) = \sum_{x_1 \in B} f_1(x_1) = \sum_{x_1 \in B} \sum_{x_2 \in R(X_2)} f(x_1, x_2).$$

Heuristically, one can think of the marginal density of $X_1$ as having been defined by "summing out" the values of $x_2$ in the bivariate PDF for $(X_1, X_2)$. Having defined $f_1(x_1)$, the probability space for the portion of the experiment involving only $X_1$, can then be defined as $\{R(X_1), \Upsilon_{X_1}, P_{X_1}\}$ where $P_{X_1}(B) = \sum_{X_1 \in B} f_1(x_1)$ for $B \in \Upsilon_{X_1}$. Note that the order in which the random variables are originally listed is immaterial to the approach taken above, and the marginal density function and probability space for $X_2$ could be defined in an analogous manner by simply reversing the roles of $X_1$ and $X_2$ in the preceding arguments. The MPDF for $X_2$ would be defined as

$$f_2(x_2) = \sum_{x_1 \in R(X_1)} f(x_1, x_2),$$

with the probability space for $X_2$ defined accordingly.

**Example 2.19**
***Marginal PDFs in a Discrete Bivariate Case***

Reexamine Example 1.16, in which an individual was to be drawn randomly from the work force of the Excelsior Corporation to receive a monthly "loyalty award." Define the bivariate random variable $(X_1, X_2)$ as

$$x_1 = \begin{Bmatrix} 0 \\ 1 \end{Bmatrix} \text{if} \begin{Bmatrix} \text{male} \\ \text{female} \end{Bmatrix} \text{is drawn,}$$

$$x_2 = \begin{Bmatrix} 0 \\ 1 \\ 2 \end{Bmatrix} \text{if} \begin{Bmatrix} \text{sales} \\ \text{clerical} \\ \text{production} \end{Bmatrix} \text{worker is drawn,}$$

so that the bivariate random variable is measuring two characteristics of the outcome of the experiment: gender and type of worker. The joint density of the bivariate random variable is represented in tabular form below, where the nonzero values of $f(x_1, x_2)$ are given in the cells formed by intersecting an $x_1$-row with a $x_2$-column.

| | | $R(X_2)$ | | | |
|---|---|---|---|---|---|
| | | 0 | 1 | 2 | $f_1(x_1)$ |
| | 0 | .165 | .135 | .150 | .450 |
| $R(X_1)$ | 1 | .335 | .165 | .050 | .550 |
| | $f_2(x_2)$ | .500 | .300 | .200 | |

The nonzero values of the marginal density of $X_2$ are given in the bottom *margin* of the table, being the definition of the marginal density

$$f_2(x_2) = \sum_{x_1 \in R(X_1)} f(x_1, x_2) = \sum_{x_1=0}^{1} f(x_1, x_2) = .5I_{\{0\}}(x_2) + .3I_{\{1\}}(x_2) + .2I_{\{2\}}(x_2).$$

The probability space for $X_2$ is thus $\{R(X_2), \Upsilon_{X_2}, P_{X_2}\}$ with $\Upsilon_{X_2} = \{A: A \subset R(X_2)\}$ and $P_{X_2}(A) = \sum_{x_2 \in A} f_2(x_2)$. If one were interested in the probability that the individual chosen was a sales or clerical worker, i.e., the event $A = \{0,1\}$, then $P_{X_2}(A) = \sum_{x_2=0}^{1} f_{X_2}(x_2) = .5 + .3 = .8$.

The nonzero values of the marginal density for $X_1$ are given in the right-hand *margin* of the table, the definition of the density being

$$f_1(x_1) = \sum_{x_2 \in R(X_2)} f(x_1, x_2) = \sum_{x_2=0}^{2} f(x_1, x_2) = .45I_{(0)}(x_1) + .55I_{(1)}(x_1).$$

The probability space for $X_1$ is thus $\{R(X_1), \Upsilon_{X_1}, P_{X_1}\}$ with $\Upsilon_{X_1} = \{A: A \subset R(X_1)\}$ and $P_{X_1}(A) = \sum_{x_1 \in A} f_1(x_1)$. If one were interested in the probability that the individual chosen was male, i.e., the event $A = \{0\}$, then $P_{X_1}(A) = \sum_{x_1=0}^{0} f_{X_1}(x_1) = .45$.□

The preceding example provides a heuristic justification for the term *marginal* in the bivariate case and reflects the historical basis for the name *marginal density function*. In particular, by summing across the rows or columns of a tabular representation of the joint PDF $f(x_1,x_2)$ one can calculate the marginal densities of $X_1$ and $X_2$ in the *margins* of the table.

We now examine the marginal density function concept for continuous random variables. Recall that the probability of event $B$ occurring for the univariate random variable $X_1$ is **identical** to the probability that the event $A = \{(x_1,x_2): x_1 \in B, (x_1,x_2) \in R(X)\}$ occurs for the bivariate random variable $\mathbf{X} = (X_1,X_2)$. Then

$$P_{X_1}(B) = P(x_1 \in B) = P(A) = \int_{(x_1,x_2) \in A} f(x_1, x_2) dx_1 dx_2.$$

Our convention of defining $f(x_1, x_2) = 0 \; \forall (x_1, x_2) \notin R(\mathbf{X})$ allows an alternative representation of $P_{X_1}(B)$ to be given by

$$P_{X_1}(B) = \int_{x_1 \in B} \int_{-\infty}^{\infty} f(x_1, x_2) dx_2 dx_1.$$

The equivalence of the two representations follows from the fact that the set of elementary events being integrated over in the latter case, $C = \{(x_1, x_2): x_1 \in B, x_2 \in (-\infty,\infty)\}$, is such that $A \subset C$, and $f(x_1, x_2) = 0 \; \forall \; (x_1, x_2) \in C - A$. The latter representation of $P_{X_1}(B)$ leads to the definition of the marginal density of $X_1$ as

$$f_1(x_1) = \int_{-\infty}^{\infty} f(x_1,x_2) dx_2.$$

This function, when integrated over the elementary events comprising the event $x_1 \in B$, yields the probability that $x_1 \in B$, i.e.,

$$P_{X_1}(B) = \int_{x_1 \in B} f_1(x_1)\, dx_1 = \int_{x_1 \in B} \int_{-\infty}^{\infty} f(x_1, x_2) dx_2 dx_1.$$

Heuristically, one might think of the marginal density of $X_1$ as having been defined by "integrating out" the values of $X_2$ in the bivariate density function for $(X_1, X_2)$.

Having defined $f_1(x_1)$, the probability space for the portion of the experiment involving only $X_1$ can then be defined as $\{R(X_1), \Upsilon_{X_1}, P_{X_1}\}$ where $P_{X_1}(A) = \int_{x_1 \in A} f_1(x_1)dx_1$ for $A \in \Upsilon_{X_1}$. Since the order in which the random variables were originally listed is immaterial, the marginal density function and probability space for $X_2$ can be defined in an analogous manner by simply reversing the roles of $X_1$ and $X_2$ in the preceding arguments. The MPDF for $X_2$ would be defined as

$$f_2(x_2) = \int_{-\infty}^{\infty} f(x_1, x_2)\, dx_1$$

with the probability space for $X_2$ defined accordingly.

**Example 2.20** ***Marginal PDFs in a Continuous Bivariate Case***

The Seafresh Fish Processing Company operates two fish processing plants. The proportion of processing capacity at which each of the plants operates on any given day is the outcome of a bivariate random variable having joint density function $f(x_1, x_2) = (x_1 + x_2)\, I_{[0,1]}(x_1)\, I_{[0,1]}(x_2)$. The marginal density function for the proportion of processing capacity at which plant 1 operates can be defined by integrating out $x_2$ from $f(x_1, x_2)$ as

$$f_1(x_1) = \int_{-\infty}^{\infty} f(x_1, x_2)\, dx_2 = \int_{-\infty}^{\infty} (x_1 + x_2)\, I_{[0,1]}(x_1)\, I_{[0,1]}(x_2)\, dx_2$$

$$= \int_0^1 (x_1 + x_2)\, I_{[0,1]}(x_1)\, dx_2 = \left(x_1\, x_2 + \frac{x_2^2}{2}\right) I_{[0,1]}(x_1)\Big|_0^1$$

$$= (x_1 + 1/2)\, I_{[0,1]}(x_1).$$

The probability space for plant 1 outcomes is given by $\{R(X_1), \Upsilon_{X_1}, P_{X_1}\}$, where $R(X_1) = [0,1]$, $\Upsilon_{X_1} = \{A: A \text{ is a Borel set} \subset R(X_1)\}$, and $P_{X_1}(A) = \int_{x_1 \in A} f_1(x_1)dx_1$, $\forall A \in \Upsilon_{X_1}$. If one were interested in the probability that plant 1 will operate at less than half of capacity on a given day, i.e., the event $A = [0, .5)$, then

$$P_{X_1}(x_1 \le .5) = \int_0^{.5} \left(x_1 + \frac{1}{2}\right) I_{[0,1]}(x_1)dx_1 = \frac{x_1^2}{2} + \frac{x_1}{2}\Big|_0^{.5} = .375. \qquad \square$$

Regarding other properties of marginal density functions, note that the significance of the term *marginal* is only to indicate the context in which the density was derived, i.e., the marginal density of $X_1$ is deduced from the joint density for $(X_1, X_2)$. Otherwise, the MPDF has no special properties that differ from the basic properties of any other PDF.

### 2.5.2 *n*-Variate Case

The concept of a discrete MPDF can be straightforwardly generalized to the $n$-variate case, in which case the marginal densities may themselves be joint density functions. For example, if we have the density function $f(x_1, x_2, x_3)$ for the trivariate random variable $(X_1, X_2, X_3)$, then we may conceive of six marginal density functions: $f_1(x_1), f_2(x_2), f_3(x_3)$, $f_{12}(x_1, x_2)$, $f_{13}(x_1, x_3)$, $f_{23}(x_2, x_3)$. In general, for an n-variate random variable, there are $(2^n - 2)$ possible MPDFs that can be defined from knowledge of $f(x_1,\ldots,x_n)$. We present the $n$-variate generalization in the following definition. We use the notation $f_{j_1 \ldots j_m}(x_{j_1}, \ldots, x_{j_m})$ to represent the MPDF of the m-variate random variable $(X_{j_1}, \ldots, X_{j_m})$ with the $j_i$'s being the indices that identify the particular random vector of interest. The motivation for the definition is analogous to the argument in the bivariate case upon identifying the equivalent events $(x_{j_1}, \ldots, x_{j_m}) \in B$ and $A = \{\mathbf{x} : (x_{j_1}, \ldots, x_{j_m}) \in B, \mathbf{x} \in R(\mathbf{X})\}$ is left to the reader as an exercise.

**Definition 2.21** ***Discrete Marginal Probability Density Functions***

Let $f(x_1,\ldots,x_n)$ be the joint discrete PDF for the $n$-dimensional random variable $(X_1,\ldots,X_n)$. Let $J = \{j_1, j_2, \ldots, j_m\}$, $1 \leq m < n$, be a set of indices selected from the index set $I = \{1, 2, \ldots, n\}$. Then the marginal density function for the $m$-dimensional discrete random variable $(X_{j_1},\ldots,X_{j_m})$ is given by

$$f_{j_1 \ldots j_m}(x_{j_1}, \ldots, x_{j_m}) = \sum_{(x_i \in R(X_i),\ i \in I-J)} \cdots \sum f(x_1, \ldots, x_n).$$

In other words, to define a MPDF in the general discrete case, we simply "sum out" the variables that are not of interest in the joint density function. We are left with the marginal density function for the random variable in which we are interested. For example, if $n = 3$, so that $I = \{1,2,3\}$, and if $J = \{j_1, j_2\} = \{1,3\}$ so that $I\text{-}J = \{2\}$, then Definition 2.21 indicates that the MPDF of the random variable $(X_1, X_3)$ is given by

$$f_{13}(x_1, x_3) = \sum_{x_2 \in R(X_2)} f(x_1, x_2, x_3).$$

Similarly, the marginal density for $x_1$ would be defined by

$$f_1(x_1) = \sum_{x_2 \in R(X_2)} \sum_{x_3 \in R(X_3)} f(x_1, x_2, x_3).$$

The concept of a continuous MPDF can be generalized to the $n$-variate case as follows:

**Definition 2.22** ***Continuous Marginal Probability Density Functions***

Let $f(x_1,\ldots,x_n)$ be the joint continuous PDF for the $n$-variate random variable $(X_1,\ldots,X_n)$. Let $J = \{j_1, j_2, \ldots, j_m\}$, $1 \leq m < n$, be a set of indices selected from the index set $I = \{1, 2, \ldots, n\}$. Then the marginal density function for the $m$-variate continuous random variable $(X_{j1}, \ldots, X_{jm})$ is given by

$$f_{j_1 \ldots j_m}(x_{j_1}, \ldots, x_{j_m}) = \int_{-\infty}^{\infty} \cdots \int_{-\infty}^{\infty} f(x_1, \ldots, x_n) \prod_{i \in I-J} dx_i.$$

In other words, to define a MPDF function in the general continuous case, we simply "integrate out" the variables in the joint density function that are not of interest. We are left with the marginal density function for the random variables in which we are interested. An example of marginal densities in the context of a trivariate random variable will be presented in Section 2.8.

### 2.5.3 Marginal Cumulative Distribution Functions (MCDFs)

*Marginal* CDFs are simply CDFs that have been derived for a subset of the random variables in $\mathbf{X} = (X_1,\ldots,X_n)$ from initial knowledge of the joint PDF or joint CDF of $\mathbf{X}$. For example, ordering the elements of a continuous random variable $(X_1,\ldots,X_n)$ so that the first $m < n$ random variables are of primary interest, the MCDF of $(X_1,\ldots,X_m)$ can be defined as

$$\begin{aligned}
F_{1\ldots m}(b_1,\ldots,b_m) &= P_{X_1\ldots X_m}(x_i \le b_i, i=1,\ldots,m)(\text{Def. of } CDF) \\
&= P(x_i \le b_i, i=1,\ldots,m; x_i < \infty, i=m+1,\ldots,n)(\textit{equivalent events}) \\
&= F(b_1,\ldots,b_m,\infty,\ldots,\infty)(\textit{Def. in terms of joint CDF}) \\
&= \int_{-\infty}^{b_1}\cdots\int_{-\infty}^{b_m}\int_{-\infty}^{\infty}\cdots\int_{-\infty}^{\infty} f(x_1,\ldots,x_n)\,dx_n\ldots dx_1\,(\textit{Def. in terms of joint PDF}) \\
&= \int_{-\infty}^{b_1}\cdots\int_{-\infty}^{b_m} f_{1\ldots m}(x_1,\ldots,x_m)\,dx_m\ldots dx_1\,(\textit{Def. in terms of marginal PDF}).
\end{aligned}$$

In the case of an arbitrary subset $(X_{j_1},\ldots,X_{j_m})$, $m < n$ of the random variables $(X_1,\ldots,X_n)$, the MCDF in terms of the joint CDF or marginal PDF can be represented as

$$F_{j_1\ldots j_m}(b_{j_1},\ldots,b_{j_m}) = F(\mathbf{b}) = \int_{-\infty}^{b_{j_1}}\cdots\int_{-\infty}^{b_{j_m}} f_{j_1\ldots j_m}(x_{j_1},\ldots,x_{j_m})\,dx_{j_m}\ldots dx_{j_1}$$

where $b_{j_i}$ is the $j_i$th entry in $\mathbf{b}$ and $b_i = \infty$ if $i \notin \{j_1,\ldots,j_m\}$.

Examples of marginal CDFs in the trivariate case are presented in Section 2.8. The discrete case is analogous, with summation replacing integration.

## 2.6 Conditional Density Functions

Suppose that we have knowledge of the probability space corresponding to an experiment involving outcomes of the $n$-dimensional random variable $\mathbf{X}_{(n)} = (X_1,\ldots,X_m, X_{m+1},\ldots,X_n)$, and we are interested in assigning probabilities to the event $(x_1,\ldots,x_m) \in C$ *given that* $(x_{m+1},\ldots,x_n) \in D$. In practical terms, this relates to an experiment in which $n$ different characteristics were recorded for each outcome and we are specifically interested in analyzing a subset of these characteristics *given that* a fixed set of possibilities will occur with certainty for the remaining characteristics. Note that this is different from asking for the *probability* of observing the event $(x_1,\ldots,x_m) \in C$ *and* $(x_{m+1},\ldots, x_n) \in D$, for we are saying that $(x_{m+1},\ldots,x_n) \in D$ *will happen with certainty.* How do we assign the appropriate probability in this case? Questions of this type can be addressed through the use of **conditional PDFs**, which can be derived from knowledge of the joint density function $f(x_1,\ldots,x_n)$.

The key to the definition of a conditional PDF is to establish the equivalence between events for the $m$-dimensional random variable $(X_1,\ldots,X_m)$ and

$(n-m)$-dimensional random variable $(X_{m+1},\ldots,X_n)$ with events for the $n$-dimensional random variable $(X_1,\ldots,X_n)$. Then conditional probabilities in the probability space for $(X_1,\ldots,X_n)$ can be used to define a conditional PDF.

### 2.6.1 Bivariate Case

Let $f(x_1, x_2)$ be the joint density function and $R(\mathbf{X})$ be the range of the bivariate random variable $(X_1, X_2)$. The event for the bivariate random variable that is equivalent to the event $x_1 \in C$ occurring for the scalar random variable $X_1$ is given by $A = \{(x_1, x_2): x_1 \in C, (x_1, x_2) \in R(X)\}$. That is, $A$ is the set of all possible outcomes for the two-tuple $(x_1, x_2)$ that result in the first coordinate $x_1$ residing in $C$. Similarly, the event for the bivariate random variable that is equivalent to the event $D$ occurring for the random variable $X_2$ is given by $B = \{(x_1, x_2): x_2 \in D, (x_1, x_2) \in R(\mathbf{X})\}$. Then the probability that $x_1 \in C$ *given that* $x_2 \in D$ can be defined by the conditional probability

$$P_{X_1|X_2}(C|D) = P(x_1 \in C | x_2 \in D) = P(A|B) = \frac{P(A \cap B)}{P(B)} \text{ for } P(B) \neq 0,$$

where $A \cap B = \{(x_1, x_2): x_1 \in C, x_2 \in D, (x_1, x_2) \in R(\mathbf{X})\}$.

In the case of a discrete random variable, the foregoing conditional probability is represented by

$$P_{X_1|X_2}(C|D) = P(A|B) = \frac{\sum_{(x_1,x_2)\in A\cap B} f(x_1,x_2)}{\sum_{(x_1,x_2)\in B} f(x_1,x_2)}$$

Given our convention that $f(x_1, x_2) = 0$ whenever $(x_1, x_2) \notin R(\mathbf{X})$, we can ignore the set-defining condition $(x_1, x_2) \in R(\mathbf{X})$ in both the sets $A \cap B$ and $B$, and represent the conditional probability as

$$P_{X_1|X_2}(C|D) = \frac{\sum_{x_1\in C}\sum_{x_2\in D} f(x_1,x_2)}{\sum_{x_1\in R(X_1)}\sum_{x_2\in D} f(x_1,x_2)} = \sum_{x_1\in C}\left[\frac{\sum_{x_2\in D} f(x_1,x_2)}{\sum_{x_2\in D} f_2(x_2)}\right]$$

where we have used the fact that $f_2(x_2) = \sum_{x_1\in R(X_1)} f(x_1,x_2)$. The expression in brackets is the conditional density function we seek, since it is the function that would be summed over the elements in $C$ to assign probability to the event $x_1 \in C$, given $x_2 \in D$, for any event $C$. We will denote the conditional density of $X_1$, given $x_2 \in D$, by the notation $f(x_1|x_2 \in D)$. If $D$ is a singleton set $\{d\}$, we will also represent the conditional density function as $f(x_1|x_2 = d)$.

In the case of a continuous bivariate random variable, the probability that $x_1 \in C$ given that $x_2 \in D$ would be given by (assuming $P_{X_2}(D) = P(B) \neq 0$)

$$P_{X_1|X_2}(C|D) = P(x_1 \in C | x_2 \in D) = P(A|B) = \frac{\int_{(x_1,x_2)\in A\cap B} f(x_1,x_2)\,dx_1\,dx_2}{\int_{(x_1,x_2)\in B} f(x_1,x_2)\,dx_1\,dx_2}$$

Using our convention that $f(x_1,x_2) = 0\,\forall\,(x_1, x_2) \notin R(\mathbf{X})$, we can also represent the conditional probability as

$$P_{X_1|X_2}(C|D) = \frac{\int_{x_1\in C}\int_{x_2\in D} f(x_1,x_2)\,dx_2\,dx_1}{\int_{-\infty}^{\infty}\int_{x_2\in D} f(x_1,x_2)\,dx_2\,dx_1} = \int_{x_1\in C}\left[\frac{\int_{x_2\in D} f(x_1,x_2)\,dx_2}{\int_{x_2\in D} f_2(x_2)\,dx_2}\right]dx_1,$$

where we have used the fact that $f_2(x_2) = \int_{-\infty}^{\infty} f(x_1,x_2)dx_1$. The expression in brackets is the conditional density function we seek, since it is the function that would be integrated over the elements in $C$ to assign probability to the event $x_1 \in C$, given $x_2 \in D$, for any event $C$. As in the discrete case, we will use the notation $f(x_1|x_2 \in D)$ or $f(x_1|x_2 = d)$ to represent the conditional density function. In both the discrete and continuous cases, we will eliminate the random variable subscripts on $P_{X_1|X_2}(\cdot)$when the random variable context of the probability set function is clear.

Once derived, a conditional PDF exhibits all of the standard properties of a PDF. The significance of the term *conditional* PDF is to indicate that the density of $X_1$ was derived from the joint density for $(X_1, X_2)$ conditional on a specific event for $X_2$. Otherwise, there are no special general properties of a conditional PDF that distinguishes it from any other PDF.

We provide examples of the derivation and use of discrete and continuous conditional PDF's in the following examples.

**Example 2.21** ***Conditional PDF in a Bivariate Discrete Case***

Recall the dice example, Example 2.15, where $f(x_1,x_2) = (1/36)\ I_{\{1,...,6\}}(x_1)\ I_{\{1,...,6\}}(x_2 - x_1)$. The conditional density function for $X_1$, given that $x_2 = 5$, is given by

$$f(x_1\,|\,x_2 = 5) = \frac{f(x_1,5)}{f_2(5)} = \frac{\frac{1}{36}I_{\{1,...,6\}}(x_1)\,I_{\{1,...,6\}}(5-x_1)}{\frac{6-|5-7|}{36}I_{\{2,...,12\}}(5)} = \frac{1}{4}\,I_{\{1,...,4\}}(x_1).$$

The probability of rolling a 3 or less on the red die, given that the total of the two dice will be 5, is then

$$P(x_1 \le 3|x_2 = 5) = \sum_{x_1=1}^{3} f(x_1|x_2 = 5) = \frac{3}{4}.$$

Note that the *unconditional* probability that $x_1 \le 3$ is equal to 1/2.

The conditional density function for $X_1$, given that $x_2 \in D = \{7, 11\}$, is given by

$$f(x_1\,|\,x_2 \in D) = \frac{\sum_{x_2\in D} f(x_1,x_2)}{\sum_{x_2\in D} f_2(x_2)} = \frac{\frac{1}{36}I_{\{1,...,6\}}(x_1)\left[I_{\{1,...,6\}}(7-x_1) + I_{\{1,...,6\}}(11-x_1)\right]}{\frac{8}{36}}$$
$$= \tfrac{1}{8}I_{\{1,...,4\}}(x_1) + \tfrac{1}{4}I_{\{5,6\}}(x_1)$$

The probability of rolling a 3 or less on the red die, given that the total of the two dice will be either a 7 or 11, is then

$$P(x_1 \le 3|x_2 \in D) = \sum_{x_1=1}^{3} f(x_1|x_2 \in D) = \frac{3}{8}.$$

□

**Example 2.22** ***Conditional PDF in a Bivariate Continuous Case***

Recall Example 2.20 regarding the proportion of daily capacity at which two processing plants operate. The conditional density function of plant 1's capacity, given that plant 2 operates at less than half of capacity, is given by

$$f(x_1 \mid x_2 \leq .5) = \frac{\int_{-\infty}^{.5} f(x_1, x_2)\, dx_2}{\int_{-\infty}^{.5} f_2(x_2)\, dx_2} = \frac{\int_0^{.5} (x_1 + x_2)\, I_{[0,1]}(x_1)\, dx_2}{\int_0^{.5} (x_2 + 1/2)\, dx_2}$$

$$= \frac{.5\, x_1 + .125}{.375} I_{[0,1]}(x_1) = \left(\frac{4}{3} x_1 + \frac{1}{3}\right) I_{[0,1]}(x_1)$$

The probability that $x_1 \leq .5$, given that $x_2 \leq .5$, is given by

$$P(x_1 \leq .5 \mid x_2 \leq .5) = \int_0^{.5} \left(\frac{4}{3} x_1 + \frac{1}{3}\right) dx_1 = \frac{1}{3}.$$

Recall that the *unconditional* probability that $x_1 \leq .5$ was .375. □

### 2.6.2 Conditioning on Elementary Events in Continuous Cases-Bivariate

A problem arises in the continuous case when defining a conditional PDF for $X_1$, conditional on an *elementary* event occurring for $X_2$. Namely, because all elementary events are assigned probability zero in the continuous case, with the integral over a singleton set being zero, our definition of the conditional density, as presented earlier, yields

$$f(x_1 \mid x_2 = b) = \frac{\int_b^b f(x_1, x_2)\, dx_2}{\int_b^b f_2(x_2)\, dx_2} = \frac{0}{0},$$

which is an indeterminate form. Thus $f(x_1|x_2 = b)$ is undefined so that $P(x_1 \in A \mid x_2 = b)$ is undefined as well. This is different than the discrete case, where

$$f(x_1 \mid x_2 = b) = \frac{f(x_1, b)}{f_2(b)}$$

is well-defined, provided $f_2(b) \neq 0$.

The problem is circumvented by redefining the conditional probability, $P(x_1 \in A|x_2 = b)$, in the continuous case in terms of a limit as

$$P(x_1 \in A \mid x_2 = b) = \lim_{\varepsilon \to 0^+} P(x_1 \in A \mid x_2 \in [b - \varepsilon, b + \varepsilon])$$

$$= \lim_{\varepsilon \to 0^+} \left[\frac{\int_{x_1 \in A} \int_{b-\varepsilon}^{b+\varepsilon} f(x_1, x_2)\, dx_2\, dx_1}{\int_{b-\varepsilon}^{b+\varepsilon} f_2(x_2) dx_2}\right]$$

where $\lim_{\varepsilon \to 0^+}$ means we are examining a limit for a sequence of $\varepsilon$-values that approach zero from positive values (i.e., $\varepsilon > 0$). The idea is to examine the limiting value of a sequence of probabilities that are conditioned on a corresponding sequence of events, $[b - \varepsilon, b + \varepsilon]$ for $\varepsilon \to 0^+$, that converge to the elementary event $\{b\}$. The following lemma will facilitate the identification of the limit.

**Lemma 2.2**
***Mean Value Theorem for Integrals***

If $g(x)$ is continuous $\forall x \in [c_1, c_2]$, then $\exists x_0 \in [c_1, c_2]$ such that $\int_{c_1}^{c_2} g(x)dx = g(x_0)(c_2 - c_1)$.[15]

To use the mean value theorem, and to ensure that the limit of the conditional probabilities exists, we assume that there exists a choice of $\varepsilon > 0$ such that $f_2(x_2)$ and $f(x_1, x_2)$ are continuous in $x_2$, $\forall x_2 \in [b - \varepsilon, b + \varepsilon]$, and that $f_2(b) > 0$. Then, by the mean value theorem,

$$P(x_1 \in A | x_2 = b) = \lim_{\varepsilon \to 0^+} \left[ \frac{\int_{x_1 \in A} \int_{b-\varepsilon}^{b+\varepsilon} f(x_1, x_2)\, dx_2 dx_1}{\int_{b-\varepsilon}^{b+\varepsilon} f_2(x_2) dx_2} \right] = \lim_{\varepsilon \to 0^+} \left[ \frac{2\varepsilon \int_{x_1 \in A} f(x_1, x_2^0)\, dx_1}{2\varepsilon f_2\,(x_2^*)} \right]$$

where both $x_2^0$ and $x_2^* \in [b - \varepsilon, b + \varepsilon]$, and $x_2^0$ will generally depend on the value of $x_1$.[16] The 2 $\varepsilon$'s in the numerator and denominator cancel each other, and as $\varepsilon \to 0^+$, the interval $[b - \varepsilon, b + \varepsilon]$ reduces to $[b,b] = b$, so that in the limit, both $x_2^0$ and $x_2^* = b$. The limiting value of the conditional probability is then

$$P(x_1 \in A | x_2 = b) = \int_{x_1 \in A} \frac{f(x_1, b)}{f_2(b)} dx.$$

Since the choice of event $A$ is arbitrary, it follows that the appropriate conditional probability density in this case is

$$f(x_1 \,|\, x_2 = b) = \frac{f(x_1, b)}{f_2\,(b)},$$

which is precisely of the same form as the discrete case. Thus, the definition of conditional density functions, when conditioning on elementary events, will be identical for continuous and discrete random variables, provided $f_2(b) \neq 0$.

**Example 2.23**
***PDF Conditioned on an Elementary Event***

Recall Example 2.22. The conditional PDF for plant 1's proportion of capacity $X_1$, given that plant 2's capacity proportion is $x_2 = .75$, can be defined as

$$f(x_1 \,|\, x_2 = .75) = \frac{f(x_1, .75)}{f_2\,(.75)} = \frac{(x_1 + .75) I_{[0,1]}(x_1)}{1.25} = \left(\frac{4}{5}x_1 + \frac{3}{5}\right) I_{[0,1]}(x_1)$$

The probability that $x_1 \leq .5$, given that $x_2 = .75$, is then given by $P(x_1 \leq .5 \mid x_2 = .75) = \int_0^{.5} (\frac{4}{5}x_1 + \frac{3}{5}) dx_1 = .4$ □

[15] R. Courant and F. John, *Introduction to Calculus and Analysis*, New York, John Wiley-Interscience, 1965, p. 143.

[16] In applying the mean value theorem to the numerator, we treat $f(x_1, x_2)$ as a function of the single variable $x_2$, fixing the value of $x_1$ for each application.

### 2.6.3 Conditioning on Elementary Events in Continuous Cases: *n*-Variate Case

The preceding concepts of discrete and continuous conditional PDFs in the bivariate case can be generalized to the $n$-variate case, as indicated in the following definition, which subsumes $n = 2$ as a special case:

**Definition 2.23** ***Conditional Probability Density Functions***

Let $f(x_1,...,x_n)$ be the joint density function for the $n$-dimensional random variable $(X_1,...,X_n)$. The conditional density function for the $m$-dimensional random variable$(X_1,...,X_m)$, given that $(X_{m+1},...,X_n) \in D$ and $P_{X_{m+1},...,X_n}(D)>0$, is as follows:

***Discrete Case:***

$$f(x_1,...,x_m|x_{m+1},...,x_n) \in D) = \frac{\sum_{(x_{m+1},...,x_n)\in D} f(x_1,...,x_n)}{\sum_{(x_{m+1},...,x_n)\in D} f_{m+1,...,n}(x_{m+1},...,x_n)}$$

***Continuous Case:***

$$f(x_1,...,x_m|x_{m+1},...,x_n) \in D) = \frac{\int_{(x_{m+1},...,x_n)\in D} f(x_1,...,x_n)dx_{m+1}...dx_n}{\int_{(x_{m+1},...,x_n)\in D} f_{m+1,...,n}(x_{m+1},...,x_n)dx_{m+1}...dx_n}$$

If $D$ is equal to the elementary event $(d_{m+1},...,d_n)$ then the definition of the conditional density in both the discrete and continuous cases can be represented as

$$f(x_1,...,x_m|x_i = d_i, i = m+1,...,n) = \frac{f(x_1,...,x_m,d_{m+1},...,d_n)}{f_{m+1,...,n}(d_{m+1},...,d_n)}$$

when the marginal density in the denominator is positive valued.[17]

For example, if $n = 3$, then the conditional density function of $(X_1, X_2)$, given that $x_3 \in D$, would be defined as

$$f(x_1,x_2|x_3 \in D) = \frac{\sum_{x_3\in D} f(x_1,x_2,x_3)}{\sum_{x_3\in D} f_3(x_3)}$$

in the discrete case, with integration replacing summation in the continuous case. If $D = d_3$, then for both the discrete and continuous cases,

$$f(x_1,x_2|x_3 = d_3) = \frac{f(x_1,x_2,d_3)}{f_3(d_3)}.$$

[17] In the continuous case, it is also presumed that $f$ and $f_{m+1,...,n}$ are continuous in $(x_{m+1},...,x_n)$ within some neighborhood of points around the point where the conditional density is evaluated in order to justify the conditional density definition via a limiting argument analogous to the bivariate case. Motivation for the conditional density expression when conditioning on an elementary event in the continuous case can then be provided by extending the mean-value theorem argument used in the bivariate case. See R.G. Bartle, *Real Analysis*, p. 429 for a statement of the general mean value theorem for integrals.

An example of conditional PDFs in the trivariate case will be presented in Section 2.8.

In summary, if we begin with the joint density function appropriate for assigning probabilities to events involving the $n$-dimensional random variable $(X_1,\ldots,X_n)$, we can derive a conditional probability density function that is the PDF appropriate for assigning probabilities to events for an m-dimensional subset of the random variables in $(X_1,\ldots,X_n)$, *given* (or *conditional*) on an event for the remaining $n-m$ random variables. The construction of the conditional density involves both the joint density of $(X_1,\ldots, X_n)$ and the marginal density of the $(n-m)$ dimensional random variable on which we are conditioning. In the special case where we are conditioning on an elementary event, the conditional density function simply becomes the ratio of the joint density function divided by the marginal density function, replacing the arguments of these functions with their conditioned values for those arguments corresponding to random variables on which we are conditioning (which represents *all* of the arguments of the marginal density, and a subset of the arguments of the joint density).

### 2.6.4 Conditional CDFs

We can define the concept of a **conditional CDF** by simply using a conditional density function in the definition of the CDF. For example, for the bivariate random variable $(X_1, X_2)$, we can define

$$F(b_1\,|\,x_2 \in D) = P(x_1 \leq b_1\,|\,x_2 \in D) = \int_{-\infty}^{b_1} f(x_1\,|\,x_2 \in D)\,dx_1$$

as one such conditional CDF, representing the CDF of $X_1$, conditional on $x_2 \in D$. Once defined, the conditional CDF possesses no special properties that distinguish it in concept from any other CDF. The reader is asked to contemplate the various conditional CDFs that can be defined for the $n$-dimensional random variable $(X_1,\ldots,X_n)$.

## 2.7 Independence of Random Variables

From our previous discussion of independence of events, we know that $A$ and $B$ are independent *iff* $P(A \cap B) = P(A)P(B)$. This concept can be applied directly to determine whether two particular events for the $n$-dimensional random variable $(X_1,\ldots,X_n)$ are independent. The general definition of independence of events (Definition 1.19) can also be applied to examine the independence of $k$ specific events for the random variable $(X_1,\ldots,X_n)$.

The concept of independence of events will now be extended further to the idea of **independence of random variables**, which is related to the question of whether the $n$ events (recall the abbreviated set definition notation of Definition 2.7) $\{x_i \in A_i\} \equiv \{(x_1,\ldots,x_n)\colon x_i \in A_i, (x_1,\ldots,x_n) \in R(\mathbf{X})\}$, $i = 1,\ldots,n$, are independent for *all* possible choices of the events $A_1,\ldots,A_n$. If so, the *n random variables* are said to be independent. In effect, the concept is one of **global independence of**

**events** for random variables – we define an event $A_i$ for each of the $n$ random variables in $(X_1,\ldots,X_n)$ and, *no matter how we define the events* (which is the meaning of the term "global" here), the events $\{x_i \in A_i\}$, $i = 1,\ldots,n$, are independent. Among other things, we will see that this implies that the probability assigned to *any* event $A_i$ for *any* random variable $X_i$ in $(X_1,\ldots,X_n)$ is unaffected by conditioning on *any* event $B$ for the remaining random variables (assuming $P(B) > 0$ for the existence of the conditional probability).

### 2.7.1 Bivariate Case

We seek to establish a condition that will ensure that the events $\{x_1 \in A_1\}$ and $\{x_2 \in A_2\}$ are independent for *all* possible choices of the events $A_1$ and $A_2$. This can be accomplished by applying independence conditions to events in the probability space, $\{R(\mathbf{X}),\Upsilon,P\}$ for the bivariate random variable $\mathbf{X} = (X_1, X_2)$. The events $x_1 \in A_1$ and $x_2 \in A_2$ are equivalent, respectively, to the following events for the bivariate random variable:

$$B_1 = \{(x_1,x_2) : x_1 \in A_1, (x_1,x_2) \in R(\mathbf{X})\} \text{and } B_2 = \{(x_1,x_2) : x_2 \in A_2, (x_1,x_2) \in R(\mathbf{X})\}.$$

The two events $B_1$ and $B_2$ are independent *iff* $P(B_1 \cap B_2) = P(B_1)P(B_2)$, which can also be represented using our abbreviated notation as $P(x_1 \in A_1, x_2 \in A_2) = P(x_1 \in A_1)P(x_2 \in A_2)$. Requiring the independence condition to hold for *all* choices of the events $A_1$ and $A_2$ leads to the definition of the independence condition for random variables.

**Definition 2.24** ***Independence of Random Variables: Bivariate***

The random variables $X_1$ and $X_2$ are said to be independent *iff* $P(x_1 \in A_1, x_2 \in A_2) = P(x_1 \in A_1)\, P(x_2 \in A_2)$ for all events $A_1$, $A_2$.

There is an equivalent characterization of independence of random variables in terms of PDFs that can be useful in practice and that also further facilitates the investigation of the implications of random variable independence.

**Theorem 2.6** ***Bivariate Density Factorization for Independence of Random Variables***

*The random variables $X_1$ and $X_2$ with joint PDF $f(x_1, x_2)$ and marginal PDFs $f_i(x_i)$, $i = 1, 2$, are independent iff the joint density factors into the product of the marginal densities as $f(x_1, x_2) = f_1(x_1)\, f_2(x_2)\ \forall\ (x_1, x_2)$.*[18]

[18]Technically, the factorization need not hold at points of discontinuity for the joint density function of a continuous random variable. However, if the random variables are independent, there will always exist a density function for which the factorization can be formed. This has to do with the fundamental non-uniqueness of PDFs in the continuous case, which can be redefined arbitrarily at a countable number of isolated points without affecting the assignment of any probabilities of events through integration. There are few practical benefits of this non-uniqueness, and we suppress this technical anomaly here.

**Proof** **Discrete Case** Let $A_1$ and $A_2$ be any two events for $X_1$ and $X_2$, respectively. Then if the joint density function $f(x_1, x_2)$ factors,

$$P(x_1 \in A_1, x_2 \in A_2) = \sum_{x_1 \in A_1} \sum_{x_2 \in A_2} f(x_1, x_2) = \sum_{x_1 \in A_1} f_1(x_1) \sum_{x_2 \in A_2} f_2(x_2) = P(x_1 \in A_1)P(x_2 \in A_2)$$

so that $X_1$ and $X_2$ are independent. Thus, factorization is sufficient for independence. Now assume $(X_1,X_2)$ are independent random variables. Let $A_1 = \{a_1\}$ and $A_2 = \{a_2\}$ for any choice of elementary events, $a_i \in R(X_i)$, corresponding to the random variable $X_i, i = 1,2$, respectively. Then, by independence,

$$P(x_1 = a_1, x_2 = a_2) = f(a_1, a_2) = P(x_1 = a_1)P(x_2 = a_2) = f_1(a_1)f_2(a_2)$$

If $a_i \notin R(X_i)$, then $f_i(a_i) = 0$ and $f(a_1,a_2) = 0$ for $i = 1,2$, and thus factorization will automatically hold. Thus, factorization is necessary for independence.

**Continuous case** Let $A_1$ and $A_2$ be any two events for $X_1$ and $X_2$, respectively. Then if the joint density function $f(x_1, x_2)$ factors,[19]

$$\begin{aligned} P(x_1 \in A_1, x_2 \in A_2) &= \int_{x_2 \in A_2} \int_{x_1 \in A_1} f(x_1, x_2) dx_1 dx_2 \\ &= \int_{x_1 \in A_1} f_1(x_1) dx_1 \int_{x_2 \in A_2} f_2(x_2) dx_2 \\ &= P(x_1 \in A_1)P(x_2 \in A_2), \end{aligned}$$

so that $X_1$ and $X_2$ are independent. Thus, factorization is sufficient for independence. Now assume $(X_1,X_2)$ are independent random variables. Let $A_i = \{x_i : x_i \leq a_i\}$ for arbitrary choice of $a_i, i = 1,2$. Then by independence,

$$\begin{aligned} P(x_1 \leq a_1, x_2 \leq a_2) &= \int_{-\infty}^{a_2} \int_{-\infty}^{a_1} f(x_1, x_2)\, dx_1\, dx_2 \\ &= P(x_1 \leq a_1)P(x_2 \leq a_2) = \int_{-\infty}^{a_1} f_1(x_1)\, dx_1 \int_{-\infty}^{a_2} f_2(x_2)\, dx_2 . \end{aligned}$$

Differentiating the integrals with respect to $a_1$ and $a_2$ yields $f(a_1, a_2) = f_1(a_1) f_2(a_2)$ wherever the joint density function is continuous. Thus, the factorization condition stated in the theorem is *necessary* for independence. ■

In other words, two random variables are independent *iff* their joint PDF can be expressed equivalently as the product of their respective marginal PDFs (the condition not being required to hold at points of discontinuity in the continuous case). An important implication of independence of $X_1$ and $X_2$ is that the

[19]Any points of discontinuity can be ignored in the definitions of the probability integrals without affecting the probability assignments.

conditional and marginal PDFs of the respective random variables are identical.[20] For example, assuming independence,

$$f(x_1 \mid x_2 \in B) = \frac{\int_{x_2 \in B} f(x_1, x_2)\, dx_2}{\int_{x_2 \in B} f_2(x_2)\, dx_2} = \frac{f_1(x_1) \int_{x_2 \in B} f_2(x_2)\, dx_2}{\int_{x_2 \in B} f_2(x_2)\, dx_2} = f_1(x_1)$$

(in the discrete case, replace integration by summation). The fact that conditional and marginal PDFs are identical implies that the probability of $x_1 \in A$, for any event $A$, is unaffected by the occurrence or nonoccurrence of event $B$ for $X_2$. For example, in the continuous case,

$$P(x_1 \in A | x_2 \in B) = \int_{x_1 \in A} f(x_1 | x_2 \in B) dx_1 = \int_{x_1 \in A} f_1(x_1) dx_1 = P(x_1 \in A)$$

(replace integration by summation in the discrete case). The result holds for any events involving $X_2$ for which the conditional density function is defined. The roles of $X_1$ and $X_2$ can be reversed in the preceding discussion.

**Example 2.24** ***Independence of Bivariate Continuous RVs***

Recall Example 2.16 concerning coating flaws in the manufacture of television screens. The horizontal and vertical coordinates of the coating flaw was the outcome of a bivariate random variable with joint density function

$$f(x_1, x_2) = \tfrac{1}{12} I_{[-2,2]}(x_1) I_{[-1.5,1.5]}(x_2).$$

Are the random variables independent?

**Answer**: The marginal densities of $X_1$ and $X_2$ are given by

$$f_1(x_1) = I_{[-2,2]}(x_1) \int_{-1.5}^{1.5} \frac{1}{12} dx_2 = .25 I_{[-2,2]}(x_1)$$

$$f_2(x_2) = I_{[-1.5,1.5]}(x_2) \int_{-2}^{2} \frac{1}{12} dx_1 = \frac{1}{3} I_{[-1.5,1.5]}(x_2).$$

It follows that $f(x_1, x_2) = f_1(x_1) f_2(x_2)\ \forall (x_1, x_2)$, and the random variables are independent. Therefore, knowledge that an event for $X_2$ has occurred has no effect on the probability assigned to events for $X_1$, and vice versa. □

**Example 2.25** ***Independence of Bivariate Discrete RVs***

Recall the dice example, Example 2.15. Are $X_1$ and $X_2$ independent random variables?

**Answer**: Examine the validity of the independence condition:

$$f(x_1, x_2) \stackrel{?}{=} f_1(x_1) f_2(x_2)\ \ \forall (x_1, x_2),$$

[20] We will henceforth suppress constant reference to the fact that factorization might not hold for some points of discontinuity in the continuous case – it will be tacitly understood that results we derive based on the factorization of $f(x_1, x_2)$ may be violated at some isolated points. For example, for the case at hand, marginal and conditional densities may not be equal at some isolated points. Assignments of probability will be unaffected by this technical anomaly.

or, specifically,

$$\frac{1}{36} I_{\{1,2,\ldots,6\}}(x_1) I_{\{1,2,\ldots,6\}}(x_2 - x_1) \stackrel{?}{=} \frac{1}{6} I_{\{1,2,\ldots,6\}}(x_1)\left(\frac{6 - |x_2 - 7|}{36}\right) I_{\{2,\ldots,12\}}(x_2) \quad \forall (x_1, x_2)$$

The random variables $X_1$ and $X_2$ are *not* independent, since, for example, letting $x_1 = 2$ and $x_2 = 4$ results in $1/36 \neq 1/72$. Therefore, knowledge that an event for $X_2$ has occurred *can* affect the probability assigned to events for $X_1$, and vice versa. □

### 2.7.2 *n*-Variate

The independence concept can be extended beyond the bivariate case to the case of **independence of random variables** $X_1, \ldots, X_n$. The formal definition of independence in the $n$-variate case is as follows:

**Definition 2.25** ***Independence of Random Variables (n-Variate)***

The random variables $X_1, X_2, \ldots, X_n$ are said to be independent *iff* $P(x_i \in A_i, i = 1, \ldots, n) = \prod_{i=1}^{n} P(x_i \in A_i)$ for all choices of the events $A_1, \ldots, A_n$.

The motivation for the definition is similar to the argument used in the bivariate case. For $B_i = \{(x_1, \ldots, x_n) : x_i \in A_i, (x_1, \ldots, x_n) \in R(\mathbf{X})\}, i = 1, \ldots, n$ to be independent events, we require (recall Definition 1.19)

$$P(\cap_{j \in J} B_j) = \prod_{j \in J} P(B_j) \;\forall J \subset \{1, 2, \ldots, n\} \text{ with } N(J) \geq 2.$$

If we require this condition to hold for *all* possible choices of the events $(B_1, \ldots, B_n)$, then the totality of the conditions can be represented as

$$P(x_i \in A_i, i = 1, \ldots, n) = P\left(\bigcap_{i=1}^{n} B_i\right) = \prod_{i=1}^{n} P(B_i) = \prod_{i=1}^{n} P(x_i \in A_i)$$

for all choices of the events $A_1, \ldots, A_n$ (or, equivalently, for corresponding choices of $B_1, \ldots, B_n$). Any of the other conditions required for independence of events, i.e.,

$$P\left(\bigcap_{j \in J} B_j\right) = \prod_{j \in J} P(B_j) \text{ with } J \subset \{1, 2, \ldots, n\} \text{ and } N(J) < n,$$

are implied by the preceding condition upon letting $A_j = R(X_j)$ (or equivalently, $B_j = R(\mathbf{X})$) for $j \in \bar{J}$.

The generalization of the joint density factorization theorem is given as Theorem 2.7. The proof is a direct extension of the arguments used in proving Theorem 2.6, and is left to the reader.

**Theorem 2.7**
***Density Factorization for Independence of Random Variables (n-Variate Case)***

The random variables $X_1, X_2,\ldots,X_n$ with joint PDF $f(x_1,\ldots,x_n)$ and marginal PDFs $f_i(x_i)$, $i = 1,\ldots,n$, are independent *iff* the joint density can be factored into the product of the marginal densities as $f(x_1,\ldots,x_n) = \prod_{i=1}^{n} f_i(x_i)\ \forall(x_1,\ldots,x_n)$.[21]

An example of the application of Theorem 2.7 is given in Section 2.8.

### 2.7.3 Marginal Densities Do Not Determine an *n*-Variate Density Without Independence

If $(X_1,\ldots,X_n)$ are independent random variables, then knowing the marginal densities $f_i(x_i)$, $i = 1,\ldots,n$ is equivalent to knowing the joint density function for $(X_1,\ldots,X_n)$, since then $f(x_1,\ldots,x_n) = \prod_{i=1}^{n} f_i(x_i)$ . However, if the random variables in the collection $(X_1,\ldots,X_n)$ are *not* independent, then knowing each of the marginal densities of the $X_i$'s is generally *not* sufficient to determine the joint density function for $(X_1,\ldots,X_n)$. In fact, it can be shown that an uncountably infinite family of different joint density functions can give rise to the same collection of marginal density functions.[22] We provide the following counter example in the bivariate case to the proposition that knowledge of the marginal PDFs is sufficient for determining the *n*-variate PDF.

**Example 2.26**
***Marginal Densities Do Not Imply n-Variate Densities***

Examine the function

$$f_\alpha(x_1,x_2) = [1+\alpha(2x_1 - 1)(2x_2 - 1)]I_{[0,1]}(x_1)I_{[0,1]}(x_2).$$

The reader should verify that $f_\alpha(x_1, x_2)$ is a PDF $\forall\ \alpha\in[-1,1]$. For any choice of $\alpha\in[-1,1]$, the marginal density function for $X_1$ is given by $f_1(x_1) = \int_{-\infty}^{\infty} f_\alpha(x_1,x_2)\,dx_2 = I_{[0,1]}(x_1)$. Similarly, the marginal density of $X_2$, for any choice of $\alpha\in[-1,1]$, is given by $f_2(x_2) = \int_{-\infty}^{\infty} f_\alpha(x_1,x_2)dx_1 = I_{[0,1]}(x_2)$.

Since the *same* marginal density functions are associated with each of an uncountably infinite collection of bivariate density functions, it is clear that knowledge of $f_1(x_1)$ and $f_2(x_2)$ is insufficient to determine which is the appropriate joint density function for $(X_1, X_2)$. If we knew the marginal densities of $X_1$ and $X_2$, as stated, and if $X_1$ and $X_2$ are *independent* random variables, then we would know that $f(x_1, x_2) = I_{[0,1]}(x_1)\ I_{[0,1]}(x_2)$. □

[21] The same technical proviso regarding points of discontinuity in the case of continuous random variables hold as in the bivariate case. See Footnote 18.

[22] E.J. Gumbel (1958) *Distributions a' plusieurs variables dont les marges sont données, C.R. Acad. Sci., Paris*, 246, pp. 2717–2720.

### 2.7.4 Independence Between Random Vectors and Between Functions of Random Vectors

The independence concepts can be extended so that they apply to independence among two or more random *vectors*. Essentially, all that is required is to interpret the $X_i$'s as *multivariate* random variables in the appropriate definitions and theorems presented heretofore, and the statements are valid. Motivation for the validity of the extensions can be provided using arguments that are analogous to those used previously. For example, to extend the previous bivariate result to two random *vectors*, let $\mathbf{X}_1 = (X_{11},\ldots,X_{1m})$ be an $m$-dimensional random variable and $\mathbf{X}_2 = (X_{21},\ldots,X_{2n})$ be an $n$-dimensional random variable. Then $\mathbf{X}_1$ and $\mathbf{X}_2$ are independent *iff*

$$\begin{aligned} P(\mathbf{x}_1 \in A_1, \mathbf{x}_2 \in A_2) &= P((\mathbf{x}_{11},\ldots,x_{1m}) \in A_1, (x_{21},\ldots,x_{2n}) \in A_2) \\ &= P((x_{11},\ldots,x_{1m}) \in A_1)P((x_{21},\ldots,x_{2n}) \in A_2) = P(\mathbf{x}_1 \in A_1)P(\mathbf{x}_2 \in A_2) \end{aligned}$$

for *all* event pairs $A_1$, $A_2$. Furthermore, in terms of joint density factorization, $\mathbf{X}_1$ and $\mathbf{X}_2$ are independent *iff*

$$\begin{aligned} f(\mathbf{x}_1, \mathbf{x}_2) &= f(x_{11}, \ldots, x_{1m}, x_{21}, \ldots, x_{2n}) \\ &= f_1(x_{11}, \ldots, x_{1m}) f_2(x_{21}, \ldots, x_{2n}) \\ &= f_1(\mathbf{x}_1) f_2(\mathbf{x}_2) \ \forall(\mathbf{x}_1, \mathbf{x}_2) \end{aligned}$$

The reader can contemplate the myriad of other independence conditions that can be constructed for discrete and continuous random *vectors*.

Implications of the extended independence definitions and theorems are qualitatively similar to the implications identified previously for the case where the $X_i$'s were interpreted as scalars. For example, if $\mathbf{X}_1 = (X_{11},\ldots,X_{1m})$ and $\mathbf{X}_2 = (X_{21},\ldots,X_{2n})$ are independent random variables, then

$$P((x_{11},\ldots,x_{1m}) \in A_1 | (x_{21},\ldots,x_{2n}) \in A_2)) = P((x_{11},\ldots,x_{1m}) \in A_1),$$

i.e., conditional and unconditional probability of events for the random variable $\mathbf{X}_1$ are identical (and similarly for $\mathbf{X}_2$) for all choices of $A_1$ and $A_2$ for which the conditional probability is defined.

It is also useful to note some results concerning the independence of random variables that are defined as functions of other independent random variables. We begin with the simplest case of two independent random variables $X_1$ and $X_2$.

**Theorem 2.8** *If $X_1$ and $X_2$ are independent random variables, and if the random variables $Y_1$ and $Y_2$ are defined by $y_1 = Y_1(x_1)$ and $y_2 = Y_2(x_2)$, then $Y_1$ and $Y_2$ are independent random variables.*

**Proof** The event involving outcomes of $X_i$ that is equivalent to the event $y_i \in A_i$ is given by $B_i = \{x_i\text{: } Y_i(x_i) \in A_i,\ x_i \in R(X_i)\}$ for $i = 1, 2$. Then

$$\begin{aligned} P(y_1 \in A_1, y_2 \in A_2) &= P(x_1 \in B_1, x_2 \in B_2) \\ &= P(x_1 \in B_1)P(x_2 \in B_2) \quad \text{(by independence of } x_1, x_2\text{)} \\ &= P(y_1 \in A_1)P(y_2 \in A_2), \end{aligned}$$

and since this holds for every event pair $A_1$, $A_2$, the random variables $Y_1$ and $Y_2$ are independent. ■

**Example 2.27** ***Independence of Functions of Continuous RVs***

A large service station sells unleaded and premium-grade gasoline. The quantities sold of each type of fuel on a given day is the outcome of a bivariate random variable with density function[23]

$$f(x_1,x_2) = \frac{1}{20}e^{-(.1x_1+.5x_2)}I_{(0,\infty)}(x_1)I_{(0,\infty)}(x_2),$$

where the $x_i$'s are measured in thousands of gallons. The marginal densities are given by (reader, please verify)

$$f_1(x_1) = \frac{1}{10}e^{-.1x_1}I_{(0,\infty)}(x_1) \text{ and } f_2(x_2) = \frac{1}{2}e^{-.5x_2}I_{(0,\infty)}(x_2)$$

and so the random variables are independent. The prices of unleaded and premium gasoline are \$3.25 and \$3.60 per gallon, respectively. The wholesale cost of gasoline plus federal state and local taxes amounts to \$2.80 and \$3.00 per gallon, respectively. Other daily variable costs in selling the two products amount to $C_i(x_i) = 20\,x_i^2$, $i = 1, 2$. Are daily profits above variable costs for the two products independent random variables?

**Answer**: Yes. Note that the profit levels in the two cases are $\Pi_1 = 450x_1 - 20x_1^2$ and $\Pi_2 = 600x_2 - 20x_2^2$, respectively. Since $\Pi_1$ is only a function of $x_1$, $\Pi_2$ is only a function of $x_2$, and $X_1$ and $X_2$ are independent, then $\Pi_1$ and $\Pi_2$ are independent by Theorem 2.8. □

A more general theorem explicitly involving random *vectors* is stated as follows:

**Theorem 2.9** *Let $\mathbf{X}_1,\ldots,\mathbf{X}_n$ be a collection of n independent random vectors, and let the random vectors $\mathbf{Y}_1,\ldots,\mathbf{Y}_n$ be defined by $\mathbf{y}_i = \mathbf{Y}_i(\mathbf{x}_i)$, $i = 1,\ldots,n$. Then the random vectors $\mathbf{Y}_1,\ldots,\mathbf{Y}_n$ are independent.*

**Proof** The event involving outcomes of the random vector $\mathbf{X}_i$ that is equivalent to the event $A_i$ for the random vector $\mathbf{Y}_i$ is given by $B_i = \{\mathbf{x}_i\colon \mathbf{Y}_i(\mathbf{x}_i) \in A_i,\ \mathbf{x}_i \in R(\mathbf{X}_i)\}$, $i = 1,\ldots,n$. Then

$$\begin{aligned} P(\mathbf{y}_i \in A_i, i=1,\ldots,n) &= P(\mathbf{x}_i \in B_i, i=1,\ldots,n) \\ &= \prod_{i=1}^{n} P(\mathbf{x}_i \in B_i) \text{ (by independence of random vectors)} \\ &= \prod_{i=1}^{n} P(\mathbf{y}_i \in A_i) \end{aligned}$$

and since this holds for every collection of events $A_1,\ldots,A_n$, the random vectors $\mathbf{Y}_1,\ldots,\mathbf{Y}_n$ are independent by a vector interpretation of the random variables in Definition 2.19. ■

[23]This must be an approximation – why?

**Example 2.28** ***Independence of Functions of Discrete RVs***

Examine the experiment of independently tossing two fair coins and rolling three fair dice. Let $X_1$ and $X_2$ represent whether heads $(x_i = 1)$ or tails $(x_i = 0)$ appears on the first and second coins, respectively, and let $X_3$, $X_4$, and $X_5$ represent the number of dots facing up on each of the three dice, respectively. Since the random variables are independent, the probability density of $X_1,\ldots,X_5$ can be written as

$$f(x_1,\ldots,x_5) = \prod_{i=1}^{2} \frac{1}{2} I_{\{0,1\}}(x_i) \prod_{i=3}^{5} \frac{1}{6} I_{\{1,\ldots,6\}}(x_i)$$

Define two new random *vectors* $\mathbf{Y}_1$ and $\mathbf{Y}_2$ using the vector functions

$$\mathbf{y}_1 = \begin{bmatrix} y_{11} \\ y_{12} \end{bmatrix} = \begin{bmatrix} x_1 + x_2 \\ x_1\ x_2 \end{bmatrix} = \mathbf{Y}_1(x_1, x_2),$$

$$\mathbf{y}_2 = \begin{bmatrix} y_{21} \\ y_{22} \end{bmatrix} = \begin{bmatrix} x_3 + x_4 + x_5 \\ x_3\ x_4 / x_5 \end{bmatrix} = \mathbf{Y}_2(x_3, x_4, x_5)$$

Then since the vector $\mathbf{y}_1$ is a function of $(x_1, x_2)$, $\mathbf{y}_2$ is a function of $(x_3, x_4, x_5)$, and since the random vectors $(X_1, X_2)$ and $(X_3, X_4, X_5)$ are independent (why?), Theorem 2.9 indicates that the random vectors $\mathbf{Y}_1$ and $\mathbf{Y}_2$ are independent. This is clearly consistent with intuition, since outcomes of the vector $\mathbf{Y}_1$ obviously have nothing to do with outcomes of the vector $\mathbf{Y}_2$. The reader should note that within vectors, the random variables are *not* independent, i.e., $Y_{11}$ and $Y_{12}$ are not independent, and neither are $Y_{21}$ and $Y_{22}$. □

## 2.8 Extended Example of Multivariate Concepts in the Continuous Case

We now further illustrate some of the concepts of this chapter with an example involving a trivariate continuous random variable. Let $(X_1, X_2, X_3)$ have the PDF $f(x_1, x_2, x_3) = (3/16)\, x_1\, x_2^2\, e^{-x_3} I_{[0,2]}(x_1)\, I_{[0,2]}(x_2)\, I_{[0,\infty)}(x_3)$.

a. What is the marginal density of $X_1$? of $X_2$? of $X_3$?
**Answer:**

$$f_1(x_1) = \int_{-\infty}^{\infty}\int_{-\infty}^{\infty} f(x_1, x_2, x_3)dx_2 dx_3$$
$$= \tfrac{3}{16} x_1 I_{[0,2]}(x_1) \int_{-\infty}^{\infty} x_2^2 I_{[0,2]}(x_2)dx_2 \int_{-\infty}^{\infty} e^{-x_3} I_{[0,\infty]}(x_3)dx_3$$
$$= \tfrac{3}{16} x_1 I_{[0,2]}(x_1)\left(\tfrac{8}{3}\right)(1) = \tfrac{1}{2} x_1 I_{[0,2]}(x_1).$$

Similarly,

$$f_2(x_2) = \int_{-\infty}^{\infty}\int_{-\infty}^{\infty} f(x_1, x_2, x_3)dx_1 dx_3 = \frac{3}{8} x_2^2 I_{[0,2]}(x_2)$$

$$f_3(x_3) = \int_{-\infty}^{\infty}\int_{-\infty}^{\infty} f(x_1, x_2, x_3)dx_1 dx_2 = e^{-x_3} I_{[0,\infty]}(x_3).$$

**b.** What is the probability that $x_1 \geq 1$?

**Answer**: $P(x_1 \geq 1) = \int_1^{\infty} f_1(x_1)dx_1 = \int_1^2 \frac{1}{2}x_1 dx_1 = \left.\frac{x_1^2}{4}\right|_1^2 = .75.$

**c.** Are the three random variables independent?
**Answer**: Yes. It is clear that $f(x_1, x_2, x_3) = f_1(x_1)\, f_2(x_2)\, f_3(x_3)\ \forall\ (x_1,x_2,x_3)$.

**d.** What is the marginal cumulative distribution function for $X_1$? for $X_3$?
**Answer**: By definition,

$$F_1(b) = \int_{-\infty}^{b} f_1(x_1)\,dx_1 = \int_{-\infty}^{b} \frac{1}{2}x_1 I_{[0,2]}(x_1)\,dx_1$$

$$= \left.\frac{1}{2}\frac{x_1^2}{2}\right|_0^b I_{[0,2]}(b) + I_{(2,\infty)}(b) = \frac{b^2}{4} I_{[0,2]}(b) + I_{(2,\infty)}(b),$$

$$F_3(b) = \int_{-\infty}^{b} f_3(x_3)\,dx_3 = \int_{-\infty}^{b} e^{-x_3} I_{[0,\infty)}(x_3)\,dx_3$$

$$= \left.-e^{-x_3}\right|_0^b I_{[0,\infty)}(b) = (1 - e^{-b}) I_{[0,\infty)}(b).$$

**e.** What is the probability that $x_1 \leq 1$? that $x_3 > 1$?
**Answer**: $P(x_1 \leq 1) = F_1(1) = .25$. $P(x_3 > 1) = 1 - F_3(1) = e^{-1} = .3679$.

**f.** What is the joint cumulative distribution function for $X_1, X_2, X_3$?
**Answer**: By definition:

$$F(b_1, b_2, b_3) = \int_{-\infty}^{b_1}\int_{-\infty}^{b_2}\int_{-\infty}^{b_3} f(x_1, x_2, x_3)\,dx_3\,dx_2\,dx_1$$

$$= \int_{-\infty}^{b_1} \frac{1}{2}x_1 I_{[0,2]}(x_1)\,dx_1 \int_{-\infty}^{b_2} \frac{3}{8} x_2^2 I_{[0,2]}(x_2)\,dx_2 \int_{-\infty}^{b_3} e^{-x_3} I_{[0,\infty)}(x_3)\,dx_3$$

$$= \left[\frac{b_1^2}{4} I_{[0,2]}(b_1) + I_{(2,\infty)}(b_1)\right]\left[\frac{3b_2^3}{24} I_{[0,2]}(b_2) + I_{(2,\infty)}(b_2)\right]\left[\left(1 - e^{-b_3}\right) I_{[0,\infty)}(b_3)\right]$$

**g.** What is the probability that $x_1 \leq 1$, $x_2 \leq 1$, $x_3 \leq 10$?
**Answer**: $F(1,1,10) = (1/4)(3/24)(1 - e^{-10}) = .031$.

**h.** What is the conditional PDF of $X_1$, given that $x_2 = 1$ and $x_3 = 0$?
**Answer**: By definition, $f(x_1|x_2 = 1, x_3 = 0) = \frac{f(x_1,1,0)}{f_{23}(1,0)}$. Also,

$$f_{23}(x_2,x_3) = \int_{-\infty}^{\infty} f(x_1,x_2,x_3)dx_1 = \frac{3}{8}x_2^2 I_{[0,2]}(x_2) e^{-x_3} I_{[0,\infty)}(x_3).$$ Thus,

$$f(x_1|x_2 = 1, x_3 = 0) = \frac{\left(\frac{3}{16}\right) x_1\, I_{[0,2]}(x_1)}{\frac{3}{8}} = \frac{1}{2}x_1 I_{[0,2]}(x_1)$$

**i.** What is the probability that $x_1 \in [0, 1/2]$, given that $x_2 = 1$ and $x_3 = 0$?

**Answer:**

$$P(x_1 \in [0,\tfrac{1}{2}] \mid x_2 = 1, x_3 = 0) = \int_0^{1/2} f(x_1 \mid x_2 = 1, x_3 = 0)\, dx_1$$

$$= \int_0^{1/2} \frac{1}{2} x_1\, I_{[0,2]}(x_1)\, dx_1 = \left.\frac{x_1^2}{4}\right|_0^{1/2} = \frac{1}{16}$$

**j.** Let the two random variables $Y_1$ and $Y_2$ be defined by $y_1 = Y_1(x_1, x_2) = x_1^2 x_2$ and $y_2 = Y_2(x_3) = x_3/2$. Are the random variables $Y_1$ and $Y_2$ independent?
**Answer**: Yes, they are independent. The bivariate random variable $(X_1, X_2)$ is independent of the random variable $X_3$ since $f(x_1, x_2, x_3) = f_{12}(x_1,x_2)\, f_3(x_3)$, i.e., the joint density function factors into the product of the marginal density of $(X_1, X_2)$ and the marginal density of $X_3$. Then, since $y_1$ is a function of only $(x_1, x_2)$ and $y_2$ is a function of only $x_3$, $Y_1$ and $Y_2$ are independent random variables, by Theorem 2.9.

## Keywords, Phrases, and Symbols

[), interval, closed lower bound and open upper bound
(], interval, open lower bound and closed upper bound
[], interval, closed bounds
(), interval, open bounds
Abbreviated set notation
CDF
Classes of discrete and continuous density functions
Composite random variable
Conditional cumulative distribution function
Conditional density function
Continuous density component
Continuous joint PDF
Continuous PDF
Continuous random variable
Cumulative distribution function
Density factorization for independence
Discrete density component
Discrete joint PDF
Discrete PDF
Discrete random variable
Duality between CDFs and PDFs
∃, there exists
Equivalent events
Event A is relatively certain
Event A is relatively impossible
Event A occurs with probability one
Event A occurs with probability zero
$F(b)$
$f(x_1,\ldots,x_m|(x_{m+1},\ldots,x_n) \in B)$
$f(x_1,\ldots,x_n)$
$f_{1\ldots m}(x_1,\ldots,x_m)$
Increasing function
Independence of random variables
Induced probability space, $\{R(X), \Upsilon_X, P_X\}$
$\bar{J}$, complement of $J$
Marginal cumulative distribution function
Marginal PDF
MCDF
Mixed discrete-continuous random variables
MPDF
Multivariate cumulative distribution function
Multivariate random variable
⇔, mutual implication or *iff*
Outcome of the random variable, $x$
$P(x \leq b)$
PDF
$R(X)$
Random variable, $X$
Real-valued vector function
Truncation function
$X(w)$
$X\colon S \to R$

## Problems

**1.** Which of the following are valid PDFs? Justify your answer.

a. $f(x) = (.2)^x (.6)^{1-x} I_{\{0,1\}}(x)$

b. $f(x) = (.3)(.7)^x I_{\{0,1,2,\ldots\}}(x)$

c. $f(x) = .6\, e^{-x/4} I_{(0,\infty)}(x)$

d. $f(x) = x^{-1} I_{[1,e]}(x)$

**2.** Graph each of the *probability density* functions in Problem 1.

**3.** Sparkle Cola, Inc., manufactures a cola drink. The cola is sold in 12 oz. bottles. The probability distribution associated with the random variable whose outcome represents the actual quantity of soda place in a bottle of Sparkle Cola by the soda bottling line is specified to be

$$f(x) = 50\left[e^{-100(12-x)} I_{(-\infty,12]}(x) + e^{-100(x-12)} I_{(12,\infty)}(x)\right].$$

In order to be considered full, a bottle must contain within .25 oz. of 12 oz. of soda.

a. Define a random variable whose outcome indicates whether or not a bottle is considered full.

b. What is the range of this random variable?

c. Define a PDF for the random variable. Use it to assign probability to the event that a bottle is "considered full."

d. The PDF $f(x)$ is only an approximation. Why?

**4.** A health maintenance organization (HMO) is currently treating 10 patients with a deadly bacterial infection. The best-known antibiotic treatment is being used in these cases, and this treatment is effective 95 percent of the time. If the treatment is not effective, the patient expires.

a. Define a random variable whose outcome represents the number of patients being treated by the HMO that survive the deadly bacterial infection. What is the range of this random variable? What is the event space for outcome of this random variable?

b. Define the appropriate PDF for the random variable you defined in (a). Define the probability set function appropriate for assigning probabilities to events regarding the outcome of the random variable.

c. Using the probability space you defined in (a) and (b), what is the probability that all 10 of the patients survive the infection?

d. What is the probability that no more than two patients expire?

e. If 50 percent of the patients were to expire, the government would require that the HMO suspend operations, and an investigation into the medical practices of the HMO would be conducted. Provide an argument in defense of the government's actions in this case.

**5.** Star Enterprises is a small firm that produces a product that is simple to manufacture, involving only one variable input. The relationship between input and output levels is given by $q = x^{.5}$, where $q$ is the quantity of product produced and $x$ is the quantity of variable input used. For any given output and input prices, Star Enterprises operates at a level of production that maximizes its profit over variable cost. The possible prices in dollars facing the firm on a given day is represented by a random variable $V$ with $R(V) = \{10,20,30\}$ and PDF

$$f(v) = .2I_{\{10\}}(v) + .5I_{\{20\}}(v) + .3I_{\{30\}}(v).$$

Input prices vary independently of output prices, and input price on a given day is the outcome of $W$ with $R(W) = \{1,2,3\}$ and PDF

$$g(w) = .4I_{\{1\}}(w) + .3I_{\{2\}}(w) + .3I_{\{3\}}(w).$$

a. Define a random variable whose outcome represents Star's profit above variable cost on a given day. What is the range of the random variable? What is the event space?

b. Define the appropriate PDF for profit over variable cost. Define a probability set function appropriate for assigning probability to events relating to profit above variable cost.

c. What is the probability that the firm makes at least \$100 profit above variable cost?

d. What is the probability that the firm makes a positive profit on a given day? Is making a positive profit a certain event? Why or why not?

e. *Given* that the firm makes at least \$100 profit above variable cost, what is the probability that it makes at least \$200 profit above variable cost?

**6.** The ACME Freight Co. has containerized a large quantity of 4-gigabyte memory chips that are to be shipped to a personal computer manufacturer in California. The shipment contains 1,000 boxes of memory chips, with each box containing a dozen chips. The chip manufacturer calls and says that due to an error in manufacturing, each box contains exactly one defective chip. The defect can be detected through an easily administered nondestructive continuity test using an ohmmeter. The chip maker requests that ACME break open the container, find the defective chip in each box, discard them, and then reassemble the container for shipment. The testing of each chip requires 1 min to accomplish.

a. Define a random variable representing the amount of testing time required to find the defective chip in a box of chips. What is the range of the random variable? What is the event space?

b. Define a PDF for the random variable you have defined in (a). Define a probability set function appropriate for assigning probabilities to events relating to testing time required to find the defective chip in a box of chips.

c. What is the probability that it will take longer than 5 min to find the defective chip in a box of chips?

d. If ACME uses 28-hour-shift workers for one shift each to perform the testing, what is the probability that testing of all of the boxes in the container will be completed?

**7.** Intelligent Electronics, Inc., manufactures monochrome liquid crystal display (LCD) notebook computer screens. The number of hours an LCD screen functions until failure is represented by the outcome of a random variable $X$ having range $R(X) = [0,\infty)$ and PDF

$$f(x) = .01 \exp\left(-\frac{x}{100}\right) I_{[0,\infty)}(x).$$

The value of $x$ is measured in thousands of hours. The company has a 1-year warranty on its LCD screen, during which time the LCD screen will be replaced free of charge if it fails to function.

a. Assuming that the LCD screen is used for 8,760 hours per year, what is the probability that the firm will have to perform warranty service on an LCD screen?

b. What is the probability that the screen functions for at least 50,000 hours? *Given* that the screen has already functioned for 50,000 hours, what is the probability that it will function for at least *another* 50,000 hours?

**8.** People Power, Inc., is a firm that specializes in providing temporary help to various businesses. Job applicants are administered an aptitude test that evaluates mathematics, writing, and manual dexterity skills. After the firm analyzed thousands of job applicants who took the test, it was found that the scores on the three tests could be viewed as outcomes of random variables with the following joint density function (the tests are graded on a 0–1 scale, with 0 the lowest score and 1 the highest):

$$f(x_1, x_2, x_3) = .80(2x_1 + 3x_2)\, x_3 \prod_{i=1}^{3} I_{[0,1]}(x_i).$$

a. A job opening has occurred for an office manager. People Power, Inc., requires scores of $> .75$ on both the mathematics and writing tests for a job applicant to be offered the position. Define the marginal density function for the mathematics and writing scores. Use it to define a probability space in which probability questions concerning events for the mathematics and writing scores can be answered. What is the probability that a job applicant who has just entered the office to take the test will qualify for the office manager position?

b. A job opening has occurred for a warehouse worker. People Power, Inc., requires a score of $> .80$ on the manual dexterity test for a job applicant to be offered the position. Define the marginal density function for the dexterity score. Use it to define a probability space in which probability questions concerning events for the dexterity score can be answered. What is the probability that a job applicant who has just entered the office to take the test will qualify for the warehouse worker position?

c. Find the conditional density of the writing test score, given that the job applicant achieves a score of $> .75$ on the mathematics test. Given that the job applicant scores $> .75$ on the mathematics test, what is the probability that she scores $> .75$ on the writing test? Are the two test scores independent random variables?

d. Is the manual dexterity score independent of the writing and mathematics scores? Why or why not?

**9.** The weekly average price (in dollars/foot) and total quantity sold (measured in thousands of feet) of copper wire manufactured by the Colton Cable Co. can be viewed as the outcome of the bivariate random variable $(P,Q)$ having the joint density function:

$$f(p,q) = 5pe^{-pq}\, I_{[.1,.3]}(p)\, I_{(0,\infty)}(q).$$

a. What is the probability that total dollar sales in a week will be less than \$2,000?

b. Find the marginal density of price. What is the probability that price will exceed \$.25/ft?

c. Find the conditional density of quantity, given price = .20. What is the probability that > 5,000 ft of cable will be sold in a given week?

d. Find the conditional density of quantity, given price = .10. What is the probability that > 5,000 ft of cable will be sold in a given week? Compare this result to your answer in (c). Does this make economic sense? Explain.

**10.** A personal computer manufacturer produces both desktop computers and notebook computers. The monthly proportions of customer orders received for desktop and notebook computers that are shipped within 1 week's time can be viewed as the outcome of a bivariate random variable $(X,Y)$ with joint probability density

$$f(x,y) = (2-x-y)\, I_{[0,1]}(x)\, I_{[0,1]}(y).$$

a. In a given month, what is the probability that more than 75 percent of notebook computers and 75 percent of desktop computers are shipped within 1 week of ordering?

b. Assuming that an equal number of desktop and notebook computers are ordered in a given month, what is the probability that more than 75 percent of all orders received will be shipped within 1 week?

c. Are the random variables independent?

d. Define the conditional probability that less than 50 percent of the notebook orders are shipped within 1 week, given that $x$ proportion of the desktop orders are shipped within 1 week (the probability will be a function of the proportion $x$). How does this probability change as $x$ increases?

**11.** A small nursery has seven employees, three of whom are salespersons, and four of whom are gardeners who tend to the growing and caring of the nursery stock. With such a small staff, employee absenteeism can be critical. The number of salespersons and gardeners absent on any given day is the outcome of a bivariate random variable $(X,Y)$. The nonzero values of the joint density function are given in tabular form as:

| | | Y | | | | |
|---|---|---|---|---|---|---|
| | | 0 | 1 | 2 | 3 | 4 |
| X | 0 | .75 | .025 | .01 | .01 | .03 |
| | 1 | .06 | .03 | .01 | .01 | .003 |
| | 2 | .025 | .01 | .005 | .005 | .002 |
| | 3 | .005 | .004 | .003 | .002 | .001 |

a. What is the probability that more than two employees will be absent on any given day?

b. Find the marginal density function of the number of gardeners that are absent. What is the probability that more than two gardeners will be absent on any given day?

c. Are the number of gardener absences and the number of salesperson absences independent random variables?

d. Find the conditional density function for the number of salespersons that are absent, given that there are no gardeners absent. What is the probability that there are no salespersons absent, given that there are no gardeners absent? Is the conditional probability higher or lower given that there is at least one gardener absent?

**12.** The joint density of the bivariate random variable $(X,Y)$ is given by

$$f(x,y) = xy\, I_{[0,1]}(x)\, I_{[0,2]}(y).$$

a. Find the joint cumulative distribution function of $(X,Y)$. Use it to find the probability that $x \le .5$ and $y \le 1$.

b. Find the marginal cumulative distribution function of $X$. What is the probability that $x \le .5$?

c. Find the marginal density of $X$ from the marginal cumulative distribution of $X$.

**13.** The joint cumulative distribution function for $(X,Y)$ is given by

$$F(x,y) = \left(1 - e^{-x/10} - e^{-y/2} + e^{-(x+5y)/10}\right) I_{(0,\infty)}(x)\, I_{(0,\infty)}(y).$$

a. Find the joint density function of $(X,Y)$.

b. Find the marginal density function of $X$.

c. Find the marginal cumulative distribution function of $X$.

**14.** The cumulative distribution of the random variable X is given by

$$F(x) = (1 - p^{x+1})\, I_{\{0,1,2,\ldots\}}(x), \text{for some choice of } p \in (0,1).$$

a. Find the density function of the random variable $X$.

b. What is the probability that $x \leq 8$ if $p = .75$?

c. What is the probability that $x \leq 1$ given that $x \leq 8$?

**15.** The federal mint uses a stamping machine to make coins. Each stamping produces 10 coins. The number of the stamping at which the machine breaks down and begins to produce defective coins can be viewed as the outcome of a random variable, $X$, having a PDF with general functional form $f(x) = \alpha\,(1 - \beta)^{x-1}\, I_{\{1,\,2,\,3,\,\ldots\}}(x)$, where $\beta \in (0,1)$.

a. Are there any constraints on the choice of $\alpha$ for $f(x)$ to be a PDF? If so, precisely what are they?

b. Is the random variable $X$ a discrete or a continuous random variable? Why?

c. It is known that the probability the machine will break down on the first stamping is equal to .05. What is the specific functional form of the PDF $f(x)$? What is the probability that the machine will break down on the tenth stamping?

d. Continue to assume the results in (a–c). Derive a functional representation for the cumulative distribution function corresponding to the random variable $X$. Use it to assign the appropriate probability to the event that the machine does not break down for at least 10 stampings.

e. What is the probability that the machine does not break down for at least 20 stampings, *given* that the machine does not break down for at least 10 stampings?

**16.** The daily quantity demanded of unleaded gasoline in a regional market can be represented as $Q = 100 - 10p + E$, where $p \in [0,8]$, and $E$ is a random variable having a probability density given by $f(e) = 0.025 I_{[-20,20]}(e)$.

Quantity demanded, Q, is measured in thousands of gallons, and price, $p$, is measured in dollars.

a. What is the probability of the quantity demanded being greater than 70,000 gal if price is equal to \$4? if price is equal to \$3?

b. If the average variable cost of supplying Q amount of unleaded gasoline is given by $C(Q) = Q^{.5}/2$, define a random variable that can be used to represent the daily profit above variable cost from the sale of unleaded gasoline.

c. If price is set equal to \$4, what is the probability that there will be a positive profit above variable cost on a given day? What if price is set to \$3? to \$5?

d. If price is set to \$6, what is the probability that quantity demanded will equal 40,000 gal?

**17.** For each of the cumulative distribution functions listed below, find the associated PDFs. For each CDF, calculate $P(x \leq 6)$.

a. $F(b) = (1 - e^{-b/6})\, I_{(0,\infty)}(b)$

b. $F(b) = (5/3)\,(.6 - .6^{\text{trunc}(b)+1})I_{[0,\infty)}(b)$

**18.** An economics class has a total of 20 students with the following age distribution:

| # of students | age |
|---|---|
| 10 | 19 |
| 4 | 20 |
| 4 | 21 |
| 1 | 24 |
| 1 | 29 |

Two students are to be selected randomly, without replacement, from the class to give a team report on the state of the economy. Define a random variable whose outcome represents the average age of the two students selected. Also, define a discrete PDF for the random variable. Finally, what is the probability space for this experiment?

**19.** Let $X$ be a random variable representing the *minimum* of the two numbers of dots that are facing up after a pair of fair dice is rolled. Define the appropriate probability density for $X$. What is the probability space for the experiment of rolling the fair dice and observing the minimum of the two numbers of dots?

**20.** A package of a half-dozen light bulbs contains two defective bulbs. Two bulbs are randomly selected from the package and are to be used in the same light fixture. Let the random variable $X$ represent the number of light bulbs selected that function properly (i.e., that are not defective). Define the appropriate PDF for $X$. What is the probability space for the experiment?

**21.** A committee of three students will be randomly selected from a senior-level political science class to present an assessment of the impacts of an antitax initiative to some visiting state legislators. The class consists of five economists, eight political science majors, four business majors, and three art majors. Referring to the experiment of drawing three students randomly from the class, let the bivariate random variable $(X,Y)$ be defined by $x$ = number of economists on the committee, and $y$ = number of business majors on the committee.

a. What is the range of the bivariate random variable $(X,Y)$? What is the PDF, $f(x,y)$, for this bivariate random variable? What is the probability space?

b. What is the probability that the committee will contain at least one economist and at least one business major?

c. What is the probability that the committee will consist of only political science and art majors?

d. *On the basis of the probability space you defined in (a) above,* is it possible for you to assign probability to the event that the committee will consists entirely of art majors? Why or why not? If you answer yes, calculate this probability using $f(x,y)$ from (a).

e. Calculate the marginal density function for the random variable $X$. What is the probability that the committee contains three economists?

f. Define the conditional density function for the number of business majors on the committee, *given* that the committee contains two economists. What is the probability that the committee contains less than one business major, *given* that the committee contains two economists?

g. Define the conditional density function for the number of business majors on the committee, *given* that the committee contains at least two economists. What is the probability that the committee contains less than one business major, *given* that the committee contains at least two economists?

h. Are the random variables $X$ and $Y$ independent? Justify your answer.

**22.** The Imperial Electric Co. makes high-quality portable compact disc players for sale in international and domestic markets. The company operates two plants in the United States, where one plant is located in the Pacific Northwest and one is located in the South. At either plant, once a disc player is assembled, it is subjected to a stringent quality-control inspection, at which time the disc player is either approved for shipment or else sent back for adjustment before it is shipped. On any given day, the proportion of the units produced at each plant that require adjustment before shipping, and the total production of disc players at the company's two plants, are outcomes of a trivariate random variable, with the following joint PDF:

$$f(x,y,z) = \tfrac{2}{3}(x+y)\, e^{-x}\, I_{(0,\infty)}(x)\, I_{(0,1)}(y)\, I_{(0,1)}(z),$$

where

$x$ = total production of disc players at the two plants, measured in thousands of units,
$y$ = proportion of the units produced at the Pacific Northwest plant that are shipped without adjustment, and
$z$ = proportion of the units produced in the southern plant that are shipped without adjustment.

a. In this application, the use of a *continuous* trivariate random variable to represent proportions and total production values must be viewed as only an *approximation* to the underlying real-world situation. Why? In the remaining parts, assume the approximation is acceptably accurate, and use the approximation to answer questions where appropriate.

b. What is the probability that less than 50 percent of the disc players produced in each plant will be shipped without adjustment and that production will be less than 1,000 units on a given day?

c. Derive the marginal PDF for the total production of disc players at the two plants. What is the probability that less than 1,000 units will be produced on a given day?

d. Derive the marginal PDF for the bivariate random variable $(Y,Z)$. What is the probability that more than 75 percent of the disc players will be shipped without adjustment from each plant?

e. Derive the conditional density function for $X$, *given* that 50 percent of the disc players are shipped from the Pacific Northwest plant without adjustment. What is the probability that 1,500 disc players will be produced by the Imperial Electric Co. on a day for which 50 percent of the disc players are shipped from the Pacific Northwest plant without adjustment?

f. Answer (e) for the case where 90 percent of the disc players are shipped from the Pacific Northwest plant without adjustment.

g. Are the random variables $(X, Y, Z)$ independent random variables?

h. Are the random variables $(Y, Z)$ independent random variables?

**23.** ACE Rentals, a car-rental company, rents three types of cars: compacts, mid-size sedans, and large luxury cars. Let $(x_1, x_2, x_3)$ represent the number of compacts, mid-size sedans, and luxury cars, respectively, that ACE rents per day. Let the sample space for the possible outcomes of $(X_1, X_2, X_3)$ be given by

$$S = \{ (x_1,x_2,x_3) : x_1, x_2, \text{ and } x_3 \in (0,1,2,3) \}$$

(ACE has an inventory of nine cars, evenly distributed among the three types of cars).
The discrete PDF associated with $(X_1, X_2, X_3)$ is given by

$$f(x_1, x_2, x_3) = \left[\frac{.004(3 + 2x_1 + x_2)}{(1 + x_3)}\right] \prod_{i=1}^{3} I_{\{0,1,2,3\}}(x_i).$$

The compact car rents for \$20/day, the mid-size sedan rents for \$30/day, and the luxury car rents for \$60/day.

a. Derive the marginal density function for $X_3$. What is the probability that all three luxury cars are rented on a given day?

b. Derive the marginal density function for $(X_1, X_2)$. What is the probability of more than one compact and more than one mid-size sedan being rented on a given day?

c. Derive the conditional density function for $X_1$, given $x_2 \leq 2$. What is the probability of renting no more than one compact care, given that two or more mid-size sedans are rented?

d. Are $X_1$, $X_2$, and $X_3$ jointly independent random variables? Why or why not? Is $(X_1, X_2)$ independent of $X_3$?

e. Derive the conditional density function for $(X_1, X_2)$, given that $x_3 = 0$. What is the probability of renting more than one compact and more than one mid-size sedan given that no luxury cars are rented?

f. If it costs \$150/day to operate ACE Rentals, define a random variable that represents the daily profit made by the company. Define an appropriate density function for this random variable. What is the probability that ACE Rentals makes a positive daily profit on a given day?

**24.** If $(X_1, X_2)$ and $(X_3, X_4)$ are independent bivariate random variables, are $X_2$ and $X_3$ independent random variables? Why or why not?

**25.** The joint density function of the discrete trivariate random variable $(X_1, X_2, X_3)$ is given by

$$f(x_1, x_2, x_3) = .20\, I_{\{0,1\}}(x_1)\, I_{\{0,1\}}(x_2)\, I_{\{|x_1 - x_2|\}}(x_3) + .05\, I_{\{0,1\}}(x_1)\, I_{\{0,1\}}(x_2)\, I_{\{1-|x_1 - x_2|\}}(x_3).$$

a. Are $(X_1, X_2)$, $(X_1, X_3)$, and $(X_2, X_3)$ *each* pairwise independent random variables?

b. Are $X_1, X_2, X_3$ jointly independent random variables?

**26.** SUPERCOMP, a retail computer store, sells personal computers and printers. The number of computers and printers sold on any given day varies, with the probabilities of the various possible sales outcomes being given by the following table:

| | | Number of computers sold | | | | | |
|---|---|---|---|---|---|---|---|
| | | 0 | 1 | 2 | 3 | 4 | |
| | 0 | .03 | .03 | .02 | .02 | .01 | Probabilities of elementary events |
| Number | 1 | .02 | .05 | .06 | .02 | .01 | |
| of | 2 | .01 | .02 | .10 | .05 | .05 | |
| printers | 3 | .01 | .01 | .05 | .10 | .10 | |
| | 4 | .01 | .01 | .01 | .05 | .15 | |

a. If SUPERCOMP has a profit margin (product sales price – product unit cost) of \$100 per computer sold and \$50 per printer sold, define a random variable representing aggregate profit margin from the sale of computers and printers on a given day. What is the range of this random variable?

b. Define a discrete density function appropriate for use in calculating probabilities of all events concerning aggregate profit margin outcomes on a given day.

c. What is the probability that the aggregate profit margin is $\geq$ \$300 on a given day?

d. The daily variable cost of running the store is \$200/day. What is the probability that SUPERCOMP's aggregate profit margin on computer and printer sales will equal or exceed variable costs on a given day?

e. Assuming that events involving the number of computers and printers sold are independent from day to day, what is the probability that for any given 6-day business week, aggregate profit margins equal or exceed variable cost all 6 days?

**27.** Given the function definitions below, determine which can be used as PDFs (PDFs) and which cannot. Justify your answers.

a. $f(x) = \begin{cases} \left(\frac{1}{4}\right)^x & \text{for} \quad x = 0, 1, 2, \ldots \\ 0 & \text{otherwise} \end{cases}$

b. $f(x) = \left(\frac{1}{4}\right)^x I_{(0,\infty)}(x)$

c. $f(x,y) = \begin{cases} (2x+y)/100, & \text{for} \quad x \text{ and } y = 0, 1, 2, 3, 4, \text{ and } y \leq x \\ 0 & \text{otherwise} \end{cases}$

d. $f(x,y) = 6xy^2 I_{[0,1]}(x) I_{[0,1]}(y)$

**28.** Given the function definitions below, determine which can be used as cumulative distribution functions (CDFs) and which cannot. Justify your answers.

a. $F(c) = \dfrac{e^c}{1+e^c}$ for $c \in (-\infty, \infty)$

b. $F(c) = \begin{cases} 1 - x^{-2}, & \text{for} \quad c \in (1, \infty) \\ 0 & \text{otherwise} \end{cases}$

c. $F(c) = \begin{cases} 1 - (.5)^{floor(c)} & \text{for} \quad c \geq 1 \\ 0 & \text{otherwise.} \end{cases}$

where *floor* (c) ≡ round down the value c.

d. $F(c_1, c_2) = \begin{cases} 1 \; if \; c_1 \text{ and } c_2 \in (1, \infty) \\ c_1^3 I_{[0,1]}(c_1) if \; c_2 \in (1, \infty) \\ c_2^2 I_{[0,1]}(c_2) if \; c_1 \in (1, \infty) \\ c_1^3 c_2^2 I_{[0,1]}(c_1) I_{[0,1]}(c_2) \text{ for } c_1 \text{ and } c_2 \in (-\infty, 1) \end{cases}$

e. $F(c_1, c_2) = (1 - e^{-c_1})(1 - e^{-c_2}) I_{[0,\infty)}(c_1) I_{[0,\infty)}(c_2)$

**29.** For those functions in (28) that are actually cumulative distribution functions (CDFs), use the duality principle to derive the PDFs (PDFs) that are associated with the CDFs.

**30.** The daily quantity demanded of milk in a regional market, measured in 1,000's of gallons, can be represented during the summer months as the outcome of the following random variable:

$Q = 200 - 50p + V$,

where V is a random variable having a probability density defined by

$f(v) = 0.02 I_{[-25,25]}(v)$ and $p$ is the price of milk, in dollars per gallon.

a. What is the probability that the quantity demanded will be greater than 100,000 gal if price is equal to \$2? if price is equal to \$2.25?

b. If the variable cost of supplying $Q$ amount of milk is given by the cost function $C(Q) = 20\,Q^{.5}$, define a random variable that represents the daily profit above variable cost from the sale of milk.

c. If price is equal to \$2, what is the probability that there will be a positive profit above variable cost on a given day? What if price is set to \$2.25?

d. Is there any conceptual problem with using the demand function listed above to model quantity demanded if $p = 4$? If so, what is it?

**31.** A small locally-owned hardware store in a western college town accepts both cash and checks for purchasing merchandise from the store. From experience, the store accountant has determined that 2 percent of the checks that are written for payment are "bad" (i.e., they are refused by the bank) and cannot be cashed. The accountant defines the following probability model $(R(X), f(x))$ for the outcome of a random variable $X$ denoting the number of bad checks that occur in $n$ checks received on a given day at the store:

$$f(x) = \begin{cases} \frac{n!}{(n-x)!x!}(.02)^x(.98)^{n-x} & \text{for} \quad x \in R(X) = \{0, 1, 2, \ldots, n\} \\ 0 & \text{elsewhere} \end{cases}$$

If the store receives 10 checks for payment on a given day, what is the probability that:

a. Half are bad?

b. No more than half are bad?

c. None are bad?

d. None are bad, *given that* no more than half are bad?

**32.** Let an outcome of the random variable $T$ represent the time, in minutes, that elapses between when an order is placed at a ticket counter by a customer and when the ticket purchase is completed. The following probability model $(R(T), f(t))$ governs the behavior of the random variable $T$:

$$f(t) = \begin{cases} 3e^{-3t} \text{ for } t & \in R\{T\} = [0, \infty) \\ 0 & \text{elsewhere} \end{cases}$$

a. What is the probability that the customer waits less than 3 min to have her ticket order completed?

b. Derive the cumulative distribution function for T. Use it to define the probability that it takes longer than 10 min to have the ticket order completed.

c. Given that the customer's wait will be less than 3 min, what is the probability that it will be less than 1 min?

d. *Given* that the customer has already waited more than 3 min, what is the probability that the customer will wait *at least* another 3 min to have the ticket order completed?

**33.** Outcomes of the random variable $Z$ represent the number of customers that are waiting in a queue to be serviced at Fast Lube, a quick stop automobile lubrication business, when the business opens at 9 A.M. on any given Saturday. The probability model $(R(Z), f(z))$ for the random variable $Z$ is given by:

$$f(z) = \begin{cases} .5^{z+1} \text{ for } z & \in R(Z) = \{0, 1, 2, 3, ...\} \\ 0 & \text{elsewhere} \end{cases}$$

a. Derive the cumulative distribution function for $Z$.

b. What is the probability that there will be less than 10 people waiting?

c. What is the probability that there will be more than 3 people waiting?

d. Given that no more than two people will be waiting, what is the probability that there will be no customers when business opens at 9 A.M.?

**34.** The daily wholesale price and quantity sold of ethanol in a Midwestern regional market during the summer months is represented by the outcome of a bivariate random variable $(P, Q)$ having the following probability model $(R(P,Q), f(p,q))$:

$$f(p, q) \begin{cases} .5pe^{-pq} \text{ for } & (p, q) \in R(P, Q) = [2, 4] \times [0, \infty) \\ 0 & \text{elsewhere} \end{cases}$$

where price is measured in dollars and quantity is measured in 100,000 gal units (e.g., $q = 2$ means 200,000 gal were sold).

a. Derive the marginal probability density of price. Use it to determine the probability that price will exceed \$3.

b. Derive the marginal cumulative distribution function for price. Use it to verify your answer to part (a) above.

c. Derive the marginal probability density of quantity. Use it to determine the probability that quantity sold will be less than \$500,000 gal.

d. Let the random variable $D = PQ$ denote the daily total dollar sales of ethanol during the summer months. What is the probability that daily total dollar sales will exceed \$300,000?

e. Are $P$ and $Q$ independent random variables?

**35.** The BigVision Electronic Store sells a large 73 inch diagonal big screen TV. The TV comes with a standard 1 year warranty on parts and labor so that if anything malfunctions on the TV in the first year of ownership, the company repairs or replaces the TV for free. The store also sells an "extended warranty" which a customer can purchase that extends warranty coverage on the TV for another 2 years, for a total of 3 years of coverage. The daily numbers of TVs and extended warranties sold can be viewed as the outcome of a bivariate random variable $(T, W)$ with probability model $(R(T,W), f(t,w))$ given by

$$f(t, w) \begin{cases} (2t + w)/100, \text{ for } & t \text{ and } w = 0, 1, 2, 3, 4, \text{ and } w \leq t \\ 0 & \text{otherwise} \end{cases}$$

a. What is the probability that all of the TVs sold on a given day will be sold with extended warranties?

b. Derive the marginal density function for the number of TVs sold. Use it to define the probability that $\leq 2$ TVs are sold on a given day?

c. Derive the marginal density function for the number of warranties sold. What is the probability that $\geq 3$ warranties are sold on a given day?

d. Are $T$ and $W$ independent random variables?

**36.** The following function is proposed as a cumulative distribution function for the bivariate random variable $(X, Y)$:

$$F(x, y) = \left(1 + e^{-(x/10 + y/20)} - e^{-x/10} - e^{-y/20}\right) I_{(0,\infty)}(x) I_{(0,\infty)}(y)$$

a. Verify that the function has the appropriate properties to serve as a cumulative distribution function.

b. Derive the marginal cumulative distribution function of Y.

c. Derive the marginal PDF of $Y$.

d. Derive the joint PDF of $(X, Y)$.

e. What is the probability that $X \leq 10$ and $Y \leq 20$?

f. Are $X$ and $Y$ independent random variables?

**37.** For each of the joint PDFs listed below, determined which random variables are independent and which are not.

a. $f(x,y) = e^{-(x+y)} I_{[0,\infty)}(x) I_{[0,\infty)}(y)$

b. $f(x,y) = \frac{x(1+y)}{300} I_{\{1,2,3,4,5\}}(x) I_{\{1,2,3,4,5\}}(y)$

c. $f(x,y,z) = 8xyz I_{[0,1]}(x) I_{[0,1]}(y) I_{[0,1]}(z)$

d. $f(x_1,x_2,x_3) = \frac{.5^{x_1}.2^{x_2}.75^{x_3}}{10} \prod_{i=1}^{3} I_{\{0,1,2,\ldots\}}(x_i)$

**38.** The daily wholesale price and quantity sold of ethanol in a Midwestern regional market during the summer months is represented by the outcome of a bivariate random variable $(P, Q)$ having the following probability model $\{R(P,Q), f(p,q)\}$:

$$f(p,q) = \begin{cases} .5pe^{-pq} & \text{for} \quad (p,q) \in R(P,Q) = [2,4] \times [0,\infty) \\ 0 & \text{elsewhere} \end{cases}$$

where price is measured in dollars and quantity is measured in 100,000 gal units (e.g., $q = 2$ means 200,000 gal were sold).

a. Derive the conditional-on-p PDF for quantity sold.

b. What is the probability that quantity sold exceeds 50,000 gal if price = \$2. What is the probability that quantity sold exceeds 50,000 gal if price = \$4. Does this make economic sense?

c. What is the probability that quantity sold exceeds 50,000 gal if price is *greater than or equal to* \$3.00?

**39.** Let the random variable $X$ represent the product of the number of dots facing up on each die after a pair of fair dice is rolled. Let $Y$ represent the sum of the number of dots facing up on the pair of dice.

a. Define a probability model $(R(X), f(x))$ for the random variable $X$.

b. What is the probability that $X \geq 16$?

c. Define a probability model $(R(X,Y), f(x,y))$ for the random vector $(X, Y)$.

d. What is the probability that $X \geq 16$ and $Y \geq 8$?

e. Are $X$ and $Y$ independent random variables?

f. Define the conditional PDF of $X$ given that $Y = 7$.

g. What is the probability that $X \geq 10$ given that $Y = 7$?

**40.** The production of a certain volatile commodity is the outcome of a stochastic production function given by $Y = L^{.5} K^{.25} e^{v}$, where $v$ is a random variable having the cumulative distribution function $F(v) = \frac{1}{1+e^{-2(v-1)}}$, $L$ denotes units of labor and $K$ denotes units of capital.

a. If labor is applied at 9 units and capital is applied at 16 units, what is the probability that output will exceed 12 units?

b. Given the input levels applied in (a), what is the probability that output will be between 12 and 16 units?

c. What level of capital and labor should be applied so that the probability of producing a positive profit is *maximized* when output price is \$10, labor price is \$5, and capital price is \$10?

d. What is the value of the maximum probability of obtaining positive profit?

# 3 Expectations and Moments of Random Variables

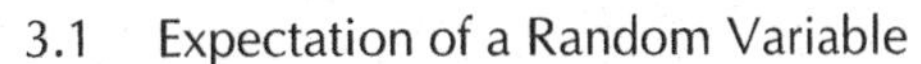



## 3.1 Expectation of a Random Variable

The definition of the expectation of a random variable can be motivated both by the concept of a weighted average and through the use of the physics concept of center of gravity, or the balancing point of a distribution of weights. We first examine the case of a discrete random variable and look at a problem involving the balancing-point concept.[1]

**Example 3.1** ***Balancing Weights*** Suppose that a weightless rod is placed on a fulcrum, a weight of 10 lb is placed on the rod exactly 4 ft to the right of the fulcrum, and a weight of 5 lb is placed on the rod exactly 8 ft to the left of the fulcrum, as shown in Figure 3.1.

[1] Readers who recollect earlier days spent on a seesaw should possess ample intuition regarding the placement of weights appropriate distances from a fulcrum so as to achieve a "balanced seesaw."

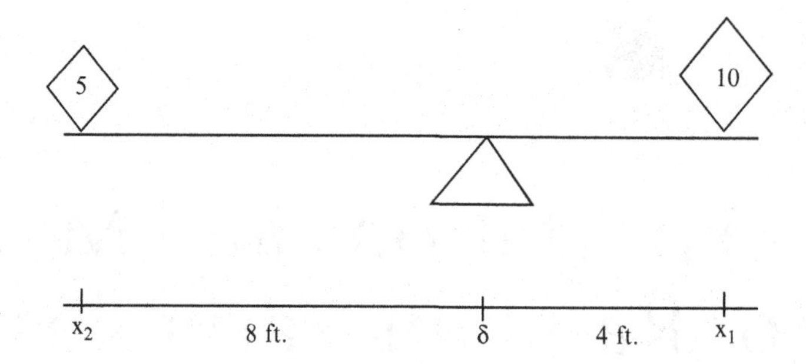


**Figure 3.1** Weights on a weightless rod.

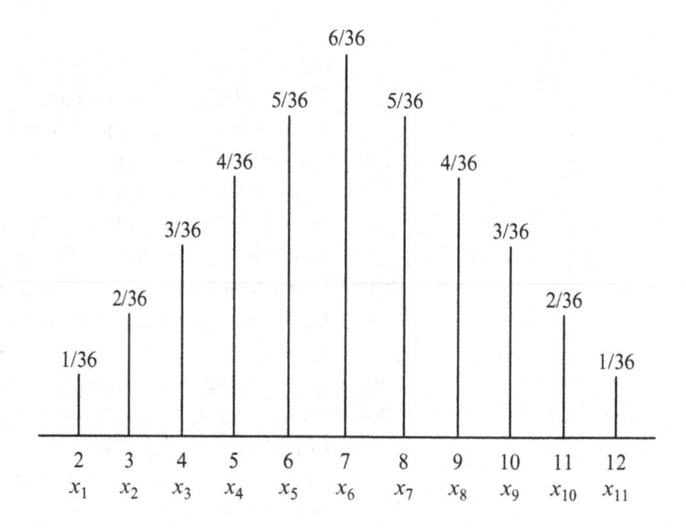


**Figure 3.2** Density function "weights" on a weightless rod.

Assume that $\delta = 0$ is the point at which the fulcrum is placed, so that the 10 lb weight is at the point $x_1 = 4$, and the 5 lb weight is at the point $x_2 = -8$. Let mass$(x)$ denote the mass placed at point $x$. The *moment* of any mass placed at a point $x$ is defined to be the product of the mass times its signed distance from the fulcrum, $[\text{mass}(x)]\,(x - \delta)$, where $\delta$ is the point at which the fulcrum is placed. Thus, the moment of the 10 lb weight is $10(4 - 0) = 40$, while the moment of the 5 lb pound weight is $5(-8 - 0) = -40$. A system of weights with fulcrum placed at $\delta$ will balance if the sum of the moments $\sum_{i=1}^{n} [\text{mass}(x_i)](x_i - \delta)$, called the *total moment of the system*, is equal to zero. Our system balances, since $40 + (-40) = 0$.

The moments concept illustrated in Example 3.1 can be used to identify the point at which a probability density "balances." Recall the dice example of Example 2.2. We place the probability "weights" on a weightless rod at the points corresponding to the outcomes with which the probabilities are associated, as shown in Figure 3.2. □

At what point should the fulcrum be placed so that the distribution of weights balances? We require that the total moment of the system be zero. Thus, we require that

$$\sum_{i=1}^{11}[\text{mass}(x_i)](x_i - \delta) = \sum_{i=1}^{11} f(x_i)(x_i - \delta) = 0,$$

which implies

$$\sum_{i=1}^{11} f(x_i)x_i = \delta\left[\sum_{i=1}^{11} f(x_i)\right] = \delta,$$

where the sum in brackets equals one because the density function $f(x)$ is being summed over the entire range of the random variable $X$. Substituting the appropriate values of $x_i$ and $f(x_i)$ in the expression obtains the result $\delta = 7$. Thus, if the fulcrum were placed at the point 7, the system of weights would balance. The quantity $\delta$ is precisely what is meant by the **expected value** of the discrete random variable $X$ with density function $f(x)$. Thus, the expected value of a discrete random variable is a measure of the center of gravity of its density function.

**Definition 3.1** ***Expectation of a Random Variable: Discrete Case***

The expected value of a discrete random variable is defined by $E(X) = \sum_{x \in R(X)} x f(x)$, provided the sum exists.[2]

Since $f(x) \geq 0 \; \forall x \in R(X)$ and $\sum_{x \in R(X)} f(x) = 1$, the expected value of a discrete random variable can also be straightforwardly interpreted as a weighted average of the possible outcomes (or range elements) of the random variable. In this context the weight assigned to a particular outcome of the random variable is equal to the probability that the outcome occurs (as given by the value of $f(x)$).

**Example 3.2** ***Expected Gain from Insurance***

A life insurance company offers a 50-year old male a \$1,000 face value, 1-year term life insurance policy for a premium of \$14. Standard mortality tables indicate that the probability a male in this age category will die within the year is .006. What is the insurance company's expected gain from issuing this policy?

**Answer**: Define a random variable $X$ having range $R(X) = \{14, -986\}$, the outcomes corresponding, respectively, to the premium of \$14 collected

[2] It is an unfortunate fact of infinite sums involving both positive and negative terms that, unless the sum is *absolutely convergent*, an appropriate reordering of terms will result in the infinite sum converging to other real numbers (Bartle, Real Analysis, p. 292). This is hardly consistent with the notion of a balancing point of the density $f(x)$. Moreover, the nonuniqueness of values to which the infinite sum can converge makes any particular convergence point arbitrary and meaningless as the expectation of $X$. Thus, to ensure the finiteness and *uniqueness* of the converged value in the countably infinite case, a technical condition can be added whereby $E(X)$ is said to exist *iff* $\sum_{x \in R(X)} |x| f(x) < \infty$ which is to say, *iff* $\sum_{x \in R(X)} x f(x)$ is *absolutely convergent*. For virtually any problem of practical interest, if the sum used in the definition of the expectation is finite, the expectation can be said to exist. It should also be noted that in many applications, random variables are nonnegative valued, in which case if the sum is convergent, it is necessarily *absolutely* convergent.

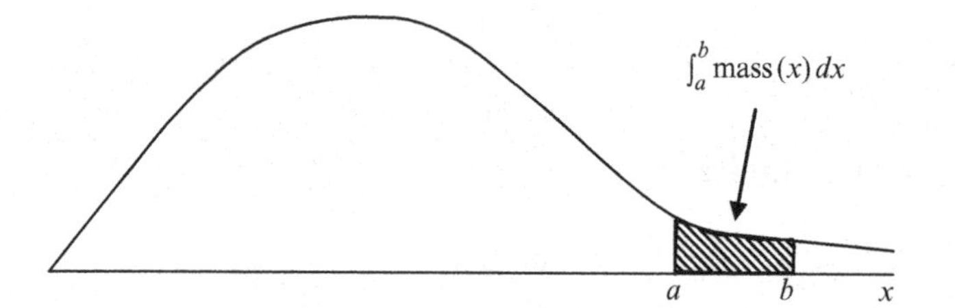


**Figure 3.3**
Continuous mass on a weightless rod.

by the company if the person lives, or the net payment of \$986 (\$1,000 minus the premium collected) to the person's estate if he dies. The probabilities of the two elementary events are .994 and .006, respectively. Then, $E(X) = (14)(.994) - (986)(.006) = 8$. □

Note that the expected value of $X$ need not be a value in the range of $X$ as the previous and following examples illustrate.

**Example 3.3**
***Expected Value not Necessarily in R(X)***

Examine the experiment of rolling a die, and recall that the density function associated with the dots facing up on the die is $f(x) = (1/6)\, I_{\{1,\ldots,6\}}(x)$. In this case $R(X) = \{1, 2, 3, 4, 5, 6\}$. The expected value of $X$ equals $E(X) = \sum_{x \in R(X)} xf(x) = \sum_{x=1}^{6} (x/6) I_{\{1,\ldots,6\}}(x) = 3.5$, and thus $E(X) \notin R(X)$. □

In the continuous case, the physics problem of balancing mass on a weightless rod can no longer be conceptualized as having weights applied to specific points on the rod. Instead, the mass is interpreted as being continuously spread out along the rod, exerting downward force along a continuum of points on the rod. The mass function, mass$(x)$, is now a density of the mass at point $x$, $\int_a^b \text{mass}(x)dx$ equals the total mass placed on the rod, and $\int_a^b \text{mass}(x)dx$ equals the mass lying between the points $a$ and $b$ (see Figure 3.3)[3]. The mass is balanced on the rod when the fulcrum is placed at the point $\delta$ such that the total moment of the mass, $\int_{-\infty}^{\infty} mass(x)(x - \delta)dx$, equals zero.

Viewing our density function as a probability mass, the continuous density "balances" with a fulcrum placed at the point $\delta$ if $\int_{-\infty}^{\infty} f(x)(x - \delta)dx = 0$, which implies that

$$\int_{-\infty}^{\infty} xf(x)dx = \delta \underbrace{\int_{-\infty}^{\infty} f(x)dx}_{1} = \delta.$$

Again, it is this balancing point or center of gravity, $\delta$, of the density that represents the expectation of the continuous random variable $X$ having density $f(x)$.

[3] The reader might notice that the mass function would exhibit properties similar to a probability density function, except the integral over the real line would not necessarily = 1, but rather equals the number reflecting the total mass placed on the rod.

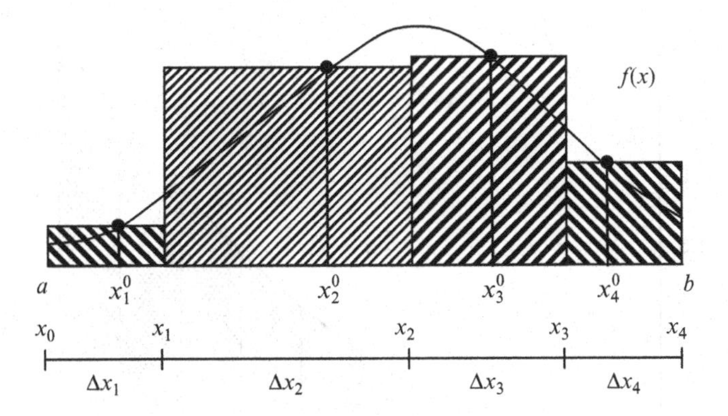


**Figure 3.4** Approximation of area under $f(x)$.

**Definition 3.2** ***Expected Value of a Random Variable: Continuous Case***

The expected value of the continuous random variable $X$ is defined by $E(X) = \int_{-\infty}^{\infty} xf(x)dx$, provided the integral exists.

The expected value of $X$ in the continuous case can also be viewed, in a limit sense, as a weighted average of the possible outcomes (or range elements) of the random variable. This interpretation follows fundamentally from the definition of the definite integral as the limit of a *Riemann sum*.[4] For the sake of exposition, we assume that the positive values of $f(x)$ all occur for $x$ within the interval $x \in [a,b]$ for finite $a$ and $b$, although a similar argument holds when $a$ and/or $b$ are infinite.

Let $x_o = a$, $x_n = b$, $x_o < x_1 < x_2 < \ldots < x_n$, $\Delta x_i = x_i - x_{i-1}$, and examine the Riemann sum $\sum_{i=1}^{n} x_i^0 f(x_i^0)\Delta x_i$, where $x_i^0$ is a value chosen such that $x_i^0 \in [x_{i-1}, x_i]$. This situation can be represented by the diagram in Figure 3.4. Thus, each $x_i^0$ is weighted by the value $f(x_i^0)\Delta x_i$, an area indicated in the diagram by a shaded rectangle, and a summation is taken over all the chosen values of the $x_i^0$. We have effectively divided the interval $[a,b]$ into a collection of subintervals of various widths $\Delta x_i$, $i = 1,\ldots,n$. The subinterval of maximum width is referred to as the *mesh* (or sometimes, the *norm*) of the collection of subintervals, i.e., mesh $= \max(\Delta x_1,\ldots,\Delta x_n)$. If $f(x)$ is continuous,[5] then as we increase without bound the number of subintervals, and in so doing decrease the mesh to zero, we have

[4] Bartle, Real Analysis, pp. 213–214.

[5] The argument can still be applied to cases where there are a finite number of discontinuities.

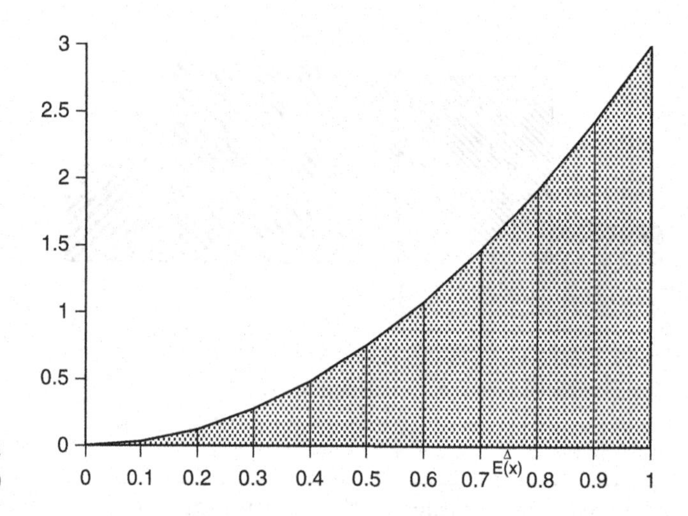


**Figure 3.5**
$f(x) = 3x^2 I_{[0,1]}(x)$

$$\lim_{n\to\infty,\,\text{mesh}\to 0} \sum_{i=1}^{n} x_i^0 f(x_i^0)\Delta x_i = \int_a^b xf(x)dx = \int_{-\infty}^{\infty} xf(x)dx$$

and

$$\lim_{n\to\infty,\,\text{mesh}\to 0} \sum_{i=1}^{n} f(x_i^0)\Delta x_i = \int_a^b f(x)dx = \int_{-\infty}^{\infty} f(x)dx = 1$$

(presuming the first limit, and hence first integral, exists—the second limit, and integral, necessarily exists since $f(x)$ is a density function). Therefore, under the assumptions of our argument, the expected value of $X$ can be viewed, in a limiting sense, as a weighted average of an infinite number of possible outcomes of the random variable.

**Example 3.4** ***Expectation of a Continuous RV***

A large domestic automobile manufacturer mails out quarterly customer satisfaction surveys to owners who have purchased new automobiles within the last 3 years. The proportion of surveys returned in any given quarter is the outcome of a random variable $X$ having density function $f(x) = 3x^2 I_{[0,1]}(x)$. What is the expected proportion of surveys returned in any given quarter?
**Answer**: By definition,

$$E(X) = \int_{-\infty}^{\infty} xf(x)dx = \int_{-\infty}^{\infty} x(3x^2 I_{[0,1]}(x))dx = \int_0^1 3x^3\,dx = \left.\frac{3x^4}{4}\right|_0^1 = .75$$

This is represented diagrammatically in Figure 3.5. □

In applications, the following result can often be a useful sufficient condition for the existence of the expected value in either the discrete or continuous case.

**Theorem 3.1** ***Existence of* E(*X*) *for Bounded* R(*X*)**

*If* $|x|< c\ \forall x \in R(X)$ *for some choice of* $c\in (0,\infty)$, *then* $E(X)$ *exists.*

**Proof** By the assumption of the theorem, $f(x) = 0\ \forall x$ such that $|x| \geq c$. Then

**Discrete:**

$\sum_{x\in R(X)} |x|f(x)< \sum_{x\in R(X)} cf(x) = c \sum_{x\in R(X)} f(x) = c<\infty$, so that $E(X)$ exists.

**Continuous:**

$\int_{-\infty}^{\infty} |x|f(x)dx< \int_{-\infty}^{\infty} cf(x)dx = c\int_{-\infty}^{\infty} f(x)dx = c<\infty$, so that $E(X)$ exists. ■

The theorem indicates that expectations exist for any random variable whose outcomes are bounded in absolute value because the respective sum or integral is absolutely convergent.

## 3.2 Stieltjes Integration and Implicit Support Convention

We will henceforth introduce a convention that serves to economize on notation whereby we tacitly assume that whenever a discrete density function is involved in a summation expression, summation is understood to occur only over the countable number of points for which $f(x) > 0$, i.e., only over points in the *range* or the *support of the random variable* (recall Definitions 2.8 and 2.9). Thus, we will generally abbreviate expressions such as $\sum_{x\in A, f(x)>0} f(x)$ or $\sum_{x\in A\cap R(X)} f(x)$ by $\sum_{x\in A} f(x)$, the condition $f(x) > 0$ or $x \in R(X)$ always being implicitly understood to apply.

We will also make occasional use of the **Stieltjes integration** operation (pronounced "Steel-yah") when it is useful in defining or proving concepts for both the discrete and continuous random variables simultaneously. This type of integration encompasses summation as a special case, when based on the differential of the cumulative distribution function of a discrete random variable. The Stieltjes integration concept actually generalizes the operation of Riemann integration in a number of important ways, but for our purposes, it suffices to think of the operation as simply defining a notation for representing either summation or ordinary Riemann integration when we have a discrete or continuous random variable case, respectively. We define the operator applied in our context below.

**Definition 3.3** ***Stieltjes Integration Notation***

Let $F(x)$ be the cumulative distribution function associated with the random variable $X$. Then the Stieltjes integral of the function $g(x)$ with respect to $F$ is defined as

$$\int_{x\in A} g(x)\, dF(x) \equiv \left\{ \begin{array}{c} \sum_{x\in A} g(x)f(x) \\ \int_{x\in A} g(x) f(x)dx \end{array} \right\} \text{when } X \text{ is } \left\{ \begin{array}{c} \text{discrete} \\ \text{continuous} \end{array} \right\}.$$

Note we have already employed our implicit support convention in the discrete case. The definition can be applied to both scalar and multivariate cases, where one simply interprets the $x$ argument in the preceding definition as a vector to define the notation for the multivariate application. Using Stieltjes integration notation, one can express the expectation of a random variable $X$, *for both the discrete and continuous cases*, as follows.

**Definition 3.4**
***Expected Value of a Random Variable (via Stieltjes Integration)***

The expected value of a random variable $X$ is defined by $\mathrm{E}(X) = \int_{-\infty}^{\infty} x\,dF(x)$, provided the integral exists.

Based on the above discussion of the Stieltjes integral notation, if $X$ is discrete,

$$\mathrm{E}(X) = \int_{-\infty}^{\infty} x\,dF(x) = \sum_{x \in R(X)} x f(x),$$

and if $X$ is continuous,

$$\mathrm{E}(X) = \int_{-\infty}^{\infty} x\,dF(x) = \int_{-\infty}^{\infty} x f(x) dx.$$

## 3.3 Expectation of a Function of Random Variables

Many cases arise in practice where one is interested in the expectation of a function of a random variable rather than the expectation of a random variable itself. For example, the profit on a stock investment will be a *function* of the difference between the per share buying and selling prices of the stock, and the net return on an advertising campaign is a *function* of consumer buying response to the campaign—both the stock selling price and consumer buying response might be viewed as random variables. How should $\mathrm{E}(Y)$ be determined when $y = g(x)$, $x \in R(X)$, and $X$ has density function $f(x)$? By definition, if we know the density of $Y$, $h(y)$, then

$$\mathrm{E}(Y) = \int_{-\infty}^{\infty} y\,dH(y) = \begin{cases} \sum_{y \in R(Y)} y h(y) & \text{(discrete)} \\ \int_{-\infty}^{\infty} y h(y) dy & \text{(continuous)} \end{cases}$$

where $H(y)$ is the cumulative distribution function. To use this expectation definition directly, one would need to establish the density of $Y$. This can be done in principle by exploiting the functional relationship between $y$ and $x$, given knowledge of the density $f(x)$, but finding $h(y)$ can sometimes be challenging (we will examine methods for deriving such densities in Chapter 6). Fortunately, one does not need to derive the density function of $y$ to obtain $\mathrm{E}(Y)$.

Since $Y$ is defined via a composition of the functions $g$ and $X$, and since the domain of $X$ is conceptually the sample space $S$, then $Y$ is defined on the elements of $S$ via the composition, i.e., an outcome of $y$ can be viewed as being given by $y = g(X(w))$ for $w \in S$, so that $y$: $S \to \mathbb{R}$. This implies that the range of $Y$ and probabilities of events for $Y$ can be represented alternatively as

$$R(Y) = \{ y{:}y{=}g(x), x \in R(X)\} = \{ y{:}y{=}g(X(w)), w \in S\}$$

and

$$P_Y(A) = P_X(\{ x{:}g(x) \in A, x \in R(X)\}) = P(\{ w{:}g(X(w)) \in A, w \in S\}).$$

Therefore, we can concentrate our attention on the $g$ function component of the composition, which has a real-valued domain $R(X)$, and conceptualize the outcomes of $Y$ as being generated by $y = g(x)$ for $x \in R(X)$, where $y$: $R(X) \to \mathbb{R}$. In so doing, we lose no information concerning the possible outcomes of $Y$ or the probabilities of events for $Y$, and we gain the convenience of being able to ignore the original probability space $\{S, \Upsilon, P\}$ and deal exclusively with a real-valued domain for the function $Y$. We will generally focus on this latter interpretation of the function $Y$ in our subsequent study, unless we make explicit reference to the domain of $Y$ as being $S$.

We now present a theorem identifying a straightforward approach for obtaining the expectation of $Y = g(X)$ by using density weightings applied to the outcomes of $X$.

**Theorem 3.2** ***Expectations of Functions of Random Variables*** *Let X be a random variable having density function $f(x)$. Then the expectation of $Y = g(x)$ is given by*[6]

$$E(g(X)) = \int_{-\infty}^{\infty} g(x)dF(x) = \begin{cases} \sum_{x \in R(X)} g(x)f(x) & \text{(discrete)} \\ \int_{-\infty}^{\infty} g(x)f(x)dx & \text{(continuous)} \end{cases}$$

**Proof** See Appendix ■

**Example 3.5** ***Expectation of Profit Function*** Let the daily profit function of a firm be given by $\Pi(X) = pq(X) - rX$, where $X$ is a random variable whose outcome represents the daily quantity of a highly perishable agricultural commodity delivered to the firm for processing, measured in hundredweights (100 lb units), $p = 5$ is the price of the processed product per pound, $r = 2$ is the cost of the raw agricultural commodity per pound, and $q(x) = x^{.9}$ is the production function indicating the relationship between raw

[6]It is tacitly assumed that the sum and integral are absolutely convergent for the expectation to exist.

and finished product measured in hundredweights. Let the density function of $X$ be $f(x) = \frac{1+2x}{110} I_{[0,10]}(x)$. What is the expected value of daily profit?
**Answer**: A direct application of Theorem 3.2 yields

$$\mathrm{E}(\Pi(X)) = \int_{-\infty}^{\infty} \Pi(x)f(x)dx = \int_{0}^{10} (5x^{.9} - 2x)\left(\frac{1+2x}{110}\right)dx = 13.77.$$

Since quantities are measured in hundredweights, this means that the expected profit is \$1,377 per day. □

**Example 3.6**
***Expectation of Per Unit Profit Function***

Your company manufactures a special 1/4-inch hexagonal bolt for the Defense Department. For the bolt to be useable in its intended application, the bolt must be manufactured within a 1 percent tolerance of the 1/4-inch specification. As part of your quality assurance program, each bolt is inspected by a laser measuring device that is 100 percent effective in detecting bolts that are not within the 1 percent tolerance. Bolts not meeting the tolerance are discarded. The actual size of a bolt manufactured on your assembly line is represented by a random variable, $X$, having a probability density $f(x) = (.006)^{-1} I_{[.247,\ .253]}(x)$, where $x$ is measured in inches. If your profit per bolt sold is \$.01, and if a discarded bolt costs your company \$.03, what is your expected profit per bolt manufactured?
**Answer**: We define a discrete random variable whose outcome represents whether a bolt provides the company with a \$.01 profit or a \$.03 loss. Specifically, $Y = g(X) = .01(I_{[.2475,\ .2525]}(X)) - .03\,(1 - I_{[.2475,\ .2525]}(X))$ is the function of $X$ that we seek, where $y = .01$ if $x \in [.2475, .2525]$ (i.e., the bolt is within tolerance) and $y = -.03$ otherwise. Then

$$\begin{aligned}\mathrm{E}(Y) = \mathrm{E}(g(X)) &= \int_{.247}^{.253} g(x)f(x)dx \\ &= .01P(.2475 \le x \le .2525) - .03[1 - P(.2475 \le x \le .2525)] \\ &= .01(.83\underline{3}) - .03(.16\underline{6}) = .0033.\end{aligned}$$

□

The reader should note that in the preceding example, while $X$ was a *continuous* random variable, $Y = g(X)$ is a *discrete* random variable. Whether $Y = g(X)$ is discrete or continuous depends on the nature of the function $g$ and whether $X$ is discrete or continuous. The reader should convince herself that if $X$ is discrete, then $Y$ must be discrete, but if $X$ is continuous, then $Y$ can be continuous or discrete (or mixed discrete-continuous).

Upon close examination of Example 3.6, the reader may have noticed that the expectation of an indicator function equals the probability of the set being indicated. In fact, any probability can be represented as an expectation of an appropriately defined indicator function.

**Theorem 3.3**
***Probabilities Expressed as Expectations***

*Let $X$ be a random variable with density function $f(x)$, and suppose $A$ is an event for $X$. Then $\mathrm{E}(I_A(X)) = P(A)$.*

**Proof** By definition,

$$E(I_A(X)) = \left\{ \begin{array}{l} \sum_{x \in R(X)} I_A(x)f(x) = \sum_{x \in A} f(x) \quad \text{(discrete)} \\ \int_{-\infty}^{\infty} I_A(x)f(x)dx = \int_{x \in A} f(x)\, dx \quad \text{(continuous)} \end{array} \right\} = P(A)$$ ■

It should be noted that the existence of E($X$) does *not* imply that E($g(X)$) exists, as the following example illustrates.

**Example 3.7** ***Existence of*** $\mathbf{E(X)} \not\Rightarrow$ ***Existence of*** $\mathbf{E(g(X))}$

Let $X$ be a random variable with density function $f(x) = (1/2)I_{\{0,1\}}(x)$. Then $E(X) = 0 \cdot 1/2 + 1 \cdot 1/2 = 1/2$. Define a new random variable $Y = g(X) = X^{-1}$. Since $|1/0|(1/2) + |1/1|(1/2) \not< \infty$, E($g(X)$) does not exist. □

The preceding example also illustrates that, in general, $E(g(X)) \neq g(E(X))$, because $E(X) = 1/2$, so that $g(E(X)) = (E(X))^{-1} = 2$, which does *not* equal E($g(X)$) since E($g(X)$) does not exist. In the special case where the function $g$ is either concave or convex,[7] there is a definite relationship between E($g(X)$) and $g$(E($X$)), as indicated by the following theorem.

**Theorem 3.4** ***Jensen's Inequalities***

*Let X be a random variable with expectation* E($X$), *and let g be a continuous function on an open interval I containing* R($X$). *Then*

**(a)** $E(g(X)) \geq g(E(X))$ *if g is convex on I, and* $E(g(X)) > g(E(X))$ *if g is strictly convex on I and X is not degenerate*[8]*;*

**(b)** $E(g(X)) \leq g(E(X))$ *if g is concave on I, and* $E(g(X)) < g(E(X))$ *if g is strictly concave on I and X is not degenerate.*

**Proof** See Appendix. ■

**Example 3.8** ***Expectation of Concave Function***

Suppose that the yield per acre of a given agricultural crop under standard cultivation practices is represented by $Y = 5X - .1X^2$, where outcomes of $Y$ are measured in bushels, and $X$ represents total rainfall during the growing season, measured in inches. If $E(X) = 20$, can you place an upper bound on the expected yield per acre for this crop?

**Answer**: Yes. Note that $Y = 5X - .1X^2$ is a *concave* function, so that Jensen's inequality applies. Then $E(Y) = E(g(X)) \leq g(E(X)) = 5E(X) - .1(E(X))^2 = 60$ is an upper bound to the expected yield. In fact, the function is strictly concave, and so the inequality can be made strict (it is reasonable to assume that rainfall is not a degenerate random variable).

[7] A continuous function, $g$, defined on a set $D$ is called concave if $\forall x \in D$, $\exists$ a line going through the point $(x, g(x))$ that lies on or above the graph of $g$. The function is convex if $\forall x \in D$, $\exists$ a line going through the point $(x, g(x))$ that lies on or below the graph of $g$. The function is strictly convex or concave if the aforementioned line has only the point $(x, g(x))$ in common with the graph of $g$.

[8] A degenerate random variable is a random variable that has one outcome that is assigned a probability of 1. More will be said about degenerate random variables in Section 3.6.

## 3.4 Expectation Properties

There are a number of properties of the expectation operation that follow directly from its definition. We first present results that involve scalar random variables. We then introduce results involving multivariate random variables.

### 3.4.1 Scalar Random Variables

**Theorem 3.5** *If c is a constant, then* $\mathrm{E}(c) = c$.

**Proof** Let $g(X) = c$. Then, by Theorem 3.2,

$$\mathrm{E}(c) = \int_{-\infty}^{\infty} c\, dF(x) = c \underbrace{\int_{-\infty}^{\infty} dF(x)}_{1} = c.$$ ■

In words, "the expected value of a constant is the constant itself."

**Theorem 3.6** *If c is a constant, then* $\mathrm{E}(cX) = c\mathrm{E}(X)$.

**Proof** Let $g(X) = cX$. Then, by Theorem 3.2,

$$\mathrm{E}(cX) = \int_{-\infty}^{\infty} cx dF(x) = c \underbrace{\int_{-\infty}^{\infty} x\, dF(x)}_{\mathrm{E}(X)} = c\mathrm{E}(X)$$ ■

In words, "the expected value of a constant times a random variable is the constant times the expected value of the random variable."

**Theorem 3.7** $\mathrm{E}\left(\sum_{i=1}^{k} g_i(X)\right) = \sum_{i=1}^{k} \mathrm{E}(g_i(X))$.

**Proof** Let $g(x) = \sum_{i=1}^{k} g_i(X)$. Then, by Theorem 3.2,

$$\mathrm{E}\left(\sum_{i=1}^{k} g_i(X)\right) = \int_{-\infty}^{\infty} \left[\sum_{i=1}^{k} g_i(x)\right] dF(x) = \sum_{i=1}^{k} \int_{-\infty}^{\infty} g_i(x)\, dF(x) = \sum_{i=1}^{k} \mathrm{E}(g_i(X))$$

In words, "the expectation of a sum is the sum of the expectations" regarding $k$ functions of the random variable $X$.

A useful Corollary to Theorem 3.7 concerns the expectation of a linear function of $X$: ■

**Corollary 3.1** *Let* $Y = a + bX$ *for real constants a and b, and let* $\mathrm{E}(X)$ *exist. Then* $\mathrm{E}(Y) = a + b\mathrm{E}(X)$.

**Proof** This follows directly from Theorem 3.7 by defining $g_1(X) = a$, $g_2(X) = bX$, and then applying Theorems 3.5 and 3.6. ■

### 3.4.2 Multivariate Random Variables

The concept of an expectation of a function of a random variable is generalizable to a function of a multivariate random variable as indicated in the following Theorem. The proof is based on an extension of the proof of Theorem 3.2 and is omitted.[9]

**Theorem 3.8** ***Expectation of a Function of a Multivariate Random Variable***

*Let $(X_1,\ldots,X_n)$ be a multivariate random variable with joint density function $f(x_1,\ldots,x_n)$. Then the expectation of $Y = g(X_1,\ldots,X_n)$ is given by*[10]

$$E(Y) = \sum_{(x_1,\ldots,x_n)\in R(X)} g(x_1,\ldots,x_n)f(x_1,\ldots,x_n) \quad \text{(discrete)}$$

$$E(Y) = \int_{-\infty}^{\infty}\cdots\int_{-\infty}^{\infty} g(x_1,\ldots,x_n)f(x_1,\ldots,x_n)dx_1\ldots dx_n \quad \text{(continuous)}.$$

We remind the reader that since $f(x_1,\ldots,x_n) = 0\,\forall(x_1,\ldots,x_n)\notin R(\mathbf{X})$, one could also sum over the points $(x_1,\ldots,x_n) \in \times_{i=1}^{n} R(X_i)$ to define E(Y) in the discrete case.

**Example 3.9** ***Expectation of a Function of a Bivariate RV***

Let the bivariate random variable $(X_1, X_2)$ represent the proportions of operating capacity at which two electricity generation plants operate on a given spring day in an east coast power grid. Assume the joint density of $(X_1, X_2)$ is given by

$$f(x_1,x_2) = 6x_1x_2^2 I_{[0,1]}(x_1)I_{[0,1]}(x_2).$$

What is the expected average proportion of operating capacity at which the two plants operate?

**Answer**: Define the average proportion of operating capacity via the function $g(X_1, X_2) = .5(X_1 + X_2)$. By Theorem 3.8,

$$E(g(X_1,X_2)) = \int_0^1\int_0^1 3x_1^2x_2^2dx_1dx_2 + \int_0^1\int_0^1 3x_1x_2^3dx_1dx_2$$
$$= \frac{1}{3}+\frac{3}{8} = \frac{17}{24} = .7083.$$

□

The expectation property in Theorem 3.7 concerning the sum of functions of a random variable $X$ can also be extended to the sum of functions of a multivariate random variable, as the following theorem indicates.

---

[9] See Steven F. Arnold, (1990), *Mathematical Statistics*, Englewood Cliffs, NJ: Prentice Hall, pp. 92, 98.

[10] It is tacitly assumed that the sum and integral are absolutely convergent for the expectation to exist.

**Theorem 3.9** $\mathrm{E}\left(\sum_{i=1}^{k} g_i(X_1,\ldots,X_n)\right) = \sum_{i=1}^{k} \mathrm{E}(g_i(X_1,\ldots,X_n))$.

**Proof** Let $g(X_1,\ldots,X_n) = \sum_{i=1}^{k} g_i(X_1,\ldots,X_n)$. Then, by Theorem 3.8, $\mathrm{E}(g(X_1,\ldots,X_n))$ is given by

$$\begin{aligned}\mathrm{E}\left(\sum_{i=1}^{k} g_i(X_1,\ldots,X_n)\right) &= \int_{-\infty}^{\infty}\cdots\int_{-\infty}^{\infty}\left(\sum_{i=1}^{k} g_i(x_1,\ldots,x_n)\right) dF(x_1,\ldots,x_n)\\ &= \sum_{i=1}^{k}\int_{-\infty}^{\infty}\cdots\int_{-\infty}^{\infty} g_i(x_1,\ldots,x_n)\, dF(x_1,\ldots,x_n)\\ &= \sum_{i=1}^{k} \mathrm{E}(g_i(X_1,\ldots,X_n))\end{aligned}$$

■

A useful corollary to Theorem 3.9 involving the sum of random variables themselves is given as follows:

**Corollary 3.2** $\mathrm{E}(\sum_{i=1}^{n} X_i) = \sum_{i=1}^{n} \mathrm{E}(X_i)$.

**Proof** This is an application of Theorem 3.9 with $g_i(X_1,\ldots,X_n) = X_i$, $i = 1,\ldots,n$, and $k=n$ ■

In words, "the expectation of a sum is equal to the sum of the expectations" regarding the $n$ random variables $X_1,\ldots,X_n$.

If the random variables $(X_1,\ldots,X_n)$ are *independent*, we can prove that "the expectation of a product is the product of the expectations."

**Theorem 3.10** *Let $(X_1,\ldots,X_n)$ be independent random variables. Then*

$$\mathrm{E}\left(\prod_{i=1}^{n} X_i\right) = \prod_{i=1}^{n} \mathrm{E}(X_i)$$

**Proof** Letting $g(X_1,\ldots,X_n) = \prod_{i=1}^{n} X_i$ in Theorem 3.8, we have

$$\begin{aligned}\mathrm{E}\left(\prod_{i=1}^{n} X_i\right) &= \int_{-\infty}^{\infty}\cdots\int_{-\infty}^{\infty}\prod_{i=1}^{n} x_i\, dF(x_1,\ldots,x_n)\\ &= \int_{-\infty}^{\infty}\cdots\int_{-\infty}^{\infty}\prod_{i=1}^{n} x_i \prod_{j=1}^{n} dF_j(x_j) \qquad \text{(by independence)}\\ &= \prod_{i=1}^{n}\int_{-\infty}^{\infty} x_i\, dF_i(x_i)\, dx_i = \prod_{i=1}^{n} \mathrm{E}(X_i).\end{aligned}$$

■

Later in our study we will find it necessary to take expectations of a vector or matrix of random variables. The following definition describes what is involved in such an operation.

**Definition 3.5**
***Expectation of a Matrix of Random Variables***

Let $\mathbf{W}$ be an $n \times k$ matrix of random variables whose $(i,j)$th element is $W_{ij}$. Then $\mathrm{E}(\mathbf{W})$, the expectation of the matrix $\mathbf{W}$, is the matrix of expectations of the elements of $\mathbf{W}$, where the $(i,j)$th element of $\mathrm{E}(\mathbf{W})$is equal to $\mathrm{E}(W_{ij})$.

If we let $k = 1$ in the above definition, we have that the expectation of a vector is the vector of expectations, i.e.,

$$\mathrm{E}(\mathbf{W}) = \mathrm{E}\begin{bmatrix} W_1 \\ W_2 \\ \vdots \\ W_n \end{bmatrix} = \begin{bmatrix} \mathrm{E}(W_1) \\ \mathrm{E}(W_2) \\ \vdots \\ \mathrm{E}(W_n) \end{bmatrix}$$

In general,

$$\underset{(n\times k)}{\mathrm{E}(\mathbf{W})} = \begin{bmatrix} \mathrm{E}(W_{11}) & \cdots & \mathrm{E}(W_{1k}) \\ \vdots & \ddots & \vdots \\ \mathrm{E}(W_{n1}) & \cdots & \mathrm{E}(W_{nk}) \end{bmatrix},$$

i.e., "the expectation of a matrix is the matrix of expectations."

Having introduced the concept of the expectation of a vector, we note that a **multivariate Jensen's inequality** (Theorem 3.4) holds true for multivariate random variables. In fact, the appropriate extension is made by letting $X$ denote an $n \times 1$ random vector and $I$ represent an open rectangle in the statement of Theorem 3.4. The reader is asked to prove the multivariate version of Theorem 3.4 (replace the line $\ell(x) = a + bx$ with the hyperplane $\ell(\mathbf{x}) = a + \sum_{i=1}^{n} b_i x_i$ in the Appendix proof).

## 3.5 Conditional Expectation

Up to this point, expectations of random variables and functions of random variables have been taken unconditionally, assuming that no additional information was available relating to the occurrence of an event for a subset of the random variables $(X_1,\ldots,X_n)$. When information is given concerning the occurrence of events for a subset of the random variables $(X_1,\ldots,X_n)$, the concept of *conditional* expectation of a random variable becomes relevant.

There is a myriad of situations that arise in which the concept of conditional expectation is relevant. For example, the expected number of housing starts calculated for planning purposes by building supply manufacturers would depend on the given level of mortgage interest assumed, or the expected sales tax revenue accruing to state government would be a function of whatever reduced level of employment was assumed due to the downsizing of a major industry in the state. More generally, we will see that conditional expectation is at the heart of *regression analysis*, whereby one attempts to explain the expected value of one random variable as a function of the values of other related random variables, e.g., the expected yield/acre of an agricultural crop is conditional on the level of rainfall, temperature, sunshine, and the degree of weed and pest infestation.

The difference between unconditional and conditional expectation is that the unconditional density function is used to weight outcomes in the former case, while a conditional density function supplies the weights in the latter case. The conditional expectation of a function of a random variable is defined in the bivariate case as follows.

**Definition 3.6** ***Conditional Expectation-Bivariate***

Let $X$ and $Y$ be random variables with joint density function $f(x,y)$. Let the conditional density of $Y$, given $x \in B$, be $f(y|x \in B)$. Let $g(Y)$ be a real-valued function of $Y$. Then the conditional expectation of $g(Y)$, given $x \in B$, is defined as

$$E(g(Y)|x \in B) = \sum_{y \in R(Y)} g(y) f(y|x \in B), \qquad \text{(discrete)}$$

$$E(g(Y)|x \in B) = \int_{-\infty}^{\infty} g(y) f(y|x \in B) dy. \qquad \text{(continuous)}$$

Note in the special case where $g(Y) = Y$, we have by Definition 3.6 that $E(Y|x \in B) = \int_{-\infty}^{\infty} y f(y|x \in B) dy$ in the continuous case, and $E(Y|x \in B) = \sum_{y \in R(Y)} y f(y|x \in B)$in the discrete case.

**Example 3.10** ***Conditional Expectation in Bivariate Case***

Let the bivariate random variable $(X, Y)$ represent the per dollar return on two investment projects. Let the joint density of $(X, Y)$ be

$$f(x,y) = \tfrac{1}{96}(x^2+2xy+2y^2) I_{[0,4]}(x) I_{[0,2]}(y).$$

What is the conditional expectation of the per dollar return on the second project, given that the per dollar return on the first project is $x = 1$?

**Answer:** To answer the question, we first need to establish the conditional density $f(y|x = 1)$. This in turn requires knowledge of the marginal density of $X$, which we find as

$$f_X(x) = \int_{-\infty}^{\infty} f(x,y) dy = \frac{1}{96} \int_0^2 (x^2 + 2xy + 2y^2) I_{[0,4]}(x) dy$$
$$= \left(\tfrac{1}{48} x^2 + \tfrac{1}{24} x + \tfrac{1}{18}\right) I_{[0,4]}(x).$$

Then

$$f(y|x=1) = \frac{f(1,y)}{f_X(1)} = \frac{\frac{1}{96}(1+2y+2y^2)\,I_{[0,2]}(y)}{\frac{17}{144}}$$
$$= \left[.088235 + .176471(y+y^2)\right] I_{[0,2]}(y).$$

Finally, by Definition 3.6,

$$\mathrm{E}(Y|x=1) = \int_{-\infty}^{\infty} y f(y|x=1)dy = \int_0^2 (.088235y + .176471(y^2+y^3))dy$$
$$= \frac{.088235y^2}{2} + .176471\left[\frac{y^3}{3}+\frac{y^4}{4}\right]\Bigg|_0^2 = 1.3529.$$

□

It is important to note that *all* of the properties of expectations derived previously apply equally well to conditional expectations. This follows from the fact that the operations of taking an unconditional or a conditional expectation are precisely the same once a PDF has been derived, and the genesis of the PDF is irrelevant to the expectation properties derived heretofore (note that the origin of a PDF in previous sections was never an issue).

Rather than specifying a particular elementary event for the outcome of $X$ when defining a conditional expectation of $g(Y)$, we might conceptualize leaving the elementary event for $X$ *unspecified*, and express the conditional expectation of $g(Y)$ as a *function of* $x$. Let $\eta(x) = \mathrm{E}(g(Y)|x)$ denote the function of $x$ whose value when $x = b$ is $\mathrm{E}(g(Y)|x = b)$. Then, by definition, we can interpret $\eta(X) = \mathrm{E}(g(Y)|X)$ as a random variable. If we take $\mathrm{E}(\eta(X)) = \mathrm{E}(\mathrm{E}(g(Y)|X))$ we obtain the *unconditional* expectation of $g(Y)$.

**Theorem 3.11** ***Iterated Expectation Theorem***

$$\mathrm{E}(\mathrm{E}(g(Y))|X) = \mathrm{E}(g(Y))$$

**Proof** (continuous) Let $f_X(x)$ be the marginal density of $X$ and $f(x,y)$ be the joint density of $X$ and $Y$. Then

$$\eta(x) = \mathrm{E}(g(Y)|x) = \int_{-\infty}^{\infty} g(y)\frac{f(x,y)}{f_X(x)}dy$$

and

$$\mathrm{E}(\eta(X)) = \mathrm{E}(\mathrm{E}(g(Y)|X)) = \int_{-\infty}^{\infty}\left[\int_{-\infty}^{\infty} g(y)\frac{f(x,y)}{f_X(x)}dy\right] f_X(x)dx$$
$$= \int_{-\infty}^{\infty}\int_{-\infty}^{\infty} g(y)f(x,y)dydx = \mathrm{E}(g(Y)).$$

The proof in the discrete case is left to the reader. ■

**Example 3.11** ***Unconditional Expectation* via *Iteration***

Suppose that the expectation of market supply for some commodity, given price $p$, is represented by $E(Q|p) = 3p + 7$ and $E(P) = 2$. Then by the iterated expectation theorem, the unconditional expectation of market supply is given by

$$E(E(Q|P)) = E(3P + 7) = 3E(P) + E(7) = 13. \qquad \square$$

In cases where one is conditioning on an elementary event $x = b$, there are useful generalizations of Definition 3.6 and Theorem 3.11, which are referred to as the *substitution theorem* and the *generalized iterated expectation theorem*, respectively.

**Theorem 3.12** ***Substitution Theorem***

$$E(g(X,Y)|x=b) = E(g(b,Y)|x=b).$$

**Proof** (Discrete Case) Let $z = g(x,y)$, and note that the PDF of $Z$, conditional on $x = b$, can be defined as

$$\begin{aligned} h(z|x=b) &= P(g(x,y) = z|x=b) = P(g(x,y) = z, x = b)/P(x=b) \\ &= \sum_{\{y:g(b,y)=z\}} f(b,y)/f_X(b) = \sum_{\{y:g(b,y)=z\}} f(y|x=b). \end{aligned}$$

It is evident that the set of $z$ values for which $h(z|x = b) > 0$ is given by $\Xi = \{z: z = g(b,y), y$ is such that $f(y|x = b) > 0\}$. Then

$$\begin{aligned} E(g(X,Y)|x=b) = E(z|x=b) &= \sum_{z\in\Xi} zh(z|x=b) \\ &= \sum_{z\in\Xi} z \sum_{\{y:g(b,y)=z\}} f(y|x=b) \\ &= \sum_{y\in R(Y)} g(b,y)f(y|x=b) = E(g(b,Y)|x=b). \end{aligned}$$

(Continuous Case) See A.F. Karr (1993) Probability, New York: Springer-Verlag, p. 230. ■

The *substitution* theorem indicates that when taking the expectation of $g(X, Y)$ conditional on $x = b$, one can substitute the constant $b$ for $X$ as $g(b,Y)$ and then take the conditional expectation with respect to the random variable $Y$. The random variable $X$ essentially acts as a constant in $g(X,Y)$ under the condition $x = b$.

**Theorem 3.13** ***Generalized Iterated Expectation Theorem***

$$E(E(g(X,Y)|X)) = E(g(X,Y))$$

**Proof** Using the substitution theorem but leaving the elementary event for X unspecified in order to express the conditional expectation as a function of the elementary event x obtains (continuous case—discrete case is analogous)

$$\eta(x) = E(g(X,Y)|x) = E(g(x,Y)|x)$$
$$= \int_{-\infty}^{\infty} g(x,y)f(y|x)dy.$$

Then

$$E(\eta(X)) = E(E(g(X,Y)|X))$$
$$= \int_{-\infty}^{\infty}\left[\int_{-\infty}^{\infty} g(x,y)f(y|x)dy\right]f_X(x)dx$$
$$= \int_{-\infty}^{\infty}\int_{-\infty}^{\infty} g(x,y)f(x,y)dydx = E(g(X,Y)).$$

Thus, taking the expectation of the conditional expectation of $g(X,Y)$ given $x$ yields the unconditional expectation of $g(X,Y)$. ■

**Example 3.12** *Unconditional Expectation of Linear Function of Bivariate RV* via *Iteration*

Let $(X,Y)$ represent the per dollar return on the two investment projects of Example 3.10. Assume \$1000 is invested in each project. What is the expected return on the portfolio, given that the per dollar return on the first project is $x = 1$?

**Answer:** The return on the portfolio can be represented as $z = g(x,y) = 1000x + 1000y$. The substitution theorem allows the conditional expectation to be defined as

$$E(Z|x=1) = E(1000X + 1000Y|x=1) = E(1000 + 1000Y|x=1)$$
$$= \int_0^2 (1000 + 1000y)f(y|x=1)dy$$
$$= 1000 + 1000\int_0^2 yf(y|x=1)dy = 2352.9.$$ □

**Example 3.13** *Unconditional Expectation of Nonlinear Function* via *Iteration*

Given the representation of expected market supply in Example 3.11, we know by the substitution theorem that the expected dollar sales of the commodity, expressed as a function of price, is $E(pQ|p) = 3p^2 + 7p$. Suppose $E(P^2) = 8$. Then, using Theorem 3.13, the (unconditional) expectation of dollar sales is given by $E(E(PQ|P)) = E(3P^2 + 7P) = 3E(P^2) + 7E(P) = 38$. □

### 3.5.1 Regression Function in the Bivariate Case

In the special case where $g(Y) = Y$, the conditional expectation of $Y$ expressed as a function of x, i.e., $E(Y|x)$, is called **the regression function of *Y* on *X*.** The regression function depicts the functional relationship between the conditional expectation of $Y$ and the potential values of $X$ on which the expectation might be conditioned. In the continuous case, the graph of the function is generally a curve in the plane, in which case $E(Y|x)$ is often referred to as the **regression curve of *Y* on *X*.**

**Definition 3.7**
***Regression Function of Y on X***

The conditional expectation of $Y$ expressed as a function of $x$, as $E(Y|x)$.

**Example 3.14**
***Regression Function of Investment Returns***

Refer to the investment return example, Example 3.10, and rather than calculate $E(Y|x = 1)$, we calculate $E(Y|x)$, the regression function of $Y$ on $X$. Letting $f(y|x)$ denote the conditional density function of $y$ expressed as a function of $x$, we have $f(y|x) = f(x,y)/f_X(x)$. Then

$$E(Y|x) = \int_{-\infty}^{\infty} y\frac{f(x,y)}{f_X(x)}dy = \int_0^2 \frac{y(x^2 + 2xy + 2y^2)\, I_{[0,4]}(x)}{(2x^2 + 4x + \frac{16}{3})\, I_{[0,4]}(x)}dy$$

$$= \left[\frac{2x^2 + \frac{16}{3}x + 8}{2x^2 + 4x + \frac{16}{3}}\right] \text{for } x \in [0,4]$$

and the regression function is undefined for $x \notin [0,4]$. The regression function represents the expected per dollar return on project 2 as a function of the various potential conditioning values for the per dollar return on project 1. Note that the regression function is a nonlinear function of $x$. The reader can verify that at the point $x = 1$, $E(Y|x) = 1.3529$ is the value of the regression function, as it should be given the answer to Example 3.10. The reader is encouraged to sketch the graph of the regression function over its domain of definition. □

The regression function has an important interpretation in terms of approximating one random variable by a function of other random variables. Examine the problem of choosing a function of $X$, say $h(X)$, whose outcomes are the minimum expected squared distance[11] from the outcome of $Y$. Assuming that $(X,Y)$ is a continuous bivariate random variable (the discrete case is analogous), we thus seek an $h(X)$ that minimizes

$$E(Y - h(X))^2 = \int_{-\infty}^{\infty}\int_{-\infty}^{\infty} (y - h(x))^2 f(x,y)dxdy$$

$$= \int_{-\infty}^{\infty}\left[\int_{-\infty}^{\infty} (y - h(x))^2 f(y|x)dy\right] f_X(x)dx.$$

If $h(x)$ could be chosen so as to minimize the bracketed integral for *each* possible $x$, then it would follow that the double integral, and thus the expected squared distance between outcomes of $Y$ and $h(X)$, would be minimized.

The optimal choice of approximating function is given by $h(x) = E(Y|x)$. To see why, note that the substitution theorem allows the preceding bracketed expression to be written as

[11] Recall that the distance between the points $a$ and $b$ is defined by $d(a,b) = |b-a|$ and thus squared distance would be given by $d^2(a, b) = (b-a)^2$.

$$\mathrm{E}\left([Y - h(x)]^2 \,|x\right) = \mathrm{E}\left([Y - \mathrm{E}(Y|x) + \mathrm{E}(Y|x) - h(x)]^2 \,|x\right)$$
$$= \mathrm{E}\left([Y - \mathrm{E}(Y|x)]^2 \,|x\right) + [\mathrm{E}(Y|x) - h(x)]^2$$

where the cross product term is zero and has been eliminated because

$$\begin{aligned}&\mathrm{E}([Y - \mathrm{E}(Y|x)][\mathrm{E}(Y|x) - h(x)]|x)\\&\quad = [\mathrm{E}(Y|x) - h(x)]\mathrm{E}(Y - \mathrm{E}(Y|x)|x)(\text{by the substitution theorem})\\&\quad = [\mathrm{E}(Y|x) - h(x)][\mathrm{E}(Y|x) - \mathrm{E}(Y|x)] = 0.\end{aligned}$$

It follows that the choice of $h(x)$ that minimizes $\mathrm{E}([Y-h(x)]^2|x)$ is given by $h(x) = \mathrm{E}(Y|x)$, since any other choice results in $[\mathrm{E}(Y|x)-h(x)]^2 > 0$.

The preceding result suggests that if one is attempting to explain or predict the outcome of one random variable from knowledge of the outcome of another random variable, and if expected squared distance (also called mean square error—to be discussed in Chapter 7) is used as the measure of closeness between actual and predicted outcomes, then the best (closest) prediction is given by values of the regression function, or equivalently by the conditional expectation of the random variable of interest. For example, in Example 3.14, if one were attempting to predict the expected dollar return on project 2 in terms of the dollar return on project 1, the regression function $\mathrm{E}(Y|x)$ presented in the example provides the predictions that minimize expected squared distance between outcomes of $Y$ and outcomes of $h(X)$. If $x = 1$, then the best prediction of $Y$'s outcome would be 1.3529.

### 3.5.2 Conditional Expectation and Regression in the Multivariate Case

The definition of conditional expectation (Definition 3.4) and the theorems involving conditional expectation extend to the case where $Y$ and/or $X$ is multivariate, in which case the reader can interpret $Y$ and $X$ as referring to random *vectors* and introduce multiple summation or integration notation appropriately when reading the definition and the theorems. Also, the notion of the *regression function of $Y$ on $X$* extends straightforwardly to the case where $\mathbf{X}$ is multivariate, in which case $\mathrm{E}(Y|\mathbf{x})$ is interpreted as a function of the *vector* $\mathbf{x}$ and would be defined by $\mathrm{E}(Y|\mathbf{x}) = \int_{-\infty}^{\infty} yf(y|x_1, \ldots, x_n)dy$ or $\sum_{y \in R(Y)} yf(y|x_1, \ldots, x_n)$ in the continuous or discrete case, respectively. An argument analogous to the bivariate case can be used to prove that $h(\mathbf{X}) = \mathrm{E}(Y|\mathbf{X})$ is the function of the multivariate $\mathbf{X}$ that is the minimum expected squared distance from $Y$. Thus, the best approximation or prediction of $Y$ outcomes via a function of the outcome of the multivariate $\mathbf{X}$ is provided by values of the regression function.

For convenience, we list below a number of general expectation results applied specifically to conditional expectations involving multivariate random variables.

**Definition 3.8**
***Conditional Expectation (General)***

Let $(X_1,\ldots,X_n)$ and $(Y_1,\ldots,Y_m)$ be random vectors having a joint density function $f(x_1,\ldots,x_n, y_1,\ldots,y_m)$. Let $g(Y_1,\ldots,Y_m)$ be a real-valued function of $(Y_1,\ldots, Y_m)$. Then the conditional expectation of $g(Y_1,\ldots,Y_m)$, given $(x_1,\ldots,x_n)\in B$, is defined as[12]:

$$E(g(Y_1,\ldots,Y_m)|(x_1,\ldots,x_n)\in B) \quad \text{(discrete)}$$
$$= \sum_{(y_1,\ldots,y_m)\in R(Y)} g(y_1,\ldots,y_m)f(y_1,\ldots,y_m|(x_1,\ldots,x_n)\in B)$$
$$E(g(Y_1,\ldots,Y_m)|(x_1,\ldots,x_n)\in B) \quad \text{(continuous)}$$
$$= \int_{-\infty}^{\infty}\cdots\int_{-\infty}^{\infty} g(y_1,\ldots,y_m)f(y_1,\ldots,y_m|(x_1,\ldots,x_n)\in B)dy_1\cdots dy_m.$$

**Theorem 3.14**
***Substitution Theorem: Multivariate***

$E(g(X_1,\ldots,X_n,Y_1,\ldots,Y_m)|\mathbf{x}=\mathbf{b}) = E(g(b_1,\ldots,b_n,Y_1,\ldots,Y_m)|\mathbf{x}=\mathbf{b})$.

**Theorem 3.15**
***Iterated Expectation Theorems: Multivariate***

$E(E(g(Y_1,\ldots,Y_m)|X_1,\ldots,X_n)) = E(g(Y_1,\ldots,Y_m))$
$E(E(g(X_1,\ldots,X_n,Y_1,\ldots,Y_m)|X_1,\ldots,X_n)) = E(g(X_1,\ldots,X_n,Y_1,\ldots,Y_m))$.

**Theorem 3.16** $E(c|(x_1,\ldots,x_n)\in B) = c$.

**Theorem 3.17** $E(cY|(x_1,\ldots x_n)\in B) = c\,E(Y|(x_1,\ldots,x_n)\in B)$.

**Theorem 3.18** $E\left(\sum_{i=1}^{k} g_i(Y_1,\ldots,Y_m)|(x_1,\ldots,x_n)\in B\right) = \sum_{i=1}^{k} E(g_i(Y_1,\ldots,Y_m)|(x_1,\ldots,x_n)\in B)$.

## 3.6 Moments of a Random Variable

The expectations of certain power functions of a random variable have uses as measures of central tendency, spread or dispersion, and skewness of the density function of the random variable, and also are important components of statistical inference procedures that we will study in later chapters. These special expectations are called **moments** of the random variable (or of the density

[12]One can equivalently sum over the points $(y_1,\ldots,y_m)\in \times_{i=1}^{m} R(Y_i)$ in defining the expectation in the discrete case.

function). There are two types of moments that we will be concerned with—**moments about the origin** and **moments about the mean.**

**Definition 3.9**
***rth Moment About the Origin***

Let $X$ be a random variable with density function $f(x)$. Then the **rth moment of** $X$ **about the origin**, denoted by $\mu'_r$, is defined for integers $r \geq 0$ as

$$\mu'_r = \mathrm{E}(X^r) = \sum_{x \in R(X)} x^r f(x) \qquad \text{(discrete)}$$

$$\mu'_r = \mathrm{E}(X^r) = \int_{-\infty}^{\infty} x^r f(x)dx. \qquad \text{(continuous)}$$

The value of $r$ in the definition of moments is referred to as the **order of the moment**, so that one would refer to $\mathrm{E}(X^r)$ as the **moment of order** $r$. Note that $\mu'_0 = 1$ for any discrete or continuous random variable, since $\mu'_0 = \mathrm{E}(X^0) = \mathrm{E}(1) = 1$.

The first moment about the origin is simply the expectation of the random variable $X$, i.e., $\mu'_1 = \mathrm{E}(X^1) = \mathrm{E}(X)$, a quantity that we have examined at the beginning of our discussion of mathematical expectation. This *balancing point* of a density function, or the weighted average of the elements in the range of the random variable, will be given a special name and symbol.

**Definition 3.10**
***Mean of a Random Variable (or Mean of a Density Function)***

The first moment about the origin of a random variable, $X$, is called the **mean of the random variable X** (or mean of the density function of $X$), and will be denoted by the symbol $\mu$.

Thus, the first moment about the origin characterizes the central tendency of a density function. Measures of spread and skewness of a density function are given by certain moments about the mean.

**Definition 3.11**
***rth Central Moment (or rth Moment About the Mean)***

Let $X$ be a random variable with density function $f(x)$. Then the $r$th **central moment of X** (or the $r$th moment of $X$ about the mean), denoted by $\mu_r$, is defined as

$$\mu_r = \mathrm{E}((X-\mu)^r) = \sum_{x \in R(X)} (x-\mu)^r f(x), \qquad \text{(discrete)}$$

$$\mu_r = \mathrm{E}((X-\mu)^r) = \int_{-\infty}^{\infty} (x-\mu)^r f(x)dx. \qquad \text{(continuous)}$$

Note that $\mu_0 = 1$ for any discrete or continuous random variable, since $\mu_0 = \mathrm{E}(X-\mu)^0 = \mathrm{E}(1) = 1$. Furthermore, $\mu_1 = 0$ for any discrete or continuous random variable for which $\mathrm{E}(X)$ exists, since $\mu_1 = \mathrm{E}(X-\mu)^1 = \mathrm{E}(X) - \mathrm{E}(\mu) = \mu - \mu = 0$. The second central moment is given a special name and symbol.

**Definition 3.12**
***Variance of a Random Variable (or Variance of a Density Function)***

The second central moment, $E((X-\mu)^2)$, of a random variable, $X$, is called the **variance of the random variable** $X$ (or the variance of the density function of $X$), and will be denoted by the symbol $\sigma^2$, or by $\text{var}(X)$.

We will also have use for the following function of the variance of a random variable.

**Definition 3.13**
***Standard Deviation of a Random Variable (or Standard Deviation of a Density Function)***

The nonnegative square root of the variance of a random variable, $X$, (i.e., $+\sqrt{\sigma^2}$) is called the **standard deviation of the random variable** $X$ (or standard deviation of the density function of $X$) and will be denoted by the symbol $\sigma$, or by $\text{std}(X)$.

The variance (and thus also the standard deviation) of $X$ is a measure of dispersion or spread of the density function $f(x)$ around its balancing point (the mean of $X$). The larger the variance, the greater the spread or dispersion of the density about its mean. In the extreme case where the entire density is concentrated at the mean of $X$ and thus has no spread or dispersion, i.e., $f(x) = I_{\{\mu\}}(x)$, then $E((X-\mu)^2) = 0$ and the variance (and standard deviation) is zero.

In order to examine the relationship between the spread of a density and the magnitude of the variance in more detail, we first present **Markov's inequality** (named after the Russian mathematician A. Markov), and we will then introduce **Chebyshev's inequality** (named after the Russian mathematician P.L. Chebyshev) as a corollary.

**Theorem 3.19**
**Markov's Inequality**

*Let $X$ be a random variable with density function $f(x)$, and let $g$ be a nonnegative-valued function of $X$. Then $P(g(x) \geq a) \leq E(g(X))/a$ for any value $a > 0$.*

**Proof**

$$
\begin{aligned}
E(g(X)) &= \int_{-\infty}^{\infty} g(x)dF(x) \\
&= \int_{\{x:g(x)\geq a\}} g(x)\,dF(x) + \underbrace{\int_{\{x:g(x)<a\}} g(x)\,dF(x)}_{\geq 0} \\
&\geq \int_{\{x:g(x)\geq a\}} g(x)\,dF(x) \\
&\geq \int_{\{x:g(x)\geq a\}} a\,dF(x)dx \text{ (since } g(x)\geq a \text{ for all } x\in\{x:g(x)\geq a\}\text{)} \\
&\geq a\int_{\{x:g(x)\geq a\}} dF(x)dx = aP(g(x)\geq a),
\end{aligned}
$$

and thus, $\dfrac{E(g(X))}{a} \geq P(g(x)\geq a)$. ■

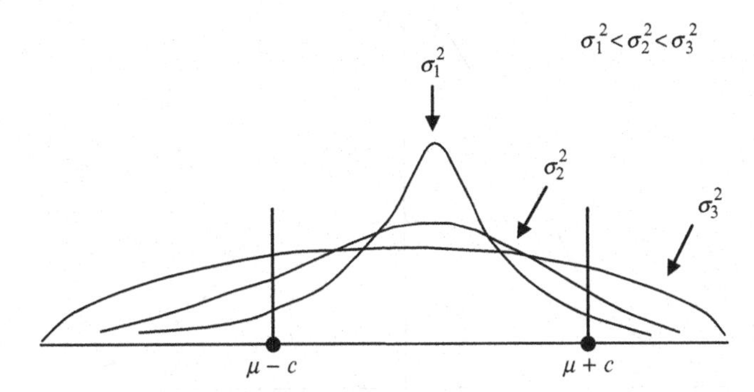


**Figure 3.6**
Density functions and variances.

**Corollary 3.3**
***Chebyshev's Inequality***

$P(|x - \mu| \geq k\sigma) \leq 1/k^2$ for $k>0$

**Proof** This follows by letting $g(X) = (X - \mu)^2$ and $a = k^2\sigma^2$ in Markov's inequality and realizing that $(x - \mu)^2 \geq k^2\sigma^2$ is equivalent to $(x - \mu) \leq -k\sigma$ or $(x - \mu) \geq k\sigma$, which is in turn equivalent to $|x - \mu| \geq k\sigma$. ■

In words, Markov's inequality states that we can *always* place an upper bound on the probability that $g(x) \geq a$ so long as $g(x)$ is nonnegative valued and $E(g(X))$ exists. Chebyshev's inequality implies that if $\mu$ and $\sigma$ are, respectively, the mean and standard deviation of the density function of $X$, then for any positive constant $k$, the probability that $X$ will have an outcome that is $k$ or more standard deviations from its mean, i.e., outside the interval $(\mu - k\sigma, \mu + k\sigma)$, is less than or equal to $1/k^2$. Note that we are able to make these probability statements *without* knowledge of the algebraic form of the density function.

Chebyshev's inequality is sometimes stated in terms of an event that is the complement of the event in Corollary 3.3.

**Corollary 3.4**
***Chebyshev's Inequality***

$P(|x - \mu| < k\sigma) \geq 1 - 1/k^2$ for $k>0$

**Proof** Follows directly from Corollary 3.3 noting that $P(|x - \mu| \geq k\sigma) = 1 - P(|x-\mu| < k\sigma) \leq 1/k^2$. ■

Markov's inequality and Chebyshev's inequalities are interesting in their own right, but at this point we will use the concepts only to further clarify our interpretation of the variance as a measure of the spread or dispersion of a density function. In Corollary 3.4, let $k\sigma = c$, where $c$ is any arbitrarily small positive number. Then $P(|x - \mu| < c) \geq 1 - \sigma^2/c^2$ where we have substituted for $k$ the value $c/\sigma$. Then note that as $\sigma^2 \rightarrow 0$, the probability inequality approaches $P(\mu - c < x < \mu + c) \rightarrow 1$, which implies that as $\sigma^2 \rightarrow 0$, the density concentrates in the interval $(\mu - c, \mu + c)$ for *any* arbitrarily small positive $c$. Diagrammatically, this can be illustrated as in Figure 3.6.

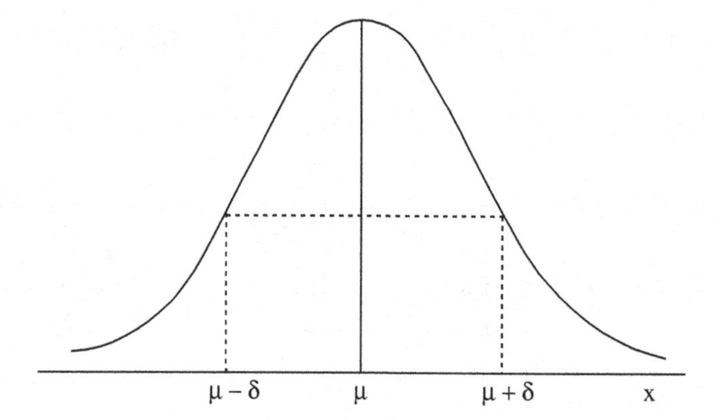


**Figure 3.7**
Density symmetric about $\mu$.

As a concrete example, let $\mu = 0$, and examine the event $B = \{x:|x| < 10\}$, where we are letting $c = 10$ in the preceding argument. Then using $P(|x - \mu| < k\sigma) \geq 1 - 1/k^2$ with $k\sigma = c = 10$, we have:

$$\text{if} \begin{bmatrix} \sigma = 5 \\ \sigma = 2 \\ \sigma = 1 \end{bmatrix}, \text{ then } \; P(B) \geq \begin{bmatrix} 1 - 1/2^2 = .75 \\ 1 - 1/5^2 = .96 \\ 1 - 1/10^2 = .99 \end{bmatrix},$$

and thus the smaller is σ (and thus the smaller the variance), the larger is the lower bound on the probability that the outcome of $X$ occurs in the interval $(-10, 10)$.

For an alternative argument in support of interpreting the variance as a measure of the spread of a density function, note that the variance of $X$ can be directly interpreted as the expected squared distance of the random variable $X$ from its mean. To see this, first recall that the distance between two points, $x$ and $y$, on the real line is defined as $d(x,y) = |x - y|$. Then $d^2(x,y) = (x - y)^2$, and letting $y = \mu$, we have $E(d^2(X,\mu)) = E((X - \mu)^2) = \sigma^2$. Therefore, the smaller is $\sigma^2$, the smaller is the expected squared distance of $X$ from its mean.

The third central moment is used as a measure of whether the density of $X$ is *skewed.*

**Definition 3.14**
***Symmetric and Skewed Densities***

A density is said to be **symmetric about** $\mu$ when $f(\mu + \delta) = f(\mu - \delta) \forall \delta > 0$. If the density is not symmetric, it is said to be a **skewed density**.

Therefore, a density is said to be **symmetric about** $\mu$ if the density is such that the graph to the right of the mean is the mirror image of the graph to the left of the mean (see Figure 3.7). If such is not the case, the density is said to be a **skewed density**. A necessary (but not sufficient) condition for a density to be symmetric about $\mu$ is that $\mu_3 = E(X - \mu)^3 = 0$.

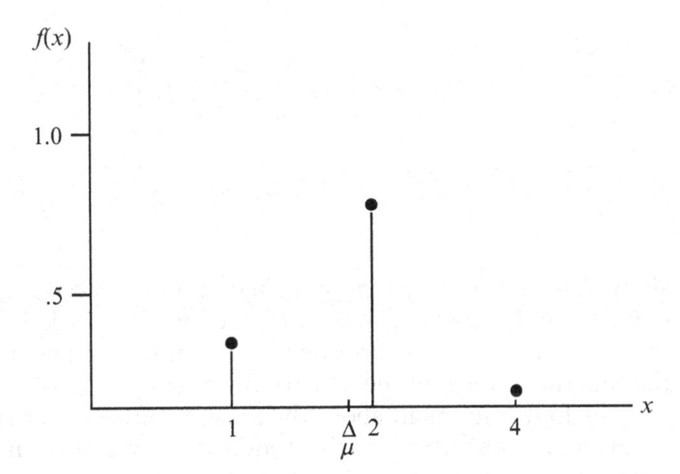


**Figure 3.8** Nonsymmetric density function of Example 3.15.

**Theorem 3.20** ***Relationship Between $\mu_3$, Symmetry, and Skewness*** *If $\mu_3$ exists, then $f(x)$ is symmetric about the mean only if $\mu_3 = 0$. If $\mu_3 \neq 0$, then $f(x)$ is skewed.*

**Proof** Note that

$$\mu_3 = \int_{-\infty}^{\infty} (x - \mu)^3 dF(x)dx = \int_{\mu}^{\infty} (x - \mu)^3 dF(x) + \int_{-\infty}^{\mu} (x - \mu)^3 dF(x).$$

By making the substitution $z = x - \mu$ in the first integral and $z = -x + \mu$ in the second integral, if the density is symmetric, then

$$\begin{aligned}\mu_3 &= \int_0^{\infty} z^3 dF(\mu + z) - \int_{\infty}^{0} (-z)^3 dF(\mu - z) \\ &= \int_0^{\infty} z^3 dF(\mu + z) - \int_0^{\infty} z^3 dF(\mu - z) = 0\end{aligned}$$

since $f(\mu + z) = f(\mu - z)\ \forall z$ by the symmetry of $f$ about $\mu$. It follows that if $\mu_3 \neq 0$, then the density function is necessarily skewed. ■

We underscore that $\mu_3 = 0$ is not *sufficient* for symmetry, i.e., a density can be skewed, and still have $\mu_3 = 0$, as the following example illustrates.

**Example 3.15** ***A Skewed Density with $\mu_3 = 0$*** Let the random variable $X$ have the density function $f(x) = .22\ I_{\{1\}}(x) + .77\ I_{\{2\}}(x) + .01\ I_{\{4\}}(x)$ (see Figure 3.8). Note that $\mu = 1.8$, and it is clear that $f(x)$ is not symmetric about $\mu$. Nonetheless, $\mu_3 = \mathrm{E}\left((X - \mu)^3\right) = 0$. □

The sign of $\mu_3$ is sometimes interpreted as indicating the direction of the skew in the density function. In particular, density functions having long "tails" to the right are called **skewed to the right**, and these densities tend to have $\mu_3 > 0$, whereas density functions with long left-hand tails are called

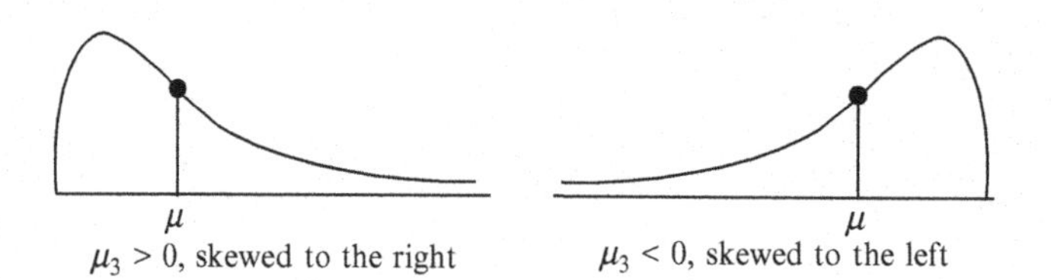


**Figure 3.9**
Skewed density functions.

**skewed to the left**, and these densities tend to have $\mu_3 < 0$ (see Figure 3.9). Unfortunately, there are exceptions to these "tendencies," and the nature of the skewness is best determined by examining the graph of the density itself if the functional form of the density is known.

The fourth moment about the mean is often used to form a measure of how "peaked" a probability density function is. Measures most often used for this purpose include a scaled version of the fourth moment, called **kurtosis**, as well as a scaled and centered (relative to a standard normal distribution, discussed in Chapter 4) version called **excess kurtosis**.

**Definition 3.15**
***Kurtosis and Excess Kurtosis***

The **kurtosis** of a probability density function $f(x)$ is given by $\mu_4/\sigma^4$. The **excess kurtosis** of $f(x)$ is given by $\mu_4/\sigma^4 - 3$.

The larger (smaller) the value of the kurtosis, the more peaked (flatter) is the density function. Excess kurtosis measures the degree to which a density function is more or less (the excess or lack thereof) peaked relative to a standard normal distribution, which has kurtosis equal to 3. The excess kurtosis measure is centered such that it equals 0 for the standard normal distribution (Chapter 4).

A probability density that has positive excess kurtosis is referred to as being **leptokurtic**. If excess kurtosis is negative the probability density is referred to as being **platykurtic**. If excess kurtosis is zero, so that the probability density has the same excess kurtosis measure as a standard normal distribution, it is referred to as being **mesokurtic**.

### 3.6.1 Relationship Between Moments About the Origin and Mean

Integer ordered moments about the origin and about the mean are functionally related. Central moments can be expressed solely in terms of moments about the origin, while moments about the origin can be expressed in terms of the mean and moments about the mean. The functional relationship is the direct result of the binomial theorem, which we review for the reader in the following lemma.

**Lemma 3.1**
***Binomial Theorem.***

Let $a$ and $b$ be real numbers. Then $(a+b)^n = \sum_{j=0}^{n} \binom{n}{j} a^j b^{n-j}$.

**Theorem 3.21** ***Central Moments as Functions of Moments About the Origin*** *If $\mu_r$ exists and r is a positive integer, then $\mu_r = \sum_{j=0}^{r} (-1)^j \binom{r}{j} \mu'_{r-j}\mu^j$.*

**Proof** By definition, $\mu_r = \mathrm{E}((X-\mu)^r)$. Substituting $(-\mu)$ for $a$, $X$ for $b$, and $r$ for $n$ in the binomial theorem (Lemma 3.1) and taking an expectation yields $\mu_r = \mathrm{E}\left(\sum_{j=0}^{r} \binom{r}{j}(-\mu)^j X^{r-j}\right)$. An application of Theorems 3.6 and 3.7 results in $\mu_r = \sum_{j=0}^{r} (-1)^j \binom{r}{j} \mu^j \mathrm{E}(X^{r-j})$. ■

**Theorem 3.22** ***Moments About the Origin as Functions of Central Moments*** *If $\mu'_r$ exists and r is a positive integer, then $\mu'_r = \sum_{j=0}^{r} \binom{r}{j} \mu_{r-j}\mu^j$.*

**Proof** By definition, $\mu'_r = \mathrm{E}(X^r) = \mathrm{E}(X - \mu + \mu)^r$. Substituting $\mu$ for $a$, $(X - \mu)$ for $b$, and $r$ for $n$ in the binomial theorem (Lemma 3.1) and taking an expectation yield $\mu'_r = \mathrm{E}\left(\sum_{j=0}^{r} \binom{r}{j} \mu^j (X-\mu)^{r-j}\right)$. An application of Theorems 3.6 and 3.7 results in $\mu'_r = \sum_{j=0}^{r} \binom{r}{j} \mu^j \mathrm{E}\left((X-\mu)^{r-j}\right)$. ■

A special case of Theorem 3.21, which we will use repeatedly in later chapters, is the case where $r = 2$, which provides a representation of the variance of a random variable in terms of moments about the origin. In particular, from Theorem 3.21, var $(X) = \mu_2 = \mu'_2 - \mu^2 = \mathrm{E}(X^2) - (\mathrm{E}(X))^2$.

**Example 3.16** ***Variance from Moments About the Origin*** Let the random variable $X$ have density function $f(x) = I_{[0,1]}(x)$. Since $\mathrm{E}(X) = 1/2$, and since $\mathrm{E}(X^2) = \int_{-\infty}^{\infty} x^2 I_{[0,1]}(x)\, dx = (x^3/3)|_0^1 = 1/3$, then $\mathrm{var}(X) = \mathrm{E}(X^2) - (\mathrm{E}(X))^2 = 1/3 - (1/2)^2 = 1/12$. □

### 3.6.2 Existence of Moments

Regarding the existence of moments, the following theorem can be useful for determining whether a series of moments of progressively higher order up to $r$ exist.

**Theorem 3.23** ***Existence of Lower Order Moments*** *If $\mathrm{E}(X^r)$ exists for a given integer $r > 0$, then $\mathrm{E}(X^s)$ exists for all integers $s \in [0,r]$.*

**Proof** Define $A_{<1} = \{x{:}|x|^s < 1\}$ and $A_{\geq 1} = \{x{:}|x|^s \geq 1\}$. Note that $\int_{-\infty}^{\infty} |x|^s dF(x) = \int_{x\in A_{<1}} |x|^s dF(x) + \int_{x\in A_{\geq 1}} |x|^s dF(x)$.
Since $f(x) \geq |x|^s f(x)\ \forall x \in A_{<1}$,

$$P(|x|^s < 1) = \int_{x \in A_{<1}} dF(x) \geq \int_{x \in A_{<1}} |x|^s dF(x).$$

Now let $r > s$, and note that $|x|^s \leq |x|^r \; \forall x \in A_{\geq 1}$. Hence it follows that

$$\int_{x \in A_{\geq 1}} |x|^r dF(x) \geq \int_{x \in A_{\geq 1}} |x|^s dF(x).$$

Finally, since $\int_{x \in A_{<1}} |x|^r dF(x) \geq 0$,

$$\int_{-\infty}^{\infty} |x|^s dF(x) \leq P(|x|^s < 1) + \int_{-\infty}^{\infty} |x|^r dF(x) < \infty,$$

where the right-most inequality is due to the fact that $P(|x|^s < 1) \in [0,1]$ and $E(X^r)$ exists, implying the absolute convergence of the improper integral defining the expectation. It follows from $\int_{-\infty}^{\infty} |x|^s dF(x) < \infty$ that $E(X^s)$ exists. ■

The theorem implies that if the existence of the $r$th-order moment about the origin can be demonstrated, then lower-order moments about the origin are known to exist. Theorem 3.23 can also be used to demonstrate the nonexistence of moments, since if $E(X^r)$ does *not* exist, then necessarily $E(X^s)$ cannot exist for $s > r$ or else Theorem 3.23 would be contradicted.

**Example 3.17** ***Nonexistence of Moments Order 2 or Higher*** Let the random variable $X$ have the density function $f(x) = 2(x+1)^{-3} I_{(0,\infty)}(x)$. Examine $E(X^\alpha) = \int_0^\infty x^\alpha \, 2(x+1)^{-3} dx$. To simplify the integral, make the substitution $y = x + 1$, so that $x = y - 1$ and $dy = dx$, to yield $E(X^\alpha) = 2 \int_1^\infty (y-1)^\alpha y^{-3} \, dy$. Note that if $\alpha = 2$, then $E(X^2) = 2\int_1^\infty (y^{-1} - 2y^{-2} + y^{-3}) dy = \lim_{y \to \infty} 2(\ln(y) + 2y^{-1} - (1/2)y^{-2})$, and since the limit diverges, $E(X^2)$ does not exist (i.e., note that $\ln(y) \to \infty$ as $y \to \infty$). This implies by Theorem 3.23 that moments of order 2 *or greater* do *not* exist for $X$. The reader can verify that $E(X)$ exists and is equal to 1. □

Existence results analogous to Theorem 3.23 can be stated for moments about the mean.

**Theorem 3.24** ***Existence of Lower Order Central Moments*** *If $E((Y-\mu)^r)$ exists for a given integer $r > 0$, then $E((Y-\mu)^s)$ exists for all integers $s \in [0,r]$.*

**Proof** This follows directly from Theorem 3.22 upon defining $X = Y - \mu$. ■

One can also infer the existence of moments about the mean from moments about the origin, and vice versa.

**Theorem 3.25** ***Existence Relationships Between Moment Types*** *If $E(X^r)$ (or $E((X-\mu)^r)$) exists for a given integer $r > 0$, then $E((X-\mu)^s)$ (or $E(X^s)$) exists for all integers $s \in [0,r]$.*

**Proof** Follows directly from Theorems 3.21–3.24. Details are left to the reader. ■

### 3.6.3 Nonmoment Measures of Probability Density Characteristics

Note that whether or not moments exist for $X$, there are other measures of probability distribution characteristics that are also of interest in applications. An alternative measure of the central tendency of a density is the median, defined as follows.

**Definition 3.16**
***Median of X***

Any number, $b$, satisfying $P(x \leq b) \geq 1/2$ and $P(x \geq b) \geq 1/2$ is called a **median of** $X$, and is denoted by med($X$).

The median is a measure of central tendency in the sense that $\geq 1/2$ of the probability mass of a density is both to the right and to the left of the median. In the continuous case, the probability inequalities in Definition 3.16 can be met with strict equalities, so that the median is a point at which exactly 1/2 the probability mass is to the left, and 1/2 to the right, as, $\int_{-\infty}^{\text{med}(X)} f(x)dx = \int_{\text{med}(X)}^{\infty} f(x)dx = .5$ . The inequalities are necessary for the concept to be generally applicable in the discrete case because the discreteness of probability assignments may prevent the conditions from being met as equalities.

Depending on how the density is defined, the median may not be unique even in the continuous case. However, if $b = \text{med}(X)$, and if in the neighborhood of the point $b$ the CDF of $X$ is continuous and strictly increasing, then med($X$) is unique (why?).

**Example 3.18**
***Median not Unique***

Let the random variable $X$ have density function $f(x) = (1/6)\, I_{\{1,2,\ldots,6\}}(x)$. Then the median of $X$ is not unique, and can be any number in the interval [3, 4], since $P(x \leq b) \geq 1/2$ and $P(x \geq b) \geq 1/2\ \forall b \in [3, 4]$. □

In practice, when there is a continuum of possible medians, as in Example 3.18, it is often the midpoint of the possible medians that is reported as "the" median. In Example 3.18 above, the value 3.5 would be reported.

**Example 3.19**
***Median of Continuous RV***

The central processing unit (CPU) used by a company that manufactures personal computers has an operating life until failure that is given by the outcome of a random variable $X$ having density function $f(x) = (1/50)e^{-x/50} I_{(0,\infty)}(x)$, where $x$ is measured in thousands of hours. What is the median operating life of the CPU?

**Answer**: We must solve the following equation for med($X$):

$$\int_{-\infty}^{\text{med}(X)} \frac{1}{50} e^{-x/50} I_{(0,\infty)}(x)dx = .5 \text{ or } 1 - e^{-\text{med}(X)/50} = .5,$$

so that med($X$) = 34.657. It is thus equally probable that the CPU will operate more or less than 34,657 hours until failure. □

Another probability density characteristic, which subsumes the median as a special case, is called a quantile.

**Definition 3.17**
*Quantile of X*

Any number, $b$, satisfying $P(x \leq b) \geq p$ and $P(x \geq b) \geq 1-p$ for $p \in (0, 1)$ is called a **quantile** of $X$ of order $p$ (or the $(100p)$th **percentile** of the distribution of $X$).

Note the median is then simply the quantile of $X$ of order .5, or the 50th percentile of the distribution of $X$. As in the case of the median, the quantile of order $p$ may not be unique for a given random variable $X$. In Example 3.18, any $b \in [4, 5]$ would be a quantile of $X$ of order 2/3, while in Example 3.19, the quantile of order 2/3 would be $b = 54.931$.

One additional characteristic of a probability distribution that will be especially useful when we study the maximum likelihood procedure of statistical inference is a **mode of the distribution of** $X$.

**Definition 3.18**
*Mode of f(x)*

Any point, $b$, at which $f(x)$ exhibits a maximum is called a **mode** of $X$, or a **mode of the distribution** of $X$, and is denoted by **mode** $(X)$.

Some density functions may not have a unique mode. Those that do are referred to as being **unimodal**. Note that the density function in Example 3.18 exhibits six modes corresponding to the points $x = 1, 2, 3, 4, 5, 6$. The density function in Example 3.17 has one mode at the point $x = 0$. The density in Example 3.19 has no mode (Why not? How might the problem be altered so that mode $(X)$ exists?).

## 3.7 Moment Generating Functions

The expectation of $e^{tX}$ results in a function of $t$ that, when differentiated with respect to the argument $t$ and then evaluated at $t = 0$, generates moments of $X$ about the origin. The function is aptly called the **moment-generating function** of $X$.

**Definition 3.19**
*Moment Generating Function (MGF)*

The expected value of $e^{tX}$ is defined to be the moment-generating function of $X$ if the expected value exists for every value of $t$ in some open interval containing 0, i.e., $\forall t \in (-h, h), h > 0$. The moment generating function of $X$ will be denoted by $M_X(t)$, and is represented by

$$M_X(t) = \mathrm{E}\left(e^{tX}\right) = \sum_{x \in R(x)} e^{tx} f(x), \qquad \text{(discrete)}$$

$$M_X(t) = \mathrm{E}\left(e^{tX}\right) = \int_{-\infty}^{\infty} e^{tx} f(x) dx, \qquad \text{(continuous)}$$

Note that $M_X(0) = \mathrm{E}(e^0) = \mathrm{E}(1) = 1$ is *always* defined, and from this property it is clear that a function of $t$ *cannot* be a MGF unless the value of the function at $t = 0$ is 1. The condition that $M_X(t)$ must be defined $\forall t \in (-h,h)$ is a technical condition that ensures $M_X(t)$ is differentiable at the point zero, a property whose importance will become evident shortly.

We now indicate how the MGF can be used to generate moments about the origin. In the following theorem, we use the notation $d^r g(a)/dx^r$ to indicate the $r$th derivative of $g(x)$ with respect to $x$ evaluated at $x = a$.

**Theorem 3.26** ***Moments from MGF*** *Let X be a Random Variable for which the MGF, $M_X(t)$, exists. Then*

$$\mu'_r = \mathrm{E}(X^r) = \frac{d^r M_X(0)}{dt^r}.$$

**Proof** The proof is facilitated by the following lemma from advanced calculus.

**Lemma 3.2**

If the function $g(t)$ defined by $g(t) = \sum_{x \in R(X)} e^{tx} f(x)$ or $\int_{-\infty}^{\infty} e^{tx} f(x) dx$ converges for $t \in (-h, h)$, $h > 0$, then $d^r g(t)/dt^r$ exists $\forall t \in (-h, h)$ and for all positive integers $r$, and the derivative can be found by *differentiating under the summation sign* or *differentiating under the integral sign*, respectively, as

$$\frac{d^r g(t)}{dt^r} = \sum_{x \in R(x)} \frac{d^r e^{tx}}{dt^r} f(x) \text{ or } \int_{-\infty}^{\infty} \frac{d^r e^{tx}}{dt^r} f(x) dx.$$

(see D.V. Widder 1961, *Advanced Calculus*, 2nd Ed., Englewood Cliffs, NJ: Prentice Hall, pp. 442–447).

If the moment generating function $M_X(t) = \mathrm{E}(e^{tx}) = \int_{-\infty}^{\infty} e^{tx} dF(x)$ exists (converges) for $t \in (-h, h)$, $h > 0$, then from Lemma 3.2,

$$\frac{d^r M_X(t)}{dt^r} = \int_{-\infty}^{\infty} \frac{d^r e^{tx}}{dt^r} dF(x) = \int_{-\infty}^{\infty} x^r e^{tx} dF(x).$$

Evaluating the $r$th derivative at $t = 0$ yields

$$\frac{d^r M_X(0)}{dt^r} = \int_{-\infty}^{\infty} x^r dF(x) dx = \mathrm{E}(X^r).$$

■

**Example 3.20** ***MGF for Continuous RV*** The random variable $X$ has the density function $f(x) = \mathrm{e}^{-x} I_{(0,\infty)}(x)$. Find the MGF, and use it to define the mean and variance of $X$.

**Answer:**

$$M_X(t) = \int_{-\infty}^{\infty} e^{tx} e^{-x} I_{(0,\infty)}(x) dx = \int_{-\infty}^{\infty} e^{x(t-1)} dx = \left. \frac{e^{x(t-1)}}{t-1} \right|_0^{\infty}$$

$$= 0 - \frac{1}{t-1} = (1-t)^{-1} (\text{provided } t<1)$$

The mean is defined as $\mu = dM_X(0)/dt = (1-0)^{-2} = 1$. For the variance, recall from Theorem 3.21 that $\sigma^2 = \mu_2' - \mu^2$. Then, $\mu_2' = d^2M_X(0)/dt^2 = 2(1-0)^{-3} = 2$, and thus, $\sigma^2 = 1$. □

There are a number of elementary results relating to moment-generating functions that can be quite useful in applications. We present these results in the next theorem.

**Theorem 3.27**
***Properties of MGFs***

Let $(X_1,\ldots,X_n)$ *be independent random variable having respective MGFs* $M_{X_i}(t)$, *for* $i = 1,\ldots,n$.

**a.** If $Y_i = a\,X_i + b$, then $M_{Y_i}(t) = e^{bt}\,M_{X_i}(at)$
**b.** If $Y = \sum_{i=1}^n X_i$, then $M_Y(t) = \prod_{i=1}^n (M_{X_i}(t))$.
**c.** If $Y = \sum_{i=1}^n a_iX_i + b$, then $M_Y(t) = e^{bt}\prod_{i=1}^n (M_{X_i}(a_it))$.

**Proof** Left to the Reader.

It is useful to note that since $X$ could be defined as a random variable that is itself a function of other random variables, a more general conceptualization of the MGF is $M_{g(X)}(t) = \mathrm{E}\,(\exp\,(g(X)t))$ where $g$ is a function of $X$.[13] The MGF $M_{g(X)}(t)$ could then be used to define moments about the origin for the random variable defined by $g(X)$. Note that because moments about the *origin* for $g(X) = X - \mu$ coincide with moments about the *mean* for $X$, the generalized MGF can be defined appropriately to generate moments about the mean directly.

**Example 3.21**
***MGF for Generating Central Moments***

Let $f(x) = e^{-x}I_{(0,\infty)}(x)$, as in Example 3.20. Recall that $\mu = 1$ in this case. We find the moment-generating function of the random variable $Y = g(X) = X - 1$, and use it to define the variance of $X$. First of all, note that Theorem 3.27 (a) is applicable with $a = 1$ and $b = -1$. Since $M_X(t) = (1-t)^{-1}$ for $t < 1$ from Example 3.20, it follows that $M_{(X-1)}(t) = e^{-t}\,(1-t)^{-1}$ for $t < 1$.

To find var $(X) = (d^2M_{(X-1)}(0))/dt^2 = \mathrm{E}(X-1)^2$:

$$\frac{dM_{(X-1)}(t)}{dt} = e^{-t}(1-t)^{-2} - (1-t)^{-1}e^{-t}$$

and

$$\frac{d^2M_{(X-1)}(t)}{dt^2} = 2e^{-t}(1-t)^{-3} - e^{-t}(1-t)^{-2} + (1-t)^{-1}e^{-t} - e^{-t}(1-t)^{-2},$$

so that

$$\frac{d^2M_{(X-1)}(0)}{dt^2} = 1 = \mathrm{var}(X).$$ □

### 3.7.1 Uniqueness of MGFs

Apart from generating moments, the MGF can be useful for identifying the density function of a given random variable. This is due to a uniqueness property

[13] Recall that $\exp(a) \equiv e^a$, where $a$ is a real number.

possessed by MGFs that essentially establishes a one-to-one correspondence between density functions and MGFs. A formal statement of the uniqueness property is given in the following theorem. The proof of the theorem relies on the fact that the MGF is a bilateral Laplace transform of the function $f(x)$, and there is a unique association between a Laplace transform and the function being transformed. These concepts are beyond the scope of our study, and the proof will be omitted. The interested reader can refer to Widder (1989) Advanced Calculus, pp. 459–460 and D.A.S. Fraser (1976) *Probability and Statistics*. North Scituate, Duxbury Press, pp. 544–546 for details.

**Theorem 3.28**
***MGF Uniqueness***

*If a moment-generating function exists for a random variable X having density function $f(x)$, then the moment generating function is unique. Conversely, the moment generating function determines the density function of X uniquely, at least up to a set of points having probability zero.*

In other words, a density function has one and only one MGF associated with it, if a MGF exists at all. Furthermore, if more than one density function is associated with a given MGF, the densities differ only on a set of points *that are irrelevant* for the purposes of assigning probabilities to events for $X$, i.e., they differ on a set of points having probability zero. Thus, if one knows the MGF for a given random variable $X$, and if one also knows of any density function that produces this MGF, then that density function suffices as the density function of the random variable $X$ for purposes of any probability assignments, or for defining expectations of any functions of $X$. The following example illustrates the logic followed in applying the uniqueness theorem.

**Example 3.22**
***Identifying a PDF by its MGF***

Examine the density function $f(x) = (b - a)^{-1} I_{[a,b]}(x)$ for $a < b$. The MGF associated with this density can be identified as follows:

$$M_X(t) = \mathrm{E}(e^{Xt}) = \int_{-\infty}^{\infty} e^{xt}(b-a)^{-1} I_{[a,b]}(x)dx = (b-a)^{-1}\int_a^b e^{xt}dx$$

$$= (b-a)^{-1}\frac{e^{xt}}{t}\bigg|_a^b = \begin{cases} \frac{e^{bt}-e^{at}}{t(b-a)} & \text{for } t \neq 0, \\ 1 & \text{for } t = 0. \end{cases}$$

Now suppose a random variable $Z$ has a MGF defined by $M_Z(t) = (e^{bt} - e^{at})/(t(b - a))$ for $t \neq 0$. Then by the uniqueness theorem, since $f(x)$ above is associated with this same MGF, the density function of $Z$ can be specified as $f(z) = (b - a)^{-1} I_{[a,b]}(z)$. □

When it exists, an MGF can be thought of as a "fingerprint" of a given density function. In Chapter 4, we will examine a collection of density functions that have been found to be useful in applications, and we will provide a list of their MGFs. Later on we will examine a number of important functions of random variables that will be used for statistical inference purposes, and in a notable number of cases, we will be able to identify the probability densities of these functions by matching their MGFs to the appropriate MGFs in the list we will have assembled.

### 3.7.2 Inversion of MGFs

The "recognition" of an MGF as a known fingerprint of some probability density function is not the only way an MGF can be used to identify the probability distribution of a random variable. There is an **inversion relationship** that allows one to integrate a function involving the MGF to identify the CDF of a random variable, from which the density function can be deduced. Unfortunately, the technique involves transform theory and generally complicated integration of expressions involving complex numbers and is beyond the scope of our study. Nonetheless, without providing the formal details, we will provide the reader with the general idea of what is involved.

If we replace the argument $t$ in an MGF by $(it)$, $i$ being the imaginary number $i = \sqrt{-1}$, then differences in the CDF values are associated with the MGF as

$$F(b) - F(a) = \frac{1}{2\pi} \int_{-\infty}^{\infty} \frac{e^{-ita} - e^{-itb}}{it} M_X(it)dt,$$

for $a < b$.[14] Then the density function can be determined from the CDF using either the differencing or derivative methods described in Chapter 2 for discrete and continuous random variables, respectively. For example, regarding the continuous case, differentiating with respect to $b$ results in the probability density function $f(x) = \int_{-\infty}^{\infty} (2\pi)^{-1} e^{-itx} M_X(it)dt$. For further information concerning this inversion property, the reader can consult M. Kendall and A. Stuart (1977), *The Advanced Theory of Statistics*, Vol. 1. New York: Macmillan, Chapter 4.

In cases where the MGF does not exist, there is an alternative function that always exists, called the **characteristic function**, which serves the same purpose as the MGF. In particular, there is a unique relationship between characteristic functions and density functions, analogous to the result stated in Theorem 3.28. The characteristic function can be inverted to obtain the density function, and the characteristic function can be used to generate any moments that exist for a random variable by differentiating the characteristic function an appropriate number of times, evaluating the derivative at the point zero, and then dividing the result by $(i)^k$, $k$ being the order of the moment sought (equivalently, $k$ is the order of the derivative). The characteristic function is defined as $\phi_X(t) = \mathrm{E}(e^{itX})$, and so complex numbers are involved in the definition of the characteristic function. When the MGF exists, the characteristic function is $\phi_X(t) \equiv M_X(it)$, i.e., the characteristic function is identically the MGF evaluated at $(it)$ rather than $t$. For example, in Example 3.22, the characteristic function of $X$ would be $(e^{bit} - \mathrm{e}^{\mathrm{ait}})/(it(\mathrm{b} - a))$ for $t \neq 0$. Despite the advantage that $\phi_X(t)$ always exists, we will not pursue the study of characteristic functions any further, in order to

[14]Phoebus Dhrymes (1989) *Topics in Advanced Econometrics*, Springer-Verlag, p. 254.

avoid the use of complex numbers. Interested readers can examine Kendall and Stuart, *Advanced Statistics*, Chapter 4, for further details.

## 3.8 Cumulant Generating Function

The natural logarithm of the moment-generating function defines a function called the **cumulant-generating function** which, when differentiated $r$ times with respect to $t$ and then evaluated at $t = 0$, defines the $r$th **cumulant** of a random variable.

**Definition 3.20** *Cumulant-Generating Function and Cumulants*

The cumulant-generating function of $X$ is defined as $\psi(t) = \ln(M_X(t))$. The $r$th cumulant of $X$ is given by $\kappa_r = \left(d^r\,\psi(0)\right)/\,dt^r$. The first four cumulants are related to moments as follows: $\kappa_1 = \mu_1'; \kappa_2 = \sigma^2; \kappa_3 = \mu_3;$ and $\kappa_4 = \mu_4 - 3\sigma^4$.

The cumulant-generating function can be used directly to generate the mean, variance, and third moment about the mean via differentiation of the function to the first, second, or third order respectively. The fourth derivative produces the fourth order cumulant that, when divided by $\sigma^4$, generates the excess kurtosis measure presented in the previous section. Thus, all of the moment measures introduced previously to measure central tendency, spread, skewness, and peakedness of a probability density function are available through use of the first four cumulants generated by the cumulant generating function.

If $X_1,\ldots,X_n$ are independent random variables, it follows from Definition 3.20 and Theorem 3.27 that the cumulant-generating function of $Y = \sum_{i=1}^n X_i$ equals the sum of the cumulant-generating functions of the $X_i$'s. It then also follows that the cumulant of the sum is the sum of the cumulants, which is the genesis of the name "cumulant." Often, the derivatives of the cumulant-generating function are easier to calculate than the derivatives of the MGF.

**Example 3.23** *Defining Moments via the Cumulant Generating Function*

Recall Examples 3.20 and 3.21 where $M_X(t) = (1 - t)^{-1}$ for $t < 1$. The cumulant-generating function of $X$ is given by $\psi_X(t) = \ln\,(M_X(t)) = -\ln(1 - t)$ for $t < 1$. Then

$$\mu = d\,\psi_X(0)/dt = (1 - t)^{-1}\,|_{t=0} = 1$$
$$\sigma^2 = d^2\,\psi_X(0)/\,dt^2 = (1 - t)^{-2}\,|_{t=0} = 1$$
$$\mu_3 = d^3\,\psi_X(0)/\,dt^3 = 2(1 - t)^{-3}\,|_{t=0} = 2$$
$$\kappa_4 = \mu_4 - 3\sigma^4 = d^4\,\psi_X(0)/\,dt^4 = 6(1 - t)^{-4}\,|_{t=0} = 6$$

□

## 3.9 Multivariate MGFs and Cumulant Generating Functions

The MGF and cumulant-generating function can be extended to the case of a multivariate random variable $\mathbf{X} = (X_1,\ldots,X_n)$, as follows:

**Definition 3.21**
***MGF and Cumulant Generating Function-Multivariate Case***

The expected value of exp $\left(\sum_{j=1}^{n} t_j X_j\right)$ is defined to be the MGF of the $n$-variate random variable $\mathbf{X} = (X_1,\ldots,X_n)$ if the expected value exists for all $t_i \in (-h,h)$, for some $h > 0$, $i = 1,\ldots,n$. The MGF will be denoted by $M_{\mathbf{X}}(\mathbf{t})$, where $\mathbf{t} = (t_1,\ldots,t_n)$. Thus,

$$M_{\mathbf{X}}(\mathbf{t}) = \sum_{(x_1,\ldots,x_n)\in R(\mathbf{X})} \exp\left(\sum_{j=1}^{n} t_j x_j\right) f(x_1,\ldots,x_n) \quad \text{(discrete)}$$

$$M_{\mathbf{X}}(\mathbf{t}) = \int_{-\infty}^{\infty} \cdots \int_{-\infty}^{\infty} \exp\left(\sum_{j=1}^{n} t_j x_j\right) f(x_1,\ldots,x_n) dx_1 \ldots dx_n \quad \text{(continuous)}.$$

The cumulant generating function of $\mathbf{X}$ is defined as $\psi_{\mathbf{X}}(\mathbf{t}) = \ln\left(M_{\mathbf{X}}(\mathbf{t})\right)$.

Letting $\mu'_r(X_i)$ denote the $r$th moment of $X_i$ about the origin, it can be shown that

$$\mu'_r(X_i) = \mathrm{E}\left(X_i^r\right) = \frac{\partial^r M_{\mathbf{X}}(0)}{\partial t_i^r}.$$

Thus, the $r$th order *partial* derivative of $M_{\mathbf{X}}(\mathbf{t})$ with respect to $t_i$, evaluated at $\mathbf{t} = \mathbf{0}$ (i.e., the *vector* $\mathbf{t}$ equal to the zero vector), equals the $r$th order moment about zero for $X_i$. Similarly, the $r$th *partial* derivative of $\psi_{\mathbf{X}}(\mathbf{t})$ with respect to $t_i$, evaluated at $\mathbf{t} = \mathbf{0}$, equals the $r$th cumulant of the random variable $X_i$, which then allows means, variances, skewness and kurtosis measures to be calculated directly for each of the $X_i$'s.

An analog of the **MFG uniqueness theorem** applies to the multivariate MGF. In fact, interpreting $X$ as a *vector* in the statement of Theorem 3.28 produces the appropriate **multivariate MGF uniqueness theorem**.

If the MGF for an $n$-variate random variable $(X_1,\ldots,X_n)$ is known, the **marginal MGF** for a subset of $m < n$ of the random variables is easily found by setting the $t_i$'s associated with the remaining $n-m$ random variables to zero, as presented in the next theorem.

**Theorem 3.29**
***Marginal MGFs from Multivariate MGFs***

*Let $\mathbf{X} = (X_1,\ldots,X_n)$ have MGF $M_{\mathbf{X}}(\mathbf{t})$, and let $\mathbf{X}_{(m)} = (X_j, j\in J)$ be any m-element subset of the random variables in $\mathbf{X}$, where $J \subset \{1, 2,\ldots,n\}$ and $N(J) = m < n$. Define $\mathbf{t}_{(m)} = (t_j, j\in J)$. Then the MGF of $\mathbf{X}_{(m)}$, referred to as the **marginal MGF** of $\mathbf{X}_{(m)}$, can be represented as $M_{X_{(m)}}\left(\mathbf{t}_{(m)}\right) = M_{\mathbf{X}}(\mathbf{t}^*)$, where the elements in $\mathbf{t}^*$ are defined by $t_j I_J(j)$.*

**Proof**

$$M_{\mathbf{X}}(\mathbf{t}^*) = \mathrm{E}\left(\exp\left(\sum_{j=1}^{n} t_j^* X_j\right)\right) = \mathrm{E}\left(\exp\left(\sum_{j\in J} t_j X_j\right)\right) \left(\text{since } t_j^* = 0 \text{ if } j \notin J\right)$$

$$= M_{\mathbf{X}_{(m)}}\left(\mathbf{t}_{(m)}\right) \qquad \text{(by definition).}$$

■

A **marginal cumulant-generating function** can be defined as the natural logarithm of a marginal MGF.

If **X** has MGF $M_{\mathbf{X}}(\mathbf{t})$, it can be shown that $X_1,\ldots,X_n$ are independent *iff* $M_X(t) = \prod_{i=1}^{n} M_{X_i}(t_i)$, or equivalently, *iff* $\psi_X(t) = \sum_{i=1}^{n} \psi_{X_i}(t)$ (see S.F. Arnold (1990) *Mathematical Statistics*, Englewood Cliffs, NJ: Prentice-Hall, pp. 118–119).

**Example 3.24** ***Marginal MGF and Cumulant Generating Functions***

Suppose the joint MGF of the bivariate random variable $(X_1, X_2)$ is given by $M_{\mathbf{X}}(\mathbf{t}) = exp\left(\sum_{i=1}^{2} \mu_i t_i + (1/2)\sum_{i=1}^{2}\sum_{j=1}^{2} \sigma_{ij} t_i t_j\right)$ (we will see in Chapter 4 that this is the MGF associated with a bivariate "normal" density function). Then the marginal MGF of $X_1$ can be defined by setting $t_2 = 0$ in $M_X(t)$ to obtain $M_{X_1}(t_1) = \exp(\mu_1 t_1 + (1/2)\sigma_{11} t_1^2)$ (which is the MGF associated with a univariate "normal" density function). The marginal cumulant-generating function is given by $\psi_{X_1}(t) = \ln(M_{X_1}(t)) = \mu_1 t_1 + (1/2)\sigma_{11} t_1^2$. □

If we take *cross* partial derivatives of the MGF of **X**, and evaluate the derivative at $\mathbf{t} = \mathbf{0}$, we obtain

$$\frac{\partial^{r+s} M_X(0)}{\partial t_i^r \, \partial t_j^s} = \mathrm{E}\left(X_i^r X_j^s\right).$$

This expectation is an example of a ***joint moment***. The cross partial derivative of the cumulant generating function given by $(\partial^2 \psi_X(\mathbf{0}))/(\partial t_i \partial t_j) = \mathrm{E}((X_i - \mathrm{E}(X_i))(X_j - \mathrm{E}(X_j)))$ is the *covariance* between $X_i$ and $X_j$. These concepts are discussed further in the next section.

## 3.10 Joint Moments, Covariance, and Correlation

In the case of multivariate random variables, the concept of **joint moments** becomes relevant. The formal definitions of joint moments about the origin and about the mean are as follows:

**Definition 3.22** ***Joint Moment About the Origin***

Let $X$ and $Y$ be two random variables having joint density function $f(x,y)$. Then the $(r,s)$th joint moment of $(X,Y)$ (or of $f(x,y)$) about the origin is defined by

$$\mu'_{r,s} = \sum_{x\in R(X)} \sum_{y\in R(Y)} x^r y^s f(x,y) \qquad \text{(discrete)}$$

$$\mu'_{r,s} = \int_{-\infty}^{\infty}\int_{-\infty}^{\infty} x^r y^s f(x,y)\,dx\,dy \qquad \text{(continuous)}$$

**Definition 3.23**
***Joint Moments About the Mean (or Central Joint Moment)***

Let $X$ and $Y$ be two random variables having joint density function $f(x,y)$. Then the $(r,s)$th joint moment of $(X,Y)$ (or of $f(x,y)$) about the mean is defined by

$$\mu_{r,s} = \sum_{x \in R(X)} \sum_{y \in R(Y)} (x - \mathrm{E}(X))^r (y - \mathrm{E}(Y))^s f(x,y) \qquad \text{(discrete)}$$

$$\mu_{r,s} = \int_{-\infty}^{\infty} \int_{-\infty}^{\infty} (x - \mathrm{E}(X))^r (y - \mathrm{E}(Y))^s f(x,y)\,dxdy \qquad \text{(continuous)}$$

### 3.10.1 Covariance and Correlation

Regarding joint moments, our immediate interest is on a particular joint moment about the mean, $\mu_{1,1}$, and the relationship between this moment and moments about the origin. The central moment $\mu_{1,1}$ is given a special name and symbol, and we will see that $\mu_{1,1}$ is useful as a measure of "linear association" between $X$ and $Y$.

**Definition 3.24**
***Covariance***

The central joint moment $\mu_{1,1} = \mathrm{E}(X - \mathrm{E}(X))(Y - \mathrm{E}(Y))$ is called the **covariance between** $X$ **and** $Y$, and is denoted by the symbol $\sigma_{XY}$, or by $\mathrm{cov}(X,Y)$.

Note that there is a simple relationship between $\sigma_{XY}$ and moments about the origin that can be used for the calculation of the covariance.

**Theorem 3.30**
***Covariance in Terms of Moments About the Origin***

$\sigma_{XY} = \mathrm{E}(XY) - \mathrm{E}(X)\mathrm{E}(Y).$

**Proof** This result follows directly from the properties of the expectation operation. In particular, by definition

$$\begin{aligned}\sigma_{XY} &= \mathrm{E}((X - \mathrm{E}(X))\,(Y - \mathrm{E}(Y))) = \mathrm{E}(XY - (\mathrm{E}(X))Y - X\mathrm{E}(Y) + \mathrm{E}(X)\mathrm{E}(Y)) \\ &= \mathrm{E}(XY) - \mathrm{E}(X)\mathrm{E}(Y)\end{aligned}$$

■

**Example 3.25**
***Covariance Calculation***

Let the bivariate random variable $(X,Y)$ have a joint density function $f(x,y) = (x + y)\, I_{[0,1]}(x) I_{[0,1]}(y)$. Find $\mathrm{cov}(X,Y)$.

**Answer**: Note that

$$\mathrm{E}(XY) = \int_0^1 \int_0^1 xy(x+y)dxdy = \int_0^1 \int_0^1 \left[x^2y + xy^2\right] dxdy = \frac{1}{3}$$

$$\mathrm{E}(X) = \int_0^1 \int_0^1 x(x+y)dxdy = \int_0^1 \int_0^1 (x^2 + xy)dxdy = \frac{7}{12}$$

$$\mathrm{E}(Y) = \int_0^1 \int_0^1 y(x+y)dxdy = \int_0^1 \int_0^1 (yx + y^2)dxdy = \frac{7}{12}.$$

Then, by Theorem 3.30, $\mathrm{cov}(X,Y) = 1/3 - (7/12)(7/12) = (-1/144)$. □

A useful corollary to Theorem 3.30 is that *the expectation of a product of two random variables is the product of the expectations iff* $\sigma_{XY} = 0$, formally stated as follows.

**Corollary 3.5** ***Expectation of Product Equals Product of Expectations***

$E(XY) = E(X)E(Y)$ *iff* $\sigma_{XY} = 0$.

**Proof** This follows directly from Theorem 3.30 upon setting $\sigma_{XY}$ to zero (sufficiency) or setting E(XY) equal to E(X) E(Y) (necessity). ■

What does $\sigma_{XY}$ measure? The covariance is a measure of the *linear association* between two random variables, where the precise meaning of linear association will be made clear shortly. Our discussion will be facilitated by observing that the value of $\sigma_{XY}$ exhibits a definite upper bound in absolute value which is expressible as a function of the variances of the two random variables involved. The bound on $\sigma_{XY}$ follows from the following inequality.

**Theorem 3.31** ***Cauchy-Schwarz Inequality***

$(E(WZ))^2 \leq E(W^2)E(Z^2)$

**Proof** The quantity $E((\lambda_1 W + \lambda_2 Z)^2)$ must be greater than or equal to 0 $\forall(\lambda_1, \lambda_2)$ since $(\lambda_1 W + \lambda_2 Z)^2$ is a random variable having only non-negative outcomes. Thus

$\lambda_1^2 E(W^2) + \lambda_2^2 E(Z^2) + 2\lambda_1\lambda_2 E(WZ) \geq 0 \; \forall(\lambda_1, \lambda_2)$, which in matrix terms can be represented as

$$[\lambda_1 \quad \lambda_2]\begin{bmatrix} E(W^2) & E(WZ) \\ E(WZ) & E(Z^2) \end{bmatrix}\begin{bmatrix} \lambda_1 \\ \lambda_2 \end{bmatrix} \geq 0 \; \forall(\lambda_1, \lambda_2).$$

The last inequality is precisely the defining property of positive semidefiniteness for the $(2 \times 2)$ matrix in brackets,[15] and the matrix in brackets will be positive semidefinite *iff* $E(W^2)E(Z^2) - (E(WZ))^2 \geq 0$ (see the Appendix Section 3.12). ■

The covariance bound we seek is stated in the following theorem.

**Theorem 3.32** ***Covariance Bound***

$|\sigma_{XY}| \leq \sigma_X \sigma_Y$.

**Proof** Let $W = (X - E(X))$ and $Z = (Y - E(Y))$ in the Cauchy-Schwarz inequality. Then $\left(E((X - E(X))(Y - E(Y)))\right)^2 \leq E((X - E(X))^2)E((Y - E(Y))^2)$, or equivalently, $\sigma_{XY}^2 \leq \sigma_X^2 \sigma_Y^2$ which holds *iff* $|\sigma_{XY}| \leq \sigma_X \sigma_Y$. ■

[15] Recall that a matrix **A** is positive semidefinite *iff* $\mathbf{t}'\mathbf{A}\mathbf{t} \geq 0 \; \forall \; \mathbf{t}$, and **A** is positive definite *iff* $\mathbf{t}'\mathbf{A}\mathbf{t} > 0 \; \forall \; \mathbf{t} \neq \mathbf{0}$.

Thus, the covariance between $X$ and $Y$ is upper-bounded in absolute value by the product of the standard deviations of $X$ and $Y$. Using this bound, we can define a useful scaled version of the covariance, called the **correlation** between $X$ and $Y$, as follows.

**Definition 3.25**
***Correlation***

The correlation between two random variables $X$ and $Y$ is defined by $corr(X, Y) = \rho_{XY} = \frac{\sigma_{XY}}{\sigma_X \, \sigma_Y}$.

**Example 3.26**
***Correlation Calculation***

Refer to Example 3.25. Note that

$$E(X^2) = \int_0^1 \int_0^1 x^2 (x + y) dx dy = \frac{5}{12}$$

$$E(Y^2) = \int_0^1 \int_0^1 y^2 (x + y) dx dy = \frac{5}{12},$$

so that

$\sigma_X^2 = E(X^2) - (E(X))^2 = 5/12 - (7/12)^2 = 11/144,$

and

$\sigma_Y^2 = E(Y^2) - (E(Y))^2 = 5/12 - (7/12)^2 = 11/144.$

Then the correlation between $X$ and $Y$ is given by

$$\rho_{XY} = \frac{\sigma_{XY}}{\sigma_X \, \sigma_Y} = \frac{-1/144}{(11/144)^{1/2} (11/144)^{1/2}} = \frac{-1}{11}. \qquad \square$$

Bounds on the correlation between $X$ and $Y$ follow directly from the bounds on the covariance between $X$ and $Y$.

**Theorem 3.33**
***Correlation Bound***

$-1 \leq \rho_{XY} \leq 1.$

**Proof** This follows directly from Theorem 3.32 via division by $\sigma_x \sigma_y$. ■

The covariance equals its upper bound value of $\sigma_X \, \sigma_Y$ *iff* the correlation equals its upper bound value of 1, and the covariance equals its lower bound value of $-\sigma_X \, \sigma_Y$ *iff* the correlation equals its lower bound value of $-1$.

Assuming that the covariance exists, a *necessary* condition for the independence of $X$ and $Y$ is that $\sigma_{XY} = 0$ (or equivalently, that $\rho_{XY} = 0$ if $\sigma_X \, \sigma_Y \neq 0$).

**Theorem 3.34**
***Relationship Between Independence and Covariance***

*If X and Y are independent, then* $\sigma_{XY} = 0$ *(assuming the covariance exists).*

**Proof** If $X$ and $Y$ are independent, then $f(x,y) = f_X(x)\,f_Y(y)$. It follows that

$$\begin{aligned}\sigma_{XY} &= \int_{-\infty}^{\infty}\int_{-\infty}^{\infty}(x - E(X))(y - E(Y))dF(x,y)\\ &= \int_{-\infty}^{\infty}(x - E(X))dF_X(x)\int_{-\infty}^{\infty}(y - E(Y))dF_Y(y)\\ &= (E(X) - E(X))(E(Y) - E(Y)) = 0 \cdot 0 = 0.\end{aligned}$$ ■

The converse of Theorem 3.34 is not true—there can be dependence between $X$ and $Y$, even *functional* dependence, and the covariance between $X$ and $Y$ could nonetheless be zero, as the following example illustrates.

**Example 3.27** ***Bivariate Function Dependence with*** $\sigma_{XY} = 0$

Let $X$ and $Y$ be two random variables having a joint density function given by $f(x,y) = 1.5\ I_{[-1,1]}(x)\ I_{[0,x^2]}(y)$. Note this density implies that $(x,y)$ points are equally likely to occur on and below the parabola represented by the graph of $y = x^2$. There is a direct functional dependence between $X$ and the range of $Y$, so that $f(y \mid x)$ will change as x changes and thus $X$ and $Y$ must be dependent random variables. Nonetheless, $\sigma_{XY} = 0$. To see this, note that

$$E(XY) = 1.5\int_{-1}^{1}\int_{0}^{x^2} xydydx = 1.5\int_{-1}^{1}(1/2)x^5dx = .75\left.\frac{x^6}{6}\right|_{-1}^{1} = 0,$$

$$E(X) = 1.5\int_{-1}^{1}\int_{0}^{x^2} xdydx = 1.5\int_{-1}^{1}x^3dx = 1.5\left.\frac{x^4}{4}\right|_{-1}^{1} = 0,$$

$$E(Y) = 1.5\int_{-1}^{1}\int_{0}^{x^2} ydydx = 1.5\int_{-1}^{1}(1/2)x^4dx = .75\left.\frac{x^5}{5}\right|_{-1}^{1} = .3.$$

Therefore, $\sigma_{XY} = E(XY) - E(X)E(Y) = 0 - 0(.3) = 0$. □

### 3.10.2 Correlation, Linear Association and Degeneracy

We now demonstrate that when the covariance takes its maximum absolute value, and thus $\rho_{XY} = +1$ or $-1$, there is a perfect positive ($\rho_{XY} = +1$) or negative ($\rho_{XY} = -1$) *linear* relationship between $X$ and $Y$ that holds with probability one (i.e., $P(y = a + bx) = 1$ or $P(y = a - bx) = 1$). The demonstration is facilitated by the following useful result.

**Theorem 3.35** ***Degeneracy when*** $\sigma^2 = 0$

*Let Z be a Random Variable for which* $\sigma_Z^2 = 0$. *Then* $P(z = E(Z)) = 1$

**Proof** Let $g(Z) = (Z - E(Z))^2$. Then

$$P(\mathrm{E}(Z) - a<z<\mathrm{E}(Z) + a) = P\left((z - \mathrm{E}(Z))^2 < a^2\right)$$
$$= 1 - P\left((z - \mathrm{E}(Z))^2 \geq a^2\right) \geq 1 - \sigma_Z^2 / a^2,$$

where the inequality is established using Markov's inequality. If $\sigma_Z^2 = 0$, then $P(\mathrm{E}(Z) - a < z < \mathrm{E}(Z) + a) = 1 \;\forall\; a > 0$, and since only $z = \mathrm{E}(Z)$ satisfies the inequality $\forall a > 0$, $P(z = \mathrm{E}(Z)) = 1$ when $\sigma_Z^2{=}0$. ■

The result on the linear relationship between $X$ and $Y$ when $\rho_{XY} = +1$ or $-1$, or equivalently, when $\sigma_{XY}$ achieves its upper and lower bound, is as follows.

**Theorem 3.36** ***Correlation Bounds and Linearity*** *If* $\rho_{XY} = +1$ *or* $-1$, *then* $P(y = a_1 + bx) = 1$ *or* $P(y = a_2 - bx) = 1$, *respectively, where* $a_1 = \mathrm{E}(Y) - (\sigma_Y/\sigma_X)\mathrm{E}(X)$, $a_2 = \mathrm{E}(Y) + (\sigma_Y/\sigma_X)\mathrm{E}(X)$, *and* $b = (\sigma_Y/\sigma_X)$.

**Proof** Define $Z = \lambda_1\ (X - \mathrm{E}(X)) + \lambda_2\ (Y - \mathrm{E}(Y))$, and note that $\mathrm{E}(Z) = 0$. It follows immediately that $\sigma_Z^2 = \mathrm{E}(Z^2) = \mathrm{E}((\lambda_1(X - \mathrm{E}(X)) + \lambda_2(Y - \mathrm{E}(Y)))^2) = \lambda_1^2\ \mathrm{E}((X - \mathrm{E}(X))^2) + \lambda_2^2\ \mathrm{E}((Y - \mathrm{E}(Y))^2) + 2\ \lambda_1\ \lambda_2\ \sigma_{XY} \geq 0 \;\forall\; \lambda_1, \lambda_2$, which can be represented in matrix terms as

$$\sigma_Z^2 = [\lambda_1 \quad \lambda_2]\begin{bmatrix} \sigma_X^2 & \sigma_{XY} \\ \sigma_{XY} & \sigma_Y^2 \end{bmatrix}\begin{bmatrix} \lambda_1 \\ \lambda_2 \end{bmatrix} \geq 0 \;\forall (\lambda_1, \lambda_2).$$

If $\rho_{XY} = +1$ or $-1$, then $\sigma_{XY}$ achieves either its (nominal) upper or lower bound, respectively, or equivalently, $\sigma_{XY}^2 = \sigma_X^2\ \sigma_Y^2$. It follows that the above $2 \times 2$ matrix is singular, since its determinant would be zero. Then the columns of the matrix are linearly dependent, so that there exist nonzero values of $\lambda_1$ and $\lambda_2$ such that

$$\begin{bmatrix} \sigma_X^2 & \sigma_{XY} \\ \sigma_{XY} & \sigma_Y^2 \end{bmatrix}\begin{bmatrix} \lambda_1 \\ \lambda_2 \end{bmatrix} = \begin{bmatrix} 0 \\ 0 \end{bmatrix}$$

and for these $\lambda$-values, the quadratic form above, and thus $\sigma_Z^2$, achieves the value 0. A solution for $\lambda_1$ and $\lambda_2$ is given by $\lambda_1 = \sigma_{XY}/\sigma_X^2$ and $\lambda_2 = -1$ which can be validated by substituting these values for $\lambda_1$ and $\lambda_2$ in the linear equations, and noting that $\sigma_Y^2 = \sigma_{XY}^2 / \sigma_X^2$ under the prevailing assumptions. Since $\sigma_Z^2 = 0$ at these values of $\lambda_1$ and $\lambda_2$, it follows from Theorem 3.35 that $P(z = 0) = 1$ (recall $\mathrm{E}(Z) = 0$).

Given the definition of $Z$, substituting the above solution values for $\lambda_1$ and $\lambda_2$ obtains an equivalent probability statement $P(y = (\mathrm{E}(Y) - (\sigma_{XY}/\sigma_X^2)\ \mathrm{E}(X)) + (\sigma_{XY}/\sigma_X^2)\ x) = 1$. If $\rho_{XY} = +1$, then $\sigma_{XY} = \sigma_X\sigma_Y$, yielding $P(y = a_1 + bx) = 1$ in the statement of the theorem, while if $\rho_{XY} = -1$, then $\sigma_{XY} = -\sigma_X\sigma_Y$, yielding $P(y = a_2 - bx) = 1$ in the statement of the theorem. ■

The theorem implies that when $\rho_{XY} = +1$ (or $-1$), the event that the outcome of $(X,Y)$ is on a straight line with positive (or negative) slope occurs with probability 1. As a diagrammatic illustration, if $(X,Y)$ is a discrete bivariate random variable, then the situation where $\rho_{XY} = +1$ would be exemplified

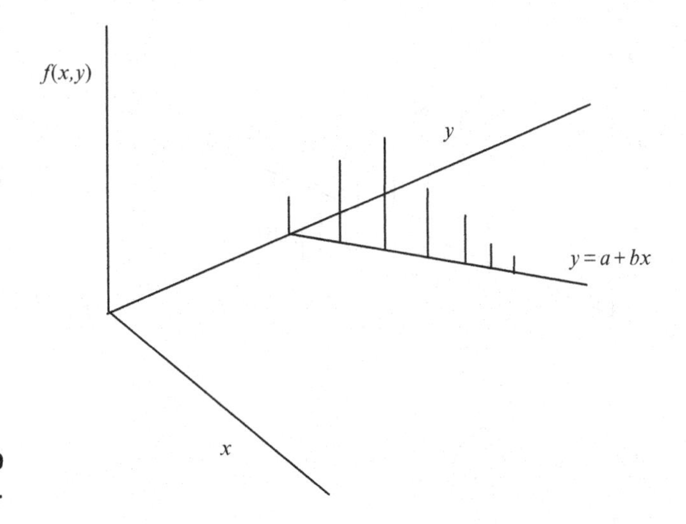


**Figure 3.10**
$\rho_{XY} = +1$, Discrete case.

by Figure 3.10. Note in Figure 3.10 that $f(x,y)$ assumes positive values only for points along the line $y = a + bx$, reflecting the fact that $P(y = a + bx) = 1$. This situation illustrates what is known as a **degenerate random variable** and a **degenerate density function**. The defining characteristic of a degenerate random variable is that it is an $n$-variate random variable $(X_1, \ldots, X_n)$ whose components satisfy one or more linear functional relationships with probability one, i.e., if $P(a_i + \sum_{j=1}^{n} b_{ij}x_j = 0) = 1$ for $i = 1,\ldots,m$, then $(X_1,\ldots,X_n)$ is a degenerate random variable.[16] A characteristic of the accompanying degenerate density function for $(X_1,\ldots,X_n)$ is that the entire mass of probability (a mass of 1) is concentrated on a collection of points that lie on a hyperplane of dimension less than $n$, the hyperplane being defined by the collection of linear functional relationships.

Degeneracy causes no particular difficulty in the discrete case—probabilities of events for the degenerate random variable $(X_1,\ldots,X_n)$ can be calculated in the usual way by summing the degenerate density function over the outcomes in the event of interest. However, degeneracy in the continuous case results in $f(x_1,\ldots,x_n)$ *not* being a density function according to our original definition of the concept. For a heuristic description of the problem, examine the diagrammatic illustration in Figure 3.11 for a degenerate bivariate random variable in the continuous case. Intuitively, because there is no volume under the graph of $f(x,y)$, $\int_{x_1}^{x_2} \int_{y_1}^{y_2} f(x,y)\, dy\, dx = 0 \; \forall\, x_1 \leq x_2$ and $\forall\, y_1 \leq y_2$, and $f(x,y)$ cannot be integrated in the usual way to assign probabilities to events for $(X, Y)$. However, there *is area* below the graph of $f(x,y)$ and above the line $y = a + bx$ representing the probability mass of 1 distributed over a segment (or perhaps, all) of this line. Since only subsets of the set $\{(x,y): y = a + bx,\, x \in R(X)\}$[17] are assigned nonzero

[16]The concept of degeneracy can be extended by calling $(X_1,\ldots,X_n)$ degenerate if the components satisfy one or more functional relationships (not necessarily linear) with probability 1. We will not examine this generalization here.

[17]Equivalently, $\{(x,y): x = b^{-1}(y-a),\, y \in R(Y)\}$.

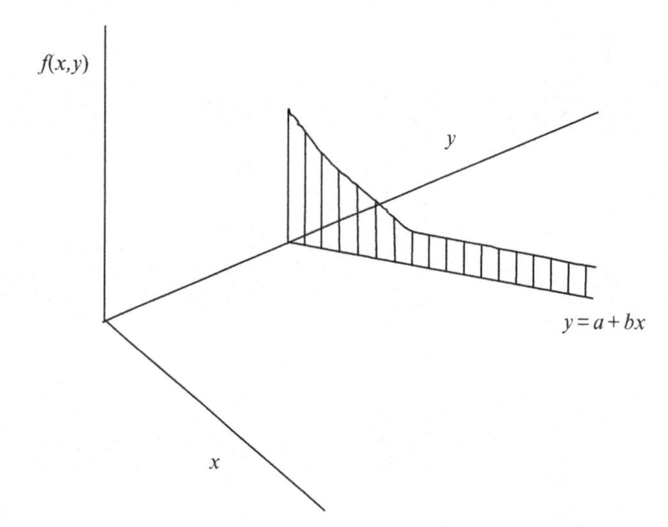


**Figure 3.11**
$\rho_{XY} = +1$, Continuous case.

probability, the degenerate density function can be used to assign probabilities to events by use of *line integrals,*[18] which essentially integrate $f(x,y)$ over subsets of points along the line $y = a + bx$. The general concept of line integrals is beyond the scope of our study, but in essence, the relevant integral in the current context is of the form $\int_{x \in A} f(x, a + bx)\, dx$. Note the linear relationship linking $y$ and $x$ is explicitly accounted for by substituting $a + bx$ for $y$ in $f(x,y)$, which converts $f(x, y)$ into a function of the single variable $x$. Then the function of $x$ is integrated over the points in the event $A$ for $x$, which determines the probability of the event $B = \{(x,y){:}y = a + bx,\ x \in A\}$ for the bivariate random variable $(X,Y)$.

Having introduced the concept of degeneracy, we can alternatively characterize $\rho_{XY} = +1$ or $-1$ as a case where the bivariate random variable $(X,Y)$, and its accompanying joint density function, are *degenerate*, with $X$ and $Y$ satisfying, respectively, a positively or negatively sloped linear functional relationship, with probability one. What can be said about the relationship between $X$ and $Y$ when $|\rho_{xy}| < 1$? The closer $|\rho_{XY}|$ is to one, the closer the relationship between $X$ and $Y$ is to being linear, where "closeness" can be interpreted as follows. Define the random variable $\hat{Y} = a + bX$ to represent predictions of $Y$ outcomes based on a linear function of $X$. We will choose the coefficients $a$ and $b$ so that $\hat{Y}$ is the *best linear* prediction of $Y$, where *best* is taken to mean "minimum expected squared distance between outcomes of $Y$ and outcomes of $\hat{Y}$."

**Theorem 3.37**
***Best Linear Prediction of Y Outcomes***

*Let $(X, Y)$ have moments of at least the second order, and let $\hat{Y} = a + bX$. Then the choices of $a$ and $b$ that minimize $\mathrm{E}(d^2(Y, \hat{Y})) = \mathrm{E}((Y - (a + b(X)))^2)$ are given by $a = \mathrm{E}(Y) - (\sigma_{XY}/\sigma_X^2)\mathrm{E}(X)$ and $b = (\sigma_{XY}/\sigma_X^2)$.*

[18]For an introduction to the concept of line integrals, see E. Kreyzig (1979) *Advanced Engineering Mathematics*, 4th ed. New York: Wiley, Chapter 9.

**Proof** Left to the reader. ■

Now define $V = Y - \hat{Y}$ to represent the deviations between outcomes of $Y$ and outcomes of the *best* linear prediction of $Y$ outcomes as defined in Theorem 3.37. Because $\mathrm{E}(Y) = \mathrm{E}(\hat{Y})$, $\mathrm{E}(V) = 0$. It follows that

$$\sigma_Y^2 = \mathrm{E}\left((Y - \mathrm{E}(Y))^2\right) = \mathrm{E}\left(\left(\hat{Y} - \mathrm{E}\left(\hat{Y}\right) + V\right)^2\right) = \sigma_{\hat{Y}}^2 + \sigma_V^2 + \sigma_{\hat{Y}V},$$

where

$$\sigma_V^2 = \mathrm{E}(V^2) = \mathrm{E}\left(d^2\left(Y, \hat{Y}\right)\right) = \mathrm{E}(d^2(Y, a + bX)) = \sigma_Y^2 - \sigma_{XY}^2/\sigma_X^2 = \sigma_Y^2[1 - \rho_{XY}^2],$$

$$\sigma_{\hat{Y}}^2 = \mathrm{E}\left(\hat{Y} - \mathrm{E}\left(\hat{Y}\right)\right)^2 = \sigma_Y^2 \rho_{XY}^2,$$

$$\sigma_{\hat{Y}V} = \mathrm{E}\left(\left(\hat{Y} - \mathrm{E}\left(\hat{Y}\right)\right)V\right) = (\sigma_{XY}/\sigma_X^2)\mathrm{E}((X - \mathrm{E}(X))V) = 0.$$

Thus, the variance of $Y$ is decomposed into a proportion $\rho_{XY}^2$ due to $\hat{Y}$ and a proportion $(1 - \rho_{XY}^2)$ due to $V$, i.e., $\sigma_Y^2 = \sigma_{\hat{Y}}^2 + \sigma_V^2 = \sigma_Y^2\, \rho_{XY}^2 + \sigma_Y^2\,(1 - \rho_{XY}^2)$.

We can now interpret values of $\rho_{XY} \in [-1, 1]$. Specifically, $\rho_{XY}^2$ is the proportion of the variance in $Y$ that is explained by the best linear prediction of the form $\hat{Y} = a + bX$, and the proportion of the variance unexplained is $(1 - \rho_{XY}^2)$. Relatedly, $\sigma_Y^2\,(1 - \rho_{XY}^2)$ is precisely the expected squared distance between outcomes of $Y$ and outcomes of the best linear prediction $\hat{Y}=a + bX$. Thus, the closer $|\rho_{XY}|$ is to 1, the more the variance in $Y$ is explained by the linear function $a + bX$, and the smaller is the expected squared distance between $Y$ and $\hat{Y}=a + bX$. It is in this sense that the higher the value of $|\rho_{XY}|$, the closer is the linear association between $Y$ and $X$.

If $\rho_{XY} = 0$, the random variables are said to be **uncorrelated**. In this case, Theorem 3.37 indicates that the best linear predictor is $\mathrm{E}(Y)$—there is effectively no linear association with $X$ whatsoever. The reader should note that $Y$ and $X$ can be interchanged in the preceding argument, leading to an analogous interpretation of the degree of linear association between $X$ and $\hat{X}=a + bY$ (for appropriate changes in the definitions of $a$ and $b$).

## 3.11 Means and Variances of Linear Combinations of Random Variables

Determining the mean and variance of random variables that are defined as linear combinations of other random variables is a problem that often arises in practice. While this determination can be accomplished from first principles by applying the basic definitions of mean and variance to the linear function of random variables, there are certain general results that facilitate and expedite the process. In particular, we will see that the mean and variance of a linear combination of random variables can be expressed as simple functions of the means, variances, and covariances of the random variables involved in the linear combination. Our first result concerns the determination of the mean.

**Theorem 3.38**
***Mean of a Linear Combination***

*Let* $Y=\sum_{i=1}^{n} a_i X_i = \mathbf{a}'\mathbf{X}$ *where the* $a_i$*'s are real constants. Then* $\mathrm{E}(Y)= \sum_{i=1}^{n} a_i E(X_i) = \mathbf{a}'\mathrm{E}(\mathbf{X})$, *where*

$$\mathbf{a} = \begin{bmatrix} a_1 \\ \vdots \\ a_n \end{bmatrix} \text{ and } \mathbf{X} = \begin{bmatrix} X_1 \\ \vdots \\ X_n \end{bmatrix}.$$

**Proof**

$$\mathrm{E}(Y) = \mathrm{E}\left(\sum_{i=1}^{n} a_i X_i\right) = \sum_{i=1}^{n} \mathrm{E}(a_i X_i) = \sum_{i=1}^{n} a_i \mathrm{E}(X_i) \quad \text{(Theorem 3.9 and 3.6)}$$ ■

Regarding the variance of the linear combination of random variables, we have the following result.

**Theorem 3.39**
***Variance of a Linear Combination***

*Let* $Y = \sum_{i=1}^{n} a_i X_i$ *where the* $a_i$*'s are real constants. Then* $\sigma_Y^2 = \sum_{i=1}^{n} a_i^2 \sigma_{X_i}^2 + 2\sum\sum_{i<j} a_i a_j \sigma_{X_i X_j} = \mathbf{a}'\mathbf{Cov}(\mathbf{X})\mathbf{a}$, *where* $\mathbf{Cov}(\mathbf{X}) = \mathrm{E}((\mathbf{X} - \mathrm{E}(\mathbf{X}))(\mathbf{X} - \mathrm{E}(\mathbf{X}))')$ *is the* ***covariance matrix*** *of* $\mathbf{X}$.

**Proof**

$$\sigma_Y^2 = \mathrm{E}\left((Y - \mathrm{E}(Y))^2\right) = \mathrm{E}\left(\sum_{i=1}^{n} a_i (X_i - \mathrm{E}(X_i))\right)^2$$

$$= \mathrm{E}\left[\sum_{i=1}^{n} a_i^2 (X_i - \mathrm{E}(X_i))^2 + 2\sum\sum_{i<j} a_i a_j (X_i - \mathrm{E}(X_i))(X_j - \mathrm{E}(X_j))\right]$$

$$= \sum_{i=1}^{n} a_i^2 \sigma_{X_i}^2 + 2\sum\sum_{i<j} a_i a_j \sigma_{X_i X_j} = \mathbf{a}'\mathrm{E}\left((\mathbf{X}-\mathrm{E}(\mathbf{X}))(\mathbf{X}-\mathrm{E}(\mathbf{X}))'\right)\mathbf{a}$$

where the penultimate equality follows from Theorems 3.9 and 3.6. ■

We formally define the notion of **covariance matrix** below and further motivate its content and meaning.

**Definition 3.26**
***Covariance Matrix***

The covariance matrix of an $n$-variate random variable $\mathbf{X}$ is the $(n \times n)$ symmetric matrix whose $(i,j)^{th}$element is the covariance between $X_i$ and $X_j$, defined as $\mathbf{Cov}(\mathbf{X}) = \mathrm{E}\left((\mathbf{X}-\mathrm{E}(\mathbf{X}))(\mathbf{X}-\mathrm{E}(\mathbf{X}))'\right)$.

Note that because the covariance between$X_i$ and $X_i$ is, by definition, the variance of $X_i$, the covariance matrix has the variances of the $X_i$'s along its diagonal. In order to appreciate the full informational content of the covariance matrix, note that, by definition,

$$\underset{(n\times n)}{\mathbf{Cov}}(\mathbf{X}) = \mathrm{E}\begin{bmatrix} X_1 - \mathrm{E}(X_1) \\ \vdots \\ X_n - \mathrm{E}(X_n) \end{bmatrix} [(X_1 - \mathrm{E}(X_1)) \cdots (X_n - \mathrm{E}(X_n))]$$

$$= \mathrm{E}\begin{bmatrix} (X_1 - \mathrm{E}(X_1))^2 & (X_1 - \mathrm{E}(X_1))(X_2 - \mathrm{E}(X_2)) & \cdots & (X_1 - \mathrm{E}(X_1))(X_n - \mathrm{E}(X_n)) \\ (X_2 - \mathrm{E}(X_2))(X_1 - \mathrm{E}(X_1)) & (X_2 - \mathrm{E}(X_2))^2 & \cdots & \vdots \\ \vdots & \cdots & \ddots & \vdots \\ (X_n - \mathrm{E}(X_n))(X_1 - \mathrm{E}(X_1)) & \cdots & & (X_n - \mathrm{E}(X_n))^2 \end{bmatrix}$$

$$= \begin{bmatrix} \sigma^2_{X_1} & \sigma_{X_1X_2} & \cdots & \sigma_{X_1X_n} \\ \sigma_{X_2X_1} & \sigma^2_{X_2} & \cdots & \vdots \\ \vdots & \cdots & \ddots & \vdots \\ \sigma_{X_nX_1} & \cdots & \cdots & \sigma^2_{X_n} \end{bmatrix}.$$

Thus, the covariance matrix has the variance of the $i$th random variable displayed in the $(i,i)$th (diagonal entry) position in the matrix, and the covariance between the $i$th and $j$th random variables displayed in the $(i,j)$th position (off-diagonal entry) in the matrix. Since $\sigma_{X_iX_j} = \sigma_{X_jX_i}$ the covariance matrix is *symmetric*, i.e., the $(i,j)$th entry is exactly equal to the $(j,i)$th entry $\forall i \neq j$.

Note that it is necessarily the case that the covariance matrix is a *positive semidefinite matrix* because $\sigma^2 = \mathbf{a}'\mathbf{Cov}(\mathbf{X})\mathbf{a} \geq 0$, $\forall \mathbf{a}$, which necessarily follows from the fact that variances cannot be negative. (Recall that a matrix $\mathbf{Z}$ is positive semidefinite *iff* $\mathbf{a}'\mathbf{Z}\mathbf{a} \geq 0$, $\forall \mathbf{a}$).

The preceding results can be extended to the case where $\mathbf{Y}$ is a vector defined by linear combinations of the $n$-variate random variable $\mathbf{X}$. We first extend the results corresponding to the mean of $\mathbf{Y}$.

**Theorem 3.40** ***Mean of a Vector of Linear Combinations (Pre Multiplication)***

*Let* $\mathbf{Y} = \mathbf{AX}$ *where* $\mathbf{A}$ *is a* $k \times n$ *constants, and* $\mathbf{X}$ *is an* $n \times 1$ *vector of random variables. Then* $\mathrm{E}(\mathbf{Y}) = \mathrm{E}(\mathbf{AX}) = \mathbf{A}\mathrm{E}(\mathbf{X})$.

**Proof** This follows straightforwardly from Theorem 3.38 and the fact that an expectation of a vector is the vector of expectations. ■

A useful corollary to Theorem 3.40 concerns the generalization where $\mathbf{X}$ is a $n \times p$ *matrix* of random variables.

**Corollary 3.6** ***Mean of a Matrix of Linear Combinations (Pre Multiplication)***

*Let* $\mathbf{Y} = \mathbf{AX}$ *where* $\mathbf{A}$ *is a* $k \times n$ *matrix of real constants, and* $\mathbf{X}$ *is an* $n \times p$ *matrix of random variables. Then* $\mathrm{E}(\mathbf{Y}) = \mathrm{E}(\mathbf{AX}) = A\mathrm{E}(\mathbf{X})$.

**Proof** This follows directly from Theorem 3.40 applied columnwise to the matrix **AX**. ■

If we *postmultiply* rather than *premultiply* a random matrix **X** by a conformable matrix of constants, we obtain a result on expectation qualitatively similar to the preceding result.

**Corollary 3.7** ***Mean of a Matrix of Linear Combinations (Post Multiplication)***

*Let* $\mathbf{Y} = \mathbf{XB}$, *where* $\mathbf{X}$ *is a* $n \times p$ *matrix of random variables and* $\mathbf{B}$ *is a* $p \times m$ *matrix of real constants. Then* $\mathrm{E}(\mathbf{Y}) = \mathrm{E}(\mathbf{XB}) = \mathrm{E}(\mathbf{X})\mathbf{B}$.

**Proof**

$$\begin{aligned}\mathrm{E}(\mathbf{XB}) &= \mathrm{E}(\mathbf{B}'\mathbf{X}')' \text{ (property of matrix transpose)}\\ &= (\mathrm{E}(\mathbf{B}'\mathbf{X}'))' \text{ (expectation is an elementwise operator)}\\ &= (\mathbf{B}'\mathrm{E}(\mathbf{X}'))' \text{ (Corollary 3.6)}\\ &= \mathrm{E}(\mathbf{X})\mathbf{B} \text{ (property of matrix transpose)}\end{aligned}$$

■

If a random matrix **X** is *both* premultiplied and postmultiplied by conformable matrices of real constants, then the previous two corollaries can be combined into the following result:

**Corollary 3.8** ***Mean of a Matrix of Linear Combinations (Pre and Post Multiplication)***

*Let* $\mathbf{A}$ *be a* $k \times n$ *matrix of real constants, let* $\mathbf{X}$ *be a* $n \times p$ *matrix of random variables, and let* $\mathbf{B}$ *be a* $p \times m$ *matrix of real constants. Then* $\mathrm{E}(\mathbf{AXB}) = \mathbf{A}\mathrm{E}(\mathbf{X})\mathbf{B}$.

**Proof** Let $\mathbf{Z} = \mathbf{XB}$. Then by Corollary 3.6, $\mathrm{E}(\mathbf{AXB}) = \mathrm{E}(\mathbf{AZ}) = \mathbf{A}\mathrm{E}(\mathbf{Z}) = \mathbf{A}\mathrm{E}(\mathbf{XB})$, which equals $A\mathrm{E}(X)B$ by Corollary 3.7. ■

When $\mathbf{Y} = \mathbf{AX}$ is a vector of two or more random variables, we can define a variance for each $Y_i$, as well as a covariance for each pair $(Y_i, Y_j)$. We are led to a generalization of Theorem 3.39 that involves the definition of the *covariance matrix* of the $(k \times 1)$ random vector $\mathbf{Y} = \mathbf{AX}$.

**Theorem 3.41** ***Covariance Matrix of Linear Combination***

*Let* $\mathbf{Y} = \mathbf{AX}$ *where* $\mathbf{A}$ *is a* $k \times n$ *matrix of real constants and* $\mathbf{X}$ *is a* $n \times 1$ *vector of random variables. Then* $\mathbf{Cov}(\mathbf{Y}) = \mathbf{Cov}(\mathbf{AX}) = \mathbf{A}\mathbf{Cov}(\mathbf{X})\mathbf{A}'$.

**Proof** By definition,

$$\begin{aligned}\mathbf{Cov}(\mathbf{Y}) &= \mathrm{E}\big((\mathbf{Y} - \mathrm{E}(\mathbf{Y}))(\mathbf{Y} - \mathrm{E}(\mathbf{Y}))'\big)\\ &= \mathrm{E}\big(\mathbf{A}(\mathbf{X} - \mathrm{E}(\mathbf{X}))(\mathbf{X} - \mathrm{E}(\mathbf{X}))'\mathbf{A}'\big) \text{ (substitution and Theorem 3.40)}\\ &= \mathbf{A}\mathrm{E}\big((\mathbf{X} - \mathrm{E}(\mathbf{X}))(\mathbf{X} - \mathrm{E}(\mathbf{X}))'\big)\,\mathbf{A}' \text{ (Corollary 3.8)}\\ &= \mathbf{A}\mathbf{Cov}(\mathbf{X})\,\mathbf{A}' \text{ (by definition)}\end{aligned}$$

■

We illustrate the use of some of the above theorems in the following example, where we also introduce the notion of a **correlation matrix** (see Example 3.28 part (*g*)).

**Example 3.28** *Calculating Means, Covariances, and Correlations of Linear Combinations*

Your company sells two brands of blank recordable DVDs: Blueray (BR) and standard (S). The price of a package of BR disks is \$4 while the standard disks sell for \$3 a package. The quantities of the disk packages sold on any given day are represented by the bivariate random variable $\mathbf{Q} = (Q_{BR}, Q_S)$, where

$$\mathrm{E}(\mathbf{Q}) = \begin{bmatrix} 10 \\ 30 \end{bmatrix} \text{ and } \mathbf{Cov}(\mathbf{Q}) = \begin{bmatrix} 2 & -3 \\ -3 & 5 \end{bmatrix}.$$

**a.** What is the expected value of the revenue obtained from the sale of DVDs on any given day?
**Answer**: Revenue (in dollars) is defined as

$$R = 4\,Q_{BR} + 3\,Q_S = [4 \quad 3]\begin{bmatrix} Q_{BR} \\ Q_S \end{bmatrix},$$

and Theorem 3.39 applies. Therefore,

$$\mathrm{E}(R) = [4 \quad 3]\begin{bmatrix} 10 \\ 30 \end{bmatrix} = 130.$$

**b.** What is the variance associated with daily revenue?
**Answer**: Theorem 3.39 applies here. We have that

$$\sigma_R^2 = [4 \quad 3]\begin{bmatrix} 2 & -3 \\ -3 & 5 \end{bmatrix}\begin{bmatrix} 4 \\ 3 \end{bmatrix} = 5.$$

**c.** Production costs per disk package are \$2.50 and \$2 for the Blueray and standard DVDs, respectively. Define the expected value of the vector $\begin{bmatrix} R \\ C \end{bmatrix}$, where $C = 2.50\,Q_{BR} + 2\,Q_S$ represents total cost of DVDs sold on any given day.
**Answer**: Theorem 3.40 can be used here (we could also apply Theorem 3.38 to obtain $\mathrm{E}(C)$, since we already know $\mathrm{E}(R)$from above).

$$\mathrm{E}\begin{bmatrix} R \\ C \end{bmatrix} = \begin{bmatrix} 4 & 3 \\ 2.5 & 2 \end{bmatrix}\begin{bmatrix} 10 \\ 30 \end{bmatrix} = \begin{bmatrix} 130 \\ 85 \end{bmatrix}$$

**d.** What is the covariance matrix of $\begin{bmatrix} R \\ C \end{bmatrix}$?
**Answer**: Using Theorem 3.41,

$$\mathbf{Cov}\left(\begin{bmatrix} R \\ C \end{bmatrix}\right) = \begin{bmatrix} 4 & 3 \\ 2.5 & 2 \end{bmatrix}\begin{bmatrix} 2 & -3 \\ -3 & 5 \end{bmatrix}\begin{bmatrix} 4 & 2.5 \\ 3 & 2 \end{bmatrix} = \begin{bmatrix} 5 & 3.5 \\ 3.5 & 2.5 \end{bmatrix}.$$

**e.** What is the expected level of profit on any given day?

**Answer**: Profit is defined as $\Pi = R - C$, and Theorem 3.38 implies that

$$\mathrm{E}(\Pi) = [1 \;\; -1]\begin{bmatrix}130\\85\end{bmatrix} = 45.$$

**f.** What is the variance of daily profit?
**Answer**: Applying Theorem 3.39 results in

$$\sigma^2_{\Pi} = [1 \;\; -1]\begin{bmatrix}5 & 3.5\\3.5 & 2.5\end{bmatrix}\begin{bmatrix}1\\-1\end{bmatrix} = .5$$

**g.** A matrix of correlations (or, **correlation matrix**) for $X = (X_1, ..., X_n)$ can be defined by pre- and post-multiplication of the covariance matrix by the inverse of the diagonal matrix of standard deviations, i.e., (reader please verify):

$$\mathbf{Cov(X)} = \begin{bmatrix}\sigma_{X_1} & & \\ & \ddots & \\ & & \sigma_{X_n}\end{bmatrix}^{-1} \mathbf{Cov(X)} \begin{bmatrix}\sigma_{X_1} & & \\ & \ddots & \\ & & \sigma_{X_n}\end{bmatrix}^{-1}$$

The $(i,j)$th entry of the correlation matrix is the correlation between $X_i$ and $X_j$. Define the correlation matrix for **Q**.
**Answer:**

$$\mathbf{Corr(Q)} = \begin{bmatrix}\sqrt{2} & 0\\0 & \sqrt{5}\end{bmatrix}^{-1}\begin{bmatrix}2 & -3\\-3 & 5\end{bmatrix}\begin{bmatrix}\sqrt{2} & 0\\0 & \sqrt{5}\end{bmatrix}^{-1} = \begin{bmatrix}1 & -.949\\-.949 & 1\end{bmatrix}$$

Note $\rho_{Q_{BR},Q_S} = -.949$, which is given by the off-diagonal elements in this $(2 \times 2)$ case, while the diagonal elements are ones because these values represent the correlation of a random variable with itself. □

## 3.12 Appendix: Proofs and Conditions for Positive Semidefiniteness

### 3.12.1 Proof of Theorem 3.2

**Discrete Case** Let $Y = g(X)$. The density function for the random variable $Y$ can be represented by:

$$h(y) = P_Y(y) = P_X(\{x : g(x) = y, x \in R(X)\}) = \sum_{\{x:g(x)=y\}} f(x).$$

That is, the probability of the outcome $y$ is equal to the probability of the equivalent event $\{x : g(x) = y\}$, which is the inverse image of $y$. Then

$$\begin{aligned}\mathrm{E}(g(X)) = \mathrm{E}(Y) &= \sum_{y \in R(Y)} yh(y) = \sum_{y \in R(Y)} y \sum_{\{x:g(x)=y\}} f(x)\\ &= \sum_{y \in R(Y)} \sum_{\{x:g(x)=y\}} g(x)f(x) = \sum_{x \in R(X)} g(x)f(x),\end{aligned}$$

where the next to last expression is true, since $g(x) = y$ for all $x \in \{x : g(x) = y\}$, and the last expression is true since $\sum_{y \in R(Y)} \sum_{\{x:g(x)=y\}} g(x)f(x)$ is equivalent to summing over all $x \in R(X)$ because the collection of all $y \in R(Y)$ (the outer sum) is the set $R(Y) = \{y : y = g(x), x \in R(X)\}$. ■

**Continuous Case** To prove the theorem for the continuous case, we first need to establish the following lemma.

**Lemma 3.1**

> For any continuous random variable Y, the expectation of Y, if it exists, can be written as
>
> $$E(Y) = \int_0^\infty P(y>z)dz - \int_0^\infty P(y \leq -z)dz.$$

**Proof of Lemma** Let $h(y)$ be the density function of Y. Then $P(y > z) = \int_z^\infty h(y)dy$, so that

$$\int_0^\infty P(y>z)dz = \int_0^\infty \int_z^\infty h(y)dydz = \int_0^\infty \left[\int_0^y dz\right] h(y)dy = \int_0^\infty yh(y)dy,$$

where the second equality was simply the result of changing the order of integration (note that the inner range of integration is a function of the outer range of integration, and the *same* set of $(y,z)$ points are being integrated over).

Similarly, $P(y \leq -z) = \int_{-\infty}^{-z} h(y)dy$, so that

$$\int_0^\infty P(y \leq -z)dz = \int_0^\infty \int_{-\infty}^{-z} h(y)dydz = \int_{-\infty}^0 \left[\int_0^{-y} dz\right] h(y)dy = -\int_{-\infty}^0 yh(y)dy.$$

Therefore,

$$\int_0^\infty P(y>z)dz - \int_0^\infty P(y \leq -z)dz = \int_0^\infty yh(y)dy + \int_{-\infty}^0 yh(y)dy = E(Y). \quad ■$$

Note that the lemma (integrals and all) also applies to discrete random variables.[19]

Using the lemma, we have

$$\begin{aligned}
E(g(X)) &= \int_0^\infty P(g(x)>z)dz - \int_0^\infty P(g(x) \leq -z)dz \\
&= \int_0^\infty \int_{\{x:g(x)>z\}} f(x)dxdz - \int_0^\infty \int_{\{x:g(x) \leq -z\}} f(x)dxdz \\
&= \int_{\{x:g(x)>0\}} \left[\int_0^{g(x)} dz\right] f(x)dx - \int_{\{x:g(x) \leq 0\}} \left[\int_0^{-g(x)} dz\right] f(x)dx \\
&= \int_{\{x:g(x)>0\}} g(x)f(x)dx + \int_{\{x:g(x) \leq 0\}} g(x)f(x)dx. \\
&= \int_{-\infty}^\infty g(x)f(x)dx.
\end{aligned}$$

■

[19] See P. Billingsley (1986) *Probability and Measure*, 2nd ed. New York: John Wiley, pp. 73–74 for the method of proof in the discrete case.

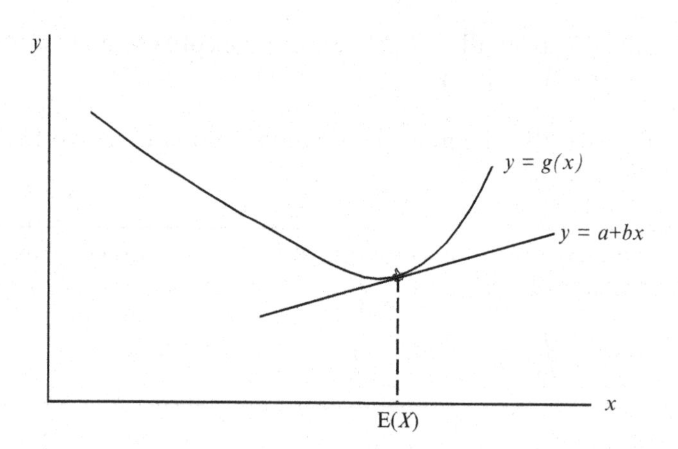


**Figure 3.12**
Convex function $g$.

### 3.12.2 Proof of Theorem 3.4 (Jensen's Inequalities)

We prove the result for the convex case. The proof of the concave case is analogous, with inequalities reversed.

If $g$ is a convex function for $x \in I$, then there exists a line going through the point E($X$), say $\ell(x) = a + bx$, such that $g(x) \geq \ell(x) = a + bx \ \forall x \in I$ and $g(\mathrm{E}(X)) = a + b\mathrm{E}(X)$ (see Figure 3.6). Now note that

$$\mathrm{E}(g(X)) = \sum_{x \in R(X)} g(x)f(x) \geq \sum_{x \in R(X)} (a + bx)f(x)$$
$$= a + b\mathrm{E}(X) = g(\mathrm{E}(X)) \qquad \text{(discrete)}$$
$$\mathrm{E}(g(X)) = \int_{-\infty}^{\infty} g(x)f(x)dx \geq \int_{-\infty}^{\infty} (a + bx)f(x)dx$$
$$= a + b\mathrm{E}(X) = g(\mathrm{E}(X)) \qquad \text{(continuous)}$$

since $g(x) \geq a + bx \ \forall x \in I$,[20] so that $\mathrm{E}(g(X)) \geq g(\mathrm{E}(X))$.

If $g$ is strictly convex, then there exists a line going through the point E($X$), say $\ell(x) = a + bx$, such that $g(x) > \ell(x) = a + bx \ \forall x \in I$ for which $x \neq \mathrm{E}(X)$, and $g(\mathrm{E}(X)) = a + b\mathrm{E}(X)$. Then, assuming that no element in $R(X)$ is assigned probability one, (i.e., $X$ is not degenerate), the previous inequality results become strict, implying $\mathrm{E}(g(X)) > g(\mathrm{E}(X))$ in either the discrete or continuous cases. ■

### 3.12.3 Necessary and Sufficient Conditions for Positive Semidefiniteness

To Prove that the symmetric matrix $\mathbf{A} = \begin{bmatrix} a_{11} & a_{12} \\ a_{21} & a_{22} \end{bmatrix}$ *is positive semidefinite iff* $a_{11} \geq 0$, $a_{22} \geq 0$, and $a_{11}\, a_{22} - a_{12}\, a_{21} \geq 0$, note that the matrix **A** will be positive semidefinite *iff* the characteristic roots of **A** are nonnegative (e.g., F.A.

[20] Recall the integral inequality that if $h(x) \geq t(x) \ \forall x \in (a,b)$, then $\int_a^b h(x)dx \geq \int_a^b t(x)dx$. Strict inequality holds if $h(x) > t(x) \ \forall x \in (a,b)$. The result holds for $a = -\infty$ and/or $b = \infty$.

Graybill (1983) Matrices with Applications in Statistics, Belmont, CA: Wadsworth, p. 397). The characteristic roots of **A** are found by solving the determinantal equation $\begin{vmatrix} a_{11}-\lambda & a_{12} \\ a_{21} & a_{22}-\lambda \end{vmatrix} = 0$ for $\lambda$, which can be represented as $(a_{11} - \lambda)(a_{22} - \lambda) - a_{12}\,a_{21} = 0$ or $\lambda^2 - (a_{11} + a_{22})\lambda + (a_{11}\,a_{22} - a_{12}\,a_{21}) = 0$.

Solutions to this equation can be found by employing the quadratic formula[21] to obtain

$$\lambda = \frac{(a_{11}+a_{22}) \pm \sqrt{(a_{11}+a_{22}\,)^2 - 4(a_{11}\,a_{22} - a_{12}\,a_{21})}}{2}$$

For $\lambda$ to be $\geq 0$, it must be the case that the numerator term is $\geq 0$. Note the term under the square root sign must be nonnegative, since it can be rewritten as $(a_{11} - a_{22})^2 + 4\,a_{12}\,a_{21}$, and because $a_{12} = a_{21}$ (by symmetry of **A**), two nonnegative numbers are being added together. If $(a_{11} + a_{22}) > 0$, then $\lambda \geq 0$ only if $a_{11}\,a_{22} - a_{12}\,a_{21} \geq 0$, since otherwise the square root term would be larger than $(a_{11} + a_{22})$, and when subtracted from $(a_{11} + a_{22})$, would result in a negative $\lambda$. Also, the term $(a_{11} + a_{22})$ cannot be negative or else at least one of the solutions for $\lambda$ would necessarily be negative. Furthermore, both $a_{11}$ and $a_{22}$ must be nonnegative, for if one were negative, then there is no value for the other that would result in both solutions for $\lambda$ being positive. Thus, necessity is proved. Sufficiency follows immediately, since $a_{11} \geq 0$, $a_{22} \geq 0$, and $a_{11}\,a_{22} - a_{12}\,a_{21} \geq 0$ imply both solutions of $\lambda$ are $\geq 0$. ■

## Keywords, Phrases, and Symbols

$\times_{i=1}^{n}$, Cartesian Product
Characteristic function
Chebyshev's inequality
Conditional expectation $E(Y|x\in B)$, $E(g(Y)|x\in B)$, $E(Y|x = b)$, $E(g(Y)|x = b)$
Correlation between two random variables $\rho_{XY}$
Correlation bound
Correlation matrix, **Corr**(**X**)
Covariance between two random variables $\sigma_{XY}$ or Cov($X$,$Y$)
Covariance bound
Covariance matrix, **Cov**(**X**)
Cumulant generating function $\psi_X(t)$
Cumulants, $\kappa_r$
Degenerate density function
Degenerate random variable
Expectation of a function of a multivariate random variable, $E(g(X_1,\ldots,X_n))$
Expectation of a function of a random variable, $E\,(g(X))$
Expectation of a matrix of random variables
Expectation of a random variable, $E(X)$
Iterated expectation theorem
Jensen's Inequality
Kurtosis, Excess Kurtosis
Leptokurtic
Marginal MGF, marginal cumulant generating function
Markov's inequality
Means and variances of linear combinations of random variables
Median, med($X$)
Mesokurtic
MGF Uniqueness theorem
Mode, mode($X$)
Moment generating function, MGF $M_X(t)$
Moments of a random variable
$\mu$, the mean of a random variable
$\mu_r$, the $r$th moment about the mean or $r$th central moment
$\mu_{r,s}$, the $(r,s)$th joint moment about the mean
$\mu'_{r,s}$, the $(r,s)$th joint moment about the origin

[21] Recall that the solutions to the quadratic equation $ax^2 + bx + c = 0$ are given by $x = \frac{-b \pm \sqrt{b^2 - 4ac}}{2a}$.

$\mu'_r$, the $r$th moment about the origin
Quantile of $X$
Platykurtic
Regression curve of $Y$ on $X$
Regression function of $Y$ on $X$
Skewed density function
Skewed to the left
Skewed to the right
Standard deviation of a random variable $\sigma$, or std($X$)
Symmetric density function
Uncorrelated
Unimodal
Variance of a random variable $\sigma^2$, or var ($X$)

## Problems

**1.** A small domestic manufacturer of television sets places a three-year warranty on its picture tubes. During the warranty period, the manufacturer will replace the television set with a new one if the picture tube fails. The time in years until picture tube failure can be represented as the outcome of a random variable $X$ with probability density function

$$f(x) = .005e^{-.005x} I_{(0,\infty)}(x).$$

The times that picture tubes operate until failure can be viewed as independent random variables. The company sells 100 television sets in a given period.

(a) What is the expected number of television sets that will be replaced due to picture tube failure?

(b) What is the expected operating life of a picture tube?

**2.** A small rural bank has two branches located in neighboring towns in eastern Washington. The numbers of certificates of deposit that are sold at the branch in Tekoa and the branch in Oakesdale in any given week can be viewed as the outcome of the bivariate random variable $(X,Y)$ having joint probability density function

$$f(x,y) = \left[\frac{x^3y^3}{(3,025)^2}\right] I_{(0,1,2,\ldots,10)}(x) I_{(0,1,2,\ldots,10)}(y).$$

(a) Are the random variables independent?

(b) What is the expected number of certificate sales by the Oakesdale Branch?

(c) What is the expected number of combined certificate sales for both branches?

(d) What is the answer to b) *given that* Tekoa branch sells four certificates?

Potentially helpful result:

$$\sum_{x=1}^{n} x^3 = \frac{n^2(n+1)^2}{4}.$$

**3.** The weekly number of luxury and compact cars sold by "Honest" Abe Smith at the Auto Mart, a local car dealership, can be represented as the outcome of a bivariate random variable $(X,Y)$ with the nonzero values of its joint probability density function given by

| | | Y | | | | |
|---|---|---|---|---|---|---|
| | | 0 | 1 | 2 | 3 | 4 |
| | 0 | .20 | .15 | .075 | .05 | .03 |
| | 1 | .10 | .075 | .04 | .03 | .02 |
| X | 2 | .05 | .03 | .02 | .01 | .01 |
| | 3 | .04 | .03 | .02 | .01 | .01 |

Al receives a base salary of \$100/week from the dealership, and also receives a commission of \$100 for every compact car sold and \$200 for every luxury car sold.

(a) What is the expected value of the weekly commission that Al obtains from selling cars? What is the expected value of his total pay received for selling cars?

(b) What is the expected value of his commission from selling compact cars? What is the expected value of his commission from selling luxury cars?

(c) Given that Al sells four compact cars, what is the expected value of his commission from selling luxury cars?

(d) If 38 percent of Al's total pay goes to federal and state taxes, what is the expected value of his pay after taxes?

**(4)** The yield, in bushels per acre, of a certain type of feed grain in the midwest can be represented as the outcome of the random variable $Y$ defined by

$$Y = 3x_l^{.30} x_k^{.45} e^U$$

where $x_l$ and $x_k$ are the per acre units of labor and capital utilized in production, and $U$ is a random variable with probability density function given by

$$f(u) = 2e^{-2u} I_{(0,\infty)}(u).$$

The price received for the feed grain is \$4/bushel, labor price per unit is \$10, and capital price per unit is \$15.

(a) What is the expected yield per acre?

(b) What is the expected level of profit per acre if labor and capital are each applied at the rate of 10 units per acre?

(c) Define the levels of input usage that maximize expected profit. What is the expected maximum level of profit?

(d) The acreage can be irrigated at a cost of \$125 per acre, in which case the yield per acre is defined by

$Y = 5x_l^{.30}x_k^{.45}e^U.$

If the producer wishes to maximize expected profit, should she irrigate?

**5.** The daily price/gallon and quantity sold (measured in *millions* of gallons) of a lubricant sold on the wholesale spot market of a major commodity exchange is the outcome of a bivariate random variable $(P,Q)$ having the joint probability density function

$f(p,q) = 2pe^{-pq}\, I_{[.5,1]}(p)\, I_{(0,\infty)}(q).$

(a) Define the regression curve of $q$ on $p$.

(b) Graph the regression curve that you have defined in (a).

(c) What is the expected value of the quantity of lubricant sold, given that price is equal to \$.75 per gallon?

(d) What is the expected value of total dollar sales of lubricant on a given day?

**6.** The short-run production function for a particular agricultural crop is critically dependent on the level of rainfall during the growing season, the relationship being $Y = 30 + 3X - .075X^2$, where $y$ is yield per acre in bushels, and $x$ is inches of rainfall during the growing season.

(a) If the expected value of rainfall is 20 inches, can the expected value of yield per acre be as high as 70 bushels per acre? Why or why not?

(b) Suppose the variance of rainfall is 40 square inches. What is the expected value of yield per acre? How does this compare to the bound placed on $E(Y)$ by Jensen's inequality?

**7.** For each of the densities below, indicate whether the mean and variance of the associated random variable exist. In addition, find the median and mode, and indicate whether or not each density is symmetric.

(a) $f(x) = 3x^2\, I_{[0,\,1]}(x)$

(b) $f(x) = 2x^{-3}\, I_{[1,\,\infty)}(x)$

(c) $f(x) = [\pi(1 + x^2)]^{-1}\, I_{(-\infty,\,\infty)}(x)$

(d) $f(x) = \binom{4}{x}(.2)^x(.8)^{4-x} I_{\{0,1,2,3,4\}}(x)$

**8.** The daily price of a certain penny stock is a random variable with an expected value of \$2. Then the probability is $\leq .20$ that the stock price will be greater than or equation to \$10. True or false?

**9.** The miles per gallon attained by purchasers of a line of pickup trucks manufactured in Detroit are outcomes of a random variable with a mean of 17 miles per gallon and a standard deviation of .25 miles per gallon. How probable is the event that a purchaser attains between 16 and 18 miles per gallon with this line of truck?

**10.** The daily quantity of water demanded by the population of a large northeastern city in the summer months is the outcome of a random variable, $X$, measured in millions of gallons and having a MGF of $M_x(t) = (1 - .5\,t)^{-10}$ for $t < 2$.

(a) Find the mean and variance of the daily quantity of water demanded.

(b) Is the density function of water quantity demanded symmetric?

**11.** The annual return per dollar for two different investment instruments is the outcome of a bivariate random variable $(X_1, X_2)$ with joint moment-generating function $M_{\mathbf{x}}(\mathbf{t}) = \exp(\mathbf{u}'\mathbf{t} + .5\mathbf{t}'\boldsymbol{\Sigma}\mathbf{t})$, where

$$\mathbf{t} = \begin{bmatrix} t_1 \\ t_2 \end{bmatrix}, \mathbf{u} = \begin{bmatrix} .07 \\ .11 \end{bmatrix} \text{ and } \boldsymbol{\Sigma} = \begin{bmatrix} .225 \times 10^{-3} & -.3 \times 10^{-3} \\ -.3 \times 10^{-3} & .625 \times 10^{-3} \end{bmatrix}.$$

(a) Find the mean annual return per dollar for each of the projects.

(b) Find the covariance matrix of $(X_1, X_2)$.

(c) Find the correlation matrix of $(X_1, X_2)$. Do the outcomes of $X_1$ and $X_2$ satisfy a linear relationship $x_1 = \alpha_1 + \alpha_2\, x_2$?

(d) If an investor wishes to invest \$1,000 in a way that maximizes her expected dollar return on the investment, how should she distribute her investment dollars between the two projects? What is the variance of dollar return on this investment portfolio?

(e) Suppose the investor wants to minimize the variance of her dollar return. How should she distribute the $1,000? What is the expected dollar return on this investment portfolio?

(f) Suppose the investor's utility function with respect to her investment portfolio is $U(M) = 5M^b$, where $M$ is the dollar return on her investment of $1,000. The investor's objective is to maximize the expected value of her utility. If $b = 1$, define the optimal investment portfolio.

(g) Repeat (f), but let $b = 2$.

(h) Interpret the investment behavior differences in (f) and (g) in terms of investor attitude toward risk.

**12.** Stanley Statistics, an infamous statistician, wants you to enter a friendly wager with him. For $1,000, he will let you play the following game. He will continue to toss a fair coin until the first head appears. Letting $x$ represent the number of times the coin was tossed to get the first heads, Stanley will then pay you $\$2^x$.

(a) Define a probability space for the experiment of observing how many times a coin must be tossed in order to observe the first heads.

(b) What is the expected payment that you will receive if you play the game?

(c) Do you want to play the game? Why or why not?

**13.** The city of Megalopolis operates three sewage treatment plants in three different locations throughout the city. The daily proportion of operating capacity exhibited by the three plants can be represented as the outcome of a trivariate random variable with the following probability density function:

$$f(x_1, x_2, x_3) = \frac{1}{3}(x_1 + 2x_2 + 3x_3) \prod_{i=1}^{3} I_{(0,1)}(x_i),$$

where $x_i$ is the proportion of operating capacity exhibited by plant $i$, $i = 1, 2, 3$.

(a) What are the expected values of the capacity proportions for the three plants, i.e., what is $E\begin{bmatrix} X_1 \\ X_2 \\ X_3 \end{bmatrix}$?

(b) What is the expected value of the average proportion of operating capacity across all three plants, i.e., what is $E\left(\frac{1}{3}\sum_{i=1}^{3} X_i\right)$?

(c) *Given* that plant 3 operates at 90 percent of capacity, what are the expected values of the proportions of capacity for plants 1 and 2?

(d) If the daily capacities of plants 1 and 2 are 100,000 gal of sewage each, and if the capacity of plant three is 250,000 gal, then what is the expected daily number of gallons of sewage treated by the city of Megalopolis?

**14.** The average price and total quantity sold of an economy brand of ballpoint pen in a large western retail market during a given sales period is represented by the outcome of a bivariate random variable having a probability density function

$$f(p, s) = 10pe^{-ps}\, I_{[.10,\,.20]}(p)\, I_{(0,\infty)}(s)$$

where $p$ is the average price, in dollars, of a single pen and $s$ is total quantity sold, measured in 10,000-pen units.

(a) Define the regression curve of $S$ on $P$.

(b) What is the expected quantity of pens sold, given that price is equal to $0.12? (You may use the regression curve if you wish.)

(c) What is the expected value of total revenue from the sale of ball point pens during the given sales period, i.e., what is $E(PS)$?

**15.** A game of chance is considered to be "equitable" or "fair" if a player's expected payoff is equal to zero. Examine the following games:

(a) The player rolls a pair of fair dice. Let $Z$ represent the amount of money that the player lets on the game outcome. If the player rolls a 7 or 11, the player payoff is $2Z$ (i.e., he gets to keep his bet of $\$Z$, plus he receives an *additional* $\$2Z$). If the player does not roll a 7 or 11, he loses the $\$Z$ that he bet on the game. Is the game fair?

(b) The player spins a spinner contained within a disk, (↗), that is segmented into five pieces as

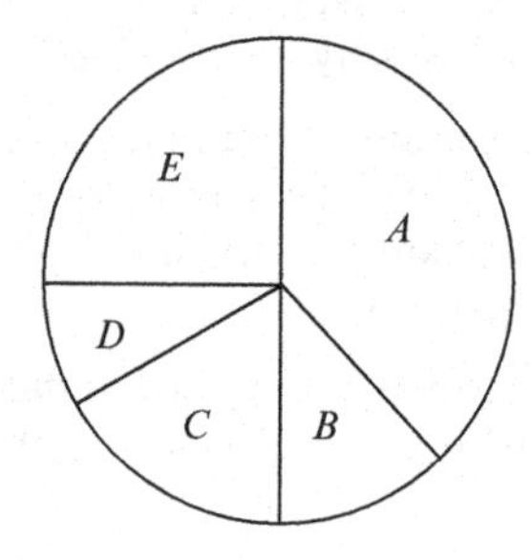

where $P(A) = 1/3$, $P(B) = 1/6$, $P(C) = 1/6$, $P(D) = 1/12$, $P(E) = 1/4$.

Each spin costs \$1. The payoffs corresponding to when the spinner lands in one of the five segments are given by:

| Segment | Payoff |
|---|---|
| A | \$.60 |
| B | \$1.20 |
| C | \$1.20 |
| D | \$2.40 |
| E | \$.80 |

Is the game fair?

(c) A fair coin will be tossed repeatedly *until* heads occurs. If the heads occurs on the $j$th toss of the coin, the player will receive $\$2^j$. How much should the player be charged to play the game if the game is to be fair? (Note: This is a trick question and represents the famous "St. Petersburg paradox" in the statistical literature.)

**16.** The manager of a bakery is considering how many chocolate cakes to bake on any given day. The manager knows that the number of chocolate cakes that will be demanded by customers on any given day is a random variable whose probability density is given by

$$f(x) = \frac{x+1}{15} I_{\{0,1,2,3\}}(x) + \frac{7-x}{15} I_{\{4,5\}}(x).$$

The bakery makes a profit of \$1.50 on each cake that is sold. If a cake is not sold on a given day, the cake is thrown away (because of lack of freshness), and the bakery loses \$1. If the manager wants to maximize expected daily profit from the sale of chocolate cakes, how many cakes should be baked?

**17.** The daily price and quantity sold of wheat in a Northwestern market during the first month of the marketing year is the outcome of a bivariate random variable $(P,Q)$ having the probability density function

$$f(p,q) = .5pe^{-pq} I_{[3,5]}(p) I_{(0,\infty)}(q)$$

where $p$ is measured in \$/bushel, and $q$ is measured in units of 100,000 bushels.

(a) Define the conditional expectation of quantity sold as a function of price, i.e., define $\mathrm{E}(Q \mid p)$ (the regression curve of $Q$ on $P$).

(b) Graph the regression curve you derived in (a). Calculate the values of $\mathrm{E}(Q \mid p = 3.50)$ and $\mathrm{E}(Q \mid p = 4.50)$.

**18.** In each case below, calculate the expected value of the random variable $Y$:

(a) $\mathrm{E}(Y|x) = 2x^2 + 3, f_X(x) = e^{-x} I_{(0,\infty)}(x)$.

(b) $\mathrm{E}(Y|x) = 3x_1x_2$, $\mathrm{E}(X_1) = 5$, $\mathrm{E}(X_2) = 7$, $X_1$ and $X_2$ are independent.

**19.** The total daily dollar sales in the ACME supermarket is represented by the outcome of the random variable $S$ having a mean of 20, where $s$ is measured in thousands of dollars.

(a) The store manager tells you the probability that sales will exceed \$30,000 on any given day is .75. Do you believe her?

(b) You are now given the information that the variance of the random variable $S$ is equal to 1.96. How probable is the sales event that $s \in (10,30)$?

**20.** The first three moments about the *origin* for the random variable $Y$ are given as follows: $\mu'_1 = .5, \mu'_2 = .5, \mu'_3 = .75$.

(a) Define the first three moments about the *mean* for Y.

(b) Is the density of Y skewed? Why or why not?

**21.** The random variable $Y$ has the PDF $f(y) = y^{-2} I_{[1,\infty)}(y)$.

(a) Find the mean of Y.

(b) Can you find the first *100* moments about the origin (i.e., $\mu'_1, \mu'_2, \ldots, \mu'_{100}$) for the random variable Y, why or why not?

**22.** The moment-generating function of the random variable $Y$ is given by $M_Y(t) = (1 - .25t)^{-3}$ for $t < 4$.

(a) Find the mean and variance of the random variable Y.

(b) Is the PDF of Y skewed? Why or why not?

(c) It is known that the moment generating function of the PDF $f(x) = \frac{1}{\beta^\alpha \Gamma(\alpha)} x^{\alpha-1} e^{-x/\beta} I_{(0,\infty)}(x)$ is given by $M_x(t) = (1 - \beta t)^{-\alpha}$ for $t < \beta^{-1}$. The $\Gamma(\alpha)$ in the preceding expression for the pdf is known as the *gamma function*, which for integer values of $\alpha$ is such that $\Gamma(\alpha) = (\alpha - 1)!$. Define the exact functional form of the probability density function for Y, if you can.

**23.** A gas station sells regular and premium fuel. The two storage tanks holding the two types of gasoline are refilled every week. The proportions of the available supplies of regular and premium gasoline that are sold during a given week in the summer is an outcome of a bivariate random variable having the joint density function $f(x,y) = \frac{2}{5}(3x + 2y)I_{[0,1]}(x)I_{[0,1]}(y)$, where $x$ = proportion of regular fuel sold and $y$ = proportion of premium fuel sold.

(a) Find the marginal density function of $X$. What is the probability that greater than 75 percent of the available supply of regular fuel is sold in a given week?

(b) Define the regression curve of $Y$ on $X$, i.e., define $\mathrm{E}(Y|x)$. What is the expected value of $Y$, given that $x = .75$? Are $Y$ and $X$ independent random variables?

(c) Regular gasoline sells for \$1.25/gal and premium gasoline sells for \$1.40/gal. Each storage tank holds 1,000 gal of gasoline. What is the expected revenue generated by the sale of gasoline during a week in the summer, *given* that $x = .75$?

**24.** Scott Willard, a famous weatherman on national TV, states that the temperature on a typical late fall day in the upper midwest, measured in terms of both the Celsius and Fahrenheit scales, can be represented as the outcome of the bivariate random variable $(C,F)$ such that

$$\mathrm{E}\begin{bmatrix} C \\ F \end{bmatrix} = \begin{bmatrix} 5 \\ 41 \end{bmatrix} \text{ and } \mathbf{Cov}(C,F) = \begin{bmatrix} 25 & 45 \\ 45 & 81 \end{bmatrix}.$$

(a) What is the correlation between $C$ and $F$?

(b) To what extent is there a linear relationship between $C$ and $F$? Define the appropriate linear relationship if it exists.

(c) Is $(C, F)$ a degenerate bivariate random variable? Is this a realistic result? Why or why not?

**25.** A fruit processing firm is introducing a new fruit drink, "Peach Passion," into the domestic market. The firm faces uncertain output prices in the initial marketing period and intends to make a short-run decision by choosing the level of production that maximize the expected value of utility:

$$\mathrm{E}(U(\pi)) = \mathrm{E}(\pi) - \alpha \mathrm{var}(\pi).$$

Profit is defined by $\pi = Pq - C(q), p$ is the price received for a unit of Peach Passion, $U$ is utility, the cost function is defined by $c(q) = .5q^2$, $\alpha \geq 0$ is a risk aversion parameter, and the probability density function of the uncertain output price is given by $f(p) = .048(5p - p^2)\, I_{[0,5]}\,(p)$.

(a) If the firm were risk neutral, i.e., $\alpha = 0$, find the level of production that maximizes expected utility.

(b) Now consider the case where the firm is risk averse, i.e., $\alpha > 0$. Graph the relationship between the optimal level of output and the level of risk aversion (i.e., the level of $\alpha$). How large does $\alpha$ have to be for optimal $q = 1$?

(c) Assume that $\alpha = 1$. Suppose that the Dept. of Agriculture were to *guarantee* a price to the firm. What *guaranteed* price would induce the firm to produce the same level of output as in the case where price was uncertain?

**26.** A Seattle newspaper intends to administer two different surveys relating to two different anti-tax initiatives on the ballot in November. The proportion of surveys mailed that will actually be completed and returned to the newspaper can be represented as the outcome of a bivariate random variable $(X,Y)$ having the density function

$$f(x,y) = \frac{2}{3}(x + 2y)I_{[0,1]}(x)I_{[0,1]}(y),$$

where $x$ is the proportion of surveys relating to initiative I that are returned, and $y$ refers to the proportion of surveys relating to initiative II that are returned.

(a) Are $X$ and $Y$ independent random variables?

(b) What is the conditional distribution of $x$, given $y = .50$? What is the probability that less than 50 percent of the initiative I surveys are returned, given that 50 percent of the initiative II surveys are returned?

(c) Define the regression curve of $X$ on $Y$. Graph the regression curve. What is the expected proportion of initiative I surveys returned, given that 50 percent of the initiative II surveys are returned?

**27.** An automobile dealership sells two types of four-door sedans, the "Land Yacht" and the "Mini-Rover." The number of Land Yachts and Mini-Rovers sold on any given day varies, with the probabilities of the various possible sales outcomes given by the following table:

| | | Number of Land Yachts sold | | | |
|---|---|---|---|---|---|
| | | 0 | 1 | 2 | 3 |
| Number of Mini-Rovers sold | 0 | .05 | .05 | .02 | .02 |
| | 1 | .03 | .10 | .08 | .03 |
| | 2 | .02 | .15 | .15 | .04 |
| | 3 | .01 | .10 | .10 | .05 |

Land Yachts sell for $22,000 each, and Mini-Rovers for $7,500 each. These cars cost the dealership $20,000 and $6,500, respectively, which must be paid to the car manufacturer.

(a) Define a random variable that represents daily profit above dealer car cost, i.e., total dollar sales—total car cost. (Let $x$ = number of Land Yachts sold and $y$ = number of Mini-Rovers sold). What is the expected value of daily profit above dealer car cost?

(b) The daily cost (other than the cost of cars) of running the dealership is equal to $4,000. What is the probability that total profit on a given day will be positive?

(c) What is the expected number of Mini-Rovers sold on a day when no Land Yachts will be sold? What is this expected number on a day when two Land Yachts are sold? Are $X$ and $Y$ independent random variables? Why or why not?

**28.** The season average price per pound, $p$, and total season quantity sold, $q$, of sweet cherries in a regional market can be represented as the outcome of a bivariate random variable $(P,Q)$ with the joint probability density function $f(p,q) = .5qe^{-q(.5+p)} I_{(0,\infty)}(q)\, I_{(0,\infty)}(p)$
where $p$ is measured in dollars, and $q$ is measured in millions of pounds.

(a) Find the marginal density of $Q$. What is the expected value of quantity sold?

(b) Define the regression curve of $P$ on $Q$. What is the expected value of $P$, given that $q = ½$?

(c) If the government receives 10 percent of the gross sales of sweet cherries every season, what is the expected value of the revenue collected by the government from the sale of sweet cherries given that $q = 1/2$?

Hint: $\int xe^{ax}dx = (e^{ax}/a^2)(ax - 1)$.

**29.** The yield/acre of wheat on a given parcel of land can be represented as the outcome of a random variable $Y$ defined by $Y = 10x^{1/3}e^{\varepsilon}$ for $x \in [8,100]$, where
$Y$ = wheat output in bushes/acre
$x$ = pounds/acre of fertilizer applied
$\varepsilon$ = is a random variable having the probability density function $f(\varepsilon) = 3e^{-3\varepsilon} I_{(0,\infty)}(\varepsilon)$.

(a) If fertilizer is applied at the rate of 27 lb/acre, what is the probability that greater than 50 bushels/acre of wheat will be produced?

(b) You sign a forward contract to sell your wheat for $3.00/bushel at harvest time. Fertilizer costs $0.20/lb. If you apply fertilizer at a rate of 27 lb/acre, what is your expected return above fertilizer cost, per acre?

(c) What is the variance of $Y$ if fertilizer is applied at the rate of 27 lb/acre? Does the variance of $Y$ change if a different level of fertilizer is applied? Why or why not?

**30.** Let $X$ have the moment generating function $M_X(t)$. Show that

(a) $M_{(X+a)}(t) = e^{at} M_X(t)$.

(b) $M_{bX}(t) = M_X(bt)$.

(c) $M_{(X+a)/b}(t) = e^{(a/b)t} M_X(t/b)$.

**31.** The AJAX Disk Co. manufactures compact disks (CDs) for the music industry. As part of its quality-control program, the diameter of each disk is measured using an electronic measuring device. Letting $X_1$ represent the actual diameter of the disk and $X_2$ represent the *measured* diameter of the disk,

$$E\begin{bmatrix} X_1 \\ X_2 \end{bmatrix} = \begin{bmatrix} 4.6775 \\ 4.6775 \end{bmatrix} \text{ and } Cov(X) = \begin{bmatrix} .00011 & .00010 \\ .00010 & .00010 \end{bmatrix},$$

where $x_1$ and $x_2$ are measured in inches.

(a) What is the correlation between the actual diameter and the measured diameter of the CDs?

(b) Assume that $P(x_1 \in [4.655, 4.700]) = 1$. Use a graph to elucidate the degree of linear association between $X_1$ and $X_2$ that is represented by the correlation value you calculated in (a).

(c) Given the characteristics of $(X_1, X_2)$ indicated above, the manager of the quality control-department states that the difference between measured and actual disk diameters is no more than .01 inches with probability $\geq .90$. Do you agree? Why or why not?

**32.** An investor wishes to invest $1,000 and is examining two investment prospects. The net dollar return per dollar invested in the two projects can be represented as the outcome of a bivariate random variable $(X_1, X_2)$ where

$$E\begin{bmatrix} X_1 \\ X_2 \end{bmatrix} = \begin{bmatrix} .15 \\ .07 \end{bmatrix} \text{ and } \mathbf{Cov(X)} = \begin{bmatrix} .04 & -.001 \\ -.001 & .0001 \end{bmatrix}.$$

(a) If the investor invests \$500 in each project, what is his/her expected net dollar return? What is the variance associated with the net dollar return?

(b) Suppose the investor wishes to invest the \$1,000 so that his/her expected utility is maximized, where $E(U(R)) = E(R) - .01\text{var}(R)$, $R = \alpha_1 X_1 + \alpha_2 X_2$ represents the total return on the investment, $\alpha_1 + \alpha_2 = 1{,}000$, and $\alpha_i \geq 0$ for $i = 1, 2$. How much money should he/she invest in each of the projects?

**33.** The length of time in minutes for an individual to be served at a local restaurant is the outcome of a random variable, $T$, having a mean of 6 and a variance of 1.5. How probable is the event that an individual will be served within 3 to 9 min?

**34.**

(a) Find the moment-generating function of a random variable $X$ having the density function

$$f(x) = \frac{1}{8}\binom{3}{x} I_{(0,1,2,3)}(x).$$

(Hint: Use of the binomial theorem may be helpful in finding a compact representation of this function.)

Use the MGF to calculate the first two moments of $X$ about the origin. Calculate the variance of $X$.

(b) Repeat (a) using the density function

$$f(x) = \frac{1}{10} e^{-x/10} I_{(0,\infty)}(x).$$

**35.** The Rockbed Insurance Company sells 1-year term life insurance policies for \$10,000 of annual coverage, where a 1-year premium is charged to put the policy in force, and then if the insured person does not live through the year, his or her estate is paid \$10,000. The mortality tables in the large market in which the company sells insurance indicates that a 25 year old person will have a .998 probability of living another year, and a 35 year old person has a .995 probability of living another year. Selling and administrative costs are \$25 per policy.

(a) What is the expected profit on a \$10,000 life insurance policy sold to a randomly chosen 25 year-old for a premium of \$50?

(b) What is the premium that would need to be set on the policy in (a) so that the policy is expected to break even (i.e., return zero profit)?

(c) What is the increment in premiums needed, above the premium charged a 25 year old, to insure a randomly chosen 35 year old with a \$10,000 life insurance policy such that the policy is expected to break even?

(d) If Rockbed had \$1,000,000 of insurance coverage to sell and was contemplating distributing it between only the 25 and the 35 year old populations of potential customers at a fixed premium of \$60 per \$10,000 policy, how should it attempt to distribute its sales so as to maximize expected profit?

**36.** The daily wholesale price and quantity sold of ethanol in a Midwestern regional market during the summer months is represented by the outcome of a bivariate random variable $(P, Q)$ having the following probability model $\{R(P,Q), f(p,q)\}$:

$$f(p,q) = \begin{cases} .5pe^{-pq} & \text{for } (p,q) \in R\{P,Q\} = [2,4] \times [0,\infty) \\ 0 & \text{elsewhere} \end{cases}$$

where price is measured in dollars and quantity is measured in 100,000 gal units (e.g., $q = 2$ means 200,000 gal were sold).

(a) Derive the conditional expectation of quantity sold, as a function of the price of ethanol (i.e., derive the regression curve of Q on P).

(b) Graph the expected conditional quantity sold as a function of price. Does the relationship make economic sense?

(c) What is the expected quantity sold if price is \$2? If price is \$4?

(d) Derive the expected value of total dollar sales, $PQ$, of ethanol.

**37.** Define the mean, median, mode, and .10 and .90 quantiles of the random variable $X$ defined in the probability models $\{R(X), f(x)\}$ below:

(a) $f(x) = \dfrac{x^2 + 4}{50} I_{\{R(X)\}}(x),\ R(X) = \{0,1,2,3,4\}$

(b) $f(x) = .5e^{-x/2} I_{\{R(X)\}}(x),\ R(X) = [0,\infty)$

(c) $f(x) = 3x^2 I_{\{R(X)\}}(x),\ R(X) = [0,1]$

(d) $f(x) = .05(.95)^{x-1} I_{\{R(X)\}}(x),\ R(X) = \{1,2,3,\ldots\}$

**38.** The regression curve of daily quantity demanded of tablet computers in a Midwestern market, measured in thousands of units and expressed as a function of price, all else held equal, is given by

$$E(Q|p) = 10p^{-.5}$$

The marginal probability density function of price is given by

$$f(p) = \frac{3}{8}p^2 I_{[0,2]}(p)$$

(a) Derive the value of quantity demanded, $E(Q)$.

(b) Derive the expected value of dollar sales, $E(PQ)$.

**39.** The yield per acre of a certain dwarf watermelon is highly dependent on the amount of rainfall that occurs during the growing season. Following standard cultivation practices, the relationship between tons per acre, $Y$, and inches of rainfall, $R$, is given by

$$Y = 25 + 2R - .05R^2 \text{ for } R \in [0, 40]$$

(a) If expected rainfall is 10, can expected yield be equal to 60?

(b) If expected rainfall is 15, define an upper bound for the value of expected yield.

(c) If expected rainfall is 15, and the variance of rainfall is 5, what is the expected value of yield?

**40.** For each probability density function below, determine the mean and variance, if they exist, and define the median and mode.

(a) $f(x) = [\pi(1 + x^2)]^{-1} I_{(-\infty,\infty)}(x)$

(b) $f(x) = 4x^3 I_{[0,1]}(x)$

(c) $f(x) = .3(.7)^{x-1} I_{\{1,2,3,\ldots\}}(x)$

(d) $f(x) = \frac{x}{55} I_{\{1,2,\ldots,10\}}(x)$

**41.** The daily dollar sales of a large retail "Big Box" store, measured in 1,000 dollar units, is a random variable, D, that has an expectation of 20.

(a) Provide an upper bound to the probability that dollar sales exceed 40,000 dollars on a given day.

(b) If the variance of D is 4, define an interval in which dollar sales will occur on any given day with probability $\geq .95$.

**42.** Given the following three moments about the origin, derive the first three moments about the mean, and determine whether the random variable has a probability density function that is skewed, if you can.

(a) $E(X) = .2, E(X^2) = .2, E(X^3) = .2$

(b) $E(X) = 1, E(X^2) = 2, E(X^3) = 5$

(c) $E(X) = 2, E(X^2) = 8, E(X^3) = 48$

**43.** The bivariate random variable $(P,Q)$ represents the weekly price, in dollars, and the quantity, in number of kegs, of an India Pale Ale beer, HopMeister, sold in the Pacific Northwest market. The moment generating function associated with this bivariate random variable is given by $M_{(P,Q)}(\mathbf{t}) = \exp(\boldsymbol{\mu}'\mathbf{t} + .5\mathbf{t}'\boldsymbol{\Sigma}\mathbf{t})$, where

$$\boldsymbol{\mu} = \begin{bmatrix} 125 \\ 500 \end{bmatrix} \text{ and } \boldsymbol{\Sigma} = \begin{bmatrix} 100 & -100 \\ -100 & 400 \end{bmatrix}.$$

(a) What is the expected weekly price of HopMeister?

(b) What is the expected weekly quantity sold of HopMeister?

(c) What is the expected weekly dollar sales of HopMeister?

**44.** Derive the moment generating function of each of the random variables below, and use it to define the mean and variance of the random variable.

(a) $f(x) = .2e^{-.2x} I_{(0,\infty)}(x)$

(b) $f(x) = 2x I_{(0,1)}(x)$

(c) $f(x) = .3^x .7^{1-x} I_{\{0,1\}}(x)$

**45.** A small manufacturing firm produces and sells a product in a market where output prices are uncertain. The owner of the firm wishes to make a short run production decision that will maximize her expected utility, defined by

$$E(U(\pi)) = E(\pi) - \alpha[\text{var}(\pi)]$$

where $U$ is utility, $\pi = Pq - c(q)$ is profit, q is measured in 1,000's of units, $P$ is the uncertain price received for a unit of the product, the cost function is defined by $c(q) = .5q + .1q^2$, $\alpha \geq 0$, is a "risk aversion" parameter, and the probability density function of the uncertain output price is given by

$$f(p) = .5e^{-.5p} I_{[0,\infty)}(p)$$

(a) If the owner were risk neutral, i.e., $\alpha = 0$, find the level of production that maximizes expected utility.

(b) Now consider the case where the owner is risk averse, i.e., $\alpha > 0$. Graph the relationship between the optimal level of output and the level of risk aversion (i.e., the level of $\alpha$). How large does $\alpha$ have to be for optimal $q = 5$?

(c) Assume that $\alpha = .1$. Suppose that the government were to *guarantee* a price to the owner. What *guaranteed* level of price would induce the owner to produce the same level of output as in the case where price was uncertain?

**46.** The weekly number of MACs and PCs sold by a salesperson at the local computer store can be represented as the outcome of a bivariate random variable $(X,Y)$ with the nonzero values of its joint probability density function given by the following table of probabilities:

| | | Y | | | | |
|---|---|---|---|---|---|---|
| | | 0 | 1 | 2 | 3 | 4 |
| | 0 | .15 | .12 | .075 | .05 | .03 |
| | 1 | .12 | .075 | .05 | .05 | .02 |
| X | 2 | .05 | .04 | .03 | .03 | .02 |
| | 3 | .01 | .03 | .02 | .02 | .01 |

(X is the number of MACs and Y is the number of PCs). The salesperson receives a salary of \$200/week, and also receives a commission of \$50 for every MAC sold and \$100 for every PC sold.

(a) What is the expected value of the weekly commission that the salesperson obtains from selling computers? What is the expected value of her total pay received from selling computers?

(b) What is the expected value of her commission from selling MACs? What is the expected value of her commission from selling PCs?

(c) *Given* that the salesperson sells three PCs, what is the expected value of her commission from selling MACs?

(d) If 40 percent of the salesperson's total pay is deducted for federal and state taxes, what is the expected value of her pay after taxes?

**47.** An investor has \$10,000 to invest between two investment projects. The rate of return per dollar invested in the two projects can be represented as the outcome of a bivariate random variable $(X_1, X_2)$where

$$\mathrm{E}\begin{bmatrix} X_1 \\ X_2 \end{bmatrix} = \begin{bmatrix} .20 \\ .05 \end{bmatrix} \text{ and } \mathbf{Cov}(\mathbf{X}) = \begin{bmatrix} .04 & .002 \\ .002 & .0001 \end{bmatrix}$$

(a) If the investor invests \$5,000 in each project, what is the expected dollar return? What is the variance associated with the dollar return?

(b) If the investor wishes to maximize expected dollar return, how should the money be invested?

(c) If the invest wishes to minimize variance of dollar returns, how should the money be invested?

(d) Suppose the investor wishes to invest the \$10,000 so that his/her expected utility is maximized, where $\mathrm{E}(U(R)) = \mathrm{E}(R) - .01\mathrm{var}(R)$, where $R = \alpha_1 X_1 + \alpha_2 X_2$ represents the total return on the investment, $\alpha_1 + \alpha_2 = 10{,}000$, and $\alpha_i \geq 0$ for $i = 1, 2$. How much money should he/she invest in each of the two projects?

**48.** The mean vector and covariance matrix of the trivariate random variable **X** is given by

$$\mathrm{E}(\mathbf{X}) = \begin{bmatrix} -2 \\ 4 \\ 2 \end{bmatrix} \text{ and } \mathbf{Cov}(\mathbf{X}) = \begin{bmatrix} 10 & 2 & 1 \\ 2 & 5 & 0 \\ 1 & 0 & 1 \end{bmatrix}.$$

The random variable **Y** is defined by $Y = \mathbf{c}'\mathbf{X}$, where $\mathbf{c}' = [5\ 1\ 3]$, and the bivariate random vector **Z** is defined by $\mathbf{Z} = \mathbf{AX}$, where $\mathbf{A} = \begin{bmatrix} 1 & 1 & 1 \\ 2 & 3 & -4 \end{bmatrix}$.

(a) Define as many of the values $\mathrm{E}(X_i X_j)$, for $i$ and $j \in \{1,2,3\}$, as you can.

(b) Define the correlation matrix for **X**.

(c) Define the mean and variance of $Y$.

(d) Define the mean vector, covariance matrix, and correlation matrix of **Z**.

**49.** The bivariate random variable $(Y,X)$ has the following mean vector and covariance matrix:

$$\mathrm{E}\begin{bmatrix} X \\ Y \end{bmatrix} = \begin{bmatrix} 10 \\ 5 \end{bmatrix} \text{ and } \mathbf{Cov}(X, Y) = \begin{bmatrix} 5 & 2 \\ 2 & 2 \end{bmatrix}$$

(a) Derive the values of $a$ and $b$ in $\hat{Y} = a + bX$ that minimize the expected squared distance between $Y$ and $\hat{Y}$, i.e., that produce the best linear predictor of $Y$ outcomes in terms of $X$ outcomes.

(b) What proportion of the variance in $Y$ is explained by the best linear predictor that you derived above?

# Parametric Families of Density Functions

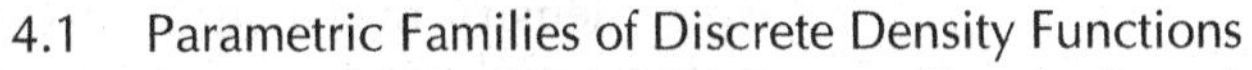



A collection of specific probability density functional forms that have found substantial use in statistical applications are examined in this chapter. The selection includes a number of the more commonly used densities, but the collection is by no means an exhaustive account of the vast array of probability densities that are available and that have been applied in the literature.[1]

Our density function definitions will actually identify **parametric families** of density functions. That is, the algebraic expressions defining the density functions will contain one or more unknowns, called **parameters**, which can be assigned values chosen from a set of admissible values called the **parameter space**. A *specific member* of a family of probability densities will be identified by choosing a specific value of the parameters contained in the parameter space. The general notation $f(x;\theta)$ will be used to distinguish elements, $x$, in the domain

---

[1]Johnson, Kotz, Balakrishnan and Kemp provide a set of volumes that provide an extensive survey of a large array of density functions that have been used in statistical applications. These include:

*Continuous Multivariate Distributions, Volume 1, Models and Applications, 2nd Edition,* by Samuel Kotz, N. Balakrishnan and Normal L. Johnson, 2000;

*Continuous Univariate Distributions, Volume 1, 2nd Edition* by Samuel Kotz, N. Balakrishnan and Normal L. Johnson, 1994;

*Continuous Univariate Distributions, Volume 2, 2nd Edition* by Samuel Kotz, N. Balakrishnan and Normal L. Johnson, 1995;

*Discrete Multivariate Distributions* by Samuel Kotz, N. Balakrishnan and Normal L. Johnson, 1997;

*Univariate Discrete Distributions, 3rd Edition* by Normal L. Johnson, Adrienne Kemp, and Samuel Kotz, 2008; all published by John Wiley and Sons, New York.

of the density function from elements,$\theta$, in the parameter space of the parametric family of functions.

For each parametric family of densities examined, we will present a particular **parameterization** of the family that identifies the parameters used in the algebraic representation of the density functions as well as the collection of admissible values for the parameters (the latter collection being the aforementioned parameter space). We will use various English and Greek letters to represent parameters. The parameter space will be generically represented by the Greek letter *capital omega*, $\Omega$. Be aware that, in general, parameterizations are *not unique*. Generally, the collection of densities in a parametric family can be **reparameterized**, meaning that an alternative set of parameters and an associated parameter space can be defined that equivalently identifies each and every density in a family of density functions. Possibilities for reparameterizations will not concern us currently, although we will revisit this issue later when we examine methods of statistical inference. And of course, the particular English and Greek letters we use to identify domain elements and elements of the parameter space are arbitrary, and can be changed.

Each parametric family has its own distinguishing characteristics that make the PDFs appropriate candidates for specifying the probability space of some experiments and inappropriate for others. Some of the main characteristics include whether the PDFs are discrete or continuous, whether the use of the PDFs are restricted to nonnegative-valued and/or integer-valued random variables, whether the densities in the family are symmetric or skewed, and the degree of flexibility with which the density can distribute probability over events in the range of a random variable. Furthermore, the functional forms of some parametric families of densities follow deductively from the characteristics of certain types of experiments. We will point out major characteristics and application contexts for each of the parametric families presented. We will also introduce procedures for assessing the adequacy of the choice of a particular family for a given application later in Chapter 10.

## 4.1 Parametric Families of Discrete Density Functions

### 4.1.1 Family Name: Uniform

*Parameterization*: $N \in \Omega = \{N\text{: } N \text{ is a positive integer}\}$
*Density Definition*: $f(x;N) = \frac{1}{N} I_{\{1,2,\ldots,N\}}(x)$
*Moments*: $\mu = (N+1)/2, \sigma^2 = (N^2-1)/12, \mu_3 = 0, \left(\frac{\mu_4}{\sigma^4} - 3\right) = -\left(\frac{6(n^2+1)}{5(n^2-1)}\right)$
*MGF*: $M_X(t) = \sum_{j=1}^{N} e^{jt}/N$
*Background and Application*: The discrete uniform density function assigns equal probability to each of $N$ possible outcomes of an experiment. The density is used to construct a probability space for any experiment having $N$ possible outcomes that are all equally likely. The outcomes are coded $1, 2, \ldots, N$, and each outcome is assigned probability $f(x;N) = 1/N$.

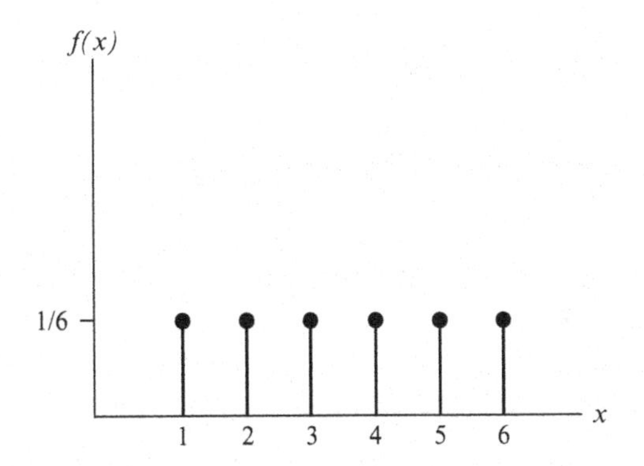


**Figure 4.1** Uniform density, $N = 6$.

**Example 4.1** ***Probabilities of Choosing Correctly at Random***

The office manager has a box of six printer cables with different pin configurations to accommodate computer hook-up of the various types of printers used by the company. A new employee needs a printer cable to hook up her printer. She randomly chooses printer cables from the box, one at a time without replacement, in an attempt to hook up her printer. Letting the outcome of $X$ denote the number of cables tried before the correct one is found, what is the probability density of $X$?

**Answer**: Let $A_i$ be the event that the correct cable is chosen on the $i$th try. Then

$$f(1) = P(A_1) = \frac{1}{6}\text{(classical probability)}$$

$$f(2) = P(A_2|\bar{A}_1)P(\bar{A}_1) = \left(\frac{1}{5}\right)\left(\frac{5}{6}\right) = \frac{1}{6}\text{(multiplication rule)}$$

$$f(3) = P(A_3|\bar{A}_1 \cap \bar{A}_2)P(\bar{A}_2|\bar{A}_1)P(\bar{A}_1)\text{(extended multiplication rule)}$$
$$= \left(\frac{1}{4}\right)\left(\frac{4}{5}\right)\left(\frac{5}{6}\right) = \frac{1}{6},$$

$$\vdots$$

$$f(6) = (1)\left(\frac{1}{2}\right)\left(\frac{2}{3}\right)\left(\frac{3}{4}\right)\left(\frac{4}{5}\right)\left(\frac{5}{6}\right) = \frac{1}{6}.$$

Thus, $X$ has the uniform distribution with $N = 6$ (see Figure 4.1). □

### 4.1.2 Family Name: Bernoulli

*Parameterization*: $p \in \Omega = \{p: 0 \leq p \leq 1\}$
*Density Definition*: $f(x; p) = p^x (1 - p)^{1-x} I_{\{0,1\}}(x)$
*Moments*: $\mu = p$, $\sigma^2 = p(1 - p)$, $\mu_3 = 2p^3 - 3p^2 + p$, $\left(\frac{\mu_4}{\sigma^4} - 3\right) = \left(\frac{1-6p(1-p)}{p(1-p)}\right)$
*MGF*: $M_X(t) = pe^t + (1 - p)$
*Background and Application*: The Bernoulli density, named after Swiss mathematician Jacques Bernoulli (1654–1705), can be used to construct a probability space for an experiment that has two possible outcomes (e.g., cure versus no cure, defective versus nondefective, success versus failure) that *may* or *may not*

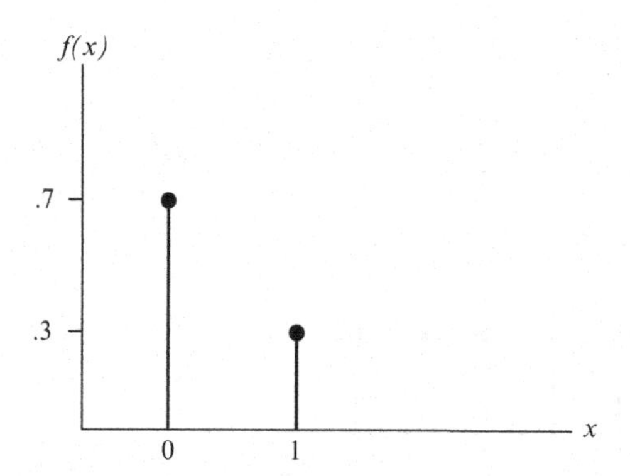


**Figure 4.2**
Bernoulli density, $p = .3$.

be equally likely to occur. The two outcomes of the experiment are coded 0 and 1, where the event $x = 0$ is assigned probability $f(0;p) = 1 - p$ and the event $x = 1$ is assigned probability $f(1; p) = p$.

**Example 4.2** ***Probability of a Dichotomous Outcome***

A shipment of DVD players to a local electronics retailer contains three defective and seven nondefective players. The players are placed on the store's shelves, and a customer randomly chooses a player to purchase. Coding the choice of a defective player as $x = 1$ and the choice of a nondefective player as $x = 0$, the Bernoulli density can be used to construct the probability space of the experiment by letting $p = .3$. (see Figure 4.2) □

### 4.1.3 Family Name: Binomial

*Parameterization*: $(n,p) \in \Omega = \{(n,p)\text{: } n \text{ is a positive integer, } 0 \leq p \leq 1\}$
*Density Definition*:

$$f(x; n, p) = \begin{cases} \frac{n!}{x!(n-x)!} & p^x (1-p)^{n-x} \text{for } x = 0, 1, 2, \ldots, n \\ 0 & \text{otherwise} \end{cases}$$

*Moments*: $\mu = np$, $\sigma^2 = np(1 - p)$, $\mu_3 = np(1 - p)(1 - 2p)$, $\left(\frac{\mu_4}{\sigma^4} - 3\right) = \left(\frac{1-6p(1-p)}{np(1-p)}\right)$
*MGF*: $M_X(t) = (1 - p + pe^t)^n$
*Background and Application*: The binomial density function is used to construct a probability space for an experiment that consists of $n$ independent repetitions (also called *Bernoulli trials)* of a given experiment of the Bernoulli type (i.e., the experiment has two possible outcomes), with the observation of interest being *how many of the n Bernoulli trials result in one of the two types of outcomes*, say type $A$ (e.g., how many successes, defectives, or cures occur in $n$ repetitions of the Bernoulli-type experiment?). The value of $x$ represents the total number of outcomes of type $A$ that occur in the $n$ Bernoulli trials. The parameters $n$ and $p$ refer respectively to the number of trials and the probability of observing the type A outcome in the underlying Bernoulli-type experiment. It is assumed that the repetitions are executed in such a way that the outcome observed on any trial does not affect the probability of occurrence of outcomes on any other

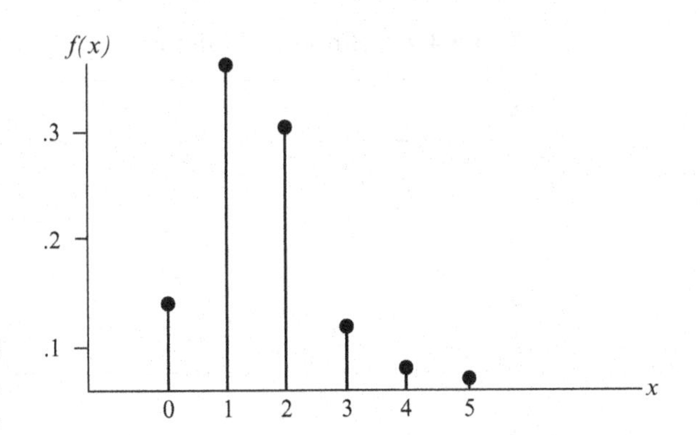


**Figure 4.3** Binomial density, $n = 5$, $p = 0.3$.

trial (which is the meaning of the phrase *independent repetitions* or *independent trials*).

The functional form of the density can be deduced directly from the characteristics of the experiment described above. Let $(X_1,\ldots,X_n)$ be a collection of $n$ independent random variables, where each $X_i$ has a Bernoulli density with the same value of $p$, as $f(x_i;p) = p^{x_i}(1-p)^{1-x_i} I_{\{0,1\}}(x_i)$. Let $x_i = 1$ indicate that outcome $A$ occurs on the *ith* Bernoulli trial. Then the random variable $X = \sum_{i=1}^{n} X_i$ represents the number of Bernoulli trials that result in outcome $A$. Since the Bernoulli random variables are independent, the probability of obtaining a *particular* sequence of $x$ outcomes of type $A$ in a sequence of $n$ trials is $p^x(1-p)^{n-x}$. The number of *different* sequences of $n$ trials that result in $x$ outcomes of type $A$ is given by $\binom{n}{x}$ (which is the number of different ways of placing $x$ outcomes of type $A$ into the $n$ positions of the sequence). Since the different sequences are mutually exclusive, it follows that the probability of observing $x$ outcomes of type $A$ is given by the sum of the probabilities of the $\binom{n}{x}$ different sequences that result in the outcome $x$, which is represented by the binomial density function defined above. The binomial density assigns the probability $f(x;n,p)$ to the outcome $x \in R(X) = \{0,1,2,\ldots,n\}$.

**Example 4.3** ***Probability of Number of Credit Card Transactions***

Upon analyzing the cash register receipts of a large department store over an extended period of time, it is found that 30 percent of the customers pay for their purchases by credit card, 50 percent pay by cash, and 20 percent pay by check. Of the next five customers that make purchases at the store, what is the probability that three of them will pay by credit card?

**Answer**: Assume that how a customer pays for her purchases is the outcome of a Bernoulli trial with $x_i = 1 \Rightarrow$ credit card, and $x_i = 0 \Rightarrow$ cash or check. Given that the respective probabilities of these outcome are .30 and .70, and assuming that customers' payment methods are independent of one another, it follows that $X = \sum_{i=1}^{5} X_i$ represents the number of customers that pay by credit card, and $X$ has a binomial density function with $n = 5$ and $p = .3$. The graph of the density is given in Figure 4.3, and the density values are displayed in Table 4.1: Thus, $P(x = 3) = .1323$. □

**Table 4.1** Binomial PDF for $n = 5$, $p = .3$

| $x$ | $f(x)$ |
|---|---|
| 0 | .1681 |
| 1 | .3602 |
| 2 | .3087 |
| 3 | .1323 |
| 4 | .0284 |
| 5 | .0024 |

### 4.1.4 Family Name: Multinomial

*Parameterization*: $(n,p_1,\ldots,p_m) \in \Omega = \{(n,p_1,\ldots,p_m)$: $n$ is a positive integer, $0 \leq p_i \leq 1, \forall i, \sum_{i=1}^m p_i = 1\}$
*Density Definition*:

$$f(x_1,\ldots,x_n;n,p_1,\ldots,p_m) = \begin{cases} \frac{n!}{\prod_{i=1}^m x_i!} \prod_{i=1}^m p_i^{x_i} \text{ for } x_i = 1,\ldots,n, \sum_{i=1}^m x_i = n, \\ 0 \quad \text{otherwise} \end{cases}$$

*Moments: For each random variable $X_i$ and random variable pair $(X_i,X_j)$* $\mu(X_i) = np_i, \sigma^2(X_i) = np_i(1-p_i), Cov(X_i,X_j) = -np_ip_j, \mu_3(X_i) = np_i(1-p_i)(1-2p_i),$ $\left(\frac{\mu_4(X_i)}{\sigma^4(X_i)} - 3\right) = \left(\frac{1-6p_i(1-p_i)}{np_i(1-p_i)}\right)$
*MGF*: $M_X(t) = \left(\sum_{i=1}^m p_i e^{t_i}\right)^n$
*Background and Application*: The multinomial density function is an extension of the binomial density function to the case where there is interest in more than two different types of outcomes for each trial of the underlying repeated experiment. In particular, the multinomial density function is used to construct a probability space for an experiment that consists of $n$ independent repetitions of a given experiment characterized by $m > 2$ different types of outcomes. The observation of interest is *how many of each of the m different types of outcomes of the experiment occur in n repetitions of the experiment.* The value of $x_i$ represents the total number of outcomes of type $A_i$ that occur in the $n$ repetitions. The parameters $n$ and $p_i$ refer, respectively, to the number of *multinomial trials* conducted and the probability of observing the type $A_i$ outcome in one trial. It is assumed that the repetitions are conducted in such a way that the outcome observed on any trial does not affect the probability of occurrence of outcomes on any other trial. The motivation for the density function definition is a direct extension of the arguments used in the binomial case upon recognizing that the number of different sequences of $n$ repetitions of the experiment that result in $x_i$-number of type $A_i$ outcomes, $i = 1,\ldots,m$, equals $\left(\frac{n!}{x_1!\ldots x_m!}\right)$. The details are left to the reader. The multinomial density assigns the probability $f(x_1,\ldots,x_m;n,p_1,\ldots,p_m)$ to the outcome $(x_1,\ldots,x_m) \in R(X) = \{(x_1,\ldots,x_m)$: $x_i \in (0,1,\ldots,n)\ \forall i, \sum_{i=1}^m x_i = n\}$.

It is useful to note that the marginal density of any of the $X_i$ variables is binomial with parameters $n$ and $p_i$. Furthermore, any subset of the random variables $(X_1,\ldots,X_m)$ has a marginal multinomial density.

**Example 4.4** ***Probability of Customer Payment Modes***

Recall Example 4.3. Of the next five customers entering the store, what is the probability that two will pay by credit card, two will pay by cash, and one will pay by check?

**Answer:** For *each* of the five experiments of observing customers' payment methods, we have *three* different types of outcomes that are of interest. In the specification of the multinomial density, we let $x_1$, $x_2$, $x_3$ refer, respectively, to the number of payments made by credit card, cash, and check. The probabilities of observing a payment by credit card, cash, or check in any given trial is $p_1 = .3$, $p_2 = .5$, and $p_3 = .2$, respectively. Then the probability we seek is given by

$$f(2,2,1;.3,.5,.2) = \left[\frac{5!}{2!2!1!}\right](.3)^2(.5)^2(.2)^1 = .135. \qquad \square$$

### 4.1.5 Family Name: Negative Binomial, and Geometric Subfamily

*Parameterization*: $(r,p)\in\Omega = \{(r,p): r \text{ is a positive integer}, 0 < p < 1\}$ (the geometric density family is defined by setting r = 1)

*Density Definition*:

$$f(x;r,p) = \begin{cases} \dfrac{(x-1)!}{(r-1)!(x-r)!}\, p^r(1-p)^{x-r} & \text{for } x = r, r+1, r+2, \ldots, \\ 0 & \text{otherwise} \end{cases}$$

*Moments*: $\mu = r/p, \sigma^2 = r(1-p)/p^2, \mu_3 = r((1-p) + (1-p)^2)/p^3$,
$\left(\frac{\mu_4}{\sigma^4} - 3\right) = \left(\frac{r(1-p)(p^2-6p+6)}{p^4}\right)$

*MGF*: $M_X(t) = e^{rt}p^r(1-(1-p)e^t)^{-r}$ for $t < -\ln(1-p)$

*Background and Application*: The negative binomial density function (also sometimes referred to as the **Pascal distribution**) can be used to construct a probability space for an experiment that consists of independent repetitions of a given experiment of the Bernoulli type, just like the case of the binomial density, *except* that the observation of interest is now *how many Bernoulli trials are necessary to obtain r outcomes of a particular type, say type A* (e.g., how many Bernoulli trials are necessary to obtain $r$ successes, defectives, or tails?). In comparing the binomial and negative binomial densities, notice that the roles of the *number of Bernoulli trials* and the *number of successes* are reversed with respect to what is the random variable and what is the parameter. For the negative binomial density, the value of $x$ represents the number of Bernoulli trials necessary to obtain $r$ outcomes of type $A$.

In order to motivate the density function definition, let $(X_1,\ldots,X_n)$ be a collection of $n$ independent random variables, where each $X_i$ has a Bernoulli density, precisely the same as in our discussion of the binomial density. Let the probability of obtaining an outcome of type $A$ be $p$ for each trial. Since the Bernoulli random variables are independent, the probability of obtaining a sequence of $x$ trials that result in $r$ outcomes of type $A$, *with the last trial being the* $r^{th}$ *such outcome*, is $p^r(1-p)^{x-r}$. The number of *different* sequences of $x$ trials that result in $r$ outcomes of type $A$, *with the* $r^{th}$ *outcome being of type* $A$, is given by $(x-1)!/((r-1)!(x-r)!)$ (which is the number of different ways of placing $r-1$ outcomes of type $A$ in the first $x-1$ positions of the sequence). Since the different sequences are mutually exclusive, it follows that the probability of needing $x$ Bernoulli trials to obtain $r$ outcome of type $A$ is given by the sum of the probabilities of the $\frac{(x-1)!}{(r-1)!(x-r)!}$ different sequences that result in the outcome $x$, this sum being represented by the negative binomial density function defined above. The negative binomial density assigns the probability $f(x;r,p)$ to the outcome $x \in R(X) = \{r, r+1, r+2, \ldots\}$.

The geometric family of densities is a subset of the family of negative binomial densities defined by setting $r = 1$. Thus, the geometric density is appropriate for assigning probability to events relating to *how many Bernoulli trials are necessary to get the first outcome of type A*. The geometric density function has a unique property in that it is the only discrete density for a nonnegative integer-valued random variable for which $P[x>i+j|x>i] = P[x>j] \forall$ $i$ and $j \in \{0,1,2,\ldots\}$. This conditional probability property is referred to as the **memoryless property**, meaning that in any experiment characterized by the geometric density, if the experiment has already resulted in $i$ trials without a type $A$ outcome, the experiment has "no memory" of this fact, since the probability that more than $j$ trials will be needed to obtain the first type $A$ outcome is precisely the same as if the first $i$ trials had never occurred. The proof that the geometric density has this property is left to the reader. The reader may wish to consult V.K. Rohatgi, (1976), *An Introduction to Probability Theory and Mathematical Statistics*, New York: John Wiley, p. 191, for a proof that the geometric density is the *only* density for nonnegative integer-valued random variables that has this property.

**Example 4.5** ***Probability of Meeting Quota in x Trials***

A salesperson has a quota of 10 sales per day that she is expected to meet for her performance to be considered satisfactory. If the probability is .25 that any given customer she contacts will make a purchase, and if purchase decisions are independent across consumers, what is the probability that the salesperson will meet her quota with no more than 30 customer contacts?

**Answer:** The negative binomial density function can be applied with $p = .25$ and $r = 10$. The event of interest is $\{x \leq 30\}$, which has the probability

$$P(x \leq 30) = \sum_{x=10}^{30} \left( \frac{(x-1)!}{9!(x-10)!} \right) (.25)^{10} (.75)^{x-10} = .1966 \qquad \square$$

**Example 4.6** ***Probability of Defect after x Trials***

A machine produces envelopes, and the probability that any given envelope will be defective is $p = .001$. The production of envelopes from the machine can be viewed as a collection of independent Bernoulli trials (i.e., each envelope is either defective or nondefective, the probability of a defective is .001, and the occurrence of a defective does not affect the probability that other envelopes will be defective or nondefective). What is the probability that the first defective envelope will occur *after* 500 envelopes have been produced? *Given that* the machine has already produced 500 envelopes without a defective, what is the probability that the first defective envelope will occur *after another* 500 envelopes have been produced?

**Answer**: The geometric density can be applied with $p = .001$. The first event of interest is $\{x > 500\}$. The probability of the event can be calculated by first noting that the cumulative distribution function for the geometric density is given by

$$F(b) = [1 - (1-p)^{\text{trunc}(b)}]\, I_{[1,\infty)}(b).$$

Then $P(x > 500) = 1 - F(500) = 1 - \left[1 - (.999)^{500}\right] = .6064$. The graph of $f(x)$ in this case has a set of spikes at $x = 1,2,3,\ldots$, that decline *very* slowly, beginning with $f(1) = .001$, and with the image of $x = 250$ still being equal to $f(250) = .00078$. As $x \to \infty$, $f(x) \to 0$.

The second probability we seek is of the form $P(x>1{,}000|x>500)$. By the memoryless property of the geometric density, we know that this probability is equal to $P(x>500) = .6064$, the same as above. □

### 4.1.6 Family Name: Poisson

*Parameterization*: $\lambda \in \Omega = \{\lambda: \lambda > 0\}$

*Density Definition*:

$$f(x;\lambda) = \begin{cases} \dfrac{e^{-\lambda}\lambda^x}{x!} & \text{for } x = 0,1,2,\ldots, \\ 0 & \text{otherwise} \end{cases}$$

*Moments*: $\mu = \lambda$, $\sigma^2 = \lambda$, $\mu_3 = \lambda$, $\left(\frac{\mu_4}{\sigma^4} - 3\right) = \left(\frac{1}{\lambda}\right)$

*MGF*: $M_X(t) = \exp(\lambda(e^t - 1))$

*Background and Application*: When the number of independent and identical Bernoulli experiments is very large and $p$ is small, the Poisson density, named after French mathematician Simeon Poisson (1781–1840), provides an approximation to the probability that $x = \sum_{i=1}^{n} x_i = c$ and thus provides an approximation to probabilities generated by the binomial density. In fact, the limit of the binomial density as $n \to \infty$ and $\lambda = np$, with $\lambda$ set to a fixed constant, is the Poisson density. We examine this situation in more detail to provide an example of how *limiting densities* arise.[2] In the discussion, we will have need for the following result.

**Lemma 4.1**

$$e^v = \lim_{n\to\infty}(1 + (v/n))^n.$$

[2]Limiting densities will be discussed further in Chapter 5.

The binomial density can be expressed alternatively as

$$f(x;n,p) = \frac{\prod_{i=1}^{x}(n-i+1)}{x!}\, p^x (1-p)^{n-x}.$$

where we have suppressed the indicator function by assuming that $x$ is a non-negative integer $\leq n$. Let $np = \lambda$ for some $\lambda > 0$, so that $p = \lambda/n$ can be substituted in the binomial density expression to yield

$$\frac{\prod_{i=1}^{x}(n-i+1)}{x!}\left(\frac{\lambda}{n}\right)^x \left(1-\frac{\lambda}{n}\right)^{n-x}.$$

Algebraically rearranging the above expression, and letting $n \to \infty$ yields

$$\begin{aligned}
&\lim_{n\to\infty} \frac{\prod_{i=1}^{x}(n-i+1)}{n^x}\frac{\lambda^x}{x!}\left(1-\frac{\lambda}{n}\right)^n\left(1-\frac{\lambda}{n}\right)^{-x} \\
&= \lim_{n\to\infty}\left(\left(\frac{n}{n}\right)\left(\frac{n-1}{n}\right)\cdots\left(\frac{n-x+1}{n}\right)\frac{\lambda^x}{x!}\left(1-\frac{\lambda}{n}\right)^n\left(1-\frac{\lambda}{n}\right)^{-x}\right) \\
&= \frac{e^{-\lambda}\lambda^x}{x!}
\end{aligned}$$

since "the limit of a product equals the product of the limits" when all of the limits exist, and since $\lim_{n\to\infty}\left(\frac{n-i}{n}\right) = 1 \;\; \forall i,\;\; \lim_{n\to\infty}\left(1-\frac{\lambda}{n}\right)^{-x} = 1,\;\; \lim_{n\to\infty}\left(1-\frac{\lambda}{n}\right)^{n} = e^{-\lambda}$ by Lemma 4.1, and $\lim_{n\to\infty}\left(\frac{\lambda^x}{x!}\right) = \frac{\lambda^x}{x!}$.

Therefore, the binomial density converges to the Poisson density for $np = \lambda$, $\lambda>0$, and $n \to \infty$. The usefulness of this result is that for *large* $n$, and thus for *small* $p = \lambda/n$, one can adopt the approximation

$$\binom{n}{x} p^x (1-p)^{n-x} \approx \left(\frac{e^{-np}(np)^x}{x!}\right),$$

that is, one can replace the parameter $\lambda$ in the Poisson density by the product of the binomial density parameters $n$ and $p$. The approximation can be quite useful since the Poisson density is relatively easy to evaluate, whereas for large $n$, dealing with the factorial expressions in the binomial density can be cumbersome. Based on two "rules of thumb", the approximation is considered reasonably good if $n \geq 20$ and $p \leq .05$, or if $n \geq 100$ and $\lambda = np \leq 10$.

**Example 4.7** ***Approximate Probability of*** $x$ ***Defects in*** $n$ ***Trials***

A publishing company is typesetting a novel that is 300 pages long and averages 1,500 typed letters per page. If typing errors are as likely to occur for one letter as another, if a typing error occurring in one place does not affect the probability of a typing error occurring in any other place, and if the probability of mistyping a letter is small, then the total number of typing errors in the book can be viewed as the outcome of a random variable having, *approximately*, a Poisson density. For example, if the probability of a typing error for any given letter is $10^{-5}$, then $\lambda = np = 4.5$ in the Poisson density. The probability of observing 10 or fewer errors in the book would be approximated as

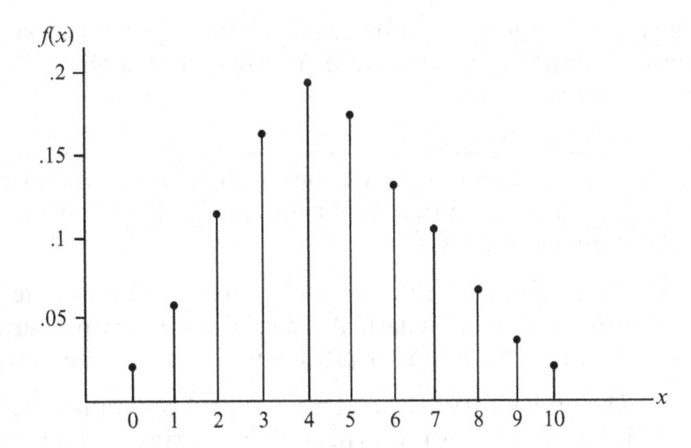


**Figure 4.4** Partial Poisson density, $\lambda = 4.5$.

**Table 4.2** Poisson Density for $\lambda = 4.5$

| *x* | *f*(*x*) |
|---|---|
| 0 | .0111 |
| 1 | .0500 |
| 2 | .1125 |
| 3 | .1687 |
| 4 | .1898 |
| 5 | .1708 |
| 6 | .1281 |
| 7 | .0824 |
| 8 | .0463 |
| 9 | .0232 |
| 10 | .0104 |

$$P(x \leq 10) \approx \sum_{x=0}^{10} e^{-4.5}(4.5)^x / x! = .9933.$$

A partial graph (truncated after $x = 10$) of the Poisson density in this case is given in Figure 4.4, and density values are exhibited in Table 4.2. □

### 4.1.7 Poisson Process and Poisson PDF

Besides serving as an approximation to the binomial density for large $n$ and small $p$, the Poisson density is important in its own right for constructing probability spaces for experiments whose outcomes are governed by the so-called **Poisson process**. The Poisson process refers to a particular type of experimental situation in which the number of occurrences of some specific type, say type $A$ (e.g., a traffic accident, a telephone call, the arrival of a customer at a checkout stand, a flaw in a length of wire) in a time, space, volume or length dimension, possesses the following general probabilistic characteristics (we state the conditions for a time dimension; the conditions can interpreted within the other dimensions of measurement as well):

**Definition 4.1**
***Poisson Process***

Let an experiment consist of observing the number of type $A$ outcomes that occur over a fixed interval of time, say $[0,t]$. The experiment is said to follow the Poisson process if:

1. The probability that precisely one type $A$ outcome will occur in a small time interval of length, $\Delta t$, is *approximately* proportional to the length of the interval, as $\gamma[\Delta t] + o(\Delta t)$, where $\gamma > 0$ is the proportionality factor;[3]
2. The probability of two or more type $A$ outcomes occurring in a small time interval of length $\Delta t$ is negligible relative to the probability that one type $A$ outcome occurs, this negligible probability being of order of magnitude $o(\Delta t)$; and
3. The numbers of type $A$ outcomes that occur in nonoverlapping time intervals are independent events.

To appreciate what is meant by the probability of two or more type $A$ outcomes occurring in a small interval of length $\Delta t$ being *negligible* relative to the probability that just one type $A$ outcome occurs, note that

$$\lim_{\Delta t \to 0}\left(\frac{P(\geq 2 \text{ typeA})}{P(1 \text{ typeA})}\right) = \lim_{\Delta t \to 0}\left(\frac{o(\Delta t)}{\gamma[\Delta t] + o(\Delta t)}\right) = \lim_{\Delta t \to o}\left(\frac{\frac{o(\Delta t)}{\Delta t}}{\frac{\gamma[\Delta t]}{\Delta t} + \frac{o(\Delta t)}{\Delta t}}\right) = 0.$$

Thus, for small enough intervals $\Delta t$, $P(\geq 2 \text{ type } A)$ is negligible *relative to* $P(1 \text{ type } A)$.

We now indicate why the Poisson process leads to the Poisson density.

**Theorem 4.1**
***Poisson Process***
***⇒ Poisson Density***

*Let X represent the number of times event A occurs in an interval of time* $[0,t]$*. If the experiment underlying X follows the Poisson process, then the density of X is the Poisson density.*

**Proof** Partition the interval $[0,t]$ into $n$ successive disjoint subintervals, each of length $\Delta t = t/n$, and denote these intervals by $I_j$, $j = 1,\ldots,n$. Let the random variable $X(I_j)$ denote the number of outcomes of type $A$ that occur within subinterval $I_j$, so that

$$X = \sum_{j=1}^{n} X(I_j).$$

[3] $o(\Delta t)$ is a generic notation applied to any function of $\Delta t$, whose values approach zero at a rate faster than $\Delta t$, so that $\lim_{\Delta t \to 0}(o(\Delta t))/(\Delta t) = 0$. The "$o(\Delta t)$" stands for "of smaller order of magnitude than $\Delta t$." For example, $h(\Delta t) = (\Delta t)^2$ is a function to which we could affix the label $o(\Delta t)$, while $h(\Delta t) = (\Delta t)^{1/2}$ is not. More will be said about orders of magnitude in Chapter 5.

Examine the event $x = k$, and note that $P(x = k) = P(A_n) + P(B_n)$, where $A_n$ and $B_n$ are the disjoint sets

$$A_n = \left\{ \sum_{j=1}^{n} x(I_j) = k, x(I_j) = 0 \text{ or } 1, \forall j \right\}$$

$$B_n = \left\{ \sum_{j=1}^{n} x(I_j) = k,\ x(I_j) \geq 2,\ \text{for 1 or more } j\text{'s} \right\}.$$

Since $B_n \subset \{x(I_j) \geq 2 \text{ for 1 or more } j\text{'s}\} \subset \cup_{j=1}^{n} \{x(I_j) \geq 2\}$ , Boole's inequality implies

$$P(B_n) \leq \sum_{j=1}^{n} P\{x(I_j) \geq 2\} = \sum_{j=1}^{n} o(\tfrac{t}{n}) = t\left[\frac{o(\frac{t}{n})}{\frac{t}{n}}\right], \text{ so that } \lim_{n\to\infty} P(B_n) = 0.$$

Now examine $P(A_n)$. For each subinterval, define a "success" as observing exactly one type $A$ outcome, and a "failure" otherwise. Then by property (1) of the Poisson process, $P(success) = \gamma[t/n] + o(t/n)$ and $P(\text{failure}) = 1 - \gamma[t/n] - o(t/n)$. Since events in one subinterval are independent of events in other subintervals by property (3) of the Poisson process, we can view the observations on the $n$ subintervals as a collection of independent Bernoulli trials, each trial yielding a success or a failure. It follows that probability can be assigned to event $A_n$ by the binomial density as

$$P(A_n) = \binom{n}{k} \left[\gamma\frac{t}{n} + o\left(\frac{t}{n}\right)\right]^k \left[1 - \gamma\frac{t}{n} - o\left(\frac{t}{n}\right)\right]^{n-k}.$$

We need the following extension of Lemma 4.1:

**Lemma 4.2**

$$e^v = \lim_{n\to\infty} (1 + (v/n) \pm o(v/n))^n.$$

Then, following a similar approach to the one used to demonstrate the convergence of the binomial to the Poisson density, it can be shown using Lemma 4.2 that $\lim_{n\to\infty} P(A_n) = \left(e^{-\gamma t}(\gamma t)^k\right)/k!$, which is the value of the Poisson density, where $\lambda = \gamma t$.

Finally, since $P(x = k) = P(A_n) + P(B_n)\ \forall n$, we have $P(x = k) = \lim_{n\to\infty} [P(A_n) + P(B_n)] = \frac{e^{-\gamma t}(\gamma t)^k}{k!}$ and so the conditions of the Poisson process lead to assignments of probability via the Poisson density function. ■

Note that when use of the Poisson density is motivated from within the context of the Poisson process, $\gamma$ can be interpreted as the *mean rate of occurrence of the type A outcome per unit of time.* This follows from the fact that $E(X) = \lambda = \gamma t$ if $X$ has a Poisson density, and then $\gamma = E(X)/t$. In applications, the Poisson density would be chosen for constructing the probability space of an experiment when either the conditions of the Poisson process hold, or when conditions exist for the Poisson density to be a reasonable approximation to the binomial density. The Poisson PDF has been used to represent the probability of such random variables as the number of machine breakdowns in a work shift, the number of customer arrivals at a checkout stand during a period of time, the number of telephone calls arriving at a switchboard during a time interval, and the number of flaws in panes of glass.

**Example 4.8**
***Probability of Number of Machine Breakdowns***

The milk bottling machine in a milk processing plant has a history of breaking down, on average, once every 2 weeks. The chief of the repair and maintenance crew is scheduling vacations for the summer months and wants to know what the probability is that the bottling machine will break down more than three times during the next 4 weeks. What is the probability?

**Answer**: When viewed in the context of ever shorter time intervals (hours, minutes, seconds), breakdowns appear to be increasingly less likely, and it seems reasonable to assume that the probability of two breakdowns within a short-enough time interval would be negligible. Assuming the repair crew returns the machine to full operating performance after each breakdown, it is reasonable to assume that the event of a breakdown in any short interval of time is independent of a breakdown occurring in other intervals of time.[4] All told, it would appear that the conditions of the Poisson process are a reasonable approximation to this situation, and we endeavor to assign probability to the event $x>3$ using the Poisson density. Since the average number of breakdowns is 1 every 2 weeks and since the chief is interested in a 4-week time interval, the Poisson density of relevance here is $(e^{-2}2^X)/x!I_{\{0,1,2,\ldots\}}(x)$, where $\lambda = \gamma t = 2$ was chosen to represent a rate of 2 breakdowns every 4 weeks ($\gamma = .5$ breakdowns per week times $t = 4$ weeks). The probability of more than 3 breakdowns in the 4-week period is then given by

$$P(x>3) = 1 - \sum_{i=0}^{3} P(x = i) = 1 - \sum_{i=0}^{3} e^{-2}\frac{2^i}{i!} = .143. \qquad \square$$

### 4.1.8 Family Name: Hypergeometric

*Parameterization*: $(M,K,n) \in \Omega = \{(M,K,n){:}\quad M = 1,2,3,\ldots; \quad K = 0,1,\ldots,M; \; n = 1,2,\ldots,M\}$

*Density Definition*:

$$f(x;M,K,n) = \begin{cases} \dfrac{\binom{K}{x}\binom{M-K}{n-x}}{\binom{M}{n}} & \text{for integer values } \max\{0,n-(M-K)\} \leq x \leq \min\{n,k\} \\ 0 & \text{otherwise} \end{cases}$$

[4] Ultimately, this assumption could be tested using a nonparametric test of hypothesis. We will examine tests of independence in our discussion of hypothesis testing procedures in Chapter 10.

*Moments*:

$$\mu = \frac{nK}{M}, \quad \sigma^2 = n\left(\frac{K}{M}\right)\left(\frac{M-K}{M}\right)\left(\frac{M-n}{M-1}\right),$$

$$\mu_3 = n\left(\frac{K}{M}\right)\left(\frac{MK}{M}\right)\left(\frac{M-2K}{M}\right)\left(\frac{M-n}{M-1}\right)\left(\frac{M-2n}{M-2}\right),$$

$$\left(\frac{\mu_4}{\sigma^4} - 3\right) = \frac{M^2(M-1)}{n(M-2)(M-3)(M-n)} \times \left[\left(\frac{M(M+1) - 6M(M-n)}{K(M-K)}\right) + \frac{3n(M-n)(M+6)}{M^2} - 6\right]$$

*MGF*: $M_X(t) = [((M-n)!(M-K)!)/M!]\, H(-n, -K, M-K-n+1, e^t)$ where $H(\cdot)$ is the hypergeometric function $H(\alpha,\beta,r,Z) = 1 + \frac{\alpha\beta}{r}\frac{Z}{1!} + \frac{\alpha\beta(a+1)(\beta+1)}{r(r+1)}\frac{Z^2}{2!} + \ldots$

(Note: this MGF is not too useful in practice since moments are defined in terms of an infinite sum; to illustrate this fact, the reader should attempt to define $\mu$ by differentiating $M_X(t)$ once with respect to $t$ and evaluating the derivative at zero.)

*Background and Application*: The hypergeometric density is used to construct a probability space for an experiment in which there are:

1. $M$ objects, of which $K$ of them are of one type, say type $A$;
2. The remaining $M$–$K$ objects are of a different type, say $B$; and
3. $n$ objects are randomly drawn *without replacement* from the original collection of $M$ objects and the number, $x$, of type $A$ outcomes in the collection of $n$ objects drawn is observed.

By **drawing randomly without replacement,** we mean that, at each draw, all of the objects that remain in the collection have an equal probability of being chosen. The hypergeometric density assigns probability to the number of objects drawn that are of type $A$ out of a total sample of $n$ objects.

To motivate the density function definition, note that the number of different ways of choosing the sample of $n$ objects is given by $\binom{M}{n}$, the number of different ways of choosing $x$ type $A$ items is $\binom{K}{x}$, and the number of different ways to choose $(n-x)$ type $B$ items is $\binom{M-K}{n-x}$. Then, since all possible sample outcomes having $x$ type $A$ and $n-x$ type $B$ outcomes are equally likely, the classical probability definition states that the probability of obtaining $x$ outcomes of type $A$ from a random sample (without replacement) of $n$ objects from the aforementioned collection is $\dfrac{\binom{K}{x}\binom{M-K}{n-x}}{\binom{M}{n}}$.

Note that the binomial and hypergeometric densities both assign probabilities to the event *"observe x type A outcomes in a sample of n observations"*. The important difference between experiments for which the binomial density or hypergeometric density applies is that in the former case, the $n$ trials are *independent* and *identical* and would correspond to randomly **drawing objects *with* replacement** (meaning once an object is drawn, and the observation made, the object is placed back in the total collection of objects so that, at each draw, *all* of the original objects in the collection are equally probable to be chosen), whereas randomly *drawing objects without replacement* characterizes the experiment to which the hypergeometric density is applied.

**Example 4.9** ***Probability of No Defectives in Random Inspection of x items without Replacement***

Suppose a shipment of 1,000 computer memory modules contains 50 defectives. What is the probability of obtaining no defective modules in a random drawing, without replacement, of five modules from the shipment for inspection?

**Answer:** The appropriate hypergeometric density for this case has $M = 1{,}000$, $K = 50$, and $n = 5$. The probability assigned to the (elementary) event $x = 0$ is

$$f(0) = \frac{\left(\frac{50!}{0!50!}\right)\left(\frac{950!}{5!945!}\right)}{\left(\frac{1000!}{5!995!}\right)} = .7734. \qquad \square$$

Regarding the graph of the hypergeometric density, $f(x;\, M,K,n)$ increases as $x$ increases until a maximum value is reached, which occurs at the largest integer value of $x$ satisfying $x \leq ((n+1)(K+1))/(M+2)$ (this can be shown by examining the values of $x$ for which the density is increasing, i.e., $f(x;\, M,K,n)/f(x-1;M,K,n) \geq 1$ ). The value of the PDF declines thereafter. Some random variables to which the hypergeometric density has been applied include observations on the number of responses of a certain type in the context of auditing, quality control, and consumer or employee attitude surveys.

### 4.1.9 Family Name: Multivariate Hypergeometric

*Parameterization*:

$$(M, K_1, \ldots, K_M, n) \in \Omega =$$
$$\left\{(M, K_1, \ldots, K_M, n) : M = 1,2,\ldots; K_i = 0,1,\ldots,M \text{ for } i = 1,\ldots,m; \right.$$
$$\left. \sum_{i=1}^{m} K_i = M; n = 1,2,\ldots,M\right\}$$

*Density Definition*:

$$f(x_1,\ldots,x_m; M,n,K_1,\ldots,K_m) = \begin{cases} \dfrac{\prod_{j=1}^{m}\binom{K_j}{x_j}}{\binom{M}{n}} & \text{for } x_i \in \{0,1,2,\ldots,n\}\ \forall\, i, \sum_{i=1}^{n} x_i = n, \\ 0 & \text{otherwise} \end{cases}$$

*Moments*: (expressed for each $X_i$)

$$\mu(X_i) = \frac{nK_i}{M}, \sigma^2(X_i) = n\left(\frac{K_i}{M}\right)\left(\frac{M-K_i}{M}\right)\left(\frac{M-n}{M-1}\right),$$

$$\mu_3(X_i) = n\left(\frac{K_i}{M}\right)\left(\frac{MK_i}{M}\right)\left(\frac{M-2K_i}{M}\right)\left(\frac{M-n}{M-1}\right)\left(\frac{M-2n}{M-2}\right)$$

$$\left(\frac{\mu_4(X_i)}{\sigma^4(X_i)} - 3\right) = \frac{M^2(M-1)}{n(M-2)(M-3)(M-n)} \times \left[\left(\frac{M(M+1) - 6M(M-n)}{K_i(M-K_i)}\right) + \frac{3n(M-n)(M+6)}{M^2} - 6\right]$$

*MGF*: not useful

*Background and Application*: The multivariate hypergeometric density is a generalization of the hypergeometric density in the same sense as the multinomial density is a generalization of the binomial density. In particular, we are considering a case where we are interested in *more than two* different types of outcomes for each object chosen from the original collection of objects. Letting $K_i$, $i = 1,\ldots,m$, refer to the number of objects of type $i$ that are in the collection, $M = \sum_{i=1}^{m} K_i$ represent the total number of objects in the collection, $n$ represent the number of objects randomly drawn *without replacement* from the collection, and $x_i$ be the number of outcome of type $i$, $i = 1,\ldots,m$, an extension of the argument used to motivate the density definition in the hypergeometric case leads to the definition of the multivariate hypergeometric density function presented above.

Note that the marginal density of each $X_i$ is hypergeometric with parameters $(M, K_i, n)$. Furthermore, any subset of the random variables $(X_1, \ldots, X_m)$ also has a multivariate hypergeometric density, as the reader can verify.

**Example 4.10** ***Probability of Quality Assurance Survey Outcome***

As part of their quality assurance program, a large northwestern bank regularly interviews a randomly selected subset of the customers who transact business at one of its branches each week. Among other questions, the customers are asked to rank the overall service they received as being "excellent," "good," "average," "below average," or "poor." In one of the smaller rural branches, there were 100 customers who entered the branch during a particular week. If the bank randomly chooses five of the 100 customers to interview, and if the 100 customers were distributed across the rating categories as 50, 30, 10, 7, 3, respectively, what is the probability that the interviews will result in 2 "excellent," 2 "good," and 1 "average rating?"

**Answer**: Use the multivariate hypergeometric density with $M = 100, K_1 = 50$, $K_2 = 30, K_3 = 10, K_4 = 7, K_5 = 3$, and $n = 5$. Then

$$P(x_1 = 2, x_2 = 2, x_3 = 1, x_4 = x_5 = 0) = \frac{\binom{50}{2}\binom{30}{2}\binom{10}{1}\binom{7}{0}\binom{3}{0}}{\binom{100}{5}} = .0708. \quad \square$$

## 4.2 Parametric Families of Continuous Density Functions

### 4.2.1 Family Name: Uniform

*Parameterization*: $(a,b) \in \Omega = \{(a,b): -\infty < a < b < \infty\}$
*Density Definition*: $f(x;a,b) = (1/(b-a))I_{[a,b]}(x)$
*Moments*: $\mu = (a+b)/2, \sigma^2 = (b-a)^2/12, \mu_3 = 0, \left(\frac{\mu_4}{\sigma^4} - 3\right) = -\frac{6}{5}$
*MGF*:

$$M_X(t) = \begin{cases} \dfrac{e^{bt} - e^{at}}{(b-a)t} \text{ for } t \neq 0, \\ 1 \text{ for } t = 0 \end{cases}$$

*Background and Application*: The continuous uniform density is used to construct probability spaces for experiments having an uncountably infinite number of possible outcomes that are all equally likely in the interval $[a,b]$, for finite $b - a$. All interval subsets of $[a,b]$ of length $k$ are assigned equal probability, $k/(b-a)$. The continuous uniform density has important applications in the computer generation of random variable outcomes for a wide array of probability distributions. We will examine these types of applications in Chapter 6, a preview of which is provided in the next example.

**Example 4.11**
***Simulating the Demand for Teller Services***

An efficiency analyst wishes to simulate the daily demand for teller services in a moderately sized branch of a regional bank. The probability that a customer will require the services of a teller is known to be .30, whereas the customer will utilize the services of a cash machine, loan officer, or investment banker with probability .70. The type of service demanded is independent across customers. The efficiency analyst concludes that of $n$ customers entering the branch on a given day, the number that utilize a teller's services is the outcome of a random variable $X$ having the binomial density $f(x) = \binom{n}{x}(.30)^x(.70)^{n-x} I_{\{0,1,\ldots,n\}}(x)$. The analyst has a computer random-number generator that produces outcomes of a random variable $Y$ having the uniform density $h(y) = I_{(0,1)}(y)$. How can the analyst simulate daily outcomes of $X$?

**Answer**: Let $F(b) = \sum_{x \leq b} f(x)$ be the CDF of $X$, and define $n + 1$ intervals as $I_0 = [0, F(0))$ and $I_j = [F(j-1), F(j))$, for $j = 1,\ldots,n$. Note that the lengths of the intervals correspond to the respective probabilities assigned to the outcomes $x \in \{0,1,\ldots,n\}$ by the aforementioned binomial density. In particular, using the relationship between CDFs and PDFs, $P(x = 0) = F(0) = f(0)$ and $P(x = j) = F(j) - F(j-1) = f(j)$ for $j \in \{1,2,\ldots,n\}$. Then an outcome of $X$ can be simulated by first generating an outcome of $Y$ and then calculating $x = \sum_{j=0}^{n} j I_{I_j}(y)$. That these outcomes follow the appropriate binomial density can be motivated by the fact that $x = j$ *iff* $y \in I_j$ and $P_X(j) = P_Y(I_j) = \int_{y \in I_j} I_{(0,1)}(y)dy = f(j)$.

For a specific illustration, let $n = 5$. Then $I_0 = [0, .1681)$, $I_1 = [.1681, .5283)$, $I_2 = [.5283, .8370)$, $I_3 = [.8370, .9693)$, $I_4 = [.9693, .9977)$, and $I_5 = [.9977, 1)$.

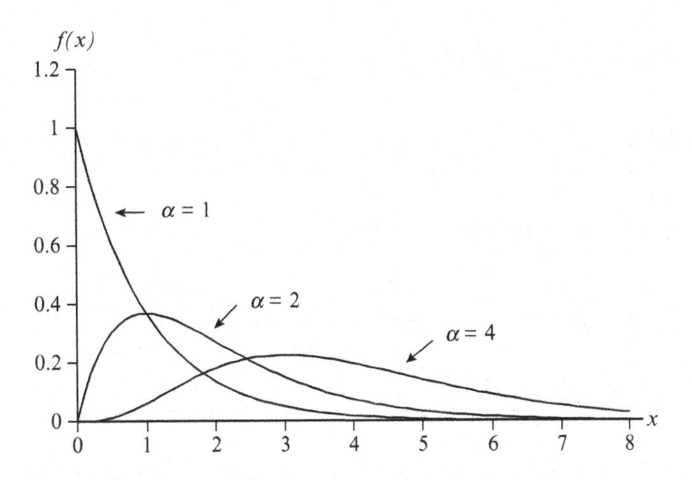


**Figure 4.5**
Gamma densities, case I, $\beta = 1$.

If the computer random-number generator were to generate an outcome $y = .6311$, then the analyst would simulate that two customers required teller's services on a day when five customers entered the branch (since $.6311 \in I_2$). □

### 4.2.2 Family Name: Gamma, and Exponential and Chi-Square Subfamilies

*Parameterization*: $(\alpha, \beta) \in \Omega = \{(\alpha, \beta): \alpha > 0, \beta > 0\}$
*Density Definition*: $f(x;\alpha,\beta) = (1/(\beta^{\alpha}\Gamma(\alpha)))\, x^{\alpha-1}\, e^{-x/\beta}\, I_{(0,\infty)}(x)$, where $\Gamma(\alpha) = \int_0^{\infty} y^{\alpha-1}e^{-y}dy$ is the **gamma function**, having the property that if $\alpha$ is a positive integer, $\Gamma(\alpha) = (\alpha - 1)!$, and if $\alpha = 1/2$, then $\Gamma(1/2) = \pi^{1/2}$. Also, for any real $\alpha > 0$, $\Gamma(\alpha + 1) = \alpha\Gamma(\alpha)$.
*Moments*: $\mu = \alpha\beta, \sigma^2 = \alpha\beta^2, \mu_3 = 2\alpha\beta^3, \left(\frac{\mu_4}{\sigma^4} - 3\right) = \frac{6}{\alpha}$
*MGF*: $M_X(t) = (1 - \beta t)^{-\alpha}$ for $t < \beta^{-1}$

*Background and Applications*: The gamma family of density functions is a versatile collection of density functions that can be used to model a wide range of experiments whose outcomes are coded as *nonnegative* real numbers and whose respective probabilities are to be assigned via a density function that is *skewed to the right*. An extensive variety of density shapes are possible by altering the parameters $\alpha$ and $\beta$, a few of which are illustrated in Figures 4.5 and 4.6. It can be shown that the gamma density is strictly decreasing when $\alpha \leq 1$. The density increases to a maximum of $((\alpha - 1)\, e^{-1})^{\alpha-1}/(\beta\, \Gamma(\alpha))$, at $x = (\alpha - 1)\beta$, for $\alpha > 1$.

While its wide variety of shapes makes the gamma family a candidate for constructing the probability space of many experiments with nonnegative outcomes, the gamma family has *specific* uses with regard to waiting times between occurrences of events based on the Poisson process. In particular, let $Y$ have a Poisson density with parameter $\lambda$, and let $y$ refer generically to the number of successes that occur in a period of time $t$, so that $\gamma = \lambda/t$ is the rate of success of the Poisson process. If $X$ measures the time that passes until the Poisson process produces the *first* success, then $X$ has a gamma density with

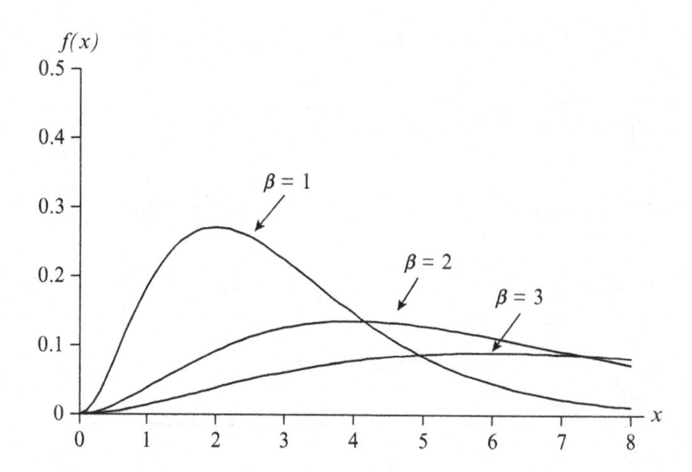


**Figure 4.6** Gamma densities, case II, $\alpha = 3$.

$\alpha = 1$ and $\beta = \gamma^{-1}$. If instead $X$ measures the time that passes until the Poisson process produces the $r^{\text{th}}$ success, then $X$ has a gamma density with $\alpha = r$ and $\beta = \gamma^{-1}$. Proofs of these propositions can be found in S.F. Arnold (1990), *Mathematical Statistics*, p. 166.

**Example 4.12** ***Probability of Time Between Machine Breakdowns***

Recall Example 4.8 regarding breakdowns of a milk bottling machine, which was assumed to be a Poisson process with rate $\gamma = \lambda/t = 2/4 = .5$. What is the probability that the machine will not breakdown in four weeks of use?

**Answer**: The event of interest is $\{x > 4\}$, that is, the first breakdown occurs after 4 weeks of operation. From the preceding discussion, we know that $X$ has a gamma density with $\alpha = 1$ and $\beta = \gamma^{-1} = 2$. Therefore, $P(x > 4) = \int_4^\infty (1/2)e^{-x/2} dx = -e^{-x/2} \big|_4^\infty = .1353$. Note this is precisely the same probability that the outcome 0 receives in the Poisson density of Example 4.8, which is as it should be, given that *no* breakdowns occurring in 4 weeks (0 outcome for the Poisson random variable) coincides with the *first* breakdown occurring after 4 weeks (an outcome greater than four for the gamma random variable). □

The gamma family of densities has an important **additivity property,** which we state in the following theorem.

**Theorem 4.2** ***Gamma Additivity***

*Let $X_1,\ldots,X_n$ be independent random variables with respective gamma densities Gamma$(\alpha_i, \beta)$, $i = 1,\ldots,n$. Then $Y = \sum_{i=1}^n X_i$ has the density Gamma $\left(\sum_{i=1}^n \alpha_i, \beta\right)$.*

**Proof** Since $M_{X_i}(t) = (1 - \beta t)^{-\alpha_i}$ for $t<\beta^{-1}$, $i = 1,\ldots,n$, and since the $X_i$'s are independent, Theorem 3.27 implies

$$M_Y(t) = \prod_{i=1}^n M_{X_i}(t) = \prod_{i=1}^n (1 - \beta t)^{-\alpha_i} = (1 - \beta t)^{-\sum_{i=1}^n \alpha_i} \text{ for } t<\beta^{-1}.$$

Thus, by the MGF uniqueness theorem, $Y$ has the density Gamma $\left(\sum_{i=1}^{n} \alpha_i, \beta\right)$. ■

Therefore, the sum of independent gamma random variables has a gamma distribution as long as the underlying gamma densities share the same $\beta$ parameter value.

Scaling a gamma random variable by a positive constant results in a random variable that also has a gamma distribution, as demonstrated in the next theorem.

**Theorem 4.3** ***Scaling of Gamma Random Variables***

*Let $X$ have a gamma density Gamma$(\alpha,\beta)$, and let $c > 0$. Then $Y = c\,X$ has a gamma density Gamma$(\alpha,\beta c)$.*

**Proof** Since $M_X(t) = (1 - \beta t)^{-\alpha}$ for $t < \beta^{-1}$, Theorem 3.27 implies that $M_Y(t) = M_{cX}(t) = M_X(ct) = (1 - \beta ct)^{-\alpha}$ for $t<(\beta c)^{-1}$, which by the MGF uniqueness theorem indicates that $Y$ has the gamma density Gamma$(\alpha,\beta c)$. ■

We will also have use for the following property of gamma PDFs.

**Theorem 4.4** ***Gamma Inverse Additivity***

*Let $Y = X_1 + X_2$, where $Y$ has density Gamma$(\alpha, \beta)$, $X_1$ has density Gamma$(\alpha_1, \beta)$, $\alpha > \alpha_1$, and $X_1$ and $X_2$ are independent. Then $X_2$ has density Gamma$(\alpha - \alpha_1, \beta)$.*

**Proof** Since $M_Y(t) = (1 - \beta t)^{-\alpha}$ and $M_{X_1}(t) = (1 - \beta t)^{-\alpha_1}$ for $t < \beta^{-1}$, Theorem 3.27 implies that

$$M_Y(t) = (1 - \beta t)^{-\alpha} = (1 - \beta t)^{-\alpha_1} M_{X_2}(t) = M_{X_1}(t)\, M_{X_2}(t),$$

which in turn implies

$$M_{X_2}(t) = (1 - \beta t)^{-\alpha}/(1 - \beta t)^{-\alpha_1} = (1 - \beta t)^{-(\alpha-\alpha_1)} \text{for } t<\beta^{-1}.$$

It follows from the MGF uniqueness theorem that $X_2$ has the gamma density Gamma$(\alpha - \alpha_1, \beta)$. ■

Its wide variety of density shapes has resulted in the gamma family's being applied to a myriad of nonnegative-valued random variables suspected of having a right-skewed PDF. Some specific applications include the waiting times between customer arrivals or machine breakdowns or telephone calls, the breaking strength of manufactured construction materials, the operating lives of electronic equipment and other objects, and the length of time required to service a customer at a store. We now examine two important subfamilies of the gamma family of densities that are defined by special choices of the parameters $\alpha$ and $\beta$.

### 4.2.3 Gamma Subfamily Name: Exponential

*Parameterization*: $\theta \in \Omega = \{\theta: \theta > 0\}$

*Density Definition*: The gamma density, with $\alpha = 1$, and $\beta = \theta$.

$$f(x;\theta) = \frac{1}{\theta} e^{-x/\theta} I_{(0,\infty)}(x)$$

*Moments*: $\mu = \theta, \sigma^2 = \theta^2, \mu_3 = 2\theta^3, \left(\frac{\mu_4}{\sigma^4} - 3\right) = 6$

*MGF*: $M_X(t) = (1 - \theta t)^{-1}$ for $t < \theta^{-1}$

*Background and Application*: The exponential density is used to construct a probability space for experiments that have a real-valued sample space given by the *nonnegative* subset of the real line, $[0,\infty)$, and in which interval events of fixed length $d > 0$ of the form $[t, t + d]$ are to be assigned probability that monotonically decreases as $t$ increases. A specific application concerns the experiment of observing the time that passes until a Poisson process with rate $\gamma = \lambda/t$ produces the *first* success, in which case the exponential density with $\theta = \gamma^{-1}$ is appropriate (recall our previous discussion regarding the relationship between the gamma density and the Poisson process).

A prominent application of the exponential density is in representing the operating lives until failure of various objects. In this regard, the following property of the exponential density is of notable importance.

**Theorem 4.5** ***Memoryless Property of Exponential Density*** *If X has an exponential density, then* $P(x > s + t \mid x > s) = P(x > t)$ $\forall$ *t and* $s > 0$.

**Proof**

$$P(x>s+t|x>s) = \frac{P(x>s+t)}{P(x>s)} = \frac{\int_{s+t}^{\infty} \frac{1}{\theta} e^{-x/\theta}\, dx}{\int_{s}^{\infty} \frac{1}{\theta} e^{-x/\theta}\, dx}$$

$$= \frac{e^{-(s+t)/\theta}}{e^{-s/\theta}} = e^{-t/\theta} = P(x>t)$$

■

Interpreted in the context of operating life, the **memoryless property** implies that *given* the object has already functioned for $s$ units of time without failing (which is the meaning of the conditioning event $x > s$), the probability that it will function for at least an *additional* $t$ units of time (which is the meaning of the event $x > s + t$) is the same as the unconditional probability that it would function for at least $t$ units of time (the meaning of the event $x > t$). In effect, the object is "**as good as new**" after functioning $s$ units of time, since the probability of functioning for at least another $t$ units of time is the same as if it had not previously functioned at all (i.e., as if it were "new").[5]

While the memoryless property certainly is not applicable to the lifetimes of all objects, the assumption is appropriate in the modeling of the lifetimes of certain electronic components, fuses, jeweled watch bearings, and other objects

[5]The exponential density is the *only* density for continuous nonnegative-valued random variables that has the memoryless property. See V.K. Rohatgi (1976) *An Introduction to Probability Theory and Mathematical Statistics.* New York: John Wiley, p. 209.

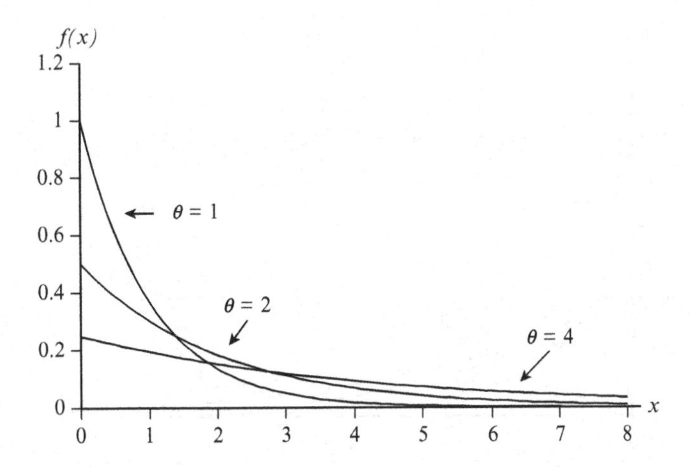


**Figure 4.7** Exponential densities.

that are not subject to significant wear and are essentially "as good as new" if they are still functioning. Furthermore, the assumption is appropriate for objects, such as machinery, that receive periodic maintenance that reconditions the object to essentially new status.

If the memoryless assumption is not appropriate, the more versatile gamma family of densities can be considered for constructing the probability space. It can be shown that members of the gamma family, other than the exponential family, exhibit "**wear-out**" effects in the sense that $P(x > s + t|x > s)$ *declines* as $s$ increases for $t > 0$ and $\alpha > 1$, that is, the probability that the object functions for at least $t$ units of time beyond the $s$ units of time for which it has already functioned declines as the value of $s$ increases. For $\alpha < 1$, the conditional probability actually increases as s increases – this is referred to in the literature as the "**work-hardening**" effect. The graphs of some exponential densities are displayed in Figure 4.7.

**Example 4.13** ***Probability of Operating Lives for "Good as New" Electronics***

The lifetime of a fuse your company manufactures is the outcome of a random variable with mean = 500 hours. The fuse is "as good as new" while functioning. What is the probability that the fuse functions for at least 1,000 hours?

**Answer:** Because of the "good as new" property, the exponential density is appropriate with $\theta = 500$. Then,

$$P(x \geq 1000) = \int_{1000}^{\infty} \frac{1}{500} e^{-x/500}\, I_{(0,\infty)}(x)dx = -e^{-x/500}\,|_{1000}^{\infty} = .1353. \qquad \square$$

### 4.2.4 Gamma Subfamily Name: Chi-Square

*Parameterization:* $v \in \Omega = \{v\colon v \text{ is a positive integer}\}$
*Density Definition:* The gamma density, with $\alpha = v/2$ and $\beta = 2$.

$$f(x; v) = \frac{1}{2^{v/2}\,\Gamma(v/2)}\, x^{(v/2)-1}\, e^{-x/2}\, I_{(0,\infty)}(x)$$

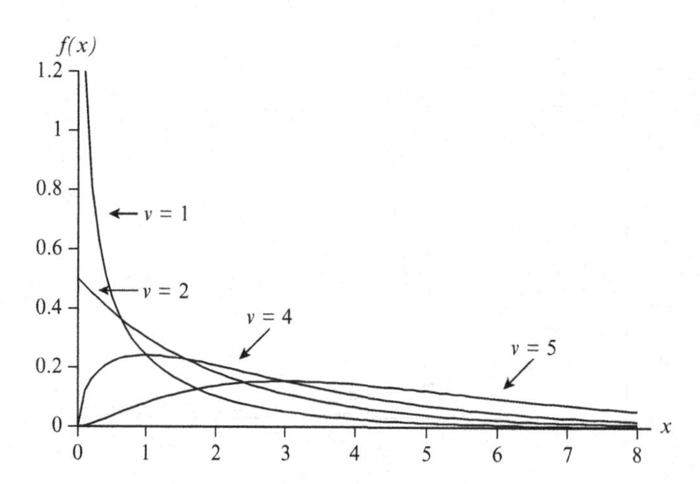


**Figure 4.8**
$\chi^2$ densities.

*Moments:* $\mu = v, \sigma^2 = 2v, \mu_3 = 8v, \left(\frac{\mu_4}{\sigma^4} - 3\right) = \frac{12}{v}$

*MGF:* $M_X(t) = (1 - 2\,t)^{-v/2}$ for $t < \frac{1}{2}$

*Background and Application*: The parameter $v$ of the chi-square density is called the **degrees of freedom**. The reason for this term will be clarified later in the chapter where we will show that the sum of the squares of $v$ independent random variables, each having a density called the standard normal (to be discussed in Section 4.3), will have a chi-square density with $v$ degrees of freedom. The chi-square density with $v$ degrees of freedom is often indicated by the notation $\chi^2_v$ or $\chi^2(v)$. We will utilize the former. The relationship between the chi-square and normal density makes the chi-square density especially important in applications concerning hypothesis testing and confidence interval estimation, which is its primary application context, as will be seen in later chapters. Note for $v = 2$, the $\chi^2$ density is equivalent to the exponential density with $\theta = 2$. Also, $\chi^2_v$ is a valid PDF even for noninteger values of $v > 0$, in which case the PDF $\chi^2_v$ is referred to as the **nonintegral chi-square density**. Our use of $\chi^2_v$ will be restricted to integer-valued $v$. Some chi-square densities are graphed in Figure 4.8.

There are two important properties of the $\chi^2$ density relating to sums of random variables that we note in the following two corollaries.

**Corollary 4.1** ***Chi-Square Additivity*** *Let $X_1,\ldots,X_k$ be independent random variables having chi-square densities with $v_1,\ldots,v_k$ degrees of freedom, respectively. Then $Y = \sum_{i=1}^k X_i$ has a chi-square density with degrees of freedom $v = \sum_{i=1}^k v_i$.*

**Proof** This follows directly from Theorem 4.2 with $\beta = 2$. ■

Thus the sum of independent chi-square random variables also has a chi-square density.

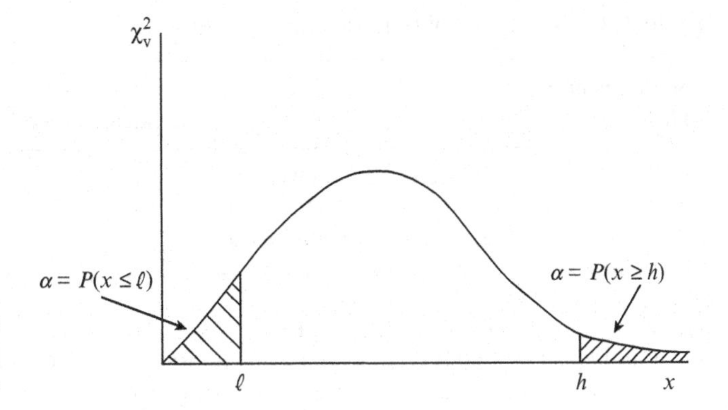


**Figure 4.9** Upper and lower $\alpha$-level tails of a $\chi^2$ density

**Corollary 4.2** ***Chi-Square Inverse Additivity*** *Let $X_1$ and $X_2$ be independent random variables, where $X_1$ has a chi-square density with $v_1$ degrees of freedom, and $Y = X_1 + X_2$ has a chi-square density with $v > v_1$ degrees of freedom. Then $X_2$ has a chi-square density with $v - v_1$ degrees of freedom.*

**Proof** This follows directly from Theorem 4.4 with $\beta = 2$, $\alpha = v/2$, and $\alpha_1 = v_1/2$. ■

Later in our study of hypothesis testing it will be useful to know the values of upper- and lower-bounds $h$ and $\ell$ for which $P(x \ge h) = \alpha$ and $P(x \le \ell) = \alpha$ are true, where $X$ has a $\chi^2$ density with $v$ degrees of freedom and $\alpha \in (0,1)$. These are, of course, **quantiles** of the chi-square distribution and are illustrated, diagrammatically, by the upper and lower tails of the $\chi^2_v$ density in Figure 4.9. Such events and their probabilities are identified in tables of the $\chi^2$ density available from many published sources, including the Appendix of this text. Typically, the events identified relate to quantiles $p = .01$, $.05$, and $.10$ in the lower tail and $p = .90$, $.95$, and $.99$ in the upper tail. Of course, the appropriate values of $h$ and $\ell$ can be found by solving the following equations given any $\alpha_\ell$ and $\alpha_h$ levels of interest $\in (0,1)$:

$$\int_h^\infty \frac{1}{2^{v/2}\Gamma\left(\frac{v}{2}\right)} x^{(v/2)-1}\, e^{-x/2}\, dx = \alpha_h \qquad \int_0^\ell \frac{1}{2^{v/2}\Gamma\left(\frac{v}{2}\right)} x^{(v/2)-1}\, e^{-x/2}\, dx = \alpha_\ell$$

Most modern statistical computer software packages can be used to identify the appropriate $h$ or $\ell$, such as STATA, SAS, MATLAB, or the GAUSS programming languages.

**Example 4.14** ***Using Chi-Square Table*** Let $X$ have a $\chi^2$ density with 10 degrees of freedom. Find the values of $h$ and $\ell$ for which $P(x \ge h) = P(x \le \ell) = .05$.

**Answer**: In the chi-square table for the row corresponding to 10 degrees of freedom, the values of $h$ and $\ell$ associated with the upper and lower .05 tails of the $\chi^2$ density are, respectively, 18.307 and 3.940. □

**Table 4.3** Summary of beta density shapes

| Conditions on $\alpha$ and $\beta$ | Behavior of $f(x)$ |
|---|---|
| $\alpha < \beta$ | Skewed to the right, $\mu_3 > 0$ |
| $\alpha > \beta$ | Skewed to the left, $\mu_3 < 0$ |
| $\alpha = \beta$ | Symmetric about $\mu = 1/2$ |
| $\alpha > 1$ and $\beta > 1$ | Maximum value when $x = (\alpha-1)/(\alpha + \beta-2)$ and $f(x) \to 0$ if $x \to 1$ or $x \to 0$ |
| $\alpha < 1$ | $f(x) \to \infty$ as $x \to 0$ |
| $\beta < 1$ | $f(x) \to \infty$ as $x \to 1$ |
| $\alpha < 1$ and $\beta < 1$ | $f(x)$ is U-shaped, having minimum value when $x = (\alpha-1)/(\alpha + \beta-2)$, and $f(x) \to \infty$ when $x \to -\infty$ or $\infty$ |
| $(\alpha-1)(\beta-1) < 0$ | J-shaped |
| $\alpha = \beta = 1$ | Uniform on (0,1) |

### 4.2.5 Family Name: Beta

*Parameterization*: $(\alpha, \beta) \in \Omega = \{(\alpha,\beta): \alpha > 0, \beta > 0\}$
*Density Definition*:

$$f(x;\alpha,\beta) = \frac{1}{B(\alpha,\beta)}\, x^{\alpha-1}(1-x)^{\beta-1}\, I_{(0,1)}(x)$$

where $B(\alpha,\beta) = \int_0^1 x^{\alpha-1}(1-x)^{\beta-1}\,dx$ is called the **beta function**. Some useful properties of the beta function include the fact that $B(\alpha,\beta) = B(\beta,\alpha)$ and $B(\alpha,\beta) = \frac{\Gamma(\alpha)\Gamma(\beta)}{\Gamma(\alpha+\beta)}$ so that the beta function can be evaluated in terms of the gamma function.
*Moments*:

$$\mu = \alpha/(\alpha+\beta), \qquad \sigma^2 = \alpha\beta/[(\alpha+\beta+1)(\alpha+\beta)^2],$$

$$\mu_3 = 2(\beta-\alpha)(\alpha\beta)/[(\alpha+\beta+2)(\alpha+\beta+1)(\alpha+\beta)^3],$$

$$\left(\frac{\mu_4}{\sigma^4}-3\right) = \frac{6\left[(\alpha+\beta)^2(\alpha+\beta+1)-\alpha\beta(\alpha+\beta+2)\right]}{\alpha\beta(\alpha+\beta+2)(\alpha+\beta+3)}$$

*MGF*: $M_X(t) = \sum_{r=0}^{\infty}(B(r+\alpha,\beta)/B(\alpha,\beta))(t^r/r!)$

*Background and Application*: The beta density is a very versatile density (i.e., it can assume a large variety of shapes) for constructing probability spaces for experiments having a continuous real-valued sample space given by the interval [0,1]. Table 4.3 provides a quick reference regarding the numerous shape characteristics of the beta density. The versatility of this density family makes it useful for representing PDFs for random variables associated with virtually any experiment whose outcomes constitute a continuum between 0 and 1. The density has obvious applications in modeling experiments whose outcomes are in the form of proportions, such as the proportion of time a certain machine is in a state of being repaired, the proportion of chemical impurities in a liquid

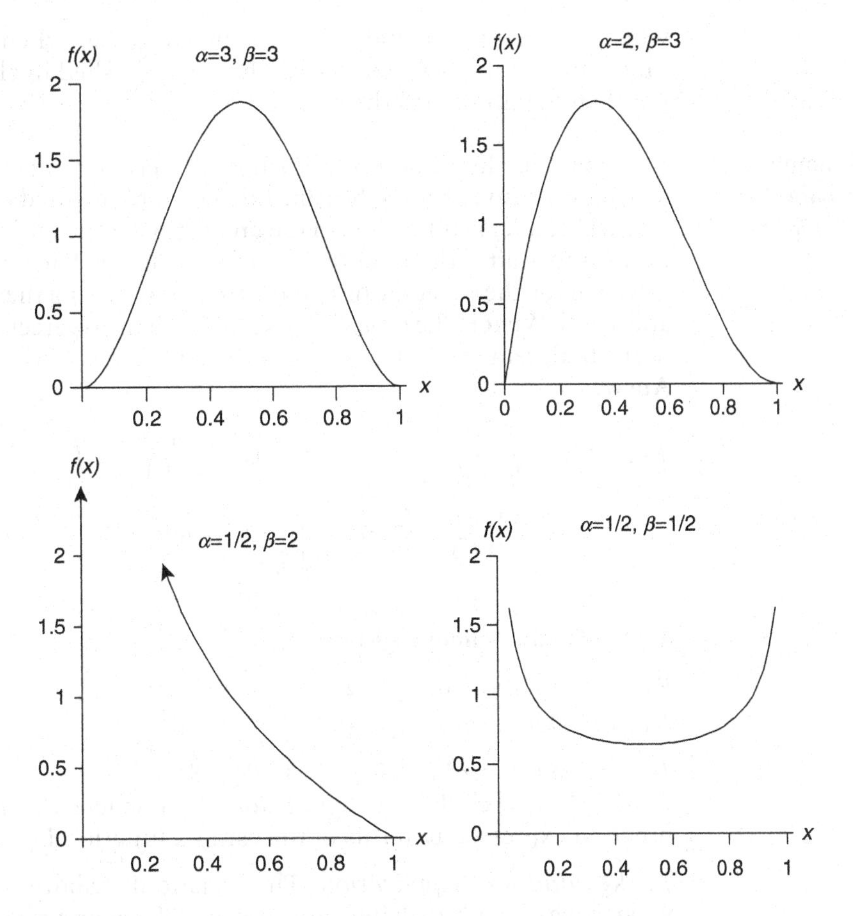


**Figure 4.10**
Beta densities.

product, or the proportion of respondents to a survey (a continuous approximation in the latter case). In Chapter 6 we will also see that the beta density has an important application in assigning probabilities to events involving so-called *order statistics*. Figure 4.10 illustrates some of the beta density shapes that are possible.

Integration of the density cannot be accomplished in closed form for noninteger values of the parameters $\alpha$ and $\beta$. Procedures are available in most modern statistical software packages for integrating the beta density. Integrals of the beta density have also been extensively tabled by Pearson.[6] When $\alpha$ and $\beta$ are integers, there is a relationship between integrals of the beta density and the binomial density that can be useful:

$$F(c) = \int_0^c \frac{1}{B(\alpha, \beta)} x^{\alpha-1} (1-x)^{\beta-1}\, dx = \sum_{i=\alpha}^{n} \binom{n}{i} c^i (1-c)^{n-i}$$

[6] K. Pearson, (1956), *Tables of the Incomplete Beta Function*, New York: Cambridge Univ. Press.

where $n = \alpha + \beta - 1$ and $c \in [0,1]$. In any case, when $\alpha$ and $\beta$ are integers, integration of the beta density can be accomplished in closed form. The following example illustrates the point.

**Example 4.15** ***Probabilities of Proportion Outcomes*** A wholesale distributor of heating oil has a storage tank that holds the distributor's inventory of heating oil. The tank is filled every Monday morning. The wholesaler is interested in the proportion of the tank's capacity that remains in inventory after the weekly sales of heating oil. The remaining proportion can be viewed as the outcome of a random variable having the beta density with $\alpha = 4$ and $\beta = 3$. What is the probability that less than 20 percent of the storage capacity of the tank remains in inventory at the end of any work week?

**Answer**:

$$P(x<.20) = \int_0^{.2} \frac{1}{B(4,3)} x^3 (1-x)^2 \, dx = \frac{\Gamma(7)}{\Gamma(4)\Gamma(3)} \int_0^{.2} (x^3 - 2x^4 + x^5) dx$$

$$= \frac{6!}{3!2!} \left[ \frac{x^4}{4} - \frac{2x^5}{5} + \frac{x^6}{6} \right] \Big|_0^{.2} = .01696. \qquad \square$$

### 4.2.6 Family Name: Logistic

*Parameterization*: $(u,s) \in \Omega = \{(u,s): -\infty < u < \infty, s > 0\}$

*Density Definition*: $f(x; u, s) = \frac{e^{-(x-u)/s}}{s(1+e^{-(x-u)/s})^2}$

*Moments*: $\mu = u, \sigma^2 = \frac{\pi^2 s^2}{3}, \mu_3 = 0\ ,\ \left(\frac{\mu_4}{\sigma^4} - 3\right) = \frac{6}{5}$

*MGF*: $M_X(t) = e^{ut} B(1 - st, 1 + st)$ *for* $|st|<1$ where $B(a,b) = \frac{\Gamma(a)\Gamma(b)}{\Gamma(a+b)}$ is the beta function expressed in terms of the gamma function $\Gamma(\cdot)$.

*Background and Application*: The logistic distribution has been applied in a wide range of fields to define probably models for growth and diffusion processes, as well as to model decision making processes where decision makers are faced with discrete alternative choices exhibiting differing characteristics. Some specific applications include how various populations of species grow in competition with each other, for characterizing the spread of epidemics, describing how learning evolves, modeling how technologies diffuse and are substituted for one another, the diffusion process relating to new product sales, and diffusion and substitution between various energy sources.

In economic applications in particular, the distribution has appeared prominently in describing the probabilities associated with dichotomous or binary choices and has been widely used in logistic regression models of the choices of decision makers in a wide variety of problems involving consumers', producers', and policy makers' decisions. Especially for this purpose, the logistic distribution has often been chosen in place of the normal distribution for defining probability models of discrete choices processes because of its tractability (e.g., closed form integrable), but also because of its similarity (symmetric, bell-shaped) with the normal distribution. In fact, the two distributions are very similar in shape, with the logistic distribution exhibiting slightly fatter tails than does the normal distribution.

Regarding the graphs of the logistic distribution, the reader can refer to the normal distribution that will be discussed in the section ahead for similar behavior in terms of shapes, and the changes that occur as the mean and scale parameters are altered. The following example provides a simple illustration of how the logistic distribution arises within a simple binary choice decision context.

**Example 4.16** ***Binary Choice and the Logistic Distribution***

Let $U_i = \mathbf{x}_i'\boldsymbol{\beta} + \varepsilon_i$ represent the utility (or net benefit) that person $i$, with a vector of personal characteristics $\mathbf{x}_i$, obtains from taking an action (as opposed to not taking the action). The person takes the action if $U_i > 0$. The unobserved residual term, $\varepsilon_i$, is assumed to have a *standard* logistic distribution[7] with $s = 1$ and $u = 0$.[8]

Then the decision process is represented as follows, where $y_i = 1$ denotes that the individual takes the action, while $y_i = 0$ indicates that the individual will not take the action:

$$y_i = \begin{Bmatrix} 1 \\ 0 \end{Bmatrix} \text{ if } U_i \begin{Bmatrix} > \\ \leq \end{Bmatrix} 0$$

where $P(y_i = 1) = P(\varepsilon_i > -\mathbf{x}_i'\boldsymbol{\beta}) = 1 - F_{\varepsilon_i}(-\mathbf{x}_i'\boldsymbol{\beta}) = \frac{1}{1+\exp(-\mathbf{x}_i'\boldsymbol{\beta})}$. □

## 4.3 The Normal Family of Densities

The **normal family** of densities is the most extensively used continuous density in applied statistics and it is for that reason that we devote an entire section to it. We will begin our discussion of the Normal family by examining the univariate case, and then we will proceed to the multivariate normal density function.

### 4.3.1 Family Name: Univariate Normal

*Parameterization*: $(a, b) \in \Omega = \{(a, b): a \in (-\infty, \infty), b > 0\}$

*Density Definition*: $f(x; a, b) = \frac{1}{\sqrt{2\pi}b}\exp\left[-\frac{1}{2}\left(\frac{x-a}{b}\right)^2\right]$.

*Moments*: $\mu = a, \sigma^2 = b^2, \mu_3 = 0, \left(\frac{\mu_4}{\sigma^4} - 3\right) = 0$

*MGF*: $M_X(t) = \exp[at + (1/2)b^2t^2]$

---

[7] There are substantive conceptual considerations underlying binary choice situations that naturally lead to the use of the logistic distribution in this context. See Train, K. (1986). Qualitative Choice Analysis: Theory, Econometrics, and an Application to Automobile Demand, MIT Press, Chapter 2.

[8] Note that this "normalization" of the logistic distribution in the current model context can be done without loss of generality since the probabilities of the decision events are unaffected thereby. This has to do with the concept of "parameter identification" in probability models, which we will discuss in later chapters when we examine statistical inference issues.

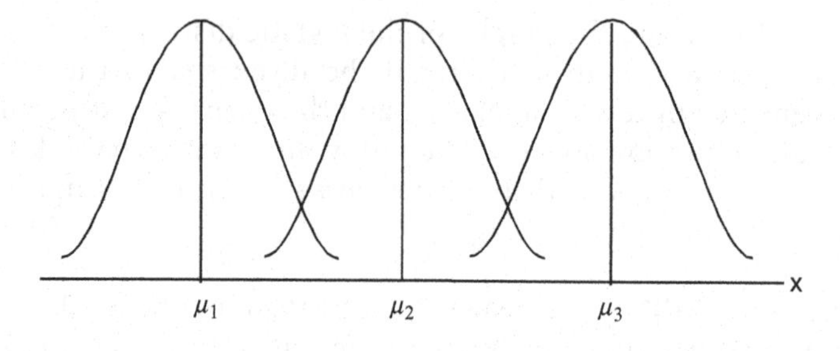


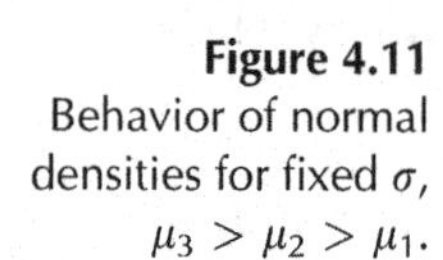

**Figure 4.11**
Behavior of normal densities for fixed $\sigma$, $\mu_3 > \mu_2 > \mu_1$.

**Figure 4.12**
Behavior of normal densities for fixed $\mu$, $\sigma_2 > \sigma_1$.

*Background and Applications*: The univariate normal family of densities is indexed by two parameters $a$ and $b$. Given the special relationship between these parameters and the mean and variance of the density, the normal density function is usually represented alternatively as

$$f(x;\mu,\sigma) = \frac{1}{\sqrt{2\pi}\sigma}\exp\left[-\frac{1}{2}\left(\frac{x-\mu}{\sigma}\right)^2\right],$$

and the moment-generating function of the normal density is given by

$$M_X(t) = \exp\left[\mu t + \frac{1}{2}\sigma^2 t^2\right]$$

The abbreviation $N(z; \mu, \sigma^2)$ is often used to signify that the random variable $Z$ has a normal distribution with mean $\mu$ and variance $\sigma^2$. When the random variable being referred to is not ambiguous, the abbreviation is often shortened to $N(\mu, \sigma^2)$. Once the mean and the variance of a normal distribution are numerically specified, then a unique member of the family of density functions is identified. The specific member of the family for which $\mu = 0$ and $\sigma^2 = 1$ is very important in applied statistics and is given the special name of the **standard normal density** or the **standard normal distribution**.

The normal density is symmetric about its mean, $\mu$, has points of inflection at $\mu - \sigma$ and $\mu + \sigma$, and has a characteristic bell shape. The bell becomes more spread out as the variance increases. We illustrate the general characteristics of a normal density in Figures 4.11 and 4.12.

A very useful property of any normally distributed random variable is that it can be easily transformed into a random variable having the *standard normal density.*

**Theorem 4.6** *Let X have the density* $N(x;\,\mu,\,\sigma^2)$. *Then* $Z = (X - \mu)/\sigma$ *has the standard normal density* $N(z;\,0,1)$.

**Proof** The MGF of $Z$ is defined by

$$M_Z(t) = E(e^{tZ}) = E\left(\exp\left(t\frac{(X-\mu)}{\sigma}\right)\right) = e^{-t\mu/\sigma} M_X\left(\frac{t}{\sigma}\right) \text{(Theorem 3.27)}$$

$$= e^{-t_*\mu} M_X(t_*) \quad \left(\text{substitute } t_* = \frac{t}{\sigma}\right)$$

and since $M_X(t_*) = \exp(\mu t_* + (1/2)\,\sigma^2\,t_*^2)$, it follows that

$$M_Z(t) = \exp(-t_*\mu)\exp(\mu t_* + (1/2)\,\sigma^2\,t_*^2) = \exp(t^2/2).$$

By the MGF uniqueness theorem, since the MGF of Z is that of a standard normal density, $Z$ has a $N(z;\,0,1)$ density. (Note that $\exp(t^2/2) = \exp(\mu t + (1/2)\sigma^2 t^2)$ with $\mu = 0$ and $\sigma^2 = 1$). ■

In applications, Theorem 4.6 implies that the probability of an event $A$, $P_X(A)$, for a random variable $X$ having a normal density $N(x;\,\mu,\sigma)$ is equal to the probability $P_Z(B)$ of the equivalent event $B = \{z\colon z = (x - \mu)/\sigma,\ x \in A\}$, for a *standard normal random variable* $Z$. This accounts for the prevalence of published tables of the standard normal CDF since, in principle, the standard normal distribution is sufficient to assign probabilities to *all* events involving normally distributed random variables.

The operation of subtracting the mean from a random variable, and then dividing by its standard deviation, is referred to as **standardizing a random variable**.

**Definition 4.2** ***Standardizing a Random Variable***

> A random variable is *standardized* by subtracting its mean and then dividing the result by its standard deviation, as $Z = \frac{X - u_X}{\sigma_X}$.

The outcome value of a *standardized* random variable can be interpreted as a measure of the distance of the outcome from its mean measured in standard deviation units, (e.g., $z = 3$ would mean the outcome of $Z$ was three standard deviations from its mean). Thus, if a random variable having a normal density is *standardized*, the standardized random variable has a standard normal density. The random variable having the density $N(0,1)$ is often referred to as a **standard normal random variable**.

**Example 4.17** ***Using Standard Normal Distribution to Assign Probabilities*** The miles per gallon (mpg) achieved by a new pickup truck produced by a Detroit manufacturer can be viewed as a random variable having a normal density with a mean of 17 mpg and a standard deviation of .5 mpg. What is the probability that a new pickup will achieve between 16 and 18 mpg?

**Answer**: Let $X$ have the density $N(x; 17, .25)$. Then,

$$P(16 \le x \le 18) = P\left(\frac{16-17}{.5} \le \frac{x-17}{.5} \le \frac{18-17}{.5}\right) = P(-2 \le z \le 2)$$
$$= F(2) - F(-2) = .9772 - .0228 = .9544,$$

where $F(\cdot)$ is the CDF of the standard normal random variable. □

An important observation should be made concerning the application of the normal density function in Example 4.17. Note that miles per gallon *cannot* be negative, and yet the normal density assigns nonzero probability to the event that $x < 0$, that is, $P(x < 0) = \int_{-\infty}^{0} N(x; \mu, \sigma^2)\, dx > 0 \;\forall\, \mu$ and $\sigma^2 > 0$. This is, in fact, illustrative of a situation that arises frequently in practice where the normal distribution is used in the construction of a probability space for an experiment whose outcomes assume only nonnegative values. The empirical justification for this apparent misuse of the normal density is that $P(x < 0)$ should be negligible in these cases, given the relevant values of $\mu$ and $\sigma^2$, in which case the anomaly of $P(x < 0) > 0$ can be ignored for all practical purposes. In Example 4.17, $P(x < 0) < 1 \times 10^{-10}$.

There is a relationship between standard normal random variables and the $\chi^2$ density that is very important for developing hypothesis-testing procedures, discussed in later chapters. We develop this relationship in the following two theorems.

**Theorem 4.7** ***Relationship Between Standard Normal and Chi-Square***

*If $X$ has the density $N(0,1)$, then $Y = X^2$ has a $\chi^2$ density with 1 degree of freedom.*

**Proof** The MGF of $Y$ is defined as

$$M_Y(t) = \mathrm{E}(e^{Yt}) = \mathrm{E}(\exp(X^2 t)) = \int_{-\infty}^{\infty} \exp(x^2 t)\frac{1}{\sqrt{2\pi}}\exp\left(-\frac{1}{2}x^2\right)dx$$
$$= \int_{-\infty}^{\infty}\frac{1}{\sqrt{2\pi}}\exp\left(-\frac{1}{2}x^2(1-2t)\right)dx$$
$$= (1-2t)^{-1/2}\int_{-\infty}^{\infty}\frac{1}{\sqrt{2\pi}(1-2t)^{-1/2}}\exp\left(-\frac{1}{2}\left(\frac{x}{(1-2t)^{-1/2}}\right)^2\right)dx$$
$$= (1-2t)^{-1/2}\underbrace{\int_{-\infty}^{\infty} N\left(x; 0, (1-2t)^{-1}\right)dx}_{1} = (1-2t)^{-1/2} \text{for } t < \frac{1}{2}.$$

Therefore, by the MGF uniqueness theorem, $Y = X^2$ has a $\chi^2$ density with 1 degree of freedom. ■

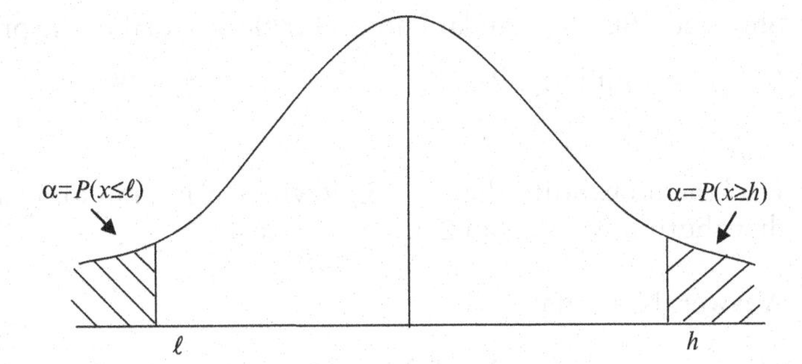


**Figure 4.13** Upper and lower α-level tails of $N(\mu,\sigma^2)$.

**Theorem 4.8 *Sums of Squares of Independent Standard Normal Random Variables is Chi-Square*** *Let $(X_1,\ldots,X_n)$ be independent random variables, each having the density N (0,1). Then $Y = \sum_{i=1}^{n} X_i^2$ has a $\chi^2$ density with n degrees of freedom.*

**Proof** The random variables $X_i^2$, $i = 1,\ldots,n$, are independent by Theorem 2.9. Then, from Theorems 3.27 and 4.7,

$$M_Y(t) = \prod_{i=1}^{n} M_{X_i^2}(t) = \prod_{i=1}^{n} (1-2t)^{-1/2} = (1-2t)^{-n/2} \text{ for } t<\frac{1}{2},$$

*which by the MGF uniqueness theorem implies that Y has a $\chi^2$ density with n degrees of freedom.* ■

In words, Theorem 4.8 is often stated as ***the sum of squares of n independent standard normal random variables has a $\chi^2$ density with n degrees of freedom.*** We can now motivate why the parameter $v$ in a $\chi^2$ density is labeled **degrees of freedom**. In particular, $v$ represents the number, or *degree*, of *freely* varying (i.e., independent) standard normal random variables whose sum of squares represents a $\chi^2_v$-distributed random variable.

It will be useful in our later study of hypothesis testing to know the values of $h$ and $\ell$ (i.e., know the appropriate quantiles) for which the following statements are true:

$$P(x \geq h) = \alpha = 1 - F(h) \text{ and } P(x \leq \ell) = \alpha = F(\ell)$$

where $F(\cdot)$ refers to the CDF of a random variable having a $N(\mu,\sigma^2)$ density and $\alpha \in (0,1)$. We diagrammatically examine the upper and lower tails of the normal family of densities in Figure 4.13. The probabilities of such events are extensively tabled for the case of the *standard* normal density, where typically the values of the cumulative standard normal density, $F(c)$, are given for numerous choices of $c$ (see Appendix). The probabilities of these events for arbitrary normal distributions can be obtained from knowledge of the standard normal distribution via standardization. For the event $x \leq \ell$, $P(x \leq \ell) = P(\frac{x-\mu}{\sigma} \leq \frac{\ell-\mu}{\sigma}) = P(z \leq \ell_*)$, where $\ell_* = (\ell - \mu)/\sigma$, and $z = (x - \mu)/\sigma$ can be interpreted as the outcome of a *standard* normal random variable using Theorem 4.6. Thus, the value of $P(x \leq \ell)$ can be obtained from the standard normal table as equal to the probability that $z \leq \ell_* = (\ell - \mu)/\sigma$. Similarly, $P(x \geq h) = P\left(\frac{x-\mu}{\sigma} \geq \frac{h-\mu}{\sigma}\right) = P(z \geq h_*)$ where $h_* = \dfrac{h-\mu}{\sigma}$.

**Example 4.18** *Using Standard Normal Table*

Find the probability that $x \geq 5.29$ where $X$ is a random variable having a normal distribution with mean 2 and variance 4.

**Answer**: Note that

$$P(x \geq 5.29) = P\left(\frac{x-2}{2} \geq \frac{5.29-2}{2}\right) = P(z \geq 1.645) = .05$$

which was found from the table of the standard normal CDF. □

Modern statistical software packages contain procedures that will numerically integrate the standard normal density over any interval $(-\infty, c]$ chosen by the user.

The normal density is often used in modeling experiments for which a symmetric, bell-shaped probability density is suspected. The normal density has been found to be a useful representation of event probabilities for literally thousands of real-world experiments, which is attributable in large part to the fact that, under general conditions, certain useful functions of a collection of independent random variables, such as sums and averages, have *approximately* normal densities when the collection is large enough, even if the original random variables in the collection do *not* have normal densities. In fact, the normal density was originally discovered by A. de Moivre (1667–1745) as an *approximation* to the binomial density (recall that the binomial density applies to the *sum* of independent Bernoulli random variables). These results are based on *central limit theorems* which will be examined in Chapter 5.

Examples of applications include fill weights of food and beverage containers, employee aptitude test scores, labor hours required to construct prefabricated homes, numbers of insurance claims filed during a time period, and weight gains of meat animals in a feedlot. In Chapter 10 we will examine statistical tests that can be used to assess the validity of the normal PDF for characterizing event probabilities in a given real-world experiment.

### 4.3.2 Family Name: Multivariate Normal Density

*Parameterization*:

$$\mathbf{a} = (a_1, \ldots, a_n)' \text{ and } \mathbf{B} = \begin{bmatrix} b_{11} & \cdots & b_{1n} \\ \vdots & \ddots & \vdots \\ b_{n1} & \cdots & b_{nn} \end{bmatrix}$$

$(\mathbf{a},\mathbf{B}) \in \Omega = \{(\mathbf{a},\mathbf{B}): \mathbf{a} \in \mathbb{R}^n, \mathbf{B}$ is a symmetric $(n \times n)$ positive definite matrix$\}$.

*Density Definition:*

$$f(\mathbf{x}; \mathbf{a}, \mathbf{B}) = \frac{1}{(2\pi)^{n/2} |\mathbf{B}|^{1/2}} \exp\left[-\frac{1}{2}(\mathbf{x}-\mathbf{a})' \mathbf{B}^{-1} (\mathbf{x}-\mathbf{a})\right]$$

*Moments*:

$$\underset{(n\times 1)}{\boldsymbol{\mu}} = \mathbf{a}, \quad \underset{(n\times n)}{\mathbf{Cov}(\mathbf{X})} = \mathbf{B}, \quad \underset{(n\times 1)}{\boldsymbol{\mu}_3} = \mathbf{0}, \left(\frac{\mu_4(X_i)}{\sigma^4(X_i)} - 3\right) = 0 \; \forall i$$

*MGF*: $M_{\mathbf{X}}(t) = \exp\left(\mathbf{a}'\mathbf{t} + (1/2)\,\mathbf{t}'\mathbf{B}\mathbf{t}\right)$, where $\mathbf{t} = (t_1, \ldots, t_n)'$.

*Background and Application*: The $n$-variate normal family of densities is indexed by $n + n(n + 1)/2$ parameters consisting of $n$ elements in the $(n \times 1)$ vector $\mathbf{a}$ and $n(n + 1)/2$ elements representing the distinct elements in the $(n \times n)$ symmetric matrix $\mathbf{B}$ (while there are $n^2$ number of elements in the matrix $\mathbf{B}$, only $n(n + 1)/2$ of these elements are distinct (or different) given the symmetry of $\mathbf{B}$, i.e., $b_{ij} = b_{ji}\,\forall i$ and $j$). Given the special relationship between the mean vector, the covariance matrix, and the parameters $\mathbf{a}$ and $\mathbf{B}$, the $n$-variate normal density is most often represented as

$$N(\mathbf{x}; \boldsymbol{\mu}, \boldsymbol{\Sigma}) = \frac{1}{(2\pi)^{n/2}\,|\boldsymbol{\Sigma}|^{1/2}} \exp\left[-\frac{1}{2}(\mathbf{x} - \boldsymbol{\mu})'\,\boldsymbol{\Sigma}^{-1}\,(\mathbf{x} - \boldsymbol{\mu})\right]$$

where $\boldsymbol{\Sigma}$ is a popular notation for the covariance matrix of $\mathbf{X}$ (The problem context will have to be relied upon to distinguish between when $\boldsymbol{\Sigma}$ designates a covariance matrix, and when $\Sigma$ signifies summation). When it is clear which random vector is being referred to, the notation $N(\mathbf{x};\boldsymbol{\mu},\boldsymbol{\Sigma})$ is often shortened to $N(\boldsymbol{\mu},\boldsymbol{\Sigma})$. The MGF of $\mathbf{X}$ is represented by $M_{\mathbf{x}}(t) = \exp(\boldsymbol{\mu}'\mathbf{t} + .5\;\mathbf{t}'\boldsymbol{\Sigma}\;\mathbf{t})$.

In order to illustrate graphically some of the characteristics of the multivariate normal density, we temporarily concentrate on the bivariate case. Thus, in the above formula, $n = 2$, $\boldsymbol{\mu}$ is a $(2 \times 1)$ column vector, and $\boldsymbol{\Sigma}$ is a $(2 \times 2)$ positive definite covariance matrix. The graph of the bivariate normal density is a three-dimensional bell of sorts, such as the illustration in Figure 4.14a. The mode of the normal density occurs at $\mathbf{x} = \boldsymbol{\mu}$, which in the bivariate case occurs at $x_1 = \mathrm{E}(X_1)$ and $x_2 = \mathrm{E}(X_2)$. **Iso-density contours** (i.e., the collection of $(x_1, x_2)$ points resulting in a fixed value of the joint density function, as $f(\mathbf{x};\boldsymbol{\mu},\boldsymbol{\Sigma}) = c$, are in the form of ellipses (ellipsoids in higher dimensions, e.g., a "football" in three dimensions) with center at $\boldsymbol{\mu}$, so that in the bivariate case the center of the ellipse is at $(\mathrm{E}(X_1), \mathrm{E}(X_2))$. One can think of these ellipses as being formed by "slicing" through the density at a certain height, removing the top portion of the density, and then projecting the exposed elliptical top onto the $(x_1,x_2)$-plane (see Figure 4.14b).

The shape and orientation of the ellipse (or ellipsoid) are determined by the elements of the covariance matrix, $\boldsymbol{\Sigma}$. The major (larger) axis of the ellipse as measured from the origin, $\boldsymbol{\mu}$, is in the direction of the characteristic vector of $\boldsymbol{\Sigma}^{-1}$ associated with the smallest characteristic root of $\boldsymbol{\Sigma}^{-1}$.[9] The length of an axis is

[9] Recall that the characteristic roots and vectors of a square matrix $\mathbf{A}$ are the scalars, $\lambda$, and associated vectors, $\mathbf{p}$, that satisfy the equation $[\mathbf{A} - \lambda\mathbf{I}]\,\mathbf{p} = \mathbf{0}$. There will be as many roots and associated vectors as there are rows (or columns) in the square matrix $\mathbf{A}$.

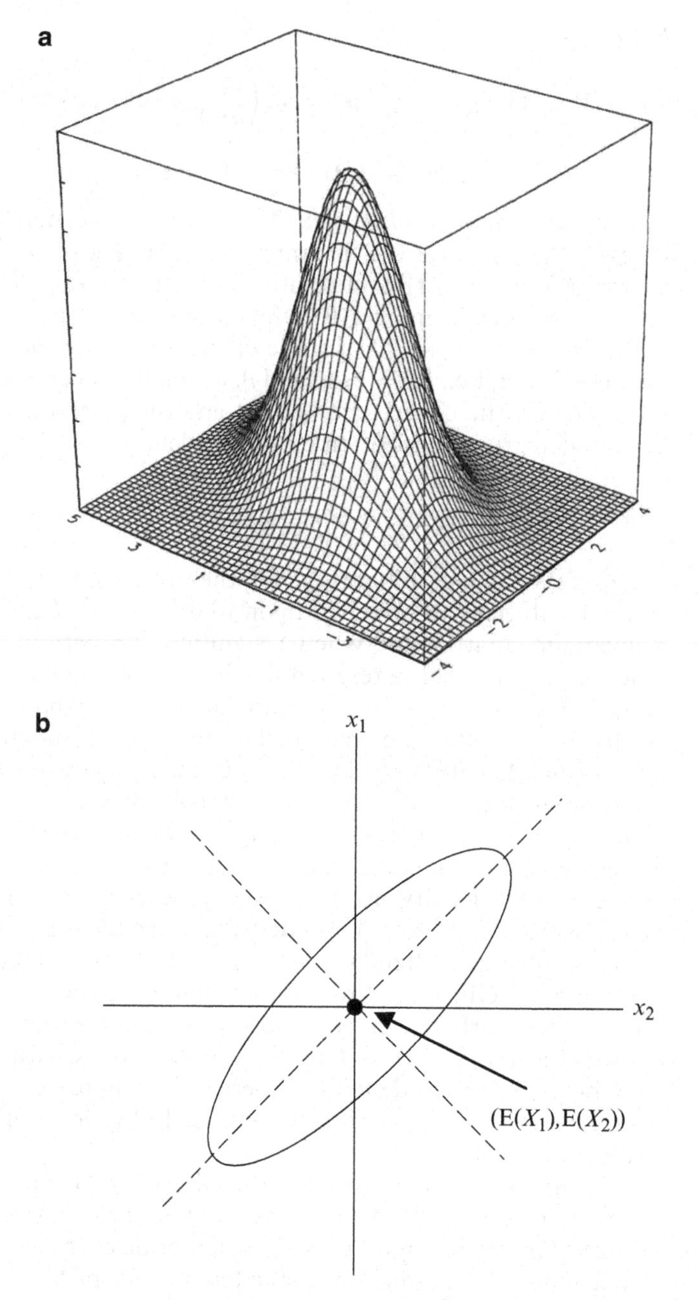


**Figure 4.14**
Bivariate normal density; (**a**) view of normal density in three-dimensions, and (**b**) an iso-density ellipse.

given by $2k/\sqrt{\lambda_i}$, where $k = \sqrt{2\pi}|\Sigma|^{.5}\, c$, $c$ is the chosen value of the density, and $\lambda_i$ is either the smallest or largest characteristic root of $\boldsymbol{\Sigma}^{-1}$.[10]

[10]These arguments extend in a natural way to higher dimensions, in which case we are examining $n$ axes of the ellipsoid. See B. Bolch, and C. Huang (1974), *Multivariate Statistical Methods for Business and Economics*, Englewood Cliffs, NJ: Prentice Hall, p. 19–23, for the matrix theory underlying the derivation of the results on axis length and orientation discussed here.

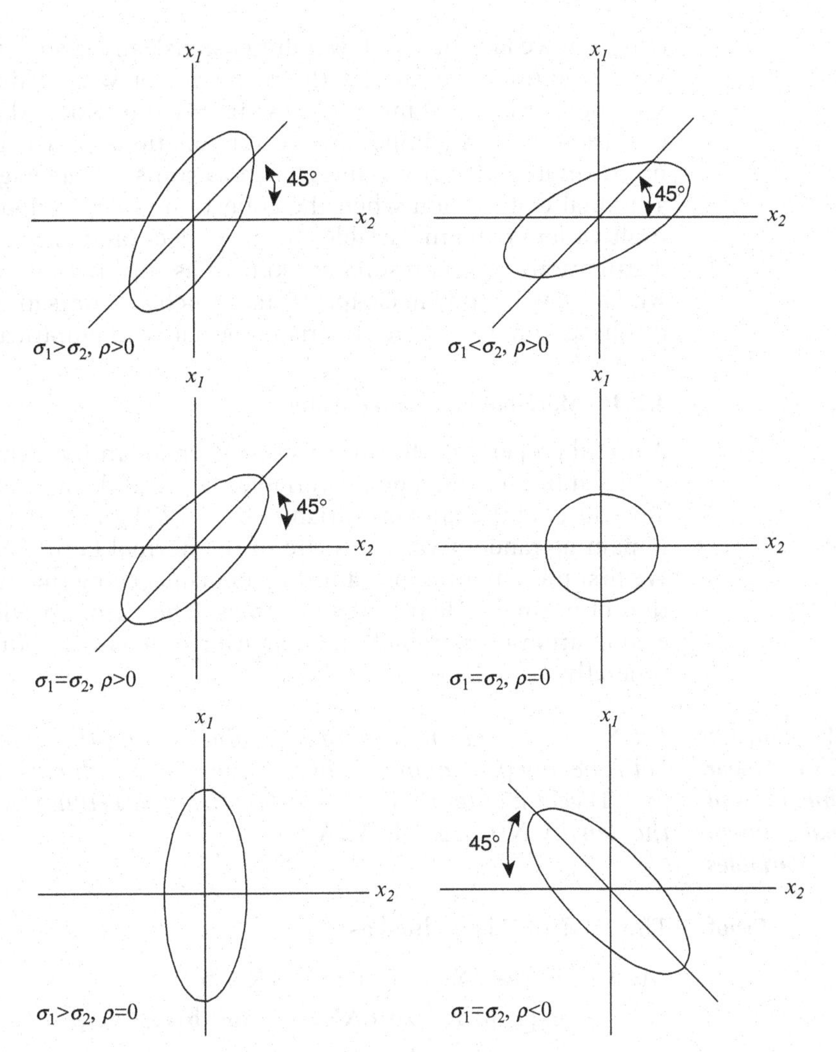


**Figure 4.15** Iso-density ellipses of bivariate normal density (all origins at $(E(X_1),E(X_2))$).

In the bivariate case, the slope, $dx_1/dx_2$, of the major axis of the ellipse is positive if $\sigma_{12} > 0$ and negative if $\sigma_{12} < 0$. The slope increases in absolute value as the ratio of the standard deviation $\sigma_1/\sigma_2$ increases, holding $\rho$ (the correlation coefficient) constant. As $|\sigma_{12}| \to \sigma_1\sigma_2$, so that $\rho \to \pm 1$, the length of the minor axis of the ellipse $\to 0$, and the ellipse concentrates on the major axis, which approaches the line $x_1 = E(X_1) \pm (\sigma_1/\sigma_2)(x_2 - E(X_2))$ (compare to Theorem 3.35 – the orientation of the slope of the line matches the sign of the correlation). If $\sigma_1 = \sigma_2$, the principal axis is given by the line $x_1 = E(X_1) \pm (x_2 - E(X_2))$, for any magnitude of the correlation.

An illustration of some of the myriad of possibilities for the graphs of the iso-density contours is given in Figure 4.15.

The multivariate normal density is often used in modeling experiments characterized by a PDF that is symmetric about its mean vector, $\boldsymbol{\mu}$, is bell-shaped (when viewed in three or fewer dimensions), and is such that the highest density weighting (mode) occurs at the mean vector with density values declining as $\mathbf{x}$ becomes more distant from $E(\mathbf{X})$ in any direction. Also, under rather general conditions, certain important vector functions of a collection of independent multivariate random variables, such as sums and averages, have approximately a normal distribution when the collection is large enough, even if the original multivariate random variables in the collection do *not* have normal probability distributions. These results are known as *multivariate central limit theorems,* which we will study in Chapter 5 and which accounts for much of the motivation for the assumption of multivariate normality in empirical work.

### 4.3.3 Marginal Normal Densities

A useful property of the multivariate normal family is that the marginal density of any subset of the random variables $(X_1,\ldots,X_n)$, and the conditional density of any subset of the random variables $(X_1,\ldots,X_n)$ *given* an elementary event for the remaining random variables, are in the normal family and are easy to identify. We first present an important theorem concerning the PDF of a linear combination of normally distributed variables. The theorem will greatly facilitate our examination of marginal and conditional densities, but it is also useful more generally.

**Theorem 4.9** ***PDF of Linear Combinations of Normal Random Variables***

*Let* $\mathbf{X}$ *be an n-variate random variable having the density function* $N(\mathbf{x};\boldsymbol{\mu},\boldsymbol{\Sigma})$. *Let* $\mathbf{A}$ *be any* $(k \times n)$ *matrix of real constants with rank* $k \leq n$, *and let* $\mathbf{b}$ *be any* $(k \times 1)$ *vector of real constants. Then the* $(k \times 1)$ *random vector* $\mathbf{Y} = \mathbf{AX} + \mathbf{b}$ *has the density* $N(\mathbf{y};\mathbf{A}\boldsymbol{\mu}+\mathbf{b},\mathbf{A}\boldsymbol{\Sigma}\mathbf{A}')$.

**Proof** The MGF of $\mathbf{Y}$ is defined as

$$\begin{aligned} M_{\mathbf{Y}}(\mathbf{t}) &= E(\exp(\mathbf{t}'\mathbf{Y})) = E\,(\exp(\mathbf{t}'(\mathbf{AX}+\mathbf{b}))) \\ &= \exp(\mathbf{t}'\mathbf{b})\;E\,(\exp(\mathbf{t}'\mathbf{AX})) = \exp(\mathbf{t}'\mathbf{b})\exp(\mathbf{t}'\mathbf{A}\boldsymbol{\mu} + (1/2)\mathbf{t}'\mathbf{A}\boldsymbol{\Sigma}\mathbf{A}'\mathbf{t}) \\ &= \exp(\mathbf{t}'(\mathbf{A}\boldsymbol{\mu}+\mathbf{b}) + (1/2)\mathbf{t}'\mathbf{A}\boldsymbol{\Sigma}\mathbf{A}'\mathbf{t}) \end{aligned}$$

where the next-to-last equality follows from the fact that $\mathbf{X}$ is normally distributed, and $E(\exp(\mathbf{t}'\mathbf{AX})) = E(\exp(\mathbf{t}_*'\mathbf{X})) = M_{\mathbf{X}}(\mathbf{t}_*) = \exp[\mathbf{t}_*'\boldsymbol{\mu} + (1/2)\mathbf{t}_*'\boldsymbol{\Sigma}\mathbf{t}_*]$, with $\mathbf{t}_*' = \mathbf{t}'A$. Thus, $M_{\mathbf{Y}}(\mathbf{t})$ identifies the multivariate normal density with mean $\mathbf{A}\boldsymbol{\mu} + \mathbf{b}$ and covariance matrix $\mathbf{A}\boldsymbol{\Sigma}\mathbf{A}'$ as the density of $\mathbf{Y} = \mathbf{AX} + \mathbf{b}$. ■

**Example 4.19** ***Probability Density of Profit and Variable Cost***

A firm uses two variable inputs in the manufacture of an output and also uses "just-in-time" production methods so that inputs arrive precisely when they are needed to produce the output. The weekly average input and selling prices, $\mathbf{r}$ and $p$, during the spring quarter can be viewed as the outcomes of a trivariate normal density function $f(r_1,r_2,p;\boldsymbol{\mu},\boldsymbol{\Sigma}) = N(\boldsymbol{\mu},\boldsymbol{\Sigma})$ with

$$\boldsymbol{\mu} = \begin{bmatrix} .50 \\ 1.25 \\ 5 \end{bmatrix}, \text{and } \boldsymbol{\Sigma} = \begin{bmatrix} .05 & .02 & .01 \\ .02 & .10 & .01 \\ .01 & .01 & .40 \end{bmatrix}.$$

What is the density function of the bivariate random variable $(\Pi,C)$, where $\Pi$ represents profit above variable cost and $C$ represents variable cost, in a week where 100 units of input 1 and 150 units of input 2 are utilized, and 100 units of output are produced and sold?

**Answer**: The random variables $\Pi$ and $C$ can be defined as

$$\mathbf{Y} = \begin{bmatrix} \Pi \\ C \end{bmatrix} = \begin{bmatrix} -100 & -150 & 100 \\ 100 & 150 & 0 \end{bmatrix} \begin{bmatrix} R_1 \\ R_2 \\ P \end{bmatrix} = \mathbf{AX} + \mathbf{b}$$

where $\mathbf{b} = \mathbf{0}$ and $\mathbf{A}$ is the bracketed $(2 \times 3)$ matrix following the second equality sign. Then, Theorem 4.9 implies that $[\Pi, C]'$ is bivariate normally distributed as $N(\boldsymbol{\mu}_*,\boldsymbol{\Sigma}_*)$, with $\boldsymbol{\mu}_* = \mathbf{A}\boldsymbol{\mu} = [262.5 \quad 237.5]'$ and $\boldsymbol{\Sigma}_* = \mathbf{A}\boldsymbol{\Sigma}\mathbf{A}' = \begin{bmatrix} 6850 & -3100 \\ -3100 & 3350 \end{bmatrix}$. □

A useful implication of Theorem 4.9 for generating or *simulating* outcomes of multivariate normally distributed random variables on the computer is that *any* such random variable can be represented in terms of linear combinations of independent random variables having the standard normal density. Specifically, if the $n \times 1$ random variable $\mathbf{Z}$ has the PDF $N(\mathbf{0},\mathbf{I})$ so that $Z_1,\ldots,Z_n$ are independent $N(0,1)$ random variables, then the $n \times 1$ random variable $\mathbf{Y}$ with PDF $N(\boldsymbol{\mu},\boldsymbol{\Sigma})$ can be represented in terms of $\mathbf{Z}$ as $\mathbf{Y} = \boldsymbol{\mu} + \mathbf{AZ}$, where $\mathbf{A}$ is chosen so that $\mathbf{A}'\mathbf{A} = \boldsymbol{\Sigma}$.[11] The utility of this representation stems from the fact that many modern statistical software packages are capable of generating independent outcomes of a random variable having the $N(0,1)$ density; such programs are referred to as "standard normal random-number generators." We will pursue the notion of simulating random variable outcomes further in Chapter 6.

We now state an important result concerning the marginal densities of subsets of the random variables $(X_1,\ldots,X_n)$ when the $n$-variate random variable has an $n$-variate normal density.

**Theorem 4.10 *Marginal Densities for* $N(\boldsymbol{\mu},\boldsymbol{\Sigma})$**

*Let* $\mathbf{Z}$ *have the density* $N(\mathbf{z}; \boldsymbol{\mu}, \boldsymbol{\Sigma})$, *where*

$$\mathbf{Z} = \begin{bmatrix} \underset{(m\times 1)}{\mathbf{Z}_{(1)}} \\ \hline \underset{(n-m)\times 1}{\mathbf{Z}_{(2)}} \end{bmatrix}, \boldsymbol{\mu} = \begin{bmatrix} \underset{(m\times 1)}{\boldsymbol{\mu}_{(1)}} \\ \hline \underset{(n-m)\times 1}{\boldsymbol{\mu}_{(2)}} \end{bmatrix}, \text{and } \boldsymbol{\Sigma} = \left[\begin{array}{c|c} \underset{(m\times m)}{\boldsymbol{\Sigma}_{11}} & \underset{(m\times(n-m))}{\boldsymbol{\Sigma}_{12}} \\ \hline \underset{((n-m)\times m)}{\boldsymbol{\Sigma}_{21}} & \underset{((n-m)\times(n-m))}{\boldsymbol{\Sigma}_{22}} \end{array}\right]$$

[11]The symmetric matrix square root $\boldsymbol{\Sigma}^{1/2}$ of $\boldsymbol{\Sigma}$ could be chosen for $\mathbf{A}$. Alternatively, there exists a lower triangular matrix, called the Cholesky decomposition of $\boldsymbol{\Sigma}$, which satisfies $\mathbf{A}'\mathbf{A} = \boldsymbol{\Sigma}$. Either choice of $\mathbf{A}$ can be calculated straightforwardly on the computer, for example, the chol(.) command in GAUSS.

*Then the marginal PDF of* $\mathbf{Z}_{(1)}$ *is* $N(\boldsymbol{\mu}_1, \boldsymbol{\Sigma}_{11})$ *and the marginal PDF of* $\mathbf{Z}_{(2)}$ *is* $N(\boldsymbol{\mu}_2, \boldsymbol{\Sigma}_{22})$.

**Proof**

Let $\mathbf{A} = \left[\underset{(m\times m)}{\mathbf{I}_m} \;\middle|\; \underset{m\times(n-m)}{\mathbf{0}}\right]$ and $\mathbf{b} = \mathbf{0}$ in Theorem 4.9, where $\mathbf{I}_m$ is the $(m \times m)$ identity matrix. It follows that $\mathbf{Z}_{(1)} = \mathbf{AZ}$ has the normal density $N(\boldsymbol{\mu}_{(1)}, \boldsymbol{\Sigma}_{11})$. The result for $\mathbf{Z}_{(2)}$ is proven similarly be letting $\mathbf{A} = \left[\underset{(n-m)\times m}{\mathbf{0}} \;\middle|\; \underset{(n-m)\times(n-m)}{\mathbf{I}_{n-m}}\right]$. ■

**Example 4.20**
***Marginal Densities for a Trivariate Normal***

Referring to Example 4.19, partition $\mathbf{X}$, $\boldsymbol{\mu}$ and $\boldsymbol{\Sigma}$ as

$$\mathbf{X} = \begin{bmatrix} \mathbf{X}_{(1)} \\ \mathbf{X}_{(2)} \end{bmatrix} = \begin{bmatrix} \mathbf{X}_1 \\ \mathbf{X}_2 \\ \hline \mathbf{X}_3 \end{bmatrix}, \boldsymbol{\mu} = \begin{bmatrix} \boldsymbol{\mu}_{(1)} \\ \boldsymbol{\mu}_{(2)} \end{bmatrix} = \begin{bmatrix} \boldsymbol{\mu}_1 \\ \boldsymbol{\mu}_2 \\ \hline \boldsymbol{\mu}_3 \end{bmatrix} = \begin{bmatrix} .50 \\ 1.25 \\ \hline 5 \end{bmatrix}$$

and

$$\boldsymbol{\Sigma} = \left[\begin{array}{c|c} \boldsymbol{\Sigma}_{11} & \boldsymbol{\Sigma}_{12} \\ \hline \boldsymbol{\Sigma}_{21} & \boldsymbol{\Sigma}_{22} \end{array}\right] = \left[\begin{array}{cc|c} .05 & .02 & .01 \\ .02 & .10 & .01 \\ \hline .01 & .01 & .40 \end{array}\right].$$

It follows from Theorem 4.10 that the marginal densities of $(X_1, X_2)$ and $X_3$ are given by

$$f_{X_1X_2}(x_1, x_2) = N\left(\begin{bmatrix} .50 \\ 1.25 \end{bmatrix}, \begin{bmatrix} .05 & .02 \\ .02 & .10 \end{bmatrix}\right), \text{and } f_{X_3}(x_3) = N(5, .4). \qquad \square$$

The reader should note that the order of the random variables in the $\mathbf{X}$ vector is arbitrary. For example, we might have $\mathbf{X} = (X_1, X_2, X_3, X_4)'$ or alternatively, the same random variables might be listed as $\mathbf{X} = (X_3, X_1, X_4, X_2)'$. The point is that Theorem 4.10 can be applied to obtain the marginal density function of *any* subset of the random variable $(X_1, \ldots, X_n)$ by simply ordering them appropriately in the definition of $\mathbf{Z}$ in the theorem. Of course, the entries in $\boldsymbol{\mu}$ and $\boldsymbol{\Sigma}$ must be correspondingly ordered so that random variables are associated with their appropriate means, variances, and covariances.

**Example 4.21**
***Marginal Normal Density* via *Reordering***

The annual percentage return on three investment instruments is the outcome of a trivariate random variable $(X_1, X_2, X_3)'$ having the density $N(\boldsymbol{\mu}, \boldsymbol{\Sigma})$, where

$$\boldsymbol{\mu} = \begin{bmatrix} 2 \\ 7 \\ 1 \end{bmatrix} \text{ and } \boldsymbol{\Sigma} = \begin{bmatrix} 4 & 1 & 0 \\ 1 & 1 & 1 \\ 0 & 1 & 3 \end{bmatrix}.$$

To identify the marginal density of the returns on investments 1 and 3 using Theorem 4.10, first reorder the random variables as $X = (X_1, X_3, X_2)'$ so that the corresponding mean vector and covariance matrix of the trivariate normal density of $\mathbf{X}$ are now

$$\boldsymbol{\mu}_* = \begin{bmatrix} 2 \\ 1 \\ 7 \end{bmatrix}, \text{and } \boldsymbol{\Sigma}_* = \begin{bmatrix} 4 & 0 & 1 \\ 0 & 3 & 1 \\ 1 & 1 & 1 \end{bmatrix}.$$

Then a straightforward application of Theorem 4.10 (with the appropriate interpretation of the symbols $\mathbf{Z}_{(1)}$ and $\mathbf{Z}_{(2)}$ used in that theorem) implies that $(X_1, X_3)'$ has the density $N\left(\begin{bmatrix} 2 \\ 1 \end{bmatrix}, \begin{bmatrix} 4 & 0 \\ 0 & 3 \end{bmatrix}\right)$. □

### 4.3.4 Conditional Normal Densities

Defining conditional densities for $n$-variate normally distributed random variables is somewhat more involved than defining marginal densities. We present the case of conditioning on an elementary event for a subset of the random variables.

**Theorem 4.11** ***Conditional Densities for $N(\mu,\Sigma)$*** *Let* $\mathbf{Z}$ *be defined as in Theorem 4.10, and let*

$$\underset{(n\times 1)}{\mathbf{z}^0} = \begin{bmatrix} \underset{(m\times 1)}{\mathbf{z}^0_{(1)}} \\ \hline \underset{(n-m)\times 1}{\mathbf{z}^0_{(2)}} \end{bmatrix}$$

*be a vector of constants. Then*

$$f(\mathbf{z}_{(1)}\,|\mathbf{z}_{(2)} = \mathbf{z}^0_{(2)}) = N(\boldsymbol{\mu}_{(1)} + \boldsymbol{\Sigma}_{12}\,\boldsymbol{\Sigma}_{22}^{-1}\,(\mathbf{z}^0_{(2)} - \boldsymbol{\mu}_{(2)}), \boldsymbol{\Sigma}_{11} - \boldsymbol{\Sigma}_{12}\,\boldsymbol{\Sigma}_{22}^{-1}\,\boldsymbol{\Sigma}_{21})$$
$$f(\mathbf{z}_{(2)}\,|\,\mathbf{z}_{(1)} = \mathbf{z}^0_{(1)}) = N(\boldsymbol{\mu}_{(2)} + \boldsymbol{\Sigma}_{21}\,\boldsymbol{\Sigma}_{11}^{-1}\,(\mathbf{z}^0_{(1)} - \boldsymbol{\mu}_{(1)}), \boldsymbol{\Sigma}_{22} - \boldsymbol{\Sigma}_{21}\,\boldsymbol{\Sigma}_{11}^{-1}\,\boldsymbol{\Sigma}_{12})$$

**Proof** We prove the result for the case of $\mathbf{Z}_{(1)}$; the case for $\mathbf{Z}_{(2)}$ can be proved analogously. By definition,

$$f\left(\mathbf{z}_{(1)}|\mathbf{z}_{(2)} = \mathbf{z}^0_{(2)}\right) = \frac{f\left(\mathbf{z}_{(1)}, \mathbf{z}^0_{(2)}\right)}{f_{\mathbf{Z}_{(2)}}\left(\mathbf{z}^0_{(2)}\right)}$$

$$= \frac{\frac{1}{(2\pi)^{n/2}|\boldsymbol{\Sigma}|^{1/2}}\exp\left(-\frac{1}{2}\begin{bmatrix} \mathbf{z}_{(1)} - \boldsymbol{\mu}_{(1)} \\ \mathbf{z}^0_{(2)} - \boldsymbol{\mu}_{(2)} \end{bmatrix}' \boldsymbol{\Sigma}^{-1} \begin{bmatrix} \mathbf{z}_{(1)} - \boldsymbol{\mu}_{(1)} \\ \mathbf{z}^0_{(2)} - \boldsymbol{\mu}_{(2)} \end{bmatrix}\right)}{\frac{1}{(2\pi)^{(n-m)/2}|\boldsymbol{\Sigma}_{22}|^{1/2}}\exp\left(-\frac{1}{2}\left(\mathbf{z}^0_{(2)} - \boldsymbol{\mu}_{(2)}\right)'\boldsymbol{\Sigma}^{-1}_{(22)}\left(\mathbf{z}^0_{(2)} - \boldsymbol{\mu}_{(2)}\right)\right)}$$

The following lemma on partitioned determinants and partitioned inversion will be useful here.

**Lemma 4.3**
***Partitioned Inversion and Partitioned Determinants***

Partition the $(n \times n)$ matrix $\mathbf{\Sigma}$ as

$$\mathbf{\Sigma} = \left[\begin{array}{c|c} \underset{(m\times m)}{\mathbf{\Sigma}_{11}} & \underset{m\times(n-m)}{\mathbf{\Sigma}_{12}} \\ \hline \underset{(n-m)\times m}{\mathbf{\Sigma}_{21}} & \underset{(n-m)\times(n-m)}{\mathbf{\Sigma}_{22}} \end{array}\right]$$

**a.** If $\mathbf{\Sigma}_{11}$ is nonsingular, then $|\mathbf{\Sigma}| = |\mathbf{\Sigma}_{11}| \cdot \left|\mathbf{\Sigma}_{22} - \mathbf{\Sigma}_{21}\mathbf{\Sigma}_{11}^{-1}\mathbf{\Sigma}_{12}\right|$.

**b.** If $\mathbf{\Sigma}_{22}$ is nonsingular, then $|\mathbf{\Sigma}| = |\mathbf{\Sigma}_{22}| \cdot \left|\mathbf{\Sigma}_{11} - \mathbf{\Sigma}_{12}\mathbf{\Sigma}_{22}^{-1}\mathbf{\Sigma}_{21}\right|$.

**c.** If $|\mathbf{\Sigma}| \neq 0, |\mathbf{\Sigma}_{11}| \neq 0$, and $|\mathbf{\Sigma}_{22}| \neq 0$, then

$$\mathbf{\Sigma}^{-1} = \left[\begin{array}{c|c} (\mathbf{\Sigma}_{11} - \mathbf{\Sigma}_{12}\mathbf{\Sigma}_{22}^{-1}\mathbf{\Sigma}_{21})^{-1} & -(\mathbf{\Sigma}_{11} - \mathbf{\Sigma}_{12}\mathbf{\Sigma}_{22}^{-1}\mathbf{\Sigma}_{21})^{-1}\mathbf{\Sigma}_{12}\mathbf{\Sigma}_{22}^{-1} \\ \hline -\mathbf{\Sigma}_{22}^{-1}\mathbf{\Sigma}_{21}(\mathbf{\Sigma}_{11} - \mathbf{\Sigma}_{12}\mathbf{\Sigma}_{22}^{-1}\mathbf{\Sigma}_{21})^{-1} & (\mathbf{\Sigma}_{22} - \mathbf{\Sigma}_{21}\mathbf{\Sigma}_{11}^{-1}\mathbf{\Sigma}_{12})^{-1} \end{array}\right]$$

**d.** The diagonal blocks in the partitioned matrix of part (c) can also be expressed as

$$\left(\mathbf{\Sigma}_{11} - \mathbf{\Sigma}_{12}\mathbf{\Sigma}_{22}^{-1}\mathbf{\Sigma}_{21}\right)^{-1} = \mathbf{\Sigma}_{11}^{-1} + \mathbf{\Sigma}_{11}^{-1}\mathbf{\Sigma}_{12}\left(\mathbf{\Sigma}_{22}^{-1} - \mathbf{\Sigma}_{21}\mathbf{\Sigma}_{11}^{-1}\mathbf{\Sigma}_{12}\right)^{-1}\mathbf{\Sigma}_{21}\mathbf{\Sigma}_{11}^{-1}$$

and

$$\left(\mathbf{\Sigma}_{22} - \mathbf{\Sigma}_{21}\mathbf{\Sigma}_{11}^{-1}\mathbf{\Sigma}_{12}\right)^{-1} = \mathbf{\Sigma}_{22}^{-1} + \mathbf{\Sigma}_{22}^{-1}\mathbf{\Sigma}_{21}\left(\mathbf{\Sigma}_{11}^{-1} - \mathbf{\Sigma}_{12}\mathbf{\Sigma}_{22}^{-1}\mathbf{\Sigma}_{21}\right)^{-1}\mathbf{\Sigma}_{12}\mathbf{\Sigma}_{22}^{-1}.$$

(see F.A. Graybill (1983), Matrices with Applications in Statistics, 2nd Ed., Belmont, CA: Wadsworth, pp. 183–186, and H. Theil, Principles of Econometrics, John Wiley and Sons, New York, pp. 16–19 for further discussion and proofs).

Utilizing Lemma 4.3, note that

$$\frac{|\mathbf{\Sigma}_{22}|^{1/2}}{|\mathbf{\Sigma}|^{1/2}} = \left|\mathbf{\Sigma}_{11} - \mathbf{\Sigma}_{12}\mathbf{\Sigma}_{22}^{-1}\mathbf{\Sigma}_{21}\right|^{-1/2}$$

and

$$\exp\left[-\frac{1}{2}\begin{bmatrix}\mathbf{z}_{(1)} - \boldsymbol{\mu}_{(1)} \\ \mathbf{z}_{(2)}^0 - \boldsymbol{\mu}_{(2)}\end{bmatrix}' \mathbf{\Sigma}^{-1} \begin{bmatrix}\mathbf{z}_{(1)} - \boldsymbol{\mu}_{(1)} \\ \mathbf{z}_{(2)}^0 - \boldsymbol{\mu}_{(2)}\end{bmatrix} + \frac{1}{2}\left(\mathbf{z}_{(2)}^0 - \boldsymbol{\mu}_{(2)}\right)'\mathbf{\Sigma}_{22}^{-1}\left(\mathbf{z}_{(2)}^0 - \boldsymbol{\mu}_{(2)}\right)\right]$$

$$= \exp\left[-\frac{1}{2}\left(\mathbf{z}_{(1)} - \left(\boldsymbol{\mu}_{(1)} + \mathbf{\Sigma}_{12}\mathbf{\Sigma}_{22}^{-1}\left(\mathbf{z}_{(2)}^0 - \boldsymbol{\mu}_{(2)}\right)\right)\right)'\Phi^{-1}\right.$$

$$\left.\times\left(\mathbf{z}_{(1)} - \left(\boldsymbol{\mu}_{(1)} + \mathbf{\Sigma}_{12}\mathbf{\Sigma}_{22}^{-1}\left(\mathbf{z}_{(2)}^0 - \boldsymbol{\mu}_{(2)}\right)\right)\right)\right]$$

where $\Phi = \left[\Sigma_{11} - \Sigma_{12}\Sigma_{22}^{-1}\Sigma_{21}\right]$. Then the expression for the conditional density function reduces to

$$f\left(\mathbf{z}_{(1)} \mid \mathbf{z}_{(2)} = \mathbf{z}_{(2)}^0\right) = \frac{1}{(2\pi)^{m/2} |\Phi|^{1/2}} \exp\left[-\frac{1}{2}(\mathbf{z}_{(1)} - \boldsymbol{\gamma})' \Phi^{-1} (\mathbf{z}_{(1)} - \boldsymbol{\gamma})\right],$$

where $\boldsymbol{\gamma} = \boldsymbol{\mu}_{(1)} + \Sigma_{12}\Sigma_{22}^{-1}(\mathbf{z}_{(2)}{}^0 - \boldsymbol{\mu}_2)$, which is the normal density $N(\boldsymbol{\gamma},\Phi)$. ■

**Example 4.22** ***Conditional Densities for a Trivariate Normal***

Let $X = (X_1, X_2, X_3)$ be the trivariate normal random variable representing the percentage returns on investments in Example 4.21. Suppose we wanted to define the conditional density of returns on investment instrument 1, *given* $x_2 = 1$ and $x_3 = 2$. By letting $z_{(1)} = \mathbf{x}_1$ and $\mathbf{z}_{(2)} = (x_2, x_3)'$ in Theorem 4.11, we know that $X_1$ will have a normal density with mean

$$E(x_1 \mid x_2 = 1, x_3 = 2) = 2 + [1 \quad 0]\begin{bmatrix} 1 & 1 \\ 1 & 3 \end{bmatrix}^{-1}\begin{bmatrix} -6 \\ 1 \end{bmatrix} = -7.5$$

and variance

$$\sigma^2_{(X_1 \mid x_2=1, x_3=2)} = 4 - [1 \quad 0]\begin{bmatrix} 1 & 1 \\ 1 & 3 \end{bmatrix}^{-1}\begin{bmatrix} 1 \\ 0 \end{bmatrix} = 2.5,$$

so that the *conditional* density of $X_1$ is $f(x_1|x_2 = 1, x_3 = 2) = N(-7.5, 2.5)$.

To find the conditional density of returns on investments 1 and 2 *given* $x_3 = 0$, Theorem 4.11, with $\mathbf{z}_{(1)} = (x_1, x_2)'$ and $z_{(2)} = x_3$, implies that the mean vector is equal to

$$E\left(\begin{bmatrix} X_1 \\ X_2 \end{bmatrix} \mid x_3 = 0\right) = \begin{bmatrix} 2 \\ 7 \end{bmatrix} + \begin{bmatrix} 0 \\ 1 \end{bmatrix}[3]^{-1}[-1] = \begin{bmatrix} 2 \\ 20/3 \end{bmatrix},$$

and the covariance matrix is

$$\mathbf{Cov}((X_1, X_2) \mid x_3 = 0) = \begin{bmatrix} 4 & 1 \\ 1 & 1 \end{bmatrix} - \begin{bmatrix} 0 \\ 1 \end{bmatrix}[3]^{-1}[0 \ 1] = \begin{bmatrix} 4 & 1 \\ 1 & 2/3 \end{bmatrix}.$$ □

In the special case where $X_1$ is a scalar and $\mathbf{X}_2$ is a $(k \times 1)$ vector, the conditional expectation of $X_1$ expressed as a *function* of $\mathbf{x}_2$ defines the **regression function** of $X_1$ on $\mathbf{X}_2$, which in this case, given the linearity of the relationship, can also be referred to as the **regression hyperplane** of $X_1$ on $\mathbf{X}_2$. If $X_2$ happens to be a scalar, the regression hyperplane is a **regression line**. The regression hyperplane, in the multivariate normal case, is thus given by

$$E\left(X_1 \mid \underset{(k\times 1)}{\mathbf{x}_2}\right) = \mu_{(1)} + \Sigma_{12}\,\Sigma_{22}^{-1}(\mathbf{x}_{(2)} - \boldsymbol{\mu}_{(2)}) = a + \mathbf{b}'\mathbf{x}_{(2)}$$

where $\mathbf{b}' = \Sigma_{12}\,\Sigma_{22}^{-1}$ and $a = \mu_{(1)} - \mathbf{b}'\boldsymbol{\mu}_{(2)}$. The special case of the regression line can be written as

$$E(X_1|\mathbf{x}_2) = \mu_1 + \frac{\rho\,\sigma_1}{\sigma_2}(x_2 - \mu_2) = a + bx_2,$$

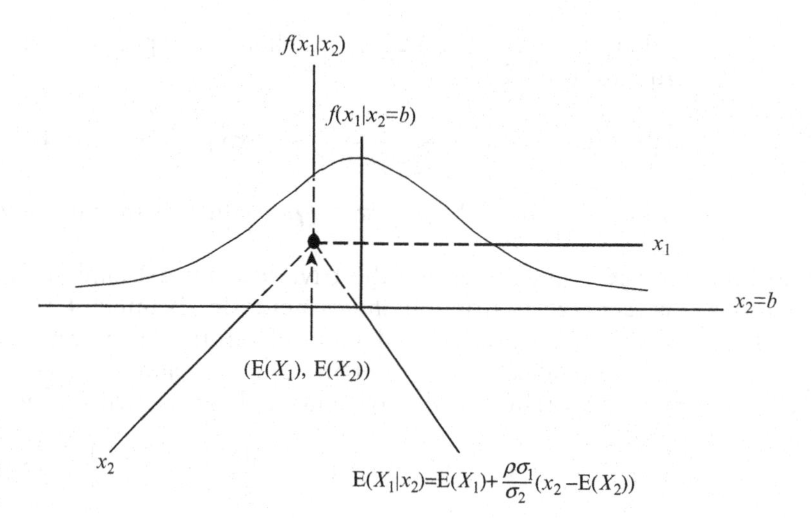


**Figure 4.16**
Regression line of $X_1$ on $X_2$.

where $\rho$ is the correlation coefficient, $b = \rho\sigma_1/\sigma_2$, and $a = \mu_1 - b\mu_2$. Figure 4.16 illustrates the regression line of $X_1$ on $X_2$.

The particular functional form of the normal density is the reason why the regression function becomes a regression *hyperplane*, and the regression curve becomes a regression *line*. In fact, the normal density belongs to a collection of families of distributions referred to as **elliptically contoured distributions**, all of which are associated with linear regression functions (S. Cambanis, et al., (1981), *On the Theory of Elliptically Contoured Distributions*, Journal of Multivariate Analysis, 11, p. 368). No other class of distributions exhibits *linear* regression functions as a general property.

While *not* true in general, in the case of a normal distribution, zero covariance implies independence of random variables, as stated in the following theorem.

**Theorem 4.12**
**Diagonal Cov(X) ⇒ *Independence* *when X has* PDF $N(\boldsymbol{\mu},\boldsymbol{\Sigma})$**

*Let* $\mathbf{X} = (X_1,\ldots,X_n)$ *have the density* $N(\boldsymbol{\mu},\boldsymbol{\Sigma})$. *Then* $(X_1,\ldots,X_n)$ *are independent iff* $\boldsymbol{\Sigma}$ *is a diagonal matrix.*

**Proof**

The *only if* (necessary) part of the theorem follows immediately from the fact that independence of random variables implies zero covariance. To see the *if* (sufficient) part, suppose $\boldsymbol{\Sigma}$ is a diagonal matrix. Then the density of $\mathbf{X}$ can be written as

$$N(\mathbf{x};\boldsymbol{\mu},\boldsymbol{\Sigma}) = \frac{1}{(2\pi)^{n/2}|\boldsymbol{\Sigma}|^{1/2}}\exp\left[-\frac{1}{2}(\mathbf{x}-\boldsymbol{\mu})'\boldsymbol{\Sigma}^{-1}(\mathbf{x}-\boldsymbol{\mu})\right]$$

$$= \prod_{i=1}^{n}\frac{1}{(2\pi)^{1/2}\sigma_i}\exp\left[-\frac{1}{2}\frac{(x_i-\mu_i)^2}{\sigma_i^2}\right] = \prod_{i=1}^{n}N(x_i;\mu_i,\sigma_i^2),$$

since

$$|\boldsymbol{\Sigma}| = \prod_{i=1}^{n} \sigma_i^2, \text{ and } \boldsymbol{\Sigma}^{-1} = \begin{bmatrix} \sigma_1^{-2} & & \\ & \ddots & \\ & & \sigma_n^{-2} \end{bmatrix}.$$

Being that the joint density factors into the product of the $n$ marginal densities, $(X_1,\ldots,X_n)$ are jointly independent. ■

A variation on the preceding theme is the case where two *vectors* $\mathbf{X}_{(1)}$ and $\mathbf{X}_{(2)}$ are independent by virtue of the covariances between the elements in $\mathbf{X}_{(1)}$ and the elements in $\mathbf{X}_{(2)}$ all being zero.

**Theorem 4.13 Block Diagonal Cov (X) ⇔ Independent Subvectors when X has PDF N(μ,Σ)**

Let $\underset{(n\times 1)}{\mathbf{Z}} = \begin{bmatrix} \underset{(m\times 1)}{\mathbf{Z}_{(1)}} \\ \hline \underset{(n-m)\times 1}{\mathbf{Z}_{(2)}} \end{bmatrix}$ *have the multivariate normal density identified in Theorem 4.10. Then the vectors* $\mathbf{Z}_{(1)}$ *and* $\mathbf{Z}_{(2)}$ *are independent iff* $\boldsymbol{\Sigma}_{12} = \boldsymbol{\Sigma}'_{21} = \mathbf{0}$.

**Proof** The *only if* part of the theorem follows immediately from the fact that the independence of random variables implies zero covariance. To see the *if* part, suppose $\boldsymbol{\Sigma}_{12} = \boldsymbol{\Sigma}'_{21} = \mathbf{0}$. Then the PDF of $\mathbf{Z}$ can be written as

$$\begin{aligned} N(\mathbf{z};\boldsymbol{\mu},\boldsymbol{\Sigma}) &= \frac{1}{(2\pi)^{n/2}|\boldsymbol{\Sigma}|^{1/2}}\exp\left[-\frac{1}{2}(\mathbf{z}-\boldsymbol{\mu})'\boldsymbol{\Sigma}^{-1}(\mathbf{z}-\boldsymbol{\mu})\right] \\ &= \frac{1}{(2\pi)^{m/2}|\boldsymbol{\Sigma}_{11}|^{1/2}}\exp\left[-\frac{1}{2}(\mathbf{z}_{(1)}-\boldsymbol{\mu}_{(1)})'\boldsymbol{\Sigma}_{11}^{-1}(\mathbf{z}_{(1)}-\boldsymbol{\mu}_{(1)})\right] \\ &\quad\times\frac{1}{(2\pi)^{(n-m)/2}|\boldsymbol{\Sigma}_{22}|^{1/2}}\exp\left[-\frac{1}{2}(\mathbf{z}_{(2)}-\boldsymbol{\mu}_{(2)})'\boldsymbol{\Sigma}_{22}^{-1}(\mathbf{z}_{(2)}-\boldsymbol{\mu}_{(2)})\right] \\ &= \prod_{i=1}^{2} N\left(\mathbf{z}_{(i)};\boldsymbol{\mu}_{(i)},\boldsymbol{\Sigma}_{ii}\right) \end{aligned}$$

because

$$|\boldsymbol{\Sigma}| = \left|\begin{array}{c|c} \boldsymbol{\Sigma}_{11} & \mathbf{0} \\ \hline \mathbf{0} & \boldsymbol{\Sigma}_{22} \end{array}\right| = |\boldsymbol{\Sigma}_{11}|\cdot|\boldsymbol{\Sigma}_{22}| \text{ and } \boldsymbol{\Sigma}^{-1} = \left[\begin{array}{c|c} \boldsymbol{\Sigma}_{11}^{-1} & \mathbf{0} \\ \hline \mathbf{0} & \boldsymbol{\Sigma}_{22}^{-1} \end{array}\right]$$

Given that the joint density of $\mathbf{Z}$ factors into the product of the marginal densities, $\mathbf{Z}_{(1)}$ and $\mathbf{Z}_{(2)}$ are independent random vectors. ■

## 4.4 The Exponential Class of Densities

The majority of the discrete and continuous density function families that we have examined in this chapter are special cases of the **exponential class** of density functions.[12] We will see in later chapters that problems of statistical inference involving experiments having probability spaces that involve density families from the *exponential* class are often easier to analyze regarding the design of statistical estimation and inference procedures.

**Definition 4.3** ***Exponential Class of Densities***

The density function $f(\mathbf{x};\boldsymbol{\Theta})$ is a member of the exponential class of density functions *iff*

$$f(\mathbf{x};\boldsymbol{\Theta}) = \begin{cases} \exp\left(\sum_{i=1}^{k} c_i(\boldsymbol{\Theta})g_i(x) + d(\boldsymbol{\Theta}) + z(\mathbf{x})\right) & \text{for } \mathrm{x} \in \mathrm{A}, \\ 0 \quad \text{otherwise}, \end{cases}$$

where $\mathbf{x} = (x_1,\ldots,x_n)'$, $\boldsymbol{\Theta} = (\Theta_1,\ldots,\Theta_k)'$; $c_i(\boldsymbol{\Theta})$, $i = 1,\ldots,k$, and $d(\Theta)$ are real-valued functions of $\boldsymbol{\Theta}$ that do not depend on $\mathbf{x}$; $g_i(\mathbf{x})$, $i = 1,\ldots,k$, and $z(\mathbf{x})$ are real-valued functions of $\mathbf{x}$ that do not depend on $\boldsymbol{\Theta}$; and $A \subset \mathbb{R}^n$ is a set of $n$-tuples contained in $n$-dimensional real space whose definition does not depend on the parameter vector $\Theta$.

In order to check whether a given density family belongs to the exponential class, one must determine whether there exists definitions of $c_i(\boldsymbol{\Theta})$, $d(\boldsymbol{\Theta})$, $g_i(\mathbf{x})$, $z(\mathbf{x})$ and $A$ such that the density can be equivalently represented in the exponential form presented in the Definition 4.2. Of the densities we have studied in this chapter, the Bernoulli, binomial and multinomial (for known values of $n$), negative binomial (for known values of $r$), Poisson, geometric, gamma (including exponential and chi-square), beta, univariate normal, and multivariate normal density families all belong to the exponential class of densities. The discrete and continuous uniform, the hypergeometric density, and the logistic families are *not* members of the exponential class.

We now present a number of examples that illustrate how membership in the exponential class of densities can be verified. In general, there is no standard method of verifying whether a density family belongs to the exponential class, and so the verification process must rely on the ingenuity of the analyst.

[12]We warn the reader that some authors refer to this collection of densities as the exponential *family* rather than the exponential *class*. The collection of densities referred to in this section is a broader concept than that of a parametric *family*, and nests a large number of probability density families within it, as noted in the text. We (and others) use the term *class* to distinguish this broader density collection from that of a parametric family, and to also avoid confusion with the exponential density family discussed in Section 4.2.

**Example 4.23**
***Normal Family ⊂ Exponential Class***

*Univariate Case.* Let $k = 2$, $n = 1$, and define

$$c_1(\Theta) = \frac{\mu}{\sigma^2}, \quad c_2(\Theta) = -\frac{1}{2\sigma^2}, \quad g_1(x) = x, \quad g_2(x) = x^2,$$

$$d(\Theta) = -\frac{1}{2}\left(\frac{\mu^2}{\sigma^2} + \ln(2\pi\sigma^2)\right), \quad z(x) = 0, \quad A = \mathbb{R}$$

Substitution into Definition 4.2 yields

$$f(x;\Theta) = \frac{1}{\sqrt{2\pi}\sigma} \exp -\left[\frac{1}{2}\left(\frac{x-\mu}{\sigma}\right)^2\right],$$

which is the univariate normal density.

*Multivariate Case.* Given an $n$-variate case, let $k = n + (n(n+1)/2)$, and note that the multivariate normal density can be written as

$$N(\mathbf{x};\boldsymbol{\mu},\boldsymbol{\Sigma}) = \frac{1}{(2\pi)^{n/2}|\boldsymbol{\Sigma}|^{1/2}} \exp\left[-\frac{1}{2}(\mathbf{x}-\boldsymbol{\mu})'\boldsymbol{\Sigma}^{-1}(\mathbf{x}-\boldsymbol{\mu})\right]$$

$$= \frac{1}{(2\pi)^{n/2}|\boldsymbol{\Sigma}|^{1/2}} \exp\left[-\frac{1}{2}\left(\Sigma_{i=1}^n x_i^2\, \boldsymbol{\Sigma}^{-1}_{(ii)} + 2\sum_{i<j}\sum x_i\, x_j\, \boldsymbol{\Sigma}^{-1}_{(ij)} - 2\boldsymbol{\mu}'\boldsymbol{\Sigma}^{-1}\mathbf{x} + \boldsymbol{\mu}'\boldsymbol{\Sigma}^{-1}\boldsymbol{\mu}\right)\right],$$

where $\boldsymbol{\Sigma}^{-1}_{(ij)}$ refers to the $(i,j)^{th}$ entry in the $\boldsymbol{\Sigma}^{-1}$ matrix. Define

$$\underset{(1\times k)}{\mathbf{c}(\Theta)'} = -\left(-\boldsymbol{\mu}'\boldsymbol{\Sigma}^{-1} \middle| \tfrac{1}{2}\boldsymbol{\Sigma}^{-1}_{(11)}, \boldsymbol{\Sigma}^{-1}_{(12)}, \ldots, \boldsymbol{\Sigma}^{-1}_{(1n)}, \tfrac{1}{2}\boldsymbol{\Sigma}^{-1}_{(22)}, \boldsymbol{\Sigma}^{-1}_{(23)}, \ldots, \boldsymbol{\Sigma}^{-1}_{(2n)}, \ldots, \tfrac{1}{2}\boldsymbol{\Sigma}^{-1}_{(nn)}\right),$$

$$\underset{(1\times k)}{\mathbf{g}(\mathbf{x})'} = \left(\mathbf{x}' \middle| x_1^2, x_1x_2, \ldots, x_1x_n, x_2^2, x_2x_3, \ldots, x_2x_n, \ldots, x_n^2\right),$$

$$d(\Theta) = -\tfrac{1}{2}\left(\boldsymbol{\mu}'\boldsymbol{\Sigma}^{-1}\boldsymbol{\mu} + \ln(2\pi)^n|\boldsymbol{\Sigma}|\right), \quad z(\mathbf{x}) = 0, \quad A = \mathbb{R}^n$$

Then it follows that $\exp[\mathbf{c}(\Theta)'\mathbf{g}(\mathbf{x}) + d(\Theta) + z(\mathbf{x})] = N(\mathbf{x};\boldsymbol{\mu},\boldsymbol{\Sigma})$. □

**Example 4.24**

Bernoulli Family ⊂ Exponential Class
Let $k = 1$, $n = 1$, and define

$$c_1(\Theta) = \ln(p/(1-p)), g_1(x) = x,\ d(\Theta) = \ln(1-p),\ z(x) = 0,\ A = \{0,1\}.$$

Substitution into Definition 4.2 yields $f(x;p) = p^x(1-p)^{1-x}\, I_{\{0,1\}}(x)$. □

**Example 4.25**

Gamma Family ⊂ Exponential Class
Let $k = 2$, $n = 1$, and define

$$c_1(\Theta) = \alpha - 1, c_2(\Theta) = (-1/\beta), g_1(x) = \ln(x), g_2(x) = x,$$

$$d(\Theta) = -\ln(\beta^\alpha \Gamma(\alpha)), z(x) = 0, A = (0, \infty).$$

Substitution into Definition 4.2 yields

$$f(x; \alpha, \beta) = \frac{1}{\beta^\alpha \Gamma(\alpha)} x^{\alpha-1} e^{-x/\beta} I_{(0,\infty)}(x).$$ □

For now, we simply acknowledge the existence of the exponential class of densities, with the recognition that it encompasses many families of densities that are commonly used in applications. Later we will examine certain general properties of the class of densities that facilitate the construction and evaluation of parameter estimation procedures and statistical hypothesis tests.

## Keywords, Phrases, and Symbols

Additivity property
Bernoulli family
Beta family
Beta function, $B(\alpha,\beta)$
Binomial family
Chi-square additivity
Chi-square sub-family, $\chi^2_v$
Continuous uniform family
Degrees of freedom
Discrete uniform family
Drawing randomly with replacement
Drawing randomly without replacement
Exponential class of densities
Exponential sub-family
Gamma family
Gamma function, $\Gamma(\alpha)$
Geometric Family
Hypergeometric family
Iso-density contours
Logistic distribution
Mean rate of occurrence
Memoryless property (or "good as new")
Multinomial family
Multivariate normal family
Negative binomial family
Parameter space
Parameters
Parametric families of densities
Pascal distribution
Poisson density approximation to binomial
Poisson family
Poisson process
Regression function (hyperplane) or curve (line) of $X_1$ on $X_2$
Reparameterized
Standard normal density
Standard normal random variable
Standardizing a random variable
Univariate normal family
Wear-out effect
Work-hardening effect

## Problems

**1.** A shipment of 100 DVDs contains $k$ defective disks. You randomly sample 20 DVDs, without replacement, from the shipment of 100 DVDs. Letting $p = k/100$, the probability that you will obtain less than three defective disks in your sample of 20 disks is then given by

$$P(x \leq 3) = \sum_{x=0}^{3} \binom{20}{x} p^x (1-p)^{20-x} I_{\{0,1,2,\ldots,20\}}(x).$$

True or False?

**2.** The daily price, $p$, and quantity, demanded, $q$, of gasoline on a European Wholesale spot market can be viewed (approximately) as the outcome of a bivariate normal random variable, where the bivariate normal density has mean vector and covariance matrix as follows:

$$\mu = \begin{bmatrix} 2.50 \\ 100 \end{bmatrix}, \ \Sigma = \begin{bmatrix} .09 & -1 \\ -1 & 100 \end{bmatrix}$$

The price is measured in U.S. dollars per gallon of gasoline, and the quantity demanded is measured in thousands of gallons.

(a) What is the probability that greater than 110,000 gallons of gasoline will be demanded on any given day?

(b) What is the probability that the price of gasoline will be between \$2.00 and \$3.00 on any given day?

(c) Define the regression function of $Q$ on $p$. Graph the regression function. What is the expected daily quantity of gasoline demanded given that price is equal to \$3.00?

(d) Given that the price equals \$3.00, what is the probability that quantity demanded will exceed 110,000 gallons? What is this probability on a day when price equals \$2.00?

**3.** Show that the following probability density functions are members of the exponential class of densities:

(a) Binomial family, for a fixed value of $n$.

(b) Poisson family.

(c) Negative binomial family for a fixed value of $r$.

(d) Multinomial family, for a fixed value of $n$.

(e) Beta family.

**4.** For each PDF family below, show whether or not the family belongs to the exponential class of densities.

(a) $f(x;\beta) = \beta x^{-(\beta+1)} I_{(1,\infty)}(x), \beta \in \Omega = (0,\infty)$. (This is a subfamily of the *Pareto family* of PDFs.)

(b) $f(x;\Theta) = (1/(2\Theta)) \exp(-|x|/\Theta), \Theta \in \Omega = (0,\infty)$. (This is a subfamily of the *double exponential family* of PDFs.)

(c) $f(x;\mu,\sigma) = \left(1/\left(x\sqrt{2\pi}\sigma\right)\right) \exp\left(-(\ln(x)-\mu)^2/(2\sigma^2)\right) I_{(0,\infty)}(x), (\mu,\sigma) \in \Omega = \{(\mu,\sigma): \mu \in (-\infty,\infty), \sigma > 0\}$ (This is the *log-normal family* of PDFs.) Hint: Expanding the square in the exponent of $e$ may be helpful as an alternative representation of the exponent.

(d) $f(x;r) = ((1-r)/r) r^x I_{\{1,2,3,\ldots\}}(x), r \in \Omega = (0,1)$.

**5.** Prove that if $X$ has the geometric density, then the "memoryless property" $P(x > s+t \;|x > s) = P(x > t)$ holds for every choice of positive integers $s$ and $t$.

**6.** Prove that the CDF of the geometric family of densities can be defined by $F(b) = [1-(1-p)^{\mathrm{trunc}(b)}] I_{[1,\infty)}(b)$.

**7.** The quantity of wheat demanded, per day, in a midwestern market during a certain marketing period is represented by

$$Q = 100{,}000 - 12{,}500\,P + V \text{ for } p \in [2,6],$$

where
$Q$ is quantity demanded in bushels;
$p$ is price/bushel; and
$V$ is *approximately normally* distributed.
You know that the expected quantity demanded is given by

$$E(Q) = 100{,}000 - 12{,}500\,p \text{ for } p \in [2,6],$$

and thus is a function of $p$, and the variance of quantity demanded is $\mathrm{var}(Q) = 16 \times 10^6$.

(a) What is the mean and variance of $V$?

(b) If $p = 4$, what is the probability that more than 50,000 bushels of wheat will be demanded?

(c) If $p = 4.50$, what is the probability that more than 50,000 bushels of wheat will be demanded?

(d) For quantity demanded to be greater than 50,000 bushels with probability .95, what does $p$ have to be?

(e) Is it possible that $V$ could actually be normally distributed instead of only *approximately* normally distributed? Explain.

**8.** An investor has \$10,000 which she intends to invest in a portfolio of three stocks that she feels are good investment prospects. During the investor's planning horizon, the weekly closing prices of the stocks can be viewed as the outcome of a trivariate normal random variable with

$$E(\mathbf{X}) = \begin{bmatrix} 27 \\ 10 \\ 18 \end{bmatrix} \text{ and } \mathbf{Cov}(\mathbf{X}) = \begin{bmatrix} 9 & 2 & 1 \\ 2 & 1 & -1 \\ 1 & -1 & 4 \end{bmatrix}.$$

The current price of the stocks are \$23, \$11, and \$19, respectively.

(a) If she invests her \$10,000 equally among the three stocks, what is the expected value of her portfolio? What is the variance of the portfolio value?

(b) What is the probability density function of the portfolio value? What is the probability that the closing value of her portfolio for a given week will exceed \$11,000?

(c) What is the probability density function of the value of stock 1? If she invests the \$10,000 entirely in stock 1, what is the probability that the closing value of her portfolio will exceed \$11,000?

(d) What is the conditional density function of the value of stock 1 *given* that stock 3 has a value of \$17? If she invests the \$10,000 entirely in stock 1, what is the conditional probability that the closing value of her portfolio will exceed \$11,000, *given* that stock 3 has a value of \$17?

(e) If she divides her \$10,000 equally among stocks 1 and 2, what is the conditional probability that this portfolio will have a closing value exceeding \$11,000 *given* that stock 3 has a value of \$17?

**9.** Let $Y$ have a chi-square distribution with 15 degrees of freedom, let $X$ have a chi-square distribution with 5 degrees of freedom, and let $Y = X + Z$, where $X$ and $Z$ are independent random variables.

(a) Calculate $P(y > 27.488)$.

(b) Calculate $P(6.262 < y < 27.488)$.

(c) Find $c$ such that $P(y > c) = .05$.

(d) Find $c$ such that $P(z > c) = .05$.

**10.** Let $Y$ have the density $N(5, 36)$, $X$ have the density $N(4, 25)$, let $Y$ and $X$ be independent random variables, and define $W = X - Y$.

(a) Calculate $P(y > 10)$.

(b) Calculate $P(-10 < y < 10)$.

(c) Calculate $P(w > 0)$.

(d) Find $c$ such that $P(w > c) = .95$.

**11.** Let $\mathbf{X}$ be a bivariate random variable having the probability density $N(\boldsymbol{\mu}, \boldsymbol{\Sigma})$, with

$$\boldsymbol{\mu} = \begin{bmatrix} 5 \\ 8 \end{bmatrix} \text{ and } \boldsymbol{\Sigma} = \begin{bmatrix} 2 & -1 \\ -1 & 3 \end{bmatrix}.$$

(a) Define the regression curve of $X_1$ on $X_2$. What is $E(X_1 \mid x_2 = 9)$?

(b) What is the conditional variance of $X_1$ *given* that $x_2 = 9$?

(c) What is the probability that $x_1 > 5$? What is the probability that $x_1 > 5$, *given* that $x_2 = 9$?

*In problems 12–20 below, identify the most appropriate parametric family of density functions from those presented in this chapter on which to base the probability space for the experiment described, and answer the questions using the probability space you define:*

**12.** WAYSAFE, a large retail supermarket, has a standard inspection policy that determines whether a shipment of produce will be accepted or rejected. Specifically, they examine 5 percent of the objects in any shipment received, and, if no defective produce is found in any of the items examined, the shipment is accepted. Otherwise, it is rejected. The items chosen for inspection are drawn randomly, one at a time, without replacement, from the objects in the shipment.
A shipment of 1,000 5-lb. bags of potatoes are received at the loading dock. Suppose that in reality, 2 percent of the 5-lb. bags have defective potatoes in them. The "objects" that are being inspected are bags of potatoes, with a bag of potatoes being defective if any of the potatoes in the bag are defective.

(a) Define an appropriate probability space for the inspection experiment.

(b) What is the probability that WAYSAFE will accept the shipment of potatoes?

(c) What is the expected number of defective bags of potatoes when choosing 5 percent of the bags for inspection in the manner described above?

(d) If the inspection policy is changed so that 10 percent of the objects will be inspected, and the shipment will be accepted only if no defectives are found, what is the probability that the shipment will be accepted?

**13.** The FLAMES-ARE-US Co. manufactures butane cigarette lighters. Your top-of-the-line lighter, which has the brand name "SURE-FLAME," costs \$29.95. As a promotional strategy, the SURE-FLAME lighter carries a guarantee that if it takes more than five attempts before the lighter actually lights, then the customer will be given \$1,000,000. The terms of the guarantee require that the demonstration of failure of a SURE-FLAME lighter to light within five attempts must be witnessed by an official of the company, and each original buyer of a new SURE-FLAME lighter is allowed only one attempt at being awarded the \$1,000,000. The lighter is such that each attempt at lighting the lighter has a probability of success (it lights) equal to .95, and the outcomes of attempts to light the lighter are independent of one another.

(a) Define the appropriate probability space for the experiment of observing the number of attempts necessary to obtain the first light with the SURE-FLAME lighter.

(b) What is the probability that a buyer who attempts to demonstrate the failure of the SURE-FLAME to light in five attempts will actually be awarded \$1,000,000?

(c) What is the expected number of attempts required to obtain a light with the SURE-FLAME lighter?

(d) What is the expected value of the award paid to any consumer who attempts to claim the $1,000,000?

**14.** An instructor in an introductory economics class has constructed a multiple-choice test for the mid-term examination. The test consists of 20 questions worth 5 points each. For each question, the instructor lists four possible answers, of which only one is correct. John Partytime, a student in the class, has not attended class regularly, and admits (to himself) that he is really not prepared to take the exam. Nonetheless, he has decided to take the exam, and his strategy is to randomly choose one of the four answers for each question. He feels very confident that this course of action will result in an exam score considerably more than zero.

(a) Define a probability space that can be used to assign probability to events involving the number of questions that John answers correctly.

(b) What is the probability that John receives a zero on the exam?

(c) What is the probability that John receives at least 25 points on the exam?

(d) What is John's expected score on the exam?

**15.** The liquid crystal display in the new Extime brand of digital watches is such that the probability it continues to function for at least $x$ hours before failure is constant (for any given choice of the number $x$), regardless of how long the display has already been functioning. The expected value of the number of hours the display functions before failure is known to be 30,000.

(a) Define the appropriate probability space for the experiment of observing the number of hours a display of the type described above functions before failure.

(b) What is the probability that the display functions for at least 20,000 hours?

(c) If the display has already functioned for 10,000 hours, what is the probability that it will continue to function for at least another 20,000 hours?

(d) The display has a rather unique guarantee in the sense that any purchaser of a Extime watch, whether the watch is new or used, has a warranty on the display of 2 years from the date of purchase, during which time if the display fails, it will be replaced free of charge. Assuming that the number of hours the watch operates in a given period of time is essentially the same for all buyers of the Extime watch, is it more likely that a buyer of a used watch will be obtaining a free display replacement than a buyer of a new watch, given an equal period of watch ownership? Explain.

**16.** The Department of Transportation in a foreign county establishes gas-mileage standards that automobiles sold in must meet or else a "gas guzzler" tax is imposed on the sale of the offending types of automobile. For the "compact, four-door" class of automobiles, the target average gas mileage is 25 miles per gallon.
Achievement of the standard is tested by randomly choosing 20 cars from a manufacturer's assembly line, and then examining the distance between the vector of 20 observed measurements of gas mileage/gallon and a $(20 \times 1)$ vector of targeted gas mileages for these cars. Letting $\mathbf{X}$ represent the $20 \times 1$ vector of observed gas mileages, and letting $\mathbf{t}$ represent the $20 \times 1$ vector of targeted gas mileages (i.e., $\mathbf{t} = (25, 25, \ldots, 25)'$), the distance measure is $D(\mathbf{x},\mathbf{t}) = [(\mathbf{x} - \mathbf{t})'(\mathbf{x} - \mathbf{t})]^{1/2}$.
If $D(\mathbf{x},\mathbf{t}) \leq 6$, then the type of automobile being tested is judged to be consistent with the standard; otherwise, the type of automobile will be taxed.
Specific Motors Company is introducing a new four-door compact into the market, and has requested that this type of automobile be tested for adherence to the gas-mileage standard. The engineers at Specific Motors know that the miles per gallon achieved by their compact four-door automobile can be represented by a normal distribution with mean 25 and variance 1.267, so that the target gas mileage is achieved *on average*.

(a) What is the probability that a car randomly chosen from Specific Motor's assembly line will be within 1 mile per gallon of the gas-mileage standard?

(b) What is the probability that Specific Motor's compact four-door will be judged as being consistent with the gas-mileage standard?

(c) A neighboring country uses a simpler test for determining whether the gas-mileage standard is met. It also has a target of 25 miles per gallon, but its test involves forming the simple average of the 20 randomly observed miles per gallon, and then simply testing whether the calculated average is within one mile per gallon of 25 miles per gallon. That is, the gas-mileage standard will be judged to have been met if

$$\frac{1}{20}\sum_{i=1}^{20} x_i \in [24, 26].$$

What is the probability that Specific Motors will pass this alternative test?

**17.** An appliance manufacturer is conducting a survey of consumer satisfaction with appliance purchases. All customers that have purchased one of the company's appliances within the last year will be mailed a customer satisfaction survey. The company is contemplating the proportion of surveys that customers will actually return. It is known from considerable past experience with these types of surveys that the expected proportion of returned surveys is equal to .40, with a variance of .04.

(a) What is the probability that there will be more than 50 percent of the surveys returned?

(b) What is the probability that less than 25 percent of the surveys will be returned?

(c) What is the median level of response to this type of survey? What is the mode?

**18.** Customers arrive at the rate of four per minute at a large bank branch in downtown Seattle. In its advertising, the bank stresses that customers will receive service promptly with little or no waiting.

(a) What is the probability that there will be more than 25 customers entering the bank in a 5-minute period?

(b) What is the expected number of customers that will enter the bank during a 5-minute period?

(c) If the bank staff can service 20 customer in a 5-minute interval, what is the probability that the customer load will exceed capacity in a 5-minute interval, so that some customers will experience delays in obtaining service?

**19.** The accounts of the Excelsior company are being audited by an independent accounting firm. The company has 200 active accounts, of which 140 are current accounts, 45 are past due 60 or more days, and 15 accounts are delinquent. The accounting firm will randomly choose five different accounts in their auditing procedure.

(a) What is the probability that none of the accounts chosen will be delinquent accounts?

(b) What is the probability that at most one of the accounts chosen will be delinquent?

(c) What is the probability that there will be three current, one past due, and one delinquent accounts chosen?

(d) What are the expected numbers of the various types of accounts that will be chosen?

**20.** The Stonebridge Tire Co. manufactures passenger car tires. The manufacturing process results in tires that are either first-quality tires, blemished tires, or defective tires. The proportions of the tires manufactured that fall in the three categories are .88, .09, and .03, respectively. The manufacturing process is such that the classification of a given tire is unaffected by the classifications of any other tires produced.

(a) A lot of 12 tires are taken from the assembly line and inspected for shipment to a tire retailer who sells only first-quality tires. What is the probability that all of the tires will be first-quality tires?

(b) For the lot of tires in part (a), what is the probability that there will be no defective tires among the 12 tires?

(c) What are the expected number of first-quality tires, blemished tires, and defective tires in the lot of 12 tires from part (a)?

(d) What is the probability that the 12 tires will contain 8 first-quality tires, 3 blemished tires, and 1 defective tire?

**21.** KoShop, a large retail department store, has a standard inspection policy that determines whether a shipment of products will be accepted or rejected. Specifically, they examine a randomly chosen sample of 10 percent of the objects in any shipment received, and, if no defectives are found in any of the items examined, the shipment is accepted. Otherwise, it is rejected. The items chosen for inspection are drawn randomly, one at a time, *without* replacement, from the objects in the shipment.
A shipment of 100 Blu-ray disk players is received at the loading dock. Suppose that in reality, unknown to the store, two of the disk players are actually defective.

(a) Define an appropriate probability model for the inspection experiment.

(b) What is the probability that KoShop will accept the shipment of disk players?

(c) What is the expected number of defective players when choosing 10 percent of the players for inspection in the manner described above?

(d) If the inspection policy is changed so that 20 percent of the objects will be inspected, and the shipment will be accepted only if no defectives are found, what is the probability that the shipment will be accepted?

**22.** Customers arrive, on average, at the rate of two per minute at a bank in downtown Portland.

(a) Define an appropriate probability model for the experiment of observing the number of customers entering the bank in a 5-minute period.

(b) What is the probability that there will be more than 20 customers entering the bank in a 5-minute period?

(c) What is the expected number of customers that will enter the bank during a 5-minute period?

(d) If the bank staff can service a maximum of 20 customers in a 5-minute interval, what is the probability that the customer load will exceed capacity in a 5-minute interval, so that some customers will experience delays in obtaining service?

**23.** The accounts of Pullman Plumbers, Inc. are being audited by an independent accounting firm. The company currently has 100 active accounts, of which 50 are current accounts, 35 are past due 60 or more days, and 15 accounts are delinquent. The accounting firm will randomly choose five different accounts in their auditing procedure.

(a) What is the probability that none of the accounts chosen will be delinquent accounts?

(b) What is the probability that at most one of the accounts chosen will be delinquent?

(c) What is the probability that there will be three current, one past due, and one delinquent accounts chosen?

(d) What are the expected numbers of the various types of accounts that will be chosen?

**24.** An instructor in an introductory economics class has a true-false section on the mid-term examination that consists of 10 questions worth 3 points each. Jeff Nostudy, a student in the class, has not attended class regularly, and knows that he is really not prepared to take the exam. Nonetheless, he decides to take the exam, and his strategy is to treat *true* and *false* as equally likely for each question, and randomly choose one as his answer to each question. He feels very confident that this course of action will result in a score for the true-false section that will be considerably more than zero.

(a) Define a probability model that can be used to assign probability to events involving the number of questions that Jeff answers correctly.

(b) What is the probability that Jeff receives zero points on the true-false section of the exam?

(c) What is the probability that Jeff receives at least 15 of the 30 points that are possible on the true-false section of the exam?

(d) What is Jeff's expected score on the true-false section of the exam?

**25.** In the board game of "Aggravation", the game begins with each player having their four game pieces in a "base" holding area and a player is unable to place one of her game pieces on the board for play until they role either a one or a six with a fair die.

(a) Define a probability model that can be used to assign probability to events involving the number of times that a player must roll the fair die to obtain a one or a six.

(b) What is the probability that it will take three or more rolls of the die for a player to get their first game piece on the board?

(c) What is the expected number of rolls of the die for a player to get their first game piece on the board?

(d) Define a new probability model, if you need to, and derive the expected number of rolls of the die for a player to get all four of their initial game pieces on the board.

**26.** The Central Processing Unit (CPU) in a laptop computer that your company manufactures is known to have the following service life characteristics:

$P(x>s+t|x>s) = P(x>t)\ \forall s \text{ and } t>0$

where outcomes, x, of the random variable $X$ measure the operating life of the CPU, measured in 100,000 hour increments, until failure of the CPU. The engineers in your company tell you that $E(X) = 3$.

(a) Define an appropriate probability model for the experiment of observing the operating life of the CPU until the point of failure.

(b) What is the probability that the CPU fails in the first 100,000 hours of use?

(c) Your company provides a warranty on the CPU stating that you will replace the CPU in any laptop in which the CPU fails within the first 5 years of use. What is the probability that you will be fixing one of the computers that has your CPU in it if the laptop were to be operating 24 hours a day?

(d) The CPU is sold to computer manufacturers at a wholesale price of $100 per CPU. The cost to repair a laptop that has a failed CPU is $200. What is your expected net revenue from the sale of one of these CPUs, taking into account the cost of warranty repair, and assuming as in (c) that the laptop were operating 24 hours a day?

**27.** The LCD screen on a popular smartphone has the following operating life characteristic

$P(x>s+t|x>s) < P(x>t)\ \forall s \text{ and } t>0$

where outcomes, x, of the random variable $X$ measure the operating life of the screen, measured in 100,000 hour increments, until failure of the screen. The engineers in your company tell you that $E(X)=1$ and $\sigma_X^2=.5$.

(a) Define an appropriate probability model for the experiment of observing the operating life of the screen until the point of failure.

(b) Does the probability model that you defined in (a) adhere to the operating life characteristic that was indicated above? Why or why not?

(c) What is the probability that the screen will fail in 1 year if the screen were operating 24 hours/day?

(d) The anticipated useful life of the smartphone itself is five calendar years. If the screen is used for 4 hours per day, what is the probability that the screen will not fail during the useful life of the smartphone?

**28.** Let $(X_1,\ldots,X_{20})$ be 20 *iid* $N(2,4)$ random variables. Also, define the random variables $Z_i=\frac{X_i-2}{2}$ for i = 1,..., 20. Answer the following questions relating to these random variables.

(a) What are the values of $E(Z_i^2)$ and $\text{var}(Z_i^2)$?

(b) What are the values of $E\left(\sum_{i=1}^{20} Z_i^2\right)$ and $\text{var}\left(\sum_{i=1}^{20} Z_i^2\right)$?

(c) Calculate the value of $P\left(\sum_{i=1}^{20} z_i^2 \le 31.4104\right)$.

(d) Calculate the value of

$$P\left(\sum_{i=1}^{10} z_i^2 \le 12.5489 \text{ and } \sum_{i=11}^{20} z_i^2 \ge 12.5489\right).$$

**29.** The grades assigned to students taking a midterm in a large principles of economics class is assumed to be normally distributed with a mean of 75 and a standard deviation of 7. The Professor teaching the class, known to be a "stringent grader", has indicated that based on this distribution of grades, she intends to "grade on the curve", whereby the top 15 percent of the students will receive A's, the next 20 percent will receive B's, the "middle" 30 percent will receive C's, the next 20 percent will receive D's, and finally, the lowest 15 percent will receive F's.

(a) Derive the grade ranges associated with each of the letter grades that will be assigned in the class.

(b) Can the grade distribution actually be normally distributed in this case, or must the normal distribution be only an approximation? Explain.

**30.** The production function for the number of toy robots that your company manufactures in a week, expressed in 100's of robots, is represented as follows:

$$Q=\mathbf{a}'\mathbf{x}-\mathbf{x}'\mathbf{B}\mathbf{x}+\varepsilon,$$

where $\mathbf{x}=\begin{bmatrix} l \\ k \end{bmatrix}$ is a 2 × 1vector of labor, $l$, and capital, $k$, applied, $\varepsilon \sim N(0,25)$, with

$$\mathbf{a}=\begin{bmatrix} 10 \\ 5 \end{bmatrix} \text{ and } \mathbf{B}=\begin{bmatrix} 2 & -1 \\ -1 & 1 \end{bmatrix}.$$

(a) Define the expected quantity produced as a function of labor and capital.

(b) If labor and capital are each applied at the level of 10 units each, what is the probability that the quantity produced will exceed 25,000 toy robots?

(c) Define the levels of labor and capital that will maximize the expected quantity produced. What is the maximum expected quantity produced?

(d) If labor and capital are each applied at the level that maximizes expected quantity produced, what is the probability that the quantity produced will exceed 25,000 toy robots?

(e) What is the probability distribution of $Q$? What is the probability distribution of $Q$ if the expected quantity produced is maximized?

(f) If all of the robots produced are sold, and they sell for $10 each, and if the prices of labor and capital per unit are $15 and $10 respectively, derive the expected value, variance, and probability distribution of the maximum weekly profit that can be made from the production and sale of toy robots.

(g) There are at least two reasons why the normal distribution can only be viewed as an approximation to the probability distribution of $\varepsilon$. What are they? Explain.

**31.** The weekly average price, in dollars per gallon, and quantity sold of organic milk, measured in thousands of gallons, in a large west coast market in the fall is represented by the outcomes of a bivariate random variable $(P, Q)$ having a multivariate normal distribution, $N(\boldsymbol{\mu}, \boldsymbol{\Sigma})$ with the following mean vector and covariance matrix:

$$\boldsymbol{\mu} = \begin{bmatrix} 3.50 \\ 100 \end{bmatrix} \text{ and } \boldsymbol{\Sigma} = \begin{bmatrix} .01 & -.7 \\ -.7 & 100 \end{bmatrix}$$

(a) Define an interval event, centered at the mean, that will contain the outcomes of the price of organic milk with .95 probability.

(b) What is the probability that quantity sold will exceed 110,000 gallons?

(c) Define the moment generating function for $(P, Q)$. Use it to define the expected value of weekly total sales of organic milk, in dollars.

(d) Define the regression function of quantity sold as a function of the price of organic milk. Does the regression function make economic sense? Use the regression function to find the expected quantity sold given a price of \$3.40.

(e) Can the bivariate distribution of price and quantity actually be multivariate normally distributed in this case, or must the normal distribution be only an approximation? Explain.

# 5 Basic Asymptotics

## 5.1 Introduction

In this chapter we establish some results relating to the probability characteristics of functions of $n$-variate random variables $\mathbf{X}_{(n)} = (X_1,\ldots,X_n)$ when $n$ is large. In particular, certain types of functions $Y_n = g(X_1,\ldots,X_n)$ of an $n$-variate random variable may converge in various ways to a constant, its probability distribution may be well-approximated by a so-called **asymptotic distribution** as $n$ increases, or the probability distribution of $g(\mathbf{X}_{(n)})$ may converge to a **limiting distribution** as $n \to \infty$.

There are important reasons why the study of the asymptotic (or large sample) behavior of $g(\mathbf{X}_{(n)})$ is an important endeavor. In practice, point estimation, hypothesis testing, and confidence-set estimation procedures (Chapters 7–10) will all be represented by functions of random variables, such as $g(\mathbf{X}_{(n)})$, where $n$ will refer to the number of sample data observations relating to the experiment being analyzed. In order to be able to evaluate, interpret, and/or compare the merits of these statistical procedures, and indeed to be able to define the latter two types of procedures at all, it is necessary to establish the probability characteristics of $g(\mathbf{X}_{(n)})$. Unfortunately, it is often the case in statistical and econometric practice that the actual probability density or distribution of $g(\mathbf{X}_{(n)})$ is difficult to derive, or even intractable to work with analytically, when $n$ is finite. Asymptotic theory often identifies methods that provide tractable approximations to the probability distribution of $g(\mathbf{X}_{(n)})$ when $n$ is sufficiently large, and thereby provides a means of evaluating, comparing, and/or defining various statistical inference procedures. Asymptotic theory also provides the principal rationale for the prevalent use of the normal probability distribution in statistical analyses.

All of the asymptotic results that will be discussed in this chapter rely on the concept of sequences *of random variables*. A fundamental difference between a non-random sequence $\{x_n\}$ and a random sequence $\{X_n\}$ is that the elements of a random sequence are random variables, as opposed to fixed numbers in the non-random case. The random variables in the random sequence are of course capable of assuming any real numbers within their respective ranges. Thus, while $\{x_n\}$ refers to only one sequence of real numbers, $\{X_n\}$ refers to a collection of possible real number sequences defined by the various possible outcomes of the random variables $X_1, X_2, X_3, \ldots$ in the sequence. Since any *particular* real number sequence associated with $\{X_n\}$ can be thought of as an outcome of the random variables involved in the sequence, it is then meaningful to define probabilities of various types of events involving outcomes of the **random sequence**. For example, one might be interested in the probability that the outcomes of a sequence of scalar random variables $\{X_n\}$ is bounded by a particular value $m > 0$, i.e., $P(|x_n| \leq m, \forall n \in N)$, or in the probability that the outcomes of the sequence converges to a limit $c$, i.e., $P(\lim_{n\to\infty} x_n = c)$. Since the sequence is random, convergence and boundedness questions cannot be verified as being unequivocally true or false on the basis of a given sequence of real numbers, but they can be assigned a probability of occurrence in the context of the probability space for the outcomes of the random variables involved.

We will examine four basic types of random variable convergence in this chapter:

1. **Convergence in distribution**;
2. **Convergence in probability**;
3. **Convergence in mean square** (or convergence in quadratic mean); and
4. **Convergence almost surely** (convergence with probability 1).

The material in this chapter is of a more advanced and technical nature, and in order to facilitate readability, the proofs of theorems have been moved to a chapter appendix, and the proofs of some of the theorems are not presented and will be deferred to a more advanced course of study. Some readers, upon first reading, may wish to skim this chapter for main results, while others may wish to move on to Chapter 6 and refer back to the results in this chapter as needed.

For those readers whose understanding of sequences, limits, continuity of functions, and orders of magnitudes of sequences is in need of refreshing, Appendix A.7 provides a review of some elements of real analysis that are particularly relevant to the concepts discussed in this chapter.

## 5.2 Convergence in Distribution

The concept of convergence in distribution involves the question of whether the sequence of random variables $\{Y_n\}$ is such that $\lim_{n\to\infty} P(y_n \in A) = P(y \in A)$ for some "limiting" random variable $Y \sim F(y)$. This is related to whether the sequence of cumulative distribution functions associated with the random variables in the sequence $\{Y_n\}$ converges to a limiting cumulative distribution function (see Definition A.30 regarding the concept of convergence of real-valued functions). The usefulness of the concept lies in establishing an *approximation* to the true CDF (i.e., using the "limiting" CDF $F(y)$, and/or its associated probability density function $f(y)$) for $Y_n$ when $n$ is large enough, where "large enough" means that the CDF of $Y_n$ is close to its limiting CDF. Such approximating CDFs and their associated PDFs can be extremely useful when the true CDF or PDF for $Y_n$ is very difficult (or impossible) to define or is intractable to work with, but the limiting CDF or PDF is easier to define and analyze.

We first characterize convergence in distribution in terms of a sequence of CDFs. We alert the reader to the fact that *all results presented henceforth can be interpreted in terms of multivariate random variables,* unless the context is explicitly defined in terms of scalar random variables.

**Definition 5.1** ***Convergence in Distribution (CDFs)***

Let $\{Y_n\}$ be a sequence of random variables, and let $\{F_n\}$ be the associated sequence of cumulative distribution functions corresponding to the random variables. If there exists a cumulative distribution function, $F$, such that $F_n(y) \to F(y)\ \forall\ y$ at which $F$ is continuous, then $F$ is called the **limiting CDF of** $\{Y_n\}$. Letting $Y$ have the distribution $F$, i.e., $Y \sim F$, we then say that $Y_n$ **converges in distribution** (or converges in law) to the random variable $Y$, and we denote this convergence by $Y_n \xrightarrow{d} Y \left(\text{or } Y_n \xrightarrow{L} Y\right)$. We also write $Y_n \xrightarrow{d} F$ as a short-hand notation for $Y_n \xrightarrow{d} Y \sim F$, which is read "$Y_n$ converges in distribution to $F$."

If $Y_n \xrightarrow{d} Y$, then as $n$ becomes large, the CDF of $Y_n$ is approximated ever more closely by its limiting CDF, the CDF of $Y$ (see Figure 5.1). Note that it is admissible in the definition that $Y = c$, i.e., the random variable can be degenerate so that $P(y = c) = 1$. When the limiting CDF is associated with a degenerate random variable, we say that the sequence of random variables **converges in distribution to a constant**, and we denote this by $Y_n \xrightarrow{d} c$.

It is not generally true that convergence in distribution will imply that the PDFs of the random variables in a sequence will also converge correspondingly.[1] However, in the cases of nonnegative integer-valued discrete random variables

[1] As an example of this situation, let $X_n$ have the discrete uniform distribution on the range $\{\frac{1}{n}, \frac{2}{n}, \ldots, \frac{n-1}{n}, 1\}$, for $n = 1, 2, 3, \ldots$, and let $X$ have the continuous uniform distribution on $[0, 1]$. The probability distribution of $X_n$ converges to the distribution of $X$ as $n \to \infty$. However, the sequence of PDFs are such that $f_n(x) \to 0$ as $n \to \infty\ \forall x \in [0, 1]$.

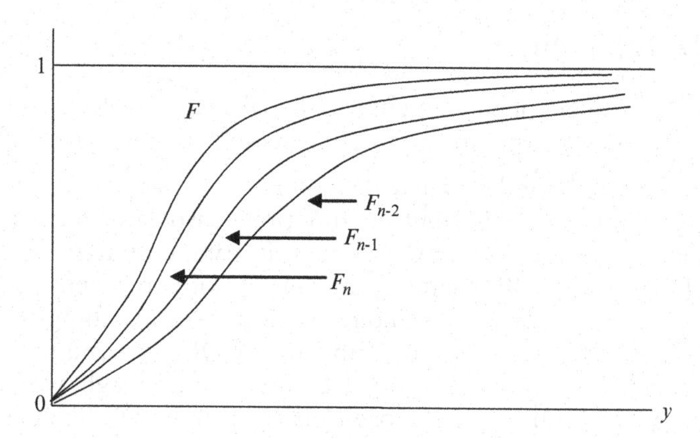


**Figure 5.1** Convergence in distribution (CDF).

and continuous random variables, convergence of a sequence of PDFs is sufficient for establishing convergence in distribution.

**Theorem 5.1** ***Convergence in Distribution (PDFs)*** *Let $\{Y_n\}$ be a sequence of either continuous or nonnegative integer-valued discrete random variables, and let $\{f_n\}$ be the associated sequence of probability density functions corresponding to the random variables. Let there exist a density function, $f$, such that $f_n(y) \to f(y)\ \forall\ y$, except perhaps on a set of points, $A$, such that $P_Y(A) = 0$ in the continuous case, where $Y \sim f$. It follows that $Y_n \xrightarrow{d} Y$ (or $Y_n \xrightarrow{L} Y$).*

If the PDFs of the elements of $\{Y_n\}$ converge to the PDF $f$ as indicated in Theorem 5.1, then $f$ is referred to as the **limiting density** of $\{Y_n\}$. The notation $Y_n \xrightarrow{d} f$ is sometimes used as a short hand for $Y_n \to Y \sim f$, where it is understood that the limiting CDF of $\{Y_n\}$ in this case coincides with the CDF $F$ associated with the PDF $f$. The term **limiting distribution** is often used generically to refer to either the limiting CDF or limiting density. If $Y_n \xrightarrow{d} Y$ and if the PDFs $\{f_n\}$ converge to a limiting density, then as $n$ becomes large, the PDF of $Y_n$ is approximated ever more closely by its limiting density, the density of $Y$ (for example, see Figure 5.2).

The following example illustrates the concept of convergence in distribution characterized through convergence of CDFs. It also provides some insight into the qualifier "such that $F$ is continuous at $y$" used in Definition 5.1.

**Example 5.1** ***Convergence of CDFs*** Let $\{Y_n\}$ be such that the cumulative distribution function for $Y_n$ is defined by $F_n(y) = \frac{n}{2}(y - \tau + n^{-1}) I_{[\tau-n^{-1},\tau+n^{-1}]}(y) + I_{(\tau+n^{-1},\infty)}(y)$, and let $Y$ have the cumulative distribution function $F(y) = I_{[\tau,\infty)}(y)$. Then $Y_n \xrightarrow{d} Y$. To see this, note that (see Figure 5.3) $\lim_{n\to\infty} F_n(y) = \frac{1}{2} I_{\{\tau\}}(y) + I_{(\tau,\infty)}(y)$ which agrees with $F(y)$ for all points at which $F$ is *continuous*, i.e., all points except $\tau$. Thus, by definition, the limiting distribution of $Y_n$ is $F(y)$. □

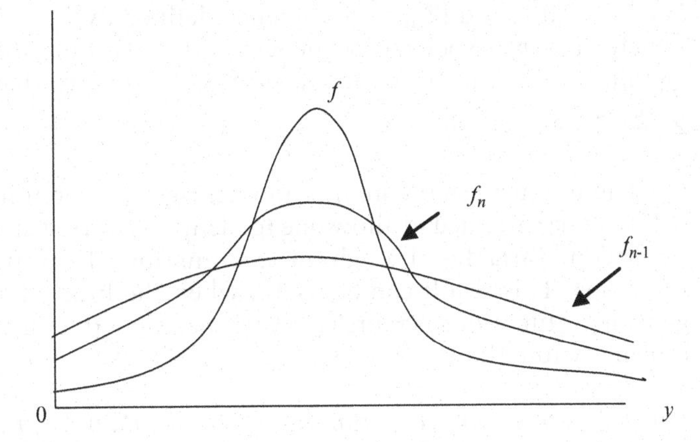


**Figure 5.2**
Convergence in distribution (PDFs).

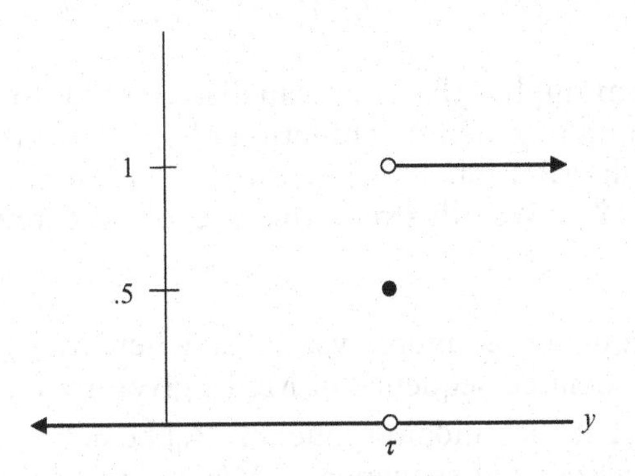


**Figure 5.3**
Graph of $\lim_{n\to\infty} F_n(y)$.

Note that $F(y)$ in Example 5.1 makes sense as a limiting distribution for $Y_n$ when interpreted in the context of its original purpose – that of providing an approximation to $F_n(y)$ for large $n$. Viewed in this way, the fact that $\lim_{n\to\infty} F_n(y)$ does not agree with $F(y)$ at the point $\tau$ is immaterial, since it is not the limit function per se, $\lim_{n\to\infty} F_n(y)$, that we wish to approximate in any case.[2] The reader will note that as $n$ increases, $F_n(y)$ is ever more closely approximated by $F(y)$. In particular, from the definition of $F_n(y)$, it follows that $P(y_n > \tau + n^{-1}) = P(y_n < \tau - n^{-1}) = 0 \;\forall\; n$, implying $P(\tau - n^{-1} \leq y_n \leq \tau + n^{-1}) = 1 \;\forall\; n$, and when $n \to \infty$, $P(y_n = \tau) \to 1$, i.e., the sequence of random variables with distribution functions $\{F_n\}$ converges in distribution to a degenerate random variable at $\tau$. Note that $F(y)$ implies this degeneracy.

In the next example, we illustrate the concept of convergence in distribution characterized through convergence of density functions.

[2]Note that $\lim_{n\to\infty} F_n(y)$ is not even a cumulative distribution function for $y \in \mathbb{R}$.

**Example 5.2** *Convergence of Densities*

Let $X \sim N(0,1)$ and $\{Z_n\}$ be a sequence defined by $Z_n = (3 + n^{-1})X + (2n/(n-1))$, which has an associated sequence of density functions $\{f_n\}$ defined by $f_n = N(2n/(n-1), (3 + n^{-1})^2)$. Since $f_n \to N(2,9) = f$ (note that $f_n(z) \to f(z)\ \forall\ z \in \mathbb{R}$), $Z_n \xrightarrow{d} Z \sim N(2,9)$. □

The uniqueness theorem of moment-generating functions (recall Theorem 3.27) can be extended to allow one to identify limiting distributions of sequences of random variables through an examination of limiting moment-generating functions. This result can be useful when the limit of a sequence of moment-generating functions is more easily defined than the limit of a sequence of CDFs or density functions.

**Theorem 5.2** *Convergence in Distribution (MGFs)*

*Let $\{Y_n\}$ be a sequence of random variables having an associated sequence of moment generating functions $\{M_{Y_n}(t)\}$. Let $Y$ have the moment generating function $M_Y(t)$. Then $Y_n \xrightarrow{d} Y$ iff $M_{Y_n}(t) \to M_Y(t)\ \forall t \in (-h, h)$, for some $h > 0$.*

The theorem implies that if we can discover that $\lim_{n\to\infty} M_{Y_n}(t)$ exists and is equal to the moment generating function $M_Y(t)\ \forall\ t$ in an open interval containing zero, then the distribution associated with $M_Y(t)$ is the limiting distribution of the sequence $\{Y_n\}$. We illustrate the use of Theorem 5.2 in the following example.

**Example 5.3** *Demonstration that* $Z_n = (X_n - n)/\sqrt{2n} \xrightarrow{d} Z \sim N(0,1)$ *if* $X_n \sim \chi^2_n$

Let $\{X_n\}$ be a sequence of random variables, where $X_n \sim \chi^2_n\ \forall n$. The sequence $\{X_n\}$ then has an associated sequence of MGFs given by $\{M_{X_n}(t)\}$, where $M_{X_n}(t) = (1 - 2t)^{-n/2} \forall n$. Let the random sequence $\{Z_n\}$ be defined by $Z_n = (X_n - n)/\sqrt{2n}$, which has an associated sequence of MFGs given by $\{M_{Z_n}(t)\}$, where $M_{Z_n}(t) = (1 - \sqrt{2/n}\ t)^{-n/2} \exp(-\sqrt{n/2}\,t)$. Note that the elements in the sequence $\{Z_n\}$ are *standardized* $\chi^2$ random variables (recall $E(X_n) = n$ and var $(X_n) = 2n$). We now show that $\ln(M_{Z_n}(t)) \to t^2/2$. To see this, first note that $\ln(M_{Z_n}(t)) = -\frac{n}{2}\ln\left(1 - \sqrt{\frac{2}{n}}t\right) - \sqrt{\frac{n}{2}}t$. Expressing the first term on the right-hand side of the equality in terms of a Taylor series expansion around $t = 0$ obtains $\ln(M_{Z_n}(t)) = \left[\sqrt{\frac{n}{2}}t + \frac{t^2}{2} + o(1)\right] - \sqrt{\frac{n}{2}}t = \frac{t^2}{2} + o(1) \to \frac{t^2}{2}$. It follows that

$$\lim_{n\to\infty}(M_{Z_n}(t)) = \lim_{n\to\infty}(\exp(\ln(M_{Z_n}(t)))) = \exp\left(\lim_{n\to\infty}(\ln(M_{Z_n}(t)))\right) = \exp\left(\frac{t^2}{2}\right)$$

since $\exp(\cdot)$ is a continuous function (recall Definition A.29, especially part (b)). Since $\exp(t^2/2)$ is the MGF of a standard normal distribution, we know by Theorem 5.2 that the sequence $\{Z_n\}$ of standardized $\chi^2_n$ random variables has a $N(0,1)$ limiting distribution, i.e., $Z_n \xrightarrow{d} Z \sim N(0,1)$. □

### 5.2.1 Asymptotic Distributions

We now introduce the concept of an **asymptotic distribution** for the elements in a random sequence. In the most general sense of the term, an *asymptotic distribution* for $Z_n$ in the sequence $\{Z_n\}$ is a distribution that provides an approximation to the true distribution of $Z_n$ for large $n$. In this general sense, if $\{Z_n\}$ has a limiting distribution, then since the limiting distribution can be interpreted as an approximation to the distribution of $Z_n$ for large $n$, the limiting distribution might be considered an *asymptotic distribution* for $Z_n$. The purpose of introducing the additional concept of asymptotic distributions is to generalize the concept of approximating distributions for large $n$ to include cases where the sequence $\{Z_n\}$ either has *no* limiting distribution, or the limiting distribution is *degenerate* (and hence not particularly useful as an approximation to the distribution of a nondegenerate $Z_n$).

We will focus on asymptotic distributions that apply to random variables defined by $g(X_n, \Theta_n)$, where $X_n \xrightarrow{d} X$ for nondegenerate $X$. This context covers most cases of practical interest.

**Definition 5.2** ***Asymptotic Distribution for $g(X_n, \Theta_n)$ When $X_n \xrightarrow{d} X$***

Let the sequence of random variables $\{Z_n\}$ be defined by $Z_n = g(X_n,\Theta_n)$, where $X_n \xrightarrow{d} X$ for *nondegenerate* $X$, and $\{\Theta_n\}$ is a sequence of real numbers, matrices, and/or parameters. Then an *asymptotic distribution for* $Z_n$ is given by the distribution of $g(X, \Theta_n)$, denoted by $Z_n \overset{a}{\sim} g(X,\Theta_n)$ and meaning "$Z_n$ **is asymptotically distributed as** $g(X,\Theta_n)$."

Before proceeding to examples, we provide some general motivation for Definition 5.2. Operationally, Definition 5.2 implies that if $Z_n$ can be defined as a function, $g(X_n,\Theta_n)$, of some random variable $X_n$ for which $X_n \xrightarrow{d} X$ and $X$ is nondegenerate, then an asymptotic distribution for $Z_n$ is given by the probability distribution associated with the same function *applied to* $X$, $g(X,\Theta_n)$. The problem of identifying asymptotic distributions is then equivalent to the problem of identifying distributions of functions of $X$ since $Z_n = g(X_n,\Theta_n)$ and $X_n \xrightarrow{d} X \Rightarrow Z_n \overset{a}{\sim} g(X,\Theta_n)$.

In order to motivate why this approach to approximating the distribution of $Z_n$ for large $n$ makes sense, first note that a correspondence between events for $Z_n$ and events for $X_n$ can be established as $z_n \in A \Leftrightarrow x_n \in B_n = \{x_n: z_n = g(x_n,\Theta_n), z_n \in A\}$. Note further that since $X_n \xrightarrow{d} X$, we can approximate the probability of the event $B_n$ for $X_n$ using the nondegenerate limiting distribution of $\{X_n\}$, as $P_{X_n}(x_n \in B_n) \approx P_X(x \in B_n)$ for large $n$. It follows from the equivalence of the events $A$ and $B_n$ that, for large $n$,

$$P_{Z_n}(z_n \in A) = P_{X_n}(x_n \in B_n) \approx P_X(x \in B_n) = P_X(g(x, \Theta_n) \in A).$$

Thus, an approximation to $P_{Z_n}(z_n \in A)$ can be calculated using the probability distribution of $g(X,\Theta_n)$ in the same sense, and with the same degree of accuracy, that $P_{X_n}(x_n \in B_n)$ can be approximated using the probability distribution of $X$.

We emphasize that a limiting distribution for $\{Z_n\} = \{g(X_n, \Theta_n)\}$ need *not* exist for the preceding approximation arguments to hold.

In the next example we illustrate a situation where the limiting distribution is degenerate, but a nondegenerate asymptotic distribution can be established.

**Example 5.4**
***Asymptotic Versus Limiting Distribution for $n^{-1}X_n$, Where $X_n \sim$ Binomial$(n, \lambda/n)$***

Let $\{X_n\}$ be such that $X_n$ has a binomial distribution with parameters $n$ and $p = \lambda/n$ for some fixed value of $\lambda > 0$, i.e., $X_n \sim$ Binomial$(n, \lambda/n)$. Note that $M_{X_n}(t) = (1 + p(e^t - 1))^n = (1 + (\lambda(e^t - 1)/n))^n \rightarrow e^{\lambda(e^t - 1)}$ (recall that $\lim_{n\rightarrow\infty}(1 + (a/n))^n = e^a$ ) which by Theorem 5.2, implies that $X_n \xrightarrow{d} X \sim (e^{-np}(np)^x/x!)I_{\{0,1,2,\ldots\}}(x)$ (Poisson distribution with parameter $\lambda = np$). Note this result is in agreement with the relationship established between the Binomial and Poisson distribution in Chapter 4.

Define the sequence $\{Z_n\}$ by $Z_n = g(X_n, n) = n^{-1}X_n$, and note that since $M_{Z_n}(t) = (1 + (\lambda/n)(e^{t/n} - 1))^n \rightarrow 1$, the *limiting distribution* of $\{Z_n\}$ exists and is *degenerate* at 0 by Theorem 5.2 (i.e., an MGF equal to the constant function 1 is associated with the density for which $P(z = 0) = 1$). On the other hand the *asymptotic distribution* of $Z_n$ as defined in Definition 5.2 is given by the distribution of $Z = g(X,n) = n^{-1}X$, so that $Z_n \overset{a}{\sim} (e^{-np}(np)^{nz}/(nz)!) I_{\{0,1/n,2/n,\ldots\}}(z)$. □

The following example illustrates a case where a limiting distribution does not exist, but a useful asymptotic distribution can nonetheless be defined.

**Example 5.5**
***Asymptotic Versus No Limiting Distribution*** $X_n \overset{a}{\sim} N(n,2n)$ if $X_n \sim \chi^2_n$

In Example 5.3, it was demonstrated that $Z_n \xrightarrow{d} Z \sim N(0,1)$, where $Z_n = (X_n - n)/\sqrt{2n}$ and $X_n \sim \chi^2_n$ . It follows from Definition 5.2, using results on linear combinations of normally distributed random variables, that $X_n = g(W_n, n) = \sqrt{2n}\, Z_n + n \overset{a}{\sim} \sqrt{2n}\, Z + n \sim N(n, 2n)$. Thus, for large $n$, a $\chi^2$-distributed random variable is approximately normally distributed with mean $n$ and variance $2n$. Note that $\{X_n\}$ does *not* have a limiting distribution. As an illustration of the use of the approximation, note that if $n = 100$, then referring to the $\chi^2$-distribution with 100 degrees of freedom, $P(x_{100} \leq 124.34) = .95$. Using the asymptotic distribution $N(100,200)$, the approximation to this probability is given by $\hat{P}(x_{100} \leq 124.34) \approx .9573$. □

### 5.2.2 Convergence in Distribution for Continuous Functions

The final result involving the concept of convergence in distribution that we will examine in this section facilitates identification of the limiting distribution of a continuous function, $g(X_n)$, when $\{X_n\}$ is such that $X_n \xrightarrow{d} X$.

**Theorem 5.3**
***Limiting Distributions of Continuous Functions*** $\{X_n\}$

*Let $X_n \xrightarrow{d} X$, and let the random variable $g(X)$ be defined by a function $g(x)$ that is continuous with probability 1 with respect to the probability distribution of $X$. Then $g(X_n) \xrightarrow{d} g(X)$.*

Thus, if $X_n \xrightarrow{d} X$, the *limiting distribution* of $g(X_n)$ is given by the distribution of $g(X)$ if $g$ is continuous with probability 1 (with respect to the probability distribution of $X$). Note that this result applies to a function of $X_n$ that does not depend on any arguments other than $X_n$ itself. In particular, the function $g$ cannot depend on $n$ or any other sequence of real numbers or matrices whose elements change with $n$, so, for example, $g(x_n) = 3 + x_n$ would be admissible, while $g(x_n,n) = n^{1/2}x_n + 2n$ would not.

**Example 5.6**
***Limiting Distribution of a Continuous Scalar Function***

Let $\{Z_n\}$ be such that $Z_n \xrightarrow{d} Z{\sim}N(0,1)$. It follows immediately from Theorem 5.3 and results on linear combinations of normally distributed random variables that $g(Z_n) = 2Z_n + 5 \xrightarrow{d} 2Z + 5 \sim N(5,4)$. One could also demonstrate via Theorem 5.3 that $g(Z_n) = Z_n^2 \xrightarrow{d} Z^2 \sim \chi_1^2$ (since the square of a standard normal random is a $\chi_1^2$ random variable). □

**Example 5.7**
***Limiting Distribution of a Continuous Vector Function***

Let $\{\mathbf{Z}_n\}$ be a sequence of bivariate random variables such that $\mathbf{Z}_n \xrightarrow{d} \mathbf{Z} \sim N(\boldsymbol{\mu},\boldsymbol{\Sigma})$, where $\boldsymbol{\mu} = [2 \;\; 3]'$ and $\boldsymbol{\Sigma} = \begin{bmatrix} 4 & 0 \\ 0 & 9 \end{bmatrix}$. Then it follows from Theorem 5.3 and results on linear combinations of normally distributed random variables that, for a matrix $\mathbf{A}$ having full row rank, $g(\mathbf{Z}_n) = \mathbf{A}\mathbf{Z}_n \xrightarrow{d} \mathbf{A}\mathbf{Z} \sim N(\mathbf{A}\boldsymbol{\mu},\mathbf{A}\boldsymbol{\Sigma}\mathbf{A}')$. For example, if

$$\mathbf{A} = \begin{bmatrix} 1 & 1 \\ 1 & -1 \end{bmatrix}, \text{ then } \mathbf{A}\mathbf{Z}_n \xrightarrow{d} \left( \begin{bmatrix} 5 \\ -1 \end{bmatrix}, \begin{bmatrix} 13 & -5 \\ -5 & 13 \end{bmatrix} \right).$$

One could also demonstrate via Theorem 5.3 that $g(\mathbf{Z}_n) = (\mathbf{Z}_n - \boldsymbol{\mu})' \, \boldsymbol{\Sigma}^{-1}(\mathbf{Z}_n - \boldsymbol{\mu}) \xrightarrow{d} (\mathbf{Z} - \boldsymbol{\mu})' \, \boldsymbol{\Sigma}^{-1}(\mathbf{Z} - \boldsymbol{\mu}) \sim \chi_2^2$, based on the fact that the two independent standard normal random variables, represented by the $(2 \times 1)$ vector $\boldsymbol{\Sigma}^{-1/2}(\mathbf{Z} - \boldsymbol{\mu})$, are squared and summed. □

## 5.3 Convergence in Probability

Referring to a random sequence of *scalars*, $\{Y_n\}$, the concept of convergence in probability involves the question of whether *outcomes* of the random variable $Y_n$ are close to the *outcomes* of some random variable $Y$ with high probability when $n$ is large enough. If so, the outcomes of $Y$ can serve as an approximation to the outcomes of $Y_n$ for large enough $n$. Stated more rigorously, the issue is whether the sequence of probabilities associated with the sequence of events $\{(y_n, y){:}|y_n - y| < \varepsilon\}$ converges to 1 for every choice of $\varepsilon > 0$, no matter how small. In the case where $\mathbf{Y}_n$ and $\mathbf{Y}$ are $(m \times k)$ *matrices*, convergence in probability of $\mathbf{Y}_n$ to $\mathbf{Y}$ requires that each element of $\mathbf{Y}_n$ converge in probability to the corresponding element of $\mathbf{Y}$. The vector case is subsumed by letting $k = 1$.

**Definition 5.3 Convergence in Probability**

The sequence of random variables, $\{Y_n\}$ *converges in probability* to the random variable, $Y$, *iff*

**a.** *Scalar Case*: $\lim_{n\to\infty} P(|y_n - y| < \varepsilon) = 1, \forall \varepsilon > 0$

**b.** *Matrix Case*:

$\lim_{n\to\infty} P(|y_n[i,j] - y[i,j]| < \varepsilon) = 1, \ \forall \varepsilon > 0, \forall i$ and $j$.

Convergence in probability will be denoted by $Y_n \xrightarrow{p} Y$, or $\text{plim}(Y_n) = Y$, the latter notation meaning the **probability limit of** $Y_n$ is $Y$.

Convergence in probability implies that as $n \to \infty$, the joint distribution of $(Y_n, Y)$ approaches a degenerate distribution defined by linear restrictions, as $P(y_n = y) = 1$. To motivate this interpretation, note that Definition 5.3 implies that for any arbitrarily small $\varepsilon > 0$, $P(y_n[i,j] \in (y[i,j] - \varepsilon, y[i,j] + \varepsilon)) \to 1, \forall i$ and $j$. It then follows from Bonferroni's probability inequality that $P(y_n[i,j] \in (y[i,j] - \varepsilon, y[i,j] + \varepsilon), \forall i \text{ and } j) \to 1$. Thus, outcomes of $Y_n$ are arbitrarily close to outcomes of $Y$ with probability approaching 1 as $n \to \infty$. Therefore, for large enough $n$, observing outcomes of $Y$ is essentially equivalent to observing outcomes of $Y_n$,[3] which motivates the idea that the random variable $Y$ can serve as an approximation to the random variable $Y_n$.

Note that the random variable $Y$ in Definition 5.3 could be *degenerate*, i.e., $Y$ could be a number or matrix of numbers. Using the notation introduced in Definition 5.3, we denote this situation by $Y_n \xrightarrow{p} c$, where $c$ is a scalar or matrix of constants. This situation is referred to as "$Y_n$ **converges in probability to a constant.**"

**Example 5.8 Convergence in Probability to a Constant**

Let $\{Y_n\}$ have an associated sequence of density functions, $\{f_n\}$, defined by $f_n(y) = n^{-1} I_{\{0\}}(y) + (1 - n^{-1}) I_{\{1\}}(y)$. Note that $\lim_{n\to\infty} P(|y_n - 1| = 0) = \lim_{n\to\infty} (1 - n^{-1}) = 1$, so that $\lim_{n\to\infty} P(|y_n - 1| < \varepsilon) = 1 \ \forall \ \varepsilon > 0$, and $\text{plim}(Y_n) = 1$. Thus, $Y_n$ converges in probability to the value 1. □

**Example 5.9 Convergence in Probability to a Random Variable**

Let $Y \sim N(0,1)$, and let $\{Z_n\}$ have an associated sequence of density functions, $\{f_n\}$, with $E(Z_n) = 0$ and $\text{var}(Z_n) = n^{-1}$. Define the random sequence $\{Y_n\}$ by $Y_n = Y + Z_n$, and assume $Y$ and $Z_n$ are independent, so that $E(Y_n) = 0$ and $\text{var}(Y_n) = 1 + n^{-1}$. Then $Y_n \xrightarrow{p} Y$ (or $\text{plim}(Y_n) = Y$) since $\forall \ \varepsilon > 0$, $\lim_{n\to\infty} P(|y_n - y| < \varepsilon) = \lim_{n\to\infty} P(|z_n| < \varepsilon) = 1$, which follows by an application of Markov's inequality. □

[3]The reader may be wondering why we do not simply replace the uncountably infinite collection of probability statements $P(|y_n - y| < \varepsilon), \forall \varepsilon > 0$, in the definition of convergence in probability with the definition $\lim_{n\to\infty} P(y_n = y) = 1$. The problem is that such a convergence definition would not be useful in the examination of any sequence of nondegenerate *continuous* random variables, since $P(y_n = y) = 0 \ \forall \ n \Rightarrow \lim_{n\to\infty} P(y_n = y) = 0$, no matter how close the outcomes of $Y_n$ become to outcomes of $Y$ as $n \to \infty$.

**Example 5.10**
***Convergence in Probability to a Constant Vector***

Let $\{\mathbf{Y}_n\}$ be such that $\mathrm{E}(\mathbf{Y}_n) = \begin{bmatrix} 2 \\ 3 \end{bmatrix}$ and $\mathbf{Cov}\,(\mathbf{Y}_n) = n^{-1}\begin{bmatrix} 2 & 1 \\ 1 & 1 \end{bmatrix}$. Note by Markov's inequality that, $\forall\, \varepsilon > 0$, $\lim_{n\to\infty} P(|y_n[1] - 2| < \varepsilon) \geq \lim_{n\to\infty}\left(1 - \frac{2/n}{\varepsilon^2}\right) = 1$,

and

$$\lim_{n\to\infty} P(|y_n[2] - 3| < \varepsilon) \geq \lim_{n\to\infty}\left(1 - \frac{n^{-1}}{\varepsilon^2}\right) = 1$$

so that $Y_n[1] \xrightarrow{p} 2$ and $Y_n[2] \xrightarrow{p} 3$. From Definition 5.3, it follows that $\mathbf{Y}_n \xrightarrow{p} \mathbf{c} = \begin{bmatrix} 2 \\ 3 \end{bmatrix}$ □

Since Definition 5.3 states that $\operatorname{plim}\left(\underset{(m\times k)}{\mathbf{Y}_n}\right) = \underset{(m\times k)}{\mathbf{Y}}$ *iff* $\operatorname{plim}(Y_n[i,j]) = Y[i,j]\ \forall\, i,j$, it follows that "**the probability limit of a matrix is the matrix of probability limits.**" Thus, just as for the expectation operator, the plim is an *element-wise* operator.

**Definition 5.4**
***Probability Limits of Matrices (and Vectors for k = 1)***

Let $\{\mathbf{Y}_n\}$ be a sequence of $(m \times k)$ random matrices. Then

$$\operatorname{p lim}\begin{pmatrix} Y_n[1,1] & \dots & Y_n[1,k] \\ \vdots & \ddots & \vdots \\ Y_n[m,1] & \dots & Y_n[m,k] \end{pmatrix} = \begin{bmatrix} \operatorname{plim}(Y_n[1,1]) & \dots & \operatorname{plim}(Y_n[1,k]) \\ \vdots & \ddots & \vdots \\ \operatorname{plim}(Y_n[m,1]) & \dots & \operatorname{plim}(Y_n[m,k]) \end{bmatrix}$$

### 5.3.1 Convergence in Probability Versus Convergence in Distribution

At this point we emphasize a fundamental difference in concept between *convergence in distribution* and *convergence in probability* (which will also apply when comparing convergence in distribution to other types of convergence we will study). For the convergence in distribution concept $Y_n \xrightarrow{d} Y$, it is immaterial whether outcomes of the random variables in the sequence $\{Y_n\}$ are related in any way to outcomes of $Y$. In particular, it makes no difference whether $Y_n$ and $Y$ are even referring to the same type of experiment. The convergence $Y_n \xrightarrow{d} Y$ states only that the random variables in the sequence $\{Y_n\}$ have a limiting distribution that is equal to the probability distribution of $Y$. There is no implication that the outcomes of $Y_n$ necessarily emulate the outcomes of $Y$ in any way as $n \to \infty$ (although they might). This is simply a result of the fact that random variables with the same probability distributions are not necessarily the same random variable. For example, $X \sim f(x) = ½\, I_{\{0,1\}}(x)$ could refer to tossing a fair coin where $x = 1$ is a head and $x = 0$ is a tail, while $Y \sim h(y) = ½\, I_{\{0,1\}}(y)$ could be referring to the rolling of a fair die, where $y = 0$ stands for a roll of 3 or less, while $y = 1$ stands for a roll of 4 or more. Clearly, the PDFs of $X$ and $Y$ are equal, i.e., $f(t) = h(t)$ $\forall\, t$. Just as clearly, outcomes of the random variables $X$ and $Y$ are not related in any sense, since whether $x = 1$ (or 0) has no bearing on whether $y = 1$ (or 0). Indeed, $X$ and $Y$ refer to two separate experiments.

The situation is quite different for convergence in probability, where the convergence being referred to involves the *outcomes* of the respective random variables $Y_n$ and $Y$ themselves, and not merely convergence of their probability distributions. That is, convergence in probability implies that, for large $n$, outcomes of the random variable $Y_n$ in the sequence $\{Y_n\}$ are close to the outcomes of $Y$ with high probability, and for large enough $n$, observing an outcome of $Y$ is a very good approximation to observing the corresponding outcome of $Y_n$. Thus, for large enough $n$, outcomes of the experiment represented by $Y_n$ can be treated essentially as outcomes of the experiment represented by $Y$.

### 5.3.2 Properties of the Plim Operator

It is useful to note that if a real-valued sequence of numbers or matrices $\{a_n\}$ converges (in the sense of real analysis) to a given number, or matrix, i.e., $a_n \to a$, then it can also be stated that $\text{plim}(a_n) = a$, since $\{a_n\}$ can be interpreted as a sequence of *degenerate* random variables.

**Theorem 5.4** ***Lim ⇒ Plim*** *Let* $\{a_n\}$ *be such that* $\underset{(m\times k)}{\mathbf{a}_n} \to \underset{(m\times k)}{\mathbf{a}}$. *Then* $\text{plim}(\mathbf{a}_n) = \mathbf{a}$.

When a sequence of random variables is defined via a *continuous* function, $g(X_n)$, of the random variables in another sequence $\{X_n\}$, the plim operator acts analogously to the lim operator of real analysis (also note the analogy to Theorem 5.3).

**Theorem 5.5** ***Plims of Continuous Functions*** *Let* $X_n \xrightarrow{p} X$, *and let the random variable* $g(X)$ *be defined by a function* $g(x)$ *that is continuous with probability 1 with respect to the probability distribution of* $X$. *Then* $g(X_n) \xrightarrow{p} g(X)$, *or equivalently,* $\text{plim}(g(X_n)) = g(\text{plim}\,(X_n))$.

Theorem 5.5 says that for functions that are continuous with probability 1 (with respect to the probability distribution of $X$), *the probability limit of the function is the function of the probability limit.* The theorem can greatly simplify finding the probability limits of complicated functions of $X_n$, especially when convergence is to a constant.

**Example 5.11** ***Plim of a Continuous Function***

**a.** Let $\{X_n\}$ be a positive-valued random variable such that $X_n \xrightarrow{p} 3$. Then $Y_n = g(X_n) = \ln(X_n) + (X_n)^{1/2}$ is such that $\text{plim}(Y_n) = \text{plim}(g(X_n)) = g(\text{plim}(X_n)) = \ln(\text{plim}(X_n)) + (\text{plim}(X_n))^{1/2} = \ln(3) + 3^{1/2} = 2.8307$.

**b.** Let $\{\mathbf{X}_n\}$ be such that $\underset{(k\times 1)}{\mathbf{X}_n} \xrightarrow{p} \underset{(k\times 1)}{\mathbf{X}} \sim N(\mathbf{0},\mathbf{I})$, and let $\mathbf{Y}_n = g(\mathbf{X}_n) = \mathbf{X}_n{}'\mathbf{X}_n$. Then $Y_n \xrightarrow{p} g(\mathbf{X}) = \mathbf{X}'\mathbf{X} \sim \chi^2_k$. □

Through examining a number of special cases of Theorem 5.5, we can establish additional useful properties of the plim operator.

**Theorem 5.6**
***Plim Properties for Addition, Multiplication, and Division***

*For conformable* $\mathbf{X}_n$, $\mathbf{Y}_n$, *and constant matrix* $\mathbf{A}$.

**a.** $\text{plim}(\mathbf{A}\mathbf{X}_n) = \mathbf{A}(\text{plim}(\mathbf{X}_n))$;
**b.** $\text{plim}\left(\sum_{i=1}^m (X_n[i])\right) = \sum_{i=1}^m \text{plim}(X_n[i])$ (*the* plim *of a sum = the sum of the* plims);
**c.** $\text{plim}\left(\prod_{i=1}^m X_n[i]\right) = \prod_{i=1}^m \text{plim}(X_n[i])$ (*the* plim *of a product = the product of the* plims);
**d.** $\text{plim}(\mathbf{X}_n \mathbf{Y}_n) = \text{plim}(\mathbf{X}_n)\text{plim}(\mathbf{Y}_n)$;
**e.** $\text{plim}(\mathbf{X}_n^{-1} \mathbf{Y}_n) = (\text{plim}(\mathbf{X}_n))^{-1}\, \text{plim}(\mathbf{Y}_n)$ (*assuming* $\text{plim}(\mathbf{X}_n)$ *is nonsingular*).

**Example 5.12**
***Plims of Scalar Additive and Multiplicative Functions***

Let $\mathbf{A} = \begin{bmatrix} 2 & 1 \\ 1 & 1 \end{bmatrix}$, and let $\{\mathbf{X}_n\}$ be such that $\text{plim}(\mathbf{X}_n) = \begin{bmatrix} 2 \\ 5 \end{bmatrix}$. Then, plim $(\mathbf{A}\mathbf{X}_n) = \mathbf{A}\,\text{plim}(\mathbf{X}_n) = \begin{bmatrix} 9 \\ 7 \end{bmatrix}$, $\text{plim}(X_n[1] + X_n[2]) = \text{plim}(X_n[1]) + \text{plim}(X_n[2]) = 2 + 5 = 7$, and $\text{plim}(X_n[1]\, X_n[2]) = \text{plim}(X_n[1])\, \text{plim}(X_n[2]) = 2 \cdot 5 = 10$. □

**Example 5.13**
***Plims of Matrix Functions to Constant Matrices***

Let $\{\mathbf{Y}_n\}$ be such that $\text{plim}(\mathbf{Y}_n) = \begin{bmatrix} 1 & 2 \\ 2 & 1 \end{bmatrix}$ and $\{\mathbf{X}_n\}$ be such that $\text{plim}(\mathbf{X}_n) = \begin{bmatrix} 3 & 1 \\ 2 & 1 \end{bmatrix}$. Then $\text{plim}(\mathbf{X}_n \mathbf{Y}_n) = \text{plim}(\mathbf{X}_n)\, \text{plim}(\mathbf{Y}_n) = \begin{bmatrix} 3 & 1 \\ 2 & 1 \end{bmatrix}\begin{bmatrix} 1 & 2 \\ 2 & 1 \end{bmatrix} = \begin{bmatrix} 5 & 7 \\ 4 & 5 \end{bmatrix}$, and $\text{plim}(\mathbf{X}_n^{-1} \mathbf{Y}_n) = (\text{plim}(\mathbf{X}_n))^{-1}\, \text{plim}(\mathbf{Y}_n) = \begin{bmatrix} 1 & -1 \\ -2 & 3 \end{bmatrix}\begin{bmatrix} 1 & 2 \\ 2 & 1 \end{bmatrix} = \begin{bmatrix} -1 & 1 \\ 4 & -1 \end{bmatrix}$. □

**Example 5.14**
***Plims of Matrix Functions to Vector Random Variables***

Let $\{\mathbf{X}_n\}$ and $\{\mathbf{Y}_n\}$ be such that $\text{plim}(\mathbf{X}_n) = \begin{bmatrix} 3 & 2 \\ 2 & 4 \end{bmatrix}$ and $\underset{(2\times1)}{\mathbf{Y}_n} \xrightarrow{p} \underset{(2\times1)}{\mathbf{Y}} \sim \mathbf{N}(\mathbf{0}, \mathbf{I})$. Then $\mathbf{X}_n \mathbf{Y}_n \xrightarrow{p} (\text{plim}(\mathbf{X}_n))\mathbf{Y} \sim \mathbf{N}\left(\mathbf{0}, \begin{bmatrix} 13 & 14 \\ 14 & 20 \end{bmatrix}\right)$, and $\mathbf{X}_n^{-1}\mathbf{Y}_n \xrightarrow{p} (\text{plim}(\mathbf{X}_n))^{-1}\mathbf{Y} \sim \mathbf{N}\left(\mathbf{0}, \begin{bmatrix} .3125 & -.2188 \\ -.2188 & .2031 \end{bmatrix}\right)$. □

### 5.3.3 Relationships Involving Both Convergence in Probability and in Distribution

There are a number of useful results on convergence that involve both convergence in distribution and convergence in probability. One such result has use in situations where it is relatively easy to show (or it is known) that $\mathbf{X}_n \xrightarrow{d} \mathbf{X}$ and $(\mathbf{X}_n - \mathbf{Y}_n) \xrightarrow{p} \mathbf{0}$ but more difficult to demonstrate directly that $\mathbf{Y}_n \xrightarrow{d} \mathbf{X}$.

**Theorem 5.7**
***Plim of Difference and Convergence in Distribution***

Let $\{(\mathbf{X}_n, \mathbf{Y}_n)\}$ be a sequence of pairs of $(m \times k)$ random matrices for which $\mathbf{X}_n \xrightarrow{d} \mathbf{X}$ and $(\mathbf{X}_n - \mathbf{Y}_n) \xrightarrow{p} \mathbf{0}$. Then $\mathbf{Y}_n \xrightarrow{d} \mathbf{X}$.

**Example 5.15**
***Defining Limiting Distribution Through Plim of Differences***

Let $\{X_n\}$ be such that $X_n \sim N(0, (n-1)/n)$, so that $X_n \xrightarrow{d} X \sim N(0,1)$. Let $\{Z_n\}$ be such that $Z_n \sim \chi^2_n$ with $X_n$ and $Z_n$ independent. Define $\{Y_n\}$ by $Y_n = (1 + n^{-1})\, X_n + n^{-1}Z_n - 1$. Then by Theorem 5.6, $Y_n \xrightarrow{d} X \sim N(0,1)$. To see this, first note that $X_n - Y_n = 1 - n^{-1}\,(X_n + Z_n)$, so that plim $(X_n - Y_n) = 1 - \text{plim}(X_n/n) - \text{plim}(Z_n/n)$. Now, note that $\text{E}(X_n/n) = 0\ \forall\ n$ and var $(X_n/n) = n^{-3}(n-1) \rightarrow 0$ implies $\text{plim}(X_n/n) = 0$ by Chebyshev's inequality. Similarly, because $\text{E}(Z_n/n) = 1\ \forall n$ and $\text{var}(Z_n/n) = (2/n) \rightarrow 0$, $\text{plim}(Z_n/n) = 1$ by Chebyshev's inequality. Then plim $(X_n - Y_n) = 0$, and since $X_n \xrightarrow{d} X \sim N(0,1)$, $Y_n \xrightarrow{d} X \sim N(0,1)$ by Theorem 5.7 with $m = k = 1$. □

The following corollary to Theorem 5.7 indicates that convergence in probability of a sequence of random variables implies convergence in distribution of the sequence.

**Corollary 5.1**
***Plim ⇒ Convergence in Distribution***

$Y_n \xrightarrow{p} Y \Rightarrow Y_n \xrightarrow{d} Y.$

Therefore, one way of discovering the limiting distribution of a sequence of scalar or multivariate random variables is to discover the probability limit of the sequence, in which case the limiting distribution is identical to the distribution of the random variable representing the probability limit.

**Example 5.16**
***Defining Limiting Distribution Through Convergence in Probability***

Let $\{Y_n\}$ be defined by $Y_n = (2 + n^{-1})X + 3$, where $X \sim N(1,2)$. Using properties of the plim operator, it follows that $\text{plim}(Y_n) = \text{plim}((2 + n^{-1})X) + \text{plim}\,(3) = 2X + 3 \sim N(5,8)$. Then, Corollary 5.1 implies that $Y_n \xrightarrow{d} N(5,8)$. □

The converse of Corollary 5.1 is not generally true, as motivated by the discussion in Section 5.3.2, i.e., convergence in distribution of two random variables does not necessarily imply that the outcomes of the two random variables are related in any way. However, in the special case where a sequence of random variables converges in distribution to a *constant*, the converse of Corollary 5.1 does hold, since then the random variable is degenerate in the limit, taking a constant value with probability 1.

**Theorem 5.8**
***Convergence in Distribution ⇒ Plim for Constants***

$Y_n \xrightarrow{d} c \Rightarrow Y_n \xrightarrow{p} c$

Thus, in the case of convergence to a constant, the notions of convergence in probability and convergence in distribution are equivalent. Otherwise, convergence in probability is the more stringent type of convergence.

The concepts of convergence in distribution and convergence in probability can be combined to produce a versatile extension of Theorem 5.3 that is useful for deriving the limiting distribution of a much wider variety of functions of $X_n$. In particular, with reference to a random sequence $\{X_n\}$ for which $X_n \xrightarrow{d} X$, the extension facilitates the discovery of the limiting distribution for a function of $X_n$ that is continuous with probability 1 (with respect to the distribution of $X$) when the function *also* depends on a convergent sequence of numbers or matrices $\{a_n\}$, and/or on a sequence of other random variables $\{Y_n\}$ that converges in probability to a *constant* matrix $\mathbf{y}$.

**Theorem 5.9**
***Limiting Distribution of Continuous Functions of $\{\mathbf{X}_n, \mathbf{Y}_n, \mathbf{a}_n\}$***

*Let $\{\mathbf{X}_n\}$, $\{\mathbf{Y}_n\}$, and $\{\mathbf{a}_n\}$ be such that $\underset{(k\times m)}{\mathbf{X}_n} \xrightarrow{d} \underset{(k\times m)}{\mathbf{X}}$, $\underset{(\ell\times q)}{\mathbf{Y}_n} \xrightarrow{p} \underset{(\ell\times q)}{\mathbf{y}}$ where $\mathbf{y}$ is a matrix of constants, and $\underset{(j\times p)}{\mathbf{a}_n} \rightarrow \underset{(j\times p)}{\mathbf{a}}$. Let the set $B$ be such that the probability distribution of $\mathbf{X}$ assigns $P(\mathbf{x}\in B) = 1$, and let the random variable $g(\mathbf{X}_n, \mathbf{Y}_n, \mathbf{a}_n)$ be defined by a (possibly vector) function $g$ that is continuous at every point in the set $B \times \mathbf{y} \times \mathbf{a}$. Then $g(\mathbf{X}_n, \mathbf{Y}_n, \mathbf{a}_n) \xrightarrow{d} g(\mathbf{X}, \mathbf{y}, \mathbf{a})$.*

Theorem 5.9 reduces the problem of identifying the limiting distribution of $g(\mathbf{X}_n, \mathbf{Y}_n, \mathbf{a}_n)$ to the problem of identifying the distribution of $g(\mathbf{X}, \mathbf{y}, \mathbf{a})$. Note that either of the arguments $\mathbf{Y}_n$ or $\mathbf{a}_n$ could be a *ghost* in the function $g(\mathbf{X}_n, \mathbf{Y}_n, \mathbf{a}_n)$, meaning that the value of the function $g$ is unaffected by $\mathbf{Y}_n$ or $\mathbf{a}_n$ (in which case $\mathbf{Y}_n$ or $\mathbf{a}_n$ could be completely ignored in Theorem 5.9).

**Example 5.17**
***Convergence in Distribution of Quadratic Form in RVs and Parameters***

Let $\{\mathbf{Z}_n\}$ be a sequence of bivariate random variables such that $\mathbf{Z}_n \xrightarrow{d} \mathbf{Z} \sim N(\boldsymbol{\mu},\boldsymbol{\Sigma})$ where $\boldsymbol{\mu} = [2\ 3]'$ and $\boldsymbol{\Sigma} = \begin{bmatrix} 4 & 0 \\ 0 & 9 \end{bmatrix}$. Let $\{\boldsymbol{\mu}_n\}$ be any sequence of $(2 \times 1)$ vectors such that $\boldsymbol{\mu}_n \rightarrow \boldsymbol{\mu}$ (e.g., $\boldsymbol{\mu}_n = (2 + n^{-1}|3/(1+\exp(-n)))'$ and let $\boldsymbol{\Sigma}_n$ be any sequence of nonsingular $(2 \times 2)$ matrices such that $\boldsymbol{\Sigma}_n \xrightarrow{p} \boldsymbol{\Sigma}$, for example, $\boldsymbol{\Sigma}_n = \begin{bmatrix} 4+n^{-1} & n^{-1} \\ n^{-1} & 9+n^{-1} \end{bmatrix}$. Then it follows from Theorem 5.9 that $g(\mathbf{Z}_n, \boldsymbol{\Sigma}_n, \boldsymbol{\mu}_n) = (\mathbf{Z}_n-\boldsymbol{\mu}_n)' \boldsymbol{\Sigma}_n^{-1} (\mathbf{Z}_n - \boldsymbol{\mu}_n) \xrightarrow{d} (\mathbf{Z} - \boldsymbol{\mu})' \boldsymbol{\Sigma}^{-1} (\mathbf{Z} - \boldsymbol{\mu}) = g(\mathbf{Z}, \boldsymbol{\Sigma}, \boldsymbol{\mu}) \sim \chi^2_2$ (based on the fact that two independent standard normal random variables represented by the $(2 \times 1)$ vector $\boldsymbol{\Sigma}^{-1/2} (\mathbf{Z} - \boldsymbol{\mu})$, are squared and summed.) Compare the generality of this result to the result found in Example 5.7. □

**Example 5.18**
***Convergence in Distribution of Continuous Function of RVs and Parameters***

Let $X_n \xrightarrow{d} X \sim \theta^{-1} \exp(-x/\theta)\, I_{(0,\infty)}(x)$ (i.e., exponential with parameter $\theta$), $Y_n \xrightarrow{p} 4$, and $a_n \rightarrow 2$. Then $Z_n = g(X_n, Y_n, a_n) = (Y_n)^2 X_n/a_n \xrightarrow{d} 4^2X/2 = 8X = g(X, 4, 2) \sim$ Gamma $(1,8\theta)$. □

A number of special cases of Theorem 5.9 provide a number of useful results that are collectively referred to as **Slutsky's theorems**. The results refer to convergence in distribution of sums and products of random matrices.

**Theorem 5.10**
***Slutsky's Theorems***

*Let* $\mathbf{X}_n \xrightarrow{d} \mathbf{X}$ *and* $\mathbf{Y}_n \xrightarrow{p} \mathbf{c}$. *Then, for conformable* $\mathbf{X}_n$ *and* $\mathbf{Y}_n$,

**a.** $\mathbf{X}_n + \mathbf{Y}_n \xrightarrow{d} \mathbf{X} + \mathbf{c}$;

**b.** $\mathbf{Y}_n \mathbf{X}_n \xrightarrow{d} \mathbf{c}\mathbf{X}$;

**c.** $\mathbf{Y}_n^{-1} \mathbf{X}_n \xrightarrow{d} \mathbf{c}^{-1}\mathbf{X}$ *(if* $\mathbf{c}^{-1}$ *exists).*

### 5.3.4 Order of Magnitude in Probability

It is sometimes useful to be able to characterize or compare random sequences and/or terms in a random sequence relative to their *order of magnitude* in addition to, or in lieu of, any examination of convergence in probability of the sequences involved. The order-of-magnitude concept is especially useful in sorting out which random terms in the definition of a random sequence make dominant contributions to the magnitude of the sequence outcome and which terms are irrelevant as $n$ increases. The order of magnitude of a random sequence, in terms of probability, is described in the following definition. The concept can be thought of as a random or probabilistic counterpart to the order-of-magnitude concept in real analysis (see the Appendix, Section A.7.5).

**Definition 5.5**
***Order of Magnitude in Probability***

Let $\{X_n\}$ be a sequence of random scalars, and let $\{\mathbf{W}_n\}$ be a real-valued random matrix sequence.

**a.** $O_p(n^k)$: The sequence $\{X_n\}$ is said to be **at most of order** $n^k$ **in probability**, denoted by $O_p(n^k)$, *iff* for every $\varepsilon > 0$ there exists a corresponding positive constant $c(\varepsilon) < \infty$ such that $P(n^{-k}|X_n| \leq c(\varepsilon)) \geq 1 - \varepsilon, \forall n$.

**b.** $o_p(n^k)$: The sequence $\{X_n\}$ is said to be **of order smaller than** $n^k$ **in probability**, denoted by $o_p(n^k)$, *iff* $n^{-k}X_n \xrightarrow{p} 0$.

**c.** If $\{W_n[i,j]\}$ is $O_p(n^k)$ (or $o_p(n^k)$) $\forall\, i$ and $j$, then the random matrix sequence $\{\mathbf{W}_n\}$ is said to be $O_p(n^k)$ (or $o_p(n^k)$).

Given Definition 5.5, a random sequence of scalars, $\{X_n\}$, is $O_p(n^k)$ *iff* one can always find a finite interval within which the outcomes of $n^{-k}X_n$ will occur with probability arbitrarily close to (but not necessarily equal to) 1 for each term in the sequence. The random sequence $\{X_n\}$ is $o_p(n^k)$ iff $n^{-k}X_n$ converges in probability to zero. If a random sequence is $O_p(1) \equiv O_p(n^o)$, the random sequence is said to be **bounded in probability**. Note that the ranges of the random variables in a sequence need *not* be finite for the sequence to be $O_p(n^k)$, or even $O_p(1)$, as the next example illustrates.

**Example 5.19**
***Random Sequence of Various Orders***

Let $\{X_n\}$ be such that $X_i \sim N(0,1)$, $\forall\, i$, with all terms in the sequence being independent random variables. Define $\{Z_n\}$ as $Z_n = \sum_{i=1}^{n} X_i$. Then $\{X_n\}$ itself is $O_p(1)$ and $\{Z_n\}$ is $O_p(n^{1/2})$. To see this, first note that since $X_i \sim N(0,1)$, there always exists a constant $c(\varepsilon) > 0$ large enough such $\int_{-c(\varepsilon)}^{c(\varepsilon)} N(x;0,1)\,dx \geq 1 - \varepsilon$ for any choice of $\varepsilon > 0$, so that $\{X_n\}$ is $O_p(1)$ (i.e., $\{X_n\}$ is bounded in probability). Now, note that $n^{-1/2} Z_n = n^{-1/2} \sum_{i=1}^{n} X_i \sim N(0,1)$, whence it follows from the preceding argument that $\{Z_n\}$ is $O_p(n^{1/2})$. It also follows that $\{X_n\}$ is $o_p(n^{\delta})$ and $\{Z_n\}$ is $o_p(n^{1/2+\delta})$ $\forall\ \delta > 0$. Finally, in the sequence defined by $Y_n = n^{-1/2}(X_n + Z_n)$, note that $n^{-1/2} X_n$ is $o_p(1)$, while $n^{-1/2} Z_n$ is $O_p(1)$, implying that as $n \to \infty$, $n^{-1/2} Z_n$ is the dominant random term in the definition of $Y_n$ while $n^{-1/2} X_n$ is stochastically irrelevant as $n \to \infty$. □

Rules for determining the order of magnitude of the sum or product of sequences of random variables are analogous to the rules introduced for the case of sequences of real numbers or matrices and will not be repeated here. In particular, Lemmas A.2 and A.3 in Appendix Section A.7.5 apply with O and o changed to $O_p$ and $o_p$, respectively. The reader should also note the following relationship between O (or o) and $O_p$ (or $o_p$) for sequences of real numbers or matrices.

**Theorem 5.11** ***Relationship Between O and $O_p$***

*Let $\{a_n\}$ be a sequence of real numbers or matrices. If $\{a_n\}$ is $O(n^k)$ (or $o(n^k)$), then $\{a_n\}$ is $O_p(n^k)$ (or $o_p(n^k)$).*

**Example 5.20** **Orders of Function Sequences**

Let $\{x_n\}$ be $O(n^1)$, $\{Y_n\}$ be $O_p(n^2)$ and $\{Z_n\}$ be $O_p(n^1)$. Then $\{x_n Y_n\}$ is $O_p(n^3)$, $\{x_n + Y_n + Z_n\}$ is $O_p(n^2)$, $\{n^{-1} x_n\}$ is $O_p(1)$, $\{n^{-2} x_n (Y_n + Z_n)\}$ is $O_p(n^1)$, $\{x_n Y_n Z_n\}$ is $O_p(n^4)$, and $\{n^{-5} x_n Y_n Z_n\}$ is $o_p(1)$. □

## 5.4 Convergence in Mean Square (or Convergence in Quadratic Mean)

Referring to a random sequence of *scalars*, $\{Y_n\}$, the concept of convergence in mean square involves the question of whether, for some random variable Y, the sequence of expectations $\{E((Y_n - Y)^2)\}$ converges to zero as $n \to \infty$. Since $E((Y_n - Y)^2)$can be interpreted as the expected squared distance between outcomes of $Y_n$ and outcomes of Y, convergence in mean square implies that outcomes of $Y_n$ and Y are "close" to one another for large enough n, and arbitrarily close when $n \to \infty$. When $\mathbf{Y}_n$ and $\mathbf{Y}$ are $m \times k$ matrices, the question of convergence in mean square concerns whether the preceding convergence of expectations to zero occurs *elementwise* between corresponding entries in $\mathbf{Y}_n$ and $\mathbf{Y}$.

**Definition 5.6** ***Convergence in Mean Square (or Convergence in Quadratic Mean)***

The sequence of random variables, $\{Y_n\}$, **converges in mean square** to the random variable, Y, *iff*

**a.** Scalar Case: $\lim_{n\to\infty} (E((Y_n - Y)^2)) = 0$

**b.** Matrix Case: $\lim_{n\to\infty} (E((Y_n[i,j] - Y[i,j])^2)) = 0,\ \forall\ i$ and $j$.

Convergence in mean square will be denoted by $Y_n \overset{m}{\to} Y$.

### 5.4.1 Properties of the Mean Square Convergence Operator

As in the case of convergence in probability, convergence in mean square imposes restrictions on the characteristics of the joint probability distribution of $(Y_n, Y)$ as $n \to \infty$. In particular, first- and second-order moments of corresponding entries in $Y_n$ and $Y$ converge to one another as indicated in the necessary and sufficient conditions for mean square convergence presented in the following theorem:

**Theorem 5.12** ***Necessary and Sufficient Conditions for Mean Square Convergence***

$\mathbf{Y}_n \overset{m}{\rightarrow} \mathbf{Y}$ *iff* $\forall i$ *and* $j$:

**a.** $\mathrm{E}(Y_n[i,j]) \to \mathrm{E}(Y[i,j])$,

**b.** $\mathrm{var}(Y_n[i,j]) \to \mathrm{var}(Y[i,j])$,

**c.** $\mathrm{cov}(Y_n[i,j], Y[i,j]) \to \mathrm{var}(Y[i,j])$.

The conditions in Theorem 5.12 simplify in the case where **Y** is a degenerate random variable equal to the constant matrix, **c**.

**Corollary 5.2** ***Necessary and Sufficient Conditions for Mean Square Convergence to c***

$\mathbf{Y}_n \overset{m}{\rightarrow} \mathbf{c}$ *iff* $\mathrm{E}(Y_n[i,j]) \to c[i,j]$ *and* $\mathrm{var}(Y_n[i,j]) \to 0$ $\forall i$ *and* $j$.

**Example 5.21** ***Convergence in Mean Square to a Constant Vector***

Let $\mathrm{E}(\mathbf{Y}_n) = \begin{bmatrix} 2+3n^{-1} \\ 1+n^{-1} \end{bmatrix}$ and $\mathbf{Cov}\,(\mathbf{Y}_n) = n^{-2}\begin{bmatrix} 2 & 1 \\ 1 & 1 \end{bmatrix}$. Since $\mathrm{E}(\mathbf{Y}_n) \to \begin{bmatrix} 2 \\ 1 \end{bmatrix}$ and $\mathrm{DIAG}(\mathbf{Cov}\,(\mathbf{Y}_n)) = \begin{bmatrix} 2/n^2 \\ 1/n^2 \end{bmatrix} \to \begin{bmatrix} 0 \\ 0 \end{bmatrix}$, it follows by Corollary 5.2 that $\mathbf{Y}_n \overset{m}{\rightarrow} \begin{bmatrix} 2 \\ 1 \end{bmatrix}$.□

In addition to the convergence of second-order moments, convergence in mean square implies that the correlation between corresponding entries in $\mathbf{Y}_n$ and $\mathbf{Y}$ converge to 1 if the entries in $\mathbf{Y}$ have nonzero variances.

**Corollary 5.3** ***Convergence in Mean Square ⇒ Correlation Convergence***

$Y_n \overset{m}{\rightarrow} Y \Rightarrow \mathrm{corr}(Y_n[i,j], Y[i,j]) \to 1$ when $\mathrm{var}(Y[i,j]) > 0$, $\forall i, j$.

The corollary indicates that if $\mathbf{Y}_n \overset{m}{\rightarrow} \mathbf{Y}$, and if none of the entries in $\mathbf{Y}$ are degenerate, then corresponding entries in $\mathbf{Y}_n$ and $\mathbf{Y}$ tend to be *perfectly positively correlated* as $n \to \infty$. It follows that as $n \to \infty$, the outcomes of $Y_n[i,j]$ and $Y[i,j]$ tend to exhibit properties that characterize a situation in which two random variables have perfect positive correlation and *equal* variances, including the fact that $P(|y_n[i,j] - y[i,j]| < \varepsilon) \to 1$ $\forall \varepsilon > 0$ (recall the discussion of correlation in Chapter 3). If $Y[i,j]$ were degenerate, then although the correlation is undefined, the preceding probability convergence result still holds since, in this case, $Y_n[i,j]$ tends toward a degenerate random variable ($\mathrm{var}(Y_n[i,j]) \to 0$) having an outcome equal to the appropriate scalar value of $Y[i,j]$. Thus, if $Y_n \overset{m}{\rightarrow} Y$, outcomes of $Y_n$ emulate outcomes of $Y$ ever more closely as $n \to \infty$, closeness being measured in terms of probability as well as expected squared distance.

### 5.4.2 Relationships Between Convergence in Mean Square, Probability, and Distribution

From the preceding discussion, the fact that mean square convergence is a sufficient condition for both convergence in probability and convergence in distribution could have been anticipated.

**Theorem 5.13** ***Relationship Between Convergence in Mean Square, Probability, and Distribution***

$Y_n \xrightarrow{m} Y \Rightarrow Y_n \xrightarrow{p} Y \Rightarrow Y_n \xrightarrow{d} Y.$

The result of Theorem 5.13 can be quite useful as a tool for establishing convergence in probability and in distribution in cases where convergence in mean square is relatively easy to demonstrate. Furthermore, the theorem allows convergence in probability or in distribution to be demonstrated, even in cases where the distributions of $Y_n$ and/or $Y$ are not fully known, so long as the appropriate convergence properties of the relevant sequences of expectations can be established.

**Example 5.22** ***Using Mean Square Convergence to Establish Convergence in Probability and Distribution***

Let $Y \sim N(0,1)$, $E(Y_n) = 0 \; \forall \, n$, var $(Y_n) \to 1$ and cov $(Y_n, Y) \to 1$. Then, since (recall the proof of Theorem 5.12. c) $E((Y_n - Y)^2) = \text{var}\,(Y_n) + \text{var}\,(Y) - 2\,\text{cov}\,(Y_n, Y) + (E(Y_n) - E(Y))^2 \to 0$, it follows that $Y_n \xrightarrow{m} Y$, which implies that $Y_n \xrightarrow{p} Y$, and $Y_n \xrightarrow{d} N(0,1)$. Note that while we did not know the forms of the probability distributions associated with the random variables in the sequence $\{Y_n\}$, we can nonetheless establish convergence in probability and in distribution via convergence in mean square. □

The following example demonstrates that convergence in mean square is *not* a necessary condition for either convergence in probability or in distribution, so that convergence in probability or in distribution does *not* imply convergence in mean square.

**Example 5.23** ***Illustration that $Y_n \xrightarrow{p} Y$ and/or $Y_n \xrightarrow{d} Y \not\Rightarrow Y_n \xrightarrow{m} Y$***

Let $\{Y_n\}$ be such that $P(y_n = 0) = 1 - n^{-2}$ and $P(y_n = n) = n^{-2}$. Then $\lim_{n\to\infty} P(y_n = 0) = 1$, so that $\text{plim}(Y_n) = 0$ and $Y_n \xrightarrow{d} 0$. However, $E((Y_n - 0)^2) = 0\,(1 - n^{-2}) + n^2(n^{-2}) = 1 \; \forall \, n$, so that $Y_n \overset{m}{\not\to} 0$. □

As a further illustration of differences between convergence in mean square and convergence in probability or in distribution, we now provide an example showing that convergence in probability and/or convergence in distribution do *not* necessarily imply convergence of the first- and second-order moments of $Y_n$ and $Y$.

**Example 5.24** ***Illustration that*** $Y_n \xrightarrow{p} Y$ ***and/or*** $Y_n \xrightarrow{d} Y \not\Rightarrow$ $E(Y_n^r) \rightarrow E(Y^r)$

Let $\{Y_n\}$ be such that $Y_n \sim f_n(y) = (1 - n^{-1})I_{\{0\}}(y) + n^{-1}I_{\{n\}}(y)$, and note that $f_n(y) \rightarrow f(y) = I_{\{0\}}(y)\ \forall\ y$, and thus $Y_n \xrightarrow{d} Y \sim f(y)$. Furthermore, since $\lim_{n\to\infty} P(|y_n - 0| < \varepsilon) = 1\ \forall \varepsilon > 0$, then $Y_n \xrightarrow{p} 0$. Now, note that $E(Y_n) = 1\ \forall n$ and $E(Y_n^2) = n\ \forall\ n$, but $E(Y) = 0$ and $E(Y^2) = 0$ when $Y \sim f(y)$. Thus, *neither* $Y_n \xrightarrow{p} Y$ *nor* $Y_n \xrightarrow{d} Y$ implies that $E(Y_n) \rightarrow E(Y)$ or $E(Y_n^2) \rightarrow E(Y^2)$. □

In summary, convergence in mean square is a *more stringent* type of convergence than either convergence in probability or convergence in distribution. In addition to implying the latter two types of convergence, convergence in mean square *also* implies convergence of first- and second-order moments about the origin and mean as well as convergence to 1 of the correlation between corresponding entries in $Y_n$ and $Y$ (when $\text{var}(Y) > 0$).

## 5.5 Almost-Sure Convergence (or Convergence with Probability 1)

Referring to a sequence of random variables $\{Y_n\}$, which could be scalars, or matrices, the concept of *almost-sure convergence* involves the question of whether, for some random variable $Y$, the limit of the outcomes of the random variable *sequence* converges to the outcome of $Y$ *with probability 1*. That is, does $P(y_n \rightarrow y) = P(\lim_{n\to\infty} y_n = y) = 1$?

**Definition 5.7** ***Almost-Sure Convergence (or Convergence with Probability 1)***

The sequence of random variables, $\{Y_n\}$, **converges almost surely** to the random variable $Y$ *iff*

**a.** Scalar case: $P(y_n \rightarrow y) = P(\lim_{n\to\infty} y_n = y) = 1$,

**b.** Matrix case: $P(y_n[i,j] \rightarrow y[i,j]) = P(\lim_{n\to\infty} y_n[i,j] = y[i,j]) = 1$, $\forall i$ and $j$.

Almost-sure convergence will be denoted by $Y_n \xrightarrow{as} Y$, or by $\textbf{aslim}(Y_n) = Y$, the latter notation meaning *the almost-sure limit of* $Y_n$ *is* $Y$.

Almost-sure convergence is the random counterpart to the non-random real-analysis concept of the limit of a sequence. Almost-sure convergence accommodates the fact that when dealing with sequences of nondegenerate random variables, more than one sequence outcome (perhaps an infinite number) is under consideration for convergence. If $Y_n \xrightarrow{as} Y$, then a real-analysis-type limit is (essentially) certain to be achieved by outcomes of the sequence $\{Y_n\}$, the limit being represented by the *outcome* of $Y$. If $Y$ is degenerate and equal to the constant $c$, then $Y_n \xrightarrow{as} c$ implies that the outcomes of $\{Y_n\}$ are (essentially) certain to converge to the value $c$.

Note that almost-sure convergence is defined in terms of an event involving an *infinite* collection of random variables contained in the sequence $\{Y_n\}$ and in $Y$, the event being that the sequence of outcomes $y_1, y_2, y_3, \ldots$ has a limit that equals the outcome $y$. This is notably different than in either the case of convergence in probability or convergence in mean square, which both relate to sequences of *marginal* probability distributions and outcomes of *bivariate* random variables $(Y_n, Y)$ for $n = 1,2,3,\ldots$. Through counterexamples, we will see that neither convergence in probability nor mean square necessarily implies that $\lim_{n\to\infty} y_n = y$ occurs with probability 1.

In order to clarify the additional restrictions imposed on the sequence $\{Y_n\}$ by almost-sure convergence relative to convergence in probability, we provide an intuitive description of why $Y_n \xrightarrow{p} c$ does *not* imply that $\lim_{n\to\infty} Y_n$ exists with probability 1. Note that for a sequence of outcomes of $\{Y_n\}$ to have a limit, $c$, it must be the case that $\forall\, \varepsilon > 0$, there exists an integer $N(\varepsilon)$ such that for $\forall n \geq N(\varepsilon)$, $|y_n - c| < \varepsilon$ (recall Definition A.25). The definition of almost-sure convergence ensures that the outcomes $\{y_n\}$ are generated in such a way that the limit is achieved *with probability 1*, and so the preceding restriction on the outcomes of $\{Y_n\}$ will be met with probability 1 if $Y_n \xrightarrow{as} c$. However, if $Y_n \xrightarrow{p} c$, so that $P(|y_n - c| < \varepsilon) \to 1, \forall \varepsilon > 0$, then all that can be said is that, for any fixed large value of $n$, the probability that a particular $y_n$ is close to $c$ is high, but *not* necessarily equal to 1. Since $Y_n \xrightarrow{p} c$ does not even imply that $P(|y_n - c| < \varepsilon) = 1$ for *any* fixed value of $n$, it certainly does not imply that there exists an $N(\varepsilon)$ for which $P(|y_n - c| < \varepsilon, \forall n \geq N(\varepsilon)) = 1$. Thus, the existence of a limit for $\{y_n\}$ with probability 1 is not implied by $Y_n \xrightarrow{p} c$.

An alternative and equivalent characterization of almost-sure convergence that follows directly from the fundamental definition of a limit (Definition A.25) is presented in the next theorem. The alternative characterization facilitates both comparisons with convergence in probability and proofs of other results involving the concept of almost-sure convergence.

**Theorem 5.14** ***Alternative Characterization of Almost-Sure Convergence (Scalar Case)***

$$P(\lim_{n\to\infty} y_n = y) = 1 \Leftrightarrow \lim_{n\to\infty} P(|y_i - y| < \varepsilon, i \geq n) = 1,\ \forall \varepsilon > 0$$

The alternative characterization can be extended to cases where $\mathbf{Y}_n$ and $\mathbf{Y}$ are matrices, by simply applying the characterization elementwise to the respective entries in $\mathbf{Y}_n$ and $\mathbf{Y}$.

### 5.5.1 Relationships Between Almost-Sure Convergence and Other Convergence Modes

Theorem 5.14 leads to a relatively straightforward demonstration that almost-sure convergence *implies* convergence in probability.

**Theorem 5.15** ***Relationship Between Almost-Sure and Plim***

$$Y_n \xrightarrow{as} Y \Rightarrow Y_n \xrightarrow{p} Y$$

The converse of Theorem 5.15 is not true, that is, convergence in probability does *not* imply almost-sure convergence. Furthermore, convergence in mean square and/or convergence in distribution also do *not* imply almost-sure convergence. These facts are demonstrated by the following counterexample.

**Example 5.25** ***Almost-Sure Not Implied by Other Convergence Modes***

$Y_n \xrightarrow{m} Y$ **and/or** $Y_n \xrightarrow{p} Y$ **and/or** $Y_n \xrightarrow{d} Y \not\Rightarrow Y_n \xrightarrow{as} Y$. Let $\{Y_n\}$ be a sequence of independent random variables such that $Y_n \sim f_n(y) = (1 - n^{-1})\, I_{\{0\}}(y) + n^{-1} I_{\{1\}}(y)$. Since $f_n(y) \to f(y) = I_{\{0\}}(y)$, $\{Y_n\}$ converges in distribution to the constant 0, and thus it is also true that $Y_n \xrightarrow{p} 0$. Furthermore, since $E(Y_n) = n^{-1} \to 0$ and var $(Y_n) = n^{-1} - n^{-2} \to 0$, it follows that $Y_n \xrightarrow{m} 0$. However, it does *not* follow that $Y_n \xrightarrow{as} 0$. To see this, note that $\forall\ \varepsilon \in (0,1)$ and $\forall$ integer $s > n$,

$$
\begin{aligned}
P(|y_i|<\varepsilon, n \leq i \leq s) &= \prod_{i=n}^{s} (1 - i^{-1}) = \prod_{i=n}^{s} \frac{i-1}{i} \\
&= \frac{n-1}{n} \cdot \frac{n}{n+1} \cdot \frac{n+1}{n+2} \cdots \frac{s-1}{s} \\
&= \frac{n-1}{s} \to 0 \text{ as } s \to \infty.
\end{aligned}
$$

Therefore, $\lim_{n\to\infty} P(|y_i|<\varepsilon, i \geq n) = 0$ because the probability value equals 0 $\forall n \geq 1$, and so the limit of the sequence of probability values equals 0 and *not 1*. Thus, $Y_n \overset{as}{\not\to} 0$. □

It is also true that almost-sure convergence does *not* imply convergence in mean square, as the following counterexample demonstrates:

**Example 5.26** ***Counter example Showing $Y_n \xrightarrow{as} Y \not\Rightarrow Y_n \xrightarrow{m} Y$***

Let $\{Y_n\}$ be a sequence of independent random variables such that $Y_n \sim f_n(y) = (1 - n^{-2})\, I_{\{0\}}(y) + n^{-2}\, I_{\{n\}}(y)$. Note that $\forall\ \varepsilon \in (0,n)$, and $\forall$ integer s $> n$,

$$
P(|y_i|<\varepsilon, n \leq i \leq s) = \prod_{i=n}^{s} (1 - i^{-2}) = \prod_{i=n}^{s} \frac{i^2-1}{i^2} = \frac{(n-1)(s+1)}{ns},
$$

so that $P(|y_i|<\varepsilon, i \geq n) = (n - 1/n)$. Then $\forall\ \varepsilon > 0$, $\lim_{n\to\infty} P(|y_i|<\varepsilon, i \geq n) = 1$ so that $Y_n \xrightarrow{as} 0$. Now, note that $E(Y_n) = n^{-1} \to 0$ but $\text{var}(Y_n) = 1 - n^{-2} \not\to 0$. Therefore, $Y_n \overset{m}{\not\to} 0$ by Corollary 5.2. □

### 5.5.2 Additional Properties of Almost-Sure Convergence

Similar to the case of convergence in probability (Theorem 5.4), it is useful to note that if a real-valued sequence of numbers, or matrices, converges (in the sense of real analysis) to a given number, or matrix, i.e., $a_n \to a$, then it can also be stated that $a_n \xrightarrow{as} a$.

**Theorem 5.16**
***Lim ⇒ Almost-Sure***

*Let* $\{\mathbf{a}_n\}$ *be such that* $\mathbf{a}_n \to \mathbf{a}$. *Then* $\mathbf{a}_n \xrightarrow{as} \mathbf{a}$.

Similar to the cases of convergence in probability and convergence in distribution, a useful result for establishing almost-sure convergence of *continuous* functions of $X_n$, when $X_n \xrightarrow{as} X$, can be stated as follows.

**Theorem 5.17**
***Almost-Sure Convergence of Continuous Functions***

*Let* $X_n \xrightarrow{as} X$, *and let the random variable* $g(X)$ *be defined by a function* $g(x)$ *that is continuous with probability 1 with respect to the probability distribution of* $X$. *Then* $g(X_n) \xrightarrow{as} g(X)$ *or, equivalently,* $\text{aslim } g(X_n) = g(\text{aslim } (X_n))$.

Note that all of the properties of the plim operator listed in Theorem 5.6 apply equally well to the aslim operator since they can all be justified as special cases of Theorem 5.17.

**Example 5.27**
***Almost-Sure Convergence of Sums and Products***

Let $\mathbf{X}_n \xrightarrow{as} \begin{bmatrix} 2 \\ 1 \end{bmatrix}$. Then $g_1(\mathbf{X}_n) = X_n[2]/X_n[1] \xrightarrow{as} 1/2$, $g_2(\mathbf{X}_n) = X_n[2] - X_n[1] \xrightarrow{as} -1$, and $g_3(\mathbf{X}_n) = g_2(\mathbf{X}_n)\, g_1(\mathbf{X}_n) \xrightarrow{as} -1/2$. □

**Example 5.28**
***Almost-Sure Convergence to RV***

Let $\underset{(2\times 1)}{\mathbf{X}_n} \xrightarrow{as} \underset{(2\times 1)}{\mathbf{X}}$, where $X[1] = 3$ and $X[2] \sim N(1,2)$. Then $g(\mathbf{X}_n) = X_n[1]\,(1 + X_n[2]) \xrightarrow{as} 3(1 + X[2]) \sim N(6,18)$. □

The final result we will present in this subsection provides a necessary and sufficient condition for almost-sure convergence *to occur*. The criterion will be useful in proving strong laws of large numbers, which we will examine shortly.

**Theorem 5.18**
***Cauchy's Necessary and Sufficient Condition for Almost-Sure Convergence***

*A sequence of random variables* $\{Y_n\}$ *converges almost surely to some (possibly degenerate) random variable iff* $\lim_{n\to\infty} P\left(\max_{m>n} |y_m - y_n| < \varepsilon\right) = 1, \forall \varepsilon > 0.$[4]

The Cauchy criterion states that for almost-sure convergence to occur, it is necessary and sufficient that the distance between the outcomes of the $n$th term in the random sequence and all subsequent terms beyond the $n$th be arbitrarily small with probability approaching 1 as $n \to \infty$. This makes intuitive sense, since for $\{y_n\}$ to converge to some value, *eventually* (i.e., for all values of $n$ large enough) all the values in the sequence must be arbitrarily close to the limit value and, thus, arbitrarily close to each other.

[4] In the event that max does not exist, max is replaced by sup (supremum, i.e., the smallest upper bound) in the statement of the theorem.

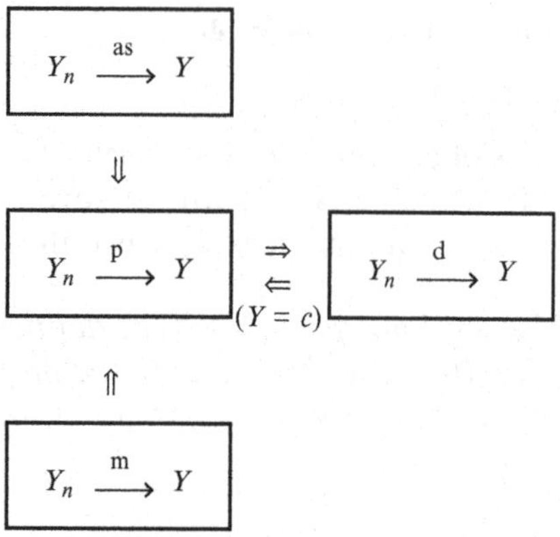

$$Y_n \xrightarrow{as} Y$$

$$\Downarrow$$

$$Y_n \xrightarrow{p} Y \quad \underset{(Y=c)}{\overset{\Rightarrow}{\Leftarrow}} \quad Y_n \xrightarrow{d} Y$$

$$\Uparrow$$

$$Y_n \xrightarrow{m} Y$$

**Figure 5.4**
General convergence mode relationships.

In summary, almost-sure convergence is a more stringent type of convergence than either convergence in probability or convergence in distribution. In addition to implying the latter two types of convergence, almost-sure convergence also implies that the sequence of outcomes of $\{Y_n\}$ *converges to a limit* represented by an outcome of $Y$ with probability 1.

## 5.6 Summary of General Relationships Between Convergence Modes

Collecting together the results presented for the four types of random variable convergence discussed in the previous sections, we can summarize the relationships between the various types of convergences in Figure 5.4.

Note that these exhaust the convergence mode relationships that can be established in terms of general implications between them. However, there are additional special cases that define additional less general interrelationships between the modes of convergence. For example, if a random sequence convergences in probability, it follows that there exists a sub-sequence that converges almost surely.[5] We leave the study of additional special cases to a more advanced course in probability theory.

## 5.7 Laws of Large Numbers

In this section we examine results concerning the convergence behavior of a specific sequence of scalar random variables defined by $\{\bar{X}_n\}$ whose $n$th term is given by $\bar{X}_n = n^{-1} \sum_{i=1}^{n} X_i$, and where $X_i$ is the $i^{\text{th}}$ element of another sequence of random variables $\{X_n\}$. Thus, $\{\bar{X}_n\}$ is a sequence whose $n$th term is given by the simple average of the first $n$ terms in the random sequence $\{X_n\}$. In the context of

[5] Gut, Allan (2005). *Probability: A Graduate Course.* Springer-Verlag, New York, Theorem 3.4.

a sample of observations from an experiment, $\bar{X}_n$ will be referred to as the *sample mean,* and a detailed examination of its properties will be presented in Chapter 6. A convergence result for $\{\bar{X}_n\}$ that uses the concept of *convergence in probability* is referred to as a **weak law of large numbers (WLLNs)**, whereas a convergence result using *almost-sure convergence* is referred to as a **strong law of large numbers (SLLNs)**.

The types of convergence we will be examining can take either of two forms. When all of the means of the random variables in $\{X_n\}$ are equal to the same number, $\mu$, we examine conditions for which $\bar{X}_n \xrightarrow{as} \mu$ (a SLLN) or $\bar{X}_n \xrightarrow{p} \mu$ (a WLLN). When the means of the random variables in $\{X_n\}$ are not necessarily equal, we examine conditions for which $\bar{X}_n - \bar{\mu}_n \xrightarrow{as} 0$ (a SLLN) or $\bar{X}_n - \bar{\mu}_n \xrightarrow{p} 0$ (a WLLN), where $\bar{\mu}_n = n^{-1}\sum_{i=1}^{n}\mu_i$ and $\mu_i = E(X_i)$ is the $i$th term in the real number sequence $\{\mu_n\}$. Thus, in the case of equal means, we are examining convergence of $\{\bar{X}_n\}$ to the common (constant) mean of the random variables in $\{X_n\}$, whereas in the case of unequal means, we are examining whether the difference between the outcomes of $\bar{X}_n$ and the average mean, $\bar{\mu}_n$, of the random variables in $\{X_n\}$ converges to zero in probability or almost surely as $n \to \infty$.[6]

The reader may wonder why the convergence behavior of such a specific sequence of random variables as $\{\bar{X}_n\}$ deserves explicit attention. The answer lies in the fact that a large number of important parameter-estimation and hypothesis-testing procedures in econometrics and statistics can be defined in terms of averages of random variables. The laws of large numbers are then useful for analyzing the asymptotic behavior of these procedures when the samples of data being analyzed are relatively large.

### 5.7.1 Weak Laws of Large Numbers (WLLNs)

There is a variety of conditions that can be placed on the random variables in the sequence $\{X_n\}$ that ensure either $\bar{X}_n \xrightarrow{p} \mu$ or $\bar{X}_n - \bar{\mu}_n \xrightarrow{p} 0$. These conditions relate in various ways to the independence, homogeneity of distribution, and/or variances and covariances of the random variables in the sequence $\{X_n\}$.

The basic idea underlying weak laws of large numbers is to have the distribution of the random variable $\bar{Y}_n = n^{-1}\sum_{i=1}^{n}(X_i - \mu_i)$ collapse and become degenerate on zero as $n \to \infty$. For this to happen, the random variable $\sum_{i=1}^{n}(X_i - \mu_i)$ must be of smaller order of magnitude (in probability) than $n$. The intuition underlying the degeneracy of $\bar{Y}_n$ in the limit is perhaps clearest in the case where the $X_i$'s are *iid* and $\text{var}(X_i) = \sigma^2$ exists. Then the distribution of $\sum_{i=1}^{n}(X_i - \mu_i)$ has a variance, or spread, given by $n\sigma^2$ which is expanding by a factor of $n$. The expanding spread of the distribution is counteracted via scaling the random variable by the factor $n^{-1}$ (actually, any $n^{-\delta}$ with $\delta > 1/2$ will do), leading to a distribution $\bar{Y}_n$ having variance $\sigma^2/n \to 0$ and implying the

[6]The concepts of SLLNs and WLLNs can be generalized to the case where $\{\mu_n\}$ is a sequence of constants that are not necessarily the means of the $X_i$'s. See Y.S. Chow and H. Teicher (1978), *Probability Theory,* p. 121.

degeneracy of $\bar{Y}_n$ at zero as $n \to \infty$. As will be seen ahead, neither the existence of $\sigma^2$ nor the *iid* condition are necessary for $\bar{Y}_n$ to be degenerate at zero in the limit.

#### 5.7.1.1 *IID* Case

The only WLLN we will examine that does *not* require the existence of the variances of the random variables in the sequence $\{X_n\}$ is Khinchin's WLLN, presented as follows.

**Theorem 5.19**
***Khinchin's WLLN***

*Let $\{X_n\}$ be a sequence of iid random variables, and suppose* $\mathrm{E}(X_i) = \mu < \infty, \forall\, i$. *Then* $\bar{X}_n \xrightarrow{p} \mu$.

**Example 5.29**
***Plim($\bar{X}_n$) for iid Gamma RVs***

Let $\{X_n\}$ be a sequence of *iid* random variables, with $X_i$ ~Gamma $(\alpha,\beta)$. It follows from Khinchin's WLLN that $\bar{X}_n \xrightarrow{p} \alpha\beta$. If $\alpha = 2$ and $\beta = 4$, it follows that $\bar{X}_n \xrightarrow{p} 8$.□

**Example 5.30**
***Plim ($\bar{X}_n$) for iid Bernoulli RVs***

Let $\{X_n\}$ be a sequence of *iid* random variables, with $X_i \sim p^x\,(1-p)^{1-x}\, I_{\{0,1\}}\,(x)$. It follows from Khinchin's WLLN that $\bar{X}_n \xrightarrow{p}$ p. If $p = .6$, it follows that $\bar{X}_n \xrightarrow{p} .6$. □

**Example 5.31**
***Illustration of WLLN When Variance Does Not Exist***

Let $f(x) = 2x^{-3} \mathrm{I}_{[1,\infty)}(x)$, and suppose that the random variables in the sequence $\{X_n\}$ are *iid*, each with density function $f(x)$. Note that if $X \sim f(x)$, then $\mathrm{E}(X) = \int_1^\infty x(2x^{-3})dx = 2$, but $\mathrm{E}(X^2) = \int_1^\infty x^2(2x^{-3})dx = \int_1^\infty 2x^{-1}dx = 2\ln(x)|_1^\infty \to \infty$, so that the $\mathrm{var}(X) = \mathrm{E}(X^2) - (\mathrm{E}(X))^2$ does *not* exist. Nonetheless, by Khinchin's theorem we know that $\bar{X}_n \xrightarrow{p} 2$. □

Khinchin's WLLN can be used to provide support for the relative-frequency definition of probability, as follows:

**Theorem 5.20**
***Convergence in Probability of Relative Frequency via WLLN***

*Let $\{S,\Upsilon,P\}$ be the probability space of an experiment, and let A be any event contained in S. Let an outcome of $N_A$ be the number of times that event A occurs in n independent and identical repetitions of the experiment. Then the relative frequency of event A occurring is such that $(N_A/n) \xrightarrow{p} P(A)$.*

Theorem 5.20 implies that as the number of independent identical repetitions of an experiment $\to \infty$, the probability that the relative frequency of event $A$ is arbitrarily close to the true probability of event $A$ approaches 1. Thus, the WLLN provides support for the notion that the relative frequency of the occurrence of an event can be used as the measure of the probability of the event as $n \to \infty$. Note, however, that we did *not* yet conclude that $P(\lim_{n\to\infty}(n_A/n) = P(A)) = 1$, that is, we cannot conclude that the *limit* of the relative frequency *exists* and *is equal to* $P(A)$ *with probability* 1. The latter result involves the notion of almost-sure convergence and will be dealt with in our subsequent discussion of SLLNs.

#### 5.7.1.2 Non-*IID* Case

WLLNs that relax the *iid* assumption of Khinchin's WLLN can be defined by imposing various other conditions on the variances and covariances of the random variables in the sequence $\{X_n\}$. The WLLN that we will present follows from the necessary and sufficient conditions for the existence of a WLLN, stated in the next theorem.

**Theorem 5.21 Necessary and Sufficient Conditions for WLLN**

*Let $\{X_n\}$ be a sequence of random variables with finite variances (not necessarily independent), and let $\{\mu_n\}$ be the corresponding sequence of their expectations. Then*

$$\lim_{n\to\infty} P(|\bar{X}_n - \bar{\mu}_n|<\varepsilon)= 1, \forall\, \varepsilon > 0 \text{ iff } E\left(\frac{(\bar{X}_n-\bar{\mu}_n)^2}{1+(\bar{X}_n-\bar{\mu}_n)^2}\right) \to 0.$$

Any condition placed on the random variables in the sequence $\{X_n\}$ that results in the convergence of $\mathrm{E}\left(\frac{(\bar{X}_n-\bar{\mu}_n)^2}{1+(\bar{X}_n-\bar{\mu}_n)^2}\right)$ to zero results in $(\bar{X}_n-\bar{\mu}_n) \xrightarrow{p} 0$ (or $\bar{X}_n \xrightarrow{p} \mu$ in the equal-means case) by Theorem 5.21. We present one such condition now.

**Theorem 5.22 WLLN for Non-IID Case**

*Let $\{X_n\}$ be a sequence of random variables with respective means given by $\{\mu_n\}$. If $\mathrm{var}(\bar{X}_n) \to 0$, then $(\bar{X}_n-\bar{\mu}_n) \xrightarrow{p} 0$.*

**Example 5.32 Plim($\bar{X}_n$) for Non-iid Gamma RVs**

Let $\{X_n\}$ be a sequence of Gamma-distributed random variables for which $\mathrm{E}(X_i) = 2^{-i}$, $\mathrm{var}(X_i) = 4$, and $\sigma_{ij} = 0, \forall\, i \neq j$. Then since $\mathrm{var}(\bar{X}_n) = 4/n \to 0$, it follows by Theorem 5.22 that $\bar{X}_n-(1-.5^n)/n \xrightarrow{p} 0$, where $\bar{\mu}_n = (1-.5^n)/n$. □

**Example 5.33 Plim($\bar{X}_n$) for Non-iid Beta RVs**

Let $\{X_n\}$ be a sequence of independent, Beta-distributed, random variables for which $\mathrm{E}(X_i) = .4\, \forall\, i$. Note that the variance of a Beta distribution exhibits a finite upper bound, say $\sigma_i^2 \leq \tau$, since $P(x\in(0,1)) = 1$. Then, *for any* variances of the random variables, $\mathrm{var}(\bar{X}_n) = n^{-2} \sum_{i=1}^n \sigma_i^2 \leq n^{-1}\tau \to 0$. By Theorem 5.22, $\bar{X}_n \xrightarrow{p} .4$. (Note: Khinchin's theorem cannot be used here since it is not known whether the $X_i$'s have identical distributions.) □

**Example 5.34 Plim($\bar{X}_n$) for Non-iid Normal RVs**

Let the sequence of random variables $\{X_n\}$ be such that $X_i \sim N(1,1 + i^{-1})$ with $\sigma_{ij} = \rho^{|i-j|}$, $\rho\in(0,1)$ and $i \neq j$. Since $\sigma_i^2 \leq 2\ \forall i$, $\sum_{i=1}^n \sigma_i^2 \leq 2n$, and thus $\sum_{i=1}^n \sigma_i^2$ is $o(n^2)$. Also, given $i$, $\sum_{j>i}^n \sigma_{ij} = \sum_{j>i}^n \rho^{|i-j|} \to \rho/(1-\rho)$, so that $\sum_{j>i}^n \sigma_{ij}$ is $o(n^1)$. Then, since $\bar{\mu}_n = 1\ \forall n$, and $\mathrm{var}(\bar{X}_n) = n^{-2}\left[\sum_{i=1}^n \sigma_i^2 + 2\sum_{i=1}^n \sum_{j>i}^n \sigma_{ij}\right] = o(1)$ implying that $\mathrm{var}(\bar{X}_n) \to 0$, it follows from Theorem 5.22 that $\bar{X}_n \xrightarrow{p} 1$. □

With reference to Example 5.33, it would seem reasonable to characterize the convergence in probability by stating that $\bar{X}_n \xrightarrow{p} 0$, because $\bar{\mu}_n = (1-.5^n)/n \to 0$. More generally, if $\bar{X}_n - \bar{\mu}_n \xrightarrow{p} 0$ and $\bar{\mu}_n \to c$, we can alternatively state that $\bar{X}_n \xrightarrow{p} c$ as indicated below.

**Theorem 5.23 WLLN via Convergence of Difference**

$\bar{X}_n - \bar{\mu}_n \xrightarrow{p} 0$ *and* $\bar{\mu}_n \to c \Rightarrow \bar{X}_n \xrightarrow{p} c$.

### 5.7.2 Strong Laws of Large Numbers (SLLNs)

As in the case of the WLLN, there is a variety of conditions that can be placed on the random variables in the sequence $\{X_n\}$ to ensure either $\bar{X}_n \xrightarrow{as} \mu$ (when

$\mathrm{E}(X_i) = \mu, \forall\, i)$ or $\bar{X}_n - \bar{\mu}_n \xrightarrow{as} 0$ (when $\mathrm{E}(X_i) = \mu_i, \forall\, i)$. The conditions relate to the independence, homogeneity of distribution, and/or variances and covariances of the random variables in the sequence. We note to the reader that the results are somewhat more difficult to establish than in the case of WLLNs.

The basic idea underlying strong laws of large numbers is to have the joint distribution of $Y_n = n^{-1}\sum_{i=1}^{n}(X_i - \mu_i)$, $n = 1,2,\ldots$, be such that the convergence event $y_n \to 0$ is assigned probability 1. For this to happen, it is known from Theorem 5.14, as well as from the basic concept of a limit itself, that the event $\{|y_i| < \varepsilon, \forall i \geq n\}$ must have a probability approaching 1 as $n \to \infty\ \forall\, \varepsilon > 0$. Thus the *marginal* distribution of the (infinite) set of random variables $\{Y_n, Y_{n+1}, Y_{n+2}, \ldots\}$ must be approaching degeneracy on a zero vector as $n \to \infty$. Through various constraints on the spread (variance) and/or degree of dependence of the underlying random variables in the sequence $\{X_i\}$, this degenerate behavior can be attained.

#### 5.7.2.1 IID Case

We begin examining SLLNs by focusing on the *iid* case, and establishing a result known as **Kolmogorov's inequality**, which can be interpreted as a *generalization of Markov's inequality.*

**Theorem 5.24** ***Kolmogorov's Inequality***

*Let $X_1,\ldots,X_n$ be independent random variables for which $\mathrm{E}(X_i) = 0$ and $\sigma_i^2 < \infty$ $\forall i$. Then $\forall\, \varepsilon > 0$, $P\left(\max_{1\leq m\leq n} |\sum_{i=1}^{m} x_i| \geq \varepsilon\right) \leq \sum_{i=1}^{n} \frac{\sigma_i^2}{\varepsilon^2}$.*

Note that with $n = 1$ and $X_1$ defined to be a nonnegative-valued random variable with zero mean, Theorem 5.24 is a statement of Markov's inequality. Kolmogorov's inequality leads to **Kolmogorov's** SLLN for *iid* random variables.

**Theorem 5.25** ***Kolmogorov's SLLN***

*Let $\{X_n\}$ be a sequence of iid random variables such that $\mathrm{E}(X_i) = \mu$ and $\mathrm{var}(X_i) = \sigma^2 < \infty$, $\forall\, i$. Then $\bar{X}_n \xrightarrow{as} \mu$.*

**Example 5.35** ***Almost-Sure Convergence of $\bar{X}_n$ from iid Experimental RVs***

Let $\{X_n\}$ be a sequence of *iid* exponentially distributed random variables, $X_i \sim \theta^{-1}\exp(-x_i/\theta)\ I_{(0,\infty)}(x_i)\ \forall i$, and assume $\theta \leq c < \infty$. Note that $\mathrm{E}(X_i) = \mu = \theta$ and $\mathrm{var}(X_i) = \sigma^2 = \theta^2 \leq c^2 < \infty$, $\forall i$. Theorem 5.25 applies, so that $\bar{X}_n \xrightarrow{as} \theta$. □

The existence of variances is not necessary for a SLLN to hold in the *iid* case. The following theorem provides necessary and sufficient conditions for a SLLN to hold.

**Theorem 5.26** ***Kolmogorov's SLLN (No Variance Existing)***

*Let $\{X_n\}$ be a sequence of iid random variables. Then the condition $\mathrm{E}(X_i) = \mu < \infty$ is necessary and sufficient for $\bar{X}_n \xrightarrow{as} \mu$.*

**Example 5.36** ***Illustration of SLLN When Variance Does Not Exist***

Recall Example 5.31, where $\{X_n\}$ was a sequence of *iid* random variables for which $\mathrm{E}(X_i) = 2$ and for which the variance of the $X_i$'s did *not* exist. Nonetheless, by Kolmogorov's SLLN, we know that $\bar{X}_n \xrightarrow{as} 2$. □

Theorem 5.26 provides stronger support for the relative-frequency definition of probability than do the WLLNs (recall Theorem 5.20), as the following theorem indicates.

**Theorem 5.27**
***Almost-Sure Convergence of Relative Frequency***

*Let $\{S,\Upsilon,P\}$ be the probability space of an experiment, and let A be any event contained in S. Let an outcome of $N_A$ be the number of times that event A occurs in n independent and identical repetitions of the experiment. Then the relative frequency of event A occurring is such that $(N_A/n) \overset{as}{\rightarrow} P(A)$.*

Theorem 5.27 implies that the relative frequency of the occurrence of event $A$ *achieves a limit* with probability 1 as $n \rightarrow \infty$. Furthermore, the value of this limit equals the probability of the event $A$. Thus, the SLLN provides strong support for the notion that the relative frequency of the occurrence of an event can be used as the measure of the probability of the event as $n \rightarrow \infty$, since we are *essentially certain* that the relative frequency of an event will converge to the probability of the event.

### *5.7.2.2 Non-IID Case*

There are many other ways that restrictions can be placed on $\{X_n\}$ so that $\bar{X}_n - \bar{\mu}_n \overset{as}{\rightarrow} 0$ or $\bar{X}_n \overset{as}{\rightarrow} \mu$. We will present a SLLN for the non-identically distributed case which can be applied whether or not the random variables in $\{X_n\}$ are independent. The theorem utilizes the concept of an *asymptotic nonpositively correlated sequence*, defined below.

**Definition 5.8**
***Asymptotic Nonpositively Correlated Sequence***

The sequence of random scalars $\{X_n\}$, where $\text{var}(X_i) = \sigma_i^2 < \infty\ \forall i$, is said to be **asymptotic nonpositively correlated** if there exists a sequence of constants $\{a_n\}$ such that $a_i \in [0,1]\ \forall\ i$, $\sum_{i=0}^{\infty} a_i < \infty$, and $\text{cov}(X_i, X_{i+t}) \leq a_t\sigma_i\sigma_{i+t}\ \forall\ t>0$.[7]

Note that for $\sum_{i=0}^{\infty} a_i$ to be finite when $a_i \in [0,1]\ \forall i$, it must be the case that $a_n \rightarrow 0$ as $n \rightarrow \infty$. Since the $a_t$'s represent upper bounds to the correlations between $X_i$ and $X_{i+t}$, the definition implies that $X_i$ and $X_{i+t}$ cannot be positively correlated when $t \rightarrow \infty$.

**Example 5.37**
***An Asymptotic Nonpositively Correlated Sequence***

Let the sequence of random variables $\{X_n\}$ adhere to the *(first-order) autocorrelation process*[8]:

(I) $X_i = \rho\, X_{i-1} + \varepsilon_i$,

[7]Some authors use the terminology "asymptotically uncorrelated" for this concept (e.g., H. White, *Asymptotic Theory*, pp. 49). However, the concept does not rule out negative correlation.

[8]This is an example of a *stochastic process* that we will revisit in our discussion of the general linear model in Chapter 8. "Stochastic process" means any collection of random variables $\{X_t;\ t \in T\}$, where $T$ is some index set that serves to order the random variables in the collection. Special cases of stochastic processes include a scalar random variable when $T = \{1\}$, an $n$-variate random vector when $T = \{1,2,\ldots,n\}$, and a random sequence when $T = \{1,2,3,\ldots\}$.

where the $\varepsilon_i$'s are *iid* with $E(\varepsilon_i) = 0$ and $var(\varepsilon_i) = \sigma^2 \in (0,\infty)$, $cov(X_{i-1}, \varepsilon_i) = 0 \; \forall i$, $X_0 = 0$, and $|\rho| < 1$. It follows that $E(X_i) = 0 \; \forall i$, and $var(X_i) = \sigma^2 \sum_{j=1}^{i} \rho^{2(j-1)}$ for $i \geq 1$. Note further that (I) implies $X_{i+t} = \rho^t X_i + \sum_{j=0}^{t-1} \rho^j \varepsilon_{t+i-j}$ for $t \geq 1$. To define the value of $corr(X_{i+t}, X_i)$, let $\sigma_i^2 = var(X_i)$, and note that

$$\begin{aligned} corr(X_{i+t}, X_i) &= E((X_{i+t} - E(X_{i+t}))(X_i - E(X_i)))/(\sigma_{i+t}\,\sigma_i) \\ &= E\left( (\rho^t (X_i - E(X_i))) \sum_{j=0}^{t-1} \rho^i \varepsilon_{t+i-j} \right) (X_i - E(X_i))/(\sigma_{i+t}\,\sigma_i) \\ &= \rho^t \sigma_i / \sigma_{i+t} \leq \begin{cases} 0 & \text{if } \rho^t \leq 0 \\ \rho^t & \text{if } \rho^t > 0 \end{cases} \end{aligned}$$

where the last inequality follows from the fact that

$$\sigma_i^2 / \sigma_{i+t}^2 = \sum_{j=1}^{i} \rho^{2(j-1)} \Big/ \sum_{j=1}^{i+t} \rho^{2(j-1)} \leq 1 \; \forall t \geq 1.$$

To demonstrate that $\{X_n\}$ is an asymptotic nonpositively correlated sequence, define $a_t = \rho^t$ if $\rho^t > 0$ and $a_t = 0$ if $\rho^t \leq 0$, so that $corr(X_{i+t}, X_i) \leq a_t \; \forall t$, and $a_t \in [0,1]$, $\forall t$. Then, since $\sum_{t=1}^{\infty} a_t \leq \sum_{t=1}^{\infty} |\rho|^t = |\rho|/(1 - |\rho|) < \infty$, the sequence $\{X_n\}$ is asymptotic nonpositively correlated. □

**Theorem 5.28**
***SLLN: Non IID Case***

*Let $\{X_n\}$ be a sequence of random variables such that $E(X_i) = \mu_i$, $var(X_i) \leq b < \infty$ $\forall i$, and $\{X_n\}$ is asymptotic nonpositively correlated. Then $\bar{X}_n - \mu_n \stackrel{as}{\rightarrow} 0$.*

The theorem indicates that a sequence of random variables will adhere to a SLLN if the variances of the random variables are bounded, and if any positive correlation between random variables in the sequence eventually dissipates when the random variables are far enough apart in the sequence.

**Example 5.38**
***Plim($\bar{X}_n$) for Asymptotically Nonpositively Correlated Sequence***

Recall Example 5.37, where it is known that $\{X_n\}$ is asymptotically nonpositively correlated. Note further that since $\rho \in [0,1)$, $\sigma_i^2 = \sigma^2 \sum_{j=1}^{i} \rho^{2(j-1)} < \infty \; \forall i$, where in fact $\sum_{j=1}^{\infty} \rho^{2(j-1)} = 1 + \rho^2 + \rho^4 + \ldots = 1/(1 - \rho^2)$, so that $\sigma_i^2 \leq 1/(1 - \rho^2) \; \forall i$. Since the $\sigma_i^2$'s are upper-bounded, the conditions of Theorem 5.28 are met, and it follows that $\bar{X}_n \stackrel{as}{\rightarrow} 0$, since $E(X_i) = \mu = 0 \; \forall i$. □

Our final result on SLLNs concerns whether $\bar{X}_n - \bar{\mu}_n \stackrel{as}{\rightarrow} 0$ and $\bar{\mu}_n \rightarrow c$ together imply that $\bar{X}_n \stackrel{as}{\rightarrow} c$. The answer is yes, as stated in the next theorem.

**Theorem 5.29**
***SLLN via Convergence of Difference***

*$\bar{X}_n - \bar{\mu}_n \stackrel{as}{\rightarrow} 0$ and $\bar{\mu}_n \rightarrow c \Rightarrow \bar{X}_n \stackrel{as}{\rightarrow} c$*

## 5.8 Central Limit Theorems

Central limit theorems (CLTs) are concerned with the conditions under which sequences of random variables *converge in distribution* to known families of distributions. We will focus primarily on results concerning convergence in distribution of sequences of random variables $\{Y_n\}$ of the following form[9]:

$$\mathbf{Y}_n = \mathbf{b}_n^{-1}(\mathbf{S}_n - \mathbf{a}_n) \xrightarrow{d} N(\mathbf{0}, \Sigma),$$

where $\{\mathbf{S}_n\}$ is a sequence of scalar or vector random variables whose $n$th term is defined by $\mathbf{S}_n = \sum_{i=1}^{n} \mathbf{X}_i$, $\{\mathbf{X}_n\}$ is a sequence of scalar or vector random variables, and $\{\mathbf{a}_n\}$ and $\{\mathbf{b}_n\}$ are suitably chosen sequences of real numbers, vectors, or matrices. A statement of conditions on $\{\mathbf{X}_n\}$, $\{\mathbf{a}_n\}$, and $\{\mathbf{b}_n\}$ for which the convergence in distribution result holds true constitutes a **central limit theorem**.

As we remarked in the introduction to the preceding section dealing with laws of large numbers, the reader may wonder why the particular problem concerning convergence in distribution defined above deserves such explicit attention. The answer lies in the fact that a large number of important parameter estimation and hypothesis-testing procedures in econometrics and statistics are defined as functions of sums of random variables (note the $\mathbf{S}_n$ term in the convergence problem above). Central limit theorems are then often useful for establishing asymptotic distributions for these procedures, as will be seen in specific examples in subsequent chapters.

In order to illustrate the general way in which the use of a CLT might arise in practice, suppose a CLT is applicable to a scalar random variable, say, as $Y_n = b_n^{-1}(S_n - a_n) \xrightarrow{d} Y \sim N(0,1)$. Then, since $S_n = g(Y_n, a_n, b_n) = b_n Y_n + a_n$, we can define an asymptotic distribution for $S_n$ using Definition 5.2 as $S_n \overset{a}{\sim} b_n Y + a_n$, or $S_n \overset{a}{\sim} N(a_n, b_n^2)$. Thus, for large $n$, $S_n = \sum_{i=1}^{n} X_i$ would have *an* asymptotic distribution that is normal, with mean $a_n$ and variance $b_n^2$. Now suppose a particular statistical procedure is based on the random variable $W_n = h(S_n; c_n) = h(g(Y_n; a_n, b_n); c_n)$. Then $W_n \overset{a}{\sim} h(g(Y; a_n, b_n); c_n)$, so that for large $n$, an asymptotic distribution for $W_n$ is given by the distribution associated with the composite function $h \circ g$ of the standard normal random variable, $Y$, or equivalently by the distribution associated with the function $h$ of the random variable $S_n$ under the assumption that $S_n \sim N(a_n, b_n^2)$. For example, if $W_n \overset{a}{\sim} h(g(Y; a_n, b_n); c_n) = c_n Y^2$, then $W_n \overset{a}{\sim} \text{Gamma}(1/2, 2c_n)$ (since $Y^2 \sim \chi_1^2$, and then $c_n Y^2$ has a gamma distribution with $\alpha = 1/2$ and $\beta = 2c_n$).

Defining asymptotic distributions for random variables of interest is most useful, and sometimes indispensable, when the *exact* distributions of the random variables are very difficult or impossible to derive. Furthermore, even if the exact distributions of random variables of interest can be defined, they

[9] The reader who wishes to read about central limit theory in its most general form can examine Chapter 5 of R.G. Laha, and V.K. Rohatgi (1979), *Probability Theory*, New York: John Wiley.

may be very difficult to work with, whereas the asymptotic distribution may be relatively easy to analyze. For both of the aforementioned reasons, central limit theorems figure prominently in the development of econometric and statistical theory and application.

We divide our presentation of CLTs into three subsections. The first subsection deals with the case of independent scalar random variables. The second subsection provides an introduction to the case where the scalar random variables in $\{X_n\}$ are *not* independent. In the final subsection, we present some CLT results relating to multivariate random variables.

### 5.8.1 Independent Scalar Random Variables

We examine three CLTs for independent random variables beginning with the simplest but least general *iid* case. We end with a CLT that presents necessary and sufficient conditions for $b_n^{-1}(S_n - a_n) \xrightarrow{d} N(0,1)$ when the $X_i$'s are independent but not necessarily identically distributed.

#### 5.8.1.1 IID Case

We begin with the simplest of all CLTs, the **Lindberg-Levy CLT.**

**Theorem 5.30**
***Lindberg-Levy CLT***

*Let $\{X_n\}$ be a sequence of iid random variables with* $\mathrm{E}(X_i) = \mu$ *and* $\mathrm{var}(X_i) = \sigma^2 \in (0,\infty)\ \forall\ i$. *Then,*

$$\left(n^{1/2}\sigma\right)^{-1}\left(\sum\nolimits_{i=1}^{n} X_i - n\mu\right) = \frac{n^{1/2}(\bar{X}_n - \mu)}{\sigma} \xrightarrow{d} N(0,1).$$

In order to enhance one's intuitive understanding of why the Lindberg-Levy CLT (LLCLT) holds, we provide some additional rationale for the result, albeit at the expense of some degree of imprecision in the mathematical details. First note that under the conditions of the LLCLT, $\sum_{i=1}^{n} X_i$ is a random variable that has a mean of $n\mu$ and a variance of $n\sigma^2$. Since both $|n\mu|$ and $n\sigma^2$ diverge to $\infty$ (assuming $\mu \neq 0$) as $n$ increases, it is clear that some form of centering and scaling of $\sum_{i=1}^{n} X_i$ will be necessary for there to be any hope of convergence to some limiting distribution. By subtracting $n\mu$ and then dividing by $n^{1/2}\sigma$, one defines random variables $Y_n = (n^{1/2}\sigma)^{-1}(\sum_{i=1}^{n} X_i - n\mu) = (n^{1/2}\sigma)^{-1}\sum_{i=1}^{n} Z_i$ which have a mean zero and variance of 1 regardless of $n$. The random variables $Z_i = X_i - \mu$, $i = 1,\ldots,n$, are *iid* random variables with zero means and variances all equal to $\sigma^2$.

Now a key observation concerning an additional effect of the aforementioned centering and scaling: when $n \to \infty$, any effect of third- and higher-order moments of $Z_i$ on the moments of $Y_n$ are "centered and scaled away" so that all probability distributions for $Z_i$ that have the same mean and variance will lead to precisely the same moments for $Y_n$ as $n \to \infty$ (we assume that all moments of the $Z_i$'s exist). To see this, first consider the third moment of $Y_n$:

$$\mathrm{E}\left(Y_n^3\right) = \left(n^{1/2}\sigma\right)^{-3}\sum_{i=1}^{n}\sum_{j=1}^{n}\sum_{k=1}^{n}\mathrm{E}(Z_i\, Z_j\, Z_k).$$

Since the $Z_i$'s are independent with zero means, it is only the case where $i = j = k$ that the expectation term is nonzero and equal to $\mu'_3$, the third moment of the distribution of the $Z_i$'s. But there are only $n$ of these terms, and thus

$E(Y_n^3) = \sigma^{-3}\, n^{-3/2}\, (n\mu'_3) \to 0$ as $n \to \infty$, and the third moment of $Y_n$ converges to zero regardless of $\mu'_3$. Following analogous logic applied to $E(Y_n^4)$, which is defined in terms of a quadruple sum of $(Z_i\, Z_j\, Z_k\, Z_\ell)$ terms premultiplied by $(n^{1/2}\sigma)^{-4}$, it can be shown that $E(Y_n^4) = \sigma^{-4}\, n^{-2}\left[n\mu'_4 + 3n(n-1)\,\sigma^4\right] \to 3$ regardless of $\mu'_4$, the fourth moment of $Z_i$. This type of argument can be continued ad infinitum to show that all higher order moments of $Y_n$ converge to known constants and that the values of higher order moments of the $Z_i$'s play no role in determining any of the moments of $Y_n$ when $n \to \infty$.

Now observe that the first four moments of $Y_n$, as defined in the previous paragraph, converge to the first four moments of the standard normal distribution, 0, 1, 0, and 3, respectively. Furthermore, all higher-order moments of $Y_n$ also converge to those of the standard normal distribution, which can in principle be verified one by one following the approach defined above, and in any case is implied by the proof of Theorem 5.30. While not true for all densities, members of the normal family of PDFs are uniquely identified by their moment sequences,[10] so that the moments of $Y_n$ are uniquely consistent with that of a standard normal density as $n \to \infty$ for all underlying probability distribution of the $X_i$'s having mean $\mu$ and variance $\sigma^2$. Thus, the centering and scaling of the $X_i$'s inherent in the definition of $Y_n$ remove any tendencies for higher-order moments of the $X_i$'s to cause the moments of $Y_n$ to deviate from those of a standard normal distribution as $n \to \infty$, leading to convergence of $Y_n$ to the $N(0,1)$ limiting distribution.

The establishment of the LLCLT provides an opportunity to revisit the relationship between *limiting distributions* and *asymptotic distributions*. Under the conditions of the LLCLT, $Y_n = (n^{1/2}\,\sigma)^{-1}\,(\sum_{i=1}^n X_i - n\mu) \xrightarrow{d} N(0,1)$. Then, following Definition 5.2, an asymptotic distribution for $S_n = \sum_{i=1}^n X_i$ can be defined by noting that $S_n = (n^{1/2}\sigma)Y_n + n\mu \overset{a}{\sim} (n^{1/2}\sigma)Y + n\mu$ where $Y \sim N(0,1)$, so that $S_n \overset{a}{\sim} N(n\mu, n\sigma^2)$. Thus, the normal distribution with mean $n\mu$ and variance $n\sigma^2$ provides an approximation to the distribution of $S_n$ when $n$ is large. Regarding $\bar{X}_n$, note that the conditions underlying the Lindberg-Levy CLT imply $\bar{X}_n \xrightarrow{p} \mu$ by Khinchins' WLLN, so that $\bar{X}_n \xrightarrow{d} \mu$. Because the *limiting distribution* of $\bar{X}_n$ is *degenerate*, it is clear that the distribution provides no information about the variability of $\bar{X}_n$ for finite $n$. The *asymptotic distribution* for $\bar{X}_n$ is more useful in this regard. Noting that $\bar{X}_n = g(Y_n,n) = (\sigma/n^{1/2})\,Y_n + \mu$, it follows from Definition 5.2 that $\bar{X}_n \overset{a}{\sim} g(Y,n) = (\sigma/n^{1/2})Y + \mu$ for $Y \sim N(0,1)$, or $\bar{X}_n \overset{a}{\sim} N(\mu, \sigma^2/n)$. Thus, for large $n$, $\bar{X}_n$ is approximately normally distributed with mean $\mu$ and variance $\sigma^2/n$. The following examples illustrate the application of the Lindberg-Levy CLT.

---

[10] M. Kendall and A. Stuart (1977), *The Advanced Theory of Statistics, Volume I*, New York: Macmillan, pp. 115. Note that any random variable for which an MGF exists is such that its moment sequence uniquely identifies its probability distribution.

**Example 5.39**
***Approximating Binomial Probabilities via the Normal Distribution***

Let $\{X_n\}$ be a sequence of *iid* Bernoulli-type random variables, i.e., $X_i \sim p^{x_i}(1-p)^{1-x_i} I_{\{0,1\}}(x_i)\ \forall\, i$ with $p \neq 0$ or 1. Then, by the Lindberg-Levy CLT,

$$\frac{\sum_{i=1}^{n} X_i - np}{n^{1/2}\left[p(1-p)\right]^{1/2}} \xrightarrow{d} N(0,1), \text{and } \sum_{i=1}^{n} X_i \overset{a}{\sim} N(np, np(1-p))$$

Since $\sum_{i=1}^{n} X_i$ has a binomial distribution under the stated conditions, we have discovered an alternative to the Poisson density for approximating the binomial distribution for large $n$. Note that we are approximating the *discrete* Binomial density with the aforementioned *continuous* Normal density. It has been found in practice that such approximations are improved, especially when $n$ is not very large, by making a **continuity correction,** whereby each outcome, $x$, in the range of the discrete random variable is associated with the interval event $(x - (1/2), x+(1/2)]$ for the purpose of assigning probability via the asymptotic normal density. For example, if $n = 40$, $p = 1/2$, and $x = 20$, then since $\mu = np = 20$ and $\sigma^2 = np(1-p) = 10$, we have, using the normal asymptotic density $N(20,10)$, $P(x = 20) \approx \int_{19.5}^{20.5} N(z; 20, 10)dz = \int_{-.5/\sqrt{10}}^{.5/\sqrt{10}} N(z; 0, 1)dz = .1272$ (from standard normal table). The actual probability assigned to the event $P(x = 20)$ by the binomial density is $P(x = 20) = \binom{40}{20}\left(\frac{1}{2}\right)^{40} = .1254.$ □

**Example 5.40**
***Approximating $\chi^2$ Probabilities via the Normal Distribution***

Let $\{X_n\}$ be a sequence of *iid* chi-square random variables with 1 degree of freedom, i.e., $X_i \sim \chi_1^2\ \forall\, i$. By the additivity property of chi-square random variables, $\sum_{i=1}^{n} X_i \sim \chi_n^2$. Also $E(X_i) = 1$ and $\text{var}(X_i) = 2\ \forall\, i$ under the prevailing assumptions. Then, by the Lindberg-Levy CLT,

$$Y_n = \frac{\sum_{i=1}^{n} X_i - n}{(2n)^{1/2}} \xrightarrow{d} N(0,1) \text{ and } \sum_{i=1}^{n} X_i \overset{a}{\sim} N(n, 2n).$$

We have thus discovered an approximation to the $\chi^2$ density function for large degrees of freedom. As an example of its use, note that (from tables of the $\chi^2$ distribution) $P\left(\chi_{30}^2 \leq 43.8\right) = .95$. We obtained our approximation of this probability by utilizing $Z_{30} \overset{a}{\sim} N(30, 60)$, and thus $P(z_{30} \leq 43.8) = P\left(\frac{z_{30}-30}{(60)^{1/2}} \leq \frac{43.8-30}{(60)^{1/2}}\right) = P(z \leq 1.783) = .9627$ (where $Z \sim N(0,1)$) □

The Lindberg-Levy CLT implies that any real-world experiment whose final outcome can be conceptualized as the result of a summation or average of the outcomes of a large number of *iid* random variables having a finite mean and variance can be treated as having approximately a normal distribution. Thus, for example, the total number of defective objects produced on an assembly line or the average miles per gallon achieved by a sample of Ford pickup trucks might be considered as approximately normally distributed to the extent that the independence and identical distribution assumptions hold true.

A natural question to ask in using asymptotic distributions is how large does $n$ have to be for the approximation to be a good one? Unfortunately, the answer depends on the characteristics of the true distribution underlying the sequence of random variables, and no general answer can be given. However, a number of inequalities have been developed that can be useful in answering the question if something is known about the moments of the underlying densities. Specifically, under the conditions of the Lindberg-Levy CLT, Van Beeck[11] has shown that the maximum absolute difference between $P(y_n \leq c)$ for the *actual* density of $Y_n = n^{1/2}(\bar{X}_n - \mu)/\sigma$ and for that of the standard normal density is $.7975\,(\zeta_3/\sigma^3)n^{-1/2}$ where $\zeta_3 = \mathrm{E}\left(|X - \mathrm{E}(X)|^3\right)$ and $\sigma$ refer to the standard deviation of the common density of the $X_i$'s. As examples of the bound, if the $X_i$'s are *iid* Bernoulli or exponential random variables, then the bounds are respectively $.7975\, n^{-1/2}[1-2p(1-p)][p(1-p)]^{-1/2}$ and $.1653n^{-1/2}$. Then, for example, if $p = .5$ and $n = 1{,}000$, the upper bounds on the errors when approximating $P(y_n \leq c)$ via the standard normal limiting distribution are .025 and .005, respectively. It should be emphasized that VanBeeck's result provides *omnibus* bounds that apply to *all* random variables having the prescribed moments, and as such the bound tends to be quite conservative, that is, the actual approximation errors are generally much smaller than the bound.

#### 5.8.1.2 Non-IID Case

While the Lindberg-Levy CLT is applicable in many experimental situations, it has the disadvantage of requiring that all of the random variables have the same mean, the same variance, and moreover, *the same probability distribution.* Various other central limit theorems can be constructed that utilize alternative conditions on the distributions of the random variables in the sequence $\{X_n\}$.[12] The most general CLT for the case of independent random variables, which subsumes the LLCLT as a special case, is the **Lindberg CLT.**

**Theorem 5.31** ***Lindberg's CLT*** *Let $\{X_n\}$ be a sequence of independent random variables with $\mathrm{E}(X_i) = \mu_i$ and $\mathrm{var}(X_i) = \sigma_i^2 < \infty\ \forall\, i$. Define $b_n^2 = \sum_{i=1}^n \sigma_i^2$, $\bar{\sigma}_n^2 = n^{-1}\sum_{i=1}^n \sigma_i^2$, $\bar{\mu}_n = n^{-1}\sum_{i=1}^n \mu_i$, and let $f_i$ be the PDF of $X_i$. If $\forall\, \varepsilon > 0$,*

$$\lim_{n\to\infty} \frac{1}{b_n^2} \sum_{i=1}^n \int_{(x_i-\mu_i)^2 \geq \varepsilon b_n^2} (x_i - \mu_i)^2 f_i(x_i)dx_i = 0 \text{ (continuous case)}$$

$$\lim_{n\to\infty} \frac{1}{b_n^2} \sum_{i=1}^n \sum_{(x_i-\mu_i)^2 \geq \varepsilon b_n^2,\, f_i(x_i)>0} (x_i - \mu_i)^2 f_i(x_i) = 0, \text{ (discrete case)}$$

$$\text{then } \frac{\sum_{i=1}^n X_i - \sum_{i=1}^n \mu_i}{\left(\sum_{i=1}^n \sigma_i^2\right)^{1/2}} = \frac{n^{1/2}(\bar{X}_n - \bar{\mu}_n)}{\bar{\sigma}_n} \xrightarrow{d} N(0,1).$$

[11] P. Van Beeck (1972), An application of Fourier methods to the problem of sharpening the Berry-Esseen Inequality, *Z. Wahrscheinlichkeits Theorie und Verw. Gebiete* 23, pp. 187–196.

[12] For example, see Y.S. Chow and H. Teicher, *Probability Theory*, (1978) Chapter 9, and R.G. Laha and V.K. Rohatgi, (1976) *Probability Theory*, Chapter 5).

It can be shown that the limit conditions in the Lindberg CLT, known as the *Lindberg conditions*, imply that $\lim_{n\to\infty}\left(\sigma_j^2/\sum_{i=1}^n \sigma_i^2\right) = 0\ \forall j$. That is, the contribution that each $X_j$ makes to the variance of $\sum_{i=1}^n X_i$ is negligible as $n \to \infty$. The Lindberg conditions can be difficult to verify in practice, and so we will present two useful special cases of the Lindberg CLT that rely on more easily verifiable conditions. The first special case essentially implies that if the random variables in the sequence $\{X_n\}$ are independent and bounded with probability 1, then $Y_n = n^{1/2}(\bar{X}_n - \bar{\mu}_n)/\bar{\sigma}_n \xrightarrow{d} N(0,1)$. It will be seen that the boundedness condition and $\sum_{i=1}^n \sigma_i^2 \to \infty$ imply the Lindberg condition.

**Theorem 5.32** ***CLT for Bounded Random Variables***

*Let $\{X_n\}$ be a sequence of independent random variables such that $P(|x_i| \leq m) = 1$ $\forall\, i$ for some $m \in (0,\infty)$, with $E(X_i) = \mu_i$ and $\mathrm{var}(X_i) = \sigma_i^2 < \infty\ \forall i$. If $\sum_{i=1}^n \mathrm{var}(X_i) = \sum_{i=1}^n \sigma_i^2 \to \infty$ as $n \to \infty$, then $n^{1/2}(\bar{X}_n - \bar{\mu}_n)/\bar{\sigma}_n \xrightarrow{d} N(0,1)$.*

The following example illustrates the discovery of an asymptotic distribution for a simple form of the *least-squares estimator*, which we will examine in more detail in Chapter 8.

**Example 5.41** ***Asymptotic Distribution for a Least Squares Estimator***

Let the sequence $\{Y_n\}$ be defined by $Y_i = z_i\,\beta + \varepsilon_i$, where:

**a.** $\beta$ is a real number,
**b.** $z_i$ is the $i$th element in the sequence of real numbers $\{z_n\}$ for which $n^{-1}\sum_{i=1}^n z_i^2 > a > 0\ \forall n$ and $|z_i| < d < \infty\ \forall\, i$, and
**c.** $\varepsilon_i$ is the $i$th element in a sequence of *iid* random variables, $\{\varepsilon_n\}$, for which $E(\varepsilon_i) = 0$, $\mathrm{var}(\varepsilon_i) = \sigma^2 \in (0,\infty)$, and $P(|\varepsilon_i| \leq m) = 1\ \forall\, i$, where $m \in (0,\infty)$.

Given the preceding assumptions, an asymptotic distribution for the least-squares estimator of $\beta$ defined by $\hat{\beta}_n = \sum_{i=1}^n z_i Y_i / \sum_{i=1}^n z_i^2$ can be defined as follows:

We transform the problem into a form that allows both an application of a CLT and an application of Definition 5.2 to define an asymptotic distribution for $\hat{\beta}_n$. Note that

$$(\hat{\beta}_n - \beta)\sum_{i=1}^n z_i(Y_i - z_i\beta)/\sum_{i=1}^n z_i^2 = \sum_{i=1}^n z_i\,\varepsilon_i/\sum_{i=1}^n z_i^2,$$

so that

$$\left(\sum_{i=1}^n z_i^2\right)^{1/2}\frac{\left(\hat{\beta}_n - \beta\right)}{\sigma} = \sum_{i=1}^n \frac{z_i\varepsilon_i}{\left(\sigma^2\sum_{i=1}^n z_i^2\right)^{1/2}} = \sum_{i=1}^n \frac{W_i}{\left(\sum_{i=1}^n \mathrm{var}(W_i)\right)^{1/2}}$$

where $W_i = z_i\varepsilon_i$. The CLT of Theorem 5.32 is applicable to this function of $\hat{\beta}_n$. To see this, observe that $E(W_i) = 0$ and $\mathrm{var}(W_i) = \sigma^2 z_i^2 \leq \sigma^2 d^2 < \infty\ \forall\, i$. Also, $P(|z_i\varepsilon_i| \leq dm) = P(|w_i| \leq dm) = 1\ \forall\, i$, and $\sum_{i=1}^n \mathrm{var}(W_i) = \sigma^2\sum_{i=1}^n z_i^2 \to \infty$

since $n^{-1} \sum_{i=1}^{n} z_i^2 > a > 0 \; \forall \, n$. Then, by Theorem 5.32, $\left(\sum_{i=1}^{n} z_i^2\right)^{1/2} \frac{\left(\hat{\beta}_n - \beta\right)}{\sigma} \xrightarrow{d}$ $N(0,1)$, and by Definition 5.2, $\hat{\beta}_n \overset{a}{\sim} N\left(\beta, \sigma^2 \left(\sum_{i=1}^{n} z_i^2\right)^{-1}\right)$. □

The **Liapounov CLT** relaxes the boundedness assumption of the previous CLT. In this case, a condition on the moments of the random variables in the sequence $\{X_n\}$ is used to imply the Lindberg conditions instead of boundedness of the random variables.

**Theorem 5.33** ***Liapounov CLT*** *Let $\{X_n\}$ be a sequence of independent random variables such that $E(X_i) = \mu_i$ and $\text{var}(X_i) = \sigma_i^2 < \infty \; \forall \, i$. If, for some $\delta > 0$,*

$$\lim_{n\to\infty} \frac{\sum_{i=1}^{n} E\left(|X_i - \mu_i|^{2+\delta}\right)}{\left(\sum_{i=1}^{n} \sigma_i^2\right)^{1+\delta/2}} = 0,$$

then

$$\frac{\sum_{i=1}^{n} X_i - \sum_{i=1}^{n} \mu_i}{\left(\sum_{i=1}^{n} \sigma_i^2\right)^{1/2}} = \frac{n^{1/2}(\bar{X}_n - \bar{\mu}_n)}{\bar{\sigma}_n} \xrightarrow{d} N(0,1).$$

An important implication of Liapounov's CLT is that $\sum_{i=1}^{n} X_i$ need not be a sum of *identically* distributed nor *bounded* random variables to have *an* asymptotic normal density. Under the conditions of the theorem, it follows that $S_n = \sum_{i=1}^{n} X_i \overset{a}{\sim} N\left(\sum_{i=1}^{n} \mu_i, \sum_{i=1}^{n} \sigma_i^2\right)$ and $\bar{X}_n \overset{a}{\sim} N(\bar{\mu}_n, n^{-1}\bar{\sigma}_n^2)$.

**Example 5.42** ***Limiting Distribution for Function of Non-IID Uniform RVs*** Let $\{X_n\}$ be a sequence of independent uniformly distributed random variables such that $X_i \sim \frac{1}{2c_i} I_{[-c_i,c_i]}(x_i) \; \forall \, i$, where $c_i \in [\tau, m]$, and $0 < \tau < m < \infty$. Then for $\delta > 0$, $E(|X_i - \mu_i|^{2+\delta}) \leq m^{2+\delta} \; \forall i$, and $\text{var}(X_i) = c_i^2/3 \geq \tau^2/3 > 0 \; \forall \, i$. Letting $\delta = 1$ in Liaponov's CLT,

$$\lim_{n\to\infty} \frac{\sum_{i=1}^{n} E|X_i - \mu_i|^3}{\left[\sum_{i=1}^{n} \sigma_i^2\right]^{3/2}} \leq \lim_{n\to\infty} \left(\frac{\sqrt{3}m}{\tau}\right)^3 n^{-1/2} = 0,$$

so that (note $\mu_i = 0 \; \forall i$)

$$Y_n = \frac{\sum_{i=1}^{n} X_i}{\left[\sum_{i=1}^{n} (c_i^2/3)\right]^{1/2}} \xrightarrow{d} N(0,1) \text{and } \bar{X}_n \overset{a}{\sim} N\left(0, n^{-2} \sum_{i=1}^{n} \left(\frac{c_i^2}{3}\right)\right).$$ □

As an illustration of how either the CLT for bounded random variables or the Liapounov CLT might be applied in practice, consider a short-run production process of a firm and focus on the effect of labor input on production. There may exist a systematic engineering relationship between the quantity of labor applied to the complement of plant equipment and the expected output of the plant, say as $y = f(L)$, for any given time period of plant operation. However, it would undoubtedly be rare that the exact quantity of production expected in an engineering sense will actually be realized. Variations from the expected quantity could occur due to variations in the health, alertness, and general performance level of each of the various employees, the extent of machine failures in any given time period and their general performance level, the varying ability of management to schedule production efficiently on any given day, weather conditions if the production process is affected thereby, and so on. Viewing the overall deviation of total production from the engineering relationship as caused by the summation of a large number of random deviations with finite absolute upper bounds caused by various uncontrolled factors results in a production relationship of the form $Y = f(L) + \varepsilon$ where $\varepsilon$, and thus $Y$, has an asymptotic normal density (assuming the summation of the random deviations can be viewed as a sum of independent, although *not* necessarily identically distributed, random variables).

As in the case of the Lindberg-Levy CLT, bounds have been established on the maximum absolute deviation between the actual value of $P(y_n \leq c)$ for $Y_n = n^{1/2}(\bar{X}_n - \bar{\mu}_n)/\bar{\sigma}_n$, and the approximated value based on the standard normal distribution. Specifically, Zolotarev[13] has shown such an upper bound to be $.9051 \left(\sum_{j=1}^{n} \xi_{3_j}\right)\left(\sum_{j=1}^{n} \sigma_{X_j}^2\right)^{-3/2}$, where $\xi_{3_j} = \mathrm{E}\left(|X_j - \mathrm{E}(X_j)|^3\right)$ and $\sigma_{X_j}^2$ refers to the variance of the $j$th random variable $X_j$. Thus, if one knew the appropriate moments of the random variables $X_1,\ldots,X_n$, a bound on the approximation error could be ascertained. If such information were not available, the researcher knows only that as $n$ increases, the accuracy of the approximation improves. As we had remarked in our discussion of similar bounds for the LLCLT, these bounds tend to be conservative and the actual approximation errors tend to be significantly less than the value of the bound.

### 5.8.2 Triangular Arrays

In analyzing the asymptotic properties of some types of econometric or statistical procedures, it is useful to be able to apply central limit theory to what is known as a **double array of random variables**. For our purposes, it will be sufficient to examine a special case of such a double array, called a **triangular array of random variables.**[14]

---

[13] M. Zolotarev (1967), *A Sharpening of the Inequality of Berry-Esseen. Z. Wahrscheinlichkeits Theorie und Verw. Gebiete*, 8 pp. 332–342.

[14] A *double array* is one where the second subscript of the random variables in the $i$th row of the array ends with the value $k_i$, rather than with the value $i$ as in the case of the triangular array, and $k_n \to \infty$ as $n \to \infty$. See Serfling, *Approximation Theorems*, pp. 31.

**Definition 5.9**
***Triangular Array of Random Variables***

The ordered collection of random variables $\{X_{11}, X_{21}, X_{22}, X_{31}, X_{32}, X_{33}, \ldots, X_{nn}, \ldots\}$, or

$$\begin{array}{llllll} X_{11}; & & & & & \\ X_{21} & X_{22}; & & & & \\ X_{31} & X_{32} & X_{33}; & & & \\ \vdots & \vdots & \vdots & \ddots & & \\ X_{n1} & X_{n2} & X_{n3} & X_{n4} & \ldots & X_{nn}; \\ \vdots & \vdots & \vdots & \vdots & \vdots & \vdots \quad \ddots \end{array}$$

is called a triangular array of random variables, and will be denoted by $\{X_{nn}\}$.

Central limit theorems that are applied to triangular arrays of random variables are concerned with the limiting distributions of appropriately defined functions of the *row averages* $\bar{X}(n) = n^{-1} \sum_{i=1}^{n} X_{ni}$. Note that all of the CLTs examined so far have dealt with functions of averages of the type $\bar{X}_n = n^{-1} \sum_{i=1}^{n} X_i$, the $X_i$'s being elements of the sequence $\{X_n\}$. It is possible that $\bar{X}(n) = \bar{X}_n \; \forall \; n$, which would occur if $x_{ij} = x_j \; \forall \; i,j$, that is, all of the elements in any given *column* of the triangular array are *identical*. Thus, the CLT results obtained heretofore apply to this special case of a triangular array. However, the triangular array $\{X_{nn}\}$ is more general than a sequence $\{X_n\}$ in the sense that the random variables in a row of the array need *not* be the same as random variables in other rows. Furthermore, previously $\bar{X}_n$ always involved *all* of the random variables in the sequence $\{X_n\}$ up to the *nth* element, whereas in the triangular array, $\bar{X}(n)$ involves only the random variables *residing in the nth row of the triangular array*. The importance of this flexibility will become apparent when we analyze the asymptotic behavior of certain statistical procedures for which central limit theory can *only* be effectively applied in the context of triangular arrays of random variables.

*All* of the CLTs presented heretofore can be extended to the case of triangular arrays. We present here the extension of the Liapounov CLT. For additional details on such extensions, the reader can refer to the book by K.L. Chung cited in the proof of the theorem. Henceforth, we will let $\bar{\mu}(n) = n^{-1} \sum_{i=1}^{n} \mu_{ni}$ and $\bar{\sigma}^2(n) = n^{-1} \sum_{i=1}^{n} \sigma_{ni}^2$.

**Theorem 5.34**
***Liapounov CLT for Triangular Arrays***

*Let $\{X_{nn}\}$ be a triangular array of random variables with independent random variables within rows. Let $\mathrm{E}(X_{ij}) = \mu_{ij}$ and $\mathrm{var}(X_{ij}) = \sigma_{ij}^2 < \infty \; \forall \; i,j$. If for some*

$$\delta > 0, \; \lim_{n \to \infty} \frac{\sum_{i=1}^{n} \mathrm{E}|X_{ni} - \mu_{ni}|^{2+\delta}}{\left(\sum_{i=1}^{n} \sigma_{ni}^2\right)^{1+\delta/2}} = 0,$$

$$\textit{then } \frac{\sum_{i=1}^{n} X_{ni} - \sum_{i=1}^{n} \mu_{ni}}{\left(\sum_{i=1}^{n} \sigma_{ni}^2\right)^{1/2}} = n^{1/2} \left(\bar{X}(n) - \bar{\mu}(n)\right) / \bar{\sigma}(n) \xrightarrow{d} N(0,1).$$

Note that the random variables *within* a row of the triangular array are assumed to be independent in the Liapounov CLT, but *no such assumption* is required for random variables in *different* rows. In fact, the random variables within a given row can be *arbitrarily dependent* on random variables in *other* rows.

The following example applies the Liapounov CLT for triangular arrays to establish the asymptotic normality of the least-squares estimator under more general conditions than those utilized in Example 5.41.

**Example 5.43**
***Asymptotic Distribution for a Least Squares Estimator with Unbounded Residuals***

Let the sequence $\{Y_n\}$ be defined by $y_i = z_i\,\beta + \varepsilon_i$, and assume the conditions of Example 5.41, except replace the boundedness assumption $P(|\varepsilon_i| \leq m) = 1\ \forall i$ with the moment assumption that $\mathrm{E}|\varepsilon_i|^{2+\delta} \leq m < \infty\ \forall i$ for some $\delta > 0$. Then the least-squares estimator $\hat{\beta}_n = \sum_{i=1}^n z_i Y_i / \sum_{i=1}^n z_i^2$ remains asymptotically normally distributed. To see this, note that

$$\left(\sum_{i=1}^{n} z_i^2\right)^{1/2} \frac{\left(\hat{\beta}_n - \beta\right)}{\sigma} = \left(\sum_{i=1}^{n} z_i \varepsilon_i\right) \Big/ \left(\sigma^2 \sum_{i=1}^{n} z_i^2\right)^{1/2} = \sum_{i=1}^{n} W_{ni},$$

where the random variables $W_{ni}$ are elements of a triangular array[15] for which $\mathrm{E}(W_{ni}) = 0$ and $\sigma_{ni}^2 = \mathrm{var}(W_{ni}) = z_i^2 / \sum_{i=1}^n z_i^2$. Because $\sum_{i=1}^n \sigma_{ni}^2 = 1$, the limit condition of the Liapounov CLT for triangular arrays is met since, for $\delta > 0$

$$\lim_{n\to\infty} \sum_{i=1}^{n} \mathrm{E}\left(|W_{ni}|^{2+\delta}\right) = \lim_{n\to\infty} \left(\sum_{i=1}^{n} |z_i|^{2+\delta} \mathrm{E}\left(|\varepsilon_i|^{2+\delta}\right) \Big/ \left(\sigma^2 \sum_{i=1}^{n} z_i^2\right)^{1+\delta/2}\right)$$

$$\leq \lim_{n\to\infty} \left(md^{2+\delta} / \left(\sigma^2 a\right)^{1+\delta/2}\right) n^{-\delta/2} = 0.$$

Therefore, $\sum_{i=1}^n W_{ni} \xrightarrow{d} N(0,1)$, which then implies that $\hat{\beta} \overset{a}{\sim} N\left(\beta, \sigma^2 \left(\sum_{i=1}^{n} z_i^2\right)^{-1}\right)$. □

### 5.8.3 Dependent Scalar Random Variables

In this subsection we provide an introduction to the notion of defining CLTs when the random variables in the sequence $\{X_n\}$ exhibit some degree of dependence. Many different CLTs can be defined by allowing different types of dependencies among the random variables in $\{X_n\}$. CLTs for dependent random variables are generally much more complicated to state and prove than CLTs for independent random variables, and the level of mathematics involved is beyond our scope of study. We will explicitly examine one useful CLT for a particular type of dependence called $m$-dependence. The reader can find additional results on CLTs for the dependent random variable case in H. White, (1980) *Asymptotic Theory*, Chapters 3 and 5 and in R.J. Serfling (1968), *Contributions to central limit theory for dependent variables, Ann. of Math. Stat.* 39 pp. 1158–1175.

---

[15]Note the $W_{ni}$, $i = 1,\ldots,n$, are independent because the $\varepsilon_i$'s are independent.

**Definition 5.10** *m-Dependence*

The sequence $\{X_n\}$ is said to exhibit $m$-dependence (or is said to be $m$-dependent) if, for $a_1 < a_2 < \ldots < a_k < b_1 < b_2 < \ldots < b_r$, the random variables $(X_{a_1}, X_{a_2} \ldots, X_{a_k})$ are independent of $(X_{b_1}, X_{b_2} \ldots, X_{b_r})$ whenever $b_1 - a_k > m$.

The definition states that $\{X_n\}$ is $m$-dependent if any two groups of random variables separated by more than $m$ positions in the sequence are independent of one another. Combining $m$-dependence with boundedness of the random variables in the sequence $\{X_n\}$ leads to the following CLT.

**Theorem 5.35** *CLT for Bounded m-Dependent Sequences*

*Let $\{X_n\}$ be an m-dependent sequence of random scalars for which* $\mathrm{E}(X_i) = \mu_i$ *and* $P(|x_i| \leq c) = 1$ *for some* $c < \infty\ \forall i$. *Let* $\sigma^2_{*n} = \mathrm{var}(\sum_{i=1}^n X_i)$. *If* $n^{-2/3}\,\sigma^2_{*n} \to \infty$, *then*

$$\left(\sigma^2_{*n}\right)^{-1/2}\left(\sum_{i=1}^n X_i - \sum_{i=1}^n \mu_i\right) \xrightarrow{d} N(0,1).$$

Regarding the variance condition stated in the theorem, note that if the $X_i$'s were *independent* and $\sigma_i^2 \geq b > 0\ \forall i$, then $\sigma^2_{*n} \geq bn \to \infty$ at a rate of $n$, so that $n^{-2/3}\,\sigma^2_{*n} \geq bn^{1/3} \to \infty$ at a rate $n^{1/3}$. If the $X_i$'s are *dependent*, then as long as the covariance terms in the determination of $var(\sum_{i=1}^n X_i)$ do not collectively *decrease* the rate at which $\sigma^2_{*n}$ increases by a factor of $n^{1/3}$ or more, the variance condition of the CLT will hold. Thus, through restricting covariance terms, the variance condition of Theorem 5.35 places restrictions on the extent of the dependence that can exist between the $X_i$'s. Note the restriction is effectively on *negative* covariances, since positive covariances actually increase the variance of $\sum_{i=1}^n X_i$.

**Example 5.44** *Asymptotic Distribution for a Least Squares Estimator with Dependent Observations*

Let the sequence $\{Y_n\}$ be defined by $Y_i = z_i\,\beta + \varepsilon_i$, and assume the conditions of Example 5.41, except replace the assumption $\mathrm{var}(\varepsilon_i) = \sigma^2 \in (0,\infty)$ with $\mathrm{var}(\varepsilon_i) = \sigma_i^2 \geq \tau > 0\ \forall i$, replace the assumption that the $\varepsilon_i$'s are *iid* with the assumption of $m$-dependence, and in addition assume that $z_i > 0\ \forall i$ and $\mathrm{cov}(\varepsilon_i, \varepsilon_j) \geq 0\ \forall i \neq j$. Applying Theorem 5.35, $\hat{\beta} \overset{a}{\sim} N\left(\beta, \sigma^2_{*n}\left(\sum_{i=1}^n z_i^2\right)^{-2}\right)$.

To see this, note that $\left(\sum_{i=1}^n z_i^2\right)\left(\hat{\beta}_n - \beta\right) = \sum_{i=1}^n z_i\varepsilon_i = \sum_{i=1}^n W_i$, where $\mathrm{E}(W_i) = 0$ and $P(|z_i\varepsilon_i| \leq db) = P(|w_i| \leq c) = 1$ for $c = db < \infty\ \forall i$. Also,

$$\sigma^2_{*n} = \mathrm{var}\left(\sum_{i=1}^n W_i\right) = \sum_{i=1}^n \sigma_i^2 z_i^2 + \sum\sum_{|i-j| \leq m,\, i \neq j} z_i z_j \mathrm{cov}(\varepsilon_i, \varepsilon_j)$$

since by $m$-dependence $\mathrm{cov}(\varepsilon_i, \varepsilon_j) = 0$ when $|i-j| > m$. Because, $z_i z_j\ \mathrm{cov}(\varepsilon_i, \varepsilon_j) \geq 0\ \forall i$ and $j$, $\sigma^2_{*n} \geq \sum_{i=1}^n \sigma_i^2 z_i^2 \geq na\tau$, so that $n^{-2/3}\ \sigma^2_{*n} \geq n^{1/3}\ a\tau \to \infty$.

Thus, by Theorem 5.35, $\left(\sum_{i=1}^n z_i^2\right)\dfrac{\left(\hat{\beta}_n - \beta\right)}{\left(\sigma^2_{*n}\right)^{1/2}} = \sum_{i=1}^n \dfrac{W_i}{\left(\sigma^2_{*n}\right)^{1/2}} \xrightarrow{d} N(0,1)$, and $\hat{\beta}_n \overset{a}{\sim} N\left(\beta, \sigma^2_{*n}\left(\sum_{i=1}^n z_i^2\right)^{-2}\right)$. □

### 5.8.4 Multivariate Central Limit Theorems

The central limit theorems presented so far are applicable to sequences of random *scalars*. Central limit theorems can be defined for sequences of random *vectors*, in which case conditions are established that ensure that an appropriate (vector) function of the random sequence converges in distribution to a multivariate normal distribution. Due to a result discovered by H. Cramer and H. Wold,[16] termed the **Cramer-Wold device**, questions of convergence in distribution for a multivariate random sequence can all be reduced to the question of convergence in distribution of sequences of random scalars, at least in principle. Thus, all of the central limit theorems discussed to this point remain highly relevant to the multivariate case.

**Theorem 5.36** ***Cramer-Wold Device*** *The sequence of* $(k \times 1)$ *random vectors* $\{\mathbf{X}_n\}$ *converges in distribution to the random* $(k \times 1)$ *vector* $\mathbf{X}$ *iff* $\boldsymbol{\ell}'\mathbf{X}_n \xrightarrow{d} \boldsymbol{\ell}'\mathbf{X}\ \forall \boldsymbol{\ell} \in \mathbb{R}^k$.

Note that in applying Theorem 5.36, $\boldsymbol{\ell}'\mathbf{X}_n \xrightarrow{d} \boldsymbol{\ell}'\mathbf{X}$ is *always* trivially true when $\boldsymbol{\ell} = \mathbf{0}$, and so the condition $\boldsymbol{\ell}'\mathbf{X}_n \xrightarrow{d} \boldsymbol{\ell}'\mathbf{X}$ need only be checked for $\boldsymbol{\ell} \neq \mathbf{0}$. We will be most concerned with convergence in distribution to members of the normal family of distributions. In this context, Theorem 5.36 implies that to establish convergence in distribution of the sequence of random $(k \times 1)$ vectors $\{\mathbf{X}_n\}$ to the random $(k \times 1)$ vector $\mathbf{X} \sim N(\boldsymbol{\mu}, \boldsymbol{\Sigma})$, it suffices to demonstrate that $\boldsymbol{\ell}'\mathbf{X}_n \xrightarrow{d} N(\boldsymbol{\ell}'\boldsymbol{\mu}, \boldsymbol{\ell}'\boldsymbol{\Sigma}\boldsymbol{\ell})\ \forall \boldsymbol{\ell} \in \mathbb{R}^k$. We formalize this observation as a corollary to Theorem 5.36.

**Corollary 5.4** ***Cramer-Wold Device for Normal Limiting Distributions*** $\mathbf{X}_n \xrightarrow{d} N(\boldsymbol{\mu}, \boldsymbol{\Sigma})$ *iff* $\boldsymbol{\ell}'\mathbf{X}_n \xrightarrow{d} N(\boldsymbol{\ell}'\boldsymbol{\mu}, \boldsymbol{\ell}'\boldsymbol{\Sigma}\boldsymbol{\ell})\ \forall \boldsymbol{\ell} \in \mathbb{R}^k$.

The Cramer-Wold device can be used to define multivariate central limit theorems. The following is a multivariate extension of the Lindberg-Levy CLT.

**Theorem 5.37** ***Multivariate Lindberg-Levy CLT*** *Let* $\{\mathbf{X}_n\}$ *be a sequence of iid* $(k \times 1)$ *random vectors with* $\mathrm{E}(\mathbf{X}_i) = \boldsymbol{\mu}$ *and* $\mathbf{Cov}(\mathbf{X}_i) = \boldsymbol{\Sigma}\ \forall i$*, where* $\boldsymbol{\Sigma}$ *is a* $(k \times k)$ *positive definite matrix. Then* $n^{1/2}\left(n^{-1}\sum_{i=1}^{n}\mathbf{X}_i - \boldsymbol{\mu}\right) \xrightarrow{d} N(\mathbf{0}, \boldsymbol{\Sigma})$.

It follows from the multivariate Lindberg-Levy CLT that $\bar{\mathbf{X}}_n \overset{a}{\sim} N(\boldsymbol{\mu}, n^{-1}\boldsymbol{\Sigma})$.

**Example 5.45** **CLT Applied to sum of Bernoulli vectors** Shipments of CPU chips from two different suppliers are to be inspected before being accepted. Chips are randomly drawn, with replacement, from each shipment and are nondestructively tested in pairs, with $(X_{1i}, X_{2i})$ representing the outcome of the tests for pair $i$. An outcome of $x_{\ell i} = 1$ indicates a faulty chip, $x_{\ell i} = 0$ indicates a nondefective chip, and the joint density of $(X_{1i}, X_{2i})$ is given by

[16]H. Cramer and H. Wold, (1936) *Some Theorems on Distribution Functions*, J. London Math. Soc., 11(1936), pp. 290–295.

$$\mathbf{X}_i = \begin{bmatrix} X_{1i} \\ X_{2i} \end{bmatrix} \sim p_1^{x_{1i}}(1-p_1)^{1-x_{1i}} I_{\{0,1\}}(x_{1i}) p_2^{x_{2i}}(1-p_2)^{1-x_{2i}} I_{\{0,1\}}(x_{2i}),$$

where $p_i \in (0,1)$ for $i = 1, 2$. Note that

$$\mathrm{E}(\mathbf{X}_i) = \begin{bmatrix} p_1 \\ p_2 \end{bmatrix} = \mathbf{p}, \text{ and}$$

$$\mathbf{Cov}(\mathbf{X}_i) = \boldsymbol{\Sigma} = \begin{bmatrix} p_1(1-p_1) & 0 \\ 0 & p_2(1-p_2) \end{bmatrix}.$$

Letting $\underset{(2\times1)}{\bar{\mathbf{X}}_n} = n^{-1} \sum_{i=1}^{n} \begin{bmatrix} X_{1i} \\ X_{2i} \end{bmatrix}$, it follows from the multivariate Lindberg-Levy CLT that $\mathbf{Z}_n = n^{1/2}[\bar{\mathbf{X}}_n - \mathbf{p}] \xrightarrow{d} \mathbf{Z} \sim N(\mathbf{0}, \boldsymbol{\Sigma})$ and also that $\bar{\mathbf{X}}_n \overset{a}{\sim} N(\mathbf{p}, n^{-1}\boldsymbol{\Sigma})$.

If one were interested in establishing an asymptotic distribution for the difference in the number of defectives observed in $n$ random pairs of CPUs from shipments 1 and 2, the random variable of interest would be $\mathbf{c}'(n\bar{\mathbf{X}}_n) = \sum_{i=1}^{n} X_{1i=1} - \sum_{i=1}^{n} X_{2i},$, where $\mathbf{c}' = [1\ -1]$. Then $\mathbf{c}'(n\bar{\mathbf{X}}_n) = g(\mathbf{Z}_n, n) = \mathbf{c}'[n^{1/2}\mathbf{Z}_n + n\mathbf{p}]$, so that by Definition 5.2, $\mathbf{c}'(n\bar{\mathbf{X}}_n) \overset{a}{\sim} \mathbf{c}'[n^{1/2}\mathbf{Z} + n\mathbf{p}] \sim N(n\mathbf{c}'\mathbf{p}, n\mathbf{c}'\boldsymbol{\Sigma}\mathbf{c})$. Thus, the asymptotic distribution for the difference in the number of defectives is given by $\mathbf{c}'(n\bar{\mathbf{X}}_n) \overset{a}{\sim} N(n(p_1 - p_2), n[p_1(1-p_1) + p_2(1-p_2)])$. □

Another useful multivariate CLT concerns the case where the elements in the sequence $\{\mathbf{X}_n\}$ are independent but not necessarily identically distributed $(k \times 1)$ random vectors that exhibit uniform (i.e., across all $n$) absolute upper bounds with probability 1.

**Theorem 5.38** ***Multivariate CLT for Independent Bounded Random Vectors*** *Let $\{\mathbf{X}_n\}$ be a sequence of independent $(k\times1)$ random vectors such that $p(|x_{1i}| \le m, |x_{2i}| \le m, \ldots, |x_{ki}| \le m) = 1\ \forall\, i$, where $m \in (0,\infty)$. Let $\mathrm{E}(\mathbf{X}_i) = \boldsymbol{\mu}_i$, $\mathbf{Cov}(\mathbf{X}_i) = \boldsymbol{\Psi}_i$, and suppose that $\lim_{n\to\infty} n^{-1}\sum_{i=1}^{n} \boldsymbol{\Psi}_i = \boldsymbol{\Psi}$, a finite positive definite $(k \times k)$ matrix. Then $n^{-1/2}\sum_{i=1}^{n}(\mathbf{X}_i - \boldsymbol{\mu}_i) \xrightarrow{d} N(\mathbf{0}, \boldsymbol{\Psi})$.*

Various other multivariate CLTs can be constructed using the Cramer-Wold device and CLTs for random scalars. In practice, one often relies on the Cramer-Wold device directly for establishing limiting distributions relating to statistical procedures of interest, and so we will not attempt to compile a list of additional multivariate CLTs here.

## 5.9 Asymptotic Distributions of Differentiable Functions of Asymptotically Normally Distributed Random Variables: The Delta Method

In this section we examine results concerning the asymptotic distributions of differentiable functions of asymptotically normally distributed random variables. General conditions will be identified for which differentiable functions of asymptotically normally distributed random variables are themselves asymptotically normally distributed. The utility of these results in practice is that once the asymptotic distribution of $X_n$ is known, the asymptotic distributions of interesting functions of $X_n$ need not be derived anew. Instead, these asymptotic distributions can generally be defined by specifying the mean and covariance matrix of a normal distribution according to well-defined and straightforwardly implemented formulas.

All of the results that we will examine in this section are based on *first-order* Taylor series expansions of the function $g(\mathbf{x})$ around a point $\boldsymbol{\mu}$. Being that the methods are based on derivatives, the methodology has come to be known as **the delta method** for deriving asymptotic distributions and associated asymptotic covariance matrices of functions of random variables. We review the Taylor series expansion concept here, paying particular attention to the nature of the *remainder term*. Recall that $d(\mathbf{x},\boldsymbol{\mu}) = [(\mathbf{x}-\boldsymbol{\mu})'(\mathbf{x}-\boldsymbol{\mu})]^{1/2}$ represents the distance between the vectors $\mathbf{x}$ and $\boldsymbol{\mu}$.

**Definition 5.11** ***First-Order Taylor Series Expansion and Remainder (Young's Form)***

Let $g$: $D \rightarrow \mathbb{R}$ be a function having partial derivatives in a neighborhood of the point $\boldsymbol{\mu} \in D$ that are continuous at $\boldsymbol{\mu}$. Let $\mathbf{G} = [\partial g(\boldsymbol{\mu})/\partial x_1, \ldots, \partial g(\boldsymbol{\mu})/\partial x_k]$ represent the $1 \times k$ gradient vector of $g(\mathbf{x})$ evaluated at the point $\mathbf{x} = \boldsymbol{\mu}$. Then for $\mathbf{x} \in D$, $g(\mathbf{x}) = g(\boldsymbol{\mu}) + \mathbf{G}(\mathbf{x}-\boldsymbol{\mu}) + d(\mathbf{x}, \boldsymbol{\mu})R(\mathbf{x})$. The remainder term $R(\mathbf{x})$ is continuous at $\mathbf{x} = \boldsymbol{\mu}$, and $\lim_{\mathbf{x}\rightarrow\boldsymbol{\mu}} R(\mathbf{x}) = R(\boldsymbol{\mu}) = 0$.

Young's form of Taylor's theorem is not prevalent in calculus texts. The reader can find more details regarding this type of expansion in G.H. Hardy (1952), *A Course of Pure Mathematics*, 10th ed., Cambridge, New York, The University Press, p. 278, for the scalar case, and T.M. Apostol (1957), *Mathematical Analysis*. Cambridge, MA: Addison-Wesley, pp. 110 and 118 for the multivariate case.

Our first result on asymptotic distributions of $g(\mathbf{x})$ concerns the case where $g(\mathbf{x})$ is a *scalar*-valued function. As will be common to all of the results we will examine, the principal requirement on the function $g(\mathbf{x})$ is that partial derivatives exist in a neighborhood of the point $\boldsymbol{\mu}$ and that they are continuous at $\boldsymbol{\mu}$ so that Lemma 5.6 can be utilized. In addition, we will also make assumptions relating to the nature of the asymptotic distribution of $\mathbf{X}$.

**Theorem 5.39** ***Asymptotic Distribution of $g(\mathbf{X}_n)$ (Scalar Function Case)***

*Let $\{\mathbf{X}_n\}$ be a sequence of $k \times 1$ random vectors such that $n^{1/2}(\mathbf{X}_n - \boldsymbol{\mu}) \xrightarrow{d} \mathbf{Z} \sim N(\mathbf{0},\boldsymbol{\Sigma})$. Let $g(\mathbf{x})$ have first-order partial derivatives in a neighborhood of the point $\mathbf{x} = \boldsymbol{\mu}$ that are continuous at $\boldsymbol{\mu}$, and suppose the gradient vector of $g(\mathbf{x})$ evaluated at $\mathbf{x} = \boldsymbol{\mu}$, $\mathbf{G}_{(1\times k)} = [\partial g(\boldsymbol{\mu})/\partial x_1 \ldots \partial g(\boldsymbol{\mu})/\partial x_k]$, is not the zero vector. Then $n^{1/2}(g(\mathbf{X}_n) - g(\boldsymbol{\mu})) \xrightarrow{d} N(0,\mathbf{G}\boldsymbol{\Sigma}\mathbf{G}')$ and $g(\mathbf{X}_n) \overset{a}{\sim} N(g(\boldsymbol{\mu}), n^{-1}\mathbf{G}\boldsymbol{\Sigma}\mathbf{G}')$.*

**Example 5.46** ***Asymptotic Distribution of $\bar{X}_n(1 - \bar{X}_n)$ for IID Bernoulli RVs***

Note from Example 5.39 that if $\{X_n\}$ is a sequence of *iid* Bernoulli-type random variables, then for $p \neq 0$ or 1, $n^{1/2}(\bar{X}_n - p) \xrightarrow{d} N(0, p(1-p))$. Consider using an outcome of $g(\bar{X}) = \bar{X}(1 - \bar{X})$ as an estimate of the variance $p(1-p)$ of the Bernoulli PDF, and consider defining an asymptotic distribution for $g(\bar{X}_n)$.

Theorem 5.39 applies with $\mu = p$ and $\Sigma = \sigma^2 = p(1-p)$. Note that $dg(p)/d\bar{X} = 1-2p$, which is continuous in $p$ and is nonzero so long as $p \neq .5$. Also, $\sigma^2 \neq 0$ if $p \neq 0$ or 1. Then for $p \neq 0$, .5, or 1, Theorem 5.39 implies that

$$\bar{X}_n(1-\bar{X}_n) \overset{a}{\sim} N\left(p(1-p), n^{-1}(1-2p)^2 p(1-p)\right).$$

An asymptotic density for $\bar{X}_n\,(1-\bar{X}_n)$ under the assumption $p = 1/2$ can be established using other methods (see Bickel and Doksum (1977), *Mathematical Statistics*, San Francisco: Holden-Day, p. 53); however, convergence is not to a normal distribution. Specifically, it can be shown that $n[\bar{X}_n(1-\bar{X}_n)-(1/4)] \overset{d}{\rightarrow} Z$, where $Z$ has the density of a $\chi_1^2$ random variable that has been multiplied by $(-1/4)$. If $p = 0$ or $p = 1$, the $X_i$'s are all degenerate random variables equal to 0 or 1, respectively, and the limiting density of $\bar{X}_n$ is then degenerate at 0 or 1 as well. □

By reinterpreting $\mathbf{g}(\mathbf{x})$ as a vector function and $\mathbf{G}$ as a Jacobian matrix, the conclusion of Theorem 5.39 regarding the asymptotic distribution of the *vector* function remains valid. The extension allows one to define the joint asymptotic distribution of the random vector $\mathbf{g}(\mathbf{X}_n)$.

**Theorem 5.40** ***Asymptotic Distribution of $\mathbf{g}(\mathbf{X}_n)$ (Vector Function Case)***

*Let $\{\mathbf{X}_n\}$ be a sequence of $k\times 1$ random vectors such that $n^{1/2}(\mathbf{X}_n - \boldsymbol{\mu}) \overset{d}{\rightarrow} \mathbf{Z} \sim N(\mathbf{0}, \Sigma)$. Let $\mathbf{g}(\mathbf{x}) = (g_1(\mathbf{x}),\ldots,g_m(\mathbf{x}))'$ be an $(m\times 1)$ vector function $(m \leq k)$ having first order partial derivatives in a neighborhood of the point $\mathbf{x} = \boldsymbol{\mu}$ that are continuous at $\boldsymbol{\mu}$. Let the Jacobian matrix of $\mathbf{g}(\mathbf{x})$ evaluated at $\mathbf{x} = \boldsymbol{\mu}$,*

$$\underset{m\times k}{\mathbf{G}} = \begin{bmatrix} \partial g_1(\boldsymbol{\mu})/\partial \mathbf{x}' \\ \vdots \\ \partial g_m(\boldsymbol{\mu})/\partial \mathbf{x}' \end{bmatrix} = \begin{bmatrix} \frac{\partial g_1(\boldsymbol{\mu})}{\partial \mathbf{x}_1} & \cdots & \frac{\partial g_1(\boldsymbol{\mu})}{\partial \mathbf{x}_k} \\ \vdots & \ddots & \vdots \\ \frac{\partial g_m(\boldsymbol{\mu})}{\partial \mathbf{x}_1} & \cdots & \frac{\partial g_m(\boldsymbol{\mu})}{\partial \mathbf{x}_k} \end{bmatrix},$$

*have full row rank. Then*
*$n^{1/2}(\mathbf{g}(\mathbf{X}_n) - \mathbf{g}(\boldsymbol{\mu})) \overset{d}{\rightarrow} N(\mathbf{0}, \mathbf{G}\Sigma\mathbf{G}')$ and $\mathbf{g}(\mathbf{X}_n) \overset{a}{\sim} N(\mathbf{g}(\boldsymbol{\mu}), n^{-1}\,\mathbf{G}\Sigma\mathbf{G}')$.*

**Example 5.47** ***Asymptotic Distribution of Products and Ratios***

Let $\{\hat{\boldsymbol{\beta}}_n\}$ be a sequence of $(2\times 1)$ random vectors such that $n^{1/2}\left(\hat{\boldsymbol{\beta}}_n - \boldsymbol{\beta}\right) \overset{d}{\rightarrow} N(\mathbf{0}, \Sigma)$, where $\boldsymbol{\beta} = [2 \quad 1]'$ and $\Sigma = \begin{bmatrix} 2 & 1 \\ 1 & 1 \end{bmatrix}$. We seek an asymptotic distribution for the vector function $\mathbf{g}\left(\hat{\boldsymbol{\beta}}\right) = \left(3\hat{\beta}[1]\hat{\beta}[2] \quad \hat{\beta}[2]/\hat{\beta}[1]\right)'$. All of the conditions of Theorem 5.40 are met, including the fact that

$$\underset{2\times 2}{\mathbf{G}} = \begin{bmatrix} dg_1(\boldsymbol{\beta})/d\boldsymbol{\beta}' \\ dg_2(\boldsymbol{\beta})/d\boldsymbol{\beta}' \end{bmatrix} = \begin{bmatrix} 3\,\beta_2 & 3\,\beta_1 \\ -\beta_2/\beta_1^2 & 1/\beta_1 \end{bmatrix} = \begin{bmatrix} 3 & 6 \\ -1/4 & 1/2 \end{bmatrix}$$

has full row rank (note that the partial derivatives exist in an open rectangle containing $\boldsymbol{\beta}$, and they are continuous at the point $\boldsymbol{\beta}$). Then, since $\mathbf{g}(\boldsymbol{\beta}) = [6 \quad ½]'$ and $\mathbf{G}\Sigma\mathbf{G}' = \begin{bmatrix} 90 & 1.5 \\ 1.5 & .125 \end{bmatrix}$, it follows from Theorem 5.40 that

$$\mathbf{g}(\hat{\boldsymbol{\beta}}_n) = \begin{bmatrix} 3\,\hat{\beta}_n[1]\,\hat{\beta}_n[2] \\ \hat{\beta}_n[2]/\,\hat{\beta}_n[1] \end{bmatrix} \overset{a}{\sim} N\left(\begin{bmatrix} 6 \\ 1/2 \end{bmatrix}, n^{-1}\begin{bmatrix} 90 & 1.5 \\ 1.5 & .125 \end{bmatrix}\right).$$

A specific distribution is obtained once $n$ is specified. For example, if $n = 20$, then $g\left(\hat{\boldsymbol{\beta}}_{20}\right) \overset{a}{\sim} N\left(\begin{bmatrix} 6 \\ 1/2 \end{bmatrix}, \begin{bmatrix} 4.5 & .075 \\ .075 & .00625 \end{bmatrix}\right)$. □

The final result that we will examine concerning the asymptotic distribution of $\mathbf{g}(\mathbf{X}_n)$ generalizes the previous two theorems to cases for which $\mathbf{V}_n^{-1/2}(\mathbf{X}_n - \boldsymbol{\mu}) \xrightarrow{d} N(\mathbf{0},\mathbf{I})$, where $\{\mathbf{V}_n\}$ is a sequence of $(m\times m)$ positive definite matrices of real numbers such that $\mathbf{V}_n \to \mathbf{0}$.[17] Note this case subsumes the previous cases upon defining $\mathbf{V}_n = n^{-1}\,\boldsymbol{\Sigma}$, in which case $\mathbf{V}_n^{-1/2}(\mathbf{X}_n - \boldsymbol{\mu}) = \boldsymbol{\Sigma}^{-1/2}\,\mathrm{n}^{1/2}\,(\mathbf{X}_n - \boldsymbol{\mu}) \xrightarrow{d} N(\mathbf{0},\mathbf{I})$ by Slutsky's theorem. The generalization allows additional flexibility in how the asymptotic distribution of $\mathbf{X}_n$ is initially established and is especially useful in the context of the least squares estimator to be discussed in Chapter 8.

**Theorem 5.41** ***Asymptotic Distribution of*** $\mathbf{g}(\mathbf{X}_n)$ ***(Generalized)***

*Let $\{\mathbf{X}_n\}$ be a sequence of $(k\times 1)$ random vectors such that $\mathbf{V}_n^{-1/2}(\mathbf{X}_n - \boldsymbol{\mu}) \xrightarrow{d} N(\mathbf{0},\mathbf{I})$, where $\{\mathbf{V}_n\}$ is a sequence of $(m\times m)$ positive definite matrices for which $\mathbf{V}_n \to \mathbf{0}$. Let $\mathbf{g}(\mathbf{x})$ be a $(m\times 1)$ vector function satisfying the conditions of Theorem 5.40. If there exists a sequence of positive real numbers $\{a_n\}$ such that $\{[a_n\mathbf{G}\mathbf{V}_n\mathbf{G}']^{-1/2}\}$ is $O(1)$ and $a_n^{1/2}\,(\mathbf{X}_n - \boldsymbol{\mu})$ is $O_p(1)$, then $(\mathbf{G}\mathbf{V}_n\mathbf{G})^{-1/2}\,[\mathbf{g}(\mathbf{X}_n) - \mathbf{g}(\boldsymbol{\mu})] \xrightarrow{d} N(\mathbf{0},\mathbf{I})$ and $\mathbf{g}(\mathbf{X}_n) \overset{a}{\sim} N(\mathbf{g}(\boldsymbol{\mu}), \mathbf{G}\mathbf{V}_n\mathbf{G}')$.*

## 5.10 Appendix: Proofs and Proof References for Theorems

**Theorem 5.1** *The discrete case is left to the reader. For the continuous case, see H. Scheffé, (1947), "A useful convergence theorem for probability distributions," Ann. Math. Stat., 18, pp. 434–438.*

**Theorem 5.2** *See E. Lukacs (1970), Characteristic Functions, London: Griffin, pp. 49–50, for a proof of this theorem for the more general case characterized by convergence of characteristic functions (which subsumes Theorem 5.2 as a special case). In the multivariate case, **t** will be a vector, and convergence of the MFG must hold $\forall t_i \in (-h,h)$, and $\forall i$.*

---

[17]Recall that $\mathbf{V}_n^{1/2}$ is the symmetric square root matrix of $\mathbf{V}_n$, and $\mathbf{V}_n^{-1/2}$ is the inverse of $\mathbf{V}_n^{1/2}$. The defining property of $\mathbf{V}_n^{1/2}$ is that $\mathbf{V}_n^{1/2}\mathbf{V}_n^{1/2} = \mathbf{V}_n$, while $\mathbf{V}_n^{-1/2}\mathbf{V}_n^{-1/2} = \mathbf{V}_n^{-1}$.

**Theorem 5.3** *See the proof of Theorem 5.17 and R. Serfling (1980), Approximation Theorems of Mathematical Statistics, New York: Wiley, pp. 24–25.*

**Theorem 5.4** *See the proof of Theorem 5.16.*

**Theorem 5.5** *See the proof of Theorem 5.17 and R. Serfling, op. cit., pp. 24–25.*

**Theorem 5.6** *All of the results follow from Theorem 5.5, and the fact that the functions being analyzed are continuous functions. Note in particular that the matrix inverse function is continuous at all points for which the matrix is nonsingular.*

**Theorem 5.7** *Y.S. Chow and H. Teicher (1978), Probability Theory, New York: Springer-Verlag, New York, p. 249.*

**Corollary 5.1** This follows immediately from Theorem 5.7 upon defining $X_n=X=Y\ \forall n$.

**Theorem 5.8: Proof** Let $\{Y_n\}$ be a sequence of scalar random variables and suppose $Y_n \xrightarrow{d} c$, so that $F_n(y) \to F(y) = I_A(y)$, where $A = \{y: y \geq c\}$. Then as $n \to \infty$, $P(|y_n - c| < \varepsilon) \geq F_n(c + \tau) - F_n(c - \tau) \to 1$, for $\tau \in (0,\varepsilon)$ and $\forall \varepsilon > 0$, which implies that $Y_n \xrightarrow{p} c$. The multivariate case can be proven similarly using marginal CDFs and the elementwise nature of the plim operator. ■

**Theorem 5.9** *This follows from the proof in V. Fabian and J. Hannon (1985), Introduction to Probability and Mathematical Statistics, New York: John Wiley, p. 159, and from Theorem 5.4.*

**Theorem 5.10: Proof** Each function on the left-hand side of (a), (b), and (c) is of the form $g(\mathbf{X}_n, \mathbf{Y}_n, \mathbf{a}_n)$ and satisfies the conditions of Theorem 5.9, with $\mathbf{a}_n$ being a ghost in the definition of the function $g$. ■

**Theorem 5.11: Proof** This follows directly from definitions of orders of magnitude in probability upon interpreting the sequence of real numbers or matrices as a sequence of degenerate random variables or random matrices. ■

**Theorem 5.12: Proof** We provide a proof for the scalar case, which suffices to prove the matrix case given the elementwise definition of mean-square convergence (Definition 5.6.b).

***Necessity*** **a.** $\mathrm{E}(Y_n) \to \mathrm{E}(Y)$ follows from the fact that

$$|\mathrm{E}(Y_n) - \mathrm{E}(Y)| = |\mathrm{E}(Y_n - Y)| \leq \mathrm{E}(|Y_n - Y|) \leq \left(\mathrm{E}\left(|Y_n - Y|^2\right)\right)^{1/2} \to 0.$$

To see this, note that the first inequality follows because $(y_n - y) \leq |y_n - y|$. Regarding the second inequality, note that $g(z) = z^2$ is a convex function on $\mathbb{R}$, and letting $Z = |Y_n - Y|$, Jensen's inequality implies $(\mathrm{E}(|Y_n - Y|))^2 \leq \mathrm{E}(|Y_n - Y|^2)$ (recall $g(\mathrm{E}(Z)) \leq \mathrm{E}(g(Z))$). Convergence to zero occurs because $\mathrm{E}(|Y_n - Y|^2) = \mathrm{E}((Y_n - Y)^2) \to 0$ by convergence in mean square.

**b.** $E(Y_n^2) = E((Y_n - Y)^2) + E(Y^2) + 2E(Y(Y_n - Y))$, and since by the Cauchy-Schwartz inequality, $|E(Y(Y_n - Y))| \leq \left[E(Y^2)E\left((Y_n - Y)^2\right)\right]^{1/2}$ it follows that

$$E\left((Y_n - Y)^2\right) + E(Y^2) - 2\left[E(Y^2)E((Y_n - Y)^2)\right]^{1/2}$$
$$\leq E(Y_n^2) \leq E\left((Y_n - Y)^2\right) + E(Y^2) + 2\left[E(Y^2)E\left((Y_n - Y)^2\right)\right]^{1/2}$$

Then since $E\left((Y_n - Y)^2\right) \to 0$ by mean-square convergence, $E(Y_n^2) \to E(Y^2)$. It follows that $\text{var}(Y_n) \to \text{var}(Y)$ since $\text{var}(Y_n) = E(Y_n^2) - (E(Y_n))^2 \to E(Y^2) - (E(Y))^2 = \text{var}(Y)$.

**c.** First note that

$$E\left((Y_n - Y)^2\right) = E(Y_n^2) - 2E(Y_nY) + E(Y^2)$$
$$= \text{var}(Y_n) + (E(Y_n))^2 - 2[\text{cov}(Y_n, Y) + E(Y_n)E(Y)] + \text{var}(Y) + (E(Y))^2$$

If $E\left((Y_n - Y)^2\right) \to 0$, then by (a) and (b), the preceding equality implies $\text{cov}(Y_n, Y) \to \text{var}(Y)$.

***Sufficiency*** From the expression defining $E\left((Y_n - Y)^2\right)$ in the preceding proof of the necessity of (c), it follows directly that (a), (b), and (c) $\Rightarrow E((Y_n - Y)^2) \to 0$, which implies $Y_n \xrightarrow{m} Y$. ■

**Corollary 5.2: Proof** This follows directly from Theorem 5.12 upon letting $\mathbf{Y} = \mathbf{c}$, and noting that $\text{var}(c[i,j]) = 0$ and $\text{cov}(Y_n[i,j], c[i,j]) = 0$. ■

**Corollary 5.3: Proof** This follows directly from Theorem 5.12 since in the scalar case, $\text{corr}(Y_n, Y) = \frac{\text{cov}(Y_n, Y)}{[\text{var}(Y_n)\text{var}(Y)]^{1/2}} \to \frac{\text{var}(Y)}{\text{var}(Y)} = 1$. The matrix case follows by applying the preceding result elementwise. ■

**Theorem 5.13: Proof** (scalar case—matrix case proved by applying the argument elementwise). Note that $(Y_n - Y)^2$ is a nonnegative-valued random variable, and letting $a = \varepsilon^2 > 0$, we have by Markov's inequality that $P((y_n - y)^2 \geq \varepsilon^2) \leq E((Y_n - Y)^2)/\varepsilon^2$. Thus,

$$P(|y_n - y| \geq \varepsilon) \leq E\left((Y_n - Y)^2\right)/\varepsilon^2 \text{ or } P(|y_n - y| < \varepsilon) \geq 1 - E\left((Y_n - Y)^2\right)/\varepsilon^2$$

By mean square convergence, $E((Y_n - Y)^2) \to 0$, so that $\lim_{n\to\infty} P(|y_n - y| < \varepsilon) = 1$ $\forall\ \varepsilon > 0$. Thus, $\text{plim}(Y_n) = Y$. Convergence in distribution follows from Corollary 5.1. ■

**Theorem 5.14: Proof**

***Necessity*** $\lim_{n\to\infty} y_n = y$ *implies* that for every $\varepsilon > 0$, there exists an integer $N(\varepsilon)$ such that $|y_i - y| < \varepsilon \; \forall i \geq N(\varepsilon)$ (recall Definition A.25). It follows that $P(\lim_{n\to\infty} y_n = y) \leq P(|y_i - y| < \varepsilon, \; i \geq N(\varepsilon)) \; \forall \varepsilon > 0$. If $Y_n \xrightarrow{as} Y$, then the left-hand side of the preceding inequality is 1, which implies that the right-hand side is also 1. It follows that, for $\forall \varepsilon > 0$, $\lim_{n\to\infty} P(|y_i - y| < \varepsilon, i \geq n) = 1$ since the values of $P(|y_i - y)| < \varepsilon, i \geq n)$ must all be ones for $n$ large enough if $\lim_{n\to\infty} y_n = y$.

***Sufficiency*** R. Serfling, Approximation Theorems, pp. 6–7. ■

**Theorem 5.15: Proof** Suppose $Y_n \xrightarrow{as} Y$. Then $\forall \varepsilon > 0$, $\lim_{n\to\infty} P(|y_i - y| < \varepsilon, i \geq n) = 1$ which follows from Theorem 5.14. Since $|y_i - y| < \varepsilon, i \geq n \Rightarrow |y_n - y| < \varepsilon$, it follows that $P(|y_i - y| < \varepsilon, i \geq n) \leq P(|y_n - y| < \varepsilon)$. Then since the left-hand side has a limiting value of 1 $\forall \varepsilon > 0$, by almost-sure convergence, it follows that the right-hand side has a limiting value of 1, $\forall \varepsilon > 0$, implying convergence in probability. ■

**Theorem 5.16: Proof** The proof is immediate from the definition of almost-sure convergence, since if $a_n[i,j] \to a[i,j] \; \forall \, i$ and $j$, then $P(\lim_{n\to\infty} a_n = a) = 1$, which implies that $a_n \xrightarrow{as} a$. ■

**Theorem 5.17: Proof** Assume that $X_n \xrightarrow{as} X$. For outcomes of $\{X_n\}$ for which $x_n \to x$ and $g$ is continuous at $x$, it must be the case that $g(x_n) \to g(x)$ (recall Definition A.29). Then defining the sets $A = \{(\{x_n\}, x)\colon x_n \to x\}$ *and* $B = \{(\{x_n\}, x)\colon g$ is continuous at $x\}$, it follows that $P(g(x_n) \to g(x)) \geq P(A \cap B) \geq 1 - P(\bar{A}) - P(\bar{B}) = 1$ since $P(\bar{A}) = 0$ by almost-sure convergence of $\{X_n\}$ to $X$, and $P(\bar{B}) = 0$ since $g$ is continuous with probability 1. Thus $g(X_n) \xrightarrow{as} g(X)$. ■

**Theorem 5.18** *Y.S. Chow and H. Teicher (1978), Probability Theory, New York: Springer-Verlag, p. 68.*

**Theorem 5.19: Proof** A general proof involving characteristic functions can be found in D.S.G. Pollock (1979), The Algebra of Econometrics, New York: John Wiley, p. 332. An alternative proof based entirely on probability inequalities is given by C.R. Rao, (1965) Statistical Inference, pp. 112–113. Our proof requires the additional assumption that $M_{X_i}(t)$ exists, but the general proof is analogous with the characteristic function replacing the MGF. The moment-generating function of $\bar{X}_n$ is given by

$$M_{\bar{X}_n}(t) = \mathrm{E}\left(\exp\left(\frac{t}{n}\sum_{i=1}^{n} X_1\right)\right) = \prod_{i=1}^{n} \mathrm{E}\left(\exp\left(X_1 \frac{t}{n}\right)\right) = \prod_{i=1}^{n} M_{X_1}\left(\frac{t}{n}\right) = \left[M_{X_1}\left(\frac{t}{n}\right)\right]^n$$

because the $X_i$'s are independent and identically distributed. It follows that

$$\lim_{n\to\infty} M_{\bar{X}_n}(t) = \lim_{n\to\infty}\left[1 + \frac{nM_{X_1}(t/n) - n}{n}\right]^n = \exp\left(\lim_{n\to\infty}\left(n\left[M_{X_1}\left(\frac{t}{n}\right) - 1\right]\right)\right)$$

by Lemma 5.1, stated below.

**Lemma 5.1**

$\lim_{n\to\infty}[1 + a_n/n]^n = \exp(\lim_{n\to\infty} a_n)$

Then applying L'Hospital's rule,

$$\lim_{n\to\infty}\left(\frac{M_{X_1}(t/n) - 1}{n^{-1}}\right) = \lim_{n\to\infty}\left((-n^2)\frac{dM_{X_1}(t/n)}{d(t/n)}\left(\frac{-t}{n^{-2}}\right)\right) = t\mu,$$

since the first derivative of $M_{X_1}(t_*) \to \mu$ as $t_* = t/n \to 0$. Then $\lim_{n\to\infty} M_{\bar{X}_1}(t) = e^{t\mu}$, which is the MGF of a random variable that is degenerate at $\mu$. Therefore, by Theorem 5.2 and Theorem 5.8 $\bar{X}_n \xrightarrow{p} \mu$. ■

**Theorem 5.20: Proof** Without loss of generality, we assume that S is a real-valued sample space. Let the $n$ *iid* random variables $Z_1,\ldots,Z_n$ represent the $n$ independent and identical repetitions of the experiment, and define $X_i = I_A(Z_i)$, for $i = 1,\ldots,n$, so that $X_1,\ldots, X_n$ are $n$ *iid* (Bernoulli) random variables for which $x_i = 1$ indicates the occurrence and $x_i = 0$ indicates the nonoccurrence of event $A$ on the $i$th repetition. Since $E(X_i) = P(A), \forall\, i$, it follows from Khinchin's WLLN that $\bar{X}_n \xrightarrow{p} P(A)$. Then, since $N_A \equiv \sum_{i=1}^n X_i$ and $(N_A/n) \equiv \bar{X}_n$, we can also conclude that $(N_A/n) \xrightarrow{p} P(A)$. ■

**Theorem 5.21: Proof**

***Sufficiency*** For any choice of $b > 0$ and $a \geq b$, $a/(a+1) \geq b/(b+1)$. It follows that $\forall \varepsilon > 0$,

$$P\left((\bar{x}_n - \bar{\mu}_n)^2 \geq \varepsilon^2\right) \leq P\left(\frac{(\bar{x}_n - \bar{\mu}_n)^2}{1 + (\bar{x}_n - \bar{\mu}_n)^2} \geq \frac{\varepsilon^2}{1+\varepsilon^2}\right) \leq \mathrm{E}\left[\frac{(\bar{X}_n - \bar{\mu}_n)^2}{1 + (\bar{X}_n - \bar{\mu}_n)^2}\right] \Big/ \left[\frac{\varepsilon^2}{1+\varepsilon^2}\right]$$

where the first inequality follows because the event on the left-hand side *implies* the event on the right, and the second inequality is an application of Markov's inequality. If the expectation of the bracketed term $\to 0$ as $n \to \infty$, then $\forall \varepsilon > 0$ $P\left((\bar{x}_n - \bar{\mu}_n)^2 < \varepsilon^2\right) = P(|\bar{x}_n - \bar{\mu}_n| < \varepsilon) \to 1$ so that $(\bar{X}_n - \bar{\mu}_n) \xrightarrow{p} 0$.

***Necessity*** See B.V. Gnedenko (1968), *The Theory of Probability*, New York: Chelsea Publishing Col, pp. 246–248. ■

**Theorem 5.22: Proof** The result follows directly from Theorem 5.21 upon recognizing that $0 \leq E\left[\frac{(\bar{X}_n - \bar{\mu}_n)^2}{1+(\bar{X}_n - \bar{\mu}_n)^2}\right] \leq \mathrm{E}\left[(\bar{X}_n - \bar{\mu}_n)^2\right] = \mathrm{var}(\bar{X}_n)$. ■

**Theorem 5.23: Proof** With reference to Theorems 5.4 and 5.5, we know that because $\bar{\mu}_n \xrightarrow{p} c$, or equivalently, $(\bar{\mu}_n - c) \xrightarrow{p} 0$, and also because $g(\bar{X}_n - \bar{\mu}_n, \bar{\mu}_n - c) = (\bar{X}_n - \bar{\mu}_n) + (\bar{\mu}_n - c) = \bar{X}_n - c$ is a continuous function of $(\bar{X}_n - \bar{\mu}_n)$ and $(\bar{\mu}_n - c)$, then plim $(\bar{X}_n - c) = \mathrm{plim}(\bar{X}_n - \bar{\mu}_n) + \mathrm{plim}(\bar{\mu}_n - c) = 0$, so that by Definition 5.3 $\bar{X}_n \xrightarrow{p} c$. ■

**Theorem 5.24: Proof** Let $W_j = \sum_{i=1}^{j} X_i$, and define the events

$$A_j = \{(x_1,\ldots,x_n): |w_j| \geq \varepsilon \, |w_i| < \varepsilon \text{ for } i<j\},\ j = 1,\ldots,n,$$

and $A = \left\{ (x_1,\ldots,x_n): \max_{1\leq m\leq n} \left| \sum_{i=1}^{m} x_i \right| \geq \varepsilon \right\}$.

The events $A_1,\ldots,A_n$ are disjoint, and $A = \cup_{i=1}^{n} A_i$, i.e., the $A_i$'s are a partition of $A$. Furthermore,

$$(I)\ \mathrm{E}(W_n^2) = \mathrm{E}\left[\sum_{i=1}^{n} X_i^2 + \sum_{i\neq j}\sum X_i X_j\right] = \sum_{i=1}^{n} \sigma_i^2$$

$$\geq \mathrm{E}(W_n^2 I_A(X)) = \sum_{i=1}^{n} \mathrm{E}(W_n^2 I_{A_i}(X))$$

where $x = (x_1,\ldots,x_n)$, and the last equality holds because $I_A(x) = \sum_{i=1}^{n} I_{A_i}(x)$. Also note that $\forall j<n, W_n - W_j = \sum_{i=j+1}^{n} X_i = f(X_{j+1},\ldots,X_n)$ is independent of $W_j = g(X_1,\ldots,X_j)$ and independent of $W_j\, I_{A_j}\ (X) = h(X_1,\ldots,X_j)$ by the independence of $(X_1,\ldots,X_j)$ and $(X_{j+1},\ldots,X_n)$ (note that $X_{j+1},\ldots,X_n$ are ghosts in the function $I_{A_j}(X)$ given the definition of $A_j$). Therefore $\mathrm{E}((W_n - W_j)W_j I_{A_j}\ (X)) = [\mathrm{E}(W_n - W_j)][\mathrm{E}(W_j I_{A_j}(X))] = 0$ since $\mathrm{E}(W_i) = 0,\ \forall\, i$. It follows that

$$\mathrm{E}(W_n^2 I_{A_j}(X)) = \mathrm{E}([W_n^2 - 2(W_n - W_j)W_j] I_{A_j}(X))$$

$$= \mathrm{E}\left(\left[W_j^2 + (W_n - W_j)^2\right] I_{A_j}(X)\right) \geq \mathrm{E}\left(W_j^2 I_{A_j}(X)\right) \geq \varepsilon^2 P(A_j)$$

where the last inequality follows from the fact that $w_j^2 \geq \varepsilon^2$ when $x \in A_j$. Using this result in $(I)$ and recalling that the $A_i$'s are a partition of $A$, $\sum_{i=1}^{n} \sigma_i^2 \geq \varepsilon^2 \sum_{i=1}^{n} P(A_i) = \varepsilon^2 P(A), \forall \varepsilon>0$. ■

**Theorem 5.25: Proof** Let $Y_i = (X_i - \mu)/i$ and note that $\mathrm{E}(Y_i) = 0$ and $\mathrm{var}(Y_i) = \sigma^2/i^2$. Define $W_k = \sum_{i=1}^{k} Y_i$, and examine the sums $W_{n+m} - W_n = \sum_{i=n+1}^{n+m} Y_i = \sum_{i=1}^{m} Y_{n+i}$, for $m = 1,2,\ldots,k$. By Kolmogorov's inequality (replace $x_i$ by $y_{n+i}$ in the statement of the inequality), it follows that $\forall \varepsilon>0$,

$$P\left(\max_{1\leq m\leq k} |w_{n+m} - w_n| \geq \varepsilon\right) = P\left(\max_{n+1\leq m\leq n+k} |w_m - w_n| \geq \varepsilon\right) \leq \sum_{i=n+1}^{n+k} \left[\frac{\sigma^2}{(i^2 \varepsilon^2)}\right].$$

Letting $k \to \infty$, we have that[18]

$$P\left(\max_{m>n} |w_m - w_n| \geq \varepsilon\right) \leq \sum_{i=n+1}^{\infty} \left[\frac{\sigma^2}{(i^2 \varepsilon^2)}\right] = \left(\frac{\sigma^2}{\varepsilon^2}\right) \sum_{i=n+1}^{\infty} i^{-2},$$

[18]If max does not exist, max is replaced by sup (the supremum).

and since $\lim_{n\to\infty}\sum_{i=n+1}^{\infty} i^{-2} = 0$,[19] it follows from the Cauchy criterion for almost-sure convergence (Theorem 5.18) that $\{W_n\}$ converges almost surely to some random variable, say $W$. Now note the Kronecker lemma from real analysis.

**Lemma 5.2** ***Kronecker's Lemma***

Let $\{a_n\}$ be a sequence of nondecreasing positive numbers, and let $\{z_n\}$ be a sequence of real numbers for which $\sum_{i=1}^{\infty} z_i/a_i$ converges. Then $\lim_{n\to\infty} a_n^{-1}\sum_{i=1}^{n} z_i = 0$. (See E. Lukacs (1968), Stochastic Convergence, Andover, MA: D.C. Heath and Co., p. 96).

Let $z_i = x_i - \mu$ and $a_i = i$, so that $y_i = z_i/a_i$ and $w_n = \sum_{i=1}^{n} z_i/a_i$. Since $w_n \to w$ with probability 1, and since by Kronecker's lemma $w_n \to w \Rightarrow n^{-1}\sum_{i=1}^{n} z_i \to 0$, then $n^{-1}\sum_{i=1}^{n} z_i = \bar{x}_n - \mu \to 0$ with probability 1. ■

**Theorem 5.26** *The proof of the theorem is somewhat difficult, and can be found in C. Rao, Statistical Inference, pp. 115–116.* ■

**Theorem 5.27: Proof** Without loss of generality, we assume that $S$ is a real-valued sample space. Let the $n$ *iid* random variables $Z_1,\ldots,Z_n$ represent the $n$ independent and identical repetitions of the experiment, and define $X_i = I_A(Z_i)$, for $i = 1,\ldots,n$, so that $X_1,\ldots,X_n$ are $n$ *iid* (Bernoulli) random variables for which $x_i = 1$ indicates the occurrence and $x_i = 0$ indicates the nonoccurrence of event $A$ on the *ith* repetition of the experiment. Note that $E(X_i) = P(A) < \infty\ \forall\ i$, so that Theorem 5.26 is applicable. Then $\bar{X}_n \overset{as}{\to} P(A)$, and since $N_A \equiv \sum_{i=1}^{n} X_i$ and $N_A/n = \bar{X}_n$, we can also conclude that $(N_A/n) \overset{as}{\to} P(A)$. ■

**Theorem 5.28** *This follows directly from Theorem 3.7.2 in W.F. Stout (1974), Almost-sure Convergence. New York: Academic Press, p. 202.*

**Theorem 5.29: Proof** The proof is based on Theorems 5.16 and 5.17 and follows the approach of Theorem 5.23. Details are left to the reader. ■

**Theorem 5.30: Proof** We prove the theorem for the case where the moment-generating function of $X_i$ exists. The proof can be made general by substituting characteristic functions in place of MGFs, and such a general proof can be found in C. Rao, Statistical Inference, p. 127.

Let $Z_i = (X_i - \mu)/\sigma$, so that $E(Z_i) = 0$ and var $(Z_i) = 1$, and define $Y_n = (n^{1/2}\sigma)^{-1}\left(\sum_{i=1}^{n} X_i - n\mu\right) = n^{-1/2}\sum_{i=1}^{n} Z_i$. Then

$$M_{Y_n}(t) = \prod_{i=1}^{n} M_{Z_i}\left(t/n^{1/2}\right) = \left[M_{Z_1}\left(t/n^{1/2}\right)\right]^n,$$

[19] $\sum_{i=1}^{n} i^{-p}$ is the so-called $p$ series that converges for $p>1$ and diverges for $p\in(0,1]$. Since the series converges for $p = 2$, it must be the case that $\sum_{i=n+1}^{\infty} i^{-2} \to 0$ as $n \to \infty$. See Bartle, The Elements of Real Analysis, 2nd Ed. pp. 290–291.

where the last equality follows because the $Z_i$'s are independent and identically distributed. Taking logarithms,

$$\lim_{n\to\infty} \ln(M_{Y_n}(t)) = \lim_{n\to\infty} \left[\ln\left(M_{Z_1}\left(t/n^{1/2}\right)/n^{-1}\right)\right],$$

which has the indeterminate form 0/0, so we can apply L'Hospital's rule. Letting $t_* = t/n^{1/2}$ and $\eta(t_*) = \ln(M_{Z_1}(t_*))$,

$$\begin{aligned}\lim_{n\to\infty} \ln M_{Y_n}(t) &= \lim_{n\to\infty} \left[\left(\frac{d\eta(t_*)}{dt_*}\right)\left(-\frac{t}{2n^{3/2}}\right)/(-n^{-2})\right] \\ &= \lim_{n\to\infty} \left[\frac{1}{2}\left[t\frac{d\eta(t_*)}{dt_*}/n^{-1/2}\right]\right],\end{aligned}$$

which remains an indeterminate form since $d\eta(t_*)/dt_* \to d\eta(0)/dt_* = \mathrm{E}(Z_1) = 0$ when $n \to \infty$. A second application of L'Hospital's rule yields

$$\lim_{n\to\infty} \ln(M_{Y_n}(t)) = (1/2)\lim_{n\to\infty}\left[t^2\frac{d^2\eta(t_*)}{dt_*^2}\right] = t^2/2,$$

since

$$\begin{aligned}\lim_{n\to\infty}\frac{d^2\eta(t_*)}{dt_*^2} = \frac{d^2\eta(0)}{dt_*^2} &= \frac{d^2M_{Z_1}(0)}{dt_*^2} - \left[\frac{dM_{Z_1}(0)}{dt_*}\right]^2 \\ &= \mathrm{E}(Z_1^2) - (\mathrm{E}(Z_1))^2 = \mathrm{var}(Z_1) = 1\end{aligned}$$

Thus, $\lim_{n\to\infty}(M_{Y_n}(t)) = \exp\left(\lim_{n\to\infty}\ln(M_{Y_n}(t))\right) = e^{t^2/2}$, which implies that $Y_n \xrightarrow{d} N(0,1)$ by Theorem 5.2. ■

**Theorem 5.31** *See Rohatgi, Mathematical Statistics, pp. 282–288 or Chow and Teicher, Probability Theory, pp. 291–293.*

**Theorem 5.32: Proof** This follows from the Lindberg CLT by first noting that since $P(|x_i - \mu_i| \le 2m) = 1$ by the boundedness assumption, then $\forall\varepsilon > 0$ (for the continuous case—the discrete case is similar),

$$\begin{aligned}\int_{(x_i-\mu_i)^2 \ge \varepsilon b_n^2} (x_i - \mu_i)^2 f_i(x_i)dx_i &\le 4m^2 \int_{(x_i-\mu_i)^2 \ge \varepsilon b_n^2} f_i(x_i)dx_i \\ &\le 4m^2 P\left((x_i - \mu_i)^2 \ge \varepsilon b_n^2\right) \\ &\le \frac{4m^2\sigma_i^2}{\varepsilon b_n^2}\end{aligned}$$

where the last inequality results from Markov's inequality. Then

$$\frac{1}{b_n^2}\sum_{i=1}^{n}\int\limits_{(x_i-\mu_i)^2\geq \varepsilon b_n^2}(x_i-\mu_i)^2 f_i(x_i)dx_i \leq \frac{1}{b_n^2}\sum_{i=1}^{n}\frac{4m^2\sigma_i^2}{\varepsilon b_n^2}$$

$$\leq \frac{4m^2}{(\varepsilon b_n^2)}$$

since $b_n^2 = \sum_{i=1}^{n}\sigma_i^2$, so that if $b_n^2 \to \infty$ when $n \to \infty$, the right-hand side above has a zero limit $\forall \varepsilon > 0$, and the Lindberg condition is satisfied. ■

**Theorem 5.33: Proof** This follows from the Lindberg CLT by first noting that, for $\delta > 0$ (for the continuous case—the discrete case is similar),

$$\int\limits_{(x_i-\mu_i)^2\geq \varepsilon b_n^2}(x_i-\mu_i)^2 f_i(x_i)dx_i = \int\limits_{(x_i-\mu_i)^2\geq \varepsilon b_n^2}|x_i-\mu_i|^{\delta}|x_i-\mu_i|^{-\delta}(x_i-\mu_i)^2 f_i(x_i)dx_i$$

$$\leq \left(\varepsilon b_n^2\right)^{-\delta/2}\int\limits_{(x_i-\mu_i)^2\geq \varepsilon b_n^2}|x_i-\mu_i|^{2+\delta}f_i(x_i)dx_i$$

$$\leq \left(\varepsilon b_n^2\right)^{-\delta/2}\mathrm{E}\left(|X_i-\mu_i|^{2+\delta}\right),$$

where the first inequality follows from the fact that over the range of integration, $(x_i-\mu_i)^2 \geq \varepsilon b_n^2$ implies $|x_i-\mu_i|^{-\delta} \leq \left(\varepsilon b_n^2\right)^{-\delta/2}$, and the second inequality results from adding the nonnegative term $\left(\varepsilon b_n^2\right)^{-\delta/2}\int_{(x_i-\mu_i)^2<\varepsilon b_n^2}|x_i-\mu_i|^{2+\delta}f_i(x_i)dx_i$ to the right-hand side of the first inequality. Then

$$\lim_{n\to\infty}\frac{1}{b_n^2}\sum_{i=1}^{n}\int\limits_{(x_i-\mu_i)^2\geq \varepsilon b_n^2}(x_i-\mu_i)^2 f_i(x_i)dx_i \leq \lim_{n\to\infty}\frac{1}{b_n^2}\sum_{i=1}^{n}\left(\varepsilon b_n^2\right)^{-\delta/2}\mathrm{E}\left(|X_i-\mu_i|^{2+\delta}\right)$$

$$\leq \varepsilon^{-\delta/2}\lim_{n\to\infty}\sum_{i=1}^{n}\left[\frac{\mathrm{E}\left(|X_i-\mu_i|^{2+\delta}\right)}{\left(b_n^2\right)^{1+\delta/2}}\right]$$

The assumptions of the Liapounov CLT state that, for some $\delta > 0$, $\sum_{i=1}^{n}\frac{\mathrm{E}\left(|X_i-\mu_i|^{2+\delta}\right)}{\left(b_n^2\right)^{1+\delta/2}} \to 0$ as $n \to \infty$, so that, $\forall\, \varepsilon > 0$ the Lindberg condition is met and the Liapounov CLT holds. ■

**Theorem 5.34** *See K.L. Chung (1974), A Course in Probability Theory, 2nd Ed., New York: Academic Press, Section 7.2.*

**Theorem 5.35** *See R.G. Laha and V.K. Rohatgi, (1979) Probability Theory, p. 355, Section 5.5.31.*

**Theorem 5.36: Proof** Sufficiency will be motivated assuming the existence of MGFs. The general proof replaces MGFs with characteristic functions. From Theorem 5.2, $\boldsymbol{\ell}'\mathbf{X_n} \xrightarrow{d} \boldsymbol{\ell}'\mathbf{X} \Rightarrow M_{\ell'\mathbf{X}_n}(t) \to M_{\ell'\mathbf{X}}(t)$ *for* $t \in (-h,h)$, $h > 0$, and $\forall \boldsymbol{\ell} \in \mathbb{R}^k$. This implies $\mathrm{E}(e^{t\ell'\mathbf{X_n}}) = \mathrm{E}(e^{\mathbf{t}'_*\mathbf{X_n}}) \to \mathrm{E}(e^{\mathbf{t}'_*\mathbf{X}}) = \mathrm{E}(e^{t\ell'\mathbf{X}}) \forall \mathbf{t}_* = t\boldsymbol{\ell}$ since $\boldsymbol{\ell}$ can be chosen arbitrarily. But this is equivalent to $M_{\mathbf{X}_n}(\mathbf{t}_*) \to M_{\mathbf{X}}(\mathbf{t}_*)$, $\forall \mathbf{t}_*$, which implies $\mathbf{X}_n \xrightarrow{d} \mathbf{X}$ by the multivariate interpretation of Theorem 5.2. Necessity follows from Theorem 5.3 since $\boldsymbol{\ell}'\mathbf{x}$ is a continuous function of $\mathbf{x}$. The general proof based on characteristic functions can be found in V. Fabian and J. Hannan (1985), Introduction to Probability and Mathematical Statistics. New York: John Wiley, p. 144. ■

**Theorem 5.37: Proof** Examine $Z_i = \boldsymbol{\ell}'\mathbf{X}_i$, where $\boldsymbol{\ell} \neq \mathbf{0}$. Note that $\mu_z = \mathrm{E}(Z_i) = \mathrm{E}(\boldsymbol{\ell}'\mathbf{X}_i) = \boldsymbol{\ell}'\boldsymbol{\mu}$ and $\sigma_z^2 = \mathrm{var}(Z_i) = \mathrm{var}(\boldsymbol{\ell}'\mathbf{X}_i) = \boldsymbol{\ell}'\Sigma\boldsymbol{\ell}\ \forall i$. Now since $\{\mathbf{X}_n\}$ is a sequence of *iid* random vectors, then $\{Z_n\} = \{\boldsymbol{\ell}'\mathbf{X}_n\}$ is a sequence of iid random scalars, and applying the Lindberg-Levy CLT for random scalars to the iid sequence $\{Z_n\}$ results in

$$\frac{\sum_{i=1}^n Z_i - n\mu_z}{n^{1/2}\sigma_z} = \frac{\sum_{i=1}^n \boldsymbol{\ell}'\mathbf{X}_i - n\boldsymbol{\ell}'\boldsymbol{\mu}}{n^{1/2}(\boldsymbol{\ell}'\boldsymbol{\Sigma}\boldsymbol{\ell})^{1/2}} = \frac{\boldsymbol{\ell}'\left[\sum_{i=1}^n \mathbf{X}_i - n\boldsymbol{\mu}\right]}{n^{1/2}(\boldsymbol{\ell}'\boldsymbol{\Sigma}\boldsymbol{\ell})^{1/2}}$$

$$= \frac{\boldsymbol{\ell}'n^{1/2}\left[n^{-1}\sum_{i=1}^n \mathbf{X}_i - \boldsymbol{\mu}\right]}{(\boldsymbol{\ell}'\boldsymbol{\Sigma}\boldsymbol{\ell})^{1/2}} \xrightarrow{d} N(0,1)$$

Then by Slutsky's theorem

$$\boldsymbol{\ell}'n^{1/2}\left[n^{-1}\sum_{i=1}^n \mathbf{X}_i - \boldsymbol{\mu}\right] \xrightarrow{d} N(0, \boldsymbol{\ell}'\boldsymbol{\Sigma}\boldsymbol{\ell})$$

which holds for any choice of the vector $\boldsymbol{\ell} \neq \mathbf{0}$. Then by the Cramer-Wold device, we can conclude that $n^{1/2}\left[n^{-1}\sum_{i=1}^n \mathbf{X}_i - \boldsymbol{\mu}\right] \xrightarrow{d} N(\mathbf{0}, \boldsymbol{\Sigma})$ ■

**Theorem 5.38: Proof** Examine $Z_i = \boldsymbol{\ell}'\mathbf{X}_i$, where $\boldsymbol{\ell} \neq \mathbf{0}$. Note that $\mathrm{E}(Z_i) = \mathrm{E}(\boldsymbol{\ell}'\mathbf{X}_i) = \boldsymbol{\ell}'\mu_i$ and $\mathrm{var}(Z_i) = \mathrm{var}(\boldsymbol{\ell}'\mathbf{X}_n) = \boldsymbol{\ell}'\psi_i\boldsymbol{\ell}$. Since $\{\mathbf{X}_n\}$ is a sequence of independent random vectors, $\{Z_n\} = \{\boldsymbol{\ell}'\mathbf{X}_n\}$ is a sequence of independent random scalars. Furthermore, since outcomes of the vector $\mathbf{X}_n$ are contained within the closed and bounded rectangle $\times_{i=1}^k [-m, m]$ in $\mathbb{R}^k$ with probability 1, then for any given nonzero vector of real numbers $\boldsymbol{\ell}$, outcomes of $Z_n = \boldsymbol{\ell}'\mathbf{X}_n$ exhibit an upper bound in absolute value with probability 1 uniformly $\forall\, n$. Thus, there exists a finite real number $\delta > 0$ such that $P(|z_i| \leq \delta) = 1\ \forall\, i$. In addition, since $n^{-1}\sum_{i=1}^n \mathrm{var}(Z_i) = n^{-1}\sum_{i=1}^n \boldsymbol{\ell}'\boldsymbol{\psi}_i\boldsymbol{\ell} \to \boldsymbol{\ell}'\boldsymbol{\psi}\boldsymbol{\ell} > 0$, $\sum_{i=1}^n \mathrm{var}(Z_i) \to \infty$. It follows from the CLT of Theorem 5.32 that

$$\frac{\sum_{i=1}^n Z_i - \sum_{i=1}^n \boldsymbol{\ell}'\boldsymbol{\mu}_i}{\left(\sum_{i=1}^n \mathrm{var}(Z_i)\right)^{1/2}} = \frac{\boldsymbol{\ell}'\left(\sum_{i=1}^n \mathbf{X}_i - \sum_{i=1}^n \boldsymbol{\mu}_i\right)}{\left(\sum_{i=1}^n \boldsymbol{\ell}'\boldsymbol{\psi}_i\boldsymbol{\ell}\right)^{1/2}}$$

$$= \frac{\boldsymbol{\ell}'n^{-1/2}\left(\sum_{i=1}^n \mathbf{X}_i - \sum_{i=1}^n \boldsymbol{\mu}_i\right)}{\left(\sum_{i=1}^n \boldsymbol{\ell}'(n^{-1}\boldsymbol{\psi}_i)\boldsymbol{\ell}\right)^{1/2}} \xrightarrow{d} N(0,1).$$

Premultiplying by the denominator term, and noting that

$$\lim_{n\to\infty}\left(\sum_{i=1}^{n}\boldsymbol{\ell}'(n^{-1}\boldsymbol{\psi}_i)\boldsymbol{\ell}\right)^{1/2}=(\boldsymbol{\ell}'\boldsymbol{\psi}\boldsymbol{\ell})^{1/2},$$

Slutsky's theorem (Theorem 5.10) then results in

$$\boldsymbol{\ell}'n^{-1/2}\left(\sum_{i=1}^{n}\mathbf{X}_i-\sum_{i=1}^{n}\boldsymbol{\mu}_i\right)\xrightarrow{d}\mathbf{N}(0,\boldsymbol{\ell}'\boldsymbol{\psi}\boldsymbol{\ell})$$

which holds for any choice of the real vector $\boldsymbol{\ell}\neq\mathbf{0}$. It follows by the Cramer-Wold device that $n^{-1/2}\left(\sum_{i=1}^{n}\mathbf{X}_i-\sum_{i=1}^{n}\boldsymbol{\mu}_i\right)\xrightarrow{d}N(\mathbf{0},\boldsymbol{\psi})$. ■

**Theorem 5.39: Proof** Representing $g(\mathbf{X}_n)$ in terms of a first-order Taylor series expansion around the point $\boldsymbol{\mu}$ and the remainder term, as in Lemma 5.6, yields $g(\mathbf{X}_n)=g(\boldsymbol{\mu})+\mathbf{G}(\mathbf{X}_n-\boldsymbol{\mu})+d(\mathbf{X}_n,\boldsymbol{\mu})\ R(\mathbf{X}_n)$. Multiplying by $n^{1/2}$ and rearranging terms obtains $n^{1/2}(g(\mathbf{X}_n)-g(\boldsymbol{\mu}))=G[n^{1/2}(\mathbf{X}_n-\boldsymbol{\mu})]+n^{1/2}d(\mathbf{X}_n,\boldsymbol{\mu})R(\mathbf{X}_n)$.

The last term converges to 0 in probability. To see this, first note by Slutsky's theorem that $\mathbf{X}_n-\boldsymbol{\mu}\xrightarrow{d}\text{plim}(n^{-1/2})\mathbf{Z}=0\cdot\mathbf{Z}=\mathbf{0}$, so that $\text{plim}(\mathbf{X}_n)=\boldsymbol{\mu}$ and then $\text{plim}(R(\mathbf{X}_n))=R(\boldsymbol{\mu})=0$ by the continuity of $R(\mathbf{x})$ at the point $\mathbf{x}=\boldsymbol{\mu}$. Also, by Theorem 5.3, $n^{1/2}d(\mathbf{X}_n,\boldsymbol{\mu})=\left([n^{1/2}(\mathbf{X}_n-\boldsymbol{\mu})]'[n^{1/2}(\mathbf{X}_n-\boldsymbol{\mu})]\right)^{1/2}\xrightarrow{d}(\mathbf{Z}'\mathbf{Z})^{1/2}$ so that by Slutsky's theorem $n^{1/2}d(\mathbf{X}_n,\boldsymbol{\mu})R(\mathbf{X}_n)\xrightarrow{d}(\mathbf{Z}'\mathbf{Z})^{1/2}\text{plim}(R(\mathbf{X}_n))=(\mathbf{Z}'\mathbf{Z})^{1/2}0=0$ and thus $n^{1/2}d(\mathbf{X}_n,\boldsymbol{\mu})\ R(\mathbf{X}_n)\xrightarrow{p}0$.

Given the previous result, it follows by another application of Slutsky's theorem, that the limiting distribution of $n^{1/2}(g(\mathbf{X}_n)-g(\boldsymbol{\mu}))$ is the same as that of $\mathbf{G}[n^{1/2}(\mathbf{X}_n-\boldsymbol{\mu})]\xrightarrow{d}\mathbf{G}\mathbf{Z}\sim N(\mathbf{0},\mathbf{G}\boldsymbol{\Sigma}\mathbf{G}')$. The asymptotic distribution of $g(\mathbf{X}_n)$ is then as stated in the theorem. ■

**Theorem 5.40: Proof** Following the proof of Theorem 5.39, a first-order Taylor series expansion applied to each coordinate function in $\mathbf{g}(\mathbf{X}_n)$ results in $n^{1/2}(\mathbf{g}(\mathbf{X}_n)-\mathbf{g}(\boldsymbol{\mu}))=\mathbf{G}[n^{1/2}(\mathbf{X}_n-\boldsymbol{\mu})]+n^{1/2}\ d(\mathbf{X}_n,\boldsymbol{\mu})\ \mathbf{R}(\mathbf{X}_n)$, where now $\mathbf{R}(\mathbf{X}_n)=(R_1(\mathbf{X}_n),\ldots,R_m(\mathbf{X}_n))'$ is an $m\times 1$ vector of remainder terms. The approach of the proof of Theorem 5.39 can be applied elementwise to conclude that $\text{plim}[n^{1/2}d(\mathbf{X}_n,\boldsymbol{\mu})\ \mathbf{R}(\mathbf{X}_n)]=\mathbf{0}$. Then, by Slutsky's theorem, the limiting distribution of $n^{1/2}(\mathbf{g}(\mathbf{X}_n)-\mathbf{g}(\boldsymbol{\mu}))$ is the same as the limiting distribution of $\mathbf{G}[n^{1/2}(\mathbf{X}_n-\boldsymbol{\mu})]$, *and* $\mathbf{G}[n^{1/2}(\mathbf{X}_n-\boldsymbol{\mu})]\xrightarrow{d}\mathbf{G}\mathbf{Z}\sim N(\mathbf{0},\mathbf{G}\boldsymbol{\Sigma}\mathbf{G}')$. The asymptotic distribution of $\mathbf{g}(\mathbf{X}_n)$ follows directly. ■

**Theorem 5.41: Proof**

***Sketch*** Represent $\mathbf{g}(\mathbf{X}_n)$ in terms of a first-order Taylor series expansion plus remainder, as in the proof of Theorem 5.40, to obtain $[\mathbf{G}\mathbf{V}_n\mathbf{G}']^{-1/2}[\mathbf{g}(\mathbf{X}_n)-\mathbf{g}(\boldsymbol{\mu})]=[\mathbf{G}\mathbf{V}_n\mathbf{G}']^{-1/2}[\mathbf{G}(\mathbf{X}_n-\boldsymbol{\mu})+d(\mathbf{X}_n,\boldsymbol{\mu})\ \mathbf{R}(\mathbf{X}_n)]$.

The first term to the right of the equality is such that $(\mathbf{G}\mathbf{V}_n\mathbf{G}')^{-1/2}\ \mathbf{G}(\mathbf{X}_n-\boldsymbol{\mu})=(\mathbf{G}\mathbf{V}_n\mathbf{G}')^{-1/2}\mathbf{G}\ \ \mathbf{V}_n^{1/2}\mathbf{V}_n^{-1/2}\ (\mathbf{X}_n-\boldsymbol{\mu})\xrightarrow{d}N(\mathbf{0},\mathbf{I})$ which follows from H. White,

Asymptotic Theory, Lemma 4.23, p. 66, upon recognizing that $\{\mathbf{A}_n\}=\{(\mathbf{G}\mathbf{V}_n\mathbf{G}')^{-1/2}\mathbf{G}\mathbf{V}_n^{1/2}\}$ is a O(1) sequence of matrices (note that $\mathbf{A}_n\mathbf{A}_n'=\mathbf{I}\ \forall n$) and $\mathbf{V}_n^{-1/2}(\mathbf{X}_n-\boldsymbol{\mu})\xrightarrow{d}N(\mathbf{0},\mathbf{I})$.

The second term converges in probability to the zero matrix since

$$[\mathbf{G}\mathbf{V}_n\mathbf{G}']^{-1/2}d(\mathbf{X}_n,\boldsymbol{\mu})=[a_n\mathbf{G}\mathbf{V}_n\mathbf{G}']^{-1/2}\left[a_n^{1/2}(\mathbf{X}_n-\boldsymbol{\mu})'(\mathbf{X}_n-\boldsymbol{\mu})a_n^{1/2}\right]^{1/2}$$

is $O_p(1)$ and $\mathbf{R}(\mathbf{X}_n)\xrightarrow{p}\mathbf{0}$. Then the convergence in distribution result of the theorem follows from Slutsky's theorem, and the asymptotic distribution of $\mathbf{g}(\mathbf{X}_n)$ follows subsequently. ■

## Keywords, Phrases, and Symbols

$\{y_n\}$ is bounded, unbounded
Almost sure convergence, $Y_n \xrightarrow{as} Y$
Asymptotic distribution of functions of asymptotically normal random variables
Asymptotic distribution, $Z_n \overset{a}{\sim} g(X, \Theta_n)$
Asymptotic nonpositively correlated
Bounded in probability
Central limit theorems, CLTs
Continuity correction
Converge to $y$
Convergence in distribution, $Y_n \xrightarrow{d} Y$ or $Y_n \xrightarrow{d} F$
Convergence in mean square, $Y_n \xrightarrow{m} Y$
Convergence in probability, $Y_n \xrightarrow{p} Y$
Convergence in quadratic mean
Convergence of a sequence $y_n \to y$, $\lim_{n\to\infty} y_n = y$
Converges in distribution to a constant, $Y_n \xrightarrow{d} c$
Converges in probability to a constant
Cramer-Wold device
Delta method
iid, independent and identically distributed
Khinchin's WLLN
Kolmogorov's inequality
Kolmogorov's SLLN
Laws of large numbers
Liapounov CLT
Limit of a real number sequence
Limit of the sequence $\{y_n\}$
Limiting density of $\{Y_n\}$, limiting CDF of $\{Y_n\}$
Limiting distribution
Limiting function of $\{f_n\}$
Lindberg CLT
Lindberg-Levy CLT
$m$ - dependent sequence
Multivariate Lindberg-Levy CLT
Natural numbers, $N$
$O(n^k)$, at most of order $n^k$
$o(n^k)$, of order smaller than $n^k$
$O_p(n^k)$, at most of order $n^k$ in probability
$o_p(n^k)$, of order smaller than $n^k$ in probability
Order of magnitude in probability
Order of magnitude of a random sequence
Order of magnitude of a sequence
Probability limit, plim
Sequence
Sequence of probability density functions
Sequence of random variables
Slutsky's theorems
Strong laws of large numbers, SLLN
Symmetric matrix square root
Triangular array of random variables
Weak laws of large numbers, WLLN
$Y \sim f(y)$, $Y$ has probability density $f(y)$, or $Y$ is distributed as $f(y)$

## Problems

**1.** Given current technology, the production of active matrix color screens for notebook computers is a difficult process that results in a significant proportion of defective screens being produced. At one company the daily proportion of defective 9.5″ and 10.4″ screens is the outcome of a bivariate random variable, $X$, with joint density function $f(x_1,x_2;\alpha)=(\alpha x_1+(2-\alpha)x_2)\,I_{[0,1]}(x_1)\,I_{[0,1]}(x_2)$, where $\alpha\in(0,2)$. The daily proportions of defectives are independent from day to day. A collection of $n$ *iid* outcomes of $X$ will be used to generate an estimate of the $(2\times1)$ vector of mean daily proportions of defectives, $\mu$, for the two types of screens being produced, as $\underset{(2\times1)}{\bar{\mathbf{X}}_n}=\sum_{i=1}^n \mathbf{X}_{(i)}/n$, where $\mathbf{X}_{(i)}=\begin{bmatrix}X_{1i}\\X_{2i}\end{bmatrix}$.

(a) Does $\bar{\mathbf{X}}_n\xrightarrow{as}\boldsymbol{\mu}$? Does $\bar{\mathbf{X}}_n\xrightarrow{p}\boldsymbol{\mu}$? Does $\bar{\mathbf{X}}_n\xrightarrow{d}\boldsymbol{\mu}$?

(b) Define an asymptotic distribution for the bivariate random variable $\bar{\mathbf{X}}_n$. If $\alpha = 1$ and $n = 200$, what is the approximate probability that $\bar{X}_n[1] > .70$, given that $\bar{X}_n[2] = .60$?

(c) Consider using an outcome of the function $g(\bar{\mathbf{X}}_n) = \bar{X}_n[1]/\bar{X}_n[2]$ to generate an estimate of the relative expected proportions of defective 9.5″ and 10.4″ screens, $\mu_1 / \mu_2$. Does $g(\bar{\mathbf{X}}_n) \xrightarrow{as} \mu_1 / \mu_2$ ? Does $g(\bar{\mathbf{X}}_n) \xrightarrow{p} \mu_1 / \mu_2$? Does $g(\bar{\mathbf{X}}_n) \xrightarrow{d} \mu_1 / \mu_2$?

(d) Define an asymptotic distribution for $g(\bar{\mathbf{X}}_n)$. If $\alpha = 1$ and $n = 200$, what is the approximate probability that the outcome of $g(\bar{\mathbf{X}}_n)$ will exceed 1?

**2.** Central limit theorems have important applications in the area of quality control. One such application concerns so-called control charts, and in particular, $\bar{X}$ charts, which are used to monitor whether the variation in the calculated mean levels of some characteristics of a production process are within acceptable limits. The actual chart consists of plotting calculated mean levels (vertical axis) over time (horizontal axis) on a graph that includes horizontal lines for the actual mean characteristic level of the process, $\mu$, and for *upper and lower control limits* that are usually determined by adding and subtracting two or more standard deviations, $\sigma_{\bar{X}}$, to the actual mean level. If, at a certain time period, the outcome of the calculated mean lies outside the control limits, the production process is considered to be no longer behaving properly, and the process is stopped for appropriate adjustments. For example, if a production process is designed to fill cans of soda pop to a mean level of 12 oz., if the standard deviation of the fill levels is .1, and if 100 cans of soda are randomly drawn from the packaging line to record fill levels and calculate a mean fill level $\bar{x}$, then the control limits on the daily calculated means of the filling process might be given by $12 \mp 3$ std$(\bar{x}) = 12 \mp .03$.

(a) Provide a justification for the $\bar{X}$ chart procedure described above based on asymptotic theory. Be sure to clearly define the conditions under which your justification applies.

(b) Suppose that control limits are defined by adding and subtracting three standard deviations of $\bar{X}$ to the mean level $\mu$. In light of your justification of the control chart procedure in (a), what is the probability that the production process will be inadvertently stopped at a given time period, even though the mean of the process remains equal to $\mu$?

(c) In the soda can-filling example described above, if the process were to change in a given period so that the mean fill level of soda cans became 12.05 oz. what is the probability that the control chart procedure would signal a shutdown in the production process in that period?

**3.** The lifetime of a certain computer chip that your company manufactures is characterized by the population distribution

$$f(z;\theta) = \frac{1}{\theta}\, e^{-z/\theta}\, I_{(0,\infty)}(z),$$

where $z$ is measured in thousands of hours. Let $(X_1, \ldots, X_n)$ represent *iid* random variables with the density $f(z;\theta)$. An outcome of the random variable $Y_n = \left(1 + \sum_{i=1}^n X_i\right)/n$ will to be used to provide an estimate of $\theta$.

(a) Is it true that $Y_n \xrightarrow{m} \theta$? Is it true that $Y_n \xrightarrow{p} \theta$?

(b) Define an asymptotic distribution for $Y_n$.

(c) Suppose $n = 100$ and $\theta = 10$. Use the asymptotic distribution you defined in (b) to approximate the probability that $Y_n \geq 15$.

**4.** In each case below, the outcome of some function, $T(X_{(n)})$, of $n$ *iid* random variables $X_{(n)}=(X_1,\ldots, X_n)$ is being considered for providing an estimate of some function of parameters, $q(\theta)$. Determine whether $\mathrm{E}(T(X_{(n)})) = q(\theta)$, $\lim_{n\to\infty}\mathrm{E}(T(X_{(n)})) = q(\theta)$, and $\operatorname{p\,lim}(T(X_{(n)})) = q(\theta)$.

(a) $X_i$'s ~ *iid* Gamma$(\alpha,\beta)$ and $T(X_{(n)}) = \bar{X}_n$ is being used to estimate $q(\alpha,\beta)=\alpha\beta$.

(b) $X_i$'s *iid* Gamma$(\alpha,\beta)$ and $T(X_{(n)}) = \sum_{i=1}^n (X_i - \bar{X}_n)^2 / (n-1)$ is being used to estimate $q(\alpha,\beta)=\alpha\beta^2$

(c) $X_i$'s ~ *iid* Bernoulli $(p)$ and $T(X_{(n)}) = \bar{X}_n(1 - \bar{X}_n)$ is being used to estimate $q(p) = p(1-p)$.

(d) $X_i$'s *iid* $N(\mu,\sigma^2)$ and $T(X_{(n)}) = \left(\sum_{i=1}^n X_i - n^{1/2}\right)/(n+1)$ is used to estimate $\mu$.

**5.** The daily number of customers entering a large grocery store who purchase one or more dairy products is given by the outcome of a binomial random variable $X_t$ with parameters $p$ and $n_t$ for day $t$. The number of customers who enter the grocery store on any given day, $n_t$, is itself an outcome of a random variable $N_t$ that has a discrete uniform distribution on the range $\{200,201,\ldots,300\}$. The $X_t$'s and the $N_t$'s are jointly independent. The local dairy products commission wants an estimate of the daily proportion of customers entering the store who purchase dairy products and wants you to use an outcome of $\bar{X}_d = (1/d)\sum_{t=1}^{d}(X_t/N_t)$, where $d$ is number of days, as an estimate.

(a) Does $\bar{X}_d \xrightarrow{as} p$? Does $\bar{X}_d \xrightarrow{p} p$? Does $\bar{X}_d \xrightarrow{d} p$?

(b) Define an asymptotic distribution for $\bar{X}_d$. If $p = .8$, $d = 300$, and $\sum_{i=1}^{300} n_t = 75,000$, what is the approximate probability that $\bar{X}_d \in (.78, .82)$?

**6.** Let $(X_1,\ldots,X_n)$ be *iid* random variables with $\sigma^2 < \infty$. We know from Khinchin's WLLN that

$$\bar{X} = n^{-1}\sum_{i=1}^{n} X_i \xrightarrow{p} u.$$

(a) Find a functional relationship between $n$, $\sigma$, and $\varepsilon$ such that

$$P(\bar{x} \in (\mu - \varepsilon, \mu + \varepsilon)) \geq .99.$$

(b) For what values of $n$ and $\sigma$ will an outcome of $\bar{X}$ be within $\mp.1$ of $\mu$ with probability $\geq .99$? Graph this relationship between the values of $n$ and $\sigma$.

(c) If $\sigma = 1$, what is the value of $n$ that will ensure that the outcome of $\bar{X}$ will be within $\mp.1$ of $\mu$ with probability $\geq .99$?

(d) If $\sigma = 1$, *and* the $X_i$'s are normally distributed, what is the value of $n$ that will ensure that the outcome of $\bar{X}$ will be within $\mp.1$ of $\mu$ with probability $=.99$?

**7.** Let the random variables in the sequence $\{X_n\}$ be *iid* with a gamma density having parameters $\alpha = 2$ and $\beta = 3$.

(a) What is the probability density for $\bar{X}_n$?

(b) What is the asymptotic probability density for $\bar{X}_n$?

(c) Plot the actual versus asymptotic probability density for $\bar{X}_n$ when $n = 10$.

(d) Repeat (c) for $n = 40$. Interpret the graphs in (c) and (d) in terms of asymptotic theory.

**8.** A pharmaceutical company claims that it has a drug that is 75 percent effective in generating hair growth on the scalps of balding men. In order to generate evidence regarding the claim, a consumer research agency conducts an experiment whereby a total of 1,000 men are randomly chosen and treated with the drug. Of the men treated with the drug, 621 experienced hair growth. Do the results support or contradict the company's claim?

**9.** A political candidate has hired a polling firm to assess her chances of winning a senatorial election in a large eastern state. She wants an estimate of the proportion of registered voters that would vote for her "if the election were held today." Registered voters are to be randomly chosen, interviewed, and their preferences recorded. The polling firm will use the outcome of $\bar{X}$ as an estimate of the proportion of voters in favor of the candidate. It is known that currently between 40 percent and 60 percent of the registered voters favor her in the election. She wants to have an estimate that is within 2 percentage points of the true proportion with probability $= .99$. How many registered voters must be interviewed, based on the asymptotic distribution of $\bar{X}$?

**10.** Let observations on the quantity supplied of a certain commodity be generated by $Y_i = x_i\,\beta + V_i$, where $|x_i| \in [a, b]$ $\forall i$ are scalar observations on fixed prices, $\beta$ is an unknown slope coefficient, and the $V_i$'s are *iid* random variables having a mean of zero, a variance of $\sigma^2 \in (0, \infty)$, and $P(|V_i| \leq m) = 1$ $\forall i$ ($a$, $b$, and $m$ are finite positive constants). Two functions of the $x_i$'s and $Y_i$'s are being considered for generating an estimate of the unknown value of $\beta$:

$$\hat{\beta} = (\mathbf{x}'\mathbf{x})^{-1}\mathbf{x}'\mathbf{Y} \text{ and } \hat{\beta}_r = (\mathbf{x}'\mathbf{x} + \mathrm{k})^{-1}\mathbf{x}'\mathbf{Y},$$

where $k > 0$, $\mathbf{x}$ is an $(n\times 1)$ vector of observations on the $x_i$'s and $\mathbf{Y}$ is an $(n\times 1)$ vector of the corresponding $Y_i$'s.

(a) Define the means and variances of the two estimators of $\beta$.

(b) Is it true that $\lim_{n\to\infty} \mathrm{E}(\hat{\beta}) = \beta$ and/or $\lim_{n\to\infty} \mathrm{E}(\hat{\beta}_r) = \beta$?

(c) Define the expected squared distances of the two estimators from $\beta$.

(d) Which, if either, of the estimators converges in mean square to $\beta$?

(e) Which, if either, of the estimators converges in probability to $\beta$?

(f) Define asymptotic distributions for each of the estimators.

(g) Under what circumstances would you prefer one estimator to the other for generating an estimate of the unknown value of $\beta$?

**11.** Let $X_1, \ldots, X_n$ be *iid* random variables having continuous uniform distributions of the form $f(z) = I_{(0,1)}(z)$.

(a) Define an asymptotic distribution for $\bar{X}_n = n^{-1}\sum_{i=1}^{n} X_i$.

(b) Using your result from (a), argue that $\left(\sum_{i=1}^{12} X_i\right) - 6 \approx Z \sim N(0,1)$.

(This approximation is very accurate, and is sometimes used for simulating N(0,1) outcomes using a uniform random number generator.)

**12.** A company produces a popular beverage product that is distributed nationwide. The aggregate demand for the product during a given time period can be represented by

$$Q = \sum_{i=1}^{n} Q_i = \sum_{i=1}^{n} (\alpha_i - \beta_i p + V_i)$$

where $Q_i$ is quantity purchased by the $i$th consumer, $\alpha_i > 0$, $\beta_i > 0$, $E(V_i) = 0$, $\text{var}(V_i) \geq c > 0$, $P(|v_i| \leq m) = 1 \; \forall i$, and $c$ and $m$ are positive constants. It can be assumed that the quantities purchased by the various consumers are jointly independent.

(a) Define an asymptotic distribution for the aggregate quantity demanded, $Q$.

(b) If $p = 2$, $E(Q) = 80$, and if $p$=5, $E(Q) = 50$. If it costs the company \$2/unit to manufacture and distribute the product, and if $p$=\$2.50, what is the asymptotic distribution of aggregate company profit during the time period?

(c) Define an interval around the mean of aggregate company profit that will contain the actual outcome of aggregate profit with (approximate) probability .95 when $p$=\$2.50.

**13.** The daily tonnage of garbage handled by the Enviro-Safe Landfill Co. is represented as the outcome of a random variable having some triangular distribution, as

$$X \sim f(x;a) = [(.5 - .25a) + .25x] I_{[a-2,a]}(x) + [(.5 + .25a) - .25x] I_{(a,a+2]}(x)$$

This distribution is represented graphically as follows:

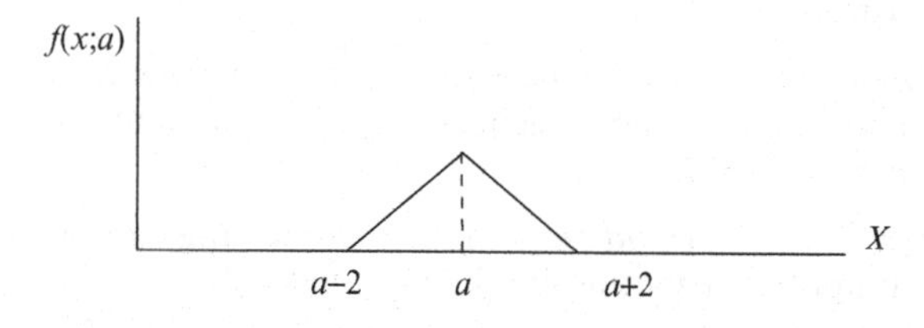


Enviro-Safe is in the process of analyzing whether or not they need to expand their facilities. It wants an estimate of the expected daily tonnage of garbage that it handles. It has collected 4 years of daily observations on tonnage handled ($n$ = 1,460 observations) and provides you with the following summary statistic:

$$\sum_{i=1}^{n} x_i = 29{,}200.$$

You may treat the observations as outcomes of *iid* random variables.

(a) Use $\bar{x} = n^{-1} \sum_{i=1}^{n} x_i$ to provide an estimate of $a$, the expected tonnage of garbage handled.

(b) Based on the LLCLT, define an asymptotic distribution for $\bar{x}$. You should be able to identify a *numerical value* for the variance of the asymptotic distribution.

(c) Using the asymptotic distribution, how probable is it that the outcome of $\bar{x}$ will be within .05 tons of the *actual* expected value of daily garbage tonnage handled?

(d) Use Van Beeck's inequality to provide an upper bound to the approximation error in the probability value that you assigned in part (c). Are you reasonably confident that you provided Enviro-Safe with an accurate "guess" of the expected daily tonnage? (Enviro-Safe management said that they would be satisfied if they could get an estimate that was "within ∀ ±1 tons of the actual expected daily tonnage.") Explain.

(e) Using your estimate $\hat{a} = \bar{x}$ to estimate the density function, as $f(x;\hat{a})$, what is your estimate of the probability that tonnage handled by Enviro-Safe will exceed 21 tons on any given day?

**14.** A statistician wants to use *iid* outcomes from some exponential distribution $f(x;\theta) = (1/\theta)\, e^{-x/\theta}\, I_{(0,\infty)}(x)$ to generate an estimate of the variance of the exponential density, $\theta^2$. She wants to use the outcome of $\bar{X}_n^2$ where $\bar{X}_n = n^{-1} \sum_{i=1}^{n} X_i$ to generate an estimate of $\theta^2$.

(a) Does $E(\bar{X}_n^2) = \theta^2$? Does $\lim_{n\to\infty} E(\bar{X}_n^2) = \theta^2$?

(b) Does plim $(\bar{X}_n^2) = \theta^2$?

(c) Define an asymptotic distribution for $\bar{X}_n^2$.

**15.** We have shown that if $\{Y_n\}$ is a sequence of $\chi^2$ random variables, where $Y_n \sim \chi_n^2$, then $(Y_n - n)/\sqrt{2n} \xrightarrow{d} N(0,1)$. Since $Y_n \sim \chi_n^2$, we know that $P(y_{25} \leq 34.3816) = P(y_{50} \leq 63.1671) = P(y_{100} \leq 118.498) = .90$. Assign (approximate) probabilities to the three events using asymptotic distributions. How good are the approximations? Does $\text{plim}(Y_n/n) = 1$? Why or why not?

**16.** Let $\{X_n\}$ be a sequence of random variables having binomial densities, where $X_n$ has a binomial density with parameters $n$ and $p$, i.e.,

$$X_n \sim \binom{n}{x} p^x (1-p)^{n-x} I_{\{0,1,2,\ldots n\}}(x).$$

(a) Show that $(X_n - np)/\left(\sqrt{n}(p(1-p))^{1/2}\right) \xrightarrow{d} N(0,1)$.

(b) Define an asymptotic distribution for $X_n$. Use Van Beeck's inequality to provide a bound on the error in approximating the probability $P(x_n \le c)$ using the asymptotic distribution for $X_n$. Calculate the numerical value of this bound when $p = .3$.

(c) Using the binomial density for $X_n$, and letting $p = .3$, it follows that

| $n$ | $k$ | $P(x_n \le k)$ |
|---|---|---|
| 15 | 6 | .8689 |
| 25 | 9 | .8106 |
| 100 | 34 | .8371 |

Assign (approximate) probabilities to the three events using asymptotic distributions based on your answer for (b). How good are the approximations?

(d) In using the CLT to approximate probabilities of events for discrete random variables whose range consists of equally spaced points, it has been found that a **continuity correction** improves the accuracy of the approximation. In particular, letting

$$R(X) = \{x_1, x_2, x_3, \ldots\} \text{ where } x_{i+1} - x_i = 2h > 0\ \forall i,$$

the continuity correction involves treating each elementary event $x_j$ for the discrete random variable $X$ as the interval event $(x_j - h, x_j + h]$ for the normal asymptotic distribution of the random variable $X$. For example, if $R(X) = \{0,1,2,3,\ldots\}$ then $P(x \in [1,2]) = \sum_{x=1}^{2} f(x) \approx \hat{P}(x \in (.5, 2.5])$, where the latter (approximate) probability is assigned using the appropriate asymptotic normal distribution for $X$. Use the continuity correction to approximate the probabilities of the three events in (c).

**17.** The Nevada Gaming Commission has been directed to check the fairness of a roulette wheel used by the WINBIG Casino. In particular, a complaint was lodged stating that a "red" slot occurs more frequently than a "black" slot for the roulette wheel used by WINBIG, whereas red and black should occur with probability .5 if the wheel is fair. The wheel is spun 100,000 times, and the number of red and black outcomes were recorded. The outcomes can be viewed as *iid* from some Bernoulli population distribution: $X \sim p^x (1-p)^{1-x} I_{\{0,1\}}(x)$ for $p \in (0,1)$. It was found that $\sum_{i=1}^{n} x_i = 49{,}873$, where $x_i = 1$ indicates that the $i$th spin resulted in a red outcome.

(a) Use $\bar{X}_n$ to provide an estimate of $p$, the probability of observing a red outcome.

(b) Define an asymptotic distribution for $\bar{X}_n$.

(c) Using the outcome of $\bar{X}_n$ mean as an estimate of $p$ in the asymptotic distribution for $\bar{X}_n$, how probable is it that an outcome of $\bar{X}_n$ is within $\pm.005$ of the true probability? Use Chebyshev's inequality to argue that the estimate $\bar{X}_n$ for $p$ should be very accurate in the sense that outcomes of $\bar{X}_n$ are very close to $p$ with high probability.

(d) Use the Van Beeck's inequality to provide an upper bound to the approximation error that can occur in assigning probability to events like $P(\bar{X}_n \le c)$ using the asymptotic distribution for $\bar{X}_n$. Your bound will unfortunately depend on the unknown value of $p$. Estimate a value for the bound using your outcome of $\bar{X}_n$ as an estimate of $p$.

(e) Define an asymptotic distribution for $g(\bar{X}_n) = \bar{X}_n(1 - \bar{X}_n)$.

(f) Compare the asymptotic distribution of the estimator in part (e) to the asymptotic distribution of the estimator $S^2 = \sum_{i-1}^{n} (X_i - \bar{X}_n)^2/(n-1)$. If the sample size were large, would you prefer one of the estimators of $p(1-p)$ over the other? Explain?

**18.** The Elephant Memory Chip Co. (EMC for short) instructs its resident statistician to investigate the operating-life characteristics of their new 4 gigabyte memory chip in order to provide product information to potential buyers. The population distribution of operating lives can be specified as some exponential family distribution. The statistician intends to draw a random sample of 10,000 chips from EMC's production and apply a nondestructive test that will determine each chip's operating life. He then intends to use the outcome of the random sample to provide estimates of both the mean and variance of the chip's operating life. He needs your help in answering a few statistical questions.

(a) Letting $\theta$ represent the unknown parameter in the exponential population distribution, what is the distribution of the sample mean, $\bar{X}_n$? What is the mean and variance of this distribution? Does plim $(\bar{X}_n) = \theta$?

(b) The outcome of the random sample resulted in the following two outcomes:

$$\bar{X}_n = 10.03702,$$

$$n^{-1}\sum_{i=1}^{n} X_i^2 = 199.09634.$$

Operating life is measured in 1,000 h units.
(**Side Note**: These are actual outcomes based on a *simulation* of the random sample outcome using a specific value of $\theta$.)
The statistician uses $\bar{X}_n$ to estimate the mean life, $\theta$, of the chips. He is considering using either $\bar{X}_n^2$ or $S_n^2 = \sum_{i=1}^n (X_i - \bar{X}_n)^2/(n-1)$ to estimate the variance, $\theta^2$, of operating lives. She asks the following questions regarding the characteristics of $\bar{X}_n^2$ and $S_n^2$: (show your work)

(a) Does $\mathrm{E}(\bar{X}_n^2) = \theta^2$? Does $\mathrm{E}(S_n^2) = \theta^2$?

(b) Does $\lim_{n\to\infty}\mathrm{E}(\bar{X}_n^2) = \theta$? Does $\lim_{n\to\infty}\mathrm{E}(S_n^2) = \theta^2$?

(c) Does plim $(\bar{X}_n^2) = \theta^2$? Does plim $(S_n^2) = \theta^2$?

(d) Define asymptotic distributions for $(\bar{X}_n^2)$ and $(S_n^2)$. Based on their asymptotic distributions, would you recommend the use of one random variable over the other for generating an estimate of $\theta^2$? Why or why not?

(e) Calculate the outcomes of both $(\bar{X}_n^2)$ and $(S_n^2)$.
(Note: The *actual* value of $\theta = 10$, and thus the actual value of $\theta^2 = 100$.)

(f) The statistician has an idea he wants you to react to. He doesn't like the fact that $\mathrm{E}(\bar{X}_n^2) \neq \theta$ (that *is* what you found–isn't it?). He wants to define a new random variable, $Y_n = a_n(\bar{X}_n^2)$, for an appropriate sequence of numbers $\{a_n\}$, so that $\mathrm{E}(Y_n) = \theta^2, \forall n$. Can he do it? How? If he (and you) can, then use the appropriate outcome of $Y_n$ to provide another estimate of $\theta^2$? Is it true that plim $(Y_n) = \theta^2$?

**19.** Let $(X_1,\ldots,X_n)$ be a random sample from a Poisson population distribution. Derive the limiting distribution of

$$T = \frac{(\bar{X}_n - \mu)}{(S_n^2/n)^{1/2}}$$

where $S_n^2 = n^{-1}\sum_{i=1}^n (X_i - \bar{X}_n)^2$.

**20.** Liquid crystal displays (LCDs) that your wholesaling company is marketing for a large Japanese electronics firm are known to have a distribution of lifetimes of the following Gamma-distribution form:

$$f(z;\alpha) = \frac{1}{2^\alpha\,\Gamma(\alpha)} z^{\alpha-1} e^{-z/2} I_{(0,\infty)}(z),$$

where $z$ is measured in 1,000's of hours.
A set of $n$ *iid* outcomes of $Z$ will be used in an attempt to obtain information about the expected value of the lifetime of the LCD's.

(a) Define the functional form of the joint density of the *iid* random variables say $(X_1,\ldots,X_n)$, of LCD lifetimes.

(b) What is the density function of the random variable $Y_n = \sum_{i=1}^n X_i$.

(c) Supposing that $n$ were large, identify an asymptotic distribution for the random variable $Y_n$. (Note: since you don't know $\alpha$ at this point, your asymptotic distribution will depend on the unknown value of $\alpha$.)

(d) If $\alpha$ were equal to 1/2, and the sample size was $n = 20$, what is the probability that $y_n \leq 31.4104$? Compare your answer to the approximate probability obtained using the asymptotic distribution you defined in (c).

(e) If $\alpha$ were equal to 1/2, and the sample size was $n = 50$, what is the probability that $y_n \leq 67.5048$? Compare your answer to the approximate probability obtained using the asymptotic distribution you defined in (c).

**21.** In each case below, determine whether the random variable sequence $\{Y_n\}$ converges in probability and/or in mean square, and if so, define what is being converged to.

(a) $Y_j = (j+5)^{-1}\sum_{i=1}^j X_i$ for $j = 1,2,3,\ldots$ ;
$X_i$'s $\sim$ *iid* Bernoulli(p)

(b) $Y_j = j^{-1}\sum_{i=1}^{j}(X_i - \lambda)^2$ for $j = 1,2,3,\ldots$ ;
$X_i$'s $\sim$ *iid* Poisson($\lambda$)

(c) $Y_j = j^{-1} \sum_{i=1}^{j} (X_i + Z_i)$ for $j = 1, 2, 3, \ldots$ ;
$(X_i, Z_i)'s \sim iid \text{ Normal}\left(\begin{bmatrix} 0 \\ 0 \end{bmatrix}, \begin{bmatrix} 1 & 0 \\ 0 & 1 \end{bmatrix}\right)$

(d) $Y_j = j^{-1} \sum_{i=1}^{j} X_i Z_i$ for $j = 1, 2, 3, \ldots$ ;
$(X_i, Z_i)'s \sim iid \text{ Normal}\left(\begin{bmatrix} 0 \\ 0 \end{bmatrix}, \begin{bmatrix} 1 & 0 \\ 0 & 1 \end{bmatrix}\right)$

**22.** In each case below, derive the probability limit and an asymptotic distribution for $n^{-1} \sum_{i=1}^{n} X_i$ and a limiting distribution for the random variable $Y_n$, if they can be defined.

(a) $X_i$'s *iid* Bernoulli$(p)$, $Y_n = \dfrac{n^{-1} \sum_{i=1}^{n} X_i - p}{n^{-1/2}(p(1-p))^{1/2}}$

(b) $X_i$'s *iid* Gamma$(\alpha, \beta)$, $Y_n = \dfrac{n^{-1} \sum_{i=1}^{n} X_i - \alpha\beta}{n^{-1/2}\alpha^{1/2}\beta}$

(c) $X_i$'s *iid* Uniform$(a, b)$, $Y_n \dfrac{n^{-1} \sum_{i=1}^{n} X_i - .5(a+b)}{(12n)^{-1/2}(b-a)}$

(d) $X_i$'s *iid* Geometric$(p)$, $Y_n = \dfrac{n^{-1} \sum_{i=1}^{n} X_i - p^{-1}}{(np^2)^{-1/2}(1-p)^{1/2}}$

# Sampling, Sample Moments and Sampling Distributions

## 6.1 Introduction

Prior to this point, our study of probability theory and its implications has essentially addressed questions of *deduction*, being of the type: "Given a probability space, what can we deduce about the characteristics of outcomes of an experiment?" Beginning with this chapter, we turn this question around, and focus our attention on *statistical inference* and questions of the form: "Given characteristics associated with the outcomes of an experiment, what can we infer about the probability space?"

The term *statistical inference* refers to the inductive process of generating information about characteristics of a real-world population or process by analyzing a *sample* of objects or outcomes from the population or process. For example, a marketer may be interested in determining whether consumers with a certain sociodemographic profile (the *population*) would purchase a new product (the *characteristic*); an auditor would be interested in assessing the accuracy (the *characteristic*); of a firm's accounts (the *population*); and a quality control engineer would have interest in determining whether commodities are being manufactured (the *process*) to within factory specifications (the *characteristic*).

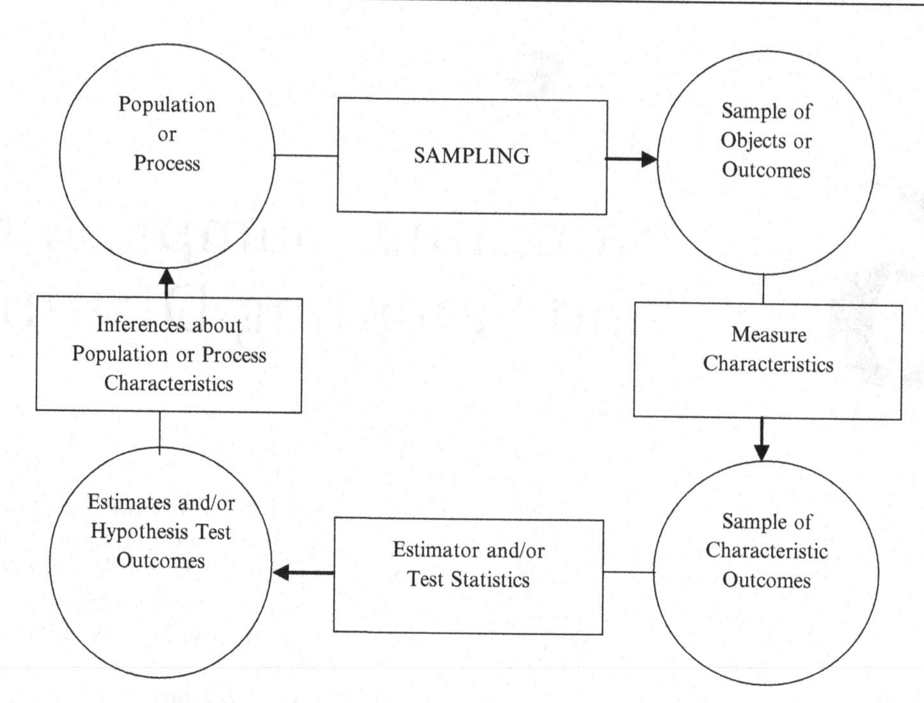


**Figure 6.1**
Overview of the statistical inference process.

The statistical inference problems would involve analyses of samples of observations from the real-world population or process to generate inferences regarding the characteristics of interest. Figure 6.1 provides a schematic overview of the process of statistical inference.

For an analysis of sample outcomes to lead to meaningful inferences about characteristics of a real-world population or process, the sample of characteristics must be in some sense representative of the incidence of characteristics in the population or process. The linkage between characteristics in the sample and in the real-world population or process is established by analyzing *probability samples* from the population or process. In particular, the objects or outcomes in a probability sample are obtained using selection procedures that establish connections between the incidence of characteristics in the population or process and the *probability* of observing the characteristics' outcomes in the sample. Then, based on probability samples and the implications of probability theory, methods of statistical inference can be devised that generate inferences about the characteristics of a population or process with a degree of accuracy or representativeness that can be measured probabilistically.

In this chapter we examine the concept of probability sampling and begin identifying methods of generating information on population or process characteristics from samples of observations. We focus primarily on sample moments and their probability distributions, and what can be learned from sample moment outcomes relative to the probability distribution associated with the population or process from which the sample was obtained. This allows us to establish foundational statistical concepts that illustrate how linkages are formed between sample and population or process characteristics, and demonstrate how knowledge about population or process characteristics can be generated from sample outcomes. A detailed and formal analysis of estimation and hypothesis-testing methodology will commence with Chapter 7 and beyond.

## 6.2 Sampling

The objective of statistical inference is to generate information about relevant characteristics of either the objects in some set of objects or the outcomes of some stochastic processes. The term **population** will henceforth refer to any set of objects with characteristics that a researcher wishes to identify, enumerate, or generally obtain information about. The term **stochastic process** will refer to any collection of experiments or measurement activities whose outcome can be interpreted as the outcome of a collection of random variables[1] and whose characteristics are of interest to the analyst. Henceforth, we will shorten *stochastic process* to simply **process** whenever the context is clear. The purpose of sampling is to obtain information about the characteristics of a population or process without having to examine each and every object or outcome relating to a population or process.

If a population has a finite number of objects, then an examination of each element of the population is conceivable, but there are reasons why the researcher might not wish to do so. An obvious reason would be if the measurement process were destructive, in which case measuring the characteristics of all of the population elements would lead to the population's destruction. This would be clearly undesirable if, for example, the objects in a population were manufactured by a firm that intended to offer them for sale (the tensile strength of steel beams is an example of a destructive measurement process, in which the beam is stressed to the breaking point). Another reason would be that the cost involved in evaluating each and every member of a population may be prohibitive. A related reason could be that the time available in which to perform the analysis is insufficient for a complete enumeration of the population. Finally, the researcher may have no choice but to analyze a sample, instead of measuring all conceivable outcomes of the characteristics of interest, as would be the case for an ongoing manufacturing (or other) process having no stipulated or identifiable end.

If we attempt to obtain information about relevant characteristics of a population or process from characteristics observed in a *sample* from that population or process, an obvious difficulty comes to mind. Namely, depending on which sample is chosen, the incidence of sample characteristics can differ from

[1] More formally, the term *stochastic process* refers to any collection of random variables indexed by some index set $T$, i.e., $\{X_t, t \in T\}$ is a stochastic process.

sample to sample, even though there exists only one fixed distribution of the characteristics in the population or process of interest. It follows that the existence or level of a characteristic in a given sample will not necessarily coincide with the existence or level of the characteristic in the population or process. The discrepancy between sample information on characteristics, and the actual state of affairs regarding the characteristics in the population or process is generically referred to as **sampling error**.

As an explicit example of sampling error, suppose a shipment contained 100 objects (the population) of which 30 were defective (the characteristic of interest). Not knowing the number of defectives in the shipment, suppose an individual attempted to infer the proportion of the population of objects that were defective by observing the proportion of defectives in a sample of 10 objects from the shipment. It is unfortunately possible, depending on the particular set of 10 objects chosen, for the observed proportion of defectives in the sample to range between 0 and 1. Thus, the actual *sampling error* in inferring the *proportion* of defectives in the population could range between a low of 0 (if there happened to be three defectives in the sample) and a high of .7 (if there happened to be 10 defectives in the sample) in absolute value, although the researcher would clearly be unaware of this range, since she is unaware that the actual proportion of defectives is .3. Without any further information about how the sample of observations from the population was obtained and/or how the sample was representative of the population, nothing can be said about the reliability of the inference. Note that sample information in this context is essentially useless – the observed proportion of defectives in the sample will lie between 0 and 1, and without any other information regarding the reliability of the inference, we know nothing more about the population than when we started the analysis.

In order for an analyst to be able to assess the reliability with which population or process characteristics are represented by sample information, it is necessary for the sample observations to have been generated in such a way that the researcher is able to establish some quantitative measures of confidence that the characteristics identified or enumerated in the sample accurately portray their incidence in the population or process. In particular, a **probability sample** will be useful for such inference.

**Definition 6.1** ***Probability Sample***

A sample obtained by a sampling mechanism by which (1) all members of a population, or all potential types of process outcomes, have some possibility of being observed in the sample, and (2) the specific sample of observations can be viewed as the outcome of some random variable $\mathbf{X}$.

In essence, a probability sample will refer to a sample whose values can be viewed as being governed by a probabilistic relationship between the distribution of characteristics inherent in the population or process and the specific method used to generate the sample observations. The linkage between the distribution of characteristics in the population or process and the nature of the joint density function of sample observations is exploited to design inference reliability measures. The terminology **random sample** and **outcome of the random sample** are used to refer to the random variable $\mathbf{X}$ and its outcome $\mathbf{x}$ in the definition of probability sample above.

**Definition 6.2**
***Random Sample***

The random variable $\mathbf{X}$ whose outcome $\mathbf{x}$ represents a probability sample is called a **random sample**, and its outcome is called the **outcome of the random sample**.

One can think of a random sample $\mathbf{X}$ as representing all of the potential outcomes of a probability sample, and it is a ***random*** sample because the sample outcome cannot be anticipated with certainty. The outcome of the random sample is then a specific set of $n$ observations that represents a probability sample. We will represent the $n$ observations comprising a sample outcome as the $n$ *rows* of $\mathbf{X}$, with the number of columns of $\mathbf{X}$ corresponding to the dimensionality of the characteristics being measured. The terminology **size of the random sample** will refer to the row dimension of $\mathbf{X}$, effectively representing the number of observations in the sample on the characteristics of interest. Thus, the size of the random sample $\mathbf{X} = [\mathbf{X}_i, i = 1, \ldots, n]'$ is $n$.

### 6.2.1 Simple Random Sampling

In this sampling design, sample observations can be conceptualized as the outcome of a collection of *iid* random variables, as in the case of observing the outcomes of a collection of *independent and identical experiments*. Therefore the random sample $\mathbf{X} = [\mathbf{X}_i, i = 1, \ldots, n]'$is such that $\mathbf{X}_i's \sim iid\ m(x)$, and the common probability distribution shared by the members of the random sample is sometimes referred to as the **population distribution**. The terminology **sampling from the population distribution** is also sometimes used to refer to this type of sampling.

There are basically two generic types of **simple random sampling**. One type involves sampling from an existent finite collection of objects, such as a shipment of personal computers or the citizens of a particular region of the county. The other type involves sampling from some ongoing stochastic process, such as a manufacturing process (e.g., measuring the net weight of potato chips in a bag of potato chips) or a market process (e.g., measuring stock market prices over time). In the former type of sampling, the ultimate objective is to obtain information about characteristics of the finite collection of objects under study, while in the latter type the objective is to obtain information about the characteristics of the stochastic process under study (e.g., the expected weight of potato chips in a bag; the probability of producing a bag containing less than the advertised quantity of chips; the variance of stock prices).

When an existent finite population is being sampled, simple random sampling from a population distribution is alternatively referred to as **random sampling with replacement**. In this case, there exists a finite set of $N$ objects having certain characteristics of interest to the analyst. The formal description of the steps involved in the sampling method is as follows.

**Definition 6.3**
***Random Sampling with Replacement***

1. An object is selected from a population in a way that gives all objects in the population an equal chance of being selected.
2. The characteristics of the object selected is observed, and the object is returned to the population prior to any subsequent selection.
3. For a sample of size $n$, (1) and (2) are performed $n$ times.

Since all members of the population are equally likely to occur at each selection, the classical probability definition is appropriate for determining the probability of observing any level of the characteristics on a given selection, as

$$m(z) = \frac{\begin{bmatrix} \text{number of population objects} \\ \text{that have characteristics level } z \end{bmatrix}}{N},$$

where $z$ can henceforth be replaced by a vector $\mathbf{f}$ anywhere in the discussion ahead depending on the number of characteristics being measured for each selection, and $N$ represents the number of objects in the population. The density function $m(z)$ is the population distribution of the characteristics.

Since all of the outcomes are independent, the probability of observing a sample of size $n$ that exhibits the collection of characteristic levels $(x_1, x_2, \ldots, x_n)$ is given by

$$f(x_1, x_2, \ldots, x_n) = \prod_{i=1}^{n} m(x_i).$$

The probability density function $f(x_1, \ldots, x_n)$ is referred to as the **joint density of the random sample.**

**Example 6.1**
***Simple Random Sampling with Replacement for Defectives***

Let a shipment of $N$ keyboards contain $J$ defective keyboards. (Note that in practice, $N$ will generally be known, while $J$ will be unknown, and information concerning $J$ would be the objective of statistical inference.) Let $z = 1$ denote a defective keyboard and $z = 0$ denote a nondefective keyboard. Then, if $n$ objects are sampled with replacement, and letting $p = J/N$, the *population distribution* of defective/nondefective keyboards is $m(z;p) = p^z(1-p)^{1-z}\, I_{\{0,1\}}(z)$, where $p$ represents the proportion of defective keyboards in the population. The joint density governing the probabilities of observing outcome $(x_1, \ldots, x_n)$ in the sample is given by

$$f(x_1, \ldots, x_n; p) = \prod_{i=1}^{n} p^{x_i} (1-p)^{1-x_i} I_{\{0,1\}}(x_i) = p^{\sum_{i=1}^{n} x_i} (1-p)^{n-\sum_{i=1}^{n} x_i} \prod_{i=1}^{n} I_{\{0,1\}}(x_i).$$

Note that the incidence of nondefectives in the population (the value of $p$) has a direct influence on the probabilities of events for the random sample outcomes.

Thus, a probabilistic linkage is established between the incidence of characteristics in the population and in the sample. □

When the population from which we are sampling is not finite, simple random sampling will refer to an ongoing stochastic process in which all of the random variables whose outcomes are being measured are *independent* and *identically* distributed. In practice, this means that whatever is the underlying random mechanism that determines the sample observations, it is unchanging from observation to observation, and observing a particular outcome for a given sample observation has no effect on the probability of outcomes of any other sample observations. Then each sample observation is the outcome of a random variable $Z \sim m(z)$, where $m(z)$ is the common PDF of characteristics outcomes (also called the *population distribution*) associated with the stochastic process. Since all of the observation experiments are identical and independent, the PDF for the $n$ random variables $(X_1,\ldots,X_n)$ characterizing the outcomes of the $n$ observation experiments is given by $f(x_1,\ldots,x_n) = \prod_{i=1}^{n} m(x_i)$. As before, $(X_1,\ldots, X_n)$ is called the *random sample*, $(x_1,\ldots,x_n)$ is the *outcome of the random sample*, and $f(x_1,\ldots,x_n)$ is the *joint density function of the random sample*.

**Example 6.2** ***Simple Random Sampling the Reliability of a Manufactured Product***

The distribution of the operating lives until failure of halogen lamps produced by a domestic manufacturer, i.e., the population distribution, is a given by a member of the gamma family of densities, as

$$m(z;\alpha,\beta) = \frac{1}{\beta^{\alpha}\Gamma(\alpha)} z^{\alpha-1} e^{-z/\beta} I_{(0,\infty)}(z).$$

The lamps are all produced using an identical manufacturing process, and $n$ lamps are arbitrarily selected from the production line for reliability testing. The $n$ measurements on operating life are interpreted as the outcome of a random sample with joint density function

$$f(x_1,\ldots,x_n;\alpha,\beta) = \prod_{i=1}^{n} m(x_i;\alpha,\beta)$$
$$= \frac{1}{\beta^{n\alpha}(\Gamma(\alpha))^{n}} \prod_{i=1}^{n} x_i^{\alpha-1} e^{-\sum_{i=1}^{n} x_i/\beta} \prod_{i=1}^{n} I_{(0,\infty)}(x_i).$$

Note that the functional form of the population distribution, and in particular the actual values of the parameters $\alpha$ and $\beta$, will have a direct influence on the probabilities of events for random sample outcomes. Thus, a probabilistic linkage is established between the incidence of characteristics in the process and in the sample. □

### 6.2.2 Random Sampling Without Replacement

**Random sampling without replacement** is relevant for a finite existent population of objects, but differs from random sampling with replacement in that once the characteristics of an object are observed, the object is removed from the population before another object is selected for observation. The sampling procedure is described as follows:

**Definition 6.4**
***Random sampling without replacement***

1. The first object is selected from the population in a way that allows all objects in the population an equal chance of being selected.
2. The characteristics of the object are observed, but the object is not returned to the population.
3. An object is selected from the remaining objects in the population in a way that gives all remaining objects an equal chance of being selected, and step (2) is repeated. For a sample of size $n$, step (3) is performed $(n-1)$ times.

In this case, the sampling process can be characterized as a collection of $n$ experiments that are neither identical nor independent. In particular, the probability of observing characteristic level $x_i$ on the $i$th selection depends on what objects were observed and removed from the population in the preceding $(i-1)$ selections. For the first selection, the probability of observing characteristics level $x_1$ is given by

$$m(x_1) = \frac{\begin{bmatrix}\text{number of population objects}\\ \text{that have characteristics level } x_1\end{bmatrix}}{N}$$

since all objects are equally likely to be selected. The density $m(x_1)$ can be thought of as the *initial* population distribution. On the $i$th selection, $i > 1$, the probability of observing characteristic level $x_i$ is conditioned on what was observed and removed from the population previously, so that

$$f(x_i|x_1,\ldots,x_{i-1}) = \frac{\begin{bmatrix}\text{number of objects remaining in the}\\ \text{population that have characteristics level}\\ x_i \text{ after the } i-1 \text{ selections } x_1,\ldots,x_{i-1}\end{bmatrix}}{(N-i+1)}$$

for $N - i + 1 > 0$. The joint density defining the probability of observing a sample of size $n$ that has the collection of characteristic levels $(x_1,\ldots,x_n)$ can then be defined by[2]

[2]This follows straightforwardly from the definitions of marginal and conditional density functions, as the reader should verify.

$$f(x_1,\ldots,x_n) = m(x_1)f(x_2\,|\,x_1)f(x_3\,|\,x_1,x_2)\ldots f(x_n\,|\,x_1,x_2,\ldots,x_{n-1})$$
$$= m(x_1)\prod_{i=2}^{n} f(x_i\,|\,x_1,\ldots,x_{i-1}).$$

**Example 6.3** ***Random Sampling Without Replacement for Defectives***

A shipment of $N$ LCD screens contains $J$ defectives. Suppose a random sample without replacement of size $n$ is drawn from the shipment. Let $x = 1$ denote that a screen is defective and $x = 0$ denote a nondefective. Then, assuming $N \geq n$ and $[(n-1)-(N-J)] \leq \sum_{i=1}^{n-1} x_i \leq J$,[3]

$$m(x_1) = \left(\frac{J}{N}\right)^{x_1}\left(\frac{(N-J)}{N}\right)^{1-x_1} I_{\{0,1\}}(x_1),$$
$$f(x_2\,|\,x_1) = \left[\frac{J-x_1}{N-1}\right]^{x_2}\left[\frac{N-J-(1-x_1)}{N-1}\right]^{1-x_2}\prod_{i=1}^{2} I_{\{0,1\}}(x_i),$$

and in general, for the $n$th selection

$$f(x_n|x_1,\ldots,x_{n-1}) = \left[\frac{J-\sum_{i=1}^{n-1}x_i}{N-(n-1)}\right]^{x_n}\left[\frac{N-J-\left(n-1-\sum_{i=1}^{n-1}x_i\right)}{N-(n-1)}\right]^{1-x_n}\prod_{i=1}^{n} I_{\{0,1\}}(x_i).$$

The joint density of the random sample $(X_1,\ldots,X_n)$ is then given by

$$f(x_1,\ldots,x_n) = m(x_1)f(x_2|x_1)\prod_{i=3}^{n} f(x_i|x_1,\ldots,x_{i-1}).$$

Note that the incidence of nondefectives in the population (the value of $J$) has a direct influence on the probabilities of events for the random sample outcomes. Thus, a probabilistic linkage is established between the incidence of characteristics in the population and in the sample. Compare this result to Example 6.1. □

As before, the $n$-variate random variable $\mathbf{X} = (X_1,\ldots, X_n)$ is the **random sample**, $\mathbf{x} = (x_1,\ldots, x_n)$ is the **outcome of the random sample**, and the joint PDF $f(x_1,\ldots,x_n)$ is the **joint density of the random sample**.

### 6.2.3 General Random Sampling

We use the term **general random sampling** to refer to any other type of sampling other than simple random sampling or sampling without replacement. In general random sampling, observations on population or stochastic process characteristics are generated by the outcomes of random variables that are not independent and/or are associated with experiments that are not performed

[3]These conditions, coupled with the condition that $x \in \{0,1\}$, ensure that the denominators in the density expressions are positive and that the numerators are nonnegative.

under identical conditions. Note that while we chose to list *random sampling without replacement* separately in Section 6.2.2 because of its prevalence in applications, it exhibits the features (i.e., non-independence and non-identical experiments) that qualify it to be included in the general random sampling category.

The joint density for the random sample in this case is some density $f(x_1,\ldots, x_n)$ that is inherent to the collection of random variables and/or non-identical experiments and their interrelationships in the sampling design. For example, if the random variables happen to be independent, but are not identically distributed, then the joint density of the random sample is given, in general, by $f(x_1,\ldots,x_n) = \prod_{i=1}^{n} f_i(x_i)$ where $f_i(x_i)$ is the probability density associated with the outcomes of the $i$th random variable or experiment.

**Example 6.4** ***General Random Sampling of Quantities Demanded Across Consumers***

Let the quantity demanded of a commodity by consumer $i$ in a given market be represented by

$$Q_i = g(p_i, y_i, \mathbf{z}_i) + V_i,$$

where $Q_i$ is quantity demanded, $p_i$ is the price of the commodity, $y_i$ is disposable income, and $\mathbf{z}_i$ is a vector of substitute/complement prices and sociodemographic characteristics for consumer $i$. The $V_i$'s are independent but not necessarily identically distributed random variables whose outcomes represent deviations of $Q_i$ from $g(p_i, y_i, \mathbf{z}_i)$ caused by errors in utility optimization, lack of information and/or inherent random human behavior. For the sake of exposition, assume that $V_i \sim N(0,\sigma_i^2)$. Then $(Q_1,\ldots,Q_n)$ is a random sample from an experiment relating to $n$ observations on a *demand process*. The *ith* experiment consists of observing the quantity demanded by consumer $i$, for which $Q_i \sim N(q_i; g(p_i, y_i, \mathbf{z}_i), \sigma_i^2)$, and the joint density of the random sample is given by a product of non-identical marginal densities as

$$f(q_1, q_2, \ldots, q_n; (p_i, y_i, \mathbf{z}_i, \sigma_i^2), i = 1, \ldots, n) = \prod_{i=1}^{n} N(q_i; g(p_i, y_i, \mathbf{z}_i), \sigma_i^2).$$

Note that the characteristics of the demand process directly influence the probabilities of events for random sample outcomes by their influence on the functional form of the joint density of the random sample. □

Again as before, the $n$-variate random variable $\mathbf{X} = (X_1,\ldots, X_n)$ is the **random sample**, $\mathbf{x} = (x_1,\ldots, x_n)$ is the **outcome of the random sample**, and the joint PDF $f(x_1,\ldots,x_n)$ is the **joint density of the random sample**.

### 6.2.4 Commonalities in Probabilistic Structure of Probability Samples

The underlying rationale leading to the joint density function associated with a simple random sample is essentially identical whether random sampling is from an existent finite population of objects or from an ongoing stochastic process. In both cases, an experiment with population distribution, $m(z)$, is independently repeated $n$ times to obtain a random sample outcome, where the joint density of the random sample is defined by $f(x_1,\ldots,x_n) = \prod_{i=1}^{n} m(x_i)$. In other words, in either case, the random sample can be thought of as a collection of *iid* random variables each having the PDF $m(z)$. In subsequent sections of this chapter we will examine certain functions of random samples that have a number of properties that will be useful in statistical inference applications and that are derived from the fact that $X_1,\ldots,X_n$ are *iid* random variables. Since the *iid* property is shared by either type of random sample from a population distribution, any property of the function $g(X_1,\ldots,X_n)$ deduced from the fact that $X_1,\ldots,$ $X_n$ are *iid* will apply regardless of whether random sampling is from a population that exists and is finite or is from an ongoing stochastic process.

There is also a commonality between how a probability sample is generated via general random sampling and how a probability sample is generated by random sampling *without* replacement. In the latter case, observations are generated by a sequence of experiments that are neither independent nor performed under identical conditions, and this characterization of the experiments is subsumed under the general description of how general random sampling occurs. Note, however, that general random sampling case is the broader concept, encompassing literally a myriad of different ways a series of experiments can be interrelated, leading to a myriad of definitions for the joint density of the random sample. On the other hand, in random sampling without replacement, there is an explicit structure to the definition of the joint density of the random sample, where the sampling procedure leads to a joint density definition based on the product of a collection of well-defined conditional density functions. In either case, it becomes somewhat difficult to establish *general* properties of functions of random samples, $g(X_1,\ldots,X_n)$, that are useful for purposes of statistical inference. In the case of random sampling without replacement, the mathematics involved in deriving properties of $g(X_1,\ldots,X_n)$ is generally more complex than in the case of simple random sampling – the *iid* assumption involved in the latter case introduces considerable simplifications. The sheer breadth of variations in the general random sampling case virtually relegates analyses of the properties of $g(X_1,\ldots,X_n)$ to analyses of special cases, and few generalizations to the entire class of general random samples can be made.

We will focus primarily on random sampling from a population distribution in the remainder of this chapter. However, Section 6.6 will present results for deriving sampling distribution for function of random sample that applies to general random sampling context. In later chapters we will examine statistical inference methods that can be applied well beyond simple random sampling contexts. We will also examine some problems of interest involving random sampling with replacement from a finite population, but our analyses of these problems will be limited. To begin further reading on sampling without replacement that parallels some of the topics discussed in the remainder of this chapter, and for a discussion of additional refinements to random sampling techniques, see M. Kendall, A. Stuart, and J. Keith Ord (1977), *The Advanced Theory of Statistics, Vol. 1,* 4th ed., New York: MacMillan, pp. 319–324, and W.G. Cochrane (1977), *Sampling Techniques,* 3rd ed., New York: Wiley.

### 6.2.5 Statistics

In statistical inference, functions of random samples will be used to map sample information into inferences regarding the relevant characteristics of a population or process. These functions, such as $T = t(X_1,\ldots,X_n)$, will be random variables whose probability densities depend on the joint density of the random sample on which they are defined. More specifically, inferential procedures will involve special functions known as **statistics**, defined as follows:

**Definition 6.5** *Statistic*

A real-valued function of observable random variables that is itself an observable random variable, and not dependent on any unknown parameters.

By *observable* random variable, we simply mean a random variable whose numerical outcomes can actually be observed in the real world. Note the following example:

**Example 6.5** *Statistics* **Versus** *Nonobservables*

Let the outcome of a Beta-distributed random variable $X$ represent the proportion of a given day's telephone orders, received by the catalogue department of a large retail store, that are shipped the same day the order is received. Define the two random variables $Y = 100(X - .5)$ and $W = a(X - b)$. The random variable $Y$ is a *statistic* representing the number of percentage points above 50 percent that are shipped the same day. The random variable $W$ is *not* a statistic. It depends on the unknown values of $a$ and $b$, and until these values are specified, the random variable is *unobservable*. □

The reason for restricting our attention to statistics when attempting statistical inference is obvious. We cannot utilize a function of a random sample outcome whose range elements are unobservable to make inferences about characteristics of the population or process from which we have sampled. In subsequent sections, we will examine a number of statistics that will be useful for statistical inference.

## 6.3 Empirical Distribution Function

There is a simple function of a random sample from a population distribution that can be used to provide an empirical characterization of the underlying population distribution from which a random sample is drawn. The function is called the **empirical distribution function** (EDF – sometimes also referred to as the *sample distribution function*), and we examine its definition and some of its properties in this section. After we have discussed common measures used to judge goodness of estimators in Chapter 7, we will see that the EDF represents a useful estimator of the underlying population's cumulative distribution function. Furthermore, the EDF can be used to test hypotheses about the appropriate parametric family of distributions to which the population distribution belongs, as we will examine in our discussion of hypothesis testing.

**Table 6.1** EDF of wheat yields.

| $t$ | $\hat{F}_n(t)$ |
|---|---|
| $(-\infty,50)$ | 0 |
| [50,55) | .1 |
| [55,60) | .2 |
| [60,67) | .3 |
| [67,71) | .4 |
| [71,75) | .5 |
| [75,78) | .6 |
| [78,81) | .8 |
| [81,90) | .9 |
| $[90,\infty)$ | 1.0 |

### 6.3.1 EDF: Scalar Case

The EDF in the scalar case is defined as follows:

**Definition 6.6**
***Empirical Distribution Function: Scalar Case***

Let the scalar random variables $X_1,\ldots,X_n$ denote a random sample from some population distribution. Then the empirical distribution function is defined, for $t \in (-\infty,\infty)$, by $F_n(t) = n^{-1}\sum_{i=1}^{n} I_{(-\infty,t]}(X_i)$, an outcome of which is defined by $\hat{F}_n(t) = n^{-1}\sum_{i=1}^{n} I_{(-\infty,t]}(x_i)$.

An outcome of the EDF can be defined alternatively using the size-of-set function $N(\cdot)$ as

$$\hat{F}_n(t) = \frac{N(\{x : x \leq t, x \in \{x_1, x_2, \ldots, x_n\}\})}{n},$$

that is, $\hat{F}_n(t)$ equals the number of $x_i$'s in the random sample outcome that have values $\leq t$, divided by the sample size.

**Example 6.6**
***EDF of Wheat Yields***

A random sample of size 10 from the population distribution of the yield per acre of a new wheat variety that a seed company has developed produced the following 10 outcomes of wheat yield, in bushels/acre: {60, 71, 55, 50, 75, 78, 81, 78, 67, 90}. Then the EDF is defined in Table 6.1, and graphed in Figure 6.2. □

Given Definition 6.6, it is apparent that the EDF defines a random variable for each value of $t\in\mathbb{R}$. In order to be able to assess the usefulness of the EDF in representing characteristics of the underlying population distribution, it will be informative to examine a number of important properties of the random variable $F_n(t)$. We begin by noting an important relationship between the binomial PDF and the PDF of $F_n(t)$.

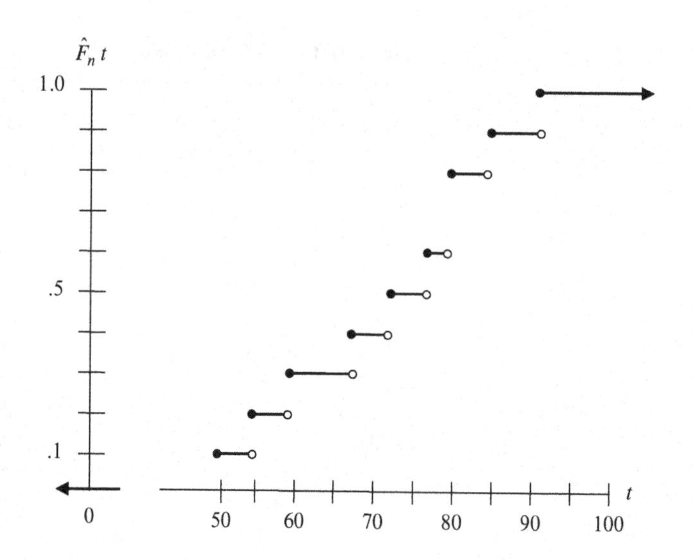


**Figure 6.2** EDF for wheat yields.

**Theorem 6.1** ***Relationship Between EDF and Binomial Distribution***

*Let $F_n(t)$ be the EDF corresponding to a random sample of size n from a population distribution characterized by $F(t)$. Then the PDF of $F_n(t)$ is defined by*

$$P\left(\hat{F}_n(t) = \frac{j}{n}\right) = \begin{cases} \binom{n}{j} & [F(t)]^j\,[1 - F(t)]^{n-j} \text{ for } j \in \{0,1,2,\ldots,n\}, \\ 0 & \text{otherwise} \end{cases}$$

**Proof** From the definition of $F_n(t)$, it follows that

$$P\left(\hat{F}_n(t) = \frac{j}{n}\right) = P\left(\sum_{i=1}^{n} I_{(-\infty,t]}(x_i) = j\right).$$

Note that $Y_i = I_{(-\infty,t]}(X_i)$ is a Bernoulli random variable with $P(y_i = 1) = P(x_i \leq t) = F(t) = p$, and $P(y_i = 0) = P(x_i > t) = 1 - F(t) = 1 - p \; \forall i = 1,\ldots,n$, and since $(X_1,\ldots,X_n)$ is a random sample from a population distribution, $Y_1,\ldots,Y_n$ are *iid* Bernoulli random variables. Then $\sum_{i=1}^{n} Y_i$ has a binomial density with parameters $n$ and $p = F(t)$ as,

$$P\left(\sum_{i=1}^{n} I_{(-\infty,t]}(x_i) = j\right) = P\left(\sum_{i=1}^{n} y_i = j\right) = \binom{n}{j}[F(t)]^j\,[1 - F(t)]^{n-j}$$

for $j = 0,1,2,\ldots,n$, with all other values of $j$ assigned probability zero. ■

The implication of the preceding theorem is that for a *given* choice of $t \in (-\infty,\infty)$, $F_n(t)$ has the same probability distribution as the random variable $n^{-1}Z_n$ where $Z_n$ has a binomial distribution with parameters $n$ and $p = F(t)$. Now that we have discovered this probability distribution for $F_n(t)$, it is rather straightforward to derive the mean, variance, probability limit and asymptotic distribution of $F_n(t)$:

**Theorem 6.2**
***EDF Properties***

*Let $F_n(t)$ be the EDF defined in Theorem 6.1. Then, $\forall t \in (-\infty,\infty)$,*

**a.** $E(F_n(t)) = F(t)$,
**b.** $\text{var}(F_n(t)) = n^{-1}[F(t)(1 - F(t))]$,
**c.** $\text{plim}(F_n(t)) = F(t)$,
**d.** $F_n(t) \overset{a}{\sim} N(F(t), n^{-1}[F(t)(1 - F(t))])$.

**Proof** Since $F_n(t)$ can be represented as $n^{-1}Z_n$, where $Z_n$ has a binomial density with $E(Z_n) = nF(t)$ and $\text{var}(Z_n) = nF(t)(1 - F(t))$, properties (a) and (b) follow immediately. Property (c) follows from the fact that $F_n(t) \overset{m}{\to} F(t)$, since $E(F_n(t)) = F(t)\ \forall n$ and $\text{var}(F_n(t)) \to 0$ as $n \to \infty$, which in turn implies $F_n(t) \overset{p}{\to} F(t)$ by mean square convergence. Property (d) follows from the fact that $F_n(t) = n^{-1}Z_n$ is the average of $n$ *iid* Bernoulli random variables, and it is known from the Lindberg-Levy CLT that this average has an asymptotic normal distribution with mean $F(t)$ and variance $n^{-1}[F(t)(1 - F(t))]$. ■

It will be seen from our discussion of estimator properties in Chapter 7 that the properties possessed by $F_n(t)$ will make it a good statistic to use in providing information about $F(t)$. In particular, the distribution of $F_n(t)$ is centered on $F(t)$; the variance, and thus "spread" of the distribution of $F_n(t)$ decreases as the sample size increases; and the probability that outcomes of $F_n(t)$ agree with $F(t)$ within any arbitrarily small nonzero tolerance converges to one as the sample size increases without bound.

The EDF can be used to generate estimates of probabilities in a way that is analogous to the way a CDF generates information on the probability that $x \in (a,b]$, namely, $\hat{P}_n(x \in (a,b]) = \hat{F}_n(b) - \hat{F}_n(a)$ provides an *empirical estimate* of the appropriate probability. Properties of the random variable $P_n(x \in (a,b]) = F_n(b) - F_n(a)$ generating these empirical estimates are established in the following theorem.

**Theorem 6.3**
***Properties of Estimated Probabilities from EDF***

*Let $F_n(t)$ be the EDF defined in Theorem 6.1. Then $\forall\, t \in (-\infty,\infty)$, and for $a < b$,*

**a.** $E(F_n(b) - F_n(a)) = F(b) - F(a)$,
**b.** $\text{var}(F_n(b) - F_n(a)) = n^{-1}\,[F(b) - F(a)]\,[1 - F(b) + F(a)]$,
**c.** $\text{plim}(F_n(b) - F_n(a)) = F(b) - F(a)$,
**d.** $F_n(b) - F_n(a) \overset{a}{\sim} N(F(b) - F(a), n^{-1}[F(b) - F(a)]\,(1 - F(b) + F(a)))$.

**Proof** Property (a) follows directly from applying the expectation operation to the linear combination of the two random variables and using Theorem 6.2, part (a). Property (c) follows from Theorem 6.2, part (c), and the fact that the probability limit of a sum is the sum of the probability limits. To prove part (b), first note that our previous results concerning variances of linear combinations of random variables (Section 3.11) implies that

$$\mathrm{var}(F_n(b) - F_n(a)) = \mathrm{var}(F_n(b)) + \mathrm{var}(F_n(a)) - 2\mathrm{cov}(F_n(b), F_n(a)).$$

Values of the variance terms are known from Theorem 6.2, part (b). To find the value of the covariance term, first note by definition,

$$\begin{aligned}\mathrm{cov}(F_n(b), F_n(a)) &= \mathrm{cov}\left(n^{-1}\sum_{i=1}^{n} I_{(-\infty,b]}(X_i), n^{-1}\sum_{j=1}^{n} I_{(-\infty,a]}(X_j)\right)\\ &= n^{-2}\sum_{i=1}^{n}\sum_{j=1}^{n}\mathrm{cov}(I_{(-\infty,b]}(X_i), I_{(-\infty,a]}(X_j))\\ &= n^{-1}\,\mathrm{cov}(I_{(-\infty,b]}(X_1)\, I_{(-\infty,a]}(X_1)).\end{aligned}$$

The last equality follows from the fact that, for $i \neq j$, $I_{(-\infty,b]}(X_i)$ and $I_{(-\infty,a]}(X_j)$ are independent random variables since $X_i$ and $X_j$ are independent, which implies the associated covariance terms are zero, and the $n$ remaining covariance terms are all identical and represented by the covariance term in the last equality involving the functions of $X_1$ (since $X_1,\ldots,X_n$ are *iid*). Then since $\mathrm{cov}(Y,Z) = \mathrm{E}(YZ) - \mathrm{E}(Y)\mathrm{E}(Z)$, by letting $Y = I_{(-\infty,b]}(X_1)$ and $Z = I_{(-\infty,a]}(X_1)$, we have that

$$\begin{aligned}\mathrm{cov}(F_n(b), F_n(a)) &= n^{-1}\left[\mathrm{E}\left(I_{(-\infty,b]}(x_1)\, I_{(-\infty,a]}(x_1)\right) - F(b)F(a)\right]\\ &= n^{-1}\left[F(a) - F(b)F(a)\right]\\ &= n^{-1}F(a)[1 - F(b)],\end{aligned}$$

where the next-to-last equality follows from the fact that $I_{(-\infty,b]}(x_1)\, I_{(-\infty,a]}(x_1) = I_{(-\infty,a]}(x_1)\ \forall\, x_1$, since $a < b$.

Having found the value of the covariance term, we finally have that

$$\begin{aligned}\mathrm{var}(F_n(b) - F_n(a)) &= n^{-1}F(b)[1 - F(b)] + n^{-1}F(a)[1 - F(a)] - 2\,n^{-1}F(a)[1 - F(b)]\\ &= n^{-1}[F(b) - F(a)][1 - F(b) + F(a)].\end{aligned}$$

To prove part (d), first note that $F_n(b)$ and $F_n(a)$ have an asymptotic bivariate normal distribution with mean vector $[F(b)\ \ F(a)]'$ and covariance matrix

$$n^{-1}\Sigma = n^{-1}\begin{bmatrix} F(b)[1 - F(b)] & F(a)[1 - F(b)] \\ F(a)[1 - F(b)] & F(a)[1 - F(a)] \end{bmatrix},$$

which follows from applying the multivariate Lindberg-Levy CLT to the sample means of the *iid* Bernoulli random variables that define the random variables $F_n(b)$ and $F_n(a)$. Now define the function $g(F_n(b), F_n(a)) = F_n(b) - F_n(a)$, $g$ being a function of the two asymptotically normal random variables, and note that Theorem 5.40 is applicable. In particular, $(\partial g/\partial F_n(b)) = 1$ and $(\partial g/\partial F_n(a)) = -1$ so that $\mathbf{G} = [1\ -1]$, and thus $n^{-1}\,\mathbf{G}\ \Sigma\ \mathbf{G}' = n^{-1}\ [F(b) - F(a)][1 - F(b) + F(a)]$. Then given $g(F(b), F(a)) = F(b) - F(a)$, Theorem 5.39 implies part (d) of the theorem. ■

The following example illustrates the use of the EDF to provide empirical estimates of probabilities.

**Example 6.7**
***Calculating Empirical Probabilities from EDF***

Referring to the EDF in the wheat yield example (Example 6.6), an empirical calculation of the probability that wheat yield is $\leq$ 76 bushels results in $\hat{P}(x \leq 76) = \hat{F}_n(76) = .6$. The probability that wheat yield is $>$ 62 bushels and $\leq$ 80 bushels is given by $\hat{P}(62<x \leq 80) = \hat{F}_n(80) - \hat{F}_n(62) = .8 - .3 = .5$. □

It can be shown that the convergence of $F_n(t)$ to $F(t)$ in probability as $n \to \infty$ can be strengthened to almost-sure convergence, i.e., $\lim_{n\to\infty} (F_n(t)) = F(t)$ occurs with probability 1 for any $t$. Furthermore, it can be shown that $\lim_{n\to\infty} (F_n(t)) = F(t)$ occurs *simultaneously* $\forall t \in (-\infty,\infty)$ with probability 1. These results are given by the celebrated **Glivenko-Cantelli theorem**, which involves convergence to zero of the supremum (or maximum if it exists) over all $t \in (-\infty,\infty)$ of the absolute difference between $F_n(t)$ and $F(t)$.

**Theorem 6.4**
***Glivenko-Cantelli's Theorem for EDF Convergence***

*Let* $D_n = \sup_t \{|F_n(t) - F(t)|\}$. *Then* $P(\lim_{n\to\infty} D_n = 0) = 1$.

**Proof** See V. Fabian and J. Hannan (1985), *Introduction to Probability and Mathematical Statistics*, New York: Wiley, pp. 80–82.

Note that the Glivenko-Cantelli Theorem provides important additional information on the use of outcomes of $F_n(t)$ as a means of providing empirical representations of $F(t)$. In particular, the theorem implies that the sequence of *functions* $\{F_n\}$ converges as $n \to \infty$ to the *function* $F$ *with probability* 1; this interpretation being supported by the fact that $F_n(t) \to F(t)\ \forall t$ with probability 1. Thus, for large enough $n$, $F_n$ represents a useful approximation to the function $F$ *over its entire domain*, and it is the Glivenko-Cantelli theorem that provides a rigorous justification for inferring the shape and functional form of $F$ from the shape and functional form of $F_n$.

### 6.3.2 EDF: Multivariate Case

The empirical distribution function can be extended to the case where the random sample from the population distribution consists of a collection of multivariate random variables, as follows.

**Definition 6.7**
***Empirical Distribution Function - Multivariate Case***

Let the $(k \times 1)$ random vectors $\mathbf{X}_1,\ldots,\mathbf{X}_n$ denote a random sample from some population distribution. Then the empirical distribution function is defined for $\mathbf{t} = [t_1,\ldots,t_k]' \in \mathbb{R}^k$ and $A(\mathbf{t}) = \times_{i=1}^{k} (-\infty, t_i]$ as $F_n(\mathbf{t}) = n^{-1} \sum_{i=1}^{n} I_{A(\mathbf{t})}(\mathbf{X}_i)$, with an outcome defined by $\hat{F}_n(\mathbf{t}) = n^{-1} \sum_{i=1}^{n} I_{A(\mathbf{t})}(\mathbf{x}_i)$.

An outcome of the EDF can be defined alternatively using the size-of-set function $N(\cdot)$ as

$$\hat{F}_n(\mathbf{t}) = \frac{N(\{\mathbf{x}:\mathbf{x} \leq \mathbf{t}, \mathbf{x} \in \{\mathbf{x}_1,\ldots,\mathbf{x}_n\}\})}{n},$$

i.e., $\hat{F}_n(\mathbf{t})$ equals the number of $\mathbf{x}_i$ *-vectors* in the outcome of the random sample that have values $\leq$ the *vector* $\mathbf{t}$, divided by the sample size.

The properties of the EDF in the multivariate case parallel those of the EDF in the scalar case. In particular, all of the previous theorems apply analogously to the multivariate case by simply reinterpreting $\mathbf{t}$, $\mathbf{a}$, and $\mathbf{b}$ as $(k \times 1)$ vectors instead of scalars in the statement of the theorems, and changing the condition $\forall t \in (-\infty, \infty)$ to $\forall \mathbf{t} \in \mathbb{R}^k$. This follows because precisely the same arguments based on the relationship between the EDF and the binomial distribution apply full well to the multivariate case. The proofs of Theorem 6.1, 6.2, and 6.3 in the multivariate case are in fact analogous to the scalar case and are left to the reader.

## 6.4 Sample Moments and Sample Correlation

Using a random sample from a population distribution (i.e., simple random sampling), which we assume is the case throughout this section, statistics called *sample moments* can be defined that represent sample counterparts to the moments of the population distribution (henceforth called **population moments**). The sample moments have properties that make them useful for estimating the values of corresponding population moments. Sample moments also form the basis for the *method-of-moments* estimation procedure which can be used to provide information on other characteristics of the population distribution besides moments. The method-of-moments procedure will be examined in Chapter 8.

The definitions of the various sample moments can all be unified through the use of the empirical distribution function concept. Specifically, applications of the *empirical substitution principle* lead to the appropriate sample moment definitions.

**Definition 6.8** ***Empirical Substitution Principle for EDFs***

Let $\mathbf{X} = [X_1,\ldots,X_n]'$ be a random sample from a population distribution having CDF $F$. Let $q = q(F)$ be any function of $F$. Then the **empirical substitution principle** representation of $q = q(F)$ is given by $\hat{q} = q(\hat{F}_n)$, where $\hat{F}_n$ is the EDF outcome based on $\mathbf{X}$ and used to estimate $F$.

### 6.4.1 Scalar Case

In order to use the empirical substitution principle to define sample moments when $(X_1,\ldots,X_n)$ is a collection of scalar random variables, first note that moments of the population distribution about the origin or mean can be expressed as functions of the CDF $F$. Specifically, letting $\mathrm{E}_F$ denote an expectation taken with respect to the probability distribution implied by $F$, we have

$\mu'_r = \mathrm{E}_F(X^r)$ and $\mu_r = \mathrm{E}_F((X - \mathrm{E}_F(X))^r)$, which are functions of $F$. Substituting the EDF outcome, $\hat{F}_n$, for $F$ when taking the expectations leads to the definition of sample moments about the origin and mean via the empirical substitution principle.

**Definition 6.9** ***Sample Moments About the Origin and Mean Based on the EDF***

Let the scalar random variables $X_1,\ldots,X_n$ be a random sample with EDF outcome $\hat{F}_n$. Then outcomes of the $r^{th}$ order sample moments about the origin and mean are defined as:

**Sample moments about the origin:** $m'_r = \mathrm{E}_{\hat{F}_n}(X^r)$

**Sample moments about the mean:** $m_r = \mathrm{E}_{\hat{F}_n}\left((X - \mathrm{E}_{\hat{F}_n}(X))^r\right)$

In effect, when defining sample moments, one proceeds as if $X$ had the CDF $\hat{F}_n$, and calculates moments associated with the probability distribution defined by $\hat{F}_n$ in precisely the same way as presented in Chapter 3. The computational difference between the moments defined in Section 3.6 and the sample moments defined here relates simply to which probability distribution is used in taking the expectations – the one implied by $F$ or by $\hat{F}_n$.

Expectations taken with respect to the probability distribution represented by the EDF outcome $\hat{F}_n$ have a common mathematical definition, regardless of the form of the underlying CDF, $F$. To see this, first recall that in *either* the discrete or continuous case $\hat{F}_n$ is a step function whose incremental value at each step equals the observed sample relative frequency of the random variable outcome corresponding to the step (recall Figure 6.2). Since by its definition $\hat{F}_n$ can always be interpreted as a CDF for some *discrete* random variable (i.e., the EDF satisfies all the properties necessary for it to behave as a genuine CDF), the value of an incremental step can be interpreted as the probability assigned to the corresponding random variable outcome by $\hat{F}_n$. In the case where an outcome value, say $x$, is observed only once in the outcome of a random sample of size $n$, it follows that the probability is assigned as $\hat{p}(x) = 1/n$. In general, the probability assigned by $\hat{F}_n$ to an outcome $x$ is given by $\hat{p}(x) = n^{-1}\sum_{i=1}^n I_{\{x\}}(x_i)$, which is the relative frequency of the occurrence of outcome $x$ in the random sample outcome $(x_1,\ldots,x_n)$.

Now let $\hat{R}(X) = \{x : \hat{p}(x) > 0\}$ be the set of $x$-values assigned a positive density weighting by $\hat{p}(x)$, i.e., $\hat{R}(x)$ is the collection of *unique* values in the sample outcome $(x_1,\ldots,x_n)$. It follows that the expectation of $g(X)$ with respect to $\hat{p}(x)$, or equivalently with respect to the probability distribution implied by $\hat{F}_n$, is

$$\mathrm{E}_{\hat{F}_n}(g(X)) = \sum\nolimits_{x\in\hat{R}(X)} g(x)\hat{p}(x) = \sum_{x\in\hat{R}(X)} g(x)\, n^{-1}\sum_{i=1}^{n} I_{\{x\}}(x_i)$$

$$= n^{-1}\sum_{x\in\hat{R}(X)}\sum_{i=1}^{n} g(x_i)\, I_{\{x\}}(x_i) = n^{-1}\sum_{i=1}^{n} g(x_i)$$

where the third equality follows because only terms for which $x_i = x$ affect the value of the inner summation term. Then, defining $g(X) = X^r$ or $g(X) = \left(X - \mathrm{E}_{\hat{F}_N}(X)\right)^r$ and taking expectations with respect to $\hat{F}_n$, we obtain the following alternative definition of sample moment outcomes.

**Definition 6.10**
***Sample Moments About the Origin and Mean Derived from the EDF***

Assume the conditions of Definition 6.7. Then sample moment outcomes can be defined as

**Sample Moments about the Origin:** $m'_r = n^{-1}\sum_{i=1}^n x_i^r$

**Sample Moments About the Mean:** $m_r = n^{-1}\sum_{i=1}^n (x_i - \bar{x}_n)^r$

where $\bar{x}_n = m'_1 = n^{-1}\sum_{i=1}^n x_i$.

We emphasize that regardless of which representation, Definition 6.9 or Definition 6.10, is used in defining sample moment outcomes, all of the previous properties of moments $\mu'_r$ and $\mu_r$ presented in Chapter 3 apply equally as well to the outcomes of sample moments $m'_r$ and $m_r$, with $F_n$ taking the place of $F$. This is so because sample moment outcomes can be interpreted as the appropriate moments of a distribution defined by the discrete CDF $\hat{F}_n$.

We now present a number of important properties of sample moments about the origin that are suggestive of the usefulness of sample moment outcomes for estimating the values of corresponding population moments.

**Theorem 6.5**
***Properties of M'$_r$***

*Let $M'_r = n^{-1}\sum_{i=1}^n X_i^r$ be the $r^{th}$ sample moment about the origin for a random sample $(X_1,\ldots,X_n)$ from a population distribution. Then, assuming the appropriate population moments exist,*

**a.** $\mathrm{E}(M'_r) = \mu'_r$,
**b.** $\mathrm{Var}(M'_r) = n^{-1}\left(\mu'_{2r} - (\mu'_r)^2\right)$,
**c.** $\mathrm{plim}(M'_r) = \mu'_r$,
**d.** $(M'_r - \mu'_r)/[\mathrm{var}(M'_r)]^{1/2} \xrightarrow{d} N(0,1)$,
**e.** $M'_r \overset{a}{\sim} N(\mu'_r, \mathrm{var}(M'_r))$.

**Proof**

(a) $\mathrm{E}(M'_r) = \mathrm{E}(n^{-1}\sum_{i=1}^n X_i^r) = n^{-1}\sum_{i=1}^n \mathrm{E}(X_i^r) = n^{-1}\sum_{i=1}^n \mu'_r = \mu'_r$, since $\mathrm{E}(X_i^r) = \mu'_r \;\forall i$ because $(X_1,\ldots,X_n)$ is a random sample with *iid* elements.

(b) $\mathrm{var}\,(M'_r) = \mathrm{var}\,(n^{-1}\sum_{i=1}^n X_i^r) = n^{-2}\,\mathrm{var}(\sum_{i=1}^n X_i^r)$, and note that the random variables $X_1^r,\ldots,X_n^r$ in the sum are *iid* since $X_i^r$ is the same real-valued function of $X_i \;\forall i$, the $X_i$'s are identically distributed, and Theorem 2.9 applies.

It follows from independence and the results on variances of linear combinations of random variables that

$$\mathrm{var}(M'_r) = n^{-2}\sum_{i=1}^n \mathrm{var}(X_i^r) = n^{-2}\sum_{i=1}^n \left[\mu'_{2r} - (\mu'_r)^2\right] = n^{-1}\left[\mu'_{2r} - (\mu'_r)^2\right].$$

(c) Since $E(M'_r) = \mu'_r \; \forall n$, and since $\text{var}(M'_r) \to 0$ as $n \to \infty$, then Corollary 5.2 with $k = 1$ implies that $M'_r \xrightarrow{m} \mu'_r$ so that $\text{plim}(M'_r) = \mu'_r$.[4]

(d) and (e) Since $(X_1^r, \ldots, X_n^r)$ are *iid* random variables with $E(X_i^r) = \mu'_r$ and $\text{var}(X_i^r) = \mu'_{2r} - (\mu'_r)^2 \; \forall i$, it follows upon substitution into the Lindberg-Levy central limit theorem that

$$Z_n = \frac{\sum_{i=1}^n X_i - n\mu'_r}{n^{1/2}\left[\mu'_{2r} - (\mu'_r)^2\right]^{1/2}} \xrightarrow{d} N(0,1).$$

Multiplying by 1 in the special form $n^{-1}/n^{-1}$ yields

$$Z_n = \frac{M'_r - \mu'_r}{n^{-1/2}\left[\mu'_{2r} - (\mu'_r)^2\right]^{1/2}} \xrightarrow{d} N(0,1),$$

which in turn implies by Definition 5.10 that

$$M'_r \overset{a}{\sim} N\left(\mu'_r, n^{-1}\left[\mu'_{2r} - (\mu'_r)^2\right]\right).$$ ■

In summary, the properties of $M'_r$ presented in Theorem 6.5 indicate that the expected value of a sample moment is equal to the value of the corresponding population moment, the variance of the sample moment monotonically decreases and converges to zero as $n \to \infty$, the sample moment converges in probability to the value of the corresponding population moment, and the sample moment is approximately normally distributed for large $n$. In the context of utilizing outcomes of $M'_r$ as estimates of $\mu'_r$, the properties indicate that the outcomes correctly estimate $\mu'_r$ *on average*, the *spread* of the estimates decreases and the outcomes of $M'_r$ become arbitrarily close to $\mu'_r$ with probability approaching one as the sample size increases, and the outcomes of $M'_r$ are approximately normally distributed around $\mu'_r$ for large enough sample sizes. The fact that outcomes of $M'_r$ are correct on average and become highly accurate individually with high probability as $n$ increase contribute to $M'_r$s being a useful *estimator* of $\mu'_r$, as will be discussed further in Chapter 7. The fact that $M'_r$ is approximately normally distributed for large enough $n$ will facilitate testing hypotheses about the value of $\mu'_r$, to be discussed in Chapters 9 and 10.

The first-order sample moment about the origin is of particular importance in a number of point estimation and hypothesis-testing situations, and it is given a special name and symbol.

**Definition 6.11**
***Sample Mean***

Let $(X_1, \ldots, X_n)$ be a random sample. The **sample mean** is defined by $\overline{X}_n = n^{-1}\sum_{i=1}^n X_i = M'_1$.

[4] An alternative proof, requiring only that moments up to the $r$th-order exist, can be based on Khinchine's WLLN. Although we will not use the property later, the reader can utilize Kolmogorov's SLLN to also demonstrate that $M'_r \xrightarrow{as} \mu'_r$ (see Chapter 5).

Based on sample moment properties, we know that $\mathrm{E}(\overline{X}_n) = \mu, \mathrm{var}(\overline{X}_n) = n^{-1}(\mu'_2 - \mu^2) = \sigma^2/n$, $\mathrm{plim}(\overline{X}_n) = \mu$, and $\overline{X}_n \stackrel{a}{\sim} N(\mu, \sigma^2/n)$. As a preview to a particular problem of statistical inference, note that $\overline{X}_n$ has properties that might be considered useful for estimating the population mean, $\mu$. In particular, the PDF of $\overline{X}_n$ is centered on $\mu$, and as the size of the random sample increases, the density of $\overline{X}_n$ concentrates within a small neighborhood of points around $\mu$ so that it becomes ever more improbable that an outcome of $\overline{X}_n$ would occur far from $\mu$. Thus, outcomes of $\overline{X}_n$ can be useful as estimates of the unknown value of a population mean.

The result on asymptotic normality extends to vectors of sample moments about the origin, in which case a multivariate asymptotic normal density is appropriate.

**Theorem 6.6** ***Multivariate Asymptotic Normality of Sample Moments About the Origin***

$$n^{1/2}\begin{bmatrix} M'_1 - \mu'_1 \\ \vdots \\ M'_r - \mu'_r \end{bmatrix} \xrightarrow{d} N\Big(\underset{r\times 1}{\mathbf{0}}, \underset{r\times r}{\boldsymbol{\Sigma}}\Big) \text{ and } \begin{bmatrix} M'_1 \\ \vdots \\ M'_r \end{bmatrix} \stackrel{a}{\sim} N((\mu'_1, \ldots, \mu'_r)', n^{-1}\boldsymbol{\Sigma}),$$

*where the nonsingular covariance matrix* $\boldsymbol{\Sigma}$ *has typical (j,k) entry equal to*

$\sigma_{jk} = \mu'_{j+k} - \mu'_j \mu'_k.$

**Proof** The proof relies on the multivariate version of the Lindberg-Levy central limit theorem. Let $\mathbf{Y}_i = (X_i^1, \ldots, X_i^r)'$. Since $X_1, \ldots, X_n$ is a random sample with *iid* elements, it follows from Theorem 2.9 that $(\mathbf{Y}_1, \ldots, \mathbf{Y}_n)$ are independent $(r \times 1)$ random vectors with $\mathrm{E}(\mathbf{Y}_i) = \boldsymbol{\mu}$ and $\mathbf{Cov}(\mathbf{Y}_i) = \boldsymbol{\Sigma}$, $\forall\, i$, where $\boldsymbol{\mu} = (\mu'_1, \ldots \mu'_r)'$. Then, given $\boldsymbol{\Sigma}$ is nonsingular, the multivariate Lindberg-Levy CLT applies, establishing the convergence in distribution result, which in turn implies the asymptotic density result.

The typical entry in $\boldsymbol{\Sigma}$ is given by

$$\begin{aligned}\sigma_{jk} &= \mathrm{cov}\left(X^j, X^k\right) = \mathrm{E}(X^j - \mu'_j)\left(X^k - \mu'_k\right) \\ &= \mathrm{E}\left(X^{j+k} - \mu'_j X^k - \mu'_k X^j + \mu'_j \mu'_k\right) = \mu'_{j+k} - \mu'_j \mu'_k.\end{aligned}$$

■

We will not study properties of sample moments *about the mean* in detail. Unlike the case of sample moments about the origin, the properties of higher-order sample moments about the mean become progressively more difficult to analyze. We will concentrate on properties of the second-order sample moment about mean, called the *sample variance.* The reader interested in the general properties of sample moments about the mean can refer to R. Serfling (1980), *Approximation Theorems of Mathematical Statistics,* New York: John Wiley, pp. 69–74.

**Definition 6.12**
***Sample Variance***

Let $X_1,\ldots,X_n$ be a random sample of size $n$. The sample variance is defined as[5] $S_n^2 = n^{-1}\sum_{i=1}^{n}(X_i - \overline{X}_n)^2 = M_2$.

Some important properties of the sample variance are presented in the following theorem.

**Theorem 6.7**
***Properties of $S_n^2$***

*Let $S_n^2$ be the sample variance for a random sample $(X_1,\ldots,X_n)$ from a population distribution. Then, assuming the appropriate population moments exist,*

**a.** $\mathrm{E}(S_n^2) = \left(\frac{n-1}{n}\right)\sigma^2$
**b.** $\mathrm{var}(S_n^2) = n^{-1}\left[((n-1)/n)^2\mu_4 - ((n-1)(n-3)/n^2)\sigma^4\right]$
**c.** $\mathrm{plim}(S_n^2) = \sigma^2$
**d.** $n^{1/2}(S_n^2 - \sigma^2) \xrightarrow{d} N(0, \mu_4 - \sigma^4)$
**e.** $S_n^2 \overset{a}{\sim} N(\sigma^2, n^{-1}(\mu_4 - \sigma^4))$.

**Proof** (a) $\mathrm{E}(S_n^2) = \mathrm{E}\left(\frac{\sum_{i=1}^{n}(X_i - \overline{X}_n)^2}{n}\right) = \mathrm{E}\left(\frac{\sum_{i=1}^{n}(X_i - \mu + \mu - \overline{X}_n)^2}{n}\right)$

$$= \mathrm{E}\left(\frac{\left[\sum_{i=1}^{n}(X_i-\mu)^2\right]}{n} - \frac{\left[n(\mu-\overline{X}_n)^2\right]}{n}\right) = \sigma^2 - \frac{\sigma^2}{n} = \left(\frac{n-1}{n}\right)\sigma^2$$

(b) The proof follows from expressing $\mathrm{var}(S_n^2) = \mathrm{E}(S_n^2 - \sigma^2)^2$ in terms of $X_i$'s and taking expectations. The proof is straightforward conceptually, but quite tedious algebraically. The details are left to the reader, or see R.G. Krutchkoff (1970), Probability and Statistical Inference, New York: Gordon and Breach, pp. 154–157.

(c) Since $\mathrm{E}(S_n^2) = ((n-1)/n)\,\sigma^2 \to \sigma^2$ as $n \to \infty$, and since $\lim_{n\to\infty} \mathrm{var}(S_n^2) = 0$, then Corollary 5.2 implies that $S_n^2 \xrightarrow{m} \sigma^2$, so that $\mathrm{plim}(S_n^2) = \sigma^2$.

(d) and (e) First note that

$$nS_n^2 = \sum_{i=1}^{n}(X_i - \overline{X}_n)^2 = \sum_{i=1}^{n}(X_i - \mu + \mu - \overline{X}_n)^2$$
$$= \sum_{i=1}^{n}(X_i - \mu)^2 + 2(\mu - \overline{X}_n)\sum_{i=1}^{n}(X_i - \mu) + n(\mu - \overline{X}_n)^2,$$

so that

$$n^{1/2}(S_n^2 - \sigma^2) = \frac{\sum_{i=1}^{n}(X_i - \mu)^2 - n\sigma^2}{n^{1/2}} + 2(\mu - \overline{X}_n)n^{-1/2}\sum_{i=1}^{n}(X_i - \mu) + n^{1/2}(\mu - \overline{X}_n)^2$$

[5] Some authors define the sample variance as $S_n^2 = (n/(n-1))M_2$, so that $\mathrm{E}(S_n^2) = \sigma^2$, which identifies $S_n^2$ as an *unbiased* estimator of $\sigma^2$ (see Section 7.2). However, this definition would be inconsistent with the aforementioned fact that $M_2$, and not $(n/(n-1))M_2$, is the second moment about the mean, and thus the *variance*, of the sample or empirical distribution function, $\hat{F}_n$.

Of the three terms added together in the previous expression, all but the first converge in probability to zero. To see this, note for the third term, $\text{p}\lim(n^{1/2})(\mu - \overline{X}_n)^2 = \text{p}\lim(n^{1/2}o_p(n^{-1/2})) = 0$. For the second term, $n^{-1/2}\sum_{i=1}^{n}(X_i - \mu) \xrightarrow{d} N(0,\sigma^2)$ by the LLCLT, and $\text{plim}(\mu - \overline{X}_n) = \text{plim}(\mu) - \text{plim}(\overline{X}_n) = 0$, so by Slutsky's theorem, the second term converges in distribution, and thus in probability, to the constant 0.

Regarding the first term, let $Y_i = (X_i - \mu)^2$, and note that $E(Y_i) = \sigma^2$ and $\text{var}(Y_i) = \mu_4 - \sigma^4$. Then since the $Y_i$'s are *iid*, the first term converges in distribution to $N(0, \mu_4 - \sigma^4)$ by the LLCLT and Slutsky's theorem. Thus, by Slutsky's theorem,

$$n^{1/2}(S_n^2 - \sigma^2) \xrightarrow{d} N(0, \mu_4 - \sigma^4), \text{so that } S_n^2 \overset{a}{\sim} N(\sigma^2, n^{-1}(\mu_4 - \sigma^4)).$$ ■

As another preview to a particular problem of statistical inference, note that $S_n^2$ has characteristics that can be useful for estimating the population variance, $\sigma^2$. Specifically, the distribution of $S_n^2$ becomes centered on $\sigma^2$ as $n \to \infty$, and as the size, $n$, of the random sample increases, the density of $S_n^2$ concentrates within a small neighborhood of points around $\sigma^2$ so that it becomes highly probable that an outcome of $S_n^2$ will occur close to $\sigma^2$.

**Example 6.8**
***Sample Measure Variance Calculation***

Calculating outcomes of the sample mean and sample variance for the wheat yield data presented in Example 6.6 yields, respectively, $\overline{x} = \sum_{i=1}^{10} \frac{x_i}{10} = 70.5$ and $s^2 = \sum_{i=1}^{10} \frac{(x_i - \overline{x})^2}{10} = 140.65$. □

### 6.4.2 Multivariate Case

When the random sample consists of a collection of $k$-variate random vectors $\mathbf{X}_1,\ldots,\mathbf{X}_n$ (i.e., where $k$ characteristics are observed for each of the $n$ sample observations), one can define sample means, sample variances, or any $r$th – order sample moment for *each* of the $k$ entries in the $\mathbf{X}_i$ – vectors. Furthermore, the concept of **joint sample moments** between pairs of entries in the $\mathbf{X}_i$ vectors becomes relevant in the multivariate case, and in particular one can define the notion of **sample covariance** and **sample correlation.**

The method of defining sample moments in the multivariate case is analogous to the approach in the scalar case – use the empirical substitution principle. All of the aforementioned moments can be defined as expectations taken with respect to the probability distribution defined by the (joint) EDF outcome, $\hat{F}_n$, just as population moments can be defined using the population distribution represented by the (joint) CDF, $F$. Using analogous reasoning to the scalar case, the PDF implied by $\hat{F}_n$ is given by $\hat{p}(\mathbf{x}) = n^{-1}\sum_{i=1}^{n} I_{\{\mathbf{x}\}}(\mathbf{x}_i)$, i.e., the probability assigned to the *vector* outcome $\mathbf{x}$ is the relative frequency of the occurrence of $\mathbf{x}$ in the random sample of *vector* outcomes $(\mathbf{x}_1,\ldots,\mathbf{x}_n)$. Then, following an argument analogous to the scalar case, expectations of $g(\mathbf{X})$ with respect to the discrete CDF $\hat{F}_n$ can be defined as

$$\mathrm{E}_{\hat{F}_n}(g(\mathbf{X})) = \sum_{\mathbf{x}\in\hat{R}(\mathbf{X})} g(\mathbf{x})\hat{p}(\mathbf{x}) = n^{-1}\sum_{i=1}^{n} g(\mathbf{x}_i).$$

By appropriate definitions of $g(\mathbf{X})$, one can define $r$th sample moments about the origin and mean for each entry in the vector $\mathbf{X}$, as well as covariances between entries in the vector $\mathbf{X}$. In particular, these function definitions would be $g(\mathbf{X}) = X[j]^r$, $g(\mathbf{X}) = (X[j] - \overline{x}[j])^r$, and $g(\mathbf{X}) = (X[i] - \overline{x}[i])(X[j] - \overline{x}[j])$, where $\overline{x}[\ell] = \mathrm{E}_{\hat{F}_n}(X[\ell])$, leading to the following definition of sample moments in the multivariate case.

**Definition 6.13**
***Sample Moments, Multivariate Case***

Let the $(k \times 1)$ vector random variables $\mathbf{X}_1,\ldots,\mathbf{X}_n$ be a random sample from a population distribution. Then the following outcomes of sample moments can be defined for $j$ and $\ell \in \{1,2,\ldots,k\}$:

**Sample moments about the origin:** $m'_r[j] = n^{-1} \sum_{i=1}^{n} x_i[j]^r$

**Sample means:** $\overline{x}[j] = m'_1[j] = n^{-1} \sum_{i=1}^{n} x_i[j]$

**Sample moments about the mean:** $m_r[j] = n^{-1} \sum_{i=1}^{n} (x_i[j] - \overline{x}[j])^r$

**Sample variances:** $s^2[j] = m_2[j] = n^{-1} \sum_{i=1}^{n} (x_i[j] - \overline{x}[j])^2$

**Sample covariance:** $s_{j\ell} = n^{-1} \sum_{i=1}^{n} (x_i[j] - \overline{x}[j])(x_i[\ell] - \overline{x}[\ell])$

The properties of each of the sample moments about the origin and each of the sample variances are precisely the same as in the scalar case. The proofs are the same as in the scalar case upon utilizing *marginal* population distributions for each $X[j]$, $j = 1,\ldots,k$. The sample covariance has no counterpart in the scalar case, and we examine its properties below.

*Sample Covariance* For clarity of exposition, we will examine the case where $k = 2$, and distinguish the two random variables involved by $X$ and $Y$. A random sample of size $n$ will then be given by the $n$ two-tuples $((X_1, Y_1),\ldots,(X_n, Y_n))$, which can be interpreted as representing observations on two characteristics, $x_i$ and $y_i$, for each of $n$ outcomes of an experiment. Applications to cases where $k > 2$ are accomplished by applying the subsequent results to any pair of random variables in the $(k \times 1)$ vector $\mathbf{X}$.

**Theorem 6.8**
***Properties of Sample Covariance***

*Let $((X_1,Y_1),\ldots,(X_n,Y_n))$ be a random sample from a population distribution and let $S_{XY}$ be the sample covariance between $X$ and $Y$. Then, assuming the appropriate population moments exist,*

**a.** $\mathrm{E}(S_{XY}) = \left(\frac{n-1}{n}\right)\sigma_{XY}$

**b.** $\mathrm{var}(S_{XY}) = n^{-1}\left(\mu_{2,2} - (\mu_{1,1})^2\right) + o(n^{-1})$

**c.** $\text{plim}(S_{XY}) = \sigma_{xy}$
**d.** $S_{xy} \overset{a}{\sim} N(\sigma_{XY}, n^{-1}(\mu_{2,2} - (\mu_{1,1})^2))$

**Proof** **(a)** Examine the $i^{th}$ term in the sum, and note that

$$\mathrm{E}((X_i - \overline{X}_n)(Y_i - \overline{Y}_n)) = \mathrm{E}\left(X_iY_i - \frac{1}{n}X_i\sum_{j=1}^{n}Y_j - \frac{1}{n}Y_i\sum_{j=1}^{n}X_j + \frac{1}{n^2}\sum_{j=1}^{n}X_j\sum_{j=1}^{n}Y_j\right).$$

Since $(X_1,Y_1),\ldots,(X_n,Y_n)$ are *iid*, $X_i$ is independent of $X_j$ and $Y_j$, for $i \neq j$, and $Y_i$ is independent of $X_j$ and $Y_j$ for $i \neq j$. Furthermore, since the $(X_i, Y_i)$'s have the same joint density function, and thus the same population moments $\forall i$, it follows that

$$\begin{aligned}\mathrm{E}((X_i - \overline{X}_n)(Y_i - \overline{Y}_n)) &= \mu'_{1,1} - \frac{2}{n}[\mu'_{1,1} + (n-1)\mu_X\mu_Y] \\ &\quad + \frac{1}{n^2}[n\mu'_{1,1} + n(n-1)\mu_X\mu_Y] \\ &= \left(\frac{n-1}{n}\right)(\mu'_{1,1} - \mu_X\mu_Y) = \left(\frac{n-1}{n}\right)\sigma_{XY}.\end{aligned}$$

Using this result, it follows that

$$\mathrm{E}(S_{XY}) = \frac{1}{n}\sum_{i=1}^{n}\left(\frac{n-1}{n}\right)\sigma_{XY} = \left(\frac{n-1}{n}\right)\sigma_{XY}.$$

**(b)** The procedure used to derive the approximation is based on Taylor series expansions, and can be found in M. Kendall and A. Stuart, The Advanced Theory of Statistics, Vol. 1, 4th ed., pp. 246–250. An alternative motivation for the approximation is given in the proof of part (d).

**(c)** Since $\mathrm{E}(S_{XY}) = ((n-1)/n)\sigma_{XY} \to \sigma_{XY}$ and $\mathrm{var}(S_{XY}) \to 0$ as $n \to \infty$, it follows by Corollary 5.2 that $S_{XY} \xrightarrow{m} \sigma_{XY}$, which in turn implies that $\text{plim}(S_{XY}) = \sigma_{XY}$.

**(d)** First note that the sample covariance can alternatively be written as $S_{XY} = \left(\sum_{i=1}^{n}\frac{X_iY_i}{n} - \overline{X}_n\overline{Y}_n\right) = (M'_{1,1} - \overline{X}_n\overline{Y}_n)$, where $M'_{1,1} = (1/n)\sum_{i=1}^{n}X_iY_i$.[6]

Now examine the *iid* random vectors

$$\begin{bmatrix} X_iY_i \\ X_i \\ Y_i \end{bmatrix}, \; i = 1,\ldots,n,$$

and note that $\forall i$

[6]This is an example of a **joint sample moment about the origin**, the general definition being given by $M'_{r,s} = (1/n)\sum_{i=1}^{n}X_i^rY_i^s$. The definition for the case of **joint sample moment about the mean** replaces $X_i$ with $X_i - \overline{X}_i$, $Y_i$ with $Y_i - \overline{Y}_i$, and $M'_{rs}$ with $M_{r,s}$.

$$\mathrm{E}\begin{bmatrix} X_i\, Y_i \\ X_i \\ Y_i \end{bmatrix} = \begin{bmatrix} \mu'_{1,1} \\ \mu_X \\ \mu_Y \end{bmatrix},$$

$$\boldsymbol{\Sigma} = \mathbf{Cov}\begin{bmatrix} X_i\, Y_i \\ X_i \\ Y_i \end{bmatrix} = \mathrm{E}\left(\begin{bmatrix} X_i\, Y_i - \mu'_{1,1} \\ X_i - \mu_X \\ Y_i - \mu_Y \end{bmatrix}\begin{bmatrix} X_i\, Y_i - \mu'_{1,1} \\ X_i - \mu_X \\ Y_i - \mu_Y \end{bmatrix}'\right)$$

$$= \begin{bmatrix} \mu'_{2,2} - (\mu'_{1,1})^2 & \mu'_{2,1} - \mu_X \mu'_{1,1} & \mu'_{1,2} - \mu_Y \mu'_{1,1} \\ & \sigma_X^2 & \sigma_{XY} \\ \text{(Symmetric)} & & \\ & & \sigma_Y^2 \end{bmatrix}$$

Then, by the multivariate Lindberg-Levy central limit theorem

$$n^{1/2}\begin{bmatrix} M'_{1,1} - \mu'_{1,1} \\ \overline{X}_n - \mu_X \\ \overline{Y}_n - \mu_Y \end{bmatrix} \xrightarrow{d} N(\mathbf{0}, \boldsymbol{\Sigma}).$$

Now let $g(M'_{1,1}, \overline{X}_n, \overline{Y}_n) = M'_{1,1} - \overline{X}_n \overline{Y}_n$, and note by Theorem 5.39 that $g(\cdot)$ has an asymptotic normal distribution given by $g(\cdot) \overset{a}{\sim} N(\mu'_{1,1} - \mu_X \mu_Y, n^{-1}\mathbf{G}\boldsymbol{\Sigma}\mathbf{G}') = N(\sigma_{XY}, n^{-1}\, \mathbf{G}\boldsymbol{\Sigma}\mathbf{G}')$, where $\mathbf{G}$ is the row vector of derivatives of g with respect to its three arguments, evaluated at the expected values of $M'_{1,1}, \overline{X}_n, \overline{Y}_n$, i.e., $\mathbf{G}_{1\times 3} = [1\; -\mu_Y\; -\mu_X]$. The variance of the asymptotic distribution of $g(\cdot)$ is then represented (after some algebraic simplification) as

$$n^{-1}\mathbf{G}\boldsymbol{\Sigma}\mathbf{G}' = n^{-1}\Big[\mu'_{2,2} - (\mu'_{1,1})^2 + 6\mu_X \mu_Y \mu'_{1,1} - 4\,\mu_X^2\,\mu_Y^2$$
$$-2\,\mu'_Y\,\mu'_{2,1} - 2\,\mu_X\,\mu'_{1,2} + \mu_Y^2\,\mu'_{2,0} + \mu_X^2\,\mu'_{0,2}\Big]$$

and it can be shown by defining $\mu_{2,2}$ and $\mu_{1,1}$ in terms of moments about the origin that the preceding bracketed expression is identically equal to $\mu_{2,2} - (\mu_{1,1})^2$. Thus $g(\cdot) \overset{a}{\sim} N(\sigma_{XY}, n^{-1}\,(\mu_{2,2} - (\mu_{1,1})^2))$. ■

Similar to the case of the sample mean and sample variance, the sample covariance has properties that can be useful for estimating its population counterpart, the covariance $\sigma_{XY}$. In particular, the distribution of $S_{XY}$ becomes centered on $\sigma_{XY}$ as $n \to \infty$ and as the sample size $n$ increases, the density of $S_{XY}$ concentrates within a small neighborhood of $\sigma_{XY}$ so that it becomes highly probable that an outcome of $S_{XY}$ will occur close to $\sigma_{XY}$.

*Sample Correlation* Having defined the concepts of sample variance and sample covariance, a rather natural definition of the **sample correlation** between random variables $X$ and $Y$ can be made as follows.

**Definition 6.14**
***Sample Correlation***

Let $((X_1,Y_1),\ldots,(X_n,Y_n))$ be a random sample from a population distribution. Then the **sample correlation** between $X$ and $Y$ is given by

$$R_{XY} = \frac{S_{XY}}{S_X\, S_Y},$$

where $S_X = \left(S_X^2\right)^{1/2}$ and $S_Y = \left(S_Y^2\right)^{1/2}$ are the **sample standard deviations** of $X$ and $Y$, respectively.

Regarding properties of $R_{XY}$, outcomes of the sample correlation are lower bounded by $-1$ and upper bounded by $+1$, analogous to their population counterparts.

**Theorem 6.9**
***Sample Correlation Bounds***

*The outcomes of $R_{XY}$ are such that $r_{XY} \in [-1,1]$.*

**Proof** This follows directly from Theorem 3.32 and 3.33 upon recognizing that $s_{XY}$, $s_X^2$ and $s_X^2$ can be interpreted as a covariance and variances associated with a probability distribution defined by $\hat{F}_n$. ■

In an analogy to the population correlation, the numerical value of the sample correlation, $r_{XY}$, can be interpreted as a measure of the degree of *linear* association between given *sample* outcomes $(x_1,\ldots,x_n)$ and $(y_1,\ldots,y_n)$. The motivation for this interpretation follows directly from Theorems 3.36 and 3.37 upon recognizing that $\hat{F}_n$ can be thought of as defining a probability distribution for $X$ and $Y$ to which the theorems can be subsequently applied. Expectations in the statements of the theorems are then interpreted as expectations based on $\hat{F}_n$, and all references to moments are then interpreted in the context of sample moments.

With the preceding interpretation of Theorem 3.36 in mind, and recalling Figure 3.10, $r_{XY} = 1$ implies that $y_i = a_1 + bx_i$, for $i = 1,\ldots,n$, where $a_1 = \bar{y} - b\bar{x}$ and $b = (s_Y/s_X)$. Thus, the sample outcomes $y_1,\ldots,y_n$ and $x_1,\ldots,x_n$ have a perfect positive linear relationship. If $r_{XY} = -1$, then $y_i = a_2 - bx_i$, for $i = 1,\ldots,n$, where $a_2 = \bar{y} + b\bar{x}$, so the $y_i$'s and $x_i$'s have a perfect negative linear relationship.

To interpret the meaning of $r_{XY} \in (-1,1)$, Theorem 3.37 and its subsequent discussion can be applied to the discrete CDF $\hat{F}_n$ and the associated random sample outcome $(x_1,y_1),\ldots,(x_n,y_n)$. The best predictor of the $y_i$'s in terms of a linear function of the associated $x_i$'s is thus given by $\hat{y}_i = a + bx_i$, where $a = \bar{y} - b\bar{x}$ and $b = s_{XY}/s_X^2$. In the current context, *best* means the choice of $a$ and $b$ that minimizes

$$\mathrm{E}_{\hat{F}_n}\left(d^2\left(Y,\hat{Y}\right)\right) = n^{-1}\sum_{i=1}^{n} (y_i - (a + bx_i))^2,$$

which is a strictly monotonically increasing function of the distance $d(\mathbf{y}, \hat{\mathbf{y}}) = \left[\sum_{i=1}^{n} (y_i - \hat{y}_i)^2\right]^{1/2}$ between the vector $(y_1,\ldots,y_n)$ and the vector $(\hat{y}_1,\ldots,\hat{y}_n)$. (This is analogous to the least-squares criterion that will be discussed in Section 8.2.) Then, since $\mathrm{E}_{\hat{F}_n}\left(d^2\left(Y, \hat{Y}\right)\right) = s_Y^2\left[1 - r_{XY}^2\right]$ (recall the discussion following Theorem 3.37), it follows that the closer $r_{XY}$ is to either $-1$ or 1, the smaller is the distance between the sample outcomes $y_1,\ldots,y_n$ and the best linear prediction of these outcomes based on the sample outcomes $x_1,\ldots,x_n$. Therefore, $r_{XY}$ is a measure of the degree of *linear* association *between sample outcomes* $y_1,\ldots,y_n$ and $x_1,\ldots,x_n$ in the sense that the larger is $|r_{XY}|$, the smaller the distance between $(y_1,\ldots,y_n)$ and $(\hat{y}_1,\ldots,\hat{y}_n)$, where $\hat{y}_i = a + bx_i$.

Also, recall from the discussion of Theorem 3.37 that $r_{XY}^2$ has an interpretation as the proportion of the (sample) variance in the $y_i$'s that is explained by the $\hat{y}_i$'s. Thus, the closer $|\, r_{XY} \,|$ is to 1, the more of the sample variance in the $y_i$'s is explained by $a + bx_i$, $i = 1,\ldots,n$. The arguments are completely symmetric in the $y_i$'s and $x_i$'s and a reversal of their roles leads to an interpretation of $r_{XY}$ (or $r_{YX}$) as a measure of the distance between $(x_1,\ldots,x_n)$ and $(\hat{x}_1,\ldots,\hat{x}_n)$ with $\hat{x}_i = a + by_i$ ($a$ and $b$ suitably redefined). Also, $r_{XY}^2$ (or $r_{XY}^2$) is the proportion of the sample variance in the $x_i$'s explained by $a + by_i$, $i = 1,\ldots,n$.

Besides its use as a measure of linear association between the *sample outcomes*, one might also inquire as to the relationship between $R_{XY}$ and its population counterpart, $\rho_{XY}$. Like the case of the sample variance and the sample covariance, it is not true that the expectation of the sample correlation equals its population counterpart, i.e., in general, $\mathrm{E}(R_{XY}) \neq \rho_{XY}$. Furthermore, because it is defined in terms of a nonlinear function of random variables, general expressions for the mean and variance of $R_{XY}$, as well as other finite sample properties, are quite complicated to state and derive, and generally depend on the particular form of the population distribution on which the random sample is based.[7] We will concentrate here on establishing the probability limit and asymptotic distribution of $R_{XY}$, for which general results can be stated.

**Theorem 6.10** ***Properties of Sample Correlation*** *Let $(X_i, Y_i)$, $i = 1,\ldots,n$, be a random sample from a population distribution and let $R_{XY}$ be the sample correlation between X and Y. Then*

**a.** $\mathrm{plim}(R_{XY}) = \rho_{XY}$,

**b.** $R_{XY} \overset{a}{\sim} N(\rho_{XY}, n^{-1}\, \boldsymbol{\tau}'\boldsymbol{\Sigma}\boldsymbol{\tau})$, *with* $\boldsymbol{\tau}$ *and* $\boldsymbol{\Sigma}$ *defined in the theorem proof.*

[7] See Kendall and Stuart (1977), *Advanced Theory, Vol.* 1, pp. 246-251, for an approach based on Taylor series expansions that can be used to approximate moments of the sample correlation.

**Proof** **(a)** Note that $R_{XY} = S_{XY}/(S_X S_Y)$ is a continuous function of $S_{XY}$, $S_X$, and $S_Y$ for all $S_X > 0$ and $S_Y > 0$ and, in particular, is continuous at the values $\sigma_{XY} = \text{plim}(S_{XY})$, $\sigma_X = \text{plim}(S_X)$, and $\sigma_Y = \text{plim}(S_Y)$. It follows from Slutsky's theorem that

$$\text{plim}(R_{XY}) = \text{plim}\left(\frac{S_{XY}}{S_X S_Y}\right) = \frac{\text{plim}(S_{XY})}{(\text{plim}(S_X)\cdot\text{plim}(S_Y))} = \frac{\sigma_{XY}}{\sigma_X \sigma_Y} = \rho_{XY}.$$

**(b)** The proof follows the approach used by Serfling, Approximation Theorems, pp. 125–126. Define the $(5 \times 1)$ vector $\mathbf{W} = (\overline{X}, \overline{Y}, n^{-1}\sum_{i=1}^n X_i^2, n^{-1}\sum_{i=1}^n Y_i^2, n^{-1}\sum_{i=1}^n X_i Y_i)'$, so that the sample correlation can be expressed as

$$R_{XY} = g(\mathbf{W}) = \frac{W_5 - W_1 W_2}{\left(W_3 - W_1^2\right)^{1/2}\left(W_4 - W_2^2\right)^{1/2}}.$$

Note that $\mathbf{Z}_i = (X_i, Y_i, X_i^2, Y_i^2, X_i Y_i)'$, $i = 1,\ldots,n$ are *iid* random vectors, so that an application of the multivariate Lindberg-Levy CLT to the $\mathbf{Z}_i$'s implies that

$$n^{1/2}(\mathbf{W} - \mathbf{EW}) \xrightarrow{d} N(\mathbf{0}, \boldsymbol{\Sigma}),$$

because $n^{-1}\sum_{i=1}^n \mathbf{Z}_i = \overline{\mathbf{Z}} = \mathbf{W}$, where $\boldsymbol{\Sigma}$ is the covariance matrix of any $\mathbf{Z}_i$.
A direct application of Theorem 5.39 to $R_{XY} = g(\mathbf{W})$ yields the statement in the theorem, where

$$\tau_1 = \frac{\partial g(\text{E}(\mathbf{W}))}{\partial w_1} = \left(\frac{\rho_{XY}\mu'_{1,0}}{\sigma_X^2}\right) - \left(\frac{\mu'_{0,1}}{\sigma_X\sigma_Y}\right)$$

$$\tau_2 = \frac{\partial g(\text{E}(\mathbf{W}))}{\partial w_2} = \left(\frac{\rho_{XY}\mu'_{0,1}}{\sigma_Y^2}\right) - \left(\frac{\mu'_{1,0}}{\sigma_X\sigma_Y}\right)$$

$$\tau_3 = \frac{\partial g(\text{E}(\mathbf{W}))}{\partial w_3} = \frac{-\rho_{XY}}{(2\sigma_X^2)}$$

$$\tau_4 = \frac{\partial g(\text{E}(\mathbf{W}))}{\partial w_4} = \frac{-\rho_{XY}}{(2\sigma_Y^2)}$$

$$\tau_5 = \frac{\partial g(\text{E}(\mathbf{W}))}{\partial w_5} = (\sigma_X\sigma_Y)^{-1}.$$

■

As we have noted for the other statistics we have examined in this section, the sample correlation can be useful for estimating its population counterpart, $\rho_{XY}$, at least in large samples. In particular, as $n \to \infty$, the density of $R_{XY}$ concentrates within a small neighborhood of $\rho_{XY}$ so that as $n$ increases, it becomes highly probable that an outcome of $R_{XY}$ will occur close to $\rho_{XY}$.

In Example 6.9 below we introduce the concepts of **sample covariance matrices** and **sample correlation matrices**.

**Example 6.9** ***Sample Means, Variances, Covariance and Correlation Matrices***

A stock analyst has 15 observations on daily average prices of three common stocks:

| Stock 1 | Stock 2 | Stock 3 |
|---|---|---|
| 1.38 | 1.66 | 4.85 |
| 3.45 | 5.95 | 2.26 |
| 4.80 | 3.02 | 4.41 |
| 4.68 | 7.08 | 4.61 |
| 9.91 | 7.55 | 9.62 |
| 6.01 | 9.49 | 9.34 |
| 8.13 | 9.43 | 8.35 |
| 8.64 | 11.96 | 10.02 |
| 12.54 | 11.32 | 13.97 |
| 11.20 | 12.09 | 10.37 |
| 15.20 | 11.85 | 13.75 |
| 13.52 | 16.98 | 14.59 |
| 15.77 | 16.81 | 16.80 |
| 16.26 | 18.03 | 18.64 |
| 18.21 | 17.07 | 16.95 |

She wants to calculate summary statistics of the data consisting of sample means, sample variances, sample covariances, and sample correlations. She also wants to assess the extent to which there are linear relationships between pairs of stock prices.

Letting $\mathbf{x}$ represent the $(15 \times 3)$ matrix of stock prices. The three sample means are given by

$$\underset{(3\times1)}{\overline{\mathbf{x}}} = \frac{1}{15}\mathbf{x}[i,\cdot]' = \begin{bmatrix} 9.98 \\ 10.69 \\ 10.57 \end{bmatrix}.$$

The **sample covariance matrix**, containing the respective sample variances on the diagonal and the sample covariances between stock $i$ and stock $j$ prices in the $(i,j)$th entry, $i \neq j$, is

$$\underset{(3\times3)}{\widehat{\mathbf{Cov}}}(\mathbf{X}) = \frac{1}{15}\sum_{i=1}^{15}\left(\mathbf{x}[i,\cdot]' - \overline{\mathbf{x}}\right)\left(\mathbf{x}[i,\cdot]' - \overline{\mathbf{x}}\right)'$$

$$= \begin{bmatrix} 25.402 & 22.584 & 23.686 \\ & 24.378 & 22.351 \\ \text{(symmetric)} & & 24.360 \end{bmatrix}.$$

The **sample correlation matrix**, containing the sample correlations between stock $i$ and stock $j$ prices in the $(i,j)$th entry, $i \neq j$, is

$$\widehat{\mathbf{Corr}}(\mathbf{X}) = \begin{bmatrix} s_{X_1} & & \\ & s_{X_2} & \\ & & s_{X_3} \end{bmatrix}^{-1} \widehat{\mathbf{Cov}}(\mathbf{X}) \begin{bmatrix} s_{X_1} & & \\ & s_{X_2} & \\ & & s_{X_3} \end{bmatrix}^{-1}$$

$$= \begin{bmatrix} 1 & .908 & .952 \\ & 1 & .917 \\ \text{(symmetric)} & & 1 \end{bmatrix}$$

The sample correlations indicate that sample observations have a pronounced tendency to follow a *positively sloped* linear relationships between pairs of stock prices. The linear relationship between stock 1 and stock 3 prices is the most pronounced with $(.952)^2 \times 100 = 90.631$ percent of the variance in either of the stock prices explained by a linear function of the other stock price. □

Note in the example above that we have used a matrix relationship between sample correlation matrices and sample covariance matrices. Specifically, letting $\mathbf{S_X}$ denote the diagonal matrix of sample standard deviations of the random variables in $\mathbf{X}$, we have that $\widehat{\mathbf{Corr}}(\mathbf{X}) = \mathbf{S}_\mathbf{X}^{-1}\widehat{\mathbf{Cov}}(\mathbf{X})\mathbf{S}_\mathbf{X}^{-1}$, and it also holds that $\widehat{\mathbf{Cov}}(\mathbf{X}) = \mathbf{S_X}\widehat{\mathbf{Corr}}(\mathbf{X})\mathbf{S_X}$.

## 6.5 Properties of $\overline{\mathbf{X}}_\mathbf{n}$ and $\mathbf{S}_\mathbf{n}^2$ when Random Sampling is from a Normal Distribution

Additional sampling properties of $\overline{X}_n$ and $S_n^2$ are available beyond the generally applicable results presented in earlier sections when random sampling is from a normal population distribution. In particular, $\overline{X}_n$ and $S_n^2$ are then independent random variables, $\overline{X}_n$ is normally distributed in *finite* samples, and $nS_n^2/\sigma^2$ has a $\chi^2$ distribution with $(n - 1)$ degrees of freedom with $S_n^2$ itself being gamma-distributed in finite samples. We establish these properties in subsequent theorems. It can also be shown that $n^{1/2}(\overline{X}_n - \mu)/\hat{\sigma}_n$, where $\hat{\sigma}_n^2 = (n/(n-1))\, S_n^2$, has a so-called ***t*-distribution**, but we defer our examination of this property until Section 6.7, where we will examine the $t$-distribution in detail.

We begin by stating a theorem on the independence of linear and quadratic forms in normally distributed random variables that has applicability in a variety of situations.

**Theorem 6.11** ***Independence of Linear and Quadratic Forms*** *Let* $\mathbf{B}$ *be a* $(q \times n)$ *matrix of real numbers,* $\mathbf{A}$ *be an* $(n \times n)$ *symmetric matrix of real numbers having rank p, and let* $\mathbf{X}$ *be an* $(n \times 1)$ *random vector such that* $\mathbf{X} \sim N(\boldsymbol{\mu}_\mathbf{X}, \sigma^2\mathbf{I})$. *Then* $\mathbf{BX}$ *and* $\mathbf{X'AX}$ *are independent if* $\mathbf{BA} = \mathbf{0}$.[8]

[8] The theorem can be extended to the case where $\mathbf{X} \sim N(\boldsymbol{\mu}_\mathbf{x}, \boldsymbol{\Sigma})$, in which case the condition for independence is that $\mathbf{B\Sigma A} = \mathbf{0}$.

**Proof** Since **A** is symmetric, it can be diagonalized by pre- and postmultiplication using the matrix of characteristic vectors of **A** as

$$\mathbf{P}'\mathbf{A}\mathbf{P} = \mathbf{\Lambda} = \left[\begin{array}{ccc|c} \lambda_1 & & & \\ & \ddots & & \mathbf{0} \\ & & \lambda_p & \\ \hline & \mathbf{0} & & \mathbf{0} \end{array}\right],$$

where **P** is the $(n \times n)$ matrix of characteristic vectors stored columnwise, and $\lambda_i$'s are the nonzero characteristic roots of **A**. It is assumed without loss of generality that the columns of **P** have been ordered so that the first p columns correspond to characteristic vectors associated with the $p$ nonzero characteristic roots. Let $\mathbf{BA} = \mathbf{0}$, so that $\mathbf{BPP}'\mathbf{AP} = \mathbf{0}$, since **P** is orthogonal (i.e., $\mathbf{PP}' = \mathbf{P}'\mathbf{P} = \mathbf{I}$). Let $\mathbf{C} = \mathbf{BP}$, so that $\mathbf{C\Lambda} = \mathbf{BPP}'\mathbf{AP} = \mathbf{0}$. Partitioning **C** and $\mathbf{\Lambda}$ appropriately, we have that

$$\mathbf{C\Lambda} = \left[\begin{array}{c|c} \underset{(q\times p)}{\mathbf{C}_1} & \underset{q\times(n-p)}{\mathbf{C}_2} \end{array}\right] \left[\begin{array}{c|c} \underset{p\times p}{\mathbf{D}} & \underset{p\times(n-p)}{\mathbf{0}} \\ \hline \underset{(n-p)\times p}{\mathbf{0}} & \underset{(n-p)\times(n-p)}{\mathbf{0}} \end{array}\right] = \underset{(q\times n)}{\mathbf{0}},$$

where

$$\mathbf{D} = \begin{bmatrix} \lambda_1 & & \\ & \ddots & \\ & & \lambda_p \end{bmatrix}$$

is the diagonal matrix of nonzero characteristic roots of **A**.

The above matrix equation implies that $\mathbf{C}_1\mathbf{D} = \mathbf{0}$, *and since* **D** *is* invertible, we have that $\mathbf{C}_1 = \mathbf{0}$. Thus

$$\mathbf{C} = \left[\begin{array}{c|c} \underset{(q\times p)}{\mathbf{0}} & \underset{q\times(n-p)}{\mathbf{C}_2} \end{array}\right].$$

Now define $\mathbf{Z}_{(n\times 1)} = \mathbf{P}'\mathbf{X} \sim N(\mathbf{P}'\boldsymbol{\mu}_{\mathbf{X}}, \sigma^2\mathbf{I})$ (recall that $\mathbf{P}'\mathbf{P} = \mathbf{I}$), and note that $(Z_1,\ldots, Z_n)$ are independent random variables. Since $\mathbf{X}'\mathbf{AX} = \mathbf{Z}'\mathbf{P}'\mathbf{APZ} = \mathbf{Z}'\mathbf{\Lambda Z} = \sum_{i=1}^{p} \lambda_i Z_i^2 = g_1(Z_1,\ldots,Z_p)$ and $\mathbf{BX} = \mathbf{BPZ} = \mathbf{CZ} = \mathbf{C}_2 \begin{bmatrix} Z_{p+1} \\ \vdots \\ Z_n \end{bmatrix} = g_2(Z_{p+1},\ldots,Z_n)$, and because $(Z_1,\ldots,Z_p)$ and $(Z_{p+1},\ldots,Z_n)$ are independent, then by Theorem 2.9, $\mathbf{X}'\mathbf{AX}$ and **BX** are independent. ■

We use the preceding theorem to prove a theorem that establishes the distribution of the random variable $nS_n^2/\sigma^2$, which we will later find to have important applications in testing hypotheses about the value of $\sigma^2$.

**Theorem 6.12** ***Independence of $\overline{X}_n$ and $S_n^2$ and $nS_n^2/\sigma^2 \sim \chi^2_{n-1}$ Under Normality***

*If $\overline{X}_n$ and $S_n^2$ are the sample mean and sample variance, respectively, of a random sample of size n from a normal population distribution with mean $\mu$ and variance $\sigma^2$, then*

**a.** *$\overline{X}_n$ and $S_n^2$ are independent;*
**b.** *$(nS_n^2/\sigma^2) \sim \chi^2_{n-1}$.*

**Proof** (a) In the context of Theorem 6.11, let $\mathbf{B}_{(1\times n)} = [n^{-1}\ldots n^{-1}]$, so that $\mathbf{BX} = \overline{X}_n$. Also, let

$$\underset{(n\times n)}{\mathbf{H}} = \begin{bmatrix}\mathbf{B}\\ \mathbf{B}\\ \vdots\\ \mathbf{B}\end{bmatrix} = \begin{bmatrix} n^{-1} & \ldots & n^{-1}\\ \vdots & \ddots & \vdots\\ n^{-1} & \ldots & n^{-1}\end{bmatrix} \text{ so that } (\mathbf{I}-\mathbf{H})\mathbf{X} = \begin{bmatrix} X_1 - \overline{X}_n\\ \vdots\\ X_n - \overline{X}_n\end{bmatrix}$$

implying $nS_n^2 = \mathbf{X}'(\mathbf{I}-\mathbf{H})'(\mathbf{I}-\mathbf{H})\mathbf{X} = \mathbf{X}'(\mathbf{I}-\mathbf{H})\mathbf{X}$ since $\mathbf{I}-\mathbf{H}$ is symmetric and $(\mathbf{I}-\mathbf{H})(\mathbf{I}-\mathbf{H}) = \mathbf{I}-\mathbf{H}$, i.e., $\mathbf{I}-\mathbf{H}$ is idempotent. Then letting $\mathbf{A} = n^{-1}(\mathbf{I}-\mathbf{H})$, we have that $\mathbf{X}'\mathbf{AX} = S_n^2$.

It follows from Theorem 6.11 that $\overline{\mathbf{X}}_n = \mathbf{BX}$ and $S_n^2 = \mathbf{X}'\mathbf{AX}$ are independent since $\mathbf{BA} = n^{-1}\mathbf{B}(\mathbf{I}-\mathbf{H}) = n^{-1}(\mathbf{B}-\mathbf{B}) = \mathbf{0}$.

(b) From the proof of part (a), it follows that

$$\frac{nS_n^2}{\sigma^2} = \frac{1}{\sigma^2}(\mathbf{X}'(\mathbf{I}-\mathbf{H})'(\mathbf{I}-\mathbf{H})\mathbf{X}).$$

Let $\boldsymbol{\mu}_\mathbf{X} = (\mu\;\mu\ldots\mu)'$, and note that $(\mathbf{I}-\mathbf{H})\,\boldsymbol{\mu}_\mathbf{X} = \boldsymbol{\mu}_\mathbf{X} - \boldsymbol{\mu}_\mathbf{X} = \mathbf{0}$. Therefore,

$$\frac{n\,S_n^2}{\sigma^2} = \frac{1}{\sigma^2}(\mathbf{X}-\boldsymbol{\mu}_\mathbf{X})'(\mathbf{I}-\mathbf{H})'(\mathbf{I}-\mathbf{H})(\mathbf{X}-\boldsymbol{\mu}_\mathbf{X}) = \frac{1}{\sigma^2}(\mathbf{X}-\boldsymbol{\mu}_\mathbf{X})'(\mathbf{I}-\mathbf{H})(\mathbf{X}-\boldsymbol{\mu}_\mathbf{X}),$$

because $\mathbf{I}-\mathbf{H}$ is symmetric and idempotent. Note that $\text{tr}[\mathbf{I}-\mathbf{H}] = n-1$, which implies that $(n-1)$ characteristic roots have value 1 because a symmetric idempotent matrix has rank equal to its trace, and its characteristic roots are a collection of 1's and 0's with the number of 1's equal to the rank of the matrix. Diagonalizing $(\mathbf{I}-\mathbf{H})$ by its orthogonal characteristic vector matrix $\mathbf{P}$ then yields

$$\mathbf{P}'(\mathbf{I}-\mathbf{H})\mathbf{P} = \left[\begin{array}{c|c}\mathbf{I} & \mathbf{0}\\ \hline \mathbf{0} & 0\end{array}\right] = \boldsymbol{\Lambda}$$

where $\mathbf{I}$ is an $(n-1)$ dimensional identity matrix. Therefore, $\mathbf{I}-\mathbf{H} = \mathbf{P\Lambda P}'$, and then $(nS_n^2/\sigma^2) = (1/\sigma^2)(\mathbf{X}-\boldsymbol{\mu}_\mathbf{X})'\mathbf{P\Lambda P}'(\mathbf{X}-\boldsymbol{\mu}_\mathbf{X}) = \mathbf{Z}'\boldsymbol{\Lambda}\mathbf{Z}$ where $\mathbf{Z} = (1/\sigma)\mathbf{P}'(\mathbf{X}-\boldsymbol{\mu}_\mathbf{X}) \sim N(\mathbf{0},\mathbf{I})$ since $\mathbf{P}'\mathbf{P} = \mathbf{I}$. Finally $(nS_n^2/\sigma^2) = \sum_{i=1}^{n-1} Z_i^2 \sim \chi^2_{n-1}$, given the definition of $\boldsymbol{\Lambda}$, i.e., we have the sum of squares of $(n-1)$ *iid* standard normal random variables. ■

It follows from part (b) of Theorem 6.12 that $S_n^2$ has a Gamma density, as stated and proved in the following theorem.

**Theorem 6.13**
***Distribution of $S_n^2$ Under Normality***

*Under the Assumptions of Theorem 6.12, $S_n^2 \sim Gamma(\alpha, \beta)$ with $\alpha = (n-1)/2$ and $\beta = (2\sigma^2/n)$.*

**Proof**

Let $Y = nS_n^2/\sigma^2$. Then it is known from Theorem 6.12 (b) that $M_Y(t) = (1 - 2t)^{-(n-1)/2}$. Note that $S_n^2 = (\sigma^2/n)Y$. Then $M_{S_n^2}(t) = \mathrm{E}(\exp(S_n^2 t)) = \mathrm{E}(\exp((\sigma^2/n)Yt)) = \mathrm{E}(\exp(Yt_*))$ where $t_* = (\sigma^2 t/n)$. But since $M_Y(t_*) = \mathrm{E}(\exp(Yt_*)) = (1-2t_*)^{-(n-1)/2}$, it follows that $M_{S_n^2}(t) = (1-2(\sigma^2 t/n))^{-(n-1)/2}$, which is associated with the Gamma density having $\alpha = (n-1)/2$ and $\beta = (2\sigma^2/n)$. ■

Since $\overline{X}_n = n^{-1} \sum_{i=1}^n X_i$ is a linear combination of *iid* $N(\mu,\sigma^2)$ random variables, $\overline{X}_n$ is also normally distributed, as indicated in the following theorem.

**Theorem 6.14**
***Distribution of $\bar{X}_n$ Under Normality***

*Under the Assumptions of Theorem 6.12, $\overline{X}_n \sim N(\mu, \sigma^2/n)$.*

**Proof**

Let $\mathbf{A}_{(1\times n)} = (n^{-1} \ldots n^{-1})$ and $b = 0$ in Theorem 4.9. Then since $\mathbf{X}_{(n\times 1)} \sim N(\boldsymbol{\mu}_\mathbf{X}, \sigma^2\mathbf{I})$, where $\mathbf{X} = (X_1,\ldots,X_n)'$ and $\boldsymbol{\mu}_\mathbf{X} = (\mu,\ldots,\mu)'$, then $\overline{\mathrm{X}} = \mathbf{AX} \sim N(\mathbf{A}\boldsymbol{\mu}_\mathbf{X}, \mathbf{A}\sigma^2\mathbf{IA}') = N(\mu, \sigma^2/n)$. ■

**Example 6.10**
***Inspection of Fill Volumes***

An inspector from the state's Department of Weights and Measures is investigating a claim that an oil company is underfilling quarts of motor oil that are sold at retail. The filling process is such that the actual volumes of oil placed in containers can be interpreted as being normally distributed. According to state specifications, a container is considered legally full if it contains $\geq 31.75$ ounces of oil. The company claims that their filling process has a mean of 32.25 and a standard deviation of .125, and that therefore the probability that a container is legally full exceeds .9999.

The inspector randomly samples 200 containers of oil produced by the oil company, measures their content in ounces, and finds that $\sum_{i=1}^{200} x_i = 6{,}462$ and $\sum_{i=1}^{200} x_i^2 = 208{,}791$, so that the sample mean and sample variance are given by $\overline{x} = \sum_{i=1}^{200} x_i/200 = 32.31$ and $s^2 = \sum_{i=1}^{200}(x_i - \overline{x})^2/200 = \left(\sum_{i=1}^{200} x_i^2 - 200\overline{x}^2\right)/200 = .0189$. Given the company's claim, $\overline{X}_n \sim N(32.25, .78125 \times 10^{-4})$ and $(200)\ S^2/(.015625) \sim \chi^2_{199}$. If the company's claim were true, then the event $\overline{x} \in [\mu - 4\sigma_{\overline{X}}, \mu + 4\sigma_{\overline{X}}] = [32.2146, 32.2854]$ would occur with probability $> .9999$. The particular outcome of $\overline{X}$ that was actually observed appears to be unusually *high* compared to the claimed mean, suggesting that $\mu$ might actually be *higher* than the company claims it to be (if the claimed standard deviation is correct). Regarding the sample variance, and *given* the company's claim, $\mathrm{E}(S^2) = (199/200)\ \sigma^2 = .015547$ and the event $s^2 \in [.01265, .01875]$ would occur with probability[9] equal to .95. Then $s^2 = .0189$ appears to be somewhat high compared to the claimed variance.

Overall, there is strong evidence to suggest that the filling process does not have the characteristics claimed by the company, but the evidence given above does not support the claim that the cans are being *underfilled*. We will examine this issue further after we derive the *t*-distribution in the next Section 6.7. □

## 6.6 Sampling Distributions of Functions of Random Variables

The statistics that we have examined in the preceding sections are useful in a number of statistical inference problems, but one needs to be concerned with a much larger variety of functions of random samples to adequately deal with the variety of inference problems that arise in practice. Furthermore, in order to assess the adequacy of statistical procedures, it will be necessary to identify probability spaces for functions of random samples that are proposed as estimator, hypothesis-testing, or confidence interval statistics in order to evaluate their performance characteristics probabilistically. In particular, we will have need for deriving the functional forms of the PDFs associated with functions of random samples, where such PDFs will be referred to as **sampling densities** or **sampling distributions**.

### 6.6.1 MGF Approach

If $Y = g(X_1,\ldots,X_n)$ is a function of interest, then one can attempt to derive the moment generating function of $Y = g(X_1,\ldots,X_n)$, i.e., $M_{g(\cdot)}(t) = \mathrm{E}(\exp(g(X_1,\ldots,X_n)t))$, and identify the density function characterized by the moment generating function. When $X_1,\ldots,X_n$ is a random sample, one refers to the density of $Y$ as the **sampling density** or **sampling distribution** of $Y$. Of course, $\mathbf{g}$ can be a vector function, in which case we would employ the multivariate moment generating function concept in the hope of identifying the joint density function of the multivariate random variable $\mathbf{Y}$.

**Example 6.11**
***Sampling Distribution of $\overline{X}_n$ when Sampling from Exponential Family***

Consider the sampling distribution of the sample mean $\overline{X}_n = n^{-1}\sum_{i=1}^{n} X_i$, when the random sample $(X_1,\ldots,X_n)$ is drawn from a population distribution represented by the exponential family of densities. Using the MGF approach, we attempt to find $\mathrm{E}(\exp(\overline{X}_n t)) = \mathrm{E}\left(\exp\left(\left(\sum_{i=1}^{n} X_i\right)t/n\right)\right)$. We know that the population distribution is given by $m(z) = \theta^{-1} e^{-z/\theta} I_{(0,\infty)}(z)$, so that the joint density of the random sample is given by

[9] This interval and associated probability was obtained by noting that if $Y \sim \chi^2_{199}$, then $P(161.83 \leq y \leq 239.96) = .95$, which was obtained via numerical integration of the $\chi^2_{199}$ density, leaving .025 probability in both the right and left tails of the density. Using the relationship $S^2 \sim (.015625/200)Y$ then leads to the stated interval.

$$f(x_1,\ldots,x_n) = \frac{1}{\theta^n} \exp\left(\frac{-\sum_{i=1}^n x_i}{\theta}\right) \prod_{i=1}^n I_{(0,\infty)}(x_i).$$

Then

$$\mathrm{E}\left(\exp\left(\left(\sum_{i=1}^n X_i\right)t/n\right)\right) = \prod_{i=1}^n \mathrm{E}\left(\exp\left(\frac{X_i t}{n}\right)\right) = \prod_{i=1}^n \left(1-\theta\frac{t}{n}\right)^{-1} = \left(1-\theta\frac{t}{n}\right)^{-n}$$

for $t < n/\theta$, which is recognized as the MGF of a Gamma($\alpha,\beta$) density with $\alpha = n$ and $\beta = \theta/n$. Thus, the *sampling distribution* of $\overline{X}$ is gamma with the aforementioned parameters. □

### 6.6.2 CDF Approach

One might also attempt to derive the sampling distribution of $Y = g(X_1,\ldots,X_n)$ by identifying the cumulative distribution function of $Y$ and then use the correspondence between cumulative distribution functions and probability density functions to derive the latter. The usefulness of this approach depends on how difficult it is to derive the cumulative distribution function for a particular problem. The advantages of this approach over the MGF approach are that it is applicable even if the MGF does not exist, and it circumvents the problem of the researcher's not knowing which density function corresponds to a particular MGF when the MGF does exist.[10]

**Example 6.12** ***Sampling Distribution of $\overline{X}_n^2$ for Random Sample from Exponential***

Refer to Example 6.11, and suppose we want the sampling distribution of $Y = g(X_1,\ldots,X_n) = \overline{X}_n^2$. Note that

$$F(c) = \begin{cases} P(y \le c) & = P(\overline{x}_n^2 \le c) = P(\overline{x}_n \in [-c^{1/2}, c^{1/2}]) = \int_0^{c^{1/2}} \frac{1}{\beta^\alpha \Gamma(\alpha)} x^{\alpha-1} e^{-x/\beta} dx \text{ for } c>0, \\ 0 & \text{otherwise.} \end{cases}$$

since $\overline{X}_n$ is Gamma distributed. The following lemma regarding differentiating an integral with respect to its bounds will be useful here and in other applications of the CDF approach for defining the PDF of $g(X)$.

[10] Although, as we have mentioned previously, the density function can always be identified in principle by an integration problem involving the MGF in the integrand.

**Lemma 6.1**
***Leibnitz's Rules***

Let $w_1(c)$ and $w_2(c)$ be functions which are differentiable at $c$, and let $h(x)$ be continuous at $x = w_1(c)$ and $x = w_2(c)$ and be integrable. Then

$$\frac{d \int_{w_1(c)}^{w_2(c)} h(x)dx}{dc} = h(w_2(c))\frac{dw_2(c)}{dc} - h(w_1(c))\frac{dw_1(c)}{dc}.$$

***Special Cases:***

**a.** Let $w_1(c) = k \;\forall\; c$, where $k$ is a constant. Then

$$\frac{d \int_{k}^{w_2(c)} h(x)dx}{dc} = h(w_2(c))\frac{dw_2(c)}{dc},$$

and the function $h(x)$ need not be continuous at $k$. Note the result is still valid if $k$ is replaced by $-\infty$.

**b.** Let $w_2(c) = k \;\forall\; c$, where $k$ is a constant. Then

$$\frac{d \int_{w_1(c)}^{k} h(x)dx}{dc} = -h(w_1(c))\frac{dw_1(c)}{dc},$$

and the function need not be continuous at $k$. Note the result is still valid if $k$ is replaced by $\infty$.

(Adapted from D.V. Widder (1961) *Advanced Calculus*, 2nd ed., Englewood Cliffs, NJ: Prentice-Hall, pp. 350–353 and R.G. Bartle (1976), *The Elements of Real Analysis*, 2nd ed., John Wiley, pp. 245–246.)

In the case at hand, $w_2(c) = c^{1/2}$, and $w_1(c) = 0$, so that Lemma 6.1 implies that the sampling distribution of $g(X_1,\ldots,X_n) = \overline{X}_n^2$ is given by

$$\frac{dF(c)}{dc} = f(c) = \begin{cases} \dfrac{1}{2\,\beta^{\alpha}\,\Gamma(\alpha)}\, c^{(\alpha/2)-1} \exp\left[\dfrac{-c^{1/2}}{\beta}\right] & \text{for } c > 0 \\ 0 & \text{otherwise} \end{cases}.$$

### 6.6.3 Event Equivalence Approach (for Discrete Random Variables)

In the case of *discrete* random variables, a third conceptually straightforward approach for deriving the density function of functions of random variables is available. In fact, without calling attention to it, we have already used the procedure, called the **equivalent-events approach**, in proving a theorem concerning the expectation of functions of random variables. Specifically, if $y = g(x)$ is a real-valued function of $x$, then $P_Y(y) = P_X(A_y)$, $A_y = \{x\colon y = g(x),\; x \in R(X)\}$, because the elementary event $y$ for $Y$ is equivalent to the event $A_y$ for $X$ in the sense that $y$ occurs *iff* $A_y$ occurs. It follows that the density function of $Y$ can be obtained from the density function of $X$ for scalar $X$ and $Y$ as

$h(y) = \sum_{(x:g(x)=y,x\in R(X))} f(x)$. The extension to the case of multivariate random variables and vector functions is straightforward. In particular, interpreting $\mathbf{y}$ and/or $\mathbf{x}$ as vectors, we have that $h(\mathbf{y}) = \sum_{\{\mathbf{x}:\mathbf{g}(\mathbf{x})=\mathbf{y},\mathbf{x}\in R(\mathbf{X})\}} f(\mathbf{x})$.

In either the scalar or multivariate case, if $\mathbf{y} = \mathbf{g}(\mathbf{x})$ is invertible, so that $\mathbf{x} = \mathbf{g}^{-1}(\mathbf{y})\ \forall \mathbf{x} \in \mathrm{R}(\mathbf{X})$, then the discrete density can be defines simply as $h(\mathbf{y}) = f(\mathbf{g}^{-1}(\mathbf{y}))$.

**Example 6.13** ***Defining Sampling Distribution via Event Equivalence***

Your company ships BluRay disks in lots of 100. As part of your quality-control program, you assure potential buyers that no more than one disk will be defective in each lot of 100. In reality, the probability that a disk manufactured by your company is defective is .001, and whether a given disk is defective is independent of whether or not any other disk is defective. Then the PDF for the number of defectives in a lot of your disks is given by the binomial density function

$$f(x) = \begin{cases} \binom{100}{x} (.001)^x (.999)^{100-x} & \text{for } x \in \{0, 1, \ldots, 100\}, \\ 0 & \text{otherwise.} \end{cases}$$

Define a new binary random variable, $Z$, such that $z = 1$ represents the event that your quality-control claim is valid (i.e., $x = 0$ or $1$) while $z = 0$ represents the event that your claim is invalid (i.e., $x = 2$ or more) on a lot of your disks. That is, $z = g(x) = I_{\{0,1\}}(x)$. Consider defining the density function of $Z$. The density is given by $h(z) = \sum_{\{x: I_{\{0,1\}}(x)=z, x\in R(X)\}} f(x)$ for $z \in R(Z) = \{0, 1\}$, and $h(z) = 0$ elsewhere. Then, referring to the binomial density function defined above,

$$h(1) = \sum_{x=0}^{1} f(x) = .995 \text{ and } h(0) = \sum_{x=2}^{100} f(x) = .005, \text{so that}$$

$$h(z) = .995\ I_{\{1\}}(z) + .005\ I_{\{0\}}(z).$$

□

### 6.6.4 Change of Variables Approach (for Continuous Random Variables)

Another very useful procedure for deriving the PDF of functions of *continuous* random variables is available if the functions involved are continuously differentiable and if $Y = g(X)$ admits an inverse function $X = g^{-1}(Y)$. We first examine the case where both $Y$ and $X$ are *scalar* random variables. For the purpose of proper interpretation of the theorem below, recall our convention established in Definition 2.13, whereby there is equivalence between the **support of the random variable** $\{x : f(x){>}0 \text{ for } x \in \mathbb{R}\}$, and the **range of the random variable** $R(X)$, i.e., $R(X) \equiv \{x : f(x){>}0 \text{ for } x \in \mathbb{R}\}$.

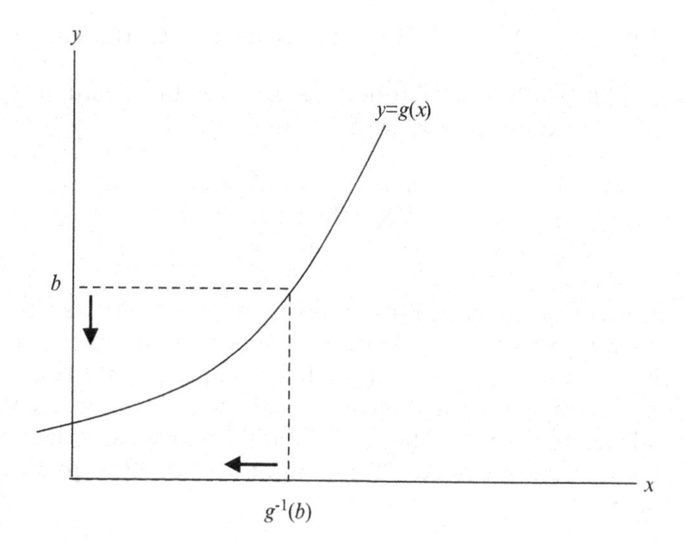


**Figure 6.3** Monotonically increasing function.

**Theorem 6.15 Change of Variables Technique: Univariate and Invertible** *Suppose the continuous random variable X has PDF f(x). Let g(x) be continuously differentiable with $dg/dx \neq 0\ \forall x$ in some open interval, $\Delta$, containing R(X). Also, let the inverse function $x = g^{-1}(y)$ be defined $\forall\ y \in R(Y)$. Then the PDF of $Y = g(X)$ is given by*

$$h(y) = f(g^{-1}(y))\left|\frac{dg^{-1}(y)}{dy}\right| \text{for } y \in R(Y), \textit{with } h(y) = 0 \textit{ elsewhere.}$$

**Proof** If $dg/dx \neq 0$ is a continuous function, then $g$ is either a monotonically increasing or monotonically decreasing function, i.e., either $dg/dx>0$ or $dg/dx<0$, respectively, $\forall\ x \in \Delta$. (Note that $dg/dx$ cannot $> 0$ for some $x$ and $< 0$ for other $x$'s since continuity of $dg/dx$ would then necessarily require that $dg/dx = 0$ at some point.)

**(a)** Case where $dg/dx > 0$: In this case, $P(y \leq b) = P(g(x) \leq b) = P(x \leq g^{-1}(b)),\ \forall\ b \in R(Y)$, since $x = g^{-1}(y)$ is monotonically increasing for $y \in R(Y)$ (see Figure 6.3). Thus the cumulative distribution function for $Y$ can be defined for all $b \in R(Y)$ by $H(b) = P(y \leq b) = P(x \leq g^{-1}(b)) = \int_{-\infty}^{g^{-1}(b)} f(x)dx$.
Then using Lemma 6.1, we can derive the density function for $Y$ by differentiation as[11]

$$h(b) = \frac{dH(b)}{db} = \begin{cases} \dfrac{d\int_{-\infty}^{g^{-1}(b)} f(x)dx}{db} = f(g^{-1}(b))\dfrac{dg^{-1}(b)}{db} & \text{for } b \in R(Y) \\ 0 & \text{otherwise} \end{cases}$$

[11]Here, and elsewhere, we are suppressing a technical requirement that $f$ be continuous at the point $g^{-1}(b)$, so that we can invoke Lemma 6.1 for differentiation of the cumulative distribution function. Even if $f$ is discontinuous at $g^{-1}(b)$, we can nonetheless *define* $h(b)$ as indicated above, since a density function can be redefined arbitrarily at a finite number of points of discontinuity without affecting the assignment of probabilities to any events.

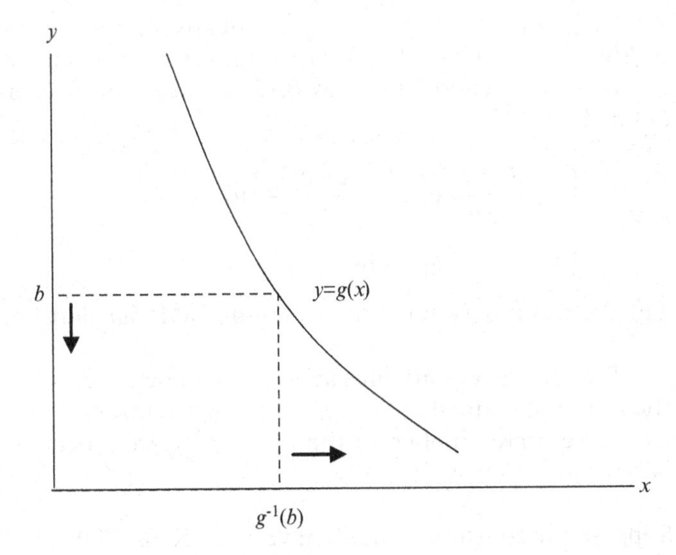


**Figure 6.4** Monotonically decreasing function.

**(b)** Case where $dg/dx < 0$: In this case, $P(y \le b) = P(g(x) \le b) = P(x \ge g^{-1}(b))$, $\forall b \in R(Y)$, since $x = g^{-1}(y)$ is monotonically decreasing for $y \in R(Y)$ (see Figure 6.4). Thus the cumulative distribution function for $Y$ can be defined for all $b \in R(Y)$ by $H(b) = P(y \le b) = P(x \ge g^{-1}(b)) = \int_{g^{-1}(b)}^{\infty} f(x)dx$.
Then, using Lemma 6.1, we can derive the density function for $Y$ by differentiation as

$$h(b) = \frac{dH(b)}{db} = \begin{cases} \dfrac{d\int_{g^{-1}(b)}^{\infty} f(x)dx}{db} & = -f(g^{-1}(b))\dfrac{dg^{-1}(b)}{db} \text{ for } b \in R(Y) \\ 0 & \text{otherwise} \end{cases}$$

Note in this latter case, since $(dg^{-1}(b)/db) < 0$, we can write $h(b)$ alternatively as

$$h(b) = \begin{cases} f(g^{-1}(b))\left|\dfrac{dg^{-1}(b)}{db}\right| \text{for } b \in R(Y) \\ = 0 \text{ otherwise} \end{cases}$$

which is also an alternative representation of the previous case where $dg/dx > 0$. ■

**Example 6.14** ***Derivation of the Log-Normal Distribution***

A stochastic form of the Cobb-Douglas production function is given by

$$Q = \beta_o \left( \prod_{i=1}^{k} x_i^{\beta_i} \right) e^W = \beta_o \left( \prod_{i=1}^{k} x_i^{\beta i} \right) V,$$

where $W \sim N(0, \sigma^2)$ and $V = e^W$. Consider deriving the PDF of $V$.

Note that $v = e^w$ is a monotonically increasing function of $w$ for which $dv/dw = e^w > 0\ \forall w$. The inverse function is given by $w = \ln(v)\ \forall v > 0$, and $dw/dv = v^{-1}$. Then Theorem 6.15 applies, and the distribution of $V$ can be defined as

$$h(v) = \begin{cases} \frac{1}{\sqrt{2\pi}\ \sigma v} \exp\left(-\frac{[\ln(v)]^2}{2\sigma^2}\right) & \text{for } v>0 \\ 0 & \text{elsewhere} \end{cases}$$

The density function is called a **log-normal distribution**. □

The change-of-variables approach can be generalized to certain cases where the function defined by $y = g(x)$ does *not* admit an inverse function. The procedure essentially applies to functions that are *piecewise invertible*, as defined below.

**Theorem 6.16 *Change-of-Variables Technique: Univariate and Piecewise Invertible***

Suppose the continuous random variable X has PDF $f(x)$. *Let $g(x)$ be continuously differentiable with $dg(x)/dx \neq 0$ for all but perhaps a finite number of x's in an open interval $\Delta$ containing the range of X. Let $R(X)$ be partitioned into a collection of disjoint intervals $D_1,\ldots,D_n$ for which $g$: $D_i \to R_i$ has an inverse function $g_i^{-1}$: $R_i \to D_i\ \forall i$.*[12] *Then the probability density of $Y = g(X)$ is given by*

$$h(y) = \begin{cases} \sum_{i \in I(y)} f(g_i^{-1}(y)) \left|\frac{dg_i^{-1}(y)}{dy}\right| & \text{for } y \in R(Y) \\ 0 & \text{elsewhere,} \end{cases}$$

*where $I(y) = \{i: \exists\ x \in D_i$ such that $y = g(x),\ i = 1,\ldots,n\}$, and $(dg_i^{-1}(y)/dy) \equiv 0$ whenever it would otherwise be undefined.*[13]

**Proof** The CDF of $Y = g(X)$ can be represented as $P(y \leq b) = \sum_{i=1}^{n} P(x{:}g(x) \leq b, x \in D_i)$

Note the following possibilities for $P(x$: $g(x) \leq b$, $x \in D_i)$:

| Case | $P(x : g(x) \leq b, x \in D_i)$: **g monotonically increasing on $D_i$** | **g monotonically decreasing on $D_i$** |
|---|---|---|
| 1. $b < \min_{x \in D_i} g(x)$ | 0 | 0 |
| 2. $\min_{x \in D_i} g(x) \leq b \leq \max_{x \in D_i} g(x)$ | $\int_{\min(D_i)}^{g_i^{-1}(b)} f(x)dx$ | $\int_{g_i^{-1}(b)}^{\max(D_i)} f(x)dx$ |
| 3. $b > \max_{x \in D_i} g(x)$ | $P(x_i \in D_i)$ | $P(x_i \in D_i)$ |

[12] These properties define a function that is **piecewise invertible** on the domain $\cup_{i=1}^{n} D_i$.

[13] Note that $I(y)$ is an index set containing the indices of all of the $D_i$ sets that have an element whose image under the function g is the value $y$.

where $\min(D_i)$ and $\max(D_i)$ refer to the minimum and maximum values in $D_i$.[14] In either case 1 or 3, $dP(x{:}g(x) \leq b,\ x \in D_i)/db = 0$, which corresponds to the case where $\nexists x$ such that $g(x) = b$. In case 2, Leibnitz's rule for differentiating the integrals yields $f(g^{-1}(b))\,[dg^{-1}(b)/db]$ and $-f(g^{-1}(b))[dg^{-1}(b)/db]$ in the monotonically increasing and decreasing cases, respectively, which can both be represented by $f(g^{-1}(b))|dg^{-1}(b)/db|$. Then the density of $Y$ can be defined as $h(b) = dP(y \leq b)/db$ for $b \in R(Y)$, with $|(dg^{-1}(y)/dy)|$ arbitrarily set to zero in the finite number of instances where $dg(b)/dx = 0$ for $b \in D_i$. ■

The following example illustrates the use of this more general change-of-variables technique.

**Example 6.15** ***Change of Variables in a Noninvertible Function Case***

The manufacturing process for a certain product is sensitive to deviations from an optimal ambient temperature of 72 °F. In a poorly air-conditioned plant owned by the Excelsior Corporation, the average daily temperature is uniformly distributed in the range [70,74], and the deviations from 72 °F are represented by the outcome of $X \sim f(x) = .25\, I_{[-2,2]}(x)$. The percentage of production lost on any given day due to temperature deviations can be represented by the outcome of $Y = g(X) = X^2$. Consider deriving the probability density for the percentage of production lost on any given day.

First note that Theorem 6.15 does not apply since $y = x^2$ is such that $dy/dx = 0$ at $x = 0$, which is in the range of X, and $y = x^2$ does not admit an inverse function $\forall y \in R(Y) = [0,4]$ (see Figure 6.5).

However, we can utilize Theorem 6.16 to derive the density function of Y. The function $g(x) = x^2$ is continuously differentiable $\forall x$, and $dg(x)/dx = 2x \neq 0$ $\forall x$ except $x = 0$. The sets $D_1 = [-2,0)$ and $D_2 = [0,2]$ are two disjoint intervals whose union equals the range of X. Also, $g\colon D_1 \to (0,4]$ has inverse function $g_1^{-1}$: $(0,4] \to D_1$ defined by $g_1^{-1}(y) = -y^{1/2}$ for $y \in (0,4]$, while $g\colon D_2 \to [0,4]$ has inverse function $g_2^{-1}{:}[0,4] \to D_2$ defined by $g_2^{-1}(y) = y^{1/2}$ for $y \in [0,4]$.[15] Finally, note that $dg_1^{-1}(y)/dy = -(1/2)y^{-1/2}$ and $dg_2^{-1}(y)/dy = (1/2)y^{-1/2}$ are defined $\forall y \in (0,4]$. Then from Theorem 6.16, it follows that

$$h(y) = \begin{cases} \frac{1}{4} I_{[-2,2]}(-y^{1/2})| -.5y^{-.5}| + \frac{1}{4} I_{[-2,2]}(y^{1/2})|.5y^{-.5}| \\ \quad = \frac{1}{8} I_{[0,4]}(y)\, y^{-.5} + \frac{1}{8} I_{[0,4]}(y)\, y^{-.5} = \frac{1}{4} y^{-.5}\, I_{[0,4]}(y) \text{ for } y \in (0,4] \\ 0 \text{ elsewhere.} \end{cases}$$

The change of variable technique can be extended to the multivariate case. We examine the case where the inverse function exists.

[14] If max or min do not exist, they are replaced with sup and inf.

[15] We are continuing to use the convention that $y^{1/2}$ refers to the positive square root of y, so that $-y^{1/2}$ refers to the negative square root.

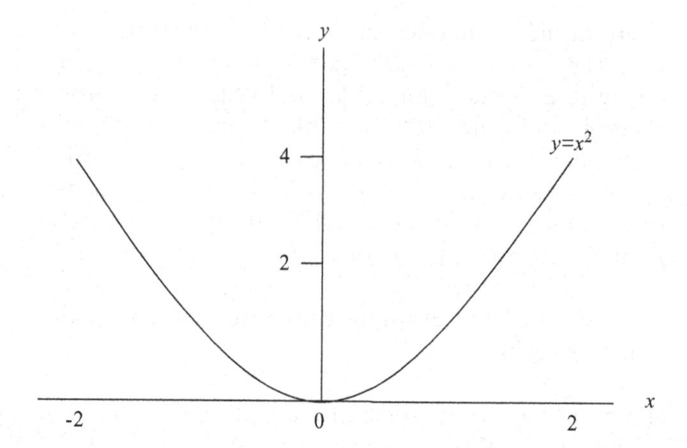


**Figure 6.5**
Function with no inverse.

**Theorem 6.17**
***Change of Variables Technique: Multivariate and Invertible***

*Suppose the continuous $(n \times 1)$ random vector $\mathbf{X}$ has joint PDF $f(\mathbf{x})$. Let $\mathbf{g}(\mathbf{x})$ be a $(n \times 1)$ real-valued vector function that is continuously differentiable $\forall\, \mathbf{x}$ vector in some open rectangle of points, $\Delta$, containing the range of $\mathbf{X}$. Assume the inverse vector function $\mathbf{x} = \mathbf{g}^{-1}(\mathbf{y})$ exists, $\forall\ \mathbf{y} \in R(\mathbf{Y})$. Furthermore, let the **Jacobian matrix** of the inverse function transformation,*

$$\mathbf{J} = \begin{bmatrix} \dfrac{\partial g_1^{-1}(\mathbf{y})}{\partial y_1} & \cdots & \dfrac{\partial g_1^{-1}(\mathbf{y})}{\partial y_n} \\ \vdots & \ddots & \vdots \\ \dfrac{\partial g_n^{-1}(\mathbf{y})}{\partial y_1} & \cdots & \dfrac{\partial g_n^{-1}(\mathbf{y})}{\partial y_n} \end{bmatrix},$$

*be such that $\det(\mathbf{J}) \neq 0$ with all partial derivatives in $\mathbf{J}$ being continuous $\forall\, \mathbf{y} \in R(\mathbf{Y})$. Then the joint density of $\mathbf{Y} = \mathbf{g}(\mathbf{X})$ is given by*

$$h(\mathbf{y}) = \begin{cases} f(g_1^{-1}(\mathbf{y}), \ldots, g_n^{-1}(\mathbf{y}))\,|\det(\mathbf{J})| \text{ for } \mathbf{y} \in R(\mathbf{Y}), \\ = 0 \text{ otherwise,} \end{cases}$$

*where $|\det(\mathbf{J})|$ denotes the absolute value of the determinant of the Jacobian.*

**Proof** The proof is quite complex, and is left to a more advanced course of study. The proof is based on the change-of-variables approach in multivariate integration problems. See T. Apostol (1974), Mathematical Analysis, 2nd ed., Reading, MA: Addison-Wesley, pp. 421. ■

We defer an illustration of the use of this theorem until the next section, where we will use the procedure to derive both the $t$-density and $F$-density. The reader should note that in Theorem 6.17, there are as many coordinate functions in the vector function $\mathbf{g}$ as there are elements in the random variable vector $\mathbf{X}$. In practice, if the researcher is interested in establishing the density of fewer than $n$ random variables defined as real-valued functions of the $(n \times 1)$ vector $\mathbf{X}$, she is obliged to define "auxiliary" random variables to obtain an invertible vector function having $n$ coordinate functions, and then later integrate out the auxiliary random variables from the $n$-dimensional joint density to arrive at the marginal density of interest. This approach will be illustrated in the next section.

A generalization of Theorem 6.17 to noninvertible functions along the lines of Theorem 6.16 is possible. See Mood, Graybill, and Boes (1974), *Introduction to the Theory of Statistics*. New York: McGraw-Hill, p. 209 for one such generalization.

## 6.7 t-and F-Densities

In this section we consider two important statistics that are defined in terms of ratios of random variables related to the normal distribution. These statistics, and their probability distributions, figure prominently in the construction of hypothesis-testing and confidence interval procedures when random sampling is from a normal probability distribution.

### 6.7.1 *t*-Density

The $t$-statistic is defined as the ratio of a standard normal random variable divided by the square root of a chisquare random variable that has been divided by its degrees of freedom. If the two random variables in the ratio are independent, then the $t$-statistic has a $t$-distribution, with a single degrees of freedom parameter that refers to the denominator chisquare random variable, as defined in the theorem below.

**Theorem 6.18** ***Definition and Derivation of the t-Density*** *Let $Z \sim N(0,1)$, $Y \sim \chi^2_v$, and let $Z$ and $Y$ be independent random variables. Then $T = Z/(Y/v)^{1/2}$ has the* ***t-density*** *with v degrees of freedom, where*

$$f(t;v) = \frac{\Gamma\left(\frac{v+1}{2}\right)}{\Gamma(v/2)\sqrt{\pi v}}\left(1+\frac{t^2}{v}\right)^{-\left(\frac{v+1}{2}\right)}.$$

**Proof** Define the $(2 \times 1)$ vector function $\mathbf{g}$ as

$$\begin{bmatrix} t \\ w \end{bmatrix} = \begin{bmatrix} g_1(z,y) \\ g_2(z,y) \end{bmatrix} = \begin{bmatrix} \dfrac{z}{(y/v)^{1/2}} \\ y \end{bmatrix},$$

where $g_2(z,y)$ is an auxiliary function of $(z,y)$ defined to allow the use of Theorem 6.17. Note $\mathbf{g}$ is continuously differentiable $\forall\, z$ and $\forall\, y > 0$, which represents an open rectangle of $(z,y)$ points containing the range of $(Z,Y)$ (note the joint density of $(Z,Y)$ in this case is the product of a standard normal density and a $\chi^2$ density, which has support $(-\infty,\infty) \times (0,\infty))$. The inverse vector function $\mathbf{g}^{-1}$ is defined by

$$\begin{bmatrix} z \\ y \end{bmatrix} = \begin{bmatrix} g_1^{-1}(t,w) \\ g_2^{-1}(t,w) \end{bmatrix} = \begin{bmatrix} t\left(\frac{w}{v}\right)^{1/2} \\ w \end{bmatrix} \forall (t,w) \in R(T,W).$$

The Jacobian of the inverse function is

$$\mathbf{J} = \left[\begin{array}{c|c} \left(\frac{w}{v}\right)^{1/2} & \frac{t}{2}\left(\frac{w}{v}\right)^{-1/2}\left(\frac{1}{v}\right) \\ \hline 0 & 1 \end{array}\right],$$

where the elements of the Jacobian are continuous functions of $(t,w)$ and $|\det(\mathbf{J})| = |(w/v)^{1/2}| = (w/v)^{1/2} \neq 0 \; \forall (t,w) \in R(T,W)$. Given the density assumptions concerning $Z$ and $Y$,

$$\begin{aligned} f(z,y) &= m_Z(z)m_Y(y) \\ &= \frac{1}{(2\pi)^{1/2}} \exp(-z^2/2) \frac{1}{2^{v/2}\Gamma(v/2)} y^{(v/2)-1} \exp(-y/2) I_{(0,\infty)}(y). \end{aligned}$$

Then, by Theorem 6.17, the joint density of $(T,W)$ is given by

$$\begin{aligned} h(t,w) &= \frac{1}{(2\pi)^{1/2}} \exp\left(-(1/2)\,t^2\left(\frac{w}{v}\right)\right) \frac{1}{2^{v/2}\Gamma(v/2)} w^{(v/2)-1} \\ &\quad \times \exp\left(-\frac{w}{2}\right) I_{(0,\infty)}(w) \left(\frac{w}{v}\right)^{1/2} \\ &= \frac{1}{\Gamma(v/2)(\pi v)^{1/2}\, 2^{(v+1)/2}} w^{(v-1)/2} \exp\left(-\frac{w}{2}\left(1+\frac{t^2}{v}\right)\right) I_{(0,\infty)}(w). \end{aligned}$$

Since our interest centers on the density of $T$, we require the marginal density of $T$, $f_T(t;v) = \int_0^\infty h(t,w)dw$. Making the substitution $p = (w/2)(1 + (t^2/v))$ in the integral, so that $w = 2p/(1 + (t^2/v))$ and $dw = 2/(1 + (t^2/v))dp$, yields

$$\begin{aligned} f_T(t;v) &= \int_0^\infty \frac{1}{\Gamma(v/2)(\pi v)^{1/2} 2^{(v+1)/2}} \left[\frac{2p}{1+\frac{t^2}{v}}\right]^{(v-1)/2} e^{-p} \left[\frac{2}{1+\frac{t^2}{v}}\right] dp \\ &= \frac{1}{\Gamma(v/2)(\pi v)^{1/2}} \left(1+\frac{t^2}{v}\right)^{-(v+1)/2} \int_0^\infty p^{(v-1)/2} e^{-p} dp \\ &= \frac{\Gamma\left(\frac{v+1}{2}\right)}{\Gamma(v/2)(\pi v)^{1/2}} \left(1+\frac{t^2}{v}\right)^{-(v+1)/2}, \end{aligned}$$

where we have used the fact that the integral in the next-to-last expression is the definition of the Gamma function $\Gamma((v+1)/2)$. The density is known as the $t$-density with $v$ degrees of freedom, the degrees of freedom referring to the denominator $\chi^2_v$ random variable in the $t$-ratio. ■

The preceding theorem facilitates the derivation of the probability distribution of $T = n^{1/2}(\overline{X}_n - \mu)/\hat{\sigma}_n$ when random sampling is from a population distribution, where $\hat{\sigma}_n = (n/(n-1))^{1/2} S_n$. We will later find that the random variable $T$ has important applications in testing hypotheses about the value of the population mean, $\mu$.

**Theorem 6.19** ***A T-Statistic for Known $\mu$***

*Under the assumptions of Theorem 6.12 and defining $\hat{\sigma}_n \equiv (n/(n-1))^{1/2} S_n$, $T = ((n^{1/2}(\overline{X}_n - \mu))/\hat{\sigma}_n)$ has the t-density with $n - 1$ degrees of freedom.*

**Proof** The proof is based on the fact that $T$ is defined as a standard normal random variable divided by the square root of a $\chi^2_v$ random variable that has been divided by its degrees of freedom $v = n - 1$, i.e.,

$$T = \frac{\dfrac{(\overline{X}_n - \mu)}{\sigma/\sqrt{n}}}{\left(\dfrac{nS^2}{\sigma^2(n-1)}\right)^{1/2}}$$

and the two random variables are independent, so Theorem 6.18 applies. Final details are left to the reader. ■

**Example 6.16** ***Assessing a Hypothesis Based on a T-Statistic***

Recall Example 6.10, regarding the question of underfilling of containers. By Theorem 6.19, we know that $T = n^{1/2}(\overline{X}_n - \mu)/\hat{\sigma}_n$ has a $t$-distribution with 199 degrees of freedom when $n = 200$. *Assuming* the company's claim of $\mu = 32.25$ to be true, the outcome of $T$ in this case is 6.1559. The probability of obtaining an outcome $\geq 6.1559$ is $P(t \geq 6.1559) = 2.022 \times 10^{-9}$ (obtained by integrating the aforementioned $t$-distribution on a computer). The evidence suggests that the company's claim is suspect – its claimed value of $\mu$ may in fact be *too low*, and there is no support for the claim that the company is underfilling their containers. □

We present some properties of the $t$-distribution below.

#### *6.7.1.1 Family Name: t*-Family

*Parameterization*: $v \in \Omega = \{v: v \text{ is a positive integer}\}$

*Density Definition*: See Theorem 6.18.

*Moments*: $\mu = 0$ for $v > 1$, $\sigma^2 = v/(v-2)$ for $v > 2$, $\mu_3 = 0$ for $v > 3$, $\left(\frac{\mu_4}{\sigma^4} - 3\right) = \frac{6}{v-4}$ for $v>4$

*MFG*: Does not exist

As Figure 6.6 shows, the graph of the $t$-density is symmetric about zero, and when compared to the standard normal density, the $t$-density tends to have a smaller mode at zero and has fatter tails. However, as $v \to \infty$, $T_v \xrightarrow{d} N(0,1)$.

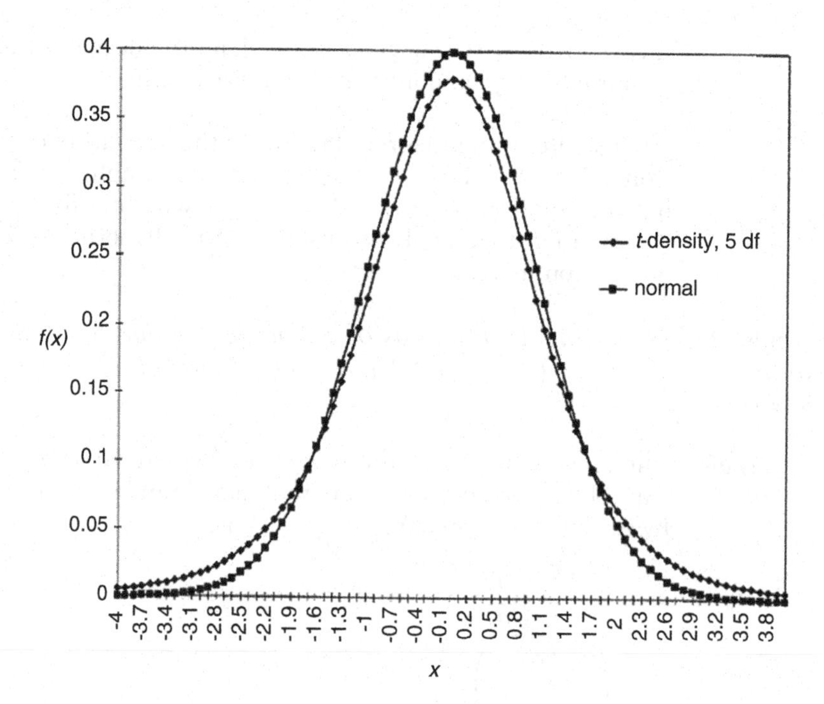


**Figure 6.6** Standard normal and $t$-density.

**Theorem 6.20** ***Convergence of t-Distribution to*** $N(0,1)$

*Let* $T_v = Z/(Y_v/v)^{1/2}$, $Z \sim N(0,1), Y_v \sim \chi^2_v$, *and let* $Z$ *and* $Y_v$ *be independent, so that* $T_v$ *has the t-density with* $v$ *degrees of freedom. Then as* $v \to \infty$, $T_v \xrightarrow{d} N(0,1)$.

**Proof** Since $Y_v \sim \chi^2_v$, then $E(Y_v) = v$ and $\text{var}(Y_v) = 2v$, and thus $E(Y_v/v) = 1$, and $\text{var}(Y_v/v) = 2v^{-1}$. It follows that $(Y_v/v) \xrightarrow{m} 1$, so that $\text{plim}(Y_v/v) = 1$. Also note that since $Z \sim N(0,1)\ \forall v$, it follows trivially that $Z \xrightarrow{d} N(0,1)$. Then, by Slutsky's theorems, $T_v = (Y_v/v)^{1/2} Z \xrightarrow{d} 1 \cdot Z \sim N(0,1)$. ■

The convergence of the $t$-density to the standard normal density is rapid, and for $v > 30$, the standard normal density provides an excellent approximation to the $t$-density. Tables of integrals of the $t$-density are widely available, generally giving the value of $c$ for which

$$P[t_v \geq c] = \int_c^\infty f_T(t; v)dt = \alpha$$

for specific choices of $\alpha$ such as .01, .025 and .05, and for selected choices of $v$, i.e., the complement of the CDF is tabled. The symmetry of the $t$-density implies that if $c$ is such that $P[t_v \geq c] = \alpha$, then $-c$ is such that $P[t_v \leq -c] = \alpha$ (see Figure 6.7). If more detail is required than what is available in the table of the $t$-density, computer programs are readily available for numerically integrating the $t$-density (e.g. GAUSS, Matlab, and SAS).

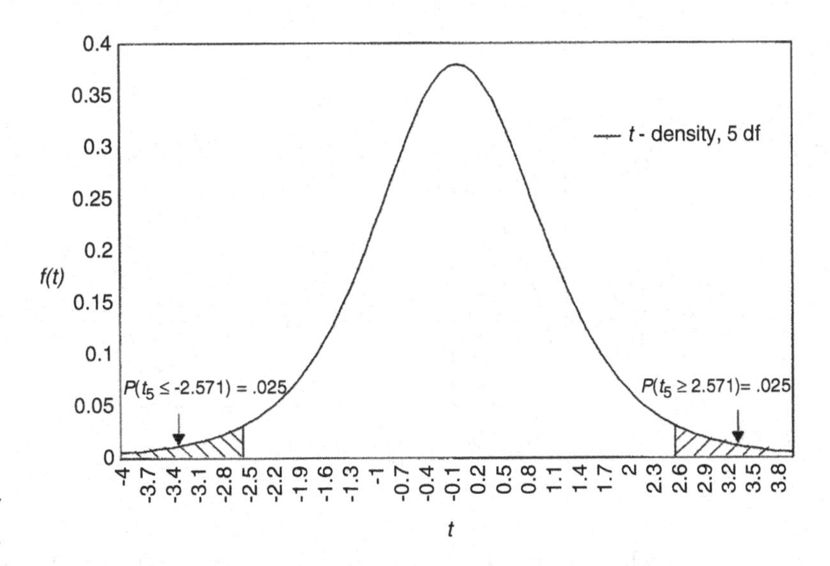


**Figure 6.7**
*t*-density symmetry.

### 6.7.2 *F*-Density

The *F*-statistic is defined as the ratio of two chisquare random variables that have each been divided by their degrees of freedom, and if the random variables are independent, the statistic has the *F*-distribution, having two parameters equal to the degrees of freedom of the chisquare random variables, as defined in the theorem below.

**Theorem 6.21**
***Definition and Derivation of the F-Density***

*Let* $Y_1 \sim \chi^2_{v_1}, Y_2 \sim \chi^2_{v_2}$, *and let* $Y_1$ *and* $Y_2$ *be independent. Then* $F = (Y_1/v_1)/(Y_2/v_2)$ *has the F-density with* $v_1$ *numerator and* $v_2$ *denominator degrees of freedom, defined as* $m(f; v_1, v_2) = \frac{\Gamma(\frac{v_1+v_2}{1})}{\Gamma(\frac{v_1}{2})\Gamma(\frac{v_2}{2})}\left(\frac{v_1}{v_2}\right)^{v_1/2} f^{(v_1/2)-1}\left(1+\frac{v_1}{v_2}f\right)^{-(1/2)(v_1+v_2)} I_{(0,\infty)}(f)$.

**Proof** Define the $(2 \times 1)$ vector function $\mathbf{g}$ as

$$\begin{bmatrix} f \\ w \end{bmatrix} = \begin{bmatrix} g_1(y_1, y_2) \\ g_2(y_1, y_2) \end{bmatrix} = \begin{bmatrix} \frac{y_1 / v_1}{y_2 / v_2} \\ y_2 \end{bmatrix},$$

where $g_2(y_1, y_2)$ is an auxiliary function defined to allow the use of Theorem 6.17. Note $\mathbf{g}$ is continuously differentiable $\forall\, y_1 > 0$ and $\forall\, y_2 > 0$, which represents an open rectangle of $(y_1, y_2)$ points containing the range of $(Y_1, Y_2)$ (their support being $X^2_{i=1}(0, \infty)$. The inverse function $\mathbf{g}^{-1}$ is defined by

$$\begin{bmatrix} y_1 \\ y_2 \end{bmatrix} = \begin{bmatrix} g_1^{-1}(f, w) \\ g_2^{-1}(f, w) \end{bmatrix} = \begin{bmatrix} \frac{v_1 f w}{v_2} \\ w \end{bmatrix} \forall (f, w) \in R(F, W).$$

The elements of the Jacobian matrix of the inverse function are continuous functions of $(f,w)$, and the absolute value of the determinant of the Jacobian is

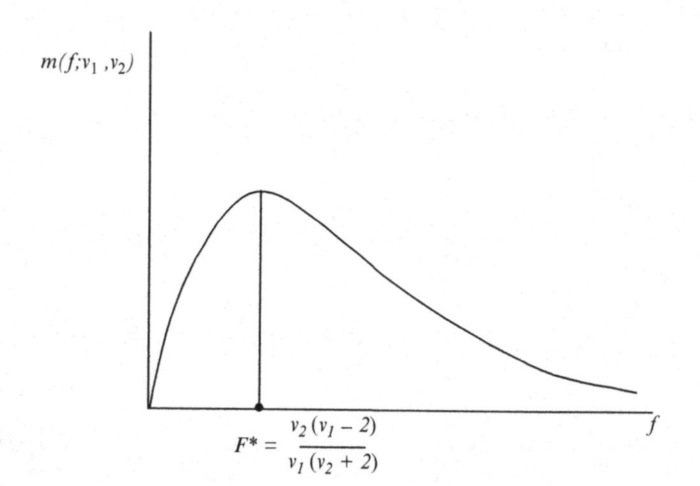


**Figure 6.8**
$F$-density for $v_1 > 2$.

such that $|\det(\mathbf{J})| = v_1 w/v_2 \neq 0 \ \forall (f,w) \in R(F,W)$. Then since $Y_1$ and $Y_2$ have independent $\chi^2$ densities with $v_1$ and $v_2$ degrees of freedom, respectively, the joint density of $(F,W)$ is given by Theorem 6.17 as

$$h(f,w) = \frac{\left(\frac{v_1}{v_2}\right)^{v_1/2} f^{(v_1/2)-1}\ w^{((v_1+v_2)/2)-1}}{2^{(v_1+v_2)/2}\Gamma(v_1/2)\Gamma(v_2/2)} \exp\left(-\frac{w}{2}\left(1+\frac{v_1 f}{v_2}\right)\right) I_{(0,\infty)}(f)\, I_{(0,\infty)}(w).$$

Substituting for $w$ using $p = (w/2)(1+(v_1 f/v_2))$, and then integrating out $p$ finally yields the marginal density of $F$, as stated in the theorem. ■

We present some useful properties of the $F$-distribution below.

### 6.7.2.1 *Family Name: F-Family*

*Parameterization*: $(v_1, v_2) \in \Omega = \{(v_1, v_2): v_1 \text{ and } v_2 \text{ are positive integers}\}$
*Density Definition*: See Theorem 6.21.
*Moments*:

$$\mu = \frac{v_2}{v_2-2} \text{ for } v_2>2, \quad \sigma^2 = \frac{2\,v_2^2\,(v_1+v_2-2)}{v_1\,(v_2-2)^2\,(v_2-4)} \text{ for } v_2>4,$$

$$\mu_3 = \left(\frac{v_2}{v_1}\right)^3 \frac{8\,v_1\,(v_1+v_2-2)(2\,v_1+v_2-2)}{(v_2-2)^3\,(v_2-4)(v_2-6)} > 0 \text{ for } v_2>6,$$

$$\left(\frac{\mu_4}{\sigma^4}-3\right) = \frac{12v_1(5v_2-22)(v_1+v_2-2)+(v_2-4)(v_2-2)^2}{v_1(v_2-6)(v_2-8)(v_1+v_2-2)} \text{ for } v_2>8$$

*MGF*: Does not exist.

The graph of the $F$-density is skewed to the right, and for $v_1 > 2$ has the typical shape shown in Figure 6.8. The mode of the density occurs at $F^* = v_2(v_1-2)/(v_1(v_2+2))$. If $v_1 = 2$, the density is monotonically decreasing, and approaches an intercept on the vertical axis equal to $\tau = 2\ v_2^{-1}\Gamma(1 + v_2/2)/\Gamma(v_2/2)$ as $F \to 0$

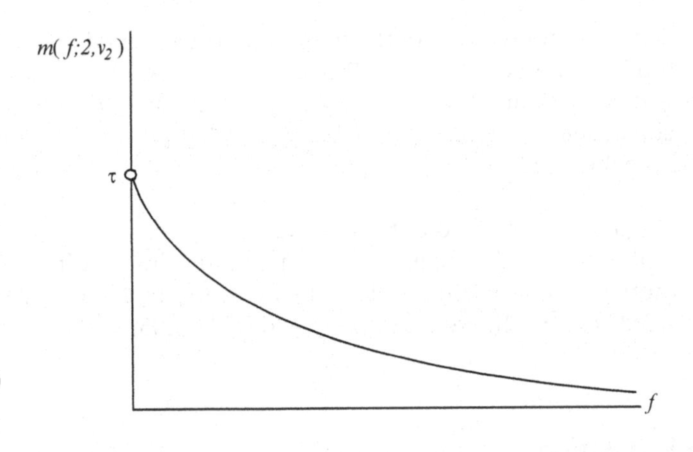


**Figure 6.9**
F-density for $v_1 = 2$.

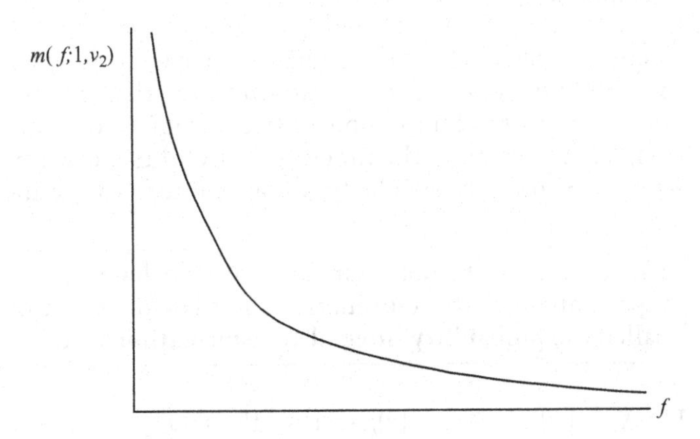


**Figure 6.10**
F-density for $v_1 = 1$.

(see Figure 6.9). If $v_1 = 1$, the density is monotonically decreasing, and approaches $\infty$ as $F \to 0$ (see Figure 6.10).

Values of $c$ for which $P[F \geq c] = \alpha$ for selected values of $v_1$, $v_2$, and $\alpha$ are available in tables of the $F$-distribution, i.e., the complement of the CDF is tabled. However, choices of $\alpha$ are very limited in these tables, generally being .05 and .01. Computer programs are available for integrating the $F$-density more generally (e.g., GAUSS, Matlab, and SAS).

It is interesting to note that if $T$ has a $t$-density with $v$ degrees of freedom, then $T^2$ has an $F$-density with 1 and $v$ degrees of freedom. The reader should verify this result, which follows directly from definitions. The reader can also verify that by letting $v_2 \to \infty$ while holding $v_1$ constant, $v_1 F \xrightarrow{d} \chi^2_{v_1}$ (a proof based on Slutsky's theorems, similar to the approach used in Theorem 6.20 can be constructed).

Finally, note that if $F_{v_1,v_2}$ denotes a random variable that has the $F$-distribution with $v_1$ (numerator) and $v_2$ (denominator) degrees of freedom, then $(F_{v_1,v_2})^{-1}$ (i.e., the *reciprocal* of $F_{v_1,v_2}$) has the $F$-distribution with $v_2$ numerator and $v_1$

denominator degrees of freedom (note the reversal of degrees of freedom), which follows immediately by definition. Therefore, if the value $c$ is such that $P(F_{v_1,v_2} \geq c) = \alpha$, then $P\left( \left(F_{v_1,v_2}\right)^{-1} \leq c^{-1}\right) = P\left( F_{v_2,v_1} \leq c^{-1}\right) = \alpha$, which allows one to construct lower-tail events having probability $\alpha$ from upper-tail events having probability $\alpha$.

**Example 6.17**
*Using F-Tables*

Suppose it is desired to find the value of $b$ for which $P(F_{2,4} \leq b) = .05$. From the tables of the $F$-density, one can obtain the result that $P(F_{4,2} \geq 19.25) = .05$ (notice the reversed order of the numerator and denominator degrees of freedom). It follows that $P(F_{2,4} \leq .0519) = .05$, where $b = (19.25)^{-1} = .0519$. □

## 6.8 Random Sample Simulation

There is a special real-valued function of a continuous random variable $X$, called the *probability integral transformation*, that is useful for simulating the outcomes of a random sample of size $n$ from a (continuous) probability distribution, $f(x)$. In addition, the function is useful in a certain goodness-of-fit test that we will examine later. The function of interest is defined as follows.

**Definition 6.15**
*Probability Integral Transformation (PIT)*

Let $X$ be a continuous random variable having a probability distribution represented by the cumulative distribution function $F$. Then $Y = F(X)$ is called the **probability integral transformation** of $X$.

**Example 6.18**
*PIT for Exponential Distribution*

Let $X \sim (1/\theta)e^{-x/\theta}\, I_{(0,\infty)}(x)$, so that the CDF of $X$ is given by $F(x) = (1 - e^{-x/\theta})\, I_{(0,\infty)}(x)$. Then the probability integral transformation of the random variable $X$ is defined by $Y = (1 - e^{-X/\theta})\, I_{(0,\infty)}(X)$. □

There is an important relationship between the probability integral transformation and the continuous uniform PDF which is identified in the following theorem.

**Theorem 6.22**
*Uniform Distribution of the PIT*

*Let $Y = F(X)$ be the probability integral transformation of the continuous random variable $X \sim f(x)$. Then $Y$ is uniformly distributed on the interval (0,1), as $Y \sim I_{(0,1)}(y)$.*

**Proof**

(a) ***Case where F is strictly increasing and continuously differentiable on an open interval containing the range of X***
Since $F$ is strictly increasing, $dF/dx = f(x) > 0\ \forall\ x$ in the range of $X$, and $x = F^{-1}(y)$ defines the inverse function $\forall y \in (0,1)$. Then, by the change-of-variables technique (Theorem 6.15), the density function for $Y$ is given by

$$h(y) = \begin{cases} f(F^{-1}(y))\left|\dfrac{dF^{-1}(y)}{dy}\right| = f(x)/f(x) = 1\ \forall y \in (0,1) \\ = 0 \text{ elsewhere} \end{cases}$$

because $dF^{-1}(y)/dy = [dF(x)/dx]^{-1} = (f(x))^{-1}$ for $y = F(x)$, so that $Y$ is uniformly distributed on the interval (0,1).

(b) ***Case where X has MGF*** $M_X(t)$
The moment generating function of the random variable defined by the probability integral transform is

$$M_Y(t) = E(e^{Yt}) = E\left(e^{F(X)t}\right) = \int_{-\infty}^{\infty} e^{F(x)t} f(x)dx.$$

Make a change of variable in the integral using $z = tF(x)$, so that $dz/dx = tf(x)$ and $dx/dz = [tf(x)]^{-1}$, yielding $M_Y(t) = \int_0^t (e^z/t)dy = t^{-1}(e^t - 1)$ for $t \neq 0$, which by the MGF uniqueness theorem identifies a uniform distribution on (0,1) (recall the MGF of the uniform density presented in Section 4.2). A general proof of the theorem along these lines can be constructed by using the characteristic function in place of the MGF above. ■

**Example 6.19** ***PIT Applied to Exponential RV***

Referring to Example 6.18, the density of $Y = F(X) = g(X)$ can be found using the change-of-variables approach. In particular, $dg/dx = (1/\theta)\, e^{-x/\theta} > 0\ \forall x \in (0,\infty)$, the inverse function $x = -\theta \ln(1 - y)$ is defined and continuously differentiable $\forall y \in (0,1)$, and $|dg^{-1}(y)/dy| = \theta/(1 - y)$. Then,

$$h(y) = \begin{cases} \dfrac{1}{\theta} \exp\left(\dfrac{\theta \ln(1-y)}{\theta}\right) \dfrac{\theta}{1-y} = 1 \text{ for } y \in (0,1) \\ 0 \text{ elsewhere} \end{cases}$$

so that $Y \sim I_{(0,1)}(y)$. □

The preceding theorem indicates how a random variable having any continuous CDF, $F$, can be transformed into a random variable having a uniform probability distribution on the interval (0,1). We now examine the converse to the result of Theorem 6.22 – a uniformly distributed random variable on the interval (0,1) can be transformed into a random variable having any continuous CDF, $F$. Specifically, if the continuous CDF $F$ admits an inverse function, then it can be shown that $X = F^{-1}(Y)$ has the CDF $F$ if $Y \sim I_{(0,1)}(y)$. Even if $F^{-1}$ does not exist, one can define a function of $Y$ that involves $F$ and that defines a random variable that has the CDF $F$. The details are provided in the next theorem.

**Theorem 6.23** ***Generating Outcomes of X ~ F using Inverse PIT and Uniform Distribution***

*Let* $Y \sim I_{(0,1)}(y)$ *and let F be a continuous CDF. If F has an inverse on the interval* (0,1), *then the inverse PIT,* $X_* = F^{-1}(Y)$, *has the CDF F. In general,* $X_* = \min_x \{x\colon F(x) \geq Y\}$ *has the CDF F.*

**Proof**

(a) ***Case where F has an inverse on* (0,1)** *Because* $F^{-1}$ exists, and since $F$ is continuous and strictly increasing, then for a given value of $y$, $\{x{:}F(x) \leq y\} = \{x\colon x \leq F^{-1}(y)\}$. The CDF of $X_* = F^{-1}(Y)$ can be represented as $G(c) = P(x_* \leq c) = P(F^{-1}(y) \leq c) = P(y \leq F(c)) = F(c)$, because $Y$ is uniformly distributed on (0,1). Therefore, $X_*$ has the CDF $F$.

**(b)** ***General*** Let $x_* = \min_x \{x: F(x) \geq y\}$. Then $P(x_* \leq c) = P(y \leq F(c)) = F(c)$ because $y \sim I_{(0,1)}(y)$, and thus $X_*$ has the CDF $F$. ■

Both of the preceding theorems have important applications. As a preview to how Theorem 6.22 is used in practice, suppose we are dealing with a random sample $(X_1,\ldots,X_n)$ and we hypothesize that the population distribution from which the sample was drawn is given by the CDF $F$. If our hypothesized distribution were the correct one, it necessarily follows from Theorem 6.22 that $Y_i = F(X_i)$, $i = 1,\ldots,n$, would constitute a random sample from a uniform population distribution on the interval (0,1). Then if the outcomes of the random sample, i.e., $y_i = F(x_i)$, $i = 1,\ldots,n$, did not exhibit behavior that would be "appropriate" for a random sample from the uniform probability distribution, our hypothesized CDF, $F$, would be suspect. We will examine how "appropriate" behavior is evaluated in this case when we examine the question of hypothesis testing.

Theorem 6.23 suggests a procedure that can be used for simulating outcomes of random samples of size $n$ from any continuous probability distribution function, $F$. Specifically, one begins with outcomes of independent uniformly distributed random variables on the interval (0,1). Computer programs are readily available that generate independent outcomes of $Y \sim I_{(0,1)}(y)$ using numerical techniques, and are commonly referred to as "uniform random number generators". Theorem 6.23 indicates that the outcome of a random sample of size $n$ from the desired population CDF $F$ can then be obtained by calculating $x_{*i} = \min_x\{x: F(x) \geq y_i\}$, $i = 1,\ldots,n$ where $(y_1,\ldots,y_n)$ are the independent outcomes of $Y \sim I_{(0,1)}(y)$ generated via the computer. It is frequently that case that the CDF $F$ has an inverse function $F^{-1}$ existing, so that the random sample can be represented more simply as $x_{*i} = F^{-1}(y_i)$, $i = 1,\ldots,n$.

**Example 6.20** ***Simulating a Random Sample from an Exponential Distribution***

Suppose we wish to simulate a random sample of size five from an exponential population distribution of computer memory chip lifetimes, where $\theta = 100$, so that $X \sim (1/100)\, e^{-x/100}\, I_{(0,\infty)}(x)$ and $F(x) = (1 - e^{-x/100}) I_{(0,\infty)}(x)$, with $x$ measured in 1,000's of hours. We use the result in Theorem 6.23, and note that for $y \in (0,1)$, $F$ has an inverse function defined by $x = -100 \ln(1 - y)$. Then, using five *iid* outcomes of $Y \sim I_{(0,1)}(y)$ generated by computer, as {.127, .871, .464, .922, .761}, we simulate a random sample outcome of five chip lifetimes as {13.582, 204.794, 62.362, 255.105, 143.129}. □

The preceding approach has wide applicability in simulating random samples from continuous probability distributions, but sometimes the procedure can be numerically complex if the inverse function is impossible to solve for explicitly, or if the solution to $\min_x\{x: F(x) \geq y\}$ is difficult to obtain. Other approaches exist for simulating outcomes of random samples, and others are being developed in the literature. For an introduction to some alternative random variable generation concepts, the interested reader can consult M. Johnson (1987), *Multivariate Statistical Simulation*. New York: John Wiley.

To this point we have not discussed a method for generating random samples from a population distribution when the distribution is discrete. In principle, the procedure in this case is relatively straightforward and, as in the continuous case, utilizes *iid* outcomes of a random variable that is uniformly distributed on (0,1). Specifically, let the range of the discrete random variable $X$ be given by $R(X) = \{x_1, x_2, x_3, \ldots\}$, where $x_1 < x_2 < x_3 < \ldots$, and let the density function of $X$ be $f(x)$. Define the function $g(y)$ as

$$g(y) = \begin{bmatrix} x_1 \\ x_2 \\ x_3 \\ \vdots \end{bmatrix} \text{if} \begin{bmatrix} 0<y\leq F(x_1) \\ F(x_1)<y\leq F(x_2) \\ F(x_2)<y\leq F(x_3) \\ \vdots \end{bmatrix}.$$

By the duality between CDFs and PDFs, this definition of $g(y)$ implies that $P(g(y) = x_i) = f(x_i)$ if $Y \sim I_{(0,1)}(y)$ since $P(y \in (F(x_{i-1}),\ F(x_i)]) = F(x_i) - F(x_{i-1}) = f(x_i)$. Thus $g(y)$, for $Y \sim I_{(0,1)}(y)$, is a discrete random variable with PDF $f$. If the range of $X$ does not exhibit a finite lower bound then all outcomes would be calculated as $g(y) = y_i$ if $F(x_{i-1})<y\leq F(X_i)$.

**Example 6.21** ***Simulating Random Samples from Bernoulli Based on Uniform***

In order to simulate a random sample of consumer buying behavior regarding a given product, assume that the appropriate probability distribution from which random sampling would occur in this case is given by the Bernoulli distribution with $p = .5$, so that the consumer is as likely to buy as not. Then define

$$g(y) = \begin{cases} 0 & \text{if } 0<y\leq .5, \\ 1 & \text{if } .5<y\leq 1, \end{cases}$$

where 1 denotes a purchase, and 0 denotes no purchase. Utilizing the numbers {.217, .766, .822, .402, .674} generated via a computer-based uniform random-number generator for *iid* outcomes of $Y \sim I_{(0,1)}(y)$, we calculate a simulated random sample of consumer purchasing decisions as {0,1,1,0,1}. □

## 6.9 Order Statistics

Situations arise in practice where the relative magnitudes of observations are of primary interest. For example, the *largest* value in a sample of observations may be of particular interest, such as if a business is deciding on the appropriate level of capacity in order to successfully service the demands of customers over a number of operating periods. The *smallest* observation may also be of interest, such as in the manufacture of a consumer electronics product that will properly function only so long as the most short-lived critical electronics component in its circuitry. One might also be interested in the median value of sample observations as a measure of central tendency of the observations, or the sample range of observations as a measure of spread. All of these statistics are examples of *order statistics* or functions of order statistics, which also play an important role in nonparametric hypothesis testing, which will be introduced in Chapter 10.

The **order statistics** corresponding to a random sample $(X_1,\ldots,X_n)$ are simply the $X_i$'s arranged in order of increasing magnitude. The formal definition is as follows.

**Definition 6.16**
***Order Statistics***

Let sort(**x**) be the $n \times 1$ vector function whose value is defined by sorting the elements of the $n \times 1$ vector $\mathbf{x} = [x_1,\ldots,x_n]'$ from the lowest to the highest values. The $n \times 1$ vector of order statistics $\mathbf{X}_o = [X_{[1]},\ldots,X_{[n]}]'$ corresponding to the random sample $\mathbf{X} = [X_1,\ldots,X_n]'$ is defined by $\mathbf{X}_o = \text{sort}(\mathbf{X})$, and has an outcome $\mathbf{x}_o = \text{sort}(\mathbf{x})$. The random variable $X_{[k]}$ is called the $k$th **order statistic**.

Note that the order statistics are indeed *statistics*, since they are defined as (vector) functions of the random sample. It is apparent from the definition that outcomes of order statistics satisfy the inequalities $x_{[1]} \leq x_{[2]} \leq \ldots \leq x_{[n]}$.

We will examine the sampling distribution of order statistics under the assumption that we are random sampling *from a population distribution*, i.e., the $X_i$'s are *iid*. In this case, the following result is available concerning the sampling distribution of the $k$th order statistic.

**Theorem 6.24**
***Sampling Distribution of the $k^{th}$ Order Statistic $X_{[k]}$***

*Let $(X_1,\ldots,X_n)$ be a random sample from a population distribution with CDF F, and let $X_{[k]}$ be the $k$th order statistic corresponding to the random sample. Then the CDF of $X_{[k]}$ is given by*

$$F_{X_{[k]}}(b) = \sum_{j=k}^{n} \binom{n}{j} F(b)^j \, [1 - F(b)]^{n-j}.$$

**Proof** For a given value of $b$, define the random variable $Y_i = I_{(-\infty,b]}(X_i)$. Note that $Y_i$ has a Bernoulli distribution with $p = P(y_i = 1) = P(x_i \leq b) = F(b)$. Since the $Y_i$'s are *iid*, it follows that $\sum_{i=1}^{n} Y_i$ has the binomial distribution with parameters $n$ and $p$.

Now note the equivalence of the following events:

$$\{(x_1,\ldots,x_n) : x_{[k]} \leq b\} = \left\{(x_1,\ldots,x_n) : \sum_{i=1}^{n} I_{(-\infty,b]}(x_i) \geq k\right\}.$$

The event to the left of the equality corresponds to the situation where the $k$th largest outcome in $(x_1,\ldots,x_n)$ is less than or equal to $b$. This event can happen *iff* at least $k$ outcomes in $(x_1,\ldots,x_n)$ are less than or equal to $b$, which is the event to the right of the equality. Hence, the events are equivalent. It then follows from the **binomial** distribution of $\sum_{i=1}^{n} Y_i$ that

$$P(x_{[k]} \leq b) = P\left(\sum_{i=1}^{n} y_i \geq k\right) = \sum_{j=k}^{n} \binom{n}{j} F(b)^j [1 - F(b)]^{n-j}. \qquad \blacksquare$$

The CDF identified in Theorem 6.24 simplifies considerably in two cases of particular relevance in applications – the smallest and the largest order statistic.

**Corollary 6.1** ***Sampling Distributions of $X_{[1]}$ and $X_{[n]}$*** *Assume the conditions of Theorem 6.24. Then $F_{X_{[1]}}(b) = 1 - [1 - F(b)]^n$ and $F_{X_{[n]}}(b) = F(b)^n$.*

**Proof** It follows from Theorem 6.24 that $F_{X_{[1]}}(b) = \sum_{j=1}^{n} \binom{n}{j} F(b)^j [1 - F(b)]^{n-j} = 1 - [1 - F(b)]^n$, where the second equality follows from the fact that $\sum_{j=0}^{n} \binom{n}{j} F(b)^j [1 - F(b)]^{n-j} = 1$ because we are summing a binomial density over all the values in the support of the random variable. Also, by direct evaluation of the CDF, $F_{X_{[n]}}(b) = \sum_{j=n}^{n} \binom{n}{j} F(b)^j (1 - F(b))^{n-j} = [F(b)]^n$. $\blacksquare$

**Example 6.22** ***Probability of Largest Waiting Time*** The waiting time between customer arrivals at the pharmacy department in a variety store is given by the exponential PDF $f(x) = .2e^{-.2x} I_{(0,\infty)}(x)$, where $x$ is measured in minutes. In a random sample of 10 customers, what is the probability that the smallest waiting time will be greater than 2 minutes? What is the probability that the largest waiting time will be no greater than 5 minutes?

The CDF of the exponential population distribution is given by $F(b) = 1 - e^{-.2b}$. Then

$$P(x_{[1]}>2) = 1 - P(x_{[1]} \leq 2) = 1 - F_{X_{[1]}}(2)$$
$$= \left[1 - \left(1 - e^{-.2(2)}\right)\right]^{10} = .0183,$$
$$P(x_{[10]} \leq 5) = F_{X_{[10]}}(5) = \left[1 - e^{-.2(5)}\right]^{10} = .0102. \qquad \square$$

The PDFs of the order statistics can be found using the duality between CDFs and PDFs. If the population distribution, $f(x)$, is discrete with support $x_1 < x_2 < \ldots < x_n$, then the PDF of the $k$th order statistic can be defined in the usual way as $f_{X_{[k]}}(x_i) = F_{X_{[k]}}(x_i) - F_{X_{[k]}}(x_{i-1})$ for $i \geq 2$ with $f_{X_{[k]}}(x_1) = F_{X_{[k]}}(x_1)$. In the continuous case, the PDF can be found after differentiation of the CDF, and some algebraic manipulation, to be

$$f_{X_{[k]}}(x) = \frac{dF_{X_k}(x)}{dx} = \frac{n!}{(k-1)!(n-k)!} f(x)F(x)^{k-1} [1 - F(x)]^{n-k}.$$

In order to assign probabilities to events involving outcomes of the **sample range**, i.e., $X_{[n]} - X_{[1]}$, or to make a joint probability statement concerning the outcomes of the largest and smallest observations in a random sample, the joint sampling distribution of $(X_{[1]}, X_{[n]})$ is needed. Furthermore, if the sample size $n$ is an even number, and one wishes to assign probabilities to events involving the **sample median** defined as $\widehat{\text{med}}(X) = (X_{[k]} + X_{[k+1]})/2$ for $k = n/2$, the joint sampling distribution of $(X_{[k]}, X_{[k+1]})$ is needed. (When $n$ is odd, the **sample median** is defined as $\widehat{\text{med}}(X) = X_{[k]}$ where $k = (n + 1)/2$, and so Theorem 6.24 covers this case.) The sampling distribution for any pair of order statistics is given as follows.

**Theorem 6.25 *Sampling Distribution of Pairs of Order Statistics* $(X_{[k]}, X_{[\ell]})$**

*Let $X_{[k]}$ and $X_{[\ell]}$, $k < \ell$, be the kth and ℓth order statistics corresponding to the random sample $X = [X_1, \ldots, X_n]'$ from a population distribution with CDF F and PDF f. Then the joint CDF of $(X_{[k]}, X_{[\ell]})$ is given by*

$$F_{X_{[k]},X_{[\ell]}}(b_k, b_\ell) = \begin{cases} F_{X_{[\ell]}}(b_\ell) \text{ for } b_k \geq b_\ell, \\ \sum_{i=k}^{n} \sum_{j=\max\{0,\ell-i\}}^{n-i} \frac{n!}{i!j!(n-i-j)!} F(b_k)^i [F(b_\ell) - F(b_k)]^j [1 - F(b_\ell)]^{n-i-j} \text{ for } b_k < b_\ell \end{cases}$$

**Proof** Given $k < \ell$, it follows by definition of the order statistics that $b_k \geq b_\ell$ implies $\{x: x_{[\ell]} \leq b_\ell\} \subset \{x: x_{[k]} \leq b_k\}$, so that

$$F_{X_{[k]},X_{[\ell]}}(b_k, b_\ell) = P(x_{[k]} \leq b_k, x_{[\ell]} \leq b_\ell) = P(x_{[\ell]} \leq b_\ell) = F_{x_{[\ell]}}(b_\ell)$$

proving the first part of the definition of the CDF.

When $b_k < b_\ell$, note that the event $\{x_{[k]} \leq b_k, x_\ell \leq b_\ell\}$ corresponds to the event that at least $k$ of the random sample outcomes $x_1, \ldots, x_n$ are less than or equal to $b_k$ and at least $\ell$ are $\leq b_\ell$. Defining the index set $I = \{(i,j)\text{: } \max\{0, \ell - i\} \leq j \leq n - i;\ k \leq i \leq n;\ i \text{ and } j \text{ are integers}\}$, we can then represent the event as

$$\{x_{[k]} \leq b_k, x_{[\ell]} \leq b_\ell\} = \bigcup_{(i,j)\in I} \{\text{exactly } i \text{ x}'_i\text{s} \leq \text{b}_\text{k}, \text{ exactly j x}'_i\text{s such that } \text{b}_\text{k} < \text{x}_\text{i} \leq b_\ell\}$$

Now note that each of the disjoint events involved in the union operation can be assigned probability via the multinomial distribution, where the outcome of each $X_i$ in the random sample is categorized into one of three types $x_i \leq b_k$, $x_i \in (b_k, b_\ell]$, and $x_i > b_\ell$ which occur with probabilities $F(b_k)$, $F(b_\ell) - F(b_k)$, and $1 - F(b_\ell)$, respectively. Then applying the multinomial distribution (Section 4.1) and summing the probabilities of all of the disjoint events in the union operation yield the second part of the definition of the CDF. ∎

As indicated following Theorem 6.24, the PDF of $(X_{[k]}, X_{[\ell]})$ can be obtained through the duality between CDFs and PDFs. In the discrete case, the PDF would be obtained by appropriate differencing of the CDF, while in the continuous case, the CDF would be differentiated (recall Theorems 2.4 and 2.5). Details are left to the reader.

The CDF simplifies considerably for the case of the joint distribution of the **extreme values** $(X_{[1]}, X_{[n]})$.

**Corollary 6.2 *Sampling Distribution of Extreme Values*** $(X_{[1]}, X_{[n]})$

*Let $k = 1$ and $\ell = n$ in Theorem 6.25. Then by the binomial theorem,*

$$F_{X_{[1]},X_{[n]}}(b_1, b_n) = \begin{cases} F(b_n)^n \text{ for } b_1 \geq b_n \\ F(b_n)^n - [F(b_n) - F(b_1)]^n \text{ for } b_1 < b_n. \end{cases}$$ □

**Example 6.23 *Probability of Minimum and Maximum Waiting Times***

Consider Example 6.22 regarding waiting time between customer arrivals. In a sample of 10 customer arrivals, what is the probability that the minimum waiting time will be $\leq 4$ minutes and the maximum waiting time will be $\leq 8$ minutes?

The probability can be calculated by evaluating the joint CDF of the extreme values identified in Corollary 6.2, yielding

$$F_{X_{[1]},X_{[10]}}(4, 8) = F(8)^{10} - [F(8) - F(4)]^{10}$$
$$= \left[1 - e^{-.2(8)}\right]^{10} - \left[e^{-.2(4)} - e^{-.2(8)}\right]^{10} = .1049$$ □

Sampling distributions for functions of order statistics, such as the sample range and the sample median when $n$ is even, can be pursued using the change-of-variable approach presented in Section 6.6. The multinomial logic of the proof of Theorem 6.25 can be extended to derive joint densities of three or more order statistics. For additional details on properties and functions of order statistics, see M. Kendall and A. Stuart, (1977) *Advanced Theory*, Vol. 1, Chapter 14, and the references therein.

## Keywords, Phrases, and Symbols

CDF approach
Change of variables technique
Empirical distribution function
Empirical substitution principle
Equivalent events approach
*F*-density
General random sampling
Glivenko-Cantelli theorem
Jacobian matrix
Joint density of the random sample
Joint sample moment about the mean
Joint sample moment about the origin
Log-normal distribution
MGF approach
Order statistics
Outcome of the random sample
Population
Population distribution
Population moment
Probability integral transformation
Random sample
Random sampling from a population distribution
Random sampling with replacement
Random sampling without replacement
$r$th sample moment about the mean, $M_r$
$r$th sample moment about the origin, $M'_r$
Sample correlation matrix
Sample correlation, $R_{XY}$
Sample covariance matrix
Sample covariance, $S_{XY}$
Sample mean, $\overline{X}_n$
Sample standard deviation
Sample variance, $S_n^2$
Sampling density or sampling distribution of a function of a random sample
Sampling error
Simple random sampling
Simulation
Statistic
Stochastic process
Support of $f(x)$
Support of the density
*t*-density

## Problems

**1.** Let $X$ be a random sample of size $n$ from a $N(\mu,\sigma^2)$ population distribution representing the weights, in ounces, of cereal placed in cereal boxes for a certain brand and type of breakfast cereal. Define $\hat{\sigma}$ as in Theorem 6.19.

(a) Show that the random variable $T = n^{1/2}(\overline{X} - \mu)/\hat{\sigma}$ has the *t*-distribution with $n$-1 degrees of freedom.

(b) Let $n = 25$. What is the probability that the *random interval* $(\overline{X} - 2.06\hat{\sigma}/n^{1/2},\ \overline{X} + 2.06\hat{\sigma}/n^{1/2})$ will have an outcome that contains the value of $\mu$? (This random interval is an example of a **confidence interval** – in this case for the population mean $\mu$. See Section 10.6.)

(c) Suppose that $\overline{x} = 16.3$ and $s^2 = .01$. Define a *confidence interval* that is designed to have a .90 probability of generating an outcome that contains the value of the population mean weight of cereal placed in the cereal boxes. Generate a confidence interval outcome for the mean weight.

**2.** Let $X$ and $Y$ be two independent random samples of sizes $n_x$ and $n_y$, respectively, from two normal population distributions that do not necessarily have the same means or variances. The two distributions refer to the miles per gallon achieved by two 1/2-ton pickup trucks produced by two rival Detroit manufacturers. Define $\hat{\sigma}$ as in Theorem 6.19.

(a) Show that the random variable $F = (\hat{\sigma}^2_X / \sigma^2_X)/(\hat{\sigma}^2_Y / \sigma^2_Y)$ has the *F*-distribution with $(n_x - 1)$ numerator and $(n_y - 1)$ denominator degrees of freedom.

(b) Let $n_x = 21$ and $n_y = 31$. What is the probability that the *random interval* $\left(.49(\hat{\sigma}^2_Y/\hat{\sigma}^2_X), 1.93(\hat{\sigma}^2_Y/\hat{\sigma}^2_X)\right)$ will have an outcome that will contain the value of the ratio of the variances $\sigma_Y{}^2/\sigma_X{}^2$? (This random interval is another example of a **confidence interval** – in this case for the ratio of the population variances $\sigma_Y{}^2/\sigma_X{}^2$.)

(c) Suppose that $s_x{}^2 = .25$ and $s_y{}^2 = .04$. Define a *confidence interval* that is designed to have a .98 probability of generating an outcome that contains the value of the ratio of population variances associated with the miles per gallon achieved by the two pickup trucks. Generate a confidence interval outcome for the ratio of variances.

**3.** Let $X$ be a random sample of size n from a $N(\mu,\sigma^2)$ population distribution representing the yield per acre, in pounds, of a new strain of hops used in the production of premium beer.

(a) Justify that the random interval $(nS^2/\chi^2_\alpha, nS^2/\chi^2_{1-\alpha})$ will have an outcome that contains the value of the population variance $\sigma^2$ with probability $\alpha$, where $\chi^2_\delta$ is a number for which $P(x > \chi^2_\delta) = \delta$ when $X$ has a $\chi^2$ distribution with $n - 1$ degrees of freedom.

(b) Suppose that $s^2 = 9$ and $n = 20$. Define a *confidence interval* that is designed to have a .95 probability of generating an outcome that contains the value of the population variance of hop yields, $\sigma^2$. Generate a confidence interval outcome for the variance.

**4.** The shipping and receiving department of a large toy manufacturer is contemplating two strategies for sampling incoming parts deliveries and estimating the proportion, $p$, of defective parts in a shipment. The two strategies are differentiated on the basis of whether random sampling will be *with* or *without* replacement. In each case, a sample mean will be calculated and used as an estimate of the proportion of defective parts in the shipment. The department wants to use the strategy that will generate estimates that are smallest in expected squared distance from $p$.

(a) Compare the means and variances of the sample mean under both sampling strategies. Which strategy should be used?

(b) Describe conditions under which there will be little difference between the two methods in terms of expected squared distance from the true proportion of defectives.

(c) Do the sample means converge in probability to $p$ in each case? Do they converge in mean square? Explain.

(d) If a shipment contains 250 parts of which 10 percent are defective, and if a random sample of 50 will be taken from the shipment, calculate the percentage reduction in expected squared distance that can be obtained by using the better strategy.

**5.** GenAG, Inc., a genetics engineering laboratory specializing in the production of better seed varieties for commercial agriculture, is analyzing the yield response to fertilizer application for a new variety of overlineley that it has developed. GenAg has planted 40 acres of the new overlineley variety and has applied a different fixed level of fertilizer to each one-acre plot. In all other respects the cultivation of the crop was identical. The GenAg scientists maintain that the relationship between observed levels of yield, in bushels per acre, and the level of fertilizer applied, in pounds per acre, will be a quadratic relationship as $Y_j = \beta_0 + \beta_1 f_j + \beta_2 f_j^{\,2} + V_j$, where $f_j$ is the level of fertilizer applied to the $j$th one-acre plot, the β's are fixed parameters, and the $V_j$'s are *iid* random variables with some continuous probability density function for which $EV_j = 0$ and $\text{var}(V_j) = \sigma^2$.

(a) Given GenAg's assumptions, is $(Y_1,\ldots,Y_{40})$ a random sample from a population distribution or more general random sampling? Explain.

(b) Express the mean and variance of the sample mean $\overline{Y}_{40}$ as a function of the parameters and $f_j$ variables. If the sample size could be increased without bound, would $\overline{Y}_n$ converge in probability to some constant? Explain.

(c) Is it true that $\left(\overline{Y}_n - \left(\beta_0 + \beta_1 n^{-1}\sum_{i=1}^{n} f_i + \beta_2 n^{-1}\sum_{i=1}^{n} f_i^2\right)\right) \xrightarrow{p} 0$? If so, interpret the meaning of this result. Based on your analysis to this point, does it appear that an outcome of $\overline{Y}_{40}$ will produce a meaningful estimate of any characteristic of the yield process?

(d) Suppose that the 40 one-acre plots were all contiguous on a given 40 acre plot of land. Might there be reasons for questioning the assumption that the $V_j$'s are *iid*? What would the outcome of $V_j$ represent in GenAg's representation of the yield process? Presuming that the $V_j$'s were not *iid*, would this change your answer to (a)?

**6.** The Always Ready Battery Co. has developed a new "Failsafe" battery which incorporates a small secondary battery that becomes immediately functional upon failure of the main battery. The operating life of the main battery is a gamma distributed random variable as $X_1 \sim \text{Gamma}(3,1)$ where $X_1$ is measured in years. The operating life of the secondary battery is also gamma distributed as $X_2 \sim \text{Gamma}(2,1)$. The operating lives of the main and secondary batteries are independent.

(a) Let $Y_1 = X_1 + X_2$ represent the total operating life of the Failsafe battery, and $Y_2 = X_1/(X_1 + X_2)$ represent the proportion of total operating life that is contributed by the main battery. Derive the joint probability distribution of $(Y_1, Y_2)$.

(b) Are $Y_1$ and $Y_2$ independent random variables?

(c) Define the marginal densities of $Y_1$ and $Y_2$ . What specific families of densities do these marginal densities belong to?

(d) What is the expected proportion of total operating life contributed by the main battery? What is the probability that the secondary battery contributes more than 50 percent of the total operating life of the Failsafe battery?

(e) What is the expected total operating life of the Failsafe battery?

**7.** The seasonal catch of a commercial fishing vessel in a certain fishery in the southern hemisphere can be represented by $Q = c(\mathbf{z})\ V$, where $\mathbf{z}$ is a vector of characteristics of the vessel relating to tonnage, length, number of crew members, holding tank size, etc., $c(\mathbf{z})$ represents maximum fishing capacity of the boat, $Q$ represents the tons of fish caught, and $V \sim \theta\ v^{\theta-1} I_{(0,1)}(v)$ represents the proportion of fishing capacity realized.

(a) Derive the density function of seasonal catch.

(b) If $\theta = 10$ and $c(\mathbf{z}) = 5{,}000$, what is the expected value of seasonal catch?

**8.** A company markets its line of products directly to consumers through telephone solicitation. Salespersons are given a base pay that depends on the number of documented phone calls made plus incentive pay for each phone call that results in a sale. It can be assumed that the number of phone calls that result in sale is a binomial random variable with parameters $p$ (probability of sale) and $n$ (number of phone calls). The base pay is \$.50 per call and the incentive pay is \$5.00 per sale.

(a) Derive the probability distribution of pay received by a salesperson making n calls in a day.

(b) Given that 100 calls are made in a day and $p = .05$, what is the expected pay of the salesperson? What is the probability that pay will be \$50 or less?

**9.** The daily quantity of a commodity that can be produced using a certain type of production technology is given by the outcome of the random variable $Q$, defined as $Q = 10 x_1^{.35} x_2^{.5} V$, where $Q$ is measured in tons/day, $x_1$ represents units of labor per day, $x_2$ represents units of capital per day, and $v = e^{\varepsilon}$, with $\varepsilon \sim N(0, \sigma^2)$.

(a) Derive the probability density function of the random variable $V$. (What you will have derived is a PDF that is a member of the "lognormal" family of densities. In general, if $X \sim N(\mu, \sigma^2)$, $Y = e^X \sim$ lognormal with mean $= \exp(\mu + \sigma^2/2)$, variance $= \exp(2\,\mu + 2\sigma^2) - \exp(2\,\mu + \sigma^2)$, and $\mu'_r = \exp[r\mu + 1/2\, r^2\sigma^2]$).

(b) Derive the density of $Q$ if $x_1 = x_2 = 2$.

(c) Define the expected value of $Q$ in terms of the levels of $x_1$, $x_2$ and $\sigma^2$. What is the expected value of $Q$ if $x_1 = x_2 = 2$ and $\sigma^2 = 1$?

(d) The above production technology is used in 1,600 plants in a country in Eastern Europe. The economy is centrally planned in that country, and all of the plants are required to use labor and capital at the levels $x_1 = 7.24579$ and $x_2 = 4$. Assume that var($\varepsilon$) = .25. An economist says that the aggregate daily production function

$$Q^* = \sum_{i=1}^{1600} Q_i = 10 x_1^{.35} x_2^{.5} \sum_{i=1}^{1600} V_i$$

is such that aggregate daily production, $Q^*$, can be considered to be approximately *normally* distributed. Do you agree? Justify or refute the economist's proposition. You may assume that $V_i$'s, and hence the $Q_i$'s, are jointly independent r.v.s.

(e) Define the appropriate Berry-Esseen inequality bound on the approximation error corresponding to a CLT applicable to part (d).

**10.** The daily price, $p$, and quantity sold, $q$, of ground beef produced by the Red Meat Co. can be represented by outcomes of the bivariate random variable $(P,Q)$ having bivariate density function

$$f(p,q) = 2pe^{-pq} I_{[.5,1]}(p) I_{(0,\infty)}(q)$$

where $p$ is measured in dollars and $q$ is measured in 1,000's of pounds.

(a) Derive the probability density function for $R = PQ$, where outcomes of $R$ represent daily revenue from ground beef sales. (Hint: define $W = P$ as an "auxiliary" random variable and use the change of variable approach).

(b) What is the expected value of daily revenue? What is the probability that daily revenue exceeds \$1,000?

**11.** Let $X = (X_1,\ldots,X_{26})$ and $Y = (Y_1,\ldots,Y_{31})$ represent two independent random samples from two normal population distributions. Let $S_X^2$ and $S_Y^2$ represent the sample variances associated with the two random samples and let $\overline{X}$ and $\overline{Y}$ represent the respective sample means. Define $\hat{\sigma}$'s as in Theorem 6.19.

(a) What is the value of $P\left(\frac{|\overline{X} - E\overline{X}|}{(\hat{\sigma}_X^2/26)^{1/2}} \leq 1.316\right)$?

(b) What is the value of $P\left(\frac{26\, s_X^2}{\sigma_X^2} > 37.652\right)$?

(c) What is the value of $P(s_X^2 > 6.02432)$ assuming $\sigma_X^2 = 4$?

(d) What is the value of $P(s_Y^2 > 1.92\, s_X^2)$, assuming that $\sigma_X^2 = \sigma_Y^2$?

(e) Find the value of c for which the following probability statement is true:

$$P\left(\frac{\sigma_Y^2\, \hat{\sigma}_X^2}{\sigma_X^2\, \hat{\sigma}_Y^2} \leq c\right) = .05.$$

**12.** The daily price, $p$, and daily quantity sold, $q$, of pink salmon produced by the AJAX Fish Packing Co. can be represented by outcomes of the bivariate random variable $(P,Q)$ with density function

$$f(p,q) = 5pe^{-pq} I_{[.2,.4]}(p) I_{(0,\infty)}(q)$$

where $p$ is measured in dollars, and $q$ is measured in 1,000's of pounds.

(a) Derive the density function for $R = PQ$, where outcomes of $R$ represent daily revenue from fish sales. (Hint: Define $W = P$ as an auxiliary random variable, and use the change of variables approach).

(b) What is the expected value of daily revenue? What is the probability that daily revenue exceeds \$300?

**13.** The probability that a customer entering an electronics store will make a purchase is equal to $p = .15$, and customers' decisions whether to purchase electronics equipment are jointly independent random variables.

(a) Simulate the buying behavior of 10 customers entering the store using the following 10 outcomes from a Uniform(0,1) computer random number generator:
(.4194,.3454,.8133,.1770,.5761,.6869,.5394,.5098,.4966,.5264).

(b) Calculate the sample mean and sample variance, and compare them to the appropriate population mean and variance.

**14.** Under the conditions of the previous problem:

(a) Of the first 10 customers that enter the store on 10 consecutive days, simulate the daily number of customers that make a purchase. Use the following 10 outcomes from a Uniform(0,1) computer random number generator:
(.0288,.7936,.8807,.4055,.6605,.3188,.6717,.2329,.1896,.8719).

(b) Calculate the sample mean and sample variance, and compare them to the appropriate population mean and variance.

**15.** The number of times that a copy machine malfunctions in a day is the outcome of a Poisson process with $\lambda = .1$.

(a) Simulate the operating behavior of the copy machine regarding the daily number of malfunctions over a 10 day period using the following 10 outcomes from a Uniform(0,1) computer random number generator:
(.5263,.8270,.8509,.1044,.6216,.9214,.1665, .5079,.1715,.1726)

(b) Calculate the sample mean and sample variance, and compare them to the appropriate population mean and variance.

**16.** The length of time that a 4 gigabyte PC computer memory module operates until failure is the outcome of an exponential random variable with mean $EX = 3.25$, where $x$ is measured in 100,000 hour units.

(a) Simulate the operating lives of 10 memory modules using the following 10 outcomes from a Uniform(0,1) computer random number generator:
(.2558,.5938,.1424,.9476,.5748,.8641,.0968, .5839,.3201,.1577).

(b) Calculate the sample mean and sample variance, and compare them to the appropriate population mean and variance.

**17.** The monthly proportion of purchases paid by check to a large grocery store that are returned because of insufficient funds can be viewed as the outcome of a Beta(1, 20) distribution.

(a) Simulate 12 monthly proportions of returned checks using the following 24 outcomes from a Uniform(0,1) computer random number generator:
(.6829,.4283,.0505,.7314,.8538,.6762,.6895,.9955,.2201, .9144,.3982,.9574, .0801,.6117,.3706,.2936,.2799,.3900,.7533,.0113,.5659, .9063,.5029,.6385)
(Hint: In a previous problem you have proven that $Y = X_1/(X_1 + X_2)$ has a beta distribution with parameters $(a,b)$ if $(X_1, X_2)$ are independent gamma-distributed random variables with parameters $(a, \beta)$ and $(b, \beta)$, respectively.)

(b) Calculate the sample mean and sample variance, and compare them to the appropriate population mean and variance.

**18.** The daily closing price for a certain stock issue on the NYSE can be represented as the outcome of $Y_t = Y_{t-1} + V_t$, where $y_t$ is the value of the stock price on day $t$, and $V_t \sim N(0,4)$ (This is an example of a stochastic process known as a **random walk.**)

(a) Use the change of variable approach to verify that if $(U_1, U_2)$ are independent and identically distributed Uniform(0,1) random variables, then

$$V_1 = [-2 \ln (U_1)]^{.5} \cos(2\pi U_2) \text{ and}$$
$$V_2 = [-2 \ln (U_1)]^{.5} \sin(2\pi U_2)$$

are independent and identically distributed $N(0,1)$ random variables.

(b) Simulate 10 days worth of stock prices $(y_1,\ldots,y_{10})$ using $y_0 = 50$, the result in (a), and the following 10 outcomes from a Uniform(0,1) computer random number generator:
(.9913,.4661,.1018,.0988,.4081,.3422,.1585,.6351,.0634, .4931).

(c) Are you simulating a random sample from a population distribution or is this more general random sampling? Calculate the sample mean and sample variance and compare them to whatever characteristics of the *random walk* process that you feel is appropriate.

**19.** A random sample of the gas mileage achieved by 20 domestic compact automobiles resulted in the following outcome:
(25.52,24.90,22.24,22.36,26.62,23.46,25.46,24.98,25.82, 26.10,21.59,22.89,27.82,22.40,23.98,27.77,23.29,24.57, 23.97,24.70).

(a) Define and graph the empirical distribution function.

(b) What is the estimated probability that gas mileage will exceed 26 miles per gallon?

(c) What is the estimated probability that gas mileage will be between 24 and 26 miles per gallon?

(d) Acting as if the EDF is the true CDF of gas mileages, calculate the expected value of gas mileage. Is the value you calculated equal to the sample mean? Why or why not?

**20.** The time between work-related injuries at the Imperial Tool and Die Co. during a given span of time resulted in the following 20 observations, where time was measured in weeks:
(9.68,6.97,7.08,.50,6.71,1.13,2.20,9.98,4.63,7.59,3.99,3.26, .92,3.07,17.96,4.69,1.80,8.73,18.13,4.02).

(a) Define and graph the empirical distribution function.

(b) What is the estimated probability that there will be at least 8 weeks between work-related injuries?

(c) What is the probability that there will be between 4 and 8 weeks between work-related injuries?

(d) Acting as if the EDF is the true CDF of time between work-related injuries, calculate the expected value of time between injuries. Is the value you calculated equal to the sample mean? Why or why not?

**21.** A realtor randomly samples homeowners who have purchased homes in the last 2 years and records their income, $y$, and home purchase price, $p$ (the population is large enough that one can consider this a random sample with replacement):

| Income | Price | Income | Price |
|---|---|---|---|
| 21,256 | 49,412 | 37,589 | 74,574 |
| 97,530 | 170,249 | 137,557 | 232,097 |
| 24,759 | 56,856 | 67,598 | 124,309 |
| 18,369 | 45,828 | 83,198 | 144,103 |
| 35,890 | 73,703 | 46,873 | 92,600 |
| 38,749 | 80,050 | 24,897 | 61,763 |
| 57,893 | 11,0658 | 36,954 | 77,971 |

(a) Calculate the sample covariance between income and home price.

(b) Calculate the sample correlation between income and home price.

(c) Calculate the linear function of income of the form $\hat{p} = a + by$ that is minimum distance from the home price observations.

(d) Discuss the extent to which there is a linear relationship between income and home price.

**22.** The proportion of the work force of a large Detroit manufacturing firm that takes at least 1 day's sick leave in a given work week is assumed to be the outcome of a random variable whose PDF is well-represented by a uniform distribution on the interval [0, .10].

(a) In a random sample of eight work weeks, what is the probability that the maximum proportion of workers who take sick leave is $\leq .05$?

(b) What is the probability that the sum of all eight sample observations will be between .25 and .75?

**23.** A large aircraft manufacturer produces a passenger jet having a navigation component consisting of three sequentially functioning redundant navigation systems that will allow the jet to be properly controlled so long as at least one of the systems remain operational. The operating life of each of the systems is the outcome of a random variable having the exponential PDF $f(x) = .1\, e^{-.1x}\, I_{(0,\infty)}(x)$ where $x$ is measured in 1,000's of hours and the operating lives are independent of one another.

(a) What is the probability that the navigation component will continue to function for at least 20,000 hours?

(b) In an economizing mode, a redesign of the jet is being considered that will reduce the navigation component from three redundant systems to two. How does this affect the probability of the event in (a)?

**24.** A news agency wants to poll the population of registered voters in the United States (over 200,000,000) to find out how many would vote for the Republican candidate, Henry Washington, if the election were held today. They intend to take a random sample, with replacement, of 1,000 registered U.S. voters, record their preferences for ($y_i = 1$) or against ($y_i = 0$) Mr. Washington, and then use the 1,000 sampled outcomes to estimate the proportion of registered voters in favor of the candidate. They conduct the random sampling, and observe that $\sum_{i=1}^{n} y_i = 593$. Given their random sampling design, they are assuming that $y_i$'s $\sim$ *iid* Bernoulli($p$), where $p$ is the proportion of registered voters in favor of Mr. Washington. They intend to use the sample mean, $\overline{X}$, as an estimator for $p$.

(a) What is the expected value of the estimator?

(b) What is the standard deviation of the estimator?

(c) Provide a lower bound to the probability that the estimate is within $\pm$ .03 of the true proportion of voters in favor of the candidate.

(d) What size of random sample would the agency need to use in order to generate an estimate, based on the sample mean, that would be within $\pm$ .01 of the true proportion of voters?

(e) What is the estimate of the proportion of voters in favor of the candidate?

(f) Would there be much gain, in the way of lower variance of the estimator, if the agency would have sampled *without* replacement instead of with replacement? Explain.

**25.** Your company manufactures LCD screens that are used in the production of a popular smartphone sold by a major wireless cell phone service provider. The engineering department suggests that the operating life of the screen "likely" exhibits the memoryless property

$P(x>s+t|x>s) = P(x>t) \forall s \text{ and } t>0$

where outcomes, $x$, of the random variable $X$ measure the operating life of the screen, measured in 100,000 hour increments, until failure of the screen. But they are not entirely sure of that as yet. A random sample of 50 of the LCD screens has been subjected to a test that accurately assesses their operating lives, and the following outcomes occurred:

| | | | | |
|---|---|---|---|---|
| 0.841 | 0.478 | 2.631 | 0.126 | 2.953 |
| 0.476 | 0.744 | 0.753 | 3.344 | 0.141 |
| 0.145 | 2.031 | 0.694 | 0.654 | 0.402 |
| 0.893 | 3.675 | 2.068 | 0.366 | 3.145 |
| 0.064 | 0.740 | 0.522 | 0.146 | 1.641 |
| 0.506 | 0.790 | 3.096 | 1.381 | 2.249 |
| 2.057 | 1.045 | 0.783 | 0.368 | 0.121 |
| 5.788 | 1.862 | 2.165 | 1.156 | 0.200 |
| 2.415 | 1.077 | 5.258 | 0.326 | 3.317 |
| 2.105 | 0.361 | 7.611 | 2.334 | 0.808 |

(a) Define the EDF estimate of the underlying CDF for the screen lifetimes.

(b) Graph the EDF that you estimated in (a).

(c) What is the mean of screen lifetimes implied by the EDF?

(d) Treating the value you obtained in (c) as if it were the true mean, plot the CDF implied by the memoryless property on top of the EDF plot in (b). Do they appear to be similar or not?

(e) What is the third moment about the mean implied by the EDF? Based on this value, would you conclude the PDF of screen lifetimes is symmetric probability distribution?

(f) Based on the EDF, estimate the probability that the screen will fail in 1 year if the screen were operating 24 hours/day.

(g) The anticipated useful life of the smartphone itself is five calendar years. If the screen is used for 4 hours per day, use the EDF to estimate the probability that the screen will not fail during the useful life of the smartphone.

**26.** The following outcomes were from a random sample of size 25 from the joint distribution of $(Y,X)$, where $Y$ denotes yield, in bushels per acre, of a new variety of overlineley and $X$ denotes average inches of rainfall during the growing season:

| $Y$ | $X$ | $Y$ | $X$ |
|---|---|---|---|
| 81.828 | 26.195 | 77.903 | 17.091 |
| 71.305 | 17.102 | 79.645 | 17.867 |
| 75.232 | 21.629 | 72.966 | 19.094 |
| 75.936 | 17.553 | 74.051 | 21.304 |
| 74.377 | 24.760 | 78.050 | 24.847 |
| 77.149 | 22.788 | 74.878 | 23.053 |
| 81.959 | 25.566 | 68.752 | 13.607 |
| 75.094 | 19.819 | 71.925 | 21.738 |
| 81.166 | 21.407 | 76.299 | 21.829 |
| 77.723 | 19.190 | 83.792 | 35.889 |
| 72.750 | 18.410 | 73.022 | 23.079 |
| 75.413 | 26.478 | 78.167 | 21.155 |
| 73.079 | 19.744 | | |

(a) Define the joint EDF for the underlying joint CDF of yield and rainfall.

(b) Define the marginal EDF for the underlying CDF of yield.

(c) Define the sample covariance and sample correlation between yield and rainfall based on the EDF you defined in (a).

(d) Treating the EDF as the true CDF of $(Y,X)$, estimate the probability that yield will be greater than 75 bushels per acres.

(e) Use the EDF to estimate the probability that yield will be greater than 75 bushels per acre, given that rainfall will be less than 20 inches.

(f) Use the EDF to estimate the expected yield.

(g) Use the EDF to estimate the expected yield, given that rainfall will be less than 20 inches.

**27.** The following 20 *iid* random variable outcomes arose from some unknown cumulative distribution $F(t)$:

| | | | |
|---|---|---|---|
| 2 | 1 | 3 | 2 |
| 1 | 2 | 1 | 3 |
| 4 | 1 | 2 | 4 |
| 2 | 2 | 1 | 1 |
| 3 | 2 | 2 | 1 |

(a) Define the EDF for the random variable.

(b) Define the range of the random variable together with the probability density function that is implied by the EDF.

(c) Use the probability density function you derived in (b) to define the mean and the variance of the random variable.

(d) Now ignore that there are "repeated observations" for some of the numbers in the table above and assign an equal weight of 1/20 to each of the 20 numbers. Repeat the calculation of the mean and the variance of the random variable acting as if each of the 20 numbers above are assigned 1/20 probability. Are the calculated mean and variance the same values? Should they be? Explain.

**In the next FOUR questions below, use the following *iid* random variable outcomes from a Uniform(0,1) probability distribution:**

| | | | | |
|---|---|---|---|---|
| **0.2957** | **0.3566** | **0.8495** | **0.5281** | **0.0914** |
| **0.5980** | **0.4194** | **0.9722** | **0.7313** | **0.1020** |
| **0.5151** | **0.6369** | **0.7888** | **0.9893** | **0.1252** |
| **0.6362** | **0.1392** | **0.1510** | **0.4202** | **0.2946** |
| **0.1493** | **0.0565** | **0.4959** | **0.8899** | **0.6343** |

**28.** The number of customers entering the lobby of a bank in any 5 minute interval during the lunch hour from noon until 1 p.m. follows a Poisson process, Poisson(10).

(a) Using the uniform random outcomes above row-wise, simulate the outcomes of the number of customer arrivals for the 12 5-minute intervals between noon and 1 p.m.

(b) Calculate the sample mean and variance, and compare them to the true mean and variance.

**29.** The operating time until failure, in 100,000 hour units, of a hard disk that your company manufactures follows an exponential distribution, Exponential(2.5).

(a) Simulate time-until-failure outcomes of 25 of your hard disks.

(b) Calculate the sample mean and variance, and compare them to the true mean and variance.

**30.** A ground-fault-protection circuit breaker is designed to "trip" and interrupt power whenever a ground fault is detected in a power line to which it attached. After it trips, it designed to be able to "reset", and restore power when a reset button is pressed. For a certain brand of circuit breaker, the number of resets that occur to get a reset failure, at which the breaker is no longer functional, can be viewed as the outcome of a geometric probability distribution, Geometric(.1).

(a) Simulate 25 observations on the number of times this brand of circuit breaker is reset, ending in a reset failure.

(b) Calculate the sample mean and variance, and compare them to the true mean and variance.

**31.** The probability density function relating to the number of seconds that individuals remain on a website until they leave is given by the Pareto distribution,

$$f(x) = \begin{pmatrix} \frac{\alpha\beta^{\alpha}}{x^{\alpha+1}} \\ 0 \end{pmatrix} \text{ for } x \begin{Bmatrix} > \\ \leq \end{Bmatrix} \beta, \ \alpha = 2 \text{ and } \beta = 5.$$

(a) Simulate observations on the number of seconds that each of 25 visitors spends visiting the website.

(b) Calculate the sample mean and variance. Compare them to the true mean and variance, if they exist.

**32.** The monthly production of beer in a large Midwestern brewery can be viewed as the outcome of the following production process:

$$Q = 100 l^{.25} k^{.5} e^{V} \text{ and } V \sim N(0, .04),$$

where Q is measured in 1,000's of gallons, $l$ is units of labor and $k$ represents units of capital applied in production.

(a) Derive the probability density function of the random variable $W = e^V$ (see the note below regarding this distribution).

(b) Derive the probability density function of quantity produced, Q.

(c) Derive the expected quantity of beer produced as a function of the levels of labor and capital applied.

(d) What is the median level of production, as a function of labor and capital?

(e) If 16 units of labor and four units of capital are applied, what are the mean and median levels of production?

(f) At the levels of labor and capital defined in part (e), what is the standard deviation of production?

(g) At the levels of labor and capital defined in part (e), what is the probability that greater than 425,000 gallons of beer will be produced?

**NOTE:** What you will have derived in part (a) is a member of the log-normal probability distribution family. (See some general properties listed in problem 9(a)).

**33.** The monthly wholesale price and quantity sold of fresh squeezed orange juice during the winter months by a small Florida growers' cooperative is represented by the following bivariate probability density function:

$$F(p,q) = .01pe^{-.01pq}I_{[1.50,2.50]}(p)I_{[0,\infty)}(q)$$

where $p$ is measured in dollars per gallon and $q$ is measured in 1,000's of gallons.

(a) Derive the probability density function of total revenue, $R = PQ$.

(b) What is the expected value of total monthly revenue?

(c) What is the probability that total monthly revenue exceeds $100,000?

# 7 Point Estimation Theory

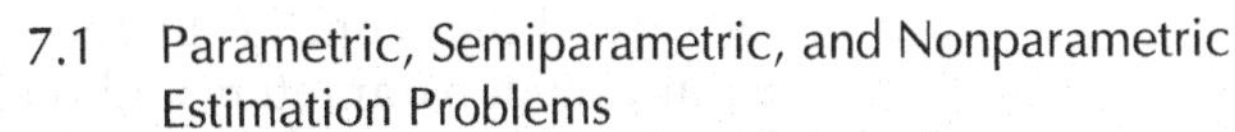



The problem of point estimation examined in this chapter is concerned with the estimation of the values of unknown parameters, or functions of parameters, that represent characteristics of interest relating to a probability model of some collection of economic, sociological, biological, or physical experiments. The outcomes generated by the collection of experiments are assumed to be outcomes of a random sample with some joint probability density function $f(x_1, \ldots, x_n; \Theta)$. The random sample need *not* be from a population distribution, so that it is *not* necessary that $X_1, \ldots, X_n$ be *iid*. The estimation concepts we will examine in this chapter can be applied to the case of general random sampling, as well as simple random sampling and random sampling with replacement, i.e., all of the random sampling types discussed in Chapter 6. The objective of point estimation will be to utilize functions of the random sample outcome to generate good (in some sense) estimates of the unknown characteristics of interest.

## 7.1 Parametric, Semiparametric, and Nonparametric Estimation Problems

The types of estimation problems that will be examined in this (and the next) chapter are problems of **parametric estimation** and **semiparametric estimation**, as opposed to **nonparametric estimation** problems. Both parametric and semiparametric estimation problems are concerned with the estimates of the values of unknown parameters that characterize **parametric probability models** or **semiparametric probability models** of the population, process, or general experiments under study. Both of these models have specific parametric functional structure to them that becomes fixed and known once values of parameters are numerically specified. The difference between the two models lies in whether a particular parametric family or class of probability distributions underlies the probability model and is fully determined by setting the values of parameters (the parametric model) or not (the semiparametric model). A **nonparametric probability model** is a model that is devoid of any specific parametric functional structure that becomes fixed when parameter values are specified. We discuss these models in more detail below.

Given the prominence of parameters in the estimation problems we will be examining, and the need to distinguish their appearance and effect in specifying parametric, semiparametric, and nonparametric probability models, we extend the scope of the term **probability model** to explicitly encompass the definition of parameters and their admissible values. Note, because it is possible that the range of the random variable can change with changing values of the parameter vector for certain specification of the joint probability density function of a random variable (e.g, a uniform distribution), we emphasize this in the definition below by including the parameter vector in the definition of the range of $\mathbf{X}$.

**Definition 7.1**
***Probability Model***

A probability model for the random variable $\mathbf{X}$ is defined by the set $\{R(\mathbf{X};\boldsymbol{\Theta}), f(\mathbf{x};\boldsymbol{\Theta}), \boldsymbol{\Theta} \in \Omega\}$, where $\Omega$ defines the admissible values of the parameter vector $\boldsymbol{\Theta}$.

In the context of point estimation problems, and later hypothesis testing and confidence interval estimation problems, $\mathbf{X}$ will refer to a random sample relating to some population, process, or general set of experiments having characteristics that are the interest of estimation, and $f(\mathbf{x};\boldsymbol{\Theta})$ will be the joint probability density function of the random sample. In our study, the parameter space $\Omega$ will generally represent all of the values of $\boldsymbol{\Theta}$ for which $f(\mathbf{x};\boldsymbol{\Theta})$ is a legitimate PDF, and thus represents all of the possible values for the unknowns one may be interested in estimating. The objective of point estimation is to increase knowledge of $\boldsymbol{\Theta}$ beyond simply knowing all of its admissible values. It can be the case that prior knowledge exists regarding the values $\boldsymbol{\Theta}$ can assume in a given empirical application, in which case $\Omega$ can be specified to incorporate that knowledge.

We note, given our convention that the range and the support of the random variable $\mathbf{X}$ are equivalent (recall Definition 2.13), that explicitly listing the range of the random variable as part of the specification of the probability model does not provide new information, per se. That is, knowing the density function and its admissible parameter values implies the range of the random variable as $R(\mathbf{X};\boldsymbol{\Theta}) \equiv \{\mathbf{x} : f(\mathbf{x};\boldsymbol{\Theta})>0\}$ for $\boldsymbol{\Theta} \in \Omega$. We will see ahead that in point estimation problems an explicit specification of the range of a random sample $\mathbf{X}$ is important for a number of reasons, including determining the types of estimation procedures that can be used in a given estimation problem, and for defining the range of estimates that are possible to generate from a particular point estimator specification. We will therefore continue to explicitly include the range of X in our specification of a probability model, but we will reserve the option to specify the probability model in the abbreviated form $\{f(\mathbf{x};\boldsymbol{\Theta}),\ \boldsymbol{\Theta}\in\Omega\}$ when emphasizing the range of the random variable is not germane to the discussion.

**Definition 7.2**
***Probability Model: Abbreviated Notation***

An abbreviated notation for the probability model of random variable $\mathbf{X}$ is $\{f(\mathbf{x};\boldsymbol{\Theta}),\ \boldsymbol{\Theta}\in\Omega\}$, where $R(\boldsymbol{X};\boldsymbol{\Theta})\equiv\{\mathbf{x}:f(\mathbf{x};\boldsymbol{\Theta})>0\}$ is taken as implicit in the definition of the model.

### 7.1.1 Parametric Models

A parametric model is one in which the functional form of the joint probability density function, $f(x_1,\ldots,x_n;\boldsymbol{\Theta})$, contained in the probability model for the observed sample data, $\mathbf{x}$, is fully specified and known once the value of the parameter vector, $\boldsymbol{\Theta}$, is given a specific numerical value. In specifying such a model, the analyst defines a collection of explicit parametric functional forms for the joint density of the random sample $\mathbf{X}$, as $f(x_1,\ldots,x_n;\boldsymbol{\Theta})$, *for* $\boldsymbol{\Theta}\in\Omega$, with the implication that if the appropriate value of the parameter vector, say $\boldsymbol{\Theta}_0$, were known, then $f(x_1,\ldots,x_n;\boldsymbol{\Theta}_0)$ would represent true probability density function underlying the observed outcome $\mathbf{x}$ of the random sample. We note that in applications the analyst may not feel fully confident in the specification of the probability model $\{f(\mathbf{x};\boldsymbol{\Theta}),\ \boldsymbol{\Theta}\in\Omega\}$, and view it as a tentative working model, in which case the adequacy of the model may itself be an issue in need of further statistical analysis and testing. However, use of parametric estimation methodology begins with, and indeed requires such a full specification of a parametric model for $\mathbf{X}$.

### 7.1.2 Semiparametric Models

A semiparametric model is one in which the functional form of the joint probability density function component of the probability model for the observed sample data, $\mathbf{x}$, is *not* fully specified and is *not* known when the value of the parameter vector of the model, $\boldsymbol{\Theta}$, is given a specific numerical value. Instead of defining a collection of explicit parametric functional forms for the joint density of the random sample $\mathbf{X}$, when defining the model, as in the parametric case, the analyst defines a number of properties that the underlying true sampling density $f(x_1,\ldots,x_n;\boldsymbol{\Theta}_0)$ is thought to possess. Such information could include parametric specifications for some of the moments that the random variables are thought to adhere to, or whether the random variables contained in the random sample exhibit independence or not. Given a numerical value for the parameter vector $\boldsymbol{\Theta}$, any parametric structural components of the model are given an explicit fully specified functional form, but other components of the model, most notably the underlying joint density function for the random sample, $f(x_1,\ldots,x_n;\Theta)$, remains unknown and not fully specified.

### 7.1.3 Nonparametric Models

A nonparametric model is one in which neither the functional form of the joint probability density function component of the probability model for the observed sample data, $\mathbf{x}$, nor any other parametric functional component of the probability model is defined and known given numerical values of parameters $\Theta$. These models proceed with minimal assumptions on the structure of the probability model, with the analyst simply acknowledging the existence of some general characteristics and relationships relating to the random variables in the random sample, such as the existence of a general regression relationship, or the existence of a population probability distribution if the sample were generated through simple random sampling.

For example, the analyst may wish to estimate the CDF $F(z)$, where $(X_1,\ldots, X_n)$ is an *iid* random sample from the population distribution $F(z)$, and no mention is made, nor required, regarding *parameters* of the CDF. We have already examined a method for estimating the CDF in the case where the random sample is from a population distribution, namely, the empirical distribution function, $F_n$, provides an estimate of $F$. We will leave the general study of nonparametric estimation to a more advanced course of study; interested readers can refer to M. Puri and P. Sen (1985) *Nonparametric Methods in General Linear Models*. New York: John Wiley, F. Hampel, E. Ronchetti, P. Rousseeuw, and W. Stahel (1986), *Robust Statistics*. New York: John Wiley; and J. Pratt and J. Gibbons (1981), *Concepts of Nonparametric Theory*. New York: Springer-Verlag and A. Pagan and A. Ullah, (1999), *Nonparametric Econometrics*, Cambridge: Cambridge University Press.[1]

We illustrate the definition of the above three types of models in the following example.

**Example 7.1 *Parametric, Semiparametric, and Nonparametric Models of Regression***

Consider the specification of a probability model underlying a relationship between a given $n \times 1$ vector of values $\mathbf{z}$, and corresponding outcomes on the $n \times 1$ random vector $\mathbf{X}$, where the $n$ elements in $\mathbf{X}$ are assumed to be independent random variables.

For a **parametric model** specification of the relationship, let the probability model $\{R(\mathbf{X};\Theta), f(\mathbf{x};\Theta), \Theta \in \Omega\}$ be defined by $\mathbf{X} = \beta_1 + \mathbf{z}\beta_2 + \boldsymbol{\varepsilon}$ and

[1] There is not universal agreement on the meaning of the terms *parametric, nonparametric,* and *distribution-free.* Sometimes nonparametric and distribution-free are used synonymously, although the case of distribution-free *parametric* estimation is pervasive in econometric work. See J.D. Gibbons, (1982), *Encyclopedia of Statistical Sciences*, Vol. 4. New York: Wiley, pp. 400–401.

$$\boldsymbol{\varepsilon} \sim \prod_{i=1}^{n} \frac{1}{\sqrt{2\pi\sigma^2}} \exp\left(-\frac{1}{2}\frac{\varepsilon_i^2}{\sigma^2}\right)$$

for $\boldsymbol{\beta} \in \mathbb{R}^2$ and $\sigma^2 > 0$, where then $R(\mathbf{X};\ \boldsymbol{\beta}, \sigma^2) = R(\mathbf{X}) = \mathbb{R}^n$ for all admissible parameter values. In this case, if the parameters $\boldsymbol{\beta}$ and $\sigma^2$ are given specific numerical values, the joint density of the random sample is fully defined and known. Specifically, $X_i \sim N(\beta_{10} + z_i\beta_{20}, \sigma_0^2)$, $i = 1, \ldots, n$ for given values of $\boldsymbol{\beta}_0$ and $\sigma_0^2$ . Moreover, the parametric structure for the mean of $\mathbf{X}$ is then fully specified and known, as $\mathrm{E}(X_i) = \beta_{10} + z_i\beta_{20}$, $i = 1, \ldots, n$. Given the fully specified probability model, the analyst would be able to fully and accurately emulate the random sampling of $\mathbf{X}$ implied by the above fully specified probability model.

For a **semiparametric model** specification of the relationship, the probability model $\{R(\mathbf{X};\boldsymbol{\Theta}), f(\mathbf{x};\boldsymbol{\Theta}),\ \boldsymbol{\Theta} \in \Omega\}$ will not be fully functionally specified. For this type of model, the analyst might specify that $\mathbf{X} = \beta_1 + \mathbf{z}\beta_2 + \boldsymbol{\varepsilon}$ with $E(\boldsymbol{\varepsilon}) = 0$ and $\mathbf{Cov}(\boldsymbol{\varepsilon}) = \sigma^2\mathbf{I}$, so that the first and second moments of the relationship have been defined as $\mathrm{E}(\mathbf{X}) = \beta_1\mathbf{1}_n + \mathbf{z}\beta_2$, $i = 1, \ldots, n$ and $\mathbf{Cov}(\mathbf{X}) = \sigma^2\mathbf{I}$. In this case, knowing the numerical values of $\boldsymbol{\beta}$ and $\sigma^2$ will fully identify the means of the random variables as well as the variances, but the joint density of the random sample will remain unknown and not fully specified. It would not be possible for the analyst to simulate random sampling of $\mathbf{X}$ given this incomplete specification of the probability model.

Finally, consider a **nonparametric model** of the relationship. In this case, the analyst might specify that $\mathbf{X} = \mathbf{g}(\mathbf{z}) + \boldsymbol{\varepsilon}$ with $\mathrm{E}(\mathbf{X}) = \mathbf{g}(\mathbf{z})$, and perhaps that $\mathbf{X}$ is a collection of independent random variables, but nothing more. Thus, the mean function, as well as all other aspects of the relationship between $\mathbf{X}$ and $\mathbf{z}$, are left completely general, and nothing is explicitly determined given numerical values of parameters. There is clearly insufficient information for the analyst to simulate random sample outcomes from the probability model, not knowing the joint density of the random sample or even any moment aspects of the model, given values of parameters. □

### 7.1.4 Scope of Parameter Estimation Problems

The objective in problems of parameter estimation is to utilize a sample outcome $[x_1, \ldots, x_n]'$ of $\mathbf{X} = [X_1, \ldots, X_n]'$ to estimate the unknown value $\boldsymbol{\Theta}_0$ or $\mathbf{q}(\boldsymbol{\Theta}_0)$, where $\boldsymbol{\Theta}_0$ denotes the value of the parameter vector associated with the joint PDF that *actually* determines the probabilities of events for the random sample outcome. That is, $\boldsymbol{\Theta}_0$ is the value of $\boldsymbol{\Theta}$ such that $\mathbf{X} \sim f(\mathbf{x};\boldsymbol{\Theta}_0)$ is a *true statement*, and for this reason $\boldsymbol{\Theta}_0$ is oftentimes referred to as the **true value of $\boldsymbol{\Theta}$**, and we can then also speak of $\mathbf{q}(\boldsymbol{\Theta}_0)$ as being the **true value of $\mathbf{q}(\boldsymbol{\Theta})$** and $f(\mathbf{x};\ \boldsymbol{\Theta}_0)$ as being the **true PDF of X**. Some examples of the many functions of $\boldsymbol{\Theta}_0$ that might be of interest when sampling from a distribution $f(z;\boldsymbol{\Theta}_0)$ include

1. $q_1(\boldsymbol{\Theta}_0) = \mathrm{E}(Z) = \int_{-\infty}^{\infty} z\, f(z;\ \boldsymbol{\Theta}_0)dz$ (mean),
2. $q_2(\boldsymbol{\Theta}_0) = \mathrm{E}(Z - \mathrm{E}(Z))^2 = \int_{-\infty}^{\infty} (z - \mathrm{E}(Z))^2 f(z;\ \boldsymbol{\Theta}_0)dz$ (variance),

3. $q_3(\Theta_0)$ defined implicitly by $\int_{-\infty}^{q_3(\theta_0)} f(z;\, \Theta_0)dz = .5$ (median),
4. $q_4(\Theta_0) = \int_a^b f(z;\, \Theta_0)dz = P(z \in [a,b])$ (probabilities)
5. $q_5(\Theta_0) = \int_{-\infty}^{\infty} z\, f(z\, |\mathbf{x};\, \Theta_0)dz =$ (regression function of z on $\mathbf{x}$)

The method used to solve a parametric estimation problem will generally depend on the degree of specificity with which one can define the family of candidates for the true PDF of the random sample, **X**. The situation for which the most statistical theory has been developed, both in terms of the actual procedures used to generate point estimates and in terms of the evaluation of the properties of the procedures, is the **parametric model** case. In this case, the density function candidates, $f(x_1,\ldots,x_n;\, \Theta)$, are assumed at the outset to belong to *specific parametric families* of PDFs (e.g., normal, Gamma, binomial), and application of the celebrated *maximum likelihood* estimation procedure (presented in Chapter 8) relies on the candidates for the distribution of **X** being members of a specific collection of density functions that are indexed, and fully algebraically specified, by the values of $\Theta$.

In the **semiparametric model** and **nonparametric model** cases, a specific functional definition of the potential PDFs for **X** is *not* assumed, although some assumptions about the lower-order moments of $f(\mathbf{x};\Theta)$ are often made. In any case, it is often still possible to generate useful point estimates of various characteristics of the probability model of **X** that are conceptually functions of parameters, such as moments, quantiles, and probabilities, even if the specific parametric family of PDFs for **X** is not specified. For example, useful point estimates (in a number of respects) of the parameters in the so-called general linear model representation of a random sample based on general random sampling designs can be made, and with only a few general assumptions regarding the lower-order moments of $f(x_1,\ldots,x_n;\Theta)$, and without any assumptions that the density is of a *specific* parametric form (see Section 8.2).

Semiparametric and nonparametric methods of estimation have an advantage of being applicable to a wide range of sampling distributions since they are defined in a distribution-nonspecific context that inherently subsumes many different functional forms for $f(\mathbf{x};\Theta)$. However, it is usually the case that superior methods of estimating $\Theta$ or $\mathbf{q}(\Theta)$ exist if a parametric family of PDFs for **X** can be specified, and if the actual sampling distribution of the random sample is subsumed by the probability model. Put another way, the more (correct) information one has about the form of $f(\mathbf{x};\Theta)$ at the outset, the more precisely one can estimate $\Theta_0$ or $\mathbf{q}(\Theta_0)$.

## 7.2 Additional Considerations for the Specification and Estimation of Probability Models

A problem of point estimation begins with either a fully or partially specified *probability model* for the random sample $\mathbf{X} = (X_1,\ldots,X_n)'$ whose outcome $\mathbf{x} = [x_1,\ldots,x_n]'$ constitutes the observed data being analyzed in a real-world problem of point estimation, or statistical inference. The probability model defines the probabilistic and parametric context in which point estimation proceeds. Once the probability model has been specified, interest centers on estimating the true values of some (or all) of the parameters, or on estimating the true values of some functions of the parameters of the problem. The specific objectives of any point estimation problem depend on the needs of the researcher, who will identify which quantities are to be estimated.

The case of parametric model estimation of $\boldsymbol{\Theta}$ or $q(\boldsymbol{\Theta})$ is associated with a fully specified probability model in which a specific parametric family of PDFs is represented by $\{f(\mathbf{x};\boldsymbol{\Theta}), \boldsymbol{\Theta}\in\Omega\}$. For example, a fully specified probability model for a random sample of miles per gallon achieved by 25 randomly chosen trucks from the assembly line of a Detroit manufacturer might be defined as $\left\{\prod_{i=1}^{25} N(x_i;\mu,\sigma^2), (\mu,\sigma^2)\in\Omega\right\}$, where $\Omega = (0,\infty)\times(0,\infty)$.

In the semiparametric model case, a specific functional form for $f(\mathbf{x};\boldsymbol{\Theta})$ is not defined and $\Omega$ may or may not be fully specified. For example, in the preceding truck mileage example, a partially specified statistical model would be $\{f(\mathbf{x};\mu,\sigma^2), (\mu,\sigma^2)\in\Omega\}$, where $\Omega = (0,\infty)\times(0,\infty)$ and $f(\mathbf{x};\mu,\sigma^2)$ is *some continuous PDF.* In this latter case, the statistical model allows for the possibility that $f(\mathbf{x};\mu,\sigma^2)$ is *any* continuous PDF having a mean of $\mu$ and variance of $\sigma^2$, with both $\mu$ and $\sigma^2$ positive, e.g., normal, Gamma, or uniform PDFs would be potential candidates.

### 7.2.1 Specifying a Parametric Functional Form for the Sampling Distribution

In specifying a probability model, the researcher presumably attempts to identify an appropriate parametric family based on a combination of experience, consideration of the real-world characteristics of the experiments involved, theoretical considerations, past analyses of similar problems, an attempt at a reasonably robust approximation to the probability distribution, and/or pragmatism. The degree of detail with which the parametric family of densities is specified can vary from problem to problem.

In some situations there will be great confidence in a detailed choice of parametric family. For example, suppose we are interested in estimating the proportion, $p$, of defective manufactured items in a shipment of $N$ items. If a random sample with replacement of size $n$ is taken from the shipment (population) of manufactured items, then

$$(X_1,\ldots,X_n) \sim f(x_1,\ldots,x_n;p) = p^{\sum_{i=1}^n x_i}(1-p)^{n-\sum_{i=1}^n x_i}\prod_{i=1}^n I_{\{0,1\}}(x_i)$$

represents the parametric family of densities characterizing the joint density of the random sample, and interest centers on estimating the unknown value of the parameter $p$.

On the other hand, there will be situations in which the specification of the parametric family is quite tentative. For example, suppose one were interested in estimating the average operating life of a certain brand of hard-disk based on outcomes of a random sample of hard-disk lifetimes. In order to add some mathematical structure to the estimation problem, one might represent the $i^{th}$ random variable in the random sample of lifetimes $(X_1,\ldots,X_n)$ as $X_i = \mu + V_i$, where $\mu$ represents the unknown mean of the population distribution of lifetimes, an outcome of $X_i$ represents the actual lifetime observed for the $i^{th}$ hard disk sampled, and the corresponding outcome of $V_i$ represents the deviation of $X_i$ from $\mu$. Since $(X_1,\ldots,X_n)$ is a random sample from the population distribution, it follows that $E(X_i) = \mu$ and $\mathrm{var}(X_i) = \sigma^2\ \forall i$ can be assumed, so that $E(V_i) = 0$ and $\mathrm{var}(V_i) = \sigma^2\ \forall i$ can also be assumed. Moreover, it is then legitimate to assume that $(X_1,\ldots,X_n)$ and $(V_1,\ldots,V_n)$ are each a collection of *iid* random variables. Then to this point, we have already specified that the parametric family of distributions associated with $\mathbf{X}$ is of the form $\prod_{i=1}^{n} m(x_i;\boldsymbol{\Theta})$, where the density $m(z;\boldsymbol{\Theta})$ has mean $\mu$ and variance $\sigma^2$ (what is the corresponding specification for $V$?).

Now, what parametric functional specification of $m(x_i;\boldsymbol{\Theta})$ can be assumed to contain the specific density that represents the actual probability distribution of $X_i$ or $V_i$? (Note, of course, that specifying a parametric family for $V_i$ would imply a corresponding parametric family for $X_i$ and vice versa). One general specification would be the collection of all continuous joint density functions $f(x_1,\ldots,x_n;\boldsymbol{\Theta})$ for which $f(x_1,\ldots,x_n;\boldsymbol{\Theta}) = \prod_{i=1}^{n} m(x_i;\boldsymbol{\Theta})$ with $E(X_i) = \mu$ and $\mathrm{var}(X_i) = \sigma^2$, $\forall i$. The advantage of such a general specification of density family is that we have great confidence that the actual density function of $\mathbf{X}$ is contained within the implied set of potential PDFs, which we will come to see as an important component of the specification of any point estimation problem. In this particular case, the general specification of the statistical model actually provides sufficient structure to the point estimation problem for a useful estimate of mean lifetime to be generated (for example, the least squares estimator can be used to estimate $\mu$ – see Chapter 8). We will see that one disadvantage of very general specifications of the probability model is that the interpretation of the properties of point estimates generated in such a general context is also usually not as specific or detailed as when the density family can be defined with greater specificity.

Consider a more detailed specification of the probability model of hard-disk operating lives. If we feel that lifetimes are symmetrically distributed around some point, $\mu$, with the likelihoods of lifetimes declining the more distant the measurement is from $\mu$, we might consider the normal parametric family for the distribution of $V_i$. It would, of course, follow that $X_i$ is then also normally distributed, and thus the normal distribution could serve only as an *approximation* since negative lifetimes are impossible. Alternatively, if we felt that the distribution of lifetimes was skewed to the right, the gamma parametric family provides a rich source of density shapes, and we might specify that the $X_i$'s have some Gamma density, and thus the $V_i$'s would have the density of a Gamma–type random variable that has been shifted to the left by $\mu$ units. Hopefully, the engineering staff could provide some guidance regarding the most defensible parametric family specification to adopt. In cases where there is considerable doubt concerning the appropriate parametric family of densities, tests of hypotheses concerning the adequacy of a given parametric family specification can be performed. Some such tests will be discussed in Chapter 10. In some problem situations, it may not be possible to provide any more than a general specification of the density family, in which case the use of semiparametric methods of parameter estimation will be necessary.

### 7.2.2 The Parameter Space for the Probability Model

Given that a parametric functional form is specified to characterize the joint density of the random sample, a parameter space, $\Omega$, must also be identified to complete the probability model. There are often natural choices for the parameter space. For example, if the Bernoulli family were specified, then $\Omega = \{p: p \in [0,1]\}$, or if the normal family were specified, then $\Omega = \{(\mu,\sigma): \mu \in (-\infty,\infty), \sigma > 0\}$. However, if only a general definition of the parametric family of densities is specified at the outset of the point estimation problem, the specification of the parameter space for the parametric family will then also be general and often incomplete. For example, a parameter space specification for the aforementioned point estimation problem involving hard-disk lifetimes could be $\Omega = \{\Theta_o: \mu \geq 0, \sigma^2 \geq 0\}$. In this case, since the specific algebraic form of $f(x_1,\ldots, x_n; \mathbf{\Theta})$ is also not specified, we can only state that the mean and variance of hard-disk lifetimes are nonnegative, possibly leaving other unknown parameters in $\Theta_o$ unrestricted depending on the relationship of the mean and variance to the parameters of the distribution, and in any case not fully specifying the functional form of the density function. Regardless of the level of detail with which $\Omega$ is specified, there are two important assumptions, presented ahead, regarding the specification of $\Omega$ that are made in the context of a point estimation problem.

*The Issue of Truth in the Parameter Space* First, it is assumed that $\Omega$ contains the true value of $\mathbf{\Theta}_0$, so that the probability model given by $\{f(\mathbf{x};\, \mathbf{\Theta}),\ \mathbf{\Theta} \in \Omega\}$ can be assumed to contain the true sampling distribution for the random sample under study. Put another way, in the context of a point estimation problem, the set $\Omega$ is assumed to represent the entire collection of possible values for $\mathbf{\Theta}_0$. The relevance of this assumption in the context of point estimation is perhaps obvious – if the objective in point estimation is to estimate the value of $\mathbf{\Theta}_0$ or $\mathbf{q}(\mathbf{\Theta}_0)$ we do not want to preclude $\mathbf{\Theta}_0$ or $\mathbf{q}(\mathbf{\Theta}_0)$ from the set of potential estimates. Note that, in practice, this may be a tentative assumption that is subjected to statistical test for verification or refutation (see Chapter 10).

*Identifiability of Parameters* The second assumption on $\Omega$ concerns the concept of the *identifiability* of the parameter vector $\mathbf{\Theta}$. As we alluded to in our discussion of parametric families of densities in Chapter 4, parameterization of density families is *not* unique. Any invertible transformation of $\mathbf{\Theta}$, say $\boldsymbol{\lambda} = \mathbf{h}(\mathbf{\Theta})$, defines an alternative parameter space $\Lambda = \{\boldsymbol{\lambda}: \boldsymbol{\lambda} = \mathbf{h}(\mathbf{\Theta}), \mathbf{\Theta} \in \Omega\}$ that can be used to specify an alternative probability model for $\mathbf{X}$ that contains the same PDF candidates as the statistical model based on $\Omega$, i.e., $\{f(\mathbf{x}; \mathbf{h}^{-1}(\boldsymbol{\lambda})), \boldsymbol{\lambda} \in \Lambda\} = \{f(\mathbf{x}; \mathbf{\Theta}), \mathbf{\Theta} \in \Omega\}$. Defining $m(\mathbf{x};\boldsymbol{\lambda}) \equiv f(\mathbf{x};\mathbf{h}^{-1}(\boldsymbol{\lambda}))$, the alternative probability model could be written as $\{m(\mathbf{x};\boldsymbol{\lambda}),\ \boldsymbol{\lambda}\in\Lambda\}$. The analyst is free to choose whatever parameterization appears to be most natural or useful in the specification of a probability model, so long as the parameters in the chosen parameterization are *identified*. In stating the definition of parameter identifiability we use the terminology **distinct PDFs** to refer to PDFs that assign *different* probabilities to at least one event for $\mathbf{X}$.

**Definition 7.3** ***Parameter Identifiability***

Let $\{f(\mathbf{x};\ \boldsymbol{\Theta}),\ \boldsymbol{\Theta}\in\Omega\}$ be a probability model for the random sample $\mathbf{X}$. The parameter vector $\boldsymbol{\Theta}$ is said to be **identified** or **identifiable** *iff* $\forall\ \boldsymbol{\Theta}_1$ and $\boldsymbol{\Theta}_2 \in \Omega$, $f(\mathbf{x};\ \boldsymbol{\Theta}_1)$ and $f(\mathbf{x};\boldsymbol{\Theta}_2)$ are *distinct* if $\boldsymbol{\Theta}_1 \neq \boldsymbol{\Theta}_2$.

The importance of parameter identifiability is related to the ability of random sample outcomes to provide discriminatory information regarding the choice of $\boldsymbol{\Theta}\in\Omega$ to be used in estimating $\boldsymbol{\Theta}_0$. If the parameter vector in a statistical model is *not* identified, then two or more different values of the parameter vector $\boldsymbol{\Theta}$, say $\boldsymbol{\Theta}_1$ and $\boldsymbol{\Theta}_2$, are associated with precisely the same sampling distribution $\mathbf{X}$. In this event, random sample outcomes cannot possibly be used to discriminate between the values of $\boldsymbol{\Theta}_1$ and $\boldsymbol{\Theta}_2$ since the probabilistic behavior of $\mathbf{X}$ under either possibility is indistinguishable. We thus insist on parameter identifiability in a point estimation problem so that different values of $\boldsymbol{\Theta}$ are associated with different probabilistic behavior of the outcomes of the random sample.

**Example 7.2** ***Parameter Identifiability***

The yield per acre of tomatoes on 20 geographically dispersed parcels of irrigated land is thought to be representable as the outcomes of $Y_i = \beta_0 + \beta_1 T_i + V_i$, $i = 1,\ldots,20$, where $\beta_0$ and $\beta_1$ are $> 0$, and $(T_i, V_i)$, $i = 1,\ldots,20$, are *iid* outcomes of a bivariate normal population distribution with mean vector $[\mu_T, 0]$ for $\mu_T > 0$, and diagonal covariance matrix with diagonal entries $\sigma_T^2$ and $\sigma_V^2$. The outcome $y_i$ represents bushels/acre on parcel $i$, and $t_i$ is season average temperature measured at the growing site. If the probability model for the random sample $\mathbf{Y} = [Y_1,\ldots,Y_{20}]'$ is specified as

$$\left\{\prod_{i=1}^{20} N(y_i;\beta_0+\beta_1\mu_T,\beta_1^2\ \sigma_T^2+\sigma_V^2),\ \text{for}\ (\beta_0,\beta_1,\mu_T,\sigma_T^2,\sigma_V^2)\in\Omega\right\}$$

where $\Omega = \times_{i=1}^{5}\ (0,\infty)$, is the parameter vector $[\beta_0,\ \beta_1,\ \mu_T,\ \sigma_T^2,\ \sigma_V^2]$ identified?

**Answer**: Define $\mu = \beta_0 + \beta_1\mu_T$ and $\sigma^2 = \beta_1^2\ \sigma_T^2+\sigma_V^2$, and examine the probability model for $\mathbf{Y}$ given by

$$\left\{\prod_{i=1}^{20} N(y_i;\mu,\sigma^2),(\mu,\sigma^2)\in\Lambda\right\},$$

where $\Lambda = \times_{i=1}^{2}(0,\infty)$. Note that any choice of positive values for $\beta_0$, $\beta_1$, $\mu_T$, $\sigma_T^2$, and $\sigma_V^2$ that result in the same given positive values for $\mu$ and $\sigma^2$ result in precisely the same sampling distribution for $\mathbf{Y}$ (there are an infinite set of such choices for each value of the vector $[\mu,\sigma^2]'$). Thus the original parameter vector is not identified. Note that the parameter vector $[\mu,\sigma^2]'$ in the latter statistical model for $\mathbf{Y}$ *is* identified since the sampling distributions associated with two different positive values of the vector $[\mu,\sigma^2]'$ are distinct. □

### 7.2.3 A Word on Estimation Phraseology

We pause here to introduce a convention regarding the interpretation of phrases such as **estimating $\Theta$** or **estimating $q(\Theta)$**, or **an estimate of $\Theta$ (or of $q(\Theta)$)**. Since $\Theta$ is simply a parameter vector that indexes a family of density functions and that can assume a range of alternative values (those specified in $\Omega$), the reader might wonder what such phrases could possibly mean? That is, *what* are we estimating if we are estimating, say, $\Theta$? The phrases are used as a shorthand or an abbreviated way of stating that one is *estimating the true value of* $\Theta$, or *estimating the true value of* $q(\Theta)$, or that one has *an estimate of the true value of* $\Theta$ *(or of* $q(\Theta)$*)*. There is widespread use of such phrases in the statistics and econometrics literature, and we will make frequent use of such phrases in this book as well. In general, one must rely on the context of the discussion to be sure whether $\Theta$ or $q(\Theta)$ refers to the quantity being estimated or merely to the indexing parameter of a family of joint density functions.

## 7.3 Estimators and Estimator Properties

Point estimation is concerned with estimating $\Theta$ or $q(\Theta)$ from knowledge of the outcome $x = [x_1, \ldots, x_n]'$ of a random sample $X$. It follows from this basic description of point estimation that *functions* are critical to the estimation problem, where inputs or domain elements are sample outcomes, $x$, and outputs or range elements are estimates of $\Theta$ or $q(\Theta)$. More formally, estimates will be generated via some function of the form $t: R(X) \rightarrow R(t)$, where $R(t)$ is the range of $t$ defined as $R(t) = \{t: t = t(x), x \in R(X)\}$. Note that $R(t)$ represents the set of all possible estimates of $\Theta$ or $q(\Theta)$ that can be generated as outcomes of $t(X)$. We will always tacitly assume that $t(X)$ is an observable random variable, and hence a *statistic*, so that estimates are observable and empirically informative.

Henceforth, when the function $t: R(X) \rightarrow R(t)$ represented by $t = t(x)$ is being utilized to generate estimates of $q(\Theta)$, we will refer to the random variable $T = t(X)$ as an **estimator** for $q(\Theta)$, and $q(\Theta)$ will be referred to as the **estimand**. An outcome, $t = t(x)$, of the estimator will be referred to as an **estimate** of $q(\Theta)$. We formalize these three terms in the following definition:

**Definition 7.4**
***Point Estimator, Estimate, and Estimand***

A statistic or vector of statistics, $T = t(X)$, whose outcomes are used to estimate the value of a scalar or vector function, $q(\Theta)$, of the parameter vector, $\Theta$, is called a **point estimator**, with $q(\Theta)$ being called the **estimand.**[2] An observed outcome of an estimator is called a **point estimate**.

[2]Note, as always, that the function $q$ can be the identity function $q(\Theta) \equiv \Theta$, in which case we could be referring to estimating the vector $\Theta$ itself. Henceforth, it will be understood that since $q(\Theta) \equiv \Theta$ is a possible choice of $q(\Theta)$, all discussion of estimating $q(\Theta)$ could be referring to estimating the vector $\Theta$ itself.

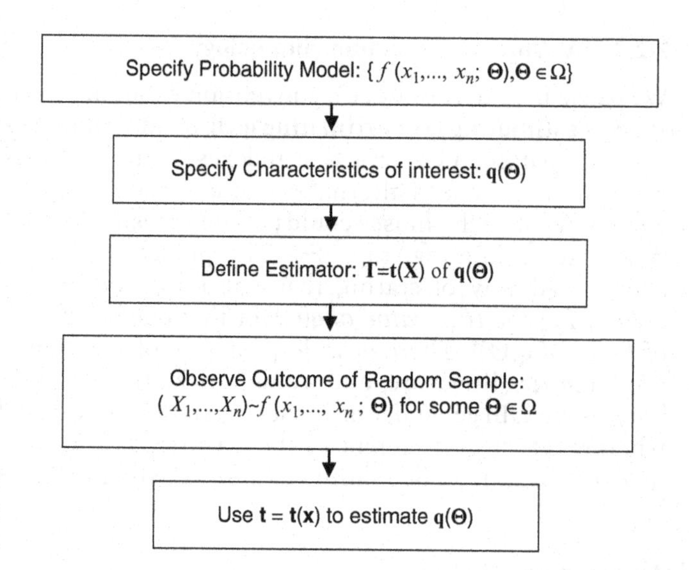


**Figure 7.1**
General point estimation procedure.

Figure 7.1 contains a schematic overview of the general context of the point estimation problem to this point.

### 7.3.1 Evaluating Performance of Estimators

Since there is literally an *uncountably infinite* set of possible functions of $\mathbf{X}$ that are potential estimators of $\mathbf{q}(\Theta)$, a fundamental problem in point estimation is the choice of a "good" estimator. In order to rank the efficacy of estimators and/ or to choose the *optimal* estimator of $\mathbf{q}(\Theta)$, an objective function that establishes an appropriate measure of "goodness" must be defined.

A natural measure to use in ranking estimators would seem to be the distance between outcomes of $\mathbf{t}(\mathbf{X})$ and $\mathbf{q}(\Theta)$, which is a direct measure of how *close* estimates are to what is being estimated. In the current context, this distance measure is $d(\mathbf{t}(\mathbf{x}), \mathbf{q}(\Theta)) = ([\mathbf{t}(\mathbf{x}) - \mathbf{q}(\Theta)]' [\mathbf{t}(\mathbf{x}) - \mathbf{q}(\Theta)])^{1/2}$, which specializes to $|t(\mathbf{x}) - q(\Theta)|$ when $k = 1$. However, this closeness measure has an obvious practical flaw for comparing alternative functions for estimating $\mathbf{q}(\Theta)$ – the estimate that would be preferred depends on the true value of $\mathbf{q}(\Theta)$, which is unknown (or else there would be no point estimation problem in the first place). This problem is clearly not the fault of the particular closeness measure chosen since any reasonable measure of closeness between the two values $\mathbf{t}(\mathbf{x})$ and $\mathbf{q}(\Theta)$ would depend on where $\mathbf{q}(\Theta)$ actually is in $\mathbb{R}^k$ vis-a-vis where $\mathbf{t}(\mathbf{x})$ is located. Thus, comparing alternative functions for estimating $\mathbf{q}(\Theta)$ on the basis of the closeness to $\mathbf{q}(\Theta)$ to an actual estimate $\mathbf{t}(\mathbf{x})$ is not tractable — we clearly need additional criteria with which to judge whether $\mathbf{t}(\mathbf{X})$ generates "good" estimates of $\mathbf{q}(\Theta)$.

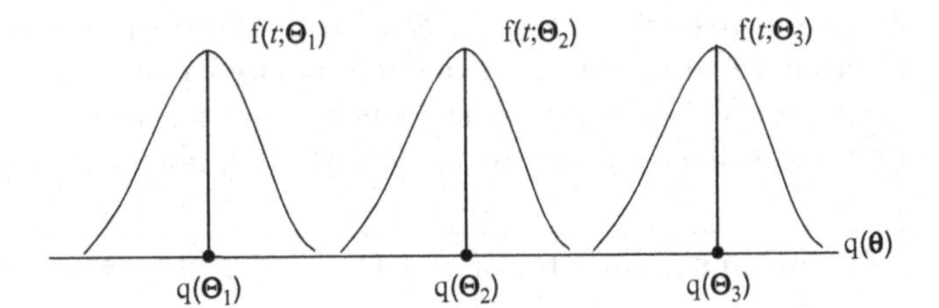


**Figure 7.2**
Scalar estimator PDFs for various values of **Θ**.

Various criteria for judging the usefulness of a given estimator **t**(**X**) for estimating **q**(**Θ**) have been presented in the literature.[3] The measures evaluate and rank estimators in terms of closeness of estimates to **q**(**Θ**) in an *expected* or *probabilistic* sense. Note that since **t**(**X**) is a function of **X**, and thus a random variable, a sampling distribution (i.e., the probability distribution of **t**(**X**)) exists on $R(\mathbf{T})$ that is induced by the probability distribution of the random sample, $\mathbf{X} = (X_1,\ldots,X_n)$. Roughly speaking, the fact that the distribution of **X** depends on **Θ** will generally result in the sampling distribution of **t**(**X**) depending on **Θ** as well, and this latter dependence can lead to changes in location, spread, and/or shape of the distribution of **t**(**X**) as **Θ** changes. If the sampling distribution of **t**(**X**) changes with **Θ** in a way that keeps the spread of potential estimates generated by **t**(**X**) narrowly focused on **q**(**Θ**) so that outcomes of **t**(**X**) occur near **q**(**Θ**) with high probability under all contingencies for $\boldsymbol{\Theta} \in \Omega$, (see Figure 7.2), then the function **T** would be useful for generating estimates of **q**(**Θ**).

We now turn our attention to specific estimator properties that have been used in practice to measure whether these objectives have been achieved. In discussing estimator properties, we will sometimes utilize a **Θ**-subscript such as $E_{\boldsymbol{\Theta}}(\cdot)$, $P_{\boldsymbol{\Theta}}(\cdot)$, or $\text{var}_{\boldsymbol{\Theta}}(\cdot)$ to emphasize that expectations or probabilities are being calculated using a particular value of **Θ** for the parameter vector of the underlying probability distribution. In cases where the parametric context of expectations and probabilities are clear or does not need to be distinguished, the subscript **Θ** will not be explicitly displayed.

### 7.3.2 Finite Sample Properties

The properties examined in this section evaluate the performance of estimators when the random sample is of fixed size, and they are therefore referred to as **finite sample properties.** This is as opposed to **asymptotic properties** that we will examine later in this section, which relate to limiting results that are established as the random sample size increases without bound (increases to infinity). All of the finite sample properties examined here are based on the first two moments of estimators, and thus relate to the central tendency of estimates as well as the spread of estimates around their central tendency. Of course, if these moments do not exist for a given estimator, then these finite sample properties cannot be used to evaluate the performance of the estimator.

[3] A concise review and comparison of a number of alternative criteria is given by T. Amemiya (1994), *Introduction to Statistics and Econometrics*, Cambridge, MA, Harvard University Press, pp. 118–121.

*Mean Square Error and Relative Efficiency* The term **mean square error** (MSE) is an alternative term for the **expected squared distance** between outcomes of an estimator $\mathbf{T} = \mathbf{t}(\mathbf{X})$, and what it is estimating, the estimand $\mathbf{q}(\boldsymbol{\Theta})$. When $T$ and $q(\boldsymbol{\Theta})$ are scalars, the following definition applies.

**Definition 7.5**
***Mean Square Error: Scalar Case***

The **mean square error** of an estimator $T$ of $q(\boldsymbol{\Theta})$ is defined as $\mathrm{MSE}_{\boldsymbol{\Theta}}(T) = \mathrm{E}_{\boldsymbol{\Theta}}\left(d^2(T, q(\boldsymbol{\Theta}))\right) = \mathrm{E}_{\boldsymbol{\Theta}}\left((T - q(\boldsymbol{\Theta}))^2\right)\ \forall \boldsymbol{\Theta} \in \Omega$.

The MSE criterion accounts for both the degree of spread in the sampling distribution of $T$ as well as the degree to which the central tendency of $T$'s distribution deviates from $q(\boldsymbol{\Theta})$. We will make this notion precise upon defining the concept of bias, as follows.

**Definition 7.6**
***Estimator Bias***

The **bias** of a scalar estimator $T$ of $q(\boldsymbol{\Theta})$ is defined as $\mathrm{bias}_{\boldsymbol{\Theta}}(T) = \mathrm{E}_{\boldsymbol{\Theta}}(T - q(\boldsymbol{\Theta}))$, $\forall \boldsymbol{\Theta} \in \Omega$. The **bias vector** of a vector estimator $\mathbf{T}$ of $\mathbf{q}(\boldsymbol{\Theta})$ is defined as $\mathbf{Bias}_{\boldsymbol{\Theta}}(\mathbf{T}) = \mathrm{E}_{\boldsymbol{\Theta}}(\mathbf{T} - \mathbf{q}(\boldsymbol{\Theta})), \forall \boldsymbol{\Theta} \in \Omega$.

Thus the bias of an estimator is the expected difference between the outcomes of the estimator and what the estimator is estimating.

The MSE of a scalar estimator $T$ can be decomposed into the sum of the variance of $T$ and the squared bias of $T$, as

$$\begin{aligned}\mathrm{MSE}_{\boldsymbol{\Theta}}(T) &= \mathrm{E}_{\boldsymbol{\Theta}}\left((T - \mathrm{E}_{\boldsymbol{\Theta}}(T) + \mathrm{E}_{\boldsymbol{\Theta}}(T) - q(\boldsymbol{\Theta}))^2\right)\\ &= \mathrm{E}_{\boldsymbol{\Theta}}(T - \mathrm{E}_{\boldsymbol{\Theta}}(T))^2 + (\mathrm{E}_{\boldsymbol{\Theta}}(T) - q(\boldsymbol{\Theta}))^2\\ &= \mathrm{var}_{\boldsymbol{\Theta}}(T) + (\mathrm{bias}_{\boldsymbol{\Theta}}(T))^2.\end{aligned}$$

The MSE criterion thus penalizes an estimator for having a high variance, a high bias, or both. It also follows that the MSE criterion allows a tradeoff between variance and bias in the ranking of estimators. In the final analysis, it is the expected squared distance between $T$ and $q(\boldsymbol{\Theta})$ implied by $\mathrm{var}_{\boldsymbol{\Theta}}(T)$ and $\mathrm{bias}_{\boldsymbol{\Theta}}(T)$, and not the variance and bias per se, that determines an estimator's relative ranking via the MSE criterion.

In the multivariate case, the MSE criterion is generalized through the use of the **mean square error matrix**.

**Definition 7.7**
***Mean Square Error Matrix***

The **mean square error matrix** of the estimator $\mathbf{T}$ of the $(k \times 1)$ vector $\mathbf{q}(\boldsymbol{\Theta})$ is defined by $\mathbf{MSE}_{\boldsymbol{\Theta}}(\mathbf{T}) = \mathrm{E}_{\boldsymbol{\Theta}}((\mathbf{T} - \mathbf{q}(\boldsymbol{\Theta}))(\mathbf{T} - \mathbf{q}(\boldsymbol{\Theta}))')\ \forall \boldsymbol{\Theta} \in \Omega$.

To appreciate the information content of $\mathbf{MSE}_{\boldsymbol{\Theta}}(\mathbf{T})$, first note that the diagonal of the MSE matrix contains the MSE of the estimator $T_i$ for $q_i(\boldsymbol{\Theta})$, $i = 1,\ldots,k$ since the $i$th diagonal entry in $\mathbf{MSE}_{\boldsymbol{\Theta}}(\mathbf{T})$ is $\mathrm{E}_{\boldsymbol{\Theta}}(T_i - q_i(\boldsymbol{\Theta}))^2$. More generally, let $\mathbf{c}$ be any $(k \times 1)$ vector of constants, and examine the MSE of the linear combination $\mathbf{c}'\mathbf{T} = \sum_{i=1}^{k} c_i T_i$ as an estimator of $\mathbf{c}'\mathbf{q}(\boldsymbol{\Theta}) = \sum_{i=1}^{k} c_i q_i(\boldsymbol{\Theta})$, defined by

$$\begin{aligned}\mathrm{MSE}_{\boldsymbol{\Theta}}(\mathbf{c}'\mathbf{T}) &= \mathrm{E}_{\boldsymbol{\Theta}}\left((\mathbf{c}'\mathbf{T}-\mathbf{c}'\mathbf{q}(\boldsymbol{\Theta}))^2\right) = \mathrm{E}_{\boldsymbol{\Theta}}(\mathbf{c}'[\mathbf{T}-\mathbf{q}(\boldsymbol{\Theta})][\mathbf{T}-\mathbf{q}(\boldsymbol{\Theta})]'\mathbf{c}) \\ &= \mathbf{c}'\mathbf{MSE}_{\boldsymbol{\Theta}}(\mathbf{T})\mathbf{c}.\end{aligned}$$

Thus, the MSEs of every possible linear combination of the $T_i$'s, used as estimators of the corresponding linear combination of the $q_i(\boldsymbol{\Theta})$'s, can be obtained from the MSE matrix. Note further that the *trace* of the MSE matrix defines the expected squared distance of the vector estimator **T** from the vector estimand $\mathbf{q}(\boldsymbol{\Theta})$, as

$$\begin{aligned}\mathbf{tr}(\mathbf{MSE}_{\Theta}(\mathbf{T})) &= \mathrm{tr}(\mathrm{E}_{\Theta}([\mathbf{T}-\mathbf{q}(\Theta)][\mathbf{T}-\mathbf{q}(\Theta)]')) \\ &= \mathrm{E}_{\Theta}([\mathbf{T}-\mathbf{q}(\Theta)]'[\mathbf{T}-\mathbf{q}(\Theta)]) = \mathrm{E}_{\Theta}\left(d^2(\mathbf{T},\mathbf{q}(\Theta))\right).\end{aligned}$$

This is the direct vector analogue to the measure of closeness of $T$ to $q(\boldsymbol{\Theta})$ that is provided by the MSE criterion in the scalar case.

The MSE matrix can be decomposed into variance and bias components, analogous to the scalar case. Specifically, **MSE(T)** is equal to the sum of the covariance matrix of **T** and the outer product of the bias vector of **T**, as

$$\begin{aligned}\mathbf{MSE}_{\boldsymbol{\Theta}}(\mathbf{T}) &= \mathrm{E}_{\boldsymbol{\Theta}}([\mathbf{T}-\mathrm{E}_{\boldsymbol{\Theta}}(\mathbf{T})+\mathrm{E}_{\boldsymbol{\Theta}}(\mathbf{T})-\mathbf{q}(\boldsymbol{\Theta})]\ [\mathbf{T}-\mathrm{E}_{\boldsymbol{\Theta}}(\mathbf{T})+\mathrm{E}_{\boldsymbol{\Theta}}(\mathbf{T})-\mathbf{q}(\boldsymbol{\Theta})]') \\ &= \mathbf{Cov}_{\boldsymbol{\Theta}}(\mathbf{T})+\mathbf{Bias}_{\boldsymbol{\Theta}}(\mathbf{T})\mathbf{Bias}_{\boldsymbol{\Theta}}(\mathbf{T})'.\end{aligned}$$

The outer product of the bias vector forms a $(k \times k)$ matrix that is called the **bias matrix.**

In the case of a scalar $q(\boldsymbol{\Theta})$, estimators with smaller MSEs are preferred. Note, however, that since the true $\boldsymbol{\Theta}$ is unknown (or else there is no point estimation problem to begin with), one must consider the performance of an estimator for all possible contingencies for the true value of $\boldsymbol{\Theta}$, which is to say, for all $\boldsymbol{\Theta}\in\Omega$. It is quite possible, and often the case, that an estimator will have lower MSEs than another estimator for some values of $\boldsymbol{\Theta}\in\Omega$ but not for others. These considerations lead to the concepts of **relative efficiency** and **relatively more efficient**.

**Definition 7.8**
***Relative Efficiency: Scalar Case***

Let $T$ and $T^*$ be two estimators of a scalar $q(\boldsymbol{\Theta})$. The **relative efficiency** of $T$ with respect to $T^*$ is given by

$$\mathrm{RE}_{\boldsymbol{\Theta}}(T,T^*) = \frac{\mathrm{MSE}_{\boldsymbol{\Theta}}(T^*)}{\mathrm{MSE}_{\boldsymbol{\Theta}}(T)} = \frac{\mathrm{E}_{\boldsymbol{\Theta}}(T^*-q(\boldsymbol{\Theta}))^2}{\mathrm{E}_{\boldsymbol{\Theta}}(T-q(\boldsymbol{\Theta}))^2}, \forall\boldsymbol{\Theta}\in\Omega.$$

$T$ is **relatively more efficient** than $T^*$ if $\mathrm{RE}_{\boldsymbol{\Theta}}(T,T^*) \geq 1\ \forall\boldsymbol{\Theta}\in\Omega$ and $> 1$ for some $\boldsymbol{\Theta}\in\Omega$.

In comparing two estimators of $q(\boldsymbol{\Theta})$, if $T$ is relatively more efficient than $T^*$, then there is no value of $\boldsymbol{\Theta}$ for which $T^*$ is preferred to $T$ on the basis of MSE, and for one or more values of $\boldsymbol{\Theta}$, $T$ is preferred to $T^*$. In this case, it is evident that $T^*$ can be discarded as an estimator of $q(\boldsymbol{\Theta})$, and in this case, $T^*$ is said to be an **inadmissible** estimator of $q(\boldsymbol{\Theta})$, as defined below.

**Definition 7.9**
***Estimator Admissibility***

Let $T$ be an estimator of $q(\Theta)$. If there exists another estimator of $q(\Theta)$ that is *relatively more efficient* than $T$, then $T$ is called **inadmissible** for estimating $q(\Theta)$. Otherwise, $T$ is called **admissible**.

It is evident that if one is judging the performance of estimators on the basis of MSE, the analyst need not consider any estimators that are *inadmissible*.

**Example 7.3**
***Two Admissible Estimators for the Mean of a Bernoulli Distribution***

Suppose $(X_1,\ldots,X_n)$ is a random sample from a Bernoulli population distribution, where $X_i$ represents whether $(X_i = 1)$ or not $(X_i = 0)$ the $i$th customer contacted by telephone solicitation purchases a product. Consider two estimators for the unknown proportion, $p$, of the consumer population who will purchase the product:

$$T = \overline{X} = n^{-1}\sum_{i=1}^{n} X_i \text{ and } T^* = (n+1)^{-1}\sum_{i=1}^{n} X_i = \left(\frac{n}{n+1}\right)\overline{X}.$$

Which estimator, if either, is the preferred estimator of $p$ on the basis of MSE given that $n = 25$? Does either estimator render the other inadmissible?
**Answer:** Note that $\text{bias}(T) = \text{E}(\overline{X}) - p = 0$, $\text{bias}(T^*) = \text{E}((n/(n+1))\overline{X}) - p = -p/(n+1) = -p/26$, $\text{var}(T) = p(1-p)/n = p(1-p)/25$, and $\text{var}(T^*) = np(1-p)/(n+1)^2 = p(1-p)/27.04$. Then the MSEs of the two estimators are given by

$$\text{MSE}(T) = \frac{p(1-p)}{25}$$

and

$$\text{MSE}(T^*) = \frac{p(1-p)}{27.04} + \frac{p^2}{676}.$$

Examine the MSE of $T^*$ relative to the MSE of $T$, as

$$\text{RE}_p(T, T^*) = \frac{\text{MSE}(T^*)}{\text{MSE}(T)} = .9246 + .0370p/(1-p).$$

Since the ratio depends on the value of $p$, which is unknown, we must consider all of the possible contingencies for $p\in[0,1]$. Note that the ratio is monotonically increasing in $p$, taking its smallest value of .9246 when $p = 0$, and diverging to infinity as $p \to 1$. The ratio of MSEs equals 1 when $p = .6708$. Thus, without constraints on the potential values of $p$, neither estimator is preferred to the other on the basis of MSE, and thus neither estimator is rendered inadmissible by the other. □

In contrast to the scalar case, a myriad of different MSE comparisons are possible when $\mathbf{q}(\Theta)$ is a $(k \times 1)$ vector. First of all, there are $k$ individual MSE comparisons that can be made between corresponding entries in the two estimators $\mathbf{T}^*$ and $\mathbf{T}$. One could also compare the expected squared distances of $\mathbf{T}^*$ and $\mathbf{T}$ from $\mathbf{q}(\Theta)$, which is equivalent to comparing the sums of the mean square errors of the entries in $\mathbf{T}^*$ and $\mathbf{T}$. Furthermore, one could contemplate estimating linear combinations of the entries in $\mathbf{q}(\Theta)$ via corresponding linear combinations of the entries in $\mathbf{T}^*$ and $\mathbf{T}$, so that MSE comparisons between the estimators $\boldsymbol{\ell}'\mathbf{T}^*$ and $\boldsymbol{\ell}'\mathbf{T}$ for $\boldsymbol{\ell}'\mathbf{q}(\Theta)$ are then of interest. *All* of the preceding MSE comparisons are accounted for simultaneously in the following **strong mean square error** (SMSE) criterion.

**Definition 7.10**
***Strong Mean Square Error Superiority***

Let $\mathbf{T}^*$ and $\mathbf{T}$ be two estimators of the $(k \times 1)$ vector $\mathbf{q}(\Theta)$. $\mathbf{T}^*$ is **strong mean square error superior** to $\mathbf{T}$ *iff* $\mathbf{MSE}_\Theta(\mathbf{T}^*) - \mathbf{MSE}_\Theta(\mathbf{T})$ is negative semidefinite $\forall \Theta \in \Omega$ and unequal to the zero matrix for some $\Theta \in \Omega$.

If $\mathbf{T}^*$ is SMSE superior to $\mathbf{T}$, it follows directly from Definition 7.10 that $MSE_\Theta(T_i^*) \leq MSE_\Theta(T_i)\ \forall i$ and $\forall \Theta \in \Omega$ because if $\mathbf{MSE}_\Theta(\mathbf{T}^*) - \mathbf{MSE}_\Theta(\mathbf{T})$ is negative semidefinite, the matrix difference necessarily has nonpositive diagonal entries.[4] It follows that

$$\mathrm{E}_\Theta\left(d^2(\mathbf{T}^*, \mathbf{q}(\Theta))\right) = \sum_{i=1}^{k} MSE_\Theta(T_i^*) \leq \sum_{i=1}^{k} MSE_\Theta(T_i) = \mathrm{E}_\Theta\left(d^2(\mathbf{T}, \mathbf{q}(\Theta))\right) \quad \forall \Theta \in \Omega$$

Furthermore, in terms of estimating $\boldsymbol{\ell}'\mathbf{q}(\Theta)$,

$$MSE_\Theta(\boldsymbol{\ell}'\mathbf{T}^*) = \boldsymbol{\ell}'\mathbf{MSE}_\Theta(\mathbf{T}^*)\boldsymbol{\ell} \leq \boldsymbol{\ell}'\mathbf{MSE}_\Theta(\mathbf{T})\boldsymbol{\ell} = MSE_\Theta(\boldsymbol{\ell}'\mathbf{T}) \ \forall \Theta \in \Omega \text{ and } \forall \boldsymbol{\ell}.$$

Thus in the sense of *all* of the MSE comparisons defined previously, $\mathbf{T}^*$ is at least as good as $\mathbf{T}$.

The fact that $\mathbf{MSE}_\Theta(\mathbf{T}^*) - \mathbf{MSE}_\Theta(\mathbf{T})$ is negative semidefinite *and* unequal to the zero matrix for some $\Theta \in \Omega$ implies that some of the weak inequalities $(\leq)$ in the aforementioned MSE comparisons become strong inequalities $(<)$ for some $\Theta$. To see this, note that a nonzero negative semidefinite symmetric matrix necessarily has one or more *negative* diagonal entries.[5] Therefore, $MSE_\Theta(T_i^*) < MSE_\Theta(T_i)$ for some $\Theta$ and $i$, so that $\mathrm{E}_\Theta(d^2(\mathbf{T}^*, \mathbf{q}(\Theta))) < \mathrm{E}_\Theta(d^2(\mathbf{T}, \mathbf{q}(\Theta)))$ for some $\Theta$ and $MSE_\Theta(\boldsymbol{\ell}'\mathbf{T}^*) < MSE_\Theta(\boldsymbol{\ell}'\mathbf{T})$ for some $\Theta$ and $\boldsymbol{\ell}$. Thus, $\mathbf{T}^*$ is superior to $\mathbf{T}$ for at least some MSE comparisons in addition to being no worse for any of the MSE comparisons. We can now define multivariate analogues to the notions of relative efficiency and admissibility.

[4] By definition, $\mathbf{A}$ is negative semidefinite *iff* $\boldsymbol{\ell}'\mathbf{A}\boldsymbol{\ell} \leq 0\ \forall \boldsymbol{\ell}$. Then the $i^{\text{th}}$ diagonal entry of $\mathbf{A}$ must be $\leq 0$ since this entry can be defined by $\boldsymbol{\ell}'\mathbf{A}\boldsymbol{\ell}$ with $\boldsymbol{\ell}$ being a zero vector except for a 1 in the $i^{\text{th}}$ position.

[5] A nonzero matrix has at least unit rank. The rank of a negative semidefinite symmetric matrix is equal to the number of negatively valued eigenvalues, and all eigenvalues of a negative semidefinite matrix are $\leq 0$. The trace of a negative semidefinite symmetric matrix is equal to the sum of its eigenvalues. Since all diagonal entries in a negative semidefinite matrix must be $\leq 0$, it follows that a nonzero negative semidefinite symmetric matrix must have one or more negative diagonal entries.

**Definition 7.11 *Relative Efficiency and Admissibility with Respect to SMSE***

Let $\mathbf{T}^*$ and $\mathbf{T}$ be estimators of the $(k \times 1)$ vector $\mathbf{q}(\boldsymbol{\Theta})$. If $\mathbf{T}^*$ is SMSE superior to $\mathbf{T}$, then $\mathbf{T}^*$ is said to be **relatively more efficient** than $\mathbf{T}$. If there exists an estimator that is relatively more efficient than $\mathbf{T}$, then $\mathbf{T}$ is said to be **inadmissible**. Otherwise, $\mathbf{T}$ is said to be **admissible**.

As in the scalar case, if MSE is being used to measure estimator performance, the analyst need not consider any estimators of $\mathbf{q}(\boldsymbol{\Theta})$ that are inadmissible when searching for good estimators of $\mathbf{q}(\boldsymbol{\Theta})$.[6]

In either the scalar or multivariate case, a natural question to ask is whether an optimal estimator exists that has the smallest MSE or MSE matrix among all estimators of $\mathbf{q}(\boldsymbol{\Theta})$. We might call such an estimator most efficient, or simply efficient. Unfortunately, no such estimator exists in general. To clarify the issues involved, consider the scalar case and note that the degenerate estimator $T^* = t^*(\mathbf{X}) = \Theta_o$ would certainly have minimum mean-square error for estimating $\Theta$ if mean-square error were evaluated *at the point* $\Theta = \Theta_o$, i.e., $\text{MSE}_\Theta(T^*) = 0$ for $\Theta = \Theta_o$. Since a similar degenerate estimator could be defined for each $\Theta \in \Omega$, then for a given estimator to have minimum mean-square error *for every potential value of* $\Theta$, (i.e., uniformly in $\Theta$) it would be necessary that $MSE_\Theta(T)=0\ \forall \Theta \in \Omega$, which would imply that $\text{var}_\Theta(T)=0\ \forall \Theta \in \Omega$, and thus, that $P_\Theta(t(\mathbf{x}) = \Theta) = 1\ \forall \Theta \in \Omega$. In order to construct an estimator $T$ that satisfies the condition $P(t(\mathbf{x})=\Theta) = 1\ \forall \Theta \in \Omega$, it would be necessary to be able to identify the true value of $\Theta$ directly upon observing the sample outcome, $\mathbf{x}$. This essentially requires that the range of the random sample be dependent on the value of $\Theta$, denoted as $R_\Theta(\mathbf{X})$, in such a way that the sets $R_\Theta(\mathbf{X})$, $\Theta \in \Omega$, are all mutually exclusive, i.e., $R_{\Theta'}(X) \cap R_{\Theta''}(X) = \emptyset$ for $\Theta' \neq \Theta''$. Then, upon observing $\mathbf{x}$, one would only need to identify the set $R_\Theta(\mathbf{X})$ to which $\mathbf{x}$ belonged, and $\Theta$ would be immediately known. This is rarely, if ever, possible in practice, and so adopting a **minimum mean-square error criterion** for choosing an estimator of $\mathbf{q}(\boldsymbol{\Theta})$ is *not* feasible. A similar argument leads to the conclusion that there is in general no estimator of a $(k \times 1)$ vector $\mathbf{q}(\boldsymbol{\Theta})$ whose MSE matrix is smallest among the MSE matrices of all estimators of $\mathbf{q}(\boldsymbol{\Theta})$.[7]

While there generally does not exist an estimator that has a uniformly (i.e., for all $\boldsymbol{\Theta} \in \Omega$) minimum MSE or MSE matrix relative to *all* other estimators of $\mathbf{q}(\boldsymbol{\Theta})$, it is often possible to find an optimal estimator if one restricts the type of estimators under consideration. Two such restrictions that have been widely used in practice are **unbiasedness** and **linearity**, which we will examine in the next two subsections ahead.

---

[6] Some analysts use a *weak mean square error* (WMSE) criterion that relates to only expected squared distance considerations. $\mathbf{T}^*$ is WMSE superior to $\mathbf{T}$ *iff* $\mathrm{E}_\Theta(d^2(\mathbf{T}^*,\mathbf{q}(\boldsymbol{\Theta})) \leq \mathrm{E}_\Theta(d^2(\mathbf{T},\mathbf{q}(\boldsymbol{\Theta}))\ \forall \boldsymbol{\Theta} \in \Omega$, and $<$ for some $\boldsymbol{\Theta} \in \Omega$. Relative efficiency and admissibility can be defined in the context of WMSE superiority and are left to the reader.

[7] By "smallest MSE matrix," we mean that $\mathbf{MSE}_\Theta(\mathbf{T}^*)$-$\mathbf{MSE}_\Theta(\mathbf{T})$ is a *negative semidefinite matrix* for all estimators $\mathbf{T}$ of $\mathbf{q}(\boldsymbol{\Theta})$ and for all $\boldsymbol{\Theta}$.

*Unbiasedness* The property of unbiasedness refers to the balancing point or expectation of an estimator's probability distribution being equal to what is being estimated.

**Definition 7.12**
***Unbiased Estimator***

An estimator $\mathbf{T}$ is said to be an **unbiased estimator** of $\mathbf{q}(\boldsymbol{\Theta})$ *iff* $\mathrm{E}_{\boldsymbol{\Theta}}(\mathbf{T}) = \mathbf{q}(\boldsymbol{\Theta})$, $\forall \boldsymbol{\Theta} \in \boldsymbol{\Omega}$. Otherwise, the estimator is said to be **biased**.

As in the case of the MSE criteria, it is important to appreciate the significance of the condition $\forall\boldsymbol{\Theta}\in\Omega$ in the above definition. In the context of the point estimation problem, we have assumed that the true value of $\boldsymbol{\Theta}$, say $\boldsymbol{\Theta}_*$, is some element of the specified parameter space, $\Omega$, but we do not know which one. Thus, the property of unbiasedness is stated for *all possible contingencies* regarding the potential values for the true value of $\boldsymbol{\Theta}$. Due to the condition $\forall\boldsymbol{\Theta}\in\Omega$, the requirement for unbiasedness essentially means that $\mathrm{E}_{\boldsymbol{\Theta}}(\mathbf{T}) = \mathbf{q}(\boldsymbol{\Theta})$ *regardless* of which value of $\boldsymbol{\Theta}\in\Omega$ is the true value. Thus, for $\mathbf{T}$ to be unbiased, its density function must be balanced on the point $\mathbf{q}(\boldsymbol{\Theta})$, whatever the true value of $\boldsymbol{\Theta}$. Whether or not $\mathbf{T}$ has the unbiasedness property depends on the functional definition of $\mathbf{T}$, and in particular, on how the function translates the density function of $\mathbf{X} \sim f(x_1,\ldots,x_n;\boldsymbol{\Theta})$ into the density function of $\mathbf{T} \sim f(\mathbf{t};\boldsymbol{\Theta})$.

An unbiased estimator has the intuitively appealing property of being equal to $\mathbf{q}(\boldsymbol{\Theta})$ *on average*, the phrase having two useful interpretations. First, since the expectation operation is inherently a weighted average of the outcomes of $\mathbf{T}$, then the outcomes of $\mathbf{T}$ have a weighted average equal to $\mathbf{q}(\boldsymbol{\Theta})$. Alternatively, if one were to repeatedly and independently observe outcomes of the random sample $\mathbf{X}$, and thus repeatedly generate estimates of $\mathbf{q}(\boldsymbol{\Theta})$ using corresponding outcomes of the vector $\mathbf{T}$, then the simple average of all of the observed estimates would converge in probability (and, in fact, converge almost surely) elementwise to $\mathbf{q}(\boldsymbol{\Theta})$ by Khinchin's WLLN (or by Kolmogorov's SLLN in the case of almost-sure convergence), provided only that $\mathbf{q}(\boldsymbol{\Theta})$ is finite.

We provide the following example of an unbiased estimator of a parameter.

**Example 7.4**
***Unbiased Estimator of the Mean of an Exponential Distribution***

Let the population distribution of hard disk operating lives for a certain brand of hard disk be a member of the exponential family of densities $f(z;\theta) = (1/\theta)\, e^{-1/\Theta} I_{(0,\infty)}(z)$, $\theta \in \Omega = (0,\infty)$. Let $T = n^{-1}\sum_{i=1}^{n} X_i$ be an estimator of the expected operating life of the hard disk, $\theta$, where $(X_1,\ldots,X_n)$ is a random sample from the exponential population distribution. Then $T$ is an unbiased estimator of $\theta$, since $\mathrm{E}(X_i) = \theta\ \forall i$, which implies that $\mathrm{E}(T) = \mathrm{E}\ (n^{-1}\sum_{i=1}^{n} X_i) = n^{-1}\sum_{i=1}^{n}\mathrm{E}(X_i) = \theta$ *regardless* of the value of $\theta > 0$. Thus, for example, if the true value of $\theta$ were 2, then $\mathrm{E}(T) = 2$, or if the true value of $\theta$ were100, then $\mathrm{E}(T) = 100$. □

*MVUE, MVLUE or BLUE, and Efficiency* The unbiasedness criterion ensures only that an estimator will have a density that has a central tendency or balancing point of $\mathbf{q}(\boldsymbol{\Theta})$. However, it is clear that we would also desire that the density not be too spread out around this balancing point for fear that an estimate could be

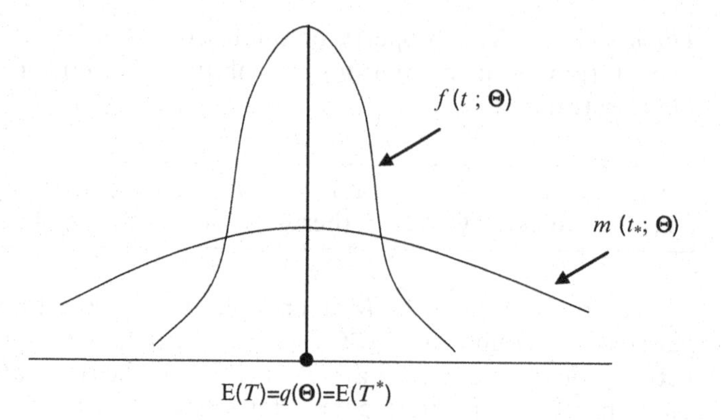


**Figure 7.3** The densities of two unbiased estimators of $q(\Theta)$.

generated that was a significant distance from $\mathbf{q}(\Theta)$ with high probability. Graphically, we would prefer the estimator $T$ to the estimator $T^*$ in Figure 7.3, where both of these estimators are unbiased estimators of $\mathbf{q}(\Theta)$.

The foregoing considerations motivate that, if one wishes to use an unbiased estimator of $\mathbf{q}(\Theta)$, one should use the unbiased estimator that also has minimum variance, or minimum covariance matrix if **T** is a vector, among all unbiased estimators of $\mathbf{q}(\Theta)$. Since $\mathbf{Bias}_\Theta(\mathbf{T}) = \mathbf{0}$ for all estimators in the unbiased class of estimators, $\text{MSE}_\Theta(T) = \text{var}_\Theta(T)$ or $\mathbf{MSE}_\Theta(\mathbf{T}) = \mathbf{Cov}_\Theta(\mathbf{T})$, and we can thus view the objective of minimizing $\text{var}(T)$ or $\mathbf{Cov}(\mathbf{T})$ equivalently as searching for the estimator with the smallest MSE or smallest MSE matrix *within the class of unbiased estimators*. In the definition below, we introduce the notation $\mathbf{A} \preceq \mathbf{B}$ to indicate that **matrix A is smaller than matrix B** by a negative semidefinite matrix, i.e., $\mathbf{A} - \mathbf{B} = \mathbf{C}$ is a negative semidefinite matrix.

**Definition 7.13**
***Minimum Variance Unbiased Estimator (MVUE)***

An estimator **T** is said to be a **minimum variance unbiased estimator** of $\mathbf{q}(\Theta)$ *iff* **T** is an unbiased estimator of $\mathbf{q}(\Theta)$, and

**(a)** (scalar case) $\text{var}_\Theta(T) \le \text{var}_\Theta(T^*)\ \forall\ \Theta \in \Omega$ and $\forall\ T^* \in U_{q(\Theta)}$;

**(b)** (vector case) $\mathbf{Cov}_\Theta(\mathbf{T}) \preceq \mathbf{Cov}_\Theta(\mathbf{T}^*)\ \forall\ \Theta \in \Omega$ and $\forall\ \mathbf{T}^* \in U_{\mathbf{q}(\Theta)}$;

where $U_{\mathbf{q}(\Theta)}$ is the set of all unbiased estimators of $\mathbf{q}(\Theta)$.[8]

[8]This is alternatively referred to in the literature by the term uniformly minimum variance unbiased estimator (UMVUE), where the adverb "uniformly" is used to emphasize the condition "$\forall\Theta\in\Omega$." In our usage of the terms, MVUE and UMVUE will be interchangeable.

Definition 7.13 implies that an estimator is a MVUE if the estimator is unbiased and if there is no other unbiased estimator that has a smaller variance or covariance matrix for any $\Theta \in \Omega$. Drawing direct analogies to the discussion of the MSE criteria, a MVUE, **T**, is such that $\mathrm{MSE}_\Theta(T_i) = \mathrm{var}_\Theta(T_i) \leq \mathrm{var}_\Theta(T_i^*) = \mathrm{MSE}_\Theta(T_i^*)\ \forall \Theta \in \Omega$ and $\forall i$, where $\mathbf{T}^*$ is any estimator *in the unbiased class of estimators*. Furthermore, $\mathrm{E}_\Theta\,(d^2(\mathbf{T},\mathbf{q}(\Theta)) \leq \mathrm{E}_\Theta\,(d^2(\mathbf{T}^*, \mathbf{q}(\Theta))\ \forall \Theta \in \Omega$ and $\mathrm{MSE}_\Theta(\ell'\mathbf{T}) = \mathrm{var}_\Theta(\ell'\mathbf{T}) \leq \mathrm{var}_\Theta(\ell'\mathbf{T}^*) = \mathrm{MSE}_\Theta(\ell'\mathbf{T}^*)\ \forall \Theta \in \Omega$ and $\forall \ell$. Thus, within the class of unbiased estimators, a MVUE of $\mathbf{q}(\Theta)$ is at least as good as any other estimator of $\mathbf{q}(\Theta)$ in terms of all of the types of MSE comparisons that we have discussed previously. If **T** is a MVUE for $\mathbf{q}(\Theta)$, then **T** is said to be **efficient** within the class of unbiased estimators.

Unfortunately, without the aid of theorems that facilitate the discovery of MVUES, finding a MVUE of $\mathbf{q}(\Theta)$ can be quite challenging even when the point estimation problem appears to be quite simple. The following example illustrates the general issues involved.

**Example 7.5** ***MVUE of the Mean of a Bernoulli Distribution***

Consider defining an MVUE for the parameter $p$ using a random sample of size 2 from the Bernoulli population distribution $f(z;p) = p^z(1-p)^{1-z} I_{\{0,1\}}(z)$. First of all, the range of the random sample is $\{(0,0), (0,1), (1,0), (1,1)\}$, which represents the domain of the estimator function $T = t(\mathbf{X})$. For $t(\mathbf{X})$ to be in the unbiased class, the following general condition must be met:

$$\mathrm{E}(t(\mathbf{X})) = t(0,0)(1-p)^2 + t(0,1)(1-p)p + t(1,0)p(1-p) + t(1,1)p^2 = p\ \forall p \in [0,1]$$

This unbiasedness conditions implies the following set of restrictions on the definition of $t(\mathbf{X})$:

$$\text{If } p \in \left\{ \begin{matrix} \{1\} \\ \{0\} \\ (0,1) \end{matrix} \right\} \text{ then}$$

$$\mathrm{E}(t(\mathbf{X})) = \left\{ \begin{matrix} t(1,1)p^2 \\ t(0,0)(1-p)^2 \\ (t(0,1)+t(1,0))(1-p)p + p^2 \end{matrix} \right\} = p \ \ iff \left\{ \begin{matrix} t(1,1) = 1 \\ t(0,0) = 0 \\ t(0,1) + t(1,0) = 1 \end{matrix} \right\}$$

Now consider the variance of $t(\mathbf{X})$, which by definition can be written as

$$\begin{aligned} \mathrm{var}(t(\mathbf{X})) &= (t(0,0)-p)^2(1-p)^2 + (t(0,1)-p)^2(1-p)p \\ &\quad + (t(1,0)-p)^2 p(1-p) + (t(1,1)-p)^2 p^2 \\ &= 2p^2(1-p)^2 + (t(0,1)-p)^2(1-p)p + (t(1,0)-p)^2 p(1-p) \end{aligned}$$

where we have used the facts that $\mathrm{E}(t(\mathbf{X}))= p$, $t(0,0) = 0$, and $t(1,1) = 1$ since $t(\mathbf{X})$ must be unbiased. Also because of the unbiasedness condition, we can substitute $t(0,1) = 1-t(1,0)$ into the variance expression to obtain

$$\mathrm{var}(t(X) = 2p^2(1-p)^2 + (1-p-t(1,0))^2(1-p)p + (t(1,0)-p)^2 p(1-p).$$

The first-order condition for a minimum of the variance is given by

$$\frac{d\,\mathrm{var}\,t(X)}{dt(1,0)} = -2(1-p)^2 p + 2t(1,0)p(1-p) + 2t(1,0)p(1-p) - 2p^2(1-p) = 0,$$

which implies that $4p(1-p)t(1,0) = 2(1-p)^2p + 2p^2(1-p)$, so that $t(1,0) = (1/2)$, which then implies $t(0,1) = (1/2)$.

We have thus defined the function $T = t(\mathbf{X})$ that represents a MVUE of $p$ by associating an appropriate outcome of $T$ with each random sample outcome. The preceding results can be represented collectively as $t(x_1, x_2) = (1/2)(x_1 + x_2) = \bar{x}$, so that the MVUE is $\bar{X}$. □

A number of general theorems that can often be used to simplify the search for a MVUE will be presented in Section 7.5.

For purposes of simplicity and tractability, as well as for cases where little can be assumed about the probability model other than conditions on low-order moments, attention is sometimes restricted to estimators that are unbiased and that have minimum variance or covariance matrix among all unbiased estimators that are *linear* functions of the sample outcome. Such an estimator is called a BLUE or MVLUE, as indicated in the following definition.

**Definition 7.14**
***Best Linear Unbiased Estimator (BLUE) or Minimum Variance Linear Unbiased Estimator (MVLUE)***

An estimator $\mathbf{T}$ is said to be a BLUE or MVLUE of $\mathbf{q}(\boldsymbol{\Theta})$ *iff*

1. $\mathbf{T}$ is a linear function, $\mathbf{T} = t(\mathbf{X}) = \mathbf{AX} + \mathbf{b}$, of the random sample $\mathbf{X}$,
2. $\mathrm{E}_{\boldsymbol{\Theta}}(\mathbf{T}) = \mathbf{q}(\boldsymbol{\Theta})\ \forall\ \boldsymbol{\Theta} \in \Omega$ ($\mathbf{T}$ is unbiased),
3. $\mathbf{T}$ has minimum variance or covariance matrix among all unbiased estimators that are also linear functions of the random sample $\mathbf{X}\ \forall\ \boldsymbol{\Theta} \in \Omega$.

A BLUE estimator of $\mathbf{q}(\boldsymbol{\Theta})$ is also referred to as an **efficient** estimator *within the class of linear unbiased estimators*. The following is an example identifying the BLUE or MVLUE of the mean of *any* population distribution with a finite mean and variance.

**Example 7.6**
***BLUE of Population Mean for Any Population Distribution***

Let $(X_1,\ldots,X_n)$ be a random sample from some population distribution $f(z;\Theta)$ having a finite mean $\mu = q_1(\Theta)$ and variance $\sigma^2 = q_2(\Theta)$. What is the BLUE of the mean of the population distribution?

**Answer:** We are examining linear estimators, and thus $t(\mathbf{X}) = \sum_{i=1}^n a_i X_i + b$. For $T$ to be unbiased, we require that $\sum_{i=1}^n a_i = 1$ and $b = 0$, since $\mathrm{E}(T)=\mathrm{E}\left[\sum_{i=1}^n a_i X_i + b\right] = \left[\sum_{i=1}^n a_i \mathrm{E}(X_i)\right] + b = \mu\left[\sum_{i=1}^n a_i\right] + b = \mu$ holds for all potential $\mu$ *iff* $\sum_{i=1}^n a_i = 1$

and $b = 0$. The variance of $T$ is simply $\sigma^2 \sum_{i=1}^n a_i^2$ because $(X_1,\ldots,X_n)$ is a random sample from $f(z;\Theta)$. Thus, to find the BLUE, we must solve the following minimization problem:

$$\min_{a_1,\ldots,a_n} \sigma^2 \sum_{i=1}^n a_i^2 \text{ subject to } \sum_{i=1}^n a_i = 1.$$

The Lagrangian form of this minimization problem is given by

$$L = \sigma^2 \sum_{i=1}^n a_i^2 - \lambda\left[\sum_{i=1}^n a_i - 1\right],$$

and the first-order conditions are

$$\frac{\partial L}{\partial a_i} = 2\sigma^2 a_i - \lambda = 0, \quad i = 1,\ldots n$$

and

$$\frac{\partial L}{\partial \lambda} = 1 - \sum_{i=1}^n a_i = 0.$$

The first $n$ conditions imply $a_1 = a_2 = \ldots = a_n$, since $a_i = \lambda/2\sigma^2 \; \forall i$, and then $\sum_{i=1}^n a_i = 1$ requires that $a_i = 1/n$, $i = 1,\ldots,n$. Thus, $t(\mathbf{X}) = \sum_{i=1}^n a_i X_i = n^{-1}\sum_{i=1}^n X_i = \overline{X}$, so that the sample mean is the BLUE (or MVLUE) of the mean of *any* population distribution having a finite mean and variance. The reader should check that the second-order conditions for a minimum are in fact met.□

In addition to estimating the means of population distributions, a prominent BLUE arises in the context of least-squares estimation of the parameters of a general linear model, which we will examine in Chapter 8.

### 7.3.3 Asymptotic Properties

When finite sample properties are intractable or else inapplicable due to the nonexistence of the appropriate expectations that define means and variances, one generally resorts to **asymptotic properties** to rank the efficiency of estimators. In addition, asymptotic properties are of fundamental interest if the analyst is interested in assessing the effects on estimator properties of an ever-increasing number of sample observations.

Asymptotic properties of estimators are essentially equivalent in concept to the finite sample properties presented heretofore, except that asymptotic properties are based on the *asymptotic distributions of estimators* rather than estimators' exact finite sampling distributions. In particular, asymptotic analogues to MSE, relative efficiency, unbiasedness, and minimum-variance unbiasedness can be defined with reference to asymptotic distributions of estimators. However, a problem of nonuniqueness of asymptotic properties arises because of the inherent nonuniqueness of asymptotic distributions.

To clarify the difficulties that can arise when using asymptotic distributions as a basis for defining estimator properties, let $T_n$ denote an estimator of the scalar $q(\Theta)$ based on $n$ sample observations, and suppose

$b_n^{-1}(T_n - q(\Theta)) \xrightarrow{d} N(0,1)$. Then one might consider defining asymptotic properties of $T_n$ in terms of the asymptotic distribution $N(q(\Theta), b_n^2)$. However, by Slutsky's theorem it follows that $(n/(n-k))^{1/2} b_n^{-1}(T_n - q(\Theta)) \xrightarrow{d} N(0,1)$ for a fixed value of $k$ since $(n/(n-k))^{1/2} \to 1$, so that an alternative asymptotic distribution could be $T_n \overset{a}{\sim} N(q(\Theta), ((n-k)/n)b_n^2)$, producing a different asymptotic variance with implications for estimator performance measures that are functions of the variance of estimators. The difficulty is that the centering and scaling required to achieve a limiting distribution is not unique, leading to both nonunique asymptotic distributions and nonunique asymptotic properties derived from them.

There are two basic ways of addressing the aforementioned nonuniqueness problem when dealing with asymptotic properties. One approach, which we will mention only briefly, is to rank estimators only on the basis of limits of asymptotic property comparisons so as to remove the effects of any arbitrary scaling or centering from the comparison. For example, referring to the previous illustration of nonuniqueness, let the asymptotic distribution of $T_n$ be $N(q(\Theta), b_n^2)$ and let $T_n^*$ have the asymptotic distribution $N(q(\Theta), ((n-k)/n)b_n^2)$. **Asymptotic MSEs (AMSEs)** based on these asymptotic distributions would be calculated as $\text{AMSE}_\Theta(T_n) = E_\Theta\left((T_n - q(\Theta))^2\right) = b_n^2$ and $\text{AMSE}_\Theta(T_n^*) = E_\Theta\left((T_n^* - q(\Theta))^2\right) = ((n-k)/n)b_n^2$. Then the **asymptotic relative efficiency** (**ARE**) of $T_n$ with respect to $T_n^*$ would be represented as

$$\text{ARE}_\Theta(T_n, T_n^*) = \frac{\text{AMSE}_\Theta(T_n^*)}{\text{AMSE}_\Theta(T_n)} = \frac{n-k}{n}.$$

Using the ARE in this form, one would be led to the conclusion that $T_n^*$ is **asymptotically relatively more efficient** than $T_n$, which in the context of the previous illustration of nonuniqueness would be absurd since $T_n$ and $T_n^*$ are the same estimator. However, $\lim_{n\to\infty}(\text{ARE}_\Theta(T_n, T_n^*)) = 1$, which leads to the conclusion that $T_n$ and $T_n^*$ are equally preferable on the basis of asymptotic MSE considerations, which of course is the correct conclusion in the current context. The limit operation removes the arbitrary scaling of the asymptotic variance. To operationalize this approach for general applications requires extensions of limit notions (limit superiors, or lim sups) which we will leave for future study (see L. Schmetterer (1974), *Introduction to Mathematical Statistics*. New York: Springer-Verlag, pp. 335–342).

An alternative approach for avoiding nonuniqueness of asymptotic properties is to restrict the use of asymptotic properties to classes of estimators for which the problem will not occur. For our purposes, it will suffice to examine the consistent asymptotically normal (CAN) class of estimators (for other possibilities, see E. Lehmann, *Point Estimation*, pp. 347–348).

Prior to identifying the CAN class of estimators, we examine the property of consistency.

*Consistency* A **consistent estimator** is an estimator that converges in probability (element-wise if $\mathbf{T}_n$ is a vector) to what is being estimated.

**Definition 7.15**
***Consistent Estimator***

$\mathbf{T}_n$ is said to be a **consistent estimator** of $\mathbf{q}(\boldsymbol{\Theta})$ *iff* $\text{plim}_{\boldsymbol{\Theta}}(\mathbf{T}_n) = \mathbf{q}(\boldsymbol{\Theta})\ \forall\ \boldsymbol{\Theta} \in \Omega$.

Thus, for large enough $n$ (i.e., for large enough sample size), there is a high probability that the outcome of a scalar estimator $T_n$ will be in the interval $(q(\boldsymbol{\Theta})-\varepsilon, q(\boldsymbol{\Theta})+\varepsilon)$ for arbitrarily small $\varepsilon > 0$ regardless of the value of $\boldsymbol{\Theta}$. Relatedly, the sampling density of $T_n$ concentrates on the true value of $q(\boldsymbol{\Theta})$ as the sample size $\rightarrow \infty$ if $T_n$ is a consistent estimator of $q(\boldsymbol{\Theta})$. Consistency is clearly a desirable property of an estimator, since it ensures that increasing sample information will ultimately lead to an estimate that is essentially certain to be arbitrarily close to what is being estimated, $q(\boldsymbol{\Theta})$.

Since $\mathbf{T}_n \xrightarrow{m} \mathbf{q}(\boldsymbol{\Theta})$ implies $\mathbf{T}_n \xrightarrow{p} \mathbf{q}(\boldsymbol{\Theta})$, we can state **sufficient conditions for consistency** of $\mathbf{T}_n$ in terms of unbiasedness and in terms of variance convergence to zero. Specifically, if $\mathbf{T}_n$ is unbiased, or if the bias vector converges to zero as $n \rightarrow \infty$, and if $\text{var}(T_n) \rightarrow 0$ as $n \rightarrow \infty$, or $\mathbf{Cov}(\mathbf{T}_n) \rightarrow \mathbf{0}$ as $n \rightarrow \infty$ if $\mathbf{T}_n$ is a vector, then $\mathbf{T}_n$ is a consistent estimator of $\mathbf{q}(\boldsymbol{\Theta})$ by mean-square convergence.

**Example 7.7**
***Sample Mean as a Consistent Estimator of a Population Mean***

Let $(X_1,\ldots,X_n)$ be a random sample from a population distribution, $f(z;\boldsymbol{\Theta})$, of the incomes of buyers of new Cadillacs, where $\text{E}(Z) = q_1(\boldsymbol{\Theta}) = \mu$ and $\text{var}(Z) = q_2(\boldsymbol{\Theta}) = \sigma^2 < \infty$. The sample mean $t_n(\mathbf{X}) = \overline{X}_n$ is a consistent estimator of the mean income, $\mu$, of Cadillac buyers since $\overline{X}_n$ is unbiased, and its variance is such that $(\sigma^2/n) \rightarrow 0$ as $n \rightarrow \infty$. □

We note for future reference that a sequence of estimators can be consistent for $\mathbf{q}(\boldsymbol{\Theta})$ even *without* $\text{E}(\mathbf{T}_n) \rightarrow \mathbf{q}(\boldsymbol{\Theta})$ even if $\text{E}(\mathbf{T}_n)$ exists $\forall n$. This at first seems counter intuitive, since if $\mathbf{T}_n$ is consistent, its density collapses on $\mathbf{q}(\boldsymbol{\Theta})$ as $n \rightarrow \infty$. Note the following counterexample.

**Example 7.8**
***Consistency without*** $\text{E}(T_n) \rightarrow q(\Theta)$

Let the sampling density of $T_n$ be defined as $f(t_n;\Theta) = (1-n^{-1/2})I_{\{\Theta\}}(t_n) + n^{-1/2}I_{\{n\}}(t_n)$. Note that as $n \rightarrow \infty$, $\lim_{n\rightarrow\infty} P[|t_n-\Theta| < \varepsilon] = 1$, for any $\varepsilon > 0$, and $T_n$ is consistent for $\Theta$. However, since $\text{E}(T_n) = \Theta(1-n^{-1/2}) + n(n^{-1/2}) = \Theta(1-n^{-1/2}) + n^{1/2}$, then as $n \rightarrow \infty$, $\text{E}(T_n) \rightarrow \infty$. □

The divergence of the expectation in Example 7.8 is due to the fact that the density function of $T_n$, although collapsing to the point $\Theta$ as $n \rightarrow \infty$, was not collapsing at a *fast enough rate* for the expectation to converge to $\Theta$. In particular, the density weighting assigned to the outcome $n$ in defining the expectation went to zero at a rate slower than $n$ went to infinity as $n \rightarrow \infty$, causing the divergence. A sufficient condition for $T_n \xrightarrow{p} q(\boldsymbol{\Theta}) \Rightarrow \lim_{n\rightarrow\infty}\text{E}(T_n) = q(\boldsymbol{\Theta})$ is provided in the following theorem:

**Theorem 7.1**
***Sufficient Condition for*** $T_n \xrightarrow{p} q(\Theta) \Rightarrow \lim_{n\rightarrow\infty} E(T_n) = q(\Theta)$

*If* $\text{E}(T_n^2)$ *exists and is bounded* $\forall n$, *so that* $\text{E}(T_n^2) \leq m < \infty\ \forall n$, *then convergence in probability implies convergence in mean.*

**Proof** Rao, Statistical Inference, pp. 121. ■

Note the sufficient condition given in Theorem 7.1 does not hold in Example 7.7.

*Consistent Asymptotically Normal (CAN) Estimators* The class of **consistent asymptotically normal (CAN)** estimators of $\mathbf{q}(\boldsymbol{\Theta})$ is defined in the statistical literature to be the collection of all estimators of $\mathbf{q}(\boldsymbol{\Theta})$ for which $n^{1/2}\,(\mathbf{T}_n - \mathbf{q}(\boldsymbol{\Theta})) \xrightarrow{d} N([\mathbf{0}], \boldsymbol{\Sigma}_T)$, where $\boldsymbol{\Sigma}_T$ is a positive definite covariance matrix that may depend on the value of $\boldsymbol{\Theta}$. We will allow this dependence to be implicit rather than utilize notation such a $\boldsymbol{\Sigma}_\mathbf{T}(\boldsymbol{\Theta})$. Note the consistency of $\mathbf{T}_n$ follows immediately, since by Slutsky's theorem $n^{-1/2}[n^{1/2}(\mathbf{T}_n - \mathbf{q}(\boldsymbol{\Theta}))] = \mathbf{T}_n - \mathbf{q}(\boldsymbol{\Theta}) \xrightarrow{d} 0 \cdot \mathbf{Z} = \mathbf{0}$, where $\mathbf{Z} \sim N(\mathbf{0}, \boldsymbol{\Sigma}_\mathbf{T})$, which implies $\mathbf{T}_n - \mathbf{q}(\boldsymbol{\Theta}) \xrightarrow{d} \mathbf{0}$ or equivalently $\mathbf{T}_n \xrightarrow{d} \mathbf{q}(\boldsymbol{\Theta})$. The CAN class contains a large number of the estimators used in empirical work.

Because all of the estimators in the CAN class utilize precisely the same sequence of centering (i.e., $\mathbf{q}(\boldsymbol{\Theta})$ is subtracted from $\mathbf{T}_n$) and scaling (i.e., $\mathbf{T}_n - \mathbf{q}(\boldsymbol{\Theta})$ is multiplied by $n^{1/2}$), the problem of nonuniqueness of asymptotic distributions and properties does not arise. Asymptotic versions of MSEs, MSE matrices, bias vectors, variances, and covariance matrices can be defined via expectations taken with respect to the unique asymptotic distribution of estimators, where $\mathbf{T}_n \overset{a}{\sim} N(\mathbf{q}(\boldsymbol{\Theta}), n^{-1}\boldsymbol{\Sigma}_\mathbf{T})$. In particular, letting the prefix $A$ denote an asymptotic property, and letting $E_A$ denote an expectation taken with respect to an asymptotic distribution, we have within the CAN class $\forall \boldsymbol{\Theta} \in \Omega$,

$$\mathbf{AMSE}(\mathbf{T}_n) = \mathbf{ACov}(\mathbf{T}_n) = E_A((\mathbf{T}_n - \mathbf{q}(\boldsymbol{\Theta}))(\mathbf{T}_n - \mathbf{q}(\boldsymbol{\Theta}))') \quad \text{(multivariate)}$$
$$= \text{Avar}(T_n) = E_A\left((T_n - q(\boldsymbol{\Theta}))^2\right) \quad \text{(scalar)}$$

and

$$\mathbf{ABIAS}(\mathbf{T}_n) = E_A(\mathbf{T}_n - \mathbf{q}(\boldsymbol{\Theta})) = \mathbf{0}.$$

The zero value of the asymptotic bias indicates that a CAN estimator of $\mathbf{q}(\boldsymbol{\Theta})$ is necessarily **asymptotically unbiased**. We pause to note that there is a lack of consensus in the literature regarding the definition of **asymptotic unbiasedness**, and Example 7.8 is useful for illustrating the issues involved. Some statisticians define asymptotic unbiasedness of an estimator sequence in terms of the limit of the expected values of the estimators in the sequence, where $\lim_{n\to\infty} E(\mathbf{T}_n) = \mathbf{q}(\boldsymbol{\Theta})\ \forall \boldsymbol{\Theta} \in \Omega$ characterizes an **asymptotically unbiased estimator**. Under this definition, the estimator in Example 7.8 would *not* be asymptotically unbiased, but rather would be **asymptotically biased**. It is clear that this definition of asymptotic unbiasedness requires that the expectations in the sequence exist, as they do in Example 7.8. Within the CAN class, the two definitions of asymptotic unbiasedness will coincide if the second order moments of the estimators in the sequence $\{\mathbf{T}_n\}$ are bounded (recall Theorem 7.1), since then $\lim_{n\to\infty} E(\mathbf{T}_n) = \mathbf{q}(\boldsymbol{\Theta}) = E_A(\mathbf{T}_n)$. Otherwise, the definitions may refer to different concepts of unbiasedness, as Example 7.8 demonstrates.

Thus, one must discern the definition of asymptotic unbiasedness being used by any analyst by the context of the discussion.

Given the preceding definition of asymptotic properties, we can now define the meaning of **asymptotic relative efficiency** and **asymptotic admissibility** uniquely for CAN estimators.

**Definition 7.16**
***Asymptotic Relative Efficiency and Asymptotic Admissibility***

Let $\mathbf{T}_n$ and $\mathbf{T}_n^*$ be CAN estimators of $\mathbf{q}(\boldsymbol{\Theta})$ such that $n^{1/2}(\mathbf{T}_n - \mathbf{q}(\boldsymbol{\Theta})) \xrightarrow{d} N(\mathbf{0}, \boldsymbol{\Sigma}_\mathbf{T})$ and $n^{1/2}(\mathbf{T}_n^* - \mathbf{q}(\boldsymbol{\Theta})) \xrightarrow{d} N(\mathbf{0}, \boldsymbol{\Sigma}_{\mathbf{T}^*})$.

**a.** If $T_n$ and $T_n^*$ are *scalars*, then the **asymptotic relative efficiency** of $T_n$ with respect to $T_n^*$ is given by

$$\text{ARE}_\Theta(T_n, T_n^*) = \frac{\text{AMSE}_\Theta(T_n^*)}{\text{AMSE}_\Theta(T_n)} = \frac{\Sigma_{\mathbf{T}^*}}{\Sigma_\mathbf{T}} \forall \Theta \in \Omega.$$

$T_n$ is **asymptotically relatively more efficient** than $T_n^*$ if $\text{ARE}_\Theta(T_n, T_n^*) \geq 1$ $\forall \boldsymbol{\Theta} \in \Omega$ and $> 1$ for some $\boldsymbol{\Theta} \in \Omega$.

**b.** $\mathbf{T}_n$ is **asymptotically relatively more efficient** than $\mathbf{T}_n^*$ *iff* $\boldsymbol{\Sigma}_\mathbf{T} - \boldsymbol{\Sigma}_{\mathbf{T}^*}$ is negative semidefinite $\forall \Theta \in \Omega$ and $\boldsymbol{\Sigma}_\mathbf{T} - \boldsymbol{\Sigma}_{\mathbf{T}^*} \neq \mathbf{0}$ for some $\boldsymbol{\Theta} \in \Omega$.

**c.** If there exists an estimator that is asymptotically relatively more efficient than $\mathbf{T}_n$, then $\mathbf{T}_n$ is **asymptotically inadmissible**. Otherwise $\mathbf{T}_n$ is **asymptotically admissible.**

A discussion of the meaning of ARE and asymptotic admissibility, as well as all of the other asymptotic properties presented to this point, would be completely analogous to the discussion presented in the finite sample case, except now all interpretations would be couched in terms of approximations based on asymptotic distributions. We leave it to the reader to draw the analogies.

**Example 7.9**
***Relative Asymptotic Efficiency of Two Estimators of Exponential Mean***

Recall Example 7.4 regarding the estimation of the expected operating lives of hard disks, $\theta$, using a random sample from an exponential population distribution. As an alternative estimator of $\theta$, consider the following:

$$T_n^* = t_n^*(\mathbf{X}) = \left[\frac{1}{2}\left(n^{-1}\sum_{i=1}^{n} X_i^2\right)\right]^{1/2} = (M'_2/2)^{1/2}.$$

Recall that in the case of the exponential probability distribution, $\mu'_2 = \text{E}(X_i^2) = 2\,\theta^2$. Since $M'_2$ is the second order sample moment based on a random sample from a probability distribution, we know that $M'_2 \xrightarrow{p} 2\theta^2$ and that

$$\frac{n^{1/2}(M'_2 - \mu'_2)}{\left[\mu'_4 - (\mu'_2)^2\right]^{1/2}} = \frac{n^{1/2}(M'_2 - 2\theta^2)}{\left[20\theta^4\right]^{1/2}} \xrightarrow{d} N(0, 1)$$

where $\mu'_4 = \left[\left(d^4(1-\theta t)^{-1}/dt^4\right)\right]_{t=0} = 24\theta^4$.

Now note that $T_n^*$ is a continuous function of $M'_2$ so that plim $(T_n^*) =$ plim$(M'_2/2)^{1/2} = (\text{plim}(M'_2)/2)^{1/2} = \theta$ by Theorem 5.5. Therefore, $T_n^*$ is a consistent estimator of $\theta$. Furthermore, $n^{1/2}(M'_2 - \theta)$ has a normal limiting distribution and is thus a CAN estimator. To see this, recall Theorem 5.39 on the asymptotic distribution of functions of asymptotically normal random variables, where in this application, $T_n^*$ is a function of the asymptotically normal random variable $M'_2 \overset{a}{\sim} N(2\theta^2, 20\theta^4/n)$. Since

$$G = \left[\frac{dT_n^*}{dM'_2}\right]_{M'_2=2\theta^2} = \left[\frac{1}{4}(M'_2/2)^{-1/2}\right]_{M'_2=2\theta^2} = (4\theta)^{-1},$$

which is $\neq 0 \;\forall\, \theta > 0$ and thus of "full row rank," it follows that $n^{1/2}(T_n^* - \theta) \xrightarrow{d} N(0, G[20\theta^4/n]G') = N(0, 1.25\,\theta^2)$.

In comparing $T_n^*$ with $\overline{X}_n$ as estimators of $\theta$, it is now clear that although both are consistent and asymptotically normal estimators of $\theta$, $\overline{X}_n$ is **asymptotically more efficient** than $T_n^*$, since in comparing the asymptotic variances of the limiting distributions of $n^{1/2}(\overline{X}_n - \theta)$ and $n^{1/2}(T_n^* - \theta)$, we have that $\theta^2 < 1.25\,\theta^2$. □

*Asymptotic Efficiency* At this point it would seem logical to proceed to a definition of *asymptotic efficiency* in terms of a choice of estimator in the CAN class that has the smallest asymptotic variance or covariance matrix $\forall \boldsymbol{\Theta} \in \Omega$ (compare to Definition 7.13). Unfortunately, LeCam (1953)[9] has shown that such an estimator does not exist without further restrictions on the class of estimators. In particular, LeCam (1953) effectively showed that for *any* CAN estimator one can always define an alternative estimator that has a smaller variance or covariance matrix for at least one $\boldsymbol{\Theta} \in \Omega$. The implication of this result is that one cannot define an achievable lower bound to the asymptotic variances or covariance matrices of CAN estimators, so that no asymptotically optimal estimator exists.

On the other hand, LeCam (1953) also showed that under mild regularity conditions, there does exist a lower bound to the asymptotic variance or covariance matrix of a CAN estimator that holds for all $\boldsymbol{\Theta} \in \Omega$ except on a set of $\boldsymbol{\Theta}$-values having *Lebesque measure zero*, which is the **Cramer-Rao Lower Bound** that will be discussed in Section 7.5. Note that the Lebesque measure of a set of $\boldsymbol{\Theta}$-values can be thought of as the volume of the set within the $k$-dimensional parameter space. A set having Lebesque measure zero is a set with zero volume in $k$-space, e.g., a collection of isolated points, or a set of points having dimension less than $k$ (such as a square and its interior in a three-dimensional space, or a line in two-dimensional space). A set of Lebesque measure zero is a nonstochastic analogue to a set having probability zero, and such a set is thus practically irrelevant relative to its complement. It is thus meaningful to speak

[9]LeCam, L., (1953) "*On Some Asymptotic Properties of Maximum Likelihood Estimates and Related Bayes Estimates*", University of California Publications in Statistics, 1:277–330, 1953.

of a lower bound on the asymptotic variance or covariance matrix of a CAN estimator of $\mathbf{q}(\Theta)$ that holds *almost everywhere* in the parameter space (i.e., except for a set of Lebesque measure zero), and then a search for an estimator that achieves this bound becomes meaningful as well.

At this point we will state a general definition of asymptotic efficiency for CAN estimators. In Section 7.5, we will be much more precise about the functional form of the asymptotic covariance matrix of an asymptotically efficient estimator.

**Definition 7.17**
***Asymptotic Efficiency***

If $\mathbf{T}_n$ is a CAN estimator of $\mathbf{q}(\Theta)$ having the smallest asymptotic covariance matrix among all CAN estimators $\forall \Theta \in \Omega$, except on a set of Lebesque measure zero, $\mathbf{T}_n$ is said to be asymptotically efficient.

As a final remark, it is possible to remove the qualifier "except on a set of Lebesque measure zero" if the CAN class of estimators is further restricted so that only estimators that converge *uniformly* to the normal distribution are considered. Roughly speaking, *uniform* convergence of a function sequence $F_n(x)$ to $F(x)$ requires that the rate at which convergence occurs is uniform across all x is the domain of $F(x)$, unlike ordinary convergence (recall Definition 5.7) which allows for the possibility that the rate is different for each $x$. The restricted class of estimators is called the Consistent Uniformly Asymptotically Normal (CUAN) class, and within the CUAN class it is meaningful to speak of an estimator that literally has the smallest asymptotic covariance matrix. The interested reader can consult C.R. Rao, (1963) "*Criteria of Estimation in Large Samples,*" Sankhya, Series A, pp. 189–206 for further details.

## 7.4 Sufficient Statistics

Sufficient statistics for a given estimation problem are a collection of statistics or, equivalently, a collection of functions of the random sample, that summarize or represent all of the information in a random sample that is useful for estimating any $\mathbf{q}(\Theta)$. Thus, in place of the original random sample outcome, it is *sufficient* to have observations on the sufficient statistics to estimate any $\mathbf{q}(\Theta)$. Of course, the random sample itself is a collection of $n$ sufficient statistics, but an objective in defining sufficient statistics is to reduce the number of functions of the random sample needed to represent all of the sample information relevant for estimating $\mathbf{q}(\Theta)$. If a small collection of sufficient statistics can be found for a given statistical model then for defining estimators of $\mathbf{q}(\Theta)$ it is sufficient to consider only functions of the smaller set of sufficient statistic outcomes as opposed to functions of all $n$ outcomes contained in the original random sample. In this way the sufficient statistics allow a data reduction step to occur in a point estimation problem. Relatedly, it will be shown that the search for estimators of $\mathbf{q}(\Theta)$ having the MVUE property or small MSEs can always be restricted to functions of the smallest collection of sufficient statistics. Finally, if the sufficient statistics have a special property, referred to as *completeness*, then an explicit procedure utilizing the complete sufficient statistics is available that is often useful in defining MVUEs. We begin by presenting a more rigorous definition of sufficient statistics.

**Definition 7.18**
***Sufficient Statistics***

Let $\mathbf{X} = [X_1, \ldots, X_n]' \sim f(\mathbf{x};\boldsymbol{\Theta})$ be a random sample, and let $s = [s_1(X), \ldots, s_r(X)]'$ be $r$ statistics. The $r$ statistics are said to be **sufficient statistics** for $f(\mathbf{x};\boldsymbol{\Theta})$ *iff* $f(\mathbf{x};\boldsymbol{\Theta}|\mathbf{s}) = h(\mathbf{x})$, i.e., the conditional density of $\mathbf{X}$, given $\mathbf{s}$, does not depend on the parameter vector $\boldsymbol{\Theta}$.[10]

An intuitive interpretation of Definition 7.18 is that once the outcomes of the $r$ sufficient statistics are observed, there is no additional information on $\boldsymbol{\Theta}$ in the sample outcome. The definition also implies that *given* the function values $\mathbf{s}(\mathbf{x}) = \mathbf{s}$, no other function of $\mathbf{X}$ provides any additional information about $\boldsymbol{\Theta}$ than that obtained from the outcomes $\mathbf{s}$. To motivate these interpretations, first note that the conditional density function $f(\mathbf{x};\boldsymbol{\Theta}|\mathbf{s})$ can be viewed as representing the probability distribution of all of the various ways in which random sample outcomes, $\mathbf{x}$, occur so as to generate the conditional value of $\mathbf{s}$. This is because the event being conditioned on requires that $\mathbf{x}$ satisfy, $\mathbf{s}(\mathbf{x}) = \mathbf{s}$. Definition 7.18 states that if $\mathbf{S}$ is a vector of sufficient statistics, then $\boldsymbol{\Theta}$ is a ghost in $f(\mathbf{x};\boldsymbol{\Theta}|\mathbf{s})$ i.e., the conditional density function really does not depend on the value of $\boldsymbol{\Theta}$ since $f(\mathbf{x};\boldsymbol{\Theta}|\mathbf{s}) = h(\mathbf{x})$. It follows that the probabilistic behavior of the various ways in which $\mathbf{x}$ results in $\mathbf{s}(\mathbf{x}) = \mathbf{s}$ *has nothing to do with* $\boldsymbol{\Theta}$, i.e., it is independent of $\boldsymbol{\Theta}$. Thus, analyzing the various ways in which a given value of $\mathbf{s}$ can occur, or examining additional functions of $\mathbf{X}$, cannot possibly provide any additional information about $\boldsymbol{\Theta}$ since the behavior of the outcomes of $\mathbf{X}$, conditioned on the fact that $\mathbf{s}(\mathbf{x}) = \mathbf{s}$, is totally unrelated to $\boldsymbol{\Theta}$.

**Example 7.10**
***Sufficient Statistic for Bernoulli Population Distribution***

Let $(X_1,\ldots,X_n)$ be a random sample from a Bernoulli population distribution representing whether phone call solicitations to potential customers results in a sale, so that $f(\mathbf{x};p) = p^{\sum_{i=1}^n x_i}(1-p)^{n-\sum_{i=1}^n x_i} \prod_{i=1}^n I_{\{0,1\}}(x_i)$ where $p \in \Omega = (0,1)$, $x_i = 1$ denotes a sale, and $x_i = 0$ denotes no sale on the $i^{\text{th}}$ call. In this case, $\sum_{i=1}^n X_i$, representing the total number of sales in the sample, is a sufficient statistic for $f(\mathbf{x};p)$. To see that this is true, first note that the appropriate conditioning event in the context of Definition 7.18, would be $s(\mathbf{x}) = \sum_{i=1}^n x_i = s$, i.e., the total number of

[10] Note that the conditional density function referred to in this definition is *degenerate* in the general sense alluded to in Section 3.10, footnote 20. That is since $(x_1,\ldots,x_n)$ satisfies the $r$ restrictions $s_i(x_1,\ldots,x_n) = s_i$, for $i = 1,\ldots,r$ by virtue of the event being conditioned upon, the arguments $x_1,\ldots,x_n$ of the conditional density are not all free to vary but rather are functionally related. If one wanted to utilize the conditional density for actually calculating *conditional* probabilities of events for $(X_1,\ldots,X_n)$, and if the random variables were continuous, then line integrals would be required as discussed previously in Chapter 3 concerning the use of degenerate densities. This technical problem is of no concern in our current discussion of sufficient statistics since we will have no need to actually calculate conditional probabilities from the conditional density.

sales equals the value $s$. It follows from the definition of conditional probability that the conditional density function can be defined as[11]

$$f(\mathbf{x};p|s) = \frac{P(x_1,\ldots,x_n,s(\mathbf{x})=s)}{P(s(\mathbf{x})=s)}.$$

The denominator probability is given directly by

$$P(s(\mathbf{x})=s) = \binom{n}{s} p^s(1-p)^{n-s} I_{\{0,1,\ldots,n\}}(s)$$

because $s(\mathbf{X}) = \sum_{i=1}^n X_i$ is the sum of *iid* Bernoulli random variables, which we know to have a binomial distribution. The numerator probability is defined by an appropriate evaluation of the joint density of the random sample, as

$$\begin{aligned} P(x_1,\ldots,x_n,s(\mathbf{x})=s) &= f(\mathbf{x};p) I_{\{\sum_{i=1}^n x_i = s\}}(\mathbf{x}) \\ &= p^{\sum_{i=1}^n x_i}(1-p)^{n-\sum_{i=1}^n x_i}\left(\prod_{i=1}^n I_{\{0,1\}}(x_i)\right) I_{\{\sum_{i=1}^n x_i = s\}}(\mathbf{x}) \end{aligned}$$

which is the probability of $x_1,\ldots,x_n$ *and* $s(\mathbf{x}) = \sum_{i=1}^n x_i = s$. Using the preceding functional representations of the numerator and denominator probabilities in the ratio defining the conditional density function, and using the fact that $s = \sum_{i=1}^n x_i$, we obtain after appropriate algebraic cancellations that

$$f(\mathbf{x};p|s) = \binom{n}{s}^{-1}\left[\prod_{i=1}^n I_{\{0,1\}}(x_i)\right] I_{\{\sum_{i=1}^n x_i = s\}}(\mathbf{x})$$

for any choice of $s \in \{0,1,\ldots,n\}$, which does *not* depend on the parameter p. Thus, $S = \sum_{i=1}^n X_i$ is a sufficient statistic for $p$.

Note the conditional density states that, *given* $\sum_{i=1}^n x_i = s$, all outcomes of $(x_1,\ldots,x_n)$ are equally likely with probability $\binom{n}{s}^{-1}$, and thus the probability of a particular pattern of sales and no sales occurring for $(x_1,\ldots,x_n)$, given that $\sum_{i=1}^n x_i = s$, has *nothing to do with the value of* $p$. It follows that only the fact that $\sum_{i=1}^n x_i = s$ provides any information about $p$ – the particular pattern of 0's and 1's in $(X_1,\ldots,X_n)$ is irrelevant. This is consistent with intuition in that it is the total number of sales in $n$ phone calls and not the particular pattern of sales that provides information in a relative frequency sense about the probability, $p$, of obtaining a sale on a phone call solicitation. Furthermore, if $Y = g(\mathbf{X})$ is any other function of the random sample, then it can provide no *additional* information about $p$ other than that already provided by $s(\mathbf{X})$. This follows from the fact

[11] The reader may wonder why we define the conditional density "from the definition of conditional probability," instead of using the rather straightforward methods for defining conditional densities presented in Chapter 2, Section 2.6. The problem is that here we are conditioning on an event that involves *all* of the random variables $X_1,\ldots,X_n$, whereas in Chapter 2 we were dealing with the usual case where the event being conditioned upon involves only a subset of the random variable $X_1,\ldots,X_n$ having fewer than n elements.

that $h(y|s(\mathbf{x}) = s)$ will not depend on $p$ because the conditional density of $Y$ will have been derived from a conditional density of $\mathbf{X}$ that is independent of $p$, i.e.,

$$h(y|s(\mathbf{x}) = s) = P(y|s(\mathbf{x}) = s) = \sum_{\{\mathbf{x}:g(\mathbf{x})=y\}} f(\mathbf{x}; p|s) = \sum_{\{\mathbf{x}:g(\mathbf{x})=y\}} h(\mathbf{x})$$

since $f(\mathbf{x};\mathrm{p}|\mathbf{s}) = h(\mathbf{x})$ if $\mathbf{s}$ is a sufficient statistic. □

In any problem of estimating $\mathbf{q}(\boldsymbol{\Theta})$, once the outcome of a set of sufficient statistics is observed, the random sample outcome $(x,\ldots,x_n)$ can effectively be ignored for the remainder of the point estimation problem since $\mathbf{s}(\mathbf{x})$ captures all of the relevant information that the sample has to offer regarding $\mathbf{q}(\boldsymbol{\Theta})$. Essentially, it is *sufficient* that the outcome of $\mathbf{s}$ be observed. For example, with reference to Example 7.10, if a colleague were to provide the information that 123 sales were observed in a total of 250 phone calls, i.e., $\sum_{i=1}^{250} x_i = 123$, we would have no need to examine any other characteristic of the random sample outcome $(x_1,\ldots,x_{250})$ when estimating $p$, or $q(p)$.

A significant practical problem in the use of sufficient statistics is knowing how to identify them. A criterion for identifying sufficient statistics which is sometimes useful is given by the Neyman Factorization Theorem:

**Theorem 7.2 Neyman Factorization Theorem** *Let $f(\mathbf{x};\boldsymbol{\Theta})$ be the density function of the random sample $(X_1,\ldots,X_n)$. The statistics $S_1,\ldots,S_r$ are sufficient statistics for $f(\mathbf{x};\boldsymbol{\Theta})$ iff $f(\mathbf{x};\boldsymbol{\Theta})$ can be factored as $f(\mathbf{x};\boldsymbol{\Theta}) = g(s_1(\mathbf{x}),\ldots,s_r(\mathbf{x});\boldsymbol{\Theta})\ h(\mathbf{x})$, where g is a function of only $s_1(\mathbf{x}),\ldots,s_r(\mathbf{x})$ and of $\boldsymbol{\Theta}$, and $h(\mathbf{x})$ does not depend on $\boldsymbol{\Theta}$.*

**Proof** The proof of the theorem in the continuous case is quite difficult, and we leave it to a more advanced course of study (see Lehmann, (1986) *Testing Statistical Hypotheses*, John Wiley, 1986, pp. 54–55). We provide a proof for the discrete case.

***Sufficiency*** Suppose the factorization criterion is met. Let $B(\mathbf{a}) = \{(X_1,\ldots,X_n):s_i(\mathbf{x}) = a_i,\ i = 1,\ldots,r;\ \mathbf{x} \in R(\mathbf{X})\}$ be such that $P(B(\mathbf{a})) > 0$, and note that

$$P(B(\mathbf{a})) = \sum_{(x_1,\ldots,x_n)\in B(\mathbf{a})} f(\mathbf{x};\boldsymbol{\Theta}) = g(a_1,\ldots,a_r;\boldsymbol{\Theta}) \sum_{(x_1,\ldots,x_n)\in B(a)} h(x_1,\ldots,x_n)$$

Therefore,

$$f(\mathbf{x};\boldsymbol{\Theta}|\mathbf{s}(\mathbf{x}) = \mathbf{a}) = \left\{ \begin{array}{ll} \dfrac{g(\mathbf{a};\boldsymbol{\Theta})h(\mathbf{x})}{g(\mathbf{a};\boldsymbol{\Theta})\sum_{\mathbf{x}\in B(\mathbf{a})} h(\mathbf{x})} = h^*(\mathbf{x}) & \text{when } \mathbf{x} \text{ satifies } \mathbf{s}(\mathbf{x}) = \mathbf{a} \\ 0 & \textit{otherwise} \end{array} \right\}$$

which does not depend on $\boldsymbol{\Theta}$ and hence $s_1,\ldots,s_r$ are sufficient statistics.

***Necessity*** Suppose $s_1,\ldots,s_r$ are sufficient statistics, and note by the definition of the conditional density that $f(\mathbf{x};\boldsymbol{\Theta}) = f(\mathbf{x}|s_i(\mathbf{x}) = a_i,\ i = 1,\ldots,r)\ P(s_i(\mathbf{x}) = a_i,\ i = 1,\ldots,r)$ where the conditional density function does not depend on $\boldsymbol{\Theta}$ by the sufficiency of $\mathbf{s}$. Then we have factored $f(\mathbf{x};\boldsymbol{\Theta})$ into the product of a function of

$s_1(\mathbf{x}),\ldots,s_r(\mathbf{x})$ and $\boldsymbol{\Theta}$ (i.e., $P(s_i(\mathbf{x}) = a_i, i = 1,\ldots,r)$ will depend on $\boldsymbol{\Theta}$), and a function that does not depend on $\boldsymbol{\Theta}$. ■

As we have alluded to previously, a practical advantage of sufficient statistics is that they can often greatly reduce the number of random variables required to represent the sample information relevant for estimating $\mathbf{q}(\boldsymbol{\Theta})$, as seen in Example 7.8 and in the following example of the use of the Neyman Factorization Theorem.

**Example 7.11**
***Sufficient Statistics* via *Neyman Factorization for Exponential***

Let $\mathbf{X} = (X_1,\ldots,X_n)$ be a random sample from the exponential population distribution $\Theta^{-1}e^{-z/\Theta}\, I_{(0,\infty)}(z)$ representing waiting times between customer arrivals at a retail store. Note that the joint density of the random sample is given $f(x_1,\ldots,x_n;\Theta) = \Theta^{-n}\exp(-\sum_{i=1}^n x_i/\Theta)\prod_{i=1}^n I_{(0,\infty)}(x_i)$.

The joint density can be factored into the form required by the Neyman Factorization theorem by defining $g(\sum_{i=1}^n x_i;\Theta) = \Theta^{-n}\exp(-\sum_{i=1}^n x_i/\Theta)$ and $h(\mathbf{x}) = \prod_{i=1}^n I_{(0,\infty)}(x_i)$. Then from the theorem, we can conclude that $S = \sum_{i=1}^n X_i$ is a sufficient statistic for $f(\mathbf{x};\Theta)$. It follows that the value of the sum of the random sample outcomes contains all of the information in the sample outcomes relevant for estimating $q(\Theta)$. □

Successful use of the Neyman Factorization Theorem for identifying sufficient statistics requires that one be ingenious enough to define the appropriate $g(\mathbf{s}(\mathbf{x});\boldsymbol{\Theta})$ and $h(\mathbf{x})$ functions that achieve the required joint probability density factorization. Since the appropriate function definitions will *not* always be readily apparent, an approach introduced by Lehmann–Scheffé[12] can sometimes be quite useful for providing direction to the search for sufficient statistics. We will discuss this useful result in the context of *minimal sufficient statistics*.

### 7.4.1 Minimal Sufficient Statistics

At the beginning of our discussion of sufficient statistics we remarked that an objective of using sufficient statistics is to reduce the number of functions of the random sample required to represent all of the information in the random sample relevant for estimating $\mathbf{q}(\boldsymbol{\Theta})$. A natural question to consider is what is the *smallest* number of functions of the random sample that can represent all of the relevant sample information in a given point estimation problem? This relates to the concept of a **minimal sufficient statistic**, which is essentially the sufficient statistic for a given $f(\mathbf{x};\boldsymbol{\Theta})$ that is defined using the fewest number of (functionally independent) coordinate functions of the random sample.

The statement of subsequent definitions and theorems will be facilitated by the concept of the **range of X over the parameter space Ω**, defined as

[12] Lehmann, E.L. and H. Scheffe' (1950). Completeness, Similar Regions, and Unbiased Estimation, *Sankhyā*, 10, pp. 305.

$R_\Omega(\mathbf{X}) = \{\mathbf{x} : f(\mathbf{x};\Theta)>0 \text{ for some } \Theta \in \Omega\}$ The set $R_\Omega(\mathbf{X})$ represents all of the values of $\mathbf{x}$ that are assigned a nonzero density weighting by $f(\mathbf{x};\Theta)$ for at least one $\Theta \in \Omega$. In other words, $R_\Omega(\mathbf{X})$ is the union of the supports of the densities $f(\mathbf{x};\Theta)$ for $\Theta \in \Omega$ and thus corresponds to the set of relevant $\mathbf{x}$-outcomes for the statistical model $\{f(\mathbf{x};\Theta), \Theta \in \Omega\}$. If the support of the density $f(\mathbf{x};\Theta)$ does not change with $\Theta$ (e.g., normal, Gamma, binomial) then $R_\Omega(\mathbf{X}) = R(\mathbf{X}) = \{\mathbf{x}: f(\mathbf{x};\Theta) > 0\}$, where $\Theta \in \Omega$ can be chosen arbitrarily and we henceforth treat the range of $\mathbf{X}$ as being synonymous with the support of its density.

**Definition 7.19** ***Minimal Sufficient Statistics***

A sufficient statistic $\mathbf{S} = \mathbf{s}(\mathbf{X})$ for $f(\mathbf{x};\Theta)$ is said to be a **minimal sufficient statistic** if for every other sufficient statistic $\mathbf{T} = \mathbf{t}(\mathbf{X})$ $\exists$ a function $\mathbf{h}_\mathbf{T}(\cdot)$ such that $\mathbf{s}(\mathbf{x}) = \mathbf{h}_\mathbf{T}(\mathbf{t}(\mathbf{x}))$ $\forall \mathbf{x} \in R_\Omega(\mathbf{X})$.

In order to motivate what is "minimal" about the sufficient statistic $\mathbf{S}$ in Definition 7.19, first note that $\mathbf{S}$ will have the fewest elements in its range compared to all sufficient statistics for $f(\mathbf{x};\Theta)$. This follows from the fact that a function can never have more elements in its range than in its domain (recall the definition of a function, which requires that there is only one range point associated with each domain element, although there can be many domain elements associated with each range element), and thus if $\mathbf{S} = \mathbf{h}_\mathbf{T}(\mathbf{T})$ for any other sufficient statistic $\mathbf{T}$, then the number of elements in $R(\mathbf{S})$ must be no more than the number of elements in $R(\mathbf{T})$, for *any* sufficient statistic $\mathbf{T}$. So, in this sense, $\mathbf{S}$ utilizes the *minimal* set of points for representing the sample information relevant for estimating $\mathbf{q}(\Theta)$.

It can also be shown that a minimal sufficient statistic can be chosen to have the fewest number of coordinate functions relative to any other sufficient statistic, i.e., the number of coordinate functions defining the minimal sufficient statistic is *minimal.* A rigorous proof of this fact is quite difficult and is deferred to a more advanced cause of study.[13] In order to at least motivate the plausibility of this fact, first note that since a minimal sufficient statistic, say $\mathbf{S}$, is a function of all other sufficient statistics, then if $\mathbf{T}$ is any other sufficient statistic, $\mathbf{t}(\mathbf{x}) = \mathbf{t}(\mathbf{y}) \Rightarrow \mathbf{s}(\mathbf{x}) = \mathbf{h}_\mathbf{T}(\mathbf{t}(\mathbf{x})) = \mathbf{h}_\mathbf{T}(\mathbf{t}(\mathbf{y})) = \mathbf{s}(\mathbf{y})$. It follows that

$$A_T = \{(\mathbf{x}, \mathbf{y}) : \mathbf{t}(\mathbf{x}) = \mathbf{t}(\mathbf{y})\} \subset \{(\mathbf{x}, \mathbf{y}) : \mathbf{s}(\mathbf{x}) = \mathbf{s}(\mathbf{y})\} = B$$

no matter which sufficient statistic, $\mathbf{T}$, is being referred to. If B is to contain the set $A_T$, then the constraints on $(\mathbf{x},\mathbf{y})$ representing the set-defining conditions of B cannot be more constraining than the constraints defining $A_T$, and in particular the number of nonredundant constraints[14] defining B cannot be more than the

[13] See E.W. Barankin and M. Katz, (1959) *Sufficient Statistics of Minimal Dimension*, Sankhya, 21:217–246; R. Shimizu, (1966) *Remarks on Sufficient Statistics*, Ann. Inst. Statist. Math., 18:49–66; D.A.S. Fraser, (1963) *On Sufficiency and the Exponential Family*, Jour. Roy. Statist. Soc., Series B, 25:115–123.

number defining $A_T$. Thus the number of nonredundant coordinate functions defining **S** must be no larger than the number of nonredundant coordinate functions defining any other sufficient statistic, so that the number of coordinate functions defining S is *minimal*. Identification of minimal sufficient statistics can often be facilitated by the following approach suggested by Lehmann and Scheffé.

**Theorem 7.3** ***Lehmann-Scheffé Minimal Sufficiency Theorem*** *Let* $\mathbf{X} \sim f(\mathbf{x};\boldsymbol{\Theta})$. *If the statistic* $\mathbf{S} = \mathbf{s}(\mathbf{X})$ *is such that* $\forall \mathbf{x}$ *and* $\mathbf{y} \in R_\Omega(\mathbf{X})$, $f(\mathbf{x};\boldsymbol{\Theta}) = \tau(\mathbf{x},\mathbf{y})\, f(\mathbf{y};\Theta)$ *iff* $(\mathbf{x},\mathbf{y})$ *satisfies* $\mathbf{s}(\mathbf{x}) = \mathbf{s}(\mathbf{y})$, *then* $\mathbf{S} = \mathbf{s}(\mathbf{X})$ *is a minimal sufficient statistic for* $f(\mathbf{x};\boldsymbol{\Theta})$.

**Proof** Define $A(\mathbf{s}) = \{\mathbf{x}: \mathbf{s}(\mathbf{x}) = \mathbf{s}\}$ and let $x_s \in A(\mathbf{s}) \cap R_\Omega(X)$, be chosen as a representative element of $A(\mathbf{s})$, $\forall \mathbf{s} \in R(\mathbf{S})$. Define $\eta(\mathbf{x}) = \mathbf{x}_\mathbf{s}\ \forall \mathbf{x} \in A(\mathbf{s})$ and $\forall \mathbf{s} \in R(\mathbf{S})$. Thus $A(\mathbf{s})$ is the set of $\mathbf{x}$-outcomes whose image $\mathbf{s}(\mathbf{x})$ is $\mathbf{s}$, and $\eta(\mathbf{x})$ is the representative element of the set $A(\mathbf{s})$ to which $\mathbf{x}$ belongs.

Assume that $(\mathbf{x},\mathbf{y}) \in \{(\mathbf{x},\mathbf{y}): \mathbf{s}(\mathbf{x}) = \mathbf{s}(\mathbf{y})\} \Rightarrow f(\mathbf{x};\boldsymbol{\Theta}) = \tau(\mathbf{x},\mathbf{y})\, f(\mathbf{x};\boldsymbol{\Theta})\ \forall\ \mathbf{x}$ and $\mathbf{y} \in R_\Omega(\mathbf{X})$. Then for $\mathbf{x} \in A(\mathbf{s}) \cap R_\Omega(\mathbf{X})$, $\mathbf{s}(\mathbf{x}) = \mathbf{s}(\mathbf{x}_\mathbf{s})$ implies

$$
\begin{aligned}
f(\mathbf{x};\Theta) &= \tau(\mathbf{x},\mathbf{x}_\mathbf{s}) f(\mathbf{x}_\mathbf{s};\Theta) \\
&= \tau(\mathbf{x},\boldsymbol{\eta}(\mathbf{x})) f(\boldsymbol{\eta}(\mathbf{x});\Theta) \begin{pmatrix} \text{substitute} \\ \boldsymbol{\eta}(x) = \mathbf{x}_\mathbf{s} \end{pmatrix} \\
&= h(\mathbf{x}) g(\mathbf{s}(\mathbf{x});\Theta)
\end{aligned}
$$

where $h(\mathbf{x}) \equiv \tau(\mathbf{x},\boldsymbol{\eta}(\mathbf{x}))$, $g(\mathbf{s}(\mathbf{x});\Theta) \equiv f(\boldsymbol{\eta}(\mathbf{x});\Theta)$, and the $g$-function in the latter identity can be defined from the fact that $\boldsymbol{\eta}(\mathbf{x}) = \mathbf{x}_\mathbf{s}$ *iff* $\mathbf{s}(\mathbf{x}) = \mathbf{s}$, so that $\boldsymbol{\eta}(\mathbf{x}) \Leftrightarrow \mathbf{s}(\mathbf{x})$. If $\mathbf{x} \in R_\Omega(\mathbf{X})$, then $f(\mathbf{x};\Theta) = h(\mathbf{x})\, g(\mathbf{s}(\mathbf{x});\Theta)$ by defining $h(\mathbf{x}) = 0$. Since Neyman factorization holds, $s(\mathbf{X})$ is a sufficient statistic.

Now assume $f(\mathbf{x};\Theta) = \tau(\mathbf{x},\mathbf{y})\, f(\mathbf{x};\Theta) \Rightarrow (\mathbf{x},\mathbf{y}) \in \{(\mathbf{x},\mathbf{y}): \mathbf{s}(\mathbf{x}) = \mathbf{s}(\mathbf{y})\}\ \forall \mathbf{x}$ and $\mathbf{y} \in R_\Omega(\mathbf{X})$. Let $\mathbf{s}_*(\mathbf{x})$ be any other sufficient statistic for $f(\mathbf{x};\Theta)$. Then by Neyman factorization, for some $g_*(\cdot)$ and $h_*(\cdot)$ functions, $f(\mathbf{x};\Theta) = g_*(\mathbf{s}_*(\mathbf{x});\Theta)\, h_*(\mathbf{x})$. If $\mathbf{s}_*(\mathbf{x}) = \mathbf{s}_*(\mathbf{y})$, then since $g_*(\mathbf{s}_*(\mathbf{x});\Theta) = g_*(\mathbf{s}_*(\mathbf{y});\Theta)$, and it follows that $f(\mathbf{x};\Theta) = [h_*(\mathbf{x})/h_*(\mathbf{y})]\, f(\mathbf{x};\Theta) = \tau(\mathbf{x},\mathbf{y})\, f(\mathbf{x};\Theta)$ whenever $h_*(\mathbf{y}) \neq 0$, so that $\mathbf{s}_*(\mathbf{x}) = \mathbf{s}_*(\mathbf{y}) \Rightarrow \mathbf{s}(\mathbf{x}) = \mathbf{s}(\mathbf{y})$. Values of $\mathbf{y}$ for which $h_*(\mathbf{y}) = 0$ are such that $f(\mathbf{x};\Theta) = 0\ \forall \Theta \in \Omega$ by Neyman Factorization, and are thus irrelevant to the minimal sufficiency of S (recall Definition 7.19). Then $s$ is a function of $\mathbf{s}_*$, as $\mathbf{s}(\mathbf{x}) = \gamma(\mathbf{s}_*(\mathbf{x}))$, $\forall \mathbf{x} \in R_\Omega(\mathbf{X})$, because for a representative $\mathbf{y}_{\mathbf{s}*} \in \{\mathbf{x}: \mathbf{s}_*(\mathbf{x}) = \mathbf{s}_*\}$, $\mathbf{s}_*(\mathbf{x}) = \mathbf{s}_*(\mathbf{y}_{\mathbf{s}*}) = \mathbf{s}_* \Rightarrow \mathbf{s}(\mathbf{x}) = \mathbf{s}(\mathbf{y}_{\mathbf{s}*}) = \mathbf{s}$, and thus $\mathbf{s}(\mathbf{x}) = \mathbf{s} = \gamma(\mathbf{s}_*) = \gamma(\mathbf{s}_*(\mathbf{x}))$. Therefore $s(\mathbf{X})$ is a minimal sufficient statistic by Definition 7.19. ■

[14]By nonredundant, we mean that none of the constraints are implied by the others. Redundant constraints are constraints that are ineffective or unnecessary in defining sets.

Before proceeding to applications of the theorem, we present two corollaries that are informative and useful in practice.

**Corollary 7.1** ***Lehmann-Scheffé Sufficiency*** Let $\mathbf{X} \sim f(\mathbf{x};\Theta)$. If the statistic $\mathbf{S} = \mathbf{s}(\mathbf{X})$ is such that $\forall\, \mathbf{x}$ and $\mathbf{y} \in R_\Omega(\mathbf{X})$, $(\mathbf{x},\mathbf{y}) \in \{(\mathbf{x},\mathbf{y})\text{: } \mathbf{s}(\mathbf{x}) = \mathbf{s}(\mathbf{y})\} \Rightarrow f(\mathbf{x};\Theta) = \tau(\mathbf{x},\mathbf{y})\, f(\mathbf{y};\Theta)$, then $\mathbf{S} = \mathbf{s}(\mathbf{X})$ is a sufficient statistic for $f(\mathbf{x};\Theta)$.

**Proof** The validity of this corollary is implied by the first part of the proof of Theorem 7.3. ■

The corollary indicates that the "only if" part of the condition in Theorem 7.3 is not required for the *sufficiency* of $\mathbf{s}(\mathbf{X})$ but it is the addition of the "only if" part that results in *minimality* of $\mathbf{s}(\mathbf{X})$.

**Corollary 7.2** ***Minimal Sufficiency when R(X) is Independent of Θ*** Let $\mathbf{X} \sim f(\mathbf{x};\Theta)$ and suppose $R(\mathbf{X})$ does not depend on $\Theta$. If the statistic $\mathbf{S} = \mathbf{s}(\mathbf{X})$ is such that $f(\mathbf{x};\Theta)/f(\mathbf{y};\Theta)$ does not depend on $\Theta$ *iff* $(\mathbf{x},\mathbf{y})$ satisfies $\mathbf{s}(\mathbf{x}) = \mathbf{s}(\mathbf{y})$ then $\mathbf{S} = \mathbf{s}(\mathbf{X})$ is a minimal sufficient statistic.

**Proof** This follows from Theorem 7.3 by dividing through by $f(\mathbf{y};\Theta)$ on the left-hand side of the *iff* condition, which is admissible for all $\mathbf{x}$ and $\mathbf{y}$ in $R(\mathbf{X}) = \{\mathbf{x}\text{: } f(\mathbf{x};\Theta) > 0\}$. Values of $\mathbf{x}$ and $\mathbf{y} \notin R(\mathbf{X})$ are irrelevant to sufficiency (recall Definition 7.19). ■

Using the preceding results for defining a minimal sufficient statistic of course still requires that one is observant enough to recognize an appropriate (vector) function $\mathbf{S}$. However, in many cases the Lehmann-Scheffé approach transforms the problem into one where a choice of $\mathbf{S}$ is readily apparent. The following examples illustrate the use of the procedure for discovering minimal sufficient statistics.

**Example 7.12** ***Lehman-Scheffé Minimal Sufficiency Approach for Bernoulli*** Let $X = (X_1,\ldots,X_n)$ be a random sample from a nondegenerate Bernoulli population distribution representing whether or not a customer contact results in a sale, so that

$$f(x;p) = p^{\sum_{i=1}^n x_i}(1-p)^{n-\sum_{i=1}^n x_i}\prod_{i=1}^n I_{\{0,1\}}(x_i) \text{ for } p \in (0,1).$$

In an attempt to define a sufficient statistic for $f(\mathbf{x};p)$, follow the Lehmann–Scheffé procedure by examining

$$\frac{f(\mathbf{x};p)}{f(\mathbf{y};p)} = \frac{p^{\sum_{i=1}^n x_i}(1-p)^{n-\sum_{i=1}^n x_i}\prod_{i=1}^n I_{\{0,1\}}(x_i)}{p^{\sum_{i=1}^n y_i}(1-p)^{n-\sum_{i=1}^n y_i}\prod_{i=1}^n I_{\{0,1\}}(y_i)}$$

for all values of $\mathbf{x}$ and $\mathbf{y} \in R(\mathbf{X}) = \times_{i=1}^n \{0,1\}$. The ratio will be independent of $p$, *iff* the constraint $\sum_{i=1}^n x_i = \sum_{i=1}^n y_i$ is imposed. A minimal sufficient statistic for $f(\mathbf{x};p)$ is thus $\mathbf{s}(\mathbf{X}) = \sum_{i=1}^n X_i$ by Corollary 7.2. □

**Example 7.13** ***Lehman-Scheffé Minimal Sufficiency Approach for Gamma***

Let $X = (X_1,\ldots,X_n)$ be a random sample from a gamma population distribution representing the operating life until failure of a certain brand and type of personal computer, so that

$$f(\mathbf{x};\alpha,\beta) = \frac{1}{\beta^{n\alpha}\Gamma^n(\alpha)}\left(\prod_{i=1}^n x_i\right)^{\alpha-1} \exp\left(-\sum_{i=1}^n x_i/\beta\right)\prod_{i=1}^n I_{(0,\infty)}(x_i).$$

Using the Lehmann–Scheffé procedure for defining a sufficient statistic for $f(\mathbf{x};\alpha,\beta)$, examine

$$\frac{f(\mathbf{x};\alpha,\beta)}{f(\mathbf{y};\alpha,\beta)} = \frac{\left(\prod_{i=1}^n x_i\right)^{\alpha-1} \exp\left(-\sum_{i=1}^n x_i/\beta\right)\prod_{i=1}^n I_{(0,\infty)}(x_i)}{\left(\prod_{i=1}^n y_i\right)^{\alpha-1} \exp\left(-\sum_{i=1}^n y_i/\beta\right)\prod_{i=1}^n I_{(0,\infty)}(y_i)}$$

for all values of $\mathbf{x}$ and $\mathbf{y} \in R(\mathbf{X}) = \times_{i=1}^n (0,\infty)$. (Note the term $(\beta^{n\alpha}\Gamma^n(\alpha))$ has been algebraically canceled in the density ratio). The ratio will be independent of both $\alpha$ and $\beta$ *iff* the constraints $\prod_{i=1}^n x_i = \prod_{i=1}^n y_i$ and $\sum_{i=1}^n x_i = \sum_{i=1}^n y_i$, are imposed. A minimal sufficient statistic for $f(\mathbf{x};\alpha,\beta)$ is then bivariate and given by $\mathbf{s}_1(\mathbf{X}) = \prod_{i=1}^n X_i$ and $s_2(\mathbf{X}) = \sum_{i=1}^n X_i$, by Corollary 7.2. □

**Example 7.14** ***Lehman-Scheffé Minimal Sufficiency Approach for Uniform***

Let $X = (X_1,\ldots,X_n)$ be a random sample from a uniform population distribution representing the number of minutes that a shipment is delivered before $(x < 0)$ or after $(x > 0)$ its scheduled arrival time, so that

$$f(\mathbf{x};a,b) = (b-a)^{-n} \prod_{i=1}^n I_{[a,b]}(x_i).$$

Unlike the previous examples, here the range of $\mathbf{X}$ depends on the parameters $a$ and $b$. Referring to the Lehmann-Scheffé procedure for defining a sufficient statistic for $f(\mathbf{x};a,b)$ as given by Theorem 7.3, examine

$$\tau(\mathbf{x},\mathbf{y},a,b) = f(\mathbf{x};a,b)/f(\mathbf{y};a,b) = \prod_{i=1}^n I_{[a,b]}(x_i)/\prod_{i=1}^n I_{[a,b]}(y_i)$$

for all values of $\mathbf{x} \in R_\Omega(\mathbf{X}) = \{\mathbf{x}: f(\mathbf{x};a,b) > 0 \text{ for some } (a,b) \text{ satisfying } -\infty < a < b < \infty\}$,[15] and for values of $\mathbf{y}$ for which the denominator is $> 0$. (Note we have algebraically canceled the $(b-a)^{-n}$ term which appears in both the numerator and the denominator of the ratio.) The $\mathbf{x}$ and $\mathbf{y}$ vectors under consideration will be $n$-element vectors, with $\mathbf{x}$ being any point in $\mathbb{R}^n$ and $\mathbf{y}$ being any point in $\times_{i=1}^n [a,b]$.

[15] It may be more appropriate to assume finite lower and upper bounds for $a$ and $b$, respectively. Doing so will not change the final result of the example.

The ratio will be independent of $a$ and $b$ *iff* min $(x_1,\ldots,x_n)$ = min $(y_1,\ldots,y_n)$ and $\max(x_1,\ldots,x_n) = \max(y_1,\ldots,y_n)$, in which case the ratio will be equal to 1. The preceding conditions also ensure that $f(\mathbf{x};a,b) = 0$ when $f(\mathbf{y};a,b) = 0$, so that $f(\mathbf{x};a,b) = \tau(\mathbf{x},\mathbf{y})f(\mathbf{y};a,b)$ holds $\forall \mathbf{x}$ and $\mathbf{y} \in R_\Omega(\mathbf{X})$. A minimal sufficient statistic for $f(\mathbf{x};a,b)$ is then bivariate and given by the *order statistics* $s_1(\mathbf{X}) = \min(X_1,\ldots,X_n)$ and $s_2(\mathbf{X}) = \max(X_1,\ldots,X_n)$ by Theorem 7.3. □

### 7.4.2 Sufficient Statistics in the Exponential Class

The exponential class of densities represent a collection of parametric families of density functions for which sufficient statistics are straightforwardly defined. Furthermore, the sufficient statistics are generally *minimal* sufficient statistics.

**Theorem 7.4** ***Exponential Class and Sufficient Statistics*** *Let $f(\mathbf{x};\Theta)$ be a member of the exponential class of density functions*

$$f(\mathbf{x};\Theta) = \exp\left[\sum_{i=1}^{k} c_i(\Theta)g_i(\mathbf{x}) + d(\Theta) + z(\mathbf{x})\right] I_A(\mathbf{x}).$$

*Then $\mathbf{s}(\mathbf{X}) = (g_1(\mathbf{X}),\ldots,g_k(\mathbf{X}))$ is a k-variate sufficient statistic, and if $c_i(\Theta)$, $i = 1,\ldots,k$ are linearly independent, the sufficient statistic is a minimal sufficient statistic.*

**Proof** That $\mathbf{s}(\mathbf{X})$ is a sufficient statistic follows immediately from the Neyman Factorization theorem by defining $g(g_1(\mathbf{x}),\ldots,g_k(\mathbf{x});\Theta) = \exp\left[\sum_{i=1}^{k} c_i(\Theta)g_i(\mathbf{x}) + d(\Theta)\right]$ and $h(\mathbf{x}) = \exp(z(\mathbf{x}))I_A(\mathbf{x})$ in the theorem.

That $\mathbf{s}(\mathbf{X})$ is a minimal sufficient statistic follows from the fact that $s(\mathbf{X})$ can be derived using the Lehmann–Scheffé approach of Corollary 7.2. To see this, note that

$$f(\mathbf{x};\Theta)/f(\mathbf{y};\Theta) = \exp\left[\sum_{i=1}^{k} c_i(\Theta)[g_i(\mathbf{x}) - g_i(\mathbf{y})] + z(\mathbf{x}) - z(\mathbf{y})\right] \frac{I_A(\mathbf{x})}{I_A(\mathbf{y})}$$

will be independent of $\Theta$ *iff* $\mathbf{x}$ and $\mathbf{y}$ satisfy $g_i(\mathbf{x}) = g_i(\mathbf{y})$ for $i = 1,\ldots,k$ assuming $c_i(\Theta), i = 1,\ldots,k$ are linearly independent.[16] ■

Note that Theorem 7.4 could be used as an alternative approach for discovering minimal sufficient statistics in the problems of random sampling

[16] If one (or more) $c_i(\Theta)$ were linearly dependent on the other $c_j(\Theta)$'s, then "only if" would not apply. To see this, suppose $c_k(\Theta) = \sum_{i=1}^{k-1} a_i c_i(\Theta)$. Then the exp term could be rewritten as $\exp\left[\sum_{i=1}^{k-1} c_i(\Theta)[g_i(\mathbf{x}) - g_i(\mathbf{y}) + a_i[g_k(\mathbf{x}) - g_k(\mathbf{y})]]\right]$ and so $g_i(\mathbf{x}) = g_i(\mathbf{y})$, $i = 1,\ldots,k$, is sufficient but not *necessary* for the term to be independent of $\Theta$, and thus $\mathbf{s}(\mathbf{X})$ would not be *minimal*.

examined in Examples 7.11 and 7.12. It could not be used in Example 7.13 since the uniform distribution is not in the exponential class.

### 7.4.3 Relationship Between Sufficiency and MSE: Rao-Blackwell

In addition to generating a condensed representation of the information in a sample relevant for estimating $q(\Theta)$, sufficient statistics can also facilitate the discovery of estimators of $q(\Theta)$ that are relatively efficient in terms of MSE. In particular, in the pursuit of estimators with low MSE, *only* functions of sufficient statistics need to be examined, which is the implication of the **Rao-Blackwell theorem**.

**Theorem 7.5** ***Rao-Blackwell Theorem - Scalar Case*** *Let* $\mathbf{S} = (S_1,\ldots,S_r)$ *be an r-variate sufficient statistic for* $f(\mathbf{x};\Theta)$, *and let* $t^*(\mathbf{X})$ *be any estimator of the scalar* $q(\Theta)$ *having finite variance. Define* $t(\mathbf{X}) = \mathrm{E}(t^*(\mathbf{X})|S_1,\ldots,S_r) = \xi(S_1,\ldots,S_r)$. *Then* $t(\mathbf{X})$ *is an estimator of* $q(\Theta)$ *for which* $MSE_\Theta(t(\mathbf{X})) \leq MSE_\Theta\,(t^*(\mathbf{X}))\ \forall\ \Theta\ \in \Omega$, *with the equality being attained only if* $P_\Theta(t(\mathbf{x}) = t^*(\mathbf{x})) = 1$.

**Proof** First note that since $\mathbf{S} = (S_1,\ldots,S_r)$ is an $r$-variate sufficient statistic, $f(\mathbf{x}|\mathbf{s})$ does not depend on $\Theta$, and thus neither does the function $t(\mathbf{X})$ (since it is defined as a conditional expectation using $f(\mathbf{x}|\mathbf{s})$), so $t(\mathbf{X})$ is a statistic that can be used as an estimator of $q(\Theta)$. Now by the iterated expectation theorem, $\mathrm{E}(t(\mathbf{X})) = \mathrm{EE}(t^*(\mathbf{X})|S_1,\ldots\ldots,S_r) = \mathrm{E}(t^*(\mathbf{X}))$, so that $t(\mathbf{X})$ and $t^*(\mathbf{X})$ have precisely the same expectation. Next examine

$$\begin{aligned}\mathrm{MSE}\,(t^*(\mathbf{X})) &= \mathrm{E}(t^*(\mathbf{X}) - q(\Theta))^2 = \mathrm{E}(t^*(\mathbf{X}) - t(\mathbf{X}) + t(\mathbf{X}) - q(\Theta))^2 \\ &= \mathrm{E}(t^*(\mathbf{X}) - t(\mathbf{X}))^2 + 2\mathrm{E}(t^*(\mathbf{X}) - t(\mathbf{X}))(t(\mathbf{X}) - q(\Theta)) \\ &\quad + \mathrm{E}(t(\mathbf{X}) - q(\Theta))^2.\end{aligned}$$

The cross–product term is zero. To see this, first note that $\mathrm{E}[(t^*(\mathbf{X}) - \mathbf{t}(\mathbf{X}))\,(\mathbf{t}(\mathbf{X}) - q(\Theta))] = \mathrm{E}[t(\mathbf{X})\,(t^*(\mathbf{X}) - t(\mathbf{X}))]$ since $\mathrm{E}(t^*(\mathbf{X}) - t(\mathbf{X}))\,q(\Theta) = 0$ because $\mathrm{E}(t^*(\mathbf{X})) = \mathrm{E}(t(\mathbf{X}))$. Now note that by definition $t(\mathbf{X})$ is a function of only sufficient statistics, so that $t(\mathbf{X})$ is a constant given $s_1,\ldots,s_r$. Therefore,

$$\mathrm{E}[t(t^*(\mathbf{X}) - t)|s_1,\ldots,s_r] = [t\ \mathrm{E}(t^*(\mathbf{X}) - t)|s_1,\ldots,s_r)] = 0$$

since $\mathrm{E}(t^*(\mathbf{X})|s_1,\ldots\ldots,s_r) = t$ by definition, so that $\mathrm{E}[t(\mathbf{X})(t^*(\mathbf{X}) - t(\mathbf{X}))] = 0$ by the iterated expectation theorem.

Then dropping the nonnegative term $\mathrm{E}(t^*(\mathbf{X}) - t(\mathbf{X}))^2$ on the right-hand side of the expression defining $\mathrm{MSE}(t^*(\mathbf{X}))$ above yields $\mathrm{MSE}_\Theta(t^*(\mathbf{X})) \geq \mathrm{E}_\Theta(t(\mathbf{X}) - q(\Theta))^2 = \mathrm{MSE}_\Theta(t(\mathbf{X}))\ \forall\ \Theta \in \Omega$. The equality is attained *iff* $\mathrm{E}_\Theta(t^*(\mathbf{X}) - t(\mathbf{X}))^2 = 0$, which requires that $P_\Theta[t^*(\mathbf{x}) = t(\mathbf{x})] = 1$. ■

The point of the theorem is that for any estimator $t^*(\mathbf{X})$ of $q(\Theta)$ there always exists an alternative estimator that is at least as good as $t^*(\mathbf{X})$ in terms of MSE and that is a function of *any* set of sufficient statistics. Thus, the Rao-Blackwell theorem suggests that the search for estimators of $q(\Theta)$ with low MSEs can always be restricted to an examination of functions of sufficient statistics, where hopefully the number of sufficient statistics required to fully represent the information about $q(\Theta)$ is substantially less than the size of the random sample itself.[17] Note that if attention is restricted to the unbiased class of estimators, so that $t^*(\mathbf{X})$ is an unbiased estimator in the statement of the theorem, then the Rao Blackwell theorem implies that the search for a minimum variance estimator within the class of unbiased estimators can also be restricted to functions of sufficient statistics. As an illustration, in Example 7.11, we know that $\sum_{i=1}^n X_i$ is a sufficient statistic for $f(\mathbf{x};\Theta)$. Thus, our search for estimators of $\Theta$ with low MSE can be confined to functions of the sufficient statistic, i.e. $t(\mathbf{X}) = \xi(\sum_{i=1}^n X_i)$. We note for future reference that $\xi(\sum_{i=1}^n X_i) = n^{-1}(\sum_{i=1}^n X_i)$is the MVUE of $\theta$.

The Rao-Blackwell theorem can be extended to the vector case as follows:

**Theorem 7.6** ***Rao-Blackwell Theorem-Vector Case*** *Let* $\mathbf{S} = (S_1,\ldots,S_r)$ *be an r-variate sufficient statistic for* $f(\mathbf{x};\boldsymbol{\Theta})$, *and let* $\mathbf{t}^*(\mathbf{X})$ *be an estimator of the* $(k \times 1)$ *vector function* $\mathbf{q}(\boldsymbol{\Theta})$ *having a finite covariance matrix. Define* $\mathbf{t}(\mathbf{X}) = \mathrm{E}(\mathbf{t}^*(\mathbf{X})|S_1,\ldots,S_r) = \mathbf{h}(S_1,\ldots,S_r)$. *Then* $\mathbf{t}(\mathbf{X})$ *is an estimator of* $\mathbf{q}(\boldsymbol{\Theta})$ *for which* $\mathrm{MSE}_{\boldsymbol{\Theta}}(\mathbf{t}(\mathbf{X})) \preceq \mathrm{MSE}_{\boldsymbol{\Theta}}(\mathbf{t}^*(\mathbf{X}))\ \forall \boldsymbol{\Theta} \in \Omega$, *the equality being attained only if* $P_{\boldsymbol{\Theta}}(\mathbf{t}(\mathbf{x}) = \mathbf{t}^*(\mathbf{x})) = 1$.

**Proof** The proof is analogous to the proof in the scalar case, except that MSE matrices are used in place of scalar MSEs in establishing that $\mathrm{MSE}(\mathbf{t}(\mathbf{X}))$ is smaller than $\mathrm{MSE}(\mathbf{t}^*(\mathbf{X}))$. The details are left to the reader. ■

The implications of Theorem 7.6 are analogous to those for the scalar case. Namely, one need only examine *vector* functions of sufficient statistics for estimating the *vector* $\mathbf{q}(\boldsymbol{\Theta})$ if the objective is to obtain an estimator with a small MSE matrix. Furthermore, the search for an MVUE of $\mathbf{q}(\boldsymbol{\Theta})$ can also be restricted to functions of sufficient statistics. As stated previously, this can decrease substantially the dimensionality of the data used in a point estimation problem if the minimal sufficient statistics for the problem are few in number.

Revisiting Example 7.12 we note for future reference that

$$\mathbf{T} = \xi\left(\sum_{i=1}^n X_i\right) = \begin{bmatrix} (\sum_{i=1}^n X_i)/n \\ \left(n(\sum_{i=1}^n X_i) - (\sum_{i=1}^n X_i)^2\right)/(n(n-1)) \end{bmatrix}$$

is the MVUE for $(p,\ p(1-p))$, the mean and variance of the Bernoulli population distribution in the example.

[17] The reader will recall that the random sample, $(X_1,\ldots,X_n)$, is by definition a set of sufficient statistics for $f(\mathbf{x};\Theta)$. However, it is clear that no improvement (decrease) in the MSE of an unbiased estimator will be achieved by conditioning on $(X_1,\ldots,X_n)$, i.e., the reader should verify that this is a case where $\mathrm{E}(t^*(\mathbf{X}) - t(\mathbf{X}))^2 = 0$ and MSE equality is achieved in the Rao–Blackwell theorem.

### 7.4.4 Complete Sufficient Statistics, Minimality, and Uniqueness of Unbiased Estimation

If a sufficient statistic, **S**, has the property of being **complete**, then it is also a **minimal sufficient statistic**. Moreover, any unbiased estimator of $q(\Theta)$ that is defined as a function of **S** is **unique.** We state the formal definition of completeness, and then provide results on minimality and motivate the uniqueness of unbiased estimators based on complete sufficient statistics.

**Definition 7.20** *Complete Sufficient Statistics*

Let $\mathbf{S} = (S_1,\ldots,S_r)$ be a sufficient statistic for $f(\mathbf{x};\Theta)$. The sufficient statistic **S** is said to be **complete** *iff* the only real valued function $\tau$ defined on the range of **S** that satisfies $E_\Theta(\tau(\mathbf{S})) = 0 \;\forall\; \Theta \in \Omega$ is the function defined as $\tau(\mathbf{S}) = 0$ with probability $1 \;\forall\; \Theta \in \Omega$.

A complete sufficient statistic condenses the random sample information as much as possible in the sense of also being a minimal sufficient statistic, as formalized below.

**Theorem 7.7** *Completeness ⇒ Minimal*

$\mathbf{s}(\mathbf{X})$ *is complete* $\Rightarrow$ $\mathbf{s}(\mathbf{X})$ *is minimal.*

**Proof** E. L. Lehmann and H. Scheffe', (1950), op. cit. ■

Thus, if one finds a set of complete sufficient statistics, one also knows there is no smaller set of sufficient statistics available for the statistical model being analyzed. Furthermore, one knows that the search for complete sufficient statistics can be limited to an examination of minimal sufficient statistics.

If a sufficient statistic, **S**, is complete, it follows that two *different* functions of **S** cannot have the same expected value. To see this, suppose that $E(t(\mathbf{S})) = E(t^*(\mathbf{S})) = q(\Theta)$, and define $\tau(\mathbf{S}) = t(\mathbf{S}) - t^*(\mathbf{S})$. Then $E(\tau(\mathbf{S})) = 0$, and since $\tau(\mathbf{S})$ is a function of the complete sufficient statistic **S**, it must be the case that $\tau(\mathbf{s}) = t(\mathbf{s}) - t^*(\mathbf{s}) = 0$ occurs with probability 1 for all $\Theta$. Thus, $t(\mathbf{s})$ and $t^*(\mathbf{s})$ are the *same* function with probability 1. An important implication of this result is that any unbiased estimator of $q(\Theta)$ that is a function of the complete sufficient statistic is *unique* – there cannot be more than one unbiased estimator of $q(\Theta)$ defined in terms of complete sufficient statistics. This uniqueness property leads to an important procedure for defining MVUEs that will be discussed in Section 7.5.

The following example illustrates the process of verifying the completeness property of a sufficient statistic. Verification of the completeness property oftentimes requires considerable ingenuity.

**Example 7.14** *Complete Sufficient Statistics for the Binomial*

Let $(X_1,\ldots,X_n)$ be a random sample from the Bernoulli population distribution $p^z(1-p)^{1-z}I_{\{0,1\}}(z)$ representing whether or not the administration of a particular drug cures the disease of a patient, and suppose we wish to estimate $q(p) = p(1-p) = \sigma^2$. Note that the joint density of the random sample in this case is given by

$$f(\mathbf{x}) = p^{\sum_{i=1}^{n} x_i}(1-p)^{n-\sum_{i=1}^{n} x_i} \prod_{i=1}^{n} I_{\{0,1\}}(x_i).$$

Using the Neyman factorization criterion, we know that $S = \sum_{i=1}^{n} X_i$ is a sufficient statistic for $f(x_1,\ldots,x_n;p)$, since the joint density can be factored into the product of

$$g(s;p) = p^s(1-p)^{n-s} \text{ and } h(x_1,\ldots,x_n) = \prod_{i=1}^{n} I_{\{0,1\}}(x_i).$$

To determine whether $S = \sum_{i=1}^{n} X_i$ is a *complete* sufficient statistic we need to determine whether the only real-valued function $h$ defined on the range of $S$ that satisfies $\mathrm{E}(\tau(S)) = 0 \; \forall p \in [0,1]$ is the function defined as $\tau(S) = 0$ with probability $1 \; \forall p \in [0,1]$. Note that since $S$ has a *binomial* density we know that for a sample size of $n$, $\mathrm{E}(\tau(S)) = 0 \; \forall p \in [0,1]$ implies

$$\mathrm{E}(\tau(S)) = \sum_{i=0}^{n} \tau(i) \binom{n}{i} p^i(1-p)^{n-i} = 0 \; \forall p \in [0,1].$$

Let $p \in (0,1)$ and note, by dividing through by $(1-p)^n$, that the preceding summation condition can be rewritten as

$$\xi(z) = \sum_{i=0}^{n} \tau(i) \binom{n}{i} z^i = 0,$$

which is a polynomial in $z = p/(1-p)$. Differentiating $n$ times with respect to $z$ yields

$$\frac{d^n \xi(z)}{dZ^n} = n!\tau(n) = 0,$$

which implies that $\tau(n) = 0$. Differentiating $\tau(z)$ only $(n-1)$ times with respect to $z$, and using the fact that $\tau(n) = 0$ yields

$$\frac{d^{n-1} \xi(z)}{dz^{n-1}} = n!\tau(n-1) = 0.$$

The process can be continued to ultimately lead to the conclusion that $\tau(i) = 0$ for $i = 0,\ldots,n$ is required. Thus, necessarily $\tau(S) = 0$ with probability 1 if $\mathrm{E}(\tau(S)) = 0$, and thus $S = \sum_{i=1}^{n} X_i$ is a *complete* sufficient statistic for $f(\mathbf{x}_1,\ldots,\mathbf{x}_n;p)$. □

### 7.4.5 Completeness in the Exponential Class

In general, the method used to verify completeness must be devised on a case-by-case basis. However, the following theorem identifies a large collection of parametric families for which complete sufficient statistics are relatively straightforward to identify.

**Theorem 7.8**
***Completeness in the Exponential Class***

*Let the joint density, $f(\mathbf{x};\boldsymbol{\Theta})$, of the random sample $\mathbf{X} = (X_1,\ldots,X_n)'$ be a member of a parametric family of densities belonging to the exponential class of densities. If the range of $\mathbf{c}(\boldsymbol{\Theta}) = (c_1(\boldsymbol{\Theta}),\ldots,c_k(\boldsymbol{\Theta}))'$, $\boldsymbol{\Theta} \in \Omega$, contains an open k-dimensional rectangle, then $\mathbf{g}(\mathbf{X}) = (g_1(\mathbf{X}),\ldots,g_k(\mathbf{X}))'$ is a set of complete sufficient statistics for $f(\mathbf{x};\boldsymbol{\Theta})$, $\boldsymbol{\Theta} \in \Omega$.*

**Proof** See Lehmann, (1986) *Testing Statistical Hypotheses*, John Wiley, 1986, pp. 142–143 and Bickel and Doksum, (1977) *Mathematical Statistics*, Holden Day, 1977, pp. 123. ■

Theorem 7.8 implies that if we are dealing with a parametric family of densities from the exponential class, once we verify that the range of $\mathbf{c}(\boldsymbol{\Theta})$ contains an open $k$-dimensional rectangle, we will have immediately identified a set of complete sufficient statistics, which are given by $\mathbf{g}(\mathbf{X})$ in the exponential class representation. Regarding the open rectangle condition, it will often be readily apparent from the definition of the range of $\mathbf{c}(\boldsymbol{\Theta})$ whether there exists an open $k$-dimensional rectangle contained in $R(\mathbf{c})$. Alternatively, it can be shown that $R(\mathbf{c})$ will contain such an open rectangle if $\mathbf{c}(\boldsymbol{\Theta})$ is continuously differentiable $\forall\ \boldsymbol{\Theta}$ in some open $k$-dimensional rectangle, $\Gamma$, contained in the parameter space $\Omega$ and $\partial\mathbf{c}/\partial\boldsymbol{\Theta}$ has full rank for at least one $\boldsymbol{\Theta} \in \Gamma$ (Bartle, Real Analysis, pp. 381).

**Example 7.16**
***Complete Sufficient Statistics for the Normal Distribution***

Let $(X_1,\ldots,X_n)$ be a random sample from a normal population distribution with mean $\mu$ and variance $\sigma^2$ representing the package weights of a certain type of cereal produced by General Mills. The joint density function for the random sample is then given by

$$f(\mathbf{x}_1,\ldots,\mathbf{x}_n;\mu,\sigma) = \frac{1}{(2\pi)^{n/2}\sigma^n}\exp\left(-(1/2)\sum_{i=1}^{n}\left(\frac{x_i-\mu}{\sigma}\right)^2\right).$$

The density is a member of the exponential class of density functions

$$\exp\left(\sum_{i=1}^{2} c_i(\mu,\sigma)\, g_i(\mathbf{x}) + d(\mu,\sigma) + z(\mathbf{x})\right) I_A(x_1,\ldots,x_n)$$

where $c_1(\mu,\sigma) = \mu/\sigma^2$, $g_1(\mathbf{x}) = \sum_{i=1}^n x_i$, $c_2(\mu,\sigma) = -1/(2\sigma^2)$, $g_2(\mathbf{x}) = \sum_{i=1}^n x_i^2$, $d(\mu,\sigma) = (-n/2)((\mu^2/\sigma^2) + \ln(2\pi\sigma^2)), z(\mathbf{x}) = 0$, and $A = \times_{i=1}^n (-\infty,\infty)$.

Now note that $\mathbf{c}(\mu,\sigma) = (c_1(\mu,\sigma), c_2(\mu,\sigma))$ has the set $(-\infty,\infty) \times (-\infty,0)$ for its range since the range of $\mathbf{c}(\mu,\sigma)$ in this case is, by definition,

$$R(\mathbf{c}) = \{(c_1,c_2) : c_1 = \mu/\sigma^2, c_2 = -1/(2\sigma^2), \mu \in (-\infty,\infty), \sigma \in (0,\infty)\}.$$

The range of $\mathbf{c}(\mu,\sigma)$ contains an open two-dimensional rectangle, i.e., there exists a set of points $\{(x_1, x_2)$: $a_i<x_i<b_i$, $i = 1,2\} \subset R(\mathbf{c})$, and so we know from Theorem 7.8 that $\{\sum_{i=1}^n X_i, \sum_{i=1}^n X_i^2\}$ is a set of *complete* sufficient statistics for the multivariate normal family of densities in this example.

As an alternative verification of the open rectangle condition, note that $\mathbf{c}(\mu,\sigma)$ is continuously differentiable for all $\mu \in (-\infty,\infty)$ and $\sigma > 0$. Furthermore, letting $\mathbf{\Theta} = (\mu, \sigma)'$,

$$\frac{\partial \mathbf{c}}{\partial \mathbf{\Theta}'} = \begin{bmatrix} \frac{\partial c_1}{\partial \mu} & \frac{\partial c_1}{\partial \sigma} \\ \frac{\partial c_2}{\partial \mu} & \frac{\partial c_2}{\partial \sigma} \end{bmatrix} = \begin{bmatrix} \frac{1}{\sigma^2} & -\frac{2\mu}{\sigma^3} \\ 0 & \frac{1}{\sigma^3} \end{bmatrix},$$

so that $\det(\partial \mathbf{c}/\partial \mathbf{\Theta}') = 1/\sigma^5 > 0, \forall \sigma > 0$. Thus $\partial \mathbf{c}/\partial \mathbf{\Theta}$ has full rank $\forall \mu \in (-\infty,\infty)$ and $\forall \sigma \in (0,\infty)$ and the open rectangle condition is verified. □

The reader might have noticed that in all of the preceding examples in which sampling was from a population distribution belonging to the exponential class, the joint density function for the random sample was also a member of the exponential class of densities. This was *not* just a coincidence, as the following theorem makes clear.

**Theorem 7.9** ***Exponential Class Population Implies Exponential Class Random Sample Distribution***

*Let $(X_1,\ldots,X_n)$ be a random sample from a population distribution that belongs to the exponential class of density functions. Then the joint density function of the random sample also belongs to the exponential class of density functions.*

**Proof** Suppose

$$X_j \sim f(x_j;\mathbf{\Theta}) = \exp\left[\sum_{i=1}^{k} c_i(\mathbf{\Theta})g_i(x_j) + d(\mathbf{\Theta}) + z(x_j)\right] I_A(x_j) \ \forall j.$$

Then

$$\begin{aligned}(X_1,\ldots,X_n) \sim \prod_{j=1}^{n} f(x_j;\mathbf{\Theta}) &= \exp\left[\sum_{i=1}^{k} c_i(\mathbf{\Theta}) \sum_{j=1}^{n} g_i(x_j) + nd(\mathbf{\Theta}) + \sum_{j=1}^{n} z(x_j)\right] \prod_{j=1}^{n} I_A(x_j) \\ &= \exp\left[\sum_{i=1}^{k} c_i(\mathbf{\Theta}) g_i^*(x_1,\ldots,x_n) + d^*(\mathbf{\Theta}) + z^*(x_1,\ldots,x_n)\right] \\ &\quad \times I_{A*}(x_1,\ldots,x_n)\end{aligned}$$

where

$$g_i^*(x_1,\ldots,x_n) = \sum\nolimits_{j=1}^{n} g_i(x_j),\ d^*(\mathbf{\Theta}) = nd(\mathbf{\Theta}),\ z^*(x_1,\ldots,x_n) = \sum\nolimits_{j=1}^{n} z(x_j),$$
$$\text{and } A^* = \times_{i=1}^{n} A.$$

Thus, the joint density function of $(X_1,\ldots,X_n)$ belongs to the exponential class.■

The importance of Theorem 7.9 is that if random sampling is from an exponential class population distribution, then Theorem 7.8 might be potentially useful for finding complete sufficient statistics, since we know that the joint density of the random sample is in fact in the exponential class of probability distributions.

### 7.4.6 Completeness, Minimality, and Sufficiency of Functions of Sufficient Statistics

Sufficient statistics are not unique. In fact any one to one (i.e., invertible) function of a sufficient statistic $\mathbf{S}$, say $\boldsymbol{\tau}(\mathbf{S})$, is also a sufficient statistic, and if $\mathbf{S}$ is complete or minimal, then it is also true that $\boldsymbol{\tau}(\mathbf{S})$ is complete or minimal, respectively. This result is perhaps not surprising given that one can invert back and forth between the two sets of statistics, using them interchangeably to represent the same sample information. We formalize this observation in the following theorem.

**Theorem 7.10** ***Sufficiency of Invertible Functions of Sufficient Statistics***

*Let* $\mathbf{S} = \mathbf{s}(\mathbf{X})$ *be a* $(r \times 1)$ *sufficient statistic for* $f(\mathbf{x};\Theta)$. *If* $\boldsymbol{\tau}(\mathbf{s}(\mathbf{X}))$ *is an* $(r \times 1)$ *invertible function of* $\mathbf{s}(\mathbf{X})$,

**a.** $\boldsymbol{\tau}(\mathbf{s}(\mathbf{X}))$ *is a* $(r \times 1)$ *sufficient statistic for* $f(\mathbf{x};\Theta)$;

**b.** *If* $\mathbf{s}(\mathbf{X})$ *is a minimal sufficient statistic, then* $\boldsymbol{\tau}(\mathbf{s}(\mathbf{X}))$ *is a minimal sufficient statistic;*

**c.** *If* $\mathbf{s}(\mathbf{X})$ *is a complete sufficient statistic, then* $\boldsymbol{\tau}(\mathbf{s}(\mathbf{X}))$ *is a complete sufficient statistic.*

**Proof**

**(a)** If $\mathbf{s}(\mathbf{X})$ is sufficient, then by Neyman factorization $f(\mathbf{x};\Theta) = g(\mathbf{s}(\mathbf{x});\Theta)\,\mathbf{h}(\mathbf{x})$. Since $\boldsymbol{\tau}(\mathbf{s}(\mathbf{X}))$ is invertible, it follows that $\mathbf{s}(\mathbf{x}) = \boldsymbol{\tau}^{-1}(\boldsymbol{\tau}(\mathbf{s}(\mathbf{x})))$, *so that* $g(\mathbf{s}(\mathbf{x});\Theta) = g(\boldsymbol{\tau}^{-1}(\boldsymbol{\tau}(\mathbf{s}(\mathbf{x}));\Theta) = g_*(\boldsymbol{\tau}(\mathbf{s}(\mathbf{x}));\Theta)$. Then by Neyman factorization, since $f(\mathbf{x};\Theta) = g_*(\boldsymbol{\tau}(\mathbf{s}(\mathbf{x}));\Theta)\,\mathrm{h}(\mathbf{x})$, $\boldsymbol{\tau}(\mathbf{s}(\mathbf{X}))$ is a sufficient statistic for $f(\mathbf{x};\Theta)$.

**(b)** If it can be shown that $\boldsymbol{\tau}(\mathbf{s}(\mathbf{X}))$ satisfies the conditions of the Lehmann–Scheffé minimal sufficiency theorem, then we know that $\boldsymbol{\tau}$ is a minimal sufficient statistic. Note that $\boldsymbol{\tau}(\mathbf{s}(\mathbf{x})) = \boldsymbol{\tau}(\mathbf{s}(\mathbf{y})) \Leftrightarrow \mathbf{s}(\mathbf{x}) = \mathbf{s}(\mathbf{y})$ by the invertibility of $\boldsymbol{\tau}$. It follows that $\{(\mathbf{x},\mathbf{y}): \boldsymbol{\tau}(\mathbf{s}(\mathbf{x})) = \boldsymbol{\tau}(\mathbf{s}(\mathbf{y}))\} = \{(\mathbf{x},\mathbf{y}): \mathbf{s}(\mathbf{x}) = \mathbf{s}(\mathbf{y})\}$ so that by Theorem 7.2 if $\mathbf{S}$ is a minimal sufficient statistic, then so is $\boldsymbol{\tau}(\mathbf{S})$.

**(c)** Suppose $\mathrm{E}(h_*(\boldsymbol{\tau}(\mathbf{S}))) = 0\ \forall\Theta \in \Omega$. Since $\boldsymbol{\tau}(\mathbf{S})$ is invertible, it follows that $\mathrm{E}(h\circ\boldsymbol{\tau}^{-1}(\boldsymbol{\tau}(\mathbf{S}))) = 0\ \forall\Theta \in \Omega$ where $h$ is such that $h_* = h\circ\boldsymbol{\tau}^{-1}$. But since $\mathbf{S}$ is complete, $P(h(\mathbf{s}) = 0) = P(h_*(\boldsymbol{\tau}(\mathbf{s})) = 0) = 1\ \forall\Theta \in \Omega$, so that $\boldsymbol{\tau}(\mathbf{S})$ is complete by Definition 7.20. ■

The implication of Theorem 7.10 is that we can transform a set of sufficient statistics, via invertible functions, in any way that is useful or convenient. For example, the minimal sufficient statistic $\sum_{i=1}^n X_i$ in Example 7.12 could be alternatively defined as $\overline{X}_n$, the minimal sufficient statistic $\left(\sum_{i=1}^n X_i, \prod_{i=1}^n X_i\right)$ in Example 7.13 could be alternatively defined as $\left(\overline{X}_n, \left(\prod_{i=1}^n X_i\right)^{1/n}\right)$, and the complete sufficient statistic $\left(\sum_{i=1}^{n} X_i, \sum_{i=1}^{n} X_i^2\right)$ in Example 7.16 could alternatively be represented as $\left(\overline{X}_n, S_n^2\right)$. All of the alternative representations are justified by Theorem 7.10, because each of the alternative representations can be defined via invertible functions of the original sufficient statistics.

## 7.5 Minimum Variance Unbiased Estimation

Results that can be helpful in the search for a minimum variance unbiased estimator (MVUE) of $\mathbf{q}(\boldsymbol{\Theta})$ are presented in this section. However, it should be noted that MVUEs do not always exist, and when they do, MVUEs may be difficult to determine even with the aid of the theorems presented in this section. If the results in this section are inapplicable or cannot be successfully applied in a given estimation setting, there are numerous alternative approaches for defining estimators of $\mathbf{q}(\boldsymbol{\Theta})$ in a wide range of problems of practical interest. In Chapter 8 we will present a number of tractable procedures for deriving estimators of $\mathbf{q}(\boldsymbol{\Theta})$ that often lead to estimators with good properties, and may also lead to MVUEs.

We begin with a theorem that identifies necessary and sufficient conditions for an unbiased estimator of a scalar $\mathbf{q}(\boldsymbol{\Theta})$ to be a MVUE. Henceforth, we tacitly restrict the class of estimators under consideration to those with *finite variances* or *finite covariance matrices* since estimators with infinite variances will clearly not minimize variance, and in fact cannot even be compared on the basis of variance.

**Theorem 7.11 Scalar MVUEs - Necessary and Sufficient Conditions** *A necessary and sufficient condition for an estimator $T = t(\mathbf{X})$ of the scalar $\mathrm{q}(\Theta)$ to be a MVUE is that $\mathrm{cov}(T,W) = 0$, $\forall\, \Theta \in \Omega$ and $\forall\, W = w(\mathrm{X}) \in \upsilon_o$, where $\upsilon_o$ is the set of all functions of $\mathbf{X}$ that have expectations equal to* 0.

**Proof**

***Necessity*** If $T$ is unbiased for $q(\boldsymbol{\Theta})$, and $\mathrm{E}(W) = 0$, then $T + \lambda W$ is unbiased for $q(\boldsymbol{\Theta})$. Given $\boldsymbol{\Theta}$ and $\lambda$, the variance of $T + \lambda W$ equals $\mathrm{var}(T + WS) = \mathrm{var}(T) + [2\lambda\ \mathrm{cov}(T,W) + \lambda^2\ \mathrm{var}(W)]$. Suppose $\mathrm{cov}(T,W) < 0$. Then the bracketed term can be made negative, and thus $\mathrm{var}(T + \lambda W) < \mathrm{var}(T)$, for $\lambda \in (0, -2\mathrm{cov}(T,W)/\mathrm{var}(W))$. Alternatively, suppose $\mathrm{cov}\ (T,W) > 0$. Then the bracketed term can be made negative, and thus $\mathrm{var}(T + \lambda W) < \mathrm{var}(T)$, for $\lambda \in (-2cov(T,W)/var(W), 0)$. Thus, $t(\mathbf{X})$ cannot be a MVUE if $\mathrm{cov}(T,W) \neq 0$.

***Sufficiency*** Let $T^*$ be any other unbiased estimator of $q(\boldsymbol{\Theta})$. *Then* $(T - T^*) \in \upsilon_o$, i.e., $(T - T^*)$ has an expectation equal to zero. Suppose that $\forall \boldsymbol{\Theta} \in \Omega$ and $T^* \in \upsilon_{q(\boldsymbol{\Theta})}$, where $\upsilon_{q(\boldsymbol{\Theta})}$ is the class of unbiased estimators of $q(\boldsymbol{\Theta})$,

$$\mathrm{cov}(T,(T - T^*)) = \mathrm{E}((T - q(\boldsymbol{\Theta}))(T - T^*)) = \mathrm{E}(T(T - T^*)) = \mathrm{E}(T^2) - \mathrm{E}(TT^*)$$
$$= 0,$$

where we have invoked the condition in the theorem upon defining $w(\mathbf{X}) = t(\mathbf{X}) - t^*(\mathbf{X})$. It follows that $\mathrm{E}(T^2) = \mathrm{E}(TT^*)$, and thus

$$\mathrm{cov}(T, T^*) = \mathrm{E}(TT^*) - \mathrm{E}(T)\mathrm{E}(T^*) = \mathrm{E}(T^2) - (q(\Theta))^2 = \mathrm{var}(T)$$

since both $T$ and $T^*$ are unbiased for $q(\Theta)$. The preceding result implies

$$[\mathrm{var}(T)]^{1/2} = \left[\frac{\mathrm{cov}(T, T^*)}{(\mathrm{var}(T))^{1/2}(\mathrm{var}(T^*))^{1/2}}\right](\mathrm{var}(T^*))^{1/2} = \rho(\mathrm{var}(T^*))^{1/2} \leq (\mathrm{var}(T^*))^{1/2}$$

since the correlation coefficient $\rho \in [0, 1]$ in this case. Thus $T$ has minimum variance. ■

The result in Theorem 7.11 facilitates the proof of the following proposition that establishes important relationships between *scalar* and *vector* MVUEs.

**Theorem 7.12 *Relationships Between Scalar and Vector MVUEs*** *Let* $\mathbf{T} = \mathbf{t}(\mathbf{X})$ *be a* $(k \times 1)$ *vector estimator of the* $(k \times 1)$ *vector function* $\mathbf{q}(\Theta)$. *Then any of the following statements implies the others:*

1. $\mathbf{T}$ *is a MVUE for* $\mathbf{q}(\Theta)$.
2. $T_i$ *is a MVUE for* $q_i(\Theta)$, $i = 1, \ldots, k$.
3. $\boldsymbol{\ell}'\mathbf{T}$ *is a MVUE for* $\boldsymbol{\ell}'\mathbf{q}(\Theta)$, $\forall \boldsymbol{\ell} \neq \mathbf{0}$.

**Proof** We prove the theorem by demonstrating that (1) $\Leftrightarrow$ (2) and (2) $\Leftrightarrow$ (3). Proof that (2) $\Leftrightarrow$ (3):

Assume (2). By Theorem 7.11, for $Z = \sum_{i=1}^k \ell_i T_i$ to be a MVUE of $\sum_{i=1}^k \ell_i q_i(\Theta)$, it is necessary and sufficient that $\mathrm{cov}(F, W) = 0 \; \forall \, W \in \upsilon_o$ and $\forall \, \Theta \in \Omega$. *Note that* $\forall \, W \in \upsilon_o$,

$$\mathrm{cov}(Z, W) = \mathrm{E}\left(\left[\sum_{i=1}^k \ell_i T_i - \sum_{i=1}^k \ell_i q_i(\Theta)\right] W\right) = \sum_{i=1}^k \mathrm{E}(\ell_i [T_i - q_i(\Theta)] W)$$
$$= \sum_{i=1}^k \ell_i \mathrm{cov}(T_i, W) = 0$$

where the last equality follows from the fact that $T_i$ is a MVUE for $q_i(\Theta)$, and thus by Theorem 7.11 $\mathrm{cov}(T_i, W) = 0$. It follows from Theorem 7.11 that $\sum_{i=1}^k \ell_i T_i$ is a MVUE of $\sum_{i=1}^k \ell_i \, q_i(\Theta)$, so that (2) $\Rightarrow$ (3).

Assume (3). Defining $\boldsymbol{\ell}$ so that $\boldsymbol{\ell}'\mathbf{T} = T_i$ (let $\boldsymbol{\ell}$ be a vector of zeros except for a 1 in the $i^{\text{th}}$ position), $T_i$ is a MVUE for $q_i(\Theta)$, $i = 1, \ldots k$. Thus, (3) $\Rightarrow$ (2).

Proof that (1) $\Leftrightarrow$ (2):

Assume (1). Note that $T_i = \boldsymbol{\ell}'\mathbf{T}$ and $\mathrm{var}(T_i) = \boldsymbol{\ell}'\mathbf{Cov}(\mathbf{T})\boldsymbol{\ell}$, where $\boldsymbol{\ell}$ is a zero vector except for a 1 in the $i^{\text{th}}$ position. Clearly $T_i$ is unbiased for $q_i(\Theta)$. Suppose $\mathrm{var}(T_i) = \boldsymbol{\ell}'\mathbf{Cov}(\mathbf{T})\,\boldsymbol{\ell} > \mathrm{var}\,(T_i^*)$ for some other unbiased estimator, $T_i^*$, of $q_i(\Theta)$. But then $\mathbf{T}$ could not be a MVUE of $\mathbf{q}(\Theta)$, since defining $\mathbf{T}^*$ equal to $\mathbf{T}$, except for $T_i^*$ replacing $\mathbf{T}_i$, would imply $\boldsymbol{\ell}'\mathbf{Cov}(\mathbf{T})\boldsymbol{\ell} > \boldsymbol{\ell}'\,\mathbf{Cov}(\mathbf{T}^*)\boldsymbol{\ell}$, contradicting the fact that

$\mathbf{Cov(T)} - \mathbf{Cov(T^*)}$ would be negative semidefinite for all other unbiased estimators $\mathbf{T}^*$ *if* $\mathbf{T}$ were a MVUE for $\mathbf{q}(\boldsymbol{\Theta})$. Thus (1) $\Rightarrow$ (2).

Assume (2). Suppose $T_i$ is a MVUE for $q_i(\boldsymbol{\Theta})$, $i = 1,\ldots,k$. Then clearly $\mathrm{E}(\mathbf{T}) = \mathbf{q}(\boldsymbol{\Theta})$, so that the vector $\mathbf{T}$ is an unbiased estimator of $\mathbf{q}(\boldsymbol{\Theta})$. Now suppose there existed another unbiased estimator, $T^*$, *of* $\mathbf{q}(\boldsymbol{\Theta})$ such that $\boldsymbol{\ell}'\mathbf{Cov}(\mathbf{T}^*)\boldsymbol{\ell} < \boldsymbol{\ell}'\mathbf{Cov}(\mathbf{T})\boldsymbol{\ell}$ for some real vector $\boldsymbol{\ell}$ and for some $\boldsymbol{\Theta}\in\Omega$. Then $\boldsymbol{\ell}'\mathbf{T}^*$ would be an unbiased estimator of $\boldsymbol{\ell}'\mathbf{q}(\boldsymbol{\Theta})$ having a smaller variance than the unbiased estimator $\boldsymbol{\ell}'\mathbf{T}$. However, the latter estimator is a MVUE of $\boldsymbol{\ell}'\mathbf{q}(\boldsymbol{\Theta})$ by the proof of (2) $\Leftrightarrow$ (3) above. This is a contradiction, so that there is no $\boldsymbol{\Theta}$ for which $\exists\, \boldsymbol{\ell}$ such that $\boldsymbol{\ell}'\mathbf{Cov}(\mathbf{T}^*)\boldsymbol{\ell} < \boldsymbol{\ell}'\mathbf{Cov}(\mathbf{T})\boldsymbol{\ell}$, i.e. $\boldsymbol{\ell}'\mathbf{Cov}(\mathbf{T})\boldsymbol{\ell} \leq \boldsymbol{\ell}'\mathbf{Cov}(\mathbf{T}^*)\boldsymbol{\ell}$ $\forall \boldsymbol{\ell}$ and $\forall\, \boldsymbol{\Theta} \in \Omega$. But this implies that $\mathbf{Cov}(\mathbf{T}) \leq \mathbf{Cov}(\mathbf{T}^*)$ $\forall\, \boldsymbol{\Theta} \in \Omega$ and for any unbiased estimator $\mathbf{T}^*$, so that $\mathbf{T}$ *is the MVUE of* $\mathbf{q}(\boldsymbol{\Theta})$. Thus (2) $\Rightarrow$ (1). ■

An important implication of Theorem 7.12 is that it allows a vector MVUE of $\mathbf{q}(\boldsymbol{\Theta})$ to be defined *elementwise* with the knowledge that once each MVUE, $T_i$, for $q_i(\boldsymbol{\Theta})$ has been *individually* defined for $i = 1,\ldots,k$, then $\mathbf{T} = (T_1,\ldots,T_k)$ is a MVUE of $\mathbf{q}(\boldsymbol{\Theta})$ in the *vector* sense. In addition, the theorem indicates that a MVUE for *any* linear combination of the entries in $\mathbf{q}(\boldsymbol{\Theta})$ is immediately known to be $\boldsymbol{\ell}'\mathbf{T}$.

Heretofore we have been referring to "a" MVUE of $\mathbf{q}(\boldsymbol{\Theta})$, rather than "the" MVUE of $\mathbf{q}(\boldsymbol{\Theta})$. The following theorem implies that if we have discovered a MVUE of $\mathbf{q}(\boldsymbol{\Theta})$, it is unique with probability one. We present the result for the scalar case, and then use it to provide the extension to vector situations in the next two theorems.

**Theorem 7.13** ***Uniqueness of Scalar MVUEs*** *If $T$ and $T^*$ are both scalar MVUEs for $q(\boldsymbol{\Theta})$, then $P[t = t^*] = 1$.*

**Proof** If $T$ and $T^*$ are both MVUE's for $q(\boldsymbol{\Theta})$, then necessarily $\mathrm{var}(T) = \mathrm{var}(T^*)\ \forall\, \boldsymbol{\Theta} \in \Omega$. From the sufficiency proof of Theorem 7.11, we know that $(\mathrm{var}\,(T))^{1/2} = \rho(\mathrm{var}\,(T^*))^{1/2}$ where $\rho$ is the correlation between $T$ and $T^*$. It also follows from the sufficiency proof of Theorem 7.11 that because $\mathrm{var}(T) = \mathrm{var}(T^*)$, $\rho = \mathrm{var}(T)/\mathrm{var}(T^*) = 1$. Then from Theorem 3.36, $\rho = 1$ implies that $P[t = a_1 + bt^*] = 1$ with $a_1 = (\mu_T\sigma_{T^*} - \mu_{T^*}\sigma_T)/\sigma_{T^*} = 0$ and $b = \sigma_T/\sigma_{T^*} = 1$, since $\mu_T = \mu_{T^*} = q(\boldsymbol{\Theta})$ and $\sigma_{T^*} = \sigma_T$, so that $P(t = t^*) = 1$. ■

**Theorem 7.14** ***Uniqueness of Vector MVUEs*** *If the vector estimators $\mathbf{T}$ and $\mathbf{T}^*$ are both MVUEs for the $(k \times 1)$ vector $\mathbf{q}(\boldsymbol{\Theta})$, then $P[\mathbf{t} = \mathbf{t}^*] = 1$.*

**Proof** Let $A_i$ represent the event that $t_i = t_i^*$ for $i = 1, \ldots,k$. Then since $P(\cap_{i=1}^k A_i) \geq 1 - \sum_{i=1}^k P(\overline{A}_i)$ by Bonferroni's inequality,

$$P[t_i = t_i^*, i = 1,\ldots,k] = P[\mathbf{t} = \mathbf{t}^*] \geq 1 - \sum_{i=1}^{k} P[t_i \neq t_i^*] = 1 - \sum_{i=1}^{k} 0 = 1$$

where the next to last inequality follows from Theorem 7.13 and the fact that both $t_i$ and $t_i^*$ are MVUEs for $q_i(\Theta)$. ■

From this point forward we will refer to *the* MVUE of $\mathbf{q}(\Theta)$ rather than *a* MVUE since a MVUE of $\mathbf{q}(\Theta)$ is unique with probability 1.

### 7.5.1 Cramer-Rao Lower Bound on the Covariance Matrix of Unbiased Estimators

We now examine a lower bound for the covariance matrix of an unbiased estimator of $\mathbf{q}(\Theta)$. If an unbiased estimator of $\mathbf{q}(\Theta)$ can be found whose covariance matrix actually attains this lower bound, then the estimator is the MVUE of $\mathbf{q}(\Theta)$. The bound is called the **Cramer-Rao Lower Bound** (CRLB), and its applicability relies on a number of so-called regularity conditions (equivalently, assumptions) on the underlying joint density function, $f(\mathbf{x};\Theta)$, of the random sample under investigation. We state the regularity conditions below. The interpretation and application of the regularity conditions can be quite challenging and the reader may wish to skim the discussion of these conditions on first reading.

**Definition 7.21 *CRLB Regularity Conditions***

1. The parameter space, $\Omega$, of the family of densities to which $f(\mathbf{x};\Theta)$ belongs is an open rectangle.[18]
2. The support of $f(\mathbf{x};\Theta)$, $A = \{\mathbf{x}: f(\mathbf{x};\Theta) > 0\}$, is the same $\forall \Theta \in \Omega$.
3. $\partial \ln(f(\mathbf{x};\Theta))/\partial\Theta_i$ exists and is finite $\forall i$, $\forall\, \mathbf{x} \in A$ and $\forall\, \Theta \in \Omega$.
4. $\forall i$ and $j$ and $\forall$ unbiased estimator $\mathbf{t}(\mathbf{X})$ of $\mathbf{q}(\Theta)$ having a finite covariance matrix, one can differentiate under the integral or summation sign as indicated below:

**Continuous Case**

a. $$\frac{\partial}{\partial\Theta_j}\int_{-\infty}^{\infty}\cdots\int_{-\infty}^{\infty} f(\mathbf{x};\Theta)dx_1\ldots dx_n = \int_{-\infty}^{\infty}\cdots\int_{-\infty}^{\infty}\frac{\partial f(\mathbf{x};\Theta)}{\partial\Theta_j}dx_1\ldots dx_n$$

b. $$\frac{\partial}{\partial\Theta_j}\int_{-\infty}^{\infty}\cdots\int_{-\infty}^{\infty} t_i(\mathbf{x})f(\mathbf{x};\Theta)dx_1\ldots dx_n$$
$$= \int_{-\infty}^{\infty}\cdots\int_{-\infty}^{\infty} t_i(\mathbf{x})\frac{\partial f(\mathbf{x};\Theta)}{\partial\Theta_j}dx_1\ldots dx_n$$

**Discrete Case**

a. $$\frac{\partial}{\partial\Theta_j}\sum_{\mathbf{x}\in A} f(\mathbf{x};\Theta) = \sum_{\mathbf{x}\in A}\frac{\partial f(\mathbf{x};\Theta)}{\partial\Theta_j}$$

b. $$\frac{\partial}{\partial\Theta_j}\sum_{\mathbf{x}\in A} t_i(\mathbf{x})\, f(\mathbf{x};\Theta) = \sum_{\mathbf{x}\in A} t_i(\mathbf{x})\frac{\partial f(\mathbf{x};\Theta)}{\partial\Theta_j}$$

[18]Recall that by open rectangle, we mean that the parameter space can be represented as $\Omega = \{(\Theta_1,\ldots,\Theta_k): a_i < \Theta_i < b_i, i = 1,\ldots,k\}$, where any of the $a_i$'s could be $-\infty$ and any of the $b_i$'s could be $\infty$. This condition can actually be weakened to requiring only that the parameter space be an open subset of $\mathbb{R}^k$, and not necessarily an open rectangle, and the CRLB would still apply.

In applications, conditions 1–3 are generally not difficult to verify, but condition 4 can be problematic. Regarding regularity condition (4a) for the continuous case, note that since $\int_{-\infty}^{\infty}\cdots\int_{-\infty}^{\infty} f(\mathbf{x};\boldsymbol{\Theta})\,\mathbf{dx} = 1$ because $f(\mathbf{x};\boldsymbol{\Theta})$ is a density function, then the left-hand side of the equality in 4a is equal to zero. Then a regularity condition that is equivalent to (4a) is

$$\int_{-\infty}^{\infty}\cdots\int_{-\infty}^{\infty} \frac{\partial f(\mathbf{x};\boldsymbol{\Theta})}{\partial \Theta_j}\mathbf{dx} = 0, \forall j.$$

Because $\partial \ln f(\mathbf{x};\boldsymbol{\Theta})/\partial\Theta_j = [f(\mathbf{x};\boldsymbol{\Theta})]^{-1}\ \partial f(\mathbf{x};\boldsymbol{\Theta})/\partial\Theta_j\ \forall\ \mathbf{x}\ \in \mathrm{A}$, another regularity condition equivalent to (4a) is given by

$$\int_{\mathbf{x}\in A} \frac{\partial \ln f(\mathbf{x};\boldsymbol{\Theta})}{\partial \Theta_j} f(\mathbf{x};\boldsymbol{\Theta})\mathbf{dx} = \mathrm{E}\left[\frac{\partial \ln f(\mathbf{x};\boldsymbol{\Theta})}{\partial \Theta_j}\right] = 0, \forall j.$$

While sometimes mathematically challenging, condition (4a) can be verified directly by demonstrating that one of the above functions of $\mathbf{x}$ equal zero, based on the functional form of the joint density function for the random sample. By replacing integration with summation, the preceding two equivalent conditions suffice to verify condition (4a) in the discrete as well.

Verification of regularity condition (4b) is complicated by the fact that it must hold true for *all* unbiased estimators of $\mathbf{q}(\boldsymbol{\Theta})$ with finite covariance matrices, and can be a challenging exercise in real analysis. We present a sufficient condition for the validity of (4b) that is based on results in advanced calculus relating to the validity of *differentiating under the integral sign.*

**Theorem 7.15** ***Sufficient Condition for CRLB Regularity Condition 4b***

*Let CRLB regularity conditions (1)–(4a) hold. Then regularity condition (4b) holds if*

$$\mathrm{E}\left[\left(\frac{\partial \ln(f(\mathbf{X};\boldsymbol{\xi}))}{\partial \xi_j}\right)^2\right] < \tau(\boldsymbol{\Theta}) < \infty$$

$\forall j, \forall \boldsymbol{\xi} \in \Gamma(\boldsymbol{\Theta})$ *and* $\forall \boldsymbol{\Theta} \in \Omega$, *where* $\Gamma(\boldsymbol{\Theta})$ *is an open rectangle containing* $\boldsymbol{\Theta}$.

**Proof** Apostol, T.M., (1974) *Mathematical Analysis, 2nd Ed.*, Addison-Wesley, Reading, MA, pp. 167. ■

Note that since $\mathrm{E}(\partial \ln(f(\mathbf{X};\boldsymbol{\xi}))/\partial \xi_j) = 0$ by condition (4a), Theorem 7.15 can be interpreted as a boundedness condition on the *variance* of the random variable $\partial \ln(f(\mathbf{X};\boldsymbol{\xi}))/\partial \xi_j\ \forall j$ and $\forall \boldsymbol{\Theta} \in \Omega$. The following example illustrates the CRLB regularity verification process.

**Example 7.17** ***Verifying CRLB Regularity in an Exponential Population***

Let $\mathbf{X} = (X_1,\ldots,X_n)'$ be a random sample from an exponential population distribution representing the waiting time between customer arrivals at a retail store, so that

$$f(\mathbf{x};\theta) = \theta^{-n} \exp\left(-\sum_{i=1}^{n} x_i/\theta\right) \prod_{i=1}^{n} I_{(0,\infty)}(x_i).$$

The parameter space, $\Omega$, is an open interval since $\theta > 0$ for the exponential density, verifying 1. The set, A, of **x**–values that satisfy $f(\mathbf{x};\theta) > 0$ is the same for all $\theta \in \Omega$, i.e., $A = \times_{i=1}^{n} (0,\infty)$, regardless of the value of $\Theta > 0$, verifying 2. Also, $\partial \ln(f(\mathbf{x};\theta))/\partial\theta = -n/\theta + \sum_{i=1}^{n} x_i/\theta^2$ exists and is finite $\forall\ \theta\ \in \Omega$ and $\mathbf{x}\ \in$ A, verifying 3.

Regarding 4, first note that

$$\mathrm{E}\left(\frac{\partial \ln(f(\mathbf{X};\theta))}{\partial\theta}\right) = \mathrm{E}\left(-\frac{n}{\theta} + \frac{\sum_{i=1}^{n} X_i}{\theta^2}\right) = \frac{-n}{\theta} + \frac{n}{\theta} = 0$$

since $\mathrm{E}(\sum_{i=1}^{n} X_i) = n\,\theta$, and thus 4a is met. Regarding 4b, note that

$$\mathrm{E}\left(\left(\frac{\partial \ln(f(\mathbf{X};\theta))}{\partial\theta}\right)^2\right) = \mathrm{E}\left(\left(-\frac{n}{\theta} + \frac{\sum_{i=1}^{n} X_i}{\theta^2}\right)^2\right) = \mathrm{E}\left(\frac{n^2}{\theta^2} - \frac{2n\sum_{i=1}^{n} X_i}{\theta^3} + \frac{\left(\sum_{i=1}^{n} X_i\right)^2}{\theta^4}\right)$$

$$= \frac{n^2}{\theta^2} - \frac{2n^2\theta}{\theta^3} + \frac{2n\theta^2 + n(n-1)\theta^2}{\theta^4} = \frac{n}{\theta^2},$$

where the next to last equality follows from the fact that $\mathrm{E}(X_i^2) = 2\,\theta^2\ \forall i$, and the fact that the $X_i$'s are *iid*. It follows that *whatever* the value of $\theta \in \Omega$, there exists an open interval $\Gamma(\theta) = (\theta - \varepsilon,\ \theta + \varepsilon)$ for some $\varepsilon > 0$, and a positive number $\tau(\theta) = n/(\theta - \varepsilon)^2$ such that

$$\mathrm{E}\left[\left(\frac{\partial \ln(f(\mathbf{X};\xi))}{\partial\theta}\right)^2\right] = \frac{n}{\xi^2} < \tau(\theta) < \infty \quad \forall \xi \in \Gamma(\theta),$$

so that CRLB condition 4b is met. Thus, the CRLB regularity conditions are met for this problem. □

In practice, there are cases where it can be quite difficult to verify all of the CRLB regularity conditions. It is useful to note that for the large collection of probability distributions in the *exponential class* discussed in Chapter 4, the regularity conditions hold quite generally, as formalized below.

**Theorem 7.16**
***CRLB Regularity in the Exponential Class of Distributions***

*Let $f(\mathbf{x};\boldsymbol{\Theta})$ be a member of the exponential class of densities, as defined in Definition 4.3 and let $\mathbf{c}(\boldsymbol{\Theta})$ and $d(\boldsymbol{\Theta})$ be continuously differentiable with, $\partial\mathbf{c}(\boldsymbol{\Theta})/\partial\boldsymbol{\Theta}$ having full rank, for $\boldsymbol{\Theta} \in \Omega$. If $\Omega$ is an open rectangle, then $f(\mathbf{x};\boldsymbol{\Theta})$ satisfies the CRLB regularity conditions.*

**Proof** A proof of the theorem can be constructed using the result in Lehmann (1986), *Testing Statistical Hypotheses*, John Wiley, 1986, pp. 59-60. ■

**Example 7.18**
***CRLB Regularity in Exponential Class: Geometric Distribution***

Let $(X_1,\ldots,X_n)$ be a random sample from a geometric population distribution representing the number of customer contacts required for a salesperson to make her first sale, so that

$$f(\mathbf{x};p) = p^n(1-p)^{\sum_{i=1}^n x_i - n}\prod_{i=1}^n I_{\{1,2,3,\ldots\}}(x_i).$$

The PDF of the random sample is a member of the exponential class of densities with $c(p) = \ln(1-p)$, $d(p) = \ln[p/(1-p)]^n$, $g(\mathbf{x}) = \sum_{i=1}^n x_i$, $z(\mathbf{x}) = 0$, and $A = \times_{i=1}^n \{1,2,3,\ldots\}$. Note that $\partial c(p)/\partial p = -1/(1-p) \neq 0$ and is continuous for all $p \in (0,1)$, and $\partial d(p)/\partial p = n/(p(1-p))$ is continuous for all $p \in (0,1)$. Then since $\Omega = (0,1)$ is an open interval, $f(\mathbf{x};p)$ satisfies the CRLB regularity conditions by Theorem 7.16. □

Based on the CRLB regularity conditions, we now present the CRLB. We remind the reader that the notation $\mathbf{A}\succeq\mathbf{B}$ indicates matrix $\mathbf{A}$ is greater than the matrix $\mathbf{B}$ by a positive semidefinite (psd) matrix, i.e., $\mathbf{A}-\mathbf{B}=\mathbf{C}$, where the matrix difference $\mathbf{C}$ is psd.

**Theorem 7.17 Cramer-Rao Lower Bound** *Let $\mathbf{t}(\mathbf{X})$ be an unbiased estimator of the $(k \times 1)$ vector function $\mathbf{q}(\boldsymbol{\Theta})$ and let $\mathbf{t}(\mathbf{X})$ have a finite covariance matrix. Let the $(k \times m)$ Jacobian matrix $\partial\mathbf{q}(\boldsymbol{\Theta})/\partial\boldsymbol{\Theta}$ exist $\forall\boldsymbol{\Theta}_{(m\times 1)} \in \Omega$. Assume the CRLB regularity conditions of Definition 7.21 hold, where $f(\mathbf{x};\boldsymbol{\Theta})$ is the joint density function of the random sample $\mathbf{X}$. Then*

$$\mathbf{Cov}_{\boldsymbol{\Theta}}(\mathbf{t}(\mathbf{X}))\succeq\left[\frac{\partial\mathbf{q}(\boldsymbol{\Theta})}{\partial\boldsymbol{\Theta}}\right]'\left[\mathrm{E}_{\boldsymbol{\Theta}}\left(\frac{\partial\ln(f(\mathbf{X};\boldsymbol{\Theta}))}{\partial\boldsymbol{\Theta}}\frac{\partial\ln(f(\mathbf{X};\boldsymbol{\Theta}))}{\partial\boldsymbol{\Theta}'}\right)\right]^{-1}\left[\frac{\partial\mathbf{q}(\boldsymbol{\Theta})}{\partial\boldsymbol{\Theta}}\right]\quad\forall\boldsymbol{\Theta}\in\Omega,$$

*provided the inverse matrix exists.*

**Proof** We prove the result for the continuous case. The discrete case is analogous, with integrals replaced by summation, and is left to the reader.

First note that CRLB regularity condition 4a implies that (the $\boldsymbol{\Theta}$-subscripts on expectations are suppressed) $\mathrm{E}[\partial\ln(f(\mathbf{X};\boldsymbol{\Theta}))/\partial\boldsymbol{\Theta}_i] = 0, i = 1,\ldots,m.$ Then defining the $(m \times 1)$ random vector

$s(\mathbf{X}) = [\partial\ln(f(\mathbf{X};\boldsymbol{\Theta}))/\partial\boldsymbol{\Theta}_1,\ldots,\partial\ln(f(\mathbf{X};\boldsymbol{\Theta}))/\partial\boldsymbol{\Theta}_m]'$,

it follows that $\mathrm{E}(\mathbf{s}(\mathbf{X})) = \mathbf{0}$ and

$$\mathbf{Cov}(\mathbf{s}(\mathbf{X})) = \mathrm{E}(\mathbf{s}(\mathbf{X})\mathbf{s}(\mathbf{X})') = \mathrm{E}\left(\frac{\partial\ln(f(\mathbf{X};\boldsymbol{\Theta}))}{\partial\boldsymbol{\Theta}}\frac{\partial\ln(f(\mathbf{X};\boldsymbol{\Theta}))}{\partial\boldsymbol{\Theta}'}\right).$$

Now, since $\mathbf{t}(\mathbf{X})$ is an unbiased estimator of $\mathbf{q}(\boldsymbol{\Theta})$, $q_i(\boldsymbol{\Theta}) = \int_{-\infty}^{\infty}\cdots\int_{-\infty}^{\infty} t_i(\mathbf{x})f(\mathbf{x};\boldsymbol{\Theta})dx_1\ldots dx_n$.

Then by CRLB regularity condition 4b,

$$\begin{aligned}\frac{\partial q_i(\boldsymbol{\Theta})}{\partial\Theta_j} &= \int_{-\infty}^{\infty}\cdots\int_{-\infty}^{\infty} t_i(\mathbf{x})\frac{\partial f(\mathbf{x};\boldsymbol{\Theta})}{\partial\Theta_j}dx_1\ldots dx_n,\\ &= \int_{-\infty}^{\infty}\cdots\int_{-\infty}^{\infty} t_i(\mathbf{x})\frac{\partial\ln(f(\mathbf{x};\boldsymbol{\Theta}))}{\partial\Theta_j}f(\mathbf{x};\boldsymbol{\Theta})dx_1\ldots dx_n\\ &= E\left(t_i(\mathbf{X})\frac{\partial\ln(f(\mathbf{X};\boldsymbol{\Theta}))}{\partial\Theta_j}\right) = \mathrm{E}(t_i(\mathbf{X})s_j(\mathbf{X}))\text{ for } i,j=1,\ldots,m.\end{aligned}$$

Examine the covariance matrix of the $(k + m) \times 1$ random vector $\mathbf{Z} = \begin{bmatrix} \mathbf{t}(\mathbf{X}) \\ \mathbf{s}(\mathbf{X}) \end{bmatrix}$.

Since $\mathrm{E}(\mathbf{Z}) = \begin{bmatrix} \mathbf{q}(\boldsymbol{\Theta}) \\ \mathbf{0} \end{bmatrix}$, it follows that

$$\mathbf{Cov}(\mathbf{Z}) = \left[\begin{array}{c|c} \mathbf{Cov}(\mathbf{t}(\mathbf{X})) & \mathrm{E}\left(\mathbf{t}(\mathbf{X})\mathbf{s}(\mathbf{X})'\right) \\ \hline \mathrm{E}\left(\mathbf{s}(\mathbf{X})\mathbf{t}(\mathbf{X})'\right) & \mathbf{Cov}(\mathbf{s}(\mathbf{X})) \end{array}\right] = \left[\begin{array}{c|c} \mathbf{Cov}(\mathbf{t}(\mathbf{X})) & \dfrac{\partial \mathbf{q}(\boldsymbol{\Theta})'}{\partial \boldsymbol{\Theta}} \\ \hline \dfrac{\partial \mathbf{q}(\boldsymbol{\Theta})}{\partial \boldsymbol{\Theta}} & \mathbf{Cov}(\mathbf{s}(\mathbf{X})) \end{array}\right]$$

Premultiply $\mathbf{Cov}(\mathbf{Z})$ by

$$\mathbf{D} = \left[\begin{array}{c|c} \underset{k\times k}{\mathbf{I}} & \underset{k\times m}{-\dfrac{\partial \mathbf{q}(\boldsymbol{\Theta})'}{\partial \boldsymbol{\Theta}}[\mathbf{Cov}(\mathbf{s}(\mathbf{X}))]^{-1}} \end{array}\right]$$

and then postmultiply by $\mathbf{D}'$ to obtain $\mathbf{D}\ \mathbf{Cov}(\mathbf{Z})\ \mathbf{D}' = \mathbf{Cov}(\mathbf{t}(\mathbf{X})) - [\partial \mathbf{q}(\boldsymbol{\Theta})/\partial \boldsymbol{\Theta}]'[\mathbf{Cov}(\mathbf{s}(\mathbf{X}))]^{-1}[\partial \mathbf{q}(\boldsymbol{\Theta})/\partial \boldsymbol{\Theta}] \succeq \mathbf{0}$ because $\mathbf{Cov}(\mathbf{Z})$ is necessarily psd, so that $\mathbf{D}\ \mathbf{Cov}(\mathbf{Z})\ \mathbf{D}'$ is positive semidefinite for any conformable matrix $\mathbf{D}$.[19] Then given the definition of $\mathbf{Cov}(\mathbf{s}(\mathbf{X}))$ above, the result of the theorem follows. ■

An **alternative form of the CRLB** that utilizes second order partial derivatives of $f(\mathbf{x};\boldsymbol{\Theta})$ with respect to $\boldsymbol{\Theta}$ is sometimes easier to use in practice. Its applicability relies on an additional regularity condition beyond the conditions assumed for the existence of the CRLB, and is presented in the next theorem.

**Theorem 7.18** ***Cramer–Rao Lower Bound: Hessian Form*** *Assume the conditions of Theorem 7.17 hold, so that the Cramer-Rao Lower Bound applies. Then if*

$$\int_{-\infty}^{\infty} \cdots \int_{-\infty}^{\infty} \frac{\partial^2 f(\mathbf{x};\boldsymbol{\Theta})}{\partial \Theta_i \partial \Theta_j} dx_1 \ldots dx_n = 0, \ \forall i \text{ and } j,$$

*the CRLB can be represented alternatively as*

$$\mathbf{Cov}_{\boldsymbol{\Theta}}(\mathbf{t}(\mathbf{X})) \succeq \left[\frac{\partial \mathbf{q}(\boldsymbol{\Theta})}{\partial \boldsymbol{\Theta}}\right]' \left[-\mathrm{E}_{\boldsymbol{\Theta}}\left(\frac{\partial \ln(f(\mathbf{X};\boldsymbol{\Theta}))}{\partial \boldsymbol{\Theta}\partial \boldsymbol{\Theta}'}\right)\right]^{-1} \left[\frac{\partial \mathbf{q}(\boldsymbol{\Theta})}{\partial \boldsymbol{\Theta}}\right] \quad \forall \boldsymbol{\Theta} \in \Omega.$$

**Proof** Comparing the two versions of the CRLB, it is clear that for the alternative version to apply, we must have

$$\mathrm{E}\left(\frac{\partial \ln f(\mathbf{X};\boldsymbol{\Theta})}{\partial \boldsymbol{\Theta}} \frac{\partial \ln f(\mathbf{X};\boldsymbol{\Theta})}{\partial \boldsymbol{\Theta}'}\right) = -\mathrm{E}\left(\frac{\partial^2 \ln f(\mathbf{X};\boldsymbol{\Theta})}{\partial \boldsymbol{\Theta}\partial \boldsymbol{\Theta}'}\right).$$

[19]This can be easily seen, since $\ell'\mathbf{D}\ \mathbf{Cov}(\mathbf{Z})\ \mathbf{D}'\ell = \ell_*'\ \mathbf{Cov}(\mathbf{Z})\ \ell_* \geq 0$, where $\ell^* = \mathbf{D}'\ell$, $\forall\ \ell$.

Note that

$$\mathrm{E}\left(\frac{\partial^2 \ln f(\mathbf{X};\boldsymbol{\Theta})}{\partial\Theta_i\partial\Theta_j}\right) = \mathrm{E}\left(\frac{\partial}{\partial\Theta_j}\left[\frac{1}{f(\mathbf{X};\boldsymbol{\Theta})}\frac{\partial f(\mathbf{X};\Theta)}{\partial\Theta_i}\right]\right)$$

$$= \mathrm{E}\left(\frac{1}{f(\mathbf{X};\boldsymbol{\Theta})}\frac{\partial^2 f(\mathbf{X};\boldsymbol{\Theta})}{\partial\Theta_i\partial\Theta_j} - \frac{1}{(f(\mathbf{X};\boldsymbol{\Theta}))^2}\frac{\partial f(\mathbf{X};\boldsymbol{\Theta})}{\partial\Theta_i}\frac{\partial f(\mathbf{X};\boldsymbol{\Theta})}{\partial\Theta_j}\right)$$

$$= \mathrm{E}\left(\frac{1}{f(\mathbf{X};\boldsymbol{\Theta})}\frac{\partial^2 f(\mathbf{X};\boldsymbol{\Theta})}{\partial\Theta_i\partial\Theta_j} - \frac{\partial \ln f(\mathbf{X};\boldsymbol{\Theta})}{\partial\Theta_i}\frac{\partial \ln f(\mathbf{X};\boldsymbol{\Theta})}{\partial\Theta_j}\right), \quad \forall i \text{ and } j.$$

Proceeding in the continuous case, if

$$\mathrm{E}\left[\frac{1}{f(\mathbf{X};\boldsymbol{\Theta})}\frac{\partial^2 f(\mathbf{X};\boldsymbol{\Theta})}{\partial\Theta_i\partial\Theta_j}\right] = \int_{-\infty}^{\infty}\cdots\int_{-\infty}^{\infty}\frac{\partial^2 f(\mathbf{x};\boldsymbol{\Theta})}{\partial\Theta_i\partial\Theta_j}dx_1\ldots dx_n = 0 \quad \forall i,\ j,$$

then we can replace the matrix being inverted in Theorem 7.17 by the expectation of the negative of the second order partial derivative matrix of ln $f(\mathbf{x};\boldsymbol{\Theta})$ with respect to $\boldsymbol{\Theta}$, justifying the alternative form of the CRLB. The discrete case follows by replacing integrals with summation. ■

Regarding the additional regularity condition stipulated in Theorem 7.18, note that if

$$\frac{\partial}{\partial\Theta_j}\int_{-\infty}^{\infty}\cdots\int_{-\infty}^{\infty}\frac{\partial f(\mathbf{x};\boldsymbol{\Theta})}{\partial\Theta_i}dx_1\ldots dx_n = \int_{-\infty}^{\infty}\cdots\int_{-\infty}^{\infty}\frac{\partial^2 f(\mathbf{x};\boldsymbol{\Theta})}{\partial\Theta_i\partial\Theta_j}dx_1\ldots dx_n \quad \forall i \text{ and } j,$$

then the integral conditions required in the theorem are met. This follows because the CRLB regularity conditions imply that the left hand side of the equality must be zero, because

$$\int_{-\infty}^{\infty}\cdots\int_{-\infty}^{\infty}\frac{\partial f(\mathbf{x};\boldsymbol{\Theta})}{\partial\Theta_i}dx_1\ldots dx_n = \frac{\partial}{\partial\Theta_i}\underbrace{\int_{-\infty}^{\infty}\cdots\int_{-\infty}^{\infty}f(\mathbf{x};\boldsymbol{\Theta})dx_1\ldots dx_n}_{1} = \frac{\partial}{\partial\Theta_i}(1) = 0.$$

Thus, the alternative CRLB applies if we can differentiate $\int_{-\infty}^{\infty}\cdots\int_{-\infty}^{\infty}f(\mathbf{x};\boldsymbol{\Theta})\,dx_1\ldots dx_n$ under the integral sign *twice*.

We note that verifying the applicability of the alternative form of the CRLB is generally easier if the joint density of the random sample is a member of the exponential class of distributions, because the functional form of the density ensures the additional regularity condition of Theorem 7.16 if some straightforward differentiability properties are met. The following theorem formalizes the result.

**Theorem 7.19 CRLB Alternative Hessian Form in the Exponential Class**

*If the joint probability density of the random sample, $f(\mathbf{x};\boldsymbol{\Theta})$, is a member of the exponential class of densities satisfying Theorem 7.16, and if $d(\boldsymbol{\Theta})$ and $\mathbf{c}(\boldsymbol{\Theta})$ are twice continuously differentiable with respect to $\boldsymbol{\Theta} \in \Omega$, then the alternative Hessian form of the CRLB presented in Theorem 7.18 is valid.*

**Proof** A proof of the theorem can be constructed using the result in Lehmann, (1986), *Testing Statistical Hypotheses*, John Wiley, pp. 59–60. ■

If $f(\mathbf{x};\boldsymbol{\Theta})$ is not in the exponential class, the admissibility of differentiating under the integral sign a second time would need to be established, or else it would need to be verified by *direct evaluation* that the integrals of the second order derivatives of $f(\mathbf{x};\boldsymbol{\Theta})$ presented in Theorem 7.18 are indeed zero. For the sake of illustration, we provide an example of how the direct evaluation process would proceed using a normal distribution, with $\sigma^2 = 1$, as the population distribution. The reader will note that since the normal distribution is a member of the exponential class satisfying Theorem 7.16, and since $\partial^2 c(\Theta)/\partial\Theta^2$ and $\partial^2 d(\Theta)/\partial\Theta^2$ exist and are continuous, we already know that the outcome of our evaluation process will result in the alternative form of the CRLB being applicable.

**Example 7.19** ***Direct Evaluation of Applicability of Hessian Form of CRLB: Normal Case***

Let $\mathbf{X} = (X_1,\ldots,X_n)$ be a random sample from a normal population distribution with $\sigma^2 = 1$, so that

$$f(\mathbf{x};\mu) = \frac{1}{(2\pi)^{n/2}} e^{-(1/2)\sum_{i=1}^{n}(x_i-\mu)^2}.$$

Note that

$$\frac{\partial f(\mathbf{x};\mu)}{\partial\mu} = f(\mathbf{x};\mu)\sum_{i=1}^{n}(x_i-\mu),$$

and

$$\frac{\partial^2 f(\mathbf{x};\mu)}{\partial\mu^2} = -nf(\mathbf{x};\mu) + \left[\sum_{i=1}^{n}(x_i-\mu)\right]^2 f(\mathbf{x};\mu).$$

It follows that

$$\begin{aligned}\int_{-\infty}^{\infty}\cdots\int_{-\infty}^{\infty}\frac{\partial^2 f(\mathbf{x};\mu)}{\partial\mu^2}\,dx_1\ldots dx_n &= -n\int_{-\infty}^{\infty}\cdots\int_{-\infty}^{\infty} f(\mathbf{x};\mu)\,dx_1\ldots dx_n\\ &\quad+\int_{-\infty}^{\infty}\cdots\int_{-\infty}^{\infty}\sum_{i=1}^{n}(x_i-\mu)^2 f(\mathbf{x},\mu)\,dx_1\ldots dx_n\\ &= -n+n=0\end{aligned}$$

since $\sigma_i^2 = 1\ \forall i$, and so the alternative form of the CRLB is applicable. Thus, in this case

$$\mathrm{E}\left(\left(\frac{\partial\ln f(\mathbf{X};\mu)}{\partial\mu}\right)^2\right) = -\mathrm{E}\left(\frac{\partial^2\ln f(\mathbf{X};\mu)}{\partial\mu^2}\right) = \mathrm{E}(n) = n.$$

If interest centers on estimating μ (as opposed to estimating some function of $\mu$), then $q(\mu) = \mu$ in the statement of the CRLB in Theorem 7.17, so that $\partial q(\mu)/\partial \mu = 1$, and the CRLB for the variance of any unbiased estimator of $\mu$ is equal to $n^{-1}$. □

Note that when the equality

$$\mathrm{E}\left(\left(\frac{\partial \ln f(\mathbf{X};\boldsymbol{\Theta})}{\partial \Theta_j}\right)^2\right) = -\mathrm{E}\left(\frac{\partial^2 \ln f(\mathbf{X};\boldsymbol{\Theta})}{\partial \Theta_j^2}\right)$$

is true, the latter expectation expression can be substituted into the previously mentioned sufficient conditions for checking the validity of CRLB regularity condition 4b (Theorem 7.15). In particular, the substitution would imply that $-\mathrm{E}\left[\partial^2 \ln f(\mathbf{x};\boldsymbol{\xi})/\partial\Theta_j^2\right] < \tau(\boldsymbol{\Theta}) < \infty\ \forall \boldsymbol{\xi}$ in an open rectangle $\Gamma(\boldsymbol{\Theta})$ containing $\boldsymbol{\Theta}$, $\forall \boldsymbol{\Theta} \in \Omega$, is the condition that would need to be verified.

Two special cases of the CRLB often arise in practice. The first concerns the case where q(Θ) is a scalar function of a scalar parameter Θ.

**Corollary 7.3 *CRLB for Scalar Function Case***

Let $k = m = 1$ in Theorem 7.17. Then the unbiased estimator $T = t(\mathbf{X})$ of the scalar function $q(\Theta)$ of the scalar parameter Θ is such that

$$\mathrm{var}_\Theta(t(\mathbf{X})) \geq \frac{(dq(\Theta)/d\Theta)^2}{\mathrm{E}_\Theta\left((d(f(\mathbf{X};\Theta))/d\Theta)^2\right)} \quad \forall \Theta \in \Omega.$$

The second special case is when $\mathbf{q}(\boldsymbol{\Theta}) = \boldsymbol{\Theta}$. Since in this case, $\partial \mathbf{q}(\boldsymbol{\Theta})/\partial \boldsymbol{\Theta} = \mathbf{I}$, we have the following result:

**Corollary 7.4 *CRLB for Estimating the Full Parameter Vector***

Let $\mathbf{q}(\boldsymbol{\Theta}) = \boldsymbol{\Theta}$ in Theorem 7.17. Then the unbiased estimator $\mathbf{T} = \mathbf{t}(\mathbf{X})$ of the vector $\boldsymbol{\Theta}$ is such that

$$\mathbf{Cov}_{\boldsymbol{\Theta}}(\mathbf{t}(\mathbf{X})) \succeq \left[\mathrm{E}_{\boldsymbol{\Theta}}\left(\frac{\partial \ln(f(\mathbf{X};\boldsymbol{\Theta}))}{\partial \boldsymbol{\Theta}} \frac{\partial \ln(f(\mathbf{X};\boldsymbol{\Theta}))}{\partial \boldsymbol{\Theta}'}\right)\right]^{-1}, \quad \forall \boldsymbol{\Theta} \in \Omega.$$

We now turn to examples of the use of the CRLB to discover MVUEs.

**Example 7.19 *Using CRLB to Define MVUES: Bernoulli Population***

Let $\mathbf{X} = (X_1,\ldots,X_n)$ be a random sample from a Bernoulli population distribution representing whether or not a vaccine developed by a pharmaceutical company prevents the flu, so that

$$f(\mathbf{x};p) = p^{\sum_{i=1}^n x_i}(1-p)^{n-\sum_{i=1}^n x_i}\prod_{i=1}^n I_{\{0,1\}}(x_i).$$

In order to find the MVUE of the probability that the vaccine prevents the flu, first note that the support of $f(\mathbf{x};p)$ does not depend on $p \in (0,1)$, and we concentrate attention on $p \in (0,1)$ to achieve the open rectangle regularity condition on the parameter space[20]. Also, $f(\mathbf{x};p)$ is continuous in $p$, and

$$\ln f(\mathbf{x};p) = \sum_{i=1}^{n} x_i \ln(p) + \left(n - \sum_{i=1}^{n} x_i\right) \ln(1-p) + \ln\left[\prod_{i=1}^{n} I_{\{0,1\}}(x_i)\right]$$

so that

$$\frac{d\ln f(\mathbf{x};p)}{dp} = \frac{\sum_{i=1}^{n} x_i}{p} - \frac{\left(n - \sum_{i=1}^{N} x_i\right)}{(1-p)}$$

exists and is finite $\forall\, p \in (0,1)$. Regarding CRLB regularity condition (4) note that since $df(\mathbf{x};p)/dp = f(\mathbf{x};p)(d\ln(f(\mathbf{x};p))/dp)$ exists $\forall\, p \in (0,1)$, and since we are dealing with finite sums when defining expectations, condition 4 is met. Note further that

$$\frac{d^2 \ln(f(\mathbf{x};p))}{dp^2} = \frac{-\sum_{i=1}^{n} x_i}{p^2} - \frac{(n - \sum_{i=1}^{n} x_i)}{(1-p)^2}$$

also exists, and since $\sum_{x_1=0}^{1} \cdots \sum_{x_n=0}^{1} df(\mathbf{x};p)/dp$ is a sum involving a *finite* number of terms, differentiation *under the summation signs* a second time is permissible, so that substitution of the alternative form $-\mathrm{E}\left(d^2 \ln(f(\mathbf{X};\Theta))/d\Theta^2\right)$ in the CRLB, as discussed above, is allowed. Note

$$\mathrm{E}\left(\frac{d^2 \ln(f(\mathbf{X};p))}{dp^2}\right) = \frac{-np}{p^2} - \frac{(n-np)}{(1-p)^2} = \frac{-n}{p} - \frac{n}{1-p} = \frac{-n(1-p)-np}{p(1-p)} = \frac{-n}{p(1-p)}$$

and thus the CRLB for the variance of an unbiased estimator of $q(p) = p$ is given by $p(1-p)/n$. Because $\mathrm{var}(\overline{X}_n) = p(1-p)/n$, and $\mathrm{E}(\overline{X}_n) = p$, $\overline{X}_n$ is the MVUE of the probability of flu prevention, $p$. □

**Example 7.21** ***Using CRLB to Define MVUES: Exponential Population***

Let $\mathbf{X} = (X_1,\ldots,X_n)$ be a random sample from an exponential population distribution representing the waiting time between customer arrivals at a retail store, so that

$$f(\mathbf{x};\theta) = \theta^{-n} e^{-\sum_{i=1}^{n} x_i/\theta} \prod_{i=1}^{n} I_{(0,\infty)}(x_i).$$

To find the MVUE of the mean waiting time between customers, $\theta$, first note that $f(\mathbf{x};\theta)$ is a member of the exponential class, i.e.,

$$f(\mathbf{x};\theta) = e^{c(\theta)g(x)+d(\theta)+z(x)} I_A(\mathbf{x}),$$

[20]Note, this rules out the degenerate cases $p = 1$ or $p = 0$, in which all sample observations would then be 1s and 0s, respectively. If such were the case for the outcome of any given random sample from the Bernoulli distribution, the best, and in fact only reasonable estimates of the parameter $p$ would be 1 and 0, respectively, since there would be no sample variability with which to conclude anything different.

with $c(\theta) = -\theta^{-1}, g(\mathbf{x}) = \sum_{i=1}^n x_i, d(\theta) = \ln(\theta^{-n}), z(\mathbf{x}) = 0, A = \times_{i=1}^n (0,\infty)$ and since $d(\theta)$ and $c(\theta)$ are *twice* continuously differentiable and $dc(\theta)/d\theta \neq 0$ for $\theta > 0$, we know by Theorem 7.16 that the CRLB regularity conditions are met, and also that the alternative form of the CRLB applies. Note

$$\ln(f(\mathbf{x};\theta)) = -n\ln(\theta) - \sum_{i=1}^n x_i/\theta + \sum_{i=1}^n \ln\left(I_{(0,\infty)}(x_i)\right),$$

so that

$$\frac{d\ln(f(\mathbf{x};\theta))}{d\theta} = \frac{-n}{\theta} + \frac{\sum_{i=1}^n x_i}{\theta^2} \text{ and } \frac{d^2\ln(f(\mathbf{x};\theta))}{d\theta^2} = \frac{n}{\theta^2} - \frac{2\sum_{i=1}^n x_i}{\theta^3},$$

and thus

$$\mathrm{E}\left(\frac{d^2\ln(f(\mathbf{x};\theta))}{d\theta^2}\right) = \frac{n}{\theta^2} - \frac{2n\theta}{\theta^3} = \frac{-n}{\theta^2}.$$

The CRLB for the variance of an unbiased estimator of $\theta$ is then $\theta^2/n$. Since $\mathrm{E}(\overline{X}_n) = \theta$ and $\mathrm{var}(\overline{X}_n) = \theta^2/n$, it follows that $\overline{X}_n$ is the MVUE of the mean waiting time, $\theta$. □

### 7.5.2 CRLB Attainment

The use of the CRLB for identifying the MVUE of $\mathbf{q}(\mathbf{\Theta})$ will only be useful if there actually exists an unbiased estimator whose covariance matrix equals the CRLB. A natural question to ask is under what conditions an unbiased estimator, $\mathbf{T}$, of $\mathbf{q}(\mathbf{\Theta})$ will have a covariance matrix that actually *achieves* the CRLB? It turns out that the CRLB is achieved only when the definition of the estimator $\mathbf{T}$ has the special form given in the following theorem.

**Theorem 7.20** ***Attainment of the CRLB*** *Assume the CRLB regularity conditions hold, and let* $\mathbf{T} = \mathbf{t}(\mathbf{X})$ *be an unbiased estimator of* $\mathbf{q}(\mathbf{\Theta})$. *Then* $\mathbf{Cov}(\mathbf{t}(\mathbf{X}))$ *equals the CRLB iff*

$$\mathbf{t}(\mathbf{X}) = \mathbf{q}(\mathbf{\Theta}) + \frac{\partial \mathbf{q}(\mathbf{\Theta})'}{\partial \mathbf{\Theta}}\left[\mathrm{E}\left(\frac{\partial\ln(f(\mathbf{X};\mathbf{\Theta}))}{\partial\mathbf{\Theta}}\frac{\partial\ln(f(\mathbf{X};\mathbf{\Theta})')}{\partial\mathbf{\Theta}}\right)\right]^{-1}\frac{\partial\ln(f(\mathbf{X};\mathbf{\Theta}))}{\partial\mathbf{\Theta}}$$

*with probability 1.*

**Proof** Referring to the proof of the CRLB, the covariance matrix of $\mathbf{t}(\mathbf{X})$ will equal the CRLB *iff* $\mathbf{Cov}(\mathbf{Y}) = \mathbf{0}$, where the random variable $\mathbf{Y}$ is defined as

$$\mathbf{Y} = \left[\mathbf{I} \;\middle|\; \frac{-\partial\mathbf{q}(\mathbf{\Theta})}{\partial\mathbf{\Theta}}[\mathbf{Cov}(\mathbf{s}(\mathbf{X}))]^{-1}\right]\begin{bmatrix}\mathbf{t}(\mathbf{X})\\ \mathbf{s}(\mathbf{X})\end{bmatrix} \text{ and } \mathbf{s}(\mathbf{X}) = \frac{\partial\ln(f(\mathbf{X};\mathbf{\Theta}))}{\partial\mathbf{\Theta}}$$

and is the random variable that has the covariance matrix $\mathbf{D}\,\mathbf{Cov}(\mathbf{Z})\,\mathbf{D}'$ in the proof of Theorem 7.17.

Since $E(\mathbf{Y}) = E(\mathbf{t}(\mathbf{X})) = \mathbf{q}(\boldsymbol{\Theta})$ *because* $E(\mathbf{s}(\mathbf{X})) = \mathbf{0}$, then setting $\mathbf{Cov}(\mathbf{Y}) = \mathbf{0}$ for the attainment of the CRLB, implies that $P(\mathbf{y} = \mathbf{q}(\boldsymbol{\Theta})) = 1$, which in turn implies $\mathbf{P}\left(\mathbf{t}(\mathbf{x}) - (\partial\mathbf{q}(\boldsymbol{\Theta})/\partial\boldsymbol{\Theta})'[\mathbf{Cov}(\mathbf{s}(\mathbf{X}))]^{-1}\mathbf{s}(\mathbf{X}) = \mathbf{q}(\boldsymbol{\Theta})\right) = 1$. Upon substitution for $[\mathbf{Cov}(\mathbf{s}(\mathbf{X}))]^{-1}$ and $\mathbf{s}(\mathbf{X})$, the result is proved. ■

The CRLB attainment theorem leads to an explicit constructive procedure for deriving the MVUE of $\mathbf{q}(\boldsymbol{\Theta})$ when it exists and when the CRLB exists. Namely, given a probability model $\{f(\mathbf{x};\boldsymbol{\Theta}), \boldsymbol{\Theta} \in \Omega\}$, define

$$\mathbf{t}(\mathbf{X}) = \mathbf{q}(\boldsymbol{\Theta}) + \frac{\partial\mathbf{q}(\boldsymbol{\Theta})'}{\partial\boldsymbol{\Theta}}\left[E\left(\frac{\partial\ln(f(\mathbf{X};\boldsymbol{\Theta}))}{\partial\boldsymbol{\Theta}}\frac{\partial\ln(f(\mathbf{X};\boldsymbol{\Theta})')}{\partial\boldsymbol{\Theta}}\right)\right]^{-1}\frac{\partial\ln(f(\mathbf{X};\boldsymbol{\Theta}))}{\partial\boldsymbol{\Theta}}$$

as a tentative estimator for $\mathbf{q}(\boldsymbol{\Theta})$, and if the expression on the right hand side of the equality does not depend on $\boldsymbol{\Theta}$, i.e. a statistic is defined, then $\mathbf{t}(\mathbf{X})$ is the MVUE for $\mathbf{q}(\boldsymbol{\Theta})$ since then $\mathbf{t}(\mathbf{X})$ is an estimator that achieves the CRLB. Note, unbiasedness of $\mathbf{t}(\mathbf{X})$ follows immediately from the fact that $E[\partial\ln(f(\mathbf{X};\boldsymbol{\Theta}))/\partial\boldsymbol{\Theta}] = \mathbf{0}$ by CRLB regularity condition 4a.

**Example 7.22** ***Using CRLB Attainment to Define MVUEs: Exponential Population***

Reexamine Example 7.21. Because we are estimating $q(\theta)$, we know that $dq(\theta)/d\theta = 1$. We also know from the example that $d\ln(f(\mathbf{X};\theta))/\partial\theta = (-n/\theta) + (\sum_{i=1}^{n} X_i/\theta^2)$, and that

$$E\left(\frac{d\ln(f(\mathbf{X};\theta))}{d\theta}\frac{d\ln(f(\mathbf{X};\theta)')}{d\theta}\right) = -E\left(\frac{d^2\ln(f(\mathbf{X};\theta))}{d\theta^2}\right) = \frac{n}{\theta^2}.$$

Then using Theorem 7.20, we define $\mathbf{t}(\mathbf{X})$ as $t(\mathbf{X}) = \theta + \frac{\theta^2}{n}\left[\frac{-n}{\theta} + \frac{\sum_{i=1}^{n} X_i}{\theta^2}\right] = \frac{\sum_{i=1}^{n} X_i}{n} = \overline{X}$,
and since $\mathbf{t}(\mathbf{X})$ is a statistic, $\mathbf{t}(\mathbf{X})$ is the MVUE of $\theta$. □

It turns out that we can be specific about the class of parametric families of density functions that will both satisfy the CRLB regularity conditions, so that the CRLB applies as a lower covariance bound to unbiased estimators, and also allow an unbiased estimator of $\mathbf{q}(\boldsymbol{\Theta})$ to *achieve* the CRLB. Specifically, the *exponential class of densities* contains the parametric families of densities for which the CRLB is attained for an unbiased estimator of some $\mathbf{q}(\boldsymbol{\Theta})$.

**Theorem 7.21** ***Exponential Class and CRLB Attainment***

*Let $f(\mathbf{x};\boldsymbol{\Theta})$ satisfy the CRLB regularity conditions, and suppose there exists an unbiased estimator of $\mathbf{q}(\boldsymbol{\Theta})$ that achieves the CRLB. Then $f(\mathbf{x};\boldsymbol{\Theta})$ belongs to the exponential class of density functions.*

**Proof** The proof for the scalar parameter case can be found in Bickel and Doksum, (1977) *Mathematical Statistics*, Holden-Day, pp. 130–131. A discussion of the multivariate case can be found in Zacks, (1971) *The Theory of Statistical*

*Inference*, John Wiley, pp. 194–201, and Cencov, C.C., (1982) *Statistical Decision Rules and Optimal Inference*, American Mathematical Society, pp. 219–225. ■

Theorem 7.21 suggests that the procedure for defining the MVUE for $\mathbf{q}(\Theta)$ implied by Theorem 7.20 has potential for success only for probability models where the joint density of the random sample belongs to the exponential class.

### 7.5.3 Asymptotic Efficiency and the CRLB

To this point, we have been using the CRLB concept in the context of examining finite sample properties of estimators. The CRLB also plays a prominent role in defining the concept of an **asymptotically efficient estimator** of $\mathbf{q}(\Theta)$, which relies on the properties of consistency and asymptotic normality as the following definition indicates.

**Definition 7.22** ***Asymptotic Efficiency or Best Asymptotically Normal (BAN)***

$\mathbf{T}_n = \mathbf{t}_n(\mathbf{X})$ is an asymptotically efficient estimator of $\mathbf{q}(\Theta)$ *iff* $\mathbf{t}_n(\mathbf{X})$ is a consistent estimator of $\mathbf{q}(\Theta)$ and

$$\left(\left[\frac{\partial \mathbf{q}(\Theta)}{\partial \Theta}\right]' \left[\mathrm{E}\left(\frac{\partial \ln f(\mathbf{X};\Theta)}{\partial \Theta}\frac{\partial \ln f(\mathbf{X};\Theta)}{\partial \Theta'}\right)\right]^{-1}\left[\frac{\partial \mathbf{q}(\Theta)}{\partial \Theta}\right]\right)^{-1/2} (\mathbf{T}_n - \mathbf{q}(\Theta)) \xrightarrow{d} N(\mathbf{0},\mathbf{I}),$$

so that

$$\mathbf{T}_n \overset{a}{\sim} N\left(\mathbf{q}(\Theta), \left[\frac{\partial \mathbf{q}(\Theta)}{\partial \Theta}\right]' \left[\mathrm{E}\left(\frac{\partial \ln f(\mathbf{X};\Theta)}{\partial \Theta}\frac{\partial \ln f(\mathbf{X};\Theta)'}{\partial \Theta}\right)\right]^{-1}\left[\frac{\partial \mathbf{q}(\Theta)}{\partial \Theta}\right]\right).$$

Thus, if $\mathbf{T}_n$ is asymptotically efficient, it has an asymptotic normal distribution with mean $\mathbf{q}(\Theta)$ and covariance matrix equal to the CRLB, and $\mathbf{T}_n$ is then *approximately* MVUE for $\mathbf{q}(\Theta)$, based on the characteristics of its asymptotic (or approximate) normal probability distribution. Also note the result by L. LeCam[21] which states that an asymptotically efficient estimator in the CAN class has an asymptotic covariance matrix which is smaller than the asymptotic covariance matrix of any other estimator of $\mathbf{q}(\Theta)$ in the CAN class, except, perhaps, on a set of $\Theta$ values having Lebesque measure zero

In order to illustrate the asymptotic efficiency concept, we demonstrate that in the case of sampling from an exponential density function (recall Example 7.21) $\overline{X}_n$ satisfies the conditions of Definition 7.22, and is thus an *asymptotically efficient* estimator of $\Theta$.

[21] LeCam, L., (1953) *On Some Asymptotic Properties of Maximum Likelihood Estimates and Related Bayesx Estimates*, University of California Publications in Statistics, pp. 1:277–330.

**Example 7.23** ***Asymptotically Efficient Estimator of Exponential Mean***

Recall Example 7.21. Since $E(\overline{X}_n) = \theta$, so that $\overline{X}_n$ is unbiased, and since $\mathrm{var}(\overline{X}_n) = \theta^2/n \to 0$ as $n \to \infty$, then $\overline{X}_n \xrightarrow{m} \theta$, which is sufficient for $\mathrm{plim}(\overline{X}_n) = \theta$ and so $\overline{X}_n$ is a consistent estimator of $\theta$. The CRLB for unbiasedly estimating $\Theta$ in this case is given by $\theta^2/n$ (again, recall Example 7.21). By the Lindberg–Levy CLT, $Z_n = (\overline{X}_n - \theta)/(n^{-1/2}\,\theta) \xrightarrow{d} N(0,1)$ and thus $\overline{X}_n$ is in the CAN class of estimators. Then by Definition 7.22, $\overline{X}_n$ is an asymptotically efficient estimator of $\theta$. The asymptotic distribution of $\overline{X}_n$ can be written as $\overline{X}_n \overset{a}{\sim} N(\theta, \theta^2/n)$. □

There is a generic way of defining asymptotic probability distributions for asymptotically efficient estimators which is inherent to the definition of asymptotic efficiency. In particular, the asymptotic distribution of an asymptotically efficient estimator can always be specified as normal, with a mean equal to whatever is being estimated, and covariance matrix equal to the CRLB.

### 7.5.4 Complete Sufficient Statistics and MVUEs

If complete sufficient statistics exist for the statistical model $\{f(\mathbf{x};\boldsymbol{\Theta}), \boldsymbol{\Theta} \in \Omega\}$, then an alternative to the CRLB approach is available to aid in the search for the MVUE of $\mathbf{q}(\boldsymbol{\Theta})$. The approach is based on the Lehmann-Scheffé completeness theorem.

**Theorem 7.22** ***Lehmann–Scheffé Completeness Theorem for MVUEs***

*Let $S_1,\ldots,S_r$ be a set of complete sufficient statistics for $f(\mathbf{x};\boldsymbol{\Theta})$. Let $\mathbf{t}(S_1,\ldots,S_r)$ be an unbiased estimator for the $(k \times 1)$ vector function $\mathbf{q}(\boldsymbol{\Theta})$. Then $\mathbf{T} = \mathbf{t}(S_1,\ldots,S_r)$ is the MVUE of $\mathbf{q}(\boldsymbol{\Theta})$.*

**Proof** Let $\mathbf{S}$ be an $r$–variate complete sufficient statistic and *let* $\mathbf{t}_1(\mathbf{S}) = E(\mathbf{t}_1{}^*(\mathbf{X})|S_1,\ldots,S_r)$ and $\mathbf{t}_2(S) = E(\mathbf{t}_2{}^*(\mathbf{X})|S_1,\ldots,S_r)$ be two unbiased estimators of $\mathbf{q}(\boldsymbol{\Theta})$ that have been defined by application of the Rao-Blackwell procedure to any two unbiased estimators of $\mathbf{q}(\boldsymbol{\Theta})$ given by $\mathbf{t}_1{}^*(\mathbf{X})$ and $\mathbf{t}_2{}^*(\mathbf{X})$. Then letting $\tau(\mathbf{S}) = \boldsymbol{\ell}'(\mathbf{t}_1(\mathbf{S}) - \mathbf{t}_2(\mathbf{S}))$ for any conformable $\boldsymbol{\ell}$-vector, it follows from unbiasedness that

$$E(\tau(\mathbf{S})) = \boldsymbol{\ell}'E(\mathbf{t}_1(\mathbf{S}) - \mathbf{t}_2(\mathbf{S})) = \boldsymbol{\ell}'(E(\mathbf{t}_1(\mathbf{S})) - E(\mathbf{t}_2(\mathbf{S}))) = \mathbf{0}\,\forall \boldsymbol{\Theta} \in \Omega.$$

But since $\mathbf{S}$ is a set of complete sufficient statistics, necessarily $\tau(\mathbf{S}) = \boldsymbol{\ell}'[\mathbf{t}_1(\mathbf{S}) - \mathbf{t}_2(\mathbf{S})] = \mathbf{0}$ with probability one because $\boldsymbol{\ell}'[\mathbf{t}_1(\mathbf{S}) - \mathbf{t}_2(\mathbf{S})]$ is a function of the set of complete sufficient statistics that has an expectation of zero $\forall\boldsymbol{\Theta} \in \Omega$. Since $\boldsymbol{\ell}$ can be chosen to have all zero entries except for a 1 in any position, it follows that the $(k \times 1)$ vectors $\mathbf{t}_1(\mathbf{S})$ and $\mathbf{t}_2(\mathbf{S})$ have the same $k$ elements, and thus are the same unbiased estimator, say $\mathbf{t}(\mathbf{S})$, of $\mathbf{q}(\boldsymbol{\Theta})$ with probability one, regardless of the choice of unbiased estimators $\mathbf{t}_1{}^*(\mathbf{X})$ and $\mathbf{t}_2{}^*(\mathbf{X})$ used in the definitions of $\mathbf{t}_1(\mathbf{S})$ and $\mathbf{t}_2(S)$, respectively. Then $\mathbf{t}(\mathbf{S}) = E(\mathbf{t}^*(\mathbf{X})|S_1,\ldots,S_r)$ must be the MVUE of $\mathbf{q}(\boldsymbol{\Theta})$, regardless of the choice of unbiased estimator $\mathbf{t}^*(\mathbf{X})$, because $\mathbf{Cov}(\mathbf{t}(\mathbf{S})) \preceq \mathbf{Cov}(\mathbf{t}^*(\mathbf{X}))$ for any choice of unbiased estimator $\mathbf{t}^*(\mathbf{X})$ by the Rao–Blackwell theorem.

We now show that regardless of the procedure used to define $\mathbf{t}(\mathbf{S})$, if $E(\mathbf{t}(\mathbf{S})) = \mathbf{q}(\boldsymbol{\Theta})$, then $\mathbf{t}(\mathbf{S})$ is the MVUE of $\mathbf{q}(\boldsymbol{\Theta})$. Let $\mathbf{t}(\mathbf{S})$ be an unbiased estimator for $\mathbf{q}(\boldsymbol{\Theta})$. Note that by definition, $E(\mathbf{t}(\mathbf{S})|\mathbf{S}) = \mathbf{t}(\mathbf{S})$. Then by the argument presented above, $\mathbf{t}(\mathbf{S})$ is the MVUE of $\mathbf{q}(\mathbf{\Theta})$ (regardless of how we were led to the definition of $\mathbf{t}(\mathbf{S})$). ■

The point of the Lehmann–Scheffé completeness theorem is that if a set of complete sufficient statistics exist for $f(\mathbf{x};\mathbf{\Theta})$, then our search for the MVUE of $\mathbf{q}(\mathbf{\Theta})$ is complete when we have found a function of the set of complete sufficient statistics whose expectation is $\mathbf{q}(\mathbf{\Theta})$. Viewed another way, Theorem 7.22 implies that if $\mathbf{S}$ is complete, then for *any* function of $\mathbf{S}$, say $\tau(\mathbf{S})$, we have that $\tau(\mathbf{S})$ is the MVUE for its own expectation, i.e., $\tau(\mathbf{S})$ is the MVUE for $E(\tau(\mathbf{S}))$.

The Lehmann–Scheffé completeness theorem then suggests a least two procedures for defining the MVUE of $\mathbf{q}(\mathbf{\Theta})$ if a set of complete sufficient statistics, $\mathbf{S}$, exists, which we delineate in the following definition of the **Lehman-Scheffé Completeness Approach** for finding MVUEs.

**Definition 7.23** ***Lehman-Scheffé Completeness Approach for Finding MVUEs***

The following methods based on complete sufficient statistics $\mathbf{S}$ lead to MVUEs for a function, $\mathbf{q}(\mathbf{\Theta})$, of a parameter vector $\mathbf{\Theta}$.

***Method 1.*** Find a statistic of the form $\mathbf{t}(\mathbf{S})$ for which $E(\mathbf{t}(\mathbf{S})) = \mathbf{q}(\mathbf{\Theta})$. Then $\mathbf{t}(\mathbf{S})$ is necessarily the MVUE of $\mathbf{q}(\mathbf{\Theta})$.

***Method 2.*** Find *any* unbiased estimator of $\mathbf{q}(\mathbf{\Theta})$, say $\mathbf{t}^*(\mathbf{X})$. Then $\mathbf{t}(\mathbf{S}) = E(\mathbf{t}^*(\mathbf{X})|\mathbf{S})$ is the MVUE of $\mathbf{q}(\mathbf{\Theta})$.

We emphasize that in either method in the preceding definition, $\mathbf{t}^*(\mathbf{X})$ and $\mathbf{q}(\mathbf{\Theta})$ can be conformable *vectors*.

**Example 7.24** ***Finding MVUEs Using Complete Sufficient Statistics***

Recall Examples 7.15 and 7.16. Since $\sum_{i=1}^n X_i$ is a complete sufficient statistic for the probability model of Example 7.15, and since $T = t(\sum_{i=1}^n X_i) = n^{-1}(\sum_{i=1}^n X_i) = \bar{x}$ is such that $E(T) = p$, we know by the Lehmann-Scheffé completeness theorem that $\bar{X}$ is the MVUE for $p$. In Example 7.16, $\sum_{i=1}^n X_i$ and $\sum_{i=1}^n X_i^2$, or alternatively by Theorem 7.10, $\bar{X}$ and $nS^2/(n-1)$ are complete sufficient statistics for the probability model. Since $E(\bar{X}) = \mu$ and $E(nS^2(n-1)) = \sigma^2$, it follows by the Lehmann-Scheffé Completeness Theorem that $(\bar{X}, nS^2/(n-1))$ is the MVUE for $(\mu, \sigma^2)$. □

## Keywords, Phrases, and Symbols

$\mathbf{A} \preceq \mathbf{B}$, Matrix $\mathbf{A}$ is no larger than matrix $\mathbf{B}$
Alternative form of the CRLB
an estimate of $\mathbf{\Theta}$ or $\mathbf{q}(\Theta)$
Asymptotic efficiency
Asymptotic MSE
Asymptotic relative efficiency
Asymptotically relatively more efficient
Asymptotically unbiased
Best Linear Unbiased Estimator, BLUE
Bias
Bias matrix
Bias vector
Biased estimator
Complete sufficient statistics
Completeness in the exponential class
Consistency
Consistent estimator
Cramer-Rao Lower bound, CRLB
CRLB regularity conditions
Distinct PDFs
Distribution-free case
Distribution-specific case

Efficient
Estimand
Estimating $\mathbf{\Theta}$ or $q(\mathbf{\Theta})$
Estimator admissibility
Estimator, MVLUE
Estimator, MVUE
Exponential class and sufficient statistics
$h \circ \tau$ (composite function)
Lehmann-Scheffé completeness theorem
Lehmann-Scheffé minimal sufficiency
Lehmann-Scheffe MVUE approach
Mean square error matrix
Mean square error, MSE
Minimal sufficient statistics
Minimum variance linear unbiased
Minimum variance unbiased
Neyman factorization theorem
Nonparametric estimation
Parameter identifiability
Parametric estimation
Point estimate
point estimator
Range of $X$ over the parameter space $\Omega$, $R_\Omega(X)$
Rao-Blackwell theorem
Relative efficiency, relatively more efficient
Statistical model
Strong means square error (SMSE) criterion
Sufficient statistics
$T_i$, $T_{(n)}$
True PDF of $\mathbf{X}$
True value of $q(\mathbf{\Theta})$
True value of $\Theta$
Unbiased estimator

## Problems

**1.** The operating life of a certain type of math coprocessor installed in a personal computer can be represented as the outcome of a random variable having an exponential density function, as

$$Z \sim f(z;\theta) = \frac{1}{\theta} e^{-z/\theta}\, I_{(0,\infty)}(z),$$

where $z$ = the number of hours the math coprocessor functions until failure, measured in thousands of hours.

A random sample, $X = (X_1,\ldots,X_n)$, of the operating lives of 200 coprocessors is taken, where the objective being is to estimate a number of characteristics of the operating life distribution of the coprocessors. The outcome of the sample mean was $\bar{x} = 28.7$.

a. Define a minimal sufficient statistic for $f(\mathbf{x};\theta)$, the joint density of the random sample.

b. Define a complete sufficient statistic for $f(\mathbf{x};\theta)$.

c. Define the MVUE for $E(Z) = \theta$ if it exists. Estimate $\theta$.

d. Define the MVUE for $\operatorname{var}(Z) = \theta^2$ if it exists. Estimate $\theta^2$.

e. Define the MVUE for $E(Z^2) = 2\theta^2$ if it exists. Estimate $2\theta^2$.

f. Define the MVUE for $q(\theta)_{(3\times 1)} = \begin{bmatrix}\theta \\ \theta^2 \\ 2\theta^2\end{bmatrix}$ if it exists. Estimate $\mathbf{q}(\theta)$.

g. Is the second sample moment about the origin, i.e., $M'_2 = \sum_{i=1}^n X_i^2/n$, the MVUE for $E(Z^2)$?

h. Is the sample variance, $S^2$, the MVUE for var($Z$)?

i. Suppose we want the MVUE for $F(b) = P(z \le b) = 1-e^{-b/\theta}$, where $F(b)$ is the probability that the coprocessor fails before 1,000 $b$ hours of use. It can be shown that

$$t(\mathbf{X}) = 1 - \left(1 - \frac{b}{\left(\sum_{i=1}^n X_i\right)}\right)^{n-1} I_{[b,\infty)}\left(\sum_{i=1}^n X_i\right)$$

is such that $E(t(\mathbf{X})) = 1-e^{-b/\theta}$. Is $t(\mathbf{X})$ the MVUE for $P(z \le b)$? Why or why not? Estimate $P(z \le 20)$.

(j) Is $t_*(\mathbf{X}) = 1 - e^{-b/\bar{X}}$ a MVUE for $F(b)$? Is $t_*(\mathbf{X})$ a consistent estimator of $F(b)$?

**2.** Use the Lehman-Scheffé minimal sufficiency theorem, or some other argument, to find a set of minimal sufficient statistics for each case below.

a. You are random sampling from a log-normal population distribution given by

$$f(z;\mu,\sigma^2) = \frac{1}{(2\pi)^{1/2}\sigma z}\exp\left(-\frac{1}{2\sigma^2}(\ln(z)-\mu)^2\right) I_{(0,\infty)}(z),$$

where $\mu \in (-\infty,\infty)$ and $\sigma^2 > 0$.

b. You are random sampling from a "power function" population distribution given by

$$f(z;\lambda) = \lambda z^{\lambda-1} I_{(0,1)}(z), \text{ where } \lambda > 0.$$

c. You are random sampling from a Poisson population distribution

$$f(x;\lambda) = \frac{e^{-\lambda}\lambda^x}{x!} I_{\{0,1,2,\ldots,\}}(x).$$

d. You are random sampling from a negative binomial density

$$f(x;r_0,p) = \frac{(x-1)!}{(r_0-1)!(x-r_0)!} p^{r_0}(1-p)^{x-r_0} I_{\{r_0,r_0+1,\ldots\}}(x)$$

where $r_0$ is a *known* positive integer.

e. You are random sampling from a $N(\mu,\sigma^2)$ population distribution.

f. You are random sampling from a continuous uniform density

$$f(x;\Theta) = \frac{1}{\Theta} I_{(0,\Theta)}(x)$$

g. You are sampling from a Beta distribution

$$f(x;\alpha,\beta) = \frac{1}{B(\alpha,\beta)} x^{\alpha-1}(1-x)^{\beta-1} I_{(0,1)}(x).$$

**3.** Identify which of the minimal sufficient statistics in (2) are *complete* sufficient statistics.

**4.** The operating life of a small electric motor manufactured by the AJAX Electric Co. can be represented as a random variable having a probability density given as

$$Z \sim f(z;\Theta) = \frac{1}{6\Theta^4} z^3 e^{-z/\Theta} I_{(0,\infty)}(z)$$

where $\Theta \in \Omega = (0,\infty)$, $E(Z) = 4\Theta$, $var(Z) = 4\Theta^2$, and $z$ is measured in thousand of hours. A random sample $(X_1,\ldots,X_{100})$ of the operating lives of 100 electric motors has an outcome that is summarized as $\bar{x} = 7.65$ and $s^2 = \sum_{i=1}^{n}(x_i - \bar{x})^2/100 = 1.73$.

a. Define a minimal, complete sufficient statistic for estimating the expected operating life of the electric motors produced by the AJAX Co.

b. Define the MVUE for estimating $E(Z) = 4\Theta$. Justify the MVUE property of your estimator. Generate an estimate of $E(Z)$ using the MVUE.

c. Is the MVUE estimator a consistent estimator of $E(Z) = 4\Theta$? Why or why not?

d. Does the variance of the estimator you defined in (b) attain the Cramer-Rao Lower Bound? (The CRLB regularity conditions hold for the joint density of the random sample. Furthermore, the alternative form of the CRLB, expressed in terms of second-order derivatives, applies in this case if you want to use it).

**5.** The number of customers that enter the corner grocery store during the noon hour has a Poisson distribution, i.e.,

$$f(z;\lambda) = \frac{e^{-\lambda}\lambda^z}{z!} I_{\{0,1,2,3,\ldots\}}(z).$$

Assume that $(X_1,X_2,\ldots,X_n)'$ is a random sample from this Poisson population distribution.

a. Show that the Cramer-Rao lower bound regularity conditions hold for the joint density of the random sample.

b. Derive the CRLB for unbiased estimation of the parameter $\lambda$. Is $\bar{X}$ the MVUE for estimating $\lambda$? Why or why not?

c. Use the CRLB attainment theorem to derive the MVUE for estimating $\lambda$. Suppose $n = 100$ and $\sum_{i=1}^{100} x_i = 283$. Estimate $\lambda$ using the MVUE.

d. Is $\bar{X}$ a member of the CAN class of estimators? Is $\bar{X}$ asymptotically efficient?

e. Define the CRLB for estimating $P(z=0) = e^{-\lambda}$. Does there exist an unbiased estimator of $e^{-\lambda}$ that achieves the CRLB? Why or why not?

**6.** Polly Pollster wants to estimate the proportion of voters in Washington State that are in favor of an anti-tax initiative. She will be using a random sample (with replacement) of 1,000 voters, and she will record their preference regarding the anti-tax initiative. She needs some statistical advice from you.

a. Define a statistical model for the problem of estimating the proportion of voters in favor of the initiative.

b. Define the MVLUE for the proportion of voters in favor of the initiative. Justify that your estimator really is a MVLUE.

c. Is the estimator that you defined in (b) a consistent estimator of the proportion of voters in favor of the initiative? Is it a CAN estimator? Is it asymptotically efficient?

d. Assume there are two million voters in the state. What is the probability that the estimator you defined in (b) generates an estimate that is within

$\pm.03$ of the true proportion of voters favoring the initiative? (Note: This may be a function of unknown parameters!)

e. Polly summarized the outcome of the random sample as $\sum_{i=1}^{1000} x_i = 670$, where $x_i = 1$ indicates that the *ith* sample voter was in favor of the initiative, and $x_i = 0$ otherwise. Estimate the proportion of voters in favor of the initiative. Given your result in (d), if the election were held today, would you predict that the initiative would pass? Why or why not?

**7.** Two economics professors are arguing about the appropriate estimator to use in estimating the mean of the population distribution of incoming freshmen's I.Q.'s. The estimators will be based on a random sample from the population distribution. One professor suggests that they simply calculate the sample mean I.Q., $\bar{x}_n$, and use it as an estimate of the mean of the population distribution. The other prefers to use an estimator of the form $t(\mathbf{X}) = \sum_{i=1}^{n} X_i/(n+k)$, where $n$ is the random sample size and $k$ is some positive integer, and she argues that her estimator has less variance than the sample mean and that for an appropriate choice of $k$, her estimator would be superior on the basis of MSE.

a. We know that $\bar{X}$ is unbiased, asymptotically unbiased, BLUE, and consistent for estimating the mean of the population distribution. Which of these properties apply to the alternative estimator?

b. Define asymptotic distributions for both estimators. On the basis of their asymptotic distributions, do you favor one estimator over the other?

c. Define the MSEs of the estimators. Is there any validity to the statement that "for an appropriate choice of $k$" $t(\mathbf{X})$ will be superior to $\bar{X}$ in terms of MSE? Explain.

d. Can you foresee any practical problems in using $t(\mathbf{X})$ to generate estimates of the population mean?

**8.** The diameters of blank compact disks manufactured by the Dandy Disk Co. can be represented as outcomes of a random variable

$$Z \sim f(z;\Theta) = \frac{1}{\Theta} I_{(4,4+\Theta)}(z), \text{ for some } \Theta > 0,$$

where $z$ is measured in inches. You will be using a random sample $(X_1, X_2, \ldots X_n)$ from the population distribution $f(z;\Theta)$ to answer the questions below.

a. Based on the random sample, define an unbiased estimator of the parameter $\Theta$.

b. Is the estimator you defined in (a) a BLUE for $\Theta$? If not, find a BLUE for $\Theta$, if it exists.

c. Is your estimator a consistent estimator for $\Theta$? Why or why not?

d. Define an asymptotic distribution for your estimator.

e. A random sample of size $n = 1{,}000$ from $f(z;\Theta)$ results in $\sum_{i=1}^{1000} x_i = 4,100$. Use your estimator to estimate the value of $\Theta$. Using your estimate of $\Theta$, what is the estimated probability that $z \in (4.05, 4.15)$?

**9.** Your company sells trigger mechanisms for air bags that are used in many modern domestic and foreign-built passenger cars. The reliability of such trigger mechanisms is obviously critical in the event that an air bag-equipped vehicle is involved in an accident. One large Detroit automobile manufacturer said that they would be willing to purchase trigger mechanisms from your company if you could provide convincing support for the statement that, in repeated simulations of an impact of 15 mph, the expected number of impacts needed to obtain the first failure of your trigger (i.e., the trigger does not signal an air bag deployment) was greater than or equal to 1,000.

Management has randomly chosen 10,000 of the trigger mechanisms for a testing program in which the number of simulated impacts needed to obtain the first failure will be observed for each mechanism. You need to estimate the expected number of impacts needed to obtain the first failure of the trigger mechanisms you manufacture. You intend to use the outcome of the sample mean as your estimate of this expected number of impacts.

a. Define an appropriate statistical model for the sampling experiment, and justify your choice.

b. Is the sample mean an unbiased estimator in this case? Why?

c. Is the sample mean an asymptotically unbiased estimator? Why?

d. Is the sample mean a consistent estimator? Why?

e. Is the sample mean a BLUE (or equivalently, a MVLUE)? Why?

f. Derive the Cramer-Rao Lower Bound for the variance of unbiased estimators of the expected number of impacts to obtain the first failure. (Hint: You may use the alternative (second-derivative) form of the bound – it might be a little easier to work with in this case.)

g. Is the sample mean a MVUE? Why?

h. Use Theorem 7.17 on the attainment of the Cramer-Rao lower bound to derive the MVUE of the expected number of impacts needed to obtain the first failure.

i. Define an appropriate asymptotic distribution for the sample mean in this case. Is the sample mean asymptotically efficient?

j. The 10,000 observations resulted in $\sum_{i=1}^{10,000} x_i = 1.5 \times 10^7$. What is your estimate of the expected number of impacts needed to obtain the first failure?

k. Is $(\overline{X}_n)^{-1}$ a consistent estimator of $p$, the probability that a trigger successfully signals deployment of the air bag on any given trial?

l. Define an asymptotic distribution for the estimator $(\overline{X}_n)^{-1}$ of $p$. Is the estimator asymptotically efficient?

m. Use the estimator $(\overline{X}_n)^{-1}$ to estimate $p$, and use this estimated value in the asymptotic distribution you defined for $\overline{X}_n$ to estimate the probability that an estimate generated by $\overline{X}_n$ would be within $\pm$ 50 units of the population mean.

n. What, if anything, can you say to the Detroit manufacturer to convince him/her to buy your trigger mechanisms?

**10.** Suppose random sampling is from an exponential population distribution representing the waiting time between customer arrivals at a bank, so that the statistical model is given by

$$f(x;\theta) = \theta^{-n} e^{-\sum_{i=1}^{n} x_i/\theta} \prod_{i=1}^{n} I_{(0,\infty)}(x_i),$$

*for* $\theta>0$.

Your objective is to use an outcome of a random sample of size $n$ ($n$ fixed and known) to estimate $q(\theta) = \theta^2$, the variance of waiting times.

a. Define a complete sufficient statistic for $f(\mathbf{x};\theta)$.

b. Define the CRLB for unbiasedly estimating $\theta^2$.

c. Does there exist an unbiased estimation of $\theta^2$ whose variance is equal to the CRLB?

d. Define the MVUE of $\theta^2$ by finding an appropriate function of the complete sufficient statistic—if you can.

e. Is the sample variance, $S_n^2$, the MVUE for the population variance $\theta^2$?

f. If $\sum_{i=1}^{100} x_i = 257$, generate a MVUE estimate of the variance of waiting times.

**11.** One hundred one-acre test plots are being used to assess the yield potential of a new variety of wheat genetically engineered by Washington State University. The actual yield-per-acre observations on the test plots can be viewed as iid observations from some log-normal population, so that the statistical model for the experiment is given by

$$f(\mathbf{y};\mu,\sigma^2) = \frac{1}{(2\pi)^{n/2}\sigma^n \prod_{i=1}^{n} y_i} \times \exp\left[-\frac{1}{2\sigma^2}\sum_{i=1}^{n}(\ln(y_i)-\mu)^2\right]\prod_{i=1}^{n} I_{(0,\infty)}(y_i)$$

for $\mu \in (-\infty,\infty)$ and $\sigma>0$.

a. Define minimal sufficient statistics for $f(\mathbf{y};\mu,\sigma^2)$.

b. Are the sufficient statistics you defined in (a) *complete* sufficient statistics?

c. Is $t_1(\mathbf{Y}) = n^{-1}\sum_{i=1}^{n}\ln(Y_i)$ the MVUE of the parameter $\mu$? Is it consistent? Why?

d. Is $t_2(\mathbf{Y}) = (n-1)^{-1}\sum_{i=1}^{n}\left(\ln(Y_i) - n^{-1}\sum_{i=1}^{n}\ln(Y_i)\right)^2$ the MVUE of the parameter $\sigma^2$? Is it consistent?

e. Define a consistent estimator of $q(\mu,\sigma) = e^{\mu+\sigma^2/2}$, which is the mean of the log-normal population. Justify your answer. (The MVUE of the mean exists, but it is quite complicated to define and calculate. See D. J. Finney (1941), "*On the distribution of a variate whose logarithm is normally distributed.*" *Roy. Statistical Society*, Series B, 7, pp. 155–161.)

f. An overworked, underpaid, gaunt-looking research assistant hands you an envelope that contains only summary information on the results of the experiment. In particular the information is that

$$\sum_{i=1}^{100}\ln(Y_i) = 375.00, \sum_{i=1}^{100}(\ln(Y_i))^2 = 1455.75.$$

Generate an estimate of $q(\mu,\sigma^2)_{(2\times1)} = \begin{bmatrix}\mu\\ \sigma^2\end{bmatrix}$ using the MVUE of $q(\mu,\sigma^2)$. Generate an estimate of the mean of the lognormal distribution using a consistent estimator.

**12.** In each case below, determine whether the estimator under consideration is unbiased, asymptotically unbiased, and/or consistent.

a. The random sample $(X_1,\ldots X_n)$ is generated from a Gamma population distribution. The estimator $t(\mathbf{X}) = \sum_{i=1}^{n} X_i/n$ will be used to estimate $E(X_i) = \alpha\beta$.

b. The random sample $(X_1,\ldots,X_n)$ is generated from an exponential population distribution. The estimator $t(\mathbf{X}) = (1/2)\sum_{i=1}^{n} X_i^2/n$ will be used to estimate $\text{var}(X_i) = \theta^2$.

c. The random sample $(X_1,\ldots,X_n)$ is generated from a geometric population distribution. The estimator $t(\mathbf{X}) = (S^2 - \overline{X})$ will be used to estimate $E(X_i^2) = p^{-2}$.

d. The random sample $(X_1,\ldots,X_n)$ is generated from a Bernoulli population distribution. The estimator $t(\mathbf{X}) = \overline{X}(1-\overline{X})$ will be used to estimate $\text{var}(X_i) = p(1-p)$. (Hint: $2\sum_{i=1}^{n}\sum_{j>i}^{n} a = n(n-1)a$).

**13.** In Problem 12(b), above, consider the alternative estimator $t^*(\mathbf{X}) = S^2$ for estimating $\theta^2$. In a mean square error sense, which estimator would you prefer for estimating $\theta^2$, $t(\mathbf{X})$ or $t^*(\mathbf{X})$? Note $\mu'_r = r!\theta^r$ for an exponential PDF.

**14.** An incoming shipment of 1,000 toys from a toy manufacturer is received by a large department store for a pre-Christmas sale. The store randomly samples 50 toys from the shipment, *without replacement*, and records whether or not the sample item is defective. The store wants to generate a MVUE estimate of the proportion of defectives in the shipment of toys. The statistical model it uses for the sampling experiment is given by the hypergeometric density with the following parameterization:

$$f(x;\Theta) = \frac{\binom{1000\Theta}{x}\binom{1000(1-\Theta)}{50-x}}{\binom{1000}{50}} I_{\{0,1,2,\ldots,50\}}(x),$$

where $\Theta \in \Omega = \{0,.001,.002,\ldots,1\}$ represents the proportion of defectives in the shipment.

a. Show that the $X \sim f(x;\Theta)$ is a minimal, complete sufficient statistic for $f(x;\Theta)$.

b. Define the MVUE for the proportion of defectives in the shipment.

c. Suppose that the outcome of $X$ was 3. Define a MVUE estimate of the proportion of defectives in the shipment.

d. Define an MVUE for the *number* of defective toys in the shipment, and provide an MVUE estimate of the this number.

**15.** The rates of return per dollar invested in two common stocks over a given investment period can be viewed as the outcome of a bivariate normal distribution $N(\boldsymbol{\mu},\boldsymbol{\Sigma})$. The rates are independent between investment periods. An investment firm intends to use a random sample from the $N(\boldsymbol{\mu},\boldsymbol{\Sigma})$ population distribution of rates of return to generate estimates of the expected rates of return, $\boldsymbol{\mu}$, as well as the variances in the rates of return, given by the diagonal of $\boldsymbol{\Sigma}$.

a. Find a minimal, complete (vector) sufficient statistic for $N(\boldsymbol{\mu},\boldsymbol{\Sigma})$.

b. Define the MVUE for $\boldsymbol{\mu}$.

c. Define the MVUE for $\boldsymbol{\Sigma}$ and for diag($\boldsymbol{\Sigma}$).

d. Define the MVUE for the vector $(\mu_1, \mu_2, \Sigma_1^2, \Sigma_2^2)$.

e. A random sample of size 50 has an outcome that is summarized by: $\bar{\mathbf{x}} = [.048\ .077]'$, $s_1^2 = .5 \times 10^{-3}$, $s_2^2 = .3 \times 10^{-4}$, and $s_{12} = .2\times10^{-4}$. Calculate the MVUE outcome for $[\mu_1, \mu_2, \Sigma_1^2, \Sigma_2^2]'$.

f. Is the MVUE of $[\mu_1, \mu_2, \Sigma_1^2, \Sigma_2^2]'$ consistent?

g. If an investor invests \$500 in each of the two investments, what is the MVUE of her expected dollar return on the investment during the investment period under consideration?

**16.** For estimating the population mean, $\mu$, based on a size $n$ iid random sample, $(X_1,\ldots,X_n)$, from some population distribution, the following three estimators are being considered

$$t_1(X) = n^{-1}\sum_{i=1}^{n} x_i,\quad t_2(X) = (n-1)\sum_{i=1}^{n-1} x_i, \text{ and}$$

$$t_3(X) = n^{-1}\left(.5\sum_{i=1}^{n/2} x_i + 1.5\sum_{i=(n/2)+1}^{n} x_i\right)$$

where $n$ is an even number.

a. Identify which of the estimators are unbiased estimators of $\mu$.

b. Identify which of the estimators are asymptotically unbiased estimators of $\mu$.

c. Define the variances of each of the estimators. Identify which estimator has the smallest variance. In comparing the variances of these particular three estimators, would you have expected one of these estimators to have the smallest variance, *a priori*?

d. Identify which of the estimators are consistent estimators of the population mean, justifying your answers.

**17.** Consider two estimators for the probability, $p$, that a tossed coin will land on "heads", based on a *iid* random sample of size $n$ from whatever Bernoulli distribution governs the probability of observing "heads" and "tails". One estimator is the sample mean $\bar{x}_n$, and the other is the estimator defined as $t(X) = \sum_{i=1}^{n} X_i/(n+k)$, where $k$ is some positive integer.

a. For each of the estimators, determine which of the following properties apply: unbiased, asymptotically unbiased, BLUE, and/or consistent?

b. Define asymptotic distributions for both estimators. On the basis of their asymptotic distributions, do you favor one estimator over the other?

c. Define the expected squared distances of the estimators from the unknown value of $p$. Is there any validity to the statement that "for an appropriate choice of $k$," $t(X)$ will be superior to $\bar{X}$ in terms of expected squared distance from $p$? Explain.

d. Can you foresee any practical problems in using $t(X)$ to generate estimates of the population mean?

e. Suggest a way of estimating when it might make sense to use $t(X)$ in place of the sample mean for estimating $p$.

**18.** In each case below, determine whether the estimator under consideration is unbiased, asymptotically unbiased, and/or consistent.

a. The *iid* random sample $(X_1,\ldots,X_n)$is generated from a Gamma population distribution. The estimator $t(\mathbf{X}) = \bar{X}$ will be used to estimate the value of $\alpha\beta$.

b. The *iid* random sample $(X_1,\ldots,X_n)$ is generated from an exponential population distribution. The estimator $t(\mathbf{X}) = (n-1)^{-1}\sum_{i=1}^{n}(X_i-\bar{X})^2$ will be used to estimate the value of $\theta^2$.

c. The *iid* random sample $(X_1,\ldots,X_n)$ is generated from a Poisson population distribution. The estimator $t(\mathbf{X}) = \bar{X}_n$ will be used to estimate the value of $\lambda$.

d. The *iid* random sample $(X_1,\ldots,X_n)$ is generated from a Poisson population distribution. The estimator $t(\mathbf{X}) = (n-1)^{-1}\sum_{i=1}^{n}(X_i-\bar{X})^2$ will be used to estimate the value of $\lambda$.

e. The *iid* random sample $(X_1,\ldots,X_n)$ is generated from a geometric population distribution. The estimator $t(\mathbf{X}) = S_n^2$ will be used to estimate $(1-p)p^{-2}$.

f. The *iid* random sample $(X_1,\ldots,X_n)$ is generated from a Bernoulli population distribution. The estimator $t(\mathbf{X}) = \bar{X}(1-\bar{X})$ will be used to estimate $p(1-p)$.

g. The random sample of size 1, $X$, is generated from a Binomial population distribution for which the value of the parameter $n$ is known. The estimator $t(X) = X/n$ will be used to estimate $p$.

**19.** If you wanted to choose the estimator with the smaller expected squared distance from the value of $\lambda$, which of the estimators in 18.c and 18.d above would you choose?

**20.** Let the size $n$ random sample $\mathbf{Y} = [Y_1, Y_2, \ldots Y_n]'$ be such that $\mathbf{Y} = \mathbf{x}\beta + \boldsymbol{\varepsilon}$, where $\mathbf{x}$ is a $n \times 1$ non-zero vector of "explanatory variable values", $\beta$ is an unknown parameter value, and $\boldsymbol{\varepsilon}$ is an $n \times 1$ random vector for which $\mathrm{E}(\boldsymbol{\varepsilon}) = \mathbf{0}$ and $\mathrm{cov}(\boldsymbol{\varepsilon}) = \sigma^2\mathbf{I}$.

a. Is $t(\mathbf{Y}) = \sum_{i=1}^{n} x_iY_i/\sum_{i=1}^{n} x_i^2$ the BLUE of $\beta$? Explain.

b. Define the mean and the variance of the estimator in a).

c. Under what conditions will the estimator in a) be a consistent estimator of $\beta$?

d. Consider an alternative estimator $t^*(\mathbf{Y}) = (\mathbf{x}'\mathbf{x}+k)^{-1}\mathbf{x}'\mathbf{Y}$ for $\beta$. (this is an example of a so-called "ridge regression" estimator). Define the mean and variance of this estimator. Is this a linear estimator? Is it unbiased? Is it asymptotically unbiased?

e. Under what conditions will the estimator in d) be a consistent estimator of $\beta$?

f. If you wanted to use the estimator that has the smaller expected squared distance from $\beta$, would you prefer one estimator over the other? Explain.

**21.** An estimate is needed of the expected number of resets it takes to get a certain brand of 15 amp ground fault protecting circuit breaker to fail to reset. A random sample of $n$ circuit breakers are each tripped and reset as many times as needed to produce a fail. The recorded numbers of resets associated with the $n$ breakers are viewed as the outcome of *iid* random variables $(X_1,\ldots,X_n)$.

a. Define the joint probability density of the random sample of the $n$ reset outcomes.

b. Define the MVLUE for the expected number of resets. Justify that your estimator really is a MVLUE of this proportion.

c. Is the estimator that you defined above a consistent estimator of the expected number of resets? Is it asymptotically normally distributed?

d. Use the CRLB attainment theorem to find a MVUE of the expected number of resets, if you can. If the MVUE exists, how does it differ from the *linear* estimator you defined in b)?

e. Does an unbiased estimator of the *variance* of reset outcomes exist that achieves the CRLB, and is thereby MVUE?

# 8 Point Estimation Methods

## 8.1 Introduction

In this chapter we examine point estimation methods that lead to specific functional forms for estimators of $\mathbf{q}(\boldsymbol{\Theta})$ that can be relied upon to define estimators that often have good estimator properties. Thus far, the only result in Chapter 7 that could be used directly to define the functional form of an estimator is the theorem on the attainment of the CRLB, which is useful only if the probability model $\{f(\mathbf{x};\boldsymbol{\Theta}), \boldsymbol{\Theta}\in\Omega\}$ and the estimand $\mathbf{q}(\boldsymbol{\Theta})$ are such that the CRLB is actually attainable. We did examine a number of important results that could be used to narrow the search for a good estimator of $\mathbf{q}(\boldsymbol{\Theta})$, to potentially improve upon an unbiased estimator that was already available, or that could verify when an unbiased estimator was actually the best in the sense of minimizing variance or in having the smallest covariance matrix. However, since the functional form of an estimator of $\mathbf{q}(\boldsymbol{\Theta})$ having good estimator properties is often not apparent even with the aid of the results assembled in Chapter 7, we now examine procedures that suggest functional forms of estimators.

Currently, there is no single procedure for generating estimators of $\mathbf{q}(\boldsymbol{\Theta})$ that will always lead to the best estimator or even to an estimator that *always* has "good" properties. However, the three general estimation methods that we will examine lead to good estimators of $\mathbf{q}(\boldsymbol{\Theta})$ for a wide range of probability models and associated random samples of data. These methods include the **least squares**, **maximum likelihood**, and the **generalized method of moments estimators**. Together, these estimators encompass a large number of the estimation procedures that are currently used in empirical practice.

In this and the remaining two chapters, we will place some proofs of theorems in the appendix to improve readability relative to the main results that need to be conveyed in estimation and inference methodology. In particular, proofs that tend to be both mathematically complex and relatively less informative, relative to enhancing statistical understanding, will be found in the appendices to chapters.

## 8.2 The Least Squares Estimator

In this section we examine the **least squares (LS) estimator** for defining the mean of a random variable **Y** conditional on the values of other random variables **X** on which the mean of **Y** potentially depends. We will investigate both the **linear LS** and **nonlinear LS** estimators, providing substantial detail on the former and an introduction to the latter. Together, these two **LS** methods account for a great deal of the empirical literature in statistics and econometrics devoted to "*regression analysis*".

### 8.2.1 Linear and Nonlinear Regression Models

In the empirical statistics and econometrics literature, a **regression model** is used to characterize how the outcomes of a random sample come about, whereby the $i^{th}$ random variable, $Y_i$, in the random sample $(Y_1, \ldots, Y_n)$ is decomposed into the sum of its expectation and the deviation from its expectation,[1] as

$$Y_i = \mu_i + \varepsilon_i, i = 1, \ldots, n.$$

In this conceptualization, $E(Y_i) = \mu_i$ and $E(\varepsilon_i) = 0$. At this level of generality, the decomposition is *always* valid so long as $E(Y_i)$ exists.

The **linear regression model** specializes this representation by further assuming that $\mu_i$ is defined via a function of $m$ explanatory variables contained in the $(m \times 1)$ vector $\mathbf{z}_i$ and that this function is known except for $k$ unknown parameters that enter the function *linearly*, leading to the representation $\mu_i = \sum_{j=1}^{k} h_j(\mathbf{z}_i)\, \beta_j, i = 1, \ldots, n,$ and $Y_i = \sum_{j=1}^{k} h_j(\mathbf{z}_i)\beta_j + \varepsilon_i,\ i = 1, \ldots, n$.

We emphasize that the adjective *linear* in *linear model* refers to the assumption that the function of the explanatory variables defining $\mu_i$ is **linear in parameters**. The representation need not be linear in the explanatory variables, as is evident in the specification above. However, it is customary to define the variables $x_{ij} \equiv h_j(\mathbf{z}_i)$ for $i = 1, \ldots, n$ and $j = 1, \ldots, k$, so that the representation of the mean $\mu_i$ is then *also* linear in the $x_{ij}$'s, as $\mu_i = \sum_{j=1}^{k} x_{ij}\, \beta_j, i = 1, \ldots, n,$ and thus

$$Y_i = \sum_{j=1}^{k} x_{ij}\, \beta_j + \varepsilon_i, i = 1, \ldots, n \text{ or } \mathbf{Y} = \mathbf{x}\boldsymbol{\beta} + \boldsymbol{\varepsilon},$$

[1] We will concentrate on models for which the mean of the random variable exists. It is possible to use this characterization in other ways when it does not, for example, using the median as the measure of the central tendency of the outcomes, such as $Y_i = \eta_i + \varepsilon_{ii}$, where $\eta_i = median(Y_i)$ and $median(\varepsilon_i) = 0$.

with the latter being a compact matrix representation of the linear relationship between the random sample $\mathbf{Y}$ and the variables explaining the mean (i.e., the **explanatory variables**), $\mathbf{x}$.

The linear regression model can be used to represent a random sample $\mathbf{Y}$ whether the random sample is from a population distribution or from a more general experiment. If the elements of $\mathbf{Y}$ are *iid* random variables from a population distribution, then $\mathbf{x}$ is simply a column vector of 1's, $\beta$ is a scalar representing the common mean of the $Y_i$'s, and the $\varepsilon_i$'s are *iid* with mean zero. More generally, if the elements of $\mathbf{Y}$ are generated by a more general sampling scheme, so that the $Y_i$'s need not be identically distributed nor independent, then linear combinations of the elements in the rows of $\mathbf{x}$ are used to represent the various mean values of the $y_i$'s, and the $\varepsilon_i$'s have a joint probability distribution with a zero mean vector and appropriate variances and covariances.

Although it is always linear in the parameters, a linear model can be more general than it might, at first, appear in terms of the types of functional relationships between $\mathbf{Y}$ and $\mathbf{x}$ that it can portray. In order to illustrate the generality of the linear model, note that

$$Y_i = \beta_1 + \beta_2 z_{i2}^2 + \beta_3 \sin(z_{i3}) + \beta_4 \left(\frac{z_{i4}}{z_{i5}}\right) + \varepsilon_i, \quad i = 1, \ldots, n$$

is consistent with the linear model characterization, and a representation that is linear in explanatory variables as well can be obtained upon defining $x_{i1} = 1$, $x_{i2} = z_{i2}^2$, $x_{i3} = \sin(z_{i3})$, and $x_{i4} = (z_{i4}/z_{i5})$, so that $Y_i = \sum_{j=1}^{4} x_{ij}\beta_j + \varepsilon_i$, $i = 1,\ldots,n$. More generally, a linear model can serve as an approximation in situations where $E(Y_i) = \mu(z_{i1},\ldots,z_{im})$ is *any continuous* function of $(z_{i1},\ldots,z_{im})$, at least in principle, since by **Weierstrass's approximation theorem** any continuous function can be approximated arbitrarily closely by a polynomial of sufficiently high degree (see Bartle, op. cit., pp. 185–186). That is, if $\mu(z_{i1},\ldots,z_{im})$ represents a continuous function, then

$$\mu(z_{i1}, \ldots, z_{im}) \approx \sum_{(j_1,\ldots,j_m)\in A} c_{j_1,\ldots,j_m}\, z_{i1}^{j_1}\, z_{i2}^{j_2} \cdots z_{im}^{j_m},$$

where $A = \{(j_1,\ldots,j_m): 0 \le j_1 + \ldots + j_m \le d,\ j_i\text{'s are nonnegative integers}\}$, and $d$ represents the degree of the polynomial. Letting the $c_{j_1,\ldots,j_m}$ values be the entries in the $\boldsymbol{\beta}$ vector, and the product terms $(z_{i1}^{j_1}\, z_{i2}^{j_2} \cdots z_{im}^{j_m})$ be the entries in the $\mathbf{x}$ matrix, it is clear that the polynomials in the Weierstrass approximation theorem can be incorporated into the linear model framework.[2]

Finally, a relationship between $\mathbf{Y}$, $\mathbf{x}$, and $\boldsymbol{\varepsilon}$ that is initially *nonlinear* in the parameters might be *transformable* into the linear model form. For example, suppose $\mathbf{Y}_i = \beta_1 k_i^{\beta_2} \ell_i^{\beta_3} e^{\varepsilon_i}$ which, for example, could represent a Cobb-Douglas production function with $y_i$ being output, and $k_i$ and $\ell_i$ being capital and labor inputs.

[2] We note, however, that for highly nonlinear functions, the degree of polynomial required to provide an adequate approximation to $\mu(z_{i1},\ldots,z_{im})$ may be so high that there will not be enough sample observations to estimate the unknown $\beta_j$'s adequately, or at all. This is related to a requirement that $\mathbf{x}$ have full column rank, which will be discussed shortly.

Applying a logarithmic transformation results in $\ln(Y_i) = \ln(\beta_1) + \beta_2 \ln(k_i) + \beta_3 \ln(\ell_i) + \ln(\varepsilon_i)$, or $y_i^* = \beta_1^* + \beta_2 k_i^* + \beta_3 \ell_i^* + \varepsilon_i^*$, which is in the linear model form for obvious definitions of the starred variables and parameters.

A **nonlinear regression model** is one in which the mean value function is a nonlinear function of the parameters, as $Y_i = g(\mathbf{x}_{i.}; \boldsymbol{\beta}) + \varepsilon_i$, and it is not transformable into linear model form, so that the model must be estimated in the nonlinear form. The computational aspects of estimating nonlinear models via a least squares objective can be challenging at times, but there are substantial similarities in the interpretation of the asymptotic behavior of both types of estimators. There are fewer similarities in the finite sample interpretation of estimator behavior. While we will leave a detailed investigation of the estimation of nonlinear regression models to a more advanced course of study, we will provide a general overview of the method of nonlinear least squares, and indicate analogies to the linear least squares estimator in Section 8.2.6 ahead.

### 8.2.2 Least Squares Estimator Under the Classical General Linear Model Assumptions

The **Classical General Linear Model** (GLM) refers to a linear model with a specific set of additional assumptions regarding the genesis of the observed random sample of data **Y**. The random variable **Y** is referred to as the **dependent variable**, **x** is called the matrix of **independent or explanatory variables**, and $\boldsymbol{\varepsilon}$ is called the **disturbance, error, or residual vector**. The main objective in analyzing the GLM will be to estimate (and later, test hypotheses about) the entries, or functions of the entries, in the parameter vector $\boldsymbol{\beta}$. In particular, we will utilize an outcome, $(y_1,\ldots,y_n)$, of the random sample, $(Y_1,\ldots,Y_n)$, together with knowledge of the values of the explanatory variables, **x**, to estimate the unknown entries in $\boldsymbol{\beta}$. Note the entries in $\boldsymbol{\varepsilon}$ are *unobservable* random variables, since they represent deviations of $(y_1,\ldots,y_n)$ from the *unknown* mean vector $\mathbf{x}\boldsymbol{\beta}$, and thus outcomes of $\boldsymbol{\varepsilon}$ will not be useful in estimating $\boldsymbol{\beta}$, per se.

The following definition delineates the set of assumptions that determine the statistical characteristics of the Classical General Linear Model.

**Definition 8.1** ***Classical General Linear Model (GLM) Assumptions***

GLM 1. $\mathrm{E}(\mathbf{Y}) = \mathbf{x}\boldsymbol{\beta}$ and $\mathrm{E}(\boldsymbol{\varepsilon}) = \mathbf{0}$

GLM 2. $\mathbf{Cov}(\mathbf{Y}) = \sigma^2\mathbf{I} = \mathbf{Cov}(\boldsymbol{\varepsilon}) = \mathrm{E}(\boldsymbol{\varepsilon}\boldsymbol{\varepsilon}')$

GLM 3. **x** is a fixed matrix of values with $\mathrm{rank}(\mathbf{x}) = k$

The first assumption simply restates what we have already asserted concerning the mean of **Y** in the linear model, namely, $\mathbf{x}\boldsymbol{\beta}$ is representing or explaining $\mathrm{E}(\mathbf{Y})$, which then necessarily implies that $\mathrm{E}(\boldsymbol{\varepsilon}) = \mathbf{0}$. In applications, $\mathrm{E}(Y_i) = \mathbf{x}_{i.}\boldsymbol{\beta}$ (recall that $\mathbf{x}_{i.}$ refers to the *i*th row of the **x** matrix) means that whatever the experiment under investigation, the expected value of the distribution associated with outcomes of $Y_i$ is a linear function of the values of the

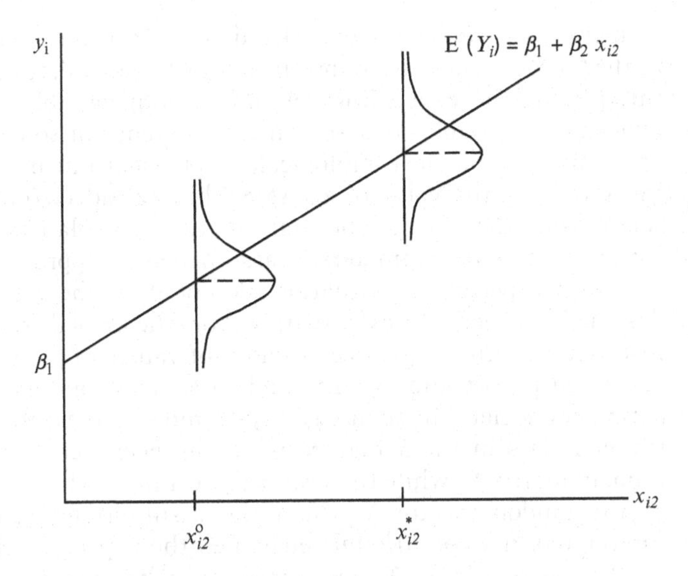


**Figure 8.1**
GLM representation of $E(Y_i)$ in bivariate case

explanatory variables in $\mathbf{x}_{i.}$. A graphical illustration, for $k = 2$ with $\mathbf{x}_{.1} = (1\ 1 \ldots 1)'$, is provided in Figure 8.1.

Note that under the Classical GLM assumptions, $\mathbf{x}$ is a fixed matrix of numbers. This implies that a sample observation $(y_i, x_{i1}, \ldots, x_{ik})$ can be interpreted as having been generated from an outcome of the vector $(Y_i, x_{i1}, \ldots, x_{ik})$, where this latter vector can be considered a random vector, pairing the (generally) *nondegenerate* random variable $Y_i$ with the *degenerate* random vector $(x_{i1}, \ldots, x_{ik})$. When observing repeated outcomes of $(Y_i, x_{i1}, \ldots, x_{ik})$, it is implied that $(x_{i1}, \ldots, x_{ik})$ remains constant, while the outcome $y_i$ that is paired with the vector $(x_{i1}, \ldots, x_{ik})$ generally varies. The distribution of $y_i$ outcomes for given $(x_{i1}, \ldots, x_{ik})$ values, when $k = 2$, is depicted in Figure 8.1 by a distribution centered at $E(Y_i)$.

It is useful to distinguish two contexts in which the $\mathbf{x}$ matrix can be considered fixed. One context is the situation where elements of the $\mathbf{x}$-matrix are controlled or fixed by the individual performing the experiment. In this case, the $\mathbf{x}$-matrix is sometimes referred to as the **design matrix**, meaning that the researcher can essentially design the vectors $(x_{i1}, \ldots, x_{ik})$, $i = 1, \ldots, n$, in a way that is of particular interest to the investigation at hand, and then she can observe the $y_i$ values associated with the chosen design matrix. As an example of this type of interpretation for $\mathbf{x}$, suppose various levels of inputs to a production process were *chosen*, and an observation was made on the output corresponding to each fixed level of the inputs. We might expect that, "on average," a certain level of output, say $E(Y_i)$, would be produced given the levels of inputs specified by $\mathbf{x}_{i.}$. However, for any given observation with input level $\mathbf{x}_{i.}$, deviations from the average level of output might occur due to a myriad of non-input type factors that are not controlled in the production process (e.g., machine function variation, labor efficiency, weather, temperature, and the like). Then observations on output levels, given the chosen (or *designed*) levels of inputs represented by the elements in $\mathbf{x}$, can be conceptualized as observations on the vector $(Y_i, x_{i1},\ldots,x_{ik})$, as the GLM classical assumptions imply. In general, the underlying rationale for interpreting the expectation of the random variable $Y_i$ as a function of the explanatory variables is based on the existence of some underlying systematic economic, sociological, biological, or physical linear relationship relating the mean of $Y_i$ to the value of $\mathbf{x}_i$. As we have alluded to previously, sometimes a relationship that is a polynomial function of explanation variables, and thus a linear function of parameters, is assumed as an approximation.

Many experimental situations do not allow the $\mathbf{x}$ matrix to be controlled or designed. In economics, business, or the social sciences in general, the researcher is often a passive observer of values of $(y_i, x_{i1},\ldots,x_{ik})$ that have been generated by consumers, entrepreneurs, the economy, by markets, or by the actions of society. In such cases, it is more natural to consider some or all of the elements in the $\mathbf{x}$ matrix as having been generated by an outcome of a random matrix $\mathbf{X}$, while the vector $\mathbf{y}$ continues to be interpreted as the outcome of the random vector $\mathbf{Y}$. Then the assumption $E(\mathbf{Y}) = \mathbf{x}\boldsymbol{\beta}$ of the GLM is interpreted in a *conditional* sense, i.e., the expected value of $\mathbf{Y}$ is *conditional on* the outcome $\mathbf{x}$ for $\mathbf{X}$, and a more revealing notation for this interpretation would be given by $E(\mathbf{Y}|\mathbf{x}) = \mathbf{x}\boldsymbol{\beta}$. The reader will recall from Chapter 3 that this conditional expectation is literally the *regression function* of $\mathbf{Y}$ on $\mathbf{x}$, and thus the GLM maintains that the regression function is actually a regression *hyperplane*.[3]

Note that since $\mathbf{Y} = \mathbf{x}\boldsymbol{\beta} + \boldsymbol{\varepsilon}$ is assumed by the GLM, it follows that in the context of operating conditionally on an outcome of $\mathbf{X}$, we must then be referring to a *conditional distribution* of the vector $\mathbf{Y}$ when we say that outcomes of Y can be decomposed into the mean vector, $\mathbf{x}\boldsymbol{\beta}$, and deviations, $\boldsymbol{\varepsilon}$, from the mean vector. We might consider using the notation $\mathbf{Y}|\mathbf{x} = \mathbf{x}\boldsymbol{\beta} + \boldsymbol{\varepsilon}|\mathbf{x}$ to emphasize this fact, implying that $E(\boldsymbol{\varepsilon}|\mathbf{x}) = \mathbf{0}$. In this situation, the entire analysis is cast in the framework of being conditional on the $\mathbf{x}$ matrix observed, and once this interpretation is affixed to the GLM together with the implicit assumption that *all* expectations are conditional on $\mathbf{x}$, the situation (and method of analysis) will be analogous to the previous case where the $\mathbf{x}$ matrix is fixed. We will generally suppress the conditional-on-$\mathbf{x}$ notation, leaving it to the reader to provide the proper context for interpreting the GLM in a given problem situation.

The assumption $\mathbf{Cov}(\mathbf{Y}) = \sigma^2\mathbf{I} = \mathbf{Cov}(\boldsymbol{\varepsilon}) = E(\boldsymbol{\varepsilon}\boldsymbol{\varepsilon}')$ implies that the covariance matrix of $\mathbf{Y}$, and of $\boldsymbol{\varepsilon}$, is a diagonal matrix with $\sigma^2$'s along the diagonal and zeroes on the off-diagonal. The fact that all of the variances of the elements in $\mathbf{Y}$ (or $\boldsymbol{\varepsilon}$) have the same value, $\sigma^2$, is referred to as the property of **homoskedasticity** (from homoskedastic, meaning "same spread"). The off-diagonal entries being zero imply that the covariance, or equivalently, the correlation between any two elements of $\mathbf{Y}$ (or $\boldsymbol{\varepsilon}$) is zero, which is called the property of **zero autocovariance** or **zero autocorrelation**. This assumption on the covariance matrix of $\mathbf{Y}$ (or $\boldsymbol{\varepsilon}$) then

[3]Note that by tradition, the previous case where $\mathbf{x}$ is designed is also referred to as a regression of $\mathbf{Y}$ on $\mathbf{x}$, where one can think of the conditional expectation in the degenerate sense, i.e., where $\mathbf{x}$ takes its observed value with probability one.

implies that the random sample $(Y_1,\ldots,Y_n)$ is a collection of uncorrelated random variables all having the same variance or measure of "spread." The reader is reminded that this assumption may be referring to the conditional distribution of $\mathbf{Y}$ (and $\boldsymbol{\varepsilon}$), if we are conditioning on an outcome of the $\mathbf{X}$ matrix.

The assumption that rank $\mathbf{x} = k$, i.e. the $\mathbf{x}$ matrix has full column rank, simply implies that there are no linear dependencies among the columns of the $\mathbf{x}$ matrix. That is, no column of the $\mathbf{x}$ matrix is representable as some linear combination of the remaining columns in the $\mathbf{x}$ matrix. This necessarily implies that $n \geq k$, i.e., we must have at least as many sample observations as there are unknown parameters in the $\boldsymbol{\beta}$-vector and columns in the $\mathbf{x}$-matrix.

To this point we still have not fully specified the *probability model* since we have not yet specified a parametric family of densities for the random vector $\mathbf{Y}$ (or $\boldsymbol{\varepsilon}$). In fact, all we have assumed, so far, is that $(Y_1,\ldots,Y_n)$ is a random sample from some experiment such that the joint density of the random sample has a mean vector $\mathbf{x}\boldsymbol{\beta}$, a covariance matrix $\sigma^2\mathbf{I}$, with $\mathbf{x}$ having full column rank. Nevertheless, it is possible to suggest an estimator for the parameter vector $\boldsymbol{\beta}$ that has a number of useful properties. We will proceed to define such an estimator of $\boldsymbol{\beta}$ in the absence of a specific parametric family specification, reserving such a specification until we have exhausted the progress that can be made in its absence. We will also define a useful estimator of the variance parameter $\sigma^2$.

*Estimator for $\boldsymbol{\beta}$ Under Classical GLM Assumptions* How should the estimator of $\boldsymbol{\beta}$ be defined? We examine two approaches to the problem that lead to the same estimator for $\boldsymbol{\beta}$ but that offer different insights into the genesis of the estimator of $\boldsymbol{\beta}$.

One approach, called the **method of least squares**, which is the principal focus of this section, defines the estimator of $\boldsymbol{\beta}$ by associating with each observation on $(\mathbf{y},\mathbf{x})$ the $(k \times 1)$ vector $\mathbf{b}$ that solves the minimization problem[4]

$$\mathbf{b} = \arg\min_{\boldsymbol{\eta}}\{(\mathbf{y}-\mathbf{x}\boldsymbol{\eta})'(\mathbf{y}-\mathbf{x}\boldsymbol{\eta})\}.$$

Note that this is equivalent to finding the vector $\hat{\mathbf{y}} = \mathbf{x}\mathbf{b}$ that is the minimum distance from $\mathbf{y}$, since the minimum of the distance $d(\mathbf{y},\hat{\mathbf{y}}) = [(\mathbf{y}-\hat{\mathbf{y}})'(\mathbf{y}-\hat{\mathbf{y}})]^{1/2}$ and the minimum of $(\mathbf{y}-\hat{\mathbf{y}})'(\mathbf{y}-\hat{\mathbf{y}})$ occur at precisely the same value of $\mathbf{b}$.[5] From the point of view of attempting to explain $\mathbf{y}$ as best one can in terms of a linear function of the explanatory variable values, $\mathbf{x}$, the least squares approach has intuitive appeal. To solve the minimization problem, note that $(\mathbf{y}-\mathbf{x}\boldsymbol{\eta})'(\mathbf{y}-\mathbf{x}\boldsymbol{\eta}) = \mathbf{y}'\mathbf{y} - 2\boldsymbol{\eta}'\mathbf{x}'\mathbf{y} + \boldsymbol{\eta}'\mathbf{x}'\mathbf{x}\boldsymbol{\eta}$, and the $k$ first order conditions for the minimum can be represented in matrix form as $-2\mathbf{x}'\mathbf{y} + 2\mathbf{x}'\mathbf{x}\mathbf{b} = \mathbf{0}$ where we have used a result contained in the following lemma concerning matrix calculus:

---

[4]Recall that $\arg\min_w\{f(w)\}$ denotes the *argument* value $w$ that *minimizes* $f(w)$.

[5]This follows because $z^{1/2}$ is a monotonic transformation of $z$ for $z \geq 0$, so that the minimum (and maximum) of $z^{1/2}$ and $z$ occur at the same values of $z$, $z \in D$, $D$ being some set of nonnegative numbers.

**Lemma 8.1**

Let $\mathbf{z}$ be a $(k \times 1)$ vector, $\mathbf{A}$ be a $(k \times j)$ matrix, and $\mathbf{w}$ be a $(j \times 1)$ vector. Then

a. $\dfrac{\partial \mathbf{z}'\mathbf{A}\mathbf{w}}{\partial \mathbf{z}} = \mathbf{A}\mathbf{w}$

b. $\dfrac{\partial \mathbf{z}'\mathbf{A}\mathbf{w}}{\partial \mathbf{w}} = \mathbf{A}'\mathbf{z}$

c. $\dfrac{\partial \mathbf{z}'\mathbf{A}\mathbf{z}}{\partial \mathbf{z}} = 2\mathbf{A}\mathbf{z}(k = j \text{ and } \mathbf{A} \text{ is symmetric})$

d. $\dfrac{\partial \mathbf{z}'\mathbf{w}}{\partial \mathbf{z}} = \mathbf{w}(\mathrm{k} = \mathrm{j})$

e. $\dfrac{\partial \mathbf{z}'\mathbf{w}}{\partial \mathbf{w}} = \mathbf{z}(\mathrm{k} = \mathrm{j})$

Since $\mathbf{x}$ has full column rank, the $(k \times k)$ matrix $(\mathbf{x}'\mathbf{x})$ necessarily has full rank and is, thus, invertible.[6] The first order conditions can then be solved for the sum of squares minimizer $\mathbf{b}$, as

$$\mathbf{b} = (\mathbf{x}'\mathbf{x})^{-1}\mathbf{x}'\mathbf{y}$$

which defines the estimate for $\boldsymbol{\beta}$ implied by the least squares method. The estimator of $\boldsymbol{\beta}$ is then defined by the $(k \times 1)$ random vector

$$\hat{\boldsymbol{\beta}} = (\mathbf{x}'\mathbf{x})^{-1}\mathbf{x}'\mathbf{Y}.$$

where for the remainder of this section, we will use $\hat{\boldsymbol{\beta}}$ to denote an estimator, and $\mathbf{b}$ will denote the associated estimate of $\boldsymbol{\beta}$. Note that the second order conditions for the minimization problem are $2\mathbf{x}'\mathbf{x}$, which is positive definite, so that the second order conditions are satisfied for a global minimum and thus $\boldsymbol{\eta} = \mathbf{b}$ does indeed minimize $(\mathbf{y} - \mathbf{x}\boldsymbol{\eta})'(\mathbf{y} - \mathbf{x}\boldsymbol{\eta})$.

*Coefficient of Determination, $R^2$* A function of the minimized value of the distance between $\mathbf{y}$ and $\mathbf{xb}$ is often used as a measure of how well the $\mathbf{y}$ outcome has been "explained" by the $\mathbf{xb}$ outcome. In particular, letting $\mathbf{1}_n$ be a $(n \times 1)$ vector of 1's and $\hat{\mathbf{y}} = \mathbf{xb}$, the measure is given by

$$R^2 = 1 - \frac{\mathrm{d}^2(\mathbf{y},\hat{\mathbf{y}})}{\mathrm{d}^2(\mathbf{y}, \mathbf{1}_n\bar{y})} = 1 - \frac{(\mathbf{y} - \hat{\mathbf{y}})'(\mathbf{y} - \hat{\mathbf{y}})}{(\mathbf{y} - \mathbf{1}_n\bar{y})'(\mathbf{y} - \mathbf{1}_n\bar{y})} = 1 - \frac{\sum_{i=1}^{n} (y_i - \hat{y}_i)^2}{\sum_{i=1}^{n} (y_i - \bar{y})^2}$$

and is called the **coefficient of determination** or "$R$-squared." It is clear that the closer $\hat{\mathbf{y}}$ is to $\mathbf{y}$ in terms of distance, the higher is $R^2$, and its *maximum* value of 1 is achieved *iff* $\mathbf{y} = \hat{\mathbf{y}}$.

If $\mathbf{x}$ contains a column of 1's, so that $\mathbf{Y} = \mathbf{x}\boldsymbol{\beta} + \boldsymbol{\varepsilon}$ contains an intercept term, then $R^2$ is lower bounded by zero and $R^2$ is identically the square of the sample

[6] A matrix $\mathbf{x}$ has full column rank *iff* $\mathbf{x}'\mathbf{x}$ has full rank. See Rao, C., op. cit., p. 30.

correlation between $\mathbf{y}$ and $\hat{\mathbf{y}}$. To see this, first define $\hat{\mathbf{e}} = \mathbf{y} - \hat{\mathbf{y}} = \mathbf{y} - \mathbf{xb}$ and note that $\mathbf{y}'\mathbf{y} = (\mathbf{xb} + \hat{\mathbf{e}})'(\mathbf{xb} + \hat{\mathbf{e}})$ can be decomposed into two components as $\mathbf{y}'\mathbf{y} = \mathbf{b}'\mathbf{x}'\mathbf{xb} + \hat{\mathbf{e}}'\hat{\mathbf{e}}$ since $\hat{\mathbf{y}}'\hat{\mathbf{e}} = \mathbf{b}'\mathbf{x}'\hat{\mathbf{e}} = \mathbf{b}'(\mathbf{x}'(\mathbf{y} - \mathbf{xb})) = \mathbf{0}$ by the first order conditions of the least squares minimization problem. Also note that $\mathbf{1}'_n\hat{\mathbf{e}} = \mathbf{1}'_n\mathbf{y} - \mathbf{1}'_n\hat{\mathbf{y}} = 0$ because $\mathbf{x}'(\mathbf{y} - \mathbf{xb}) = \mathbf{x}'\hat{\mathbf{e}} = \mathbf{0}$ and $\mathbf{x}'$ contains a row of 1's. It follows that $\bar{y} = \bar{\hat{y}}$. Subtracting $n\bar{y}^2$ from both sides of the preceding decomposition of $\mathbf{y}'\mathbf{y}$, and then dividing by $\mathbf{y}'\mathbf{y} - n\bar{y}^2$, implies

$$R^2 = 1 - \frac{\hat{\mathbf{e}}'\hat{\mathbf{e}}}{(\mathbf{y} - \mathbf{1}_n\bar{y})'(\mathbf{y} - \mathbf{1}_n\bar{y})} = \frac{\mathbf{b}'\mathbf{x}'\mathbf{xb} - n\bar{y}^2}{(\mathbf{y} - \mathbf{1}_n\bar{y})'(\mathbf{y} - \mathbf{1}_n\bar{y})} = \frac{\sum_{i=1}^{n} \left(\hat{y}_i - \bar{\hat{y}}\right)^2}{\sum_{i=1}^{n} (y_i - \bar{y})^2}.$$

Finally, since $\sum_{i=1}^{n} \left(\hat{y}_i - \bar{\hat{y}}\right)^2 = \sum_{i=1}^{n} (y_i - \bar{y})\left(\hat{y}_i - \bar{\hat{y}}\right)$,

$$R^2 = \frac{\left[\sum_{i=1}^{n} (y_i - \bar{y})\left(\hat{y}_i - \bar{\hat{y}}\right)\right]^2}{\sum_{i=1}^{n} (y_i - \bar{y})^2 \sum_{i=1}^{n} \left(\hat{y}_i - \bar{\hat{y}}\right)^2} = r^2_{Y\hat{Y}},$$

i.e., $R^2$ is the square of the sample correlation between $\mathbf{Y}$ and $\hat{\mathbf{Y}}$.

It follows from our discussion of the *sample correlation* in Chapter 6 that $R^2$ can be interpreted as the proportion of the sample variance in the $y_i$'s that is explained by the corresponding values of the $\hat{y}_i$'s. This follows from the fact that the vector of values $a + b\hat{y}_i$, $i = 1,\ldots,n$ has the smallest expected squared distance from $y_1,\ldots,y_n$ when $b = s_{Y\hat{Y}} / s^2_{\hat{Y}} = s^2_{\hat{Y}} / s^2_{\hat{Y}} = 1$ and $a = \bar{y} - b\bar{\hat{y}} = \bar{y} - \bar{y} = 0$.

*Best Linear Unbiased Estimator (BLUE) of* $\boldsymbol{\beta}$ A second approach for defining an estimator of $\boldsymbol{\beta}$ in the Classical GLM begins with the objective of defining the BLUE of $\boldsymbol{\beta}$. In order to be in the *linear* class, the estimator must be of the form $\hat{\boldsymbol{\beta}} = \mathbf{AY} + \mathbf{d}$ for some nonrandom $(k \times n)$ matrix $\mathbf{A}$ and some nonrandom $(k \times 1)$ vector $\mathbf{d}$. If the estimator is to be in the *unbiased* class, then $E(\hat{\boldsymbol{\beta}}) = \mathbf{Ax}\boldsymbol{\beta} + \mathbf{d} = \boldsymbol{\beta}$, $\forall\boldsymbol{\beta}$, which requires that $\mathbf{Ax} = \mathbf{I}$ and $\mathbf{d} = \mathbf{0}$. For the estimator to be the *best* in the class of linear unbiased estimators, its covariance matrix must be as small or smaller than the covariance matrix of any other estimator in the linear unbiased class. Since $\hat{\boldsymbol{\beta}} = \mathbf{AY}$ and $\mathbf{Cov}(\mathbf{Y}) = \sigma^2\mathbf{I}$, it follows that $\mathbf{Cov}(\hat{\boldsymbol{\beta}}) = \sigma^2\mathbf{AA}'$ is the covariance matrix that must be minimized through choice of the matrix $\mathbf{A}$ subject to the unbiasedness constraint that $\mathbf{Ax} = \mathbf{I}$. The next theorem solves the problem.

**Theorem 8.1**
***BLUE of $\boldsymbol{\beta}$ in the GLM: Gauss-Markov Theorem***

*Under the assumptions of the Classical GLM,* $\hat{\boldsymbol{\beta}} = (\mathbf{x}'\mathbf{x})^{-1}\mathbf{x}'\mathbf{Y}$ *is the best linear unbiased estimator of* $\boldsymbol{\beta}$.

**Proof** The linear estimator $\hat{\boldsymbol{\beta}} = \mathbf{AY} + \mathbf{d}$ is unbiased under the assumptions of the Classical GLM *iff* $\mathbf{Ax} = \mathbf{I}$ and $\mathbf{d} = \mathbf{0}$. The feasible choices for $\mathbf{A}$ for defining a linear unbiased estimator can be represented as $\mathbf{A} = (\mathbf{x}'\mathbf{x})^{-1}\mathbf{x}' + \mathbf{D}$, where $\mathbf{D}$ is any matrix such that $\mathbf{Dx} = \mathbf{0}$. To see this, first note that any matrix $\mathbf{A}$ can be equivalently represented as $(\mathbf{x}'\mathbf{x})^{-1}\mathbf{x}' + \mathbf{D}$ by simply choosing $\mathbf{D} = \mathbf{A} - (\mathbf{x}'\mathbf{x})^{-1}\mathbf{x}'$. Now for $\mathbf{A}$ to satisfy $\mathbf{Ax} = \mathbf{I}$, *it must be the case that* $\mathbf{Ax} = [(\mathbf{x}'\mathbf{x})^{-1}\mathbf{x}' + \mathbf{D}]\mathbf{x} = \mathbf{I} + \mathbf{Dx} = \mathbf{I}$, so that $\mathbf{Dx} = \mathbf{0}$ is implied. Substituting $(\mathbf{x}'\mathbf{x})^{-1}\mathbf{x}' + \mathbf{D}$ for $\mathbf{A}$ represents the linear unbiased estimator as $\hat{\boldsymbol{\beta}} = ((\mathbf{x}'\mathbf{x})^{-1}\mathbf{x}' + \mathbf{D})\mathbf{Y}$, which has covariance matrix $\mathbf{Cov}(\hat{\boldsymbol{\beta}}) = \sigma^2\left[(\mathbf{x}'\mathbf{x})^{-1} + \mathbf{DD}'\right]$ because $\mathbf{Dx} = \mathbf{0}$. Being that $\mathbf{DD}'$ is necessarily positive semidefinite, it follows that the covariance matrix is minimized when $\mathbf{D} = \mathbf{0}$, which implies that the BLUE estimator is $\hat{\boldsymbol{\beta}} = (\mathbf{x}'\mathbf{x})^{-1}\mathbf{x}'\mathbf{Y}$. ■

It is also true that $\boldsymbol{\xi}'\hat{\boldsymbol{\beta}} = \boldsymbol{\xi}'(\mathbf{x}'\mathbf{x})^{-1}\mathbf{x}'\mathbf{Y}$ is the BLUE of $\boldsymbol{\xi}'\hat{\boldsymbol{\beta}}$ for any choice of $\boldsymbol{\xi} \neq \mathbf{0}$. Thus, once the BLUE $\hat{\boldsymbol{\beta}}$ is known, one can calculate BLUE estimates of any linear combination of the entries in $\boldsymbol{\beta}$ by simply calculating the corresponding linear combination of the entries in $\hat{\boldsymbol{\beta}}$. We formalize this result in the following theorem.

**Theorem 8.2** ***BLUE for*** $\boldsymbol{\xi}'\hat{\boldsymbol{\beta}}$ *Under the assumptions of the Classical GLM,* $\boldsymbol{\xi}'\hat{\boldsymbol{\beta}} = \boldsymbol{\xi}'(\mathbf{x}'\mathbf{x})^{-1}\mathbf{x}'\mathbf{Y}$ *is the best linear unbiased estimator of* $\boldsymbol{\xi}'\boldsymbol{\beta}$.

**Proof** Let $\mathbf{T} = \mathbf{t}(\mathbf{Y}) = \mathbf{c}'\mathbf{Y} + d$ be any linear estimator of the scalar $\boldsymbol{\xi}'\boldsymbol{\beta}$, where $\mathbf{c}$ is an $(n \times 1)$ vector and $d$ is a scalar. For $\mathbf{T}$ to be unbiased, it is required that $\mathrm{E}(\mathbf{T}) = \mathrm{E}(\mathbf{c}'\mathbf{Y} + d) = \mathbf{c}'\mathbf{x}\boldsymbol{\beta} + d = \boldsymbol{\xi}'\boldsymbol{\beta}\,\forall\boldsymbol{\beta}$, which in turn requires that $\mathbf{x}'\mathbf{c} = \boldsymbol{\xi}$ and $d = 0$. Note the following lemma:

**Lemma 8.2**

> Let $\mathbf{x}$ be a matrix of rank $k$, and let be any $(k \times 1)$ vector. Then $\mathbf{x}'\mathbf{c} = \boldsymbol{\xi}$ *iff*
>
> $\mathbf{c} = \mathbf{x}(\mathbf{x}'\mathbf{x})^{-1}\boldsymbol{\xi} + [\mathbf{I} - \mathbf{x}(\mathbf{x}'\mathbf{x})^{-1}\mathbf{x}']\mathbf{h}$,
>
> where the $(n \times 1)$ vector $\mathbf{h}$ can be chosen arbitrarily.
> F.A. Graybill, (1983) *Matrices With Applications in Statistics, 2nd Ed.*, Wadsworth, p. 153.

Using the solution for $\mathbf{c}$ given by the lemma, note that $\mathrm{var}(\mathbf{c}'\mathbf{Y}) = \sigma^2[\boldsymbol{\xi}'(\mathbf{x}'\mathbf{x})^{-1}\boldsymbol{\xi} + \mathbf{h}'[\mathbf{I} - \mathbf{x}(\mathbf{x}'\mathbf{x})^{-1}\mathbf{x}']\mathbf{h}]$. The variance is minimized by setting $\mathbf{h} = \mathbf{0}$, and thus $\mathrm{T} = \mathbf{c}'\mathbf{Y} = \boldsymbol{\xi}'(\mathbf{x}'\mathbf{x})^{-1}\mathbf{x}'\mathbf{Y} = \boldsymbol{\xi}'\hat{\boldsymbol{\beta}}$ is the BLUE for $\boldsymbol{\xi}'\boldsymbol{\beta}$. ■

The estimator $\hat{\boldsymbol{\beta}} = (\mathbf{x}'\mathbf{x})^{-1}\mathbf{x}'\mathbf{Y}$ is referred to as the **least squares estimator** of $\boldsymbol{\beta}$ due to its definition via the least squares approach described previously. Without additional assumptions on the statistical model, the BLUE property exhausts the properties that we can attribute to $\hat{\boldsymbol{\beta}}$. In particular, we cannot demonstrate consistency of $\hat{\boldsymbol{\beta}}$, nor can we make progress towards defining a MVUE of $\boldsymbol{\beta}$.

*Estimator for* $\sigma^2$ *and* $\mathbf{Cov}(\hat{\boldsymbol{\beta}})$ *Under Classical GLM Assumptions* The classical assumptions are sufficient to allow the definition of an unbiased estimator of the common variance, $\sigma^2$, of the $Y_i$'s (or equivalently, the common variance of the $\varepsilon_i$'s), and from it, an unbiased estimator of the covariance matrix of the least squares estimator. We formalize these results in the next Theorem. Note that we use the hat notation, ^, to distinguish this unbiased estimator from the sample variance, $S^2$, introduced previously in Chapter 6.

**Theorem 8.3** ***Unbiased Estimators of $\sigma^2$ and $\text{Cov}(\hat{\boldsymbol\beta})$ in the Classical GLM***

*The estimators* $\hat{S}^2 = (\mathbf{Y} - \mathbf{x}\hat{\boldsymbol\beta})'(\mathbf{Y} - \mathbf{x}\hat{\boldsymbol\beta})/(n-k)$ *and* $\hat{S}^2(\mathbf{x}'\mathbf{x})^{-1}$ *are unbiased estimators of* $\sigma^2$ *and* $\mathbf{Cov}(\hat{\boldsymbol\beta})$, *respectively.*

**Proof** Note that substitution of the linear model representation of $\mathbf{Y} = \mathbf{x}\boldsymbol\beta + \boldsymbol\varepsilon$ for $\mathbf{Y}$, and the definition of the least squares estimator $\hat{\boldsymbol\beta} = (\mathbf{x}'\mathbf{x})^{-1}\mathbf{x}'\mathbf{Y}$ into the definition of $\hat{S}^2$ results in $\boldsymbol\varepsilon'(\mathbf{I} - \mathbf{x}(\mathbf{x}'\mathbf{x})\mathbf{x}')\boldsymbol\varepsilon/(n-k)$. Then

$$\begin{aligned}
\mathrm{E}\left(\hat{S}^2\right) &= \mathrm{E}(\boldsymbol\varepsilon'(\mathbf{I} - \mathbf{x}(\mathbf{x}'\mathbf{x})\mathbf{x}')\boldsymbol\varepsilon/(n-k)) \\
&= \mathrm{E}(\boldsymbol\varepsilon'\boldsymbol\varepsilon/(n-k)) - \mathrm{E}(tr(\mathbf{x}(\mathbf{x}'\mathbf{x})\mathbf{x}'\boldsymbol\varepsilon\boldsymbol\varepsilon')/(n-k)) \\
&= (n/(n-k))\sigma^2 - (tr(\mathbf{x}(\mathbf{x}'\mathbf{x})\mathbf{x}'\mathrm{E}(\boldsymbol\varepsilon\boldsymbol\varepsilon'))/(n-k)) \\
&= (n/(n-k))\sigma^2 - (k/(n-k))\sigma^2 = \sigma^2.
\end{aligned}$$

It follows that $\mathrm{E}\left(\hat{S}^2(\mathbf{x}'\mathbf{x})^{-1}\right) = \mathrm{E}\left(\hat{S}^2\right)(\mathbf{x}'\mathbf{x})^{-1} = \sigma^2(\mathbf{x}'\mathbf{x})^{-1}$. ■

The following is an example of the use of the estimators $\hat{\boldsymbol\beta}$ and $\hat{S}^2$.

**Example 8.1** ***Least Squares Estimation of Production Function***

Let the production of a commodity in a given period of time be represented by $Y_t = \beta_1 x_{1t}^{\beta_2} x_{2t}^{\beta_3} e^{\varepsilon_t}$ where $y_t$ = thousands of units produced in time period $t$, $(x_{1t}, x_{2t})$ = units of labor and capital applied to the production process in year $t$, and $\varepsilon_t$ = disturbance term value in year $t$. Assume that $\mathrm{E}(\varepsilon_t) = 0$ and $\mathrm{var}(\varepsilon_t) = \sigma^2$, $\forall t$, and let $\varepsilon_i$ and $\varepsilon_j$ be uncorrelated $\forall i$ and $j$.

Ten time periods' worth of observations on $y_t$, $x_{1t}$, and $x_{2t}$ were obtained, yielding

$$y = \begin{bmatrix} 47.183 \\ 53.005 \\ 43.996 \\ 38.462 \\ 54.035 \\ 59.132 \\ 79.763 \\ 67.252 \\ 55.267 \\ 38.972 \end{bmatrix}, \quad x_1 = \begin{bmatrix} 7 \\ 8 \\ 6 \\ 4 \\ 5 \\ 9 \\ 11 \\ 9 \\ 8 \\ 4 \end{bmatrix}, \quad x_2 = \begin{bmatrix} 2 \\ 3 \\ 2 \\ 5 \\ 6 \\ 4 \\ 8 \\ 5 \\ 5 \\ 4 \end{bmatrix}$$

We seek a BLUE estimate of $\ln(\beta_1)$, $\beta_2$, $\beta_3$, and of $\beta_2 + \beta_3$, the latter representing the degree of homogeneity of the function, as well as an unbiased estimate of $\sigma^2$.

Before we can apply the least squares estimator we must transform the model to be in GLM form. By taking the natural logarithm of both sides of the production function relationship, we obtain

$$\ln(Y_t) = \ln(\beta_1) + \beta_2 \ln(x_{1t}) + \beta_3 \ln(x_{2t}) + \varepsilon_t \quad \text{or} \quad Y_t^* = \beta_1^* + \beta_2\, x_{1t}^* + \beta_3\, x_{2t}^* + \varepsilon_t.$$

The transformed model satisfies the classical assumptions of the GLM. Defining $\underset{(10\times1)}{\mathbf{y}_*} = \ln(\mathbf{y})$ and $\underset{(10\times3)}{\mathbf{x}_*} = [\mathbf{1}_{10} \;\; \ln(\mathbf{x}_1) \;\; \ln(\mathbf{x}_2)]$, where $\mathbf{1}_{10}$ is a $(10 \times 1)$ vector with each element equal to 1, we can represent the observations on the transformed relationship as outcomes of $\mathbf{Y}_* = \mathbf{x}_* \boldsymbol{\beta}_* + \boldsymbol{\varepsilon}$, where $\boldsymbol{\beta}'_* = (\ln(\beta_1) \;\; \beta_2 \;\; \beta_3)$. The least squares (BLUE) estimate of $\boldsymbol{\beta}_*$ is then $\mathbf{b} = (\mathbf{x}'_*\mathbf{x}_*)^{-1}\mathbf{x}'_*\mathbf{y}_* = \begin{bmatrix} 2.638 \\ .542 \\ .206 \end{bmatrix}$.

The BLUE estimate of $\beta_2 + \beta_3$, is given by $b_2 + b_3 = .748$ by Theorem 8.2 with $\boldsymbol{\ell}' = [0\;1\;1]$. An unbiased estimate of $\sigma^2$ is given by $\hat{s}^2 = (\mathbf{y} - \mathbf{xb})'(\mathbf{y} - \mathbf{xb})/7 = .0043$.

We emphasize that in the process of transforming the production relationship to GLM form, we also transformed the parameter $\beta_1$ to $\ln(\beta_1)$, which then results in a situation where the least squares estimator is estimating *not* $\beta_1$, but $\ln(\beta_1)$. The phenomenon of transforming parameters often occurs when the overall relationship must be transformed in order to convert it into GLM form. □

*Consistency of* $\hat{\boldsymbol{\beta}}$ The following theorem demonstrates that additional conditions are needed to ensure that the least squares estimator is a consistent estimator of $\boldsymbol{\beta}$.

**Theorem 8.4** ***Consistency of*** $\hat{\boldsymbol{\beta}} = (\mathbf{x}'\mathbf{x})^{-1}\mathbf{x}'\mathbf{Y}$

*If in addition to the assumptions of the Classical GLM,* $(\mathbf{x}'\mathbf{x})^{-1} \to \mathbf{0}$ *as* $n \to \infty$, *then* $\hat{\boldsymbol{\beta}} = (\mathbf{x}'\mathbf{x})^{-1}\mathbf{x}'\mathbf{Y} \xrightarrow{p} \boldsymbol{\beta}, \forall\boldsymbol{\beta}$, *so that* $\hat{\boldsymbol{\beta}}$ *is a consistent estimator of* $\boldsymbol{\beta}$.

**Proof** Given the assumptions of the Classical GLM, we know that $\mathrm{E}(\hat{\boldsymbol{\beta}}) = \boldsymbol{\beta}$, and the covariance matrix of $\hat{\boldsymbol{\beta}}$ is given by $\mathbf{Cov}(\hat{\boldsymbol{\beta}}) = \sigma^2(\mathbf{x}'\mathbf{x})^{-1}$. It follows from Corollary 5.2 that if $\mathbf{Cov}(\hat{\boldsymbol{\beta}}) = \sigma^2(\mathbf{x}'\mathbf{x})^{-1} \to \mathbf{0}$ then $\hat{\boldsymbol{\beta}}$ converges in mean square to $\boldsymbol{\beta}$. But since convergence in mean square implies convergence in probability (Theorem 5.13), it follows that $\mathrm{plim}(\hat{\boldsymbol{\beta}}) = \boldsymbol{\beta}$, so that $\hat{\boldsymbol{\beta}}$ is a consistent estimator of $\boldsymbol{\beta}$. ■

As a practical matter, what is required of the $\mathbf{x}$ matrix for $\lim_{n\to\infty}(\mathbf{x}'\mathbf{x})^{-1} = \mathbf{0}$? The convergence to the zero matrix will occur *iff* each *diagonal* entry of $(\mathbf{x}'\mathbf{x})^{-1}$ converges to zero, the necessity of this condition being obvious. The sufficiency follows from the fact that $(\mathbf{x}'\mathbf{x})^{-1}$ is positive definite since $\mathbf{x}'\mathbf{x}$ is positive definite,[7] and then the $(i,j)$th entry of $(\mathbf{x}'\mathbf{x})^{-1}$ is upper-bounded in absolute value by the product of the square roots of the $(i,i)$th and $(j,j)$th (diagonal) entries of $(\mathbf{x}'\mathbf{x})^{-1}$. To see the boundedness of the $(i,j)$th entry of $(\mathbf{x}'\mathbf{x})^{-1}$, note that for any symmetric positive definite matrix, all principal submatrices formed by deleting one or more diagonal entries together with the rows and columns in which they appear

[7]The inverse of a symmetric positive definite matrix is necessarily a symmetric positive definite matrix. This can be shown by noting that a symmetric matrix is positive definite *iff* all of its characteristic roots are positive, and the characteristic roots of $\mathbf{A}^{-1}$ are the reciprocals of the characteristic roots of $\mathbf{A}$.

must also be positive definite.[8] In particular, retaining only the $i$th and $j$th diagonal entries results in a $(2 \times 2)$ principal submatrix of the form $\begin{bmatrix} a_{ii} & a_{ij} \\ a_{ji} & a_{jj} \end{bmatrix}$ where $a_{ij}$ is used to denote the $(i,j)$th entry in $(\mathbf{x}'\mathbf{x})^{-1}$. The conditions for the submatrix to be positive definite are the standard ones, i.e., $a_{ii} > 0$ and $a_{ii}a_{jj} > a_{ij}a_{ji}$, the latter condition implying $a_{ii}a_{jj} > (a_{ij})^2$ by the symmetry of $(\mathbf{x}'\mathbf{x})^{-1}$, and thus $|a_{ij}| < (a_{ii})^{1/2}(a_{jj})^{1/2}$ which is the boundedness result mentioned above. Therefore, if $a_{ii} \to 0\ \forall i$, then $a_{ij} \to 0\ \forall i$ *and* $j$.

Regarding the convergence of the diagonal entries of $(\mathbf{x}'\mathbf{x})^{-1}$ to zero, examine the (1,1) entry of $(\mathbf{x}'\mathbf{x})^{-1}$. We will again have use for the concept of partitioned inversion (see Lemma 4.3), where we partition the $\mathbf{x}$ matrix as $\mathbf{x} = [\mathbf{x}_{.1}\ \mathbf{x}_*]$, where $\mathbf{x}_{.1}$ denotes the first column of $\mathbf{x}$ and $\mathbf{x}_*$ denotes the remaining $(k-1)$ columns of the $\mathbf{x}$ matrix. Then

$$\mathbf{x}'\mathbf{x} = \begin{bmatrix} \mathbf{x}'_{.1}\mathbf{x}_{.1} & \mathbf{x}'_{.1}\mathbf{x}_* \\ \mathbf{x}'_*\mathbf{x}_{.1} & \mathbf{x}'_*\mathbf{x}_* \end{bmatrix}$$

and using partitioned inversion, the (1,1) entry in $(\mathbf{x}'\mathbf{x})^{-1}$ is represented by

$$(\mathbf{x}'\mathbf{x})^{-1}_{1,1} = \left(\mathbf{x}_{.1}'\mathbf{x}_{.1} - \mathbf{x}_{.1}'\mathbf{x}_*(\mathbf{x}_*'\mathbf{x}_*)^{-1}\mathbf{x}_*'\mathbf{x}_{.1}\right)^{-1}.$$

Thus, for the (1,1) entry in $(\mathbf{x}'\mathbf{x})^{-1}$ to have a limit of zero as $n \to \infty$, it is clear that the expression in parentheses on the right-hand side of the preceding equality must $\to \infty$ as $n \to \infty$. Because $(\mathbf{x}_*'\mathbf{x}_*)$ is positive definite (recall footnote 18), this necessarily requires that $\mathbf{x}_{.1}'\mathbf{x}_{.1} \to \infty$ as $n \to \infty$ since a nonnegative quantity, $\mathbf{x}_{.1}'\mathbf{x}_*(\mathbf{x}_*'\mathbf{x}_*)^{-1}\mathbf{x}_*'\mathbf{x}_{.1}$, is being subtracted from $\mathbf{x}_{.1}'\mathbf{x}_{.1}$.

Now note further that $(\mathbf{x}_{.1}'\mathbf{x}_{.1} - \mathbf{x}_{.1}'\mathbf{x}_*(\mathbf{x}_*'\mathbf{x}_*)^{-1}\mathbf{x}_*'\mathbf{x}_{.1})$ is the sum of the squared deviations of $\mathbf{x}_{.1}$ from a vector of predictions of $\mathbf{x}_{.1}$ generated by a least squares-estimated linear explanation of $\mathbf{x}_{.1}$ (the dependent variable $\mathbf{x}_{.1}$ is being predicted by a linear function of $\mathbf{x}_*$, which are the explanatory variables in this context). To see this, note that the vector $\mathbf{x}_*\hat{\mathbf{b}}$ is the least distance from $\mathbf{x}_{.1}$ when we choose $\hat{\mathbf{b}} = (\mathbf{x}_*'\mathbf{x}_*)^{-1}\mathbf{x}_*'\mathbf{x}_{.1}$, as we have argued previously with regard to the least squares procedure. Then the deviations between entries in $\mathbf{x}_{.1}$ and $\mathbf{x}_*\hat{\mathbf{b}}$ are represented by the vector $\mathbf{e}_1 = \mathbf{x}_{.1} - \mathbf{x}_*(\mathbf{x}_*'\mathbf{x}_*)^{-1}\mathbf{x}_*'\mathbf{x}_{.1}$, and straightforward matrix multiplication then demonstrates that $\mathbf{e}_1'\mathbf{e}_1 = \mathbf{x}_{.1}'\mathbf{x}_{.1} - \mathbf{x}_{.1}'\mathbf{x}_*(\mathbf{x}_*'\mathbf{x}_*)^{-1}\mathbf{x}_*'\mathbf{x}_{.1}$ which is the *reciprocal* of the (1,1) entry of $(\mathbf{x}'\mathbf{x})^{-1}$. Thus, in order for the (1,1) entry of $(\mathbf{x}'\mathbf{x})^{-1}$ to converge to zero, we require that $\mathbf{e}_1'\mathbf{e}_1 \to \infty$ as $n \to \infty$. We thus specifically rule out the possibility that $\mathbf{e}_1'\mathbf{e}_1 < \delta < \infty$ i.e., that the sum of squared deviations is bounded as $n \to \infty$. As a practical matter, it is *sufficient* that the average squared error, $n^{-1}\mathbf{e}_1'\mathbf{e}_1$, in predicting entries in $\mathbf{x}_{.1}$ via linear combinations of corresponding entries in the remaining columns of the $\mathbf{x}$ matrix exhibits some

[8] Suppose $\mathbf{B}$ is a $(k \times k)$ symmetric positive definite matrix. Let $\mathbf{C}$ be a $(r \times k)$ matrix such that each row consists of all zeros except for a 1 in one position, and let the $r$-rows of $\mathbf{C}$ be linearly independent. Then $\mathbf{CBC}'$ defines an $(r \times r)$ principal submatrix of $\mathbf{B}$. Now note that $\ell'\mathbf{CBC}'\ell = \ell_*'\mathbf{B}\ell_* > 0\ \forall\ \ell \neq 0$ since $\ell_* = \mathbf{C}'\ell \neq \mathbf{0}\ \forall\ \ell \neq \mathbf{0}$ by the definition of $\mathbf{C}$, and $\mathbf{B}$ is positive definite. Therefore, the principal submatrix $\mathbf{CBC}'$ is positive definite.

positive lower bound and thus does not converge to zero as $n \to \infty$. Roughly speaking, it is sufficient to assume that a linear dependence between $\mathbf{x}_{\cdot 1}$ and the column vectors in $\mathbf{x}_*$ never develops regardless of sample size.

Note that the column of $\mathbf{x}$ that we utilize in applying the partitioned inversion result is arbitrary—we can always rearrange the columns of the $\mathbf{x}$ matrix to place whichever explanatory variable we choose in the "first position." Thus, our discussion above applies to each diagonal entry in the $(\mathbf{x}'\mathbf{x})^{-1}$ matrix. Then in summary, our sufficient condition for the consistency of $\hat{\boldsymbol{\beta}}$ is that $\lim_{n\to\infty}((\mathbf{x}'\mathbf{x})^{-1}) = \mathbf{0}$,[9] which holds if $(\mathbf{x}_{\cdot i}'\mathbf{x}_{\cdot i}) \to \infty \forall i$, and if $\exists$ some positive number $\delta > 0$ such that no column of $\mathbf{x}$ can be represented via a linear combination of the remaining columns of $\mathbf{x}$ with an average squared prediction error less than $\delta$ as $n \to \infty$.

*Consistency of $\hat{S}^2$* Additional conditions beyond those underlying the Classical GLM are necessary if $\hat{S}^2$ is to be a consistent estimator $\sigma^2$, as the next two theorems indicate.

**Theorem 8.5** ***Consistency of $\hat{S}^2$-iid Residuals*** *If in addition to the assumptions of the Classical GLM, the elements of the residual vector, $\boldsymbol{\varepsilon}$, are iid, then $\hat{S}^2 \xrightarrow{p} \sigma^2$, so that $\hat{S}^2$ is a consistent estimator of $\sigma^2$.*

**Proof** Recall from our discussion of the unbiasedness of $\hat{S}^2$ that $\hat{S}^2 = \left((\mathbf{Y}-\mathbf{x}\hat{\boldsymbol{\beta}})'(\mathbf{Y}-\mathbf{x}\hat{\boldsymbol{\beta}})\right)/(n-k) = \left(\boldsymbol{\varepsilon}'(\mathbf{I}-\mathbf{x}(\mathbf{x}'\mathbf{x})^{-1}\mathbf{x}')\boldsymbol{\varepsilon}\right)/(n-k)$. Examine

$$\text{plim}\left(\hat{S}^2\right) = \text{plim}\left(\frac{\boldsymbol{\varepsilon}'\boldsymbol{\varepsilon}}{n-k}\right) - \text{plim}\left(\frac{\boldsymbol{\varepsilon}'\mathbf{x}(\mathbf{x}'\mathbf{x})^{-1}\mathbf{x}'\boldsymbol{\varepsilon}}{n-k}\right)$$

and focus on the last term first. Note that $\left(\boldsymbol{\varepsilon}'\mathbf{x}(\mathbf{x}'\mathbf{x})^{-1}\mathbf{x}'\boldsymbol{\varepsilon}\right)/(n-k)$ is a nonnegative-valued random variable since $(\mathbf{x}'\mathbf{x})^{-1}$ is positive definite. Also note that under the Classical GLM assumptions $\mathrm{E}(\boldsymbol{\varepsilon}\boldsymbol{\varepsilon}') = \sigma^2\mathbf{I}$, so that

$$\mathrm{E}\left((n-k)^{-1}\boldsymbol{\varepsilon}'\mathbf{x}(\mathbf{x}'\mathbf{x})^{-1}\mathbf{x}'\boldsymbol{\varepsilon}\right) = (n-k)^{-1}\,\mathrm{tr}\left(\mathbf{x}(\mathbf{x}'\mathbf{x})^{-1}\mathbf{x}'\mathrm{E}(\boldsymbol{\varepsilon}\boldsymbol{\varepsilon}')\right)$$

$$= \sigma^2(n-k)^{-1}\mathrm{tr}\left((\mathbf{x}'\mathbf{x})^{-1}\mathbf{x}'\mathbf{x}\right) = \sigma^2\,(k/(n-k))$$

Then by Markov's inequality, $\forall c > 0$,

$$P\left((n-k)^{-1}\boldsymbol{\varepsilon}'\mathbf{x}(\mathbf{x}'\mathbf{x})^{-1}\mathbf{x}'\boldsymbol{\varepsilon} \geq c\right) \leq \frac{\sigma^2\,(k/(n-k))}{c}$$

and since $\lim_{n\to\infty}\sigma^2(k/(n-k)) = 0$, it follows that $\text{plim}\left((n-k)^{-1}\boldsymbol{\varepsilon}'\mathbf{x}(\mathbf{x}'\mathbf{x})^{-1}\mathbf{x}'\boldsymbol{\varepsilon}\right) = 0$.

---

[9]Consistency of $\hat{\boldsymbol{\beta}}$ can be proven under alternative conditions on $\mathbf{x}$. Judge, et al., (1982) *Introduction to the Theory and Practice of Econometrics*, John Wiley, pp. 269–269, prove the result using the stronger condition that $\lim_{n\to\infty} n^{-1}\,\mathbf{x}'\mathbf{x} = Q$, where Q is a finite, positive definitive matrix. Halbert White, (1984) *Asymptotic Theory for Econometricians*, Academic Press, p. 20, assumes that $n^{-1}\,\mathbf{x}'\mathbf{x}$ is bounded and uniformly positive definite, which is also a stronger condition than the one we use here.

Given the preceding result, it is clear that for $\text{plim}(\hat{S}^2) = \sigma^2$, it is necessary and sufficient that $\text{plim}(\boldsymbol{\varepsilon}'\boldsymbol{\varepsilon}/(n-k)) = \sigma^2$. Recalling that plim(WZ) = plim(W) plim(Z)) by Theorem 5.6 and since $((n-k)/n) \to 1$ as $n \to \infty$, an equivalent necessary and sufficient condition is that $\text{plim}(\boldsymbol{\varepsilon}'\boldsymbol{\varepsilon}/n) = \sigma^2$. If the $\varepsilon_i$'s are not only uncorrelated, but also *iid*, then Khinchin's WLLN (Theorem 5.19) allows us to conclude the preceding plim result, because $\text{E}(\varepsilon_i^2) = \sigma^2$, $\forall i$ the $\varepsilon_i^2$'s are *iid*, and $n^{-1}\boldsymbol{\varepsilon}'\boldsymbol{\varepsilon} = n^{-1}\sum_{i=1}^n \varepsilon_i^2$, which is precisely in the form of Khinchin's theorem applied to the sequence of *iid* random variables $\{\varepsilon_1^2, \varepsilon_2^2, \varepsilon_3^2, \ldots\}$. Thus, $\text{plim}(\boldsymbol{\varepsilon}'\boldsymbol{\varepsilon}/n) = \sigma^2$, and $\hat{S}^2$ is a consistent estimator of $\sigma^2$. ■

Consistency of $\hat{S}^2$ for estimating $\sigma^2$ can be demonstrated without requiring that the $\varepsilon_i$'s are *iid*.

**Theorem 8.6 *Consistency of $\hat{S}^2$: Non-iid Residuals*** *Assume the Classical Assumptions of the GLM. Then $\hat{S}^2$ is a consistent estimator of $\sigma^2$ if $\text{E}(\varepsilon_i^4) \leq \tau < \infty$, $\forall i$ and either of the following conditions hold:*

**a.** $\sum_{j=1, j\neq i}^n \text{Cov}\left(\varepsilon_i^2, \varepsilon_j^2\right) = o(n^1)$, $\forall i$.

**b.** $\{\varepsilon_n\}$ *is an m-dependent sequence.*

**Proof** See Appendix. ■

Regarding the conditions in Theorem 8.6, first note that the boundedness of the moments of the residuals, i.e., $\text{E}(\varepsilon_i^4) \leq \tau < \infty$, $\forall i$ for some $\tau > 0$, is a relatively weak assumption. Note, for example, that these moment boundedness assumptions are satisfied for *every* parametric family of density functions that we examined in Chapter 4. The remaining conditions on the residuals relate to the degree of association between outcomes of $\varepsilon_i$ and $\varepsilon_j$ as $i$ and $j$ become further and further apart. Essentially, (a) implies that the average covariance between $\varepsilon_i^2$ and all of the remaining squared residuals, $\varepsilon_j^2$ for $j > i$, converges to zero $\forall i$; and (b) states that $\varepsilon_i$ and $\varepsilon_j$ become independent of one another when $|i-j|$ exceeds a fixed finite value. If the random sample refers to a time series of observations, then $|i-j|$ refers to separation of $\varepsilon_i$ and $\varepsilon_j$ in the time dimension. If the random sample refers to a cross-section of observations, then one must search for an ordering of the elements in the random sample to which the conditions of Theorem 8.6 can be applied.

*Asymptotic Normality of $\hat{\beta}$* We establish the asymptotic normality of $\hat{\boldsymbol{\beta}}$ by initially stipulating that the $\varepsilon_i$'s are *iid*, that $|x_{ij}| < \xi < \infty$ $\forall i, j$, i.e., the explanatory variable values are bounded in absolute value, and that $P(|\varepsilon_i| < m) = 1$ $\forall i$, for $m < \infty$, i.e., the disturbance terms are bounded in absolute value with probability 1. As a practical matter, it is often reasonable to assume that real-world explanatory variables in a linear model do not have values that increase without bound; e.g., all published data on the economy, census data, samples of socio-demographic information and the like contain finite numbers, no matter how vast the collection. Indeed it is safe to say that all of the numerical data ever measured and recorded can be bounded in absolute value by some large enough finite number, $\xi$. The boundedness assumption on the explanatory variables thus has wide applicability.[10] Similar reasoning suggests that, in practice, boundedness of the disturbance terms in a linear model is often a reasonable assumption. That is, if $y_i$ is an observation on real-world economic or social data and thus bounded in magnitude, it follows that $\varepsilon_i = y_i - \mathrm{E}(Y_i)$ would also be bounded, so that $P(|\varepsilon_i| < m) = 1$.

We finally assume that $\lim_{n\to\infty}(n^{-1}\mathbf{x}'\mathbf{x}) = \mathbf{Q}$, where $\mathbf{Q}$ is a finite, positive definite matrix. This of course implies that $n^{-1}\mathbf{x}'\mathbf{x}$ has full rank, not just for a finite number of observations which would be ensured by rank$(\mathbf{x}) = k$, but also in the limit as $n \to \infty$. It also requires that the entries in $n^{-1}\mathbf{x}'\mathbf{x}$ be bounded (or else the limit would not exist). Note that this assumption poses no conceptual difficulty if $\mathbf{x}$ is fixed and designed by the researcher (or someone else), since the $\mathbf{x}$ matrix can then always be designed to behave in the required manner. If $\mathbf{x}$ is not designed, it must be assumed that the $x_{ij}$'s are generated in a way that allows the limit of $n^{-1}\mathbf{x}'\mathbf{x}$ to be finite and positive definite.

**Theorem 8.7** ***Asymptotic Normality of $\hat{\boldsymbol{\beta}}$–iid Residuals*** *Assume the Classical Assumptions of the GLM. In addition, assume that $\{\varepsilon_n\}$ is a collection of iid random variables and that $P(|\varepsilon_i| < m) = 1$ for $m < \infty$ and $\forall i$. Finally, assume that the explanatory variables are such that $|x_{ij}| < \xi < \infty$ $\forall i$ and $j$, and that $\lim_{n\to\infty}(n^{-1}\mathbf{x}'\mathbf{x}) = \mathbf{Q}$, where $\mathbf{Q}$ is a finite positive definite matrix. Then*

$$n^{1/2}(\hat{\boldsymbol{\beta}} - \boldsymbol{\beta}) \xrightarrow{d} N(\mathbf{0}, \sigma^2\mathbf{Q}^{-1}) \text{ and } \hat{\boldsymbol{\beta}} \overset{a}{\sim} N(\boldsymbol{\beta},\ n^{-1}\sigma^2\mathbf{Q}^{-1}).$$

**Proof** Recall that the least squares estimator can be expressed as $\hat{\boldsymbol{\beta}} = \boldsymbol{\beta} + (\mathbf{x}'\mathbf{x})^{-1}\,\mathbf{x}'\,\boldsymbol{\varepsilon}$. Then $n^{1/2}(\hat{\boldsymbol{\beta}} - \boldsymbol{\beta}) = n^{1/2}\,(\mathbf{x}'\mathbf{x})^{-1}\,\mathbf{x}'\,\boldsymbol{\varepsilon} = (n^{-1}\,\mathbf{x}'\mathbf{x})^{-1}\,n^{-1/2}\mathbf{x}'\boldsymbol{\varepsilon}$. Note that $n^{-1/2}\mathbf{x}'\boldsymbol{\varepsilon}$ can be written as the sum of $n$ $k$-vectors as $n^{-1/2}\mathbf{x}'\boldsymbol{\varepsilon} = n^{-1/2}\sum_{i=1}^{n}\mathbf{x}'_{i.}\varepsilon_i$ where $\mathbf{x}_{i.}$ is the $(1 \times k)$ vector representing the $i$th row of $\mathbf{x}$. The expectation of the $(k \times 1)$ random vector $\mathbf{x}'_{i.}\varepsilon_i$ is $\mathbf{0}$ since $\mathrm{E}(\varepsilon_i) = 0$. The covariance matrix of the random vector is given by $\mathbf{Cov}(\mathbf{x}'_{i.}\varepsilon_i) = \mathrm{E}((\mathbf{x}'_{i.}\varepsilon_i)(\varepsilon_i\mathbf{x}_{i.})) = \mathrm{E}(\varepsilon_i^2)\mathbf{x}'_{i.}\mathbf{x}_{i.} = \sigma^2\mathbf{x}'_{i.}\mathbf{x}_{i.}$.

The random vectors $\mathbf{x}'_{i.}\varepsilon_i$, $i = 1,\ldots,n$, are independent because the $\varepsilon_i$'s are independent, and the entries in the random vector $\mathbf{x}'_{i.}\varepsilon_i$ are bounded in absolute value with probability 1, $\forall i$, since $P(|\varepsilon_i| < m) = 1$ and $|x_{ij}| < \xi$, so that $P(|x_{ij}\varepsilon_i| < m\xi) = 1$ $\forall i,j$. Given that

[10] We should note, however, that in the specification of some linear models, certain proxy variables might be used to explain $\mathbf{y}$ that literally violate the boundedness assumption. For example, a linear "time trend" $t = (1,2,3,4,\ldots)$ is sometimes used to explain an apparent upward or downward trend in $\mathrm{E}(Y_t)$and the trend clearly violates the boundedness constraint. In such cases, one may wonder whether it is really to be believed that $t \to \infty$ is relevant in explaining $\mathbf{y}$, or whether the time trend is just an artifice relevant for a certain range of observations, but for which extrapolation ad infinitum is not appropriate.

$$\lim_{n\to\infty}\left(n^{-1}\sum_{i=1}^{n}\mathbf{Cov}(\mathbf{x}'_{i.}\varepsilon_i)\right)=\lim_{n\to\infty}\left(n^{-1}\sum_{i=1}^{n}\sigma^2(\mathbf{x}'_{i.}\mathbf{x}_{i.})\right)$$
$$=\lim_{n\to\infty}\left(n^{-1}\sum_{i=1}^{n}\sigma^2(\mathbf{x}'\mathbf{x})\right)=\sigma^2\,\mathbf{Q},$$

and since $\mathbf{Q}$ is a finite positive definite matrix, it follows directly from Theorem 5.38 concerning a multivariate CLT for independent bounded random vectors that $n^{-1/2}\sum_{i=1}^{n}\mathbf{x}'_{i.}\varepsilon_i = n^{-1/2}\mathbf{x}'\boldsymbol{\varepsilon}\xrightarrow{d} N(\mathbf{0},\sigma^2\mathbf{Q})$.

Then by Slutsky's theorem, since $\lim_{n\to\infty}(n^{-1}\mathbf{x}'\mathbf{x}) = \text{plim}(n^{-1}\mathbf{x}'\mathbf{x}) = \mathbf{Q}$, and thus

$$\lim_{n\to\infty}\left((n^{-1}\mathbf{x}'\mathbf{x})^{-1}\right)=\text{plim}\left((n^{-1}\mathbf{x}'\mathbf{x})^{-1}\right)=(\text{plim}(n^{-1}\mathbf{x}'\mathbf{x}))^{-1}=\mathbf{Q}^{-1},$$

we have that

$$n^{1/2}(\hat{\boldsymbol{\beta}}-\boldsymbol{\beta})=n^{1/2}(\mathbf{x}'\mathbf{x})^{-1}\mathbf{x}'\boldsymbol{\varepsilon}=(n^{-1}\mathbf{x}'\mathbf{x})^{-1}n^{-1/2}\mathbf{x}'\boldsymbol{\varepsilon}\xrightarrow{d}\text{N}(\mathbf{0},\sigma^2\mathbf{Q}^{-1})$$

which proves the limiting distribution part of the theorem. The asymptotic distribution then follows directly from Definition 5.2. □

While the previous theorem establishes the asymptotic normality of the least squares or BLUE estimator $\hat{\boldsymbol{\beta}}$, the asymptotic distribution is expressed in terms of the generally unobservable limit matrix $\mathbf{Q}$. We now present a corollary to Theorem 8.6 that replaces $n^{-1}\mathbf{Q}^{-1}$ with the observable matrix $(\mathbf{x}'\mathbf{x})^{-1}$, leading to a covariance matrix for the asymptotic distribution that is identical to the finite sample covariance matrix of $\hat{\boldsymbol{\beta}}$.

**Corollary 8.1** ***Alternative Normal Asymptotic Distribution Representation for*** $\hat{\boldsymbol{\beta}}$***–iid Residuals***

*Under the conditions of Theorem 8.6,* $(\mathbf{x}'\mathbf{x})^{1/2}(\hat{\boldsymbol{\beta}}-\boldsymbol{\beta})\xrightarrow{d}N(\mathbf{0},\sigma^2\mathbf{I})$ *and* $\hat{\boldsymbol{\beta}}\overset{a}{\sim}N(\boldsymbol{\beta},\sigma^2(\mathbf{x}'\mathbf{x})^{-1})$.

**Proof** By Theorem 8.7 and Slutsky's theorem, it follows that $\mathbf{Q}^{1/2}\,n^{1/2}(\hat{\boldsymbol{\beta}}-\boldsymbol{\beta})\xrightarrow{d}N(\mathbf{0},\sigma^2\,\mathbf{I})$. Since $(n^{-1}\mathbf{x}'\mathbf{x})^{1/2}\mathbf{Q}^{-1/2}\to\mathbf{Q}^{1/2}\,\mathbf{Q}^{-1/2}=\mathbf{I}$ because $n^{-1}\,\mathbf{x}'\mathbf{x}\to\mathbf{Q}$,[11] Slutsky's theorem also implies that

$$(n^{-1}\,\mathbf{x}'\mathbf{x})^{1/2}\,\mathbf{Q}^{-1/2}\,\mathbf{Q}^{1/2}\,n^{1/2}(\hat{\boldsymbol{\beta}}-\boldsymbol{\beta})=(\mathbf{x}'\mathbf{x})^{1/2}(\hat{\boldsymbol{\beta}}-\boldsymbol{\beta})\xrightarrow{d}N(\mathbf{0},\sigma^2\,\mathbf{I}),$$

so that $\hat{\boldsymbol{\beta}}\overset{a}{\sim}N(\boldsymbol{\beta},\sigma^2(\mathbf{x}'\mathbf{x})^{-1})$. ■

[11] Regarding the symmetric square root matrix $\mathbf{A}^{1/2}$, note that $\mathbf{P}'\mathbf{A}\mathbf{P}=\boldsymbol{\Lambda}$, where $\mathbf{P}$ is the orthogonal matrix of characteristic vectors of the symmetric positive semidefinite matrix $\mathbf{A}$, and $\boldsymbol{\Lambda}$ is the diagonal matrix of characteristic roots. Then $\mathbf{A}^{1/2}=\mathbf{P}\boldsymbol{\Lambda}^{1/2}\mathbf{P}'$, where $\boldsymbol{\Lambda}^{1/2}$ is the diagonal matrix formed from $\boldsymbol{\Lambda}$ by taking the square root of each diagonal element. Note that since $\mathbf{P}'\mathbf{P}=\mathbf{I}$, $\mathbf{A}^{1/2}\mathbf{A}^{1/2}=\mathbf{P}\boldsymbol{\Lambda}^{1/2}\mathbf{P}'\mathbf{P}\boldsymbol{\Lambda}^{1/2}\mathbf{P}'=\mathbf{P}\boldsymbol{\Lambda}^{1/2}\boldsymbol{\Lambda}^{1/2}\mathbf{P}'=\mathbf{P}\boldsymbol{\Lambda}\mathbf{P}'=\mathbf{A}$. The matrix square root is a continuous function of the elements of $\mathbf{A}$. Therefore, $\lim_{n\to\infty}(\mathbf{A}_n)^{1/2}=(\lim_{n\to\infty}\mathbf{A}_n)^{1/2}$.

The asymptotic normality of the least squares estimator can be demonstrated more generally without reliance on the boundedness of the disturbance terms, without the *iid* assumption on the error terms and without assuming the existence of a finite positive definite limit of $n^{-1}\,\mathbf{x}'\mathbf{x}$ as $n \to \infty$. The demonstration relies on somewhat more complicated central limit arguments. We state one such result below.

**Theorem 8.8 Asymptotic Normality of $\hat{\boldsymbol{\beta}}$–non-iid Residuals** *Assume the Classical Assumptions of the GLM. Then* $(\mathbf{x}'\mathbf{x})^{1/2}(\hat{\boldsymbol{\beta}}-\boldsymbol{\beta}) \xrightarrow{d} N(\mathbf{0}, \sigma^2\mathbf{I})$ *and* $\hat{\boldsymbol{\beta}} \overset{a}{\sim} N(\boldsymbol{\beta}, \sigma^2(\mathbf{x}'\mathbf{x})^{-1})$ *if the elements of the sequence* $\{\varepsilon_n\}$ *are independent random variables such that* $\mathrm{E}(\varepsilon_i^4) < \tau < \infty\ \forall i$, *and the explanatory variables are such that* $|\mathbf{x}_{ij}| < \xi < \infty\ \forall i$ *and* $j$, *with* $\det(n^{-1}\,\mathbf{x}'\mathbf{x}) > \eta > 0$ *and* $\{n^{-1}\,\mathbf{x}'\mathbf{x}\}$ *being* $O(1)$.

**Proof** See Appendix. ■

The difference in assumptions between our previous result on the asymptotic normality of $\hat{\boldsymbol{\beta}}$ and the result in Theorem 8.8 is that the existence of moments of the $\varepsilon_i$'s of order four replaces the assumption that the errors are bounded with probability 1, the $\varepsilon_i$'s are no longer required to be identically distributed (although first and second order moments are assumed to be the same by the classical GLM assumptions), and bounds on the determinant and elements of $n^{-1}\,\mathbf{x}'\mathbf{x}$ replace the assumption that $\lim_{n\to\infty}(n^{-1}\,\mathbf{x}'\mathbf{x})$ exists and is positive definite. The boundedness assumptions are less restrictive than the assumption that $\lim_{n\to\infty} n^{-1}\,\mathbf{x}'\mathbf{x}$ exists and is a nonsingular positive definite matrix, especially in cases where there is no control over how the $\mathbf{x}$ matrix was generated and no knowledge that the process generating the $\mathbf{x}$ matrix would inherently lead to the existence of a limit for $n^{-1}\,\mathbf{x}'\mathbf{x}$.

Regarding the moment condition on the $\varepsilon_i$'s, note that our previous assumption $P(|\varepsilon_i| < m) = 1\ \forall i$ implies boundedness of the moments and, thus, implies the moment condition in Theorem 8.7. However, the moment conditions in Theorem 8.7 also allow situations in which there is no absolute upper bound to the value of the error term that holds with probability one, and thus, for example, allows the $\varepsilon_i$'s to have a normal distribution or a mean-shifted (to zero) gamma distribution whereas the former assumption would not.

In summary, the least squares estimator $\hat{\boldsymbol{\beta}}$ can be approximately normally distributed, even if $\boldsymbol{\varepsilon}$ is not multivariate normally distributed, given certain conditions on the values of the explanatory variables and certain assumptions regarding distributional characteristics of the disturbance vector. Note that asymptotic normality of $\hat{\boldsymbol{\beta}}$ can be demonstrated using still weaker assumptions—the interested reader is directed to Chapter 5 of the book by White, H., *Asymptotic Theory for Econometricians*, Academic Press, 1984, for further details.

*Asymptotic Normality of* $\hat{S}^2$ If in addition to the classical GLM assumption it is assumed that the $\varepsilon_i$'s are *iid* and have bounded fourth-order moments about the origin, then $\hat{S}^2$ will be asymptotically normally distributed.

**Theorem 8.9 Asymptotic Normality of $\hat{S}^2$-iid Residuals** *Under the Classical Assumptions of the GLM, if the elements of the residual vector,* $\boldsymbol{\varepsilon}$, *are iid, and if* $\mathrm{E}(\varepsilon_i^4) \le \tau < \infty$, *then*

$$n^{1/2}\left(\hat{S}^2 - \sigma^2\right) \xrightarrow{d} N(0, \mu'_4 - \sigma^4) \text{ and } \hat{S}^2 \overset{a}{\sim} N(\sigma^2, n^{-1}(\mu'_4 - \sigma^4)).$$

**Proof** Note that

$$n^{1/2}\left(\hat{S}^2 - \sigma^2\right) = \frac{n^{1/2}}{(n-k)}\boldsymbol{\varepsilon}'\left(\mathbf{I} - \mathbf{x}(\mathbf{x}'\mathbf{x})^{-1}\mathbf{x}'\right)\boldsymbol{\varepsilon} - n^{1/2}\sigma^2$$
$$= \left[\frac{n^{1/2}\boldsymbol{\varepsilon}'\boldsymbol{\varepsilon}}{(n-k)} - n^{1/2}\sigma^2\right] - \frac{n^{1/2}}{n-k}\boldsymbol{\varepsilon}'\mathbf{x}(\mathbf{x}'\mathbf{x})^{-1}\mathbf{x}'\boldsymbol{\varepsilon}.$$

Based on Markov's inequality,

$$P\left(\left(\frac{n^{1/2}}{n-k}\right)\boldsymbol{\varepsilon}'\mathbf{x}(\mathbf{x}'\mathbf{x})^{-1}\mathbf{x}'\boldsymbol{\varepsilon} \geq c\right) \leq \frac{\sigma^2\, n^{1/2}\, k}{c(n-k)}$$

(recall the proof of Theorem 8.4), and since $\sigma^2\, n^{1/2}\, k/(c(n-k)) \to 0$ as $n \to \infty$, $\text{plim}\ (n^{1/2}/(n-k))\boldsymbol{\varepsilon}'\mathbf{x}(\mathbf{x}'\mathbf{x})^{-1}\mathbf{x}'\boldsymbol{\varepsilon} = 0$. Then it follows from Slutsky's theorem that the limiting density of $n^{1/2}\,(\hat{S}^2 - \sigma^2)$ will depend only on the bracketed term in its definition above. Also by Slutsky's theorem, the limiting density is unaffected if we multiply the bracketed term by $(n-k)/n$, since $(n-k)/n \to 1$ as $n \to \infty$, or if we then also add the term $-k\sigma^2/n^{1/2}$ since $-k\sigma^2/n^{1/2} \to 0$ as $n \to \infty$. Thus to establish the limiting density of $n^{1/2}\,(\hat{S}^2 - \sigma^2)$, it suffices to examine the limiting density of

$$\frac{\boldsymbol{\varepsilon}'\boldsymbol{\varepsilon}}{n^{1/2}} - \frac{(n-k)\sigma^2}{n^{1/2}} - \frac{k\sigma^2}{n^{1/2}} = \frac{\boldsymbol{\varepsilon}'\boldsymbol{\varepsilon} - n\sigma^2}{n^{1/2}}.$$

Since $\mathrm{E}\left(\varepsilon_i^4\right) \leq \tau < \infty \Leftrightarrow \mathrm{var}\left(\varepsilon_i^2\right) \leq \tau - \sigma^4 < \infty$, a direct application of the Lindberg-Levy CLT to the sequence of *iid* random variables $\{\varepsilon_1^2, \varepsilon_2^2, \varepsilon_3^2, \ldots\}$ yields

$$\frac{\boldsymbol{\varepsilon}'\boldsymbol{\varepsilon} - n\sigma^2}{n^{1/2}(\mu'_4 - \sigma^4)^{1/2}} \xrightarrow{d} N(0,1)$$

and thus by Slutsky's theorem

$$\frac{\boldsymbol{\varepsilon}'\boldsymbol{\varepsilon} - n\sigma^2}{n^{1/2}} \xrightarrow{d} N(0, \mu'_4 - \sigma^4).$$

Therefore,

$$n^{1/2}\left(\hat{S}^2 - \sigma^2\right) \xrightarrow{d} N(0, \mu'_4 - \sigma^4) \text{ and } \hat{S}^2 \overset{a}{\sim} N\left(\sigma^2, \tfrac{\mu'_4 - \sigma^4}{n}\right).$$ ■

The asymptotic normality of $\hat{S}^2$ can be established without assuming that the disturbance terms are *iid*. We present one such result in the following theorem.

**Theorem 8.10 Asymptotic Normality of $\hat{S}^2$–Non-iid Residuals** *Assume the Classical Assumptions of the GLM. Also assume that the $\varepsilon_i$'s are independent and $\mathrm{E}\left(|\varepsilon_i^2 - \sigma^2|^{2+\delta}\right) \leq \tau < \infty\ \forall i$ for some $\delta > 0$. Then letting $\xi_n = n^{-1}\sum_{i=1}^n \mathrm{var}(\varepsilon_i^2) \geq \eta > 0\ \forall n,$, $n^{1/2}\left(\hat{S}^2 - \sigma^2\right)/\xi_n^{1/2} \xrightarrow{d} N(0,1)$ and $\hat{S}^2 \overset{a}{\sim} N(\sigma^2, n^{-1}\xi_n)$.*

**Proof** See Appendix. □

*Summary of Least Squares Estimator Properties for the Classical General Linear Model*

Table 8.1 summarizes assumptions and the resultant properties of $\hat{\boldsymbol{\beta}}$ and $\hat{S}^2$ as estimators of $\boldsymbol{\beta}$ and $\sigma^2$ in the GLM context.

**Example 8.2** ***Least Squares Estimator When Random Sampling from a Population Distribution***

Let $(Y_1,\ldots,Y_n)$ be a random sample from a population distribution having finite mean $\beta$ and finite variance $\sigma^2$ (*both* scalars). The GLM representation of the vector **Y** is given by

$$\mathbf{Y} = \begin{bmatrix} Y_1 \\ Y_2 \\ \vdots \\ Y_n \end{bmatrix} = \mathbf{x}\beta + \boldsymbol{\varepsilon} = \begin{bmatrix} 1 \\ 1 \\ \vdots \\ 1 \end{bmatrix} \beta + \begin{bmatrix} \varepsilon_1 \\ \varepsilon_2 \\ \vdots \\ \varepsilon_n \end{bmatrix},$$

where $\mathrm{E}(\boldsymbol{\varepsilon}) = \mathbf{0}$, and since the $\varepsilon_i$'s are *iid*, $\mathrm{E}(\boldsymbol{\varepsilon}\boldsymbol{\varepsilon}') = \sigma^2\mathbf{I}$. Then note that $\hat{\beta} = (\mathbf{x}'\mathbf{x})^{-1}\mathbf{x}'\mathbf{Y} = n^{-1}\sum_{i=1}^{n} Y_i = \bar{Y}$, so that the least squares estimator of β is the sample mean. Furthermore, $\hat{S}^2 = (\mathbf{Y} - \mathbf{x}\hat{\beta})'(\mathbf{Y} - \mathbf{x}\hat{\beta})/(n - k) = \sum_{i=1}^{n} (Y_i - \bar{Y})^2/(n - 1)$ is used as an estimator of $\sigma^2$. The reader can verify that cases 2 and 3 in Table 8.1 are met, and thus $\hat{\beta}$ is BLUE and consistent for $\beta$, while $\hat{S}^2$ is an unbiased and consistent estimator for $\sigma^2$. Under additional assumptions such as cases 5–8 in Table 8.1, or recall those discussed in Chapter 6 concerning the asymptotic normality of the sample mean and sample variance, $\hat{\beta}$ and $\hat{S}^2$ are both asymptotically normally distributed. □

**Example 8.3** ***Empirical Application of Least Squares to a GLM Specification***

A realtor is analyzing the relationship between the number of new one-family houses sold in a given year in the United States and various explanatory variables that she feels had important impacts on housing sales during the decade of the 1980s. She has collected the following information:

| | # New homes sold (1,000s) | Conventional mortgage interest rate | Medium family income | New home purchase price (1,000's) | CPI (1982–1984 = 100) |
|---|---|---|---|---|---|
| 1980 | 545 | 13.95 | 21,023 | 83.2 | 82.4 |
| 1981 | 436 | 16.52 | 22,388 | 90.3 | 90.9 |
| 1982 | 412 | 15.79 | 23,433 | 94.1 | 96.5 |
| 1983 | 623 | 13.43 | 24,580 | 93.9 | 99.6 |
| 1984 | 639 | 13.80 | 26,433 | 96.8 | 103.9 |
| 1985 | 688 | 12.28 | 27,735 | 105.0 | 107.6 |
| 1986 | 750 | 10.07 | 29,458 | 119.8 | 109.6 |
| 1987 | 671 | 10.17 | 30,970 | 137.2 | 113.6 |
| 1988 | 676 | 10.30 | 32,191 | 150.5 | 118.3 |
| 1989 | 650 | 10.21 | 34,213 | 160.1 | 124.0 |
| $\bar{z}$ | 609 | 12.65 | 27,242 | 113.1 | 104.6 |

Source: Statistical Abstracts of the United States, (1991), *U.S. Dept. of Commerce, Washington D.C., Tables 1272, 837, 730, 825, and 771.*

**Table 8.1** Assumptions and properties of $\hat{\boldsymbol{\beta}}$ and $\hat{S}^2$ in the GLM point estimation problem

| Case | Classical GLM assumptions | Additional x-assumptions | Additional ε-assumptions | Properties $\hat{\boldsymbol{\beta}}$ | Properties $S^2$ |
|---|---|---|---|---|---|
| 1. | Hold | – | – | BLUE | Unbiased |
| 2. | Hold | $(\mathbf{x}'\mathbf{x})^{-1} \to \mathbf{0}$ | – | Consistent | – |
| 3. | Hold | – | $\varepsilon_i$'s are *iid* | – | Consistent |
| 4. | Hold | – | $E(\varepsilon_i^4) \leq \tau < \infty$ <br> $\left\{ \begin{array}{l} \{\varepsilon_n\} \text{ is } m-\text{dependent or} \\ \text{cov}\left(\varepsilon_i^2, \varepsilon_j^2\right) \to 0, \lvert i-j \rvert \to \infty \end{array} \right\}$ | – <br> – | Consistent |
| 5. | Hold | $\lvert x_{ij} \rvert < \xi < \infty$ <br> $n^{-1}\,\mathbf{x}'\mathbf{x} \to \mathbf{Q}$ positive definite | $\varepsilon_i$'s are *iid* <br> $P(\lvert \varepsilon_i \rvert < m) = 1$ | $\overset{a}{\sim} N(\beta, \sigma^2(\mathbf{x}'\mathbf{x})^{-1})$ | – |
| 6. | Hold | $\lvert x_{ij} \rvert < \xi < \infty$ <br> $\det(n^{-1}\,\mathbf{x}'\mathbf{x}) > \eta > 0$ <br> $\{n^{-1}\mathbf{x}'\mathbf{x}\}$ is O(1) | $\varepsilon_i$'s are independent <br> $E(\varepsilon_i^4) \leq \tau < \infty$ | | |
| 7. | Hold | – | $\varepsilon_i$'s are *iid*, $E(\varepsilon_i^4) \leq \tau < \infty$ | – | $\overset{a}{\sim} N(\sigma^2, n^{-1}(\mu'_4 - \sigma^4))$ |
| 8. | Hold | – | $\varepsilon_i$'s are independent <br> $E(\lvert \varepsilon_i^4 - \sigma^2 \rvert^{2+\delta}) \leq \tau < \infty, \delta > 0$ <br> $\xi_n = n^{-1} \sum_{i=1}^{n} \text{var}(\varepsilon_i^4) \geq \eta > 0$ | – | $\overset{a}{\sim} N(\sigma^2, n^{-1}\,\xi_n)$ |

She specifies the dependent variable, $\mathbf{Y}$, as the number of new homes sold, and the explanatory variable matrix, $\mathbf{x}$, contains a column of 1's (for the intercept), followed by the remaining four variables in the table above. That is, $\mathbf{x}[.,1] = 1$, $\mathbf{x}[.,2]$ = Conventional Mortgage Interest Rates, $\mathbf{x}[.,3]$ = Median Family Incomes, $\mathbf{x}[.,4]$ = New Home Purchase Prices, $\mathbf{x}[.,5]$ = CPIs, and the GLM is given by $\mathbf{Y} = \sum_{i=1}^{5} \beta_i \mathbf{x}[.,i] + \boldsymbol{\varepsilon}$. She calculates the following values on her personal computer:

$$\mathbf{b} = (\mathbf{x}'\mathbf{x})^{-1}\mathbf{x}'\mathbf{y} = \begin{bmatrix} 1074.82 \\ -46.01 \\ .0376 \\ -6.16 \\ -2.03 \end{bmatrix}, \quad \hat{s}^2 = (\mathbf{y} - \mathbf{x}\mathbf{b})'(\mathbf{y} - \mathbf{x}\mathbf{b})/5 = 905.32, \quad R^2 = .96.$$

The linear model thus explains 96 percent of the observed sample variation in new homes sold in terms of the values of mortgage interest rates, median family incomes, home prices, and the general cost of living. An estimate of the covariance matrix of the entries in $\hat{\boldsymbol{\beta}}$ is calculated as

$$\hat{s}^2(\mathbf{x}'\mathbf{x})^{-1} = \begin{bmatrix} 79586.35 & -3389.91 & -5.9323 & 273.0789 & 896.8833 \\ & 191.4445 & .4094 & -14.5792 & -81.6034 \\ & & .0013 & -.0538 & -.2665 \\ & \text{(symmetric)} & & 3.3401 & 9.5606 \\ & & & & 60.3537 \end{bmatrix}.$$

The entries in $\mathbf{b}$ divided by their respective estimated standard deviations are given as (3.80, −3.33, 1.05, −3.37, −.26). Thus, the parameters associated with the intercept, interest rates, and new home prices are each more than three standard deviations away from zero (further insight into the significance of this observation will be provided in Chapter 10).

The realtor calculates estimates of elasticities of home sales with respect to each of the explanatory variables, other than the intercept, evaluated at the means of the observations, as

$$e_{\text{sales,rate}} = b_2 \frac{\bar{x}_2}{\bar{y}} = -46.01\left(\frac{12.65}{609}\right) = -.96$$

$$e_{\text{sales,income}} = b_3 \frac{\bar{x}_3}{\bar{y}} = .0376\left(\frac{27,242}{609}\right) = 1.68,$$

$$e_{\text{sales,price}} = b_4 \frac{\bar{x}_4}{\bar{y}} = -6.16\left(\frac{113.1}{609}\right) = -1.14,$$

$$e_{\text{sales,CPI}} = b_5 \frac{\bar{x}_5}{\bar{y}} = -2.03\left(\frac{104.6}{609}\right) = -.35.$$

Based on the estimated elasticities, home sales appear to be quite responsive to changes in interest rates, income levels, and home prices. Response to changes in the general cost of living is notably inelastic. □

### 8.2.3 Violations of Classic GLM Assumptions

The Classical assumptions of the GLM given in Definition 8.1 form essentially the base-level set of assumptions on which useful properties of the least squares estimator depend. It is instructive to examine the effect that a violation in each of the Classical assumptions has on the basic estimator properties of unbiasedness, BLUE, and consistency.

*Assumption Violation:* $E(\boldsymbol{\varepsilon}) \neq \mathbf{0}$ In this section we address implications of $E(\boldsymbol{\varepsilon}) \neq \mathbf{0}$, but we retain the other classical GLM assumptions. Note the covariance assumption must then be restated to accommodate the nonzero mean of $\boldsymbol{\varepsilon}$, as $\mathbf{Cov}(\mathbf{Y}) = \sigma^2\mathbf{I} = \mathbf{Cov}(\boldsymbol{\varepsilon}) = E(\boldsymbol{\varepsilon} - E(\boldsymbol{\varepsilon}))(\boldsymbol{\varepsilon} - E(\boldsymbol{\varepsilon}))'$.

$\hat{\boldsymbol{\beta}}$ Properties

If $E(\boldsymbol{\varepsilon}) = \boldsymbol{\phi} \neq \mathbf{0}$, then since $\hat{\boldsymbol{\beta}} = \boldsymbol{\beta} + (\mathbf{x}'\mathbf{x})^{-1}\mathbf{x}'\boldsymbol{\varepsilon}$, $E(\hat{\boldsymbol{\beta}}) = \boldsymbol{\beta} + (\mathbf{x}'\mathbf{x})^{-1}\mathbf{x}'\boldsymbol{\phi}$. Except for the special case where $\mathbf{x}'\boldsymbol{\phi} = \mathbf{0}$, the estimator $\hat{\boldsymbol{\beta}}$ is a *biased* estimator of $\boldsymbol{\beta}$ since $(\mathbf{x}'\mathbf{x})^{-1}\mathbf{x}'\boldsymbol{\phi} = \mathbf{0}$ *iff* $\mathbf{x}'\boldsymbol{\phi} = (\mathbf{x}'\mathbf{x}) \cdot \mathbf{0} = \mathbf{0}$. If $\mathbf{x}'\boldsymbol{\phi} \neq \mathbf{0}$, then $\hat{\boldsymbol{\beta}}$ is clearly *not* BLUE since $\hat{\boldsymbol{\beta}}$ is not even unbiased. In the special case where $\mathbf{x}'\boldsymbol{\phi} = \mathbf{0}$, $\hat{\boldsymbol{\beta}}$ retains the BLUE property, as can be verified by repeating the previous derivation of the BLUE for $\boldsymbol{\beta}$ using the conditions $E(\boldsymbol{\varepsilon}) = \boldsymbol{\phi}$ and $\mathbf{x}'\boldsymbol{\phi} = \mathbf{0}$.

Regarding consistency, it follows from $\hat{\boldsymbol{\beta}} = \boldsymbol{\beta} + (\mathbf{x}'\mathbf{x})^{-1}\mathbf{x}'\boldsymbol{\varepsilon}$, that plim $\hat{\boldsymbol{\beta}} = \boldsymbol{\beta}$ *iff* $\text{plim}((\mathbf{x}'\mathbf{x})^{-1}\mathbf{x}'\boldsymbol{\varepsilon}) = \mathbf{0}$. If we assume as before that $(\mathbf{x}'\mathbf{x})^{-1} \to \mathbf{0}$ when $n \to \infty$ (Table 8.1, Condition 2), then since $E((\mathbf{x}'\mathbf{x})^{-1}\mathbf{x}'\boldsymbol{\varepsilon} - (\mathbf{x}'\mathbf{x})^{-1}\mathbf{x}'\boldsymbol{\phi}) = \mathbf{0}$ and $\mathbf{Cov}((\mathbf{x}'\mathbf{x})^{-1}\mathbf{x}'\boldsymbol{\varepsilon}) = \sigma^2(\mathbf{x}'\mathbf{x})^{-1} \to \mathbf{0}$ as $n \to \infty$ it follows from mean square convergence that $\text{plim}((\mathbf{x}'\mathbf{x})^{-1}\mathbf{x}'\boldsymbol{\varepsilon} - (\mathbf{x}'\mathbf{x})^{-1}\mathbf{x}'\boldsymbol{\phi}) = \mathbf{0}$. Thus, if $\lim_{n\to\infty}((\mathbf{x}'\mathbf{x})^{-1}\mathbf{x}'\boldsymbol{\phi}) = \boldsymbol{\xi} \neq \mathbf{0}$, or if $(\mathbf{x}'\mathbf{x})^{-1}\mathbf{x}'\boldsymbol{\phi}$ does not converge to a limit, then $\hat{\boldsymbol{\beta}}$ would *not* be consistent for $\boldsymbol{\beta}$ since $\text{plim}((\mathbf{x}'\mathbf{x})^{-1}\mathbf{x}'\boldsymbol{\varepsilon}) \neq \mathbf{0}$ and $\text{plim}(\hat{\boldsymbol{\beta}}) \neq \boldsymbol{\beta}$.

$\hat{S}^2$ Properties

Since $\hat{S}^2 = (n-k)^{-1}\,\text{tr}((\mathbf{I} - \mathbf{x}(\mathbf{x}'\mathbf{x})^{-1}\mathbf{x}')\boldsymbol{\varepsilon}\boldsymbol{\varepsilon}')$, $\mathbf{Cov}(\boldsymbol{\varepsilon}) = \sigma^2\mathbf{I}$, and $E(\boldsymbol{\varepsilon}\boldsymbol{\varepsilon}') = \sigma^2\mathbf{I} + \boldsymbol{\phi}\boldsymbol{\phi}'$, the expectation of $\hat{S}^2$ is given by

$$E\left(\hat{S}^2\right) = \sigma^2 + \frac{\boldsymbol{\phi}'\left(\mathbf{I} - \mathbf{x}(\mathbf{x}'\mathbf{x})^{-1}\mathbf{x}'\right)\boldsymbol{\phi}}{n-k}.$$

Unless $\boldsymbol{\phi}'(\mathbf{I} - \mathbf{x}(\mathbf{x}'\mathbf{x})^{-1}\mathbf{x}')\boldsymbol{\phi} = \mathbf{0}$, the estimator $\hat{S}^2$ is a *biased* estimator of $\sigma^2$ and the bias is nonnegative, since the matrix $\mathbf{I} - \mathbf{x}(\mathbf{x}'\mathbf{x})^{-1}\mathbf{x}'$ is symmetric and idempotent, and hence positive semidefinite.[12] Note that if $E(\boldsymbol{\varepsilon}) = \boldsymbol{\phi} \neq \mathbf{0}$, it is *impossible* for both $\hat{\boldsymbol{\beta}}$ and $\hat{S}^2$ to remain unbiased. To see this, first note that if $\mathbf{x}'\boldsymbol{\phi} = \mathbf{0}$, so that $\hat{\boldsymbol{\beta}}$ is unbiased, then the bias in $\hat{S}^2$ is $(E(\hat{S}^2) - \sigma^2) = \boldsymbol{\phi}'\boldsymbol{\phi}/(n-k) > 0$. Alternatively, if $\hat{S}^2$ is unbiased, then it must be the case that $\boldsymbol{\phi}'\boldsymbol{\phi} = \boldsymbol{\phi}'\mathbf{x}(\mathbf{x}'\mathbf{x})^{-1}\mathbf{x}'\boldsymbol{\phi}$

[12] Positive semidefiniteness can be deduced from the fact that the characteristic roots of a symmetric idempotent matrix are all nonnegative, being a collection of 0's and 1's.

$> 0$, and since $\boldsymbol{\phi}'\mathbf{x}(\mathbf{x}'\mathbf{x})^{-1}\mathbf{x}'\boldsymbol{\phi}$ is a positive *definite* quadratic form in the vector $\mathbf{x}'\boldsymbol{\phi}$ (i.e., $(\mathbf{x}'\mathbf{x})^{-1}$ is positive definite), it necessarily follows that $\mathbf{x}'\boldsymbol{\phi} \neq \mathbf{0}$ if $\boldsymbol{\phi}'\mathbf{x}(\mathbf{x}'\mathbf{x})^{-1}\mathbf{x}'\boldsymbol{\phi}$ is positive, and thus $\hat{\boldsymbol{\beta}}$ is biased. Thus at least one, and generally both of the estimators $\hat{\boldsymbol{\beta}}$ and $\hat{S}^2$ will be biased if $\mathrm{E}(\boldsymbol{\varepsilon}) \neq \mathbf{0}$.

Regarding consistency, define $\mathbf{V} = \boldsymbol{\varepsilon} - \boldsymbol{\delta}$, and note that $\hat{S}^2 = (n-k)^{-1}(\boldsymbol{\phi} + \mathbf{V})'(\mathbf{I} - \mathbf{x}(\mathbf{x}'\mathbf{x})^{-1}\mathbf{x}')(\boldsymbol{\phi} + \mathbf{V})$ where $\mathrm{E}(\mathbf{V}) = \mathbf{0}$ and $\mathbf{Cov}(\mathbf{V}) = \sigma^2\mathbf{I}$. Letting $\mathbf{M} = \mathbf{I} - \mathbf{x}(\mathbf{x}'\mathbf{x})^{-1}\mathbf{x}'$, it follows that $\hat{S}^2 = (n-k)^{-1}[\boldsymbol{\phi}'\mathbf{M}\boldsymbol{\phi} - 2\boldsymbol{\phi}'\mathbf{M}\mathbf{V} + \mathbf{V}'\mathbf{M}\mathbf{V}]$, so that $\hat{S}^2$ is consistent *iff* the expression on the right hand side of the equality has a probability limit of $\sigma^2$. To illustrate that this is generally not true, assume that the $\mathbf{V}_i$'s are *iid*, which is tantamount to assuming that the $\varepsilon_i$'s are independent and identically distributed once they are transformed to have mean zero, as $\varepsilon_i - \phi_i$. Then $\mathrm{plim}((n-k)^{-1}\mathbf{V}'\mathbf{M}\mathbf{V}) = \sigma^2$, which follows from an argument identical to that used in the proof of Theorem 8.5. It follows that $\hat{S}^2$ will be consistent for $\sigma^2$ *iff* $(n-k)^{-1}[\boldsymbol{\phi}'\mathbf{M}\boldsymbol{\phi} - 2\boldsymbol{\phi}'\mathbf{M}\mathbf{V}] \xrightarrow{p} 0$.

Now assume $|\phi_i| \leq c < \infty$, so that the means of the residuals are finite and $\boldsymbol{\phi}'\boldsymbol{\phi} = O(n^1)$, or at least assume that $\boldsymbol{\phi}'\boldsymbol{\phi}$ is $o(n^2)$, implying that the sum of the squared means approaches infinity at a rate less than $n^2$. It follows that $\boldsymbol{\phi}'\mathbf{M}\boldsymbol{\phi}$ is $o(n^2)$, since $0 \leq \boldsymbol{\phi}'\mathbf{M}\boldsymbol{\phi} = \boldsymbol{\phi}'\boldsymbol{\phi} - \boldsymbol{\phi}'\mathbf{x}(\mathbf{x}'\mathbf{x})^{-1}\mathbf{x}'\boldsymbol{\phi} \leq \boldsymbol{\phi}'\boldsymbol{\phi}$. Then since $\mathrm{E}(2(n-k)^{-1}\boldsymbol{\phi}'\mathbf{M}\mathbf{V}) = 0$ and $\mathrm{var}(2(n-k)^{-1}\boldsymbol{\phi}'\mathbf{M}\mathbf{V}) = 4\sigma^2\boldsymbol{\phi}'\mathbf{M}\boldsymbol{\phi}/(n-k)^2 \leq 4\sigma^2 o(n^2)/(n-k)^2 \rightarrow 0$ as $n \rightarrow \infty$, $\mathrm{plim}(2(n-k)^{-1}\boldsymbol{\delta}'\mathbf{M}\mathbf{V}) = 0$ follows from mean square convergence. Thus, except for the special case $\lim_{n\rightarrow\infty}(n-k)^{-1}\boldsymbol{\delta}'\mathbf{M}\boldsymbol{\phi} = 0$, $\hat{S}^2$ will *not* be consistent for $\sigma^2$.

*Assumption Violation:* $\mathrm{E}(\boldsymbol{\varepsilon}\boldsymbol{\varepsilon}') \neq \boldsymbol{\sigma}^2\mathbf{I}$ In this section we examine the impacts of having $\mathrm{E}(\boldsymbol{\varepsilon}\boldsymbol{\varepsilon}') = \boldsymbol{\Phi} \neq \sigma^2\mathbf{I}$, so that the error terms are **heteroskedastic** (unequal variances) and/or **autocorrelated** (nonzero correlations between some $\varepsilon_i$ and $\varepsilon_j$'s, $i \neq j$)

$\hat{\boldsymbol{\beta}}$ Properties

The estimator $\hat{\boldsymbol{\beta}}$ remains *unbiased* for $\boldsymbol{\beta}$ regardless of heteroskedasticity or autocorrelation so long as the remaining Classical assumptions of the GLM remain true. This follows because the representation $\hat{\boldsymbol{\beta}} = \boldsymbol{\beta} + (\mathbf{x}'\mathbf{x})^{-1}\mathbf{x}'\boldsymbol{\varepsilon}$ is unaffected, and taking expectations still yields $\mathrm{E}(\hat{\boldsymbol{\beta}}) = \boldsymbol{\beta} + (\mathbf{x}'\mathbf{x})^{-1}\mathbf{x}'\mathrm{E}(\boldsymbol{\varepsilon}) = \boldsymbol{\beta}$.[13]

To examine consistency, note that now $\mathbf{Cov}(\hat{\boldsymbol{\beta}}) = (\mathbf{x}'\mathbf{x})^{-1}\mathbf{x}'\boldsymbol{\phi}\mathbf{x}(\mathbf{x}'\mathbf{x})^{-1}$. If it were true that $\mathrm{var}(\hat{\beta}_i) \rightarrow \mathbf{0}$ as $n \rightarrow \infty\ \forall i$, then since $\mathrm{E}(\hat{\boldsymbol{\beta}}) = \boldsymbol{\beta}$, we could conclude by convergence in mean square that $\mathrm{plim}(\hat{\boldsymbol{\beta}}) = \boldsymbol{\beta}$. Under mild conditions on the $\boldsymbol{\phi}$ matrix, and assuming as before that $(\mathbf{x}'\mathbf{x})^{-1} \rightarrow \mathbf{0}$, convergence of $\mathrm{var}(\hat{\beta}_i)$ to zero $\forall i$ will be achieved. The argument is facilitated by the following matrix theory lemma:

---

[13]We are making the tacit assumption that x contains no lagged values of the dependent variable, which is as it must be if it is presumed that $\mathbf{x}$ can be held fixed. In the event that $\mathbf{x}$ contains lagged values of the dependent variable and the error terms are autocorrelated, then in general $\mathrm{E}((\mathbf{X}'\mathbf{X})^{-1}\mathbf{X}'\boldsymbol{\varepsilon}) \neq \mathbf{0}$, and $\hat{\boldsymbol{\beta}}$ is biased. Issues related to this case are discussed in subsection 8.2.3.

**Lemma 8.3**

Let $A$ and $B$ be symmetric positive semi-definite matrices of order $(n \times n)$. Then $\text{tr}(\mathbf{AB}) \leq \lambda_L(\mathbf{A})\ \text{tr}(\mathbf{B})$, where $\lambda_L(\mathbf{A})$ represents the value of the largest characteristic root of the matrix $\mathbf{A}$.

**Proof**: Let $\mathbf{P}$ be the $(n \times n)$ characteristic vector matrix of the symmetric matrix $\mathbf{A}$, so that $\mathbf{P}'\mathbf{AP} = \mathbf{\Lambda}$, where $\mathbf{\Lambda}$ is the diagonal matrix of characteristic roots of $\mathbf{A}$. Since $\mathbf{PP}' = \mathbf{I}$ by the orthogonality of $\mathbf{P}$, it follows by a property of the trace operator $(\text{tr}(\mathbf{DFG}) = \text{tr}(\mathbf{GDF}))$ that $\text{tr}(\mathbf{AB}) = \text{tr}(\mathbf{APP'BPP'}) = \text{tr}\ (\mathbf{P'APP'BP}) = \text{tr}\ (\mathbf{\Lambda C})$, where $\mathbf{C} = \mathbf{P'BP}$. But then $\text{tr}\ (\mathbf{\Lambda C}) = \sum_{i=1}^{n} \lambda_i C_{ii} \leq \lambda_L(\mathbf{A}) \sum_{i=1}^{n} C_{ii}$ where $\lambda_i$ is the $i$th diagonal entry in the diagonal matrix of characteristic roots $\mathbf{\Lambda}$, $C_{ii}$ represents the $(i,i)$th entry in $\mathbf{C}$ and $\lambda_i \geq 0$ and $C_{ii} \geq 0\ \forall i$ by the positive semidefiniteness of $\mathbf{A}$, $\mathbf{B}$, and $\mathbf{P'BP}$. But since $\sum_{i=1}^{n} C_{ii} = \text{tr}(\mathbf{C}) = \text{tr}(\mathbf{P'BP}) = \text{tr}(\mathbf{BPP'}) = \text{tr}(\mathbf{B})$, we have that $\text{tr}(\mathbf{AB}) \leq \lambda_L(\mathbf{A})\ \text{tr}(\mathbf{B})$. ■

Using the lemma, and properties of the trace operator, note that

$$\text{tr}(\mathbf{x}'\mathbf{x})^{-1}\mathbf{x}'\mathbf{\Phi}\mathbf{x}(\mathbf{x}'\mathbf{x})^{-1} = \text{tr}\ \mathbf{\Phi}\mathbf{x}(\mathbf{x}'\mathbf{x})^{-2}\mathbf{x}' \leq \lambda_L(\mathbf{\Phi})\ \text{tr}\ \mathbf{x}(\mathbf{x}'\mathbf{x})^{-2}\mathbf{x}' = \lambda_L(\mathbf{\Phi})\ \text{tr}(\mathbf{x}'\mathbf{x})^{-1}.$$

Thus, if $\lambda_L(\mathbf{\Phi}) < \tau < \infty$, i.e. if the largest characteristic root of $\mathbf{\Phi}$ is bounded, then since $\text{tr}(\mathbf{x}'\mathbf{x})^{-1} \to 0$ as $n \to \infty$, $\text{tr}(\mathbf{x}'\mathbf{x})^{-1}\ \mathbf{x}'\ \mathbf{\Phi}\ \mathbf{x}(\mathbf{x}'\mathbf{x})^{-1} \to 0$ as $n \to \infty$. It follows that all of the diagonal entries in $(\mathbf{x}'\mathbf{x})^{-1}\ \mathbf{x}'\ \mathbf{\Phi}\ \mathbf{x}(\mathbf{x}'\mathbf{x})^{-1}$ must converge to zero, since the $(k \times k)$ matrix is (at least) positive semidefinite and thus has nonnegative diagonal entries, and the sum of these $k$ nonnegative numbers converging to zero requires that each diagonal entry converge to zero. Thus, by convergence in mean square, $\text{plim}\ (\hat{\boldsymbol{\beta}}) = \boldsymbol{\beta}$ even if $\text{E}(\boldsymbol{\varepsilon\varepsilon}') = \mathbf{\Phi} \neq \sigma^2\mathbf{I}$.

As a practical matter, the assumption that $\lambda_L(\mathbf{\Phi})$ is bounded is not very restrictive. The following matrix theory lemma will be useful for establishing sufficient conditions for the boundedness of $\lambda_L(\mathbf{\Phi})$ and the consistency of $\hat{\boldsymbol{\beta}}$.

**Lemma 8.4**

The absolute value of any characteristic root of a matrix, $\mathbf{A}$, is less than or equal to the sum of the absolute values of the elements in the row of $\mathbf{A}$ for which the sum is largest.

See Hammarling, S.J. (1970), *Latent Roots and Latent Vectors*, Univ. of Toronto Press, p. 9.

Using Lemma 8.4, we present three different sufficient conditions on the covariance matrix $\mathbf{\Phi}$ that ensure the boundedness of $\lambda_L(\mathbf{\Phi})$ and thus the consistency of $\hat{\boldsymbol{\beta}}$. Assume that $\text{var}(\varepsilon_i) = \sigma_{ii} < \xi < \infty\ \forall i$, so that the variances of the disturbance terms exhibit some (perhaps very large) upper bound.

***Consistency Sufficient Condition 1:*** As a first sufficient condition, if the disturbance vector is *heteroskedastic*, but there is *zero autocorrelation*, then it is immediate that $\lambda_L(\mathbf{\Phi}) < \xi$, since $\mathbf{\Phi}$ is a diagonal matrix of variances, and the characteristic roots of a diagonal matrix are directly the diagonal elements. Thus, $\lambda_L(\mathbf{\Phi})$ is bounded and $\text{plim}\ (\hat{\boldsymbol{\beta}}) = \boldsymbol{\beta}$.

***Consistency Sufficient Condition 2:*** If the disturbance vector is either *heteroskedastic* or *homoskedastic* and exhibits *nonzero autocorrelation*, then $\lambda_L(\mathbf{\Phi})$ will be bounded if the covariance terms in $\mathbf{\Phi}$ decline sufficiently fast as $|i-j|$ increases. The most straightforward sufficient condition to examine in this case is $\sigma_{ij}=0$ for $|i-j|>m$, where $m$ is some (perhaps very large) positive integer $m$, which is true if the $\varepsilon_i$'s are $m$-dependent (recall Definition 5.10). Then let row $i$ of $\mathbf{\Phi}$ be the row for which the sum of the absolute values of the elements in the row is largest in comparison with the other rows of $\mathbf{\Phi}$. Because $\sigma_{ij}=0$ when $|i-j|>m$, the *maximum* number of nonzero entries in row $i$ is $2\ m+1$. Also by the bound on covariances, $|\sigma_{ij}|\leq|\sigma_{ii}|^{1/2}|\sigma_{jj}|^{1/2}\ \forall i,j$, it follows that the sum of the absolute values of the entries in the $i$th row of the covariance matrix $\boldsymbol{\phi}$ is upper-bounded by $(2m+1)\ \xi$, and thus by Lemma 8.4, $\lambda_L(\mathbf{\Phi})<(2\ m+1)\xi$. Then since $\lambda_L(\mathbf{\Phi})$ is bounded, $\text{plim}(\hat{\boldsymbol{\beta}})=\boldsymbol{\beta}$.

***Consistency Sufficient Condition 3:*** Suppose that $\sigma_{ij}\neq 0$ and there is no value of $|i-j|$ beyond which the covariance is assumed to be zero. Then again using Lemma 8.4, a third sufficient condition for $\lambda_L(\mathbf{\Phi})<\tau<\infty$ is that $\sum_{j=1}^{n}|\sigma_{ij}|<\tau<\infty\ \forall i$ and $\forall n$. Since we are assuming $\sigma_{ii}<\xi\ \forall i$, the sufficient condition can alternatively be stated in terms of the boundedness of $\sum_{j\neq i}|\sigma_{ij}|\ \forall i$. Thus, if covariances decline sufficiently fast so that $\sum_{j\neq i}|\sigma_{ij}|<\eta<\infty\ \forall i$, then $\text{plim}\ (\hat{\boldsymbol{\beta}})=\boldsymbol{\beta}$.

We should note that the consistency of $\hat{\boldsymbol{\beta}}$ does *not require* that $\lambda_L(\mathbf{\Phi})$ be bounded. It is sufficient that $\lambda_L(\mathbf{\Phi})$ increase at a rate slower than the rate at which $\text{tr}(\mathbf{x}'\mathbf{x})^{-1}\rightarrow 0$, for then it would still be true that $\lambda_L(\mathbf{\Phi})\ \text{tr}(\mathbf{x}'\mathbf{x})^{-1}\rightarrow 0$. For additional results on the consistency of $\hat{\boldsymbol{\beta}}$ the reader is referred to the book by H. White, *Asymptotic Theory.*

Having seen that $\hat{\boldsymbol{\beta}}$ is unbiased and consistent for $\boldsymbol{\beta}$ under general conditions on $\text{E}(\boldsymbol{\varepsilon}\boldsymbol{\varepsilon}')=\mathbf{\Phi}$, one may wonder whether $\hat{\boldsymbol{\beta}}$ retains the BLUE property. In the general (and usual) case where $\mathbf{\Phi}$ is unknown, *no* linear estimator exists that has the BLUE property. To see this, note that if $\text{E}(\boldsymbol{\varepsilon}\boldsymbol{\varepsilon}')=\mathbf{\Phi}$, then

$$\mathbf{\Phi}^{-1/2}\mathbf{Y}=\mathbf{\Phi}^{-1/2}\mathbf{x}'\boldsymbol{\beta}+\mathbf{\Phi}^{-1/2}\boldsymbol{\varepsilon}\ \text{or}\ \mathbf{Y}_*=\mathbf{x}_*'\boldsymbol{\beta}+\boldsymbol{\varepsilon}_*$$

satisfies the Assumptions of the Classical GLM (assuming all of the classical assumptions *except* $\mathbf{Cov}(\mathbf{Y})=\mathbf{Cov}(\boldsymbol{\varepsilon})=\sigma^2\mathbf{I}$ apply to $\mathbf{Y}=\mathbf{x}\boldsymbol{\beta}+\boldsymbol{\varepsilon}$). In particular, note that $\text{E}(\boldsymbol{\varepsilon}_*\boldsymbol{\varepsilon}_*')=\mathbf{\Phi}^{-1/2}\ \text{E}(\boldsymbol{\varepsilon}\boldsymbol{\varepsilon}')\mathbf{\Phi}^{-1/2}=\mathbf{\Phi}^{-1/2}\ \mathbf{\Phi}\ \mathbf{\Phi}^{-1/2}=\mathbf{I}$ and $\text{E}(\boldsymbol{\varepsilon}_*)=\text{E}(\mathbf{\Phi}^{-1/2}\ \boldsymbol{\varepsilon})=\mathbf{\Phi}^{-1/2}\ \text{E}(\boldsymbol{\varepsilon})=\mathbf{0}$. Then the Gauss-Markov theorem applied to the transformed linear model implies that

$$\hat{\boldsymbol{\beta}}=(\mathbf{x}_*'\mathbf{x}_*)^{-1}\mathbf{x}_*'\mathbf{Y}_*=(\mathbf{x}'\ \mathbf{\Phi}^{-1}\mathbf{x})^{-1}\mathbf{x}'\ \mathbf{\Phi}^{-1}\mathbf{Y}$$

would be the BLUE for $\boldsymbol{\beta}$. However, the definition of the BLUE estimator depends on the unknown value of $\mathbf{\Phi}$, and because there does not then exist a fixed choice of $\mathbf{A}$ and $\mathbf{b}$ such that $\mathbf{AY}+\mathbf{d}$ is BLUE for all potential values of $\mathbf{\Phi}$, no BLUE estimator of $\boldsymbol{\beta}$ exists when $\mathbf{\Phi}$ is unknown.

In the *very* special case where $\mathbf{\Phi}$ is known up to a scalar multiple, i.e., $\mathbf{\Phi}=\sigma^2\ \mathbf{\Omega}$ with $\mathbf{\Omega}$ *known*, then the linear estimator $\hat{\boldsymbol{\beta}}^*=(\mathbf{x}'\ \mathbf{\Phi}^{-1}\mathbf{x})^{-1}\mathbf{x}'\ \mathbf{\Phi}^{-1}\mathbf{Y}=(\mathbf{x}'\mathbf{\Omega}^{-1}\mathbf{x})^{-1}\ \mathbf{x}'\mathbf{\Omega}^{-1}\mathbf{Y}$ is the BLUE estimator. This special estimator is referred to in the literature as the **generalized least squares estimator** of $\boldsymbol{\beta}$.

### $\hat{S}^2$ Properties

The estimator $\hat{S}^2$ under the condition $E(\boldsymbol{\varepsilon}\boldsymbol{\varepsilon}') = \boldsymbol{\Phi} \neq \sigma^2\mathbf{I}$ is only useful if there exists a counterpart to the parameter $\sigma^2$ which can be estimated. Two such situations arise when $\boldsymbol{\Phi} = \sigma^2\,\boldsymbol{\Omega}$, and either $\boldsymbol{\Omega}$ is a known positive definite symmetric matrix, or else $\boldsymbol{\varepsilon}$ is homoskedastic with common variance $\sigma^2$. In either case,

$$E\left(\hat{S}^2\right) = (n-k)^{-1}E\left(\boldsymbol{\varepsilon}'\left(\mathbf{I}-\mathbf{x}(\mathbf{x}'\mathbf{x})^{-1}\mathbf{x}'\right)\boldsymbol{\varepsilon}\right) = (n-k)^{-1}\mathrm{tr}\left(\left(\mathbf{I}-\mathbf{x}(\mathbf{x}'\mathbf{x})^{-1}\mathbf{x}'\right)E(\boldsymbol{\varepsilon}\boldsymbol{\varepsilon}')\right)$$
$$= \sigma^2\mathrm{tr}\left(\frac{\boldsymbol{\Omega}\left(\mathbf{I}-\mathbf{x}(\mathbf{x}'\mathbf{x})^{-1}\mathbf{x}'\right)}{(n-k)}\right)$$

and thus $\hat{S}^2$ is generally *biased* as an estimator of $\sigma^2$, since the trace of the matrix in the preceding expression will generally not be equal to 1. Of course, in the *very* special where $\boldsymbol{\Omega}$ were known, $\hat{S}^2$ could be scaled by the known value of the trace to define an unbiased estimator of $\sigma^2$.

In certain cases, the bias in $\hat{S}^2$ converges to 0 as $n \to \infty$. Note that Lemma 8.3 implies that

$$\sigma^2\mathrm{tr}\left(\frac{\left(\boldsymbol{\Omega}\mathbf{x}(\mathbf{x}'\mathbf{x})^{-1}\mathbf{x}'\right)}{(n-k)}\right) \leq \sigma^2\lambda_L(\boldsymbol{\Omega})\left(\frac{k}{(n-k)}\right),$$

and if $\lambda_L(\boldsymbol{\Phi})$ is bounded or at least $o(n^1)$ as we have argued above, then $\sigma^2\lambda_L(\boldsymbol{\Omega}) = \lambda_L(\boldsymbol{\Phi})$[14] is bounded and $\sigma^2\,\lambda_L(\boldsymbol{\Omega})(k/(n-k)) \to 0$ as $n \to \infty$. Then since $\mathrm{tr}(\boldsymbol{\Omega}\mathbf{x}(\mathbf{x}'\mathbf{x})^{-1}\mathbf{x}') \geq 0$,[15] it must be the case that $\sigma^2\,\mathrm{tr}(\boldsymbol{\Omega}\mathbf{x}(\mathbf{x}'\mathbf{x})^{-1}\mathbf{x}')/(n-k) \to 0$ as $n \to \infty$. It follows that $E(\hat{S}^2) \to \sigma^2$ as $n \to \infty$ *iff* $\mathrm{tr}(\boldsymbol{\Omega}/(n-k)) \to 1$ as $n \to \infty$, which *does* occur if $\boldsymbol{\varepsilon}$ is homoskedastic ($\boldsymbol{\varepsilon}$ may still exhibit nonzero autocorrelation) with $\sigma^2$ representing the common variance of the $\varepsilon_i$'s, for then $\mathrm{tr}(\boldsymbol{\Omega}) = n$, since the diagonal entries of $\boldsymbol{\Omega}$ would all be equal to the number 1, and then $n/(n-k) \to 1$ as $n \to \infty$. In the heteroskedastic case, there is no reason to expect that the preceding condition will hold.

Regarding consistency of $\hat{S}^2$ for $\sigma^2$, Markov's inequality can be used as in the proof of Theorem 8.5 to show that $\mathrm{plim}((n-k)^{-1}\,\boldsymbol{\varepsilon}'\mathbf{x}(\mathbf{x}'\mathbf{x})^{-1}\mathbf{x}'\,\boldsymbol{\varepsilon}) = 0$, assuming $\lambda_L(\boldsymbol{\Phi})$ is bounded. If $\boldsymbol{\varepsilon}$ is *homoskedastic* with variance $\sigma^2$, and if the conditions of Theorem 8.6 other than $E(\boldsymbol{\varepsilon}\boldsymbol{\varepsilon}') = \sigma^2\mathbf{I}$ can be assumed to hold, then the weak law of large numbers given by Theorem 5.22 implies that $\mathrm{plim}((n-k)^{-1}\,\boldsymbol{\varepsilon}'\boldsymbol{\varepsilon}) = \mathrm{plim}\,(n^{-1}\boldsymbol{\varepsilon}'\boldsymbol{\varepsilon}) = \sigma^2$, and thus $\hat{S}^2$ would be consistent for $\sigma^2$ (consistency under alternative conditions on $\boldsymbol{\varepsilon}$ can be demonstrated as well—recall Table 8.1). If $\boldsymbol{\varepsilon}$ is

[14] The characteristic roots of $\tau\mathbf{A}$ are equal to the characteristic roots of $\mathbf{A}$ times the scalar $\tau$.

[15] Note $\mathrm{tr}(\boldsymbol{\Omega}\mathbf{x}(\mathbf{x}'\mathbf{x})^{-1}\mathbf{x}') = \mathrm{tr}(\boldsymbol{\Omega}^{1/2}\mathbf{x}(\mathbf{x}'\mathbf{x})^{-1}\mathbf{x}'\boldsymbol{\Omega}^{1/2})$, and since $(\mathbf{x}'\mathbf{x})^{-1}$ is positive definite, $\boldsymbol{\Omega}^{1/2}\,\mathbf{x}(\mathbf{x}'\mathbf{x})^{-1}\mathbf{x}'\,\boldsymbol{\Omega}^{1/2}$ is at least positive semidefinite and its trace must be nonnegative.

heteroskedastic with $\mathbf{Cov}(\boldsymbol{\varepsilon}) = \sigma^2\boldsymbol{\Omega}$, while maintaining the remaining preceding assumptions, then by Theorem 5.22

$$\text{plim}\left((n-k)^{-1}\left[\boldsymbol{\varepsilon}'\boldsymbol{\varepsilon} - \sum_{i=1}^{n} \sigma^2 \omega_{ii}\right]\right) = \text{plim}\left(n^{-1}\left[\boldsymbol{\varepsilon}'\boldsymbol{\varepsilon} - \sigma^2 \sum_{i=1}^{n} \omega_{ii}\right]\right) = 0$$

where $\omega_{ij}$ denotes the elements in $\boldsymbol{\Omega}$, and except for the very special case $\lim_{n\to\infty} n^{-1} \sum_{i=1}^{n} \omega_{ii} = 1$, $\hat{S}^2$ will not be consistent for $\sigma^2$.

*Assumption Violation: Rank* $(\mathbf{x}) < \mathbf{k}$, or $|\mathbf{x}'\mathbf{x}| \approx 0$ In this subsection we assume that either $\mathbf{x}$ is less than full column rank so that rank $\mathbf{x} < k$, or else there is a "near" linear dependency among the columns of $\mathbf{x}$, so that $|\mathbf{x}'\mathbf{x}| \approx 0$. Note that this latter assumption is *not* a violation of the assumptions of the Classical GLM per se, but rather "nearly" a violation. We retain the other assumptions of the classical GLM.

$\hat{\boldsymbol{\beta}}$ Properties

If the classical assumption concerning rank$(\mathbf{x}) = k$ is violated, then the least squares estimator $\hat{\boldsymbol{\beta}} = (\mathbf{x}'\mathbf{x})^{-1}\mathbf{x}'\mathbf{Y}$ does not exist, since $(\mathbf{x}'\mathbf{x})^{-1}$ does not exist. In this case, there are an *infinite* number of solutions to the problem of minimizing $(\mathbf{y} - \mathbf{xb})'(\mathbf{y} - \mathbf{xb})$ through choice of the vector $\mathbf{b}$. To see this, recall that the first order conditions for the minimization problem are given by $(\mathbf{x}'\mathbf{x})\,\mathbf{b} = \mathbf{x}'\mathbf{y}$, which is a system of $k$ equations in the $k$ unknowns represented by the vector $\mathbf{b}$. If $\mathbf{x}$ is less than full column rank, so that $(\mathbf{x}'\mathbf{x})$ is less than full rank, then this system of linear equations effectively contains one or more redundant equations, so that there are essentially more unknowns than equations. Then there are an infinite number of solution values for $\mathbf{b}$. In this case, the parameter vector $\boldsymbol{\beta}$ cannot be estimated uniquely. This problem is referred to in the literature as **perfect multicollinearity**.

More frequent in applications is the case where rank$(\mathbf{x}) = k$, but $\mathbf{x}'\mathbf{x}$ is *nearly singular*, i.e., its determinant is near zero. In such cases, $(\mathbf{x}'\mathbf{x})^{-1}$ tends to have large diagonal entries, implying that the variances of elements of $\hat{\boldsymbol{\beta}}$, given by the diagonal elements of $\sigma^2(\mathbf{x}'\mathbf{x})^{-1}$, are very large. This follows from our previous application of Lemma 8.2 on partitioned inversion to the matrix $(\mathbf{x}'\mathbf{x})^{-1}$. In particular, the lemma implies that the variance of $\hat{\beta}_i$(i.e., the $(i,i)$th entry in $\sigma^2(\mathbf{x}'\mathbf{x})^{-1}$) is given by $\sigma^2\ (\mathbf{x}_{\cdot i}'\mathbf{x}_{\cdot i} - \mathbf{x}_{\cdot i}'\ \mathbf{x}_{*}(\mathbf{x}_{*}'\mathbf{x}_{*})^{-1}\ \mathbf{x}_{*}'\ \mathbf{x}_{\cdot i})^{-1} = \sigma^2(\mathbf{e}_i'\mathbf{e}_i)^{-1}$, where $\mathbf{e}_i$ represents a vector of deviations of the $i$th column of $\mathbf{x}$, represented here by $\mathbf{x}_{\cdot i}$, from a least squares prediction of $\mathbf{x}_{\cdot i}$ based on a linear combination of the columns of $\mathbf{x}$ other than $\mathbf{x}_{\cdot i}$ (these remaining $(k-1)$ columns being represented by $\mathbf{x}_{*}$). The more closely $\mathbf{x}_{\cdot i}$ can be approximated by $\mathbf{x}_{*}\hat{\mathbf{b}}_{*}$ (where $\hat{\mathbf{b}}_{*} = (\mathbf{x}_{*}'\mathbf{x}_{*})^{-1}\ \mathbf{x}_{*}'\mathbf{x}_{\cdot i})$, and thus the more closely that $\mathbf{x}_{\cdot i}$ is linearly related to the remaining column vectors in the $\mathbf{x}$ matrix, the smaller is $\mathbf{e}_i$ and $\mathbf{e}_i'\mathbf{e}_i$, and thus the larger is var$(\hat{\beta}_i) = \sigma^2(\mathbf{e}_i'\mathbf{e}_i)^{-1}$. Note that so long as rank$(\mathbf{x}) = k$, and the appropriate other assumptions hold, $\hat{\boldsymbol{\beta}}$ remains BLUE and consistent as an estimator of $\boldsymbol{\beta}$. However, the large variances associated with the $\hat{\beta}_i$'s imply that in small samples, outcomes of $\hat{\boldsymbol{\beta}}$ can be quite distant from $\boldsymbol{\beta}$ with high probability.

### $\hat{S}^2$ Properties

In the case of perfect multicollinearity, $\hat{S}^2$ can nonetheless be used to generate estimates of $\sigma^2$, and with minor modifications to the derivations used previously to establish properties of $\hat{S}^2$, it can be shown that $S^2$ is unbiased and consistent for $\sigma^2$. Of course, the definition of $\hat{S}^2$ cannot be expressed in terms of $(\mathbf{Y} - \mathbf{x}\hat{\boldsymbol{\beta}})'(\mathbf{Y} - \mathbf{x}\hat{\boldsymbol{\beta}})/(n - k)$ since $\hat{\boldsymbol{\beta}} = (\mathbf{x}'\mathbf{x})^{-1}\mathbf{x}'\mathbf{Y}$ does not exist in this case. However, $\hat{S}^2 = (\mathbf{e}'\mathbf{e})/(n - k)$ can nonetheless be calculated, where $\mathbf{e}'\mathbf{e}$ is the minimum sum of squared errors calculated from any choice of $\hat{\boldsymbol{\beta}}_{*}$ which solves the first order conditions $(\mathbf{x}'\mathbf{x})\,\hat{\boldsymbol{\beta}}_{*} = \mathbf{x}'\mathbf{y}$. A rigorous demonstration of the properties of $\hat{S}^2$ under perfect multicollinearity relies on the notion of generalized inverses of matrices (e.g., see Graybill, F., (1976), *Theory and Application of the Linear Model*, Duxbury Press, pp. 23–39), and given that perfect multicollinearity is by far the exception rather than the rule, we will not pursue the details here.

In the case where rank($\mathbf{x}$) = $k$, but $(\mathbf{x}'\mathbf{x})$ is *nearly singular*, the proofs of unbiasedness and consistency of $\hat{S}^2$ nonetheless apply *exactly* as stated previously, and thus these properties are attained by $\hat{S}^2$ under the assumptions introduced heretofore. It is also useful to note that unlike the variances of the $\hat{\beta}_i$'s, which can increase without bound as the multicollinearity becomes increasingly severe, the variance of $\hat{S}^2$ exhibits a finite bound, and this bound is *unaffected* by the degree of multicollinearity. To derive such a bound, assume the Classical GLM Assumptions hold, assume the $\varepsilon_i$'s are *iid*, and let $\mathrm{E}(\varepsilon_i^4) = \mu'_4 < \infty$ exist. Recall that $\mathrm{var}(\hat{S}^2) = \mathrm{E}(\hat{S}^4) - (\mathrm{E}(\hat{S}^2))^2 = \mathrm{E}(\hat{S}^4) - \sigma^4$ where we have used the fact that $\mathrm{E}(\hat{S}^2) = \sigma^2$. Now note that

$$\hat{S}^2 = (n-k)^{-1}(\boldsymbol{\varepsilon}'\boldsymbol{\varepsilon} - \boldsymbol{\varepsilon}'\mathbf{x}(\mathbf{x}'\mathbf{x})^{-1}\mathbf{x}'\boldsymbol{\varepsilon}) \leq (n-k)^{-1}\boldsymbol{\varepsilon}'\boldsymbol{\varepsilon},$$

since $\mathbf{x}(\mathbf{x}'\mathbf{x})^{-1}\mathbf{x}'$ is positive semidefinite, which implies

$$\mathrm{E}\left(\hat{S}^4\right) \leq (n-k)^{-2}\mathrm{E}(\boldsymbol{\varepsilon}'\boldsymbol{\varepsilon})^2 \leq (n-k)^{-2}\mathrm{E}\left[\sum_{i=1}^{n}\varepsilon_i^4 + 2\sum\sum_{i<j}\varepsilon_i^2\,\varepsilon_j^2\right]$$

$$\leq (n-k)^{-2}\left[n\mu'_4 + n(n-1)\sigma^4\right] = \tau(\mathrm{n})$$

where $\tau(n) \geq \sigma^4$, $\tau(n)$ is monotonically decreasing in $n$ $\forall n > k$, and $\lim_{n\to\infty}(\tau(n)) = \sigma^4$.

Then the variance of $\hat{S}^2$ is bounded as $\mathrm{var}(\hat{S}^2) \leq \tau(n) - \sigma^4$ regardless of the severity of the multicollinearity, so long as rank($\mathbf{x}$) = $k$. Other bounds can be derived in cases where the $\varepsilon_i$'s are not *iid*.

*Assumption Violation: Stochastic* $\mathbf{X}$ *with* $\mathrm{E}((\mathbf{X}'\mathbf{X})^{-1}\mathbf{X}'\boldsymbol{\varepsilon}) \neq \mathbf{0}$ *and plim* $(\mathbf{X}'\mathbf{X})^{-1}\mathbf{X}'\boldsymbol{\varepsilon} \neq \mathbf{0}$ In this section we focus explicitly on the case where the explanatory variable matrix $\mathbf{X}$ is a nondegenerate random matrix. Note the condition

$E((\mathbf{X}'\mathbf{X})^{-1}\mathbf{X}'\boldsymbol{\varepsilon}) \neq \mathbf{0}$ necessarily implies (by the iterated expectation theorem) that $E(\boldsymbol{\varepsilon}|\mathbf{x}) \neq \mathbf{0}$ with positive probability, and so the conditional mean of $\boldsymbol{\varepsilon}$ is *not* independent of $\mathbf{X}$. This in turn implies that $E(\mathbf{Y}|\mathbf{x}) \neq \mathbf{x}\boldsymbol{\beta}$ with positive probability, and thus assumption 1 of the Assumptions of the Classical GLM, applied explicitly to the stochastic $\mathbf{X}$ case, is violated. Furthermore, the dependence between $\mathbf{X}$ and $\boldsymbol{\varepsilon}$ persists in the limit, as indicated by $\text{plim}((\mathbf{x}'\mathbf{x})^{-1}\mathbf{X}'\boldsymbol{\varepsilon}) \neq \mathbf{0}$. The interdependence of $\mathbf{X}$ and $\boldsymbol{\varepsilon}$ is typically caused by either *measurement error* in the $\mathbf{x}$ outcomes or by *simultaneous determination* of $\mathbf{Y}$ and $\mathbf{X}$ outcomes, and is discussed in the econometric literature under the categories of *errors in variables* and *simultaneous equations*, respectively.

### $\hat{\boldsymbol{\beta}}$ Properties

Since $E(\hat{\boldsymbol{\beta}}) = \boldsymbol{\beta} + E((\mathbf{X}'\mathbf{X})^{-1}\mathbf{X}'\boldsymbol{\varepsilon}) \neq \boldsymbol{\beta}$ because $E((\mathbf{X}'\mathbf{X})^{-1}\mathbf{X}'\boldsymbol{\varepsilon}) \neq \mathbf{0}$, it follows that the estimator is a *biased* estimator of $\boldsymbol{\beta}$, so that $\hat{\boldsymbol{\beta}}$ is also *not* BLUE. Regarding consistency, $\text{plim}(\hat{\boldsymbol{\beta}}) = \boldsymbol{\beta} + \text{plim}((\mathbf{X}'\mathbf{X})^{-1}\mathbf{X}'\boldsymbol{\varepsilon}) \neq \boldsymbol{\beta}$ because $\text{plim}((\mathbf{X}'\mathbf{X})^{-1}\mathbf{X}'\boldsymbol{\varepsilon}) \neq \mathbf{0}$ and so $\hat{\boldsymbol{\beta}}$ is also *not* consistent.

### $\hat{S}^2$ Properties

Letting $\boldsymbol{\delta}(\mathbf{x}) = E(\boldsymbol{\varepsilon}|\mathbf{x})$, the expectation of $(n-k)\hat{S}^2$ can be written via the iterated expectation theorem as

$$\begin{aligned}
E(n-k)\hat{S}^2 &= E(\boldsymbol{\varepsilon}'\boldsymbol{\varepsilon}) - E\left(E\left(\boldsymbol{\varepsilon}'\mathbf{X}(\mathbf{X}'\mathbf{X})^{-1}\mathbf{X}'\boldsymbol{\varepsilon}|\mathbf{X}\right)\right)\\
&= n\sigma^2 - E\left[\text{tr}\left(\mathbf{X}(\mathbf{X}'\mathbf{X})^{-1}\mathbf{X}'E(\boldsymbol{\varepsilon}\boldsymbol{\varepsilon}'|\mathbf{X})\right)\right]\\
&= n\sigma^2 - E\left(\text{tr}\left(\mathbf{X}(\mathbf{X}'\mathbf{X})^{-1}\mathbf{X}'(\mathbf{Cov}(\boldsymbol{\varepsilon}|\mathbf{X}) + \boldsymbol{\delta}(\mathbf{X})\boldsymbol{\delta}(\mathbf{X})')\right)\right)\\
&= (n-k)\sigma^2 - E\left(\boldsymbol{\delta}(\mathbf{X})'\mathbf{X}(\mathbf{X}'\mathbf{X})^{-1}\mathbf{X}'\boldsymbol{\delta}(\mathbf{X})\right)
\end{aligned}$$

assuming $\mathbf{Cov}(\boldsymbol{\varepsilon}|\mathbf{x}) = \sigma^2\mathbf{I}, \forall \mathbf{x}$.[16] Thus $E(\hat{S}^2) < \sigma^2$ in general, so that $\hat{S}^2$ is *biased*.

Regarding the consistency of $\hat{S}^2$, first note that $\hat{S}^2 = (n-k)^{-1}\boldsymbol{\varepsilon}'\mathbf{M}\boldsymbol{\varepsilon} = n^{-1}\boldsymbol{\varepsilon}'\mathbf{M}\boldsymbol{\varepsilon} + o_p(1) = n^{-1}\boldsymbol{\varepsilon}'\boldsymbol{\varepsilon} - n^{-1}\boldsymbol{\varepsilon}'\mathbf{X}(\mathbf{X}'\mathbf{X})^{-1}\mathbf{X}'\boldsymbol{\varepsilon} + o_p(1)$, and since $(n^{-1}\boldsymbol{\varepsilon}'\boldsymbol{\varepsilon} - \sigma^2) \xrightarrow{p} 0$,

$$\left(\hat{S}^2 - \sigma^2\right) = -\boldsymbol{\varepsilon}'\mathbf{X}(\mathbf{X}'\mathbf{X})^{-1}(n^{-1}\mathbf{X}'\mathbf{X})(\mathbf{X}'\mathbf{X})^{-1}\mathbf{X}'\boldsymbol{\varepsilon} + o_p(1).$$

If $n^{-1}\mathbf{X}'\mathbf{X} \xrightarrow{p} \mathbf{Q}$, a positive definite symmetric matrix, and if $\text{plim}((\mathbf{X}'\mathbf{X})^{-1}\mathbf{X}'\boldsymbol{\varepsilon}) = \boldsymbol{\xi} \neq \mathbf{0}$, then $(\hat{S}^2 - \sigma^2) \xrightarrow{p} -\boldsymbol{\xi}'\mathbf{Q}\boldsymbol{\xi} < 0$, and $\hat{S}^2$ is not consistent. Even if neither $n^{-1}\mathbf{X}'\mathbf{X}$ nor $(\mathbf{X}'\mathbf{X})^{-1}\mathbf{X}'\boldsymbol{\varepsilon}$ converge at all, so long as $\det(n^{-1}\mathbf{x}'\mathbf{x}) > \eta > 0\ \forall n$, as in Table 8.1, and $(\mathbf{X}'\mathbf{X})^{-1}\mathbf{X}'\boldsymbol{\varepsilon} \not\xrightarrow{p} \mathbf{0}$, it follows that $(\hat{S}^2 - \sigma^2) \not\xrightarrow{p} 0$ and $\hat{S}^2$ is not consistent.

[16] Recall that $E(\boldsymbol{\varepsilon}\boldsymbol{\varepsilon}') = \mathbf{Cov}(\boldsymbol{\varepsilon}) + E(\boldsymbol{\varepsilon})E(\boldsymbol{\varepsilon}')$.

**Table 8.2** General Least Squares Estimator Properties Under GLM Assumption Violations

| Violation | $E(\varepsilon) = \phi \neq 0$ | $E(\varepsilon\varepsilon') = \Phi \neq \sigma^2 I$ | $\det(x'x) \approx 0$ | Random X, $E((X'X)^{-1}X'\varepsilon) \neq 0$, $\text{Plim}((X'X^{-1})X'\varepsilon) \neq 0$ |
|---|---|---|---|---|
| Terminology | Specification error | Heteroskedasticity and/or autocorrelation | Multicollinearity | Errors in variables, simultaneity |
| $\hat{\beta}$ *Properties* | | | | |
| Unbiased | Generally no yes if $x'\phi = 0$ | Yes | Yes | No |
| BLUE | Generally no yes if $x'\phi = 0$ | No | Yes | No |
| Consistent | Generally no yes if $(x'x)^{-1}x'\phi \to 0$ | Generally yes if $\lambda_L(\Phi)\ \text{tr}(x'x)^{-1} \to 0$ | Yes | No |
| $S^2$ *Properties* | | | | |
| Unbiased | Generally no yes if $\phi'(I - x(x'x)^{-1}x')\phi \to 0$ | Generally no | Yes | No |
| Consistent | Generally no yes if $\phi'(I - x(x'x)^{-1}x')\phi \to 0$ | Generally no Yes if $\lambda_L(\Phi)$is $o(n^1)$ and $\varepsilon_i$'s are homoskedastic | Yes | No |

### 8.2.4 GLM Assumption Violations: Property Summary and Epilogue

The basic properties of the OLS-based estimators of $\beta$ and $\sigma^2$ under the various violations of the Classical GLM Assumptions are summarized in Table 8.2. The row of the table identified as "terminology" indicates the label under which the problem is generally discussed in the econometric literature. When reading Table 8.2, it is assumed that the conditions in Table 8.1 hold, except those that are noted.

A major focus of research based on the GLM concerns the detection and remedies for violations of the assumptions needed to achieve the desirable properties of the estimators of $\beta$ and $\sigma^2$. Indeed, fields such as **Econometrics**, **Psychometrics**, and **Sociometrics** were necessitated by the fact that, unlike disciplines in which the experiments under investigation can generally be designed and controlled to achieve the conditions needed for the least squares estimator to have optimal properties, many experiments in economics, business, psychology or sociology, and other social sciences, are oftentimes not under the control of the researcher. Variations on the least squares estimator, such as *restricted least squares*, the *feasible generalized least squares estimator*, *instrumental variables estimators, two and three stage least squares estimators*, *generalized method of moments*, and *limited and full-information maximum likelihood estimators* represent procedures for remedying violations of the GLM classical assumptions that the reader will encounter in her subsequent studies. We will examine some hypothesis tests for detecting GLM assumption violations in Chapter 10.

### 8.2.5 Least Squares Under Normality

Before examining the implications of making specific parametric family assumptions for the probability model underlying sample data (in the next section), we emphasize that the type of random sample envisioned for $(Y_1,\ldots,Y_n)$ in the GLM context is generally of the general experimental-type, i.e. note that $E(Y_i)$ does not necessarily equal $E(Y_j)$ for $i \neq j$, nor is it necessarily assumed that $(Y_1,\ldots,Y_n)$, or $(\varepsilon_1,\ldots,\varepsilon_n)$, are *iid*. We now examine the assumption of normality for $(Y_1,\ldots,Y_n)$ and, hence, for $(\varepsilon_1,\ldots,\varepsilon_n)$.

*Finite Sample Distributions of* $\hat{\boldsymbol{\beta}}$ *and* $\hat{S}^2$ In the GLM, under the classical GLM assumptions, let $\mathbf{Y} \sim N(\mathbf{x}\boldsymbol{\beta}, \sigma^2\mathbf{I})$. Under the normality assumption, the classical GLM assumptions *necessarily* imply the $\varepsilon_i$'s are *iid* normal, with mean zero and variance $\sigma^2$. Furthermore, *all* of the assumptions in Table 8.1 corresponding to the disturbance vector hold, *except* for $P(|\varepsilon_i| < m) = 1\ \forall i$, which does not hold. Of course, our entire preceding discussion concerning the properties of $\hat{\boldsymbol{\beta}}$ and $\hat{S}^2$ then applies equally well to the case where $\mathbf{Y}$ is multivariate normally distributed. The question addressed in this section is "what *additional* properties can be attributed to the estimators when $\mathbf{Y}$ is multivariate normally distributed?"

One immediate property is that $\hat{\boldsymbol{\beta}}$ is multivariate normally distributed for every $n \geq k$, and not just *asymptotically* normally distributed. This follows straightforwardly from the fact that $\hat{\boldsymbol{\beta}}$ is a linear function of the entries in $\mathbf{Y}$, so that (by Theorem 4.9) $\hat{\boldsymbol{\beta}}$ is normally distributed with mean $E(\hat{\boldsymbol{\beta}}) = (\mathbf{x}'\mathbf{x})^{-1}\mathbf{x}'(\mathbf{x}\boldsymbol{\beta}) = \boldsymbol{\beta}$, and covariance matrix $\mathbf{Cov}(\hat{\boldsymbol{\beta}}) = (\mathbf{x}'\mathbf{x})^{-1}\mathbf{x}'[\sigma^2\mathbf{I}]\mathbf{x}(\mathbf{x}'\mathbf{x})^{-1} = \sigma^2(\mathbf{x}'\mathbf{x})^{-1}$, i.e., $\hat{\boldsymbol{\beta}} \sim N(\boldsymbol{\beta}, \sigma^2(\mathbf{x}'\mathbf{x})^{-1})$.

Another property is that $(n-k)\,\hat{S}^2/\sigma^2 \sim \chi^2_{n-k}$. To see this, first note that

$$\frac{(n-k)\hat{S}^2}{\sigma^2} = \frac{\boldsymbol{\varepsilon}'[\mathbf{I} - \mathbf{x}(\mathbf{x}'\mathbf{x})^{-1}\mathbf{x}']\boldsymbol{\varepsilon}}{\sigma^2} = \sigma^{-1}\,\boldsymbol{\varepsilon}'\mathbf{P}\boldsymbol{\Lambda}\mathbf{P}'\boldsymbol{\varepsilon}\,\sigma^{-1},$$

where $\mathbf{I} - \mathbf{x}(\mathbf{x}'\mathbf{x})^{-1}\mathbf{x}' = \mathbf{P}\boldsymbol{\Lambda}\mathbf{P}'$, and $\boldsymbol{\Lambda}$ and $\mathbf{P}$ are, respectively, the diagonal matrix of characteristic roots and the matrix of characteristic vectors (stored columnwise) associated with the symmetric and idempotent matrix $(\mathbf{I} - \mathbf{x}(\mathbf{x}'\mathbf{x})^{-1}\mathbf{x}')$ (recall the proof of Theorem 6.12b). Examine the probability distribution of $\mathbf{Z} = \mathbf{P}'\boldsymbol{\varepsilon}\,\sigma^{-1}$. Since $\boldsymbol{\varepsilon} \sim N(\mathbf{0}, \sigma^2\mathbf{I})$, then $\mathbf{Z} \sim N(\sigma^{-1}\,\mathbf{P}'\mathbf{0}, \sigma^{-1}\,\mathbf{P}'\sigma^2\mathbf{I}\mathbf{P}\sigma^{-1}) = N(\mathbf{0}, \mathbf{I})$, because $\mathbf{P}'\mathbf{P} = \mathbf{I}$ by the orthogonality of $\mathbf{P}$. Then

$$\frac{(n-k)\hat{S}^2}{\sigma^2} = \mathbf{Z}'\boldsymbol{\Lambda}\mathbf{Z} = \sum_{i=1}^{n-k}\mathbf{Z}_i^2 \sim \chi^2_{n-k},$$

i.e., we have the sum of the squares of $(n-k)$ *iid* standard normal random variables which has a Chisquare distribution with $n-k$ degrees of freedom.

Since by the preceding result $\hat{S}^2$ can be characterized as a $\chi^2$ random variable that has been multiplied by the constant $\sigma^2/(n-k)$, it follows that $\hat{S}^2 \sim$

Gamma$((n-k)/2,\ 2\sigma^2/(n-k))$, as can be shown by deriving the MGF of $\hat{S}^2$. Using properties of the gamma density, this in turn implies that

$$\mathrm{E}\left(\hat{S}^2\right) = ((n-k)/2)(2\sigma^2/(n-k)) = \sigma^2 \text{ and}$$

$$\mathrm{var}\left(\hat{S}^2\right) = ((n-k)/2)(2\sigma^2/(n-k))^2 = (2\sigma^4/(n-k)).$$

Still another property is that $\hat{\boldsymbol{\beta}}$ and $\hat{S}^2$ are *independent* random variables. This follows from an application of Theorem 6.11 along the lines of the proof of Theorem 6.12a, and is left as an exercise for the reader.

*MVUE Property of $\hat{\boldsymbol{\beta}}$ and $S^2$* Perhaps the most important additional property that results when $\mathbf{Y} \sim N(\mathbf{x}\boldsymbol{\beta}, \sigma^2\mathbf{I})$ is that $\begin{bmatrix} \hat{\boldsymbol{\beta}} \\ \hat{S}^2 \end{bmatrix}$ is then the MVUE for $\begin{bmatrix} \boldsymbol{\beta} \\ \sigma^2 \end{bmatrix}$.

**Theorem 8.11** ***MVUE Property of $(\hat{\boldsymbol{\beta}}, \hat{S}^2)$ Under Normality*** *Assume the Assumptions of the Classical GLM, and assume that $\mathbf{Y} \sim N(\mathbf{x}\boldsymbol{\beta}, \sigma^2\mathbf{I})$. Then $(\hat{\boldsymbol{\beta}}, \hat{S}^2)$ is the MVUE for $(\boldsymbol{\beta}, \sigma^2)$.*

**Proof** Note that the multivariate normal density belongs to the exponential class of densities

$$\exp\left(\sum_{i=1}^{k+1} c_i(\boldsymbol{\beta},\sigma^2)\, g_i(\mathbf{y}) + d(\boldsymbol{\beta},\sigma^2) + z(\mathbf{y})\right) I_A(\mathbf{y}) = \exp(\mathbf{c}(\boldsymbol{\beta},\sigma^2)'\mathbf{g}(\mathbf{y}) + d(\boldsymbol{\beta},\sigma^2) + z(\mathbf{y}))I_A(\mathbf{y})$$

where

$$\mathbf{c}(\boldsymbol{\beta},\sigma^2) = \begin{bmatrix} c_1(\boldsymbol{\beta},\sigma^2) \\ \vdots \\ c_k(\boldsymbol{\beta},\sigma^2) \\ c_{k+1}(\boldsymbol{\beta},\sigma^2) \end{bmatrix} = \begin{bmatrix} \beta_1/\sigma^2 \\ \vdots \\ \beta_k/\sigma^2 \\ \dfrac{-1}{2\sigma^2} \end{bmatrix} = \begin{bmatrix} \dfrac{1}{\sigma^2}\boldsymbol{\beta} \\ \dfrac{-1}{2\sigma^2} \end{bmatrix},\ \mathbf{g}(\mathbf{y}) = \begin{bmatrix} g_1(\mathbf{y}) \\ \vdots \\ g_{k+1}(\mathbf{y}) \end{bmatrix} = \begin{bmatrix} \mathbf{x}'\mathbf{y} \\ \mathbf{y}'\mathbf{y} \end{bmatrix}$$

$d(\boldsymbol{\beta},\sigma^2) = -((\boldsymbol{\beta}'\mathbf{x}'\mathbf{x}\boldsymbol{\beta})/2\sigma^2) - \ln((2\pi\sigma^2)^{n/2})$, $z(\mathbf{y}) = 0$, and $A = \times_{i=1}^{n} (-\infty,\infty) = \mathbb{R}^n$. Then by Theorem 7.4, $\begin{bmatrix} \mathbf{x}'\mathbf{Y} \\ \mathbf{Y}'\mathbf{Y} \end{bmatrix}$ is a vector of minimal sufficient statistics for estimating $\beta$ and $\sigma^2$. These sufficient statistics are also complete sufficient statistics, since the range of the vector function represented by $\mathbf{c}(\boldsymbol{\beta},\boldsymbol{\sigma}^2)$ contains an open $(k+1)$ dimensional rectangle (recall Theorem 7.8). In particular, note that the range is given by $R(c) = \{(c_1,\ldots,c_{k+1}): -\infty < c_i < \infty, i = 1,\ldots,k,\ c_{k+1} < 0\}$ since $\beta_i \in (-\infty,\infty)$ for $i = 1,\ldots,k$, and $\sigma^2 > 0$, so that any open rectangle of the form $A = \{(c_1,\ldots,c_{k+1}): a_i < c_i < b_i,\ i = 1,\ldots,(k+1)\}$, for $a_i < b_i$, and $b_{(k+1)} < 0$

[17]Note that $\beta_i \in (-\infty,\infty)$, $\forall i$, and $\sigma^2 > 0$ are the *admissible* parameter values for $\mathbf{Y} \sim N(\mathbf{x}\boldsymbol{\beta}, \sigma^2\mathbf{I})$. It may be the case that only a subset of these values for the parameters are deemed to be relevant in a given estimation problem (e.g., a price effect may be restricted to be of one sign, or the realistic magnitude of the effect of an explanatory variable may be bounded). Restricting the parameter space comes under the realm of *prior information* models, which we do not pursue here. So long as the admissible values of $\boldsymbol{\beta}$ and $\sigma^2$ form an open rectangle themselves, the range of $\mathbf{c}(\boldsymbol{\beta}, \sigma^2)$ will contain an open rectangle.

is a $(k+1)$ dimensional open rectangle subset of $R(\mathbf{c})$.[17] Then because $\hat{\boldsymbol{\beta}} = (\mathbf{x}'\mathbf{x})^{-1}\mathbf{x}'\mathbf{Y}$ and $\hat{S}^2 = (n-k)^{-1}\ (\mathbf{Y} - \mathbf{x}\,\hat{\boldsymbol{\beta}}\,)'(\mathbf{Y} - \mathbf{x}\,\hat{\boldsymbol{\beta}}\,) = (\mathbf{Y}'\mathbf{Y} - \mathbf{Y}'\mathbf{x}(\mathbf{x}'\mathbf{x})^{-1}\mathbf{x}'\mathbf{Y})/(n-k)$ are functions of the complete sufficient statistics., and since we have shown previously that $\hat{\boldsymbol{\beta}}$ and $\hat{S}^2$ are unbiased for $\boldsymbol{\beta}$ and $\sigma^2$ under the assumptions of the Classical GLM, it follows from the Lehmann-Scheffé completeness theorem (Theorem 7.22) that $\begin{bmatrix} \hat{\boldsymbol{\beta}} \\ \hat{S}^2 \end{bmatrix}$ is the MVUE for $\begin{bmatrix} \boldsymbol{\beta} \\ \sigma^2 \end{bmatrix}$. ■

Note that the CRLB is

$$\mathrm{E}\left[\frac{\partial \ln f(\mathbf{Y};\boldsymbol{\Theta})}{\partial \boldsymbol{\Theta}} \frac{\partial \ln f(\mathbf{Y};\boldsymbol{\Theta})}{\partial \boldsymbol{\Theta}'}\right]^{-1} = \left[\begin{array}{c|c} \sigma^2(\mathbf{x}'\mathbf{x})^{-1} & \mathbf{0} \\ \hline \mathbf{0} & \frac{2\sigma^4}{n} \end{array}\right]$$

where $\boldsymbol{\Theta} = \begin{pmatrix} \boldsymbol{\beta} \\ \sigma^2 \end{pmatrix}$. Then because $\mathbf{Cov}(\hat{\boldsymbol{\beta}}) = \sigma^2(\mathbf{x}'\mathbf{x})^{-1}$, $\hat{\boldsymbol{\beta}}$ achieves the CRLB for unbiased estimators of $\boldsymbol{\beta}$ providing an alternative demonstration that $\hat{\boldsymbol{\beta}}$ is the MVUE for $\boldsymbol{\beta}$. However, since $\mathrm{var}(\hat{S}^2) = 2\sigma^4/(n-k) > 2\sigma^4/n$, $\hat{S}^2$ does *not* achieve the CRLB for unbiased estimators of $\sigma^2$, and the CRLB approach would have left the question unanswered regarding whether $\hat{S}^2$ was the MVUE for $\sigma^2$. The approach using complete sufficient statistics used above demonstrates that no unbiased estimator of $\sigma^2$ can achieve the CRLB in this case, for indeed $\hat{S}^2$ *is* the MVUE.

*On the Assumption of Normality* The reader might wonder what considerations would lead to a specification of the normal family of densities for the probability distribution of $\mathbf{Y}$. Of course, if there is an underlying theoretical or physical rationale for why $\mathbf{Y}$ is multivariate normally distributed, the assumption is obviously supported. However, the underlying rationale for normality is often not clear. In these cases, normality is sometimes rationalized by an appeal to a central limit theorem. In particular, it is often argued that the elements in the disturbance vector are themselves defined as the summation of a large number of random variables, i.e., $\varepsilon_i = \sum_{j=1}^{m} V_{ij}$ for large $m$. The $V_{ij}$'s may represent a myriad of neglected explanatory factors which affect $\mathrm{E}(Y_i)$, $i = 1,\ldots,n$, which, because of the need for tractability, parsimony, or because of a lack of data, are not explicitly represented in the specification of $\mathrm{E}(Y_i) = \mathbf{x}_{i.}\boldsymbol{\beta}$. Alternatively, the $V_{ij}$'s may be intrinsic to the error term specification if the GLM is being used to represent a summation of individual micro relations. For example, an aggregate short-run supply function in an aggregate economic analysis might be represented as

$$\sum_{j=1}^{m} Q_{ij} = \sum_{j=1}^{m} (\beta_{1j} + \beta_{2j} p_i + V_{ij}) \Rightarrow Q_i^* = \beta_1^* + \beta_2^* p_i + \varepsilon_i,$$

where $Q_i^* = \sum_{j=1}^{m} Q_{ij}$ is the $i$th observation on aggregate supply, $\beta_1^* = \sum_{j=1}^{m} \beta_{1j}$ is the intercept term of the aggregate supply relationship, $\beta_2^* = \sum_{j=1}^{m} \beta_{2j}$ is the aggregate price effect, $p_i$ is the $i$th observation on supply price, and $\varepsilon_i = \sum_{j=1}^{m} V_{ij}$ is the disturbance term for the $i$th observation, defined as the sum of the individual disturbance terms of the $m$ micro supply functions. Our investigation of CLT's in Chapter 5 have suggested that under a variety of conditions, sums of large numbers of random variables are asymptotically normally distributed, i.e., $\boldsymbol{\varepsilon} = \sum_{j=1}^{m} \mathbf{V}_{.j} \overset{a}{\sim} N(\boldsymbol{\mu}, \boldsymbol{\Sigma})$.

A prudent approach to the assumption of normality is to view arguments such as those presented above as *suggestive* of a normal approximation to the true distribution of $\boldsymbol{\varepsilon}$, but the assumption should be tested for acceptability whenever possible. We will investigate testing the hypothesis of normality (and indeed, hypotheses of other parametric families as well) in Chapter 10.

### 8.2.6 Overview of Nonlinear Least Squares Estimation

We provide an introduction to the nonlinear least squares (NLS) estimator in this subsection. The nonlinear least squares estimator generalizes the ordinary least squares estimator by allowing for more functional flexibility in the way explanatory variables, $\mathbf{x}$, and parameters, $\boldsymbol{\beta}$, can affect values of $E[\mathbf{Y}|\mathbf{x}]$. In particular, the relationship between $\mathbf{Y}$ and $\mathbf{x}$ is specified as $\boldsymbol{Y} = \mathbf{g}(\mathbf{x}, \boldsymbol{\beta}) + \boldsymbol{\varepsilon}$ where $\mathbf{Y} = (Y_1, Y_2, \ldots, Y_n)'$ is a $(n \times 1)$ vector of observable random variables, $\mathbf{x}$ is a $(n \times k)$ matrix representing $n$ fixed values of $k$ explanatory variables, $\boldsymbol{\beta}$ is a $k$-dimensional fixed vector of unknown parameters, $\boldsymbol{\varepsilon}$ is a $(n \times 1)$ vector of unobservable random residuals, and $\mathbf{g}$ is a $(n \times 1)$ nonlinear vector-valued function of both $\mathbf{x}$ and $\boldsymbol{\beta}$ representing the systematic component of the model.

*The Nonlinear Least Squares Estimator* The NLS estimator minimizes the sums of squared residuals function, or squared distance between $\mathbf{Y}$ and $\mathbf{g}(\mathbf{x}, \boldsymbol{\beta})$,

$$d^2(\mathbf{Y}, \mathbf{g}(\mathbf{x}, \boldsymbol{\beta})) = (\mathbf{Y} - \mathbf{g}(\mathbf{x}, \boldsymbol{\beta}))'(\mathbf{Y} - \mathbf{g}(\mathbf{x}, \boldsymbol{\beta}))$$

and the NLS estimator is defined by

$$\hat{\boldsymbol{\beta}} = \arg\min_{\boldsymbol{\beta}} \left[d^2(\mathbf{Y}, \mathbf{g}(\mathbf{x}, \boldsymbol{\beta}))\right] = \arg\min_{\boldsymbol{\beta}} \{(\mathbf{Y} - \mathbf{g}(\mathbf{x}, \boldsymbol{\beta}))'(\mathbf{Y} - \mathbf{g}(\mathbf{x}, \boldsymbol{\beta}))\}.$$

Regarding motivation for the squared error metric used in defining the estimator, note that

$$\begin{aligned} E(n^{-1} d^2(\mathbf{Y}, \mathbf{g}(\mathbf{x}, \boldsymbol{\beta}))) &= E(n^{-1}(\mathbf{Y} - \mathbf{g}(\mathbf{x}, \boldsymbol{\beta}))'(\mathbf{Y} - \mathbf{g}(\mathbf{x}, \boldsymbol{\beta}))) \\ &= E(n^{-1}(\mathbf{Y} - \mathbf{g}(\mathbf{x}, \boldsymbol{\beta}_0) + \mathbf{g}(\mathbf{x}, \boldsymbol{\beta}_0) - \mathbf{g}(\mathbf{x}, \boldsymbol{\beta}))'(\mathbf{Y} - \mathbf{g}(\mathbf{x}, \boldsymbol{\beta}_0) \\ &\quad + \mathbf{g}(\mathbf{x}, \boldsymbol{\beta}_0) - \mathbf{g}(\mathbf{x}, \boldsymbol{\beta}))) \\ &= \sigma^2 + \frac{1}{n}[\mathbf{g}(\mathbf{x}, \boldsymbol{\beta}_0) - \mathbf{g}(\mathbf{x}, \boldsymbol{\beta})]' [\mathbf{g}(\mathbf{x}, \boldsymbol{\beta}_0) - \mathbf{g}(\mathbf{x}, \boldsymbol{\beta})] \end{aligned}$$

where $\boldsymbol{\beta}_0$ denotes the true value of the parameter vector. If $\mathbf{g}(\mathbf{x}, \boldsymbol{\beta}_0) \neq \mathbf{g}(\mathbf{x}, \boldsymbol{\beta})$ $\forall \boldsymbol{\beta} \neq \boldsymbol{\beta}_0$, then it is clear that the minimum occurs uniquely at $\boldsymbol{\beta} = \boldsymbol{\beta}_0$. Thus, if the preceding expectation could actually be calculated, minimizing it would recover $\boldsymbol{\beta}_0$ exactly. In practice the expectation is unknown and so the scaled sample analog given by $d^2(\mathbf{y}, \mathbf{g}(\mathbf{x}, \boldsymbol{\beta}))$ is used instead. Under an appropriate law of large numbers, $n^{-1}d^2(\mathbf{Y}, \mathbf{g}(\mathbf{x}, \boldsymbol{\beta}))$ will converge to the preceding expectation and thus one would expect that $\hat{\boldsymbol{\beta}}$ as defined above should result in an estimate close to $\boldsymbol{\beta}_0$ for large enough $n$, justifying the choice of the squared error metric.

An alternative rationale based on a prediction criterion is analogous to the rationale given in the linear model case. One need only replace $\hat{\mathbf{Y}} = \mathbf{x}\hat{\boldsymbol{\beta}}$ with $\hat{\mathbf{Y}} = \mathbf{g}\left(\mathbf{x}, \hat{\boldsymbol{\beta}}\right)$, and the arguments are identical.

Unlike the LS estimator in the general linear model, there is generally no analytical or closed form solution for the NLS estimator. The solution for the minimum sum of squares problem is most often found by using a computer and software designed to find minimums of nonlinear objective functions. A popular minimization algorithm used for solving NLS problems is the so-called Gauss-Newton method. We do not pursue computational methodology here, and the reader is directed to the nonlinear regression and econometrics literature for discussion of computational approaches (e.g., Mittelhammer, Judge, and Miller, (2000) *Econometric Foundations*, Cambridge University Press).

*Sampling Properties of the NLS Estimator* An overview of sampling properties of the NLS estimator is provided in this subsection. The discussion begins immediately with asymptotic properties simply because there is little that can be said about the NLS estimator regarding generally applicable finite sample properties.

As we noted at the outset of Section 8.2, in the process of motivating asymptotic properties, striking analogies to the linear model case will be observed. These are based on first order approximations to the conditional expectation function of the model.

We can expand $\mathbf{g}(\mathbf{x}, \boldsymbol{\beta})$ in a first order Taylor series around $\boldsymbol{\beta}_0$ as

$$\mathbf{g}(\mathbf{x}, \boldsymbol{\beta}) \approx \mathbf{g}(\mathbf{x}, \boldsymbol{\beta}_0) + \frac{\partial \mathbf{g}(\mathbf{x}, \boldsymbol{\beta}_0)}{\partial \boldsymbol{\beta}'}(\boldsymbol{\beta} - \boldsymbol{\beta}_0) = \mathbf{g}(\mathbf{x}, \boldsymbol{\beta}_0) + \mathbf{x}(\boldsymbol{\beta}_0)(\boldsymbol{\beta} - \boldsymbol{\beta}_0).$$

where

$$\mathbf{x}(\boldsymbol{\beta}_0) \equiv \frac{\partial \mathbf{g}(\mathbf{x}, \boldsymbol{\beta}_0)}{\partial \boldsymbol{\beta}'} \equiv \left.\frac{\partial \mathbf{g}(\mathbf{x}, \boldsymbol{\beta})}{\partial \boldsymbol{\beta}'}\right|_{\boldsymbol{\beta}_0}.$$

Substituting into the NLS objective function results in an approximate sum of squares function defined by

$$d^2(\mathbf{y}, \mathbf{g}(\mathbf{x}, \boldsymbol{\beta})) \approx [\mathbf{Z} - \mathbf{x}(\boldsymbol{\beta}_0)\boldsymbol{\theta}]'[\mathbf{Z} - \mathbf{x}(\boldsymbol{\beta}_0)\boldsymbol{\theta}]$$

where $\mathbf{Z} \equiv \mathbf{Y} - \mathbf{g}(\mathbf{x}, \boldsymbol{\beta}_0) = \boldsymbol{\varepsilon}$ and $\boldsymbol{\theta} \equiv \boldsymbol{\beta} - \boldsymbol{\beta}_0$.

Minimizing the first order approximation is analogous to finding the least squares estimator of $\boldsymbol{\theta}$ in the linear model $\mathbf{z} = \mathbf{x}(\boldsymbol{\beta}_0)\boldsymbol{\theta} + \mathbf{v}$, and results in

$$\hat{\boldsymbol{\theta}} = \left[\mathbf{x}(\boldsymbol{\beta}_0)'\mathbf{x}(\boldsymbol{\beta}_0)\right]^{-1}\mathbf{x}(\boldsymbol{\beta}_0)'\mathbf{z}.$$

It follows that the NLS estimator is characterized by the approximation

$$\hat{\boldsymbol{\beta}} \approx \boldsymbol{\beta}_0 + \left[\mathbf{x}(\boldsymbol{\beta}_0)'\mathbf{x}(\boldsymbol{\beta}_0)\right]^{-1}\mathbf{x}(\boldsymbol{\beta}_0)'\boldsymbol{\varepsilon}.$$

Therefore, *to the first order of approximation*, the NLS estimator is unbiased and its covariance matrix is $\sigma^2\left[\mathbf{x}(\boldsymbol{\beta}_0)'\mathbf{x}(\boldsymbol{\beta}_0)\right]^{-1}$, which follows from results on the means and covariance matrices of linear combinations of random variables.

Assuming regularity conditions on $\mathbf{x}(\boldsymbol{\beta}_0)$ and $\boldsymbol{\varepsilon}$ analogous to those that were applied to $\mathbf{x}$ and $\boldsymbol{\varepsilon}$ in the linear model case to establish asymptotic properties of the LS estimator, it can be shown that

$$n^{-1}\mathbf{x}(\boldsymbol{\beta}_0)'\mathbf{x}(\boldsymbol{\beta}_0) \to \boldsymbol{\Xi},\ n^{-1}\mathbf{x}(\boldsymbol{\beta}_0)'\boldsymbol{\varepsilon} \xrightarrow{p} \mathbf{0},\ \text{ and } n^{-1/2}\mathbf{x}(\boldsymbol{\beta}_0)'\boldsymbol{\varepsilon} \xrightarrow{d} N(\mathbf{0}, \sigma^2\boldsymbol{\Xi}),$$

where $\Xi$ is a finite, symmetric, positive definite matrix. It then follows to the first order of approximation, and analogous to the linear LS estimator application, that

$$n^{1/2}\left(\hat{\boldsymbol{\beta}} - \boldsymbol{\beta}_0\right) \xrightarrow{d} N\left(\mathbf{0}, \sigma^2\boldsymbol{\Xi}^{-1}\right) \text{ and } \hat{\boldsymbol{\beta}} \overset{a}{\sim} N\left(\boldsymbol{\beta}_0, \sigma^2\left[\mathbf{x}(\boldsymbol{\beta}_0)'\mathbf{x}(\boldsymbol{\beta}_0)\right]^{-1}\right).$$

Moreover, it follows from the limiting distribution result above that $n^{1/2}\left(\hat{\boldsymbol{\beta}} - \boldsymbol{\beta}_0\right)$ is $O_p(1)$, so that $\left(\hat{\boldsymbol{\beta}} - \boldsymbol{\beta}_0\right)$ is $o_p(1)$, $\hat{\boldsymbol{\beta}} - \boldsymbol{\beta}_0 \xrightarrow{p} \mathbf{0}$, and thus $\hat{\boldsymbol{\beta}}$ is a consistent estimator of $\boldsymbol{\beta}_0$.

The preceding approximations are accurate within small neighborhoods of the true parameter vector $\boldsymbol{\beta}_0$, derived from the accuracy of the first order Taylor series in such neighborhoods. It can be shown that $\hat{\boldsymbol{\beta}}$ is almost certain to be in a neighborhood of $\boldsymbol{\beta}_0$ as $n \to \infty$ under general regularity conditions. In applications, the asymptotic normality and asymptotic covariance results provide the basis for hypothesis testing and confidence region estimation, as we will see in later chapters. Operational versions of the covariance matrix are obtained by replacing $\boldsymbol{\beta}_0$ and $\sigma^2$ by consistent NLS estimator outcomes, as will be discussed further ahead.

*Estimating* $\sigma^2$ *and* $\mathbf{Cov}\left(\hat{\boldsymbol{\beta}}\right)$ Estimates of random residual vector outcomes are provided by the estimator

$$\hat{\boldsymbol{\varepsilon}} = \mathbf{Y} - \mathbf{g}(\mathbf{x}, \hat{\boldsymbol{\beta}}) = \boldsymbol{\varepsilon} + \left[\mathbf{g}(\mathbf{x}, \boldsymbol{\beta}_0) - \mathbf{g}\left(\mathbf{x}, \hat{\boldsymbol{\beta}}\right)\right]$$

where $\boldsymbol{\beta}_0$ is again being used to denote the true value of the parameter vector. Given the consistency of the NLS estimator, $\hat{\boldsymbol{\beta}} \xrightarrow{p} \boldsymbol{\beta}_0$, it follows under mild regularity conditions (continuity of the $\mathbf{g}$ function) that $\hat{\boldsymbol{\varepsilon}} - \boldsymbol{\varepsilon} = \left[\mathbf{g}(\mathbf{x}, \boldsymbol{\beta}_0) - \mathbf{g}\left(\mathbf{x}, \hat{\boldsymbol{\beta}}\right)\right] \xrightarrow{p} \mathbf{0}$ element-wise. Therefore, as $n$ increases, the outcomes of $\hat{\varepsilon}_i$ will eventually become indistinguishable from the outcomes of $\varepsilon_i$ with probability converging to 1, and thus the distributions of $\hat{\varepsilon}_i$ and $\varepsilon_i$ coincide as well (recall that convergence in probability implies convergence in distribution).

We now show that $\hat{\boldsymbol{\varepsilon}}$ is asymptotically linear in $\boldsymbol{\varepsilon}$, which emulates the relationship between the estimated and actual residuals in the linear model case. Expand $\hat{\varepsilon}_i$ in a Taylor series around the point $\boldsymbol{\beta}_0$ based on the definition for $\hat{\varepsilon}_i$ above, to obtain

$$\hat{\varepsilon}_i = y_i - g(\mathbf{x}_{i\cdot}, \boldsymbol{\beta}_0) - \frac{\partial g(\mathbf{x}_{i\cdot}, \boldsymbol{\beta})}{\partial \boldsymbol{\beta}}\Big|_{\boldsymbol{\beta}_*} \left(\hat{\boldsymbol{\beta}} - \boldsymbol{\beta}_0\right)$$

where $\boldsymbol{\beta}_* = \lambda\hat{\boldsymbol{\beta}} + (1-\lambda)\boldsymbol{\beta}_0$ for some $\lambda \in [0,1]$ by the Mean Value Theorem. Using the relationship $\hat{\boldsymbol{\beta}} = \boldsymbol{\beta}_0 + \left[\mathbf{x}(\boldsymbol{\beta}_0)'\mathbf{x}(\boldsymbol{\beta}_0)\right]^{-1}\mathbf{x}(\boldsymbol{\beta}_0)'\boldsymbol{\varepsilon} + o_p(n^{-1/2})$ from before, and substituting $\hat{\varepsilon}_i$ for $y_i - g(\mathbf{x}_{i\cdot}, \boldsymbol{\beta}_0)$ obtains

$$\hat{\varepsilon}_i = \varepsilon_i - \mathbf{x}_i(\boldsymbol{\beta}_*)\left[\mathbf{x}(\boldsymbol{\beta}_0)'\mathbf{x}(\boldsymbol{\beta}_0)\right]^{-1}\mathbf{x}(\boldsymbol{\beta}_0)'\boldsymbol{\varepsilon} + o_p\left(n^{-1/2}\right).$$

Given that $g(\mathbf{x}_{i\cdot}, \boldsymbol{\beta})$ is continuously differentiable, so that $\mathbf{x}_i(\boldsymbol{\beta})$ is a continuous function of $\boldsymbol{\beta}$, and given that $\hat{\boldsymbol{\beta}} \xrightarrow{p} \boldsymbol{\beta}_0$ so that necessarily $\boldsymbol{\beta}_* \xrightarrow{p} \boldsymbol{\beta}_0$, it follows that $\mathbf{x}_i(\boldsymbol{\beta}_*) \xrightarrow{p} \mathbf{x}_i(\boldsymbol{\beta}_0)$. Substituting $\mathbf{x}_i(\boldsymbol{\beta}_0)$ for $\mathbf{x}_i(\boldsymbol{\beta}_*)$, and recognizing that an analogous argument holds $\forall i$, finally obtains

$$\begin{aligned}\hat{\boldsymbol{\varepsilon}} &= \left[\mathbf{I}_n - \mathbf{x}(\boldsymbol{\beta}_0)\left[\mathbf{x}(\boldsymbol{\beta}_0)'\mathbf{x}(\boldsymbol{\beta}_0)\right]^{-1}\mathbf{x}(\boldsymbol{\beta}_0)'\right]\boldsymbol{\varepsilon} + o_p\left(n^{-1/2}\right)\\ &= \mathbf{m}(\boldsymbol{\beta}_0)\boldsymbol{\varepsilon} + o_p\left(n^{-1/2}\right)\end{aligned}$$

where now $o_p(n^{-1/2})$ denotes an $(n \times 1)$ vector whose elements are each $o_p(n^{-1/2})$ and $\mathbf{m}(\boldsymbol{\beta}_0)$ denotes the outer-bracketed matrix in the expression for $\hat{\boldsymbol{\varepsilon}}$ above.

Note that $\mathbf{m}(\boldsymbol{\beta}_0)$ is an $(n \times n)$ idempotent matrix of rank $(n-k)$. Comparing this result to the associated estimator for residuals in the linear model, $\hat{\boldsymbol{\varepsilon}} = \left[\mathbf{I}_n - \mathbf{x}(\mathbf{x}'\mathbf{x})^{-1}\mathbf{x}'\right]\boldsymbol{\varepsilon}$, establishes the asymptotically valid analogy between the linear and nonlinear cases. It can be shown that $\hat{\boldsymbol{\varepsilon}}$ has the smallest limiting covariance matrix of all estimators that are asymptotically linear and consistent for $\boldsymbol{\varepsilon}$.

Turning our attention to the issue of estimating the value of $\sigma^2$, examine the inner product of the estimator $\hat{\boldsymbol{\varepsilon}}$ and use the preceding representation of it to conclude that $\hat{\boldsymbol{\varepsilon}}'\hat{\boldsymbol{\varepsilon}} = \boldsymbol{\varepsilon}'\mathbf{m}(\boldsymbol{\beta}_0)\boldsymbol{\varepsilon} + 2\boldsymbol{\varepsilon}'\mathbf{m}(\boldsymbol{\beta}_0)o_p(n^{-1/2}) + o_p(1)$. Regarding the order of magnitude of the trailing term above, note that $o_p(n^{-1/2})'o_p(n^{-1/2})$ is the sum of $n$ terms of order $o_p(n^{-1})$, which then is of order of magnitude $o_p(1) = n\, o_p(n^{-1})$. Then define the estimator for $\sigma^2$ by

$$\hat{S}^2 = \frac{\left(\mathbf{Y} - \mathbf{g}\left(\mathbf{x}, \hat{\boldsymbol{\beta}}\right)\right)'\left(\mathbf{Y} - \mathbf{g}\left(\mathbf{x}, \hat{\boldsymbol{\beta}}\right)\right)}{n-k}.$$

The estimator is a consistent estimator for $\sigma^2$, the proof of which follows closely the proof of the consistency of $\hat{S}^2$ in the linear model case (Theorems 8.5 and 8.6) upon noting that

$$\hat{S}^2 = \frac{\boldsymbol{\varepsilon}'\mathbf{m}(\boldsymbol{\beta}_0)\boldsymbol{\varepsilon}}{n-k} + o_p\left(n^{-1/2}\right)$$

and ignoring the asymptotically irrelevant $o_p(n^{-1/2})$ terms. Note that dividing $\hat{\boldsymbol{\varepsilon}}'\hat{\boldsymbol{\varepsilon}}$ by $(n - k)$ in the definition of the estimator produces an estimator of $\sigma^2$that has the advantage of being unbiased up to terms of order $o_p(n^{-1/2})$, in another analogy to the linear model case (see Theorem 8.3). Note that had we divided $\hat{\boldsymbol{\varepsilon}}'\hat{\boldsymbol{\varepsilon}}$ by $n$ instead of $(n - k)$, we would still have arrived at a consistent estimator of $\sigma^2$.

It can also be shown that $S^2$ is asymptotically normally distributed under additional noise component moment and convergence assumptions.

Having defined an estimator for information on the value of $\sigma^2$, we can now address the issue of generating an operational version of the asymptotic covariance matrix of the NLS estimator,

$$\mathbf{cov}\left(\hat{\boldsymbol{\beta}}\right) = \sigma^2\left(\frac{\partial \mathbf{g}(\mathbf{x}, \boldsymbol{\beta}_0)}{\partial \boldsymbol{\beta}} \frac{\partial \mathbf{g}(\mathbf{x}, \boldsymbol{\beta}_0)}{\partial \boldsymbol{\beta}'}\right)^{-1} = \sigma^2(\mathbf{x}(\boldsymbol{\beta}_0)'\mathbf{x}(\boldsymbol{\beta}_0))^{-1}$$

defined above. In particular, one replaces $\sigma^2$and $\boldsymbol{\beta}_0$ with consistent estimators, yielding

$$\hat{\mathbf{cov}}\left(\hat{\boldsymbol{\beta}}\right) = \hat{S}^2\left(\frac{\partial \mathbf{g}\left(\mathbf{x}, \hat{\boldsymbol{\beta}}\right)}{\partial \boldsymbol{\beta}} \frac{\partial \mathbf{g}\left(\mathbf{x}, \hat{\boldsymbol{\beta}}\right)}{\partial \boldsymbol{\beta}'}\right)^{-1} = \hat{S}^2\left(\mathbf{x}\left(\hat{\boldsymbol{\beta}}\right)'\mathbf{x}\left(\hat{\boldsymbol{\beta}}\right)\right)^{-1}$$

Under the assumptions leading to the consistency, asymptotic normality, and asymptotic linearity of the NLS estimator, it can be shown that $n\left(\hat{\mathbf{cov}}\left(\hat{\boldsymbol{\beta}}\right)\right)$ consistently estimates the covariance matrix of the limiting distribution of $n^{1/2}\left(\hat{\boldsymbol{\beta}} - \boldsymbol{\beta}_0\right)$.

*Summary Remarks on NLS Estimation* The preceding overview of the NLS estimator provides a basic introduction to nonlinear estimation of unknown parameters in nonlinear regression models. It was shown that asymptotically, at least to the first order of approximation, there are striking analogies to the least squares estimator applied in the linear model context in so far as statistical properties are concerned. Indeed, most of the derivations of properties of the NLS estimator can follow by analogy to the linear model derivations upon replacing $\mathbf{x}$ with $\mathbf{x}(\boldsymbol{\beta}_0)$. Other results that can be demonstrated this way include asymptotic efficiency of the NLS estimator and an associated asymptotically-valid MVUE property.

We will see ahead that the analogy can be continued regarding asymptotically valid hypotheses testing and confidence interval or region generation. Of course, as we noted at the outset, calculation of the actual NLS estimates is another matter, and must often be done numerically on a computer. For additional reading on both the theory and application of nonlinear estimation methodology, the reader can begin by consulting A. R. Gallant's *Nonlinear Statistical Models*, John wiley and Sons, and R. Mittelhammer, G. Judge, and D. Miller's *Econometric Foundations*, Cambridge University Press.

## 8.3 The Method of Maximum Likelihood

The method of Maximum Likelihood (ML) can be used to estimate the unknown parameters, or functions of unknown parameters, corresponding to the joint density function of a random sample. The procedure leads to an estimate of $\mathbf{\Theta}$ or $\mathbf{q}(\mathbf{\Theta})$ by maximizing the so-called *likelihood function* of the parameters, given the observed outcome of the random sample. We will focus initially on the problem of estimating $\mathbf{\Theta}$ itself. It will be seen later that maximum likelihood estimation of $\mathbf{q}(\mathbf{\Theta})$ is easily implemented through the so-called **invariance principle of ML** once the problem of estimating $\mathbf{\Theta}$ has been solved.

The **likelihood function** is *identical* in functional form to the joint density function of the random sample. However, there is an important interpretational difference between the two functions. The joint density function is interpreted as a function of the values of the random variable outcomes $(x_1,\ldots,x_n)$, *given* values of the parameters $(\Theta_1,\ldots,\Theta_k)$. The interpretation of the likelihood function is just the reverse—it is a function of the values of the parameters $(\Theta_1,\ldots,\Theta_k)$, *given* values of the random variable outcomes $(x_1,\ldots,x_n)$. Thus, $L(\mathbf{\Theta};\mathbf{x}) \equiv f(\mathbf{x};\mathbf{\Theta})$ defines the functional form of the likelihood function $L(\mathbf{\Theta};\mathbf{x})$, but now $\mathbf{\Theta}$ *precedes* the semicolon and $\mathbf{x}$ *follows* the semicolon to denote that the likelihood function is a function of $\mathbf{\Theta}$, for given values of $\mathbf{x}$.

The **maximum likelihood estimate** of $\mathbf{\Theta}$ is the solution of the maximization problem $\max_{\mathbf{\Theta}\in\Omega}\{L(\mathbf{\Theta};\mathbf{x})\}$ where $\Omega$ is the appropriate parameter space. The maximum likelihood estimate is thus defined as $\hat{\mathbf{\theta}} = \hat{\mathbf{\Theta}}(\mathbf{x}) = \arg\max_{\mathbf{\Theta}\in\Omega}\{L(\mathbf{\Theta};\mathbf{x})\}$ .[18] The **maximum likelihood estimator (MLE)** is the random variable or vector that generates the estimates above, and is defined as $\hat{\mathbf{\Theta}} = \hat{\mathbf{\Theta}}(\mathbf{X}) = \arg\max_{\mathbf{\Theta}\in\Omega}\{L(\mathbf{\Theta};\mathbf{X})\}$. The maximum likelihood procedure can be interpreted as choosing, from among all feasible candidates, the value of the parameter vector $\mathbf{\Theta}$ that identifies the joint density function $f(\mathbf{x};\mathbf{\Theta})$ assigning the highest probability (discrete case) or highest density weighting (continuous case) to the random sample outcome, $\mathbf{x}$, actually observed. Put another way, the maximum likelihood procedure chooses a parameter vector value so as to identify a particular member of a parametric family of densities, $L(\mathbf{\Theta};\mathbf{x}) \equiv f(\mathbf{x};\mathbf{\Theta})$, $\mathbf{\Theta}\in\Omega$, that assigns the highest "likelihood" to generating the random sample outcome $\mathbf{x}$ actually observed. Of course, whether the estimator of $\mathbf{\Theta}$ implied by this procedure is a "good one" depends on the statistical properties of the maximum likelihood estimator $\hat{\mathbf{\Theta}}$.

### 8.3.1 MLE Implementation

In order to implement the ML procedure, the functional form of $f(\mathbf{x};\mathbf{\Theta})$ and hence $L(\mathbf{\Theta};\mathbf{x})$ must be specified along with the feasible choices of $\mathbf{\Theta}$, represented by the parameter space, $\Omega$. This is generally accomplished in practice by specifying a

[18]Recall that $\arg\max_w\{f(w)\}$denotes the *argument* value of $f(w)$ that *maximizes* $f(w)$, where *argument* value means *the value of* $w$. Also, $\arg\max_{w\in\Omega}\{f(w)\}$ denotes the value of $w \in \Omega$ that maximizes $f(w)$.

parametric family of density functions to provide a representation of the probability model for the random sample under investigation. Once the parametric family is identified, the estimation problem focuses on the maximization of the likelihood function. In many cases, the likelihood function will be differentiable with respect to the parameter vector and will possess a maximum which will be interior to the parameter. In these cases, the classical calculus-based approach to maximization, in which first order conditions are used to solve for the likelihood-maximizing value of the $\boldsymbol{\Theta}$ vector, can be used to solve the MLE problem. That is, the ML estimate, $\hat{\boldsymbol{\theta}}$, can be found as the solution to the vector equation

$$\underset{(k\times 1)}{\frac{\partial L(\boldsymbol{\Theta};\mathbf{x})}{\partial \boldsymbol{\Theta}}} = \begin{bmatrix} \frac{\partial L(\boldsymbol{\Theta};\mathbf{x})}{\partial \boldsymbol{\Theta}_1} \\ \vdots \\ 0\frac{\partial L(\boldsymbol{\Theta};\mathbf{x})}{\partial \boldsymbol{\Theta}_k} \end{bmatrix} = \underset{(k\times 1)}{\mathbf{0}}.$$

The solution will be a function of the random sample outcome $\mathbf{x}$, as $\hat{\boldsymbol{\theta}} = \hat{\boldsymbol{\Theta}}(\mathbf{x}) = \arg_{\boldsymbol{\Theta}\in\Omega}\{\partial L(\boldsymbol{\Theta};\mathbf{x})/\partial\boldsymbol{\Theta} = \mathbf{0}\}$.[19]

Note that it may not be possible to *explicitly* solve the first order conditions for $\hat{\boldsymbol{\theta}}$ in terms of a function of $\mathbf{x}$. If not, numerical methods are used to find the value of $\hat{\boldsymbol{\theta}}$ that satisfied the first order conditions. More generally, even if the classical maximization approach is not appropriate (e.g., there is no interior maximum, or $L(\boldsymbol{\Theta};\mathbf{x})$ is not differentiable with respect to $\boldsymbol{\Theta}$), a value of $\hat{\boldsymbol{\theta}}$ that solves $\max_{\boldsymbol{\Theta}\in\Omega}\{L(\boldsymbol{\Theta};\mathbf{x})\}$ is a ML estimate of $\boldsymbol{\Theta}$, *no matter how it is derived.*

The estimator function $\hat{\boldsymbol{\theta}} = \hat{\boldsymbol{\Theta}}(\mathbf{x})$ will either be explicitly derivable, or else it will be an *implicit* function implied by the functional dependence of $\hat{\boldsymbol{\theta}}$ on $\mathbf{x}$, and in either case $\hat{\boldsymbol{\Theta}} = \hat{\boldsymbol{\Theta}}(\mathbf{X})$ is referred to as a maximum likelihood estimator of $\boldsymbol{\Theta}$.[20]

The following examples illustrate the derivation of MLEs. Note that in some problem situations the calculations are considerably simplified by maximizing $\ln(L(\boldsymbol{\Theta};\mathbf{x}))$ as opposed to $L(\boldsymbol{\Theta};\mathbf{x})$. Since the logarithmic transformation is strictly monotonically increasing, $\hat{\boldsymbol{\theta}}$ maximizes $L(\boldsymbol{\Theta};\mathbf{x})$ *iff* it also maximizes $\ln(L(\boldsymbol{\Theta};\mathbf{x}))$, i.e.,

$$L\left(\hat{\boldsymbol{\theta}},\mathbf{x}\right) = \max_{\boldsymbol{\Theta}\in\Omega}\{L(\boldsymbol{\Theta};\mathbf{x})\} \Leftrightarrow \ln\left(L\left(\hat{\boldsymbol{\theta}};\mathbf{x}\right)\right) = \max_{\boldsymbol{\Theta}\in\Omega}\{\ln(L(\boldsymbol{\Theta};\mathbf{x}))\}$$

and

$$\hat{\boldsymbol{\theta}} = \arg\max_{\boldsymbol{\Theta}\in\Omega}\{L(\boldsymbol{\Theta};\mathbf{x})\} = \arg\max_{\boldsymbol{\Theta}\in\Omega}\{\ln(L(\boldsymbol{\Theta};\mathbf{x}))\}.$$

---

[19] Recall that $\arg_{\boldsymbol{\Theta}\in\Omega}\{\mathbf{g}(\boldsymbol{\Theta}) = \mathbf{c}\}$ represents the value of $\boldsymbol{\Theta}\in\Omega$ that satisfies or solves $\mathbf{g}(\boldsymbol{\Theta}) = \mathbf{c}$.

[20] We are suppressing the fact that a maximum of $L(\boldsymbol{\Theta};\mathbf{x})$ may not be attainable. For example, if the parameter space is an open interval and if the likelihood function is strictly monotonically increasing, then no maximum can be stated. If a maximum of $L(\boldsymbol{\Theta};\mathbf{x})$ for $\boldsymbol{\Theta}\in\Omega$ does not exist, then the MLE of $\boldsymbol{\Theta}$ does not exist.

Thus the objective function of the ML estimation problem can be chosen to be $L(\boldsymbol{\Theta};\mathbf{x})$ or $\ln(L(\boldsymbol{\Theta};\mathbf{x}))$, whichever is more convenient.[21] In applying the logarithmic transformation, we *define* $\ln(0) \equiv -\infty$ to accommodate points where the likelihood function is zero-valued.

**Example 8.4** ***MLE for Exponential Distribution***

Let $\mathbf{X} = (X_1,\ldots,X_n)$ be a random sample from an exponential population distribution representing the operating time until a work stoppage occurs on an assembly line, so that $X_i \sim \theta^{-1} \exp(-x_i/\theta) I_{(0,\infty)}(x_i)\ \forall i$. We seek the MLE for $\theta$, the mean operating time until a work stoppage.

The functional form for the joint density function of the random sample, and hence the functional form of the likelihood function, is given by

$$L(\theta; x_1,\ldots,x_n) \equiv f(x_1,\ldots,x_n;\theta) = \theta^{-n} \exp\left(-\sum_{i=1}^{n} x_i/\theta\right) \prod_{i=1}^{n} I_{(0,\infty)}(x_i),$$

where $\theta \in \Omega = (0,\infty)$. Then,

$$\ln(L(\theta; x_1,\ldots,x_n)) = -n\ln(\theta) - \frac{\sum_{i=1}^{n} x_i}{\theta} + \ln\left(\prod_{i=1}^{n} I_{(0,\infty)}(x_i)\right).$$

The first order condition for maximizing $\ln(L(\theta;\mathbf{x}))$ with respect to $\theta$ is given by

$$\frac{d\ln L}{d\theta} = -\frac{n}{\theta} + \frac{\sum_{i=1}^{n} x_i}{\theta^2} = 0.$$

Thus the solution $\hat{\theta} = \hat{\Theta}(\mathbf{x}) = \sum_{i=1}^{n} x_i/n = \bar{x}$ is the ML estimate of $\theta$, and $\hat{\Theta} = \hat{\Theta}(\mathbf{X}) = \sum_{i=1}^{n} X_i/n = \bar{X}_n$ is the MLE of $\Theta$. (The second order conditions for a maximum are met.) □

**Example 8.5** ***MLE for Normal Distribution***

Let $\mathbf{X} = (X_1,\ldots,X_n)$ be a random sample from a normal population distribution representing the actual fill volumes of 1 liter bottles of liquid laundry detergent so that

$$X_i \sim \frac{1}{(2\pi)^{1/2}\sigma} \exp\left[-\frac{1}{2\sigma^2}(x_i - \mu)^2\right],\ \forall i.$$

---

[21] In the case where the classical first order conditions are applicable, note that if $L(\boldsymbol{\Theta},\mathbf{x}) > 0$ (which will necessarily be true at the maximum value), then

$$\frac{\partial \ln(L(\boldsymbol{\Theta};\mathbf{x}))}{\partial\boldsymbol{\Theta}} = \frac{1}{L(\boldsymbol{\Theta};\mathbf{x})}\frac{\partial L(\boldsymbol{\Theta};\mathbf{x})}{\partial\boldsymbol{\Theta}},$$

and thus any $\boldsymbol{\Theta}$ for which $\partial L(\boldsymbol{\Theta};\mathbf{x})/\partial\boldsymbol{\Theta} = \mathbf{0}$ also satisfies $\partial \ln(L(\boldsymbol{\Theta};\mathbf{x}))/\partial\boldsymbol{\Theta} = \mathbf{0}$. Regarding second order conditions, note that if $\boldsymbol{\Theta}$ satisfies the first order conditions, then

$$\frac{\partial^2 \ln(L(\boldsymbol{\Theta};\mathbf{x}))}{\partial\boldsymbol{\Theta}\partial\boldsymbol{\Theta}'} = \frac{1}{L(\boldsymbol{\Theta};\mathbf{x})}\frac{\partial^2 L(\boldsymbol{\Theta};\mathbf{x})}{\partial\boldsymbol{\Theta}\partial\boldsymbol{\Theta}'} - \frac{\partial L(\boldsymbol{\Theta};\mathbf{x})}{\partial\boldsymbol{\Theta}}\frac{\partial L(\boldsymbol{\Theta};\mathbf{x})}{\partial\boldsymbol{\Theta}'} = \frac{1}{L(\boldsymbol{\Theta};\mathbf{x})}\frac{\partial^2 L(\boldsymbol{\Theta};\mathbf{x})}{\partial\boldsymbol{\Theta}\partial\boldsymbol{\Theta}'}$$

since $\partial L(\boldsymbol{\Theta};\mathbf{x})/\partial\boldsymbol{\Theta} = \mathbf{0}$. Then since $L(\boldsymbol{\Theta};\mathbf{x}) > 0$ at the maximum, $\partial^2 \ln(L(\boldsymbol{\Theta};\mathbf{x}))/\partial\boldsymbol{\Theta}\partial\boldsymbol{\Theta}'$ is negative definite *iff* $\partial^2(L(\boldsymbol{\Theta};\mathbf{x}))/\partial\boldsymbol{\Theta}\partial\boldsymbol{\Theta}'$ is negative definite.

We seek the MLE for the mean and variance of fill volumes.

The functional form for the joint density of the random sample, and hence the functional form of the likelihood function, is given by

$$L(\mu,\sigma^2;\mathbf{x}) \equiv f(\mathbf{x};\mu,\sigma^2) = \frac{1}{(2\pi\sigma^2)^{n/2}} \exp\left[-\frac{1}{2\sigma^2}\sum_{i=1}^{n}(x_i-\mu)^2\right],$$

where $\mu \geq 0$ and $\sigma^2 > 0$. Then

$$\ln L(\mu,\sigma^2;\mathbf{x}) = -\frac{n}{2}\ln(2\pi) - \frac{n}{2}\ln(\sigma^2) - \frac{1}{2\sigma^2}\sum_{i=1}^{n}(x_i-\mu)^2.$$

The first order conditions for the maximum of $\ln(L(\mu,\sigma^2;\mathbf{x}))$are

$$\frac{\partial \ln(L)}{\partial \mu} = \sigma^{-2}\sum_{i=1}^{n}(x_i-\mu) = 0$$

$$\frac{\partial \ln(L)}{\partial \sigma^2} = \frac{-n}{2\sigma^2} + \frac{1}{2\sigma^4}\sum_{i=1}^{n}(x_i-\mu)^2 = 0.$$

The solution $\hat{\mu}(\mathbf{x}) = \sum_{i=1}^{n} x_i/n$ and $\hat{\sigma}^2(\mathbf{x}) = \sum_{i=1}^{n}(x_i-\bar{x})^2/n$ defines the ML estimate of $\mu$ and $\sigma^2$, and the MLE is given by $\hat{\mu}(\mathbf{X})$ and $\hat{\sigma}^2(\mathbf{X})$. (The second order conditions for a maximum are met.) □

**Example 8.6** ***Maximum Likelihood Estimation in the GLM***

Let $\mathbf{Y} = \mathbf{x}\boldsymbol{\beta} + \boldsymbol{\varepsilon}$ with $\mathbf{Y} \sim N(\mathbf{x}\boldsymbol{\beta},\sigma^2\mathbf{I})$ represent the relationship between a random sample of family expenditures on consumer durables, $\mathbf{Y}$, and the respective levels of disposable income and other sociodemographic factors for the families, $\mathbf{x}$. We seek the MLE for the marginal effects of sociodemographic variables on consumption, i.e., for $\partial \mathrm{E}(Y_i)/\partial \mathbf{x}'_{i.} = \boldsymbol{\beta}$, and for the variance of consumption, $\sigma^2$. The likelihood function in this case is

$$L(\boldsymbol{\beta},\sigma^2\mathbf{y}) = N(\mathbf{x}\boldsymbol{\beta},\sigma^2\mathbf{I}) = \frac{1}{(2\pi\sigma^2)^{n/2}}\exp\left[-\frac{1}{2\sigma^2}(\mathbf{y}-\mathbf{x}\boldsymbol{\beta})'(\mathbf{y}-\mathbf{x}\boldsymbol{\beta})\right],$$

where $\boldsymbol{\beta} \in \mathbb{R}^k$ and $\sigma^2 > 0$.[22] Then

$$\ln(L(\boldsymbol{\beta},\sigma^2;\mathbf{y})) = -(n/2)\ln(2\pi) - (n/2)\ln(\sigma^2) - (1/2\sigma^2)(\mathbf{y}-\mathbf{x}\boldsymbol{\beta})'(\mathbf{y}-\mathbf{x}\boldsymbol{\beta}).$$

The first order conditions for the maximum of $\ln(L(\boldsymbol{\beta},\sigma^2;\mathbf{y}))$ are given by (recall Lemma 8.1)

$$\frac{\partial \ln(L(\boldsymbol{\beta},\sigma^2;\mathbf{y}))}{\partial \boldsymbol{\beta}} = \frac{\partial\left[-\frac{1}{2\sigma^2}(\mathbf{y}'\mathbf{y} - 2\boldsymbol{\beta}'\mathbf{x}'\mathbf{y} + \boldsymbol{\beta}'\mathbf{x}'\mathbf{x}\boldsymbol{\beta})\right]}{\partial \boldsymbol{\beta}} = -\frac{1}{2\sigma^2}(-2\mathbf{x}'\mathbf{y} + 2\mathbf{x}'\mathbf{x}\boldsymbol{\beta}) = \mathbf{0},$$

$$\frac{\partial \ln(L(\boldsymbol{\beta},\sigma^2;\mathbf{y}))}{\partial \sigma^2} = -\frac{n}{2\sigma^2} + \frac{1}{2\sigma^4}(\mathbf{y}-\mathbf{x}\boldsymbol{\beta})'(\mathbf{y}-\mathbf{x}\boldsymbol{\beta}) = 0$$

[22] Economic theory may suggest constraints on the signs of some of the entries in $\boldsymbol{\beta}$ (e.g., the effect of income on durables consumption will be positive), in which case $\boldsymbol{\beta} \in \Omega_{\beta} \subset \mathbb{R}^k$ may be more appropriate.

The solution to the first order conditions implies that the ML estimate is

$$\hat{\boldsymbol{\beta}}(\mathbf{y}) = (\mathbf{x}'\mathbf{x})^{-1}\mathbf{x}'\mathbf{y} \text{ and } \hat{\sigma}^2(\mathbf{y}) = \frac{\left(\mathbf{y} - \mathbf{x}\hat{\boldsymbol{\beta}}(\mathbf{y})\right)'\left(\mathbf{y} - \mathbf{x}\hat{\boldsymbol{\beta}}(\mathbf{y})\right)}{n},$$

with the MLE given by $\hat{\boldsymbol{\beta}}(\mathbf{Y})$ and $\hat{\sigma}^2(\mathbf{Y})$. (The second order conditions for a maximum are met.) □

The following examples illustrate cases where the standard calculus approach *cannot* be utilized.

**Example 8.7** ***MLE for Uniform Distribution***

Let $\mathbf{X} = (X_1,\ldots,X_n)$ be a random sample from a uniform population distribution representing measurements of the hardness of steel, based on the Rockwell scale, produced by a certain foreign manufacturer, so that $X_i \sim (1/(b-a))I_{[a,b]}(x_i)\forall i$. We seek an MLE for the lower and upper bounds to the hardness measurements, $a$ and $b$, respectively.

The likelihood function is given by

$$L(a,b;\mathbf{x}) \equiv f(\mathbf{x};a,b) = (b-a)^{-n}\prod_{i=1}^{n} I_{[a,b]}(x_i).$$

where $a < b$. It is clear that for $L(a,b;\mathbf{x})$ to be maximized, $a$ and $b$ must be chosen so as to make $(b-a)$ as small as possible while still maintaining $\prod_{i=1}^{n} I_{[a,b]}(x_i) = 1$. The smallest choice for $b$ is given by $\max(x_1,\ldots,x_n)$, while the largest choice for $a$ is given by $\min(x_1,\ldots,x_n)$, yielding the smallest $(b-a) = \max(x_1,\ldots,x_n) - \min(x_1,\ldots,x_n)$. Thus, the MLE estimates are given by outcomes of the smallest and largest *order statistics*, as $\hat{a}(\mathbf{x}) = \min(x_1,\ldots,x_n)$ and $\hat{b}(\mathbf{x}) = \max(x_1,\ldots,x_n)$, and the MLEs are then $\hat{a}(\mathbf{X})$ and $\hat{b}(\mathbf{X})$. □

**Example 8.8** ***MLE for Hypergeometric Distribution***

Let $X \sim$ Hypergeometric $(x;M,n,k)$ represent the number of defective parts found in a random sample without replacement of $n$ parts taken from a shipment of $M$ parts, where $M$ and $n$ are known. We seek an MLE of the number of defective parts, $k$, in the shipment of $M$ parts.

The likelihood function is given by

$$L(k;x) \equiv f(x;k) = \frac{\binom{k}{x}\binom{M-k}{n-x}}{\binom{M}{n}} \quad \text{for } x = 0,\ 1,\ldots,k,$$

where $k \in \{0, 1, 2,\ldots,M\}$. Finding the solution for $k$ that maximizes $L(k;x)$ is essentially an *integer programming problem*. Specifically, note that

$$\frac{L(k;x)}{L(k-1;x)} = \frac{\dfrac{k!}{x!(k-x)!}}{\dfrac{(k-1)!}{x!(k-x-1)!}} \; \frac{\dfrac{(M-k)!}{(n-x)!(M-k-(n-x))!}}{\dfrac{(M-k+1)!}{(n-x)!(M-k-(n-x)+1)!}} = \frac{k(M-k-(n-x)+1)}{(k-x)(M-k+1)}.$$

Then $L(k;x)/L(k-1;x) \geq 1$ *iff* $k(M-k-(n-x)+1) \geq (k-x)(M-k+1)$, which after algebraic simplification is equivalent to $k \leq n^{-1}x(M+1)$. The implication of the preceding result is that the likelihood function increases as the integer value of $k$ increases so long as $k \leq n^{-1}x(M+1)$. Therefore, the ML estimate of $k$ equals the largest integer not exceeding $n^{-1}x(M+1)$, which we can represent as $\hat{k}(x) = \text{trunc}(n^{-1}x(M+1))$, where recall that trunc($w$) is the truncation function that truncates the decimal part of $w$, e.g., trunc(2.76) = 2. The MLE of $k$ would then be $\hat{k}(X) = \text{trunc}(n^{-1}X(M+1))$. □

### 8.3.2 MLE Properties: Finite Sample

As we stated at the beginning of this section, whether estimates produced by the ML procedure are "good ones" depends on the statistical properties that an MLE possesses. It turns out that there are a number of reasons why we might expect that the ML procedure would lead to good estimates of $\boldsymbol{\Theta}$. First of all, if an unbiased estimator of $\boldsymbol{\Theta}$ exists that achieves the CRLB, the MLE will be this estimator if the MLE is defined by solving first order conditions for maximizing the likelihood function.

**Theorem 8.12** ***MLE Attainment of the CRLB*** *If an unbiased estimator,* $\mathbf{T} = \mathbf{t}(\mathbf{X})$, *of* $\boldsymbol{\Theta}$ *exists whose covariance matrix equals the CRLB, and if the MLE is defined by solving first order conditions for maximizing the likelihood function, then the MLE is equal to* $\mathbf{T} = \mathbf{t}(\mathbf{X})$ *with probability* 1.

**Proof** If there exists an unbiased estimator, $\mathbf{t}(\mathbf{X})$, whose covariance matrix equals the CRLB, then regardless of the value of $\boldsymbol{\Theta} \in \Omega$, the CRLB attainment theorem (Theorem 7.20) implies that the estimator has outcomes defined by

$$\mathbf{t}(\mathbf{x}) = \boldsymbol{\Theta} + \left[ \mathrm{E}\left( \frac{\partial \ln(L(\boldsymbol{\Theta};\mathbf{X}))}{\partial \boldsymbol{\Theta}} \frac{\partial \ln(L(\boldsymbol{\Theta};\mathbf{X}))}{\partial \boldsymbol{\Theta}'} \right) \right]^{-1} \frac{\partial \ln(L(\boldsymbol{\Theta};\mathbf{x}))}{\partial \boldsymbol{\Theta}}$$

with probability 1, where we have expressed the result of the CRLB attainment theorem using likelihood function notation. Substituting the ML estimate, $\hat{\boldsymbol{\theta}} \in \Omega$, for $\boldsymbol{\Theta}$ in the preceding equality implies that $\mathbf{t}(\mathbf{x}) = \hat{\boldsymbol{\theta}}$, since $\hat{\boldsymbol{\theta}}$ would satisfy the first order conditions $\partial \ln(L(\hat{\boldsymbol{\theta}};\mathbf{x})/\partial \boldsymbol{\Theta}) = \mathbf{0}$. Thus outcomes of the MLE and $\mathbf{t}(\mathbf{X})$ coincide with probability 1. ■

Therefore, under the conditions of Theorem 8.12, the MLE will also be the MVUE for $\boldsymbol{\Theta}$.

We can also show that if the MLE is uniquely defined, then the MLE can be equivalently represented as a function of *any* sufficient statistics for $f(\mathbf{x};\boldsymbol{\Theta})$, and in particular, a function of *complete* sufficient statistics when complete sufficient statistics exist.

**Theorem 8.13 Unique MLEs Are Functions of Any Sufficient Statistics for $f(\mathbf{x};\Theta)$**

*Assume that the MLE $\hat{\Theta}$ of $\Theta$ is uniquely defined in terms of $\mathbf{X}$. If $\mathbf{S} = (S_1,\ldots,S_r)'$ is any vector of sufficient statistics for $f(\mathbf{x};\Theta) \equiv L(\Theta;\mathbf{x})$, then there exists a function of $\mathbf{S}$, say $\boldsymbol{\tau}(\mathbf{S})$, such that $\hat{\boldsymbol{\theta}} = \boldsymbol{\tau}(\mathbf{s})$.*

**Proof** The Neyman Factorization theorem states that $L(\Theta;\mathbf{x}) \equiv f(\mathbf{x};\Theta) = g(s_1,\ldots,s_r;\Theta)\,h(\mathbf{x})$ where $(s_1,\ldots,s_r)$ are sufficient statistics, which can be chosen as complete sufficient statistics if they exist. Now because $L(\Theta;\mathbf{x}) \geq 0$, $g$ and $h$ can always be defined as nonnegative-valued functions, in which case for a given value of $\mathbf{x}$, $L(\Theta;\mathbf{x}) \propto g(s_1,\ldots,s_r;\Theta)$, where "$\propto$" means "proportional to" and the proportionality constant is $h(\mathbf{x})$. It follows that for a given value of $\mathbf{x}$, if the MLE is unique, then

$$\hat{\boldsymbol{\theta}} = \underset{\Theta\in\Omega}{\arg\max}\{L(\Theta;\mathbf{x})\} = \underset{\Theta\in\Omega}{\arg\max}\,\{g(s_1,\ldots,s_r;\Theta)\}.$$

Thus $\hat{\Theta}$ maximizes $L(\Theta;\mathbf{x})$ *iff* $\hat{\boldsymbol{\theta}}$ maximizes $g(s_1,\ldots,s_r;\Theta)$. But the latter maximization problem implies that the unique maximizing choice of $\Theta$ is then a function of the values $(s_1,\ldots,s_r)$, so that $\hat{\Theta} = \boldsymbol{\tau}(s_1,\ldots,s_r)$.[23] ■

If the sufficient statistics $(S_1,\ldots,S_r)$ used in the Neyman Factorization theorem are complete, then the unique MLE $\hat{\Theta} = \boldsymbol{\tau}(S_1,\ldots,S_r)$ is a function of the complete sufficient statistics by Theorem 8.13. It follows from the Lehmann-Scheffe' completeness theorem that if $\hat{\Theta}$ is also unbiased (or if $\hat{\Theta}$ can be transformed so as to be unbiased) then the MLE (or the **bias-adjusted MLE**) is the MVUE for $\Theta$. We formalize this observation in the following theorem:

**Theorem 8.14 MVUE Property of Unique Unbiased or Bias-Adjusted MLEs**

*Assume that the MLE $\hat{\Theta}$ of $\Theta$ is uniquely defined in terms of $\mathbf{X}$, and that a vector of complete sufficient statistics, $\mathbf{S}$, exists for $f(\mathbf{x};\Theta) \equiv L(\Theta;\mathbf{x})$. If $\hat{\Theta}$ or $\boldsymbol{\eta}(\hat{\Theta})$ is an unbiased estimator of $\Theta$, then $\hat{\Theta}$ or $\boldsymbol{\eta}(\hat{\Theta})$ is the* MVUE of $\Theta$.

**Proof** From Theorem 8.13 it follows that the MLE is a function of the complete sufficient statistics as $\hat{\Theta} = \boldsymbol{\tau}(\mathbf{S})$. It follows from the Lehmann-Scheffe' Completeness Theorem (Theorem 7.22) that if $\hat{\Theta}$ is unbiased, then $\hat{\Theta}$ is the MVUE of $\Theta$. Alternatively, if $\hat{\Theta}$ is biased, but the function $\boldsymbol{\eta}(\hat{\Theta})$ of the MLE is an unbiased estimator of $\Theta$, then because $\boldsymbol{\eta}(\hat{\Theta}) = \boldsymbol{\eta}(\boldsymbol{\tau}(\mathbf{S}))$ is a (composite) function of the complete sufficient statistics, $\boldsymbol{\eta}(\hat{\Theta})$ is the MVUE of $\Theta$ by the Lehmann-Scheffe' Completeness Theorem. ■

[23]In the event that a MLE is *not* unique, then the *set* of MLE's is a function of any set of sufficient statistics. However, a particular MLE within the set of MLE's need not necessarily be a function of $(s_1,\ldots,s_r)$, although it is always possible to choose an MLE that *is* a function of $(s_1,\ldots,s_r)$. See Moore, D.S., (1971), "*Maximum Likelihood and Sufficient Statistics*", American Mathematical Monthly, January, pp. 50–52.

In Example 8.4, $\sum_{i=1}^{n} X_i$ is a complete sufficient statistic for $f(\mathbf{x};\boldsymbol{\Theta}) \equiv L(\boldsymbol{\Theta};\mathbf{x})$, and the MLE is unique and unbiased, and thus the MLE is MVUE. In Example 8.5, $\sum_{i=1}^{n} X_i^2$ and $\sum_{i=1}^{n} X_i$ are complete sufficient statistics, and the MLEs $\hat{\mu}$ and $\hat{\sigma}^2$ are unique, so that they are functions of the complete sufficient statistics. However, $\hat{\sigma}^2$ is *not* unbiased. The bias can be removed by multiplying $\hat{\sigma}^2$ by $n/(n-1)$. Then the bias-adjusted MLE of $\mu$ and $\sigma^2$, namely $(\hat{\mu}, n\hat{\sigma}^2/(n-1))$, is the MVUE by the Lehmann-Scheffe' completeness theorem. The case of Example 8.6 was discussed in the previous section on the least squares estimator, where $(\hat{\boldsymbol{\beta}}, n\hat{\sigma}^2/(n-k))$ is also the bias-adjusted MLE and was found to be the MVUE of $(\boldsymbol{\beta},\sigma^2)$. In Example 8.7, it can be shown that the MLE $\hat{b}(\mathbf{X}) = \max(X_1,\ldots,X_n)$ and $\hat{a}(\mathbf{X}) = \min(X_1,\ldots,X_n)$ is unique and is itself a complete sufficient statistic and that $\hat{\alpha}(\mathbf{X}) = ((n+1)/n)\hat{a}(\mathbf{X})$, and $\hat{\beta}(\mathbf{X}) = ((n+1)/n)\hat{b}(\mathbf{X})$ are unbiased estimators of $a$ and $b$, respectively. Thus, the bias-adjusted MLE $\left(\hat{\alpha}(\mathbf{X}), \hat{\beta}(\mathbf{X})\right)$ is the MVUE for $(a,b)$. (See Bickel and Doksum, (1969) op. cit., pp. 125–126, and Johnson and Kotz, *Discrete Distributions*, John Wiley, pp. 146–148). Finally, in Example 8.8, it can be shown that $X$ is a complete sufficient statistic, and the unique MLE can be transformed as $t(X) = (M/n)X$ to define an unbiased estimator of $k$. Then the bias-adjusted MLE, $t(X) = (M/n)X$, is the MVUE for $k$ (see Bickel and Doksum, (1969) op. cit., pp. 122–123).

As we have seen above, the maximum likelihood procedure is a rather straight-forward approach to defining estimators of unknown parameters, and in many cases the estimator, or a simple transformation of it, will be unbiased and the MVUE. And if not an MVUE itself, a unique MLE will always be a function of complete sufficient statistics whenever the latter exist, so that the unique MLE is a reasonable starting point in the search for the MVUE. On the other hand, the MLE need not be unbiased nor be the MVUE, and there may be no apparent transformation of the MLE that achieves the MVUE property. It is useful to examine asymptotic properties of MLE's since, even if an MLE does not possess the finite sample properties that one might desire, MLE's possess desirable large sample properties under general conditions, and these latter properties can still serve to rationalize the use of an MLE for estimating the parameters of a particular statistical model.

### 8.3.3 MLE Properties: Large Sample

There are two basic approaches that one can follow in establishing asymptotic properties of an MLE. First of all, if a MLE can be explicitly solved for, so that one can analyze an explicit real-valued function, $\hat{\boldsymbol{\Theta}} = \hat{\boldsymbol{\Theta}}(\mathbf{X})$, of the random sample $\mathbf{X}$, then it might be possible to apply laws of large numbers and/or central limit theorems directly to $\hat{\boldsymbol{\Theta}}(\mathbf{X})$ to investigate the asymptotic properties of the MLE. Note the following example:

**Example 8.9** ***Asymptotics of Exponential MLE via Direct Evaluation***

Recall Example 8.4, where it was found that the MLE of $\theta$, when random sampling from an exponential population distribution, is given by $\hat{\Theta} = n^{-1}\sum_{i=1}^{n} X_i = \bar{X}_n$. Through direct evaluation of the MLE function definition, we can establish that $\hat{\Theta}$ is a consistent, asymptotically normal, and asymptotically efficient estimator of $\theta$. In fact, the procedure for establishing these asymptotic properties has already been carried out in Example 7.23, where it was demonstrated that $\hat{\Theta} = \bar{X}_n$ was a consistent and asymptotically efficient estimator of $\theta$ having an asymptotically normal distribution given by $\hat{\Theta} \overset{a}{\sim} N(\theta, \theta^2/n)$.□

At times, the function defining a MLE cannot be defined explicitly, even though the MLE estimates can be calculated, or else the explicit definition of an MLE may be so complicated as to make it unclear how laws of large numbers or central limit theorems could be applied. For these cases, regularity conditions on the likelihood functions have been presented in the literature that ensure the MLE is consistent, asymptotically normal, and asymptotically efficient. As an illustration of a situation in which direct evaluation of the asymptotic properties (and finite sample properties) of an MLE is not possible, consider the following example in which maximum likelihood estimation of the parameters of a gamma density is being pursued.

**Example 8.10** ***MLE for Parameters of a Gamma Distribution***

Let $(X_1,\ldots,X_n)$ be a random sample from a gamma population distribution representing the time between breakdowns of a certain type of refrigeration equipment used in the frozen foods section of a major supermarket chain, so that

$$X_i \sim \frac{1}{\beta^{\alpha}\Gamma(\alpha)} x_i^{\alpha-1} e^{-x_i/\beta} I_{(0,\infty)}(x_i)\ \forall i.$$

The likelihood function is given by

$$L(\alpha,\beta;\mathbf{x}) = \frac{1}{\beta^{n\alpha}[\Gamma(\alpha)]^n} \prod_{i=1}^{n} x_i^{\alpha-1} \exp\left(-\sum_{i=1}^{n} x_i/\beta\right) \prod_{i=1}^{n} I_{(0,\infty)}(x_i)$$

where $\alpha > 0$ and $\beta > 0$. The log-likelihood is given by

$$\ln(L(\alpha,\beta;\mathbf{x})) = -n\alpha(\ln(\beta)) - n(\ln(\Gamma(\alpha))) + (\alpha-1)\sum_{i=1}^{n}\ln(x_i) - \sum_{i=1}^{n} x_i/\beta + \ln\left(\prod_{i=1}^{n} I_{(0,\infty)}(x_i)\right).$$

The first-order conditions characterizing the maximum of the log-likelihood function are given by

$$\frac{\partial \ln(L)}{\partial \alpha} = -n(\ln(\beta)) - \frac{n}{\Gamma(\alpha)}\frac{d\Gamma(\alpha)}{d\alpha} + \sum_{i=1}^{n}\ln(x_i) = 0$$

$$\frac{\partial \ln(L)}{\partial \beta} = -n(\alpha/\beta) + \left(\sum_{i=1}^{n} x_i\right)/\beta^2 = 0.$$

Note that the second condition implies that $\alpha\beta = \bar{X}_n$, or $\beta = \bar{X}_n/\alpha$. Substituting this result for $\beta$ in the first condition implies

$$\ln(\alpha) - \frac{d\Gamma(\alpha)}{d\alpha}/\Gamma(\alpha) = \ln(\bar{x}_n) - \left(\sum_{i=1}^{n} \ln(x_i)\right)/n,$$

and there is *no* explicit solution for $\alpha$ in terms of $(x_1,\ldots,x_n)$, although $\alpha$ is an *implicit* function of $(x_1,\ldots,x_n)$. A unique value of $\alpha$ satisfying the above equality can be solved for numerically on a computer, which can then be used to solve for $\beta$ using the equation $\beta = \bar{X}_n/\alpha$.[24] Thus, the ML estimates for $(\alpha,\beta)$ can be calculated. However, since an explicit functional form for the MLE is not identifiable, an analysis of the estimator's finite sample and asymptotic properties is quite difficult.

Regarding finite sample properties of the MLE, the reader can verify by an appeal to Theorem 7.6 that $(\sum_{i=1}^{n} X_i, \sum_{i=1}^{n} \ln(X_i))$ is a set of complete sufficient statistics for this problem, so that the MLE $\left(\hat{\alpha}(\mathbf{X}), \hat{\beta}(\mathbf{X})\right)$ is a (implicit) function of complete sufficient statistics. However, the MLE is biased, and no MVUE estimator for $(\alpha,\beta)$ has been presented in the literature. Bowman and Shenton[25] have obtained expressions for low order moments of $\left(\hat{\alpha}(\mathbf{X}), \hat{\beta}(\mathbf{X})\right)$ that are accurate to terms having order of magnitude $n^{-6}$. For example, when $\alpha \geq 1$ and $n \geq 4$, they found that

$$\mathrm{E}(\hat{\alpha}(\mathbf{X})) \approx \alpha + \left[3\alpha - \frac{2}{3} + \frac{1}{9}\alpha^{-1} + \frac{13}{405}\alpha^{-2}\right]/(n-3),$$

and they suggest, as an *approximately unbiased* estimator of $\alpha$, the following function of the MLE for $\alpha$:

$$\hat{\alpha}^*(\mathbf{X}) = \left[(n-3)\hat{\alpha}(\mathbf{X}) + \frac{2}{3}\right]/n$$

Since $\hat{\alpha}^*(\mathbf{X})$ is a function of complete sufficient statistics (because the MLE $\hat{\alpha}^*(\mathbf{X})$ is), and since $\hat{\alpha}^*(\mathbf{X})$ is approximately unbiased, $\hat{\alpha}^*(\mathbf{X})$ can be interpreted as being *approximately MVUE* for $\alpha$. For further details on finite sample properties of the MLE $\left(\hat{\alpha}(\mathbf{X}), \hat{\beta}(\mathbf{X})\right)$, see Bowman and Shenton. An analysis of the asymptotic properties of $\left(\hat{\alpha}(\mathbf{X}), \hat{\beta}(\mathbf{X})\right)$ will be developed in a subsequent example. □

[24] Alternatively, the solution for $\alpha$ can be determined by consulting tables generated by Chapman which were constructed specifically for this purpose (Chapman, D.G. "Estimating Parameters of a Truncated Gamma Distribution," *Ann. Math. Stat.*, 27, 1956, pp. 498–506).

[25] Bowman, K.O. and L.R. Shenton. *Properties of Estimators for the Gamma Distribution*, Report CTC-1, Union Carbide Corp., Oak Ridge, Tennessee.

A varied collection of regularity conditions on likelihood functions have been presented in the literature that represent sufficient conditions for MLE's to possess desirable asymptotic properties. Most of these conditions apply specifically to the case of random sampling from a population distribution, so that the random sample $(X_1,\ldots,X_n)$ must be a collection of *iid* random variables. For a survey of alternative types of regularity conditions, the reader can refer to the article by Norden.[26] We concentrate on regularity conditions that do not cover all cases, but that are relatively simple to comprehend and apply and that focus attention on key assumptions that lead to good asymptotic properties of MLEs. The conditions we present do *not* require that the random sample be a collection of *iid* random variables, and so the conditions can also be applied to cases other than random sampling from a population distribution.

Henceforth we focus on the case where an MLE is the unique global maximizer of the likelihood function, which covers the majority of applications. An examination of the multiple local optima case is more complicated and is best left to a more advanced course of study.[27]

*Consistency* We first examine conditions that ensure the consistency of the MLE of $\Theta$ in the scalar case.

**Theorem 8.15 *MLE Consistency-Sufficient Conditions for Scalar $\Theta$***

*Let $\{f(\mathbf{x};\Theta), \Theta\in\Omega\}$ be the statistical model for the random sample $\mathbf{X}$, where $\Theta$ is a scalar.[28] Assume the following regularity conditions:*

1. *The PDFs $f(\mathbf{x};\Theta)$, $\Theta \in \Omega$, have common support, $\Xi$*
2. *The parameter space, $\Omega$ is an open interval;*
3. *$\ln(L(\Theta;\mathbf{x}))$ is continuously differentiable with respect to $\Theta \in \Omega$, $\forall\, \mathbf{x} \in \Xi$;*
4. *$\partial \ln(L(\Theta;\mathbf{x}))/\partial\Theta = 0$ has a unique solution for $\Theta \in \Omega$, and the solution defines the unique maximum likelihood estimate, $\hat{\Theta}(\mathbf{x})$, $\forall\, \mathbf{x} \in \Xi$;*
5. *$\lim_{n\to\infty} P(\ln(L(\Theta_o;\mathbf{x})) > \ln(L(\Theta;\mathbf{x}))) = 1$ for $\Theta \neq \Theta_o$, where $\Theta_o$ is the true value of $\Theta \in \Omega$.[29] Then $\hat{\Theta} \xrightarrow{p} \Theta_o$, and thus the MLE is consistent for $\Theta$.*

**Proof** Let $h > 0$ be such that $\Theta_o - h \in \Omega$ and $\Theta_o + h \in \Omega$, where such an $h$ exists by condition (2), and define the events

[26] Norden, R.H. (1972; 1973) "A Survey of Maximum Likelihood Estimation," *International Statistical Revue*, (40): 329–354 and (41): 39–58.

[27] See Lehmann, E. (1983) *Theory of Point Estimation*, John Wiley and Sons, pp. 420–427.

[28] We remind the reader of our tacit assumption that $\Theta$ is identified (Definition 7.2).

[29] By *true value* of $\Theta$, we again mean that $\Theta_o$ is the value of $\Theta \in \Omega$ for which $f(\mathbf{x};\Theta_o) \equiv L(\Theta_o;\mathbf{x})$ is the actual joint density function of the random sample $\mathbf{X}$. The value of $\Theta_o$ is generally unknown, and in the current context is the objective of point estimation.

$$A_n = \{\mathbf{x} : \ln L(\Theta_o;\mathbf{x}) > \ln L(\Theta_o - h;\mathbf{x})\},$$

$$B_n = \{\mathbf{x} : \ln L(\Theta_o;\mathbf{x}) > \ln L(\Theta_o + h;\mathbf{x})\},$$

and $H_n = A_n \cap B_n$. As $n \to \infty$, $P(H_n) \to 1$, since $P(H_n) = P(A_n) + P(B_n) - P(A_n \cup B_n)$, $P(A_n)$ and $P(B_n)$ both converge to 1 as $n \to \infty$ by Assumption (5), and $P(A_n \cup B_n)$ converges to 1 since $P(A_n \cup B_n) \geq P(A_n)$ and $P(A_n \cup B_n) \geq P(B_n)$.

*Now note that* $\mathbf{x} \in H_n \Rightarrow L(\Theta;\mathbf{x})$ exhibits its unique maximum for some value $\Theta_*$ such that $\Theta_o - h < \Theta_* < \Theta_o + h$ because by the differentiability of $L(\Theta;\mathbf{x})$, $\Theta_*$ solves $\partial \ln(L(\Theta;\mathbf{x}))/\partial\Theta = 0$ and is thus the ML estimate, $\hat{\Theta}(\mathbf{x})$, of $\Theta$ given $\mathbf{x}$. Note further that $H_n \subset \{\mathbf{x}: \Theta_o - h < \hat{\Theta}(\mathbf{x}) < \Theta_o + h\}$ because $\mathbf{x} \in H_n$ implies $\Theta_o - h < \hat{\Theta}(\mathbf{x}) < \Theta_o + h$. Then $P(H_n) \to 1$ as $n \to \infty$ implies $P(H_n) \leq P(\Theta_o - h < \hat{\Theta}(\mathbf{x}) < \Theta_o + h) \to 1$ as $n \to \infty$. Since the foregoing result remains true if we decrease the values of $h$ to be arbitrarily close to zero (but still positive-valued), $\hat{\Theta} \xrightarrow{p} \Theta_o$. ■

In some cases the verification of condition 5 in Theorem 8.15 can be challenging. If random sampling is such that the random sample $X_1,\ldots,X_n$ is a collection of *iid* random variables, then condition 5 is *not* needed.

**Theorem 8.16** ***MLE Consistency: iid Random Sampling and Scalar*** $\Theta$

*Assume conditions 1–4 of Theorem 8.15, and assume further that the random sample* $X_1,\ldots,X_n$ *is a collection of iid random variables. Then* $\hat{\Theta} \xrightarrow{p} \Theta_o$.

**Proof** Let $\Theta_\ell = \Theta_o - \varepsilon$ and $\Theta_h = \Theta_o + \varepsilon$ for any $\varepsilon > 0$ s.t. $\Theta_\ell$ and $\Theta_h \in \Omega$ (such $\varepsilon$'s exists by assumption 2). Define $H(\varepsilon) = \{\mathbf{x}: \ln(L(\Theta_o;\mathbf{x})) > \ln(L(\Theta_\ell;\mathbf{x}))$ and $\ln(L(\Theta_o;\mathbf{x})) > \ln(L(\Theta_h;\mathbf{x}))\}$ and note that $\mathbf{x} \in H(\varepsilon) \Rightarrow \hat{\Theta} \in (\Theta_o - \varepsilon, \Theta_o + \varepsilon)$ because the MLE is unique and is defined via $\partial \ln(L(\Theta;\mathbf{x}))/\partial\Theta = 0$.

Now define $A(\Theta) = \{\mathbf{x}: \ln(L(\Theta_o;\mathbf{x})) > \ln(L(\Theta;\mathbf{x}))\}$ *for* $\Theta \neq \Theta_o$, and note that the event $A(\Theta)$ can be equivalently represented as

$$\tau_n(\mathbf{x}) = n^{-1}\sum_{i=1}^{n} \ln(f(x_i,\Theta)/f(x_i;\Theta_o)) < 0.$$

Because $\tau_n(\mathbf{x})$ can be interpreted as the sample mean of $n$ *iid* random variables of the form $\ln(f(X_i;\Theta)/f(X_i;\Theta_o))$, Khinchin's WLLN implies that $t_n(\mathbf{X}) \xrightarrow{p} \mathrm{E}(\ln[f(X_i;\Theta)/f(X_i;\Theta_o)])$. Also, $\ln(z)$ is strictly concave over its domain so that, Jensen's inequality implies $\mathrm{E}(\ln(f(X_i;\Theta)/f(X_i;\Theta_o))) < \ln(\mathrm{E}(f(X_i;\Theta)/f(X_i;\Theta_o)))$. Then since $\mathrm{E}(f(X_i;\Theta)/f(X_i;\Theta_o)) = 1$,[30] the right-hand side of the preceding inequality is zero. Thus, $\tau_n(\mathbf{X})$ converges in probability to a negative number, which implies that $\lim_{n\to\infty} P(A(\Theta)) = 1$ when $\Theta \neq \Theta_o$. This in turn implies that $\lim_{n\to\infty} P(H(\varepsilon))$

[30] If $X_i$ is a continuous random variable, then since $\Theta_o$ is the true value of $\Theta$,

$$\mathrm{E}[f(X_i;\Theta)/f(X_i;\Theta_0)] = \int_{-\infty}^{\infty} \frac{f(x_i;\Theta)}{f(x_i;\Theta_o)} f(x_i,\Theta_o)dx_i = \int_{-\infty}^{\infty} f(x_i;\Theta)dx_i = 1$$

because $f(x_i;\Theta)$ is a probability density function. The discrete case is analogous.

$= 1\ \forall\varepsilon>0$ since $H(\varepsilon) = A(\Theta_\ell) \cap A(\Theta_h)$, and thus $\lim_{n\to\infty} P\left(\hat{\Theta} \in (\Theta_o - \varepsilon, \Theta_o + \varepsilon)\right)$ $= 1\ \forall\varepsilon>0$ and $\hat{\Theta} \xrightarrow{p} \Theta_o$. ■

**Example 8.11**
***Consistency of MLE for Exponential Distribution***

Reexamine the case of estimating the value of $\theta$ using the MLE when random sampling is from an exponential density function (recall Examples 8.4, and 8.9). In this case the joint density function of the random sample is given by

$$f(\mathbf{x};\theta) = \theta^{-n} e^{-\sum_{i=1}^n x_i/\theta} \prod_{i=1}^n I_{(0,\infty)}(x_i).$$

The parameter space is an open interval since $\Omega = (0, \infty)$. The log of the likelihood function is continuously differentiable for $\theta \in \Omega$ and $\partial \ln(L(\theta;\mathbf{x}))/\partial\theta = (-n/\theta) + (\sum_{i=1}^n x_i)/\theta^2 = 0$ has the unique solution $\hat{\theta} = \sum_{i=1}^n x_i/n$, which is the unique maximum likelihood estimate $\forall \mathbf{x} \in \mathbb{R}_+^n$.[31] Therefore, it follows from Theorem 8.16 that $\hat{\Theta} \xrightarrow{p} \Theta_o$, so that the MLE is a consistent estimator. □

We now examine sufficient conditions for MLE consistency when $\mathbf{\Theta}$ is $k$-dimensional. We present two sets of sufficient conditions. One set allows unbounded parameter spaces but is generally more difficult to apply (1–4a). The other set(1–3, 4b) is generally more tractable but requires the parameter space to be a bounded and closed rectangle (or more generally, a closed and bounded *set*). As a practical matter, one can often state (perhaps very large) absolute bounds for the parameters of a statistical model based on real-world or theoretical considerations relating to an experiment so that the boundedness of the parameter space may not represent a serious restriction in practice.

**Theorem 8.17**
***MLE Consistency-Sufficient Conditions for Parameter Vectors***

*Let $\{f(\mathbf{x};\mathbf{\Theta}),\ \mathbf{\Theta} \in \Omega\}$ be the probability model for the random sample $\mathbf{X}$. Let $N(\varepsilon) = \{\mathbf{\Theta}: d(\mathbf{\Theta},\mathbf{\Theta}_o) < \varepsilon\}$ be an open $\varepsilon$-neighborhood of $\mathbf{\Theta}_o$, where $\mathbf{\Theta}_o$ is the true value of $\mathbf{\Theta}$.[32] Assume the following regularity conditions:*

1. *The PDFs $f(\mathbf{x};\mathbf{\Theta})$, $\mathbf{\Theta}\in\Omega$, have common support, $\Xi$;*
2. *$\ln(L(\mathbf{\Theta};\mathbf{x}))$ has continuous first-order partial derivatives with respect to $\mathbf{\Theta}\in\Omega$, $\forall\mathbf{x}\in\Xi$;*[33]
3. *$\partial \ln(L(\mathbf{\Theta};\mathbf{x}))/\partial\mathbf{\Theta} = \mathbf{0}$ has a unique solution that defines the unique maximum likelihood estimate $\hat{\mathbf{\Theta}}(\mathbf{x}) = \arg\max_{\mathbf{\Theta}\in\Omega}\{L(\mathbf{\Theta};\mathbf{x})\}\ \forall\mathbf{x}\in\Xi$;*

---

[31] That this unique solution is a maximum can be demonstrated by noting that $\partial^2 \ln(L(\Theta;\mathbf{x}))/\partial\Theta^2 = n/\Theta^2 - 2\sum_{i=1}^n x_i/\Theta^3$, which when evaluated at the maximum likelihood estimate $\hat{\Theta} = \sum_{i=1}^n x_i/n$, yields $\left(\left(\partial^2 \ln\left(L\left(\hat{\theta};\mathbf{x}\right)\right)\right)/(\partial\Theta^2)\right) = -n^3/(\sum_{i=1}^n x_i)^2 < 0$.

[32] $N(\varepsilon)$ is an open interval, the interior of a circle, the interior of a sphere, and the interior of a hypersphere in 1,2,3, and $\geq 4$ dimension, respectively.

[33] It is allowable that conditions (2) and (3) be violated on a set of $\mathbf{x}$-values having probability zero.

**4a.** $\lim_{n\to\infty} P(\ln(L(\Theta_o;\mathbf{x}))>\max_{\Theta\in\overline{N(\varepsilon)}} \ln(L(\Theta;\mathbf{x}))) = 1\ \forall\ \varepsilon > 0$ *with* $\Omega$ *being an open rectangle containing* $\Theta_o$;[34]

**4b.** $\lim_{n\to\infty} P(\max_{\Theta\in\Omega}\{|n^{-1}\ln(L(\Theta_o;\mathbf{x})) - G(\Theta)|\}<\varepsilon) = 1\ \forall\varepsilon>0$ *with* $G(\Theta)$ *being a continuous function that is uniquely globally maximized at* $\Theta = \Theta_o$ *and* $\Omega$ *is a bounded closed rectangle containing* $\Theta_o$.

*Then* $\hat{\Theta} \xrightarrow{p} \Theta_o$.

**Proof** See Appendix. ■

We now return to the case of random sampling from the gamma density where there is no closed form solution for the MLE, although $\hat{\Theta}$ is *implicitly* defined by first order conditions.

**Example 8.12** ***Consistency of MLE for Gamma Distribution*** Reexamine the case of estimating the value of α and β using the MLE when random sampling is from a gamma population distribution (recall Example 8.10). In an attempt to utilize Theorem 8.16 for demonstrating the consistency of the MLE, first note that $f(\mathbf{x};\Theta) > 0$ for $\mathbf{x} \in \mathbb{R}^n_+$, and so condition (1) is satisfied. Recalling Example 8.10, it is evident that $\ln(L(\Theta;\mathbf{x}))$ is continuously differentiable with respect to $\alpha$ and $\beta\ \forall\mathbf{x} \in \mathbb{R}^n_+$ and for all $\alpha > 0$ and $\beta > 0$, validating condition (2). Also, the first order conditions have a unique solution for $(\alpha, \beta)$ $\forall\mathbf{x} \in \mathbb{R}^n_+$, which defines the unique maximum likelihood estimate$\left(\hat{\alpha}, \hat{\beta}\right)$ (again recall Example 8.10) satisfying condition (3).

Assume that bounds can be placed on the parameter values of $\alpha$ and $\beta$, and examine the validity of condition (4b) of Theorem 8.17. Note that (suppressing the indicator function)

$$n^{-1}\ln(L(\alpha,\beta;\mathbf{x})) = -\alpha\ln(\beta) - \ln(\Gamma(\alpha)) + (\alpha-1)n^{-1}\sum_{i=1}^{n}\ln(x_i) - \bar{x}/\beta.$$

Then because[35]

$$\text{p}\lim\left(n^{-1}\sum_{i=1}^{n}\ln(X_i)\right) = ((d\Gamma(\alpha_o)/d\alpha)/(\alpha_o)) + \ln(\beta_o),$$

it follows that

$$\text{plim}\,(n^{-1}\ln(L(\alpha,\beta;\mathbf{x}))) = G(\alpha,\beta) = \alpha\ln\left(\frac{\beta_o}{\beta}\right) + (\alpha-1)\left[\frac{d\Gamma(\alpha_o)/d\alpha}{\Gamma(\alpha_o)}\right] - \ln(\Gamma(\alpha)) - \frac{\alpha_o\beta_o}{\beta} - \ln(\beta_o)$$

uniformly in $(\alpha,\beta)$. Furthermore, the maximum of $G(\alpha,\beta)$ occurs at $\alpha = \alpha_o$ and $\beta = \beta_o$. To see this, note that $\alpha_o$ and $\beta_o$ solve the first order conditions for the maximization problem given by

[34] Change max to sup if max does not exist.

[35] This follows from Khinchin's WLLN upon recognizing that the right hand side expression represents $E(\ln(x_i))$.

$$\frac{\partial G(\alpha,\beta)}{\partial \alpha} = \ln\left(\frac{\beta_o}{\beta}\right) + \frac{d\Gamma(\alpha)/d\alpha}{\Gamma(\alpha_o)} - \frac{d\Gamma(\alpha)/d\alpha}{\Gamma(\alpha)} = 0$$

and

$$\frac{\partial G(\alpha,\beta)}{\partial \beta} = -\alpha/\beta + \alpha_o\beta_o/\beta^2 = 0.$$

The hessian matrix is negative definite, so that the second order conditions for a maximum are met (see Example 8.15 for the explicit representation of the hessian matrix), and thus condition (4b) holds. Therefore, the estimator $\hat{\boldsymbol{\Theta}}$, defined *implicitly* by the first order conditions for maximizing the likelihood function, is consistent, i.e., $\hat{\boldsymbol{\Theta}} \xrightarrow{p} \boldsymbol{\Theta}_o$, so that both $\hat{\alpha} \xrightarrow{p} \alpha_o$ and $\hat{\beta} \xrightarrow{p} \beta_o$. □

In the next example, we revisit Example 8.6, in which the MLE of $(\boldsymbol{\beta},\sigma^2)$ for the GLM was defined for the case where the random sample from a general sampling experiment had a joint density of the multivariate normal form, $N(\mathbf{x}\boldsymbol{\beta}, \sigma^2\mathbf{I})$. In this case, the MLE can be explicitly solved for, and its consistency property evaluated directly, as we have done in Section 8.2, but for the sake of illustration we reexamine consistency using Theorem 8.17.

**Example 8.13**
***Consistency of MLE in GLM with $\mathbf{Y} \sim N(\mathbf{x}\boldsymbol{\beta}, \sigma^2\mathbf{I})$***

Reexamine Example 8.6. To demonstrate consistency of the MLE using Theorem 8.17, first note $f(\mathbf{y};\boldsymbol{\beta},\sigma^2) > 0$ for $\mathbf{y} \in \mathbb{R}^n$, and so condition (1) is satisfied.

It is evident from Example 8.6 that $L(\boldsymbol{\beta},\sigma^2;\mathbf{y})$ is continuously differentiable with respect to $\boldsymbol{\Theta} \in \Omega$ for all $\mathbf{y} \in \mathbb{R}^n$, so that condition (2) is satisfied. Also, $\partial \ln(L(\boldsymbol{\beta},\sigma^2;\mathbf{y}))/\partial\boldsymbol{\Theta} = \mathbf{0}$ has a unique solution (assuming $\mathbf{x}'\mathbf{x}$ is of full rank), and this solution defines the unique MLE of $(\boldsymbol{\beta},\sigma^2)$. We assume that $\sigma^2 \in \left[\sigma^2_\ell, \sigma^2_h\right]$ and $\boldsymbol{\beta} \in \times_{i=1}^k[\beta_{i\ell}, \beta_{ih}]$ and verify condition (4b). Note that

$$\mathbf{Z}_n = n^{-1}\ln(L(\boldsymbol{\beta},\sigma^2;\mathbf{y})) = -(1/2)\ln(2\pi\sigma^2) - (1/2)n^{-1}\left[(\mathbf{y}-\mathbf{x}\boldsymbol{\beta})'(\mathbf{y}-\mathbf{x}\boldsymbol{\beta})/\sigma^2\right].$$

Because $\mathbf{Y} = \mathbf{x}\boldsymbol{\beta}_o + \boldsymbol{\varepsilon}$, $\mathbf{Z}_n$ can be expressed equivalently as

$$\mathbf{Z}_n = -(1/2)\left[\ln(2\pi\sigma^2) + (\boldsymbol{\beta}_o - \boldsymbol{\beta})'(n^{-1}\mathbf{x}'\mathbf{x})(\boldsymbol{\beta}_o - \boldsymbol{\beta})/\sigma^2 + n^{-1}\boldsymbol{\varepsilon}'\boldsymbol{\varepsilon}/\sigma^2 + 2(\boldsymbol{\beta}_o - \boldsymbol{\beta})'(n^{-1}\mathbf{x}'\boldsymbol{\varepsilon})/\sigma^2\right].$$

Since the $\varepsilon_i$'s are *iid* because $\mathbf{Y} \sim N(\mathbf{x}\boldsymbol{\beta}_o,\sigma^2\mathbf{I})$, it follows from Khinchin's WLLN that $\text{plim}\,(n^{-1}\boldsymbol{\varepsilon}'\boldsymbol{\varepsilon}) = \sigma_0^2$. Also note that $\mathrm{E}(n^{-1}\mathbf{x}'\boldsymbol{\varepsilon}) = \mathbf{0}$ and $\mathbf{Cov}(n^{-1}\mathbf{x}'\boldsymbol{\varepsilon}) = \sigma^2 n^{-2}\mathbf{x}'\mathbf{x}$, so that assuming $\mathbf{x}'\mathbf{x}$ is $o(n^2)$, so that $n^{-2}\mathbf{x}'\mathbf{x} \to \mathbf{0}$ as $n \to \infty$, then $n^{-1}\mathbf{x}'\boldsymbol{\varepsilon} \xrightarrow{p} \mathbf{0}$, and thus $n^{-1}\mathbf{x}'\boldsymbol{\varepsilon} \xrightarrow{p} \mathbf{0}$ and $(\boldsymbol{\beta}_o - \boldsymbol{\beta})'(n^{-1}\mathbf{x}'\boldsymbol{\varepsilon})/\sigma^2 \xrightarrow{p} 0$. Assuming further that $n^{-1}\mathbf{x}'\mathbf{x} \to \mathbf{Q}$, a positive definite symmetric matrix, then

$$\mathbf{Z}_n \xrightarrow{p} G(\boldsymbol{\beta},\sigma^2) = -\frac{1}{2}\left[\ln(2\pi\sigma^2) + (\boldsymbol{\beta}_o - \boldsymbol{\beta})'\mathbf{Q}(\boldsymbol{\beta}_o - \boldsymbol{\beta})/\sigma^2 + \sigma_o^2/\sigma^2\right]$$

uniformly in $\boldsymbol{\beta}$ and $\sigma^2$.

The function $G(\boldsymbol{\beta},\sigma^2)$ is maximized when $\boldsymbol{\beta} = \boldsymbol{\beta}_o$ and $\sigma^2 = \sigma_o^2$. To see this, note that $\boldsymbol{\beta}_o$ and $\sigma_o^2$ satisfy the first order conditions for maximizing $G(\boldsymbol{\beta},\sigma^2)$ given by

$$\frac{\partial G(\boldsymbol{\beta},\sigma^2)}{\partial \boldsymbol{\beta}} = \mathbf{Q}(\boldsymbol{\beta}_o - \boldsymbol{\beta}) = \mathbf{0}$$

$$\frac{\partial G(\boldsymbol{\beta},\sigma^2)}{\partial \sigma^2} = -\frac{1}{2\sigma^2} + \frac{(\boldsymbol{\beta}_o - \boldsymbol{\beta})'\mathbf{Q}(\boldsymbol{\beta}_o - \boldsymbol{\beta})}{2\sigma^2} + \frac{\sigma_o^2}{2\sigma^2} = 0$$

The hessian matrix for checking second order conditions, evaluated at $\boldsymbol{\beta} = \boldsymbol{\beta}_o$ and $\sigma^2 = \sigma_o^2$, is given by

$$\begin{bmatrix} \dfrac{\partial^2 G(\boldsymbol{\beta}_o,\sigma_o^2)}{\partial \boldsymbol{\beta} \partial \boldsymbol{\beta}'} & \dfrac{\partial^2 G(\boldsymbol{\beta}_o,\sigma_o^2)}{\partial \boldsymbol{\beta} \partial \sigma^2} \\ \dfrac{\partial^2 G(\boldsymbol{\beta}_o,\sigma_o^2)}{\partial \sigma^2 \partial \boldsymbol{\beta}'} & \dfrac{\partial^2 G(\beta_o,\sigma_o^2)}{\partial (\sigma^2)^2} \end{bmatrix} = \left[\begin{array}{c|c} -\mathbf{Q} & \mathbf{0} \\ \hline \mathbf{0} & -\dfrac{1}{2\sigma_o^4} \end{array}\right],$$

which is negative definite, and thus $G(\boldsymbol{\beta},\sigma^2)$ is indeed maximized at $\boldsymbol{\beta}_o, \sigma_o^2$. Therefore, by Theorem 8.17, and under the preceding assumptions on $\mathbf{x}$, $\hat{\boldsymbol{\Theta}} \xrightarrow{p} \boldsymbol{\Theta}_o$ so that $\text{plim}\left(\hat{\boldsymbol{\beta}}\right) = \boldsymbol{\beta}_o$ and $\text{plim}(\hat{\sigma}^2) = \sigma_o^2$. □

*Asymptotic Normality and Asymptotic Efficiency* In order that the MLE be asymptotically normally distributed, additional regularity conditions on the maximum likelihood estimation problem are needed. We present a collection of sufficient conditions below, and note that there exist a variety of alternative sufficient conditions in the literature (see Norden, op. cit.; Amemiya, T., *Advanced Econometrics*, pp. 111–112 for conditions related to those presented here).

**Theorem 8.18** ***MLE Asymptotic Normality-Sufficient Conditions*** *In addition to conditions (1–4) of Theorem 8.17, assume that:*

1. $\partial^2 \ln(L(\boldsymbol{\Theta};\mathbf{x}))/\partial\boldsymbol{\Theta}\partial\boldsymbol{\Theta}'$ *exists and is continuous in* $\boldsymbol{\Theta}$ $\forall \boldsymbol{\Theta} \in \Omega$ *and* $\forall \mathbf{x} \in \boldsymbol{\Xi}$,
2. $\text{plim}(n^{-1}(\partial^2 \ln(L(\boldsymbol{\Theta}^*;\mathbf{X}))/\partial\boldsymbol{\Theta}\partial\boldsymbol{\Theta}')) = \mathbf{H}(\boldsymbol{\Theta}_o)$ *is a nonsingular matrix for any sequence of random variables* $\{\boldsymbol{\Theta}_n^*\}$ *such that* $\text{plim}(\boldsymbol{\Theta}_n^*) = \boldsymbol{\Theta}_o$,
3. $n^{-1/2}(\partial \ln(L(\boldsymbol{\Theta}_o;\mathbf{X})))/\partial\boldsymbol{\Theta} \xrightarrow{d} N(\mathbf{0},\mathbf{M}(\boldsymbol{\Theta}_o))$ *where* $\mathbf{M}(\boldsymbol{\Theta}_o)$ *is a positive definite symmetric matrix.*

*Then the MLE,* $\hat{\boldsymbol{\Theta}}$, *is such that*

$$n^{1/2}\left(\hat{\boldsymbol{\Theta}} - \boldsymbol{\Theta}_o\right) \xrightarrow{d} N\left(\mathbf{0}, \mathbf{H}(\boldsymbol{\Theta}_o)^{-1}\mathbf{M}(\boldsymbol{\Theta}_o)\mathbf{H}(\boldsymbol{\Theta}_o)^{-1}\right) \quad \textit{and}$$

$$\hat{\boldsymbol{\Theta}} \overset{a}{\sim} N\left(\boldsymbol{\Theta}_o, n^{-1}\mathbf{H}(\boldsymbol{\Theta}_o)^{-1}\mathbf{M}(\boldsymbol{\Theta}_o)\mathbf{H}(\boldsymbol{\Theta}_o)^{-1}\right).$$

**Proof** See Appendix. ■

We present one further condition that leads to the asymptotic efficiency of the MLE.

**Theorem 8.19** ***MLE Asymptotic Efficiency: Sufficient Conditions*** *In addition to the assumptions in Theorem 8.17 and Theorem 8.18, assume that*

$$\mathbf{M}(\Theta_o) = \lim_{n\to\infty}\left(n^{-1}\mathrm{E}\left(\frac{\partial \ln(L(\Theta_o;\mathbf{X}))}{\partial \Theta}\frac{\partial \ln(L(\Theta_o;\mathbf{X}))}{\partial \Theta'}\right)\right) = -\mathbf{H}(\Theta_o)$$

*Then $\hat{\Theta}$ is asymptotically efficient.*

**Proof** From Theorem 8.18, if $\mathbf{M}(\Theta_o)$ and $\mathbf{H}(\Theta_o)$ *are as defined above, then* $n^{1/2}\left(\hat{\Theta}-\Theta_o\right) \xrightarrow{d} \mathbf{Z} \sim N\left(\mathbf{0}, \mathbf{M}(\Theta_o)^{-1}\right)$. Referring to the definition of asymptotic efficiency (Definition 7.22), it follows that[36]

$$n^{-1/2}\left[\mathrm{E}\left(\frac{\partial \ln L(\Theta_o;\mathbf{X})}{\partial \Theta}\frac{\partial \ln L(\Theta_o;\mathbf{X})}{\partial \Theta'}\right)\right]^{1/2} n^{1/2}\left(\hat{\Theta}-\Theta\right) \xrightarrow{d} \mathbf{M}(\Theta_o)^{1/2}\mathbf{Z} \sim N(\mathbf{0},\mathbf{I})$$

so that $\hat{\Theta} \overset{a}{\sim} N\left(\Theta_o, \mathbf{M}(\Theta_o)^{-1}\right)$ is asymptotically efficient. ■

In applications, since the value of $\Theta_o$ is unknown, the asymptotic covariance of $\hat{\Theta}$ is also unknown. However, under the assumptions of Theorems 8.18 and 8.19, a consistent estimate of $\mathbf{M}(\Theta_o)^{-1}$ is obtained by using an outcome of $-\left[n^{-1}\left(\partial^2 \ln\left(L\left(\hat{\Theta};\mathbf{X}\right)\right)/\partial\Theta\partial\Theta'\right)\right]^{-1}$ (recall assumption (2)) of Theorem 8.18 and the statement of Theorem 8.19).

We now revisit Examples 8.11 and 8.12 and illustrate the use of Theorems 8.18 and 8.19 for establishing the asymptotic normality and asymptotic efficiency of the MLE.

**Example 8.14** ***Asymptotic Normality and Efficiency of MLE - Exponential Distribution*** Reexamine Example 8.11. Note that $\partial^2 \ln(L(\theta;\mathbf{x}))/\partial\theta^2 = (n/\theta^2) - \left(2\sum_{i=1}^n x_i/\theta^3\right)$ exists and is continuous for $\forall\theta > 0$ and $\forall \mathbf{x} \in \Xi$, so assumption (1) of Theorem 8.18 is satisfied. Furthermore, assuming $\mathrm{plim}(\Theta_*) = \theta_o$, it follows by Slutsky's theorem that

$$\mathrm{plim}\left(n^{-1}\frac{\partial^2 \ln(L(\Theta_*;\mathbf{X}))}{\partial\theta^2}\right) = \mathrm{plim}(\Theta_*^{-2} - 2\bar{X}_n/\Theta_*^3) = (\theta_o^{-2} - 2\theta_o/\theta_o^3) = -\theta_o^{-2} = \mathrm{H}(\theta_o)$$

so that assumption (2) of Theorem 8.18 applies. Also note that

[36] Recall that the matrix square root is a *continuous* function of its arguments, so that $\mathrm{plim}\left(\mathbf{A}_n^{1/2}\right) = (\mathrm{plim}\,\mathbf{A}_n)^{1/2}$. Letting $\mathbf{A}_n = n^{-1}\mathrm{E}\left(\frac{\partial \ln L(\Theta_o;\mathbf{X})}{\partial\Theta}\frac{\partial \ln L(\Theta_o;\mathbf{X})}{\partial\Theta'}\right)$ leads to $\mathrm{plim}\left(\mathbf{A}_n^{1/2}\right) = \mathbf{M}(\Theta_o)^{1/2}$.

$$n^{-1/2}\frac{\partial \ln(L(\theta_o;\mathbf{X}))}{\partial\theta} = -n^{1/2}\theta_o^{-1} + n^{-1/2}\left(\sum_{i=1}^{n}\frac{X_i}{\theta_o^2}\right)$$
$$= \frac{1}{\theta_o}\left[\frac{n^{1/2}(\bar{X}_n - \theta_o)}{\theta_o}\right] \xrightarrow{d} N(0,\theta_o^{-2}) = N(0,\mathrm{M}(\theta_o))$$

where $\mathrm{M}(\theta_o) = \theta_o^{-2}$, which follows by a direct application of the Lindberg-Levy CLT to the bracketed expression and by Slutsky's theorem, since $\mathrm{E}(X_i) = \theta_o$ and $\mathrm{var}(X_i) = \theta_o^2 \forall i$. Then, from Theorem 8.18, it follows that

$$n^{1/2}\left(\hat{\Theta} - \theta_o\right) \xrightarrow{d} N(0,\theta_o^2),\ \text{and}\ \hat{\Theta} \overset{a}{\sim} N(\theta_o, n^{-1}\theta_o^2).$$

Regarding asymptotic efficiency, recall from Example 7.21 that

$$\mathrm{E}\left(\frac{\partial \ln L(\theta_o;\mathbf{X})}{\partial\theta}\frac{\partial \ln L(\theta_o;\mathbf{X})}{\partial\theta'}\right) = n/\theta_o^2$$

Asymptotic efficiency of $\hat{\Theta}$ follows immediately from Theorem 8.19 since the results above demonstrate the equality

$$\mathrm{M}(\theta_o) = -\mathrm{H}(\theta_o) = \lim_{n\to\infty}\left(\mathrm{E}\left(n^{-1}\left[\frac{\partial \ln L(\theta_o;\mathbf{X})}{\partial\theta}\frac{\partial \ln L(\theta_o;\mathbf{X})}{\partial\theta'}\right]\right)\right) = \theta_o^{-2}.$$ □

**Example 8.15** ***Asymptotic Normality and Efficiency of MLE: Gamma Distribution***

Reexamine Example 8.12. The second order derivatives of $\ln(L(\alpha,\beta;\mathbf{x}))$, divided by $n$, are given by

$$\frac{1}{n}\frac{\partial^2 \ln(L(\alpha,\beta;\mathbf{x}))}{\partial\alpha^2} = -\left[\Gamma(\alpha)\frac{d^2\Gamma(\alpha)}{d\alpha^2} - \left(\frac{d\Gamma(\alpha)}{d\alpha}\right)^2\right]/\Gamma(\alpha)^2,$$
$$\frac{1}{n}\frac{\partial^2 \ln(L(\alpha,\beta;\mathbf{x}))}{\partial\alpha\partial\beta} = \frac{1}{n}\frac{\partial^2 \ln(L(\alpha,\beta;\mathbf{x}))}{\partial\beta\partial\alpha} = -\beta^{-1},$$
$$\frac{1}{n}\frac{\partial^2 \ln(L(\alpha,\beta;\mathbf{x}))}{\partial\beta^2} = \frac{\alpha}{\beta^2} - 2\frac{\bar{x}}{\beta^3},$$

and the derivatives themselves are continuous functions of $(\alpha,\beta)$ for $\alpha > 0$ and $\beta > 0$ and $\forall \mathbf{x} \in \Xi$, so that assumption (1) of Theorem 8.18 is satisfied.[37] Letting $\boldsymbol{\Theta} = \begin{pmatrix}\alpha\\ \beta\end{pmatrix}$, $\boldsymbol{\Theta}_o = \begin{pmatrix}\alpha_o\\ \beta_o\end{pmatrix}$, , and $\boldsymbol{\Theta}_*$ be such that $\boldsymbol{\Theta}_* \xrightarrow{p} \boldsymbol{\Theta}_o$, it follows from continuity that

[37] The gamma function, $\Gamma(\alpha)$, is continuous in $\alpha$, and its first two derivatives are continuous in $\alpha$, for $\alpha > 0$. $\Gamma(\alpha)$ is in fact strictly convex, with its second order derivative strictly positive for $\alpha > 0$. See Bartle, op. cit., p. 282.

$$\text{plim}\left(n^{-1}\frac{\partial^2 \ln(L(\mathbf{\Theta}_*;\mathbf{X}))}{\partial\mathbf{\Theta}\partial\mathbf{\Theta}'}\right) = -\left[\begin{array}{c|c}\left[\Gamma(\alpha_o)\dfrac{d^2\Gamma(\alpha_o)}{d\alpha^2}-\left(\dfrac{d\Gamma(\alpha_o)}{d\alpha}\right)^2\right]\Big/\Gamma(\alpha_o)^2 & \beta_o^{-1} \\ \hline \beta_o^{-1} & \dfrac{\alpha_o}{\beta_o^2}\end{array}\right] = \mathbf{H}(\mathbf{\Theta}_o)$$

(recall that $\text{plim}(\bar{X}) = \alpha\beta$ in this case), so that the convergence condition in (2) of Theorem 8.18 applies.

Regarding condition (3), note that

$$n^{-1/2}\frac{\partial \ln(L(\mathbf{\Theta}_o;\mathbf{X}))}{\partial\mathbf{\Theta}} = \begin{bmatrix} -n^{1/2}\left[\ln(\beta_o)+\left[\frac{d^2\Gamma(\alpha_o)}{d\alpha^2}\right]/\Gamma(\alpha_o)^2\right]+n^{-1/2}\sum_{i=1}^{n}\ln(X_i) \\ -n^{1/2}\left[\frac{\alpha_o}{\beta_o}\right]+n^{1/2}\left[\frac{\bar{X}_n}{\beta_o^2}\right]\end{bmatrix}$$

To establish the bivariate normal limiting density of this (2 × 1) random vector, first note that $(\mathbf{Z}_1,\ldots,\mathbf{Z}_n)$, where $\mathbf{Z}_i = \begin{bmatrix}\ln(X_i)\\ X_i/\beta_o^2\end{bmatrix}$, is a collection of *iid* random variables with[38]

$$\text{E}(\mathbf{Z}_i) = \boldsymbol{\mu} = \begin{bmatrix}\frac{d\Gamma(\alpha_o)}{d\alpha}/\Gamma(\alpha_o)+\ln(\beta_o)\\ \alpha_o/\beta_o\end{bmatrix},$$

$$\mathbf{Cov}(\mathbf{Z}_i) = \mathbf{\Phi} = \left[\begin{array}{c|c}\left[\Gamma(\alpha_o)\frac{d^2\Gamma(\alpha_o)}{d\alpha^2}-\left(\frac{d\Gamma(\alpha_o)}{d\alpha}\right)^2\right]/\Gamma(\alpha_o)^2 & \beta_o^{-1}\\ \hline \beta_o^{-1} & \alpha_o/\beta_o^2\end{array}\right].$$

The Multivariate Lindberg-Levy CLT then implies that

$$n^{1/2}(\bar{\mathbf{Z}}-\boldsymbol{\mu})\xrightarrow{d}N(\mathbf{0},\mathbf{\Phi}),\text{where } \bar{\mathbf{Z}} = \begin{bmatrix}n^{-1}\sum_{i=1}^{n}\ln(X_i)\\ \bar{X}/\beta_o^2\end{bmatrix}.$$

Noting that

$$n^{-1/2}(\partial\ln(L(\mathbf{\Theta}_o;\mathbf{X}))) = n^{1/2}(\bar{\mathbf{Z}}-\boldsymbol{\mu})\xrightarrow{d}N(\mathbf{0},\mathbf{M}(\mathbf{\Theta}_o)),$$

[38] A way of deriving the expectations involving $\ln(X_i)$ that is conceptually straightforward, albeit somewhat tedious algebraically, is first to derive the MGF of $\ln(X_i)$, which is given by $\beta^t\Gamma(\alpha+t)/\Gamma(\alpha)$. Then using the MGF in the usual way establishes the mean and variance of $\ln(X_i)$. The covariance between $(X_i/\beta_o^2)$ and $\ln(X_i)$ can be established by noting that

$$\beta_o^{-2}\text{E}(X_i\ln(X_i)) = \beta_o^{-2}\left[\frac{1}{\beta_o^{\alpha_o}\Gamma(\alpha_o)}\right]\int_0^\infty(\ln(x_i))x_i^{\alpha_o}e^{-x_i/\beta_o}dx_i$$
$$= \alpha_o\beta_o^{-1}\text{E}_*(\ln(X_i))$$

where $\text{E}_*$ denotes an expectation of $\ln(X_i)$ using a gamma density having parameter values $\alpha_o+1$ and $\beta_o$. Then $\text{cov}((X_i/\beta_o^2),\ln(X_i))$ is equal to

$$\alpha_o\beta_o^{-1}\text{E}_*(\ln(X_i)) - (\text{E}(\ln(X_i))\text{E}(X_i/\beta_o^2)) = \alpha_o\beta_o^{-1}(\text{E}_*(\ln(X_i))-\text{E}(\ln(X_i))) = \beta_o^{-1}.$$

where $\mathbf{M}(\boldsymbol{\Theta}_o) = \boldsymbol{\Phi}$, it follows that convergence to a bivariate normal density is established, and thus condition (3) of Theorem 8.18 is met. Finally, note that

$$\mathbf{M}(\boldsymbol{\Theta}_o) = \lim_{n\to\infty}\left(n^{-1}\mathrm{E}\left(\frac{\partial \ln L(\boldsymbol{\Theta}_o;\mathbf{X})}{\partial \boldsymbol{\Theta}}\frac{\partial \ln L(\boldsymbol{\Theta}_o;\mathbf{X})}{\partial \boldsymbol{\Theta}'}\right)\right) = -\mathbf{H}(\boldsymbol{\Theta}_o)$$

which follows immediately upon taking the expectation and recalling that $\mathrm{E}(\bar{X}) = \alpha\beta$. Thus, the condition of Theorem 8.19 is met, and we can conclude that $n^{1/2}\left(\hat{\boldsymbol{\Theta}} - \boldsymbol{\Theta}_o\right) \xrightarrow{d} N\left(\mathbf{0}, \mathbf{M}(\boldsymbol{\Theta}_o)^{-1}\right)$, $\hat{\boldsymbol{\Theta}} \overset{a}{\sim} N\left(\boldsymbol{\Theta}_o, n^{-1}\mathbf{M}(\boldsymbol{\Theta}_o)^{-1}\right)$, and $\hat{\boldsymbol{\Theta}}$ is asymptotically efficient. □

We emphasize that although we were unable to explicitly solve for the function defining the MLE in the case of random sampling from a general gamma density, our theorems on the asymptotic properties of the MLE still allowed us to establish the asymptotic properties of the MLE which were *implicitly* defined by first order conditions. In particular, we now know that $\hat{\boldsymbol{\Theta}}$ is consistent, asymptotically normal, and asymptotically efficient as an estimator of $(\alpha,\beta)$.

The verification of the MLE's asymptotic properties can be somewhat complicated at times. Regarding the density functions commonly used in practice, it can be said that the MLE is quite generally consistent, asymptotically normal, and asymptotically efficient when the random sample is a collection of *iid* random vectors. In particular, the MLE is consistent, asymptotically normal, and asymptotically efficient when random sampling from *any* of the exponential class densities we have presented in Chapter 4, and these asymptotic properties hold quite generally for exponential class densities (see Lehmann, op. cit., pp. 417–418 and pp. 438–439). In more general situations involving general experiments, one must typically verify the asymptotic properties of the MLE on a case-by-case basis.

### 8.3.4 MLE Invariance Principle and Density Reparameterization

Our discussion of MLEs to this point has concentrated on the estimation of the parameter vector $\boldsymbol{\Theta}$ itself. Fortunately, the maximum likelihood procedure has an *invariance property* that renders our preceding discussion entirely and directly relevant for estimating *functions* of the parameter vector. Before stating the property formally, we provide some additional background discussion concerning the **parameterization of density functions**.

The reader will recall that the members of parametric families of densities are indexed, or identified, by values of a parameter vector whose admissible values are identified by a set of values represented by the parameter space, $\Omega$. Upon a moment's reflection, it is evident that there is nothing sacrosanct about a particular indexing of the members of a parametric family of densities. That is, a collection of density functions, $f(\mathbf{x};\boldsymbol{\Theta})$ for $\boldsymbol{\Theta} \in \Omega$, could be equally well represented by $f(\mathbf{x};\mathbf{q}^{-1}(\boldsymbol{\xi}))$, where $\boldsymbol{\xi} = \mathbf{q}(\boldsymbol{\Theta})$ is any *invertible* function of $\boldsymbol{\Theta} \in \Omega$.

Such a parameter transformation is referred to as a **reparameterization of the density family**. As a specific example, consider the exponential family $f(\mathbf{x},\theta) = (1/\theta)e^{-x/\theta}I_{(0,\infty)}(\mathbf{x})$ for $\theta \in \Omega$, where $\Omega = (0,\infty)$. This same family of densities could be represented equally well by $h(\mathbf{x};\xi) = f(\mathbf{x}; q_1^{-1}(\xi)) = \xi e^{-\xi x}I_{(0,\infty)}(\mathbf{x})$ for $\xi \in \Omega_\xi$, where $\xi = q_1(\theta) = \theta^{-1}$ for $\theta \in \Omega$, so that $\Omega_\xi = (0,\infty)$. Another parameterization of this same family of densities is given by $m(\mathbf{x};\tau) = f(\mathbf{x}; q_2^{-1}(\tau)) = \tau^{-1/2}\exp(-\mathbf{x}\tau^{-12})I_{(0,\infty)}(\mathbf{x})$ for $\tau \in \Omega_\tau$, where now $\tau = q_2(\theta) = \theta^2$ for $\theta \in \Omega$, so that $\Omega_\tau = (0,\infty)$. In all cases the exact same collection of densities is identified. Of course, the interpretation of the parameter differs in each case, where $\theta$ is the mean of $X$, $\xi$ is the reciprocal of the mean of $X$, and $\tau$ is the variance of $X$ (recall $\sigma^2 = \theta^2$ in the exponential family). The fact that a density family can be **reparameterized** provides flexibility for redefining the maximum likelihood problem in terms of a parameter or parameter vector having an interpretation that may be of more fundamental interest to research objectives. The fact that the actual reparameterization process is *unnecessary* for obtaining the maximum likelihood estimate of $\mathbf{q}(\Theta)$ is presented in the following theorem.

**Theorem 8.20 *MLE Invariance Principle: Invertible Case***

*Let $\hat{\Theta}$ be a MLE of $\Theta$, and suppose that the function $\mathbf{q}(\Theta)$, for $\Theta \in \Omega$, is invertible. Then $\mathbf{q}(\hat{\Theta})$ is a MLE of $\mathbf{q}(\Theta)$.*

**Proof** The maximum likelihood estimate of $\Theta$ is given by $\hat{\boldsymbol{\theta}} = \arg\max_{\Theta\in\Omega}\{L(\Theta;\mathbf{x})\}$. Suppose $\boldsymbol{\xi} = \mathbf{q}(\Theta)$ is invertible, so the likelihood function can be reexpressed as a function of $\boldsymbol{\xi}$ by substituting $\Theta = \mathbf{q}^{-1}(\boldsymbol{\xi})$, $\boldsymbol{\xi} \in \Omega_\xi = \{\boldsymbol{\xi} : \boldsymbol{\xi} = \mathbf{q}(\Theta),\ \Theta \in \Omega\}$, in $L(\Theta;\mathbf{x})$. Then the maximum likelihood estimate for $\boldsymbol{\xi}$ would be $\hat{\boldsymbol{\xi}} = \arg\max_{\boldsymbol{\xi}\in\Omega_\xi}\{L(\mathbf{q}^{-1}(\boldsymbol{\xi});\mathbf{x})\}$. But since $\hat{\boldsymbol{\theta}}$ maximizes $L(\Theta;\mathbf{x})$, the value of $\hat{\boldsymbol{\xi}}$ that solves the latter likelihood maximization problem satisfies $\mathbf{q}^{-1}\left(\hat{\boldsymbol{\xi}}\right) = \hat{\boldsymbol{\theta}}$, which implies by the invertibility of $\mathbf{q}$ that $\hat{\boldsymbol{\xi}} = \mathbf{q}\left(\hat{\boldsymbol{\theta}}\right)$. ■

The invariance principle in Theorem 8.20 implies that once a MLE, $\hat{\Theta}$, of $\Theta$ has been found, one can immediately define the MLE of *any* invertible function $\mathbf{q}(\Theta)$ of $\Theta$ as $\mathbf{q}(\hat{\Theta})$. The invariance property is obviously a very convenient feature of the maximum likelihood estimation approach. In fact, the invariance principle of maximum likelihood estimation extends to more general functions of $\Theta$, *invertible or not*, although the intuitive interpretation of invariance in terms of reparameterizing a family of density functions no longer follows. The more general invariance principle utilizes the notion of maximizing an **induced likelihood function**, and the concept of the **parameter space induced by $\mathbf{q}(\Theta)$**, as defined in the proof of the following generalization of Theorem 8.20.

**Theorem 8.21 *MLE Invariance Principle: General Case***

*Let $\hat{\Theta}$ be a MLE of the $(k \times 1)$ parameter vector $\Theta$, and let $\mathbf{q}(\Theta)$ be a $(r \times 1)$ real-valued vector function of $\Theta \in \Omega$, where $r \leq k$. Then $\mathbf{q}(\hat{\Theta})$ is a MLE of $\mathbf{q}(\Theta)$.*

**Proof** See Appendix. ■

Note from the proof of Theorem 8.21 that a maximizing value of $\boldsymbol{\tau}$ then maintains the intuitively appealing property of associating with the sample outcome, $\mathbf{x}$, the highest probability (discrete case) or highest density weighting (continuous case) that is possible within the parametric family $f(\mathbf{x};\boldsymbol{\Theta})$, $\boldsymbol{\Theta} \in \Omega$. The following examples illustrate both the versatility and the simplicity of Theorems 8.20 and 8.21.

**Example 8.16** ***MLE of Bernoulli Variance via the Invariance Principle***

Let $\mathbf{X} = (X_1,\ldots,X_n)'$ be a random sample from the Bernoulli population distribution $f(z;p) = p^z(1-p)^{1-z}I_{\{0,1\}}(z)$ representing whether or not a customer contact results in a sale. The MLE of $p$ is given by $\hat{p} = \bar{X}$. Then by the invariance principle, the MLE of the variance of the Bernoulli population distribution, $\sigma^2 = q(p) = p(1-p)$, is given by $\hat{\sigma}^2 = \bar{X}(1-\bar{X})$. □

**Example 8.17** ***MLEs of Parameter Sums and Ratios, and Residual Standard Deviation in the GLM via Invariance Principle***

Reexamine the case of maximum likelihood estimation in the GLM (Example 8.6), where the MLE of $\boldsymbol{\beta}$ and $\sigma^2$ is given by $\hat{\boldsymbol{\beta}} = (\mathbf{x}'\mathbf{x})^{-1}\mathbf{x}'\mathbf{Y}$ and $\hat{\sigma}^2 = (\mathbf{Y} - \mathbf{x}\hat{\boldsymbol{\beta}})'(\mathbf{Y} - \mathbf{x}\hat{\boldsymbol{\beta}})/n$. Then by the invariance principle, the MLE of $\sum_{i\in I}\beta_i$ is given by $\sum_{i\in I}\hat{\beta}_i$, the MLE of $\beta_i/\beta_j$is given by $\hat{\beta}_i/\hat{\beta}_j$, and the MLE of $\sigma$ is given by $(\hat{\sigma}^2)^{1/2}$.□

While the invariance principle allows convenient and straightforward definitions of MLEs for functions of $\boldsymbol{\Theta}$, the invariance principle does *not* imply that the estimator properties attributed to $\hat{\boldsymbol{\Theta}}$ transfer to the estimator $\mathbf{q}(\hat{\boldsymbol{\Theta}})$ of $\mathbf{q}(\boldsymbol{\Theta})$. In particular, the fact that $\hat{\boldsymbol{\Theta}}$ may be unbiased, MVUE, consistent, asymptotically normal, and/or asymptotically efficient does not necessarily imply that the MLE $\mathbf{q}(\hat{\boldsymbol{\Theta}})$ possesses any of the same properties. The small sample properties of $\mathbf{q}(\hat{\boldsymbol{\Theta}})$ as an estimator of $\mathbf{q}(\boldsymbol{\Theta})$ need to be checked on a case-by-case basis. Useful generally applicable results are not available that delineate problem conditions under which small sample properties of $\hat{\boldsymbol{\Theta}}$ are transferred to a general function of $\hat{\boldsymbol{\Theta}}$, $\mathbf{q}(\hat{\Theta})$, used to estimate $\mathbf{q}(\boldsymbol{\Theta})$. However, asymptotic properties of $\hat{\boldsymbol{\Theta}}$ transfer to $\mathbf{q}(\hat{\boldsymbol{\Theta}})$ under fairly general conditions, as the following two theorems indicate:

**Theorem 8.22** ***Consistency of MLE Defined via the Invariance Principle***

*Let $\hat{\boldsymbol{\Theta}}$ be a consistent MLE of $\boldsymbol{\Theta}$, and let $\mathbf{q}(\boldsymbol{\Theta})$ be a continuous function of $\boldsymbol{\Theta} \in \Omega$. Then the MLE $\mathbf{q}(\hat{\boldsymbol{\Theta}})$ is a consistent estimator of $\mathbf{q}(\boldsymbol{\Theta})$.*

**Proof** This follows directly from Theorem 5.5, since if $\mathbf{q}(\boldsymbol{\Theta})$ is a continuous function of $\boldsymbol{\Theta}$, and if $\text{plim}\left(\hat{\boldsymbol{\Theta}}\right) = \boldsymbol{\Theta}$, then $\text{plim}(\mathbf{q}(\hat{\boldsymbol{\Theta}}_n)) = \mathbf{q}(\text{plim}(\hat{\boldsymbol{\Theta}}_n)) = \mathbf{q}(\boldsymbol{\Theta})$. ■

**Theorem 8.23**
***Asymptotic Normality and Efficiency of MLE Defined via the Invariance Principle***

*Let $\hat{\Theta}$ be a consistent, asymptotically normal, and asymptotically efficient MLE of $\Theta$, satisfying the conditions of Theorems 8.18 and 8.19 so that $n^{1/2}\left(\hat{\Theta} - \Theta_n\right) \xrightarrow{d} N\left(\mathbf{0}, \mathbf{M}(\Theta_o)^{-1}\right)$. Let $\mathbf{q}(\Theta)$ be continuously differentiable with respect to $\Theta$ and let $\partial \mathbf{q}(\Theta_o)/\partial \Theta$ have full row rank. Then $\mathbf{q}(\hat{\Theta})$ is a consistent, asymptotically normal and asymptotically efficient MLE of $\mathbf{q}(\Theta)$.*

**Proof** It follows directly from Theorem 5.40 and Slutsky's theorem that

$$\left[\frac{\partial \mathbf{q}(\Theta_o)'}{\partial \Theta} \mathbf{M}(\Theta_o)^{-1} \frac{\partial \mathbf{q}(\Theta_o)}{\partial \Theta}\right]^{-1/2} n^{1/2}\left(\mathbf{q}\left(\hat{\Theta}\right) - \mathbf{q}(\Theta_o)\right) \xrightarrow{d} N(\mathbf{0}, \mathbf{I}).$$

Using Theorems 8.18 and 8.19 and Slutsky's theorems, it then also follows that

$$\left[\frac{\partial \mathbf{q}(\Theta_o)'}{\partial \Theta}\left[\mathrm{E}\left(\frac{\partial \ln(L(\Theta_o; \mathbf{X}))}{\partial \Theta} \frac{\partial \ln(L(\Theta_o; \mathbf{X}))}{\partial \Theta'}\right)\right]^{-1} \frac{\partial \mathbf{q}(\Theta_o)}{\partial \Theta}\right]^{-1/2}\left(\mathbf{q}\left(\hat{\Theta}\right) - \mathbf{q}(\Theta_o)\right) \xrightarrow{d} N(\mathbf{0}, \mathbf{I}),$$

which by Definition 7.22 justifies the conclusion of the theorem. ■

**Example 8.18**
***MLE Consistency, Asymptotic Normality, and Asymptotic Efficiency for Bernoulli Variance***

Recall Example 8.16. It is clear from results on sample moment properties that $\hat{P} = \bar{X}$ is consistent for $p$, $n^{1/2}\left(\hat{P} - p\right) \xrightarrow{d} N(0, p(1-p))$, and $\hat{P}$ is asymptotically efficient as an estimator of $p$. Examine the MLE of $\sigma^2 = p(1-p)$, which by the invariance principle is given by $\hat{\sigma}^2 = \bar{X}(1-\bar{X})$.

Because $p(1-p)$ is continuous for $p \in [0,1]$, $\hat{\sigma}^2$ is consistent for $p(1-p)$ by Theorem 8.20. Now note that $dq(p)/dp = 1 - 2p$ is continuous and is $\neq 0$ (and thus full rank) provided $p \neq .5$. Then $\bar{X}(1-\bar{X})$ is asymptotically normal and asymptotically efficient by Theorem 8.23, where $\hat{\sigma}^2 \overset{a}{\sim} N\left(\sigma^2, (1-2p)^2 p\,(1-p)/n\right)$. In the case where $p = .5$, the asymptotic distribution of $\hat{\sigma}^2$ can be shown to be equal to the distribution of the random variable $.25\,(1 - Y/n)$, where $Y \sim \chi_1^2$. See Bickel and Doksum, op. cit., p. 53. □

**Example 8.19**
***MLE Consistency, Asymptotic Normality, and Asymptotic Efficiency for Parameter Sums and Standard Deviation in the GLM***

Recall Example 8.17. Assuming $n^{-1}\,\mathbf{x}'\mathbf{x} \to \mathbf{Q}$, a symmetric positive definite matrix, it can be shown that the MLE $\begin{pmatrix}\hat{\boldsymbol{\beta}} \\ \hat{\sigma}^2\end{pmatrix}$ is consistent, asymptotically normal, and asymptotically efficient as an estimator of $\begin{pmatrix}\boldsymbol{\beta} \\ \sigma^2\end{pmatrix}$ where

$$n^{1/2}\left(\begin{bmatrix}\hat{\boldsymbol{\beta}} \\ \hat{\sigma}^2\end{bmatrix} - \begin{bmatrix}\boldsymbol{\beta}_o \\ \sigma_o^2\end{bmatrix}\right) \xrightarrow{d} N\left(\mathbf{0}, \left[\begin{array}{c|c}\sigma^2\mathbf{Q}^{-1} & \mathbf{0} \\ \hline \mathbf{0} & 2\sigma^4\end{array}\right]\right).$$

Examine the MLE $q_1\left(\hat{\boldsymbol{\beta}}, \hat{\sigma}^2\right) = \sum_{i=1}^k \hat{\beta}_i$ of $q_1(\beta, \sigma^2) = \sum_{i=1}^k \beta_i$. Since $q_1(\boldsymbol{\beta}, \sigma^2)$ is continuous in $\boldsymbol{\beta}$ and $\sigma^2$, $q_1\left(\hat{\boldsymbol{\beta}}, \hat{\sigma}^2\right)$ is a consistent estimator of $q_1(\boldsymbol{\beta}, \sigma^2)$ by Theorem 8.22. Also, since $\partial q_1/\partial \Theta = (1\ 1\ \ldots\ 1\ 0)' \neq \mathbf{0}$ $\left(\text{where } \Theta = \begin{bmatrix}\boldsymbol{\beta} \\ \sigma^2\end{bmatrix}\right)$,

$q_1\left(\hat{\boldsymbol{\beta}}, \hat{\sigma}^2\right)$ is asymptotically normal and asymptotically efficient by Theorem 8.22, where $\sum_{i=1}^{k} \beta_i \overset{a}{\sim} N\left(\boldsymbol{\iota}'\boldsymbol{\beta}_o, \boldsymbol{\iota}'\sigma^2(\mathbf{x}'\mathbf{x})^{-1}\boldsymbol{\iota}\right)$ with $\boldsymbol{\iota}$ a $(k \times 1)$ vector of ones.

Now examine the MLE $q_2\left(\hat{\boldsymbol{\beta}}, \hat{\sigma}^2\right) = \left(\hat{\sigma}^2\right)^{1/2} = \hat{\sigma}$ of $q_2(\boldsymbol{\beta},\sigma^2) = (\sigma^2)^{1/2} = \sigma$. Since $q_2(\boldsymbol{\beta},\sigma^2)$ is continuous in $\boldsymbol{\beta}$ and $\sigma^2 > 0$, $\hat{\sigma}$ is a consistent estimator of $\sigma$ by Theorem 8.22. Also, since $\partial q_2/\partial \boldsymbol{\Theta} = [0 \ldots 0 \ ((1/2)(\sigma^2)^{-1/2})]' \neq \mathbf{0}$ and the derivative is continuous for $\sigma^2 > 0$, $\hat{\sigma}$ is asymptotically normal and asymptotically efficient by Theorem 8.23, where $\hat{\sigma} \overset{a}{\sim} N(\sigma, \sigma^2/2n)$. □

**Example 8.20** ***MLE Consistency, Asymptotic Normality, and Asymptotic Efficiency for Exponential Variance***

Recall Example 8.4. We know that the MLE of $\theta$ in the exponential population distribution is given by $\hat{\Theta} = \bar{X}$, and the MLE is consistent, asymptotically normal, and asymptotically efficient, where $n^{1/2}\left(\hat{\Theta} - \theta\right) \xrightarrow{d} N(0, \theta^2)$. Examine the MLE of the variance of the exponential population distribution, $q(\theta) = \theta^2$. By the invariance principle, the MLE is given by $q(\hat{\Theta}) = \bar{X}^2$. Since $q(\theta)$ is continuous in $\theta$, $\bar{X}^2$ is consistent for $\theta^2$ by Theorem 8.22. Since $\partial q(\theta)/\partial \theta = 2\theta \neq 0$ and the derivative is continuous $\forall \theta > 0$, $q(\hat{\Theta})$ is asymptotically normal and asymptotically efficient by Theorem 8.23, where $\bar{X}^2 \overset{a}{\sim} N(\theta^2, 4\theta^4/n)$. □

### 8.3.5 MLE Property Summary

Overall, the maximum likelihood procedure is a relatively straightforward approach for defining point estimates of $\boldsymbol{\Theta}$ or a myriad of functions of $\boldsymbol{\Theta}$. The procedure can sometimes lead to a MVUE for $\boldsymbol{\Theta}$ or $\mathbf{q}(\boldsymbol{\Theta})$, and very often in practice, the MLE will possess good asymptotic properties. Specifically, properties of a MLE that we have examined in this section include:

1. A MLE is not necessarily unbiased.
2. If an unbiased estimator of $\boldsymbol{\Theta}$ exists that achieves the CRLB, and if the MLE is defined by solving first order conditions for $\hat{\boldsymbol{\Theta}}$, then the MLE will be unique, unbiased, and achieve the CRLB.
3. If a MLE is unique, then the MLE is a function of any set of sufficient statistics, including complete sufficient statistics if they exist.
4. If a MLE of $\boldsymbol{\Theta}$ is unique, and complete sufficient statistics exist, then

   (a) If the MLE is unbiased for $\boldsymbol{\Theta}$, it is MVUE for $\boldsymbol{\Theta}$;
   (b) If a function of the MLE is unbiased for $\boldsymbol{\Theta}$, then this function is the MVUE for $\boldsymbol{\Theta}$.

5. Under general regularity conditions on the estimation problem, the MLE is consistent, asymptotically normal, and asymptotically efficient.
6. If $\hat{\boldsymbol{\Theta}}$ is the MLE of the $(k \times 1)$ parameter vector $\boldsymbol{\Theta}$, then $\mathbf{q}(\hat{\boldsymbol{\Theta}})$ is the MLE of the $(r \times 1)$ vector $\mathbf{q}(\boldsymbol{\Theta})$, $r \leq k$ (MLE invariance principle).

7. Under general regularity conditions, if the MLE $\hat{\boldsymbol{\Theta}}$ is consistent, asymptotically normal, and asymptotically efficient for estimating $\boldsymbol{\Theta}$, then the MLE $\mathbf{q}(\hat{\boldsymbol{\Theta}})$ is also consistent, asymptotically normal, and asymptotically efficient for estimating $\mathbf{q}(\boldsymbol{\Theta})$.

## 8.4 Method of Moments and Generalized Method of Moments Estimation

Both the method of moments (MOM) and the Generalized Method of Moments (GMM) approaches to point estimation begin with the specification of a vector of *moment conditions* that involve the random variables in a random sample, $\mathbf{y}$, and the value of the parameter vector in the probability model for the random sample, $\{f(\mathbf{y};\boldsymbol{\Theta}),\ \boldsymbol{\Theta} \in \Omega\}$. The moment conditions take the general form $\mathrm{E}(\mathbf{g}(\mathbf{Y},\boldsymbol{\Theta})) = \mathbf{0}$, where it is understood that the expectation is taken with respect to $f(\mathbf{y};\boldsymbol{\Theta})$, i.e., $\int_{\mathbf{y}\in\mathbb{R}^n} \mathbf{g}(\mathbf{y},\boldsymbol{\Theta})\, dF(\mathbf{y};\boldsymbol{\Theta}) = \mathbf{0}$. Both the MOM and GMM approaches to estimation are based on sample-based estimates of the moment equations, $\hat{\mathrm{E}}(\mathbf{g}(\mathbf{Y},\boldsymbol{\Theta})) = \mathbf{0}$, in order to generate estimates of the parameters.

The traditional MOM approach, originally suggested long ago by Karl Pearson,[39] is based on *iid* sampling and moment conditions that are equal in number to the number of unknown parameters. The moment conditions are represented empirically via sample moments, and then the system of estimated sample moments is solved for the parameters to produce a MOM estimate. A general representation of the MOM estimate can be specified as $\hat{\boldsymbol{\Theta}} = \arg_{\boldsymbol{\Theta}}\left\{\hat{\mathrm{E}}(\mathbf{g}(\mathbf{Y},\boldsymbol{\Theta})) = \mathbf{0}\right\}$.

The GMM approach applies in more general settings where either sampling is not necessarily *iid*, or the number of sample moment conditions is larger than the number of unknown parameters, or both. If the number of moment equations equals the number of unknown parameters, a corresponding system of sample-based estimates of the moments are solved for the parameter values, as in the MOM approach, producing a GMM estimate. If the number of moment equations exceeds the number of unknown parameters, then the GMM estimate is the value of the parameter vector that solves the sample moments as closely as possible in terms of weighted squared Euclidean distance. A general representation of the GMM estimate can be specified as

$$\hat{\boldsymbol{\Theta}} = \underset{\boldsymbol{\Theta}}{\operatorname{argmin}} \left\{ \left[\hat{\mathrm{E}}(\mathbf{g}(\mathbf{Y},\boldsymbol{\Theta}))\right]' \mathbf{W} \left[\hat{\mathrm{E}}(\mathbf{g}(\mathbf{Y},\boldsymbol{\Theta}))\right] \right\},$$

where $\mathbf{W}$ is some positive definite weighting matrix, which may also depend on the sample size.

Details regarding implementation and estimator properties of the MOM and GMM approaches are discussed ahead.

---

[39]Pearson, K.P. (1894), *Contributions to the Mathematical Theory of Evolution*, Phil. Trans. R. Soc. London A, 185: pp. 71–110.

### 8.4.1 Method of Moments

The basic rationale underlying MOM estimation is rooted in the laws of large numbers. If $\mathrm{E}(\mathbf{g}(\mathbf{Y}_t,\boldsymbol{\Theta})) = \mathbf{0}$ defines a vector of $k$ moment conditions involving the $(k \times 1)$ vector $\boldsymbol{\Theta}$ that hold $\forall t$ when $\mathbf{Y}_t \sim iid\, f(\mathbf{Y};\boldsymbol{\Theta})$, and if a law of large numbers applies to the random variables $\mathbf{g}(\mathbf{Y}_t,\boldsymbol{\Theta})$, $t = 1,\ldots,n$, then for a size $n$ random sample of data from $f(\mathbf{Y};\boldsymbol{\Theta})$, $n^{-1}\sum_{t=1}^{n}\mathbf{g}(\mathbf{Y}_t,\boldsymbol{\Theta}) \xrightarrow{\mathrm{p}} \mathbf{0}$ (weak law) or $n^{-1}\sum_{t=1}^{n}\mathbf{g}(\mathbf{Y}_t,\boldsymbol{\Theta}) \xrightarrow{\mathrm{as}} \mathbf{0}$ (strong law). Moreover, if in a given experiment, the population distribution (unknown to the analyst) is actually given by $f(\mathbf{Y};\boldsymbol{\Theta}_{\mathrm{o}})$and $\mathrm{E}(\mathbf{g}(\mathbf{Y}_t,\boldsymbol{\Theta})) = \boldsymbol{\xi}(\boldsymbol{\Theta}) \neq \mathbf{0}$ when $\boldsymbol{\Theta} \neq \boldsymbol{\Theta}_{\mathrm{o}}$ so that $n^{-1}\sum_{t=1}^{n}\mathbf{g}(\mathbf{Y}_t,\boldsymbol{\Theta}) \rightarrow \boldsymbol{\xi}(\boldsymbol{\Theta})$ in probability or almost surely, then intuitively we would expect that solving $n^{-1}\sum_{t=1}^{n}\mathbf{g}(\mathbf{y}_t,\boldsymbol{\Theta}) = \mathbf{0}$ for $\boldsymbol{\Theta}$ should produce an estimate value, $\hat{\boldsymbol{\theta}}$, that becomes very close to $\boldsymbol{\Theta}_{\mathrm{o}}$ as $n$ increases. If not, the preceding convergence results would be ultimately contradicted. In fact under appropriate regularity conditions, this is precisely true and the MOM estimator $\hat{\boldsymbol{\Theta}}$ converges in probability, and also almost surely, to $\boldsymbol{\Theta}_{\mathrm{o}}$. It is also possible that $\hat{\boldsymbol{\Theta}}$ is asymptotically normal and efficient based on an application of central limit theory. Related arguments establishing asymptotic properties of GMM estimators under *iid* sampling can be applied in the more general context where the dimension of $\mathbf{g}$ exceeds $\boldsymbol{\Theta}$, in which case some weighted squared distance between $n^{-1}\sum_{t=1}^{n}\mathbf{g}(\mathbf{y}_t,\boldsymbol{\Theta})$ and $\mathbf{0}$ is minimized, as

$$\hat{\boldsymbol{\theta}} = \underset{\boldsymbol{\Theta}}{\operatorname{argmin}}\left\{\left[n^{-1}\sum_{t=1}^{n}\mathbf{g}(\mathbf{y}_t,\boldsymbol{\Theta})\right]'\mathbf{W}\left[n^{-1}\sum_{t=1}^{n}\mathbf{g}(\mathbf{y}_t,\boldsymbol{\Theta})\right]\right\}$$

with $\mathbf{W}$ being positive definite. We examine details of the GMM case in Section 8.4.2.

As an explicit illustration of the MOM approach, consider a statistical model for a random sample of size $n$ from a Bernoulli population distribution, so that the probability model is $\{f(\mathbf{y};p),\ p \in \Omega\}$ where $f(\mathbf{y};p) = p^{\sum_{t=1}^{n} y_i}(1-p)^{n-\sum_{t=1}^{n} y_t}\prod_{t=1}^{n} I_{\{0,1\}}(y_t)$ and $\Omega = [0,1]$. A moment condition that must hold for each $y_i$ when $p = p_{\mathrm{o}}$ in this probability model is given by $\mathrm{E}(\mathbf{g}(Y_t,p_{\mathrm{o}})) = \mathrm{E}(Y_t - p_{\mathrm{o}}) = 0,\ t = 1,\ldots,n$. The sample moment counterpart to this moment condition is $n^{-1}\sum_{t=1}^{n}\mathbf{g}(y_t,p) = n^{-1}\sum_{t=1}^{n}(y_t - p) = 0$, which upon solving for $p$ yields the MOM estimate $\hat{p} = \bar{y}$. In this situation, the MOM estimator is unbiased, BLUE, MVUE, consistent, asymptotically normal, and asymptotically efficient for estimating $p$.

*MOM Implementation* In applying the MOM procedure, it is assumed that the random sample $Y_1,\ldots,Y_n$ is a collection of *iid* random variables. We further assume that the $Y_t$'s are scalar random variables, although this is not necessary to apply the general methodology. Given a probability model for the random sample, $\{f(\mathbf{y};\boldsymbol{\Theta}),\ \boldsymbol{\Theta} \in \Omega\}$, our initial interest will focus on estimating $\boldsymbol{\Theta}$. We have seen in our study of parametric families of densities in Chapter 4 that the moments of density functions are functions of the parameters that characterize the parametric family. For example, in the exponential density case, $\mathrm{E}(Y_t) = \theta$

and $E(Y_t^2) = 2\theta^2$, or in the normal density case, $E(Y_t) = \mu$ and $E(Y_t^2) = \sigma^2 + \mu^2$. In general, $\mu'_r = E(Y_t^r) = h_r(\boldsymbol{\Theta})$, i.e., the $r$th moment about the origin for a parametric family of densities $f(\mathbf{y};\boldsymbol{\Theta})$, $\boldsymbol{\Theta} \in \Omega$, is a function of the parameter vector $\boldsymbol{\Theta}$.

A typical MOM implementation for estimating the $(k \times 1)$ vector $\boldsymbol{\Theta}$ is to first define a $(k \times 1)$ vector of invertible moment conditions of the form

$$E(\mathbf{g}(Y_t, \boldsymbol{\Theta})) = E\left(\begin{bmatrix} Y_t - h_1(\boldsymbol{\Theta}) \\ Y_t^2 - h_2(\boldsymbol{\Theta}) \\ \vdots \\ Y_t^k - h_k(\boldsymbol{\Theta}) \end{bmatrix}\right) = \mathbf{0}, \quad t = 1, \ldots, n.$$

The sample moment counterparts to the moment conditions are then specified as

$$n^{-1}\sum_{t=1}^{n} \mathbf{g}(y_t, \boldsymbol{\Theta}) = \begin{bmatrix} m'_1 - h_1(\boldsymbol{\Theta}) \\ m'_2 - h_2(\boldsymbol{\Theta}) \\ \vdots \\ m'_k - h_k(\boldsymbol{\Theta}) \end{bmatrix} = \mathbf{0}$$

and the solution for $\boldsymbol{\Theta}$ defines the MOM estimate via the inverse function $\mathbf{h}^{-1}$ as $\hat{\theta}_j = h_j^{-1}(m'_1, m'_2, \ldots, m'_k), j = 1, \ldots, k$.

We emphasize that while it is typical in practice to use the first $k$ population moments when defining the needed moment conditions, other moments about the origin, or even moments about the mean, can be used in defining the moment conditions. The key requirement is that the moment conditions be invertible so that $\boldsymbol{\Theta}$ can be solved in terms of whatever sample moments are utilized.

Note that an alternative motivation for the MOM estimate is to define the sample moment conditions in terms of expectations taken with respect to the empirical distribution function of the $y_i$'s, and then solve for $\boldsymbol{\Theta}$ as $\hat{\boldsymbol{\theta}} = \arg_{\boldsymbol{\Theta}}\left[E_{\hat{F}}(\mathbf{g}(\mathbf{Y}, \boldsymbol{\Theta})) = \mathbf{0}\right]$. Some examples of the procedure will illustrate its relative simplicity.

**Example 8.21** ***MOM Estimates of Gamma Distribution Parameters***

Let $(Y_1, \ldots, Y_n)$ be a random sample from a gamma population distribution of waiting times between customer arrivals in minutes, and suppose it is desired to estimate the parameter vector $(\alpha, \beta)$. In this case the moment conditions can be specified as,

$$E(\mathbf{g}(Y_t, \alpha, \beta)) = E\left(\begin{bmatrix} Y_t - \alpha\beta \\ Y_t^2 - \alpha\beta^2(1 + \alpha) \end{bmatrix}\right) = \mathbf{0}, \quad t = 1, \ldots, n.$$

The sample moment counterpart to the moment conditions is given by

$$n^{-1}\sum_{t=1}^{n}\mathbf{g}(y_t,\alpha,\beta)=\begin{bmatrix} m'_1-\alpha\beta \\ m'_2-\alpha\beta^2(1+\alpha)\end{bmatrix}=\mathbf{0}$$

so that the MOM estimate is defined by

$$\begin{bmatrix}\hat{\alpha}\\ \hat{\beta}\end{bmatrix}=\begin{bmatrix}(m'_1)^2/\left[m'_2-(m'_1)^2\right]\\ \left[m'_2-(m'_1)^2\right]/m'_1\end{bmatrix}.$$

For example, if $m'_1 = 1.35$ and $m'_2 = 2.7$, then the MOM estimate is given by $\hat{\alpha} = 2.0769$ and $\hat{\beta} = .65$. □

**Example 8.22** ***MOM Estimates of Normal Distribution Parameters*** Let $(Y_1,\ldots,Y_n)$ be a random sample from a normal population distribution representing the weights, in hundredweights, of steers fed a certain ration of feed for 6 months, and suppose it is desired to estimate the parameter vector $(\mu,\sigma^2)$. The moment conditions in this case are

$$\mathrm{E}(\mathbf{g}(Y_t,\mu,\sigma^2))=\mathrm{E}\left(\begin{bmatrix}Y_t-\mu\\ Y_t^2-(\sigma^2+\mu^2)\end{bmatrix}\right)=\mathbf{0},\ t=1,\ldots,n.$$

The sample moment counterpart to the moment conditions is given by

$$n^{-1}\sum_{t=1}^{n}\mathbf{g}(y_t,\mu,\sigma^2)=\begin{bmatrix}m'_1-\mu\\ m'_2-(\sigma^2+\mu^2)\end{bmatrix}=\mathbf{0},$$

so that the MOM estimate is defined by

$$\begin{bmatrix}\hat{\mu}\\ \hat{\sigma}^2\end{bmatrix}=\begin{bmatrix}m'_1\\ m'_2-(m'_1)^2\end{bmatrix}.$$

For example, if $m'_1 = 12.3$ and $m'_2 = 155.2$, then the MOM estimate is given by $\hat{\mu} = 12.3$ and $\hat{\sigma}^2 = 3.91$. □

*MOM Estimator Properties* It is very often the case that the inverse function $\hat{\boldsymbol{\Theta}} = \mathbf{h}^{-1}(M'_1,\ldots,M'_k)$ is continuous, in which case the MOM estimator inherits consistency from the consistency of the sample moments $M'_r$ for $\mu'_r$, $\forall r$.

**Theorem 8.24** ***Consistency of MOM Estimator of*** $\boldsymbol{\Theta}$ *Let the MOM estimator* $\hat{\boldsymbol{\Theta}}_{(k\times 1)} = \mathbf{h}^{-1}(M'_1,\ldots,M'_k)$ *be such that the inverse function* $\mathbf{h}^{-1}(\mu'_1,\ldots,\mu'_k)$ *is continuous* $\forall(\mu'_1,\ldots,\mu'_k)\in\Gamma=\{(\mu'_1,\ldots,\mu'_k):\mu'_i = h_i(\boldsymbol{\Theta}), i=1,\ldots,k, \boldsymbol{\Theta}\in\Omega\}$. *Then* $\hat{\boldsymbol{\Theta}}\xrightarrow{p}\boldsymbol{\Theta}$.

**Proof** This is a direct consequence of Theorem 5.5 and the probability limit properties of sample moments about the origin. In particular, given the assumed continuity property of $\mathbf{h}^{-1}$, $\mathrm{plim}\left(\hat{\boldsymbol{\Theta}}\right) = \mathbf{h}^{-1}(\mathrm{plim}(M'_1),\ldots,\mathrm{plim}(M'_k)) = \mathbf{h}^{-1}(\mu'_1,\ldots,\mu'_k) = \boldsymbol{\Theta}$ by the invertibility of $\mathbf{h}$. ■

**Example 8.23**
**MOM Consistency for Gamma and Normal Distribution Parameters**

In Example 8.21, $(\alpha,\beta)$ is a continuous function of $\mu'_1$ and $\mu'_2$ for all $(\mu'_1,\mu'_2) \in \Gamma = \left\{(\mu'_1,\mu'_2) : \mu'_1>0, \mu'_2>0, \mu'_2>(\mu'_1)^2\right\}$, which constitutes all relevant values of $(\mu'_1,\mu'_2)$. (Recall $\sigma^2 = \mu'_2 - (\mu'_1)^2 = \alpha\beta^2>0$ in the Gamma family.) Thus by Theorem 8.23, the MOM estimator of $(\alpha,\beta)$ is consistent. In Example 8.22, $(\mu,\sigma^2)$ is a continuous function of $(\mu'_1,\mu'_2)$ so that the MOM estimator of $(\mu,\sigma^2)$ is consistent. □

In addition to consistency, the MOM estimator will have an asymptotic normal density if the inverse function $\boldsymbol{\Theta} = \mathbf{h}^{-1}(\mu'_1,\ldots,\mu'_k)$ is continuously differentiable and its Jacobian matrix has full rank.

**Theorem 8.25**
**Asymptotic Normality of MOM Estimator of $\boldsymbol{\Theta}$**

*Let the MOM estimator $\hat{\boldsymbol{\Theta}} = \mathbf{h}^{-1}(M'_1,\ldots,M'_k)$ be such that $\mathbf{h}^{-1}(\mu'_1,\ldots,\mu'_k)$ is differentiable $\forall\, (\mu'_1,\ldots,\mu'_k) \in \Gamma = \{(\mu'_1,\ldots,\mu'_k):\mu'_i = h_i(\boldsymbol{\Theta}),\ i = 1,\ldots,k,\ \boldsymbol{\Theta} \in \Omega\}$, and let the elements of*

$$\mathbf{A}(\mu'_1,\ldots,\mu'_k) = \begin{bmatrix} \dfrac{\partial h_1^{-1}(\mu'_1,\ldots,\mu'_k)}{\partial \mu'_1} & \cdots & \dfrac{\partial h_1^{-1}(\mu'_1,\ldots,\mu'_k)}{\partial \mu'_k} \\ \vdots & \ddots & \vdots \\ \dfrac{\partial h_k^{-1}(\mu'_1,\ldots,\mu'_k)}{\partial \mu'_1} & \cdots & \dfrac{\partial h_k^{-1}(\mu'_1,\ldots,\mu'_k)}{\partial \mu'_k} \end{bmatrix}$$

*be continuous functions with $\mathbf{A}(\mu'_1,\ldots,\mu'_k)$ having full rank $\forall(\mu'_1,\ldots,\mu'_k) \in \Gamma$. Then*

$$n^{1/2}\left(\hat{\boldsymbol{\Theta}} - \boldsymbol{\Theta}\right) \xrightarrow{d} N(\mathbf{0}, \mathbf{A}\boldsymbol{\Sigma}\mathbf{A}'), \text{ and } \hat{\boldsymbol{\Theta}} \overset{a}{\sim} N(\boldsymbol{\Theta}, n^{-1}\mathbf{A}\boldsymbol{\Sigma}\mathbf{A}'), \text{ where } \boldsymbol{\Sigma} = \mathbf{Cov}(M'_1,\ldots, M'_k).$$

**Proof** Recall that the sample moments converge to a normal limiting distribution as

$$n^{1/2}\left[\begin{bmatrix} M'_1 \\ \vdots \\ M'_k \end{bmatrix} - \begin{bmatrix} \mu'_1 \\ \vdots \\ \mu'_k \end{bmatrix}\right] \xrightarrow{d} N(\mathbf{0}, \boldsymbol{\Sigma}).$$

Then since the partial derivatives contained in $\mathbf{A}$ are continuous and since $\mathbf{A}$ has full rank, the result follows directly from Theorem 5.40. ■

**Example 8.24**
**MOM Asymptotic Normality for Gamma and Normal Distribution Parameters**

Reexamine Examples 8.21 and 8.22. In Example 8.21,

$$\mathbf{A}(\mu'_1, \mu'_2) = \begin{bmatrix} \dfrac{2\mu'_2\mu'_1}{\left[\mu'_2 - (\mu'_1)^2\right]^2} & \dfrac{-(\mu'_1)^2}{\left[\mu'_2 - (\mu'_1)^2\right]^2} \\ \dfrac{-\left[\mu'_2 + (\mu'_1)^2\right]}{(\mu'_1)^2} & \dfrac{1}{(\mu'_1)} \end{bmatrix},$$

which exists with continuous elements $\forall\, (\mu'_1, \mu'_2) \in \Gamma$ . Furthermore, $\mathbf{A}$ has full rank, because $\det(\mathbf{A}) = \left(\mu'_2 - (\mu'_1)^2\right)^{-1} > 0\ \forall (\mu'_1, \mu'_2) \in \Gamma$ . Therefore, by Theorem 8.25, the MOM estimator of $(\alpha, \beta)$ is asymptotically normally distributed. In Example 8.22,

$$\mathbf{A}(\mu'_1, \mu'_2) = \begin{bmatrix} 1 & 0 \\ -2\mu'_1 & 1 \end{bmatrix},$$

which exists with continuous elements and has full rank $\forall (\mu'_1, \mu'_2)$ so that the MOM estimator is asymptotically normally distributed by Theorem 8.25. □

The reader should be warned that although the MOM estimator of $\boldsymbol{\Theta}$ is conceptually straightforward to define, and is often quite straightforward to compute, in some cases the inverse function may be difficult to define explicitly, if an explicit representation exists at all. A computer algorithm for solving systems of nonlinear equations might be needed to compute the inverse function values and thereby calculate outcomes of the MOM estimator. As an example of more complicated cases, the reader should consider MOM estimation of the parameter vector $\boldsymbol{\Theta}$ when sampling from a Beta family of densities.

MOM estimators for a function of $\boldsymbol{\Theta}$, say $\mathbf{q}(\boldsymbol{\Theta})$, can be defined by the corresponding function of the MOM estimator of $\boldsymbol{\Theta}$, as $\mathbf{q}\left(\hat{\boldsymbol{\Theta}}\right) = \mathbf{q}\left(\mathbf{h}^{-1}(M'_1, \ldots, M'_k)\right)$. Under appropriate conditions on the function $\mathbf{q}$, and if the MOM estimator of $\boldsymbol{\Theta}$ satisfies the conditions of Theorems 8.24 and 8.25, then $\mathbf{q}(\hat{\boldsymbol{\Theta}})$ is consistent and asymptotically normal as an estimator of $\mathbf{q}(\boldsymbol{\Theta})$.

**Theorem 8.26** ***Consistency and Asymptotic Normality of MOM Estimator of*** $\mathbf{q}(\boldsymbol{\Theta})$

*Let the MOM estimator* $\hat{\boldsymbol{\Theta}} = \mathbf{h}^{-1}(M'_1, \ldots, M'_k)$ *of* $\boldsymbol{\Theta}$ *satisfy the conditions of Theorems 8.24 and 8.25. If the function* $\mathbf{q}(\boldsymbol{\Theta})$ *is continuous for* $\boldsymbol{\Theta} \in \boldsymbol{\Omega}$, *then the MOM estimator* $\mathbf{q}(\hat{\boldsymbol{\Theta}})$ *is consistent for* $\mathbf{q}(\boldsymbol{\Theta})$. *If* $\partial \mathbf{q}(\boldsymbol{\Theta})/\partial \boldsymbol{\Theta}'$ *exists, has full row rank, and its elements are continuous functions of* $\boldsymbol{\Theta}$ for $\boldsymbol{\Theta} \in \Omega$,

$$n^{1/2}\left(\mathbf{q}\left(\hat{\boldsymbol{\Theta}}\right) - \mathbf{q}(\boldsymbol{\Theta})\right) \xrightarrow{d} N\left(\mathbf{0}, \frac{\partial \mathbf{q}(\boldsymbol{\Theta})'}{\partial \boldsymbol{\Theta}} \boldsymbol{\Phi} \frac{\partial \mathbf{q}(\boldsymbol{\Theta})}{\partial \boldsymbol{\Theta}}\right)$$

*and*

$$\mathbf{q}\left(\hat{\boldsymbol{\Theta}}\right) \overset{a}{\sim} N\left(\mathbf{q}(\boldsymbol{\Theta}), n^{-1} \frac{\partial \mathbf{q}(\boldsymbol{\Theta})'}{\partial \boldsymbol{\Theta}} \boldsymbol{\Phi} \frac{\partial \mathbf{q}(\boldsymbol{\Theta})}{\partial \boldsymbol{\Theta}}\right),$$

*where* $\boldsymbol{\Phi}$ *is the covariance matrix of the limiting distribution of* $n^{1/2}(\hat{\boldsymbol{\Theta}} - \boldsymbol{\Theta})$ *in Theorem 8.25.*

**Proof** The proof is analogous to the proofs of Theorem's 8.24 and 8.25 and is left to the reader. ■

Overall, the MOM approach to point estimation of $\boldsymbol{\Theta}$ or $\mathbf{q}(\boldsymbol{\Theta})$ is conceptually simple, computation of estimates is often straightforward, and the MOM estimator is consistent and asymptotically normal under fairly general conditions. However, the MOM estimator is not necessarily unbiased, BLUE, MVUE, or asymptotically efficient. It has been most often applied in cases where the definition or computation of other estimators, such as the MLE, are extremely complex, or when the sample size is quite large so that the consistency of the estimator can be relied upon to assure a reasonably accurate estimate. Further discussion of asymptotic efficiency of MOM estimators will be undertaken in the next subsection in the context of GMM estimators.

### 8.4.2 Generalized Method of Moments (GMM)

In the case of the GMM, the random sample is not restricted to *iid* random variables, and the moment conditions can be greater in number than the number of parameters being estimated. However, the general estimation principles remain the same, namely, moment conditions $\mathrm{E}(\mathbf{g}(\mathbf{Y},\boldsymbol{\Theta})) = \mathbf{0}$ are specified pertaining to the probability model $\{f(\mathbf{y};\boldsymbol{\Theta}), \boldsymbol{\Theta} \in \Omega\}$, sample moment counterparts are specified as $\hat{\mathrm{E}}(\mathbf{g}(\mathbf{Y},\boldsymbol{\Theta})) = \mathbf{0}$, and then the GMM estimate is defined as the value of $\boldsymbol{\Theta}$ that satisfies the sample moment conditions as closely as possible—exactly if $\mathbf{g}$ is the same dimension as $\boldsymbol{\Theta}$.

The following two examples demonstrate how the GMM estimator can be formulated to subsume the least squares estimator and the MLE as special cases.

**Example 8.25** ***Least Squares Estimator as a GMM Estimator***

Let the probability model of a random sample of the yields/acre of a particular type of corn be (incompletely) specified by the linear model $\mathbf{Y} = \mathbf{x}\boldsymbol{\beta} + \boldsymbol{\varepsilon}$, where the classical GLM assumptions are assumed to apply and $\boldsymbol{\beta}$ is a $(k \times 1)$ vector of parameters indicating the responsiveness of yield to various inputs. Moment conditions for the statistical model can be specified by the $(k \times 1)$ vector function

$$\mathrm{E}(\mathbf{g}(Y_t,\boldsymbol{\beta})) = \mathrm{E}(\mathbf{x}'_{t.}(Y_t - \mathbf{x}_{t.}\boldsymbol{\beta})) = \mathrm{E}(\mathbf{x}'_{t.}\varepsilon_t) = \mathbf{0},\ \ t = 1,\ldots,n,$$

which reflect the orthogonality of the residuals from the explanatory variables. Corresponding sample moment conditions can be defined as

$$n^{-1}\sum_{t=1}^{n}\mathbf{g}(y_t,\boldsymbol{\beta}) = n^{-1}\sum_{t=1}^{n}\mathbf{x}'_{t.}(y_t - \mathbf{x}_{t.}\boldsymbol{\beta}) = n^{-1}[\mathbf{x}'\mathbf{y} - \mathbf{x}'\mathbf{x}\boldsymbol{\beta}] = \mathbf{0},$$

which when solved for $\boldsymbol{\beta}$ defines the GMM estimate as $\mathbf{b} = (\mathbf{x}'\mathbf{x})^{-1}\mathbf{x}'\mathbf{y}$. Thus, the least squares estimator is a GMM estimator. □

**Example 8.26** ***MLE as a GMM Estimator***

Let $\{f(\mathbf{y};\boldsymbol{\Theta}),\boldsymbol{\Theta} \in \boldsymbol{\Omega}\}$ be the probability model for a random sample for which the $Y_i$'s are not necessarily *iid*, and represent the joint density of the random sample as $f(\mathbf{y};\boldsymbol{\Theta}) = m(y_1;\boldsymbol{\Theta})\prod_{t=2}^{n} f(y_t|y_{t-1},\ldots,y_1;\boldsymbol{\Theta})$. Assume that differentiation under the integral or summation sign is possible (recall CRLB regularity condition (4) of Definition 7.21) so that moment conditions can be defined as[40]

$$\mathrm{E}(\mathbf{g}(Y_1,\boldsymbol{\Theta})) = \mathrm{E}\left(\frac{\partial \ln(m(Y_1;\boldsymbol{\Theta}))}{\partial \boldsymbol{\Theta}}\right) = \mathbf{0},$$

$$\mathrm{E}(\mathbf{g}(Y_t,\boldsymbol{\Theta})) = \mathrm{E}\left(\frac{\partial \ln(f(Y_t|y_{t-1},\ldots,y_1;\boldsymbol{\Theta}))}{\partial \boldsymbol{\Theta}}\right) = \mathbf{0} \text{ for } t = 2,\ldots,n$$

Then specify the sample moment conditions as

$$n^{-1}\sum_{t=1}^{n}\mathbf{g}(y_t,\boldsymbol{\Theta}_o) = n^{-1}\left[\frac{\partial \ln(m(y_1;\boldsymbol{\Theta}))}{\partial \boldsymbol{\Theta}} + \sum\nolimits_{t=2}^{n}\frac{\partial \ln(f(y_t|y_{t-1},\ldots,y_1;\boldsymbol{\Theta}))}{\partial \boldsymbol{\Theta}}\right]$$
$$= n^{-1}\frac{\partial \ln(L(\boldsymbol{\Theta};\mathbf{y}))}{\partial \boldsymbol{\Theta}} = \mathbf{0},$$

which when solved for $\boldsymbol{\Theta}$ defines the GMM estimate $\hat{\boldsymbol{\theta}} = \arg_{\boldsymbol{\Theta}}\{\partial \ln(L(\boldsymbol{\Theta};\mathbf{y}))/\partial\boldsymbol{\Theta} = \mathbf{0}\}$. Thus assuming the MLE of $\boldsymbol{\Theta}$ is defined as the solution to first order conditions, the MLE is a GMM estimator. □

The reader will come to find in her later studies of econometrics or statistics that the GMM subsumes a large number of the estimation procedures used in practice. For example, *instrumental variable techniques, two and three-stage least squares*, and *quasi- or pseudo-maximum likelihood estimation techniques* can all be interpreted as GMM procedures. The interested reader is referred to A.R. Gallant, (1987), *Nonlinear Statistical Models*, NY, John Wiley, Chapter 4, and to R. Mittelhammer, G. Judge, and D. Miller, (2000), *Econometric Foundations*, Cambridge University Press, Cambridge, Chapters 16 and 17, for additional details on applying the GMM approach to more general statistical models.

It is possible to apply the GMM procedure in cases where the vector function defining the moment conditions is of larger dimension than the dimension of the vector $\boldsymbol{\Theta}$. In these cases the GMM estimator is defined as the value of $\boldsymbol{\Theta}$ that minimizes a weighted measure of the squared distance between the sample moments and the zero vector, where the weights can depend on the sample size, as

$$\hat{\boldsymbol{\Theta}} = \arg\min_{\boldsymbol{\Theta}\in\boldsymbol{\Omega}}\left\{\left[\hat{\mathrm{E}}(\mathbf{g}(\mathbf{Y},\boldsymbol{\Theta}))\right]'\mathbf{W}_n(\mathbf{y})\left[\hat{\mathrm{E}}(\mathbf{g}(\mathbf{Y},\boldsymbol{\Theta}))\right]\right\}.$$

The matrix $\mathbf{W}_n(\mathbf{y})$ will be a symmetric, positive definite conformable weighting matrix which may or may not depend on $\mathbf{y}$, and which is such that $\mathbf{W}_n \xrightarrow{p} \mathbf{W}$ with $\mathbf{W}$ being a nonrandom, symmetric, and positive definite matrix.

[40]Under the assumed conditions, for each density (conditional or not),

$$\mathrm{E}_{\boldsymbol{\Theta}_o}\left(\frac{\partial \ln f(\mathbf{Z};\boldsymbol{\Theta}_o)}{\partial\boldsymbol{\Theta}}\right) = \int_{-\infty}^{\infty}\frac{\partial f(\mathbf{z};\boldsymbol{\Theta}_o)}{\partial\boldsymbol{\Theta}}dz = \frac{\partial\int_{-\infty}^{\infty}f(\mathbf{z};\boldsymbol{\Theta}_o)dz}{\partial\boldsymbol{\Theta}} = 0$$

in the continuous case, and likewise in the discrete case.

The previous case where $\mathbf{g}$ and $\boldsymbol{\Theta}$ are of equal dimension is subsumed within the current context because in this case the minimum of the distance measure occurs at the value of $\boldsymbol{\Theta}$ for which $\hat{\mathrm{E}}(\mathbf{g}(\mathbf{Y}, \boldsymbol{\Theta})) = \mathbf{0}$. The GMM estimator is consistent and asymptotically normal quite generally with respect to choices of the weighting matrix, although asymptotic efficiency considerations require a specific choice of $\mathbf{W}_n$. These issues will be considered in the next section.

*General GMM Estimator Properties* As was the case for the MOM procedure, general statements regarding the properties of GMM estimators are relegated to the asymptotic variety. We will examine a set of *sufficient* conditions for the consistency and asymptotic normality of the GMM estimator, and a sufficient condition for asymptotic efficiency within the class of estimators based on a particular set of moment conditions. The reader can refer to Gallant, op. cit., and L.P. Hansen, (1982), "Large Sample Properties of Generalized Method of Moments Estimators," *Econometrica*, pp. 1029–1054 for alternative conditions that lead to asymptotic properties of the GMM estimator. Throughout the remainder of our discussion we will be referring to the GMM estimator defined in terms of the minimization of the weighted squared distance measure presented at the end of the previous section, which subsumes the case where the dimension of $\mathbf{g}$ and $\boldsymbol{\Theta}$ are equal.

*GMM Consistency* Sufficient conditions under which the GMM estimator is consistent are given below.

**Theorem 8.27 *Consistency of GMM Estimator*** *Let $\{f(\mathbf{y};\boldsymbol{\Theta}), \boldsymbol{\Theta} \in \boldsymbol{\Omega}\}$ be the probability model for the random sample $\mathbf{Y}$, where $\boldsymbol{\Theta}$ is $(k \times 1)$. Let $\mathrm{E}(\mathbf{g}_t(\mathbf{Y}_t,\boldsymbol{\Theta})) = \mathbf{0}$ be an $(m \times 1)$ moment condition for $t = 1,\ldots,n$ with $m \geq k$. Define the GMM estimator*

$$\hat{\boldsymbol{\Theta}} = \arg\min_{\boldsymbol{\Theta}\in\boldsymbol{\Omega}}\{Q_n(\mathbf{Y}, \boldsymbol{\Theta})\} = \arg\min_{\boldsymbol{\Theta}\in\boldsymbol{\Omega}}\left\{\left[n^{-1}\sum_{t=1}^{n}\mathbf{g}_t(\mathbf{Y}_t, \boldsymbol{\Theta})\right]' \mathbf{W}_n(\mathbf{Y})\left[n^{-1}\sum_{t=1}^{n}\mathbf{g}_t(\mathbf{Y}_t, \boldsymbol{\Theta})\right]\right\}$$

*where $\mathrm{plim}(\mathbf{W}_n(\mathbf{Y})) = \mathbf{W}$, a nonrandom positive definite symmetric matrix. Suppose that $\boldsymbol{\Theta}_o$ is the true parameter vector such that $\mathbf{Y} \sim f(\mathbf{y}; \boldsymbol{\Theta}_o)$ and assume that*

**a.** *$\boldsymbol{\Omega}$ is a closed and bounded rectangle,*
**b.** *$\hat{\boldsymbol{\Theta}}$ is unique,*
**c.** $\mathrm{plim}\left[n^{-1}\sum_{t=1}^{n}\mathbf{g}_t(\mathbf{Y}_t, \boldsymbol{\Theta})\right] = \mathbf{G}(\boldsymbol{\Theta})$, *a continuous nonstochastic $(m \times 1)$ vector function of $\boldsymbol{\Theta}$ for which $\mathbf{G}(\boldsymbol{\Theta}) = \mathbf{0}$ iff $\boldsymbol{\Theta} = \boldsymbol{\Theta}_o$,*
**d.** $\lim_{n\to\infty}(P(\max_{\boldsymbol{\Theta}\in\boldsymbol{\Omega}}|Q_n(\mathbf{y}, \boldsymbol{\Theta}) - h(\boldsymbol{\Theta})| < \varepsilon)) = 1, \forall\varepsilon > 0$, *where* $h(\boldsymbol{\Theta}) = \mathbf{G}(\boldsymbol{\Theta})'\mathbf{W}\,\mathbf{G}(\boldsymbol{\Theta})$.

*Then $\hat{\boldsymbol{\Theta}} \xrightarrow{p} \boldsymbol{\Theta}_o$.*

**Proof** See Appendix. ■

In practice, when the GMM estimator is represented by an *explicit* function of the random sample, it is often more straightforward to verify consistency by taking probability limits directly. Sufficient conditions, such as those in Theorem 8.26, are generally needed when the GMM estimator is only an *implicit* function of the random sample. We illustrate both approaches to verifying consistency in the following example.

**Example 8.27**
***Consistency of GMM Instrumental Variables Estimator***

Let the aggregate demand for a given commodity be approximated by some linear function $\mathbf{Y} = \mathbf{X}\boldsymbol{\beta} + \mathbf{V}$ for some appropriate value of $\boldsymbol{\beta}$, where the $V_t$'s are *iid* with $\mathrm{E}(V_t) = 0$ and $\mathrm{var}(V_t) = \sigma^2$, $\forall t$. Suppose that the $(n \times k)$ matrix $\mathbf{X}$ contains factors affecting demand, including some commodity prices that are simultaneously determined with quantity demanded, $\mathbf{Y}$, so that $\mathrm{E}((\mathbf{X}'\mathbf{X})^{-1}\,\mathbf{X}'\mathbf{V}) \neq \mathbf{0}$ (recall Table 8.2). Thus, the OLS estimator of $\boldsymbol{\beta}$, $\hat{\boldsymbol{\beta}} = (\mathbf{X}'\mathbf{X})^{-1}\mathbf{X}'\mathbf{Y}$, will be biased and inconsistent.

Suppose there existed a conformable $(n \times k)$ matrix of variables, $\mathbf{Z}$, (called *instrumental variables* in the literature) such that

$$\mathrm{E}(\mathbf{g}_t(Y_t, \boldsymbol{\beta})) = \mathrm{E}(\mathbf{Z}'_{t.}(Y_t - \mathbf{X}_{t.}\boldsymbol{\beta})) = \mathrm{E}(\mathbf{Z}'_{t.}V_t) = \mathbf{0} \text{ for } t = 1, \ldots, n,$$

where also $n^{-1}\mathbf{Z}'\mathbf{V} \xrightarrow{p} \mathbf{0}$, $n^{-1}\mathbf{Z}'\mathbf{Z} \xrightarrow{p} \mathbf{A}_{zz}$, a finite positive definite symmetric matrix, and $n^{-1}\mathbf{Z}'\mathbf{X} \xrightarrow{p} \mathbf{A}_{zx}$, a finite nonsingular square matrix. The sample moment conditions are specified by

$$n^{-1}\sum_{t=1}^{n} \mathbf{g}_t(y_t, \boldsymbol{\beta}) = n^{-1}\sum_{t=1}^{n} \mathbf{z}_{t.}(y_t - \mathbf{x}_{t.}\boldsymbol{\beta}) = n^{-1}[\mathbf{z}'\mathbf{y} - \mathbf{z}'\mathbf{x}\boldsymbol{\beta}] = \mathbf{0}.$$

Assuming that $\mathbf{z}'\mathbf{x}$ is nonsingular, the sample moment conditions have a *unique* solution defining the GMM estimate as $\hat{\mathbf{b}} = (\mathbf{z}'\mathbf{x})^{-1}\mathbf{z}'\mathbf{y}$, which is also referred to in the literature as the **instrumental variables estimator**. Thus condition (b) of Theorem 8.27 is satisfied.

Letting $\boldsymbol{\beta}_o$ denote the true (and unknown) value of the parameter vector, so that $\mathbf{Y} = \mathbf{X}\boldsymbol{\beta}_o + \mathbf{V}$ is the correct linear model, note that

$$\mathrm{plim}\left[n^{-1}\sum_{t=1}^{n} \mathbf{g}_t(Y_t, \boldsymbol{\beta})\right] = \mathbf{A}_{zx}[\boldsymbol{\beta}_o - \boldsymbol{\beta}] = \mathbf{G}(\boldsymbol{\beta}),$$

which equals $\mathbf{0}$ *iff* $\boldsymbol{\beta} = \boldsymbol{\beta}_o$. Thus condition (c) of Theorem 8.27 is satisfied.

Now assume that $\Omega$ is a closed and bounded rectangle, so that the GMM estimate can be represented equivalently as

$$\hat{\boldsymbol{\beta}} = \arg\min_{\beta}\{Q_n(\mathbf{y}, \boldsymbol{\beta})\} = \arg\min_{\boldsymbol{\beta}\in\Omega}\{n^{-1}[\mathbf{z}'\mathbf{y} - \mathbf{z}'\mathbf{x}\boldsymbol{\beta}]'\mathbf{I}[\mathbf{z}'\mathbf{y} - \mathbf{z}'\mathbf{x}\boldsymbol{\beta}]n^{-1}\}.$$

Note that $\max_{\boldsymbol{\beta}\in\Omega}|Q_n(\mathbf{y}, \boldsymbol{\beta}) - h(\boldsymbol{\beta})|$ will exist by Weierstrass's theorem, where $h(\boldsymbol{\beta}) = (\boldsymbol{\beta}_o - \boldsymbol{\beta})'\mathbf{A}'_{zx}\mathbf{A}_{zx}(\boldsymbol{\beta}_o - \boldsymbol{\beta})$, and the maximum is a continuous function of the elements in $n^{-1}\mathbf{z}'\mathbf{y}$ and $n^{-1}\mathbf{z}'\mathbf{x}$ by the theorem of the maximum. It follows that

$$\mathrm{plim}\left(\max_{\boldsymbol{\beta}\in\Omega}\{|Q_n(\mathbf{Y}, \boldsymbol{\beta}) - h(\boldsymbol{\beta})|\}\right) = \max_{\boldsymbol{\beta}\in\Omega}\{|\mathrm{p}\lim(Q_n(\mathbf{Y}, \boldsymbol{\beta})) - h(\boldsymbol{\beta})|\} = \mathbf{0},$$

so that condition (d) of Theorem 8.27 is satisfied. Therefore, $\hat{\boldsymbol{\beta}} \xrightarrow{p} \boldsymbol{\beta}_o$.

Given the explicit functional definition of $\hat{\boldsymbol{\beta}}$, the consistency of the GMM estimator can be proven more directly by noting that

$$\text{plim}\left(\hat{\boldsymbol{\beta}}\right) = \text{plim}\left(n^{-1}\mathbf{Z}'\mathbf{X}\right)^{-1}\left(n^{-1}\mathbf{Z}'\mathbf{Y}\right),$$

$$= \text{plim}\left(n^{-1}\mathbf{Z}'\mathbf{X}\right)^{-1}\text{plim}\left(n^{-1}(\mathbf{Z}'\mathbf{X}\boldsymbol{\beta}_o + \mathbf{Z}'\mathbf{V})\right) = \mathbf{A}_{zx}^{-1}(\mathbf{A}_{zx}\boldsymbol{\beta}_o) = \boldsymbol{\beta}_o$$

where $n^{-1}\mathbf{Z}'\mathbf{V} \xrightarrow{p} \mathbf{0}.$ □

*Asymptotic Normality and Efficiency in the GMM Class* Under additional assumptions relating to the moment conditions, such as those in the theorem below, the GMM estimators are asymptotically normally distributed.

**Theorem 8.28** ***Asymptotic Normality of GMM Estimator*** *Assume*[41] *the conditions of Theorem 8.27. In addition, assume that*

**a.** $\partial^2 Q_n(\mathbf{y},\boldsymbol{\Theta})/\partial\boldsymbol{\Theta}\partial\boldsymbol{\Theta}'$ *exists and is continuous* $\forall \boldsymbol{\Theta} \in \boldsymbol{\Omega}$;

**b.** *For any sequence* $\{\boldsymbol{\Theta}_n^*\}$ *for which*

$$\text{plim}(\boldsymbol{\Theta}_n^*) = \boldsymbol{\Theta}_o,\ \partial^2\boldsymbol{\Theta}_n(\mathbf{Y},\boldsymbol{\Theta}_n^*)/\partial\boldsymbol{\Theta}\partial\boldsymbol{\Theta}' \xrightarrow{p} \mathbf{D}(\boldsymbol{\Theta}_o),$$

*a finite symmetric nonsingular matrix; and*

**c.** $n^{1/2}(\partial Q_n(\mathbf{Y},\boldsymbol{\Theta}_o)/\partial\boldsymbol{\Theta}) \xrightarrow{d} \mathbf{Z} \sim N(\mathbf{0},\mathbf{C}(\boldsymbol{\Theta}_o)).$

*Then* $n^{1/2}\left(\hat{\boldsymbol{\Theta}} - \boldsymbol{\Theta}_o\right) \xrightarrow{d} N\left(\mathbf{0},\mathbf{D}(\boldsymbol{\Theta}_o)^{-1}\mathbf{C}(\boldsymbol{\Theta}_o)\mathbf{D}(\boldsymbol{\Theta}_o)^{-1}\right)$, *and*

$$\hat{\boldsymbol{\Theta}} \overset{a}{\sim} N\left(\boldsymbol{\Theta}_o, n^{-1}\mathbf{D}(\boldsymbol{\Theta}_o)^{-1}\mathbf{C}(\boldsymbol{\Theta}_o)\mathbf{D}(\boldsymbol{\Theta}_o)^{-1}\right).$$

**Proof** By a first-order Taylor series representation,

$$\frac{\partial Q_n\left(\mathbf{Y},\hat{\boldsymbol{\Theta}}\right)}{\partial\boldsymbol{\Theta}} = \frac{\partial Q_n(\mathbf{Y},\boldsymbol{\Theta}_o)}{\partial\boldsymbol{\Theta}} + \frac{\partial^2 Q_n(\mathbf{Y},\boldsymbol{\Theta}_*)}{\partial\boldsymbol{\Theta}\partial\boldsymbol{\Theta}'}\left(\hat{\boldsymbol{\Theta}}-\boldsymbol{\Theta}_o\right)$$

where $\boldsymbol{\Theta}_* = \lambda\hat{\boldsymbol{\Theta}} + (1-\lambda)\boldsymbol{\Theta}_o$ and $\lambda \in [0,1]$.[42] By the first order conditions to the minimization problem defining the GMM estimator, $\partial Q_n\left(\mathbf{y};\hat{\boldsymbol{\Theta}}\right)/\partial\boldsymbol{\Theta} = \mathbf{0}$, so that by Slutsky's theorems and the consistency of $\hat{\boldsymbol{\Theta}}$

$$n^{1/2}\left(\hat{\boldsymbol{\Theta}} - \boldsymbol{\Theta}_o\right) = -\left[\frac{\partial^2 Q_n(\mathbf{Y},\boldsymbol{\Theta}_*)}{\partial\boldsymbol{\Theta}\partial\boldsymbol{\Theta}'}\right]^{-1} n^{1/2}\frac{\partial Q_n(\mathbf{Y},\boldsymbol{\Theta}_o)}{\partial\boldsymbol{\Theta}}$$

$$\xrightarrow{d} -\mathbf{D}(\boldsymbol{\Theta}_o)^{-1}\mathbf{Z} \sim N\left(\mathbf{0},\mathbf{D}(\boldsymbol{\Theta}_o)^{-1}\mathbf{C}(\boldsymbol{\Theta}_o)\mathbf{D}(\boldsymbol{\Theta}_o)^{-1}\right).$$

and the asymptotic distribution result follows directly. ■

[41] See T. Amemiya, op. cit., pp. 111–112 for a closely related theorem relating to extremum estimators. The assumptions of Theorem 8.26 can be weakened to assuming uniqueness and consistency of $\hat{\boldsymbol{\Theta}}$ for use in this theorem.

[42] We are suppressing the fact that each row of the matrix $\partial^2 Q_n(Y,\boldsymbol{\Theta}_*)/\partial\boldsymbol{\Theta}\partial\boldsymbol{\Theta}'$ will generally require a different value of $\boldsymbol{\Theta}_*$ defined by a different value of $\lambda$ for the representation to hold. The conclusions of the argument will remain the same.

**Example 8.28**
***Asymptotic Normality of GMM Instrumental Variables Estimator***

Revisit the instrumental variable estimator of Example 8.27. Note that $\partial^2 Q_n(\mathbf{y}, \boldsymbol{\beta})/\partial\boldsymbol{\beta}\partial\boldsymbol{\beta}' = 2n^{-2}\mathbf{x}'\mathbf{z}\mathbf{z}'\mathbf{x}$, which exists and is (trivially) continuous for $\boldsymbol{\beta} \in \Omega$. Also, $\partial^2 Q_n(\mathbf{Y}, \boldsymbol{\beta}_*)/\partial\boldsymbol{\beta}\partial\boldsymbol{\beta}' = 2(n^{-1}\mathbf{X}'\mathbf{Z})(\mathbf{Z}'\mathbf{X}n^{-1}) \xrightarrow{p} \mathbf{D}(\boldsymbol{\beta}_o) = 2\mathbf{A}'_{zx}\mathbf{A}_{zx}$, which is a finite nonsingular matrix regardless of the sequence $\{\boldsymbol{\beta}_*\}$ for which $\boldsymbol{\beta}_* \xrightarrow{p} \boldsymbol{\beta}_o$. Finally, note that

$$n^{1/2}\partial Q_n(\mathbf{Y}, \boldsymbol{\beta}_o)/\partial\boldsymbol{\beta} = -2(n^{-1}\mathbf{X}'\mathbf{Z})\left(n^{-1/2}\mathbf{Z}'\mathbf{V}\right) \xrightarrow{d} -2\mathbf{A}'_{zx}\mathbf{T} \sim N(\mathbf{0}, 4\sigma^2\mathbf{A}'_{zx}\mathbf{A}_{zz}\mathbf{A}_{zx})$$

where $n^{-1/2}\mathbf{Z}'\mathbf{V} \xrightarrow{d} \mathbf{T} \sim N(\mathbf{0}, \sigma^2\mathbf{A}_{zz})$ by an application of the multivariate Lindberg-Levy CLT, assuming that $[\mathbf{X}_{i.}, \mathbf{Z}_{i.}, \mathbf{V}_i]$ are *iid* (or else, a different central limit theorem would be used). Then from Theorem 8.28,

$$n^{1/2}\left(\hat{\boldsymbol{\beta}} - \boldsymbol{\beta}_o\right) \xrightarrow{d} N\left(\mathbf{0}, \sigma^2(\mathbf{A}_{zx})^{-1}\mathbf{A}_{zz}(\mathbf{A}'_{zx})^{-1}\right), \text{ and } \hat{\boldsymbol{\beta}} \overset{a}{\sim} N\left(\boldsymbol{\beta}_o, (\sigma^2/n)(\mathbf{A}_{zx})^{-1}\mathbf{A}_{zz}(\mathbf{A}_{zx})^{-1}\right).$$

An *estimate* of the asymptotic covariance matrix can be defined as

$$\hat{\sigma}^2(\mathbf{z}'\mathbf{x})^{-1}\mathbf{z}'\mathbf{z}(\mathbf{x}'\mathbf{z})^{-1}, \text{where } \hat{\sigma}^2 = n^{-1}\left(\mathbf{y} - \mathbf{x}\hat{\mathbf{b}}\right)'\left(\mathbf{y} - \mathbf{x}\hat{\mathbf{b}}\right).$$

Regarding the efficiency of the GMM estimator, L.P. Hansen, *Moments Estimators,* has provided conditions that characterize the optimal choice of $\mathbf{W}_n(\mathbf{y})$ in the definition of the GMM estimator

$$\hat{\boldsymbol{\beta}}_{\mathbf{W}_n} = \arg\min_{\boldsymbol{\Theta}}\left\{\left[n^{-1}\sum_{t=1}^{n}\mathbf{g}_t(\mathbf{Y}_t, \boldsymbol{\Theta})\right]'\mathbf{W}_n(\mathbf{Y})\left[n^{-1}\sum_{t=1}^{n}\mathbf{g}_t(\mathbf{Y}_t, \boldsymbol{\Theta})\right]\right\}.$$

In particular, setting $\mathbf{W}_n(\mathbf{Y})$ equal to the inverse of the asymptotic covariance matrix of $n^{-1/2}\sum_{i=1}^{n}\mathbf{g}_t(\mathbf{Y}_t, \boldsymbol{\Theta})$ is the choice that defines the GMM estimator that is asymptotically most efficient relative to all choices of the weighting matrix. Note that this result establishes asymptotic efficiency within the class of GMM estimators based on a *particular* set of moment conditions. Results relating to the optimal choice of moment conditions with which to define the GMM estimator in the general case are not available, although a number of special cases have been investigated. The interested reader can consult MacKinnon and Davidson, *Estimation and Inference,* Chapter 17 for additional reading and references.

**Example 8.29**
***Asymptotic Efficiency of GMM Instrumental Variables Estimator***

Revisit Example 8.28 concerning the instrumental variable estimator. Note that

$$n^{-1/2}\sum_{t=1}^{n}\mathbf{g}_t(Y_t, \boldsymbol{\beta}_o) = n^{-1/2}[\mathbf{Z}'\mathbf{Y} - \mathbf{Z}'\mathbf{X}\boldsymbol{\beta}_o] = n^{-1/2}[\mathbf{Z}'\mathbf{V}] \xrightarrow{d} N(\mathbf{0}, \sigma^2\mathbf{A}_{zz}),$$

as presented in Example 8.28. Thus, the optimal GMM estimator, which is based on $\mathbf{W}_n(\mathbf{Y}) = (\sigma^2\mathbf{A}_{zz})^{-1}$, is defined by

$$\hat{\boldsymbol{\beta}} = \arg\min_{\boldsymbol{\beta}}\left\{n^{-2}(\mathbf{Z}'\mathbf{Y} - \mathbf{Z}'\mathbf{X}\boldsymbol{\beta})'\left(\sigma^2\mathbf{A}_{zz}\right)^{-1}(\mathbf{Z}'\mathbf{Y} - \mathbf{Z}'\mathbf{X}\boldsymbol{\beta})\right\} = (\mathbf{Z}'\mathbf{X})^{-1}\mathbf{Z}'\mathbf{Y}. \quad \square$$

In practice, if the solution for $\hat{\mathbf{\Theta}}$ depends on a weighting matrix that contains unknown elements (in Example 8.29, the fact that $\mathbf{A}_{zz}$ was unknown is irrelevant—only its positive definiteness mattered), an estimator that converges in probability to the asymptotically efficient GMM estimator can be constructed using any weighting matrix that converges in probability to the appropriate asymptotic covariance matrix of $n^{-1/2}\sum_{t=1}^{n}\mathbf{g}_t(\mathbf{Y}_t,\mathbf{\Theta})$.

Also, unlike Example 8.29, if the number of instrumental variables, and thus equivalently the number of moment conditions, exceeds the number of parameters, the weight matrix will appear in the solution for the GMM estimator, as the following example illustrates.

**Example 8.30**
***Asymptotic Efficiency of GMM Two-Stage Least Squares Estimator***

Revisit Examples 8.27–8.29, except assume that the instrumental variable matrix $\mathbf{Z}$ is $m \times k$, with $m > k$. Following arguments analogous to those used in the preceding examples, the moment conditions would be represented by

$$n^{-1}\sum_{t=1}^{n}\mathbf{g}_t(y_t,\boldsymbol{\beta}) = n^{-1}\sum_{t=1}^{n}\mathbf{z}_{t.}(y_t - \mathbf{x}_{t.}\boldsymbol{\beta}) = n^{-1}[\mathbf{z}'\mathbf{y} - \mathbf{z}'\mathbf{x}\boldsymbol{\beta}] = \mathbf{0}.$$

To pursue an optimal GMM estimation approach in this case, recall that

$$n^{-1/2}\sum_{t=1}^{n}\mathbf{g}_t(Y_t,\boldsymbol{\beta}_o) = n^{-1/2}[\mathbf{Z}'\mathbf{Y} - \mathbf{Z}'\mathbf{X}\boldsymbol{\beta}_o] = n^{-1/2}[\mathbf{Z}'\mathbf{V}]\xrightarrow{d} N(\mathbf{0},\sigma^2\mathbf{A}_{zz})$$

where now $\mathbf{A}_{zz} = \text{plim}(n^{-1}\mathbf{Z}'\mathbf{Z})$ is $(m \times m)$, and thus the optimal weighting matrix in the GMM procedure is given by $\mathbf{W} = \sigma^{-2}\mathbf{A}_{zz}^{-1}$. Note that the scalar $\sigma^{-2}$ can be ignored since it will have no effect on the choice of $\boldsymbol{\beta}$ that would minimize the weighted squared distance measure that defines the GMM. The remaining matrix component of the optimal weight matrix is not observable, but a consistent estimator of it is given by $\hat{\mathbf{A}}_{zz}^{-1} = \left[n^{-1}\mathbf{Z}'\mathbf{Z}\right]^{-1}$.

Suppressing the $n$ value in the estimated optimal weighting matrix (which is irrelevant to the minimization problem), the estimated optimal GMM estimator is finally represented by

$$\hat{\boldsymbol{\beta}}_{GMM} = \arg\min_{\boldsymbol{\beta}}\left\{[n^{-1}\mathbf{Z}'(\mathbf{Y}-\mathbf{X}\boldsymbol{\beta})]'[\mathbf{Z}'\mathbf{Z}]^{-1}[n^{-1}\mathbf{Z}'(\mathbf{Y}-\mathbf{X}\boldsymbol{\beta})]\right\}.$$

Differentiating the objective function with respect to $\boldsymbol{\beta}$ and then solving the resultant first order conditions, it is found that the optimal GMM estimator in this case of more instrumental variables, and thus more moment conditions, than unknown parameters is equivalent to the so-called **Two Stage Least Squares Estimator** found in the econometrics literature. The solution is given by $\hat{\boldsymbol{\beta}}_{GMM} = \hat{\boldsymbol{\beta}}_{2SLS} = \left[\mathbf{X}'\mathbf{Z}[\mathbf{Z}'\mathbf{Z}]^{-1}\mathbf{Z}'\mathbf{X}\right]^{-1}\mathbf{X}'\mathbf{Z}[\mathbf{Z}'\mathbf{Z}]^{-1}\mathbf{Z}'\mathbf{Y}$. □

## 8.5 Appendix: Proofs and Proof References for Theorems

**Theorem 8.6**

**Proof** **a.** Under the stated conditions, the sequence of squared residuals $\{\varepsilon_n^2\} = \{\varepsilon_1^2, \varepsilon_2^2, \ldots\}$ satisfies the assumptions of the WLLN stated in Theorem 5.22. To see this, note that

$$\operatorname{var}(\varepsilon_i^2) = \mathrm{E}(\varepsilon_i^4) - \left(\mathrm{E}(\varepsilon_i^2)\right)^2 = \mathrm{E}(\varepsilon_i^4) - \sigma^4 \leq \tau < \infty \;\forall i,$$

so that

$$\operatorname{var}(n^{-1}\boldsymbol{\varepsilon}'\boldsymbol{\varepsilon}) = n^{-2}\left[\sum_{i=1}^{n} \operatorname{var}(\varepsilon_i^2) + \sum_{i \neq j} \operatorname{Cov}(\varepsilon_i^2, \varepsilon_j^2)\right] \leq \left[\tau + o(n^1)\right]/n \to 0$$

as $n \to \infty$. It follows immediately that plim $(n^{-1}\boldsymbol{\varepsilon}'\boldsymbol{\varepsilon}) = \sigma^2$. Also, given the Assumptions of the Classical GLM, it can be shown that plim $\left((n-k)^{-1}\boldsymbol{\varepsilon}'\mathbf{x}(\mathbf{x}'\mathbf{x})^{-1}\mathbf{x}'\boldsymbol{\varepsilon}\right) = 0$ using Markov's inequality in exactly the same way as in the proof of Theorem 8.5. It follows that

$$\operatorname{plim}\left(\hat{S}^2\right) = \underbrace{\operatorname{plim}(n^{-1}\boldsymbol{\varepsilon}'\boldsymbol{\varepsilon})}_{\sigma^2} - \underbrace{\operatorname{plim}\left[(n-k)^{-1}\boldsymbol{\varepsilon}'\mathbf{x}(\mathbf{x}'\mathbf{x})^{-1}\boldsymbol{\varepsilon}\right]}_{0} = \sigma^2.$$

**b.** The conditions in part (b) imply the conditions in part (a), since $\operatorname{cov}(\varepsilon_i^2, \varepsilon_j^2) = 0$ when $|i-j| > m$ by $m$-dependence. Thus, $\hat{S}^2 \xrightarrow{p} \sigma^2$. ■

**Theorem 8.8**

**Proof** The proof is based on the Liapounov CLT for triangular arrays (Theorem 5.34) and the Cramer-Wold device (Corollary 5.4). Define $\mathbf{V}_n = n^{-1}\,\mathbf{x}'\mathbf{x}$ and note that

$$(\mathbf{x}'\mathbf{x})^{1/2}\left(\hat{\boldsymbol{\beta}} - \boldsymbol{\beta}\right) = n^{-1/2}\mathbf{V}_n^{-1/2}\mathbf{x}'\boldsymbol{\varepsilon} = n^{-1/2}\sum_{t=1}^{n}\mathbf{V}_n^{-1/2}\mathbf{x}'_{t.}\varepsilon_t.$$

Examine

$$\boldsymbol{\ell}'(\mathbf{x}'\mathbf{x})^{1/2}\left(\hat{\boldsymbol{\beta}} - \boldsymbol{\beta}\right) = n^{-1/2}\sum_{t=1}^{n}\boldsymbol{\ell}'\mathbf{V}_n^{-1/2}\mathbf{x}'_{t.}\varepsilon_t = n^{-1/2}\sum_{t=1}^{n}W_{nt}$$

where $\boldsymbol{\ell}$ is any conformable vector such that $\boldsymbol{\ell}'\boldsymbol{\ell} = 1$. Note that $\mathrm{E}(W_{nt}) = 0$ and $\operatorname{var}(W_{nt}) = \sigma^2\,\boldsymbol{\ell}'\,\mathbf{V}_n^{-1/2}\,\mathbf{x}'_{t.}\,\mathbf{x}_{t.}\,\mathbf{V}_n^{-1/2}\boldsymbol{\ell} < \infty\;\forall\, n$ and $t$, where the finiteness of $\operatorname{var}(W_{nt})$ follows from the boundedness of $\mathrm{E}(\,W_{nt}^4\,)$, which is shown below. Because $\mathrm{E}(W_{nt}) = 0$, it follows that $\mathrm{E}(|W_{nt} - \mathrm{E}(W_{nt})|^4) = \mathrm{E}(W_{nt}^4)$, and then

$$\mathrm{E}(W_{nt}^4) = E(\varepsilon_i^4)\left[\boldsymbol{\ell}'\mathbf{V}_n^{-1/2}\mathbf{x}'_{t.}\,\mathbf{x}_{t.}\,\mathbf{V}_n^{-1/2}\boldsymbol{\ell}\right]^2 < \tau\left[\boldsymbol{\ell}'\mathbf{V}_n^{-1/2}\mathbf{x}'_{t.}\,\mathbf{x}_{t.}\,\mathbf{V}_n^{-1/2}\boldsymbol{\ell}\right]^2.$$

Now note that $\boldsymbol{\ell}'\mathbf{V}_n^{-1/2}\,\mathbf{x}'_{t.}\,\mathbf{x}_{t.}\,\mathbf{V}_n^{-1/2}\boldsymbol{\ell} < c < \infty\;\forall n$ and $t$. This follows from the fact that $|\mathbf{V}_n| > \eta > 0\;\forall n$ with $\{\mathbf{V}_n\}$ being 0(1) $\Rightarrow \{\mathbf{V}_n^{-1/2}\}$ is O(1) (H. White, (1982),

"Instrumental Variables Regression with Independent Observations," *Econometrica*, pp. 484–485), and since $\mathbf{x}'_{t.}\mathbf{x}_{t}$ is a positive semidefinite matrix all of whose elements are bounded in absolute value by $\xi^2$, it follows that $\mathbf{V}_n^{-1/2}\mathbf{x}'_{t.}\mathbf{x}_{t.}\mathbf{V}_n^{-1/2}$ is a O(1) positive semidefinite matrix, so that $\boldsymbol{\ell}'\mathbf{V}_n^{-1/2}\mathbf{x}'_{t.}\mathbf{x}_{t.}\mathbf{V}_n^{-1/2}\boldsymbol{\ell}$ is bounded as claimed. Then $\mathrm{E}(W_{nt}^4)<\tau c^2=\gamma<\infty\ \forall n$ and $t$.

Given the preceding results, Theorem 5.34 is applicable. Note that the $W_{nt}$'s can be represented in the form of a triangular array (Definition 5.9) with typical row $(W_{n1},\ldots,W_{nn})$, $\forall n$, and the $W_{nt}$'s are independent within rows since the $\varepsilon_t$'s are independent. The limit condition of Theorem 5.34 is met, since

$$0\leq\lim_{n\to\infty}\left[\frac{\sum_{t=1}^{n}\mathrm{E}(W_{nt}^4)}{\left[\sum_{t=1}^{n}\mathrm{Var}(W_{nt})\right]^2}\right]\leq\gamma\lim_{n\to\infty}\left[\frac{n}{n^2\sigma^4}\right]=0.$$

Then because $n^{-1}\sum_{t=1}^{n}\mathrm{Var}(W_{nt})=\sigma^2$, it follows from Theorem 5.34 that

$$\boldsymbol{\ell}'(\mathbf{x}'\mathbf{x})^{1/2}\left(\hat{\boldsymbol{\beta}}-\boldsymbol{\beta}\right)=n^{1/2}\left[n^{-1}\sum_{t=1}^{n}W_{nt}\right]\xrightarrow{d}N(0,\sigma^2),$$

and by the Cramer-Wold device $(\mathbf{x}'\mathbf{x})^{1/2}\left(\hat{\boldsymbol{\beta}}-\boldsymbol{\beta}\right)\xrightarrow{d}N(\mathbf{0},\sigma^2\mathbf{I})$ , so that $\hat{\boldsymbol{\beta}}\overset{a}{\sim}N\left(\boldsymbol{\beta},\sigma^2(\mathbf{x}'\mathbf{x})^{-1}\right)$. ■

**Theorem 8.10**

**Proof** Similar to the proof of Theorem 8.9, we can show that the limiting distributions of $n^{1/2}\left(\hat{S}^2-\sigma^2\right)/\xi_n^{1/2}$ and of $(\sum_{i=1}^{n}\varepsilon_i^2-n\sigma^2)/\left(n^{1/2}\xi_n^{1/2}\right)$ coincide. To see this, note by Markov's inequality that

$$P\left(\frac{n^{1/2}}{(n-k)\xi_n^{1/2}}\boldsymbol{\varepsilon}'\mathbf{x}(\mathbf{x}'\mathbf{x})^{-1}\mathbf{x}\boldsymbol{\varepsilon}\geq c\right)\leq\frac{\sigma^2n^{1/2}k}{c(n-k)\xi_n^{1/2}},$$

and if $\xi_n\geq\eta>0$, $\forall n$, then the right hand side (RHS) of the inequality $\to 0$ as $n\to\infty$, so that $\mathrm{plim}\left(n^{1/2}/\left((n-k)\xi_n^{1/2}\right)\right)\boldsymbol{\varepsilon}'\mathbf{x}(\mathbf{x}'\mathbf{x})^{-1}\mathbf{x}'\boldsymbol{\varepsilon}=0$. The remaining steps for showing the equivalence of the limiting distributions are analogous to those in the proof of Theorem 8.9.

Liapounov's CLT (Theorem 5.33) can now be applied to

$$\left(\sum_{i=1}^{n}\varepsilon_i^2-n\sigma^2\right)/\left(n^{1/2}\xi_n^{1/2}\right).$$

Note that the $\varepsilon_i^2$'s are independent, with $\mathrm{E}(\varepsilon_i^2)=\sigma^2$ and $\mathrm{var}(\varepsilon_i^2)=\mathrm{E}\left(\left(\varepsilon_i^2-\sigma^2\right)^2\right)<\infty\ \forall i$ (the existence of $\mathrm{var}(\varepsilon_i^2)$ follows from the existence of $\mathrm{E}\left(|\varepsilon_i^2-\sigma^2|^{2+\delta}\right)$ for $\delta>0$; recall Theorem 3.23). Furthermore, the limit condition of the Liapounov CLT is met, because

$$\lim_{n\to\infty} \frac{\sum_{i=1}^{n} \mathrm{E}\left(|\varepsilon_i^2 - \sigma^2|^{2+\delta}\right)}{\left[\sum_{i=1}^{n} \mathrm{var}(\varepsilon_i^2)\right]^{1+\delta/2}} \le \lim_{n\to\infty}\left(\frac{n\tau}{(n\eta)^{1+\delta/2}}\right) = \lim_{n\to\infty}\left(\tau/\eta^{1+\delta/2}\right)n^{-\delta/2} = 0$$

Therefore, by the Liapounov CLT,

$$\left(\sum_{i=1}^{n} \varepsilon_i^2 - n\sigma^2\right)/(n\,\xi_n)^{1/2} \xrightarrow{d} N(0,1),$$

which in turn implies that

$$\frac{n^{1/2}\left(\hat{S}^2 - \sigma^2\right)}{\xi_n^{1/2}} \xrightarrow{d} N(0,1) \text{ and } \hat{S}^2 \overset{a}{\sim} N(\sigma^2, n^{-1}\xi_n). \quad \blacksquare$$

**Theorem 8.17**

**Proof** ***Sufficiency of* 1-4a:** Define $H(\varepsilon) = \{\mathbf{x}: \ln(L(\Theta_o;\mathbf{x})) > \max_{\Theta\in\overline{N(\varepsilon)}}(L(\Theta;\mathbf{x}))\}$, and note that $\mathbf{x} \in H(\varepsilon) \Rightarrow \hat{\Theta} = \arg\max_{\Theta\in\Omega}(L(\Theta;\mathbf{x})) \in N(\varepsilon)$, where $\hat{\Theta}$ is unique by 3. Assumption (4a) implies that $P(\mathbf{x} \in H(\varepsilon)) \longrightarrow 1$ *as* $n \longrightarrow \infty$ $\forall\varepsilon > 0$, which in turn implies that $P(\hat{\Theta} \in N(\varepsilon)) \longrightarrow 1$ *as* $n \longrightarrow \infty$ $\forall\varepsilon > 0$. It follows from the definition of $N(\varepsilon)$ that $\hat{\Theta} \xrightarrow{p} \Theta_o$.

***Sufficiency of 1-3, 4b*:**[43] Define $\xi(\varepsilon) = G\,\Theta_o) - \max_{\Theta\in\Omega\cap\overline{N(\varepsilon)}}(G(\Theta))$ (*because* $\Omega$ is closed and bounded, the feasible space to the maximization problem is closed and bounded by the definition of $N(\varepsilon)$, and thus the maximum exists by Weierstrass's theorem). Let the event $A_n(\varepsilon)$ be defined as

$$A_n(\varepsilon) = \left\{\mathbf{x} : \max_{\Theta\in\Omega}\left\{|n^{-1}\ln(L(\Theta;\mathbf{x})) - G(\Theta)|\right\} < \xi(\varepsilon)/2\right\}.$$

Letting $\hat{\Theta}$ represent the unique MLE, it follows from $\mathbf{x} \in A_n(\varepsilon)$ that

1. $G(\hat{\Theta}) > n^{-1}\ln(L(\hat{\Theta};\mathbf{x})) - \xi(\varepsilon)/2$
2. $n^{-1}\ln(L(\Theta_o;\mathbf{x})) > G(\Theta_o) - \xi(\varepsilon)/2.$

By definition of the MLE, $L(\hat{\Theta};\mathbf{x}) \ge L(\Theta_o;\mathbf{x})$, so it follows from (1) that

3. $G(\hat{\Theta}) > n^{-1}\ln(L(\Theta_o;\mathbf{x})) - \xi(\varepsilon)/2.$

Then substituting the inequality (2) into the right side of the inequality in (3) yields

4. $G(\hat{\Theta}) > G(\Theta_o) - \xi(\varepsilon),$

which upon substituting the definition of $\xi(\varepsilon)$ yields

5. $G(\hat{\Theta}) > \max_{\Theta\in\Omega\cap\overline{N(\varepsilon)}}(G(\Theta)) \Rightarrow \hat{\Theta} \in N(\varepsilon).$

Then since

$$\lim_{n\to\infty} P(A_n(\varepsilon)) = 1\ \forall\varepsilon>0,\ \lim_{n\to\infty} P\left(\hat{\Theta} \in N(\varepsilon)\right) = 1\ \forall\varepsilon>0, \text{ so that } \hat{\Theta} \xrightarrow{p} \Theta_o. \blacksquare$$

[43]This proof is related to a proof by T. Amemiya,(1985), *Advanced Econometrics*, Harvard University Press, p. 107, dealing with the consistency of extremum estimators.

**Theorem 8.18**

**Proof** The first order derivative function of $\ln(L(\hat{\Theta};\mathbf{x}))$ can be represented by a Taylor series expansion around $\Theta_o$ as[44]

$$\frac{\partial \ln\left(L\left(\hat{\Theta};\mathbf{X}\right)\right)}{\partial \Theta} = \frac{\partial \ln(L(\Theta_o;\mathbf{X}))}{\partial \Theta} + \frac{\partial^2 \ln(L(\Theta^*;\mathbf{X}))}{\partial \Theta \partial \Theta'}\left(\hat{\Theta} - \Theta\right),$$

where $\Theta^* = \lambda(\mathbf{X})\hat{\Theta} + (1 - \lambda(\mathbf{X}))\Theta_o$ for $\lambda(\mathbf{X}) \in [0,1]$. Since $\hat{\Theta}_n$ is the ML estimator, the value of the first derivative vector, $\partial \ln\left(L\left(\hat{\Theta};\mathbf{X}\right)\right)/\partial\Theta$, is $\mathbf{0}$ by the first order conditions for the maximum likelihood problem. Then premultiplying the Taylor series expansion by $n^{-1/2}$ obtains

$$-n^{-1}\frac{\partial^2 \ln\left(L\left(\hat{\Theta}^*;\mathbf{X}\right)\right)}{\partial \Theta \partial \Theta'} n^{1/2}\left(\hat{\Theta} - \Theta_o\right) = n^{-1/2}\frac{\partial \ln(L(\Theta_o;\mathbf{X}))}{\partial \Theta} = \mathbf{Z} \xrightarrow{d} N(\mathbf{0}, \mathbf{M}(\Theta_o))$$

where convergence to the $N(\mathbf{0}, \mathbf{M}(\Theta_o))$ limiting density follows from assumption (3) of the theorem.

Now note that $\text{plim}(\Theta^*) = \Theta_o$ since $\forall \varepsilon > 0$,

$$\begin{aligned}\lim_{n\to\infty} P(\|\Theta^* - \Theta_o\| < \varepsilon) &= \lim_{n\to\infty} P\left(\left\|\lambda(\mathbf{x})\hat{\Theta}(\mathbf{x}) + (1 - \lambda(\mathbf{x}))\Theta_o - \Theta_o\right\| < \varepsilon\right) \\ &= \lim_{n\to\infty} P\left(\left\|\lambda(\mathbf{x})\left(\hat{\Theta}(\mathbf{x}) - \Theta_o\right)\right\| < \varepsilon\right) \\ &= \lim_{n\to\infty} P\left(\lambda(\mathbf{x})\left\|\left(\hat{\Theta}(\mathbf{x}) - \Theta_o\right)\right\| < \varepsilon\right)\end{aligned}$$

where the last equality follows from the fact that $\lambda(\mathbf{x}) \in [0,1]$ and therefore $\lambda(\mathbf{x})$ is nonnegative. Noting that $\left\{\mathbf{x} : \left\|\hat{\Theta}(\mathbf{x}) - \Theta_o\right\| < \varepsilon\right\} \subset \left\{\mathbf{x} : \lambda(\mathbf{x})\left\|\hat{\Theta}(\mathbf{x}) - \Theta_o\right\| < \varepsilon\right\}$ (again since $\lambda(\mathbf{x}) \in [0,1]$), it follows that $P\left(\left\|\hat{\Theta} - \Theta_o\right\| < \varepsilon\right) \le P\left(\lambda(\mathbf{x})\left\|\hat{\Theta} - \Theta_o\right\| < \varepsilon\right)$, and since $\hat{\Theta} \xrightarrow{p} \Theta_o$, the left hand side of the above inequality converges to 1 as $n \to \infty$, implying that the right hand side converges to 1 as $n \to \infty$, $\forall \varepsilon > 0$. Therefore, $\lim_{n\to\infty} P(\|\Theta^* - \Theta_o\| < \varepsilon) = 1$, or $\text{plim}(\Theta^*) = \Theta_o$.

By assumption (2) of the theorem, and by Slutsky's theorem, it follows from the preceding results that

$$\left[\mathbf{H}(\Theta_o)^{-1} n^{-1}\left[\frac{\partial^2 \ln L(\Theta^*;\mathbf{X})}{\partial \Theta \partial \Theta'}\right]\right] n^{1/2}\left(\hat{\Theta} - \Theta_o\right) \xrightarrow{d} n^{1/2}\left(\hat{\Theta} - \Theta_o\right)$$

$$\xrightarrow{d} -\mathbf{H}(\Theta_o)^{-1}\mathbf{Z} \sim N\left(\mathbf{0}, \mathbf{H}(\Theta_o)^{-1}\mathbf{M}(\Theta_o)\mathbf{H}(\Theta_o)^{-1}\right)$$

since the probability limit of the bracketed expression is equal to the identity matrix. The statement in the theorem concerning the asymptotic distribution of $\hat{\Theta}$ then follows immediately from Definition 5.2. ■

[44]See Bartle, R.G.,(1976) *The Elements of Real Analysis*, 2nd Edition, John Wiley, p. 371.

**Theorem 8.21**

**Proof** Let $\Upsilon(\boldsymbol{\tau}) = \{\boldsymbol{\Theta} : \mathbf{q}(\boldsymbol{\Theta}) = \boldsymbol{\tau}, \boldsymbol{\Theta} \in \Omega\}$ $\forall \boldsymbol{\tau} \in R(\mathbf{q})$ i.e., $\Upsilon(\boldsymbol{\tau})$ is the set of $\boldsymbol{\Theta}$-values having the image value $\boldsymbol{\tau}$ based on the function $\mathbf{q}$. The collection of $\boldsymbol{\tau}$-values represented by the range of the function, $R(\mathbf{q})$, is be called the **parameter space induced by $\mathbf{q}(\boldsymbol{\Theta})$**, and note that because there can be more than one $\boldsymbol{\Theta} \in \Omega$ that satisfy $\mathbf{q}(\boldsymbol{\Theta}) = \boldsymbol{\tau}$, there can be more than one density and likelihood function, $f(\mathbf{x};\boldsymbol{\Theta}) \equiv L(\boldsymbol{\Theta};\mathbf{x})$, associated with $\boldsymbol{\tau}$, which are identified by the $\boldsymbol{\Theta}$-values that satisfy $\mathbf{q}(\boldsymbol{\Theta}) = \boldsymbol{\tau}$. Define the **likelihood function induced by $\mathbf{q}(\boldsymbol{\Theta})$** as

$$L^*(\boldsymbol{\tau};\mathbf{x}) = \max_{\boldsymbol{\Theta}\in\Upsilon(\boldsymbol{\tau})} \{L(\boldsymbol{\Theta};\mathbf{x})\} = \max_{\boldsymbol{\Theta}\in\Upsilon(\boldsymbol{\tau})} \{f(\mathbf{x};\boldsymbol{\Theta})\}$$

i.e., $L^*(\boldsymbol{\tau};\, \mathbf{x})$ is the largest likelihood consistent with $\mathbf{q}(\boldsymbol{\Theta}) = \boldsymbol{\tau}$, or equivalently, $L^*(\boldsymbol{\tau};\, \mathbf{x})$ is the highest density weighting assigned to the random sample outcome, $\mathbf{x}$, by the family of densities $f(\mathbf{x};\boldsymbol{\Theta})$, $\boldsymbol{\Theta} \in \Omega$, given that $\mathbf{q}(\boldsymbol{\Theta}) = \boldsymbol{\tau}$. A maximum likelihood estimate of $\boldsymbol{\tau} = \mathbf{q}(\boldsymbol{\Theta})$ is then given by a value of $\boldsymbol{\tau} \in R(\mathbf{q})$ that maximizes the induced likelihood function, as

$$\hat{\boldsymbol{\tau}} = \arg\max_{\boldsymbol{\tau}\in R(\mathbf{q})} \{L^*(\boldsymbol{\tau};\mathbf{x})\} = \arg\max_{\boldsymbol{\tau}\in R(\mathbf{q})} \left\{ \max_{\boldsymbol{\Theta}\in\Upsilon(\boldsymbol{\tau})} \{L(\boldsymbol{\Theta};\mathbf{x})\} \right\} = \arg\max_{\boldsymbol{\tau}\in R(\mathbf{q})} \left\{ \max_{\boldsymbol{\Theta}\in\Upsilon(\boldsymbol{\tau})} \{f(\mathbf{x};\boldsymbol{\Theta})\} \right\}$$

Since $\hat{\boldsymbol{\theta}}$ maximizes $L(\boldsymbol{\Theta};\mathbf{x}) \equiv f(\mathbf{x};\boldsymbol{\Theta})$, it follows that $\hat{\boldsymbol{\tau}} = \mathbf{q}(\hat{\boldsymbol{\theta}})$ maximizes the induced likelihood function $L^*(\boldsymbol{\tau};\mathbf{x})$, *so that* $\mathbf{q}(\hat{\boldsymbol{\Theta}})$ is the MLE of $\mathbf{q}(\boldsymbol{\Theta})$. (See Zehna, P.W., (1966) "Invariance of Maximum Likelihood Estimation," *Ann. Math. Stat.*, 37, p. 755, for further discussion of the induced likelihood function concept). ■

**Theorem 8.27**

**Proof** The function $h(\boldsymbol{\Theta}) = \mathbf{G}(\boldsymbol{\Theta})'\mathbf{W}\mathbf{G}(\boldsymbol{\Theta})$ is a continuous function of $\boldsymbol{\Theta}$ that is uniquely minimized at $\boldsymbol{\Theta} = \boldsymbol{\Theta}_o$ since $\mathbf{W}$ is positive definite and $\mathbf{G}(\boldsymbol{\Theta}) = \mathbf{0}$ *iff* $\boldsymbol{\Theta} = \boldsymbol{\Theta}_o$. Define $N(\boldsymbol{\varepsilon}) = \{\boldsymbol{\Theta}: d(\boldsymbol{\Theta},\boldsymbol{\Theta}_o) < \varepsilon\}$, which is an open ε-neighborhood of the true parameter value $\boldsymbol{\Theta}_o$, and let $\xi(\varepsilon) = \min_{\boldsymbol{\Theta}\in\Omega\cap\overline{N(\varepsilon)}}\{h(\boldsymbol{\Theta}) - h(\boldsymbol{\Theta}_o)\}$ (where the minimum exists by Weierstrass's theorem because $h(\boldsymbol{\Theta})$ is continuous and $\Omega \cap \overline{N(\boldsymbol{\varepsilon})}$ is closed and bounded). Let the event $A_n(\varepsilon)$ be defined as

$$A_n(\varepsilon) = \{\mathbf{y} : \max_{\boldsymbol{\Theta}\in\Omega} | Q_n(\mathbf{y},\boldsymbol{\Theta}) - h(\boldsymbol{\Theta})| < \xi(\varepsilon)/2\} \text{for } \varepsilon > 0.$$

Letting $\hat{\boldsymbol{\Theta}}$ represent the unique GMM estimate, it follows that for $\mathbf{y} \in A_n(\varepsilon)$:

1. $h(\hat{\boldsymbol{\Theta}}) < Q_n(\mathbf{y}, \hat{\boldsymbol{\Theta}}) + \xi(\varepsilon)/2$,
2. $Q_n(\mathbf{y},\boldsymbol{\Theta}_o) < h(\boldsymbol{\Theta}_o) + \xi(\varepsilon)/2$.

By the definition of the GMM, $Q_n(\mathbf{y},\hat{\boldsymbol{\Theta}}) \leq Q_n(\mathbf{y},\boldsymbol{\Theta}_o)$, so it follows from (1) that

3. $h(\hat{\boldsymbol{\Theta}}) < Q_n(\mathbf{y},\boldsymbol{\Theta}_o) + \xi(\varepsilon)/2$.

Substituting (2) into the right side of the inequality in (3) yields

4. $h(\hat{\boldsymbol{\Theta}}) < h(\boldsymbol{\Theta}_o) + \xi(\varepsilon)$.

Then substituting the definition of $\xi(\varepsilon)$ into (4) yields

5. $h(\hat{\boldsymbol{\Theta}}) < \min_{\boldsymbol{\Theta}\in\Omega\cap\overline{N(\varepsilon)}}\{h(\boldsymbol{\Theta})\} \Rightarrow \hat{\boldsymbol{\Theta}} \in N(\varepsilon)$.

Since by condition (d) of the Theorem 8.28 $\lim_{n\to\infty}(P(A_n(\varepsilon))) = 1\ \forall\varepsilon>0$, then $\lim_{n\to\infty}\left(P\left(\hat{\Theta} \in N(\varepsilon)\right)\right) = 1\ \forall\varepsilon>0$, and thus $\hat{\Theta} \xrightarrow{p} \Theta_o$. ■

## Keywords, Phrases, and Symbols

$\propto$(proportional to)
Autocorrelated
Bias-adjusted MLE
Classical Assumptions of the GLM
Coefficient of determination, $R^2$
Dependent variable
Design matrix
Disturbance, error, or residual vector
Econometrics, sociometrics, psychometrics
Explanatory (or independent) variables
Gauss-Markov theorem
General linear model (GLM)
Generalized least-squares estimator
Generalized method of moments (GMM)
Heteroskedastic
Homoskedasticity
Instrumental variable estimation
Invariance principle
Least squares under normality
Least-squares estimator
Likelihood function induced by $\mathbf{q}(\Theta)$
Likelihood function, $L(\Theta;\mathbf{x})$
Maximum likelihood (ML) estimate
Maximum likelihood estimator (MLE)
Method of least squares
Method of moments (MOM)
Nonlinear least squares
Parameter space induced by $\mathbf{q}(\Theta)$
Perfect multicollinearity
Reparameterization of the density family
Specification error
Zero autocovariance or zero autocorrelation

## Problems

**1.** The daily production of electricity generated by a coal-fired power plant operating in the midwest can be represented as

$$Y_i = \beta_1\, l_i^{\beta_2}\, m_i^{\beta_3}\, e^{\varepsilon_i}$$

where
$Y_i$ = quantity of electricity produced on day $i$, measured in megawatts;
$l_i$ = quantity of labor input used on day $i$, measured in 100's of hours;
$m_i$ = units of fossil fuel input on day $i$; and
$\varepsilon_i \sim N(0,\sigma^2)$, the $\varepsilon_i$'s being *iid*.

One hundred days of observations on the levels of inputs used and the quantity of electricity generated were collected. The following information is provided to you by the utility company, where $\mathbf{y}_* = \ln(\mathbf{y})$, and $\mathbf{x}_*$ is a matrix consisting of a vector of 1's followed by column vectors corresponding to the natural logarithms of labor and fuel levels.

$$\mathbf{y}'_*\mathbf{y}_* = 288.93382,\ \mathbf{x}'_*\mathbf{y}_* = \begin{bmatrix} 165.47200 \\ 180.32067 \\ 122.90436 \end{bmatrix},$$

$$(\mathbf{x}'_*\mathbf{x}_*)^{-1} = \begin{bmatrix} .06156 & -.05020 & -.00177 \\ & .09709 & -.07284 \\ & \text{(symmetric)} & .11537 \end{bmatrix}$$

(a) Transform the production function into a form in which parameters can be estimated using the least squares estimator. Estimate the parameters of the transformed model.

(b) Is the estimator you used in part (a) to estimate $\mathbf{q}(\boldsymbol{\beta})$ (1) unbiased, (2) asymptotically unbiased, (3) BLUE, (4) MVUE, (5) consistent, and/or (6) normally distributed?

(c) Is the estimator you used to estimate $\sigma^2$ (1) unbiased, (2) asymptotically unbiased (3) BLUE, (4) MVUE, (5) consistent, and/or (6) gamma-distributed?

(d) Define the MVUE for the degree of homogeneity of the production function (i.e., define the MVUE for $q_*(\boldsymbol{\beta}) = \beta_2 + \beta_3$. Estimate the degree of homogeneity of the production function using the MVUE. Is the MVUE a consistent estimator of the degree of homogeneity? Is the MVUE normally distributed?

(e) Define an MVUE estimator for the covariance matrix of the least-squares estimator. Estimate the covariance matrix of the least-squares estimator.

**2.** Smith's Dairy is contemplating the profitability of utilizing a new bovine growth hormone for increasing the milk production of its cows. Smith's randomly selects cows and administers given dosages of growth hormone at regular intervals. All animals are cared for identically except for the levels of hormone administered. A random sample of size $n$ from the composite experiment measuring the total milk production over the lactation of each animal is represented by

$$Y_t = \beta_1 + \beta_2 \mathbf{x}_t + \varepsilon_t, t = 1, \ldots, n,$$

where the $\varepsilon_t$'s are presumed to be *iid* with a common marginal density function

$$f(z; a, b) = \frac{1}{b-a} I_{(a,b)}(z)$$

with $a = -b$, and $b$ is some positive number.
The $\mathbf{x}_t$ values are the dosages of the growth hormone, measured in cc units, at levels defined by

$$\mathbf{x}_t = t - 10 \text{ trunc}\left(\frac{t-1}{10}\right), t = 1, 2, 3, \ldots$$

and $y_t$ is the milk production of the $t$th dairy cow, measured in hundredweights.

(a) Is the least-squares estimator of $\boldsymbol{\beta}$ (1) unbiased, (2) BLUE, (3) consistent, and/or (4) asymptotically normal? Justify your answer.

(b) Is the estimator $S^2$ of $\sigma^2$ (1) unbiased, (2) consistent, and/or (3) asymptotically normal? Justify your answer. (Hint: The $x_t$'s occur in a repeating sequence of the numbers 1,2,3,...,10. It follows that $\sum_{t=1}^{n} \mathbf{x}_t/n \to 5.5$ and $\sum_{t=1}^{n} \mathbf{x}_t^2/n \to 38.5$.)

(c) In the outcome of a random sample of size 100, Smith's found that $\mathbf{y}'\mathbf{y} = 4{,}454{,}656.3$ and $\mathbf{x}'\mathbf{y} = \begin{bmatrix} 21097.673 \\ 117570.78 \end{bmatrix}$. Estimate the parameters of the linear model using $\hat{\boldsymbol{\beta}}$ and $S^2$.

(d) Given that raw milk sells for \$10 per hundredweight, define a BLUE for the expected marginal revenue per cow obtained from administering the growth hormone. If the total cost of the hormone treatment per cow over the entire lactation is \$12 per cc, and if the maximum allowable dose is 10 cc, is it profitable for Smith's to utilize the hormone, based on the BLUE estimate of marginal revenue calculated above? Why or why not? If so, determine what level should be administered, and provide a BLUE estimate of the gain in profitability.

(e) Regarding Smith's linear model assumptions, are there other distributional assumptions that you would suggest for consideration besides the uniform distribution? Would this change any of your answers above? Would you suggest examining alternative functional forms for the relationship between milk production and hormone treatment? If so, would this affect your answers to the questions above?

**3.** Suppose $\mathbf{Y}_{(n\times 1)} = \mathbf{x}_{(n\times k)}\boldsymbol{\beta}_{(k\times 1)} + \boldsymbol{\varepsilon}_{(n\times 1)}$, where $\boldsymbol{\varepsilon} \sim N(\mathbf{0}, \sigma^2\mathbf{I})$ and $\mathbf{x}$ has full column rank.

(a) Show that

$$T = \frac{\boldsymbol{\ell}'\left(\hat{\boldsymbol{\beta}} - \boldsymbol{\beta}\right)}{\left(S^2\, \boldsymbol{\ell}'(\mathbf{x}'\mathbf{x})^{-1}\boldsymbol{\ell}\right)^{1/2}},$$

for any conformable $\boldsymbol{\ell} \neq \mathbf{0}$, has a $t$-distribution with $(n-k)$ degrees of freedom. (Hint: Transform $T$ so that it is expressed as a ratio of two independent random variables, with an $N(0,1)$ random variable in the numerator, and the square root of a $\chi^2$ random variable divided by its degrees of freedom in the denominator.)

(b) Using the fact that $T$ has a $t$-distribution with $(n-k)$ degrees of freedom, define a random interval $(Z_1, Z_2)$ which satisfies $P(\boldsymbol{\ell}'\boldsymbol{\beta} \in (z_1, z_2)) = .95$ when $n - k = 25$. (Hint: Define your $z_1$ and $z_2$ variables as appropriate functions of $\left(\hat{\boldsymbol{\beta}}, S^2\right)$, which will be suggested by a transformation of $P(t \in (t_\ell, t_h)) = .95$), for outcomes $t$ of $T$.

**4.** An economist is analyzing the relationship between disposable income and food expenditure for the citizens of a developing country. A survey of the citizens has been taken, and data have been collected on income and food expenditures. The following statistical model is postulated for the data:

$$Y_i = \beta_1 z_i^{\beta_2} e^{\varepsilon_i},$$

where
$y_i$ = expenditure on food for the $i$th household;
$Z_i$ = disposable income for the $i$th household; and
$\varepsilon_i$'s ~ *iid* $N(0,\sigma^2)$

(a) Transform the model into the GLM form. What parameters or functions of parameters are being estimated by the least-squares estimator applied to the transformed model?

(b) Is the least-squares estimator unbiased, the BLUE, and/or the MVUE for the parameters or functions of parameters being estimated?

(c) The actual survey consisted of 5,000 observations, and the following summary of the data is available:

$$(\mathbf{x}'\mathbf{x})^{-1} = \begin{bmatrix} .17577542 & -.019177442 \\ -.019177442 & .0020946798 \end{bmatrix},$$

$$\mathbf{x}'\mathbf{y} = \begin{bmatrix} 10579.646 \\ 98598.324 \end{bmatrix},$$

$$\mathbf{y}'\mathbf{y} = 14965.67,$$

where the **x**-matrix is a (5,000 × 2) matrix consisting of a column of 1's and a column representing the natural logarithms of the observations on income, and the **y**-vector refers to the corresponding natural logarithms of food expenditures. Calculate the least-squares estimate of the parameters of the transformed model. What is your estimate of the elasticity of food expenditure with respect to income?

(d) What is the probability distribution of $\hat{\beta}_2$, the least-squares estimator of $\beta_2$? Generate the MVUE estimates of the mean and variance of this distribution. (You might find it useful to know that $(\mathbf{y}'\mathbf{y} - \mathbf{y}'\mathbf{x}(\mathbf{x}'\mathbf{x})^{-1}\mathbf{x}'\mathbf{y})/(4,998) = 25.369055$). Given the assumptions of the model, and using the MVUE estimates of mean and variance, what is the (estimated) probability that the income elasticity estimated by the least-squares approach will be within $\mp.2$ of the true income elasticity?

(e) Discuss any alterations to the specification of the relationship between food expenditure and disposable income that you feel is appropriate for this problem.

**5.** The research department of the *Personal Computer Monthly* magazine is analyzing the operating life of computer chips that are produced by a major manufacturer located in the Silicon Valley. The research staff postulates that a random sample of lifetimes being analyzed adheres to the statistical model

$$\mathbf{X} \sim \theta^{-n} e^{-\sum_{i=1}^{n} x_i/\theta} \prod_{i=1}^{n} I_{(0,\infty)}(x_i), \quad \text{where } \theta>0.$$

The magazine publishes an index defined by $\beta = 1/\theta$ to measure the quality of computer chips, where the closer $\beta$ is to zero, the better the computer chip. The joint density of **X** is reparameterized so that the density function is parameterized by $\beta$, as

$$\mathbf{X} \sim \beta^{n} e^{-\beta \sum_{i=1}^{n} x_i} \prod_{i=1}^{n} I_{(0,\infty)}(x_i), \quad \text{where } \beta>0.$$

(a) Does the reparameterized family of density functions belong to the exponential class of density functions?

(b) Define a set of minimal sufficient statistics for the reparameterized density function. Are the minimal sufficient statistics complete sufficient statistics?

(c) Does there exist an unbiased estimator of $\beta$ whose variance achieves the Cramer-Rao lower bound?

(d) Define the maximum likelihood estimator for the parameter $\beta$. Is the MLE a function of the complete sufficient statistic? Is the MLE a consistent estimator of $\beta$?

(e) Is the MLE the MVUE of $\beta$? Is the MLE asymptotically normally distributed? Asymptotically efficient?

**6.** A large commercial bank intends to analyze the accuracy with which their bank tellers process cash transactions. In particular, it desires an estimate of the expected proportion of daily cash transactions that the bank tellers process correctly. It plans to analyze 200 past observations on the daily proportion of correct cash transactions by the tellers, and specify the statistical model underlying the daily observations as:

$$f(z;\alpha) = \alpha z^{\alpha-1} I_{(0,1)}(z), \quad \alpha \in (0,\infty)$$

(i.e., $f(z;\alpha)$ is a beta density function with $\beta = 1$).

(a) Define the maximum likelihood estimator of $\alpha$.

(b) Show that the MLE is a function of the complete sufficient statistic for this problem.

(c) Is the MLE of $\alpha$ a consistent estimator?

(d) It can be shown (*you* don't have to) that

$$\mathrm{E}\left(\sum_{i=1}^{n} \ln(\mathbf{X}_i)\right)^{-1} = -\alpha/(n-1).$$

(See W.C. Guenther (1967), "A best statistic with variance Not Equal to the Cramer-Rao lower bound," *American Mathematical Monthly*, 74, pp. 993–994, or else you can derive the density of $\left(\sum_{i=1}^{n} \ln(\mathbf{X}_i)\right)^{-1}$ and find its expectations—the density of $\sum_{i=1}^{n} \ln(\mathbf{X}_i)$ is the mirror image (around the vertical axis at zero) of a Gamma density). Is the MLE the MVUE for $\alpha$? If not, is there a function of the MLE that is MVUE for $\alpha$?

(e) Show that the MLE is asymptotically normal and asymptotically efficient, where $n^{1/2}(\hat{\alpha} - \alpha) \xrightarrow{d} N(0, \alpha^2)$.

(f) Define the MLE of $q(\alpha) = \alpha/(\alpha + 1)$, which is the expected proportion of correct cash transactions.

(g) Is the MLE of $q(\alpha)$ a consistent estimator?

(h) Is the MLE of $q(\alpha)$ asymptotically normal and asymptotically efficient? If so, define the asymptotic distribution of the MLE estimator.

(i) The outcome of the sufficient statistic was $\sum_{i=1}^{n} \ln(\mathbf{x}_i) = -9.725$.
Calculate the maximum likelihood estimate $\alpha$. Calculate the estimate of $\alpha$ using the MVUE of $\alpha$.

(j) Calculate the maximum likelihood estimate of $q(\alpha) = \alpha/(\alpha+1)$, the expected proportion of correct transactions.

(k) Calculate the maximum likelihood estimate of the probability that there will be greater than or equal to 97 percent correct cash transactions on a given day. (Hint: Determine the appropriate function of $\alpha$ in this case, and use the invariance principle.)

**7.** The Personnel Department of the ACME Textile Co. administers an aptitude test to all prospective assembly line employees. The average number of garments per hour that an employee can produce is approximately proportional to the score received on the aptitude test. In particular, the relationship is represented by

$$Y_i = x_i\beta + \varepsilon_i$$

where
$y_i$ = average number of garments/hour produced by employee $i$,
$x_i$ = score of employee $i$ on aptitude test,
$\beta$ = proportionality factor,
$\varepsilon_i$ = error term, representing the deviation between actual average number of garments per hour produced by employee $i$ and the production level implied by $x_i\beta$.

You may assume that the $\varepsilon_i$'s are independent, and you may also assume that $E(\varepsilon_i) = 0$ and $E(\varepsilon_i^2) = \sigma^2$, $\forall i$. Suppose you had the outcome, $\{y_1, y_2, \ldots, y_n\}$, of a random sample of average production rates for $n$ employees, together with their associated scores, $\{x_1, x_2, \ldots, x_n\}$, on the aptitude test.

(a) Should the random sample $\{Y_1, \ldots, Y_n\}$ be interpreted as a random sample from some population distribution, or should it be interpreted as a random sample generated from a general experiment? (Note: It cannot be expected that the aptitude scores will all be the same.)

(b) Derive the functional form of the least-squares estimator of the proportionality factor, $\beta$. Is the least squares estimator BLUE in this case? Is it the MVUE of $\beta$?

(c) Presuming that you could increase the sample size without bound, is the estimator you derived in (b) a consistent estimator of $\beta$? Is it asymptotically normally distributed? Justify your answer, being explicit about any assumptions you have made about the behavior of the $x_i$ values and/or the $\varepsilon_i$ values.

(d) From a sample of size $n = 100$, the sample outcome resulted in

$$\sum_{i=1}^{100} x_i y_i = 92{,}017 \text{ and } \sum_{i=1}^{n} x_i^2 = 897{,}235$$

Use the estimator you derived in (b) to generate an estimate of $\beta$.

**8.** A business consultant to the ACME Textile Co. suggests that the estimator

$$\hat{\beta}^* = (\mathbf{x}'\mathbf{x} + k)^{-1}\mathbf{x}'\mathbf{Y}$$

might be useful to consider as an alternative to the least-squares estimator of $\beta$ in the preceding problem (the estimator $\hat{\beta}^*$ is a special case of the so-called "ridge regression" estimator in the statistics literature). In this case,

$$\mathbf{Y} = \begin{bmatrix} Y_1 \\ Y_2 \\ \vdots \\ Y_n \end{bmatrix} \text{ and } \mathbf{x} = \begin{bmatrix} x_1 \\ x_2 \\ \vdots \\ x_n \end{bmatrix},$$

and $k$ is some positive constant.

(a) Is the estimator unbiased? If the estimator is not unbiased, derive an expression for the bias.

(b) Derive an expression for the variance of this estimator.

(c) Is the estimator a consistent estimator of $\beta$? Justify your answer, being explicit about any assumptions you have made about the behavior of the $x_i$ values.

(d) Compare the mean square errors of the least-squares estimator and the estimator $\hat{\beta}^*$. Is one estimator superior in MSE to the other $\forall\beta$ and $\sigma^2$? If not, can you characterize the problem conditions under which each estimator would be superior in terms of MSE?

**9.** Henri Theil, a famous economist/econometrician, analyzed the demand for textiles in the Netherlands during the period 1923–1939, using a general linear model framework. In particular, he specified the relationship between per-capita textile consumption, real price of textiles, and per capita real income as

$Y_t = \beta_1 p_t^{\beta_2} i_t^{\beta_3} e^{\varepsilon_t}, t = 1923, \ldots, 1939$

where

$Y_t$ = per capita textile consumption in year $t$, represented as an index with base year 1925;

$p_t$ = retail price index of clothing divided by a general cost-of-living index in year $t$, represented as an index with base year 1925,

$i_t$ = real income per capita in year $t$, defined as the total money income of private consumers divided by the population size and a general cost of living index, represented as an index with base year 1925,

$\varepsilon_t \sim iid\ N(0,\sigma^2)$, $t = 1923,\ldots,1939$.

The data used by Theil in estimating the demand relationship is presented in the following table.

| YEAR | Y | i | p |
|---|---|---|---|
| 1923 | 99.2 | 96.7 | 101.0 |
| 1924 | 99.0 | 98.1 | 100.1 |
| 1925 | 100.0 | 100.0 | 100.0 |
| 1926 | 111.6 | 104.9 | 90.6 |
| 1927 | 122.2 | 104.9 | 86.5 |
| 1928 | 117.6 | 109.5 | 89.7 |
| 1929 | 121.1 | 110.8 | 90.6 |
| 1930 | 136.0 | 112.3 | 82.8 |
| 1931 | 154.2 | 109.3 | 70.1 |
| 1932 | 153.6 | 105.3 | 65.4 |
| 1933 | 158.5 | 101.7 | 61.3 |
| 1934 | 140.6 | 95.4 | 62.5 |
| 1935 | 136.2 | 96.4 | 63.6 |
| 1936 | 168.0 | 97.6 | 52.6 |
| 1937 | 154.3 | 102.4 | 59.7 |
| 1938 | 149.0 | 101.6 | 59.5 |
| 1939 | 165.5 | 103.8 | 61.3 |

(a) Present the statistical model in a form that is consistent with the general linear model framework (variables should be measured in a way that coincides with the way they are used in the GLM specification). Can the random sample $(Y_1, \ldots, Y_n)$ be interpreted as a random sample from a population distribution? Why or why not?

(b) Define complete (and minimal) sufficient statistics for the parameters $(\boldsymbol{\beta},\sigma)$.

(c) Define the BLUE estimator of $(\ln(\beta_1), \beta_2, \beta_3)$? Generate a BLUE estimate for this vector.

(d) Define the MVUE for the vector $(\ln(\beta_1), \beta_2, \beta_3)$. Justify that your estimator is in fact the MVUE. Generate a MVUE estimate for the vector.

(e) What is the MVUE for $(\beta_2 + \beta_3)$, i.e., what is the MVUE for the degree of homogeneity of the demand function in terms of relative prices and real income? Justify that your estimator is the MVUE for $(\beta_2 + \beta_3)$. Generate an MVUE estimate of $(\beta_2 + \beta_3)$.

(f) Define the probability distribution of the MVUE for $(\beta_2 + \beta_3)$. What is the probability distribution of the MVUE for $(\ln(\beta_1), \beta_2, \beta_3)$?

(g) Present conditions under which the MVUE of $(\ln((\beta_1), \beta_2, \beta_3))$ would be: (1) a consistent estimator, and (2) an asymptotically normally distributed estimator.

(h) Define the MVUE for $\sigma^2$. Justify that the estimator is, in fact, the MVUE. Generate an MVUE estimate of $\sigma^2$.

(i) Present conditions under which the MVUE for $\sigma^2$ would be a consistent estimator.

(j) Estimate the probability that your MVUE estimator of the price elasticity of demand will generate an estimate that is within $\pm.15$ of the true price elasticity. You may use estimates of unknown parameters in generating this probability estimate.

(k) Is $\hat{E}(Y_t) = e^{\hat{\beta}_1^*} p_t^{\hat{\beta}_2} i_t^{\hat{\beta}_3}$, where $(\hat{\beta}_1^*, \hat{\beta}_2, \hat{\beta}_3)$ is the BLUE estimator of $(\ln(\beta_1), \beta_2, \beta_3)$, a consistent estimator of $E(Y_t)$ for given values of $i$ and $i_t$?

(l) Is the estimator in (k) BLUE? MVUE?

**10.** In each case below, determine whether, and under what assumptions, the stated relationship between Y and x can be represented in general linear model form such that the least squares estimator will provide a BLUE estimator of the $\boldsymbol{\beta}$ parameters.

(a) $Y_i = \exp\left(\sum_{j=1}^n x_{ij}\beta_j + \varepsilon_i\right)$

(b) $Y_i = \beta_0 + \beta_1 x_i + \beta_2 x_i^2 + \varepsilon_i$

(c) $Y_i = \beta_0 \prod_{j=1}^k x_{ij}^{b_j} + \varepsilon_i$

(d) $Y_i = \dfrac{x_{2i}}{1 + e^{x_{1i}\beta + \varepsilon_i}}$

**11.** The following statistical model is postulated for representing the relationship between real aggregate disposable income and real aggregate expenditure on nondurable goods:

$Y_i = \exp(\beta_1 + \beta_2 x_i + \varepsilon_i)$ where

$y_i$ = real aggregate expenditure on nondurables in period $i$ measured in billions of dollar;

$x_i$ = real aggregate disposable income in period $i$, measured in billions of dollars; and $\varepsilon_i$'s are *iid*, with

$\varepsilon_i + ab \sim \dfrac{1}{b^a \Gamma(a)} \varepsilon_i^{a-1} e^{-\varepsilon_i/b} I_{(0,\infty)}(\varepsilon_i)$.

There are 100 observations $(y_1,x_1),\ldots,(y_{100},x_{100})$ available to estimate the values of $\beta_1$, $\beta_2$, and $\sigma^2 = \text{var}(\varepsilon_i)$. Assuming the model specification is correct, answer the following questions:

(a) Transform the model into GLM form. What parameters or functions of parameters are being estimated by the least-squares estimator applied to the transformed model?

(b) Is the least-squares estimator unbiased? BLUE? Asymptotically unbiased?

(c) Letting

$$\mathbf{x} = \begin{bmatrix} 1 & x_1 \\ 1 & x_2 \\ \vdots & \vdots \\ 1 & x_n \end{bmatrix}, \text{ if } (\mathbf{x}'\mathbf{x})^{-1} \to \mathbf{0} \text{ and } n^{-1}\mathbf{x}'\mathbf{x} \to \mathbf{Q}$$

(a symmetric, positive definite matrix) as $n \to \infty$, would it follow that the least-squares estimator is consistent and asymptotically normally distributed? Why or why not?

(d) Letting $\mathbf{y}_* = \begin{bmatrix} \ln(y_i) \\ \vdots \\ \ln(y_n) \end{bmatrix}$ and $\hat{\boldsymbol{\beta}}$ be the BLUE of $\boldsymbol{\beta}$, is it true that

$$S^2 = \left(\mathbf{Y}_* - \mathbf{x}\hat{\boldsymbol{\beta}}\right)'\left(\mathbf{Y}_* - \mathbf{x}\hat{\boldsymbol{\beta}}\right)/(n-2)$$

is an unbiased and consistent estimator of $ab^2$?

**12.** In each case, indicate whether the statement regarding the relationship $\mathbf{Y} = \mathbf{x}\boldsymbol{\beta} + \boldsymbol{\varepsilon}$ is true or false, and justify your answer.

(a) Let the random $(n \times 1)$ vector $\mathbf{Y}$ represent a random sample from some composite experiment, where $\text{E}(\boldsymbol{\varepsilon}) = \mathbf{0}$, and $\text{E}(\boldsymbol{\varepsilon}\boldsymbol{\varepsilon}') = \sigma^2\mathbf{I}$. Suppose the $\mathbf{x}$-matrix has full column rank, but that the first and second columns of $\mathbf{x}$ are nearly linearly dependent and, as a result, the determinant of $\mathbf{x}'\mathbf{x}$ is near zero, equaling $.273 \times 10^{-7}$. In this case, although $\hat{\boldsymbol{\beta}} = (\mathbf{x}'\mathbf{x})^{-1}\mathbf{x}'\mathbf{Y}$ is still an unbiased estimator, it is no longer BLUE (i.e., it loses its "best" property of having the smallest covariance matrix in the linear unbiased class of estimators).

(b) The $\varepsilon_i$'s are homoskedastic and jointly independent with $\text{E}(\varepsilon_i) = \delta \neq 0\ \forall i$. Also, $\mathbf{x}\boldsymbol{\beta} = \beta_1\boldsymbol{\iota} + \beta_2\mathbf{Z}$ where $\boldsymbol{\iota}$ is an $(n \times 1)$ column vector of 1's, and $\mathbf{Z}$ is a $(n \times 1)$ column vector of explanatory variable values. Then if $\hat{\boldsymbol{\beta}}$ is the least-squares estimator of $\boldsymbol{\beta}$, $\hat{\beta}_2$ is the BLUE of $\beta_2$.

(c) The disturbance terms are related as $\varepsilon_t = \rho\varepsilon_{t-1} + V_t$, where the $V_t$'s are *iid* with $\text{E}(V_t) = 0$ and $\text{var}(V_t) = \sigma^2$ $\forall t$, and $|\rho| < 1$. The least squares estimator is both BLUE and consistent.

**13.** Your company markets a disposable butane lighter called "surelight." In your product advertising, you use the slogan "lights on the first try-everytime!" As a quality check, you intend to examine a random sample of 10,000 lighters from the assembly line and observe for each lighter the number of trials required for the lighter to light. Your assistant obtains the random sample outcome and reports to you that a total of 10,118 trials were required to get all of the lighters to light. She did *not* record the 10,000 individual outcomes of how many trials were required for each lighter to light. You are interested in estimating both the expected number of trials needed for a lighter to light and the probability, $p$, that the lighter lights on any given trial.

(a) Define an appropriate statistical model for the 10,000 outcomes of how many trials were required for each lighter to light.

(b) Define the MLE for the expected number of trials needed for a lighter to light. Is the estimator the MVUE? Is it consistent? Is it asymptotically normal? Is asymptotically efficient?

(c) Define the MLE for the probability that the lighter lights on any given trial. Is the estimator the MVUE? Is it consistent? Is it asymptotically normal? Is it asymptotically efficient? (Hint: Can you show that $t(\mathbf{X}) = (n-1)/((\sum_{i=1}^{n} X_i) - 1)$ has an expectation equal to $p$?)

(d) Provide MLE estimates and MVUE estimates of both the expected number of trials needed for the first light and the probability that the lighter lights on any given trial.

**14.** A regional telephone company is analyzing the number of telephone calls that are connected to wrong numbers at its telephone exchange. It collects the number of wrong telephone connections on each of 200 days, and treats the observations as the outcome of a random sample of size 200 from a Poisson population distribution:

$$f(z;\lambda) = \frac{e^{-\lambda}\lambda^z}{z!} I_{\{0,1,2,\ldots\}}(z)$$

(a) Define the MLE of $\lambda$, the expected number of wrong connections per day.

(b) Is the MLE the MVUE for $\lambda$? Is it consistent? Asymptotically normal? Asymptotically efficient?

(c) If $\sum_{i=1}^{200} \mathbf{x}_i = 4{,}973$, what is the ML estimate of the expected number of wrong connections? If each wrong connection costs the company \$070, define a MLE for the expected daily cost of wrong connections, and generate a ML estimate of this cost.

(d) Define a MLE for the standard deviation of the daily number of wrong connections. Is the MLE consistent? Is it asymptotically normal? Generate a ML estimate of the standard deviation.

**15.** The number of minutes past the scheduled departure time that jets with no mechanical problems leave the terminal in an overcrowded airport in the northeast are *iid* outcomes from a uniform population distribution of the form $f(z;\Theta) = \Theta^{-1} I_{(0,\Theta)}(z)$. A random sample of 1,000 departures is to be used to estimate the parameter $\Theta$ and the expected number of minutes past the scheduled departure time that a jet will leave the terminal. Summary statistics from the outcome of the random sample include $\min(\mathbf{x}) = .1$, $\max(\mathbf{x}) = 13.8$, $\bar{\mathbf{x}} = 6.8$, $s^2 = 15.9$.

(a) Define a MLE for $\Theta$ and for expected number of minutes past the scheduled departure time that a jet will leave the terminal. Are these MLEs functions of minimal sufficient statistics?

(b) Use the MLEs you defined above to generate ML estimates of the respective quantities of interest.

(c) Are the estimators in (a) unbiased? consistent? (Hint: $\mathrm{E}(\max(X)) = \Theta[n/(n+1)]$ and $\mathrm{E}((\max(X))^2) = \Theta^2[n/(n+2)]$)

(d) Are the estimators in (a) MVUES?

**16.** Define MLEs for the following problems:

(a) Estimating the $p_i$'s based on a random sample from a multinomial population distribution.

(b) Estimating the unknown parameters in the mean vector $\boldsymbol{\mu}$ and covariance matrix $\Sigma$ based on a random sample from a bivariate normal distribution.

(c) Estimating $\beta$ in the population distribution

$$f(x;\beta) = \frac{2x}{\beta} \exp\left[\frac{-\mathbf{x}^2}{\beta}\right] I_{(0,\infty)}(x)$$

(d) Estimating the parameter $p$ in a negative binomial population distribution where $r$ is known.

**17.** Define MOM estimators in each of the following cases. Are estimators consistent? Are they asymptotically normal?

(a) Estimating $\mu$ and $\sigma^2$ based on a random sample from a normal population distribution.

(b) Estimating $\alpha$ and $\beta$ based on a random sample from a Beta population distribution.

(c) Estimating $p$ based on a random sample from a geometric population distribution.

(d) Estimating $a$ and $b$ based on a random sample from a continuous uniform population distribution.

**18.** (*Generalized Least Squares Estimator*) Consider the linear model $\mathbf{Y} = \mathbf{x}\boldsymbol{\beta} + \boldsymbol{\varepsilon}$ in which all of the classical assumptions apply, except that $\mathrm{E}(\boldsymbol{\varepsilon}\boldsymbol{\varepsilon}') = \boldsymbol{\Phi} \neq \sigma^2\mathbf{I}$. Consider a GMM estimator of the parameter vector $\boldsymbol{\beta}$ based on the moment conditions $\mathrm{E}(g_t(Y_t,\boldsymbol{\beta})) = \mathrm{E}(\mathbf{z}_{t\cdot}'(Y_t - \mathbf{x}_{t\cdot}\boldsymbol{\beta})) = \mathbf{0}$ where $\mathbf{z} = \boldsymbol{\Phi}^{-1}\mathbf{x}$.

(a) Assuming temporarily that $\boldsymbol{\Phi}$ were known, identify the sample moment conditions that would be used to define the GMM estimator. Solve the moment conditions to provide an explicit functional representation of the estimator (called the *generalized least squares estimator* in the literature).

(b) Discuss conditions under which the estimator you have defined in (a) is consistent, asymptotically normal, and asymptotically efficient.

(c) For the general case where $\boldsymbol{\Phi}$ is unknown, discuss how you would define an operational version of the GMM estimator, and discuss its relationship with the estimator in (a).

**19.** (*Linear and Nonlinear Models*) Determine which of the following models can be transformed into linear models and which cannot, and identify the transformation in the cases where it is possible.

(a) $Y_t = \left(\prod_{j=1}^{k} X_{jt}^{\beta_j}\right) V_t$, where $V_t \sim \mathrm{Gamma}(a,b)$

(b) $\frac{Y_t^\lambda - 1}{\lambda} = \beta_1 + \beta_2\left(\frac{X_t^\delta - 1}{\delta}\right) + \varepsilon_t$, where $\varepsilon_t \sim N(0,\sigma^2)$

(c) $Y_t = \beta_0 + \beta_1 X_t^{\beta_2} + \varepsilon_t$ where $\varepsilon_t \sim N(0,\sigma^2)$

(d) $Y_t = \beta_1\left(\beta_2 L_t^{-\beta_3} + (1-\beta_2)K_t^{-\beta_3}\right)^{-\beta_4/\beta_1} \exp(\varepsilon_t)$, where $\varepsilon_t \sim f(e)$

(e) $Y_t = \left(1 + \exp\left(\mathbf{X}_t'\boldsymbol{\beta} + \varepsilon_t\right)\right)^{-1}$, where $\varepsilon_t \sim \mathrm{Logistic}(0,s)$

**20.** (*Nonlinear Least Squares Estimator*) Consider the nonlinear regression model

$$\mathbf{Y} = \mathbf{g}(\mathbf{x},\boldsymbol{\beta}) + \boldsymbol{\varepsilon} = \beta_1\mathbf{1}_n + \beta_2\mathbf{x} + \beta_3\mathbf{x}^{\beta_4} + \boldsymbol{\varepsilon}, \text{ for } \boldsymbol{\beta} \in \Omega$$

where $\mathbf{Y}$ is $n \times 1$, $\mathbf{1}_n$ is an $n \times 1$ vector of 1's, the $n \times 1$ vector $\mathbf{x}$ is not a vector of identical constants, $E(\boldsymbol{\varepsilon}|\mathbf{x}) = \mathbf{0}$ and $\mathbf{Cov}(\boldsymbol{\varepsilon}|\mathbf{x}) = \sigma^2\mathbf{I}$.

(a) Is a parameter vector with $\beta_4 = 0$ identifiable in this model?

(b) Is a parameter vector with $\beta_4 = 1$ identifiable in this model?

(c) Define an admissible parameter space, $\boldsymbol{\Omega}$, for this model

(d) Given a random sample of data $(\mathbf{Y},\mathbf{x})$, describe how you would go about estimating this model based on the least squares principle.

(e) Identify an asymptotic distribution for the least squares estimator of $\boldsymbol{\beta}$.

(f) Suppose that $\beta_4$ was "in a close neighborhood", but not exactly equal to one of the values 0 or 1. Could there be any issues relating to "multicollinearity" affecting the least squares estimator of the parameter vector? Explain.

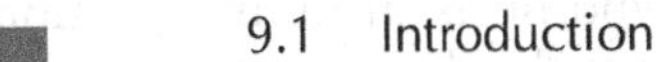

# 9 Hypothesis Testing Theory



## 9.1 Introduction

A primary goal of scientific research often concerns the verification or refutation of assertions, conjectures, currently accepted laws, or descriptions relating to a given economic, sociological, psychological, physical, or biological process or population. Statistical hypothesis testing concerns the use of probability samples of observations from processes or populations of interest, together with probability and mathematical statistics principles, to judge the validity of stated assertions, conjectures, laws, or descriptions in such a way that the probability of falsely rejecting a correct hypothesis can be controlled, while the probability of rejecting false hypotheses is made as large as possible. The precise nature of the types of errors that can be made, how the probabilities of such errors can be controlled, and how one designs a test so that the probability of rejecting false hypotheses is as large as possible is the subject of this chapter.

We will henceforth use the word **hypothesis** generically to refer to any statement relating to a process or population under study for which an analyst wishes verification or refutation. Examples of the myriad of hypotheses that might be tested include statements such as:

1. The demand for tablet PCs in the United States is price inelastic,
2. A new medication is more effective than the currently accepted treatment for treating high blood pressure,

3. A new LED light bulb has an expected usable life span of over 30,000 hours, or
4. Less than 1 in 10,000 college-educated individuals engage in criminal activity in a given year in North America.

## 9.2 Statistical Hypotheses

In order for probability and mathematical statistics principles to be applied to a problem of hypothesis testing, hypotheses must be translated into statements about characteristics of the probability space associated with a probability sample from the process or population being analyzed. The point of doing so is to translate a hypothesis into implications for expected behavior of sample outcomes when the hypothesis is true versus when it is false. If sample outcomes behave differently under a true and false hypothesis, the differences in behavior can be used to help identify when observed sampling behavior is refuting a stated hypothesis. Relatedly, if sample outcome behavior is no different whether a given hypothesis is true or false, then it is clear that sample data cannot possibly be used to identify when a hypothesis is false.

In the most general meaning of the concept, a **statistical hypothesis** is a set of potential probability distributions for a sample $\mathbf{X}$ associated with some process or population. The statistical hypothesis is defined by statements identifying the characteristics asserted to be true about the probability space associated with $\mathbf{X}$. In parametric and semiparametric models, these statements are often presented in the form of assertions about the values of parameters or functions of parameters relating to the underlying probability distribution of $\mathbf{X}$. For nonparametric models, and for the components of semiparametric models that do not involve parameters, hypothesized characteristics of the probability space for $\mathbf{X}$ are stated in more general terms. Our primary focus in this chapter will be on tests of parameters of functions of parameters, but the general concepts of statistical hypothesis testing presented ahead extend to nonparametric contexts as well.

A hypothesis is **statistically testable** only if it can be represented in terms of a statistical hypothesis, as defined below.

**Definition 9.1**
***Statistical Hypothesis***

A set of potential probability distributions for a probability sample from a process or population.

If a statistical hypothesis consists of a single element that defines a specific probability distribution for $\mathbf{X}$ that is completely and uniquely identified, the statistical hypothesis is called a **simple hypothesis**. If the statistical hypothesis is not simple, then it is called a **composite hypothesis**. A composite hypothesis is a statistical hypothesis containing two or more (possibly infinite) potential probability distributions for the outcomes of the random sample.

It is customary to represent the set of probability distributions that defines a statistical hypothesis by the capital letter $H$, and to use subscripts, when needed, to distinguish between various statistical hypotheses under investigation, such as $H_0$, $H_a$, or $H_i$. Regarding the set defining conditions that are used to represent a statistical hypothesis, the analyst chooses the characteristics of the probability space of $\mathbf{X}$ that are of interest in any given problem setting. The following examples illustrate the representation of statistical hypotheses.

**Example 9.1**
***Composite Statistical Hypothesis for Bernoulli Process***

It is asserted by a manufacturer of metal hardware that the percentage of defective bolts in a shipment of 1,000 bolts is no more than 2 percent. The receiver of the shipment intends to use a random sample, with replacement, of size $n$ from the bolt shipment to assess the validity of the manufacturer's assertion. The hypothesis of the manufacturer, translated into a statistical hypothesis stated in terms of a random sample of size $n$ from the bolt population, is then (letting $x_i = 1$ indicate a defective bolt, and $x_i = 0$ represent a nondefective bolt)

$$H = \left\{ f(\mathbf{x}; p) = \prod_{i=1}^{n} p^{x_i}(1-p)^{1-x_i} I_{\{0,1\}}(x_i),\ \ p \in [0, .02] \right\}.$$

This statistical hypothesis is a *composite* hypothesis and specifies that the probability distribution of the random sample of size $n$ from the bolt population is the product of $n$ Bernoulli distributions, with the probability of observing a defective bolt in any Bernoulli trial being $p \leq .02$. The statistical hypothesis implies an infinite set of potential probability distributions for defective bolts. □

**Example 9.2**
***Composite Statistical Hypothesis for Exponential Population***

It is hypothesized that the expected life of a computer chip manufactured by a certain manufacturing process exceeds the industry average of 20,000 hours, and it is further hypothesized that the lifetimes of computer chips manufactured by this process are distributed according to the exponential family of distributions. The statistical hypothesis, stated in terms of an *iid* random sample of size $n$ from the population distribution of computer chip lifetimes, is then (assuming elements of $\mathbf{X}$ are measured in 1,000's of hours):

$$H = \left\{ f(\mathbf{x}; \theta) = \theta^{-n} \exp\left( -\sum_{i=1}^{n} x_i/\theta \right) \prod_{i=1}^{n} I_{(0,\infty)}(x_i), \theta \in (20, \infty) \right\}.$$

This statistical hypothesis is *composite* since a single probability distribution is not completely and uniquely defined by $H$—the mean life measured in 1,000's of hours, $\theta$, is any number larger than 20, according to $H$. Thus, the statistical hypothesis implies an infinite set of potential probability distributions for the random sample of computer chip lifetimes. □

**Example 9.3**
***Simple Statistical Hypothesis for Bernoulli Process***

Suppose in Example 9.1 that the hypothesis was that there are *no* defective bolts in the 1,000 bolt shipment. Then the statistical hypothesis could be specified in terms of a random sample of size $n$ from the bolt population as

$$H = \left\{ f(\mathbf{x}) = \prod_{i=1}^{n} I_{\{0\}}(x_i) \right\}.$$

This is a *simple* statistical hypothesis indicating that the probability distribution is degenerate with the outcome $\mathbf{x} = \mathbf{0}$ (all $x_i$'s are nondefective) having probability 1. The statistical hypothesis could have been equivalently represented as in Example 9.1, but with $p = 0$ instead of $p \in [0, .02]$. □

**Example 9.4**
***Composite Statistical Hypothesis for non-iid Normally-Distributed Sample***

The average yield per acre for a particular variety of wheat grown in the Pacific Northwest is represented by

$$Y_i = \beta_0 + \beta_1 f_i + \beta_2 r_i + \varepsilon_i, \quad \varepsilon_i \sim N(0, \sigma^2),$$

for appropriate ranges of $f_i$ and $r_i$, where $y_i$ is average yield per acre measured in bushels in year $i$, $f_i$ is average pounds of fertilizer applied per acre, and $r_i$ is average rainfall per acre measured in inches. Independent observations on 30 years worth of values for $f_i$ and $r_i$ and corresponding outcomes of $Y_i$ are available. An analyst wishes to assess the hypothesis that a 1 pound per acre increase in fertilizer applied to the crop generates an expected wheat yield increase of .25 bushels per acre. The observations on wheat yields can be conceptualized as the vector outcome of a random sample generated via a general experiment with the joint probability distribution of $\mathbf{Y} = (Y_1, \ldots, Y_{30})'$ being

$$\mathbf{Y} \sim \prod_{i=1}^{30} N(y_i; \beta_0 + \beta_1 f_i + \beta_2 r_i, \sigma^2).$$

Translating the hypothesis of the researcher into a statistical hypothesis yields

$$H = \{ f(\mathbf{y}; \boldsymbol{\beta}, \sigma^2) = \prod_{i=1}^{30} N(y_i; \beta_0 + \beta_1 f_i + \beta_2 r_i, \sigma^2), \beta_1 = .25 \}.$$

The statistical hypothesis is *composite*, since a unique probability distribution for $\mathbf{Y}$ has not been completely identified because $\beta_0$, $\beta_2$, and $\sigma^2$ were left unspecified and thus can take any values that result in a legitimate normal density function. (Normal distributions are clearly only approximations in this case since yields cannot be negative; it is assumed that $P(y_i < 0)$ is negligible for the relevant $f_i$ and $r_i$ values being analyzed.) □

**Example 9.5**
***Composite Statistical Hypothesis for a Semiparametric Model with iid Random Sampling***

The average salary of accountants in a certain region of the county is \$48,500. An allegation is made that male accountants in this region have lower than average salaries, and an analyst wishes to assess the allegation. It is not known which parametric family of PDFs encompasses the distribution of male accountants' salary in the region. Translating the allegation into a statistical hypothesis relating to a random sample of size $n$ from the population distribution of male accountants in the region yields

$$H = \left\{ f(\mathbf{x};\boldsymbol{\Theta}) = \prod_{i=1}^{n} m(x_i;\boldsymbol{\Theta}),\ \ \mu = \mathrm{E}(X_i) = \mathrm{h}(\boldsymbol{\Theta}) < 48{,}500\ \ \forall i \right\}.$$

The statistical hypothesis is *composite*, since $H$ contains *all* PDFs for $\mathbf{X}$ that have $\mathrm{E}(X_i) < 48{,}500\ \forall i$. □

When defining $H$, choices of functional forms for $f(\mathbf{x})$ are sometimes suggested by the nature of the random process or population being analyzed. For example, in Example 9.1, the Bernoulli family of density functions was indicated by the fact that the random sample was defined via random sampling, with replacement, from a finite population of defective/nondefective bolts. In Example 9.2, the exponential family of distributions might be motivated by certain physical features of the computer chips under study (nonnegativity of lifetimes is obvious; the density might be suggested by a "good as new while functioning" characteristic of the chip—recall the "memoryless property" of the exponential family of distributions). In Example 9.3, the stated hypothesis implied that only one outcome of $\mathbf{X}$ was possible, so that the probability distribution of $\mathbf{X}$ must be degenerate. In Example 9.4, the normal family of distributions might be motivated via an appeal to a central limit theorem, where the error term, $\varepsilon_i$, would be assumed to represent the additive effect of a large number of random influences not specifically accounted for in the simple linear relationship specified to represent $\mathrm{E}(Y_i)$. In Example 9.5, there was no indication of the functional form of the probability distribution except for the range of possible values for its mean.

In representing statistical hypotheses, a simplification is often used when the functional form of the joint density of $\mathbf{X}$ is taken to be known except for the value of some parameter or parameter vector $\boldsymbol{\Theta}$, as in Example 9.1–9.4. In this case, $H$ is often represented as simply a set of *parameter values*, the idea being that if $f(\mathbf{x};\boldsymbol{\Theta})$ is known except for the value of $\boldsymbol{\Theta}$, then $\boldsymbol{\Theta} \in H$ will characterize a set of potential probability distributions for $\mathbf{X}$ perfectly well and provide an equivalent representation of a statistical hypothesis.

**Example 9.6** ***Statistical Hypotheses in Examples 9.1–9.4 Stated in Abbreviated Form***

Assuming the functional form for the joint density of $\mathbf{X}$ to be given as stated in each example, the statistical hypotheses can be represented in *abbreviated* form as $H = \{p\colon 0 \le p \le .02\}$, $H = \{\theta\colon\ \theta > 20\}$, $H = \{p\colon\ p = 0\}$, and $H = \{(\boldsymbol{\beta},\sigma^2)\colon \beta_1 = .25\}$, respectively. □

The representation of statistical hypotheses is often abbreviated still further, using general notation of the form **H: set defining conditions**, meaning that $H$ is a set of parameter values defined by whatever set defining conditions are stated following the colon. This further abbreviation of the representation of $H$ is illustrated below for the case of Examples 9.1–9.4.

**Example 9.7**
***Statistical Hypotheses in Terms of Set-Defining Conditions for Examples 9.1–9.4***

Alternative abbreviated representations for the statistical hypotheses of Examples 9.1–9.4 are given by $H$: $0 \leq p \leq .02$, $H$:$\theta > 20$, $H$: $p = 0$, and $H$: $\beta_1 = .25$, respectively. □

In general, the use of abbreviated representations of $H$ is acceptable so long as the context of the problem being analyzed results in no ambiguity regarding the definition of the set of potential probability distributions for $\mathbf{X}$ implied by the statistical hypothesis. The notation is prevalent in cases of *parametric hypothesis testing*, where it is assumed at the outset that the probability distribution of $\mathbf{X}$ is a member of a given parametric family of PDFs.

For later reference, note that the specification of a statistical hypothesis, $H$, specifies a set of potential outcomes for the random sample. That is, each PDF in $H$ will imply a particular range of possible outcomes for the random sample $\mathbf{X}$, and the union of all of these respective ranges of $\mathbf{X}$ is logically the full set of possible random sample outcomes implied by $H$. The idea is that since any of the PDFs in $H$ is asserted to be a candidate for the true PDF of $\mathbf{X}$, then any of the associated ranges of $\mathbf{X}$ are also candidates for $R(\mathbf{X})$. Recalling that the range of $\mathbf{X}$ is synonymous with the support of $\mathbf{X}$'s PDF, we can define the set of potential random sample outcomes implied by $H$ as

$$R(\mathbf{X}|H) = \{\mathbf{x} : f(\mathbf{x};\boldsymbol{\Theta})>0 \text{ and} f(\mathbf{x};\boldsymbol{\Theta}) \in H\}.$$

We will refer to $R(\mathbf{X}|H)$ as **the range of X over H.**[1] Analogous definitions for the range of $\mathbf{X}$ over $\bar{H}$ and over $H \cup \bar{H}$ can be given as

$$R(\mathbf{X}|\bar{H}) = \{\mathbf{x} : f(\mathbf{x};\boldsymbol{\Theta})>0 \text{ and } f(\mathbf{x};\boldsymbol{\Theta}) \in \bar{H}\} \text{ and}$$

$$R(\mathbf{X}|H \cup \bar{H}) = \{\mathbf{x} : f(\mathbf{x};\boldsymbol{\Theta})>0 \text{ and } f(\mathbf{x};\boldsymbol{\Theta}) \in H \cup \bar{H}\}.$$

For now, it will suffice to interpret the complement of $H$, $\bar{H}$, as the collection of all probability distributions not specified in H that are potential candidates for the true PDF of the probability sample under study. Thus, the complement operation is being applied in the context of a universal set of all potential probability distributions relevant for the probability sample being examined. Then $R(\mathbf{X}|H \cup \bar{H})$ can be interpreted as the set of *all* potential outcomes of the random sample.

It can be the case that $R(\mathbf{X}|H) = R(\mathbf{X}|\bar{H}) = R(\mathbf{X}|H \cup \bar{H})$, which occurs when all of the supports of the PDFs in $H$ and $\bar{H}$ are the same. For example, such is the case in Example 9.1 where all three sets equal $\times_{i=1}^{n} \{0, 1\}$. Alternatively, the sets

[1]Compare this set to the *range of* $\mathbf{X}$ *over* $\Omega$ introduced in our discussion of minimal sufficient statistics, Section 7.4.

can differ, as in Example 9.3 where these sets are respectively **0**, $\times_{i=1}^{n}\{0,1\}$, and $\times_{i=1}^{n}\{0,1\}$. The nature of the ranges of $\mathbf{X}$ over $H$, $\bar{H}$, and $H \cup \bar{H}$ will play important roles in determining whether an ideal statistical test of $H$ exists (i.e., a test which makes no errors in deciding the validity of $H$) and are also useful for motivating why one cannot generally expect statistical tests of $H$ to be error-free. We examine these ideas in the next section.

## 9.3 Basic Hypothesis Testing Concepts

In this section we define the concept of a **statistical hypothesis test**, we identify two types of errors that such a test can make, and we indicate in what sense such errors can be controlled. We also point out the general fact that the incidence of one type of test error can be lessened only at the expense of increasing the incidence of the other if the size of the probability sample on which the test is based is held constant.

### 9.3.1 Statistical Hypothesis Tests

The objective of a *test* of a statistical hypothesis is to assess the validity of the statistical hypothesis. In practice, a statistical hypothesis is tested by first obtaining the outcome of a probability sample from the process or population being analyzed and then using the observed outcome of the sample to assess whether it is reasonable to conclude that one of the probability distributions for the probability sample implied by the statistical hypothesis could have governed the outcome behavior of the sample. Deciding when an outcome of a sample will cause rejection of $H$ and when it will not is accomplished by a rule or procedure that partitions the range of potential sample outcomes into two subsets. The hypothesis is then rejected or not depending on which subset the outcome of the sample belongs to. In its most general sense, a **test of a statistical hypothesis** can be defined as follows.

**Definition 9.2** ***Test of a Statistical Hypothesis***

A rule or procedure, based on the outcome of a probability sample from the process or population under study, used to decide whether to reject a statistical hypothesis.

A test of a statistical hypothesis is often referred to as simply a **statistical test**. It follows from Definition 9.2 that a valid functional representation of a statistical test would be in the form of an indicator function whose domain consists of the potential outcomes of the sample, and whose range indicates whether or not the statistical hypothesis is rejected. A statistical test is fully defined when the set of potential sample outcomes is partitioned into two disjoint subsets, one called the **rejection region** or (**critical region**) and the

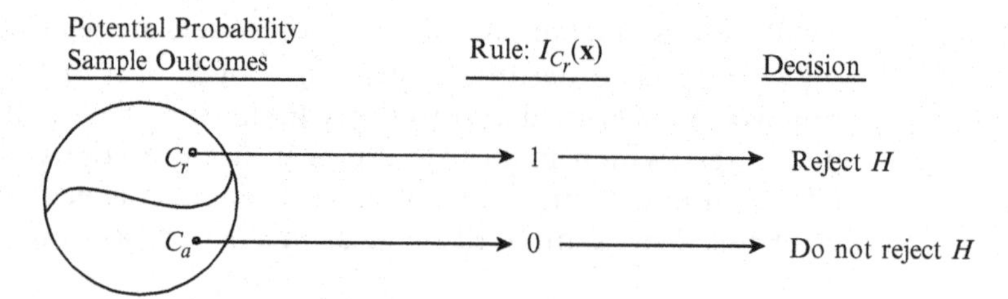


**Figure 9.1** Test of a statistical hypothesis.

other called the **acceptable region.**[2] Letting the set of potential sample outcomes be represented by $C_r \cup C_a$, where $C_r$ is the rejection region, $C_a$ is the acceptable region, and $C_r \cap C_a = \emptyset$, then

$$I_{C_r}(\mathbf{x}) = \begin{bmatrix} 1 \\ 0 \end{bmatrix} \Rightarrow \begin{bmatrix} \text{reject } H \\ \text{do not reject } H \end{bmatrix}$$

is a fully defined rule for deciding whether to reject the statistical hypothesis (see Figure 9.1).

In the remainder of the discussion of hypothesis testing concepts and methods, it will always be tacitly assumed that $C_a = \bar{C}_r$, if an explicit definition of the acceptable region, $C_a$, is not given in any hypothesis testing situation. The complement operation will be interpreted to occur within a universal set defined by $R(\mathbf{X}|H \cup \bar{H})$, which represents all possible outcomes of the random sample under PDFs belonging to either $H$ or $\bar{H}$. With this convention in mind, a statistical test can be defined completely in terms of a rejection region, $C_r$, as was the case in our indicator function representation above. Furthermore, we will sometimes use the phrase **the statistical test defined by $C_r$** to mean a statistical test defined as $\mathbf{x} \in C_r \Rightarrow$ reject $H$, $\mathbf{x} \notin C_r$ (or equivalently $\mathbf{x} \in \bar{C}_r = C_a$) $\Rightarrow$ do not reject $H$.

A schematic overview of the general context of a hypothesis testing problem is given in Figure 9.2.

The design or choice of a statistical test is seen to be equivalent to the design or choice of the partition of the set of potential probability sample outcomes into rejection and acceptable regions. This begs the question of how the partition should be chosen so as to design a test that is "good" in some appropriate sense. Operationally, this will amount to designing tests that control the probability of incorrectly rejecting correct hypotheses while making the rejection of a false hypothesis as large as possible. The types of incorrect decisions that can be made using statistical tests are examined next.

[2]The terminology "acceptable region" is sometimes replaced by "*acceptance* region" in the literature. We will see later that while the behavior of sample outcomes may be "acceptable" to $H$ given the characteristics of the probability space implied by $H$, there are statistical reasons why one might not want to literally conclude *acceptance* of $H$ on the basis of this "acceptable" behavior. We will clarify this subtle but important distinction with additional rigorous rationale later, which will motivate further why we choose to use the terminology "acceptable".

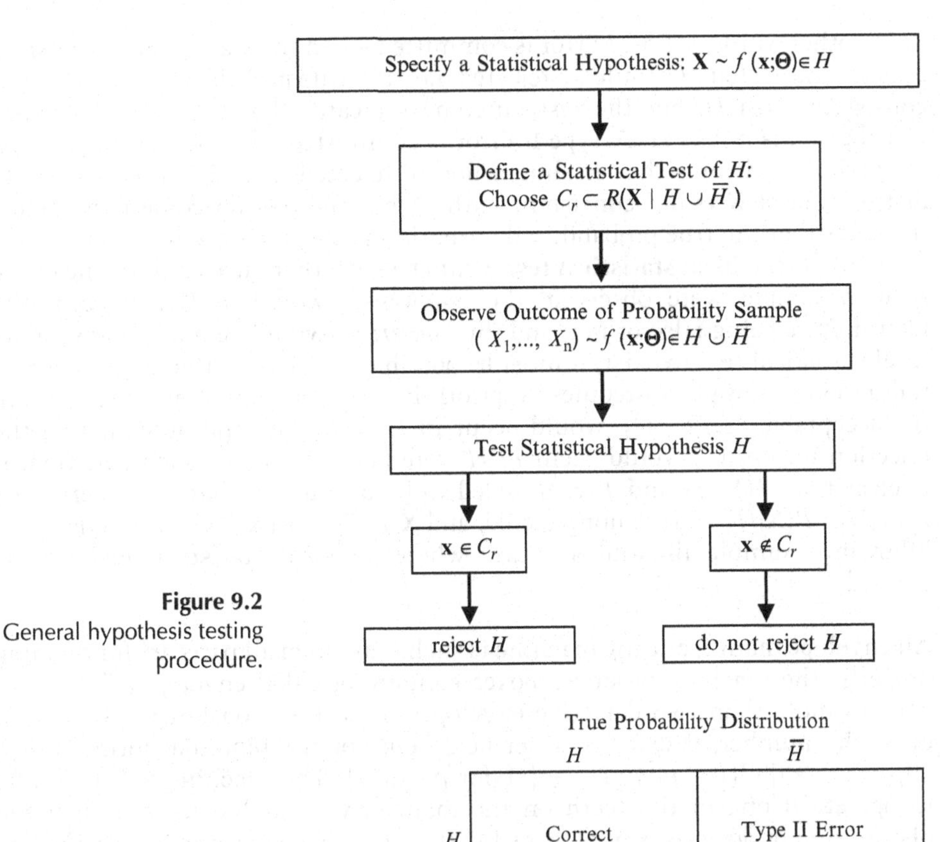


**Figure 9.2** General hypothesis testing procedure.

| Test Decision | True Probability Distribution: $H$ | True Probability Distribution: $\bar{H}$ |
|---|---|---|
| $H$ | Correct | Type II Error |
| $\bar{H}$ | Type I Error | Correct |

**Figure 9.3** Potential outcomes of statistical test relative to true probability distribution.

### 9.3.2 Type I Error, Type II Error, and General Non-Existence of Ideal Statistical Tests

Given a situation where the analyst must decide whether to reject a statistical hypothesis, $H$, and given that in reality the statistical hypothesis is either true or false, there are four possible states of affairs with respect to test decisions and their validity, as summarized in Figure 9.3.

The two potential error situations that can occur when deciding to reject $H$ are given distinct, albeit somewhat uninspired, names so that they can be clearly distinguished.

**Definition 9.3** ***Type I and Type II Error***

Let $H$ be a statistical hypothesis being tested for rejection. Then the two types of errors that can be made by the statistical test are:

1. **Type I Error**: rejecting $H$ when $H$ is true.
2. **Type II Error**: not rejecting $H$ when $H$ is false.

In other words, a Type I Error is committed when the statistical test *mistakenly* indicates that $H$ should be rejected, i.e., the true probability distribution is consistent with $H$, but the test outcome indicates that the true probability distribution is not in $H$. A Type II Error is committed when the statistical test *mistakenly* indicates that $H$ should not be rejected, i.e., the true probability distribution of $\mathbf{X}$ is not consistent with $H$, but the test outcome nonetheless indicates that the true probability distribution is consistent with $H$.

Clearly, the **ideal statistical test** would be such that once an outcome of the random sample were observed, the hypothesis would *always* be correctly identified as being false or not, and thus *no errors* would be made. For such an ideal statistical test to exist, it must be possible to partition the range of potential random sample outcomes a priori in such a way that outcomes in the acceptable region, $C_a$, would occur *iff* $H$ were true and outcomes in the rejection region, $C_r$, would occur *iff* $H$ were false. To define such partition, it is clear that $R(\mathbf{X}|H)$ and $R(\mathbf{X}|\bar{H})$ need to be disjoint so that, *with certainty*, $\mathbf{x} \in C_a = R(\mathbf{X}|H) \Rightarrow H$ is not rejected, and $\mathbf{X} \in C_r = R(\mathbf{X}|\bar{H}) \Rightarrow H$ is false. The following example illustrates a case where an ideal statistical test can be defined.

**Example 9.8**
***An Ideal Statistical Test***

An envelope manufacturing machine is such that when all parts are functioning properly, the machine produces boxes containing 1,000 envelopes that never contain more than two defective envelopes, with the probability distribution of $X$, the number of defectives per box, being of the binomial form $f(x;p) = (2!/[x!(2-x)!])p^x(1-p)^{2-x} I_{\{0,1,2\}}(x)$, for $p \in (0,1)$. The machine will continue to operate if one of the teeth on the main drive gear breaks, but then the distribution of defectives per box of 1,000 envelopes changes such that the box will always contain between 6 and 8 defective envelopes, with the probability distribution of the number of defectives per box changing to the distribution of $Y = X + 6$, i.e.,

$$g(y;p) = (2!/[(y-6)!(8-y)!])p^{y-6}(1-p)^{8-y} I_{\{6,7,8\}}(y)$$

for $p \in (0,1)$. Any other machine problem will cause the machine to shut down, so that no envelopes will be produced.

A quality control engineer wants to test the hypothesis that all parts of the envelope machine are working properly. Her statistical hypothesis, in terms of a random sample of 1 box of envelopes, is $H = \{f(x;p),\ p \in (0,1)\}$. The engineer designs the statistical test as follows

$$x \in C_a = \{0,1,2\} \Rightarrow \text{do not reject } H \text{ and } x \in C_r = \{6,7,8\} \Rightarrow \text{reject } H.$$

Note that the test is an **ideal statistical test,** which produces no decision error of either type. For example, if the engineer were to observe that $x = 1$ in a box of envelopes she randomly chose to examine, she would conclude that $H$ is true; all parts of the machine are working properly. She is certain the assessment is correct, since the statistical test used in making the decision is an ideal statistical test. □

More generally, if $R(\mathbf{X}|H) \cap R(\mathbf{X}|\bar{H}) \neq \emptyset$, which is virtually always the case in practice, then there are potential outcomes of the sample that reside simultaneously in the supports of some PDFs in $H$ and some PDFs in $\bar{H}$. In this case, a sample outcome equal to one of the points in $R(\mathbf{X}|H) \cap R(\mathbf{X}|\bar{H})$ cannot be used to resolve *with certainty* whether or not $H$ is true since such points are not uniquely associated with either $H$ or $\bar{H}$. It follows that there would not exist a dichotomous partition of $R(\mathbf{X}|H \cup \bar{H})$ that would define a statistical test which, with a priori certainty, could always correctly decide the truth of $H$, since any dichotomous partition would include some points that were in both $R(\mathbf{X}|H)$ and $R(\mathbf{X}|\bar{H})$.

If it were true that $R(\mathbf{X}|H) \neq R(\mathbf{X}|\bar{H})$, then a *trichotomous* partition of $R(\mathbf{X}|H \cup \bar{H})$ can be devised that leads to a decision rule that makes no errors in deciding the truth of $H$, but the rule will also result in situations in which some sample outcomes lead to no decision whatsoever. The rule could be based on the sets $C_n = R(\mathbf{X}|H) \cap R(\mathbf{X}|\bar{H})$, $C_a = R(\mathbf{X}|H) - C_n$, and $C_r = R(\mathbf{X}|\bar{H}) - C_n$, as

$$\mathbf{x} \in \left\{ \begin{matrix} C_a \\ C_r \\ C_n \end{matrix} \right\} \Rightarrow \left\{ \begin{matrix} \text{do not reject } H \\ \text{reject } H \\ \text{no decision} \end{matrix} \right\}.$$

While making no errors, such a rule can be wasteful of sample information, especially when outcomes in the set $C_n$ have a high probability of occurring. Furthermore in the majority of hypothesis testing applications $R(\mathbf{X}|H) = R(\mathbf{X}|\bar{H})$. So not only is an ideal statistical test not available, the preceding trichotomous partition of $R(\mathbf{X}|H \cup \bar{H})$ is not available either. In practice, it is therefore generally the case that error-free statistical tests *cannot* be defined. Instead, one must seek to define statistical tests that control the incidence of errors. We examine this idea next.

### 9.3.3 Controlling Type I and II Errors

In applications, a statistical test can often be designed to *control* the propensity for Type I and Type II errors to occur. By "controlling" the errors, we mean that $C_a$ and $C_r$ are chosen in a way such that the error *probabilities* are known, and thus *controlled*, in advance of performing the test and are at levels that are, in some sense, acceptable to the researcher.

In order to illustrate how a statistical test can be designed to "control" errors recall Example 9.3. In this case, $R(\mathbf{X}|H) = 0$, $R(\mathbf{X}|\bar{H}) = \times_{i=1}^{n} \{0,1\}$, and thus $R(\mathrm{X}|H \cup \bar{H}) = \times_{i=1}^{n} \{0,1\}$, where the definition of $R(\mathbf{X}|\bar{H})$ follows from the fact that if $\bar{H}$ is true, then the population distribution for the problem is a Bernoulli density with $p > 0$ and so both 0 and 1 are potential outcomes for each $X_i$. Also note that $R(\mathbf{X}|H) \cap R(\mathbf{X}|\bar{H}) = 0 \neq \emptyset$, so that no ideal statistical test exists for this problem.

Suppose a statistical test of $H$ is defined by the following rejection region for a random sample of size 100: $C_r = \{\mathbf{x}\colon \sum_{i=1}^{100} x_i > 0\}$. In other words, if a random

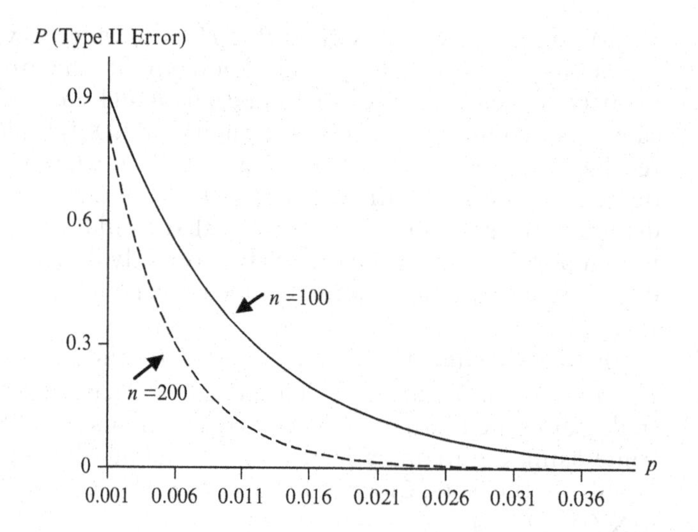


**Figure 9.4**
Probabilities of type II error for sample sizes: $n = 100$ and $n = 200$.

sample of size 100 results in no defectives, we will decide $H$ is not rejected, while if any defectives are observed in the outcome of the random sample, $H$ will be declared false. Using this test, how probable are Type I and Type II errors? Determining the probability of Type I Error is straightforward in this case, since if $H$ is true, then $P(\mathbf{x} \notin C_r) = P(\mathbf{x} = \mathbf{0}) = 1$, which follows from the fact that $P(x_i = 0) = 1\ \forall i$, so that $P(\mathbf{x} = \mathbf{0}) = P(x_i = 0, \forall i) = \prod_{i=1}^{100} P(x_i = 0) = 1$ (recall that the $X_i$'s are independent random variables). It follows that if $H$ is true, the probability of rejecting $H$ is $P(\text{Type I Error}) = P(\mathbf{x} \notin C_a) = P(\mathbf{x} \in C_r) = 0$. Note that having "controlled" the Type I Error to have probability zero, we have effectively complete confidence that $H$ *is false* if the test *rejects* $H$, since the probability is 0 that the test would mistakenly reject $H$.

Now examine the probability of Type II Error. Assuming $H$: $p = 0$ is false, then $X_i$ has a Bernoulli distribution with probability, $\overline{H}:\ p \in \{.001,.002,\ldots,1\}$. Thus, the probability of Type II Error depends on *which* probability distribution in $\bar{H}$ is the correct one. Since we do not know which is correct (or else we would have no need to test a statistical hypothesis in the first place), we examine the range of possibilities for Type II Error as a function of $p$. In this case the parameter $p$ serves as an index for the 1,000 possible alternative probability distributions for $\mathbf{X}$ that are contained in $\bar{H}$. Figure 9.4 provides a partial graph of the functional relationship between $p$ and the probability of Type II Error (see graph labeled $n = 100$).[3]

Note that in this case that probabilities of Type II Error are equal to $P(\mathbf{x} \notin C_r) = P(\sum_{i=1}^{100} x_i = 0)$ calculated on the basis of a binomial distribution for $\sum_{i=1}^{100} X_i$ for the various values of $p \in \bar{H}$, with $n = 100$. This definition of Type II Error probabilities is motivated from the fact that the event $\sum_{i=1}^{n} x_i = 0$ results in

[3]For convenience, we have chosen to "connect the dots" and display the graph as a continuous curve. We will continue with this practice wherever it is convenient and useful.

non-rejection of $H$, and $p \in \bar{H}$ indicates that $H$ is false, so that together these two conditions define a situation where a false $H$ is not rejected, i.e., a Type II Error is committed. The explicit functional relationship between $p$ and the level of Type II Error in this example, which indicates the probability levels at which Type II Error has been controlled for the various values of $p \in \bar{H}$, is given by

$$P(\text{Type II Error}) = h(p) = (1-p)^{100} \text{for} \, p \in \{.001, .002, \ldots, 1\}.$$

Figure 9.4 indicates that, in this problem, the probability of committing a Type II Error declines rapidly as $p$ increases. When $p$ is only .03, the probability of Type II Error is less than .05. Note, however, that when $p$ is very close to its hypothesized value in $H$, i.e., $p = 0$, the probability of Type II Error can be quite high. For example, if $p = .001$, then $P(\text{Type II Error}) = .905$, or if $p = .002$, then $P(\text{Type II Error}) = .819$. Thus, in practice, if the shipment has very few defectives (implying a very small $p$ associated with the *iid* Bernoulli random sample from the shipment), then the proposed statistical test will not be very effective in rejecting the assertion that the shipment has no defectives, i.e., the probability of rejection will not be high.

In general, the probabilities of Type I Error are equal to the values of $P(\mathbf{x} \in C_r)$ for each of the PDFs contained in $H$. The probabilities of Type II Error equal the values of $P(\mathbf{x} \notin C_r)$ for each of the PDFs contained in $\bar{H}$. If the error probability characteristics of a given statistical test are unacceptable, the researcher generally has two options, which can also be pursued simultaneously—she can *choose a different* rejection region, $C_r$, and thus define a different statistical test, or she can *alter the size of the probability sample* on which to base the test. We continue examining the hypothesis in Example 9.3 to illustrate these options. First suppose we increase the sample size to 200 but otherwise use a test rule analogous to the preceding rule to test the hypothesis that there are no defectives in the 1,000 bolt shipment. Stated in terms of the parameter $p$, the statistical hypothesis, $H$, remains precisely as before, $H\!: p = 0$. The probability of Type I Error remains equal to zero, since if $H$ were true, then $P(\text{Type I Error}) = 1 - P(\mathbf{x} = \mathbf{0}) = 1 - P(\sum_{i=1}^{200} x_i = 0) = 0$, which can be demonstrated following the same reasoning as before except that now we are dealing with $n = 200$ independent Bernoulli random variables as opposed to $n = 100$.

The probabilities of Type II Error as a function of $p$ are now given by the value of $P(\sum_{i=1}^{200} x_i = 0)$ calculated using the binomial distribution of $\sum_{i=1}^{200} X_i$ for the various values of $p > 0$, except now $n = 200$. The explicit functional relationship between $p$ and the level of Type II Error is now represented by

$$P(\text{Type II Error}) = h(p) = (1-p)^{200} \text{for } p \in \{.001, .002, \ldots, 1\}$$

and a partial graph of $P(\text{Type II Error})$ as a function of $p$ is given in Figure 9.4 (see the graph labeled $n = 200$). It is seen that the probability of Type II Error has been uniformly lowered for all $p > 0$. For example, the probability of Type II Error is less than .05 when p is only .015. The probability of Type II Error is appreciably reduced even for very small values of $p$, although the probability remains high for values of $p$ very close to the hypothesized value of 0(when $p = .001$ or $.002$, the probability is still .819 and .670, respectively).

If the sample size were increased further, the probabilities of Type II Error would continue to decline uniformly as a function of $p$. This becomes clear upon examining the functional relationship between $p$ and the probability of Type II Error for an unspecified sample size of $n$, as

$$P(\text{Type II Error}) = h(p,n) = (1-p)^n \text{for}\, p \in \{.001, .002, \ldots, 1.000\}.$$

Note that $dh(p,n)/dn = (1-p)^n \ln(1-p) < 0 \ \forall \ p \in (0,1)$ since $\ln(1-p) < 0$, and so the probability of Type II Error is a decreasing function of $n \ \forall \ p \in (0,1)$ (when $p = 1$, the probability of Type II Error is 0 $\forall n$). Note also that $P$(Type II Error) $= h(p,n) \to 0 \ \forall \ p \in (0,1)$ as $n \to \infty$. These results illustrate a generally valid principle for "good" statistical tests that the more sample information one has, the more accurate a statistical test will be.

Suppose it were too costly, or impossible, to increase the sample size beyond $n = 100$. Consider an alternative rejection region in order to alter the error probabilities of the associated statistical test. In particular, suppose we define $C_r = \{\mathbf{x}\colon \sum_{i=1}^{100} x_i \geq 2\}$ and make test decisions accordingly. Then the probability of Type I Error is still zero, since if $H$ were true, then $P(\sum_{i=1}^{100} x_i = 0) = 1$, and so long as the point $\mathbf{x} = \mathbf{0} \notin C_r$, $P(\mathbf{x} \notin C_r) = 1$ when $H$ is true, and so the probability of Type I Error remains zero. The probability of Type II Error as a function of $p$, based on the binomial distribution of $\sum_{i=1}^{n} X_i$, is now given by

$$\begin{aligned} \text{P(Type II Error)} &= P\left(\sum_{i=1}^{100} x_i = 0 \text{ or } 1\right) \\ &= (1-p)^{100} + 100p(1-p)^{99} \text{for } p \in \{.001, .002, \ldots, 1\} \end{aligned}$$

It is evident that the probabilities of Type II Error are *uniformly higher* for this statistical test as compared to the previous test (both tests based on $n = 100$).

In fact the original statistical test that we examined is, for any given sample size, the best one can do with respect to minimizing the probability of Type II Error while maintaining the probability of Type I Error at its optimal value of *zero*. This follows from the fact that the Type I Error probability will be zero *iff* $C_r$ does not contain the element $\mathbf{x} = \mathbf{0}$, and any element removed from $C_r$, while not affecting the probability of Type I Error, will increase the probability of Type II Error $\forall \ p \in (0,1)$. Note further that the only other choice of Type I Error probability in this statistical hypothesis testing situation is the value 1, which would be the case for any rejection region (i.e., for any statistical test) for which $\mathbf{0} \in C_r$. A statistical test having $P$(Type I Error) $= 1$ would clearly be absurd since *if H were* true, one would be essentially certain to *reject H*! Thus, for the case in Example 9.3 the statistical test based on $C_r = \{\mathbf{x}\colon \sum_{i=1}^{n} x_i > 0\}$ is the best one can do with respect to controlling Type I and Type II Error probabilities for any sample size.

In general hypothesis testing applications, a primary concern is the choice of a statistical test that provides acceptable levels of control on both Type I and Type II errors. Unfortunately, these are most often conflicting objectives, as we discuss next.

### 9.3.4 Type I Versus Type II Error Tradeoff and Protection Against Errors

The choice of values at which Type I Error probability can be controlled is generally not so limited as in the example discussed in the previous section. It is more typical for there to be a large number, or even a continuum, of possible choices for the Type I Error probability of a statistical test of a simple hypothesis. Furthermore, there is also then typically a trade-off between the choice of Type I and Type II Error probabilities such that the probability of one type of error can be decreased (increased) only at the expense of increasing (decreasing) the other. When $H$ is composite, there is also a range of Type I Error probabilities to consider which results from the fact that $H$ then contains a range of potential probability distributions for $\mathbf{X}$. We will encounter a large number of examples of these more typical situations in the remainder of this chapter, and in Chapter 10.

In order to illustrate some of the more general characteristics of statistical hypothesis tests, reexamine the hypothesis described in Example 9.1. In this case

$$R(\mathbf{X}|H \cup \bar{H}) = R(\mathbf{X}|H) = R(\mathbf{X}|\bar{H}) = \times_{i=1}^{n}\{0, 1\},$$

and thus no ideal statistical test exists for the statistical hypothesis that the probability distribution of the random sample is defined via a product of identical Bernoulli densities with $p \leq .02$. How then should we choose a statistical test of $H$? Suppose we want the probability of Type I Error to equal 0. In the current testing situation, this would mean that if the true PDF for the random sample were a product of identical Bernoulli densities having any value of $p \in \{0,.001,.002,\ldots, .020\}$, we want our test to $H$ with probability 1. Unfortunately, the only choice for $C_r \subset R(\mathbf{X}|H \cup \bar{H})$ that will ensure that $P(\mathbf{x} \notin C_r) = 1$ when $H$ is true is $C_r = \emptyset$. This follows because $\forall\, p \in \{.001,.002,\ldots,.020\}$, each and every point in $\times_{i=1}^{n}\{0,1\}$ is assigned a positive probability value. Then the statistical test would have the characteristic that $P(\mathbf{x} \in C_r) = 0$, *whether or not H is true.* The test would *always* accept $H$, no matter what the outcome of the random sample, and thus to obtain a probability of Type I Error equal to 0, we have to accept a probability of Type II Error equal to 1. This is clearly unacceptable—we would never reject $H$, even if it were profoundly false (even if *all* of the bolts in the shipment were defective).

Given the preceding results, it is clear that to choose a useful statistical test in the current situation requires that one be willing to accept some positive level of Type I Error probability. Suppose the maximum probability of Type I Error that one is willing to accept is .05, i.e., we are implicitly stating that $P$(Type I Error) $\leq .05$ provides sufficient **protection against Type I Error**. The interpretation of "protection" is that if $P$(Type I Error) $\leq .05$, then we know that the test will *mistakenly* reject a true $H$ no more than 1 time in 20, on average, in a repeated sampling context. Then if the test actually rejects $H$ for a given random sample outcome, it is much more likely that the test is rejecting $H$ because $H$ is false than because the test has mistakenly rejected a true $H$. The level of confidence we have that $H$ is false when $H$ is rejected by a statistical test is thus derived from the level of protection against Type I Error that the test provides.

How should $C_r$ be chosen in order to define a test with the desired control of Type I Error? We need to define a rejection region $C_r \subset R(\mathbf{X}|H \cup \bar{H})$ such that $P(\mathbf{x} \in C_r) \leq .05$ no matter which probability distribution in $H$ were true, i.e., no matter which $p \in \{0,.001,...,.020\}$ were true. Subject to the previous inequality constraint on the probability of Type I Error, we want $C_r$ to be such that $P$(Type II Error) is as small as possible no matter which probability distribution in $\bar{H}$ were true, i.e., no matter which $p > .02$ were true. Intuitively, it would make sense that if $P$(Type II Error) is to be minimized for a given level of control for Type I Error, we should attempt to choose points for the rejection region, $C_r$, that would have their *highest* probability of occurrence if $\bar{H}$ were true. In the example at hand, this corresponds to choosing sample outcomes for inclusion in $C_r$ that represent a higher number of defectives than lower, since $\bar{H}$ corresponds to bolt populations with more defectives than is the case for $H$.

Suppose that a random sample of size $n = 200$ will be used. Consider the rejection region $C_r = \{\mathbf{x}\colon \sum_{i=1}^{200} x_i > 7\}$ contained in $R(\mathbf{X}|H \cup \bar{H}) = \times_{i=1}^{200}\{0,1\}$. In order to verify that the statistical test implied by $C_r$ has a probability of Type I Error that does not exceed .05, one must verify that $P(\mathbf{x} \in C_r) \leq .05$ no matter which probability model in $H$ were true. In other words, we must verify that the outcome of the binomial random variable $\sum_{i=1}^{200} X_i$ is such that

$$P\left(\sum_{i=1}^{200} x_i > 7\right) = \sum_{j=8}^{200} \frac{200!}{j!(200-j)!} p^j (1-p)^{200-j} \leq .05 \ \forall \ p \in \{0, .001, \ldots, .020\}.$$

One can show that $P(\sum_{i=1}^{200} x_i > 7)$ achieves a maximum of .049 when $p = .02$, and so the inequality is met for all choices of $p$ implied by $H$. Thus, the probability of Type I Error is upper-bounded by .05. We emphasize that in this case, since $H$ was composite, the probability of Type I Error is a function of $p$, with the range of probabilities upper-bounded by .05. The largest value of Type I Error probability over all PDFs in $H$ is generally referred to as the **size of the statistical test**, and is a measure of the minimum degree of protection against Type I Error provided by the test. We will examine this concept in more detail in Section 9.4.

Now examine the probability of Type II Error as a function of $p$. The various probabilities of Type II Error that are possible are given by $P(\mathbf{x} \notin C_r) = P(\sum_{i=1}^{200} x_i \leq 7)$ evaluated with respect to binomial distributions for $\sum_{i=1}^{200} X_i$ that have $p > .02$ (which are the values of $p$ implied by $\overline{H}$). Figure 9.5 presents a partial graph of the probability of Type II Error as a function of $p$ for this statistical test.

It is seen from the graph that the probability of Type II Error decreases rapidly as $p$ increases, with $P$(Type II Error) $\leq .05$ when $p \geq .065$. However, as we have seen previously, $P$(Type II Error) is quite high for values of $p$ close to those

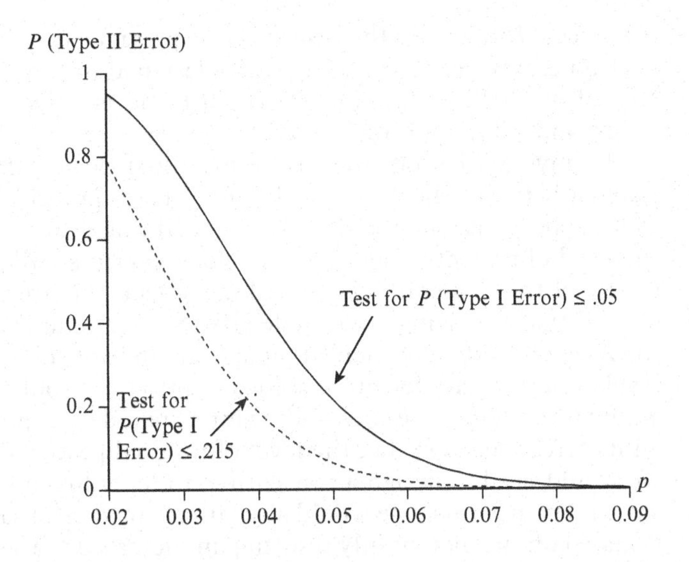


**Figure 9.5**
Probabilities of type II error for tests of H: $p \leq .02$.

implied by $H$, e.g., if $p = .025$ or $.03$, then $P$(Type II Error) = .87 or .75, respectively. Thus, if the test does *not* reject $H$, the behavior of P(Type II Error) as a function of $p$ suggests that we would have substantial confidence that $p < .065$, say, since otherwise it would have been highly probable that the test would have rejected $H$ (with a probability of 1 − $P$(Type II Error) = .95 or greater). However, we do not have a great deal of confidence that $H$ is *literally true,* since the test does not have a high probability of rejecting $H$ if $p > .02$ but $p$ is near .02. In fact, we will see that this situation is typical of statistical tests in practice so that it is advisable to conclude that non-rejection of $H$ does not necessarily mean that $H$ is *accepted,* but rather that there is insufficient statistical evidence to reject it.

If the relationship depicted in Figure 9.5 for the case where $P$(Type I Error) $\leq .05$ were deemed unacceptable, the researcher has two basic options, as we discussed previously. One option would be to increase the sample size to some value larger than 200 and apply an analogous test rule to the larger random sample. This would decrease $P$(Type II Error) uniformly for all $p > .02$ assuming $C_r$ were chosen to maintain $P$(Type I Error) $\leq .05$. The other option would be to accept a larger probability bound on Type I Error (i.e., define a different test based on the same sample size), which would also result in a uniform reduction in $P$(Type II Error). To illustrate the tradeoff between the two types of errors that occurs when pursuing the latter option, suppose we redesign the statistical test using the following rejection region: $C_r = \{\mathbf{x}\colon \sum_{i=1}^{200} x_i > 5\}$. Using the binomial distribution for $\sum_{i=1}^{200} X_i$, it can be shown that $P(\sum_{i=1}^{200} x_i > 5) \leq .215$ $\forall\ p \in \{0,.001,\ldots,.02\}$, and so $P$(Type I Error) is now upper-bounded by .215. Thus, if $H$ were true, then at most 21.5 percent of the time, on average in a repeated sampling context, the test would nonetheless reject $H$. Note, using the redesigned test, there is notably *less* confidence that $H$ is actually false when the test rejects $H$ than was the case with the previous test. However, $P$(Type II Error) is also notably decreased using the redesigned test, as illustrated in Figure 9.5. Now if $p = .025$ or $.03$, then $P$(Type II Error) is reduced to .62 and .44, respectively, and $P(\text{Type II Error}) \leq .062 \; \forall \; p > .05$.

In any application, the researcher must decide the appropriate degree of tradeoff between the Type I and Type II Error probabilities of the test, as well as the appropriate sample size to use in defining the test. An acceptable choice of the level of protection against test decision errors will generally depend on the nature of the consequences of making a decision error, and on the cost and/or feasibility of changing the sample size. For example, if the hypothesis was that the level of residual pesticide that appears in fruit sold in supermarkets exceeds levels that are safe for human consumption, it would seem that Type I Error would be extremely serious—we want a great deal of protection against making an incorrect decision that the level of residual pesticide is at a safe level. Thus, we would need to design a test with extremely low probability of Type I Error. Type II Error in this case would be a situation where a safe level of pesticide was declared unsafe, potentially resulting in increased processing and growing costs, loss of sales by the pesticide manufacturer, and the loss of an input to the agricultural sector that could potentially lower the cost of food production. Thus, one would also need to strive for a low $P$(Type II Error) in fairness to the pesticide manufacturer, farmers, and consumers. The precise choice of the Type I/Type II Error tradeoff must be addressed on a case-by-case basis.[4]

### 9.3.5 Test Statistics

It should be noted that the rejection regions of the statistical tests defined in the previous two subsections could all be defined in terms of outcomes of a scalar statistic. In particular, the statistic used was of the form $T = t(\mathbf{X}) = \sum_{i=1}^{n} X_i$, and the set defining conditions for the rejection regions were of the general form $t(\mathbf{x}) > c$. A scalar statistic whose outcomes are partitioned into rejection and acceptable regions to define statistical tests is called a **test statistic**.

**Definition 9.4** ***Test Statistic and Rejection Region***

Let $C_r$ define the rejection region associated with a statistical test of the hypothesis $H$ versus $\bar{H}$. If $T = t(\mathbf{X})$ is a scalar statistic such that $C_r = \{\mathbf{x} : t(\mathbf{x}) \in C_r^T\}$, so that the rejection region can be defined in terms of outcomes, $C_r^T$, of the statistic $T$, then $T$ is referred to as a **test statistic** for the hypothesis $H$ versus $\bar{H}$. The set $C_r^T$ will be referred to as the **rejection (or critical) region of the test statistic, *T*.**

[4]If the magnitudes of the costs or losses incurred when errors are omitted can be expressed in terms of a loss function, then a formal analysis of expected losses can lead to a choice of type I and type II error probabilities. For an introduction to the ideas involved, see Mood, A., F. Graybill, and D. Boes, (1974). *Introduction to the Theory of Statistics*, 3rd Ed., New York: McGraw-Hill, pp. 414–418.

The use of test statistics can simplify the problem of testing statistical hypotheses in at least two significant ways. First of all, it allows one to check whether a sample outcome is an element of the rejection region of a statistical test by examining whether a scalar outcome of the test statistic resides in a set of scalars (the set $C_r^T$ in Definition 9.4). This eliminates the need for dealing with $n$-dimensional outcomes and $n$-dimensional rejection regions. In effect, the rejection region of the statistical test is alternatively represented as a unidimensional set of real numbers. Secondly, test statistics can facilitate the evaluation of Type I and Type II Error probabilities of statistical tests if the PDF of the test statistic can be identified and if the PDF can be tractably analyzed. Relatedly, if an asymptotic distribution for a test statistic can be identified, then it may be possible to approximate Type I and Type II Error probabilities based on this asymptotic distribution even when exact error probabilities for the statistical test cannot be readily established. We will examine the notion of **asymptotic tests** in Chapter 10.

In practice, the large majority of statistical hypothesis testing is conducted using test statistics, and the statistical test itself is generally defined in terms of the rejection region of the test statistic. That is, a statistical test of $H$ versus $\bar{H}$ will generally be defined as

$$t(\mathbf{x}) \in \left\{ \begin{array}{c} C_r^T \\ C_a^T \end{array} \right\} \Rightarrow \left\{ \begin{array}{c} \text{reject} \\ \text{do not reject} \end{array} \right\} H$$

We will henceforth endeavor to express statistical tests in terms of test statistics whenever it is useful to do so.

It should be noted that test statistics are *not* unique. For example, the rejection regions of the statistical tests in the preceding two subsections can be represented in terms of the test statistics $\sum_{i=1}^n X_i$, $\bar{X}$, or $\ln(\bar{X})$ (among a myriad of other possibilities), with corresponding rejection regions for the test statistics given by $(c,n]$, $(c/n,1]$, and $(\ln(c/n),0]$, respectively. The choice of $\sum_{i=1}^n X_i$ was particularly attractive in the preceding application because its PDF (binomial) was easily identified and tractable to work with. In general, the choice of functional form for a test statistic is motivated by the ease with which the outcomes of the statistic can be obtained *and* by the tractability of the test statistic's PDF.

### 9.3.6 Null and Alternative Hypotheses

The terminology *null* and *alternative* hypotheses is often used in the context of testing statistical hypotheses. Historically, a **null hypothesis** was a hypothesis interpreted as characterizing no change, no difference, or no effect. For example, if the average life of a "typical" 25-watt compact fluorescent light bulb sold in the U.S. is 10,000 hours, then a test of the hypothesis that a bulb manufactured by General Electric Co. has a life equal to the average life, i.e., $H$: $\mu = 10{,}000$, could be labeled a null hypothesis, and interpreted as representing a situation where there is "no difference" between General Electric's bulb and the average life of a typical bulb. The hypothesis that is accepted if the null hypothesis is rejected is referred to as the **alternative hypothesis**. In current usage, the term null hypothesis is used more generally to refer to any hypothesis that, if rejected by mistake, characterizes a Type I Error situation. Thus, Type I Error refers to the mistaken rejection of a null hypothesis, while a Type II Error refers to a mistaken non-rejection of the null hypothesis.

Having examined a number of basic conceptual issues related to statistical hypothesis testing, we henceforth focus our attention on two specific classes of hypothesis testing situations: (1) parametric hypothesis testing, and (2) testing distributional assumptions. The next two sections examine parametric hypothesis testing in some detail. We will examine testing of distributional assumptions in Chapter 10.

## 9.4 Parametric Hypothesis Tests and Test Properties

As the phrase inherently implies, **parametric hypothesis testing** concerns the testing of statistical hypotheses relating to the values of parameters or functions of parameters contained in a probability model for a sample of data. In a parametric model, it is assumed at the outset that the probability distributions in both $H$ and $\bar{H}$ are characterized by members of a known or given parametric family or class of distributions that are indexed by the parameters in some way. In a semiparametric model, the parameters refer to the parametric component of the model, and $H$ and $\bar{H}$ define separate and distinct sets of probability models distinguished by the specific parametric components defined by $H$ and $\bar{H}$. Examples 9.1 and 9.3 are illustrations of parametric hypothesis tests in the context of parametric models, where the known parametric family of distributions underlying the definitions of the statistical hypotheses and their complements is the Bernoulli distribution. Once it is understood that the probability models are defined in terms of a Bernoulli distribution with parameter $p$, then the statistical hypotheses under consideration can be compactly represented in terms of statements about the value of $p$ or about the value of a function of $p$. Example 9.5 is an illustration of parametric hypothesis tests in the context of a semiparametric model, where no explicit family of parametric distributions is identified by the hypotheses, but a parametric component of the model is identified. In particular, $H$ relates to all distributions with $\mu < 48,500$, whereas $\bar{H}$ relates to distributions for which $\mu \geq 48,500$.

In the parametric hypothesis testing case, we will most often utilize the abbreviated notation *H: Set Defining Conditions* introduced in Section 9.2 to denote the statistical hypothesis. In this case, the *set defining conditions* will denote that a parameter vector, or a function of the parameter vector, is contained in a certain set of values (recall Example 9.7). We will adopt a further simplification that is made possible by the parametric context and interpret $H$ itself as a set of parameter values, and thus the notation $\Theta \in H$ will mean that the parameter (vector) is an element of the set of hypothesized values. We emphasize that in the parametric hypothesis testing context, $H$ still concurrently identifies, at least to some degree of specificity,[5] a set of probability distributions for **X**.

The real-world meaning of a statistical hypothesis concerning the values of parameters or functions of parameters depends on the characteristics of the real-world population or process being analyzed, and on the interpretation of what the parameters represent or measure in the context of the probability model being analyzed. For example, the exponential family of densities used in Example 9.2 leads to the interpretation of $H\colon \Theta > 20$ as being an assertion about the *mean* of the distribution of computer lifetimes. The Bernoulli family of densities used in Example 9.1 leads to the interpretation of $H\colon p \leq .02$ as an assertion about the *proportion* of defectives in a shipment of bolts.

### 9.4.1 Maintained Hypothesis

A parametric hypothesis testing situation will be characterized by two major components. One component is the statement of the **statistical hypothesis**, as described and exemplified earlier. The other is the **maintained hypothesis**, as defined below.

**Definition 9.5** ***Maintained Hypothesis***

All facts, assertions, or assumptions about the probability model of a probability sample that are in common to $H$ and $\bar{H}$, and that are maintained to be true regardless of the outcome of a statistical test of $H$ versus $\bar{H}$.

For example, in a parametric hypothesis testing situation within the context of a parametric model, whatever parametric family of distributions is assumed to characterize the probability distributions in $H$ and $\bar{H}$ is part of the maintained hypothesis. In Example 9.1, the Bernoulli family of distributions is part of the maintained hypothesis, while in Example 9.2, the exponential family of distributions could be part of the maintained hypothesis if the distributions in $\bar{H}$ were also assumed to be in the exponential family. Besides an assertion about the parametric family of distributions if the model is parametric, a number of other "facts" about the probability space might be contained in the maintained hypothesis. For example, with reference to Example 9.4, in examining the hypothesis $H\colon \beta_1 = .25$ versus $\bar{H}\colon \beta_1 \neq .25$, the maintained hypothesis includes assertions (explicitly or implicitly) that the normal family of distributions characterize the probability distributions for the $\varepsilon_i$'s, the $\varepsilon_i$'s are independent random variables with $\mathrm{E}(\varepsilon_i) = 0$ and $\mathrm{var}(\varepsilon_i) = \sigma^2 \,\forall\, i$, and the expected value of wheat yield in any given time period is a linear function of both the level of fertilizer applied and the level of rainfall.

The facts contained in the maintained hypothesis can be thought of as contributing to the definition of the specific context in which a statistical test

[5] In parametric models, sets of fully specified probability distributions will be identified, whereas in semiparametric models, only a subset of moments or other characteristics of the underlying probability distributions are generally identified by the hypotheses.

of any hypothesis is to be applied and interpreted. The probabilities of Type I and Type II Error for a given statistical test of an assertion like $H$: $\Theta > 20$, and in fact the fundamental meaning of a statistical hypothesis itself, will generally depend on the context of the parametric hypothesis testing problem defined by the maintained hypothesis.

Regarding the genesis of the maintained hypothesis, the facts that are included in the maintained hypothesis might have been supported by the results of previous statistical analyses. Alternatively, the facts may be derived from laws or accepted theories about the process or population being studied. Finally, the facts may be largely unsubstantiated conjectures about the population or process under study that are *tentatively* accepted as truths for the purpose of conducting a parametric hypothesis test. In any case, the interpretation of a statistical test can depend critically on facts stated in the maintained hypothesis, and the result of a statistical test must be interpreted as being conditional on these facts. If one or more components of the maintained hypothesis are actually false, the interpretation of the Type I and Type II Error probabilities of a given statistical test, and the interpretation of a statistical hypothesis itself, can be completely changed and/or disrupted. Thus, in a parametric hypothesis testing situation, one must be careful that the facts stated in the maintained hypothesis are defensible. Facts that are "tentative" will generally require statistical testing themselves when convincing supporting evidence is lacking. At the least, the analyst is obliged to point out any tentative assumptions contained in the maintained hypothesis when reporting the results of a statistical test.

### 9.4.2 Power Function

The **power function** of a statistical test provides a complete summary of all of the operating characteristics of a statistical test with respect to probabilities of making correct/incorrect decisions about $H$. The power function is particularly useful in comparing alternative statistical tests of a particular parametric hypothesis. The power function is defined as follows.

**Definition 9.6**
***Power Function of a Statistical Test***

Let a parametric statistical hypothesis be defined by $H$: $\boldsymbol{\Theta} \in \boldsymbol{\Omega}_H$ and let its complement be defined by $\bar{H}$: $\boldsymbol{\Theta} \in \boldsymbol{\Omega}_{\bar{H}}$ . Let the rejection region $C_r$ define a statistical test of $H$. Finally, let the CDF $F(\mathbf{x};\boldsymbol{\Theta})$, $\boldsymbol{\Theta} \in \boldsymbol{\Omega} = H \cup \bar{H}$, represent the parametric family of probability distributions encompassed by $H$ and $\bar{H}$. Then the **power function** of the statistical test is defined by $\pi(\boldsymbol{\Theta}) = P(\mathbf{x} \in C_r; \boldsymbol{\Theta}) \equiv \int_{x \in C_r} dF(\mathbf{x}; \boldsymbol{\Theta})$ for $\boldsymbol{\Theta} \in H \cup \bar{H}$.

In words, the power function indicates the probability of rejecting $H$ for every possible value of $\boldsymbol{\Theta}$. The value of the function $\pi$ at a particular value of the parameter vector, $\boldsymbol{\Theta}$, is called the **power of the test at $\boldsymbol{\Theta}$**, which is the probability of rejecting $H$ if $\boldsymbol{\Theta}$ were the true value of the parameter vector. From the definition of the power function, it follows that $\pi(\boldsymbol{\Theta})$ is identically a probability of Type I Error if $\boldsymbol{\Theta} \in H$, or else a probability of *not* making a Type II Error if $\boldsymbol{\Theta} \in \bar{H}$.

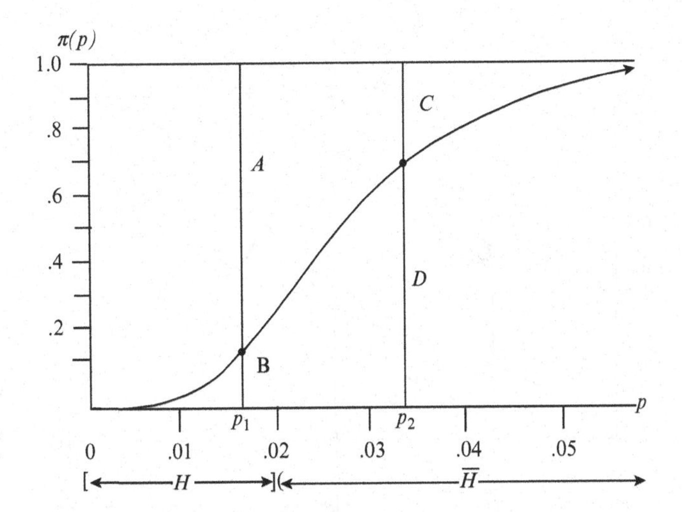


**Figure 9.6**
Power function of test for $H$: $p \leq .02$ in Example 9.1.

Then $[1 - \pi(\boldsymbol{\Theta})]$ is identically the probability of Type II Error if $\boldsymbol{\Theta} \in \bar{H}$ and the probability of *not* making a Type I Error if $\boldsymbol{\Theta} \in H$. The graph in Figure 9.6 represents a partial graph of the power function for the test of the hypothesis $H$: $p \leq .02$ from Example 9.1 using the statistical test based on $C_r = \{\mathbf{x}: \sum_{i=1}^{200} x_i > 5\}$. We know from before that $P(\text{Type I Error}) \leq .215$ for this test. The lines A, B, C, and D in the figure illustrate the probabilities of not making a Type I error, making a Type I error, making a Type II error, and not making a Type II error respectively.

The closer the power function of a test is to the theoretical **ideal power function** for the hypothesis testing situation, the better the test. The ideal power function can be compactly defined as $\pi_*(\boldsymbol{\Theta}) = I_{\bar{H}}(\boldsymbol{\Theta})$ and would correspond to the *ideal statistical test* for a given $H$, were such a test to exist. When comparing two statistical tests of a given H, a test is better if it has higher power for $\boldsymbol{\Theta} \in \bar{H}$ and lower power for $\boldsymbol{\Theta} \in H$, which implies that the better test will have lower probabilities of both Type I and Type II Error. Figure 9.7 provides graphs of hypothetical power functions for two tests of the hypothesis that $H$: $\Theta \leq c$, and a graph of the ideal power function for the test situation, where $\Omega = H \cup \bar{H} = [0,k]$.

From Figure 9.7, it is apparent that the statistical test associated with power function $\pi_2(\boldsymbol{\Theta})$ is the better test, having probabilities of Type I and Type II Error everywhere less than or equal to the corresponding probabilities associated with the alternative test. The power function $\pi_2(\boldsymbol{\Theta})$ is also seen to be closer to the ideal power function, $\pi_*(\boldsymbol{\Theta})$.

### 9.4.3 Properties of Statistical Tests

There are a number of test properties that can be examined to assess whether a given statistical test is appropriate in a given problem context, and/or to compare a test rule to competing test rules.[6] The power function of a statistical test

[6]The properties we will examine do not exhaust the possibilities. See E. Lehmann, (1986), *Testing Statistical Hypotheses*, John Wiley, NY.

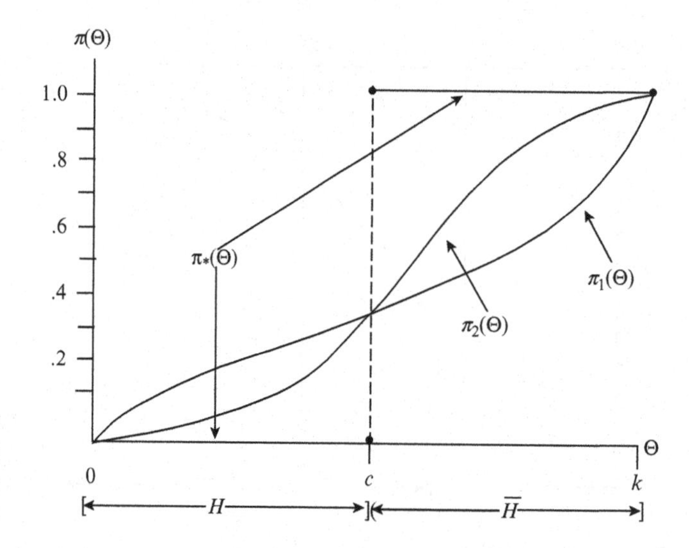


**Figure 9.7** Two power functions and the ideal power function for testing $H: \Theta \leq c$.

defined in the previous section, can be used to define properties of statistical tests, including the **size of the test** and the **significance level of the test**, and whether a test is **unbiased**, **consistent**, and/or **uniformly most powerful**. We examine each of these test properties below.

*Test Size and Level* One such property, called the **size of the test**, is a measure of the minimum level of protection against Type I Error that a given statistical test provides.

**Definition 9.7** ***Size and Level of a Statistical Test***

Let $\pi(\mathbf{\Theta})$ be the power function for a statistical test of $H$. Then $\alpha = \sup_{\mathbf{\Theta} \in H} \{\pi(\mathbf{\Theta})\}$ is called the **size of the statistical test.**[7] The test is a **level $\alpha$ test** if $\pi(\mathbf{\Theta}) \leq \alpha \ \forall \mathbf{\Theta} \in H$.

The size of the statistical test is essentially the maximum probability of Type I Error associated with a given statistical test. The lower the size of the test, the lower the maximum probability of mistakenly rejecting $H$. Tests of $H$ having **level $\alpha$** are tests for which $P_{\mathbf{\Theta}}$(Type I Error) $\leq \alpha \ \forall \mathbf{\Theta} \in H$. Thus, the level is an *upper bound* to the Type I Error probability of statistical tests. The key difference between the concepts of *size* and *level* of statistical tests is that *size* represents the maximum (or at least the supremum) value of $P_{\mathbf{\Theta}}$(Type I Error), $\mathbf{\Theta} \in H$, for a given test, while *level* is only some stated bound on *size* that might not equal $P_{\mathbf{\Theta}}$(Type I Error) for any $\mathbf{\Theta} \in H$ or the supremum of $P_{\mathbf{\Theta}}$(Type I Error) for $\mathbf{\Theta} \in H$ for a particular test. A statistical test of $H$ that has size $\gamma$ is an $\alpha$-level test for any $\alpha \geq \gamma$.

[7]Recall that $\sup_{\mathbf{\Theta} \in H}\{\pi(\mathbf{\Theta})\}$ denotes the smallest upper bound to the values of $\pi(\mathbf{\Theta})$ for $\mathbf{\Theta} \in H$ (i.e., the supremum). If the maximum of $\pi(\mathbf{\Theta})$ for $\mathbf{\Theta} \in H$ exists, then sup is the same as max.

In applications when a hypothesis $H$ is (not) rejected, terminology that is often used to indicate the protection against Type I Error used in the test is "**H is (not) rejected at the $\alpha$ – level.**" Moreover, *level* is often taken to be synonymous in meaning with *size*, although in statistical concept this need not be the case. Care should be taken to eliminate ambiguity regarding whether the stated *level* is equal to the *size* of a statistical test used. A phrase that eliminates the potential ambiguity is "**H is (not) rejected using a size-$\alpha$ test.**" However, the use of the term *level* is ubiquitous in the statistical testing applications literature, and we simply caution the reader to be careful to understand how the term *level* is used in any given application, i.e., synonymous with *size* of the test or not.

As an illustration of terminology, the power function in Figure 9.6 refers to a test that has a *size* of .215, and a *level* synonymous with size is also .215, whereas any number greater than .215 is also a legitimate upper bound to the size of the test, and is thus a *level*. If the hypothesis $H$: $p \leq .02$ is (not) rejected, then one could state that $H$: $p \leq .02$ is (not) rejected at the .215 level, or that H: $p \leq .02$ is (not) rejected using a size .215 test.

*Unbiasedness* The concept of *unbiasedness* within the context of hypothesis testing refers to a statistical test that has a smaller probability of rejecting the null hypothesis when it is true compared to when it is false.

**Definition 9.8** ***Unbiasedness of a Statistical Test***

Let $\pi(\Theta)$ be the power function of a statistical test of $H$. The statistical test is called unbiased *iff* $\sup_{\Theta \in H} \{\pi(\Theta)\} \leq \inf_{\Theta \in \bar{H}} \{\pi(\Theta)\}$.[8]

The property of unbiasedness makes intuitive sense as a desirable property of a test rule since we would generally prefer to have a higher probability of rejecting H when H is false than when H is true. As indicated in Definition 9.8, whether a test rule is unbiased can be determined from the behavior of its power function. In particular, if the height of the power function graph is everywhere lower for $\Theta \in H$ than for $\Theta \in \bar{H}$, the statistical test is *unbiased.* The tests associated with the power functions in Figures 9.6 and 9.7 are unbiased tests.

*More Powerful, Uniformly Most Powerful, Admissibility* Another desirable property of a statistical test is that, for a given level, $\alpha$, the test exhibits the highest probability of rejecting $H$ when $H$ is false compared to all other competing tests of $H$ having level $\alpha$. Such a test is called **uniformly most powerful** of level $\alpha$, where the term *uniformly* refers to the test being the most powerful (i.e., having highest power) for *each and every* $\Theta$ in $\bar{H}$ (i.e., *uniformly* in $\bar{H}$). In the formal definition of the concept presented below, we introduce the notation $\pi_{Cr}(\Theta)$ to indicate the power function of a test of $H$ when the test is defined by the rejection region $C_r$.

[8] Recall the previous footnote, and the fact that $\inf_{\Theta \in \bar{H}} \{\pi(\Theta)\}$ denotes the largest lower bound to the values of $\pi(\Theta)$ for $\Theta \in \bar{H}$ (i.e., the infimum). The sup and inf of $\pi(\Theta)$ are equivalent to max and min, respectively, when the maximum and/or minimum exists.

**Definition 9.9**
***Uniformly Most Powerful (UMP) Level $\alpha$ Test***

Let $\Xi = \{C_r\colon \sup_{\Theta \in H} \{\pi(\Theta)\} \leq \alpha\}$ be the set of all rejection regions of level $\alpha$ for testing the hypothesis $H$ based on a sample $\mathbf{X} \sim f(\mathbf{x};\Theta)$ of size $n$. The rejection region $C_r^* \in \Xi$, and the associated statistical test of $H$ that $C_r^*$ defines, are **uniformly most powerful of level $\alpha$** *iff* $C_r^*$ has level $\alpha$ and $\pi_{C_r^*}(\Theta) \geq \pi_{C_r}(\Theta)$ $\forall \Theta \in \bar{H}$ and $\forall C_r \in \Xi$.

In the case where $\bar{H}$ is a *simple hypothesis*, the rejection region, $C_r^*$, and the associated test are also referred to as being **most powerful** (the adverb "uniformly" being effectively redundant in this case). The UMP test of $H$ is thus the test of $H$, among all possible level $\alpha$ tests, that has the highest probability of rejecting $H$ when $H$ is false (equivalently, it is the level $\alpha$ test having the most power uniformly in $\Theta$ for $\Theta \in \bar{H}$). Such a test of $H$ is effectively the "best one can do" with respect to minimizing the probability of Type II Error, *given* that protection against Type I Error is at level $\alpha$. As implied by Definition 9.9, if one could plot the power function graphs of all tests of level $\alpha$, the power function of the UMP level $\alpha$ test of $H$ would lie on or above the power functions of all other level $\alpha$ tests of $H$ $\forall \Theta \in \bar{H}$. We will examine general methods for finding UMP tests of $H$ (when such tests exist) in Section 9.5. Once we have introduced the Neyman-Pearson Lemma in Section 9.5, we will also show that a UMP level $\alpha$ test is also generally an unbiased test.

Unfortunately, in a significant number of cases of practical interest, UMP tests do not exist. We will see that it is sometimes possible to restrict attention to the unbiased class of tests and define a UMP test within this class, but even this approach sometimes fails. In practice, one then often resorts to the use of statistical tests that have at least acceptable (to the analyst and to those she wishes to convince regarding the validity of her hypotheses) power function characteristics. Assuming that a test of level $\alpha$ is deemed to provide sufficient protection against Type I Error, a test that is **more powerful** is preferred, as defined below.

**Definition 9.10**
***More Powerful Level $\alpha$ Test***

Let $C_r$ and $C_r^*$ represent two level $\alpha$ statistical tests of $H$. The test based on $C_r^*$ is said to be *more powerful* than the test based on $C_r$ if $\pi_{C_r^*}(\Theta) \geq \pi_{C_r}(\Theta)\ \forall \Theta \in \bar{H}$, with strict inequality for at least one $\Theta \in \bar{H}$.

The reason for preferring a more *powerful test* is clear—if both tests provide the same desired level of protection against Type I error, then the one that provides more protection against Type II error is better.

While in comparing level $\alpha$ tests the more powerful test is preferred, the less powerful test may have some redeeming qualities. In particular, the latter test may provide better protection against Type I Error for some or all $\Theta \in H$, and the test may then be appropriate in cases where a smaller test level is desired. Tests with no such possible redeeming qualities are *inadmissible* tests, formally defined as follows.

**Definition 9.11**
***Admissibility of a Statistical Test***

Let $C_r$ represent a statistical test of $H$. If there exists a rejection region $C_r^*$ such that

$$\pi_{C_r^*}(\Theta)\left\{\begin{matrix}\geq\\ \leq\end{matrix}\right\}\pi_{C_r}(\Theta)\ \forall\Theta\in\left\{\begin{matrix}\bar{H}\\ H\end{matrix}\right\},$$

with strict inequality holding for some $\Theta \in H \cup \bar{H}$, then the test based on $C_r$ is **inadmissible**. Otherwise, the test is **admissible.**

From the definition, it follows that an inadmissible test is one that is dominated by another test in terms of protection against both types of test errors. Inadmissible tests can be eliminated from consideration in any hypothesis testing application.

*Consistency* The limiting behavior of the power functions of a sequence of level $\alpha$ tests as the size of the probability sample on which the tests are based increases without bound relates to the property of *consistency*. In particular, a sequence of level $\alpha$ tests of $H$ for which the probability of Type II Error $\to 0$ as $n \to \infty$, will be called a **consistent sequence of level $\alpha$ tests**. Since the definition of the rejection region of a test will generally change as the sample size, $n$, changes, we will introduce the notation $C_{rn}$ to indicate the dependence of the rejection region on sample size.

**Definition 9.12**
***Consistent Sequence of Level $\alpha$ Tests***

Let $\{C_{rn}\}$ represent a sequence of tests of $H$ based on a random sample $(X_1,\ldots,X_n) \sim f(x_1,\ldots,x_n;\Theta)$ of increasing size, $n$, and let the level of the test defined by $C_{rn}$ be $\alpha\ \forall\ n$. Then the sequence of tests is **consistent** *iff* $\lim_{n\to\infty}(\pi_{C_{rn}}(\Theta)) = 1$ $\forall\Theta \in \bar{H}$.

Thus, a consistent sequence of level $\alpha$ tests is such that, in the limit, the probability is one that $H$ will be rejected whenever $H$ is false. Then, for a large enough sample size $n$, the $n$th test in a consistent sequence of tests provides a level of protection against Type I Error given by a probability $\leq \alpha$ and is essentially certain to reject $H$ if $H$ is false.

*P-Values* Different analysts and/or reviewers of statistical analyses do not always agree on what the appropriate level of statistical tests should be when testing statistical hypotheses. In order to accommodate such differences of opinion, it is frequent in applications to report ***P*-values** (an abbreviation for "probability values") for each statistical test conducted. A $P$-value is effectively the **minimum size test** at which a given hypothesis would still be rejected based on the observed outcome of the sample.

In order for the idea of a *minimum size* to make sense, there must exist a prespecified family of *nested* rejection regions, and thus an associated family of statistical tests, that corresponds to increasing Type I Error probability values, $\alpha$, for testing the null hypothesis under consideration. By *nested* rejection regions corresponding to increasing values of $\alpha$, we mean that $C_r(\alpha_1) \subset C_r(\alpha_2)$ for $\alpha_1 < \alpha_2$, where $C_r(\alpha)$ denotes the rejection region corresponding to $P(\text{Type I Error}) = \alpha$. Then upon observing $\mathbf{x}$, one calculates the $P$-value as equal to the smallest $\alpha$ value that identifies the smallest rejection region that contains $\mathbf{x}$, i.e., $P$-value $= \arg\min_\alpha\{\mathbf{x} \in C_r(\alpha)\} = \min_\alpha\{\sup_{\Theta \in H}\{P(\mathbf{x} \in (C_r(\alpha); \Theta))\}\}$ such that $\mathbf{x} \in C_r(\alpha)\}$. The $P$-value can thus be interpreted as the smallest size at which the null hypothesis would be rejected based on the nested set of rejection regions or statistical tests.

Assuming the nested set of rejection regions and associated statistical tests to be used for testing a particular $H$ are well-defined, then once the analyst reports the $P$-value, a reviewer of the statistical analysis will know whether she would reject $H$ based on her own choice of test size within the set of possible size values.[9] In particular, if the reviewer's choice of size were less than (greater than) the reported $P$-value, then she would not (would) reject $H$ since her rejection region would be smaller than (larger than) the one associated with the reported *P-value* and would not (would) contain $\mathbf{x}$.

The $P$-value is also often interpreted as indicating the **strength of evidence** against the null hypothesis, where the smaller the $P$-value, the greater the evidence against $H$. The idea is that if the $P$-value is small, then the protection against Type I Error can be set high and $H$ would still be rejected by the test.

A well-defined and accepted set of nested rejection regions and statistical tests necessary for calculating $P$-values generally exist in practice, especially when the rejection regions can be defined in terms of a test statistic. We illustrate both the use of $P$-values and other test properties in the following example.

**Example 9.9** ***Illustration of Power Function, Size, Unbiasedness, Consistency, and P-Values***

A random sample of $n = 100$ observations on the miles per gallon achieved by a new model pickup truck manufactured by a major Detroit manufacturer is going to be used to test the null hypothesis that the expected miles per gallon achieved by the truck model in city driving is less than or equal to 15, the alternative hypothesis being that expected miles per gallon $> 15$. It can be assumed that miles per gallon measurements are independent, and that the population is (approximately) normally distributed with a standard deviation equal to .1.

The null and alternative hypotheses in this parametric hypothesis testing problem can be represented by $H_0$: $\mu \leq 15$ and $H_a$: $\mu > 15$, respectively. Suppose we desire to assess the validity of the null hypothesis using a statistical test of size .05. Examine a test based on the following rejection region: $C_{rn} = \{\mathbf{x} : 10n^{1/2}(\bar{x}_n - 15) \geq 1.645\}$. To verify that the rejection region implies a test of size .05, first note that

$$\frac{\bar{X}_n - 15}{\sigma/\sqrt{n}} = \frac{(\bar{X}_n - 15)}{(.1/\sqrt{n})} = 10n^{1/2}(\bar{X}_n - 15)$$

$$= \underbrace{10n^{1/2}(\bar{X}_n - \mu)}_{\sim N(0,1)} + 10n^{1/2}(\mu - 15) \sim N\left(10n^{1/2}(\mu - 15),\ 1\right).$$

[9] In the case of continuous $\mathbf{X}$, the choice of size is generally a continuous interval contained in [0,1]. If $\mathbf{X}$ is discrete, the set of choices for size is generally finite, as previous examples have illustrated.

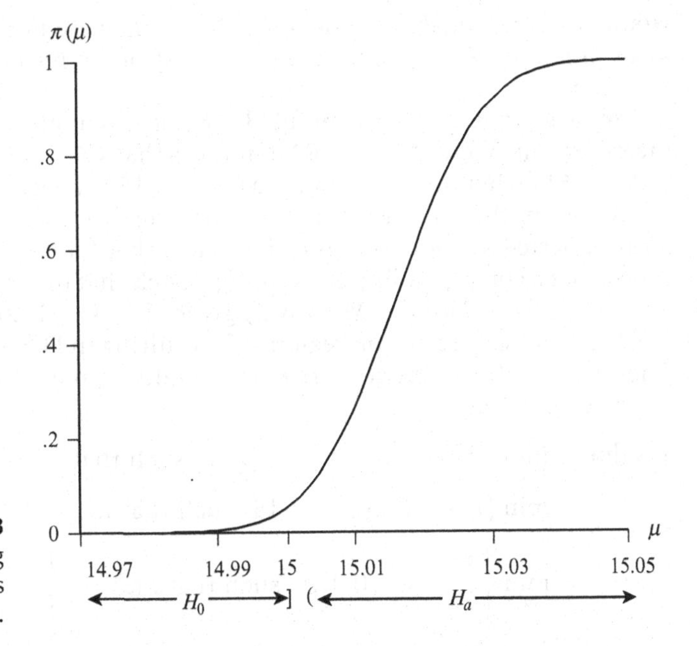


**Figure 9.8**
Power function for testing $H_0$: $\mu \leq 15$ versus $H_a$: $\mu > 15$, $n = 100$.

It follows that the *power function* of the test can be defined as

$$\pi_n(\mu) = \int_{1.645}^{\infty} N\left(z; 10n^{1/2}(\mu - 15),\ 1\right)dz = \int_{1.645-10n^{1/2}(\mu-15)}^{\infty} N(z; 0, 1)dz,$$

so that $\sup_{\mu \leq 15} \pi_n(\mu) = .05$ and the test is of size .05.[10] The test is also *unbiased,* since it is evident that $\sup_{\mu \leq 15}\{\pi_n(\mu)\} \leq \inf_{\mu > 15}\{\pi_n(\mu)\} = .05$.[11] The *consistency* of the test can be demonstrated by noting that $\forall\ \mu \in H_a$,

$$\lim_{n\to\infty}\{\pi_n(\mu)\} = \lim_{n\to\infty} \int_{1.645-10n^{1/2}(\mu-15)}^{\infty} N(z; 0, 1)dz = \int_{-\infty}^{\infty} N(z; 0, 1)dz = 1,$$

and so the test is a consistent test.

The test is also a *uniformly most powerful* test for testing the null hypothesis $H_0$: $\mu \leq 15$ versus $H_a$: $\mu > 15$, i.e., there is no other test rule with $\pi(\mu) \leq .05$ $\forall\ \mu \in H_0$ that has higher power for any $\mu \in H_a$. The rationale in support of the UMP property will be provided in Section 9.5.

A graph of the power function for this test is given in Figure 9.8.

Upon examination of the power function of the test, it is clear that protection against Type II Error increases rapidly for values of $\mu > 15$. In particular, it

[10]The maximum is achievable in this case, and equals .05 when $\mu = 15$.

[11]Note that $\min_{\mu > 15}\{\pi_n(\mu)\}$ does not exist in this case. The largest possible lower bound (i.e., the infimum) is .05, which is $< \pi_n(\mu)$, $\forall\ \mu > 15$.

would seem reasonable to conclude that if $H_0$ were not rejected, then the true value of $\mu \leq 15.03$, say, since the probability of not rejecting $H_0$ when $\mu > 15.03$ is $< .088$.

Suppose that the outcome of the random sample of 100 miles per gallon measurements yielded $\bar{x} = 15.02$. It follows that $\mathbf{x} \in C_{rn}$ since $10n^{1/2}(15.02 - 15) = 2 \geq 1.645$ when $n = 100$, and thus $H$: $\mu \leq 15$ is rejected using a size .05 test.

Regarding the $P$-value for this test, first note that $C_{rn}(\alpha) = \{\mathbf{x}: 100(\bar{\mathbf{x}} - 15) \geq k(\alpha)\}$ forms a nested set of rejection regions such that for each value of size $\alpha \in (0,1)$, there exist a corresponding constant $k(\alpha)$ such that $\max_{\mu \leq 15}\{P(\mathbf{x} \in C_{rn}(\alpha); \mu)\} = P(\mathbf{x} \in C_{rn}(\alpha); \mu = 15) = \alpha$. Moreover, given that $\bar{x} = 15.02$, the largest value of $k(\alpha)$ that defines a rejection region still resulting in H being rejected is $k(\alpha) = 2$. Then the smallest size test that still results in rejecting the null hypothesis can be calculated as

$$
\begin{aligned}
P\text{-value} &= \min_{\alpha}\left\{\max\nolimits_{\mu \leq 15}\{P(\mathbf{x} \in C_{rn}(\alpha); \mu) \text{ such that } k(\alpha) \leq 2\}\right\} \\
&= \min_{\alpha}\{P(\mathbf{x} \in C_{rn}(\alpha); \mu = 15) \text{ such that } k(\alpha) \leq 2\} \\
&= \min_{\alpha}\left\{\int_{k(\alpha)}^{\infty} N(z; 0, 1)dz \text{ such that } k(\alpha) \leq 2\right\} \\
&= \int_{2}^{\infty} N(z; 0, 1)dz = .023.
\end{aligned}
$$

It follows that $H_0$ would *not* be rejected at $\alpha = .01$ and would be rejected at $\alpha = .05$. □

*Asymptotic Tests* It is sometimes difficult, intractable or impossible (e.g., in semiparametric models) to determine the exact probabilities of Type I and Type II errors for a given statistical test. Thus the determination of the exact power function of a test may also be difficult, intractable, or impossible. Similar to the case of point estimation, one might then rely on asymptotic properties to assess the efficacy of a given test rule.

In the majority of hypothesis testing applications, the rejection regions of statistical tests are defined in terms of test statistics. It is often the case that laws of large numbers and central limit theory can be utilized to establish various convergence properties of a test statistic. In particular, asymptotic distributions of test statistics can be used to approximate power functions of tests, in which case the efficacy of a given test procedure can be assessed based on the power function approximation. In practice, the choice of a particular statistical test for a given hypothesis is often motivated by its adequacy in terms of Type I and Type II Error protection that are assessed through an examination of its *approximate* rather than its *exact* power function characteristics. We will examine the asymptotics of some important general classes of statistical tests, including generalized likelihood ratio, Lagrange multiplier, and Wald tests, in Chapter 10.

## 9.5 Classical Hypothesis Testing Theory: UMP and UMPU Tests

The focus of this section is on the discovery of uniformly most powerful (UMP) and uniformly most powerful unbiased (UMPU) tests of statistical hypotheses. The theory in this area is classical in nature requiring a variety of regularity conditions to be applicable, it is not widely applicable in multiparameter settings, and it applies primarily to only parametric models directly, although in semiparametric models it can be applied in an asymptotically-valid approximative sense. Moreover, it should be emphasized at the outset that UMP tests of hypotheses do not always exist. Whether a UMP test exists depends on the type of hypothesis being tested and on the characteristics of the joint density of the probability sample under study. In a case where the classical results of this section cannot be applied, or in cases where UMP tests simply do not exist, it is generally still possible to define statistical tests with good exact or asymptotic power function characteristics, and we will examine procedures for defining such tests in Chapter 10. Readers interested in more generally applicable, albeit not necessarily optimal, methods and who on first reading prefer to avoid a number of rather complex and involved proofs may prefer to skim this section at first, or else move directly on to Chapter 10.

We examine four basic approaches for finding UMP tests: Neyman-Pearson, monotone likelihood, exponential class, and conditioning approaches. While these approaches have distinct labels, they are all interrelated to some degree. Much of our discussion will focus on scalar parameter cases, although some important multiparameter results will also be examined. This scalar orientation is a practical one—UMP tests simply do not exist for most multiparameter hypothesis testing contexts. In cases where a UMP test does not exist, it is sometimes still possible to define a UMP test within the class of *unbiased* tests, as will be seen ahead.

As we noted at the beginning of Chapter 8, and for the reasons stated there, we will place some of the Proofs of Theorems in the Appendix to this chapter to facilitate readability of this section.

### 9.5.1 Neyman-Pearson Approach

While we will examine cases involving composite alternative hypotheses, we begin with the case where both the null and alternative hypotheses are *simple*. This situation is not representative of the typical hypothesis testing problem where at least one of the hypotheses is composite (often, the alternative hypothesis), but it provides a simplified context for motivating some of the basic principles underlying the definition of a UMP test. We note that because the alternative hypothesis consists of a single point in this case, we could drop the adverb "uniformly," and characterize our objective as finding the *most powerful* test of the null hypothesis versus the alternative hypothesis.

*Simple Hypotheses* The two simple hypotheses can be represented as $H_0$: $\boldsymbol{\Theta} = \boldsymbol{\Theta}_0$ and $H_a$: $\boldsymbol{\Theta} = \boldsymbol{\Theta}_a$, where $\boldsymbol{\Theta}_0$ and $\boldsymbol{\Theta}_a$ are parameter vectors (or scalars). A most powerful test of $H_0$ can be motivated by the following theorem, which is referred to as the *Neyman-Pearson Lemma.* [12] Recall from Def. 9.6 that the notation $P(\mathbf{x} \in A;\boldsymbol{\Theta})$ represents the probability that $\mathbf{x} \in A$ when $\boldsymbol{\Theta}$ is the value of the parameter vector.

**Theorem 9.1 *Neyman-Pearson Lemma*** *A rejection region of the form* $C_r(k) = \{\mathbf{x} : f(\mathbf{x};\boldsymbol{\Theta}_0) \leq kf(\mathbf{x};\boldsymbol{\Theta}_a)\}$, *for* $k>0$, *is a most powerful rejection region of* level $\alpha = P(\mathbf{x} \in C_r(k);\boldsymbol{\Theta}_0) \in (0, 1)$ *for testing the hypothesis* $H_0$: $\boldsymbol{\Theta} = \boldsymbol{\Theta}_0$ versus $H_a$: $\boldsymbol{\Theta} = \boldsymbol{\Theta}_a$. *Furthermore,* $C_r(k)$ *is the unique (with probability 1) most powerful rejection region of size* $\alpha$.

**Proof** See Appendix. ■

Application of the Neyman-Pearson approach for constructing most powerful tests of $H$: $\boldsymbol{\Theta} = \boldsymbol{\Theta}_0$ versus $H_a$: $\boldsymbol{\Theta} = \boldsymbol{\Theta}_a$ is relatively straightforward, at least in principle. In particular, one chooses the value of $k$ that defines the rejection region in Theorem 9.1 having the desired size $\alpha = P(\mathbf{x} \in C_r(k);\boldsymbol{\Theta}_0)$, and then the most powerful statistical test of level $\alpha$ is defined by

$$\mathbf{x} \in \left\{ \begin{array}{c} C_r(k) \\ \bar{C}_r(k) \end{array} \right\} \Rightarrow \left\{ \begin{array}{c} \text{reject} \\ \text{do not reject} \end{array} \right\} H_0.$$

Note that since the probability calculation for determining the size of the rejection region $C_r(k)$ is based on the density $f(\mathbf{x};\boldsymbol{\Theta}_o)$, values of $\mathbf{x}$ for which $f(\mathbf{x};\boldsymbol{\Theta}_o) = 0$ are irrelevant and can be ignored. Then an equivalent method of finding an $\alpha$-size most powerful test is to find the value of $k$ for which $P(f(\mathbf{x};\boldsymbol{\Theta}_a)/f(\mathbf{x};\boldsymbol{\Theta}_o) \geq k^{-1}) = \alpha$, and define $C_r(k)$ accordingly. Likewise, if the supports[13] of the densities $f(\mathbf{x};\boldsymbol{\Theta}_o)$ and $f(\mathbf{x};\boldsymbol{\Theta}_a)$ are the same, then another equivalent method is to find the value of $k$ for which $P(f(\mathbf{x};\boldsymbol{\Theta}_o)/f(\mathbf{x};\boldsymbol{\Theta}_a) \leq k) = \alpha$, and use it in the definition of $C_r(k)$. We illustrate the procedure in the following examples. Note that in the case where $\mathbf{X}$ is discrete, the feasible choices of $\alpha$ can be somewhat limited, as will also be illustrated below.[14]

**Example 9.10 *UMP Test of Bernoulli Means Using Neyman-Pearson*** Your company manufactures a hair-restoring treatment for middle aged balding men and claims that the treatment will stimulate hair growth in 80 percent of the men who use it. A competitor of the company claims that your scientists must be referring to the complementary event by mistake—the counter claim is

---

[12] Neyman, J. and E.S. Pearson, "On the Problem of the Most Efficient Tests of Statistical Hypotheses," *Phil. Trans.*, A, vol. 231, 1933, p. 289.

[13] Recall that the support of a density function is the set of $\mathbf{x}$-values for which $f(\mathbf{x};\boldsymbol{\Theta}_0) > 0$, i.e., $\{\mathbf{x}: f(\mathbf{x};\boldsymbol{\Theta}_0) > 0\}$ is the support of the density $f(\mathbf{x};\boldsymbol{\Theta}_0)$.

[14] This limitation can be overcome, in principle, by utilizing what are known as **randomized tests**. Essentially, the test rule is made to depend not only on the outcomes of $\mathbf{X}$ but also on auxiliary random variables that are independent of $\mathbf{X}$, so as to allow any level of test size to be achieved. However, the fact that the test outcome can depend on random variables that are independent of the experiment under investigation has discouraged its use in practice. For an introduction to the ideas involved, see Kendall, M. and A. Stuart, (1979) *The Advanced Theory of Statistics*, Vol. 2, 4th Edition, New York: MacMillan, 1979, p. 180–181. Also, see problem 9.8.

that your treatment stimulates hair growth in only 20 percent of the men who use it.

A government regulatory agency steps in to settle the dispute. They intend to test the hypothesis that the treatment is 20 percent effective versus the alternative hypothesis that it is 80 percent effective. A random sample of 20 middle aged balding men received your company's hair-restoring treatment. The success ($x_i = 1$) or failure ($x_i = 0$) of the chemical in stimulating hair growth for any given individual is viewed as the outcome of a Bernoulli trial, where $X_i \sim p^{x_i}(1-p)^{1-x_i} I_{\{0,1\}}(x_i)$ *for* $p \in \Omega = \{.2,\ .8\}$.

The collection of 20 trials is viewed as a random sample from a Bernoulli population distribution, and letting $X = (X_1,\ldots,X_{20})$ represent the random sample of size 20 from a Bernoulli distribution, we then have that

$$f(\mathbf{x};p) = p^{\sum_{i=1}^{20} x_i}(1-p)^{20-\sum_{i=1}^{20} x_i}\prod_{i=1}^{20} I_{\{0,1\}}(x_i), \quad p \in \Omega = \{.2,\ .8\}.$$

In order to define the most powerful level $\alpha$ test of the hypothesis $H_0$: $p = .2$ versus the hypothesis $H_a$: $p = .8$, use the Neyman-Pearson Lemma and examine

$$C_r(k) = \{\mathbf{x} : f(\mathbf{x};.20)/f(\mathbf{x};.80) \le k\} = \left\{\mathbf{x} : (.25)^{\sum_{i=1}^{20} x_i}(4)^{20-\sum_{i=1}^{20} x_i} \le k\right\}$$

$$= \left\{\mathbf{x} : \left(\sum_{i=1}^{20} x_i\right)\ln(.25) + \left(20 - \sum_{i=1}^{20} x_i\right)\ln(4) \le \ln(k)\right\}$$

$$= \left\{\mathbf{x} : \sum_{i=1}^{20} x_i \ge 10 - .36067\ln(k)\right\} \quad \text{(to five decimal places)}.$$

Since $Z = \sum_{i=1}^{20} x_i$ has a binomial distribution with parameters $p$ and $n = 20$, the relationship between the choice of $k$ and the size of the test implied by $C_r(k)$ is given by

$$\alpha(k) = P(x \in C_r(k); p = .20) = \sum_{z \ge 10 - .36067\ln(k)} \binom{20}{z}(.20)^z(.80)^{20-z} I_{\{0,1,\ldots,20\}}(z).$$

Some possible choices of test size are given as follows:

| ln(*k*) | 10−.36067 ln(*k*) | $\alpha(k)$ |
|---|---|---|
| 24.95356 | 1 | .9885 |
| 22.18094 | 2 | .9308 |
| 19.40832 | 3 | .7939 |
| 16.63571 | 4 | .5886 |
| 13.86309 | 5 | .3704 |
| 11.09047 | 6 | .1958 |
| 8.31785 | 7 | .0867 |
| 5.54524 | 8 | .0321 |
| 2.77261 | 9 | .0100 |

Note because of the discrete nature of the random variables involved, there are no other choices of $\alpha$ within the range [.01, .9885] other than the ones displayed above. The most powerful statistical test of $H_0$: $p = .20$ versus $H_a$: $p = .80$ having level .01, say, is thus given by

$\mathbf{x} \in C_r \Rightarrow$ reject $H_0 : p = .20$ and $\mathbf{x} \notin C_r \Rightarrow$ do not reject $H_0 : p = .20$

where we use $k = \exp(2.77261) = 16$ to define

$C_r = \{\mathbf{x} : f(\mathbf{x}; .20)/f(\mathbf{x}; .80) \leq 16\}$.

The rejection region can be represented more simply using the test statistic $t(\mathbf{X}) = \sum_{i=1}^{n} X_i$ as

$$C_r = \left\{\mathbf{x} : \sum_{i=1}^{20} x_i \geq 9\right\}.$$

Suppose upon observing the successes and failures in the 20 treatments, the value of the test statistic was calculated to be $\sum_{i=1}^{20} x_i = 5$. It follows that $\mathbf{x} \notin C_r$, and the hypothesis $H_0$: $p = .2$ is *not* rejected. The government concludes that your competitor's claim has merit, and instructs your company to either provide further proof that their claim is correct, or else refrain from claiming that your hair-restoring treatment is 80 percent effective. (Issue to consider: Would it be better if the *null* hypothesis were $H_0$: $p = .8$ and the *alternative* hypothesis were $H_a$: $p = .2$ in performing the statistical test? Why or why not?). □

The reader may have noticed that in Example 9.10, the choice of $k$ used in the Neyman-Pearson Lemma to define a UMP level $\alpha$ test of $H_0$: $\Theta = \Theta_0$ versus $H_a$: $\Theta = \Theta_a$ was not unique. For example, any choice of $k$ such that $2.77261 \leq \ln(k) < 5.54524$, would correspond to a rejection region $C_r^* = \left\{\mathbf{x} : \sum_{i=1}^{20} x_i \geq k_*\right\}$ for $k_* \in (8,9]$ that would have defined a UMP rejection region of level .01. This is due to the fact that none of the values of $\sum_{i=1}^{20} x_i \in (8,9)$, or equivalently, none of the points in $\{\mathbf{x}$: $2.77261 < \ln(f(\mathbf{x};\Theta_0)/f(\mathbf{x};\Theta_a)) < 5.54524\}$ are in the support of $f(\mathbf{x};\Theta_0)$, i.e., the event $\sum_{i=1}^{20} x_i \in (8,9)$ has probability zero under $f(\mathbf{x};\Theta_0)$. Henceforth, we will always assume that when applying the Neyman-Pearson Lemma, the value of $k$ is chosen so that $\mathbf{x}$-values in the event $\{\mathbf{x}$: $f(\mathbf{x};\Theta_0) = kf(\mathbf{x};\Theta_a)\}$ are in the support of $f(\mathbf{x};\Theta_0)$. Nothing of any practical consequence is lost by the assumption, and it allows one to avoid an irrelevant ambiguity in the definition of UMP tests.

**Example 9.11**
***UMP Test of Two Exponential Means Using Neyman-Pearson***

Your company is about to ship ten LCD screens to a notebook computer manufacturing firm. The manufacturer requested that the screens have an expected operating life of 10,000 hours. Your company manufactures two types of screens—one having a 10,000-hour mean life, and one having a 50,000-hour mean life. The label on the batch of ten screens you are about to send to the manufacturer is missing, and it is not known which type of screen you are about to send (although they were all taken from one of the screen production lines, so it is known they are all of the same type). The screens with a mean life of 50,000 hours cost considerably more to produce, and so you would rather not send the batch of ten screens if they were the more expensive variety. The lifetimes of the screens can be nondestructively and inexpensively determined by a test which you perform. Having observed the lifetime of the screens, you wish to test the hypothesis that the mean lifetime of the screens is 10,000 hours versus the hypothesis that their mean lifetime is 50,000 hours. The lifetime distribution of each type of screen belongs to the exponential family of distributions as

$$X_i \sim \theta^{-1} e^{-x_i/\theta} I_{(0,\infty)}(x_i), \theta \in \Omega = \{1, 5\}$$

where $x_i$ is measured in 10,000 hour units.

Let $X = (X_1, \ldots, X_{10})$ be the random sample of size 10 from the appropriate (unknown) exponential population distribution, so that

$$f(\mathbf{x}, \theta) = \theta^{-10} \exp\left( -\sum_{i=1}^{10} x_i/\theta \right) \prod_{i=1}^{10} I_{(0,\infty)}(x_i), \theta \in \Omega = \{1, 5\}.$$

In order to define the most powerful level $\alpha$ test of the hypothesis $H_0$: $\theta = 1$ versus the hypothesis $H_a$: $\theta = 5$, use the Neyman-Pearson Lemma and examine

$$\begin{aligned} C_r(k) &= \{\mathbf{x} : f(\mathbf{x}; 1)/f(\mathbf{x}; 5) \le k\} \\ &= \left\{ \mathbf{x} : (.2)^{-10} \exp\left( -.8 \sum_{i=1}^{10} x_i \right) \le k \right\} \\ &= \left\{ \mathbf{x} : -10 \ln(.2) - .8 \sum_{i=1}^{10} x_i \le \ln(k) \right\} \\ &= \left\{ \mathbf{x} : \sum_{i=1}^{10} x_i \ge 20.11797 - 1.25 \ln(k) \right\} \text{(to five decimal places).} \end{aligned}$$

Since $Z = \sum_{i=1}^{10} X_i$ has a gamma distribution with parameters $\alpha = 10$ and $\beta = \theta$, the relationship between the choice of $k$ and the size of the test implied by $C_r(k)$ is given by

$$\alpha(k) = P(\mathbf{x} \in C_r(k); \theta = 1) = \int_{20.11797 - 1.25 \ln(k)}^{\infty} \frac{1}{\Gamma(10)} z^9 \exp(-z) dz,$$

where we are integrating the gamma distribution with $\alpha = 10$ and $\beta = 1$. Given the continuity of the random variables involved, any choice for the size of the test, $\alpha$, within the range (0,1) is possible. For example, to define the most powerful statistical test of $H_0$: $\theta = 1$ versus $H_a$: $\theta = 5$ having level .05, the value of $k$ would be found by solving the integral equation $\alpha(k) = .05$ for $k$ (recall the definition of $\alpha(k)$ above). Through the use of numerical integration on a computer, it can be shown that

$$\int_{15.70522}^{\infty} \frac{1}{\Gamma(10)} z^9 \exp(-z) dz = .05 \text{(to five decimal places).}$$

Therefore, $k = 34.13079$ (to five decimal places), and the most powerful statistical test of level .05 is given by

$\mathbf{x} \in C_r \Rightarrow$ reject $H_0 : \theta = 1$ and $\mathbf{x} \notin C_r \Rightarrow$ do not reject $H_0 : \theta = 1$

where

$C_r = \{\mathbf{x} : f(\mathbf{x}; 1)/f(\mathbf{x}; 5) \leq 34.13079\}$.

The rejection region can be represented more simply using the test statistic $t(\mathbf{X}) = \sum_{i=1}^{n} X_i$ as $C_r = \left\{\mathbf{x} : \sum_{i=1}^{10} x_i \geq 15.70522\right\}$.

Suppose the sum of the lifetimes of the ten screens ultimately had an outcome equal to $\sum_{i=1}^{10} x_i = 54.4$. It follows that $\mathbf{x} \in C_r$, and the hypothesis $H_0$: $\theta = 1$ is *rejected* at the .05 level. You conclude that the batch of ten screens are the type having a mean life of 50,000 hours. (Issue to consider: Would it be better if the *null* hypothesis were $H_0$: $\theta = 5$ and the *alternative* hypothesis were $H_a$: $\theta = 1$ in performing the statistical test? Why or why not?) □

The density ratio implied in the statement of the Neyman-Pearson Lemma can itself be viewed as a test statistic, i.e. $t(\mathbf{X}) = f(\mathbf{X};\boldsymbol{\Theta}_a)/f(\mathbf{X};\boldsymbol{\Theta}_0)$ is a test statistic for given (hypothesized) values of $\boldsymbol{\Theta}_0$ and $\boldsymbol{\Theta}_a$. The test implied by the lemma can then be conducted by determining whether $t(\mathbf{x}) \geq c = k^{-1}$. We also note that the Neyman-Pearson Lemma can be (and often is) stated in terms of likelihood functions instead of probability density functions. The alternative statement of the lemma then utilizes

$C_r(k) = \{\mathbf{x} : L(\boldsymbol{\Theta}_0; \mathbf{x}) \leq kL(\boldsymbol{\Theta}_a; \mathbf{x})\}$, for k>0

in the statement of Theorem 9.1. Of course, the proof of the restated lemma would be identical with that of Theorem 9.1 with $f(\mathbf{x};\boldsymbol{\Theta}_0)$ and $f(\mathbf{x};\boldsymbol{\Theta}_a)$ replaced by $L(\boldsymbol{\Theta}_0;\mathbf{x})$ and $L(\boldsymbol{\Theta}_a;\mathbf{x})$. When the rejection region is expressed in terms of *likelihood ratio values* as $C_r = \{\mathbf{x}\text{: } L(\boldsymbol{\Theta}_a;\mathbf{x})/L(\boldsymbol{\Theta}_0;\mathbf{x}) \geq k^{-1}\}$, the test implied by the lemma is then referred to as a **likelihood ratio test**.

A most powerful test of $H_0$: $\boldsymbol{\Theta} = \boldsymbol{\Theta}_0$ versus $H_a$: $\boldsymbol{\Theta} = \boldsymbol{\Theta}_a$ defined via the Neyman-Pearson Lemma is also an unbiased test, as we state formally below.

**Theorem 9.2 *Unbiasedness of Neyman-Pearson Most Powerful Test of $H_0 : \boldsymbol{\Theta} = \boldsymbol{\Theta}_0$ versus $H_a : \boldsymbol{\Theta} = \boldsymbol{\Theta}_a$***

*Let $C_r$ represent a most powerful level $\alpha$ rejection region defined by the Neyman-Pearson Lemma for testing $H_0$: $\boldsymbol{\Theta} = \boldsymbol{\Theta}_0$ versus $H_a$: $\boldsymbol{\Theta} = \boldsymbol{\Theta}_a$. Then the test implied by $C_r$ is unbiased.*

**Proof** Let $Z$ represent a Bernoulli random variable that is independent of the random sample $\mathbf{X}$ and for which $p(z = 1) = \alpha$ and $p(z = 0) = 1 - \alpha$. Suppose that regardless (i.e., independent) of the outcome of $\mathbf{X}$, the hypothesis $H_0$ is subsequently rejected or accepted based on whether the outcome of $Z$ is in the rejection region $C_r^* = \{1\}$. Let $\mathrm{E}_{\mathbf{X}}$ and $\mathrm{E}_Z$ represent expectations taken with respect to the probability distributions of $\mathbf{X}$ and $Z$, respectively. It follows by the independence of $\mathbf{X}$ and $Z$ that $E_Z(I_{\{1\}}(Z)|\mathbf{x}) = \alpha$ is the probability of rejecting $H_0$ conditional on any given outcome of $\mathbf{X}$, regardless of the value of $\boldsymbol{\Theta} \in \{\boldsymbol{\Theta}_0, \boldsymbol{\Theta}_a\}$. Then the (unconditional) probability of rejecting $H_0$ is given by an application of the iterated expectation theorem as

$$\pi_{C_r^*}(\boldsymbol{\Theta}) = P_*(\text{reject } H_0; \boldsymbol{\Theta}) = E_X E_Z(I_{(1)}(Z)|\mathbf{X}) = \alpha, \quad \boldsymbol{\Theta} \in \{\boldsymbol{\Theta}_0, \boldsymbol{\Theta}_a\}.$$

Now let $C_r$ represent a most powerful level $\alpha$ test of $H_0$: $\boldsymbol{\Theta} = \boldsymbol{\Theta}_0$ versus $H_a$: $\boldsymbol{\Theta} = \boldsymbol{\Theta}_a$ defined via the Neyman-Pearson Lemma. Using an argument analogous to that used in the sufficiency proof of the lemma (replacing $I_{C_r^*}(\mathbf{x})$ with $I_{\{1\}}(z)$),

$$[I_{C_r}(x) - I_{\{1\}}(z)]f(\mathbf{x}; \boldsymbol{\Theta}_a)h(z) \geq k^{-1}[I_{C_r}(x) - I_{\{1\}}(z)]f(\mathbf{x}; \boldsymbol{\Theta}_0)h(z), \quad \forall \mathbf{x}, z,$$

where $h(z)$ represents the Bernoulli density function for $Z$. Integrating both sides of the inequality over $\mathbf{x} \in \mathbb{R}^n$ if $\mathbf{X}$ is continuous or else summing over all $\mathbf{x}$-values for which $f(\mathbf{x};\boldsymbol{\Theta}_0) > 0$ or $f(\mathbf{x};\boldsymbol{\Theta}_a) > 0$ if $\mathbf{X}$ is discrete, and then summing over the range of $Z$ obtains

$$P(\mathbf{x} \in C_r; \boldsymbol{\Theta}_a) - \alpha \geq k^{-1}[P(\mathbf{x} \in C_r; \boldsymbol{\Theta}_0) - \alpha] = 0$$

where the right hand side of the inequality equals zero because $C_r$ is an $\alpha$-size rejection region. Then $P(\mathbf{x} \in C_r;\boldsymbol{\Theta}_a) \geq \alpha$, so that the test implied by $C_r$ is unbiased. ■

The significance of Theorem 9.2 for applications is that once a most powerful level $\alpha$ test of $H_0$: $\boldsymbol{\Theta} = \boldsymbol{\Theta}_0$ versus $H_a$: $\boldsymbol{\Theta} = \boldsymbol{\Theta}_a$ has been derived via the Neyman-Pearson Lemma, there is no need to check whether the test is unbiased, since such tests are *always* unbiased. As an illustration, the tests in Examples 9.10–9.11 are unbiased because they were most powerful Neyman-Pearson tests. We note, however, that unbiased tests are not necessarily most powerful.

*Composite Hypotheses* In some cases, the Neyman-Pearson Lemma can also be used to identify UMP tests when the alternative hypotheses is composite. The basic idea is to show via the Neyman-Pearson Lemma that the rejection region, $C_r$, of the most powerful test of $H_0$: $\boldsymbol{\Theta} = \boldsymbol{\Theta}_0$ versus $H_a$: $\boldsymbol{\Theta} = \boldsymbol{\Theta}_a$ is the *same* $\forall\ \boldsymbol{\Theta}_a \in \Omega_a$, where $\Omega_a$ is the set of alternative hypothesis values. It would then follow that the rejection region defines a *uniformly* most powerful test of $H_0$: $\boldsymbol{\Theta} = \boldsymbol{\Theta}_0$ versus $H_a$: $\boldsymbol{\Theta} \in \Omega_a$, since the *same* rejection region would be most powerful for each and every $\boldsymbol{\Theta} \in \Omega_a$, i.e., uniformly in $\boldsymbol{\Theta} \in \Omega_a$. We formalize the approach in the following theorem.

**Theorem 9.3** ***UMP Test of $H_0: \boldsymbol{\Theta} = \boldsymbol{\Theta}_0$ versus $H_a: \boldsymbol{\Theta} \in \Omega_a$ Using Neyman-Pearson Lemma***

*The given rejection region, $C_r$, of a statistical test of $H_0$: $\boldsymbol{\Theta} = \boldsymbol{\Theta}_0$ versus $H_a$: $\boldsymbol{\Theta} \in \Omega_a$ defines a UMP level $\alpha$ test if $P(\mathbf{x} \in C_r;\boldsymbol{\Theta}_0) = \alpha$ and $\exists\ k(\boldsymbol{\Theta}_a) \geq 0$ such that the given rejection region can be defined equivalently by $C_r = \{\mathbf{x}: f(\mathbf{x};\boldsymbol{\Theta}_0)/f(\mathbf{x};\boldsymbol{\Theta}_a) \leq k(\boldsymbol{\Theta}_a)\}$, $\forall\ \boldsymbol{\Theta}_a \in \Omega_a$. Furthermore, $C_r$ is then the unique (with probability 1) UMP rejection region of size $\alpha$.*

**Proof** Because $C_r$ is defined via the Neyman-Pearson Lemma, it represents the most powerful level $\alpha$ test of $H_0$: $\mathbf{\Theta} = \mathbf{\Theta}_0$ *versus* $H_a$: $\mathbf{\Theta} = \mathbf{\Theta}_a$. Since the same and unchanged rejection region, $C_r$, applies for any $\mathbf{\Theta}_a \in \Omega_a$, the test defined by $C_r$ is also most powerful for any choice of $\mathbf{\Theta}_a \in \Omega_a$, and is thus a UMP level $\alpha$ test of $H_0$: $\mathbf{\Theta} = \mathbf{\Theta}_0$ versus $H_a$: $\mathbf{\Theta} \in \Omega_a$. Finally, $C_r$ is the unique size $\alpha$ rejection region (with probability 1) since the most powerful rejection region of size $\alpha$ is unique (with probability 1) by the Neyman-Pearson Lemma $\forall \mathbf{\Theta}_a$. ■

**Example 9.12** ***UMP Test of Bernoulli Mean Against Composite Alternative***

Recall Example 9.10 in which hypotheses regarding the effectiveness of a hair restoring treatment were being analyzed. Examine the problem of testing the simple hypothesis $H_0$: $p = .2$ versus the composite alternative hypothesis $H_a$: $p > .2$ using a UMP test having level .01. Let $p_a$ represent any choice of $p \in (.2,1]$, and define the rejection region of the test of $H_0$: $p = .2$ versus $H_a$: $p = p_a$ using the Neyman-Pearson Lemma:

$$\begin{aligned} C_r &= \{\mathbf{x} : f(\mathbf{x}; .2)/f(\mathbf{x}; p_a) \leq k(p_a)\} \\ &= \left\{\mathbf{x} : (.2/p_a)^{\sum_{i=1}^{20} x_i} (.8/(1-p_a))^{20-\sum_{i=1}^{20} x_i} \leq k(p_a)\right\} \\ &= \left\{\mathbf{x} : \sum_{i=1}^{20} x_i \geq \frac{\ln(k(p_a)) - 20\ln(.8/(1-p_a))}{\ln(.25(1-p_a)/p_a)}\right\}. \end{aligned}$$

Using reasoning identical to that used in Example 9.10, $C_r$ will define a size .01 test *iff*

$$C_r = \left\{\mathbf{x} : \sum_{i=1}^{20} x_i \geq 9\right\}$$

*regardless* of the value of $p_a$, since under $H_0$ the statistic $\sum_{i=1}^{20} X_i$ has a binomial distribution with $n = 20$ and $p = .2$. The nonnegative values of $k(p_a)\ \forall p_a \in (.2,1]$ can be solved for accordingly, and thus by Theorem 9.3 $C_r = \left\{\mathbf{x} : \sum_{i=1}^{20} x_i \geq 9\right\}$ defines a UMP level .01 test of $H_0$: $p = .2$ versus $H_a$: $p > .2$. □

**Example 9.13** ***UMP Test of Exponential Mean Against Composite Alternative***

Recall Example 9.11 in which hypotheses regarding the operating lives of screens for notebook computers were being analyzed. Examine the problem of testing the simple hypothesis $H_0$: $\theta = 1$ versus the composite hypothesis $H_a$: $\theta > 1$ using a UMP test having level .05. Let $\theta_a$ represent any choice of $\theta > 1$, and define the rejection region of the test of $H_0$: $\theta = 1$ versus $H_a$: $\theta = \theta_a$ using the Neyman-Pearson Lemma:

$$\begin{aligned} C_r &= \{\mathbf{x} : f(\mathbf{x}; 1)/f(\mathbf{x}; \theta_a) \leq k(\theta_a)\} \\ &= \left\{\mathbf{x} : \theta_a^{10} \exp\left(-\sum_{i=1}^{10} x_i\left[1-\theta_a^{-1}\right]\right) \leq k(\theta_a)\right\} \\ &= \left\{\mathbf{x} : \sum_{i=1}^{10} x_i \geq \frac{10\ln(\theta_a) - \ln(k(\theta_a))}{(1-\theta_a^{-1})}\right\} \end{aligned}$$

Using reasoning identical to that used in Example 9.11, $C_r$ will define a size .05 test *iff*

$$C_r = \left\{ \mathbf{x} : \sum_{i=1}^{10} x_i \geq 15.70522 \right\}$$

*regardless* of the value of $\theta_a$, because under $H_0$ the statistic $\sum_{i=1}^{10} X_i$ has a gamma distribution with parameters $\alpha = 10$ and $\beta = 1$. The nonnegative values of $k(\theta_a)$ $\forall \theta_a \in (1,\infty)$ can be solved for accordingly, and thus by Theorem 9.3, $C_r = \left\{ \mathbf{x} : \sum_{i=1}^{10} x_i \geq 15.70522 \right\}$ defines a UMP level .05 test of $H_0$: $\theta = 1$ versus $H_a$: $\theta > 1$. □

The uniqueness result of Theorem 9.3 can also be used to demonstrate that a UMP test of $H_0$: $\boldsymbol{\Theta} = \boldsymbol{\Theta}_0$ versus $H_a$: $\boldsymbol{\Theta} \in \Omega_a$ having size $\alpha$ does *not* exist, as in the following example.

**Example 9.14** ***Demonstrating Nonexistence of a UMP Test***

An untrusting gambler wishes to use a size .10 test that is also a UMP level .10 test of the hypothesis that a roulette wheel being used for betting purposes in an Atlantic City Casino is fair. The gambler suggests that the wheel be spun 100 times, and a random sample of red and black outcomes be used to test the "fairness" hypothesis. The joint density of the random sample in this case is given by

$$f(\mathbf{x}; p) = p^{\sum_{i=1}^{100} x_i} (1-p)^{100 - \sum_{i=1}^{100} x_i} \prod_{i=1}^{100} I_{\{0,1\}}(x_i)$$

assuming the red/black outcomes are *iid* Bernoulli trials, where $x_i = 1$ denotes red and $x_i = 0$ denotes black. The null hypothesis to be tested is $H_0$: $p = .5$ versus the alternative hypothesis that $H_a$: $p \neq .5$. In the notation of Theorem 9.43, $\Omega_a = [0,1] - \{.5\}$. Following Theorem 9.3, define a rejection region as

$$\begin{aligned} C_r &= \left\{ \mathbf{x} : \frac{f(\mathbf{x}; .5)}{f(\mathbf{x}; p_a)} \leq k(p_a) \right\} \\ &= \left\{ \mathbf{x} : (.5)^{100} / \left[ p_a^{\sum_{i=1}^{100} x_i} (1-p_a)^{100 - \sum_{i=1}^{100} x_i} \right] \leq k(p_a) \right\} \\ &= \left\{ \mathbf{x} : \ln(p_a/(1-p_a)) \sum_{i=1}^{100} x_i \geq \gamma(p_a) \right\}, \end{aligned}$$

where $\gamma(p_a) = 100 \ln(.5/(1 - p_a)) - \ln(k(p_a))$. The rejection region depends on $p_a$ (i.e., $C_r$ is *not* the same $\forall\, p_a \in \Omega_a$), and thus there is *no* UMP level .10 statistical test of the "fairness" hypothesis that has size .10.

To show how $C_r$ depends on $p_a$, note that for $p_a > .5$, $\ln(p_a/(1 - p_a)) > 0$, while for $p_a < .5$, $\ln(p_a/(1 - p_a)) < 0$. It follows that

$$C_r = \left\{ \mathbf{x} : \sum_{i=1}^{100} x_i \geq \eta(p_a) \right\} \text{ if } p_a > .5$$

or

$$C_r = \left\{ \mathbf{x} : \sum_{i=1}^{100} x_i \leq \eta(p_a) \right\} \text{ if } p_a < .5,$$

where $\eta(p_a) = \gamma(p_a)/\ln(p_a/(1-p_a))$. Given the test size of .10, the reader can show by using the binomial distribution of $\sum_{i=1}^{n} X_i$ with parameters $n = 100$ and $p = .5$ that $P\left(\sum_{i=1}^{100} x_i \geq 57; p = .5\right) = P\left(\sum_{i=1}^{100} x_i \leq 43; p = .5\right) = .10$ (to two decimal places). Thus, for $P(\mathbf{x} \in C_r; 5) = .10$, *two different* rejection regions are defined as

$$C_r = \left\{ \mathbf{x} : \sum_{i=1}^{100} x_i \geq 57 \right\} \text{ for } p_a > .5$$

or

$$C_r = \left\{ \mathbf{x} : \sum_{i=1}^{100} x_i \leq 43 \right\} \text{ for } p_a < .5.$$

For values of $p_a > .5$, the first rejection region is UMP of level .10 by Theorem 9.3, while for values of $p_a < .5$ the second rejection region is UMP of level .10 by Theorem 9.3. Since the UMP level .10 rejection regions for $p_a > .5$ and $p_a < .5$ are not the same, no UMP level .10 rejection region exists for testing the fairness hypothesis $H_0$: $p = .5$ versus $H_a$: $p \neq .5$. □

A UMP level $\alpha$ test of $H_0$: $\boldsymbol{\Theta} = \boldsymbol{\Theta}_0$ versus $H_a$: $\boldsymbol{\Theta} \in \Omega_a$ defined using Theorem 9.3 is also an unbiased test, as indicated in the following extension of Theorem 9.2.

**Theorem 9.4 Unbiasedness of Uniformly Most Powerful Test of** $H_0 : \boldsymbol{\Theta} = \boldsymbol{\Theta}_0$ *versus* $H_a : \boldsymbol{\Theta} \in \Omega_a$

*Let $C_r$ represent a UMP level $\alpha$ rejection region for testing $H_0$: $\boldsymbol{\Theta} = \boldsymbol{\Theta}_0$ versus $H_a$: $\boldsymbol{\Theta} \in \Omega_a$ defined using the Neyman-Pearson Lemma as indicated in Theorem 9.3. Then the test implied by $C_r$ is unbiased.*

**Proof** Let $\boldsymbol{\Theta}_a$ be any choice of $\boldsymbol{\Theta} \in \Omega_a$. Using an argument analogous to the proof of Theorem 9.2, it can be shown that $P(\mathbf{x} \in C_r; \boldsymbol{\Theta}_a) - \alpha \geq k(\boldsymbol{\Theta}_a)^{-1} [P(\mathbf{x} \in C_r; \boldsymbol{\Theta}_0) - \alpha] = 0$ where, again, the right hand side of the inequality is zero because $C_r$ is a size $\alpha$ rejection region. Since this holds $\forall\, \boldsymbol{\Theta}_a \in \Omega_a$, it follows that $P(\mathbf{x} \in C_r; \boldsymbol{\Theta}_a) \geq \alpha\ \forall\, \boldsymbol{\Theta}_a \in \Omega_a$, so that the test implied by $C_r$ is unbiased. ■

The theorem implies that once a UMP level $\alpha$ test of $H_0$: $\boldsymbol{\Theta} = \boldsymbol{\Theta}_0$ versus $H_a$: $\boldsymbol{\Theta} \in \Omega_a$ has been found via the Neyman-Pearson approach, one need not check to see whether the test is unbiased, since such tests are *always* unbiased. Examples 9.12 and 9.13 illustrate this fact, where both UMP level $\alpha$ tests are also unbiased tests. However, we underscore that an unbiased test is not necessarily a UMP test.

*One-Sided and Two-Sided Alternative Hypotheses* The previous three examples illustrate the concept of **one-sided** and **two-sided alternative hypotheses**. In the current context, a **one-sided alternative hypothesis** is such that *either* $\boldsymbol{\Theta}_a > \boldsymbol{\Theta}_0$ $\forall\, \boldsymbol{\Theta}_a \in \Omega_a$, *or* $\boldsymbol{\Theta}_a < \boldsymbol{\Theta}_0\, \forall\, \boldsymbol{\Theta}_a \in \Omega_a$, i.e., all values of $\boldsymbol{\Theta}_a \in \Omega_a$ are larger than $\boldsymbol{\Theta}_0$, or else they are all smaller than $\boldsymbol{\Theta}_0$, so that the values of $\boldsymbol{\Theta}_a$ are all on "one-side" of $\boldsymbol{\Theta}_0$. A **two-sided alternative hypothesis** is such that $\exists\, \boldsymbol{\Theta}_a \in \Omega_a$ for which $\boldsymbol{\Theta}_a > \boldsymbol{\Theta}_0$ *and* $\exists\, \boldsymbol{\Theta}_a \in \Omega_a$ for which $\boldsymbol{\Theta}_a < \boldsymbol{\Theta}_0$, i.e., some values of $\boldsymbol{\Theta}_a \in \Omega_a$ are larger than $\boldsymbol{\Theta}_0$ *and* some values of $\boldsymbol{\Theta}_a \in \Omega_a$ are smaller than $\boldsymbol{\Theta}_0$, so there are values of $\boldsymbol{\Theta}_a$ on "both-sides" of $\boldsymbol{\Theta}_0$. In practice, UMP level $\alpha$ tests of simple hypotheses versus *one-sided alternative hypotheses* often exist when $\boldsymbol{\Theta}$ is a scalar, but such is not the case when the alternative hypothesis is two-sided. In the latter case, one must generally resort to seeking a UMP test within a smaller class of tests, such as the class of unbiased tests. We will examine this case in more detail later.

### 9.5.2 Monotone Likelihood Ratio Approach

Results in this section will be useful for constructing UMP level $\alpha$ tests of the *composite* null hypothesis $H_0$: $\boldsymbol{\Theta} \leq \boldsymbol{\Theta}_0$ (or $H_0$: $\boldsymbol{\Theta} \geq \boldsymbol{\Theta}_0$) versus the composite *one-sided alternative* hypothesis $H_a$: $\boldsymbol{\Theta} > \boldsymbol{\Theta}_0$ (or $H_a$: $\boldsymbol{\Theta} < \boldsymbol{\Theta}_0$). The results in this section apply equally well to the case where $H_0$: $\boldsymbol{\Theta} = \boldsymbol{\Theta}_0$, and thus, this section can be interpreted as also providing additional results for testing $H_0$: $\boldsymbol{\Theta} = \boldsymbol{\Theta}_0$ versus the composite *one-sided alternative* $H_a$: $\boldsymbol{\Theta} > \boldsymbol{\Theta}_0$ (or $H_a$: $\boldsymbol{\Theta} < \boldsymbol{\Theta}_0$).

The procedure for defining UMP level $\alpha$ tests that we will present relies on the concept of a **monotone likelihood ratio in the statistic** $T = t(\mathbf{X})$.

**Definition 9.13** ***Monotone Likelihood Ratio in the Statistic*** $T = t(\mathbf{X})$

Let $f(\mathbf{x};\Theta)$, $\Theta \in \Omega$, be a family of probability density functions indexed by a scalar parameter $\Theta$. The family of density functions is said to have a **monotone likelihood ratio in the statistic** $T = t(\mathbf{X})$ *iff* $\forall\, \Theta_1, \Theta_2 \in \Omega$ for which $\Theta_1 > \Theta_2$, the likelihood ratio $L(\Theta_1;\mathbf{x})/L(\Theta_2;\mathbf{x})$ is a nondecreasing function of $t(\mathbf{x})\ \forall\, \mathbf{x} \in \{\mathbf{x}: f(\mathbf{x};\Theta_1) > 0 \text{ and/or } f(\mathbf{x};\Theta_2) > 0\}$.

Verification of whether a family of density functions has a monotone likelihood ratio in some statistic $t(\mathbf{X})$ generally requires some ingenuity. However, if the family of density functions belongs to the exponential class of densities the verification process is often simplified by the following result.

**Theorem 9.5 Monotone Likelihood Ratio and the Exponential Class of Densities** *Let $f(\mathbf{x};\Theta)$, $\Theta \in \Omega$, be a one-parameter exponential class family of densities*

$$f(\mathbf{x};\Theta) = \exp(c(\Theta)g(\mathbf{x}) + d(\Theta) + z(\mathbf{x}))I_A(\mathbf{x}), \Theta \in \Omega.$$

*If $c(\Theta)$ is a nondecreasing function of $\Theta$, then $f(\mathbf{x};\Theta)$, $\Theta \in \Omega$ has a monotone likelihood ratio in the statistic $g(\mathbf{X})$.*

**Proof** Let $\Theta_1 > \Theta_2$, and $\forall \mathbf{x} \in A$ examine the likelihood ratio

$$\begin{aligned} L(\Theta_1;\mathbf{x})/L(\Theta_2;\mathbf{x}) &= \exp([c(\Theta_1) - c(\Theta_2)]g(\mathbf{x}) + d(\Theta_1) - d(\Theta_2)) = \eta(g(\mathbf{x})) \\ &= \exp(c^* g(\mathbf{x}) + d^*) \end{aligned}$$

where $c_* = c(\Theta_1) - c(\Theta_2)$ and $d_* = d(\Theta_1) - d(\Theta_2)$. Since $c(\Theta)$ is a nondecreasing function of $\Theta$, $c_* \geq 0$ so that the likelihood ratio can be expressed as a nondecreasing function of $g(\mathbf{x})$, $\forall\, \mathbf{x} \in A$. ■

The following examples illustrate the use of Theorem 9.5 for verifying the monotone likelihood ratio property.

**Example 9.15 Monotone Likelihood Ratio for Bernoulli Distribution** Let $(X_1,\ldots,X_n)$ be a random sample from a Bernoulli distribution representing the population of television viewers in a certain region who can $(x_i = 1)$ or cannot $(x_i = 0)$ recall seeing a certain television commercial. Then

$$\begin{aligned} f(\mathbf{x};p) &= p^{\sum_{i=1}^n x_i}(1-p)^{n-\sum_{i=1}^n x_i}\prod_{i=1}^n I_{\{0,1\}}(x_i) \\ &= \exp(c(p)g(\mathbf{x}) + d(p) + z(\mathbf{x}))I_A(x), \quad p \in (0,1), \end{aligned}$$

where $c(p) = \ln(p/(1-p))$, $g(\mathbf{x}) = \sum_{i=1}^n x_i$, $d(p) = n\ln(1-p)$, $z(\mathbf{x}) = 0$, and $A = \times_{i=1}^n \{0,1\}$. Then because $dc(p)/dp = [p(1-p)]^{-1} > 0$, $c(p)$ is strictly increasing in $p$, and $f(\mathbf{x};p)$, $p \in (0,1)$ has a monotone likelihood ratio in the statistic $g(\mathbf{X}) = \sum_{i=1}^n X_i$. □

**Example 9.16 Monotone Likelihood Ratio for the Gamma Distribution** Let $(X_1,\ldots,X_n)$ be a random sample from a gamma population distribution with $\beta = 2$ representing the survival time of cancer patients treated with a new form of chemotherapy. Then

$$\begin{aligned} f(\mathbf{x};\alpha) &= \frac{1}{2^{n\alpha}[\Gamma(\alpha)]^n}\left(\prod_{i=1}^n x_i^{\alpha-1}\right)e^{-\sum_{i=1}^n x_i/2}\prod_{i=1}^n I_{(0,\infty)}(x_i) \\ &= \exp(c(\alpha)g(\mathbf{x}) + d(\alpha) + z(\mathbf{x}))I_A(\mathbf{x}), \quad \alpha>0, \end{aligned}$$

where $c(\alpha) = \alpha - 1, g(\mathbf{x}) = \sum_{i=1}^{n} \ln(x_i), d(\alpha) = -n\ln(2^{\alpha}\Gamma(\alpha)), z(\mathbf{x}) = -(1/2)\sum_{i=1}^{n} x_i$, and $A = \times_{i=1}^{n}(0,\infty)$. Then because $dc(\alpha)/d\alpha = 1 > 0$, $c(\alpha)$ is strictly increasing in $\alpha$ and $f(\mathbf{x};\alpha)$, $\alpha > 0$, has a monotone likelihood ratio in the statistic $g(\mathbf{X}) = \sum_{i=1}^{n} \ln(X_i)$. □

In the next example, we illustrate verification of the monotone likelihood ratio property for a family of density functions that does *not* belong to the exponential class of densities.

**Example 9.17** ***Monotone Likelihood Ratio for the Hypergeometric Distribution***

Let $X$ have a hypergeometric probability density function in which the parameters $n$ and $M$ have known values $n_0$ and $M_0$, respectively, with $X$ representing the number of defectives found in a random sample, without replacement, of size $n_0$ from a shipment of $M_0$ DVDs. The density function for $X$ can then be represented as

$$f(x;K) = \frac{\binom{K}{x}\binom{M_0 - K}{n_0 - x}}{\binom{M_0}{n_0}} I_{\{0,1,2,\ldots,K\}}(x) \text{ for } K \in \{0,1,2,\ldots,M_0\}.$$

Examine the likelihood ratio

$$\frac{L(K;x)}{L(K-1;x)} = \left[\frac{K}{(K-x)}\right]\left[\frac{(M_0 - K - n_0 + 1 + x)}{(M_0 - K + 1)}\right]\left[\frac{I_{\{0,1,\ldots,K\}}(x)}{I_{\{0,1,\ldots,K-1\}}(x)}\right]$$

(recall Example 8.8), and note that the ratio is a nondecreasing function of $x \in \{0,1,\ldots,K\} = \{x\colon f(x;K) > 0 \text{ and/or } f(x;\ K-1) > 0\}$, because the ratio is a product of three nondecreasing functions (in brackets) of $x$. Now let $K_1 > K_2$, and note that the likelihood ratio

$$\begin{aligned}\frac{L(K_1;x)}{L(K_2;x)} &= \prod_{i=0}^{K_1-K_2-1}\left[\frac{L(K_1-i;x)}{L(K_1-(i+1);x)}\right] \\ &= \prod_{i=0}^{K_1-K_2-1}\left(\left[\frac{K_1-i}{K_1-i-x}\right]\left[\frac{(M_0-(K_1-i)-n_0+1+x)}{(M_0-(K_1-i)+1)}\right]\right) \text{for } x \in \{0,1,\ldots,K_2\} \\ &= \infty \text{ for } x \in \{K_2+1,\ldots,K_1\}\end{aligned}$$

is a nondecreasing function of $x \in \{0,1,\ldots,K_1\} = \{x\colon f(x;K_1) > 0 \text{ and/or } f(x;K_2) > 0\}$, because it is strictly increasing for $x \in \{0,1,\ldots,K_2\}$, and equals $\infty$ (and hence nondecreasing) for $x \in \{K_2 + 1,\ldots,K_1\}$. Then the hypergeometric family of densities $f(\mathbf{x};K)$, $K \in \{0,1,\ldots,M_0\}$, has a monotone likelihood ratio in the statistic $t(X) = X$. □

If the joint density function of the random sample can be shown to have a monotone likelihood ratio in some statistic, then UMP level $\alpha$ tests will exist for $H_0$: $\Theta \leq \Theta_0$ versus the one-sided alternative $H_a$: $\Theta > \Theta_0$, or for $H_0$: $\Theta \geq \Theta_0$ versus the one-sided alternative $H_a$: $\Theta < \Theta_0$, *or* for $H_0$: $\Theta = \Theta_0$ versus the two-sided alternative of *either* $H_a$: $\Theta > \Theta_0$ *or* $H_a$: $\Theta < \Theta_0$. In fact, it will be seen below that a monotone likelihood ratio is effectively a sufficient condition for the Neyman-Pearson approach of Theorem 9.3 to be applicable to a testing problem.

**Theorem 9.6 *Monotone Likelihood Ratios and UMP Level α Tests***

*Let $f(\mathbf{x};\Theta)$, $\Theta \in \Omega$, be a family of density functions having a monotone likelihood ratio in the statistic $t(\mathbf{X})$. Let $C_r = \{\mathbf{x}: f(\mathbf{x};\Theta_0) \leq k\, f(\mathbf{x};\Theta_a)\}$ define a size α rejection region for testing $H_0$: $\Theta = \Theta_0$ versus $H_a$: $\Theta = \Theta_a$.*

1. *If $\Theta_0 < \Theta_a$, $C_r$ is UMP level α for testing either*

   a. $H_0$: $\Theta = \Theta_0$ versus $H_a$: $\Theta > \Theta_0$, *or*
   b. $H_0$: $\Theta \leq \Theta_0$ versus $H_a$: $\Theta > \Theta_0$;

2. *If $\Theta_0 > \Theta_a$, then $C_r$ is UMP level α for testing either*

   a. $H_0$: $\Theta = \Theta_0$ versus $H_a$: $\Theta < \Theta_0$, *or*
   b. $H_a$: $\Theta \geq \Theta_0$ versus $H_a$: $\Theta < \Theta_0$.

**Proof** See Appendix. ■

The UMP level α rejection region defined by Theorem 9.6 can be defined alternatively in terms of the statistic $t(\mathbf{X})$ for which the likelihood ratio is monotone.

**Corollary 9.1 *UMP Level α Tests in Terms of the t(X) of a Monotone Likelihood Ratio***

*Let $f(\mathbf{x};\Theta), \Theta \in \Omega$, be a family of density functions having a monotone likelihood ratio in $t(\mathbf{X})$. Then*

1. $C_r = \{\mathbf{x}: t(\mathbf{x}) \geq c\}$, *for choice of c such that $P(t(\mathbf{x}) \geq c;\Theta_0) = \alpha$, is a UMP level α rejection region for testing $H_0$: $\Theta = \Theta_0$ versus $H_a$: $\Theta > \Theta_0$,* or *$H_0$: $\Theta \leq \Theta_0$ versus $H_a$: $\Theta > \Theta_0$.*
2. $C_r = \{\mathbf{x}: t(\mathbf{x}) \leq c\}$, *for choice of c such that $P(t(\mathbf{x}) \leq c;\Theta_0) = \alpha$, is a UMP level α rejection region for testing $H_0$: $\Theta = \Theta_0$ versus $H_a$: $\Theta < \Theta_0$,* or *$H_0$: $\Theta \geq \Theta_0$ versus $H_a$: $\Theta < \Theta_0$.*

**Proof** The $C_r$'s are equivalent representations of the Neyman-Pearson rejection regions referred to in Theorem 9.6. ■

It is also true that the UMP level α tests based on the monotone likelihood ratio procedure are unbiased tests. We state this fact as a second corollary to that theorem.

**Corollary 9.2 Unbiasedness of UMP Tests Based on Monotone Likelihood Ratios**

*The rejection region, $C_r$, defined in Theorem 9.6 and Corollary 9.1 defines an unbiased size $\alpha$ test of the respective hypotheses stated in the theorem.*

**Proof** This follows immediately from the characteristics of the power function of the test established in the proof of Theorem 9.6. ■

**Example 9.18 UMP Test in Exponential Distribution Based on Monotone Likelihood Ratio**

Your company manufactures personal computers and is in the process of evaluating the purchase of hard disk controllers from various input suppliers. Among other considerations, a necessary condition for a disk controller to be used in the manufacture of your PCs is that the controller have a minimum expected life of more than 50,000 operating hours. The lifetimes of all of the various brands of disk controllers are characterized by exponential densities, as

$$X_i \sim f(x_i;\theta) = \theta^{-1}\exp[-x_i/\theta]I_{(0,\infty)}(x),\ \ \theta>0,$$

where $x_i$ is measured in 10,000's of operating hours. Consider defining a UMP level .01 test of $H_0$: $\theta \leq 5$ versus $H_a$: $\theta > 5$ based on a random sample of 100 lifetimes of a given brand of disk controller.

The joint density of the random sample is given by $f(\mathbf{x};\theta) = \theta^{-100}\exp\left[-\sum_{i=1}^{100} x_i/\theta\right]\prod_{i=1}^{100} I_{(0,\infty)}(x_i),\ \ \theta>0$. The density is in the exponential class with $c(\theta) = -\theta^{-1}$, which is a nondecreasing function of $\theta$, so that $f(\mathbf{x};\theta)$ has a monotone likelihood ratio by Theorem 9.6. In particular, the likelihood ratio $L(\theta_1;\mathbf{x})/L(\theta_2;\mathbf{x}) = \exp\left[-\sum_{i=1}^{100} x_i(\theta_1^{-1} - \theta_2^{-1})\right]$ is a *strictly increasing* function of $t(\mathbf{x}) = \sum_{i=1}^{100} x_i$ for $\theta_1 > \theta_2\ \forall \mathbf{x} \in \times_{i=1}^{100}(0,\infty)$. Then Theorem 9.6 and its corollaries are applicable so that the UMP rejection region of level .01 is given by $C_r = \{\mathbf{x}\text{: } \sum_{i=1}^{100} x_i \geq c\}$ for $c$ chosen so that $P(\mathbf{x} \in C_r;\ \theta = 5) = .01$.

To calculate the appropriate value of $c$, first note that $Z = \sum_{i=1}^{100} X_i$ has a gamma distribution, with parameters $\alpha = 100$ and $\beta = \theta$.[15] Then c is the solution to the integral equation

$$\int_c^{\infty} \frac{1}{\theta^{100}\Gamma(100)} z^{99}\exp(-z/\theta)dz = .01$$

with $\theta = 5$, which can be found with the aid of a computer to be $c = 623.61$. The UMP level .01 and unbiased test of $H_0$: $\theta \leq 5$ versus $H_a$: $\theta > 5$ is then defined by the rejection region for the test statistic $T = t(\mathbf{X}) = \sum_{i=1}^{n} X_i$ given by $C_r^T = [623.61,\infty)$. An alternative test statistic for performing this UMP level .01

[15]This can be shown via the MGF approach, since the MGF of $\sum_{i=1}^{100} X_i = \prod_{i=1}^{100} M_{X_i}(t) = \prod_{i=1}^{100}(1-\theta t)^{-1} = (1-\theta t)^{-100}$ for $t<\theta^{-1}$, which is of the gamma form with $\beta = \theta$, $\alpha = 100$.

and unbiased test is the sample mean so that the test can also be conducted using the rejection region for the test statistic $\bar{X}$ given by $C_r^{\bar{x}} = [6.2361,\infty)$. □

**Example 9.19** ***UMP Test in Hypergeometric Distribution Based on Monotone Likelihood Ratio***

A shipment of 300 blank recordable DVDs arrives at your video store. You intend to randomly sample, *without replacement*, 25 DVDs and observe whether or not they are defective. On the basis of the sample outcome, you wish to test the hypothesis that the shipment of DVDs contains no more than 5 percent defectives. We seek a UMP and unbiased test of the hypothesis having level .05 and a size as close to .05 as possible.[16]

The density function associated with the number of defectives observed in a random sample, without replacement, of size 25 from the population of 300 DVDs can be defined as

$$f(x;K) = \frac{\binom{K}{x}\binom{300-K}{25-x}}{\binom{300}{25}} I_{\{0,1,2,\ldots,25\}}(x),$$

i.e., hypergeometric with $M = 300$, $n = 25$, and $K \in \{0,1,\ldots,300\}$. As we have shown in Example 9.17, this hypergeometric density has a monotone likelihood ratio in the statistic $t(X) = X$. Note that the likelihood ratio $L(K_1;x)/L(K_2;x)$, $K_1 > K_2$, is in fact *strictly increasing* in $t(x) = x$ for all values of $x$ in the support of $f(x;K_2)$, as was shown in Example 9.17. Theorem 9.6 and its corollaries are applicable so that the UMP rejection region of size $\alpha$ for testing $H_0$: $K \leq 15$ versus $H_a$: $K > 15$ is represented by $C_r = \{x\colon\ x \geq c\}$ for $c$ chosen so that $P(x \in C_r;\ K = 15) = \alpha$. Computing the values of the hypergeometric CDF on the computer reveals that $\alpha = .027$, when $c = 4$. This is the choice of $c$ that generates a value of $\alpha$ closest to .05 without exceeding it. Thus, a UMP level .05 and unbiased size .027 test of $H_0$: $K \leq 15$ versus $H_a$: $K > 15$ in a random sample, without replacement, of size 25 from the population of 300 DVDs is defined by the rejection region $C_r = [4, 25]$ for x. □

There does *not* generally exist a UMP level $\alpha$ test of the hypothesis $H_0$: $\Theta = \Theta_0$ versus the *two-sided* alternative $H_a$: $\Theta \neq \Theta_0$ in the case of monotone likelihood ratios. This follows from the fact that the UMP level $\alpha$ rejection region of the test of such a hypothesis is different, depending on "which side" of $\Theta_0$ the alternative value $\Theta$ is on. We demonstrate this phenomenon for the case where the monotone likelihood ratio statistic is a continuous random variable.

[16] As we noted previously, it is possible to use a *randomized test* to achieve a size of .05 exactly, but the test can depend on the outcome of a random variable that has nothing to do with the experiment being analyzed. See problem 9.8 for an example of this approach. Randomized tests are not often used in practice.

**Theorem 9.7 Nonexistence of UMP Level $\alpha$ Test of $H_0: \Theta = \Theta_0$ versus $H_a: \Theta \neq \Theta_0$ for Monotone Likelihood Ratios**

*Let $f(\mathbf{x};\Theta)$, $\Theta \in \Omega$, be a family of PDFs having a monotone likelihood ratio in $t(\mathbf{x})$ where $t(\mathbf{X})$ is a continuous random variable. Then there does not exist a UMP level $\alpha \in (0, 1)$ rejection region for testing $H_0$: $\Theta = \Theta_0$* versus *$H_a$: $\Theta \neq \Theta_0$.*

**Proof** From Theorem 9.6 and Corollary 9.1, the UMP size $\alpha$ rejection region for testing $H_0$: $\Theta = \Theta_0$ versus $H_a$: $\Theta > \Theta_0$ is of the form $C_r = \{\mathbf{x}: t(\mathbf{x}) \geq c\}$ with $P(t(\mathbf{x}) \geq c) = \alpha$ while the UMP size $\alpha$ rejection region for testing $H_0$: $\Theta = \Theta_0$ versus $H_a$: $\Theta < \Theta_0$ is of the form $C_r = \{\mathbf{x}: t(\mathbf{x}) \leq c\}$ with $P(t(\mathbf{x}) \leq c) = \alpha$. Appropriate choices of $c$ for either set of hypotheses exist $\forall \alpha \in (0,1)$ since $t(\mathbf{X})$ is a continuous random variable. It follows that there is no rejection region, $C_r^*$, that is UMP level $\alpha$ for both $\Theta < \Theta_0$ and $\Theta > \Theta_0$. ■

An implication of Theorem 9.7 is that we must consider alternative criteria than the UMP property when defining statistical tests of $H_0$: $\Theta = \Theta_0$ versus $H_0$: $\Theta \neq \Theta_0$ and the sampling density has a monotone likelihood ratio. A similar, albeit somewhat more complicated, argument can also be made when $t(\mathbf{X})$ is a discrete random variable. In the next subsection we will examine situations in which UMP level $\alpha$ tests of $H_0$: $\Theta = \Theta_0$ versus $H_a$: $\Theta \neq \Theta_0$ exist *within the unbiased class of tests.*[17] A UMP test within the class of unbiased tests will be referred to as a **Uniformly Most Powerful Unbiased (UMPU) test.**

### 9.5.3 UMP and UMPU Tests in the Exponential Class of Densities

For cases where the joint density of the probability sample belongs to the exponential class of densities with a scalar parameter $\Theta$, our preceding discussion of monotone likelihood ratios can be used to justify the general existence of UMP level $\alpha$ tests for the case of the one-sided alternatives

$H_0 : \Theta = \Theta_0$ versus $H_a : \Theta > \Theta_0$,
$H_0 : \Theta = \Theta_0$ versus $H_a : \Theta < \Theta_0$,
$H_0 : \Theta \leq \Theta_0$ versus $H_a : \Theta > \Theta_0$,
$H_0 : \Theta \geq \Theta_0$ versus $H_a : \Theta < \Theta_0$,

and the general *nonexistence* of a UMP level $\alpha$ test for the case of the two-sided alternative

$H_0 : \Theta = \Theta_0$ versus $\mathrm{H}_a : \Theta \neq \Theta_0$.

[17] To this point, we have established UMP tests in the class of *all* tests of a certain level $\alpha$. The reader should note that in all cases examined heretofore, we have shown that UMP level $\alpha$ tests were also unbiased. This is clearly different than examining *only unbiased* tests of a certain level $\alpha$, and within this restricted set of tests, attempting to find one that is UMP.

We now reconsider the problem of defining a test of the latter hypothesis. We will find that if the joint density of the probability sample belongs to the exponential class with a scalar parameter $\Theta$, then it will generally be possible to find a rejection region that is UMP level $\alpha$ within the class of *unbiased* rejection regions for testing $H_0$: $\Theta = \Theta_0$ versus $H_a$: $\Theta \neq \Theta_0$. We will also establish UMPU rejection regions for testing $H_0$: $\Theta_1 \leq \Theta \leq \Theta_2$ versus $H_a$: $\Theta \notin [\Theta_1, \Theta_2]$.[18] Our discussion will be facilitated by the following result concerning the differentiability of the expectation of a function taken with respect to an exponential class density.

**Lemma 9.1**
***Differentiability of $\gamma(\Theta) = \mathrm{E}_\Theta(\phi(\mathbf{X}))$ in the Case of an Exponential Class Density***

Let $f(\mathbf{x};\Theta) = \exp(c(\Theta)g(\mathbf{x}) + d(\Theta) + z(\mathbf{x}))I_A(\mathbf{x})$, $\Theta \in \Omega$, be an exponential class density with $\Omega$ defined as an open interval contained in $\mathbb{R}$, and with $c(\Theta)$ and $d(\Theta)$ being differentiable functions of $\Theta \in \Omega$. Let $\mathrm{E}_\Theta$ denote an expectation taken with respect to $f(\mathbf{x};\Theta)$.

If the function $\phi(\mathbf{x})$ is such that $\gamma(\Theta) = \mathrm{E}_\Theta(\phi(\mathbf{X}))$ exists $\forall\ \Theta \in \Omega$, then $\gamma(\Theta) = \mathrm{E}_\Theta(\phi(\mathbf{X}))$ is a differentiable function of $\Theta \in \Omega$. Furthermore,

$$\text{(continuous)} \frac{\partial^r}{\partial\Theta^r} \int_{-\infty}^{\infty} \cdots \int_{-\infty}^{\infty} \phi(\mathbf{x}) f(\mathbf{x};\Theta) d\mathbf{x} = \int_{-\infty}^{\infty} \cdots \int_{-\infty}^{\infty} \phi(\mathbf{x}) \frac{\partial^r f(\mathbf{x};\Theta)}{\partial\Theta^r} d\mathbf{x}$$

$$\text{(discrete)} \frac{\partial^r}{\partial\Theta^r} \sum_{\mathbf{x}\in R(\mathbf{x})} \phi(\mathbf{x}) f(\mathbf{x};\Theta) = \sum_{\mathbf{x}\in R(\mathbf{x})} \phi(\mathbf{x}) \frac{\partial^r f(\mathbf{x};\Theta)}{\partial\Theta^r},$$

i.e., differentiation can occur under the integral or summation sign.

See E.L. Lehmann, (1986), *Testing Statistical Hypotheses*, 2nd Ed., John Wiley, New York, pp. 59–60, or D.V. Widder, (1946), *The LaPlace Transform*, Princeton University Press, Princeton, N.J., pp. 240–241.

The lemma allows one to establish the differentiability (and thus also continuity) of the power function associated with the rejection region of a hypothesis test when the joint density of the probability sample is an exponential class distribution. In particular, by defining $\phi(\mathbf{x}) = I_{C_r}(\mathbf{x})$, Lemma 9.1 becomes a result concerning the differentiability of power functions.

**Theorem 9.8**
***Differentiability and Continuity of Power Functions for Exponential Class Densities***

*Let the joint density function of a probability sample be a member of the exponential class of densities $f(\mathbf{x};\Theta) = \exp(c(\Theta)g(\mathbf{x}) + d(\Theta) + z(\mathbf{x}))I_A(\mathbf{x})$, $\Theta \in \Omega$, with $\Omega$ being an open interval contained in $\mathbb{R}$, and with $c(\Theta)$ and $d(\Theta)$ being differentiable functions of $\Theta \in \Omega$. Let $C_r$ be any rejection region for testing some hypothesis $H_0$: $\Theta \in \Omega_0$ versus $H_a$: $\Theta \in \Omega_a$.*[19] *Then the power function $\pi_{C_r}(\Theta) = P(\mathbf{x} \in C_r;\Theta)$ is differentiable, and hence continuous, with respect to $\Theta \in \Omega$.*

---

[18]Results are available for a more general class of densities referred to as Polya distributions, which subsumes the exponential class densities as a special case. However, the mathematics involved in analyzing the more general distributions is beyond the scope of our study. Interested readers can consult the work of S. Karlin, (1957), "Polya Type Distributions II," *Ann. Math. Stat.*, 28, pp. 281–308.

[19]We are assuming that $C_r$ is such that a power function is defined, i.e., $C_r$ can be assigned probability by $f(\mathbf{x};\Theta)$, $\Theta \in \Omega$.

**Proof** Let $\phi(\mathbf{x}) \equiv I_{Cr}(\mathbf{x})$ in Lemma 9.1. Then since $E_\Theta(\phi(\mathbf{X})) = E_\Theta(I_{Cr}(\mathbf{X})) = \pi_{Cr}(\Theta) = P(\mathbf{x} \in C_r;\Theta)\ \forall\ \Theta \in \Omega$, the power function $\pi_{Cr}(\Theta)$ is differentiable $\forall\ \Theta \in \Omega$. Continuity of $\pi_{Cr}(\Theta)$ follows from differentiability of $\pi_{Cr}(\Theta)$. ■

We now turn to the main result concerning the definition of UMPU level $\alpha$ *two-sided* tests when the random sample has an exponential class density.

**Theorem 9.9** ***UMPU Level $\alpha$ Two-Sided Tests for Exponential Class Densities*** *Let the joint density of the random sample be given by* $f(\mathbf{x};\Theta) = \exp(c(\Theta)g(\mathbf{x}) + d(\Theta) + z(\mathbf{x}))I_A(\mathbf{x})$, $\Theta \in \Omega$, *with* $\Omega$ *being an open interval contained in* $\mathbb{R}$, $c(\Theta)$ *and* $d(\Theta)$ *being differentiable functions of* $\Theta \in \Omega$ *and* $c(\Theta)$ *being strictly monotonic (either increasing or decreasing) in* $\Theta$. *Define a rejection region as* $C_r = \{\mathbf{x} : g(\mathbf{x}) \leq c_1 \text{ or } g(\mathbf{x}) \geq c_2\}$ *where* $c_1 < c_2$.

1. *A size* $\alpha$ $C_r$ *defines a UMPU level* $\alpha$ *test of* $H_0$: $\Theta = \Theta_0$ versus $H_a$: $\Theta \neq \Theta_0$ *iff* $\Theta_0 = \arg\min_{\Theta \in H_0 \cup H_a}\{\pi_{C_r}(\Theta)\} = \arg[d\pi_{C_R}(\Theta)/d\Theta = 0]$, i.e., *the power of the test is minimized at* $\Theta_0$.
2. *A size* $\alpha$ $C_r$ *defines a UMPU level* $\alpha$ *test of* $H_0$: $\Theta_1 \leq \Theta \leq \Theta_2$ versus $H_a$: $\Theta \notin [\Theta_1, \Theta_2]$, $\Theta_1 < \Theta_2$, *iff* $\pi_{C_r}(\Theta_1) = \pi_{C_R}(\Theta_2) = \alpha$.

**Proof** **Necessity:** For result (1), if the power function is not minimized at $\Theta_0$, then $C_r$ is not an $\alpha$-size unbiased test because then $\exists\ \Theta \in H_a$ such that $\pi_{C_r}(\Theta) < \pi_{C_r}(\Theta_0) = \alpha$. The power function is differentiable by Theorem 9.8, and can be shown (see reference below) to be strictly convex, so that the minimum occurs at $\Theta_0$ *iff* $d\pi_{C_r}(\Theta_0)/d\Theta = 0$.

For result (2), since the power function is continuous by Theorem 9.8, it is necessary that $\pi_{C_r}(\Theta_1) = \pi_{C_r}(\Theta_2) = \alpha$, for consider the contrary that $\pi_{C_r}(\Theta_1) < \alpha$ and/or $\pi_{C_r}(\Theta_2) < \alpha$ (note that $\pi_{C_r}(\Theta_i) > \alpha$ for $i = 1,2$ is ruled out since the test has size $\alpha$). Then there exists a value of $\Theta \in H_a$ close to $\Theta_1$ *or* $\Theta_2$ such that $\pi_{C_r}(\Theta) < \alpha$ by the continuity of $\pi_{C_r}(\Theta)$, contradicting unbiasedness.

**Sufficiency:** E. Lehmann, *Testing Statistical Hypotheses – 2nd Edition*, pp. 135–137. ■

The application of Theorem 9.9 to find UMPU level $\alpha$ tests of $H_0$: $\Theta = \Theta_0$ versus $H_a$: $\Theta \neq \Theta_a$ or of $H_0$: $\Theta_1 \leq \Theta \leq \Theta_2$ versus $H_a$: $\Theta \in [\Theta_1,\Theta_2]$ is conceptually straightforward although a computer will often be needed as an aid in making the necessary calculations. Essentially, one searches for the appropriate $c_1$ and $c_2$ values in $C_r = \{\mathbf{x}: g(\mathbf{x}) \leq c_1 \text{ or } g(\mathbf{x}) \geq c_2\}$ that produce an $\alpha$-size rejection region and that also satisfy either $d\pi_{C_r}(\Theta_0)/d\Theta = 0$ or $\pi_{C_r}(\Theta_1) = \pi_{C_r}(\Theta_2)$, for the respective pairs of hypotheses. The following examples illustrate the process.

**Example 9.20**
***UMPU Test of $H_0$: $p = p_0$ versus $H_a$: $p \neq p_0$ for Bernoulli Population Distribution***

Recall Example 9.14 where an untrusting gambler desired to test whether a roulette wheel used by an Atlantic City Casino was fair. A random sample of 100 spins of the wheel was to be used for the test. The joint density of the random sample was in the form of a product of Bernoulli densities, which can be represented as a member of the exponential class $f(\mathbf{x};p) = \exp[c(p)\ g(\mathbf{x}) + d(p) + z(\mathbf{x})]\ I_A\ (\mathbf{x})$ where $c(p) = \ln(p/(1-p))$, $g(\mathbf{x}) = \sum_{i=1}^{100} x_i$, $d(p) = 100\ln(1-p)$, $z(\mathbf{x}) = 0$, and $A = \times_{i=1}^{100}\{0,1\}$, with $p \in \Omega = (0,1)$. We demonstrated in Example 9.14 that there does *not* exist a UMP level .10 test of the "fairness" hypothesis $H_0$: $p = .5$ versus $H_a$: $p \neq .5$ having size .10. We now use Theorem 9.9 to show that a UMPU level and size $\alpha$ test of the fairness hypothesis does exist. The rejection region we seek has the form

$$C_r = \left\{\mathbf{x} : \sum_{i=1}^{100} x_i \leq c_1 \text{ or } \sum_{i=1}^{100} x_i \geq c_2\right\}$$

for choices of $c_1$ and $c_2$ that produce an $\alpha$-size rejection region satisfying $d\pi_{C_r}(.5)/dp = 0$. Since $Z = \sum_{i=1}^{100} X_i$ has a binomial distribution, $C_r$ will have size $\alpha$ when

$$1 - \pi_{C_r}(.5) = P(\mathbf{x} \in \overline{C_r}; p = .5) = \sum_{z=c_1+1}^{c_2-1} \binom{100}{z} (.5)^z (.5)^{100-z} I_{\{0,1,\ldots,100\}}(z) = 1 - \alpha$$

The derivative of the power function can be represented as

$$\frac{d\pi_{C_r}(p)}{dp} = -\sum_{z=c_1+1}^{c_2-1} \binom{100}{z} p^{z-1}(1-p)^{99-z}[z - 100p] I_{\{0,1,\ldots,100\}}(z),$$

which when evaluated at $p = .5$ and then set equal to 0 results in the appropriate condition on the choices of $c_1$ and $c_2$, as

$$d\pi_{C_r}(.5)/dp = -\sum_{z=c_1+1}^{c_2-1} \binom{100}{z} (.5)^{98}[z - 50] = 0.$$

Thus $c_1$ and $c_2$ must be chosen so that $c_1 < z < c_2$ is a symmetric interval around the value 50, which is a direct consequence of the binomial density being a symmetric density around the point $z = 50$ when $p = .5$. The possible choices of $c_1$ and $c_2$ are given by the finite set of two-tuples $A = \{(c_1, c_2) : c_1 = 50 - i, c_2 = 50 + i, i = 1, \ldots, 50\}$, which restricts the admissible choices of test sizes to be $B = \{\alpha : \alpha = P(z \leq c_1 \text{ or } z \geq c_2; p = .5), (c_1, c_2) \in A\}$. Some specific possibilities for the size of the test are given as follows (calculated using a computer):

| $c_1, c_2$ | $\alpha = P(\mathbf{x} \in C_r; p = .5)$ |
|---|---|
| 37,63 | .012 |
| 38,62 | .021 |
| 39,61 | .035 |
| 40,60 | .057 |
| 41,59 | .089 |
| 42,58 | .133 |

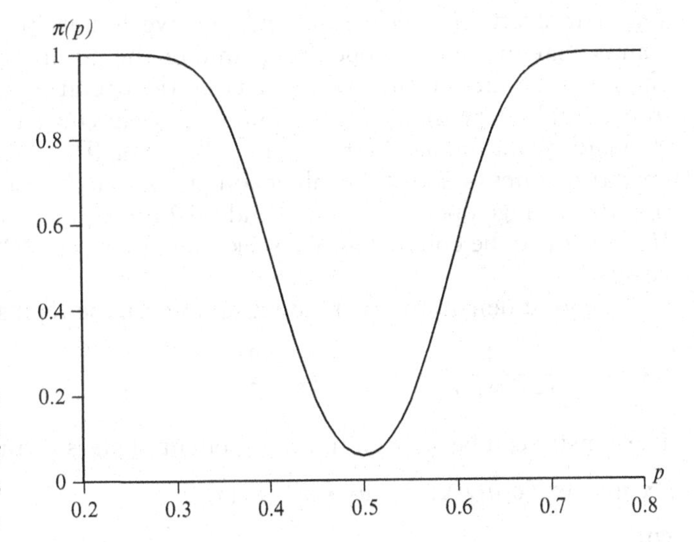


**Figure 9.9** Power function for the UMPU level .057 test of $H_0$: $p = .5$ versus $H_a$: $p \neq .5$.

Suppose a test size of $\alpha = .057$ is acceptable to the concerned gambler. Then the rejection region of the UMPU level .057 test of the fairness hypothesis is given by

$$C_r = \left\{ \mathbf{x} : \sum_{i=1}^{100} x_i \leq 40 \text{ or } \sum_{i=1}^{100} x_i \geq 60 \right\}$$

and the hypothesis that the roulette wheel was fair would be rejected if red ($x_i = 1$) occurred 40 or less times, or 60 or more times. Note that the values of the power function (see Figure 9.9) for $p \in (0,1)$ are given by

$$\pi_{C_r}(\mathrm{p}) = 1 - \sum_{z=41}^{59} \binom{100}{z} p^z (1-\mathrm{p})^{100-z} I_{\{0,1,\ldots,100\}}(z).$$

Some selected values of the power function are given below.

| $p$ | $\pi_{C_r}(p)$ |
|---|---|
| .5 | .057 |
| .45,.55 | .185 |
| .4, .6 | .543 |
| .3, .7 | .988 |
| .2, .8 | ≈1 |

Issue to Consider: Do you think the power function given in Figure 9.9 would be acceptable to the gambler? Why or why not? If not, what could you do to alter the power function to make it more acceptable to the gambler?) □

**Example 9.21** ***UMPU Test of $H_0$: $\mu = \mu_0$ versus $H_a$: $\mu \neq \mu_0$ for Normal Population Distribution***

The manufacturer of a certain inexpensive battery-powered tablet computer claims that the average operating time of the computer between full battery charges is 2 hours. It can be assumed that the operating time, in hours, obtained from a full battery charge is a random variable having a normal distribution with standard deviation equal to .2, i.e., $X_i \sim N(\mu, .04)$. A random sample of the operating times of 200 of the tablet computers is to be used to test the hypothesis that the average operating time is indeed 2 hours, i.e., a test of $H_0$: $\mu = 2$ versus $H_a$: $\mu \neq 2$ is to be conducted. We seek a test that is UMPU level $\alpha$ for testing $H_0$ versus $H_a$.[20]

The joint density of the random sample can be represented in the form

$$f(\mathbf{x};\mu) = \frac{1}{(2\pi)^{100}(.2)^{200}} \exp\left[-12.5 \sum_{i=1}^{200} (x_i - \mu)^2\right].$$

The density can be written in the exponential class form

$$f(\mathbf{x};\mu) = \exp[c(\mu)g(\mathbf{x}) + d(\mu) + z(\mathbf{x})]I_A(\mathbf{x})$$

with

$$c(\mu) = 25\mu, \ g(\mathbf{x}) = \sum_{i=1}^{200} x_i, \ d(\mu) = -2500\mu^2,$$

$$z(\mathbf{x}) = -\ln\left((2\pi)^{100}(.2)^{200}\right) - 12.5\sum_{i=1}^{200} x_i^2, \text{ and } A = \mathbb{R}^n.$$

From Theorem 9.9, we seek a rejection region of the form

$$C_r = \left\{\mathbf{x} : \sum_{i=1}^{200} x_i \leq c_1 \text{ or } \sum_{i=1}^{200} x_i \geq c_2\right\}$$

for choices of $c_1$ and $c_2$ that produce an $\alpha$-size rejection region satisfying $d\pi_{C_r}(2)/d\mu = 0$. Using results for linear combinations of normally distributed random variables we know that $\sum_{i=1}^{200} X_i \sim N(200\mu, 8)$, and thus $C_r$ will have size $\alpha$ when

$$1 - \pi_{C_r}(2) = P(\mathbf{x} \in \bar{C}_r; \mu = 2) = \int_{c_1}^{c_2} \frac{1}{\sqrt{2\pi}\sqrt{8}} \exp\left[-\frac{1}{16}(z - 400)^2\right] dz = 1 - \alpha.$$

The derivative of the power function can be represented as

$$\frac{d\pi_{C_r}(\mu)}{d\mu} = -\int_{c_1}^{c_2} \frac{25[z - 200\mu]}{\sqrt{2\pi}\sqrt{8}} \exp\left[\frac{1}{16}(z - 200\mu)^2\right] dz,$$

which when evaluated at $\mu = 2$ and then set equal to 0 results in the condition

$$\frac{d\pi_{C_r}(2)}{d\mu} = -25\int_{c_1}^{c_2} (z - 400)N(z; 400, 8)dz = 0.$$

[20] One can show using the monotone likelihood ratio approach that a UMP level $\alpha$ test of $H_0$ versus $H_a$ does *not* exist. Recall Theorem 9.9.

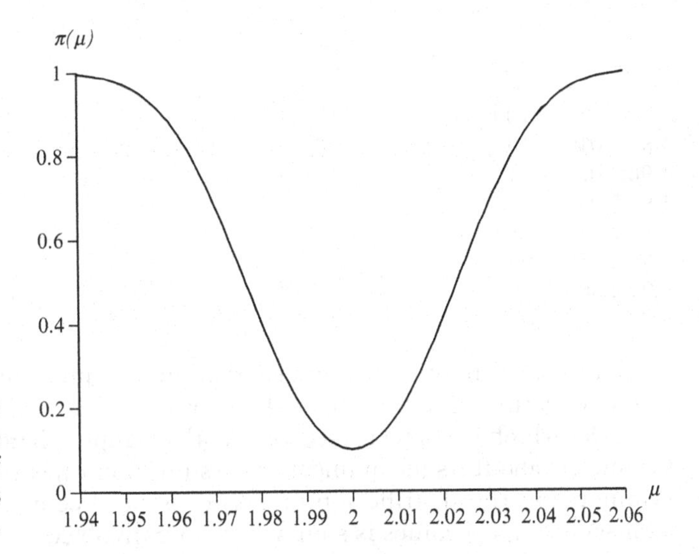


**Figure 9.10** Power function of the UMPU level .10 Test of $H_0$: $\mu = 2$ versus $H_a$: $\mu \neq 2$.

Thus $c_1$ and $c_2$ must be chosen so that $c_1 < z < c_2$ is a symmetric interval around the value 400, which is a direct consequence of the symmetry of the normal distribution around the point $z = 400$. The possible choices of $c_1$ and $c_2$ are given by the uncountably infinite set of two-tuples $A = \{(c_1, c_2) : c_1 = 400 - c, c_2 = 400 + c, c \in (0, \infty)\}$, which implies that the choice of test size can be any value $\in (0,1)$, since

$$B = \{\alpha : \alpha = P(z \leq c_1 \text{ or } z \geq c_2; \mu = 2), (c_1, c_2) \in A\} = (0, 1).$$

Suppose $\alpha = .10$ is chosen to be the size of the test. The appropriate values of $c_1$ and $c_2$ can be found by solving for $c$ in the integral equation $\int_{400-c}^{400+c} N(z; 400, 8)dz = \int_{-c/\sqrt{8}}^{c/\sqrt{8}} N(z; 0, 1)dz = .90$ and then setting $c_1 = 400 - c$ and $c_2 = 400 + c$. Using the computer, or the table of the standard normal distribution, we find that $c/\sqrt{8} = 1.645$, or $c = 4.653$. Then $c_1 = 395.347$ and $c_2 = 404.653$, and the rejection region of the UMPU size .10 test is given by

$$C_r = \left\{ \mathbf{x} : \sum_{i=1}^{200} x_i \leq 395.347 \text{ or } \sum_{i=1}^{200} x_i \geq 404.653 \right\}.$$

The rejection region can be equivalently expressed using the sample mean as a test statistic, in which case $C_r = \{\mathbf{x} : \bar{x} \leq 1.977 \text{ or } \bar{x} \geq 2.023\}$. Thus, the hypothesis is rejected if the average operating time of the 200 sampled notebook computers is $\leq 1.977$ hours or $\geq 2.023$ hours.

Note that the values of the power function (see Figure 9.10) for $\mu \in (-\infty, \infty)$ are given by

$$\pi_{C_r}(\mu) = 1 - \int_{395.347}^{404.653} N(z; 200\mu, 8)dz = 1 - \int_{(395.347 - 200\mu)/\sqrt{8}}^{(404.653 - 200\mu)/\sqrt{8}} N(z; 0, 1)dz.$$

Some selected values of the power function are given below.

| $\mu$ | $\pi_{C_r}(\mu)$ |
|---|---|
| 2 | .100 |
| 1.99,2.01 | .183 |
| 1.98,2.02 | .410 |
| 1.97,2.03 | .683 |
| 1.96,2.04 | .882 |
| 1.95,2.05 | .971 |
| 1.94,2.06 | .995 |

The reader may have noticed that in the preceding two examples, the characterization of admissible values for $c_1$ and $c_2$ was simplified by the fact that $g(\mathbf{X})$, which was a test statistic in both examples, had a distribution that was symmetric about its mean under the assumption $H_0$ is true. In cases where this symmetry in the distribution of $g(\mathbf{X})$ does not occur, the characterization of admissible $(c_1, c_2)$ values is somewhat more involved.

**Example 9.22** ***UMPU Test for $H_0$: $\Theta = \Theta_0$ versus $H_a$: $\Theta \neq \Theta_0$ in the Exponential Population Distribution***

The operating life of a 8 gigabyte memory chip manufactured by the Elephant Computer Chip Co. can be viewed as a random variable with a density function belonging to the exponential family of densities. A random sample of the lifetimes of 500 memory chips is to be used to test the hypothesis that the mean life of the chip is 50,000 hours. With the $X_i$'s measured in 10,000's of hours, the joint density of the random sample is given by

$$f(\mathbf{x};\theta) = \theta^{-500} \exp\left(-\sum_{i=1}^{500} x_i/\theta\right) \prod_{i=1}^{500} I_{(0,\infty)}(x_i) \text{ for } \theta \in \Omega = (0,\infty).$$

We seek a UMPU level .05 test of $H_0$: $\theta = 5$ versus $H_a$: $\theta \neq 5$.

The joint density can be written in the exponential class form

$$f(\mathbf{x};\theta) = \exp[c(\theta)g(\mathbf{x}) + d(\theta) + z(\mathbf{x})]I_A(\mathbf{x})$$

with

$$c(\theta) = -\theta^{-1}, \quad g(\mathbf{x}) = \sum_{i=1}^{500} x_i, \quad d(\theta) = \ln(\theta^{-500}), \quad z(\mathbf{x}) = 0, \text{ and } A = \times_{i=1}^{500}(0,\infty).$$

From Theorem 9.9, we seek a rejection region of the form

$$C_r = \left\{\mathbf{x} : \sum_{i=1}^{500} x_i \leq c_1 \text{ or } \sum_{i=1}^{500} x_i \geq c_2\right\}$$

for choices of $c_1$ and $c_2$ that produce an $\alpha$-size rejection region satisfying $d\pi_{C_r}(5)/d\theta = 0$. Using the MGF approach, it can be established that $Z = \sum_{i=1}^{500} X_i$ has a Gamma$(z;\ 500,\theta)$ distribution i.e., $Z \sim (1/(\theta^{500}\Gamma(500)))z^{499}\exp(-z/\theta)I_{(0,\infty)}(z)$ (recall Example 9.18).

Then $C_r$ will have size $\alpha$ when

$$1 - \pi_{C_r}(5) = \int_{c_1}^{c_2} \frac{1}{5^{500}\Gamma(500)} z^{499} \exp(-z/5) I_{(0,\infty)}(z) = 1 - \alpha. \tag{1}$$

The derivative of the power function can be represented as

$$d\pi_{C_r}(\theta)/d\theta = -\int_{c_1}^{c_2} \left[(z/\theta^2) - 500\theta^{-1}\right] \text{ Gamma}(z; 500, \theta)dz,$$

which when evaluated as $\theta = 5$ and then set equal to 0 results in the condition

$$d\pi_{C_r}(5)/d\theta = -\int_{c_1}^{c_2} [(z/25) - 100] \text{ Gamma}(z; 500, 5)dz = 0. \tag{2}$$

Unlike the previous two examples, the distribution of $Z$ is not symmetric about its mean, and the preceding condition does *not* imply that $(c_1, c_2)$ is a symmetric interval around $E(Z) = 2{,}500$. Nonetheless, (2) defines an implicit functional relationship between $c_2$ and $c_1$, say $c_2 = \gamma(c_1)$, that determines admissible values of $c_1$ and $c_2$. Using the computer, the simultaneous equations (1) and (2) can be numerically solved for $c_1$ and $c_2$, given a choice of the test size $\alpha \in (0,1)$. The table below provides $c_1$ and $c_2$ values for selected test sizes:

| $\alpha$ | $c_1$ | $c_2$ |
|---|---|---|
| .01 | 2222.920 | 2799.201 |
| .05 | 2287.190 | 2725.618 |
| .10 | 2320.552 | 2688.469 |
| .20 | 2359.419 | 2646.057 |

To define a UMPU level .05 test of $H_0$: $\theta = 5$ versus $H_a$: $\theta \neq 5$, the rejection region would be defined as

$$C_r = \left\{ \mathbf{x} : \sum_{i=1}^{500} x_i \leq 2287.190 \text{ or} \sum_{i=1}^{500} x_i \geq 2725.618 \right\}.$$

One could also use the sample mean as a test statistic, in which case the rejection region could be defined alternatively as $C_r = \{\mathbf{x} : \bar{x} \leq 4.574 \text{ or } \bar{x} \geq 5.451\}$. Thus, the hypothesis $H_0$: $\theta = 5$ will be rejected if the average life of the 500 sampled chips is $\leq$ 4574 hours or $\geq$ 5451 hours.

Power function values for $\theta > 0$ are given by (see Figure 9.11)

$$\pi_{C_r}(\theta) = 1 - \int_{2287.190}^{2725.618} \frac{1}{\theta^{500}\Gamma(500)} z^{499} \exp(-z/\theta)\text{dz}.$$

Some selected values of the power function are given below.

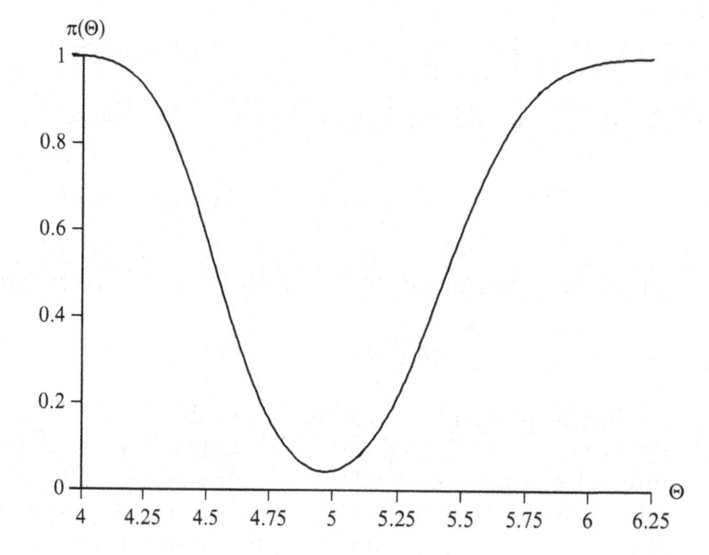


**Figure 9.11** Power function of UMPU level .05 test of $H_0$: $\theta = 5$ versus $H_a$: $\theta \neq 5$.

| $\theta$ | $\pi_{Cr}(\theta)$ |
|---|---|
| 4 | .999 |
| 4.25 | .954 |
| 4.50 | .649 |
| 4.75 | .206 |
| 4.90 | .073 |
| 5.00 | .050 |
| 5.10 | .073 |
| 5.25 | .196 |
| 5.50 | .573 |
| 5.75 | .878 |
| 6.00 | .981 |
| 6.25 | .999 |

Unlike the previous two examples, the power function is *not* symmetric about the point $\theta_0 = 5$. However, it is "nearly" symmetric because $Z_n = \sum_{i=1}^{n} X_i$ has an asymptotic normal distribution, and a sample size of $n = 500$ is sufficient for the asymptotic distribution to provide a good approximation to $Z_n$'s gamma distribution. □

**Example 9.23** ***UMPU Test of $H_0$: $\mu \in [\mu_1, \mu_2]$ versus $H_a$: $\mu \notin [\mu_1, \mu_2]$ for Normal Population Distribution***

Recall Example 9.21 regarding the operating times of tablet computers between full battery charges. Suppose instead of testing the hypothesis $H_0$: $\mu = 2$, a *range* of values for the mean operating time of the computer is to be tested. In particular, suppose we wish to test the hypothesis $H_0$: $1.75 \leq \mu \leq 2.25$ versus $H_a$: $\mu < 1.75$ or $\mu > 2.25$. We know from Example 9.21 that $f(\mathbf{x};\mu)$ belongs to the exponential class of densities and satisfies the other conditions of Theorem 9.9. Using Theorem 9.9, we seek an $\alpha$-size rejection region defined by

$$C_r = \left\{ \mathbf{x} : \sum_{i=1}^{200} x_i \leq c_1 \text{ or } \sum_{i=1}^{200} x_i \geq c_2 \right\}$$

for choices of $c_1$ and $c_2$ that satisfy

$$1 - \pi_{C_r}(350) = \int_{c_1}^{c_2} N(z; 350, 8)dz = \int_{c_1}^{c_2} N(z; 450, 8)dz = 1 - \pi_{C_r}(450) = 1 - \alpha.$$

Choosing a specific value of $\alpha$, the preceding condition becomes a system of *two* integral equations which can be solved simultaneously (using the computer) for the two unknowns $c_1$ and $c_2$.

Suppose the size of the test is chosen to be $\alpha = .05$. Using a nonlinear simultaneous equations solver,[21] the solution to the preceding two-equation system was found to be $c_1 = 345.348$ and $c_2 = 454.652$. Thus, the rejection region of the UMPU level-.05 test defined by Theorem 9.9 is

$$C_r = \left\{ \mathbf{x} : \sum_{i=1}^{200} x_i \leq 345.348 \text{ or } \sum_{i=1}^{200} x_i \geq 454.652 \right\}.$$

One could also use the sample mean for the test statistic, in which case

$C_r = \{\mathbf{x} : \bar{x} \leq 1.727 \text{ or } \bar{x} \geq 2.273\}$.

Power function values are given by

$$\pi_{C_r}(\mu) = 1 - \int_{345.348}^{454.652} N(z; 200\mu, 8)dz$$

(see Figure 9.12).

Some selected values of the power function are given below.

| $\mu$ | $\pi_{Cr}(\mu)$ |
|---|---|
| 2 | ≈0 |
| 1.76,2.24 | .01 |
| 1.75,2.25 | .05 |
| 1.74,2.26 | .17 |
| 1.73,2.27 | .41 |
| 1.72,2.28 | .68 |
| 1.71,2.29 | .88 |
| 1.70,2.30 | .97 |
| 1.69,2.31 | ≈1 |

The power function verifies that the test defined by $C_r$ is unbiased and has size .05. The hypothesis $H_0$: $1.75 \leq \mu \leq 2.25$ is then rejected by the UMPU level .05 test if the average operating time of the 200 sampled tablet computers is $\leq 1.727$ hours or $\geq 2.273$ hours. □

[21]The algorithm actually used was the NLSYS procedure in the GAUSS Matrix language.

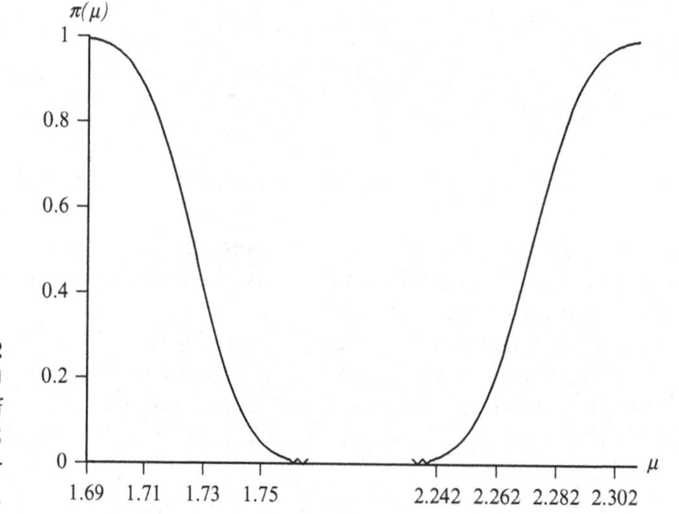


**Figure 9.12** Power function of UMPU level .05 test of $H_0$: $1.75 \le \theta \le 2.25$ versus $H_a$: $\theta < 1.75$ or $\theta > 2.25$.

**Example 9.24** ***UMPU Test of $H_0$: $\theta \in [\theta_1, \theta_2]$ versus $H_a$: $\theta \notin [\theta_1, \theta_2]$ in Exponential Population Distribution***

Recall Example 9.22 regarding the operating life of memory chips. Suppose instead of testing the hypothesis $H_0$: $\theta = 5$, a range of values for the mean life of the memory chips is to be tested. In particular, suppose we wish to test the hypothesis $H_0$: $4.9 \le \theta \le 5.1$ versus $H_a$: $\theta < 4.9$ or $\theta > 5.1$. We know from Example 9.22 that $f(\mathbf{x}; \theta)$ belongs to the exponential class of densities and adheres to the other conditions of Theorem 9.9. Using Theorem 9.9, we seek a size $\alpha$ rejection region of the form

$$C_r = \left\{ \mathbf{x} : \sum_{i=1}^{500} x_i \le c_1 \text{ or } \sum_{i=1}^{500} x_i \ge c_2 \right\}$$

for choices of $c_1$ and $c_2$ that satisfy

$$1 - \pi_{C_r}(4.9) = \int_{c_1}^{c_2} \text{Gamma}(z; 500, 4.9) dz = \int_{c_1}^{c_2} \text{Gamma}(z; 500, 5.1) dz = 1 - \pi_{C_r}(5.1) = 1 - \alpha.$$

Choosing a specific value of test size $\alpha$, the preceding condition becomes a system of *two* integral equations which can be solved simultaneously (using the computer) for the two unknowns $c_1$ and $c_2$.

Suppose the size of the test is chosen to be $\alpha = .05$. Using a nonlinear equation solver[22] the solution to the two-equation system was found to be $c_1 = 2268.031$ and $c_2 = 2746.875$. Thus, the rejection region of the UMPU level .05 test defined by Theorem 9.9 is given by

$$C_r = \left\{ \mathbf{x} : \sum_{i=1}^{500} x_i \le 2268.031 \text{ or } \sum_{i=1}^{500} x_i \ge 2746.875 \right\}.$$

[22]The algorithm actually used was the NLSYS procedure in the GAUSS matrix language.

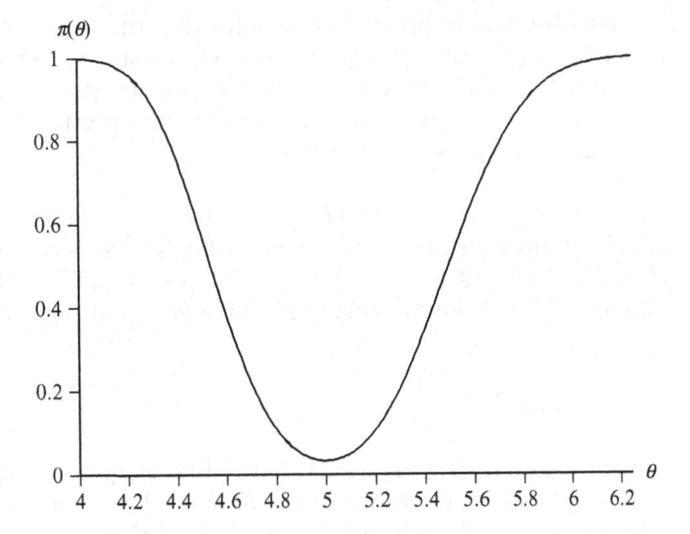


**Figure 9.13** Power function of UMPU level .05 test of $H_0$: $4.9 \le \theta \le 5.1$ *versus* $H_a$: $\theta < 4.9$ or $\theta > 5.1$.

One could also use the sample mean for the test statistic, in which case $C_r = \{\mathbf{x} : \bar{x} \le 4.536 \text{ or } \bar{x} \ge 5.494\}$.

Power function values are given

$$\pi_{C_r}(\theta) = 1 - \int_{2268.031}^{2746.875} Gamma(z; 500, \theta)dz$$

(see Figure 9.13).

Some selected values of the power function are given below

| $\theta$ | $\pi_{Cr}(\theta)$ |
|---|---|
| 4.00 | .998 |
| 4.25 | .932 |
| 4.50 | .576 |
| 4.75 | .157 |
| 4.90 | .050 |
| 5.00 | .032 |
| 6.00 | .973 |
| 5.75 | .841 |
| 5.50 | .504 |
| 5.25 | .151 |
| 5.10 | .050 |

The power function verifies that the test defined by $C_r$ is a size .05 unbiased test. The hypothesis $H_0$: $4.9 \le \theta \le 5.1$ is then rejected by the UMPU level .05 test if the average life of the 500 sampled memory chips is $\le$ 4,536 hours or $\ge$ 5,494 hours. □

We observed in previous examples that the definition of the UMPU level $\alpha$ rejection region for testing $H_0$: $\Theta = \Theta_0$ versus $H_a$: $\Theta \neq \Theta_0$ was substantially simplified if the test statistic had a density that was symmetric about its mean. This is a general property of the construction of UMPU tests based on Theorem 9.9 that we now formalize.

**Theorem 9.10** ***Two Sided UMPU Tests of $H_0 : \Theta = \Theta_0$ versus $H_a : \Theta \neq \Theta_0$ for Symmetric PDFs***

*Assume the conditions of Theorem 9.9 and let the density of the test statistic $g(\mathbf{X})$ be symmetric about its mean $\gamma = \mathrm{E}(g(\mathbf{X}))$ when $\Theta = \Theta_0$. If $c_1$ and $c_2$ are such that $P(g(\mathbf{x}) \geq c_2;\Theta_0) = \alpha/2$ and $c_1 = 2\gamma - c_2$, then $C_r = \{\mathbf{x}\text{: } g(\mathbf{x}) \leq c_1 \text{ or } g(\mathbf{x}) \geq c_2\}$ defines a UMPU level $\alpha$ test of $H_0$: $\Theta = \Theta_0$ versus $H_a$: $\Theta \neq \Theta_0$.*

**Proof** See Appendix. ■

**Example 9.25** ***UMPU Two-Sided Tests Using Symmetry of Binomial and Normal Distributions***

Revisit Examples 9.20 and 9.21. In the first example, since the binomial density of $Z = \sum_{i=1}^{100} X_i$ is symmetric about $\mathrm{E}(Z) = 50$, we know from Theorem 9.10 that choosing $c_1$ and $c_2$ such that $P(z \geq c_2; .5) = \alpha/2$ and $c_1 = 2(50) - c_2$ will define a UMPU level $\alpha$ rejection region $C_r = \{\mathbf{x}\text{: } \sum_{i=1}^{100} x_i \leq c_1 \text{ or } \sum_{i=1}^{100} x_i \geq c_2\}$. In particular, $P(z \geq 60; .5) = .0285$, $c_1 = 100 - 60 = 40$, and then $C_r = \{\mathbf{x}\text{: } \sum_{i=1}^{100} x_i \leq 40 \text{ or } \sum_{i=1}^{100} x_i \geq 60\}$ defines the appropriate size .057 rejection region.

In the second example, since the normal distribution of $Z = \sum_{i=1}^{100} X_i$ is symmetric about $\mathrm{E}(Z) = 400$, we know from Theorem 9.10 that choosing $c_1$ and $c_2$ such that $P(z \geq c_2;400) = \alpha/2$ and $c_1 = 2(400) - c_2$ will define a UMPU level $\alpha$ rejection region $C_r = \{\mathbf{x}\text{: } \sum_{i=1}^{200} x_i \leq c_1 \text{ or } \sum_{i=1}^{200} x_i \geq c_2\}$. In particular, $P(z \geq 404.653) = .05$, $c_1 = 800 - 404.653 = 395.347$, and then $C_r = \{\mathbf{x}\text{: } \sum_{i=1}^{200} x_i \leq 395.347 \text{ or } \sum_{i=1}^{200} x_i \geq 404.653\}$ defines the appropriate size .10 rejection region. □

It should also be noted that in the discrete case, it may not be possible to define an unbiased test, let alone a UMPU test, without resorting to so-called randomized tests (see Example 9.8). The reader should revisit Example 9.20 and examine alternative choices of $p$ to explore the difficulty. The problem does not arise in the continuous case.

### 9.5.4 Conditioning in the Multiple Parameter Case

In the multiple parameter case, the probability sample, $\mathbf{X}$, has a joint distribution, $f(\mathbf{x};\boldsymbol{\Theta})$, that belongs to a family of distributions indexed by a $(k \times 1)$ parameter vector $\boldsymbol{\Theta} \in \Omega$, where $k > 1$. We will examine hypotheses of the form $H_0$: $\Theta_i \in \Omega_0^i$ versus $H_a$: $\Theta_i \in \Omega_a^i$, where $\Omega^i = \Omega_0^i \cup \Omega_a^i$ represents the admissible values of the $i$th parameter $\Theta_i$. Note that for both $H_0$ and $H_a$, the values of $\Theta_j$, $j \neq i$, are left unspecified, and so it is tacitly understood that $\Theta_1, \Theta_2, \ldots, \Theta_{i-1}, \Theta_{i+1}, \ldots, \Theta_k$ can assume any admissible values for which $\Theta \in \Omega$. In this hypothesis testing context the parameters $\Theta_j$, $j \neq i$, are often referred to as **nuisance parameters**, since we are not interested in them from the standpoint of our stated hypotheses, but they must nonetheless be dealt with in defining a statistical test of $H_0$ versus $H_a$. We will also examine tests of the more general hypotheses $H_0$: $\gamma(\mathbf{\Theta}) \in \Omega_0^\gamma$ versus $H_a$: $\gamma(\mathbf{\Theta}) \in \Omega_a^\gamma$, where $\gamma(\cdot)$ is a scalar function of the parameter vector $\mathbf{\Theta}$.

The approach we will use for defining tests in the multiparameter case will allow us to use the results we have previously established concerning tests for the single parameter case. In particular, we will seek to transform the problem into a single parameter situation by conditioning on sufficient statistics, and then apply previous results for the single parameter case to the transformed problem. In many cases of practical interest, the approach can be used to define UMP level $\alpha$ tests, or at least UMP level $\alpha$ tests within the unbiased class of tests. We will examine a number of other useful and more versatile test procedures for the multiparameter case in Chapter 10, although the procedures will not necessarily be UMP or UMPU tests.

There is a substantial literature on UMP and UMPU tests of statistical hypotheses in multiparameter situations that is more advanced than what we present here and that also applies to a wider array of problems. For additional reading, and a multitude of references, see E. Lehmann, (1986), *Testing Statistical Hypotheses*, 2nd ed., pp. 134–281.

*UMPU Level $\alpha$ Tests via Conditioning* The basic idea of conditioning in order to transform a multiparameter problem into a problem involving only a single parameter, or scalar function of parameters, is as follows. Let $S_1,\ldots,S_r$ be a set of sufficient statistics for $f(\mathbf{x};\mathbf{\Theta})$. Suppose that the elements in the parameter vector $\mathbf{\Theta}$ have been ordered so that $\Theta_1$ is the parameter of interest in the hypothesis test, and $\Theta_2,\ldots,\Theta_k$ are nuisance parameters. Suppose further that, for each *fixed* value of $\Theta_1^0 \in \Omega^1$, $S_j,\ldots,S_r$ are sufficient statistics for the $k-1$ parameter density $f(\mathbf{x};\Theta_1^0, \Theta_2,\ldots,\Theta_k)$, $(\Theta_1^0, \Theta_2,\ldots,\Theta_k) \in \Omega$. It follows from the definition of sufficiency (recall Definition 7.18) that

$$f_*(\mathbf{x};\Theta_1^0) = f(\mathbf{x};\Theta_1^0,\Theta_2,\ldots,\Theta_k|s_j,\ldots,s_r), \Theta_1^0 \in \Omega^1,$$

i.e., the *conditional* distribution of $\mathbf{X}$, given $s_j,\ldots,s_r$, does *not* depend on the nuisance parameters $\Theta_2,\ldots,\Theta_k$. In effect, by conditioning on $s_j,\ldots,s_r$, the problem will have been converted into one involving only the parameter $\Theta_1$. So long as we are concerned only with results obtained in the context of the conditional problem, single parameter UMP or UMPU hypothesis testing results would apply.

In order to extend the previous argument to functions of parameters, consider an analogous argument in terms of a *reparameterized* version of $f(\mathbf{x};\mathbf{\Theta})$. Suppose the joint density function of $\mathbf{X}$ was reexpressed in terms of the parameter vector $\mathbf{c}$, where $c_i = c_i(\Theta)$ for $i = 1,\ldots,k$, so that $f_*(\mathbf{x};\mathbf{c})$, for $\mathbf{c} \in \Omega_\mathbf{c} = \{\mathbf{c}:\ c_i = c_i(\Theta),\ i = 1,\ldots,k,\ \Theta \in \Omega\}$, is an alternative representation of the parametric family of densities $f(\mathbf{x};\Theta)$, $\Theta \in \Omega$. Then if, for each fixed $c_1^0 \in \Omega_c^1$, $S_j^*, \ldots, S_r^*$ is a set of sufficient statistics for $f_*(\mathbf{x};c_1^0, c_2,\ldots,c_k)$, $(c_1^0, c_2,\ldots,c_k) \in \Omega_c$, it would follow that

$$f_*(\mathbf{x};c_1^0) = f(\mathbf{x};c_1^0, c_2, \ldots, c_k | s_j^*, \ldots, s_r^*), c_1^0 \in \Omega,$$

i.e., the conditional distribution of $\mathbf{X}$, given $s_j^*, \ldots, s_r^*$, would not depend on $c_2,\ldots, c_k$. Then single parameter UMP and UMPU results could be applied to testing hypotheses about $c_1(\Theta)$, *conditional* on $s_j^*, \ldots, s_r^*$.

For the conditioning procedure to have practical value, the optimal properties possessed by the rejection region, $C_r$, in the *conditional* problem need to carry over to the context of the original *unconditional* problem. In particular, we will focus on determining when a $C_r$ that is UMPU level $\alpha$ for the conditional problem also possesses this property in the unconditional problem. The exponential class of densities represents an important set of cases in which the conditioning procedure works well. We will have use for the following generalizations of Lemma 9.1 and Theorem 9.8.

**Lemma 9.2** ***Differentiability of $\gamma(\Theta) = \mathrm{E}_\Theta(\phi(\mathbf{X}))$ in the Case of a Multiparameter Exponential Class Density***

Let $f(\mathbf{x};\Theta) = \exp\left(\sum_{i=1}^k c_i(\Theta)g_i(\mathbf{x}) + d(\Theta) + z(\mathbf{x})\right)I_A(\mathbf{x}), \Theta \in \Omega$, be an exponential class density with $\Omega$ defined as an open rectangle contained in $\mathbb{R}^k$ and with $c_1(\Theta),\ldots,c_k(\Theta)$ and $d(\Theta)$ being differentiable functions of $\Theta \in \Omega$. If the function $\phi(\mathbf{x})$ is such that $\gamma(\Theta) = \mathrm{E}_\Theta(\phi(\mathbf{X}))$ exists $\forall\, \Theta \in \Omega$, then $\gamma(\Theta) = \mathrm{E}_\Theta(\phi(\mathbf{X}))$ is a differentiable function of $\Theta \in \Omega$. Furthermore,

$$(\text{continuous})\frac{\partial^{r+s}}{\partial\Theta_i^r\partial\Theta_j^s}\int_{-\infty}^{\infty}\cdots\int_{-\infty}^{\infty}\phi(\mathbf{x})f(\mathbf{x};\Theta)d\mathbf{x} = \int_{-\infty}^{\infty}\cdots\int_{-\infty}^{\infty}\phi(\mathbf{x})\frac{\partial^{r+s}f(\mathbf{x};\Theta)}{\partial\Theta_i^r\partial\Theta_j^s}d\mathbf{x}$$

$$(\text{discrete})\frac{\partial^{r+s}}{\partial\Theta_i^r\partial\Theta_j^s}\sum_{\mathbf{x}\in R(X)}\phi(\mathbf{x})f(\mathbf{x};\Theta) = \sum_{\mathbf{x}\in R(\mathbf{X})}\phi(\mathbf{x})\frac{\partial^{r+s}f(\mathbf{x};\Theta)}{\partial\Theta_i^r\partial\Theta_j^s}$$

for $i$ and $j = 1,\ldots,k$, i.e., differentiation can occur under the integral or summation sign.

**Proof** See references listed for Lemma 9.1. ■

**Theorem 9.11** ***Differentiability and Continuity of Power Functions for Multiparameter Exponential Class Densities***

*Let the joint density of a random sample be a member of the exponential class of densities*

$$f(\mathbf{x};\Theta) = \exp\left(\sum_{i=1}^k c_i(\Theta)g_i(\mathbf{x}) + d(\Theta) + z(\mathbf{x})\right)I_A(\mathbf{x}),$$

*$\boldsymbol{\Theta} \in \Omega$, with $\Omega$ being an open rectangle contained in $\mathbb{R}^k$ and with $c_1(\boldsymbol{\Theta}),\ldots,c_k(\boldsymbol{\Theta})$, $d(\boldsymbol{\Theta})$ being differentiable functions of $\boldsymbol{\Theta} \in \Omega$. Let $C_r$ be any rejection region for testing the hypothesis $H_0$: $\boldsymbol{\Theta} \in \Omega_0$ versus $H_a$: $\boldsymbol{\Theta} \in \Omega_a$.*[23] *Then the power function $\pi_{Cr}(\boldsymbol{\Theta}) = P(\mathbf{x} \in C_r;\boldsymbol{\Theta})$ is differentiable, and hence continuous, with respect to $\boldsymbol{\Theta} \in \Omega$.*

**Proof** Follows from Lemma 9.2 with a proof analogous to the proof of Theorem 9.8. ■

We will state our main results more generally in terms of hypotheses concerning the value of the $c_1(\boldsymbol{\Theta})$ function in the definition of the exponential class density.[24] There is no loss of generality in stating our results this way. First note that the ordering of the functions $c_1(\boldsymbol{\Theta}),\ldots,c_k(\boldsymbol{\Theta})$ in the exponential class definition is arbitrary, and so the results we present can be interpreted as referring to any of the originally-specified $c_i(\boldsymbol{\Theta})$ functions, given an appropriate reordering and relabeling of the functions $c_1(\boldsymbol{\Theta}),\ldots,c_k(\boldsymbol{\Theta})$. Furthermore, since the parameterization of a density function is not unique, the $c_1(\boldsymbol{\Theta})$ function can be defined to represent various functions of $\boldsymbol{\Theta}$ that might be of interest to the analyst. In particular, a parameterization of the density of $\mathbf{X}$ that results in $c_1(\boldsymbol{\Theta}) = \Theta_i$ places hypotheses about $\Theta_i$ under consideration. We will also be able to extend our results to the case of testing hypotheses concerning linear combinations of the functions $c_1(\boldsymbol{\Theta}),\ldots,c_k(\boldsymbol{\Theta})$. Thus, types of hypotheses that can be tested using the following theorem are much more general than they might at first appear.

**Theorem 9.12 *Hypothesis Testing in the Multiparameter Exponential Class*** *Let the density of the random sample $\mathbf{X}$ be given by*

$$f(\mathbf{x};\Theta) = \exp\left(\sum_{i=1}^{k} c_i(\boldsymbol{\Theta})g_i(\mathbf{x}) + d(\boldsymbol{\Theta}) + z(\mathbf{x})\right)I_A(\mathbf{x}),$$

$\boldsymbol{\Theta} \in \Omega \subset \mathbb{R}^k$, where $\Omega$ is an open rectangle, $s_i(\mathbf{X}) = g_i(\mathbf{X})$, $i = 1,\ldots,k$, represents a set of complete sufficient statistics for $f(\mathbf{x};\boldsymbol{\Theta})$, and $c(\boldsymbol{\Theta}) = (c_1(\boldsymbol{\Theta}),\ldots,c_k(\boldsymbol{\Theta}))$ and

[23] We are assuming that $C_r$ is such that a power function is defined, i.e., $C_r$ can be assigned probability by $f(\mathbf{x};\boldsymbol{\Theta})$, $\boldsymbol{\Theta} \in \Omega$.

[24] Note that $c = (c_1,\ldots,c_k)$ could be viewed as an alternative parameterization of the exponential class of densities, where

$$f_*(\mathbf{x};\mathbf{c}) = \exp\left(\sum_{i=1}^{k} c_i g_i(\mathbf{x}) + d_*(\mathbf{c}) + z(\mathbf{x})\right)I_A(\mathbf{x}),\quad c \in \Omega_c,$$

with $d_*(c) = \ln\left(\int_{-\infty}^{\infty}\cdots\int_{-\infty}^{\infty}\exp\left(\sum_{i=1}^{k} c_i g_i(\mathbf{x}) + z(\mathbf{x})\right)I_A(\mathbf{x})d\mathbf{x}\right)^{-1}$ (use summation in the discrete case). This parameterization is referred to as the **natural parameterization of the exponential class** of densities. Note that the definition of $d_*(\mathbf{c})$ is a direct result of the fact that the density must integrate (or sum) to 1.

$d(\Theta)$ are differentiable functions of $\Theta \in \Omega$. If the size $\alpha$ rejection region $C_r$ is unbiased, then the following relationships between $H_0$, $H_a$ and $C_r$ hold:

| Case | $H_0$ | $H_a$ | $C_r$ for UMPU Level $\alpha$ Test |
|---|---|---|---|
| 1. | $c_1(\Theta) = c_1^0$<br>$c_1(\Theta) \leq c_1^0$ | $c_1(\Theta) > c_1^0$<br>$c_1(\Theta) > c_1^0$ | $C_r = \{\mathbf{x}: s_1(\mathbf{x}) \geq h(s_2,\ldots,s_k)\}$ such that<br>$P(\mathbf{x} \in C_r; c_1^0 \mid s_2,\ldots,s_k) = \alpha \ \forall (s_2,\ldots,s_k)^{25}$ |
| 2. | $c_1(\Theta) = c_1^0$<br>$c_1(\Theta) \geq c_1^0$ | $c_1(\Theta) < c_1^0$<br>$c_1(\Theta) < c_1^0$ | $C_r = \{\mathbf{x}: s_1(\mathbf{x}) \leq h(s_2,\ldots,s_k)\}$ such that<br>$P(\mathbf{x} \in C_r; c_1^0 \mid s_2,\ldots,s_k) = \alpha, \ \forall (s_2,\ldots,s_k)^{25}$ |
| 3. | $c_1(\Theta) = c_1^0$ | $c_1(\Theta) \neq c_1^0$ | $C_r = \{\mathbf{x}: s_1(\mathbf{x}) \leq h_1(s_2,\ldots,s_k)$, or<br>$s_1(\mathbf{x}) \geq h_2(s_2,\ldots,s_k)\}$ such that<br>$P(\mathbf{x} \in C_r; c_1^0 \mid s_2,\ldots,s_k) = \alpha, \ \forall (s_2,\ldots,s_k)^{25}$ |
| 4. | $c_1(\Theta) \in [c_1^\ell, c_1^h]$ | $c_1(\Theta) \notin [c_1^\ell, c_1^h]$ | $C_r = \{\mathbf{x}: s_1(\mathbf{x}) \leq h_1(s_2,\ldots,s_k)$ or<br>$s_1(\mathbf{x}) \geq h_2(s_2,\ldots,s_k)\}$ such that<br>$P(\mathbf{x} \in C_r; c_1^\ell \mid s_2,\ldots,s_k) = \alpha,$<br>$P(\mathbf{x} \in C_r; c_2^h \mid s_2,\ldots,s_k) = \alpha, \ \forall (s_2,\ldots,s_k)^{25}$ |

**Proof** See Appendix. ■

### 9.5.5 Examples of UMPU Tests in the Multiparameter Exponential Class

We now examine examples illustrating how Theorem 9.12 can be used in practice to define UMPU level $\alpha$ tests of statistical hypotheses. Some of the examples will illustrate the use of *reparameterization* for transforming a problem into a form that allows Theorem 9.12 to be used for defining a statistical test. We also identify general hypothesis testing contexts in which the results of the examples can be applied.

**Example 9.26** ***Testing Inequality Hypotheses About the Variance of a Normal Population Distribution***

The number of miles that a certain brand of automobile tire can be driven before the tread-wear indicators are visible, and the tires need to be replaced, is a random variable having a normal distribution with unknown mean *and* variance, i.e., $X_i \sim N(\mu, \sigma^2)$. A random sample of 28 tires is used to obtain information about the mean and variance of the population distribution of mileage obtainable from this brand of tire. We wish to define a level .05 UMPU test of the hypotheses $H_0$: $\sigma^2 \leq 4$ versus $H_a$: $\sigma^2 > 4$, where the mileage measurement, $x_i$, is measured in 1,000's of miles, and given that $\bar{x} = 45.175$ and $s^2 = 2.652$, we wish to test the hypothesis.

First note that the joint density of the random sample, $f(\mathbf{x};\mu,\sigma^2) = \left((2\pi)^{n/2}\sigma^n\right)^{-1} \exp(-(1/2)\sum_{i=1}^n (x_i-\mu)^2/\sigma^2)$, for $n = 28$, can be written in exponential class form as

$$f(\mathbf{x};\mu,\sigma^2) = \exp\left[\sum_{i=1}^{2} c_i(\mu,\sigma^2) g_i(\mathbf{x}) + d(\mu,\sigma^2) + z(\mathbf{x})\right]$$

[25]Except, perhaps, on a set having probability zero.

where

$$c_1(\mu,\sigma^2) = -(2\sigma^2)^{-1}, \qquad g_1(\mathbf{x}) = \sum_{i=1}^{n} x_i^2,$$

$$c_2(\mu,\sigma^2) = (\mu/\sigma^2), \qquad g_2(\mathbf{x}) = \sum_{i=1}^{n} x_i,$$

$$d(\mu,\sigma^2) = -\frac{n}{2}(\mu^2/\sigma^2) - \ln\left((2\pi)^{n/2}\sigma^n\right), \text{ and } z(\mathbf{x}) = 0.$$

Letting $s_i(\mathbf{x}) = g_i(\mathbf{x})$ for $i = 1,2$, $s_1(\mathbf{X})$ and $s_2(\mathbf{X})$ are complete sufficient statistics for $f(\mathbf{x};\mu,\sigma^2)$. Examining the parameterization of the exponential class density, it is evident that hypotheses concerning $\sigma^2$ can be framed in terms of hypotheses about $c_1$. In particular, $H_0$: $\sigma^2 \le 4$ versus $H_a$: $\sigma^2 > 4$ can be alternatively expressed as $H_0$: $c_1 \le -.125$ versus $H_a$: $c_1 > -.125$.

Given result (1) of Theorem 9.12, we seek $h(s_2)$ such that for $c_1^0 = -.125$,

$$\pi_{C_r}(c_1^0) = P(s_1(\mathbf{x}) \ge h(s_2); c_1^0|s_2) = P\left(\sum_{i=1}^{n} x_i^2 \ge h\left(\sum_{i=1}^{n} x_i\right); c_1^0|\sum_{i=1}^{n} x_i\right) = \alpha = .05$$

$\forall\, s_2 = \sum_{i=1}^{n} x_i$. We can simplify this problem substantially by first recalling (Theorem 6.12) that the two random variables $\sum_{i=1}^{n}(X_i - X)^2 = \sum_{i=1}^{n} X_i^2 - n^{-1}\left(\sum_{i=1}^{n} X_i\right)^2$ and $\sum_{i=1}^{n} X_i$ are *independent* random variables, given that the $X_i$'s are *iid* $N(\mu,\sigma^2)$. Then the preceding probability equality can be written alternatively as

$$P\left(\sum_{i=1}^{n}(x_i - \bar{x})^2 \ge h_*\left(\sum_{i=1}^{n} x_i\right); c_1^0\right) = \alpha = .05, \text{where } h_* = h - \mathrm{n}^{-1}\left(\sum_{\mathrm{i}=1}^{\mathrm{n}} \mathrm{x_i}\right)^2.$$

Note that we have eliminated the conditional probability notation since the probability distribution of the random variable $\sum_{i=1}^{n}(X_i - \bar{X})^2$ is *unaffected* by the value of $\sum_{i=1}^{n} x_i$ given the aforementioned independence property. The value of $\sum_{i=1}^{n} x_i$ only serves to determine the lower bound value $h_*\left(\sum_{i=1}^{n} x_i\right)$ on outcomes of the random variable $\sum_{i=1}^{n}(X_i - \bar{X})^2$ but has no effect on the random variable's distribution.

We can simplify the problem still further by first recalling that $\sum_{i=1}^{n}(X_i - \bar{X})^2/\sigma^2 \sim \chi^2_{n-1}$ when the $X_i$'s are *iid* $N(\mu,\sigma^2)$ (recall Theorem 6.12). Now rewrite the preceding probability equality as

$$P\left(\sum_{i=1}^{n}\frac{(x_i - \bar{x})^2}{\sigma_0^2} \ge h_*\left(\sum_{i=1}^{n} x_i\right)/\sigma_0^2; c_1^0\right) = \alpha = .05, \text{where } \sigma_0^2 = 4.$$

Note the probability is being calculated using the parameter value $c_1^0 = -.125$, which coincides with $\sigma_0^2 = 4$. In other words, the probability is being calculated assuming that $X_i \sim N(\mu,4)\ \forall i$, so that $\sum_{i=1}^{n}(X_i - X)^2/\sigma_0^2 = \sum_{i=1}^{n}(X_i - X)^2/4 \sim \chi^2_{n-1}$. It follows that the value of $h_*\left(\sum_{i=1}^{n} x_i\right)/\sigma_0^2$ that solves the probability

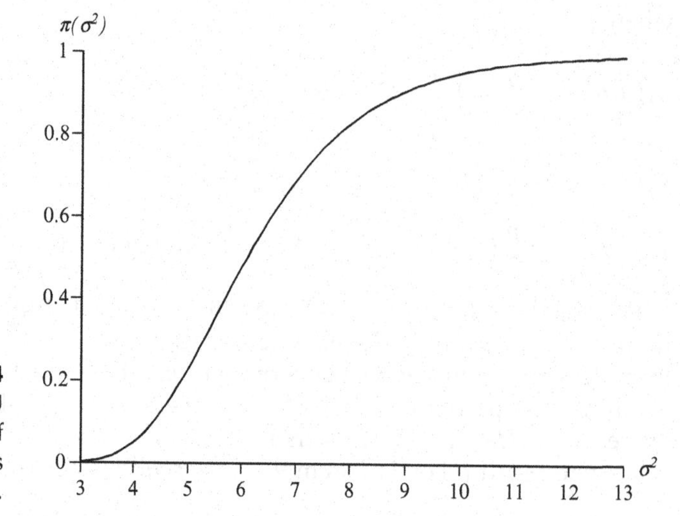


**Figure 9.14** Power function of UMPU level .05 unbiased test of $H_0$: $\sigma^2 \leq 4$ versus $H_a$: $\sigma^2 > 4$.

equality can be obtained from the table of the $\chi^2$ distribution, with $n - 1 = 27$ degrees of freedom in this case. The tabled value of 40.1 corresponding to $\alpha = .05$ implies that

$$h_*\left(\sum_{i=1}^{n} x_i\right)/\sigma_0^2 = \left[h\left(\sum_{i=1}^{n} x_i\right) - n^{-1}\left(\sum_{i=1}^{n} x_i\right)^2\right]/\sigma_0^2$$

$$= \left[h\left(\sum_{i=1}^{28} x_i\right) - \frac{1}{28}\left(\sum_{i=1}^{28} x_i\right)^2\right]/4 = 40.1$$

so that $h\left(\sum_{i=1}^{28} x_i\right) = 160.4 + \left(\sum_{i=1}^{28} x_i\right)^2/28$.

Given that $\sum_{i=1}^{28} x_i = 28(45.175) = 1264.9$, then $h(1264.9) = 57{,}302.258$. By Theorem 9.12 the rejection region becomes $C_r = \left\{\mathbf{x} : \sum_{i=1}^{28} x_i^2 \geq 57{,}302.258\right\}$. The rejection region can also be expressed in terms of the sample variance as $C_r = \{\mathbf{x} : 28s^2/4 \geq 40.1\}$.Then since $28s^2/4 = 28(2.652)/4 = 18.564 \ngeq 40.1$, the null hypothesis $\sigma^2 \leq 4$ is *not* rejected.

The power function of the test can be defined in terms of the parameter $\sigma^2$ as:

$$\pi_{C_r}(\sigma^2) = P\left(\frac{28s^2}{\sigma^2}\left(\frac{\sigma^2}{4}\right) \geq 40.1; \sigma^2\right) = P\left(\frac{28s^2}{\sigma^2} \geq \frac{160.4}{\sigma^2}; \sigma^2\right)$$

where again, assuming the value of $\sigma^2$ is true, $\frac{28s^2}{\sigma^2} \sim \chi^2_{27}$.

The power function could also have been expressed as a function of $c_1$, but power expressed in terms of $\sigma^2$ is more natural given the original statement of the hypothesis. We can use the CDF of the $\chi^2_{27}$ distribution to plot a graph of the power function for various potential values of $\sigma^2$: (see Figure 9.14)

| $\sigma^2$ | $P(28s^2/\sigma^2 \geq 160.4/\sigma^2; \sigma^2) = P(\chi^2_{27} \geq 160.4/\sigma^2; \sigma^2)$ |
|---|---|
| 4 | .050 |
| 3.9, 4.1 | .040, .062 |
| 3.75, 4.25 | .028, .082 |
| 3.50, 4.50 | .013, .123 |
| 3.00, 5.00 | .002, .229 |
| 6.00 | .478 |
| 7.00 | .700 |
| 8.00 | .829 |
| 10.00 | .952 |
| 13.00 | .993 |

Note that even though $C_r$ defines a *UMPU level .05 test* (by Theorem 9.12), the test is not very powerful for values of $\sigma^2 \in (4,7)$, say. To increase the power of the test in this range, one would either need to increase the size of the random sample on which the test is based, or else use a test size larger than .05, although the latter approach would, of course, increase the Type I Error probability associated with the test. □

We can generalize the results of the previous example as follows:

**Definition 9.14**
***UMPU Level $\alpha$ Statistical Test of $H_0$: $\sigma^2 \leq \sigma_0^2$ versus $H_a$: $\sigma^2 > \sigma_0^2$ when Sampling from a Normal Population Distribution***

Let $X_1,\ldots,X_n$ be a random sample from $N(\mu, \sigma^2)$. The rejection region denoted by

$$C_r = \left\{ \mathbf{x} : \frac{ns^2}{\sigma_0^2} \geq \chi^2_{n-1;\alpha} \right\},$$

where $\chi^2_{n-1;\alpha}$ solves $\int_{\chi^2_{n-1;\alpha}}^{\infty} Chisquare(z; n-1)dz = \alpha$, defines a UMPU level $\alpha$ statistical test of $H_0$: $\sigma^2 \leq \sigma_0^2$ versus $H_0$: $\sigma^2 > \sigma_0^2$.

The next example illustrates that a great deal of ingenuity may be required, concomitant with a rather high degree of mathematical sophistication, in order to implement the results of Theorem 9.12.

**Example 9.27**
***Testing One and Two-Sided Hypotheses about the Mean of a Normal Population Distribution Having Unknown Variance***

Recall Example 9.26 regarding mileage obtained from automobile tires. Define UMPU level .05 tests of $H_0$: $\mu = 40$ versus $H_a$: $\mu \neq 40$, and of $H_0$: $\mu \leq 40$ versus $H_a$: $\mu > 40$.

The parameterization used in Example 9.26 is not directly useful for implementing the results of Theorem 9.12 to define statistical tests of the hypotheses under consideration. In considering alternative parameterizations, the following lemma is useful.

**Lemma 9.3** ***Reparameterizing Exponential Class Densities* via *Linear Transformations***

Let an exponential class density be given by

$$f(\mathbf{x};\mathbf{c}) = \exp\left[\sum_{i=1}^{k} c_i g_i(\mathbf{x}) + d(\mathbf{c}) + z(\mathbf{x})\right] I_A(\mathbf{x}).$$

Let $c_1^* = \sum_{i=1}^{k} a_i c_i$, with $a_1 \neq 0$, and let $\mathbf{c}^* = (c_1^*, c_2, \ldots, c_k)'$. Then $f(\mathbf{x};\mathbf{c})$ can be alternatively represented as

$$f(\mathbf{x};\mathbf{c}^*) = \exp\left[c_1^* g_1^*(\mathbf{x}) + \sum_{i=2}^{k} c_i g_i^*(\mathbf{x}) + d^*(\mathbf{c}^*) + z(\mathbf{x})\right] I_A(\mathbf{x}),$$

where

$$g_1^*(\mathbf{x}) = \frac{g_1(\mathbf{x})}{a_1};\quad g_i^*(\mathbf{x}) = g_i(\mathbf{x}) - \left(\frac{a_i}{a_1}\right) g_1(\mathbf{x}),\quad i = 2,\ldots,k;$$

and

$$d^*(\mathbf{c}^*) = \ln\left(\left[\int_{-\infty}^{\infty}\cdots\int_{-\infty}^{\infty} \exp\left[c_1^* g_1^*(\mathbf{x}) + \sum_{i=2}^{k} c_i g_i^*(\mathbf{x}) + z(\mathbf{x})\right] I_A(\mathbf{x}) d\mathbf{x}\right]^{-1}\right).$$

The proof is immediate upon substituting the definitions of $c_1^*$ and $g_i^*(\mathbf{x})$, $i = 1,\ldots k$, into the expression for $f(\mathbf{x};\mathbf{c}^*)$, and defining $d(\mathbf{c}^*)$ so that the density integrates to 1. Replace integration with summation in the discrete case. ■

The point of the lemma is that any linear combination of the $c_i(\cdot)$ functions in the definition of an exponential class density can be used as a parameter in a reparameterization of the density. Note that the $c_i(\cdot)$ functions can always be reordered and relabeled so that the condition $a_1 \neq 0$ is not a restriction in practice. Then Theorem 9.12 can be used to test hypotheses about $c_1^*$, or equivalently, about $\sum_{i=1}^{k} a_i c_i$.

In the case at hand, *reorder* the $c_i(\cdot)$ functions and $g_i(\mathbf{x})$ functions so that

$$c_1^* = \sum_{i=1}^{2} a_i c_i(\mu,\sigma^2) = a_1\left(\frac{\mu}{\sigma^2}\right) + a_2\left(-\frac{1}{2\sigma^2}\right).$$

Without knowledge of the value of $\sigma^2$, it is clear that for $c_1^* = c_1^0$ to imply a unique value of $\mu$, the condition $c_1^0 = 0$ is required, in which case $\mu = a_2/(2a_1)$. Then letting $c_1^0 = 0$ and $a_2/(2a_1) = \mu_0 = 40$, it follows that $c_1^* = 0$ *iff* $\mu = 40$, and $c_1^* \leq 0$ *iff* $\mu_0 \leq 40$, assuming $a_1 > 0$. This establishes an equivalence between the hypotheses concerning $\mu$ and hypotheses concerning $c_1^*$ as follows:

$$\left\{\begin{matrix} H_0: \mu = 40 \\ \text{versus} \\ H_a: \mu \neq 40 \end{matrix}\right. \Leftrightarrow \left.\begin{matrix} H_0: c_1^* = 0 \\ \text{versus} \\ H_a: c_1^* \neq 0 \end{matrix}\right\} \text{ and } \left\{\begin{matrix} H_0: \mu \leq 40 \\ \text{versus} \\ H_a: \mu > 40 \end{matrix}\right. \Leftrightarrow \left.\begin{matrix} H_0: c_1^* \leq 0 \\ \text{versus} \\ H_a: c_1^* > 0 \end{matrix}\right\}$$

**Test of $H_0$:$\mu \leq 40$ versus $H_a$:$\mu > 40$**: The reparameterized exponential class density can be written as

$$f(\mathbf{x};\mathbf{c}^*) = \exp\left(c_1^*\left(\sum_{i=1}^n \mathbf{x}_i/a_1\right) + c_2\left(\sum_{i=1}^n x_i^2 - \frac{a_2}{a_1}\sum_{i=1}^n x_i\right) + d^*(\mathbf{c}^*)\right) I_A(\mathbf{x})$$

(recall Lemma 9.3, and the fact that $z(\mathbf{x}) = 0$ in this case). Our approach will be simplified somewhat if we rewrite $f(\mathbf{x};\mathbf{c}^*)$ as

$$f(\mathbf{x};\mathbf{c}^*) = \exp\left(c_1^*\sum_{i=1}^n \frac{(x_i - \mu_0)}{a_1} + c_2\sum_{i=1}^n (x_1 - \mu_0)^2 + d^0(\mathbf{c}^*)\right) I_A(\mathbf{x})$$
$$= \exp\left(c_1^* s_1^* + c_2 s_2^* + d^0(\mathbf{c}^*)\right) I_A(\mathbf{x})$$

where $d^0(\mathbf{c}^*) = d^*(\mathbf{c}^*) - n\mu_0^2 c_2 + n\mu_0 c_1^*/a_1$, $\mu_0 = 40$, and we have used the fact that $-a_2/a_1 = -2\ \mu_0$.

Result (1) of Theorem 9.19 indicates that a UMPU level $\alpha$ test of $H_0$: $c_1^* \leq 0$ versus $H_a$: $c_1^* > 0$ can be defined by first finding a function $h(s_2^*)$ such that

$$P(s_1^*(\mathbf{x}) \geq h(s_2^*); c_1^0|s_2^*) = P\left(\sum_{i=1}^n \frac{(x_i - \mu_0)}{a_1} \geq h\left(\sum_{i=1}^n (x_i - \mu_0)^2\right); c_1^0|s_2^*\right) = \alpha$$
$$= .05 \quad (*)$$

$\forall s_2^* = \sum_{i=1}^n (x_i - \mu_0)^2$, with $c_1^0 = 0$. Letting $\mathbf{c}^0 = [c_1^0, c_2]' = [0, c_2]'$, it is useful to recognize that $f(\mathbf{x};\mathbf{c}^0) = \exp\left[c_2 s_2^*(\mathbf{x}) + d^0(\mathbf{c}^0)\right] I_A(\mathbf{x})$, which indicates that the value of $f(\mathbf{x};\mathbf{c}^0)$ changes *only* when the value of $s_2^*(\mathbf{x})$ changes. This observation then clarifies the nature of the conditional distribution of $\mathbf{x}$ *given* $s_2^*$, namely, the conditional distribution of $\mathbf{X}$ is a *uniform* distribution on the range $R(\mathbf{X}|s_2^*) = \left\{\mathbf{x} : \sum_{i=1}^n (x_i - \mu_0)^2 = s_2^*\right\}$, where $R(\mathbf{X}|s_2^*)$ defines the boundary of an $n$-dimensional hypersphere having radius $(s_2^*)^{1/2}$. Specifically,[26]

$$f(\mathbf{x};\mathbf{c}^0|s_2^*) = \left[\frac{n\left(s_2^*\right)^{(n-1)/2}\pi^{n/2}}{\Gamma\left(\frac{n}{2}+1\right)}\right]^{-1} I_{R(\mathbf{X}|s_2^*)}(\mathbf{x}).$$

It follows from the definition of the support of the density of $\mathbf{X}$ that $f(\mathbf{x};\mathbf{c}^0|s_2^*)$ is a *degenerate* density function, since all of the probability mass for the $(n \times 1)$ vector $\mathbf{X}$ is concentrated on the surface $R(\mathbf{X}|s_2^*)$ which has dimension $n-1$.

---

[26]The surface area of an n-dimensional hypersphere is given by $A = (nr^{n-1}\pi^{n/2})/\Gamma((n/2)+1)$, where $r$ is the radius of the hypersphere. See R.G. Bartle, (1976), *The Elements of Real Analysis*, 2nd Ed., John Wiley, pp. 454–455, and note that the surface area can be defined by differentiating the volume of the hypersphere with respect to $r$. For $n = 2$, the bracketed expression becomes simply $2\pi(s_2^*)^{1/2}$, which is the familiar $2\pi r$.

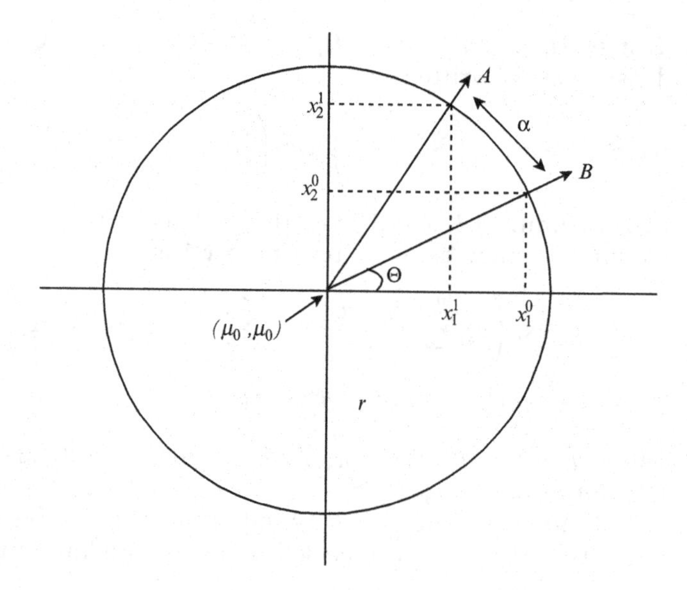


**Figure 9.15**
Graph of $R(X|r) = \left\{\mathbf{x} : \sum_{i=1}^{2} (x_i - \mu_0)^2 \le r^2\right\}$.

We can now define the function $h(s_2^*)$ that satisfies (*). Let $h(s_2^*) = \gamma_\alpha (s_2^*)^{1/2}$, where $\gamma_\alpha$ is a positive constant. Multiplying both sides of the inequality in (*) by $a_1$ defines

$$P\left(\sum_{i=1}^{n} (x_i - \mu_0) \ge \gamma_\alpha^* (s_2^*)^{1/2}; c_1^0 | s_2^*\right) = \alpha = .05 \quad **$$

where $\gamma_\alpha^* = a_1 \gamma_\alpha$ is chosen so that (**) is true. We motivate the choice of $\gamma_\alpha^*$ by examining the case where $n = 2$, although an analogous argument applies to arbitrary finite $n$. Examine a circle with center $(\mu_0, \mu_0)$ and radius $r = (s_2^*)^{1/2}$, as in Figure 9.15. Recall from basic trigonometry that with $(\mu_0, \mu_0)$ as the origin, the point $(x_1^0 - \mu_0, x_2^0 - \mu_0)$ can be represented in terms of polar coordinates as $(r\cos(\Theta), r\sin(\Theta))$. The function of $\Theta$ defined by $\sum_{i=1}^{2} (x_i - \mu_0) = r(\cos\Theta + \sin\Theta)$ is strictly concave for $\Theta \in (-135, 225)$, attains a maximum at $\Theta = 45$ (the maximized value being $1.4142r$ to 4 decimals), and is symmetric around $\Theta = 45$. It follows that an $\alpha$-proportion of the points $(x_1, x_2)$ on the circumference of the circle can be defined by appropriately choosing $\gamma_\alpha^* < 1.4142$ as

$$\begin{aligned} D_\alpha &= \{(x_1, x_2) : (x_1 - \mu_1) = r\cos\Theta, (x_2 - \mu_2) = r\sin(\Theta), r(\cos(\Theta) + \sin(\Theta)) \ge \gamma_\alpha^* r\} \\ &= \left\{(x_1, x_2) : \sum_{i=1}^{2} (x_i - \mu_0) \ge \gamma_\alpha^* (s_2^*)^{1/2}\right\}, \end{aligned}$$

and since $(x_1, x_2)$ is distributed uniformly on the circumference, it follows that $P(D_\alpha | s_2^*) = \alpha$.

Now note that the preceding argument applies for *arbitrary* $r = (s_2^*)^{1/2} > 0$, so that $\gamma_\alpha^*$ is the *same* value $\forall r > 0$. An analogous argument leading to the constancy of $\gamma_\alpha^*$ can be applied to the case of an $n$-dimensional hypersphere.

Thus the rejection region for the test will have the form

$$C_r = \left\{\mathbf{x} : \sum_{i=1}^{n} (x_i - \mu_0) \geq \gamma_\alpha^* (s_2^*)^{1/2}\right\},$$

regardless of the value of $s_2^*$. The remaining task is then to find the appropriate value of $\gamma_\alpha^*$ that defines a test of size $\alpha$.

It turns out that there is an alternative representation of the rejection region in terms of the so-called **t-statistic**, which is the form in which the test is usually defined in practice. To derive this alternative representation, first note that[27]

$$s_2^* = \sum_{i=1}^{n} (x_i - \mu_0)^2 = \sum_{i=1}^{n} (x_i - \bar{x})^2 \left[1 + \frac{n(\bar{x} - \mu_0)^2}{\sum_{i=1}^{n} (x_i - \bar{x})^2}\right].$$

Then the rejection region can be represented as

$$C_r = \left\{\mathbf{x} : \frac{n(\bar{x} - \mu_0)}{\left[\sum_{i=1}^{n} (x_i - \bar{x}^2)\right]^{1/2}} \geq \gamma_\alpha^* \left[1 + \frac{n(\bar{x} - \mu_0)^2}{\left[\sum_{i=1}^{n} (x_i - \bar{x})^2\right]}\right]^{1/2}\right\}.$$

Multiplying both sides of the inequality by $[(n-1)/n]^{1/2}$ and defining $t = n^{1/2}(\bar{x} - \mu_0)/\left[\sum_{i=1}^{n} (x_i - \bar{x})^2/(n-1)\right]^{1/2}$, the representation of $C_r$ becomes

$$C_r = \left\{\mathbf{x} : \frac{t}{(t^2 + n - 1)^{1/2}} \geq \gamma_\alpha^0\right\}$$

where $\gamma_\alpha^0 = \gamma_\alpha^*/n^{1/2}$. Since $t/(t^2 + n - 1)^{1/2} = \left(1 + (n-1)/t^2\right)^{-1/2}$ is strictly monotonically increasing in $t$, there exists a value $t_{n-1;\alpha}$ such that

$$\frac{t}{(t^2 + n - 1)^{1/2}} \geq \gamma_\alpha^0 \Leftrightarrow t \geq t_{n-1;\alpha},$$

so that $C_r$ can be substantially simplified to

$$C_r = \{\mathbf{x} : t \geq t_{n-1;\alpha}\}.$$

Assuming $\mu = \mu_0$ to be true, $T$ has the student $t$-distribution with $n-1$ degrees of freedom so that the value of $t_{n-1;\alpha}$ can be found as the value that makes $P(t \geq t_{n-1;\alpha}; \mu_0) = \alpha = .05$ true. From the table of the student $t$-distribution, with degrees of freedom equal to $n - 1 = 27$, it is found that $t_{27;\alpha} = 1.703$. Given that $\bar{x} = 45.175$ and $s^2 = 2.652$, it follows that for $\mu_0 = 40$, $t = (28)^{1/2}(45.175 - 40)/(2.75)^{1/2} = 16.513$. Because $t > 1.703$ we *reject* $H_0$: $\mu \leq 40$ in favor of $H_a$: $\mu > 40$.

[27]Expanding $\sum_{i=1}^{n} [(x_i - \bar{x}) + (\bar{x} - \mu_0)]^2$ leads to the result.

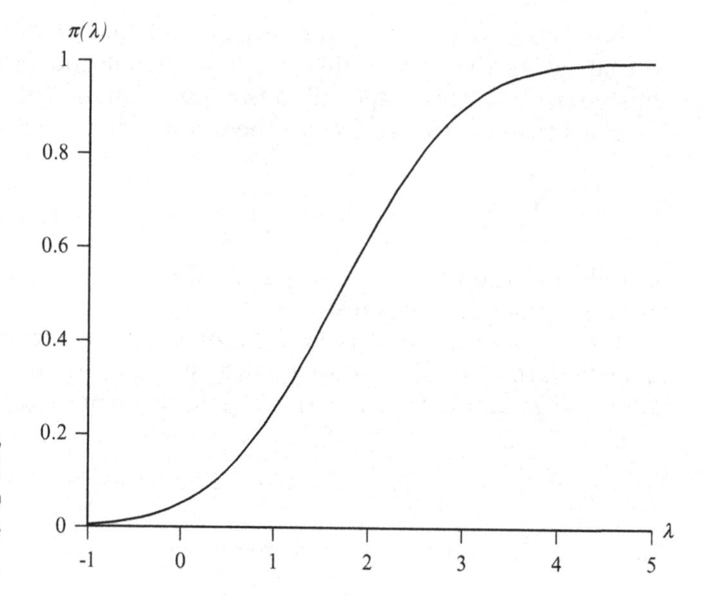


**Figure 9.16**
Power function of UMPU level .05 test of $H_0$: $\mu \leq 40$ versus $H_a$: $\mu > 40$, where $\lambda = (\mu - 40)/(\sigma/\sqrt{28})$.

In order to graph the power function of the test (See Figure 9.16), we must calculate values of $\pi(\mu) = P(t \geq 1.703;\ \mu)$, which requires that we identify the distribution of $T$ when $\mu \neq \mu_0$. To see what is involved, note that we can represent $T$ as

$$T = \frac{(\bar{X} - \mu) + (\mu - \mu_0)}{\sigma/\sqrt{n}} \bigg/ \left[\frac{nS^2}{\sigma^2}/(n-1)\right]^{1/2},$$

so that if $\mu \neq \mu_0$ we have

$$T = \left[Z + \frac{\mu - \mu_0}{\sigma/\sqrt{n}}\right] \bigg/ \left[\frac{Y}{(n-1)}\right]^{1/2},$$

where $Z \sim N(0,1)$, $Y \sim \chi^2_{n-1}$, and $Z$ and $Y$ are independent. Letting $\lambda = (\mu - \mu_0)/(\sigma/\sqrt{n})$, the random variable $T$ has a **noncentral *t*-distribution,** with **noncentrality parameter $\boldsymbol{\lambda}$**, and degrees of freedom $n - 1$, which we denote by $T_{n-1}(\lambda)$. When $\lambda = 0$, the ordinary **central student t-distribution** is defined. We discuss properties of the noncentral $t$-distribution in Section 9.6 of this chapter. For now, note that $\lambda$ depends not only on $\mu$, but also on $\sigma$, which effectively implies that we must either plot the power of the test in two dimensions (in terms of values of $\mu$ and $\sigma$), or else in a single dimension as a function of the noncentrality parameter. The latter is the approach most often followed in practice, and we follow it here. Integrals of the noncentral $t$-distribution can be evaluated by many modern statistical software packages, such as SAS, GAUSS, and MATLAB.

| $\lambda$ | $\pi(\lambda) = P(t \geq 1.703;\lambda)$ where $T \sim T_{27}(\lambda)$ |
|---|---|
| 0 | .050 |
| −.1,.1 | .041,.061 |
| −.5,.5 | .016,.124 |
| −1,1 | .004,.251 |
| 1.5 | .428 |
| 2 | .620 |
| 3 | .900 |
| 4 | .988 |
| 5 | .999 |

**Test of** $H_0$: $\mu = 40$ **versus** $H_a$: $\mu \neq 40$: In order to define a test of $H_0$: $\mu = 40$ versus $H_a$: $\mu \neq 40$, result 3 of Theorem 9.12 suggests that the rejection region can be defined as

$$C_r = \{\mathbf{x} : s_1^*(\mathbf{x}) \leq h_1(s_2^*) \text{ or } s_1^*(\mathbf{x}) \geq h_2(s_2^*)\}$$

for $h_1$ and $h_2$ chosen such that $P(\mathbf{x} \in C_r; c_1^0|s_2^*) = \alpha = .05\ \forall\, s_2^*$. Using an argument along the lines of the previous discussion (see T. Ferguson, op. cit., p. 232) it can be shown that the rejection region can be specified as

$$C_r = \{\mathbf{x} : t \leq t_1 \text{ or } t \geq t_2\}$$

where $t$ is defined as before, and $t_1$ and $t_2$ are constants chosen so that when $\mu = \mu_0 = 40$, $P(\mathbf{x} \in C_r; \mu_0) = .05$. The additional condition in result 3 of Theorem 9.12 is met by choosing $t_1 = -t_2$ since $T$ has the symmetric student t-distribution when $\mu = \mu_0$ and Theorem 9.10 applies. In particular, $C_r$ is defined by

$$C_r = \{\mathbf{x} : t \leq -t_{n-1;\alpha/2} \text{ or } t \geq t_{n-1;\alpha/2}\},$$

where $t_{n-1;\alpha/2}$ solves $\int_{t_{\alpha/2}}^{\infty} \text{Tdist}(z; n-1)dz = \alpha/2$, where Tdist (Z,v) is the $t$-distribution with v degrees of freedom. For a $t$-distribution with 27 degrees of freedom, $t_{.025} = 2.052$. Since $t = 16.513 > 2.052$, $H_0$: $\mu = 40$ is rejected in favor of $H_a$: $\mu \neq 40$.

The power function of the test is defined by values of $P(\mathbf{x} \in C_r;\mu)$. As before, $T$ has a noncentral $t$-distribution (see Section 9.6) with noncentrality parameter $\lambda = (\mu - 40)/(\sigma/\sqrt{28})$ (see Figure 9.17). Defining the power function in terms of $\lambda$ yields

| $\lambda$ | $\pi(\lambda) = P(t \leq -2.052 \text{ or } t \geq 2.052;\lambda)$ where $T \sim T_{27}(\lambda)$ |
|---|---|
| 0 | .050 |
| −.1,.1 | .051 |
| −.5,.5 | .077 |
| −1,1 | .161 |
| −2,2 | .488 |
| −3,3 | .824 |
| −4,4 | .971 |
| −5,5 | .998 |

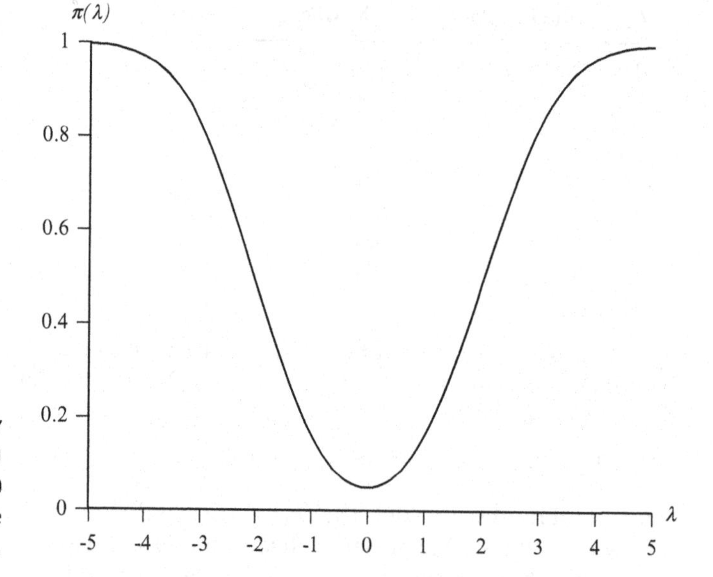


**Figure 9.17** Power function of UMPU level .05 test of $H_0$:$\mu = 40$ versus $H_a$: $\mu \neq 40$, where $\lambda = (\mu - 40)/(\sigma/\sqrt{28})$.

We can generalize the results of this example as follows.

**Definition 9.15 *UMPU Level $\alpha$ Statistical Test of $H_0$: $\mu \leq \mu_0$ Versus $H_a$: $\mu > \mu_0$, and of $H_0$: $\mu = \mu_0$ Versus $H_a$: $\mu \neq \mu_0$ When Sampling From a Normal Population Distribution***

Let $X_1,\ldots,X_n$ be a random sample from $N(\mu,\sigma^2)$, and define $T = (n-1)^{1/2}(\bar{x} - \mu_0)/S$. Then,

1. $C_r = \{\mathbf{x}: t \geq t_{n-1,\alpha}\}$ defines a UMPU level $\alpha$ statistical test of $H_0$: $\mu \leq \mu_0$ versus $H_a$: $\mu > \mu_0$, where $\int_{t_{n-1,\alpha}}^{\infty} Tdist(t;v)dt = \alpha$.
2. $C_r = \{\mathbf{x}: t \leq -t_{n-1;\alpha/2} \text{ or } t \geq t_{n-1;\alpha/2}\}$ defines a UMPU level $\alpha$ statistical test of $H_0$: $\mu = \mu_0$ versus $H_a$: $\mu \neq \mu_0$, where $\int_{t_{n-1,\alpha/2}}^{\infty} Tdist(t;v)dt = \alpha/2$.

### 9.5.6 Concluding Remarks

In this section we examined some classical statistical theory and procedures for defining either UMP or UMPU level $\alpha$ tests for various types of hypotheses. As is evident from the proofs of the theorems and the illustrative examples, attempting to define UMP or UMPU level $\alpha$ tests can sometimes be a substantial challenge. While the results provided in this section provide direction to the search for a statistical test, application of the results still requires considerable ingenuity, and the procedures are not always tractable or even applicable.

In Chapter 10 we will examine well-defined procedures that will lead to specific statistical tests requiring relatively less mathematical effort than the procedures introduced so far. While leading to the definition of statistical tests that *generally work well in practice*, the procedures do not explicitly incorporate the objective of finding a UMP or UMPU level $\alpha$ test. As a result, they cannot generally be relied upon to produce tests with these properties. However, the power functions of such tests are nonetheless often quite adequate for hypothesis testing applications, and the tests frequently do possess optimal properties.

## 9.6 Noncentral *t*-Distribution

**Family Name: Noncentral *t*-Distribution**

*Parameterization:*

$v$ (degrees of freedom) and $\lambda$ (noncentrality parameter).

$(v,\lambda) \in \Omega = \{(v,\lambda): v \text{ is a positive integer}, \lambda \in (-\infty,\infty)\}$

*Density Definition:*

$$f(t;v,\lambda) = \frac{v^{v/2}}{\pi^{1/2}\Gamma(v/2)2^{(v-1)/2}}\left(t^2+v\right)^{-(v+1)/2}\exp\left[-\frac{\lambda^2 v}{2(t^2+v)}\right]\times$$

$$\int_0^\infty \exp\left[-\frac{1}{2}\left(x-\frac{\lambda t}{\left(t^2+v\right)^{1/2}}\right)^2\right]x^v dx$$

*Moments:*

$$\mu = a_1\lambda \text{ for v} > 1, \sigma^2 = a_2\lambda^2 + a_3 \text{ for v} > 2, \mu_3 = a_4\lambda^3 + a_5\lambda \text{ for v} > 3,$$

where:

$$a_1 = (v/2)^{1/2}\Gamma\left(\frac{v-1}{2}\right)/\Gamma\left(\frac{v}{2}\right), a_2 = \frac{v}{v-2} - a_1^2, a_3 = \frac{v}{v-2},$$

$$a_4 = a_1\left[\frac{v(7-2v)}{(v-2)(v-3)} + 2a_1^2\right], \text{and } a_5 = \left[\frac{3v}{((v-2)(v-3))}\right]a_1.$$

Note that the expression for $(\mu_4/\sigma^4 - 3)$ is quite complicated, but it can be determined using the general result on higher order moments presented below.

*MGF:* Does not exist. However, higher order moments about the origin (from which moments about the mean can also be derived) can be defined directly as follows:

$$\mu'_k = \mathrm{E}\left(T^k\right) = \begin{cases} \left(\frac{v}{2}\right)^{\frac{k}{2}}\frac{\Gamma\left(\frac{v-k}{2}\right)}{\Gamma\left(\frac{v}{2}\right)}\exp\left(-\frac{\lambda^2}{2}\right)\frac{d^k}{d\lambda^k}\left(\exp\left(\frac{\lambda^2}{2}\right)\right) & \text{if } v > k; \\ \text{Does not exist} & \text{if } v \le k. \end{cases}$$

*Background and Application:* The noncentral *t*-distribution is the distribution of the random variable

$$T = \frac{Z}{\sqrt{Y/v}},$$

where $Z \sim N(\lambda, 1)$, $Y \sim \chi^2_v$, and $Z$ and $Y$ are independent. Note that when $\lambda = 0$, $T$ will have the ordinary (*central*) student *t*-distribution derived in Section 6.7. The density definition can be found by the change of variables approach, similar to the previous derivation of the student (central) *t*-density.

When $\lambda = 0$, the $t$-distribution is symmetric around zero. When $\lambda > 0$ ($\lambda < 0$) the mean of the distribution moves to the right (left) of zero, and the distribution becomes skewed to the right (left).

Our principal use of the noncentral $t$-distribution is in analyzing the power function of statistical tests based on the *t-statistic*. We have already encountered a case where the statistic $T = (n-1)^{1/2}(\bar{X} - \mu_0)/S$ was used to test hypotheses about the mean of a normal distribution, such as $H_0$: $\mu = \mu_0$ versus $H_a$: $\mu \neq \mu_0$, or $H_0$: $\mu \leq \mu_0$ versus $H_a$: $\mu > \mu_0$ (recall Example 9.27 and Definition 9.17). In particular, letting

$$T = \frac{\bar{X} - \mu_0}{\sigma/\sqrt{n}} \Big/ \left(\frac{nS^2}{\sigma^2}/(n-1)\right)^{1/2} = \frac{Z}{(Y/(n-1))^{1/2}},$$

where $Z \sim N((\mu - \mu_0)/(\sigma/\sqrt{n}), 1)$, $Y \sim \chi^2_{n-1}$, and $Z$ and $Y$ are independent (as in the context of Example 9.27), it follows from the definition of the noncentral $t$-density that $T \sim f(t;n-1,\lambda)$, with $\lambda = (\mu - \mu_0)/(\sigma/\sqrt{n})$. Then the power function of any test defined by $C_r = \{\mathbf{x} : t \in C_r^t\}$ can be expressed as a function of $\lambda$ as

$$\pi(\lambda) = \int_{t \in C_r^t} f(t; n-1, \lambda) dt.$$

The noncentral $t$-distribution can be integrated numerically, or the integral can be approximated via series expansions. Integration of the noncentral $t$-distribution adequate for performing power calculations is readily available in software packages such as GAUSS, SAS, and MATLAB. Regarding power function calculations, it is useful to note that

$$\int_c^\infty f(t; v, \lambda) dt \text{ increases as } \lambda \text{ increases,}$$

$$\int_{-\infty}^c f(t; v, \lambda) dt \text{ increases as } \lambda \text{ decreases,}$$

and

$$\int_{-c}^c f(t; v, \lambda) dt \text{ decreases as } |\lambda| \text{ increases.}$$

For further information, see the excellent survey article by D.B. Owen, (1968), "A Survey of Properties and Applications of the Noncentral $t$-Distribution," *Technometrics*, pp.445–478.

## 9.7 Appendix: Proofs and Proof References for Theorems

### Theorem 9.1

**Proof** (**Most Powerful Level α**) Let $C_r^*$ refer to any other rejection region for which $P(\mathbf{x} \in C_r^*;\Theta_0) \leq P(\mathbf{x} \in C_r(k);\Theta_0) = \alpha$, so that the test of $H_0$ versus $H_a$ based on $C_r^*$

has a size no larger than the size of the statistical test based on $C_r$. Note by definition that

$$I_{C_r(k)}(\mathbf{x}) - I_{C_r^*}(\mathbf{x}) = \begin{bmatrix} 1 - I_{C_r^*}(\mathbf{x}) \geq 0, \forall \mathbf{x} \in C_r(k) \\ 0 - I_{C_r^*}(\mathbf{x}) \leq 0, \forall \mathbf{x} \notin C_r(k) \end{bmatrix}.$$

Also note that

$$f(\mathbf{x};\boldsymbol{\Theta}_a)\begin{Bmatrix} \geq \\ < \end{Bmatrix} k^{-1} f(\mathbf{x};\boldsymbol{\Theta}_0) if \begin{bmatrix} \mathbf{x} \in C_r(k) \\ \mathbf{x} \notin C_r(k) \end{bmatrix},$$

which follows directly from the definition of $C_r(k)$. The preceding results together imply that

$$[I_{C_r(k)}(\mathbf{x}) - I_{C_r^*}(\mathbf{x})] f(\mathbf{x};\boldsymbol{\Theta}_a) \geq k^{-1} [I_{C_r(k)}(\mathbf{x}) - I_{C_r^*}(\mathbf{x})] f(\mathbf{x};\boldsymbol{\Theta}_0), \quad \forall \mathbf{x}.$$

Integrating both sides of the inequality over $\mathbf{x} \in \mathbb{R}^n$ if $\mathbf{X}$ is continuous, or summing over all $\mathbf{x}$-values for which $f(\mathbf{x};\boldsymbol{\Theta}_0) > 0$ *or* $f(\mathbf{x};\boldsymbol{\Theta}_a) > 0$ if $\mathbf{X}$ is discrete, obtains

$$P(\mathbf{x} \in C_r(k);\boldsymbol{\Theta}_a) - P(\mathbf{x} \in C_r^*;\boldsymbol{\Theta}_a) \geq k^{-1} [P(\mathbf{x} \in C_r(k);\boldsymbol{\Theta}_0) - P(\mathbf{x} \in C_r^*;\boldsymbol{\Theta}_0)] \geq 0.$$

The right hand side of the first inequality is nonnegative because $k^{-1} > 0$ and because the bracketed probability difference multiplying $k^{-1}$ is nonnegative (recall that the size of the test based on $C_r^*$ is no larger than that of the test based on $C_r(k)$). Since the probabilities on the left hand side of the first inequality represent the powers of the respective tests when $\boldsymbol{\Theta} = \boldsymbol{\Theta}_a$, the test based on $C_r(k)$ is the most powerful test.

(**Unique Most Powerful Size $\alpha$**): We discuss the proof for the discrete case. The continuous case is proven similarly by replacing summation with integration. Let the rejection region C define any other most powerful size $\alpha$ test of $H_0$: $\boldsymbol{\Theta} = \boldsymbol{\Theta}_0$ versus $H_a$: $\boldsymbol{\Theta} = \boldsymbol{\Theta}_a$. Letting $E_{\boldsymbol{\Theta}_0}$ and $E_{\boldsymbol{\Theta}_a}$ denote expectations taken using the densities $f(\mathbf{x};\boldsymbol{\Theta}_0)$ *and* $f(\mathbf{x};\boldsymbol{\Theta}_a)$, respectively, note that $\alpha = E_{\boldsymbol{\Theta}_0}(I_{C_r(k)}(\mathbf{X}))$ $= E_{\boldsymbol{\Theta}_0}(I_{C_r^*}(\mathbf{X}))$, and $E_{\boldsymbol{\Theta}_a}(I_{C_r(k)}(\mathbf{X})) = E_{\boldsymbol{\Theta}_a}(I_{C_r^*}(\mathbf{X}))$, since by assumption both tests have size $\alpha$, and both tests are most powerful tests. It follows that

$$\sum_{\mathbf{x} \in R(\mathbf{x})} [I_{C_r(k)}(\mathbf{x}) - I_{C_r^*}(\mathbf{x})][f(\mathbf{x};\boldsymbol{\Theta}_0) - kf(\mathbf{x};\boldsymbol{\Theta}_a)]$$
$$= E_{\boldsymbol{\Theta}_0}(I_{C_r(k)}(\mathbf{X}) - I_{C_r^*}(\mathbf{X})) - kE_{\boldsymbol{\Theta}_a}(I_{C_r(k)}\mathbf{X}) - I_{C_r^*}(\mathbf{X})) = 0.$$

Note further that the summand in the preceding summation is nonpositive valued $\forall \mathbf{x}$, as was shown in the proof of sufficiency above. Then the sum itself can be zero-valued only if the summands equals zero $\forall \mathbf{x}$. This in turn implies that $I_{C_r(k)}(\mathbf{x}) = I_{C_r^*}(\mathbf{x})\ \forall \mathbf{x} \in \Gamma = \{\mathbf{x} : f(\mathbf{x};\boldsymbol{\Theta}_0) \neq kf(\mathbf{x};\boldsymbol{\Theta}_a)\}$. Thus both $C_r(k)$ and $C_r^*$ contain the set of $\mathbf{x}$'s, $\Gamma$, for which $f(\mathbf{x};\boldsymbol{\Theta}_o) \neq kf(\mathbf{x};\boldsymbol{\Theta}_a)$.

Regarding when $f(\mathbf{x};\boldsymbol{\Theta}_o) = kf(\mathbf{x};\boldsymbol{\Theta}_a)$ note that $\mathbf{x} \in C_r(k)\, \forall\, \mathbf{x} \in K = \{\mathbf{x}: f(\mathbf{x};\boldsymbol{\Theta}_0) = kf(\mathbf{x};\boldsymbol{\Theta}_a)\}$. Let $A = K - C_r^*$ be the set of $\mathbf{x}$-values satisfying $f(\mathbf{x};\boldsymbol{\Theta}_0) = kf(\mathbf{x};\boldsymbol{\Theta}_a)$ that are not in $C_r^*$. Then $P(\mathbf{x} \in A;\boldsymbol{\Theta}_0) = P(\mathbf{x} \in A;\boldsymbol{\Theta}_a) = 0$, for consider the contrary. If $P(\mathbf{x} \in A;\boldsymbol{\Theta}_0) > 0$, then since $C_r(k)$ is an $\alpha$ level test and $C_r^* = C_r(k) - A$, it follows that $P(\mathbf{x} \in C_r^*;\boldsymbol{\Theta}_0) < \alpha = P(\mathbf{x} \in C_r(k);\boldsymbol{\Theta}_0)$, contradicting that $C_r^*$ is an

$\alpha$-size rejection region. If $P(\mathbf{x} \in A;\boldsymbol{\Theta}_a) > 0$, then since $C_r^*$ does not contain $A$, $P(\mathbf{x} \in C_r^*;\, \boldsymbol{\Theta}_a) < P(\mathbf{x} \in C_r(k);\boldsymbol{\Theta}_a)$, contradicting that $C_r^*$ is a most powerful rejection region. Thus, the most powerful rejection region for testing $H_0$: $\boldsymbol{\Theta} = \boldsymbol{\Theta}_0$ versus $H_a$: $\boldsymbol{\Theta} = \boldsymbol{\Theta}_a$ has the form given in the statement of the theorem with probability 1. ■

**Theorem 9.6**

**Proof** We prove part 1. The proof of part 2 is analogous with appropriate inequality reversals.

Since $f(\mathbf{x};\Theta_a)/f(\mathbf{x};\Theta_0)$ has a monotone likelihood ratio in $t(\mathbf{x})$, $f(\mathbf{x};\Theta_a)/f(\mathbf{x};\Theta_0) \geq k^{-1}$ *iff* $g(t(\mathbf{x})) \geq k^{-1}$ for some nondecreasing function $g(\cdot)$. *Moreover,* $g(t(\mathbf{x})) \geq k^{-1}$ *iff* $t(\mathbf{x}) \geq c$ where $c$ is chosen to satisfy[28] $\min_c\{g(c)\} \geq k^{-1}$. A Neyman-Pearson most powerful level $\alpha$ test of $H_0$: $\Theta = \Theta_0$ versus $H_a$: $\Theta = \Theta_a$, *for* $\Theta_a > \Theta_0$, is defined by choosing $c$ so that $P(t(\mathbf{x}) \geq c;\Theta_0) = \alpha$. Note that this probability depends only on $\Theta_0$, so that the choice of $c$ does not depend on $\Theta_a$ and thus the same rejection region $C_r = \{\mathbf{x}\colon t(\mathbf{x}) \geq c\}$ defines a Neyman-Pearson level $\alpha$ test of $H_0$: $\Theta = \Theta_0$ versus $H_a$: $\Theta = \Theta_a$, regardless of the value of $\Theta_a > \Theta_0$. Thus, by Theorem 9.3, $C_r$ is UMP level $\alpha$ for testing $H_0$: $\Theta = \Theta_0$ versus $H_a$: $\Theta > \Theta_0$.

Now examine $\Theta_* < \Theta_0$ and let $\alpha_* = P(\mathbf{x} \in C_r;\Theta_*)$. From the preceding argument, it is known that $C_r$ represents a UMP level $\alpha_*$ test of $H_0$: $\Theta = \Theta_*$ versus $H_a$: $\Theta > \Theta_*$, and because the test is then unbiased (Theorem 9.4), we know that $P(\mathbf{x} \in C_r;\Theta_*) \leq P(\mathbf{x} \in C_r;\, \Theta_0) = \alpha$, which holds $\forall \Theta_* < \Theta_0$. Thus, $C_r$ represents a level $\alpha$ test of $H_0$: $\Theta \leq \Theta_0$ versus $H_a$: $\Theta > \Theta_0$.

Finally, let $C_r^*$ represent any other level $\alpha$ test for testing $H_0$: $\Theta \leq \Theta_0$ versus $H_a$: $\Theta > \Theta_0$. Then by definition $\alpha_o = P(\mathbf{x} \in C_r^*;\Theta_o) \leq \sup_{\Theta \leq \Theta_0}\{P(\mathbf{x} \in C_r^*;\Theta)\} \leq \alpha$, so that $C_r^*$ has size $\alpha_o \leq \alpha$ for testing $H$: $\Theta = \Theta_0$ versus $H_a$: $\Theta > \Theta_0$. Because $C_r$ is UMP level $\alpha$ for testing this hypothesis, $P(\mathbf{x} \in C_r;\Theta) \geq P(\mathbf{x} \in C_r^*;\Theta)\ \forall\ \Theta > \Theta_0$, and thus $C_r$ is also a UMP level $\alpha$ test of $H_0$: $\Theta \leq \Theta_0$ versus $H_a$: $\Theta > \Theta_0$. ■

**Theorem 9.10**

**Proof** The random variable $z = g(\mathbf{X}) - \gamma$ is symmetric about zero, and thus

$$\begin{aligned} P(g(\mathbf{x}) \geq c_2;\Theta_0) = P(z \geq c_2 - \gamma;\Theta_0) &= P(z \leq \gamma - c_2;\Theta_0) \\ &= P(g(\mathbf{x}) \leq 2\gamma - c_2;\Theta_0) \\ &= P(g(\mathbf{x}) \leq c_1;\Theta_0) = \alpha/2. \end{aligned}$$

It follows that $C_r$ defines an $\alpha$-size test.

To see that $d\pi_{C_r}(\Theta_0)/d\Theta = 0$ is satisfied, first note that since $\pi_{C_r}(\Theta) = E_\Theta(I_{C_r}(\mathbf{X}))$, it follows from the exponential class representation of the density of $\mathbf{X}$ and Lemma 9.1 that (continuous case-discrete case is analogous)

---

[28] If the likelihood ratio is strictly increasing in $t(\mathbf{x})$, then c can be chosen to satisfy $g(c) = k^{-1}$.

$$d\pi_{C_r}(\Theta)/d\Theta = \frac{d\int_{-\infty}^{\infty}\cdots\int_{-\infty}^{\infty} I_{C_r}(\mathbf{x})\exp[c(\Theta)g(\mathbf{x})+d(\Theta)+z(\mathbf{x})]I_A(\mathbf{x})}{d\Theta}$$
$$= \mathrm{E}_\Theta(I_{C_r}(\mathbf{X})g(\mathbf{X}))\frac{dc(\Theta)}{d\Theta}+\frac{d(d(\Theta))}{d\Theta}P(\mathbf{x}\in C_r;\Theta).$$

Therefore,

$$d\pi_{C_r}(\Theta_0)/d\Theta = 0 \; iff \; \mathrm{E}_{\Theta_0}(I_{C_r}(\mathbf{X})g(\mathbf{X}))\frac{dc(\Theta_0)}{d\Theta} = -\alpha\frac{d(d(\Theta_0))}{d\Theta}.$$

The right-hand side of the *iff* statement holds under the current assumptions, since

$$\mathrm{E}_{\Theta_0}(I_{C_r}(\mathbf{X})g(\mathbf{X}))\frac{dc(\Theta_0)}{d\Theta} = \mathrm{E}_{\Theta_0}\left((g(\mathbf{X})-\gamma)I_{C_r}(\mathbf{x})\frac{dc(\Theta_0)}{d\Theta}\right)+\mathrm{E}(\mathrm{I}_{C_r}(\mathbf{X}))\gamma\frac{dc(\Theta_0)}{d\Theta}$$
$$= \alpha\gamma\frac{dc(\Theta_0)}{d\Theta} = -\alpha\frac{d(d(\Theta_0))}{d\Theta}$$

where the last equality follows from the fact that $\mathrm{E}_{\Theta_0}(g(\mathbf{X}))(dc(\Theta_0)/d\Theta) = -(d(d(\Theta_0))/d\Theta)$, because

$$0 = \frac{d\int_{-\infty}^{\infty}\cdots\int_{\infty}^{\infty}\exp(c(\Theta)g(\mathbf{x})+d(\Theta)+Z(\mathbf{x}))I_A(\mathbf{x})d\mathbf{x}}{d\Theta} = \mathrm{E}(g(\mathbf{X}))\frac{dc(\Theta)}{d\Theta}+\frac{d(d(\Theta))}{d\Theta}.$$

■

**Theorem 9.12**

**Proof** We focus on the case where the density function of the random sample is discrete. The proofs in the continuous case can be constructed using a similar approach, except the conditional densities involved would be degenerate, so that line integrals would be needed to assign conditional probabilities to events (recall the footnote to our definition of sufficient statistics (Definition 7.18) regarding degenerate conditional distribution $f(\mathbf{x}|s)$.

Given the definition of $f(\mathbf{x};\boldsymbol{\Theta})$, examine the distribution of $\mathbf{X}$, conditional on $s_2,\ldots,s_k$:

$$f(\mathbf{x};\boldsymbol{\Theta}|s_2,\ldots,s_k) = \frac{P(\{\mathbf{x}\}\cap\{\mathbf{x}: s_i(\mathbf{x})=s_i, i=2,\ldots,k\})}{P(s_2,\ldots,s_k)}$$
$$= \frac{\exp\left(c_1(\boldsymbol{\Theta})s_1(\mathbf{x})+\sum_{i=2}^k c_i(\boldsymbol{\Theta})s_i+d(\boldsymbol{\Theta})+z(\mathbf{x})\right)I_{A_*}(\mathbf{x})}{\exp\left(\sum_{i=2}^k c_i(\boldsymbol{\Theta})s_i+d(\boldsymbol{\Theta})\right)\sum_{(\mathbf{x}:s_i(\mathbf{x})=s_i,i=2,\ldots,k)}\exp(c_1(\boldsymbol{\Theta})s_1(\mathbf{x})+z(\mathbf{x}))I_A(\mathbf{x})}$$
$$= \exp(c_1(\boldsymbol{\Theta})s_1(\mathbf{x})+d_*(c_1)+z(\mathbf{x}))I_{A_*}(\mathbf{x})$$

where $A_* = A \cap \{\mathbf{x}: s_i(\mathbf{x}) = s_i,\ i = 2,\ldots,k\}$, and

$$d_*(c_1) = \ln\left(\sum_{x\in A_*}\exp(c_1(\boldsymbol{\Theta})s_1(\mathbf{x})+z(\mathbf{x}))\right)^{-1}.$$

Thus, $f_*(\mathbf{x};c_1(\Theta)|s_2,\ldots,s_k)$ is a one-parameter exponential class density with parameter $c_1(\boldsymbol{\Theta})$ and parameter space $\Omega_{c_1} = \{c_1 : c_1 = c_1(\boldsymbol{\Theta}), \boldsymbol{\Theta}\in\Omega\}$. Note that $c_1$

is differentiable with respect to $c_1$, and it is also true that $d_*(c_1)$ is differentiable. The latter result follows from the chain rule of differentiation, since the natural logarithmic function is differentiable for all positive values of its argument, and the positive-valued bracketed term in the definition of $d_*(c_1)$ is differentiable by Lemma 9.1 with $\phi(\mathbf{x}) = 1$.

Since the assumptions of Theorem 9.11 are met under the stated conditions, all power functions for testing $H_0$ versus $H_a$ are continuous and differentiable. It follows that, conditional, on $(s_2,\ldots,s_k)$, all of the previous results regarding UMP and UMPU tests for the case of a single parameter exponential class density apply. In particular, the definitions of the $C_r$'s in cases (1–4) are all conditionally UMPU level $\alpha$ based on Theorem's 9.5–9.6 and corollaries 9.1–9.2 for results (1) and (2), and Theorem 9.9 for results (3) and (4).

Examine result (1). Let $C_r^*$ be any other unbiased level $\alpha$ test, so that $\pi_{C_r^*}(c_1^0) = \alpha_* \leq \alpha$. Then it must be the case that $\pi_{C_r^*}(c_1^0|S_2,\ldots,S_k) \leq \alpha$ with probability 1. To see this, note that by the iterated expectation theorem

$$\pi_{C_r^*}(c_1^0) - \alpha_* = \mathrm{E}_{c_1^0}(I_{C_r^*}(\mathbf{X})) - \alpha_* = \mathrm{E}\left[\mathrm{E}_{c_1^0}(I_{C_r^*}(\mathbf{X})|S_2,\ldots,S_k) - \alpha_*\right] = 0,$$

and since the bracketed expression is a function of the complete sufficient statistics $(S_2,\ldots,S_k)$ it follows by definition that the bracketed expression must equal 0 with probability 1, and thus $\mathrm{E}_{c_1^0}(I_{C_r^*}(\mathbf{X})|S_2,\ldots,S_k) = \alpha_* \leq \alpha$ with probability 1 (recall Definition 7.20). Thus, $C_r^*$ must be conditionally unbiased of level $\alpha$.

Now since $C_r$ has maximum power for $c_1 > c_1^0$ among all unbiased level $\alpha$ tests conditional on $(s_2,\ldots,s_k)$, it follows from the iterated expectation theorem that for $c_1 > c_1^0$,

$$\pi_{C_r}(c_1) = \mathrm{E}_c(\mathrm{E}_{c_1}(I_{C_r}(\mathbf{X})|S_2,\ldots,S_k)) \geq \mathrm{E}_c(\mathrm{E}_{c_1}(I_{C_r^*}(\mathbf{X})|S_2,\ldots,S_k)) = \pi_{C_r^*}(c_1)$$

so that $C_r$ defines a UMPU level $\alpha$ test of the hypotheses in case (1).

The proof of case (2) follows directly from the proof of case (1) via appropriate inequality reversals. Moderate extensions of the preceding argument can be used to prove results (3) and (4) (see T. Ferguson (1967), *Mathematical Statistics*, Academic Press, New York, pp. 230–232). ■

## Keywords, Phrases and Symbols

$\pi_{C_r}(\Theta)$ power function
Acceptable region
Alternative hypothesis
Composite hypothesis
Conditioning in Multiparameter case
Consistent test sequence of level $\alpha$
Controlling Type I and Type II errors
Critical (or rejection) region
$H$ is (not) rejected at the $\alpha$-level of significance
$H$ is (not) rejected using a size-$\alpha$ test
$H$: set defining conditions
$H_0$, $H_a$
Ideal power function
Ideal statistical test
Likelihood ratio test
Maintained hypothesis
Monotone likelihood ratio and the exponential class of densities
Monotone likelihood ratio in the statistic $t(\mathbf{X})$
Monotone likelihood ratios and UMP level $\alpha$ tests
More powerful level $\alpha$ test
Most Powerful level $\alpha$ test
Multiparameter exponential class and hypothesis testing

Neyman-Pearson lemma
Noncentral $t$-distribution
Noncentrality parameter
Nuisance parameters
Null hypothesis
One-sided and two-sided alternative hypotheses
Operating characteristic function of a statistical test
Parametric hypothesis testing
Power function of a statistical test
Power of the test at $\boldsymbol{\Theta}$
Protection against Type I error
$P$-value
Range of $\mathbf{X}$ over $H$, $R(\mathbf{X}|H)$
Rejection (or critical) region
Reparameterizing exponential class densities via linear transformation
Significance level of the test
Simple hypothesis
Size of test
Statistical hypothesis
Statistical test defined by $C_r$
Sufficient statistic representation of likelihood ratio test
Sufficient statistic representation of test of $H_0$: $\boldsymbol{\Theta} = \boldsymbol{\Theta}_o$ versus $H_a$: $\boldsymbol{\Theta} \in \Omega_a$
Test of a statistical hypothesis
Test statistic
Type I and Type II errors
Type I/Type II error tradeoff
UMP level $\alpha$ test of $H_0$: $\boldsymbol{\Theta} = \boldsymbol{\Theta}_o$ versus $H_a$: $\boldsymbol{\Theta} \neq \boldsymbol{\Theta}_o$, nonexistence in case of MLR
UMP test of $H_0$: $\boldsymbol{\Theta} = \boldsymbol{\Theta}_o$ versus $H_a$: $\boldsymbol{\Theta} \in \Omega_a$, Neyman-Pearson approach
UMPU level $\alpha$ test of $H_0$: $\boldsymbol{\Theta} = \boldsymbol{\Theta}_o$ versus $H_a$: $\boldsymbol{\Theta} \neq \boldsymbol{\Theta}_o$ in case of exponential class density
UMPU level $\alpha$ test of $H_0$: $\boldsymbol{\Theta}_1 \leq \boldsymbol{\Theta} \leq \boldsymbol{\Theta}_2$ versus $H_a$: $\boldsymbol{\Theta} < \boldsymbol{\Theta}_1$ or $\boldsymbol{\Theta} > \boldsymbol{\Theta}_2$ in case of exponential class density
UMPU level $\alpha$ test of $H_0$: $\mu \leq \mu_0$ versus $H_a$: $\mu > \mu_0$, and of $H_0$: $\mu = \mu_0$ versus $H_a$: $\mu \neq \mu_0$, normal population
UMPU level $\alpha$ test of $H_0$: $\sigma^2 \leq \sigma_0^2$ versus $H_a$: $\sigma^2 > \sigma_0^2$ when sampling from normal distribution
Unbiasedness of a test rule
Unbiasedness of most powerful test of $H_0$: $\boldsymbol{\Theta} = \boldsymbol{\Theta}_o$ versus $H_a$: $\boldsymbol{\Theta} = \boldsymbol{\Theta}_a$
Unbiasedness of UMP level $\alpha$ tests using monotone likelihood ratios
Uniformly most powerful level $\alpha$ test

## Problems

**1.** A shipment of 20 projection screen television sets is at the receiving dock of a large department store. The department store has a policy of not accepting shipments that contain more than 10 percent defective merchandise. The receiving clerk is instructed to have a quality inspection done on two sets that are randomly drawn, without replacement, from the shipment. Letting $k$ represent the unknown number of defective sets in the shipment of 20 sets, the null hypothesis $H_0{:}k \leq 2$ will be rejected iff both sets that are inspected are found to be defective.

(a) Calculate the probabilities of committing type I errors when $k = 0, 1$, or 2.

(b) Calculate the probabilities of committing type II errors when $k \geq 3$.

(c) Plot the power function of this testing procedure. Interpret the implications of the power function from the standpoint of both the department store and the television manufacturer.

**2.** A pharmaceutical company is analyzing the effectiveness of a new drug that it claims can stimulate hair growth in balding men. For the purposes of an advertising campaign, the marketing department would like to be able to claim that the drug will be effective for at least 50 percent of the balding men who use it. To test the claim, a random sample of 25 balding men are given the drug treatment, and it is found that 10 applications were effective in stimulating hair growth. The population of balding men is sufficiently large that you may treat this as a problem of random sampling *with* replacement.

(a) Test the null hypothesis $H_0$: $p \geq .50$ using as close to a .10-size test as you can.

(b) Plot the power function for this test. Interpret the power function from the standpoint of both the pharmaceutical company and the consuming public.

(c) Is the test you used in (a) a UMP test? Is it a UMPU test? Is it an unbiased test? Is it a consistent test?

(d) Calculate and interpret the *P-value* for the test.

**3.** In each case below, identify whether the null and the alternative hypotheses are simple or composite hypotheses.

(a) You are random sampling from a gamma population distribution and you are testing $H_0$: $\alpha \leq 2$ versus $H_a$: $\alpha > 2$.

(b) You are random sampling from a geometric population distribution and you are testing $H_0$: $p = .01$ versus $H_a$: $p > .01$.

(c) The joint density of the random sample $\mathbf{Y} = \mathbf{x}\boldsymbol{\beta} + \boldsymbol{\varepsilon}$ is $N(\mathbf{x}\boldsymbol{\beta}, \sigma^2\mathbf{I})$ and you are testing whether $\boldsymbol{\beta} = \mathbf{0}$.

(d) You are random sampling from a poisson population distribution and you are testing $H_0$: $\lambda = 2$ versus $H_a$: $\lambda = 3$.

**4.** A large metropolitan branch of a savings and loan is examining staffing issues and wants to test the hypothesis that the expected number of customers requiring the services of bank personnel during the midweek (Tuesday-Thursday) noon hour is $\leq 50$. The bank has obtained the outcome of a random sample consisting of 100 observations on the number of noon hour customers requiring service from bank personnel. It was observed that $\bar{\mathbf{x}} = 54$. You may assume that the population distribution is Poisson in this case.

(a) Design a UMP level .05 test of the null hypothesis having size as close to .05 as possible. Test the hypothesis.

(b) Plot the power function for the test. Interpret the power function both from the standpoint of management's desire for staff reductions and the need to provide quality customer service to bank customers.

**5.** The annual proportion of new restaurants that survive in business for at least 1 year in a U.S. city with population $\geq 500{,}000$ people is assumed to be the outcome of some Beta population distribution. Part of the maintained hypothesis is that $b = 1$ in the Beta distribution, so that the population distribution is assumed to be *Beta*$(a,1)$. A random sample of size 50 from the beta population distribution results in the geometric mean $\bar{x}_g = .84$.

(a) Define a UMP level .05 test of the hypothesis that less than three-quarters of new restaurants are expected to survive at least 1 year in business in U.S. cities of size $\geq 500{,}000$. Test the hypothesis.

(b) Plot the power function for the test. Interpret the power function both from the standpoint of a potential investor in a restaurant and from the perspective of the managing director of a chamber of commerce.

(c) Calculate and interpret the $p$-value for the test.

**6.** A complaint has been lodged against a major domestic manufacturer of potato chips stating that their 16 oz bags of chips are being underfilled. The manufacturer claims that their filling process produces fill weights that are normally distributed with a mean of 16.1 oz and a standard deviation of $\leq .05$ so that over 97 percent of their product has a weight of $\geq 16$ oz. They suggest that their product be randomly sampled and their claims be tested for accuracy. Two *independent* random samples of observations on fill weights, each of size 250, resulted in the following summary statistics $\bar{x}_1 = 16.05, \bar{x}_2 = 16.11$, $s_1^2 = .0016$, and $s_2^2 = .0036$.

(a) Define a UMPU level .05 test of $H_0$: $\mu = 16.1$ versus $H_a$: $\mu \neq 16.1$ based on a random sample of size 250. Test the hypothesis using the statistics associated with the first random sample outcome. Plot and interpret the power function of this test.

(b) Define a UMPU level .05 test of $H_0$: $\sigma \leq .05$ versus $H_a$: $\sigma > .05$ based on a random sample of size 250. Test the hypothesis using the statistics associated with the second random sample outcome. Plot and interpret the power function of this test.

(c) Calculate and interpret the $p$-values of the tests in (a) and (b). (Hint: It might be useful to consider Bonferroni's inequality for placing an upper bound on the probability of Type I Error.)

(d) Treating the hypotheses in (a) and (b) as a joint hypothesis on the parameter vector of the normal population distribution, what is the probability of Type I Error for the joint hypothesis $H_0$: $\mu = 16.1$ and $\sigma \leq .05$ when using the outcome of the two test statistics above to determine acceptance or rejection of the joint null hypothesis? Does the complaint against the company appear to be valid?

(e) Repeat (a–c) using a pooled sample of 500 observations.

**7.** In a random sample of size 10 from a Bernoulli population distribution, how many (nonrandomized) critical regions can you define that have size $\leq .10$ and that are also unbiased for testing the null hypothesis $H_0$: $p = .4$ versus $H_a$: $p \neq .4$ ?

**8.** ***Randomized Test*** It was demonstrated in Example 9.10 that the choices of size for most powerful tests of the hypothesis $H_0$: $p = .2$ versus $H_a$: $p = .8$ was quite limited. Suppose that a .05 level test of the null hypothesis was desired and that you were willing to utilize a *randomized* test. In particular, examine the following randomized test rule:

$x \geq 8 \Rightarrow$ reject $H_0$

$x = 7 \Rightarrow$ reject $H_0$ with probability $\tau$, do not reject $H_0$ with probability $(1 - \tau)$

$x \leq 6 \Rightarrow$ do not reject $H_0$

To implement the rule when $x = 7$ occurs, a uniform random number $z$ with range (0,1) could be drawn, and if $z \leq \tau$, $H_0$ would be rejected, and if $z > \tau$, $H_0$ would not be rejected.

(a) Find a value of $\tau$ that defines a .05 test of the null hypothesis.

(b) It is possible that two analysts, using exactly the same random sample outcome and using exactly the same test rule could come to different conclusions regarding the validity of the null hypothesis. Explain. (This feature of randomized tests has discouraged their use.)

(c) Is the test you defined in (a) an unbiased size .05 test of the null hypothesis?

(d) Is the test you defined in (a) a most powerful size .05 test of the null hypothesis?

**9.** The number of work-related injuries per week that occur at the manufacturing plant of the Excelsior Corporation is a Poisson-distributed random variable with mean $\lambda \geq 3$, according to company analysts. In an attempt to lower insurance costs, the Corporation institutes a program of intensive safety instruction for all employees. Upon completion of the program, a 12-week period produced an average of two accidents per week.

(a) Design a uniformly most powerful level .10 test of the null hypothesis $H_0$: $\lambda \geq 3$ versus the alternative hypothesis $H_a$: $\lambda < 3$ having size as close to .10 as possible without exceeding .10.

(b) Test the null hypothesis using the test you defined in part (a). What can you say about the effectiveness of the safety program?

**10.** Being both quality and cost conscious, a major foreign manufacturer of compact disk players is contemplating their warranty policy. The standard warranty for compact disk players sold by competing producers is 1 year. The manufacturer is considering a 2 year warranty. The operating life until failure of your compact disk player has the density

$$f(x;\beta) = (x/\beta^2)\exp(-x/\beta)\,I_{(0,\infty)}(x)$$

for some value of $\beta > 0$, where x is measured in years.

(a) The manufacturer wants their exposure to warranty claims to be, on average, no more than 5 percent of the units sold. Find the values of $\beta$ for which $P(x \leq 2;\beta) = \int_0^2 f(x;\beta)dx \leq .05$.

(b) Based on a random sample of size 50, design a uniformly most powerful level .10 test of the null hypothesis that $\beta$ will be in the set of values you identified in (a).

(c) A nondestructive test of the disk players that determines their operating life until failure is applied to 50 players that are randomly chosen from the assembly line. The measurements resulted in $\bar{x} = 4.27$. Test the null hypothesis in (b).

(d) Plot the power curve of this test, and interpret its meaning to the management of the manufacturing firm.

**11.** Referring to Definition 9.14, state the form of the UMPU level $\alpha$ test of the null hypothesis $H_0$: $\sigma^2 \geq \sigma_0^2$ versus $H_a$: $\sigma^2 < \sigma_0^2$. Justify your answer.

**12.** Referring to Definition 9.15, state the form of the UMPU level $\alpha$ test of the null hypothesis $H_0$: $\mu \geq \mu_0$ versus $H_0$: $\mu < \mu_0$. Justify your answer.

**13.** Your company supplies an electronic component that is critical to the navigational systems of large jet aircraft. The operating life of the component has an exponential distribution with some mean value v, where operating life is measured in 100,000 hour units. You are seeking a contract to supply these components to a major aircraft manufacturer on the West Coast. The contract calls for a minimum mean operating life of 750,000 hours for the component, and you must provide evidence on the reliability of your product. You have a random sample of observations on tests of 300 of your components that provide measurements on the components' operating lives. According to the tests performed, the mean operating life of the components was 783,824 hours.

(a) In designing a UMP level $\alpha$ test in this situation, should the null hypothesis be defined as $\theta \geq 7.5$ or $\theta \leq 7.5$? Base your discussion on the characteristics of the power function of each of the tests.

(b) Design a UMP size $\alpha$ test of whichever null hypothesis you feel is appropriate based on you discussion in (a). Choose whatever size test you feel is appropriate, and discuss your choice of size.

(c) Conduct the hypothesis test. Should your company get the contract? Why or why not?

**14.** The *Pareto* distribution

$f(\mathbf{x}; \theta, c) = c^\theta \theta x^{-(1+\theta)} I_{(c,\infty)}(x)$

for $\theta > 1$ and $c > 0$ has been used to model the distribution of incomes in a given population of individuals, where c represents the minimum level of income in the population. In a certain large state on the east coast the office of fiscal management is investigating a claim that professors at teaching colleges have average annual salaries that exceed the state average annual salary for middle management white collar workers, which is known to be $62,471. A random sample of 250 professors' salaries were obtained, and the *geometric* mean of the observations was found to be $\left(\prod_{i=1}^n x_i\right)^{1/n} = 61.147$ where the $x_i$'s are measured in 1,000's of dollars. As part of the maintained hypothesis, the value of $c$ is taken to be 30.

(a) Express the mean level of income as a function of the parameter $\theta$.

(b) Define a test statistic on which you can base a UMP level $\alpha$ test of the null hypothesis $H_0$: $\mu \leq 62.471$ versus $H_a$: $\mu > 62.471$.

(c) Define a UMP size .05 test of the null hypothesis $H_0$: $\mu \leq 62.471$ versus $H_a$: $\mu > 62.471$. (Hint: Use a test statistic for which you can apply an asymptotic normal distribution and use the normal approximation).

(d) Test the hypothesis. Are university professors paid more than white collar middle management workers in this state?

**15.** Suppose that a random sample of size $n$ is drawn from a normal population distribution for which $\sigma^2$ is assumed to be *known* and equal to the given value $\sigma_*^2$. Define UMPU level $\alpha$ tests of the following null hypotheses:

(a) $H_0$: $\mu \leq \mu_0$ versus $H_a$: $\mu > \mu_0$

(b) $H_0$: $\mu \geq \mu_0$ versus $H_a$: $\mu < \mu_0$

(c) $H_0$: $\mu = \mu_0$ versus $H_a$: $\mu \neq \mu_0$

**16.** Suppose that a random sample of size $n$ is drawn from a normal population distribution for which $\mu$ is assumed to be *known* and equal to the given value $\mu_*$. Define UMPU level $\alpha$ tests of the following null hypotheses:

(a) $H_0$: $\sigma^2 \leq \sigma_0{}^2$ versus $H_a$: $\sigma^2 > \sigma_0{}^2$

(b) $H_0$: $\sigma^2 \geq \sigma_0{}^2$ versus $H_a$: $\sigma^2 < \sigma_0{}^2$

(c) $H_0$: $\sigma^2 = \sigma_0{}^2$ versus $H_a$: $\sigma^2 \neq \sigma_0{}^2$

**17.** ***Control Charting:*** A large mail-order house has initiated a quality control program. They randomly sample 100 of each day's orders and monitor whether or not the mail order-taking process is "under control" in the sense that errors in order-taking are at minimum levels. The number of orders per day is sufficiently large that one can assume that the sampling is done with replacement. The daily error proportion prior to the initiation of this program has been 3.2 percent.

(a) Define a UMPU level $\alpha$ test of the hypothesis $H_0$: $p = .032$ versus $H_a$: $p \neq .032$. You may use the asymptotic normal distribution of the test statistic in defining the critical region.

(b) Conduct a size .05 UMPU test of the null hypothesis on a day where $\bar{x} = 3.4$. Is the order process under control?

(c) After quality control training and closer monitoring of clerical workers, a daily random sample resulted in $\bar{x} = 2.6$. Is there evidence that quality has increased over what it has been in the past? Why or why not?

(The mail order company of Alden's Inc. was one of the earliest companies to use control charting techniques for monitoring clerical work. See Neter, John, (1952), "Some Applications of Statistics for Auditing", *Journal of the American Statistical Association*, March, pp. 6–24.

**18.** Revisit Example 9.27 and the power function graph in Figure 9.15 and consider the implications of the power function graph in the two dimensional parameter space $(\mu,\sigma)$.

(a) Plot the power surface in three dimensions, the axes referring to power, the value of $\mu$, and the value of $\sigma$. (This is probably best done with the aid of a computer).

(b) Plot the *isopower contour* in the $(\mu,\sigma)$-plane for a power level of .90. (An *isopower contour* is the set of $(\mu,\sigma)$ points that result in the same level of power, which in the case at hand is equivalent to the set of $(\mu,\sigma)$ points that result in the same value of the noncentrality parameter $\lambda$). Interpret the isopower contour with respect to the ability of the test to detect deviations from the null hypothesis.

**19.** The Gibralter Insurance Co. is reevaluating the premiums it charges on car insurance and is analyzing classifications of cars into high risk, average risk, and low risk on the basis of frequency of accidents. They are currently examining an imported mid-size four-door sedan and wish to examine whether its frequency of claims history is significantly different than a range of values considered to be consistent with the expected frequency of claims of the average risk class, the range being between 2 percent and 4 percent of the vehicles insured. A random sample with replacement of 400 insured vehicles of the type in question resulted in a claims percentage of 7 percent.

(a) Design a UMPU level $\alpha$ test of the null hypothesis $H_0$: $\mu \in [2,4]$ versus $H_a$: $\mu \notin [2,4]$. You can base your test on the asymptotic normal distribution of the test statistic.

(b) Test the null hypothesis using a size .05 UMPU test. Does the outcome contradict classifying the vehicle in the average risk class? Why or why not?

(c) Supposing you rejected the hypothesis, what would be your conclusion? Is further statistical analysis warranted?

**20.** A certain business uses national telephone solicitation to sell its product. Its sales staff have individual weekly sales quotas of 10 sales that they must meet or else their job performance is considered to be unsatisfactory and they receive only base pay and no sales commission. In hiring sales staff, the company has claimed that the proportion of customers solicited that will ultimately buy the company's product is .05, so that on average, 200 phone calls per week should produce the required 10 sales. The company requires that a salesperson keep a record of how many phone solicitations were made, and when the 10th sale is made, the salesperson must indicate the number of phone calls that were made to obtain the 10 sales. The data on the last 200 weekly quotas that were met by various salespersons indicated that 289 phone calls were needed, on average, to meet the quota. A disgruntled employee claims that the company has overstated the market for the product, and wants the quota lowered.

(a) Define a UMP level $\alpha$ test of the hypothesis $H_0$: $p = .05$ versus $H_a$: $p < .05$. You may use an asymptotic normal distribution for the test statistic, if it has one.

(b) Test the hypothesis with a UMP size .10 test. Does the disgruntled employee have a legitimate concern?

(c) Examine the asymptotic power function of the test (i.e., construct a power function based on the asymptotic normal distribution of the test statistic). Interpret the implications of the power function for the test you performed, both from the perspective of the company and from the perspective of the employee. If you were primarily interested in worker's rights, might you design the test differently and/or would you consider testing a different null hypothesis? Explain.

(d) Suppose there was a substantial difference in the abilities of salespersons to persuade consumers to purchase the company's product. Would this have an impact on your statistical analysis above? Explain.

# 10 Hypothesis Testing Methods and Confidence Regions

## 10.1 Introduction

In this chapter we examine general methods for defining tests of statistical hypotheses and associated confidence intervals and regions for parameters or functions of parameters. In particular, the likelihood ratio, Wald, and Lagrange multiplier methods for constructing statistical tests are widely used in empirical work and provide well-defined procedures for defining test statistics, as well as for generating rejection regions via a duality principle that we will examine. In addition, it is possible to define useful test statistics and confidence regions based entirely on heuristic principles of test construction. None of these four methods is guaranteed to produce a statistical test with optimal properties in all cases. In fact, no method of defining statistical tests can provide such a guarantee. The virtues of these methods are that they are relatively straightforward to apply (in comparison to direct implementation of many of the theorems in Section 9.5), they are applicable to a wide class of problems that are relevant in applications, they generally have excellent asymptotic properties, they often have good power in finite samples, they are sometimes unbiased and/or UMP, and they have intuitive appeal.

The four methods of defining test rules presented in this chapter, together with the results on UMP and UMPU testing in Section 9.5, by no means exhaust the ways in which statistical tests can be defined. (See E. Lehmann, *Testing Statistical Hypotheses*, for further reading). However, the methods we present cover the majority of approaches to defining test rules used in practice. Ultimately, regardless of the genesis of a statistical test, whether the test is useful in an application will depend on its properties, as discussed in Chapter 9.

We will also introduce in this chapter a number of nonparametric tests that have been designed to assess assumptions relating to the functional form of the joint density of a probability sample, as well to assess the ubiquitous *iid* assumption underlying simple random samples.

As we noted at the beginning of the previous two chapters, we will choose to move the proofs of some theorems to the Appendix of this chapter to enhance readability.

## 10.2 Heuristic Approach to Statistical Test Construction

In the heuristic approach to defining statistical tests, one attempts to implement the following general **heuristic principle of test construction**: "*Discover a test statistic whose probabilistic behavior is different under $H_0$ and $H_a$, and exploit the difference in defining a rejection region for a statistical test.*" For the choice of an appropriate test statistic $T=t(\mathbf{X})$, one might examine a good estimator of $\boldsymbol{\Theta}$ or $\mathbf{q}(\boldsymbol{\Theta})$, such as a maximum likelihood, least-squares, or generalized method of moments estimator. Alternatively, a (minimal and/or complete) sufficient statistic for $\boldsymbol{\Theta}$ or $\mathbf{q}(\boldsymbol{\Theta})$ might be useful for defining a test statistic.

If the range of a statistic $T$ over $H_0 \cup H_a$ can be partitioned as $C_r^T \cup C_a^T$ so that $P(t \in C_r^T; \boldsymbol{\Theta}) \leq \alpha \; \forall \boldsymbol{\Theta} \in H_0$, a level $\alpha$ test of $H_0$ will have been defined. If it is also true that $P(t \in C_r^T; \boldsymbol{\Theta}) \geq \alpha \; \forall \; \boldsymbol{\Theta} \in H_a$, the test would be unbiased. If the power function $\pi_{C_r}^T(\boldsymbol{\Theta}) = P(t \in C_r^T; \boldsymbol{\Theta})$ is acceptable to the analyst, the test represents at least a useful, if not optimal, statistical test of $H_0$ versus $H_a$. The heuristic approach can be a substantially less complicated method of defining a test statistic than were the classical **UMP** and UMPU approaches, which were discussed in Section 9.5. The following examples illustrate the heuristic method:

**Example 10.1**
***Heuristic Test of $H_0$: $\lambda \leq \lambda_0$ Versus $H_a$: $\lambda > \lambda_0$ in Poisson Distribution***

At the Union Bank a debate is raging over the expected rate at which customers arrive at the teller windows. If the expected rate is $\leq 2$ per minute, management feels that staffing can be reduced. A random sample of 100 observations from the (assumed) Poisson($\lambda$) population of customer arrivals/minute was obtained, and the outcome of the complete sufficient statistic, and MLE estimator of $\lambda$, $\bar{X}$, was 2.48.

In order to test $H_0$: $\lambda \leq 2$ versus $H_a$: $\lambda > 2$, the joint density of the random sample is given by $f(\mathbf{x};\lambda) = \left[ e^{-n\lambda} \lambda^{\sum_{i=1}^{n} x_i} \Big/ \prod_{i=1}^{n} x_i! \right] \prod_{i=1}^{n} I_{\{0,1,2,\ldots\}}(x_i)$, with $n = 100$.

Note that the statistic $n\bar{X} = \sum_{i=1}^{n} X_i \sim \text{Poisson}(n\lambda)$ (straightforwardly shown via the MGF approach), and thus the event $100\bar{x} \geq c$ will be more probable for higher values of $\lambda$ than for lower values. Thus heuristically, it seems reasonable to reject $H_0$: $\lambda \leq 2$ for large values of $100\,\bar{x}$ and not reject for small values. Furthermore, the probability of rejection, $P(100\,\bar{x} \geq c;\ 100\lambda)$, *increases* as $\lambda$ increases, and so to identify a value of $c$ that defines a level $\alpha$ test of $H_0$ it suffices to choose $c$ so that $P(100\bar{x} \geq c;\ 200) = \alpha$, since then $P(100\bar{x} \geq c;\ 100\lambda) < \alpha\ \forall\ \lambda < 2$. The preceding observation also implies that the rejection region defined by $100\,\bar{x} \geq c$ defines an *unbiased* $\alpha$ level test.

With the aid of the computer, we identify the following potential test sizes:

| $c$ | $P(100\bar{x} \geq c;\ 200) = 1 - \sum_{j=0}^{c-1} e^{-200}200^j/j!$ |
|---|---|
| 219 | .097 |
| 220 | .086 |
| 221 | .075 |
| 222 | .066 |
| 223 | .058 |
| 224 | .050 |
| 225 | .044 |

Assuming the labor union and bank management agree that a level .05 test is acceptable the rejection region expressed in terms of the test statistic $\bar{X}$ is $C_r^{\bar{X}} = \{\bar{x} : \bar{x} \geq 2.24\}$. Because the outcome $\bar{x} = 2.48 \in C_r^{\bar{X}}$, $H_0$ is *rejected* at the .05 level. Using the results of Section 9.5, it can be further shown that $C_r^{\bar{X}}$ defines the UMP level .05 test of $H_0$: $\lambda \leq 2$ versus $H_a$: $\lambda > 2$. Note that the $p$-value for this test is $P(\bar{x} \geq 2.48;\ 200) = P(n\bar{x} \geq 248;\ 200) < .001$, which indicates there is strong evidence against $H_0$. □

The next example deals with a complicated case involving nuisance parameters, and yet the heuristic principle of test construction leads rather straightforwardly to a test with acceptable power characteristics. In fact, the test is a UMPU level $\alpha$ test.

**Example 10.2** ***Heuristic Test for Significance of a Regression Parameter***

Revisit Example 8.3 relating to the estimation of a linear relationship between the number of new homes sold and determinants of home sales. Assume that the disturbance term of the linear model is approximately normally distributed (why *must* this be only an approximation?), so that (approximately) $\hat{\boldsymbol{\beta}} = (\mathbf{x}'\mathbf{x})^{-1}\mathbf{x}'\mathbf{Y} \sim N\left(\boldsymbol{\beta}, \sigma^2(\mathbf{x}'\mathbf{x})^{-1}\right)$. The realtor wishes to test whether mortgage interest rates actually impact home sales, and so she specifies the hypothesis $H_0$: $\beta_2 = 0$ versus $H_a$: $\beta_2 \neq 0$, where $\beta_2$ is the parameter associated with the mortgage interest rate variable.

Testing the hypothesis at the .05 level under the prevailing assumptions, note that $\hat{\beta}_2 \sim N\left(\beta_2, \text{var}\left(\hat{\beta}_2\right)\right)$, where $\text{var}(\hat{\beta}_2)$ is the (2,2) entry in the $(2 \times 2)$ covariance matrix $\sigma^2(\mathbf{x}'\mathbf{x})^{-1}$. Consider the statistic $T = \hat{\beta}_2/\widehat{std}\left(\hat{\beta}_2\right)$ for defining a statistical test of the hypothesis, where $\widehat{std}\left(\hat{\beta}_2\right)$ is the square root of the (2,2) variance entry $\widehat{\text{var}}\left(\hat{\beta}_2\right)$ in the covariance matrix estimator $\hat{S}^2(\mathbf{x}'\mathbf{x})^{-1}$. We know

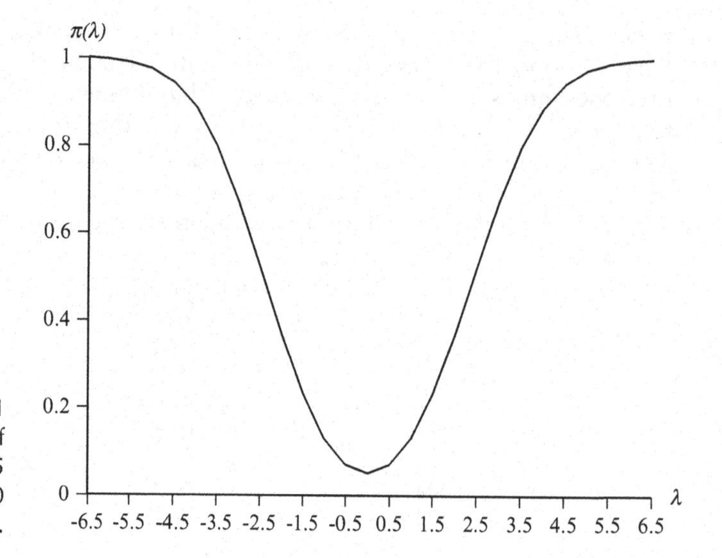


**Figure 10.1** Power function of UMPU size .05 test of $H_0$: $\beta_2 = 0$ versus $H_a$: $\beta_2 \neq 0$.

that $\hat{\beta}_2$ and the estimator of $\sigma^2$, $\hat{S}^2$, are independent random variables under the normality assumption, and in the case at hand $\left(\hat{\beta}_2 - \beta_2\right)/\left[\text{var}\left(\hat{\beta}_2\right)\right]^{1/2} \sim N(0,1)$ and $(n-k)\hat{S}^2/\sigma^2 \sim \chi^2_{n-k}$. It follows that

$$\frac{\left(\hat{\beta}_2 - \beta_2\right)/\left[\text{var}\left(\hat{\beta}_2\right)\right]^{1/2}}{\left[\hat{S}^2/\sigma^2\right]^{1/2}} = \frac{\hat{\beta}_2 - \beta_2}{\left[\widehat{\text{var}}(\hat{\beta}_2)\right]^{1/2}}$$

has a (central) $t$-distribution with $n$-$k$ degrees of freedom. Then under $H_0$: $\beta_2 = 0$, $T = \hat{\beta}_2/\widehat{std}\left(\hat{\beta}_2\right)$ has a central $t$-distribution with $n$-$k$ degrees of freedom, whereas under $H_a$: $\beta_2 \neq 0$, $T$ has a noncentral $t$-distribution with $n$-$k$ degrees of freedom and noncentrality parameter $\lambda = \beta_2/[\text{var}(\hat{\beta}_2)]^{1/2}$ (recall Section 9.6). We then know that $P(|t| \geq c;\lambda)$ increases monotonically as $|\lambda| \to \infty$, so that the event $|t| \geq c$ becomes ever more probable as $|\beta_2| \to \infty$. Thus heuristically, it seems reasonable to reject $H_0$: $\beta_2 = 0$ for large values of $|t|$ and not reject for small values. Therefore, the rejection region for the test statistic $T$ will be $C_r^T = (-\infty,-c]\cup[c,\infty)$. Given the aforementioned behavior of $P(|t|\geq c;\lambda)$ as a function of $\lambda$, it follows that $C_r^T$ will define an unbiased $\alpha$ level test of $H_0$.

Referring to the table of the (central) $t$-distribution, we find that $P(t\geq c) = .025$ when $c = 2.571$ and degrees of freedom equal $n-k = 10$–$5 = 5$. Then, by the symmetry of the (central) $t$-distribution, it follows that $P(t\in C_r^T) = .05$ where $C_r^T = (-\infty,-2.571]\cup[2.571,\infty)$. Since the outcome of $T$ is $t = -3.33\in C_r^T$, we *reject* $H_0$ and conclude that mortgage interest rates do impact home sales. The $p$ value of the test is given by $P(|t| \geq 3.33) = .021$, which suggests that the evidence against $H_0$ is substantial.

The power function of the test can be calculated with the aid of a computer (we use the GAUSS procedure CDFTNC here). Selected values of the power function are displayed in the following table, and a graph of the power function as a function of $\lambda = \beta_2/[\text{var}(\hat{\beta}_2)]^{1/2}$ is given in Figure 10.1.

| $\lambda = \beta_2/[\text{var}(\hat{\beta}_2)]^{1/2}$ | $\pi(\lambda) = P(t \in C_r^T; \lambda)$ |
|---|---|
| 0 | .05 |
| .5, −.5 | .07 |
| 1, −1 | .13 |
| 1.5, −1.5 | .23 |
| 2, −2 | .37 |
| 3, −3 | .67 |
| 4, −4 | .89 |
| 5, −5 | .98 |

Note that $\lambda$ effectively measures deviations of $\beta_2$ from 0 in terms of standard deviation units (i.e., $\beta_2$ is expressed relative to the standard deviation $[\text{var}(\hat{\beta}_2)]^{1/2}$). The power function suggests that $\beta_2$ must be a number of standard deviations away from zero before the test has appreciable power in rejecting $H_0$. This is typical for power functions associated with tests of the significance of parameters estimated by least squares (or many other procedures for that matter), and suggests that to be able to detect small departures of $\beta_2$ from zero, $\beta_2$ must be estimated quite accurately (i.e., $\text{var}(\hat{\beta}_2)$ must be small).

In this application, it might be argued, based on considerations of economic theory, that $\beta_2$ must be nonpositively valued. If this view is adopted as part of the maintained hypothesis, then a one-sided alternative hypothesis can be considered as $H_a$: $\beta_2 < 0$. The reader is invited to reexamine the problem of testing $H_0$: $\beta_2 = 0$ using the one-sided alternative hypothesis and a .05 level test. In such a case, the power of the test for $\beta_2 \in H_a$ is increased. □

In the next example we revisit Example 10.1 and illustrate how the asymptotic distribution of a test statistic can be used to define an asymptotic test of $H_0$. In the case at hand, this will allow the test to be conducted using a standard normal distribution. It also allows one to circumvent the limited choice of test sizes in this discrete case, albeit in an approximate sense.

**Example 10.3** ***Heuristic Test of $H_0$: $\lambda \leq \lambda_0$ Versus $H_a$: $\lambda > \lambda_0$ for Poisson Based on Asymptotic Normality***

Revisit Example 10.1 and consider using the asymptotic normal distribution of the test statistic $\bar{X}$ to conduct the test of $H_0$: $\lambda \leq 2$ versus $H_a$: $\lambda > 2$. In this case, where we are random sampling from the Poisson population distribution, we know that $\bar{X} \overset{a}{\sim} N(\lambda, n^{-1}\lambda)$, and thus $Z = n^{1/2}(\bar{X} - \lambda)/\lambda^{1/2} \overset{a}{\sim} N(0,1)$. Using heuristic reasoning analogous to that used in Example 10.1, we should reject $H_0$ for large values of $\bar{x}$, and thus for large values of $z$. This suggests a rejection region of the form $C_r^Z = [c, \infty)$ for outcomes of $Z$. Referring to the standard normal table assuming a .05 level test is desired, we choose $c = 1.645$, so that $P(z \geq 1.645) = .05$ (as an asymptotic approximation).

The reader may have suspected a serious practical flaw in the use of $Z$ as a test statistic—namely, $\lambda$ is unknown so that $Z$ is *not* a statistic. However, this problem is overcome by assigning $\lambda$ a numerical value. Setting $\lambda = 2$ in the definition of $Z$ is the logical choice under the current circumstances. This follows because if $\bar{X} \sim N(\lambda, n^{-1}\lambda)$, then $Z = n^{1/2}(\bar{X} - 2)/2^{1/2} \sim N\left((n/2)^{1/2}(\lambda - 2), \lambda/2\right)$, and thus (asymptotically)

$$\pi_{C_r}(\lambda) = P(z \geq 1.645)\left\{\begin{matrix} \leq \\ > \end{matrix}\right\}.05, \quad \forall\lambda\left\{\begin{matrix} \leq \\ > \end{matrix}\right\} 2,$$

so that $C_r^Z$ then defines a .05 level unbiased test of $H_0$: $\lambda \leq 2$ versus $H_a$: $\lambda > 2$.

Restating the test in terms of a rejection region for the outcome of $\bar{X}$ itself, and recalling that $n = 100$ in this case, we have that $z \geq 1.645$ *iff* $\bar{x} \geq 2.23$, so that $C_r^{\bar{X}} = [2.23,\infty)$, which is virtually the same rejection region as in Example 10.1 based on the actual Poisson population distribution. The near equivalence of the two rejection regions is due to the accuracy of the asymptotic normal distribution of $\bar{X}$. Since $\bar{x} = 2.48 \in C_r^{\bar{X}}$, we reject $H_0$, as before. An asymptotic $p$-value for the test can be calculated based on the observed value of $z = 10(2.48 - 2)/2^{1/2} = 3.394$, so that $p$ value $= \int_{3.394}^{\infty} N(z; 0,1)\, dz < .001$ provides strong evidence for the rejection of $H_0$. Also, under the assumption of normality for $\bar{X}$, it could be argued, using the results of Section 9.5, that the preceding test is a UMP level .05 test. In this case, we then state that the test is *asymptotically* UMP level .05. The reader might consider plotting the power function of the test. □

In each of the preceding examples, a specific random sample size was involved. One might also consider whether sequences of level .05 unbiased tests defined analogously to the preceding tests have the property of **consistency** as sample sizes increase without bound. Using asymptotic theory applied to both the respective test statistics and the definitions of rejection regions, one can argue that the respective test sequences are indeed consistent.

**Example 10.4** ***Consistency of Some Heuristic Tests***

Revisit Examples 10.1–10.3, and consider whether the respective sequences of .05-level unbiased tests are *consistent*. Regarding Example 10.1, the $n$th element in the appropriate sequence of rejection regions will be of the form $C_{rn}^{\bar{X}} = [c_n,\infty)$ such that $P(\bar{x}_n \geq c_n) \leq .05\ \forall n$, where $\lambda = 2$ is used when defining the probability of the rejection region for $\bar{X}_n$ and $c_n$ is chosen as small as possible in order to maximize power for $\lambda \in H_a$. The preceding sequence of probability inequalities can be written alternatively as

$$P\left(n^{1/2}(\bar{x}_n - \lambda)/\lambda^{1/2} \geq n^{1/2}(c_n - \lambda)/\lambda^{1/2}\right) \leq .05,\ \forall n,$$

and assuming $\lambda$ were the true Poisson mean, $Z_n = n^{1/2}(\bar{X}_n - \lambda)/\lambda^{1/2} \xrightarrow{d} N(0,1)$ by the LLCLT. Therefore, $n^{1/2}(c_n - \lambda)/\lambda^{1/2} \rightarrow 1.645$ when defining the rejection region, which implies $c_n \rightarrow \lambda$, and given that $\lambda = 2$ was used in defining the rejection region, then $c_n \rightarrow 2$. By the consistency of $\bar{X}_n$ for $\lambda$, we also know that $\bar{X}_n \xrightarrow{p} \lambda$, and if the true Poisson mean is $\lambda > 2$, so that $H_a$ is true, then $\bar{X}_n - c_n \xrightarrow{p} \lambda - 2 > 0$ so that $H_0$ is rejected with probability converging to 1 when $H_a$ is true. Thus the sequence of tests is consistent, and a similar argument can be applied to demonstrate consistency in the case of Example 10.3.

Regarding Example 10.2, note that the rejection region for the $T$-statistic will ultimately converge to $C_m^T \rightarrow C_r^T = (-\infty, -1.96] \cup [1.96, \infty)$ since the (central) $T_n$ random variable is such that $T_n \xrightarrow{d} Z \sim N(0,1)$ and $P(z \in C_r^T) = .05$. Assuming $\sigma^2(\mathbf{x}'\mathbf{x})^{-1} \rightarrow \mathbf{0}$ so that $\text{var}(\hat{\beta}_2) \rightarrow 0$ and $\hat{\beta} \xrightarrow{p} \beta$, and assuming that $\hat{S}^2 \xrightarrow{p} \sigma^2$ (e.g., assume that the error terms in the linear model are *iid*), it follows that the noncentrality parameter of the $t$-distribution associated with $T_n = \hat{\beta}_2 / \widehat{std}(\hat{\beta}_2)$, which is given by $\lambda = \beta_2/[\text{var}(\hat{\beta}_2)]^{1/2}$, diverges to infinity as $n \rightarrow \infty$ whenever $\beta_2 \neq 0$. Then $P(t_n \in C_m^T; \lambda) \rightarrow P(t_n \in C_r^T; \lambda) \rightarrow 1$ as $n \rightarrow \infty$ when $H_a$ is true, and thus the test is consistent. □

## 10.3 Generalized Likelihood Ratio Tests

As its name implies, a generalized likelihood ratio (GLR) test of a statistical hypothesis is based on a test rule that is defined in terms of a ratio of likelihood function values. The adjective *generalized* is used here to distinguish the GLR test from the simple ratio of two likelihood function values presented in our discussion of the Neymann-Pearson Lemma (Theorem 9.1). In the latter context, the likelihood function was evaluated at two distinct values of the parameter vector, $\boldsymbol{\Theta}$. In the current context, the likelihood ratio is "generalized" by forming the ratio of likelihood function suprema[1] (or maxima if maximums exist), where the two suprema are taken with respect to two different feasible sets–the set of null hypothesis values for $\boldsymbol{\Theta}$, and the set of values of $\boldsymbol{\Theta}$ represented by the union of the null and alternative hypotheses. The GLR test is a natural procedure to use for testing hypotheses about $\boldsymbol{\Theta}$ or functions of $\boldsymbol{\Theta}$ when maximum likelihood estimation is being used in a statistical analysis.

**Definition 10.1** ***Generalized Likelihood Ratio (GLR) Test of Size $\alpha$***

Let the probability sample $(X_1,\ldots,X_n)$ have the joint probability density function $f(x_1,\ldots,x_n;\boldsymbol{\Theta})$ and associated likelihood function $L(\boldsymbol{\Theta};x_1,\ldots,x_n)$. The **generalized likelihood ratio** (GLR) is defined as

$$\lambda(\mathbf{x}) = \frac{\sup_{\boldsymbol{\Theta} \in H_0}\{L(\boldsymbol{\Theta}; x_1, \ldots x_n)\}}{\sup_{\boldsymbol{\Theta} \in H_0 \cup H_a}\{L(\boldsymbol{\Theta}; x_1, \ldots, x_n)\}},$$

and a generalized likelihood ratio test for testing $H_0$ versus $H_a$ is given by the following test rule: reject $H_0$ *iff* $\lambda(\mathbf{x}) \leq c$, or equivalently, reject $H_0$ *iff* $I_{[0,c]}(\lambda(\mathbf{x})) = 1$.

For a size $\alpha$ test, the constant $c$ is chosen to satisfy

$$\sup_{\boldsymbol{\Theta} \in H_0}\{\pi(\boldsymbol{\Theta})\} = \sup_{\boldsymbol{\Theta} \in H_0}\{P(\lambda(\mathbf{x}) \leq c; \boldsymbol{\Theta})\} = \alpha.$$

[1] Recall that the supremum of $g(w)$ for $w \in A$ is the smallest upper bound for the value of $g(w)$ when $w \in A$. If the supremum is attainable for some value $w \in A$, the supremum is the maximum.

In order to provide some intuitive rationale for the GLR test, first note that the numerator of the GLR is essentially the largest likelihood value that can be associated with the sample outcome $(x_1,\ldots,x_n)$ when we are able to choose only among probability distributions that are contained in the null hypothesis, $H_0$. The numerator can then be interpreted as the likelihood function evaluated at the *constrained* maximum likelihood estimate of $\mathbf{\Theta}$, the constraint being $\mathbf{\Theta}\in H_0$. The denominator of the GLR is the largest likelihood value that can be associated with $(x_1,\ldots,x_n)$ when we can choose any probability distribution contained in $H_0$ and/or $H_a$. In most applications $H_0 \cup H_a$ will be the entire parameter space, $\Omega$, and the denominator is the likelihood function evaluated at the maximum likelihood estimate of $\mathbf{\Theta}$. Since likelihood functions are nonnegative valued, and since the feasible space for the numerator supremum problem is contained in the feasible space for the denominator supremum problem, we can infer that $\lambda\in[0,1]$.

To see why the critical region of the test statistic $\lambda(\mathbf{x})$ is defined in terms of the *lower tail*, $[0,c]$, of the range of $\lambda$, note that the smaller the value of $\lambda$, the larger are the maximum likelihood for values of $\mathbf{\Theta}\in H_0 \cup H_a$ relative to the maximum likelihood for $\mathbf{\Theta}\in H_0$. Intuitively, this means that when $\lambda$ is small, there is a value of $\mathbf{\Theta}\in H_a$ that is notably "more likely" to have characterized the true density $f(\mathbf{x};\mathbf{\Theta})$ associated with the sample outcome $\mathbf{x}$ than any other value of the parameter $\mathbf{\Theta}\in H_0$. When $f(\mathbf{x};\mathbf{\Theta})$ is a discrete density function, we could also infer that there is a value of $\mathbf{\Theta}\in H_a$ that implies a notably higher *probability* of observing $\mathbf{x}$ than does any value of $\mathbf{\Theta}\in H_0$. Thus, for small values of $\lambda$, say $c$ or less, it appears *reasonable* to reject $H_0$ as containing the true probability distribution of $\mathbf{x}$, and to conclude instead that a better representation of the true probability distribution of $\mathbf{x}$ resides in the set $H_a$.

### 10.3.1 GLR Test Properties: Finite Sample

Intuition aside, whether the GLR test represents a good statistical test for a given statistical hypothesis ultimately depends on the statistical properties of the test. There is no guarantee that a GLR test is UMP, or even unbiased, in finite samples. Finite sample properties of the GLR test must be established on a case-by-case basis and depend on both the characteristics of $f(\mathbf{x};\mathbf{\Theta})$ and the definition of the sets underlying $H_0$ and $H_a$. Nonetheless, before we proceed to asymptotic properties for which general results do exist, we point out some parallels with the results presented in Chapter 9.

#### *10.3.1.1 Simple Hypotheses*

In the case where both $H_0$ and $H_a$ are simple hypotheses, the size $\alpha$ GLR test and the most powerful level $\alpha$ test based on the Neyman-Pearson lemma (Theorem 9.1) will be equivalent.

**Theorem 10.1**
***Equivalence of GLR Test and Neyman-Pearson Most Powerful Test When $H_0$ and $H_a$ are Simple Hypotheses***

*Suppose a size $\alpha$ GLR test of $H_0$: $\boldsymbol{\Theta} = \boldsymbol{\Theta}_0$ versus $H_a$: $\boldsymbol{\Theta} = \boldsymbol{\Theta}_a$ exists with critical region*

$$C_r^{\mathrm{GLR}} = \{\mathbf{x} : \lambda(\mathbf{x}) \le c\}, \text{where } \mathrm{P}(\mathbf{x} \in \mathrm{C}_r^{\mathrm{GLR}}; \boldsymbol{\Theta}_0) = \alpha \in (0,1).$$

*Furthermore, suppose a Neyman-Pearson most powerful level $\alpha$ test also exists with critical region*

$$C_r = \{\mathbf{x} : L(\boldsymbol{\Theta}_0; x) \le kL(\boldsymbol{\Theta}_a; \mathbf{x})\}, \text{ where } \mathrm{P}(\mathbf{x} \in \mathrm{C}_r; \boldsymbol{\Theta}_0) = \alpha.$$

*Then the GLR test and the Neymann-Pearson most powerful test are equivalent.*

**Proof** Since $\lambda(\mathbf{x}) \in [0,1]$, and given that $\alpha \in (0,1)$, it follows that $P(\lambda(\mathbf{x}) \le c; \boldsymbol{\Theta}_0) = \alpha$ only if $c < 1$. Now let $\hat{\boldsymbol{\theta}} = \hat{\boldsymbol{\Theta}}(\mathbf{x}) = \arg\max_{\boldsymbol{\Theta} \in \{\boldsymbol{\Theta}_0, \boldsymbol{\Theta}_a\}}\{L(\boldsymbol{\Theta}; \mathbf{x})\}$, so that the GLR can be represented as $\lambda(\mathbf{x}) = \mathrm{L}(\boldsymbol{\Theta}_0; \mathbf{x})/\mathrm{L}(\hat{\boldsymbol{\theta}}; \mathbf{x})$. Note the following relationship implied by the maximization process, and leading to a dichotomous partition of the range of $\mathbf{X}$:

$$\mathbf{x} \in A = \left\{\mathbf{x}\colon \hat{\boldsymbol{\Theta}}(\mathbf{x}) = \boldsymbol{\Theta}_a\right\} \Rightarrow \frac{L(\boldsymbol{\Theta}_0; \mathbf{x})}{L(\boldsymbol{\Theta}_a; \mathbf{x})} = \frac{L(\boldsymbol{\Theta}_0; \mathbf{x})}{L\left(\hat{\boldsymbol{\theta}}; \mathbf{x}\right)} \le 1,$$

$$\mathbf{x} \in B = \left\{\mathbf{x}\colon \hat{\boldsymbol{\Theta}}(\mathbf{x}) = \boldsymbol{\Theta}_0\right\} \Rightarrow \frac{L(\boldsymbol{\Theta}_0; \mathbf{x})}{L(\boldsymbol{\Theta}_a; \mathbf{x})} \ge \frac{L(\boldsymbol{\Theta}_0; \mathbf{x})}{L\left(\hat{\boldsymbol{\theta}}; \mathbf{x}\right)} = 1.$$

It follows that $L(\boldsymbol{\Theta}_0; \mathbf{x})/L(\boldsymbol{\Theta}_a; \mathbf{x}) \le c < 1$ only if $\lambda(\mathbf{x}) = L(\boldsymbol{\Theta}_0; \mathbf{x})/L\left(\hat{\boldsymbol{\theta}}; \mathbf{x}\right) \le c < 1$. When both preceding inequalities hold, $\hat{\boldsymbol{\theta}} = \boldsymbol{\Theta}_a$ and $L(\boldsymbol{\Theta}_0; \mathbf{x})/L(\boldsymbol{\Theta}_a; \mathbf{x}) = L(\boldsymbol{\Theta}_0; \mathbf{x})/L\left(\hat{\boldsymbol{\theta}}; \mathbf{x}\right)$. Thus, for $c < 1$ and $\alpha \in (0,1)$ if $P(L(\boldsymbol{\Theta}_0; \mathbf{x})/L(\hat{\boldsymbol{\theta}}; \mathbf{x}) \le c;\, \boldsymbol{\Theta}_0) = \alpha$, and $P(L(\boldsymbol{\Theta}_0; \mathbf{x})/L(\boldsymbol{\Theta}_a; \mathbf{x}) \le c;\, \boldsymbol{\Theta}_0) = \alpha$, then $C_r = C_r^{\mathrm{GLR}}$ and the GLR test is equivalent to the Neyman-Pearson most powerful test of size $\alpha$. ■

Since the GLR test is equivalent to the Neymann-Pearson most powerful test when $H_0$ and $H_a$ are simple, we also know by Theorem 9.2 that the GLR test is *unbiased*. We revisit a previous example, in which a most powerful test was found via the Neyman-Pearson Lemma to illustrate the GLR approach to the problem.

**Example 10.5**
***GLR Test of Simple Hypotheses for an Exponential Distribution***

Recall Example 9.11, in which a decision is to be made regarding the population mean life of a type of computer screen. The likelihood function for the parameter $\theta$ (the mean life of the screens) given the 10 observations on screen lifetimes represented by $(x_1,\ldots,x_{10})$ is given by (suppressing the indicator functions)

$$L(\theta, x) = \theta^{-10} \exp\left(-\sum_{i=1}^{10} x_i/\theta\right).$$

The null and alternative hypotheses under consideration are $H_0$: $\theta = 1$ and $H_a$: $\theta = 5$, respectively.

The GLR for this problem is given by

$$\lambda(\mathbf{x}) = \frac{L(1;\mathbf{x})}{\max\limits_{\theta\in\{1,5\}}\{L(\Theta,\mathbf{x})\}} = \min\left\{1, \frac{\exp\left(-\sum_{i=1}^{10} x_i\right)}{(5)^{-10}\exp\left(-\sum_{i=1}^{10} x_i/5\right)}\right\}$$

$$= \min\left\{1, (.2)^{-10}\exp\left(-.8\sum_{i=1}^{10} x_i\right)\right\}.$$

It follows that, for $c < 1$, the probability that $\lambda(\mathbf{x}) \leq c$ is given by

$$P(\lambda(x) \leq c) = P\left((.2)^{-10}\exp\left(-.8\sum_{i=1}^{10} x_i\right) \leq c\right)$$

$$= P\left(\sum_{i=1}^{10} x_i \geq 20.11797 - 1.25\ln(c)\right) \quad \text{(to five decimal places).}$$

A size $\alpha$ GLR test with critical region $[0,c]$ is defined by choosing $c$ so that $P(\lambda(\mathbf{x}) \leq c) = \alpha$, where the probability value can be determined by utilizing the fact that $\sum_{i=1}^{10} X_i \sim \text{Gamma}(10, \theta)$ and $\theta = 1$ under $H_0$. Comparing this result to the definition of the most powerful rejection region given in Example 9.11, it is evident that the two rejection regions are identical, so that the GLR test is both unbiased and the most powerful test of $H_0$: $\theta = 1$ versus $H_a$: $\theta = 5$. For $\alpha = .05$, the critical region of the GLR test would be given by $[0, 34.13079]$, which can be transformed into a rejection region stated in terms of the test statistic $\sum_{i=1}^{10} X_i$ as $[15.70522, \infty)$. ■

*10.3.1.2 Composite Hypotheses*

Similar to the extension of the Neyman-Pearson Lemma to the case of defining UMP tests for testing simple null versus composite alternative hypotheses (Theorem 9.3), the result of Theorem 10.1 can be extended to the case of testing simple null versus composite alternative hypotheses as follows:

**Theorem 10.2** ***UMP Level $\alpha$ GLR Test of $H_0$: $\Theta = \Theta_0$ versus $H_a$: $\Theta \in \Omega_a$ When $C_r$ Is Invariant***

*Suppose the given rejection region, $C_r^{GLR} = \{\mathbf{x} : \lambda(\mathbf{x}) \leq c\}$, of the GLR test of $H_0$: $\Theta = \Theta_0$ versus $H_a$: $\Theta \in \Omega_a$ defines a size $\alpha$ test and $\forall \Theta_a \in H_a\ \exists c_{\Theta_a} \geq 0$ such that $C_r^{GLR} = \{\mathbf{x} : \lambda_{\Theta_a}(\mathbf{x}) \leq c_{\Theta_a}\}$ where*

$$\lambda_{\Theta_a}(\mathbf{x}) = \frac{L(\Theta_0;\mathbf{x})}{\max_{\Theta\in\{\Theta_0,\Theta_a\}}\{L(\Theta;\mathbf{x})\}} \text{ and } P(\lambda_{\Theta_a}(\mathbf{x}) \leq c_{\Theta_a};\Theta_0) = \alpha.$$

*Furthermore, suppose a Neyman-Pearson UMP test of $H_0$ versus $H_a$ having size $\alpha$ exists. Then $C_r^{GLR}$ defines a UMP level $\alpha$ test of $H_0$ versus $H_a$.*

**Proof** Given that a Neyman-Pearson UMP test having size $\alpha$ exists, it is the Neyman-Pearson most powerful size $\alpha$ test for every pair $(\Theta_0, \Theta_a)$, by Theorem 9.3, where $C_r$ is invariant to the choice of $\Theta_a \in H_a$. Because $\forall(\Theta_0,\Theta_a)$ pair, both the Neyman-Pearson and GLR size $\alpha$ tests exist, the tests are equivalent $\forall(\Theta_0, \Theta_a)$ by Theorem 10.1. Then $C_r^{GLR}$ defines a UMP level $\alpha$ test of $H_0$ versus $H_a$. ■

The theorem implies that if the rejection region of a size $\alpha$ GLR test of $H_0$: $\Theta = \Theta_0$ versus $H_a$: $\Theta = \Theta_a$ is the *same* $\forall\Theta_a\in\Omega_a$ i.e., $C_r^{GLR}$ is **invariant**, then if a Neyman-Pearson UMP test of $H_0$: $\Theta = \Theta_0$ versus $H_a$: $\Theta\in\Omega_a$ having size $\alpha$ exists, it is given by the GLR test. The UMP test would also be *unbiased* (Theorem 9.4).

**Example 10.6** ***GLR Test of Simple Null and Composite Alternative for Exponential Distribution***

Recall Examples 9.11, 9.13, and 10.1 regarding the operating lives of computer screens. Examine the problem of defining a size .05 GLR test of $H_0$: $\theta = 1$ versus $H_a$: $\theta \in\Omega_a$, where $\Omega_a = (1,\infty)$. The GLR for this problem is given by

$$\lambda(\mathbf{x}) = \frac{L(1;\mathbf{x})}{\sup_{\theta\in[1,\infty)}\{L(\theta;\mathbf{x})\}} = \frac{\exp\left(-\sum_{i=1}^{10} x_i\right)}{\sup_{\theta\in[1,\infty)}\left\{\theta^{-10}\exp\left(-\sum_{i=1}^{10} x_i/\theta\right)\right\}}.$$

The maximum of $L(\theta;\mathbf{x})$ for $\theta\in[1,\infty)$ can be defined by first solving the first-order condition

$$\frac{d\ln(L(\theta;x))}{d\theta} = \frac{-10}{\theta} + \frac{\sum_{i=1}^{10} X_i}{\theta^2} = 0,$$

which yields $\theta = \sum_{i=1}^{10} x_i/10$ as the choice of $\theta$ that maximizes $\ln(L(\theta;\mathbf{x}))$, and hence maximizes $L(\theta;\mathbf{x})$, when there are *no constraints* on $\theta$. Then, if the constraint $\theta \geq 1$, is recognized, $\lambda(\mathbf{x})$ can be defined ultimately as[2]

$$\lambda(\mathbf{x}) = \begin{cases} \left(\sum_{i=1}^{10} x_i/10\right)^{10} \exp\left(10 - \sum_{i=1}^{10} x_i\right), & \text{for } \sum_{i=1}^{10} x_i>10, \\ 1 & \text{otherwise} \end{cases}$$

It follows that for $c < 1$, the probability that $\lambda(\mathbf{x}) \leq c$ is given by

$$\begin{aligned} P(\lambda(\mathbf{x}) \leq c) &= P\left(\left[\left(\sum_{i=1}^{10} x_i/10\right)^{10} \exp\left(10 - \sum_{i=1}^{10} x_i\right)\right] \leq c\right) \\ &= P\left(\left[\sum_{i=1}^{10} x_i - 10\ln\left(\sum_{i=1}^{10} x_i\right)\right] \geq [-13.0259 - \ln(c)].\right) \quad (*) \end{aligned}$$

Recall that $\sum_{i=1}^{10} X_i \sim \text{Gamma}(10,1)$ when $\theta = 1$. Also, from Example 9.11 it is known that $P\left(\sum_{i=1}^{10} x_i \geq 15.70522\right) = .05$. Note that $\sum_{i=1}^{10} x_i - 10\ln\left(\sum_{i=1}^{10} x_i\right)$ is strictly monotonically increasing in the value of $\sum_{i=1}^{10} x_i$ for values of $\sum_{i=1}^{10} x_i>10$, so that there will exist a value of $c$ in (*) that defines the event $\sum_{i=1}^{10} x_i \geq 15.70522$.

[2] A more elegant solution procedure for this inequality constrained maximization problem could be formulated in terms of Kuhn-Tucker conditions.

In particular, the appropriate value is $c = .30387$ (to five decimals places). Thus, the rejection region for the GLR is [0, .30387], and the associated rejection region for $\mathbf{x}$ is $C_r^{\text{GLR}} = \left\{\mathbf{x} : \sum_{i=1}^{10} x_i \geq 15.70522\right\}$, which we know (from Example 9.13) defines the UMP and unbiased test of $H_0$: $\theta = 1$ versus $H_a$: $\theta > 1$.

To see that the GLR test satisfies Theorem 10.2, so that we can declare the test to be UMP and unbiased independently of knowing the result of Example 9.13, the reader can retrace the steps followed in Example 10.1, replacing $H_a$: $\theta = 5$ with $H_a$: $\theta = \theta_a$, where $\theta_a$ is an arbitrary choice of $\theta \in (1,\infty)$. The rejection region will invariably have the form $C_r^{\text{GLR}} = \left\{\mathbf{x} : \sum_{i=1}^{10} x_i \geq 15.70522\right\}$ and will agree with the Neyman-Pearson most powerful rejection region. □

The GLR test can also lead to UMP and unbiased tests of the simple or *composite* null hypothesis $H_0$: $\Theta \in \Omega_0$ versus the composite alternative hypothesis $H_a$: $\Theta \in \Omega_a$ when $\Theta$ is a scalar, the alternative hypothesis is one-sided, and the problem is characterized by a monotone likelihood ratio.

**Theorem 10.3** ***UMP and Unbiased GLR Level $\alpha$ Test of $H_0$: $\Theta \in \Omega_0$ versus One-Sided $H_a$: $\Theta \in \Omega_a$ in Case of Monotone Likelihood Ratio***

*Let the sampling density of $\mathbf{X}$, given by $f(\mathbf{x};\Theta)$ for scalar $\Theta \in \Omega$, be a family of density functions having a monotone likelihood ratio in the statistic $T = t(\mathbf{X})$. Then the GLR test of $H_0$: $\Theta \in \Omega_0$ versus one-sided $H_a$: $\Theta \in \Omega_a$ is a UMP and unbiased level $\alpha$ test if $P(\lambda(\mathbf{x}) \leq c) = \alpha$ and either*

1. $H_0$: $\Theta = \Theta_0$ or $\Theta \leq \Theta_0$, $H_a$: $\Theta > \Theta_0$, and $\lambda(\mathbf{x}) \leq c$ *iff* $t(\mathbf{x}) \geq c_*$,
   or
2. $H_0$: $\Theta = \Theta_0$ or $\Theta \geq \Theta_0$, $H_a$: $\Theta < \Theta_0$, and $\lambda(\mathbf{x}) \leq c$ *iff* $t(\mathbf{x}) \leq c_*$.

**Proof** The proof follows immediately from Corollaries 9.1 and 9.2 upon recognition that $\lambda(\mathbf{x}) \leq c$ is simply an alternative representation of the UMP and unbiased level $\alpha$ rejection region for the respective hypothesis tests based on the properties of monotone likelihood ratios. ■

**Example 10.7** ***UMP Test of $H_0$: $p \leq p_0$ versus $H_a$: $p > p_0$ in Bernoulli Population Using GLR***

A personal computer manufacturer claims that $\geq$ 80 percent of its new computers are shipped to its customers without any defects whatsoever. A random sample, with replacement, of 20 purchases of the company's products resulted in 6 reports of initial defects upon delivery of computers. The responses of the purchasers are viewed as outcomes of *iid* Bernoulli random variables $X_i \sim p^{x_i}(1-p)^{1-x_i} I_{\{0,1\}}(x_i)$, where $x_i = 1 \Rightarrow$ defect reported and $x_i = 0 \Rightarrow$ no defect reported. We seek to construct a size .01 GLR test of the hypothesis that $H_0$: $p \leq .20$ versus $H_a$: $p > .20$, i.e., the null hypothesis is that the proportion of computers shipped that are defective is less than or equal to .20 versus the alternative that the proportion is greater than .20.

The GLR for this problem is

$$\lambda(\mathbf{x}) = \frac{\sup_{p \le .20}\left\{p^{\sum_{i=1}^{20} x_i}(1-p)^{20-\sum_{i=1}^{20} x_i}\right\}}{\sup_{p\in[0,1]}\left\{p^{\sum_{i=1}^{20} x_i}(1-p)^{20-\sum_{i=1}^{20} x_i}\right\}}.$$

The value of $p$ that solves the denominator supremum problem is simply the MLE outcome for $p$, $\hat{p} = \sum_{i=1}^{20} x_i/20$. The value of $p$ that solves the numerator supremum problem is defined as

$$\hat{p}_0 = \left\{ \begin{array}{ll} \sum_{i=1}^{20} x_i/20 & \text{if } \sum_{i=1}^{20} x_i/20 \le .20 \\ .20 & \text{otherwise} \end{array} \right\}.$$

Therefore, $\lambda(\mathbf{x})$ can be represented as

$$\lambda(\mathbf{x}) = \begin{cases} \dfrac{(.20)^z(.80)^{20-z}}{(z/20)^z(1-z/20)^{20-z}} & \text{if } \sum_{i=1}^{20} x_i > 4 \\ 1 & \text{otherwise} \end{cases}$$

where $Z = \sum_{i=1}^{20} X_i \sim$ Binomial $(20, p)$.

Note that $\lambda(\mathbf{x})$ is a strictly decreasing function of $z$ for $z > 4$. Because $P(z \ge 9) = .01$ it follows that $P(\lambda(\mathbf{x}) \le .04172) = P(z \ge 9) = .01$ so that the rejection region for the GLR test is

$$C_r^{\text{GLR}} = \{\mathbf{x} : \lambda(\mathbf{x}) \le .04172\} = \left\{\mathbf{x} : \sum_{i=1}^{20} x_i \ge 9\right\}.$$

Note the statistic $t(\mathbf{X})$ of the monotone likelihood ratio can be specified as $t(\mathbf{X}) = \sum_{i=1}^{20} X_i$. It follows from Theorem 10.3 that the GLR test is UMP and unbiased with level .01. Given that $\sum_{i=1}^{20} x_i = 6$, the hypothesis $H_0$: $p \le .20$ *cannot* be rejected at the .01 level. □

At this point we focus the reader's attention on a common procedure that was used in the preceding examples when assigning probabilities to events of the form $\lambda(\mathbf{x}) \le c$ for the GLR. Namely, in each case we were able to find a strictly monotonically increasing or decreasing function of $\lambda(\mathbf{x})$, say $h(\lambda(\mathbf{x}))$, whose PDF had a known tractable form. Then probability was assigned using the PDF of $h(\lambda(\mathbf{X}))$ as $P(\lambda(\mathbf{x}) \le c) = P(h \le k)$ or $P(h \ge k)$ for $h(\cdot)$ monotonically increasing or decreasing respectively. It is often the case that the probability density of the GLR, $\lambda(\mathbf{X})$, is difficult to define or intractable to work with for defining a size $\alpha$ critical region. In applications of GLR tests one must often seek a *test statistic*, $h(\lambda(\mathbf{X}))$, having a tractable probability density in order to be able to both define a size $\alpha$ critical region and to investigate the finite sample properties of the GLR test. Unfortunately, in practice, it is not always possible to define such a test statistic.

We will refrain from attempting to provide additional results concerning finite sample properties of GLR tests mainly because there are few additional generalizations that can be made. Typically, a GLR test is constructed in a given problem context, and then an attempt is made to assess its properties. It is sometimes the case that little can be definitively established regarding the finite sample properties of a GLR test. Fortunately, it is typically the case that the large sample properties of GLR tests are very good, and the results apply quite generally to simple or composite hypotheses involving scalar or multidimensional parameters for problems with or without nuisance parameters. We examine this topic next.

### 10.3.2 GLR Test Properties: Asymptotics

The GLR test is generally a consistent test, and in cases where $H_0$ is defined by functional restrictions on the parameter space, the asymptotic distribution of $-2\ln(\lambda(\mathbf{X}))$ is generally a $\chi^2$ distribution. In many cases, the verification of consistency and the identification of the asymptotic distribution of the GLR test can best be accomplished by analyzing either the properties of the random variable $\lambda(\mathbf{X})$ directly or else the properties of a test statistic $h(\lambda(\mathbf{X}))$ that is a function of the GLR statistic. In other cases, there exist regularity conditions that ensure the consistency and asymptotic distribution of the GLR test. We present some results for the asymptotic properties of the GLR test below. Additional results can be found in S. Wilks, (1962), *Mathematical Statistics*, New York: John Wiley, p. 419; and R.J. Serfling, (1980), *Approximation Theorems of Mathematical Statistics*, New York: John Wiley, pp. 151–160.

#### *10.3.2.1 Consistency of GLR Tests*

We begin with the property of consistency and introduce a sufficient condition that applies quite generally and is often not difficult to establish.

**Theorem 10.4 *Consistency of the GLR Test*** *Assume the conditions for consistency of the maximum likelihood estimator (MLE) given by Theorems 8.16–8.18. Let* $[0, c_n]$ *for* $n=1,2,3,\ldots,$ *represent level* $\alpha$ *rejection regions of the GLR statistic for testing* $H_0$: $\boldsymbol{\Theta}\in\Omega_0$ *versus* $H_a$: $\boldsymbol{\Theta}\in\Omega_a$ *based on increasing sample size n. Then* $\lim_{n\to\infty}(P(\lambda(\mathbf{x}) \leq c_n;\boldsymbol{\Theta})) = 1\ \forall\boldsymbol{\Theta}\in H_a$, *so that the sequence of GLR tests is consistent if either of the following conditions hold:*

**(a)** *The GLR statistic is bounded below 1 with probability* $\to 1$ *as* $n \to \infty$ $\forall\boldsymbol{\Theta}\in H_a$, *i.e.,* $\lim_{n\to\infty}(P(\lambda(\mathbf{x}) \leq \tau;\ \boldsymbol{\Theta})) = 1\ \forall\boldsymbol{\Theta}\in H_a$, *where* $\tau < 1$.

**(b)** $\text{plim}(\lambda(\mathbf{X})) = \delta(\boldsymbol{\Theta}) \leq \tau < 1\ \forall\boldsymbol{\Theta}\in H_a$.

**Proof** **(a)** Assuming that the true $\boldsymbol{\Theta}_0\in H_0$, it follows that $\text{plim}(\lambda(\mathbf{X})) = 1$. To see this, let $\hat{\Theta}_0$ and $\hat{\Theta}$ represent the MLEs for $\boldsymbol{\Theta}\in H_0$ and $\boldsymbol{\Theta}\in H_0 \cup H_a$, respectively, and expand the logarithm of the likelihood function $L(\hat{\boldsymbol{\Theta}}_0;\mathbf{x})$ in a Taylor series around the point $\hat{\boldsymbol{\Theta}}$ to obtain $\ln(\lambda(\mathbf{x})) = \ln L(\hat{\boldsymbol{\Theta}}_0;\mathbf{x}) - \ln L(\hat{\boldsymbol{\Theta}};\mathbf{x}) =$

$(\partial \ln L(\Theta_*;\mathbf{X})/\partial\Theta)'(\hat{\Theta}_0 - \hat{\Theta})$, where $\Theta_* = \eta\hat{\Theta}_0 + (1-\eta)\hat{\Theta}$ and $\eta \in [0,1]$.[3] Since both $\hat{\Theta}_0 \xrightarrow{p} \Theta_0$ and $\hat{\Theta} \xrightarrow{p} \Theta_0$ (recall $\Theta_0 \in H_0$), then $\Theta_* \xrightarrow{p} \Theta_0$, and it follows that both $\partial \ln L(\Theta_*;\mathbf{x})/\partial\Theta$ and $(\hat{\Theta}_0 - \hat{\Theta}) \xrightarrow{p} \mathbf{0}$. Therefore, $\ln(\lambda(\mathbf{X})) \xrightarrow{p} 0$, so that $\lambda(\mathbf{X}) \xrightarrow{p} 1$ and if the GLR test is to be of size $\leq \alpha\ \forall n$, so that $\forall n$ and $\forall \Theta \in H_0$, $P(\lambda(\mathbf{x}) \leq c_n;\Theta) \leq \alpha$, it follows that $c_n \to 1$ as $n \to \infty$. Finally, if $\lim_{n\to\infty}(P(\lambda(\mathbf{x}) \leq \tau;\Theta)) = 1$ with $\tau < 1\ \forall \Theta \in H_a$, then $\lim_{n\to\infty}(P(\lambda(\mathbf{x}) \leq c_n;\Theta)) = 1\ \forall \Theta \in H_a$.

**(b)** $\forall \Theta \in H_a$, plim $(\lambda(\mathbf{X})) = \delta(\Theta) \leq \tau < 1 \Rightarrow \lim_{n\to\infty}(P(\lambda(\mathbf{x}) \leq \tau;\Theta)) = 1$. ■

**Example 10.8** ***Consistency of GLR Test for Exponential Distribution***

Revisit Example 10.6 regarding the operating lives of computer screens. Assume that the true $\theta_0 \in H_a = (1,\infty)$ and examine the behavior of the GLR $\lambda(\mathbf{X})$. Letting $Z_n = \bar{X}_n \exp(n(1-\bar{X}_n))$, the GLR can be represented as $\lambda(\mathbf{X}) = I_{[0,1]}(\bar{X}_n) + Z_n\, I_{(1,\infty)}(\bar{X}_n)$, and if $\theta_0 \in H_a$, then $I_{[0,1]}(\bar{X}_n) \xrightarrow{p} 0$ and $I_{(1,\infty)}(\bar{X}_n) \xrightarrow{p} 1$. Then $\text{plim}(\lambda(\mathbf{X})) = \text{plim}(Z_n)$ if the latter probability limit exists. Note that $n^{-1}\ln(Z_n) = n^{-1}\ln(\bar{X}_n) + (1-\bar{X}_n) \xrightarrow{p} 1 - \theta_0 = \xi < 0$ since $\bar{X}_n \xrightarrow{p} \theta_0 > 1$. It follows from the definition of convergence in probability that $\forall \varepsilon > 0$,

$$P(n^{-1}\ln(z_n) < \xi + \varepsilon) \geq P(\xi - \varepsilon < n^{-1}\ln(z_n) < \xi + \varepsilon) \to 1$$

and then choosing $\varepsilon > 0$ small enough so that $\xi + \varepsilon < 0$,

$$P(n^{-1}\ln(z_n) < \xi + \varepsilon) = P(\ln(z_n) < n(\xi + \varepsilon)) = P(z_n < \exp(n(\xi + \varepsilon))) \to 1.$$

Since $\exp(n(\xi + \varepsilon)) \to 0$ and $z_n \geq 0$, $\lim_{n\to\infty}(P(z_n \in [0,\tau))) = 1\ \forall \tau > 0$, so that $\text{plim}(Z_n) = 0$, and thus $\text{plim}(\lambda(\mathbf{X})) = 0$. Then the sequence of GLR tests of $H_0$: $\theta = 1$ versus $H_a$: $\theta > 1$ is consistent by Theorem 10.4.b. □

**Example 10.9** ***Consistency of GLR Test for Binomial Distribution***

Recall Example 10.7 regarding the claim that $\geq 80$ percent of new computers are shipped defect free. Assume that the true $p_0 \in H_a = (.20, 1]$ and examine the behavior of the GLR $\lambda(\mathbf{X})$. Letting $w_n = (.20)^{n\bar{x}}(.80)^{n(1-\bar{x})}/\left[\bar{x}^{n\bar{x}}(1-\bar{x})^{n(1-\bar{x})}\right]$, the GLR can be represented as $\lambda(\mathbf{X}) = I_{[0,.20]}(\bar{X}_n) + W_n\, I_{(.20,\infty)}(\bar{X}_n)$, and if $p_0 \in H_a$, then $I_{[0,.20]}(\bar{X}_n) \xrightarrow{p} 0$ and $I_{(.20,\infty)}(\bar{X}_n) \xrightarrow{p} 1$. Then $\text{plim}(\lambda(\mathbf{X})) = \text{plim}(W_n)$ if the latter probability limit exists. Note that

$$(W_n)^{1/n} = \frac{(.20)^{\bar{X}}(.80)^{(1-\bar{X})}}{\left[\bar{X}^{\bar{X}}(1-\bar{X})^{(1-\bar{X})}\right]} \xrightarrow{p} \frac{(.20)^p(.80)^{1-p}}{p^p(1-p)^{1-p}} = \left(\frac{.20}{p}\right)^p\left(\frac{.80}{(1-p)}\right)^{1-p} = \xi < 1\ \forall p \in (.2, 1].$$

Following reasoning similar to the previous example one can establish that $P(w_n < (\xi + \varepsilon)^n) \to 1$ for $\varepsilon > 0$ small enough such that $\xi + \varepsilon$ is positive and less than 1. It follows from $(\xi+\varepsilon)^n \to 0$ and $w_n \geq 0$ that $\lim_{n\to\infty} P(w_n \in [0,\tau)) = 1\ \forall \tau > 0$,

[3] R.G. Bartle, *Real Analysis*, p. 371.

so that $W_n \xrightarrow{p} 0$, and thus $\lambda(\mathbf{X}) \xrightarrow{p} 0$. Then the sequence of GLR tests of $H_0$: $p \leq .20$ versus $H_a$: $p > .20$ is consistent by Theorem 10.4.b. □

We have shown in the preceding examples that sequences of statistical tests were consistent, so that for large enough sample sizes, one is essentially certain to reject the null hypothesis if it is false. It is useful to note that consistent tests are effectively **asymptotically unbiased tests** in the sense that for large enough $n$, the rejection probability will eventually exceed whatever level $\alpha < 1$ is associated with the consistent test sequence since the probability of rejecting a false $H_0 \to 1$. Thus, the tests in the sequence eventually become unbiased (if they weren't unbiased to begin with) as $n \to \infty$.

*10.3.2.2 Asymptotic Distribution of GLR Statistics Under the Null Hypothesis* In the examples of statistical tests that have been presented heretofore, the probability density of either the GLR statistic or a function of the GLR statistic has been readily identifiable and tractable to work with. In cases where the identification and/or tractability of the probability distributions of GLR test statistics is problematic, there are asymptotic results relating to the GLR test that can be helpful so long as sample sizes are not too small. We will examine one such result relating to the asymptotic distribution of the natural logarithm of the GLR statistic for a special but important and prevalent form of null hypothesis. A discussion of additional results, and further readings, can be found in D.R. Cox and D.V. Hinkley, (1979), *Theoretical Statistics*, Chapman and Hall, London, pp. 311–342.

Regarding the asymptotic distribution of the GLR, it will prove useful to focus attention on the function $-2\ln(\lambda(\mathbf{x}))$ rather than on $\lambda(\mathbf{x})$ itself. A general result regarding the asymptotic distribution of $-2\ln(\lambda(\mathbf{X}))$ can be obtained when $H_0$ is defined via functional restrictions on the parameter space. In particular, we will be examining null hypotheses of the form $H_0 = \{\boldsymbol{\Theta}: \mathbf{R}(\boldsymbol{\Theta}) = \mathbf{r}, \boldsymbol{\Theta} \in \Omega\}$ where $\mathbf{R}(\boldsymbol{\Theta})$ is a $(q \times 1)$ differentiable vector function and $\mathbf{R}(\boldsymbol{\Theta}) = \mathbf{r}$ places linear and/or nonlinear constraints on the elements of the parameter vector $\boldsymbol{\Theta}$. It will be assumed that *none* of the $q$ coordinate functions in $\mathbf{R}(\boldsymbol{\Theta})$ are *redundant*. In this case, we can show that when the null hypothesis is true, $-2\ln(\lambda(\mathbf{X})) \overset{a}{\sim} \chi^2_q$ so that an asymptotically valid size $\alpha$ GLR test of $H_0$: $\mathbf{R}(\boldsymbol{\Theta}) = \mathbf{r}$ versus $H_a$: $\mathbf{R}(\boldsymbol{\Theta}) \neq \mathbf{r}$ can be conducted as

$$-2\ln(\lambda(\mathbf{X})) \geq \chi^2_{q;\alpha} \Rightarrow \text{reject } H_0,$$

or in terms of the GLR statistic itself,

$$\lambda(\mathbf{x}) \leq \exp\left(-\frac{1}{2}\chi^2_{q;\alpha}\right) \Rightarrow \text{reject } H_0,$$

where $\chi^2_{q;\alpha}$ is the value of a $\chi^2$-random variable with $q$ degrees of freedom such that the event $(\chi^2_{q;\alpha}, \infty)$ is assigned probability $\alpha$. Furthermore, $-2\ln(\lambda(\mathbf{X}))$ will have a *noncentral* $\chi^2$ asymptotic distribution (see Section 10.9) when $H_0$ is false (and when we examine so-called *local* alternative hypotheses, to be discussed shortly), allowing asymptotically valid power functions to be constructed. The formal result on the asymptotic distribution of the GLR test when $H_0$ is true is given below.

**Theorem 10.5** ***Asymptotic Distribution of GLR Test of $H_0$: $\mathbf{R}(\boldsymbol{\Theta}) = \mathbf{r}$ versus $H_a$: $\mathbf{R}(\boldsymbol{\Theta}) \neq \mathbf{r}$ When $H_0$ is True***

*Assume the conditions for the consistency, asymptotic normality, and asymptotic efficiency of the MLE of the $(k\times 1)$ vector $\boldsymbol{\Theta}$ as given in Theorem 8.19.* Let $\lambda(\mathbf{x}) = \sup_{\boldsymbol{\Theta}\in H_0}\{L(\boldsymbol{\Theta};\mathbf{x})\}/\sup_{\boldsymbol{\Theta}\in H_0\cup H_a}\{L(\boldsymbol{\Theta};\mathbf{x})\}$ *be the GLR statistic for testing* $H_0$: $\mathbf{R}(\boldsymbol{\Theta}) = \mathbf{r}$ versus $H_a$: $\mathbf{R}(\boldsymbol{\Theta}) \neq \mathbf{r}$, *where* $\mathbf{R}(\boldsymbol{\Theta})$ is a $(q\times 1)$ *continuously differentiable vector function having nonredundant coordinate functions and* $(q\leq k)$. *Then* $-2\ln(\lambda(\mathbf{X})) \xrightarrow{d} \chi^2_q$ *when $H_0$ is true.*

**Proof** See Appendix. ■

In Examples 10.5 and 10.6, which were based on random sampling from an exponential population distribution, we know from our study of the MLE in Chapter 8 that the MLE adheres to the conditions of Theorem 8.19 and is consistent, asymptotically normal, and asymptotically efficient. It follows by Theorem 10.5 that the GLR statistic for testing the simple null hypothesis $H_0$: $\theta = 1$ is such that $-2\ln(\lambda(\mathbf{X})) \overset{a}{\sim} \chi^2_1$. The asymptotic result would also apply to any *simple* hypotheses tested in the context of Example 10.7 where sampling was from a Bernoulli population distribution. While neither of these previous cases presented significant difficulties in determining rejection regions of the test in terms of the distribution of the GLR test statistic, it is useful to note that the asymptotic result for the GLR test provides an approximate method for circumventing the inherently limited choices of test sizes in the discrete case (recall Example 9.10 and Problem 9.7). In particular, since the $\chi^2$ distribution is continuous, *any* size test can be defined in terms of the *asymptotic* distribution of $-2\ln(\lambda(\mathbf{X}))$, whether or not $f(\mathbf{x};\boldsymbol{\Theta})$ is continuous, albeit the size will be an approximation.

*10.3.2.3 Asymptotic Distribution of GLR Statistics Under Local Alternatives*

In order to establish an asymptotically valid method of investigating the power of GLR tests, we now consider the asymptotic distribution of the GLR when the *alternative* hypothesis is true. We will analyze so-called **local alternatives** to the null hypothesis $H_0$: $\mathbf{R}(\boldsymbol{\Theta}) = \mathbf{r}$. In particular, we will focus on alternatives of the form $\mathbf{R}(\boldsymbol{\Theta}) = \mathbf{r} + n^{-1/2}\,\boldsymbol{\phi}$. In this context, the vector $\boldsymbol{\phi}$ specifies in what direction alternative hypotheses will be examined, and since $n^{-1/2}\,\boldsymbol{\phi} \to \mathbf{0}$ as $n \to \infty$, we are ultimately analyzing alternatives that are close or *local* to $\mathbf{R}(\boldsymbol{\Theta}) = \mathbf{r}$ for large enough $n$. These types of local alternatives are also referred to by the term *Pittman drift*.

A primary reason for examining *local alternatives* as opposed to fixed alternative hypotheses is that in the latter case, power will always be 1 for consistent tests, and thus no further information is gained about the operating characteristics of the test from asymptotic considerations other than what is already known about consistent tests, namely, that one is sure to reject a false null hypothesis when $n \to \infty$. There is an analog to degenerate limiting distributions, which are equally uninformative about the characteristics of random variables other than that they converge in probability to a constant. In the latter case, the random variable sequence was centered and scaled to obtain a nondegenerate limiting distribution that was more informative about the random variable characteristics of interest, e.g., variance. In the current context, the alternative hypotheses are scaled to establish "non-degenerate" power function behavior.

**Theorem 10.6 *Asymptotic Distribution of GLR Test of $H_0$: $\mathbf{R}(\Theta) = \mathbf{r}$ versus $H_a$: $\mathbf{R}(\Theta) \neq \mathbf{r}$ for Local Alternatives When $H_0$ Is False***

*Consider the GLR test of $H_0$: $\mathbf{R}(\Theta) = \mathbf{r}$ versus $H_a$: $\mathbf{R}(\Theta) \neq \mathbf{r}$ under the conditions and notation of Theorem 10.5, and let a sequence of local alternatives be defined by $H_{an}$: $\mathbf{R}(\Theta) = \mathbf{r} + n^{-1/2}\,\boldsymbol{\phi}$. Assume further that $\partial\mathbf{R}(\Theta_0)/\partial\Theta$ has full row rank. Then the limiting distribution of $-2\ln(\lambda(\mathbf{X}))$ under the sequence of local alternatives is noncentral $\chi^2$ as*

$$-2\ln(\lambda(\mathbf{X})) \xrightarrow{d} \chi^2_q(\lambda) \quad where\ \lambda = \frac{1}{2}\boldsymbol{\phi}' \left[\frac{\partial\mathbf{R}(\Theta_0)'}{\partial\Theta} M(\Theta_0)^{-1} \frac{\partial\mathbf{R}(\Theta_0)}{\partial\Theta}\right]^{-1} \boldsymbol{\phi}.$$

**Proof** See Appendix. ■

The GLR statistic, and the LM and Wald statistics discussed in subsequent sections, all share the same limiting distribution under sequences of local alternatives of the type identified here and are thus referred to as **asymptotically equivalent tests**, which will be established when we examine the alternative testing procedures. For related readings on the relationships between the triad of tests, see S.D. Silvey, (1959), "The Lagrangian Multiplier Test," *Annals of Mathematical Statistics*, (30), and R. Davidson and J.G. MacKinnon, (1993) *Estimation and Inference in Econometrics*, Oxford Univ. Press, NY, pp. 445–449.

**Example 10.10 *Asymptotic Power of GLR Test of $H_0$: $\theta=\theta_0$ versus $H_a$: $\theta > \theta_0$ in Exponential Population***

Revisit Example 10.6 and consider the asymptotic power of the GLR test of $H_0$: $\theta = \theta_0$ versus $H_a$: $\theta > \theta_0$ for the sequence of local alternatives $\theta_n = \theta_0 + n^{-1/2}\delta$, where in this application $\theta_0 = 1$. For this case we know that $-2\ln(\lambda(\mathbf{X})) \overset{a}{\sim} \chi^2_1(\lambda)$ with $\lambda = \delta'\delta/2 = \delta^2/2$, and $n = 10$ according to Example 10.6 (which is a bit small for the asymptotic calculations to be accurate). The asymptotic power function can be plotted in terms of the noncentrality parameter, or in this single parameter case, in terms of $\delta$. It is more conventional to graph power in terms of noncentrality, and we do this for a size .05 test in Figure 10.2, where $\pi(\lambda) = \int_{\chi^2_{1;.05}}^{\infty} f(w;1,\lambda)dw = \int_{3.841}^{\infty} f(w;1,\lambda)dw$ and $f(w;1,\lambda)$ is a noncentral $\chi^2$ density with 1 degree of freedom and noncentrality parameter $\lambda$ (the GAUSS procedure CDFCHINC was used to calculate the integrals).

As is typical of other tests that we have examined, the closer $\delta$ is to zero, and thus the closer $H_a$: $\theta = \theta_0 + n^{-1/2}\,\delta$ is to $H_0$: $\theta = \theta_0$, the less power there is for detecting a false $H_0$. □

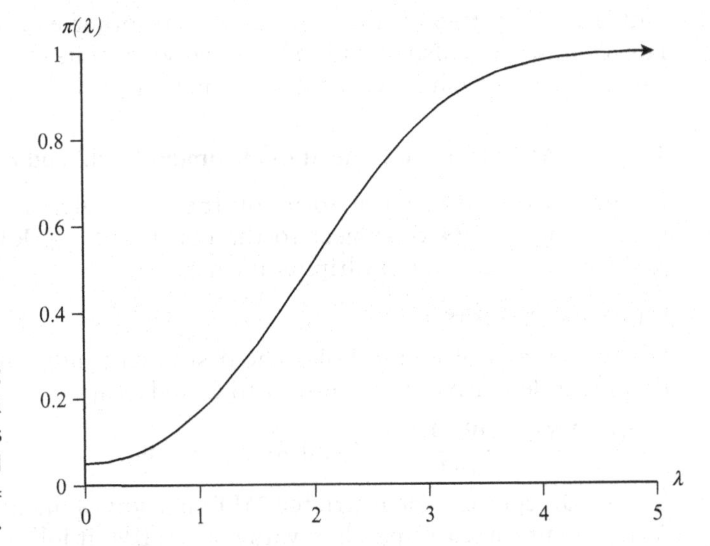


**Figure 10.2**
Asymptotic power of GLR test for $H_0$: $\theta = \theta_0$ versus $H_a$: $\theta > \theta_0$, local alternatives $\theta_n = \theta_0 + n^{-1/2}\,\delta$, $\lambda = \delta^2/2$.

## 10.4 Lagrangian Multiplier Tests

The Lagrangian multiplier (LM) test of a statistical hypothesis utilizes the size of Lagrangian multipliers as a measure of the discrepancy between restricted (by $H_0$) and unrestricted estimates of the parameters of a probability model. The approach can be applied to various types of estimation objectives, such as maximum likelihood, least squares, or to the minimization of the quadratic forms that define generalized method of moments estimators. The Lagrange multipliers measure the marginal changes in the optimized estimation objective that are caused by imposing constraints on the estimation problem defined by $H_0$. The intuition for this approach is that large values of Lagrangian multipliers indicate that large increases in likelihood function values, large decreases in sum of squared errors, or substantial reductions in moment discrepancies (in the GMM approach) are possible from constraint relaxation. If the LM values are substantially different from zero, the indication is that the estimation function can be substantially improved by examining parameter values contained in $H_a$, suggesting that $H_0$ is false and that it should be rejected.

Like the GLR test, the LM test is a natural testing procedure to use for testing hypotheses about $\boldsymbol{\Theta}$ or functions of $\boldsymbol{\Theta}$ when restricted maximum likelihood estimation is being used in statistical analysis. It has a computational advantage relative to the GLR approach in that *only* the restricted maximum likelihood estimates are needed to perform the test, whereas the GLR approach requires the unrestricted ML estimates as well. We will focus on the asymptotic properties of the test,[4] and on its application in maximum likelihood settings,

[4]Excellent references for additional details include L.G. Godfrey, (1988), *Misspecification Tests in Econometrics*, Cambridge Univ. Press, New York, pp. 5–20; and R.F. Engle, (1984), "Wald, Likelihood Ratio and Lagrange Multiplier Tests," in *Handbook of Econometrics* vol. 2, Z. Giliches and M. Intriligata, Amsterdam: North Holland, pp. 775–826.

which is indicative of how it is applied in other estimation settings as well. However, we will also provide a brief overview of its extensions to other estimation objectives such as least squares and GMM.

### 10.4.1 LM Tests in the Context of Maximum Likelihood Estimation

In order to establish the form of the test rule, examine the problem of maximizing $\ln(L(\boldsymbol{\Theta};\mathbf{x}))$ subject to the constraint $H_0$: $\mathbf{R}(\boldsymbol{\Theta}) = \mathbf{r}$. We express the problem in Lagrangian multiplier form as

$$\ln(L(\boldsymbol{\Theta};\mathbf{x})) - \boldsymbol{\lambda}'[\mathbf{R}(\boldsymbol{\Theta}) - \mathbf{r}]$$

where $\boldsymbol{\lambda}$ is $(q\times 1)$ vector of LMs. The first order conditions for this problem (i.e., first order derivatives with respect to $\boldsymbol{\Theta}$ and $\boldsymbol{\lambda}$) are

$$\frac{\partial \ln(L(\boldsymbol{\Theta};\mathbf{x}))}{\partial \boldsymbol{\Theta}} - \frac{\partial \mathbf{R}(\boldsymbol{\Theta})}{\partial \boldsymbol{\Theta}}\boldsymbol{\lambda} = \mathbf{0} \text{ and } \mathbf{R}(\boldsymbol{\Theta}) - \mathbf{r} = \mathbf{0}.$$

Letting $\hat{\boldsymbol{\Theta}}_r$ represent the restricted MLE that solves the first order conditions and $\boldsymbol{\Lambda}_r$ represent the corresponding value of the LM, it follows that

$$\frac{\partial \ln\left(L\left(\hat{\boldsymbol{\Theta}}_r;\mathbf{X}\right)\right)}{\partial \boldsymbol{\Theta}} - \frac{\partial \mathbf{R}\left(\hat{\boldsymbol{\Theta}}_r\right)}{\partial \boldsymbol{\Theta}}\boldsymbol{\Lambda}_r = 0 \text{ and } \mathbf{R}\left(\hat{\boldsymbol{\Theta}}_r\right) - \mathbf{r} = \mathbf{0}.$$

We now establish the LM test and its asymptotic distribution under $H_0$.

**Theorem 10.7** ***The LM Test of*** $H_0$: $\mathbf{R}(\boldsymbol{\Theta}_0)=\mathbf{r}$ ***versus*** $H_a$: $\mathbf{R}(\boldsymbol{\Theta}_0)\neq\mathbf{r}$

*Assume the conditions and notation of Theorem 8.19 ensuring the consistency, asymptotic normality, and asymptotic efficiency of the MLE,* $\hat{\boldsymbol{\Theta}}$, *of* $\boldsymbol{\Theta}$. *Let* $\hat{\boldsymbol{\Theta}}_r$ *and* $\boldsymbol{\Lambda}_r$ *be the restricted MLE and the value of the Lagrangian multiplier that satisfy the first order conditions of the restricted ML problem specified in Lagrange form, respectively, where* $\mathbf{R}(\boldsymbol{\Theta})$ *is a continuously differentiable* $(q\times 1)$ *vector function containing no redundant coordinate functions. If* $\mathbf{G} = \partial\mathbf{R}(\boldsymbol{\Theta}_0)/\partial\boldsymbol{\Theta}'$ *has full row rank, then under* $H_0$ *it follows that*[5]*:*

1. $W = \boldsymbol{\Lambda}'_r \dfrac{\partial \mathbf{R}\left(\hat{\boldsymbol{\Theta}}_r\right)'}{\partial \boldsymbol{\Theta}} \left[-\dfrac{\partial^2 \ln L\left(\hat{\boldsymbol{\Theta}}_r;\mathbf{X}\right)}{\partial\boldsymbol{\Theta}\partial\boldsymbol{\Theta}'}\right]^{-1} \dfrac{\partial \mathbf{R}\left(\hat{\boldsymbol{\Theta}}_r\right)}{\partial \boldsymbol{\Theta}}\boldsymbol{\Lambda}_r \xrightarrow{d} \chi^2_q;$

2. *An asymptotic size* $\alpha$ *and consistent test of* $H_0$: $\mathbf{R}(\boldsymbol{\Theta}_0) = \mathbf{r}$ *versus* $H_a$: $\mathbf{R}(\boldsymbol{\Theta}_0) \neq \mathbf{r}$ *is given by*

$$w\begin{bmatrix}\geq\\<\end{bmatrix}\chi^2_{q;\alpha} \Rightarrow \begin{bmatrix}\text{reject } H_0\\ \text{do not reject } H_0\end{bmatrix}; \text{and}$$

3. *An alternative and equivalent representation of W is given by*

$$W = \frac{\partial \ln\left(L\left(\hat{\boldsymbol{\Theta}}_r;\mathbf{X}\right)\right)'}{\partial\boldsymbol{\Theta}}\left[-\frac{\partial^2 \ln\left(L\left(\hat{\boldsymbol{\Theta}}_r;\mathbf{X}\right)\right)}{\partial\boldsymbol{\Theta}\partial\boldsymbol{\Theta}}\right]^{-1}\frac{\partial \ln\left(L\left(\hat{\boldsymbol{\Theta}}_r;\mathbf{X}\right)\right)}{\partial\boldsymbol{\Theta}}.$$

[5] $\boldsymbol{\Lambda}_r$ is the random vector whose outcome is $\boldsymbol{\lambda}_r$.

*(The test based on this alternative form of W was called the* ***score test*** *by C. R. Rao, (1948), "Large Sample Tests of Statistical Hypotheses," Proceedings of the Cambridge Philosophical Society, (44), pp. 50–57.)*

**Proof** See Appendix. ■

In certain applications the LM test can have a computational advantage over the GLR test since the latter involves *both* the restricted and unrestricted estimates of $\boldsymbol{\Theta}$ whereas the former requires *only* the restricted estimates of $\boldsymbol{\Theta}$. In the following example, the relative convenience of the LM test is illustrated in the case of testing a hypothesis relating to the gamma distribution. Note the GLR approach in this case would be complicated by the fact that the unrestricted MLE cannot be obtained in closed form. Obtaining the restricted MLE in the case below is relatively straightforward.

**Example 10.11** ***LM Test of $H_0$: $\alpha = 1$ (Exponential Family) versus $H_0$: $\alpha \neq 1$ in Gamma Population Distribution***

The operating life of a new MP3 player produced by a major manufacturer is considered to be gamma distributed as

$$z \sim \frac{1}{\beta^{\alpha}\Gamma(\alpha)} z^{\alpha-1} e^{-z/\beta} I_{(0,\infty)}(z).$$

The marketing department is contemplating a warranty policy on the MP3 player, and wants to test the hypothesis that the player is "as good as new while operating," i.e., does the player's operating life adhere to an exponential population distribution? To test the null hypothesis $H_0$: $\alpha = 1$ versus $H_a$: $\alpha \neq 1$, consider performing a size .05 LM test using a sample of 50 observations on the operating lives of the new MP3 player.

The likelihood function is (suppressing the indicator function)

$$L(\alpha, \beta; \mathbf{x}) = \frac{1}{\beta^{n\alpha}\Gamma^{n}(\alpha)} \prod_{i=1}^{n} x_i^{\alpha-1} e^{-\sum_{i=1}^{n} x_i/\beta},$$

with $n = 50$. We know from Example 8.20 that the conditions of Theorem 8.19 apply and so Theorem 10.7 is applicable. Under the constraint $\alpha = 1$, the restricted ML estimate is the value of $\hat{\beta}$ such that

$$\hat{\beta} = \arg\max_{\beta} \left\{ \frac{1}{\beta^{50}} e^{-\sum_{i=1}^{50} x_i/\beta} \right\},$$

which we know to be the sample mean, $\bar{x}$.

To implement the LM test, consider calculating the LM statistic as indicated in Theorem 10.7.3. The second order derivatives of $\ln(L(\alpha,\beta;\mathbf{x}))$ are given in Example 8.15. In order to evaluate these derivatives at $\alpha = 1$ and $\beta = \bar{x}$, note that (to 5 decimal places)

$$\frac{d\Gamma(\alpha)}{d\alpha}\Big|_{\alpha=1} = -.57722 \text{ and } \frac{d^2\Gamma(\alpha)}{d\alpha^2}\Big|_{\alpha=1} = 1.64493$$

(see M. Abramowitz and I. Stegun, (1970) *Handbook of Mathematical Functions*, Dover Publications, New York, pp. 258–260, or else calculate numerically on a computer). Then letting $\boldsymbol{\Theta} = (\alpha, \beta)'$,

$$\frac{\partial^2 \ln L(\boldsymbol{\Theta};\mathbf{x})}{\partial\boldsymbol{\Theta}\partial\boldsymbol{\Theta}'}\Big|_{\alpha=1,\beta=\bar{x}} = n\left[\begin{array}{c|c} -1.31175 & -\bar{x}^{-1} \\ \hline -\bar{x}^{-1} & -\bar{x}^{-2} \end{array}\right],$$

where in this application, $n = 50$.

The first derivatives of $\ln(L(\alpha,\beta;\mathbf{x}))$ are given in Example 8.10, and when evaluated at $\alpha = 1$ and $\beta = \bar{x}$ yield

$$\frac{\partial \ln(L(\alpha,\beta;\mathbf{x}))}{\partial\alpha}\Bigg|_{\substack{\alpha=1\\ \beta=\bar{x}}} = -n\ln(\bar{x}) + .57722n + \sum_{i=1}^{n}\ln(x_i),$$

$$\frac{\partial \ln(L(\alpha;\beta;\mathbf{x}))}{d\beta}\Bigg|_{\substack{\alpha=1\\ \beta=\bar{x}}} = -n\bar{x}^{-1} + n\bar{x}^{-1} = 0.$$

The LM test statistic can then be written in the **score test form** (note that $\hat{\boldsymbol{\theta}}_r = [1,\bar{x}]'$)

$$w = \frac{\partial\ln\left(L\left(\hat{\boldsymbol{\theta}}_r;\mathbf{x}\right)\right)'}{\partial\boldsymbol{\Theta}}\left[-\frac{\partial^2\ln\left(L\left(\hat{\boldsymbol{\theta}}_r;\mathbf{x}\right)\right)}{\partial\boldsymbol{\Theta}\partial\boldsymbol{\Theta}'}\right]^{-1}\frac{\partial\ln\left(L\left(\hat{\boldsymbol{\theta}}_r;\mathbf{x}\right)\right)}{\partial\boldsymbol{\Theta}}$$

$$= \frac{n(\ln(\bar{x}_G) - \ln(\bar{x}) + .57722)}{(.31175)},$$

where $\bar{x}_G = \left(\prod_{i=1}^{n} x_i\right)^{1/n}$ is the geometric mean of the $x_i$'s. The test rule is then

$$w = 160.38492\left(\ln\left(\frac{\bar{x}_G}{\bar{x}}\right) + .57722\right)\left[\begin{array}{c}\geq\\ <\end{array}\right]3.84146 \Rightarrow \left[\begin{array}{c}\text{reject } H_0\\ \text{do not reject } H_0\end{array}\right]$$

where $\chi^2_{1;.05} = 3.84146$.

Suppose the 50 observations on MP3 player lifetimes yielded $\bar{x} = 3.00395$ and $\bar{x}_G = 2.68992$. The value of the LM test statistic is $w = 74.86822$, and since $w \geq 3.84146$, $H_0$ is rejected. We conclude the population distribution is *not* exponential on the basis of a level .05 test. (In actuality, $\bar{x}$ and $\bar{x}_G$ were simulated from a random sample of 50 observations from a Gamma(2, 10) population distribution, and so the test made the *correct* decision in this case).□

Analogous to the limiting distribution of $-2\ln(\lambda(\mathbf{X}))$ under local alternatives in the case of the GLR test, the limiting distribution of the LM statistic is noncentral $\chi^2$ under general conditions.

**Theorem 10.8 Asymptotic Distribution of LM Test of $H_0$: $\mathbf{R}(\boldsymbol{\Theta}) = \mathbf{r}$ versus $H_a$: $\mathbf{R}(\boldsymbol{\Theta}) \neq \mathbf{r}$ for Local Alternatives When $H_0$ is False**

*Consider the LM test of $H_0$: $\mathbf{R}(\boldsymbol{\Theta}) = \mathbf{r}$ versus $H_a$: $\mathbf{R}(\boldsymbol{\Theta}) \neq \mathbf{r}$ under the conditions and notation of Theorem 10.7, and let a sequence of local alternatives be defined by $H_{an}$: $\mathbf{R}(\boldsymbol{\Theta}) = \mathbf{r} + n^{-1/2}\,\boldsymbol{\phi}$. Then the limiting distribution of the LM statistic is noncentral $\chi^2$ as*

$$\mathbf{W} = \boldsymbol{\Lambda}'_r \frac{\partial \mathbf{R}\left(\hat{\boldsymbol{\Theta}}_r\right)'}{\partial \boldsymbol{\Theta}} \left[ -\frac{\partial^2 \ln\left(L\left(\hat{\boldsymbol{\Theta}}_r; \mathbf{X}\right)\right)}{\partial \boldsymbol{\Theta} \partial \boldsymbol{\Theta}'} \right]^{-1} \frac{\partial \mathbf{R}\left(\hat{\boldsymbol{\Theta}}_r\right)}{\partial \boldsymbol{\Theta}} \boldsymbol{\Lambda}_r \xrightarrow{d} \chi^2_q(\lambda),$$

where the noncentrality parameter equals

$$\lambda = \frac{1}{2} \boldsymbol{\phi}' \left[ \frac{\partial \mathbf{R}(\boldsymbol{\Theta}_0)'}{\partial \boldsymbol{\Theta}} \mathbf{M}\left(\hat{\boldsymbol{\Theta}}_r\right)^{-1} \frac{\partial \mathbf{R}(\boldsymbol{\Theta}_0)}{\partial \boldsymbol{\Theta}} \right]^{-1} \boldsymbol{\phi}.$$

**Proof** See Appendix. ■

Given Theorem 10.8, the asymptotic power of the LM test as a function of $\lambda$ can be calculated and graphed. The asymptotic power function is monotonically increasing in the noncentrality parameter $\lambda$. Thus the LM test is also *asymptotically unbiased* for testing $\mathbf{R}(\boldsymbol{\Theta}) = \mathbf{r}$ versus $H_a$: $\mathbf{R}(\boldsymbol{\Theta}) \neq \mathbf{r}$. Being that the LM statistic has the same limiting distribution under local alternatives as the statistic $-2[\ln(L(\hat{\boldsymbol{\Theta}}_r; \mathbf{X})) - \ln(L(\hat{\boldsymbol{\Theta}}; \mathbf{x}))]$ used in the GLR test, the two procedures cannot be distinguished on the basis of these types of asymptotic power considerations. In applications, a choice of whether to use a GLR or LM test is often made on the basis of convenience, or when both are tractable to compute, one can consider both tests in assess $H_0$. Comparisons of the finite sample properties of the tests continue to be researched.

Historically, the LM test procedure is the most recent of the GLR, Wald, and LM triad of tests, listed in order of discovery. Recently, substantial progress has been made in applying the LM approach to a wide range of testing contexts in the econometrics literature. Interested readers can consult the article by R.F. Engle, op. cit., and the book by L.G. Godfrey to begin further reading. We provide a brief overview of LM extensions beyond ML in the next section.

### 10.4.2 LM Tests in Other Estimation Contexts

In general, Lagrange multiplier (LM) tests are based on the finite or asymptotic probability distribution of the Lagrange multipliers associated with functional constraints defined by a null hypothesis $H_0$, within the context of a constrained estimation problem that has been expressed in Lagrange multiplier form. The form of the estimation problem depends on the objective function that is being optimized to define the estimator. The objective function could involve maximizing a likelihood function to define a ML estimator as discussed above, minimizing a sum of squares of model residuals to define a linear or nonlinear least squares estimator, minimizing the weighted Euclidean distance of a vector of moment constraints from the zero vector to define a generalized method of moments estimator, or in general optimizing any measure of fit with the data to define an estimator.

When functional constraints defined by $H_0$ are added to an optimization problem that defines an estimator, a constrained estimator is defined together with Lagrange multipliers associated with the functional constraints. Under general regularity conditions, the Lagrange multipliers have asymptotic normal distributions, and appropriately weighted quadratic forms in these multipliers have asymptotic chisquare distributions that can be used to define asymptotically valid tests of the functional restrictions, and thus a test of $H_0$. In some fortunate cases, the Lagrange multipliers will be functions of the data that have a tractable finite sample probability distribution, in which case exact size, level, and power considerations can be determined for the LM tests. Establishing finite sampling behavior will occur on a case by case basis, and we concentrate here on the more generally applicable results that are possible using asymptotic considerations.

Suppose that an estimator under consideration is defined as the function $\hat{\boldsymbol{\Theta}} = \arg\max_{\boldsymbol{\Theta}}\{h(\mathbf{Y},\boldsymbol{\Theta})\}$ of the probability sample $\mathbf{Y}$. For example, in maximum likelihood estimation $h(\mathbf{Y},\boldsymbol{\Theta}) = \ln(L(\boldsymbol{\Theta};\mathbf{Y}))$, or in the case of applying the least squares criteria in a linear model framework, $h(\mathbf{Y},\boldsymbol{\Theta}) = -(\mathbf{Y}-\mathbf{x}\boldsymbol{\Theta})'(\mathbf{Y}-\mathbf{x}\boldsymbol{\Theta})$. Let $\mathbf{R}(\boldsymbol{\Theta}) = \mathbf{r}$ define $q$ functionally independent restrictions on the parameter vector $\boldsymbol{\Theta}$. Then the constrained estimator is defined by $\hat{\boldsymbol{\Theta}} = \arg\max_{\boldsymbol{\Theta}}\{h(\mathbf{Y},\boldsymbol{\Theta}) -\boldsymbol{\lambda}'(\mathbf{R}(\boldsymbol{\Theta}) - \mathbf{r})\}$where $\boldsymbol{\lambda}$ is a $q \times 1$ vector of Lagrange multipliers.

Assuming appropriate differentiability and second order conditions, the first order conditions to the problem define the constrained estimator as the solution to

$$\frac{\partial h(\mathbf{Y},\boldsymbol{\Theta})}{\partial\boldsymbol{\Theta}} - \frac{\partial\mathbf{R}(\boldsymbol{\Theta})}{\partial\boldsymbol{\Theta}}\boldsymbol{\lambda} = \mathbf{0} \text{ and } \mathbf{R}(\boldsymbol{\Theta}) - \mathbf{r} = \mathbf{0}.$$

The estimators for $\boldsymbol{\Theta}$ and $\boldsymbol{\lambda}$ then take the general form $\hat{\boldsymbol{\Theta}}_r = \hat{\boldsymbol{\Theta}}_r(\mathbf{Y})$ and $\hat{\boldsymbol{\Lambda}}_r = \hat{\boldsymbol{\Lambda}}_r(\mathbf{Y})$, respectively.

In seeking an asymptotic normal distribution for $\hat{\boldsymbol{\Lambda}}_r$ and subsequently deriving an asymptotic chisquare-distributed test statistic in terms of $\hat{\boldsymbol{\Lambda}}_r$ , one generally appeals to some central limit theorem argument, and attempts to find some transformation of $(\partial h(\mathbf{Y},\boldsymbol{\Theta}_0))/\partial\boldsymbol{\Theta}$ that has an asymptotic normal distribution, where $\boldsymbol{\Theta}_0$ denotes the true value of the parameter vector. For example, it is often argued in statistical and econometric applications, that $n^{-1/2}(\partial h(\mathbf{Y},\boldsymbol{\Theta}_0))/\partial\boldsymbol{\Theta} \xrightarrow{d} N(\mathbf{0},\boldsymbol{\Sigma})$. In the case of an MLE with $h(\mathbf{Y},\boldsymbol{\Theta}_0) = \ln(L(\boldsymbol{\Theta}_0;\mathbf{Y}))$, this is the typical result that $n^{-1/2}(\partial\ln(L(\mathbf{Y},\boldsymbol{\Theta}_0)))/\partial\boldsymbol{\Theta} \xrightarrow{d} N(\mathbf{0},\boldsymbol{\Sigma})$. In the case of the classical linear model, where $h(\mathbf{Y},\boldsymbol{\Theta}_0) = -(\mathbf{Y}-\mathbf{x}\boldsymbol{\Theta}_0)'(\mathbf{Y}-\mathbf{x}\boldsymbol{\Theta}_0)$, this is the standard result that $-2n^{-1/2}\mathbf{x}'(\mathbf{Y}-\mathbf{x}\boldsymbol{\Theta}_0) = -2n^{-1/2}\mathbf{x}'\boldsymbol{\varepsilon} \xrightarrow{d} N(\mathbf{0},\boldsymbol{\Sigma})$.

Assuming that the limiting distribution results can be argued, it follows that $n^{-1/2}\Sigma^{-1/2}(\partial h(\mathbf{Y},\Theta_0))/\partial\Theta \xrightarrow{d} N(\mathbf{0},\mathbf{I})$. Then following an approach that is analogous to that used in the proof of Theorem 10.7, one can show that

$$n^{-1/2}\Lambda_r \stackrel{d}{=} \left[\frac{\partial\mathbf{R}(\Theta_0)'}{\partial\Theta}\Sigma^{-1}\frac{\partial\mathbf{R}(\Theta_0)}{\partial\Theta}\right]^{-1}\frac{\partial\mathbf{R}(\Theta_0)'}{\partial\Theta}\Sigma^{-1/2}\left[n^{-1/2}\Sigma^{-1/2}\frac{\partial h(\mathbf{Y},\Theta_0)}{\partial\Theta}\right].$$

It follows that under $H_0$ that

$$n^{-1/2}\Lambda'_r \xrightarrow{d} N\left(\mathbf{0},\left[\frac{\partial\mathbf{R}(\Theta_0)'}{\partial\Theta}\Sigma^{-1}\frac{\partial\mathbf{R}(\Theta_0)}{\partial\Theta}\right]^{-1}\right).$$

Then a test statistic based on the Lagrange multiplier vector can be defined as

$$W = n^{-1}\Lambda'_r\left[\frac{\partial\mathbf{R}(\Theta_0)'}{\partial\Theta}\Sigma^{-1}\frac{\partial\mathbf{R}(\Theta_0)}{\partial\Theta}\right]\Lambda_r \stackrel{d}{=} \Lambda'_r\frac{\partial\mathbf{R}\left(\hat{\Theta}_r\right)'}{\partial\Theta}\hat{\Sigma}^{-1}\frac{\partial\mathbf{R}\left(\hat{\Theta}_r\right)}{\partial\Theta}\Lambda_r,$$
$$\xrightarrow{d} \chi^2_q \text{ under } H_0.$$

where $\hat{\Sigma}$ is a consistent estimator of the limiting covariance matrix under the null hypothesis, and $\stackrel{d}{=}$ denotes equivalence in limiting distribution.

Regarding the rejection region for a test based on the LM statistic above, recall that large values of Lagrange multipliers indicate that their associated restriction on parameters are substantially reducing the optimized value of the objective function, which indicates a situation where the hypothesized restrictions $H_0$: $\mathbf{R}(\Theta) = \mathbf{r}$ are in considerable conflict with the data in terms of a maximum likelihood, least squares, or other estimation criterion for defining an estimator. Since the LM statistic is a positive definite quadratic form in the value of the Lagrange multiplier vector, it then makes sense that $H_0$ should be rejected for *large* values of LM, and not rejected for small values. Thus, a size $\alpha$ test is based on a rejection region of the form $C_r^W = \left[\chi^2_{q;\alpha}, \infty\right)$.

Power considerations for LM tests in other estimation contexts are analogous to the approach discussed in Section 10.4 for the ML context, where the power was analyzed in terms of local alternatives and a noncentral chisquare distribution was used to calculate the appropriate rejection probabilities. Demonstrations of the consistency of such tests also follows along the lines demonstrated in the previous section.

## 10.5 Wald Tests

The Wald test for testing statistical hypotheses, named after mathematician and statistician Abraham Wald, utilizes a third alternative measure of the discrepancy between restricted (by $H_0$) and unrestricted estimates of the parameters of the joint density of a probability sample in order to define a test procedure. In particular, the Wald test assesses the significance of the difference between the unrestricted estimate of $\mathbf{R}(\boldsymbol{\Theta})$ and the value of $\mathbf{R}(\boldsymbol{\Theta})$ that is specified in $H_0$, $\mathbf{r}$. Significantly large values of $|\mathbf{R}(\hat{\boldsymbol{\Theta}}) - \mathbf{r}|$ indicate significant discrepancies between the hypothesized value of $\mathbf{R}(\boldsymbol{\Theta})$ and the unrestricted estimate of $\mathbf{R}(\boldsymbol{\Theta})$ provided by the data, suggesting that $H_0$ is false.

Note that it is not necessary that the estimator $\hat{\boldsymbol{\Theta}}_n$ be an MLE for the Wald test to be applicable. In fact, one need not specify a likelihood function at all to perform the test, and in this respect, the Wald test is more general than the GLR test presented heretofore. Moreover, unlike the GLR testing context, but like the LM context, one does not need *both* the unrestricted and restricted estimates of the parameters in order to conduct the test. The Wald test requires only the outcome of the *unrestricted* estimator, whereas the LM test requires only the outcome of the *restricted* estimator, as we have seen in Section 10.4.

Similar to the GLR and LM contexts, we will concentrate on providing some asymptotic results on the behavior of the Wald test in this section because the results are relatively general, and apply to a wide scope of applications. Finite sample properties can sometimes be established as well, but these proceed case by case, and require additional specific assumptions regarding the probability model in order to be applicable. We examine some of these results in Section 10.6, where we discuss statistical tests in the context of the General Linear Model. The asymptotic results presented ahead can also be made to apply even more generally if more advanced limiting distribution arguments are invoked, and an excellent beginning reference for such extensions is provided by H. White, *Asymptotic Theory for Econometricians*, Academic Press, Orlando.

**Theorem 10.9** ***Wald Test of Asymptotic Size α When*** $\hat{\boldsymbol{\Theta}} \overset{a}{\sim} N(\boldsymbol{\Theta}_0, n^{-1}\boldsymbol{\Sigma})$

*Let the probability sample* $(X_1,\ldots,X_n)$ *have the joint probability density function* $f(\mathbf{x};\boldsymbol{\Theta}_0)$, *let* $\hat{\boldsymbol{\Theta}}$ *be a consistent estimator for* $\boldsymbol{\Theta}_0$ *such that* $n^{1/2}(\hat{\boldsymbol{\Theta}} - \boldsymbol{\Theta}_0) \xrightarrow{d} N(\mathbf{0}, \boldsymbol{\Sigma})$ *and let* $\hat{\boldsymbol{\Sigma}}_n$ *be a consistent estimator of the positive definite* $\boldsymbol{\Sigma}$. *Furthermore, let the null and alternative hypotheses be of the form* $H_0$: $\mathbf{R}(\boldsymbol{\Theta}) = \mathbf{r}$ *and* $H_a$: $\mathbf{R}(\boldsymbol{\Theta}) \neq \mathbf{r}$, *where* $\mathbf{R}(\boldsymbol{\Theta})$ *is a* $(q\times 1)$ *continuously differentiable vector function of* $\boldsymbol{\Theta}$ *for which* $(q\leq k)$ *and* $\mathbf{R}(\boldsymbol{\Theta})$ *contains no redundant coordinate functions. Finally, let* $\partial\mathbf{R}(\boldsymbol{\Theta}_0)/\partial\boldsymbol{\Theta}'$ *have full row rank. Then a consistent and asymptotic size* α *test of* $H_0$ *versus* $H_a$ *is given by*

$$w\left\{\begin{matrix}\geq\\<\end{matrix}\right\}\chi^2_{q;\alpha} \Rightarrow \left\{\begin{matrix}\text{reject } H_0\\ \text{do not reject } H_0\end{matrix}\right\}$$

where

$$W = \left[\mathbf{R}(\hat{\boldsymbol{\Theta}}) - \mathbf{r}\right]' \left[\frac{\partial\mathbf{R}(\hat{\boldsymbol{\Theta}})'}{\partial\boldsymbol{\Theta}}\left(n^{-1}\hat{\boldsymbol{\Sigma}}_n\right)\frac{\partial\mathbf{R}(\hat{\boldsymbol{\Theta}})}{\partial\boldsymbol{\Theta}}\right]^{-1}\left[\mathbf{R}(\hat{\boldsymbol{\Theta}}) - \mathbf{r}\right]$$

**Proof** Under the stated assumptions, Theorem 5.40 is applicable so that

$$n^{1/2}\left[\mathbf{R}(\hat{\boldsymbol{\Theta}}) - \mathbf{r}\right] \xrightarrow{d} N\left(\mathbf{0}, \frac{\partial\mathbf{R}(\boldsymbol{\Theta}_0)'}{\partial\boldsymbol{\Theta}}\boldsymbol{\Sigma}\frac{\partial\mathbf{R}(\boldsymbol{\Theta}_0)}{\partial\boldsymbol{\Theta}}\right)$$

under $H_0$. Letting $\boldsymbol{\Sigma}_r$ represent the covariance matrix of this limiting distribution, it follows that

$$n^{1/2}\,\boldsymbol{\Sigma}_r^{-1/2}\left[\mathbf{R}(\hat{\boldsymbol{\Theta}})-\mathbf{r}\right]\overset{d}{\to}N(\mathbf{0},\mathbf{I}_q)$$

so that by Theorem 5.3

$$n\left[\mathbf{R}(\hat{\boldsymbol{\Theta}})-\mathbf{r}\right]'\boldsymbol{\Sigma}_r^{-1}\left[\mathbf{R}(\hat{\boldsymbol{\Theta}})-\mathbf{r}\right]\overset{d}{\to}\chi^2_q.$$

Now note that

$$\left[\frac{\partial\mathbf{R}(\hat{\boldsymbol{\Theta}})'}{\partial\boldsymbol{\Theta}}\hat{\boldsymbol{\Sigma}}_n\frac{\partial\mathbf{R}(\hat{\boldsymbol{\Theta}})}{\partial\boldsymbol{\Theta}}\right]^{-1}\overset{p}{\to}\boldsymbol{\Sigma}_r^{-1}$$

by the continuity of $\partial\mathbf{R}(\boldsymbol{\Theta})/\partial\boldsymbol{\Theta}$ and the continuity of the inverse matrix function, in which case it follows from Theorem 5.9 that $W\overset{d}{\to}\chi^2_q$, so that the test as defined above has size $\alpha$ asymptotically.

Regarding consistency, assume $\alpha > 0$ and thus $c = \chi^2_{q;\alpha} < \infty$. Since $\hat{\boldsymbol{\Theta}}\overset{d}{\to}\boldsymbol{\Theta}_0$ and $\mathbf{R}(\boldsymbol{\Theta})$ is continuous, $\mathbf{R}(\hat{\boldsymbol{\Theta}})-\mathbf{r}\overset{p}{\to}\boldsymbol{\phi}\neq\mathbf{0}$ assuming $H_a$: $\mathbf{R}(\boldsymbol{\Theta})\neq\mathbf{r}$ is true. Then $n^{-1}W\overset{p}{\to}\boldsymbol{\phi}'\boldsymbol{\Sigma}_r^{-1}\boldsymbol{\phi}>0$, so that $\lim_{n\to\infty}(P(w\geq c))=1$ for any $c<\infty$, and the test is consistent. ■

**Example 10.12** ***Wald Test of Variance in Exponential Population***

Revisit Example 10.5, and suppose in addition to a hypothesis regarding the mean life of the computer screen a hypothesis regarding the variance of the operating life was to be investigated. In particular, suppose $H_0$: $\theta^2 = 1$ versus $H_a$: $\theta^2 \neq 1$ was under consideration. We know in this case that for $\hat{\Theta} = \bar{X}$, $n^{1/2}(\bar{X}-\mu)\overset{d}{\to}N(0,\sigma^2)$, and $\hat{\Sigma}_n = S^2 \overset{p}{\to}\sigma^2$ so that Theorem 10.9 is applicable with $\mathbf{R}(\theta) = \theta^2$. Because $d\mathbf{R}(\theta)/d\theta = 2\theta$, and $n=10$, the Wald statistic becomes

$$w=\left(\hat{\theta}^2-1\right)'\left[\frac{4\hat{\theta}^2s^2}{10}\right]^{-1}\left(\hat{\theta}^2-1\right)=\frac{10(\bar{x}^2-1)^2}{(4\bar{x}^2s^2)}.$$

and a Wald test of asymptotic size .05 can be defined as

$$w\begin{Bmatrix}\geq\\<\end{Bmatrix}3.84\Rightarrow\begin{Bmatrix}\text{reject } H_0\\ \text{do not reject } H_0\end{Bmatrix}$$

where $P(w\geq 3.84)=.05$ when $W\sim\chi^2_1$.

Supposing that $\bar{x}=1.4$ and $s^2=2.1$, then $w=.78<3.84$, and thus $H_0$ would *not* be rejected using the Wald test. □

The asymptotic distribution of the Wald statistic when $H_0$: $\mathbf{R}(\boldsymbol{\Theta})=\mathbf{r}$ is false can be established based on *local alternatives*, and leads to a noncentral $\chi^2$ distribution, as in the case of the LM and GLR procedures.

**Theorem 10.10**
***Asymptotic Distribution of Wald Statistic Under*** $H_a$: $\mathbf{R}(\mathbf{\Theta}) \neq \mathbf{r}$

*Let the conditions and notation of Theorem 10.9 hold for testing* $H_0$: $\mathbf{R}(\mathbf{\Theta}) = \mathbf{r}$ *versus* $H_a$: $\mathbf{R}(\mathbf{\Theta}) \neq \mathbf{r}$. *Then for a sequence of local alternatives* $H_{an}$: $\mathbf{R}(\mathbf{\Theta}) = \mathbf{r} + n^{-1/2}\,\boldsymbol{\phi}$, *it follows that* $W \xrightarrow{d} \chi^2_q(\lambda)$ *with noncentrality parameter*

$$\lambda = \frac{1}{2}\boldsymbol{\phi}'\left[\frac{\partial \mathbf{R}(\mathbf{\Theta}_0)'}{\partial \mathbf{\Theta}}\mathbf{\Sigma}\frac{\partial \mathbf{R}(\mathbf{\Theta}_0)}{\partial \mathbf{\Theta}}\right]^{-1}\boldsymbol{\phi}.$$

**Proof** It follows from Theorem 5.40 that $n^{1/2}(\mathbf{R}(\hat{\mathbf{\Theta}}) - \mathbf{R}(\mathbf{\Theta}_0)) \xrightarrow{d} N(\mathbf{0}, \mathbf{G\Sigma G}')$ where $\mathbf{G} = \partial \mathbf{R}(\mathbf{\Theta}_0)/\partial \mathbf{\Theta}'$. Under the sequence of local alternatives, $\mathbf{R}(\mathbf{\Theta}) = \mathbf{r} + n^{-1/2}\,\boldsymbol{\phi} \to \mathbf{r}$ and $n^{1/2}\left(\mathbf{R}(\hat{\mathbf{\Theta}}) - (\mathbf{r} + n^{-1/2}\boldsymbol{\phi})\right) \xrightarrow{d} N(\mathbf{0}, \mathbf{G\Sigma G}')$, in which case $n^{1/2}\left(\mathbf{R}(\hat{\mathbf{\Theta}}) - \mathbf{r}\right) \xrightarrow{d} N(\boldsymbol{\phi}, \mathbf{G\Sigma G}')$ and $[\mathbf{G\Sigma G}']^{-1/2}\, n^{1/2}\left(\mathbf{R}\left(\hat{\mathbf{\Theta}}\right) - \mathbf{r}\right) \xrightarrow{d} N\left([\mathbf{G\Sigma G}']^{-1/2}\boldsymbol{\phi},\, \mathbf{I}\right)$. Because $\partial \mathbf{R}\left(\hat{\mathbf{\Theta}}\right)/\partial \mathbf{\Theta}' \xrightarrow{p} \mathbf{G}$ and $\hat{\mathbf{\Sigma}}_n \xrightarrow{p} \mathbf{\Sigma}$, it follows from Theorem 5.9 and properties of the noncentral $\chi^2$ distribution that

$$W = n^{1/2}\left(\mathbf{R}\left(\hat{\mathbf{\Theta}}\right) - \mathbf{r}\right)'\left[\frac{\partial \mathbf{R}(\hat{\mathbf{\Theta}})'}{\partial \mathbf{\Theta}}\hat{\mathbf{\Sigma}}_n\frac{\partial \mathbf{R}(\hat{\mathbf{\Theta}})}{\partial \mathbf{\Theta}}\right]^{-1}(\mathbf{R}(\hat{\mathbf{\Theta}}) - \mathbf{r})n^{1/2}$$

$$\stackrel{d}{=} n^{1/2}(\mathbf{R}(\hat{\mathbf{\Theta}}) - \mathbf{r})'[\mathbf{G\Sigma G}']^{-1}(\mathbf{R}(\hat{\mathbf{\Theta}}) - \mathbf{r})n^{1/2} \xrightarrow{d} \chi^2_q(\lambda),$$

with $\lambda = \frac{1}{2}\boldsymbol{\phi}'(\mathbf{G\Sigma G}')^{-1}\boldsymbol{\phi}$. ■

**Example 10.13**
***Wald Test Based on a GMM Estimator***

Revisit Example 8.28 concerning GMM estimation of $\boldsymbol{\beta}$ in the demand function $\mathbf{Y} = \mathbf{X}\boldsymbol{\beta} + \mathbf{V}$ when $\mathbf{X}$ is random and $\mathrm{E}((\mathbf{X}'\mathbf{X})^{-1}\,\mathbf{X}'\mathbf{V}) \neq \mathbf{0}$. The GMM estimator $\hat{\boldsymbol{\beta}}_G = (\mathbf{Z}'\mathbf{X})^{-1}\mathbf{Z}'\mathbf{Y}$ as defined in the example is a consistent and asymptotically normal estimator of $\boldsymbol{\beta}$, where $\mathrm{n}^{1/2}\,(\hat{\boldsymbol{\beta}}_G - \boldsymbol{\beta}_0) \xrightarrow{d} N(\mathbf{0},\, \sigma^2\,(\mathbf{A}'_{ZX}\mathbf{A}_{ZZ}^{-1}\mathbf{A}_{ZX})^{-1}$. A consistent estimator of the covariance matrix of the limiting distribution is given by $n\hat{\sigma}^2\left(\mathbf{X}'\mathbf{Z}(\mathbf{Z}'\mathbf{Z})^{-1}\mathbf{Z}'\mathbf{X}\right)^{-1}$ where $\hat{\sigma}^2 = \left(\mathbf{Y} - \mathbf{X}\hat{\boldsymbol{\beta}}_G\right)'\left(\mathbf{Y} - \mathbf{X}\hat{\boldsymbol{\beta}}_G\right)/n$. Let $H_0$: $\mathbf{R}\boldsymbol{\beta} = \mathbf{r}$ be any hypothesis concerning linear combinations of the parameters of the demand function where $\mathbf{R}$ has full row rank. Then a Wald test of $H_0$ having asymptotic size $\alpha$ can be based on outcomes of

$$W = \left(\mathbf{R}\hat{\boldsymbol{\beta}}_G - \mathbf{r}\right)'\left[\hat{\sigma}^2\mathbf{R}\left(\mathbf{X}'\mathbf{Z}(\mathbf{Z}'\mathbf{Z})^{-1}\mathbf{Z}'\mathbf{X}\right)^{-1}\mathbf{R}'\right]^{-1}\left(\mathbf{R}\hat{\boldsymbol{\beta}}_G - \mathbf{r}\right),$$

where the rejection region is defined by $C_r^W = \left[\chi^2_{q;\alpha}, \infty\right)$. How would the test change if the null hypothesis were *nonlinear* in $\boldsymbol{\beta}$, i.e., $H_0$: $\mathbf{R}(\boldsymbol{\beta}) = \mathbf{r}$? □

The asymptotic power function of the Wald test is monotonically increasing in the noncentrality parameter $\lambda$ since $P(w \geq c;\, \lambda)$ for $c < \infty$ is monotonically increasing in $\lambda$ if $W \sim \chi^2_q(\lambda)$. It follows that the Wald test is also approximately unbiased for testing $H_0$: $\mathbf{R}(\mathbf{\Theta}_0) = \mathbf{r}$ versus $H_0$: $\mathbf{R}(\mathbf{\Theta}_0) \neq \mathbf{r}$ based on asymptotic power considerations.

As we had noted at the end of our LM discussion, each of the tests in the GLR, LM, Wald triad share the same limiting distribution under sequences of local alternatives when all of the tests are based on the unrestricted and/or restricted MLEs of $\boldsymbol{\Theta}$. As such, the tests cannot be distinguished on the basis of these asymptotic power comparisons. If convenience dictates the choice of test procedure, then the Wald procedure will be most useful when the unrestricted estimate (maximum likelihood, least square, generalized method of moments) is relatively easy to obtain. Consideration of small sample properties requires a case-by-case analysis. The Wald test has flexibility in being quite generally applicable to a wide array of estimation contexts other than ML estimation. We emphasize, however, that this flexibility is not exclusive to the Wald test. In particular, as we note earlier, the LM approach has been gradually expanding into contexts more general than maximum likelihood estimation (see the Engle and Godfrey references).

## 10.6 Tests in the Context of the Classical GLM

In this section, we examine some statistical tests that are often applied in the context of the Classical General Linear Model (GLM). The collection of tests that are presented here will allow testing of statistical hypotheses relating to values of the $\boldsymbol{\beta}$ and $\sigma^2$ parameters. We begin in the context of multivariate normality of the residual term of $\mathbf{Y} = \mathbf{x}\boldsymbol{\beta} + \boldsymbol{\varepsilon}$ and later examine asymptotically valid procedures that do not depend on the normality of $\boldsymbol{\varepsilon}$, and that are appropriate to semiparametric model applications.

### 10.6.1 Tests When $\boldsymbol{\varepsilon}$ Is Multivariate Normal

In this subsection we begin by assuming that the classical assumptions of the GLM hold and in addition that $\boldsymbol{\varepsilon} \sim N(\mathbf{0}, \sigma^2 \mathbf{I})$ in $\mathbf{Y} = \mathbf{x}\boldsymbol{\beta} + \boldsymbol{\varepsilon}$. We concentrate here on testing hypotheses relating to the value of *linear* combinations of $\boldsymbol{\beta}$, i.e. $\mathbf{R}\boldsymbol{\beta}$, and the value of $\sigma^2$. We will discuss hypotheses concerning nonlinear functions of $\boldsymbol{\beta}$ when we examine the more general non-normal semiparametric cases.

*10.6.1.1 Testing $H_0$: $R\beta = r$ Versus $H_a$: $R\beta \neq r$ When $R$ Is $(q\times k)$: F-Tests* Under the prevailing assumptions, $\mathbf{Y} \sim N(\mathbf{x}\boldsymbol{\beta}, \sigma^2\mathbf{I})$ and $\hat{\boldsymbol{\beta}} \sim N(\boldsymbol{\beta}, \sigma^2(\mathbf{x}'\mathbf{x})^{-1})$. Examine a Wald-type test of the null hypothesis $H_0$: $\mathbf{R}\boldsymbol{\beta} = \mathbf{r}$ versus $H_a$: $\mathbf{R}\boldsymbol{\beta} \neq \mathbf{r}$ where $\mathbf{R}$ is $(q\times k)$ with full row rank. We know that $n^{1/2}(\hat{\boldsymbol{\beta}} - \boldsymbol{\beta}) \sim N(\mathbf{0}, \sigma^2(n^{-1}\mathbf{x}'\mathbf{x})^{-1})$, and thus $n^{1/2}(\hat{\boldsymbol{\beta}} - \boldsymbol{\beta}) \xrightarrow{d} N(\mathbf{0}, \sigma^2\mathbf{Q}^{-1})$ assuming $n^{-1}\mathbf{x}'\mathbf{x} \to \mathbf{Q}$, a positive definite matrix. Also, $\hat{S}^2(n^{-1}\mathbf{x}'\mathbf{x})^{-1}$ is a consistent estimator of the covariance matrix of the limiting distribution for $n^{1/2}(\hat{\boldsymbol{\beta}} - \boldsymbol{\beta})$. Then, Theorem 10.10 suggests that outcomes of the Wald statistic

$$W = (\mathbf{R}\hat{\boldsymbol{\beta}} - \mathbf{r})' \left[\hat{S}^2\mathbf{R}(\mathbf{x}'\mathbf{x})^{-1}\mathbf{R}'\right]^{-1} \left(\mathbf{R}\hat{\boldsymbol{\beta}} - \mathbf{r}\right)$$

can be used to test $H_0$, where the rejection region $C_r^W = \left[\chi^2_{q;\alpha}\infty\right)$ defines an asymptotic size $\alpha$, asymptotically unbiased and consistent test.

In this case, we have sufficient information to establish the small sample distribution of $W$, and we will also be able to define an exact size $\alpha$ test of $H_0$. In particular, rewrite $W$ in the following form:

$$W = \frac{\left(\mathbf{R}\hat{\boldsymbol{\beta}} - \mathbf{r}\right)' \left[\sigma^2 \mathbf{R}(\mathbf{x}'\mathbf{x})^{-1}\mathbf{R}'\right]^{-1} \left(\mathbf{R}\hat{\boldsymbol{\beta}} - \mathbf{r}\right)}{\left[(n-k)\hat{S}^2/(\sigma^2(n-k))\right]} = \frac{Y_q}{(Z_{n-k}/(n-k))}.$$

Note that the numerator random variable is the sum of the squares of $q$ independent standard normal variates since $\mathbf{V} = [\sigma^2\ \mathbf{R}(\mathbf{x}'\mathbf{x})^{-1}\ \mathbf{R}']^{-1/2}\ (\mathbf{R}\hat{\boldsymbol{\beta}}-\mathbf{r}) \sim N(\mathbf{0}, \mathbf{I}_q)$ under $H_0$, and the numerator can be represented as $\mathbf{V}'\mathbf{V}$. We know from Section 8.2 that $(n\text{-}k)\ \hat{S}^2/\sigma^2 \sim \chi^2_{n-k}$, and thus the denominator random variable is a $\chi^2_{n-k}$ random variable divided by its degrees of freedom. Finally, recall that $\hat{\boldsymbol{\beta}}$ and $\hat{S}^2$ are independent random variables (Section 8.2) so that $Y_q$ and $Z_{n\text{-}k}$ are independent. It follows from Section 6.7 that $W/q \sim F(q, (n-k))$ under $H_0$,[6] so that $W$ itself is distributed as an $F$ random variable multiplied by $q$, where the $F$-distribution has $q$ numerator and $(n-k)$ denominator degrees of freedom.

An exact size $\alpha$ test of $H_0$: $\mathbf{R}\boldsymbol{\beta} = \mathbf{r}$ versus $H_a$: $\mathbf{R}\boldsymbol{\beta} \neq \mathbf{r}$ is defined by

$$w\left\{\begin{matrix} \geq \\ < \end{matrix}\right\} qF_\alpha(q,\ n-k) \Rightarrow \left\{\begin{matrix} \text{reject } H_0 \\ \text{do not reject } H_0 \end{matrix}\right\}$$

where the value $F_\alpha(q, (n-k))$ is such that $P(f \geq F_\alpha(q, (n-k))) = \alpha$ when $F \sim F(q, (n-k))$. This test is *unbiased* and is also *consistent* if $(\mathbf{x}'\mathbf{x})^{-1} \to \mathbf{0}$ as $n \to \infty$. To motivate unbiasedness, note that if $\mathbf{R}\boldsymbol{\beta} - \mathbf{r} = \boldsymbol{\phi} \neq \mathbf{0}$, so that $H_a$ is true, then $[\sigma^2\ \mathbf{R}(\mathbf{x}'\mathbf{x})^{-1}\mathbf{R}']^{-1/2}\ (\mathbf{R}\ \hat{\boldsymbol{\beta}} - \mathbf{r} - \boldsymbol{\phi}) \sim N(\mathbf{0}, \mathbf{I})$ so that $[\sigma^2\ \mathbf{R}(\mathbf{x}'\mathbf{x})^{-1}\ \mathbf{R}']^{-1/2}(\mathbf{R}\ \hat{\boldsymbol{\beta}} - \mathbf{r}) \sim N([\sigma^2\ \mathbf{R}(\mathbf{x}'\mathbf{x})^{-1}\ \mathbf{R}']^{-1/2}\ \boldsymbol{\phi}, \mathbf{I})$ and thus the numerator of the previous representation of $W$ is a random variable having a *noncentral* $\chi^2$ distribution, i.e., $Y_q \sim \chi^2_q(\lambda)$, where $\lambda = (1/2)\ \boldsymbol{\phi}'[\sigma^2\ \mathbf{R}(\mathbf{x}'\mathbf{x})^{-1}\ \mathbf{R}']^{-1}\ \boldsymbol{\phi}$. It follows that $W/q$ is the ratio of a *noncentral* $\chi^2_q$ random variate divided by an independent central $\chi^2_{n-k}$ variate, each divided by their respective degrees of freedom, and thus $W/q$ has a **noncentral F-distribution**, $F(q, (n-k), \lambda)$ (see Section 10.8). Then because $P(w/q \geq c; \lambda)$ is monotonically increasing in $\lambda\ \forall\ c < \infty$, it follows that the power function of the F test is strictly monotonically increasing in the noncentrality parameter, $\lambda$, so that the test is an unbiased test.

To motivate consistency, first note that if $(\mathbf{x}'\mathbf{x})^{-1} \to \mathbf{0}$, so that $\mathbf{R}(\mathbf{x}'\mathbf{x})^{-1}\ \mathbf{R}' \to \mathbf{0}$, then all of the characteristic roots, $\lambda_1,\ldots,\lambda_q$, of $\mathbf{R}(\mathbf{x}'\mathbf{x})^{-1}\mathbf{R}'$ limit to 0 as $n \to \infty$ since $\sum_{i=1}^q \lambda_i = \mathrm{tr}(\mathbf{R}(\mathbf{x}'\mathbf{x})^{-1}\ \mathbf{R}') \to 0$ and $\lambda_i \geq 0\ \forall i$ by the positive semi-definiteness of $\mathbf{R}(\mathbf{x}'\mathbf{x})^{-1}\ \mathbf{R}'$. Since the characteristic roots of $[\mathbf{R}(\mathbf{x}'\mathbf{x})^{-1}\mathbf{R}']^{-1}$ are the reciprocals

[6]Note that under $H_0$ $W/q$ is the ratio of two independent central $\chi^2$ random variables each divided by their respective degrees of freedom.

of the characteristic roots of $\mathbf{R}(\mathbf{x}'\mathbf{x})^{-1}\mathbf{R}'$, all of the characteristic roots of $[\mathbf{R}(\mathbf{x}'\mathbf{x})^{-1}\mathbf{R}']^{-1}$, say $\xi_1,\ldots,\xi_q$, will then increase without bound as $n \to \infty$. If $\mathbf{R\beta} - \mathbf{r} = \boldsymbol{\phi} \neq \mathbf{0}$, it follows from results on the extrema of quadratic forms (see C.R. Rao, *Statistical Inference* p. 62) that

$$\lambda = \frac{1}{2}\boldsymbol{\phi}'\left[\sigma^2\mathbf{R}(\mathbf{x}'\mathbf{x})^{-1}\mathbf{R}'\right]^{-1}\boldsymbol{\phi} \geq \frac{1}{2\sigma^2}\xi_s\boldsymbol{\phi}'\boldsymbol{\phi},$$

where $\xi_s$ is the smallest characteristic root of $[\mathbf{R}(\mathbf{x}'\mathbf{x})^{-1}\mathbf{R}']^{-1}$, and thus if $\xi_s \to \infty$ and $\boldsymbol{\phi}'\boldsymbol{\phi} > 0$, then $\lambda \to \infty$. Finally, note that because $\hat{S}^2 \xrightarrow{p} \sigma^2$, it follows that $W \xrightarrow{d} \chi^2_q$ under $H_0$, and $q\, F_\alpha(q,(n-k)) \to \chi^2_{q;\alpha} = c < \infty$. Therefore, since $\lim_{\lambda\to\infty}(P(w \geq c;\lambda)) = 1$ when $W \sim \chi^2_q(\lambda)$, the test sequence is consistent.

**Example 10.14**
***Hypothesis Test of Joint Explanatory Variable Significance in Representing* E(Y)**

In this example we examine a size $\alpha$ test of the null hypothesis that *none* of the explanatory variables, other than the intercept, are significant in explaining the expected value of the dependent variable in a GLM, $\mathbf{Y} = \mathbf{x\beta} + \boldsymbol{\varepsilon}$ with $\boldsymbol{\varepsilon} \sim N(\mathbf{0}, \sigma^2\mathbf{I})$. We suppose that the first column of $\mathbf{x}$ is a vector of 1's, so that $\beta_1$ represents the intercept. Then from the foregoing discussion, a test of the hypothesis that none of the nonintercept explanatory variables are significant, i.e., $H_0$: $\beta_i = 0$, $i=2,\ldots,k$, versus $H_a$: not $H_0$ (i.e., *at least one* explanatory variable is significant) can be performed in terms of the $F$-statistic $W = \left(\mathbf{R}\hat{\boldsymbol{\beta}} - \mathbf{r}\right)'\left(\hat{S}^2\mathbf{R}(\mathbf{x}'\mathbf{x})^{-1}\mathbf{R}'\right)^{-1}\left(\mathbf{R}\hat{\boldsymbol{\beta}} - \mathbf{r}\right)$, where $\mathbf{r}_{(k-1)\times 1} = \mathbf{0}$ and $\mathbf{R}_{(k-1)\times k} = [\mathbf{0}|\mathbf{I}_{k-1}]$. In the current context, the test statistic can be represented alternatively as $W = \hat{\boldsymbol{\beta}}'_*\left(\widehat{\mathbf{Cov}}\left(\hat{\boldsymbol{\beta}}_*\right)\right)^{-1}\hat{\boldsymbol{\beta}}_*$, where $\hat{\boldsymbol{\beta}}_* = \left(\hat{\beta}_2,\ldots,\hat{\beta}_k\right)'$ and $\widehat{\mathbf{Cov}}\left(\hat{\boldsymbol{\beta}}_*\right)$ is the sub-matrix of $\hat{S}^2(\mathbf{x}'\mathbf{x})^{-1}$ referring to the estimated covariance matrix of $\hat{\boldsymbol{\beta}}_*$. The test is defined by

$$w\left\{\begin{matrix}\geq\\<\end{matrix}\right\}(k-1)F_\alpha(k-1,n-k) \Rightarrow \left\{\begin{matrix}\text{reject } H_0\\ \text{do not reject } H_0\end{matrix}\right\}.$$

As a numerical illustration, suppose $\mathbf{Y} = \mathbf{x\beta} + \boldsymbol{\varepsilon}$ represented a linear relationship that was hypothesized to explain quarterly average stock price levels for a certain composite of stocks sensitive to international market conditions. Let

$$\mathbf{b} = [1.2 \quad 2.75 \quad 3.13]' \text{ and } \hat{s}^2(\mathbf{x}'\mathbf{x})^{-1} = \begin{bmatrix} .25 & .15 & .31\\ & .9 & .55\\ \text{(symmetric)} & & 1.2\end{bmatrix}$$

represent least squares-based estimates of $\boldsymbol{\beta}$ and $\sigma^2(\mathbf{x}'\mathbf{x})^{-1}$, where n $= 30$, $b_1 = 1.2$ represents the estimated intercept of the relationship, and $b_2 = 2.75$ and $b_3 = 3.13$ represents estimated coefficients on quarterly average composite exchange rates and western country aggregate real income. The value of the $F$-statistic for testing the significance of the explanatory variables other than the intercept is given by

$$w = \mathbf{b}'_*\left(s^2\mathbf{R}(\mathbf{x}'\mathbf{x})^{-1}\mathbf{R}'\right)^{-1}\mathbf{b}_* = [2.75 \quad 3.13]\begin{bmatrix} .9 & .55 \\ .55 & 1.2 \end{bmatrix}^{-1}\begin{bmatrix} 2.75 \\ 3.13 \end{bmatrix} = 10.8347.$$

Then $H_0$: $\beta_2 = \beta_3 = 0$ is *rejected* using a size .05 unbiased and consistent test, since $w = 10.8347 \geq 6.7082 = 2F_{.05}(2, 27)$. The interpretation of the outcome of the statistical test is that *at least one* of the two explanatory variables is significant in explaining the expected value of the stock price composite. □

It can be shown that there does *not* exist a UMP or UMPU test of $H_0$: $\mathbf{R\beta} = \mathbf{r}$ versus $H_a$: $\mathbf{R\beta} \neq \mathbf{r}$ when $\mathbf{R}$ has two or more rows. However, we will see next that for the case where $\mathbf{R}$ is a single row vector, a UMPU test does exist.

*10.6.1.2 Testing* $H_0$*:* $\mathbf{R\beta} = r$*, or* $H_0$*:* $\mathbf{R\beta} \leq r$*, or* $H_0$*:* $\mathbf{R\beta} \geq r$ *versus* $H_a = \bar{H}_0$ *When* **R** *Is* $(1 \times k)$*: T-Tests*

In order to test $H_0$: $\mathbf{R\beta} = r$ versus $H_a$: $\mathbf{R\beta} \neq r$ when $\mathbf{R}$ is a *row* vector, the *F*-statistic in the preceding subsection can be used without modification. However, since $\mathbf{R\beta}$ is a scalar, an alternative representation of *W* is possible as

$$W = \left(\mathbf{R}\hat{\boldsymbol{\beta}} - r\right)^2 / \left[\hat{S}^2\mathbf{R}(\mathbf{x}'\mathbf{x})^{-1}\mathbf{R}'\right] \sim F(1, n-k) \text{ under } H_0.$$

An alternative test statistic, whose square is equal to *W*, can be based on the t-distribution as

$$T = (\mathbf{R}\hat{\boldsymbol{\beta}} - r)/[\hat{S}^2\mathbf{R}(\mathbf{x}'\mathbf{x})^{-1}\mathbf{R}']^{1/2} \sim T_{(n-k)} \text{under } H_0$$

and is referred to as the *T*-statistic. Justification for the *t*-distribution of *T* can be provided from first principles (recall Problem 3(a) of Chapter 8), or else from the fact that the square of a $T_{(n-k)}$ random variable is a random variable having a $F(1, n-k)$ distribution. In terms of the *t*-statistic, the test is defined by

$$t\left\{\begin{matrix} \in \\ \notin \end{matrix}\right\}(-\infty, -t_{\alpha/2}(n-k)] \cup [t_{\alpha/2}(n-k), \infty) \Rightarrow \left\{\begin{matrix} \text{reject } H_0 \\ \text{do not reject } H_0 \end{matrix}\right\},$$

where $t_{\alpha/2}(n-k)$ denotes the value of a *t*-distribution with $n-k$ degrees of freedom for which $P(t \geq t_{\alpha/2}(n-k)) = \alpha/2$.

It can be shown using the principles in Section 9.5 that the *t*-test (and thus the *F*-test) is a UMPU level $\alpha$ test of $H_0$: $\mathbf{R\beta} = r$ versus $H_a$: $\mathbf{R\beta} \neq r$. We will defer the demonstration of the UMPU property, as it is quite involved. Unbiasedness can be demonstrated from the equivalence of the *t*-test and *F*-test or directly from the power function of the *t*-test. Regarding the power function, when $\mathbf{R\beta} - r = \phi \neq 0$, the aforementioned *T*-statistic has a **noncentral *t*-distribution** with noncentrality parameter $\lambda = \phi / [\sigma^2 \mathbf{R}(\mathbf{x}'\mathbf{x})^{-1}\mathbf{R}']^{1/2}$. The power function, $\pi_{Cr}(\lambda) = 1 - \int_{-t_{\alpha/2}}^{t_{\alpha/2}} f(t; n-k, \lambda)dt$ achieves its minimum value when $\lambda = 0$ (i.e., $H_0$ is true) and is strictly monotonically increasing as $\lambda \to \infty$ or $\lambda \to -\infty$ (see Section 9.6). Thus the test is *unbiased*.

Consistency of the $t$-test sequence follows from consistency of the $F$-test sequence. As an alternative motivation for consistency, first note that because $T_{n-k} \xrightarrow{d} N(0,1)$ under $H_0$, then $t_{\alpha/2}(n-k) \to z_{\alpha/2}$, where $\int_{z_{\alpha/2}}^{\infty} N(z;0,1)\,dz = \alpha/2$. Based on the $t$-test, the null hypothesis will be rejected *iff* $\left|\frac{(\mathbf{Rb}-\mathbf{r})}{\left[\hat{s}^2\mathbf{R}(\mathbf{x}'\mathbf{x})^{-1}\mathbf{R}'\right]^{1/2}}\right| \geq t_{\alpha/2}(n-k)$. Since the $\varepsilon_i$'s are *iid*, $\hat{S}^2 \xrightarrow{p} \sigma^2$, and assuming $(\mathbf{x}'\mathbf{x})^{-1} \to \mathbf{0}$, then $\hat{\boldsymbol{\beta}} \xrightarrow{p} \boldsymbol{\beta}$. All of the preceding results imply collectively that the condition for rejecting $H_0$ becomes $|\phi/\sigma| > 0$ in the limit. Thus when $\phi \neq 0$, $H_0$ is rejected with probability $\to 1$ as $n \to \infty$, and the test sequence is consistent.

**Example 10.15** ***Hypothesis Tests of Individual Explanatory Variable Significance in Representing*** **E(Y)**

Revisit Example 10.14, and consider testing the individual null hypotheses that each of the non-intercept explanatory variables are significant in explaining the daily composite stock value. That is, we wish to test $H_0$: $\beta_i = 0$ versus $H_a$: $\beta_i \neq 0$, for $i$=2,3.

Using a size .05 test for each test, in order to test the first null hypothesis, set $\mathbf{R} = [0\ 1\ 0]$ and $r = 0$, and calculate the $t$-statistic as

$$t = \frac{\mathbf{Rb}-\mathbf{r}}{\left(\hat{s}^2\mathbf{R}(\mathbf{x}'\mathbf{x})^{-1}\mathbf{R}'\right)^{1/2}} = \frac{2.75}{(.9)^{1/2}} = 2.8988.$$

Then since

$$t = 2.8988 \in C_r^T = (-\infty, -t_{.025}(27)] \cup [t_{.025}(27)) = (-\infty, -2.052] \cup [2.052, \infty)$$

we *reject* $H_0$: $\beta_2 = 0$. To test the second null hypothesis, set $\mathbf{R} = [0\ 0\ 1]$ and $\mathrm{r} = 0$, and calculate the $t$-statistic as

$$t = \frac{\mathbf{Rb}-\mathbf{r}}{\left(\hat{s}^2\mathbf{R}(\mathbf{x}'\mathbf{x})^{-1}\mathbf{R}'\right)^{1/2}} = \frac{3.13}{(1.2)^{1/2}} = 2.8573.$$

The rejection region is the same as above, so that $t \in C_r^T$ and $H_0$: $\beta_3 = 0$ is *rejected*. It appears that *both* explanatory variables are significant in explaining the composite stock value, but see the next subsection for the appropriate interpretation of this *joint* conclusion in terms of the probability of Type I Error. □

From the previous example it is clear that a test of the significance of any individual explanatory variable in the GLM will be based on a statistic whose value is equal to the explanatory variable's associated parameter estimate, $b_i$, divided by the estimated standard error of the estimate, $\hat{s}_{(\hat{\beta}_i)}$, as $t = b_i/\hat{s}_{(\hat{\beta}_i)}$. The statistic will be $t$-distributed with $(n-k)$ degrees of freedom, and the rejection region will be of the form $(-\infty, -t_{\alpha/2}(n-k)] \cup [t_{\alpha/2}(n-k), \infty)$.

The $t$-statistic can also be used to define level $\alpha$ UMPU tests of $H_0$: $\mathbf{R\beta} \leq \mathbf{r}$ versus $H_a$: $\mathbf{R\beta} > \mathbf{r}$ or of $H_0$: $\mathbf{R\beta} \geq \mathbf{r}$ versus $H_a$: $\mathbf{R\beta} < \mathbf{r}$. The rejection regions defined in terms of the $t$-statistic are as follows:

| Case | $H_0$ | $H_a$ | $C_r^T$ |
|---|---|---|---|
| 1. | $\mathbf{R\beta} \leq r$ | $\mathbf{R\beta} > r$ | $t \in [t_\alpha(n\text{-}k), \infty)$ |
| 2. | $\mathbf{R\beta} \geq r$ | $\mathbf{R\beta} < r$ | $t \in (-\infty, -t_\alpha(n\text{-}k)]$ |

The proof of the UMPU property can be based on the results of Section 9.5, but it is quite involved and is deferred. Unbiasedness can be verified from an investigation of the power functions of the tests. Examine Case 1, and note that in terms of the noncentrality parameter $\lambda = (\mathbf{R}\,\boldsymbol{\beta}\text{-}r)/[\sigma^2\,\mathbf{R}(\mathbf{x}'\mathbf{x})^{-1}\,\mathbf{R}']^{1/2}$ of the noncentral t-distribution, an equivalent representation of $H_0$ and $H_a$ is given by $H_0$: $\lambda \leq 0$ versus $H_a$: $\lambda > 0$. Then since the power function of the test

$$\pi_{C_r}(\lambda) = \int_{t_{\alpha/2}}^{\infty} f(t; n-k, \lambda)dt$$

is strictly monotonically increasing in $\lambda$, the test is an unbiased test. Unbiasedness of the test in Case 2 follows from an analogous argument with inequality reversals and is left to the reader.

Regarding consistency of the respective test sequences, again examine case 1 and recall that $t_\alpha(n-k) \to z_\alpha$ where $\int_{z_\alpha}^{\infty} N(z; 0,1)dz = \alpha$. Based on the $t$-test, the null hypothesis will be rejected *iff* $\left|\frac{(\mathbf{Rb}-r)}{[\hat{s}^2\mathbf{R}(\mathbf{x}'\mathbf{x})^{-1}\mathbf{R}']^{1/2}}\right| \geq t_{\alpha/2}(n-k)$. Given that $\hat{S}^2 \xrightarrow{p} \sigma^2$, $(\mathbf{x}'\mathbf{x})^{-1} \to \mathbf{0}$, and thus $\hat{\boldsymbol{\beta}} \xrightarrow{p} \boldsymbol{\beta}$, the condition for rejecting $H_0$ becomes $\phi/\sigma > 0$ in the limit. Then if $H_a$ is true, so that $\phi > 0$, the null hypothesis will be rejected with probability $\to 1$ as $n \to \infty$ and the test sequence is consistent.

**Example 10.16** ***Testing Various Hypotheses About Elasticities in a Constant Elasticity GLM***

A constant elasticity quarterly demand relationship is specified as $\ln(\mathbf{Y}) = \beta_1 \mathbf{1_n} + \beta_2 \ln(\mathbf{p}) + \beta_3 \ln(\mathbf{p}_s) + \beta_4 \ln(\mathbf{m}) + \boldsymbol{\varepsilon}$ where $\mathbf{Y}$ is quantity demanded of a commodity, $\mathbf{p}$ is its price, $\mathbf{p}_s$ is an index of substitute prices, $\mathbf{m}$ is disposable income, $\mathbf{1_n}$ is a vector $(n\times 1)$ of 1's, and $\boldsymbol{\varepsilon} \sim N(\mathbf{0}, \sigma^2\mathbf{I})$. Forty quarterly observations were used to calculate the least squares-based estimates

$$\mathbf{b} = [10.1 \; -.77 \;\; .41 \;\; .56]'$$

and

$$\hat{s}^2(\mathbf{x}'\mathbf{x})^{-1} = \begin{bmatrix} 3.79 & .05 & .02 & .01 \\ & .04 & .9\times 10^{-3} & .1\times 10^{-3} \\ & & .03 & .1\times 10^{-3} \\ \text{(symmetric)} & & & .02 \end{bmatrix}.$$

We seek answers to the following statistical testing questions:

**(a)** Test the significance of the matrix of explanatory variable values other than the intercept using a size .05 test.

**Table 10.1** Statistical tests of the value of $\mathbf{R}\boldsymbol{\beta}$ relative to $\mathbf{r}$

| Test | $R$ | $\begin{pmatrix} \le \\ = \\ \ge \end{pmatrix}$ | r | w or t | $C_r$ | Outcome |
|---|---|---|---|---|---|---|
| a. | $\begin{bmatrix} 0100 \\ 0010 \\ 0001 \end{bmatrix}$ | $=$ | $\begin{bmatrix} 0 \\ 0 \\ 0 \end{bmatrix}$ | $w = 36.6237$ | $w > 8.5988$ $> 3F_{.05}(3,36)$ | Reject $H_0$ |
| b. | [0 1 0 0] | $=$ | 0 | $t = -3.85$ | $t \notin (-2.028, 2.028)$ $\notin (-t_{.025}(36), t_{.025}(36))$ | Reject $H_0$ |
| | [0 0 1 0] | $=$ | 0 | $t = 2.3671$ | | Reject $H_0$ |
| | [0 0 0 1] | $=$ | 0 | $t = 3.9598$ | | Reject $H_0$ |
| c. | [0 1 1 1] | $=$ | 0 | $t = .6587$ | $t \notin (-2.028, 2.028)$ $\notin (-t_{.025}(36), t_{.025}(36))$ | Do not reject $H_0$ |
| d. | [0 1 0 0] | $\le$ | $-1$ | $t = 1.15$ | $t \in [1.688, \infty)$ $\in [t_{.05}(36), \infty)$ | Do not reject $H_0$ |
| e. | [0 0 0 1] | $\ge$ | 1 | $t = -3.1113$ | $t \in (-\infty, -1.688]$ $\in (-\infty, -t_{.05}(36)]$ | Reject $H_0$ |

**(b)** Test the individual significance of the effect of price, of the substitute price index, and of disposable income each at the .05 level of significance.
**(c)** Test whether the demand equation is homogenous degree zero in prices and income.
**(d)** Test whether the own price elasticity is $\le -1$.
**(e)** Test whether the income elasticity is unitary elastic or larger.

For each of the hypotheses being tested, we list the specification of $\mathbf{R}$, the value of $\mathbf{r}$, the relationship hypothesized between $\mathbf{R}\boldsymbol{\beta}$ and $\mathbf{r}$, whether the statistic used is a $F$-statistic ($w$) or a $t$-statistic ($t$), the value of the calculated statistic, the rejection region of the test statistic with degrees of freedom and distribution indicated, and the outcome of the test (see Table 10.1).

Test (a) indicates that at least one of the explanatory variables is significant at the .05 level; test (b) indicates that each explanatory variable is individually significant at the .05 level (but see the next subsection on the interpretation of the collective outcome of the three tests); test (c) indicates that homogeneity of degree zero cannot be rejected; test (d) indicates that the price elasticity being $\le -1$ cannot be rejected, and test (e) indicates that the conjecture that the income elasticity is unitary elastic or greater is rejected suggesting that the income response is inelastic. Regarding the two hypotheses that were not rejected, the reader should contemplate the power function of the tests used in order to temper ones enthusiasm for *literal acceptance* of the respective hypotheses. □

*10.6.1.3 Bonferroni Joint Tests of* $\mathbf{R}_i \boldsymbol{\beta} = r_i$, $\mathbf{R}_i \boldsymbol{\beta} \leq r_i$, or $\mathbf{R}_i \boldsymbol{\beta} \geq r_i$, $i=1,\ldots,m$

Based on the Bonferroni probability inequality, it is possible to *individually* test a number of equality and/or inequality hypotheses concerning various linear combinations of $\boldsymbol{\beta}$ and still provide an upper bound to the probability of making a Type I Error with respect to the null hypotheses taken *collectively* or *simultaneously*. To see what is involved, suppose a collection of $m$ tests based on the aforementioned $t$-statistics are performed that relate to any combination of hypotheses of the form $H_{0i}$: $\mathbf{R}_i\boldsymbol{\beta} = r_i$ or $H_{0i}$: $\mathbf{R}_i\boldsymbol{\beta} \leq r_i$ or $H_{0i}$: $\mathbf{R}_i\boldsymbol{\beta} \geq r_i$ versus respectively $H_{ai}$: $\mathbf{R}_i\boldsymbol{\beta} \neq \mathbf{r}_i$ or $H_{ai}$: $\mathbf{R}_i\boldsymbol{\beta} > \mathbf{r}_i$ or $H_{ai}$: $\mathbf{R}_i\boldsymbol{\beta} < r_i$, where $\mathbf{R}_i$'s and $r_i$'s are $(1\times k)$ vectors and scalars, respectively. Let $T_i$ and $C_r^i$, $i=1,\ldots,m$, represent the associated $t$-statistics and rejection regions, and suppose the sizes of the tests were $\alpha_i$, $i = 1,\ldots,m$. It follows by Bonferroni's inequality that if all of the null hypotheses are true,

$$P(t_i \notin C_r^i, i = 1,\ldots,m) \geq 1 - \sum_{i=1}^{m} P(t_i \in C_r^i) = 1 - \sum_{i=1}^{m} \alpha_i. \; .$$

It follows that the overall level of protection against Type I Error provided by the collection of tests is $\leq \sum_{i=1}^{m} \alpha_i$.

For example, if four different tests are performed, each of size .025, then treating the four different null hypotheses collectively as a joint hypothesis regarding the parameter vector $\boldsymbol{\beta}$, the size of the joint test is $\leq .10$. In performing the four tests, one knows that the probability one or more individual hypotheses will be rejected by mistake is $\leq .10$. In practice, one generally sets the overall bound on Type I Error, say $\alpha$, that is desired for the joint test comprised of the $m$ individual tests, and then sets the size of each individual test to $\alpha/m$. Alternatively, any distribution of test sizes across the $m$ individual tests that sum to $\alpha$ will afford the same overall level of Type I Error protection. The Bonferroni approach can of course be applied in contexts other than the GLM.

While the individual tests in the collection of tests may have optimal properties such as being UMPU, the properties do not generally transfer to the Bonferroni-type joint test. However, asymptotically the situation is favorable in the sense that if the individual test sequences are consistent, the joint test sequence will also be consistent. This follows from the fact that the joint test will reject the collection of $H_0$'s as being simultaneously true with probability

$$P(\cup_{i=1}^{m}\{\mathbf{y} : t_i(\mathbf{y}) \in C_r^i) \geq P(\{\mathbf{y} : t_i(\mathbf{y}) \in C_r^i)\forall i,$$

and if $P(\{\mathbf{y}: t_i(\mathbf{y}) \in C_r^i) \to 1$ as $n \to \infty$ when $H_{ai}$ is true, then $P(\cup_{i=1}^{m}\{\mathbf{y} : t_i(\mathbf{y}) \in C_r^i\}) \to 1$ as $n \to \infty$. Thus the collection of null hypothesis $\cup_{i=1}^{m} H_{0i}$ is rejected with probability $\to$ 1, and the joint test sequence is consistent.

A useful feature of the Bonferroni approach to joint testing of hypotheses is the ability to "look inside" a joint vector hypothesis $\mathbf{R}\boldsymbol{\beta} = \mathbf{r}$, or any other collection of individual equality and/or inequality hypotheses, to analyze which of the hypotheses in the collection are causing the rejection of the overall joint hypothesis. For example, if the joint equality test $\mathbf{R}\boldsymbol{\beta} = \mathbf{r}$ is rejected, say by a joint $F$-test, one can consider performing a series of tests of the individual null hypotheses $H_{0i}$: $\mathbf{R}_i\boldsymbol{\beta} = r_i$, $i = 1,\ldots,m$, to attempt to see *which* of these linear combinations hypotheses are to be rejected and which are not. By choosing an overall level of Type I Error of $\alpha$, and then distributing $\alpha/m$ of this Type I Error probability to each of the individual tests, the collection of conclusions remains protected at the level of Type I Error $\alpha$.

**Example 10.17**
***Bonferroni Approach for Testing Components of a Joint Hypothesis***

Revisit Examples 10.14–10.16. In each case, an $F$-test rejected a joint null hypothesis of the form $H_0$: $\beta_i = 0$, $i \in I$ and then individual $t$-tests were performed on each of the null hypotheses $H_{0i}$: $\beta_i = 0$ separately. In the case of 10.14 and 10.15, since each of the two tests had size .05, the overall conclusion that both explanatory variables are significant is protected against a Type I Error at level .10 by the Bonferroni approach. Likewise, the conclusion that each of the three explanatory variables is significant in Example 10.16 is protected against of Type I Error at level .15. □

*10.6.1.4 Tests for $\sigma^2$ Under Normality: $\chi^2$-Tests*

In order to test hypotheses concerning the magnitude of $\sigma^2$, consider the GLR approach. The likelihood function for $(\boldsymbol{\beta},\sigma^2)$ is given by

$$L(\boldsymbol{\beta},\sigma^2|\mathbf{y}) = \frac{1}{(2\pi\sigma^2)^{n/2}} \exp\left(-\frac{1}{2}(\mathbf{y}-\mathbf{x}\boldsymbol{\beta})'(\mathbf{y}-\mathbf{x}\boldsymbol{\beta})/\sigma^2\right).$$

**Case I**: $H_0$: $\sigma^2 = d$ versus $H_a$: $\sigma^2 \neq d$

Let $H_0$: $\sigma^2 = d$ and $H_a$: $\sigma^2 \neq d$ . We have already seen in Section 9.3 that the MLE for $\boldsymbol{\beta}$ and $\sigma^2$ is given by $\hat{\boldsymbol{\beta}} = (\mathbf{x}'\mathbf{x})^{-1}\mathbf{x}'\mathbf{Y}$ and $\hat{\sigma}^2(\mathbf{Y}) = \left(\mathbf{Y}-\mathbf{x}\hat{\boldsymbol{\beta}}\right)'\left(\mathbf{Y}-\mathbf{x}\hat{\boldsymbol{\beta}}\right)/n$, so that

$$\max_{\beta,\sigma^2} L(\boldsymbol{\beta},\sigma^2|\mathbf{y}) = \left(2\pi\hat{\sigma}^2(\mathbf{y})\right)^{-n/2} \exp\left(-\frac{1}{2}(\mathbf{y}-\mathbf{x}\mathbf{b})'(\mathbf{y}-\mathbf{x}\mathbf{b})/\hat{\sigma}^2(\mathbf{y})\right)$$
$$= \left(2\pi\hat{\sigma}^2(\mathbf{y})\right)^{-n/2} \exp(-n/2).$$

Now consider the MLE subject to the restriction $\sigma^2 = d$. It is evident that $L(\boldsymbol{\beta},d|\mathbf{y})$ is maximized when $(\mathbf{y}-\mathbf{x}\boldsymbol{\beta})'(\mathbf{y}-\mathbf{x}\boldsymbol{\beta})$ is minimized, so that $\hat{\boldsymbol{\beta}}$ remains the optimum choice of $\boldsymbol{\beta}$ under the restriction. The GLR statistic is then given by

$$\lambda(\mathbf{Y}) = \left(\frac{\hat{\sigma}^2(\mathbf{Y})}{d}\right)^{n/2} \exp\left(\frac{n}{2}\left(1-\frac{\hat{\sigma}^2(\mathbf{Y})}{d}\right)\right).$$

In order to define a size $\alpha > 0$ test, we need to find a positive number $c < 1$ such that $P(\lambda \leq c) = \alpha$ when $H_0$ is true. The GLR statistic has a complicated probability distribution and is therefore not convenient for solving this problem. We seek an alternative test statistic for determining the rejection region of the GLR test. Letting $z = \hat{\sigma}^2(\mathbf{y})/d$, note that $\ln(\lambda) = (n/2)\ln(z) + (n/2)(1-z)$ is such that $d\ln(\lambda)/dz = (n/2)[(1/z)-1]$ and $d^2\ln(\lambda)/dz^2 = -n/2z^2 < 0$ for $z > 0$, which indicates that $\ln(\lambda)$, and thus $\lambda$ itself, attains its maximum when $z = 1$, and is strictly monotonically decreasing as $z$ increases in value above 1 or decreases in value below 1 and moves toward zero. It follows that

$\lambda \leq c$ *iff* $\hat{\sigma}^2(\mathbf{y})/d \leq \tau_1$ or $\geq \tau_2$ for appropriate choices of $\tau_1 < \tau_2$.

The problem can be simplified further via multiplying $\hat{\sigma}^2(\mathbf{Y})/d$ by the constant $n$, resulting in $n\hat{\sigma}^2(\mathbf{Y})/d = (n-k)\hat{S}^2/d \sim \chi^2_{(n-k)}$ under $H_0$. We thus seek appropriate values of $\tau_1 < \tau_2$ that satisfy both the size $\alpha$ requirement $1 - \int_{n\tau_1}^{n\tau_2} f(w;(n-k))\,dz = \alpha$, where $f(w;(n-k))$ is a $\chi^2_{n-k}$ density, and the requirement that the GLR $\lambda$ have the same value, $c$, at $\tau_1$ and $\tau_2$, implying that $\ln(\tau_1) + (1-\tau_1) = \ln(\tau_2) + (1-\tau_2)$, or $\ln(\tau_1/\tau_2) = \tau_1 - \tau_2$. The two simultaneous equations must be solved numerically to obtain the appropriate values of $\tau_1$ and $\tau_2$ that define the rejection region $(-\infty, n\tau_1] \cup [n\tau_2, \infty)$ for the $\chi^2$ test statistic outcome $(n-k)\hat{s}^2/d$. While obtaining the solution is highly feasible on personal computers, it is probably safe to say that most practitioners nevertheless approximate the generalized likelihood ratio test by using an "equal tails" test that can be defined by consulting a table of the $\chi^2$ CDF. The equal tails test would be

$$\frac{(n-k)\hat{s}^2}{d}\left\{\begin{matrix}\in\\ \notin\end{matrix}\right\}\left(0,\chi^2_{n-k;1-\alpha/2}\right]\cup\left[\chi^2_{n-k;\alpha/2}\infty\right)\Rightarrow\left\{\begin{matrix}\text{reject } H_0\\ \text{do not reject } H_0\end{matrix}\right\}.$$

Note that the GLR test is a UMPU level $\alpha$ test of $H_0$, but the equal tails level $\alpha$ test is neither UMPU nor unbiased.

Both the GLR test and "equal tails" test are both consistent. To motivate consistency in the case of the equal tails test, note that $H_0$ is rejected *iff* $\hat{S}^2/d \notin \left((n-k)^{-1}\chi^2_{n-k;1-\alpha/2}, (n-k)^{-1}\chi^2_{n-k;\alpha/2}\right)$. Because $(n-k)^{-1}\chi^2_{n-k} \xrightarrow{p} 1$, in the limit the test amounts to deciding whether or not plim $\hat{S}^2/d = 1$, and since $\hat{S}^2 \xrightarrow{p} \sigma_0^2$ the test then amounts to whether $\sigma_0^2/d = 1$. Thus if $d \neq \sigma_0^2$, so that $H_a$ is true, $H_0$ is rejected with probability $\to 1$ as $n \to \infty$, and the test sequence is consistent.

The consistency of the GLR test sequence can be established by showing that Theorem 10.4 applies. To see this, recall that $\ln(\lambda) = (n/2)\ln(\hat{\sigma}^2(\mathbf{y})/d) + (n/2)(1 - \hat{\sigma}^2(\mathbf{y})/d)$ achieves its maximum value, equal to 0, when $\hat{\sigma}^2(\mathbf{y})/d = 1$. Now note that $\text{plim}(n^{-1}(2\ln\lambda(\mathbf{Y}))) = \ln(\sigma_0^2/d) + 1 - \sigma_0^2/d < 0\ \forall \sigma_0^2 \neq d$, which implies that $\lambda(\mathbf{Y}) \xrightarrow{p} 0$ (recall Example 10.8). Then Theorem 10.4.b applies, and the GLR test sequence is consistent.

**Case II:** $H_0$: $\sigma^2 \leq d$ *versus* $H_a$: $\sigma^2 > d$ or $H_0$: $\sigma^2 \geq d$ *versus* $H_a$: $\sigma^2 < d$

In order to test the hypothesis $H_0$: $\sigma^2 \leq d$ versus $H_a$: $\sigma^2 > d$ the GLR approach of the previous subsection can be applied except now the constrained MLE is found by maximizing $L(\boldsymbol{\beta}, \sigma^2|\mathbf{y})$ subject to the inequality constraint $\sigma^2 \leq d$. The solution for $\boldsymbol{\beta}$ remains $\mathbf{b} = (\mathbf{x}'\mathbf{x})^{-1}\mathbf{x}'\mathbf{y}$, while the solution for $\sigma^2$ is $\hat{\sigma}^2_*(\mathbf{y}) = \hat{\sigma}^2(\mathbf{y}) = n^{-1}(\mathbf{y} - \mathbf{xb})'(\mathbf{y} - \mathbf{xb})$ if $\hat{\sigma}^2(\mathbf{y}) \leq d$ and $\hat{\sigma}^2_*(\mathbf{y}) = d$ otherwise. The resulting GLR statistic is therefore defined to be

$$\lambda = \left(\frac{\hat{\sigma}^2(\mathbf{Y})}{\hat{\sigma}^2_*(\mathbf{Y})}\right)^{n/2} \exp\left[\frac{n}{2}\left(1 - \frac{\hat{\sigma}^2(\mathbf{Y})}{\hat{\sigma}^2_*(\mathbf{Y})}\right)\right].$$

Note that the definition of $\hat{\sigma}_*^2$ is such that $\hat{\sigma}^2/\hat{\sigma}_*^2 \geq 1$. Given our previous discussion of the behavior of $\lambda = (z)^{n/2} \exp((n/2)(1-z))$ indicating that $\lambda$ achieves its maximum value of 1 when $z = 1$ and $\lambda$ is strictly monotonically decreasing as $z$ increases in value above 1, it follows that

$$\lambda \leq c \text{ iff } \hat{\sigma}^2(\mathbf{y})/\hat{\sigma}_*^2(\mathbf{y}) \geq \tau$$

for some appropriate choice of $\tau > 1$ for a size $\alpha > 0$ test. Furthermore, since $\hat{\sigma}^2(\mathbf{y})/\hat{\sigma}_*^2(\mathbf{y}) > 1$ only if $\hat{\sigma}_*^2(\mathbf{y}) = d$, it follows that

$$\lambda \leq c \text{ iff } \hat{\sigma}^2(\mathbf{y})/d \geq \tau.$$

Finally, by multiplying $\hat{\sigma}^2(\mathbf{y})/d$ by $n$ as before, we have

$$\lambda \leq c \text{ iff } (n-k)\hat{s}^2/d \geq n\tau.$$

To define the rejection region of the size $\alpha$ GLR test, we need to find the value of $n\tau$ that satisfies

$$\int_{n\tau}^{\infty} f(z; n-k)dz = \alpha,$$

where $f(w;n-k)$ is a $\chi^2$ density function with $n-k$ degrees of freedom. The appropriate value of $n\tau$ can be found in tables of the $\chi^2$ CDF for the typical choices of $\alpha = .01, .05$, or $.10$, or else they can be found through use of the computer. The GLR test is then

$$\frac{(n-k)\hat{s}^2}{d} \begin{Bmatrix} \in \\ \notin \end{Bmatrix} \left[\chi^2_{n-k;\alpha}, \infty\right) \Rightarrow \begin{Bmatrix} \text{reject } \mathrm{H}_0 \\ \text{do not reject } \mathrm{H}_0 \end{Bmatrix}.$$

It can be shown that the GLR test is a UMPU level $\alpha$ test of $H_0$: $\sigma^2 \leq d$ versus $H_a$: $\sigma^2 > d$. Consistency of the GLR test sequence can be demonstrated by showing that Theorem 10.4 applies following a similar argument to the one used in the previous subsection. In order to test the hypothesis $H_0$: $\sigma^2 \geq d$ versus $H_a$: $\sigma^2 < d$, the GLR approach can be followed with appropriate inequality reversals, and can be shown to be

$$\frac{(n-k)\hat{s}^2}{d} \begin{Bmatrix} \in \\ \notin \end{Bmatrix} \left[0, \chi^2_{n-k;1-\alpha}\right) \Rightarrow \begin{Bmatrix} \text{reject } H_0 \\ \text{do not reject } H_0 \end{Bmatrix}.$$

The test is also UMPU level $\alpha$ and consistent.

**Example 10.18** ***Testing Hypotheses About Variances in the GLM Under Normality***

Suppose that in Example 10.15, $\hat{s}^2 = .13$. Consider testing the null hypothesis that (a) $\sigma^2 = .10$, (b) $\sigma^2 \leq .10$, and (c) $\sigma^2 \geq .25$. Use a size .05 test in each case.

In each case, $n-k = 40-4 = 36$. We use the *equal tails* test for (a), and the GLR tests for (b) and (c). Noting that $\chi^2_{36;.975} = 21.335$, $\chi^2_{36;.025} = 54.437$, $\chi^2_{36;.05} = 50.998$, and $\chi^2_{36;.95} = 23.268$, then

**(a)** $(n-k)\hat{s}^2/d = 36(.13)/.10 = 46.8 \notin [0, 21.335] \cup [54.437, \infty) \Rightarrow$ do not reject $H_0$;
**(b)** $(n-k)\hat{s}^2/d = 36(.13)/.10 = 46.8 \notin [50.998, \infty) \Rightarrow$ do not reject $H_0$;
**(c)** $(n-k)\hat{s}^2/d = 36(.13)/.25 = 18.72 \in [0, 23.268] \Rightarrow$ reject $H_0$.

If one were to investigate the power function of the test used in (a) and (b), one would gain perspective on the danger of accepting the null hypothesis in the *literal sense* in these cases. □

### 10.6.2 Tests in Semiparametric Models

In order to test hypotheses about the parameters of the GLM in a semiparametric context where the family of probability distributions underlying $\boldsymbol{\varepsilon}$ is not known, we resort to asymptotic distributions of test statistics to establish asymptotically valid size $\alpha$ consistent tests. We assume that the classical GLM assumptions apply and that the $\boldsymbol{\varepsilon}$ is distributed according to some unknown PDF in such a way that $n^{1/2}\left(\hat{\boldsymbol{\beta}} - \boldsymbol{\beta}\right) \xrightarrow{d} N(\mathbf{0}, \sigma^2\mathbf{Q}^{-1})$ and $n^{1/2}\left(\hat{S}^2 - \sigma^2\right) \xrightarrow{d} N(0, \tau)$ (e.g., see Table 8.1 for some such conditions). Asymptotic tests can be devised for more general cases where $n^{-1}\,\mathbf{x}'\mathbf{x} \not\to \mathbf{Q}$ and for cases where the limiting distribution result for the variance estimator does not hold if one resorts to more advanced central limit theorems. An excellent source for beginning one's reading about these extensions is H. White, (1984), *Asymptotic Theory for Econometricians*, Academic Press, Orlando.

#### 10.6.2.1 *Testing* $\mathbf{R}\boldsymbol{\beta} = r$ *or* $\mathbf{R}(\boldsymbol{\beta}) = r$ *When* $\mathbf{R}$ *Has q Rows: Asymptotic* $\chi^2$ *Tests*

We first reexamine the use of the Wald statistic for testing the linear restrictions $H_0$: $\mathbf{R}\boldsymbol{\beta} = \mathbf{r}$ versus $H_a$: $\mathbf{R}\boldsymbol{\beta} \neq \mathbf{r}$ where $\mathbf{R}$ is $(q\times k)$ with full row rank. Assume that $n^{1/2}\left(\hat{\boldsymbol{\beta}} - \boldsymbol{\beta}\right) \xrightarrow{d} N(\mathbf{0}, \sigma^2\mathbf{Q}^{-1})$, and that a consistent estimator of $\sigma^2\mathbf{Q}^{-1}$ is given by $\hat{S}^2\left(n^{-1}\mathbf{x}'\mathbf{x}\right)^{-1}$. It follows directly from Theorem 10.9 that the Wald test of $H_0$: $\mathbf{R}\boldsymbol{\beta} = \mathbf{r}$ versus $H_a$: $\mathbf{R}\boldsymbol{\beta} \neq \mathbf{r}$ defined by

$$w = (\mathbf{Rb} - \mathbf{r})'\left[\hat{s}^2\mathbf{R}(\mathbf{x}'\mathbf{x})^{-1}\mathbf{R}\right]^{-1}(\mathbf{Rb} - \mathbf{r})\left\{\begin{array}{c}\geq\\<\end{array}\right\}\chi^2_{q;\alpha} \Rightarrow \left\{\begin{array}{c}\text{reject } H_0\\ \text{do not reject } H_0\end{array}\right\}$$

is an asymptotically valid size $\alpha$, asymptotically unbiased, and consistent test. It is thus seen that under assumptions that ensure the asymptotic normality of the estimator $\hat{\boldsymbol{\beta}}$ and the consistency of $\hat{S}^2$, the Wald test of $H_0$: $\mathbf{R}\boldsymbol{\beta} = \mathbf{r}$ versus $H_a$: $\mathbf{R}\boldsymbol{\beta} \neq \mathbf{r}$ can be applied when the family of distributions for $\boldsymbol{\varepsilon}$ is unknown.

A minor extension of the preceding argument allows the conclusion that the preceding test rule is appropriate for testing nonlinear hypotheses $H_0$: $\mathbf{R}(\boldsymbol{\beta}) = \mathbf{r}$ versus $H_a$: $\mathbf{R}(\boldsymbol{\beta}) \neq \mathbf{r}$. In particular, assuming that $\mathbf{R}(\boldsymbol{\beta})$ is continuously differentiable, contains no redundant constraints, and $\partial\mathbf{R}(\boldsymbol{\beta}_0)/\partial\boldsymbol{\beta}'$ has full row rank, it follows directly from Theorem 10.9 that an asymptotic size $\alpha$, asymptotically unbiased and consistent test of $H_0$: $\mathbf{R}(\boldsymbol{\beta}) = \mathbf{r}$ versus $H_a$: $\mathbf{R}(\boldsymbol{\beta}) \neq \mathbf{r}$ is defined by

$$w = (\mathbf{R}(\mathbf{b}) - \mathbf{r})' \left( \hat{s}^2 \frac{\partial \mathbf{R}(\mathbf{b})'}{\partial \boldsymbol{\beta}} (\mathbf{x}'\mathbf{x})^{-1} \frac{\partial \mathbf{R}(\mathbf{b})}{\partial \boldsymbol{\beta}} \right)^{-1} (\mathbf{R}(\mathbf{b}) - \mathbf{r}) \left\{ \begin{array}{c} \geq \\ < \end{array} \right\} \chi^2_{q;\alpha}$$

$$\Rightarrow \left\{ \begin{array}{c} \text{reject } H_0 \\ \text{do not reject } H_0 \end{array} \right\}.$$

The asymptotically valid power function of the Wald test under sequences of local alternatives can be defined in terms of the noncentral $\chi^2$ distribution based on Theorem 10.10. We provide an example of this Wald test later in Example 10.19.

*10.6.2.2 Testing $R(\boldsymbol{\beta}) = r$, $R(\boldsymbol{\beta}) \leq r$, or $R(\boldsymbol{\beta}) \geq r$ When R Is a Scalar Function: Asymptotic Normal Tests*

In the case where $R(\boldsymbol{\beta})$ is a scalar function of the vector $\boldsymbol{\beta}$, the Wald statistic of the preceding subsection specializes to

$$W = \left(R\left(\hat{\boldsymbol{\beta}}\right) - r\right)^2 / \left[ \hat{S}^2 \frac{\partial R\left(\hat{\boldsymbol{\beta}}\right)'}{\partial \boldsymbol{\beta}} (\mathbf{x}'\mathbf{x})^{-1} \frac{\partial R\left(\hat{\boldsymbol{\beta}}\right)}{\partial \boldsymbol{\beta}} \right] \xrightarrow{d} \chi^2_1 \text{ under } H_0.$$

An alternative test statistic whose square is $W$ is given by

$$Z = \left(R\left(\hat{\boldsymbol{\beta}}\right) - r\right) / \left[ \hat{S}^2 \frac{\partial R\left(\hat{\boldsymbol{\beta}}\right)'}{\partial \boldsymbol{\beta}} (\mathbf{x}'\mathbf{x})^{-1} \frac{\partial R\left(\hat{\boldsymbol{\beta}}\right)}{\partial \boldsymbol{\beta}} \right]^{1/2} \xrightarrow{d} N(0,1) \text{ under } H_0.$$

The latter limiting distribution can be established via an application of Theorem 5.39 applied to the scalar function, $R(\hat{\boldsymbol{\beta}})$, of the asymptotically normally distributed random vector $\hat{\boldsymbol{\beta}}$ and through Slutsky's theorem. An asymptotic size $\alpha$, unbiased and consistent test of $H_0$: $R(\boldsymbol{\beta}) = r$ versus $H_a$: $R(\boldsymbol{\beta}) \neq r$ in terms of a rejection region for the $Z$-statistic is given by

$$z \left\{ \begin{array}{c} \in \\ \notin \end{array} \right\} (-\infty, -z_{\alpha/2}] \cup [z_{\alpha/2}, \infty) \Rightarrow \left\{ \begin{array}{c} \text{reject } H_0 \\ \text{do not reject } H_0 \end{array} \right\}.$$

This and other asymptotic size $\alpha$, unbiased, and consistent tests based on the $Z$-statistic are summarized in the table ahead (Table 10.2).

**Table 10.2** Asymptotic level $\alpha$ tests on $R(\boldsymbol{\beta})$ using a $Z$-statistic

| Case | $H_0$ | $H_a$ | $C_r$ |
|---|---|---|---|
| 1. | $R(\boldsymbol{\beta}) \leq r$ | $R(\boldsymbol{\beta}) > r$ | $[z_\alpha, \infty)$ |
| 2. | $R(\boldsymbol{\beta}) \geq r$ | $R(\boldsymbol{\beta}) < r$ | $(-\infty, -z_\alpha]$ |
| 3. | $R(\boldsymbol{\beta}) = r$ | $R(\boldsymbol{\beta}) \neq r$ | $(-\infty, -z_{\alpha/2}] \cup [z_{\alpha/2}, \infty)$ |

In order to test the *linear* hypothesis $H_0$: $\mathbf{R\beta} = r$ versus $H_a$: $\mathbf{R\beta} \neq r$, the test is performed with $\mathbf{R}$ replacing $\partial\mathbf{R}(\hat{\boldsymbol{\beta}})/\partial\boldsymbol{\beta}'$ and $\mathbf{R}\hat{\boldsymbol{\beta}}$ replacing $R(\hat{\boldsymbol{\beta}})$ in the definition of the $z$-statistic, and thus the linear case is subsumed.

Note that in practice, the $t$-distribution is sometimes used in place of the standard normal distribution to define the rejection regions of the aforementioned tests. In the limit, there is no difference between the two procedures, since $t_v \to N(0,1)$ as $v \to \infty$. However, in small samples, the $t$-distribution has fatter tails, so that $t_\alpha > z_\alpha$ and thus the *actual size* of the aforementioned tests is smaller when based on $t_\alpha$ and $t_{\alpha/2}$ in place of $z_\alpha$ and $z_{\alpha/2}$, respectively. The use of the $t$-distribution in this context can be viewed as a conservative policy towards rejection of $H_0$ in the sense that the probability of Type I Error is less, and stronger data evidence will be required for rejection of a null hypothesis. On the other hand, the *actual power* of the test is also reduced when the $t$-distribution is utilized.

Asymptotically valid power calculations for local alternatives when one of the $z$-tests is used can be based on the standard normal distribution. In particular, for local alternatives of the form $R(\boldsymbol{\beta}) - r = n^{-1/2}\phi$, it follows that

$$\left(R\left(\hat{\boldsymbol{\beta}}\right) - r - n^{-1/2}\phi\right) / \left[\hat{S}^2 \frac{\partial R\left(\hat{\boldsymbol{\beta}}\right)'}{\partial \boldsymbol{\beta}} (\mathbf{x}'\mathbf{x})^{-1} \frac{\partial R\left(\hat{\boldsymbol{\beta}}\right)}{\partial \boldsymbol{\beta}}\right]^{1/2} \xrightarrow{d} N(0,1).$$

Then assuming that $n^{-1}\mathbf{x}'\mathbf{x} \to \mathbf{Q}$, Slutsky's theorems can be used to show that

$$Z \xrightarrow{d} N\left(\phi / \left[\sigma^2 \frac{\partial R(\boldsymbol{\beta})'}{\partial \boldsymbol{\beta}} \mathbf{Q}^{-1} \frac{\partial R(\boldsymbol{\beta})}{\partial \boldsymbol{\beta}}\right]^{1/2}, 1\right) = N\left(\frac{\phi}{\xi}, 1\right),$$

where

$$\xi = \left[\sigma^2 \frac{\partial R(\boldsymbol{\beta})'}{\partial \boldsymbol{\beta}} \mathbf{Q}^{-1} \frac{\partial R(\boldsymbol{\beta})}{\partial \boldsymbol{\beta}}\right]^{1/2},$$

so that the asymptotic power function of the $z$-test is represented as

$$\pi_{C_r}(\phi) = \int_{x \in C_r} N\left(z; \frac{\phi}{\xi}, 1\right) dz.$$

Note that the Bonferroni approach to hypothesis testing can be applied as before using either Wald statistics or $Z$-statistics except that now the bound on the level of Type I Error is only asymptotically valid.

**Example 10.19**
***Tests of β Parameters in a Semiparametric GLM***

A quadratic production function representing output in terms of two variable inputs is specified as

$$y_t = \beta_1 x_{1t} + \beta_2 x_{2t} + [x_{1t}\ x_{2t}] \begin{bmatrix} \beta_3 & \beta_5 \\ \beta_5 & \beta_4 \end{bmatrix} \begin{bmatrix} x_{1t} \\ x_{2t} \end{bmatrix} + \varepsilon_t$$

for $t = 1,2,\ldots,n$, where $y_t$ is output in period $t$, $x_{it}$ is quantity of input $i$ used in period $t$, and the $\varepsilon_t$'s are *iid* with $E(\varepsilon_t) = 0$ and $E(\varepsilon_t^2) = \sigma^2$, $\forall t$. Representing the production relationship in standard GLM form, we have

$$y_t = \begin{bmatrix} x_{1t} & x_{2t} & x_{1t}^2 & x_{2t}^2 & 2x_{1t}x_{2t} \end{bmatrix} \begin{bmatrix} \beta_1 \\ \beta_2 \\ \beta_3 \\ \beta_4 \\ \beta_5 \end{bmatrix} + \varepsilon_t, \text{ for t} = 1,2,\ldots,n; \text{ or}$$

$\mathbf{Y} = \mathbf{x}_*\boldsymbol{\beta} + \boldsymbol{\varepsilon}$ in compact matrix form.

Assuming that observations on $\mathbf{x}_*$produces a matrix with full column rank, the classical GLM assumptions hold. Assume further that $n^{-1}\mathbf{x}_*'\mathbf{x}_* \rightarrow \mathbf{Q}$, so that $n^{1/2}(\hat{\boldsymbol{\beta}} - \boldsymbol{\beta}) \xrightarrow{d} N(\mathbf{0}, \sigma^2\mathbf{Q}^{-1})$, and note that $\hat{S}^2(n^{-1}\mathbf{x}'_*\mathbf{x}_*)^{-1}$ is a consistent estimator of $\sigma^2\mathbf{Q}^{-1}$.

Suppose 30 observations are used to generate the following least squares-based estimates:

$$\mathbf{b} = [7.1 \;\; 5.2 - .11 - .07 - .02]', \quad \hat{s}^2 = .6 \times 10^{-3},$$

$$(\mathbf{x}'_*\mathbf{x}_*)^{-1} = \begin{bmatrix} 9.571 & 6.463 & 7.187 & 4.400 & 8.247 \\ & 9.686 & 4.124 & 7.364 & 8.800 \\ & & 5.766 & 2.814 & 6.087 \\ & & & 5.945 & 6.914 \\ \text{(symmetric)} & & & & 11.254 \end{bmatrix}$$

We will test the following hypotheses relative to expected output:

**(a)** $H_0$ : the production function is concave. Use a level .10 test.
**(b)** $H_0$ : the marginal products of inputs $x_1$ and $x_2$ are identical at equal input levels. Use a size .05 test.

Regarding hypothesis (a), note that a necessary and sufficient condition for the production function to be concave is that $\beta_3 \leq 0$ and $\beta_3\,\beta_4 - \beta_5^2 \geq 0$. We will test each hypothesis using a $Z$ test of size .05, so that based on the Bonferroni approach, the joint test will have size $\leq .10$ and thus level .10. For $H_0$: $\beta_3 \leq 0$, the Z-statistic outcome is

$$z = \frac{\mathbf{Rb} - r}{\left[\hat{s}^2\mathbf{R}(\mathbf{x}'\mathbf{x})^{-1}\mathbf{R}'\right]^{1/2}} = \frac{-.11 - 0}{.0588} = -1.8707,$$

where $\mathbf{R} = [0\,0\,1\,0\,0]$ and $r = 0$. The rejection region for the test is $C_r = [z_{.05}, \infty)$ $= [1.645, \infty)$, and since $w \notin C_r$, $H_0$ is not rejected. For $H_0$: $\beta_3\beta_4 - \beta_5^2 \geq 0$, first note that $\partial R(\boldsymbol{\beta})/\partial\boldsymbol{\beta} = \partial(\beta_3\,\beta_4 - \beta_5^2)/\partial\boldsymbol{\beta} = [0 \quad 0 \quad \beta_4 \quad \beta_3 \quad -2\beta_5]'$. Then

$\hat{s}^2\left(\partial R\left(\hat{\boldsymbol{\beta}}\right)/\partial\boldsymbol{\beta}'\right)(\mathbf{x}_*'\mathbf{x}_*)^{-1}\left(\partial R\left(\hat{\boldsymbol{\beta}}\right)/\partial\boldsymbol{\beta}\right) = .3996\times 10^{-4}$, so that the $Z$-statistic outcome is

$$z = \frac{.0073 - 0}{(.3996\times 10^{-4})^{1/2}} = 1.1548.$$

Because the rejection region is $C_r = [-\infty, -z_{.05}] = (-\infty, -1.645)$ and $w \notin C_r$, $H_0$ is not rejected. Overall, the joint null hypothesis of concavity of the production function is not rejected at the .10 level.

Regarding hypothesis (b), note that:

$$\frac{\partial E(Y_t)}{\partial x_{t1}} = \beta_1 + 2\beta_3 x_{t1} + 2\beta_5 x_{t2},$$

$$\frac{\partial E(Y_t)}{\partial x_{t2}} = \beta_2 + 2\beta_5 x_{t1} + 2\beta_4 x_{t2}.$$

The marginal products will be identical at equal input levels *iff*

$$\frac{\partial E(Y_t)}{\partial x_{t1}} - \frac{\partial E(Y_t)}{\partial x_{t2}} = \beta_1 - \beta_2 + 2(\beta_3 - \beta_4)\xi = 0 \;\; \forall \xi = x_{t1} = x_{t2}.$$

Thus we must test $H_0$: $\beta_1 - \beta_2 = 0$ and $\beta_3 - \beta_4 = 0$ versus $H_a$: not $H_0$. The Wald statistic appropriate for this problem has the form

$$w = (\mathbf{Rb} - \mathbf{r})'\left[\hat{s}^2\mathbf{R}(\mathbf{x}_*'\mathbf{x}_*)^{-1}\mathbf{R}'\right]^{-1}(\mathbf{Rb} - \mathbf{r}).$$

where

$$\mathbf{Rb} = \begin{bmatrix} b_1 - b_2 \\ b_3 - b_4 \end{bmatrix}, \quad \mathbf{r} = \begin{bmatrix} 0 \\ 0 \end{bmatrix}, \quad \text{and } \mathbf{R} = \begin{bmatrix} 1 & -1 & 0 & 0 & 0 \\ 0 & 0 & 1 & -1 & 0 \end{bmatrix}.$$

The Wald statistic outcome is 17,390.828, and because the rejection region of the test is $C_r = \left[\chi^2_{2;.05}, \infty\right) = [5.991, \infty)$ and $17{,}390.828 \in C_r$, $H_0$ is rejected at the .05 level. □

*10.6.2.3 Tests for $\sigma^2$: Asymptotic $\chi^2$ and Normal Tests*

In order to test hypotheses concerning the magnitude of $\sigma^2$, consider using a Wald statistic. Assuming $n^{1/2}\left(\hat{S}^2 - \sigma^2\right) \xrightarrow{d} N(0, (\mu'_4 - \sigma^4))$(e.g., see Table 8.1), it follows directly from Theorem 10.9 that a consistent, asymptotically unbiased, and asymptotic size $\alpha$ Wald test of $H_0$: $\sigma^2 = d$ versus $H_a$: $\sigma^2 \neq d$ is given by

$$w = \frac{n\left(\hat{s}^2 - d\right)^2}{\hat{\xi}} \begin{Bmatrix} \in \\ \notin \end{Bmatrix} \left[\chi^2_{1;\alpha/2}, \infty\right) \Rightarrow \begin{Bmatrix} \text{reject } H_0 \\ \text{do not reject } H_0 \end{Bmatrix}$$

**Table 10.3** Asymptotic level $\alpha$ tests on $\sigma^2$ using a $Z$-statistic

| Case | $H_0$ | $H_a$ | $C_r$ |
|---|---|---|---|
| 1. | $\sigma^2 \leq d$ | $\sigma^2 > d$ | $[z_\alpha, \infty)$ |
| 2. | $\sigma^2 \geq d$ | $\sigma^2 < d$ | $(-\infty, -z_\alpha]$ |
| 3. | $\sigma^2 = d$ | $\sigma^2 \neq d$ | $(-\infty, -z_{\alpha/2}] \cup [z_{\alpha/2}, \infty)$ |

where $\hat{\xi}$ is a consistent estimate of $\text{var}(\varepsilon_i^2) = \mu'_4 - \sigma^4$, such as $\hat{\xi} = n^{-1}\sum_{i=1}^n \hat{e}_i^4 - \hat{s}^4$, with $\hat{\mathbf{e}} = \mathbf{y} - \mathbf{xb}$.

This and other asymptotic size $\alpha$, unbiased, and consistent tests for hypotheses about $\sigma^2$based on the use of the asymptotically valid $Z$-statistic, $z = n^{1/2}\left(\hat{s}^2 - d\right)/\hat{\xi}^{1/2}$, are summarized in Table 10.3.

The reader will be asked to justify the asymptotic size and consistency of these tests in the problem section.

**Example 10.20**
***Tests of $\sigma^2$ Parameter in a Semiparametric GLM***

Recall Example 10.19, and test the null hypothesis $H_0$: $\sigma^2 \leq .25 \times 10^{-3}$ versus $H_a$: $\sigma^2 > .25 \times 10^{-3}$ at the .05 level.

Assuming the appropriate conditions for the asymptotic normality of $\hat{S}^2$ and consistency of $n^{-1}\sum_{i=1}^n \hat{e}_i^4 - \hat{S}^4$ for $\mu'_4 - \sigma^4$, suppose $n^{-1}\sum_{i=1}^n \hat{e}_i^4 = .87 \times 10^{-6}$, so that the $z$-statistic equals

$$z = \frac{n^{1/2}\left(\hat{s}^2 - d\right)}{\hat{\xi}^{1/2}} = \frac{(40)^{1/2}(.6 \times 10^{-3} - .25 \times 10^{-3})}{\left(.51 \times 10^{-6}\right)^{1/2}} = 3.0997.$$

The rejection region of the test is $C_r = [z_{.05},\infty) = [1.645, \infty)$, and because $3.0997 \in C_r$, $H_0$ is rejected. □

Note that in the preceding three subsections we have made no explicit assumptions about the functional forms of the PDFs associated with the residual terms of the GLM and proceeded in a semiparametric context, relying on asymptotic properties of estimators to design statistical tests. If an explicit parametric functional form for the PDF of the residual term is assumed (other than the normal distribution previously assumed), it might be possible to use the GLR and/or LM hypothesis testing procedures, in addition to the Wald methodology, to provide alternative test statistics and statistical tests for the preceding hypotheses. Such applications are generally a case by case affair, but the general approaches presented in this and the previous chapter for defining tests and tests statistics can serve as a guide for how to pursue such definitions.

## 10.7 Confidence Intervals and Regions

In this section we provide an introduction to the concept of *confidence intervals* and *confidence regions.* We provide motivation both from the perspective of the **duality** between hypothesis testing and confidence region or interval estimation, and from the concept of **pivotal quantities**. The study of confidence region/interval estimation encompasses its own theory and practice that would require considerably more space if treated at a general level. In fact there is an entire body of theory relating to the properties of confidence regions/intervals and methods of estimating optimal confidence regions/intervals that parallels the theories developed for hypothesis testing. Moreover, there is a full duality between certain optimality properties of hypothesis tests and their counterpart for confidence region or interval estimation. We will note some of these parallels in the presentation ahead, but we will not examine them in detail. Readers interested in more depth can begin their readings by examining the work of M. Kendall and A. Stuart, (1979), *Advanced Statistics*, Chapter 20, and E. Lehmann, (1986), *Testing Statistical Hypotheses*, Chapter 5.

In the case of a single scalar parameter ,$\Theta$, or a scalar function of parameters, R($\Theta$), a **confidence interval** (CI) is a *random interval* whose outcomes have a known probability of containing the true value of the parameter or function of parameters. In practice, a confidence interval will be defined by random variables that represent the upper and lower bounds of the interval, and thus outcomes of the random variables defining the bounds also define outcomes of the random interval. It is possible for one of the random variables defining the bounds of the confidence interval to be replaced by a constant, in which case the confidence interval is a **one-sided confidence interval** or a **confidence bound**. We provide a formal definition of what is meant by a confidence interval and confidence bound below.

**Definition 10.2**
***Confidence Intervals and Confidence Bounds with Confidence Level*** $\gamma$ ***and Confidence Coefficient*** $1-\alpha$

Let $f(\mathbf{x};\Theta)$ be the joint density of the probability sample $\mathbf{X}$ for some $\Theta\in\Omega$, and let $R(\Theta)$ be some scalar function of the parameter vector $\Theta$. The following are confidence intervals or confidence bounds for $R(\Theta)$ that have **confidence level** $\gamma$ and **confidence coefficient** $1-\alpha$.[7]

**(a)** *Two-Sided CI:* $(\ell(\mathbf{X}), u(\mathbf{X}))$ such that $P(\ell(\mathbf{x}) < R(\Theta) < u(\mathbf{x});\Theta) \geq \gamma,\ \forall\Theta\in\Omega$ and $\inf_{\Theta\in\Omega}\{P(\ell(\mathbf{x})<R(\Theta)<u(\mathbf{x});\Theta)\} = 1-\alpha$.

**(b)** *One-Sided CI or Lower Confidence Bound:* $(\ell(\mathbf{X}), \infty)$ such that $P(\ell(\mathbf{x}) < R(\Theta);\Theta) \geq \gamma,\ \forall\Theta\in\Omega$ and $\inf_{\Theta\in\Omega}\{P(\ell(\mathbf{x})<R(\Theta);\Theta)\} = 1-\alpha$.

**(c)** *One-Sided CI or Upper Confidence Bound:* $(-\infty, u(\mathbf{X}))$ such that $P(R(\Theta) < \mu(\mathbf{x});\Theta) \geq \gamma,\ \forall\Theta\in\Omega$ and $\inf_{\Theta\in\Omega}\{P(R(\Theta)<\mu(\mathbf{x});\Theta)\} = 1-\alpha$.

The random variables $\ell(\mathbf{X})$ and $u(\mathbf{X})$ are called the **lower and upper confidence limits**, respectively.

[7]Recall that inf denotes infimum, which is the largest lower bound to the set of values under consideration. The infimum is the minimum if the minimum is contained in the set of values.

Note that the confidence level, $\gamma$, is simply some stated lower bound on the probability of the event that the outcome of the CI will contain the true value of $R(\boldsymbol{\Theta})$. The confidence coefficient, $1-\alpha$, is effectively the minimum probability of the event that the CI will contain the true value of $R(\boldsymbol{\Theta})$. Note the reason we express the confidence coefficient in terms of the expression $1-\alpha$ as opposed to referring to it as a standalone letter or symbol has to do with the duality between confidence intervals and hypothesis tests, where we will see that there is a direct connection between the size of a hypothesis test, $\alpha$, and the confidence coefficient, $1-\alpha$. Note that because $\gamma \leq 1-\alpha$, a CI with confidence coefficient $1-\alpha$ is also a CI with level $\gamma$, although in using the terminology, the confidence level and the confidence coefficient do not need to be equal.

From the definition, it is seen that the random variables representing the lower and/or upper confidence limits are chosen so that they have outcomes defining an interval containing the value of $R(\boldsymbol{\Theta})$ with probability bounded as $\geq \gamma$ (a stated level) whatever the value of $\boldsymbol{\Theta}$, i.e. $\forall \boldsymbol{\Theta} \in \Omega$. Note that for any given outcome of a confidence interval, either $R(\boldsymbol{\Theta})$ *is* or *is not* in the interval, and so in practice one does not know whether $R(\boldsymbol{\Theta})$ is really contained in a *given* confidence interval outcome or not. However, since the probability of the event that the confidence interval outcome "covers" or contains $R(\boldsymbol{\Theta})$ is known to be $\geq \gamma$, we do know that in repeated sampling $\geq 100\gamma$ percent of the intervals generated *will* contain the value of $R(\boldsymbol{\Theta})$, on average. *It is in this sense that we have "confidence" that a confidence interval contains the value of* $R(\boldsymbol{\Theta})$.

The concept of a *confidence region* generalizes the concept of confidence intervals to the case where (1) the random set designed to contain a scalar $R(\boldsymbol{\Theta})$ with known probability is not necessarily in the form of an interval, or else (2) $\mathbf{R}(\boldsymbol{\Theta})$ is a *vector* function so that the random set used to contain $\mathbf{R}(\boldsymbol{\Theta})$ with a given probability is inherently not in interval form. The random set defining a confidence region will be designed so the probability that outcomes of the confidence region contain the value of $\mathbf{R}(\boldsymbol{\Theta})$ is known or is at least lower bounded. Note that confidence regions can be interpreted as subsuming confidence intervals as a special case. We provide the formal definition of the concept of a confidence region below.

**Definition 10.3** ***Confidence Region with Confidence Level*** $\gamma$ ***and Confidence Coefficient*** $1-\alpha$

Let $f(\mathbf{x};\boldsymbol{\Theta})$ be the joint density of the probability sample $\mathbf{X}$, and let $\mathbf{R}(\boldsymbol{\Theta})$ be a vector (or scalar) function of the parameter vector $\boldsymbol{\Theta}$. Then a confidence region for $\mathbf{R}(\boldsymbol{\Theta})$ with confidence level $\gamma$ and confidence coefficient $1-\alpha$ is defined by a random set $A(\mathbf{X})$ for which $P(\mathbf{R}(\boldsymbol{\Theta}) \in A(\mathbf{X});\boldsymbol{\Theta}) \geq \gamma$, $\forall \boldsymbol{\Theta} \in \Omega$, and $\inf_{\boldsymbol{\Theta}\in\Omega} \{P(\mathbf{R}(\boldsymbol{\Theta}) \in A(\mathbf{X});\boldsymbol{\Theta})\} = 1-\alpha$.

In practice the random set $A(\mathbf{X})$ will be defined using random variables that appear in the set-defining conditions of the set, and different outcomes of the random set are defined as these random variables assume different outcomes in repeated sampling. For example, one possibility is that $A(\mathbf{X})$ is a random open rectangle defined as $A(\mathbf{X}) = \{\mathbf{R}(\boldsymbol{\Theta}): \ell_i(\mathbf{X}) < R_i(\boldsymbol{\Theta}) < u_i(\mathbf{X}),\ i=1,\ldots,m\}$ so that a given outcome of the confidence region for $\mathbf{R}(\boldsymbol{\Theta})$ would be $A(\mathbf{X}) = \times_{i=1}^{m} (\ell_i(\mathbf{x}), u_i(\mathbf{x}))$. Other examples of confidence sets will be presented subsequently.

Note that confidence intervals or regions can be thought of as an alternative or supplement to point estimation, where instead of, or in addition to only generating a best (in some sense) *point* estimate of the value of some unknown function of the parameters of a probability model, a *set* of values is generated that, a priori, has a given probability of containing the unknown function of the parameters. When one is interested in knowing a probable range of values for the unknown $\mathbf{R}(\boldsymbol{\Theta})$, the confidence interval or region concept has obvious appeal. As further motivation for the use of confidence regions, recall that point estimators generally have a high probability of generating point estimates that are literally wrong, and in fact the probability that a continuous point estimator will generate the true value of $\mathbf{R}(\boldsymbol{\Theta})$ is *zero*. Alternatively, a random confidence interval or region has a high probability of generating an outcome that contain the true $\mathbf{R}(\boldsymbol{\Theta})$ by construction.

### 10.7.1 Defining Confidence Intervals and Regions via Duality with Rejection Regions

How does one go about defining confidence intervals or confidence regions? There is a duality between confidence intervals and regions for $\mathbf{R}(\boldsymbol{\Theta})$ and rejection regions for hypothesis tests about $\mathbf{R}(\boldsymbol{\Theta})$ in the sense that if a rejection region has been defined for the hypothesis testing problem, one has in effect already defined a confidence region for $\mathbf{R}(\boldsymbol{\Theta})$. The specific nature of this duality is presented in the following theorem. Note, when we henceforth refer to confidence regions, the concept of confidence intervals is tacitly assumed to be subsumed within the confidence region concept.

**Theorem 10.11** ***Duality Between Confidence and Rejection Regions***

*Let the probability sample* $\mathbf{X}$ *have joint density function* $f(\mathbf{x};\boldsymbol{\Theta})$ *for some* $\boldsymbol{\Theta}\in\Omega$, *and let* $C_r(\boldsymbol{\tau})$ *be a level (or size)* $\alpha$ *rejection region for testing* $H_0$: $\mathbf{R}(\boldsymbol{\Theta}) = \boldsymbol{\tau}$ *versus one of the alternatives (1)* $H_a$: $\mathbf{R}(\boldsymbol{\Theta}) \neq \boldsymbol{\tau}$, *(2)* $H_a$: $\mathbf{R}(\boldsymbol{\Theta}) < \boldsymbol{\tau}$, or *(3)* $H_a$: $\mathbf{R}(\boldsymbol{\Theta}) > \boldsymbol{\tau}$, *for* $\boldsymbol{\tau}\in \Omega_{\mathbf{R}(\Theta)} = \{\boldsymbol{\tau}: \boldsymbol{\tau} = \mathbf{R}(\boldsymbol{\Theta}), \boldsymbol{\Theta}\in\Omega\}$. *Then the random set represented by* $A(\mathbf{X}) = \{\boldsymbol{\tau}: \mathbf{X}\in\bar{C}_r(\boldsymbol{\tau}), \boldsymbol{\tau}\in\Omega_{\mathbf{R}(\Theta)}$ *is a confidence region for* $\mathbf{R}(\boldsymbol{\Theta})$ *having confidence level (or coefficient)* $1-\alpha$.

**Proof** Given the definitions of $C_r(\boldsymbol{\tau})$ and $A(\mathbf{X})$, $\mathbf{x}\in\bar{C}_r(\boldsymbol{\tau}) \Leftrightarrow \boldsymbol{\tau}\in A(\mathbf{x})$. To motivate this relationship, first note that if $\mathbf{x}\in\bar{C}_r(\tau)$, then hypothesis $\mathbf{R}(\boldsymbol{\Theta}) = \boldsymbol{\tau}$ would not be rejected by the test defined by $C_r(\boldsymbol{\tau})$. It follows that $\boldsymbol{\tau}$ is in the set of null hypotheses, $A(\mathbf{X})$, that would not be rejected on the basis of the sample outcome $\mathbf{x}$, so that $\mathbf{x}\in\bar{C}_r(\boldsymbol{\tau}) \Rightarrow \boldsymbol{\tau}\in A(\mathbf{X})$. Alternatively, if $\boldsymbol{\tau}\in A(\mathbf{X})$, then $\mathbf{x}$ is a sample outcome that results in $H_0$: $\mathbf{R}(\boldsymbol{\Theta}) = \boldsymbol{\tau}$ not being rejected on the basis of $C_r(\boldsymbol{\tau})$, so that $\boldsymbol{\tau}\in A(\mathbf{X}) \Rightarrow \mathbf{x}\in\bar{C}_r(\boldsymbol{\tau})$, which completes the motivation for the relationship $\mathbf{x}\in\bar{C}_r(\boldsymbol{\tau}) \Leftrightarrow \boldsymbol{\tau}\in A(\mathbf{x})$.

From the equivalence of the two sets $\{\mathbf{x}: \mathbf{x}\in\bar{C}_r(\boldsymbol{\tau})\}$ and $\{\mathbf{x}: \boldsymbol{\tau}\in A(\mathbf{x})\}$, it follows that

$$P(\boldsymbol{\tau} \in A(\mathbf{x})) = P(\mathbf{x} \in \bar{C}_r(\boldsymbol{\tau})) \geq 1 - \alpha,$$

where the inequality holds because $C_r(\boldsymbol{\tau})$ defines a level $\alpha$ test, so that a true $\boldsymbol{\Theta}$ would be rejected with $\leq \alpha$ probability. Thus, $A(\mathbf{X})$ is a confidence region for $\boldsymbol{\tau} = \mathbf{R}(\boldsymbol{\Theta})$ with confidence level $1-\alpha$. If the size of the test defined by $C_r(\boldsymbol{\tau})$ is $\alpha$ so that the maximum probability of rejecting a true $\boldsymbol{\Theta}$ equals $\alpha$, then it follows that the smallest value of $P(\boldsymbol{\tau}\in A(\mathbf{X}))$ is $1-\alpha$, which equals the confidence coefficient. ■

Theorem 10.11 states that to generate an outcome of a confidence region for $\mathbf{R}(\boldsymbol{\Theta})$, one can begin with a statistical test for the null hypothesis $H_0$: $\mathbf{R}(\boldsymbol{\Theta}) = \boldsymbol{\tau}$ against a one or two-sided alternative alternative. Then a confidence region outcome is the collection of all possible values of $\boldsymbol{\tau}$ (i.e., values in $\Omega_{\mathbf{R}(\boldsymbol{\Theta})}$) that represent null hypotheses that are *not rejected* on the basis of the testing procedure applied to the sample outcome, **x**. Some examples presented ahead will clarify the mechanics of the procedure. Note, as will be seen, that one-sided and two-sided tests can be used to define one-sided and two-sided confidence intervals, respectively.

**Example 10.21** ***Confidence Interval for $\mu$ in a Normal Distribution via Duality***

Let **X** be a random sample of size 50 from a $N(\mu,\sigma^2)$ population distribution representing observations on fill levels of 16 oz. bottles of a certain brand of liquid detergent. A sample outcome resulted in $\bar{x} = 16.02$ and $s^2 = .0001$. We will calculate the outcome of a two-sided confidence interval having confidence coefficient .95 for the mean fill level of the bottles.

From Definition 9.15, we know that a UMPU level and size .05 test of $H_0$: $\mu{=}\mu_0$ versus $H_a$: $\mu{\neq}\mu_0$ is given by

$$\frac{\bar{x}-\mu_0}{\left(s^2/(n-1)\right)^{1/2}}\left\{\begin{matrix}\in\\ \notin\end{matrix}\right\}(-\infty,-t_{.025}]\cup[t_{.025},\infty)\Rightarrow\left\{\begin{matrix}\text{reject } H_0\\ \text{do not reject } H_0\end{matrix}\right\}$$

where $t_{.025}$ refers to the upper tail of the $t$-distribution with $n$-1 degrees of freedom. Then on the basis of a given outcome **x** for the random sample, the collection of null hypotheses, $\mu_0$, that would not be rejected on the basis of this test procedure is given by

$$A(\mathbf{X}) = \{\mu_0 : \bar{x} - t_{.025}(s/\sqrt{n-1}<\mu_0<\bar{x} + t_{.025}(s/\sqrt{n-1})\}.$$

Because $t_{.025} = 2.262$ for a $t$-distribution with 9 degrees of freedom, it follows that the confidence interval outcome is given by (16.013, 16.027). The actual mean fill level is then a number in the interval with confidence level (and coefficient) .95. □

**Example 10.22** ***Confidence Interval for $R\beta$ in the GLM via Duality***

Let $Y_t = \beta_0 \prod_{i=1}^{3} x_{ti}^{\beta_i} \exp(\varepsilon_t)$, $t = 1,\ldots,25$, represent a probability sample of observations on a production function based on three inputs $(x_{t1}, x_{t2}, x_{t3})$, and assume that the classical GLM assumptions apply to the natural logarthmic transformation of the production function,

$$\ln(Y_t) = \beta_0^* + \sum_{i=1}^{3} \beta_i \ln(x_{ti}) + \varepsilon_t.$$

Furthermore, let $\varepsilon_t \sim iid\ N(0,\sigma^2)$. An outcome of the sample resulted in the following estimator outcomes corresponding to the transformed production function:

$$\mathbf{b} = \begin{bmatrix} 2.33 \\ .17 \\ .60 \\ .18 \end{bmatrix} \text{ and } \hat{s}^2(\mathbf{x}_*'\mathbf{x}_*)^{-1} = \begin{bmatrix} .64 & .011 & .0001 & .0008 \\ & .0025 & .23 \times 10^{-6} & .3 \times 10^{-5} \\ & & .0003 & .47 \times 10^{-6} \\ \text{(symmetric)} & & & .0036 \end{bmatrix}.$$

We will calculate a two-sided confidence interval having confidence level .95 for the degree of homogeneity of the production function, $\mathbf{R\beta} = [0\ 1\ 1\ 1]\,\boldsymbol{\beta} = \sum_{i=1}^{3} \beta_i$. We will also calculate a one-sided *upper* confidence bound with confidence level .95 for the output elasticity with respect to input 1, $\mathbf{R\beta} = [0\ 1\ 0\ 0]\,\boldsymbol{\beta} = \beta_1$.

Examine the problem of the two-sided confidence interval. We know from Section 10.6 that a UMPU level and size $\alpha$ test of $H_0$: $\mathbf{R\beta} = \tau$ versus the *two-sided* alternative $H_a$: $\mathbf{R\beta} \neq \tau$ is defined by

$$\frac{\mathbf{Rb} - \tau}{[\hat{s}^2\mathbf{R}(\mathbf{x}_*'\mathbf{x}_*)^{-1}\ \mathbf{R}']^{1/2}} \begin{Bmatrix} \in \\ \notin \end{Bmatrix} (-\infty, -t_{\alpha/2}] \cup [t_{\alpha/2}, \infty) \Rightarrow \begin{Bmatrix} \text{reject } H_0 \\ \text{do not reject } H_0 \end{Bmatrix}.$$

For a given outcome $\mathbf{b}$ and $\hat{s}^2$, the set of null hypothesis values, $\boldsymbol{\tau}$, that would not be rejected on the basis of this test procedure is given by

$$A(\mathbf{y}) = \left\{\tau : \mathbf{Rb} - t_{\alpha/2}\left(\hat{s}^2\mathbf{R}(\mathbf{x}_*'\mathbf{x}_*)^{-1}\mathbf{R}'\right)^{1/2} < \tau < \mathbf{Rb} + t_{\alpha/2}\left(\hat{s}^2\mathbf{R}(\mathbf{x}_*'\mathbf{x}_*)^{-1}\mathbf{R}'\right)^{1/2}\right\}.$$

Since $t_{.025} = 2.08$ for a $t$-distribution with 21 degrees of freedom, $\mathbf{Rb} = .95$, and $\hat{s}^2\ \mathbf{R}(\mathbf{x}_*'\mathbf{x}_*)^{-1}\mathbf{R}' = .00641$, it follows that the two-sided confidence interval for the homogeneity of the production function is $(.78, 1.12)$. We have level of confidence (and confidence coefficient) .95 that the degree of homogeneity is in the interval.

Regarding the one-sided *upper* confidence bound for $\mathbf{R\beta} = \boldsymbol{\beta}_1$, we know from Section 10.6 that a UMPU size $\alpha$ test of $H_0$: $\mathbf{R\beta} = \tau$ versus the (*lower*) one-sided alternative hypothesis $H_a$: $\mathbf{R\beta} < \tau$ is given by

$$\frac{\mathbf{Rb} - \tau}{[\hat{s}^2\mathbf{R}(\mathbf{x}_*'\mathbf{x}_*)^{-1}\ \mathbf{R}']^{1/2}} \begin{Bmatrix} \in \\ \notin \end{Bmatrix} (-\infty, -t_{\alpha}] \Rightarrow \begin{Bmatrix} \text{reject } H_0 \\ \text{do not reject } H_0 \end{Bmatrix}.$$

The set of null hypothesis values, $\tau$, that would not be rejected on the basis of a given sample outcome and this test procedure is given by

$$A(\mathbf{y}) = \{\tau : \tau < \mathbf{Rb} + t_{\alpha}(\hat{s}^2\mathbf{R}(\mathbf{x}_*'\mathbf{x}_*)^{-1}\mathbf{R}')^{1/2}\}$$

Since $t_{.05} = 1.721$ for a $t$-distribution with 21 degrees of freedom, $\mathbf{Rb} = .17$, and $\hat{s}^2\mathbf{R}(\mathbf{x}_*'\mathbf{x}_*)^{-1}\mathbf{R}') = .0025$, it follows that the one-sided upper confidence bound for the output elasticity of input 1 is $(-\infty, .26)$. We have level of confidence (and confidence coefficient) .95 that the elasticity is in the interval. □

**Example 10.23**
***Confidence Interval for $\theta$ in an Exponential Distribution via Duality***

Recall Example 10.6 in which operating lives of computer screens were being analyzed, and a UMPU level and size .05 test of $H_0$: $\theta = 1$ versus $H_a$: $\theta > 1$ on the basis of a random sample of size 10 was found. Retracing the development of the statistical test, it is seen that the UMPU level .05 GLR test of $H_0$: $\theta = \theta_0$ versus $H_a$: $\theta > \theta_0$, for *arbitrary* choice of $\theta_0 > 0$, can be represented in the form $\ln(\lambda(\mathbf{x})) = 10 \ln(\bar{x}/\theta_0) + 10(1 - \bar{x}/\theta_0) \leq \ln(c)$. Under $H_0$: $\theta = \theta_0$, and assuming $X_i$'s ~ *iid* $\theta_0^{-1} \exp(-x_i/\theta_0)$, we have that $\bar{X}$ ~ Gamma$(10,\theta_0/10)$ and then $\bar{X}/\theta_0$ ~ Gamma(10, .1). Note that the value of $\ln(\lambda(\mathbf{x}))$ attains its maximum value of zero when $\bar{x}/\theta_0 = 1$, and $\ln(\lambda(\mathbf{x}))$ strictly decreases for movements of $\bar{x}/\theta_0$ away from 1 in either direction. Since the one-sided nature of the GLR test is such that $c < 1 \Leftrightarrow \bar{x}/\theta_0 > 1$ (recall Example 10.6), the test can be performed in terms of the test statistic $\bar{x}/\theta_0$, and we seek a value of $d$ such that $P(\bar{x}/\theta_0 \geq d) = .05$. Using the Gamma(10, .1) distribution of $\bar{X}/\theta_0$, the value of $d$ is found to be 1.57052 (compare to Example 10.6), and the UMPU level .05 test of $H_0$: $\theta = \theta_0$ versus $H_a$: $\theta > \theta_0$ is

$$\bar{x}/\theta_0 \left\{ \begin{matrix} \geq \\ < \end{matrix} \right\} 1.57052 \Rightarrow \left\{ \begin{matrix} \text{reject } H_0 \\ \text{do not reject } H_0 \end{matrix} \right\}.$$

In order to calculate a one-sided *lower* confidence bound for $\theta$ having confidence level .95, note that the set of null hypothesis values, $\theta_0$, not rejected by the preceding UMPU level .05 testing procedure for a given sample outcome is

$$A(\mathbf{x}) = \left\{ \theta : \theta > \frac{\bar{x}}{1.57052} \right\}.$$

Suppose that $\bar{x} = 1.37$. Then an outcome for the one-sided lower confidence bound is $(.87232, \infty)$. We have level of confidence (and confidence coefficient) .95 that the true mean operating life of the computer screens is in the interval. □

Confidence intervals or regions can be based on hypothesis testing procedures that are only asymptotically valid, in which case the confidence regions inherit asymptotic validity from the duality result of Theorem 10.11.[8] The use of asymptotic procedures can be especially convenient for simplifying cases where the joint density of the random sample is discrete. Of course, the simplification comes at the price of the confidence region's confidence level being only an approximation to the true confidence level.

**Example 10.24**
***Asymptotic Confidence Interval for p of Bernoulli Distribution via Duality***

A food processor has developed a new fat-free butter substitute and intends to use a random sample of consumers to determine the proportion of U.S. food consumers that prefer the taste of the new product to that of butter. Of 250 consumers who sampled the product, 97 preferred the taste of the new butter substitute. We will calculate a confidence interval outcome with confidence

[8] We are suppressing a technical condition for this inheritance in that convergence of the test statistic's probability distribution to a limiting distribution should be uniform in $\Theta \in \Omega$. This will occur for the typical PDFs used in practice. For further details, see C.R. Rao, op. cit., pp. 350–351.

level .90 for the proportion of consumers who prefer the taste of the butter substitute. Given the small sample size *relative to* the size of the food consuming public, we will treat the sample as having occurred with replacement.

Recall that $n^{1/2}(\bar{X}-p)/(p(1-p))^{1/2} \xrightarrow{d} N(0,1)$ by the LLCLT when random sampling from the Bernoulli population distribution. Since the sample size $n = 250$ is large, we use a Wald test based on a $\chi_1^2$ limiting distribution, or equivalently, a $Z$-statistic based on the $N(0,1)$ distribution to construct an asymptotic level .10 consistent test of $H_0$: $p = p_0$ versus $H_a$: $p \neq p_0$. Adopting the latter approach, and realizing that $\bar{X}(1-\bar{X}) \xrightarrow{p} p(1-p)$, an asymptotically valid level .10 consistent test is given by

$$\frac{(250)^{1/2}(\bar{x}-p_0)}{[\bar{x}(1-\bar{x})]^{1/2}} \left\{ \begin{matrix} \in \\ \notin \end{matrix} \right\} (-\infty, -1.645] \cup [1.645, \infty) \Rightarrow \left\{ \begin{matrix} \text{reject } H_0 \\ \text{do not reject } H_0 \end{matrix} \right\}$$

where $\int_{1.645}^{\infty} N(z;0,1)\, dz = .05$.

Now consider using duality to define a confidence interval for $p$ that has an asymptotic confidence level of .90. Based on a given outcome of the random sample, the set of null hypothesis values, $p_0$, that would not be rejected on the basis of the preceding test procedure is given by

$$A(\mathbf{x}) = \left\{ p_0 : \bar{x} - 1.645 \left( \frac{\bar{x}(1-\bar{x})}{250} \right)^{1/2} < p_0 < \bar{x} + 1.645 \left( \frac{\bar{x}(1-\bar{x})}{250} \right)^{1/2} \right\}.$$

Since $\bar{x} = .388$, the confidence interval outcome is (.337, .439). We have confidence at approximately (asymptotic) level (and approximate confidence coefficient) .90 that the true proportion of individuals preferring the butter substitute is in the interval. □

Although in principle a confidence region for two or more functions of parameters may have a myriad of shapes, the typical shape of a confidence region that is derived from duality with a hypothesis test based on the Wald statistic is an ellipse (2 dimensions) or ellipsoid ($\geq$ 3 dimensions). A typical application is in the GLM.

**Example 10.25** ***Ellipsoid Confidence Region in the GLM via Duality***

Revisit the production function problem in Example 10.22. We construct a confidence region having confidence level .95 for the three output elasticities $\beta_1, \beta_2, \beta_3$.

From Section 10.6, we know that a level .05 and consistent test of $H_0 : \mathbf{R}\boldsymbol{\beta} = \boldsymbol{\tau}$ versus $H_a$: $\mathbf{R}\boldsymbol{\beta} \neq \boldsymbol{\tau}$ can be defined in terms of the Wald statistic $W = (\mathbf{R}\hat{\boldsymbol{\beta}}-\boldsymbol{\tau})'(\hat{s}^2\mathbf{R}(\mathbf{x}_*'\mathbf{x}_*)^{-1}\mathbf{R}')^{-1}(\mathbf{R}\hat{\boldsymbol{\beta}}-\boldsymbol{\tau})$ as

$$w \left\{ \begin{matrix} \geq \\ < \end{matrix} \right\} qF_\alpha(q, n-k) \Rightarrow \left\{ \begin{matrix} \text{reject } H_0 \\ \text{do not reject } H_0 \end{matrix} \right\}$$

where in the current application $q = 3$, $n-k = 21$, $\alpha = .05$, $F_{.05}(3, 21) = 3.07$, and

**Table 10.4** Relationships between hypothesis tests and confidence regions

| Hypothesis test property | ⇔ | Confidence region property |
|---|---|---|
| Significance level, $\alpha$ | | Confidence level, $1-\alpha$ |
| Size, $\alpha$ | | Confidence coefficient, $1-\alpha$ |
| Unbiased | | Unbiased |
| UMP (uniformly most powerful) | | UMA (uniformly most accurate) |
| UMPU (uniformly most powerful unbiased) | | UMAU (uniformly most accurate unbiased) |
| Consistent | | Consistent |

$$\mathbf{R} = \begin{bmatrix} 0 & 1 & 0 & 0 \\ 0 & 0 & 1 & 0 \\ 0 & 0 & 0 & 1 \end{bmatrix}.$$

Based on the sample outcome reported in Example 10.22, the set of null hypotheses, $\boldsymbol{\tau}$, not rejected by the preceding test procedure is given by

$$A(\mathbf{y}) = \left\{ \boldsymbol{\tau} : \begin{bmatrix} .17-\tau_1 \\ .60-\tau_2 \\ .18-\tau_3 \end{bmatrix}' \begin{bmatrix} .0025 & .23\times10^{-6} & .3\times10^{-5} \\ & .0003 & .47\times10^{-6} \\ (\text{symmetric}) & & .0036 \end{bmatrix}^{-1} \begin{bmatrix} .17-\tau_1 \\ .60-\tau_2 \\ .18-\tau_3 \end{bmatrix} < 9.21 \right\}.$$

The confidence region is a three-dimensional ellipsoid with center at (.17, .60, .18), and its shape resembles a football. We have confidence at level .95 that the true values of the output elasticities $\beta_1$, $\beta_2$, $\beta_3$ are in the ellipsoid. ■

### 10.7.2 Properties of Confidence Regions

Given the duality between confidence regions and hypothesis tests, one might expect that there is also a duality between properties of hypothesis tests and properties of confidence regions. Such is indeed the case, and we will briefly discuss properties of confidence regions and their relationship with properties of hypothesis tests.

Recall that in our discussion of hypothesis tests we examined the properties of significance level, size, unbiasedness, uniformly most powerful, uniformly most powerful unbiased, and consistency. Each of these has a counterpart with respect to properties of confidence regions (see Table 10.4)

In deriving confidence regions via duality with hypothesis tests, each of the properties possessed by the hypothesis testing procedure is transferred to the corresponding property of the confidence region. We now examine definitions for the latter four confidence region properties.

A confidence region is said to be **unbiased** if the probability that the confidence region contains the *true* $\mathbf{R}(\boldsymbol{\Theta})$ is greater than or equal to the probability that it contains a *false* $\mathbf{R}(\boldsymbol{\Theta})$. The formal definition is as follows, where $P(\mathbf{R}(\boldsymbol{\Theta}) \in A(\mathbf{X});\boldsymbol{\Theta}_*)$ denotes the probability that $\mathbf{R}(\boldsymbol{\Theta}) \in A(\mathbf{X})$ when $\mathbf{X} \sim f(\mathbf{x}; \boldsymbol{\Theta}_*)$.

**Definition 10.4** ***Unbiased Confidence Region***

A confidence region $A(\mathbf{X})$ for $\mathbf{R}(\boldsymbol{\Theta})$ is unbiased *iff* $P(\mathbf{R}(\boldsymbol{\Theta}) \in A(\mathbf{X});\boldsymbol{\Theta}) \geq P(\mathbf{R}(\boldsymbol{\Theta}) \in A(\mathbf{X});\boldsymbol{\Theta}_*)\ \forall \mathbf{R}(\boldsymbol{\Theta}) \neq \mathbf{R}(\boldsymbol{\Theta}_*)$.

Unbiasedness is a reasonable property for a confidence region to possess, since one would certainly desire a confidence region to contain the true value of $\mathbf{R}(\boldsymbol{\Theta})$ more often, or with higher probability, than false values.

A confidence region for $\mathbf{R}(\boldsymbol{\Theta})$ is **uniformly most accurate (UMA)** at confidence level $\gamma$ if it has the lowest probability of containing false values of $\mathbf{R}(\boldsymbol{\Theta})$ relative to any other confidence region for $\mathbf{R}(\boldsymbol{\Theta})$ with confidence level $\gamma$. Formally,

**Definition 10.5** ***Uniformly Most Accurate (UMA) Level $\gamma$ Confidence Region***

A confidence region $A(\mathbf{X})$ for $\mathbf{R}(\boldsymbol{\Theta})$ having confidence level $\gamma$ is uniformly most accurate *iff* $P(\mathbf{R}(\boldsymbol{\Theta}) \in A(\mathbf{X});\ \boldsymbol{\Theta}_*) \leq P(\mathbf{R}(\boldsymbol{\Theta}) \in A_*(\mathbf{X}));\ \boldsymbol{\Theta}_*)\ \forall\ \mathbf{R}(\boldsymbol{\Theta}) \neq \mathbf{R}(\boldsymbol{\Theta}_*)$ and $\forall\ A_*(\mathbf{X})$ having confidence level $\gamma$.

Essentially, a UMA confidence region "filters out" false values of $\mathbf{R}(\boldsymbol{\Theta})$ with higher probability than any other confidence region of like confidence level, which is clearly a desirable property.

We noted in our discussion of hypothesis testing in Section 9.5 that UMP tests, and thus now UMA confidence regions, do not exist with any degree of generality. However, UMPU tests exist much more frequently, and thus so do UMAU confidence regions. A UMAU confidence region is simply a confidence region that exhibits the UMA property when compared *only* to other unbiased confidence regions.

**Definition 10.6** ***Uniformly Most Accurate Unbiased (UMAU) Confidence Regions***

A confidence region $A(\mathbf{X})$ for $R(\boldsymbol{\Theta})$ having confidence level $\gamma$ is uniformly most accurate unbiased *iff* $A(\mathbf{X})$ is UMA within the class of unbiased confidence regions having confidence level $\gamma$.

In comparing confidence *intervals* for $R(\boldsymbol{\Theta})$, it can be shown that a UMAU confidence interval with confidence level $\gamma$ has the **smallest expected length** of any other confidence interval for $R(\boldsymbol{\Theta})$ with confidence level $\gamma$. In terms of narrowing ignorance of the value of $R(\boldsymbol{\Theta})$, this is clearly a desirable property.

Finally, *consistency* of a confidence region sequence means that as the sample size $n \to \infty$, the length or volume of the confidence regions shrinks to zero at any confidence level, and the true $\mathbf{R}(\boldsymbol{\Theta})$ becomes the only remaining point in the confidence region with probability $\to 1$. Formally,

**Definition 10.7** ***Consistent Level $\gamma$ Confidence Region Sequence***

A sequence of level $\gamma$ confidence regions $\{A_n(\mathbf{X})\}$ for $\mathbf{R}(\boldsymbol{\Theta})$ is a consistent confidence region sequence *iff* $\lim_{n\to\infty} (P(\mathbf{R}(\boldsymbol{\Theta}) \in A_n(\mathbf{x}); \boldsymbol{\Theta}_*)) = 0, \forall \mathbf{R}(\boldsymbol{\Theta}) \neq \mathbf{R}(\boldsymbol{\Theta}_*)$.

A consistent confidence region sequence is such that as $n \to \infty$, all false values of $\mathbf{R}(\boldsymbol{\Theta})$ are ultimately "filtered out" with probability $\to 1$.

Based on the duality between hypothesis tests and confidence regions, we can state that the confidence interval for $\mu$ in Example 10.21, for $\sum_{i=1}^{3} \beta_i$ and $\beta_1$ in Example 10.22, and the confidence interval for $\theta$ in Example 10.23 are all unbiased, UMAU, and consistent. The confidence interval for $p$ in Example 10.24 is approximately (asymptotically) unbiased and UMAU, and is also consistent. Finally, the ellipsoid confidence region for $(\beta_1, \beta_2, \beta_3)$ in Example 10.25 is unbiased and consistent.

### 10.7.3 Confidence Regions from Pivotal Quantities

A method of defining a confidence region for $\mathbf{R}(\boldsymbol{\Theta})$ without invoking duality with a hypothesis test involves so-called *pivotal quantities*, defined as follows.

**Definition 10.8** ***Pivotal Quantities***

Let the probability sample $\mathbf{X}$ have PDF $f(\mathbf{x};\boldsymbol{\Theta})$. A function of $\mathbf{X}$ and $\mathbf{R}(\boldsymbol{\Theta})$, $Q = q(\mathbf{X},\mathbf{R}(\boldsymbol{\Theta}))$, is a **pivotal quantity** for $\mathbf{R}(\boldsymbol{\Theta})$ if the probability distribution of $Q$ does not depend on the value of $\boldsymbol{\Theta}\in\Omega$ .

The fact that the pivotal quantity has a fixed probability distribution that does not depend on the parameter vector $\boldsymbol{\Theta}$ allows a confidence region for $\mathbf{R}(\boldsymbol{\Theta})$ to be defined via the following method.

**Theorem 10.12** ***Pivotal Quantity Method of Confidence Region Construction***

*Let $Q = q(\mathbf{X},\mathbf{R}(\boldsymbol{\Theta}))$ be a pivotal quantity for $\mathbf{R}(\boldsymbol{\Theta})$. Define the values $\ell$ and $\mu$ so that $P(\ell < q(\mathbf{x},\mathbf{R}(\boldsymbol{\Theta})) < \mu; \boldsymbol{\Theta}) = \gamma$. Then $A(\mathbf{X}) = \{\mathbf{R}(\boldsymbol{\Theta}): \ell < q(\mathbf{X},\mathbf{R}(\boldsymbol{\Theta})) < \mu\}$ defines a confidence region for $\mathbf{R}(\boldsymbol{\Theta})$ having confidence level (and coefficient) $\gamma$.*

**Proof** Since Q is a pivotal quantity, $P(\ell < q(\mathbf{x},\mathbf{R}(\boldsymbol{\Theta})) < \mu; \boldsymbol{\Theta}) = \gamma$ holds for every $\boldsymbol{\Theta}\in\Omega$, and in particular, for the true $\boldsymbol{\Theta}_0\in\Omega$. Also note that for a given outcome of $\mathbf{x}$, $\ell < q(\mathbf{X},\mathbf{R}(\boldsymbol{\Theta})) < \mu \Leftrightarrow \mathbf{R}(\boldsymbol{\Theta})\in A(\mathbf{X})$. It follows that $P(\mathbf{R}(\boldsymbol{\Theta}) \in A(\mathbf{X}); \boldsymbol{\Theta}) = \gamma\ \forall\boldsymbol{\Theta}\in\Omega$, and thus by Definition 10.3, $A(\mathbf{X})$ defines a confidence region for $\mathbf{R}(\boldsymbol{\Theta})$ with confidence level (and coefficient) $\gamma$. ■

Historically, the reason why the random variable $Q = q(\mathbf{X};\mathbf{R}(\boldsymbol{\Theta}))$ of Definition 10.8 and Theorem 10.12 was called a "pivotal" quantity is because when $R(\boldsymbol{\Theta})$ is a scalar many such random variables were such that the $\mathbf{x}$ argument in $\ell < q(\mathbf{x},R(\boldsymbol{\Theta})) < \mu$ could be "pivoted" (or better, inverted) out of the center term to yield the alternative inequality representation $t_\ell(\mathbf{x}) < R(\boldsymbol{\Theta}) < t_\mu(\mathbf{x})$. The latter inequality defines the confidence interval $(t_\ell(\mathbf{X}),t_\mu(\mathbf{X}))$ for $R(\boldsymbol{\Theta})$ which is the

**Table 10.5** Pivotal quantities and some associated confidence regions

| Ex | R(Θ) | Pivotal quantity, Q | PDF for Q, ∀Θ∈Ω | Event for Q | Level 1−α confidence region, A |
|---|---|---|---|---|---|
| 10.21 | $\mu$ | $\dfrac{\bar{X}-\mu}{S/\sqrt{n-1}}$ | *t*-distribution, *n*−1 *df* | $-t_{\alpha/2} < \mathbf{Q} < t_{\alpha/2}$ | $(\bar{x} - t_{\alpha/2}\, S/\sqrt{n-1}, \bar{x} + t_{\alpha/2} S/\sqrt{n-1})$ |
| 10.22 | $\mathbf{R\beta}$ | $\dfrac{\mathbf{R}\hat{\boldsymbol\beta} - \mathbf{R\beta}}{[\hat{S}^2\mathbf{R}(\mathbf{x}'\mathbf{x})\mathbf{R}']^{1/2}}$ | *t*-distribution, *n-k df* | $-t_{\alpha/2} < Q < t_{\alpha/2}$ | $\left(\mathbf{R}\hat{\boldsymbol\beta} - t_{\alpha/2}D,\ \mathbf{R}\hat{\boldsymbol\beta} + t_{\alpha/2}D\right)$ |
| | | | | $-t_\alpha < Q < \infty$ | $(-\infty, \mathbf{R}\hat{\boldsymbol\beta} + t_\alpha\, d)$ |
| | | | | | Where $D = \left[\hat{S}^2\mathbf{R}(\mathbf{x}'\mathbf{x})^{-1}\mathbf{R}'\right]^{1/2}$ |
| 10.23 | $\theta$ | $\bar{X}/\theta$ | Gamma$(n,n^{-1})$ | $Q < g_\alpha$ | $(\bar{x}/g_\alpha, \infty)$[a] |
| 10.24 | $p$ | $\dfrac{n^{1/2}(\bar{X}-p)}{[\bar{X}(1-\bar{X})]^{1/2}}$ | $N(0,1)$ (asymptotically) | $-z_{\alpha/2} < Q < z_{\alpha/2}$ | $\left(\bar{x} - z_{\alpha/2}D, \bar{x} + z_{\alpha/2}D\right)$ Where $D = \left[\dfrac{\bar{x}(1-\bar{x})}{n}\right]^{1/2}$ |
| 10.25 | $\mathbf{R\beta}$ | $q^{-1}(\mathbf{R}\hat{\boldsymbol\beta} - \mathbf{R\beta})'\,[\hat{S}^2\mathbf{R}(\mathbf{x}'_*\mathbf{x}_*)^{-1}\mathbf{R}']^{1/2}(\mathbf{R}\hat{\boldsymbol\beta} - \mathbf{R\beta})$ | *F*-distribution, *q* and *n-k df* | $Q < F_\alpha(q, n\text{-}k)$ | Ellipsoid with center $\mathbf{R}\hat{\boldsymbol\beta}$ |

[a] $g_\alpha$ *is such that* $\int_{g_\alpha}^{\infty}$ *Gamma*$(x;\, n, n^{-1})\, dx = \alpha$, df: *degrees of freedom*

confidence region $A(\mathbf{X})$ of Theorem 10.9. In practice, while it is often the case that such "pivoting" can be accomplished so that a confidence *interval* is defined, the characteristics of a pivotal quantity in Definition 10.8 do *not* guarantee that $q(\mathbf{x},R(\boldsymbol\Theta))$ can be pivoted, in which case $A(\mathbf{X})$ may *not* be an *interval*. Of course, if $\mathbf{R}(\boldsymbol\Theta)$ is a $(j\times 1)$ vector and $j\geq 2$, then obtaining a confidence *interval* will be neither possible nor relevant. The reader may find it interesting to know that *every* example of confidence regions examined heretofore can be motivated within the context of the pivotal quantity method, as will be seen in the next example.

**Example 10.26**
***Confidence Regions from Pivotal Quantities***

Revisit Examples 10.21–10.25. For each of the examples, Table 10.5 identifies the pivotal quantity that can be used to derive the confidence region for the respective $\mathbf{R}(\boldsymbol\Theta)$ as defined in the example. In addition, the table identifies (1) the fixed PDF for each pivotal quantity that applies regardless of the value of $\boldsymbol\Theta\in\Omega$, (2) the event for each pivotal quantity from which the confidence region is defined, and (3) the resultant explicit form of the confidence region for $\mathbf{R}(\boldsymbol\Theta)$. Note that in the case of Example 10.24, the table identifies an **asymptotic pivotal quantity**, meaning a random variable of the form $q(\mathbf{X},\mathbf{R}(\boldsymbol\Theta))$ whose limiting distribution does not depend on $\boldsymbol\Theta\in\Omega$. In practice, such asymptotic pivotal quantities are motivated via central limit theory. □

A notable practical difficulty in using pivotal quantities to define confidence regions is finding a pivotal quantity for a given $\mathbf{R}(\boldsymbol\Theta)$. There is no general method for defining pivotal quantities, and a pivotal quantity need not exist for a given probability model. However, for certain special but important classes of problems, pivotal quantities are readily available.

**Theorem 10.13 Pivotal Quantities for Location-Scale Parameter Families of PDFs**

*Let* $\mathbf{X}$ *be a random sample from a population distribution* $f(z;\boldsymbol{\Theta})$. *Let* $\hat{\boldsymbol{\Theta}}$ *denote the MLE of* $\boldsymbol{\Theta}$. *The following relationships exist between the functional form of* $f(z;\boldsymbol{\Theta})$ *and pivotal quantities for the elements of* $\boldsymbol{\Theta}$, *where* $f_0(y)$ *denotes a PDF whose values do not depend on unknown parameters:*

1. ***Location Parameter Family of PDFs***: $f(z;\Theta) = f_0(z-\Theta) \Rightarrow Q = \hat{\Theta} - \Theta$ *is a pivotal quantity if* $\Theta$ *is a scalar.*
2. ***Scale Parameter Family of PDFs:*** $f(z;\Theta) = \Theta^{-1} f_0(z/\Theta) \Rightarrow Q = \hat{\Theta}/\Theta$ *is a pivotal quantity if* $\Theta$ *is a scalar.*
3. ***Location-Scale Family of PDFs:*** $f(z;\boldsymbol{\Theta}) = \Theta_2^{-1} f_0((z-\Theta_1)/\Theta_2) \Rightarrow Q_1 = (\hat{\Theta}_1 - \Theta_1)/\hat{\Theta}_2$ *and* $Q_2 = \hat{\Theta}_2/\Theta_2$ *are pivotal quantities for* $\Theta_1$ *and* $\Theta_2$ if $\boldsymbol{\Theta}$ is a $(2\times 1)$ *parameter vector.*

**Proof** Antle, C.E. and L.J. Bain, (1969), "A Property of Maximum Likelihood Estimators of Location and Scale Parameters," *SIAM Review*, 11, p. 251. ■

The theorem indicates that so long as one is random sampling from a population distribution, and that population distribution is a location, scale, or location-scale family of PDFs, there are then specific known functions of the MLEs that define pivotal quantities for the parameters of the population distribution.

A more general result which applies to random sampling from *any* continuous PDF having a *scalar* parameter $\Theta$ is as follows.

**Theorem 10.14 Pivotal Quantities for Continuous Population PDFs**

*Let* $\mathbf{X}$ *be a random sample from a population distribution having the continuous PDF* $f(z;\Theta)$, *where* $\Theta$ *is a scalar. Then*

$$-2\sum_{i=1}^{n} \ln(F(X_i;\Theta)) \sim \chi^2_{2n}$$

*is a pivotal quantity for* $\Theta$, *where* $F(z;\Theta)$ *is the common CDF for the* $X_i$*'s in the random sample.*

**Proof** If $Z \sim f(z;\Theta)$, then the probability integral transform of $Z$, $F(Z;\Theta)$, is distributed uniform (0,1) (recall Theorem 6.22). It follows that $W = -\ln(F(Z;\Theta)) \sim$ Exponential(1). Then $Q = q(X,\Theta) = -2\sum_{i=1}^{n} \ln(F(X_i;\Theta))$ is 2 times the sum of $n$ independent Exponential(1) random variables, which has *a* $\chi^2$ distribution with $2n$ degrees of freedom. Because $Q = q(X,\Theta) \sim \chi^2_{2n}\ \forall\Theta\in\Omega$, it follows that $Q$ is a pivotal quantity for $\Theta$. ■

**Table 10.6** Relationships between events for pivotal quantity and confidence intervals

| Pivotal quantity event | Montonicity of $q(\mathbf{x},R(\Theta))$ in $R(\Theta)$ | Confidence interval |
|---|---|---|
| $\ell < q < \mu$ | Increasing | $t_1(\mathbf{x}) < R(\Theta) < t_2(\mathbf{x})$ |
| | Decreasing | $t_1(\mathbf{x}) < R(\Theta) < t_2(\mathbf{x})$ |
| $-\infty < q < \mu$ | Increasing | $-\infty < R(\Theta) < t(\mathbf{x})$ |
| | Decreasing | $t(\mathbf{x}) < R(\Theta) < \infty$ |
| $\ell < q < \infty$ | Increasing | $t(\mathbf{x}) < R(\Theta) < \infty$ |
| | Decreasing | $-\infty < R(\Theta) < t(\mathbf{x})$ |

Theorem 10.14 implies that pivotal quantities always exist for a scalar $\Theta$ when random sampling is from a continuous PDF $f(x;\Theta)$. Note that one could also demonstrate that $-2\sum_{i=1}^{n}\ln[1-F(X_i;\Theta)] \sim \chi^2_{2n}$ is an alternative pivotal quantity for $\Theta$ (see Problem 10.21).

**Example 10.27** ***Confidence Interval for the Mean of a Power Distribution via Pivotal Quantities***

The proportion of the work day that a particular assembly line is stopped because of malfunctions on the line is the outcome of a random variable with power distribution PDF $f(z;\Theta) = \Theta z^{\Theta-1} I_{(0,1)}(z)$, for $\Theta>0$. Based on a random sample of $n$ daily observations from $f(z;\Theta)$, we define a confidence region for $\Theta$ having confidence coefficient .95.

To use Theorem 10.14, first note that the CDF of $Z$ is $F(b;\Theta) = b^{\Theta} I_{[0,1]}(b) + I_{(1,\infty)}(b)$. Then $-2\ln(F(X_i;\Theta)) = -2\Theta\ln(X_i)$, and thus $Q = -2\Theta\sum_{i=1}^{n}\ln(X_i) \sim \chi^2_{2n}$ is a pivotal quantity for $\Theta$. The probability of the event $\left\{q : \chi^2_{2n;1-\alpha/2} < q < \chi^2_{2n;\alpha/2}\right\}$ is $1-\alpha$, and given a random sample outcome $\mathbf{x}$, the event can be pivoted to define a $1-\alpha$ level confidence interval for $\Theta$ as

$$\frac{\chi^2_{2n;1-\alpha/2}}{-2\sum_{i=1}^{n}\ln(x_i)} < \Theta < \frac{\chi^2_{2n;\alpha/2}}{-2\sum_{i=1}^{n}\ln(x_i)}.$$

Note when "pivoting" that $-2\sum_{i=1}^{n}\ln(x_i) > 0$ in this case, and so the sense of the inequalities do not reverse in the pivot operation. □

Regarding the choice of the pivotal quantity event from which the confidence region is defined, if the pivotal quantity is a monotonic function of $R(\Theta)$ and a confidence interval or bound is desired, the relationships are useful, where $t_i(\mathbf{x})$ denotes functions of the sample outcomes ultimately used to define lower and/or upper bounds for the confidence intervals (Table 10.6).

Also note that there are often many choices of $\ell$ and $\mu$ available that are such that $P(\ell < q < \mu) = \gamma$, e.g., an infinite number in the case when $q$ is a continuous random variable and $\gamma \in (0,1)$. If $q(\mathbf{x}, R(\Theta))$ is a monotonic function of $R(\Theta)$, then it is desirable to choose $\ell$ and $\mu$ so that the length, or expected length, of the resultant confidence *interval* is minimized. However, in practice, a rule of thumb for two-sided confidence intervals is often followed whereby $\ell$ and $\mu$ are chosen so as to define "equal tail probabilities" in the definition of a $\gamma$-level confidence interval. That is, $\ell$ and $\mu$ are chosen such that $P(q < \ell) = P(q > \mu) = \alpha/2$, leading to a $\gamma = 1-\alpha$ level confidence interval. Note that this principle was followed when defining the confidence interval in Example 10.27. The expected length of the confidence interval could have been reduced slightly by choosing the lower and upper $\chi^2$ values, say $\chi^2_{\alpha_\ell}$ and $\chi^2_{\alpha_\mu}$, so that $\chi^2_{\alpha_\mu} - \chi^2_{\alpha_\ell}$ is minimized subject to the confidence level condition $P\left(\chi^2_{\alpha_\ell} < q < \chi^2_{\alpha_\mu}\right) = \gamma$, which can be accomplished on a computer. Note that if the distribution for the pivotal quantity is symmetric, then choosing $\ell$ and $\mu$ so as to have "equal tail probabilities" is *equivalent* to choosing the confidence bounds so that the length, or expected length, of the confidence interval is minimized.

The pivotal quantity method does not guarantee that any optimal properties will apply to the confidence regions derived from it. However, the method is relatively straightforward, it applies in many cases of practical importance, and the confidence regions defined by the method are often quite adequate for their intended purpose.

### 10.7.4 Using Confidence Regions to Conduct Statistical Tests

We briefly note that because there is a duality between confidence regions and statistical tests of hypotheses, one can utilize the duality in the reverse of what we have done heretofore and define statistical testing procedures from confidence regions. Specifically, a null hypothesis is either *not rejected* or *rejected* depending on whether the hypothesized value of $\mathbf{R}(\boldsymbol{\Theta})$ *is contained* or *is not contained* in the confidence region, respectively. It follows that the preceding pivotal quantity method can be used to define statistical tests. A hypothesis test is unbiased, UMP, UMPU, and/or consistent according to whether the confidence region is unbiased, UMA, UMAU, and/or consistent, respectively.

We will not pursue this "reverse duality" approach to defining statistical tests from confidence regions any further because for the approach to be of substantive practical importance, we would need to establish additional methods of deriving UMA, UMAU, and/or consistent critical regions independent of the statistical tests to which they are dual. This requires further study of the theory of confidence region estimation, which the reader can begin by referring to the readings suggested at the beginning of this section.

## 10.8 Nonparametric Tests of Distributional Assumptions

We provide an introduction to **nonparametric testing** of the distributional assumptions underlying a point estimation, hypothesis testing, or confidence region estimation problem in this section. By nonparametric we mean here that the hypotheses under consideration are not defined in terms of the values of parameters, per se, as has been the case heretofore. Rather, the hypotheses under consideration will be more general and refer to functional forms of probability density functions, and whether random variables contained in a sample are *iid*. The tests we will introduce address questions such as "could the probability sample have come from a normal or exponential, or beta, or. . .distribution?" and "can it be assumed that the outcomes observed in the sample are outcomes of an *iid* random sample from some population distribution?"

Note that to a large degree the methods of statistical inference that we have examined heretofore required certain basic assumptions to hold, collectively representing the *maintained hypothesis*, before any analysis could proceed. It is important to be able to rigorously assess the validity of assumptions that are held only tentatively and/or that one does not have substantial confidence are true. Two of the more frequent assumptions made to this point have been the assumption of a specific functional form for the probability density of a random sample and the assumption that random variables in the sample are *iid*. In this section we focus on some nonparametric tests of these assumptions that have been useed in practice, and concentrate on scalar random variables. The field of nonparametric analysis is vast and growing. A useful place to begin additional reading is J.D. Gibbons, (1985), *Nonparametric Methods for Quantitative Analysis (2nd Ed)*, American Science Press, Columbus, Ohio, and A. Pagan and A. Ullah, (1999), *Nonparametric Econometrics*, Cambridge University Press, Cambridge.

### 10.8.1 Nonparametric Tests of Functional Forms of Distributions

There are a number of testing procedures available for testing hypotheses regarding the functional form of the joint density of the probability sample (see M. Kendall and A. Stuart, Vol. 2, op. cit., Chapter 30; and C. Huang and B. Bolch, (1974), "On Testing of Regression Disturbances for Normality," *JASA*, 1974, pp. 330–335 for alternatives and references). We will examine the $\chi^2$ goodness of fit test, because of its versatility and wide applicability, the Kolmogonov-Smirnov test and its refinement to the Lilliefors test, because of their refinement in analyzing continuous distributions, and the Shapiro-Wilks and Jarque-Bera tests because of their specific use in testing the ubiquitous normality assumption.

*10.8.1.1 $\chi^2$ Goodness of Fit Test* The $\chi^2$ goodness of fit test is used to test the null hypothesis that a random sample is from a population distribution of the form $f(z;\boldsymbol{\Theta})$ where $\boldsymbol{\Theta}$ is a $(k\times 1)$ vector. In particular, the hypotheses under consideration are $H_0$: $\mathbf{X} \sim \prod_{i=1}^{n} f(x_i;\boldsymbol{\Theta})$ versus $H_a$: $\mathbf{X} \nsim \prod_{i=1}^{n} f(x_i;\boldsymbol{\Theta})$. The test procedure differs depending on whether the null hypothesis is simple, meaning that the values of any parameters in $f(z;\boldsymbol{\Theta})$ are fully specified, or the null hypothesis is composite in the sense that $\boldsymbol{\Theta}$ in $f(z;\boldsymbol{\Theta})$ is left unspecified so that an entire *family* of density functions is implied by the null hypothesis.

We first examine the case where the null hypothesis is simple. Given the null hypothesis $H_0$: Z ~ $f(z;\boldsymbol{\Theta}_0)$ for the population distribution, where $\boldsymbol{\Theta}_0$ is its fixed and known value, and assuming the range of $Z$ has been partitioned into $m$ subintervals $\Delta_i$, $i = 1,\ldots,m$, it follows that under $H_0$, $P(\Delta_i; \Theta_0) = \int_{z\in\Delta_i} f(z;\Theta_0)\,dz = p_i(\Theta_0)$ for $i = 1,\ldots,m$. For a random sample of size $n$, let $n_i$ represent the number of times an outcome occurs in interval $\Delta_i$, $i = 1,\ldots,m$. Then the probability distribution of $(N_1,\ldots,N_m)$ is multinomial with parameters $(p_1,\ldots,p_m)$, $\sum_{i=1}^{m} p_i = 1$ (recall Section 4.1.4).

Using the multivariate version of the LLCLT, it follows that $n^{1/2}(\bar{\mathbf{X}}_* - \mathbf{p}_*)) \xrightarrow{d} N(\mathbf{0}, \boldsymbol{\Sigma}_*)$, where $\bar{\mathbf{X}}_* = n^{-1}(N_1,\ldots,N_{m-1})'$, $\mathbf{p}_* = (p_1,\ldots,p_{m-1})'$, and $\Sigma_*$is a $(m-1)\times(m-1)$ covariance matrix with $p_i(1 - p_i)$, $i$=1,…,$(m-1)$ along the diagonal and $-p_ip_j$ in the off-diagonal entries. (Note that $n_m$ is determined by $n_m = n - \sum_{i=1}^{m-1} n_i$ and thus $(n_m/n) = 1 - \sum_{i=1}^{m-1} \bar{x}_{*i}$). Then from Theorem 5.9,

$$W = n(\bar{\mathbf{X}}_* - \mathbf{p}_*)'\boldsymbol{\Sigma}_*^{-1}(\bar{\mathbf{X}}_* - \mathbf{p}_*) = (\mathbf{N}_* - n\mathbf{p}_*)'(n\boldsymbol{\Sigma}_*)^{-1}(\mathbf{N}_* - n\mathbf{p}_*) \xrightarrow{d} \chi^2_{m-1},$$

where $\mathbf{N}_* = (N_1,\ldots,N_{m-1})'$. It follows that a size $\alpha$ test of $H_0$: $Z \sim f(z;\Theta_0)$ versus $H_0$: $Z \nsim f(z;\Theta_0)$ is given by

$$w \begin{Bmatrix} \geq \\ < \end{Bmatrix} \chi^2_{m-1,\alpha} \Rightarrow \begin{Bmatrix} \text{reject } H_0 \\ \text{do not reject } H_0 \end{Bmatrix}$$

where $(p_1,\ldots,p_{m-1})$ in $\mathbf{p}_*$ and in $\boldsymbol{\Sigma}_*$ are defined by their values under $H_0$, i.e., $p_i = p_i(\Theta_0)$, $i = 1,\ldots,m-1$. The test is seen to reject $H_0$ when the probabilities of the subinterval events $\Delta_i$, $i = 1,\ldots,m$ that are hypothesized under $H_0$ are in conflict with their estimated values, given by the observed relative frequencies, $n_i/n$.

In practice, a statistic that is algebraically and numerically equivalent to $W$ is generally used to calculate the test outcome. In particular, some tedious but conceptually straightforward matrix algebra leads to the simplified expression (see Kendall and Stuart, vol. 2, op. cit., p. 381):

$$w = \sum_{i=1}^{m} \frac{(n_i - np_i)^2}{np_i}.$$

Now suppose that the null hypothesis is *composite* as $H_0$: $Z \sim f(z;\Theta)$, $\Theta\in\Omega$ versus $H_a$: $Z \nsim f(z;\Theta)$, $\Theta\in\Omega$. How do we proceed in this case when the specific numerical value of $\Theta$ is left unspecified? The unknown parameters are estimated from the data, but how this is done has a substantial impact on the form of the test procedure. Consider two different forms of MLEs, the difference being the choice of likelihood function.

For one approach, let the likelihood function be the multinomial distribution parameterized via $\Theta$ under $H_0$, i.e.,

$$L(\Theta;\mathbf{x}) = \left[\frac{n!}{n_1!n_2!\ldots n_m!}\right]\prod_{i=1}^{m} p_i(\Theta)^{n_i} \text{ for } n_i \in \{0,1,\ldots,n\}, \forall i, \sum_{i=1}^{m} n_i = n$$

where $p_i(\Theta) = \int_{z\in\Delta_i} f(z;\Theta)\,dx\ \forall i$. After one obtains the maximum likelihood estimate of $\Theta$, $\hat{\theta}$, one proceeds by replacing the $p_i$'s by $\hat{p}_i = p_i(\hat{\theta})$ in the calculation of the $w$-statistic, obtaining

$$w = \sum_{i=1}^{m} \frac{(n_i - n\hat{p}_i)^2}{n\hat{p}_i}$$

Finally, the test rule is defined as (note the reduced degrees of freedom compared to the previous test)

$$w \left\{ \begin{matrix} \geq \\ < \end{matrix} \right\} \chi^2_{m-1-k,\alpha} \Rightarrow \left\{ \begin{matrix} \text{reject } H_0 \\ \text{do not reject } H_0 \end{matrix} \right\}.$$

The reason that one degree of freedom is lost for each parameter estimated is somewhat involved. A detailed proof can be found in H. Cramer, (1946), *Mathematical Methods of Statistics*, Princeton Univ. Press.

An alternative ML procedure is to estimate $\boldsymbol{\Theta}$ using the ML method applied to the likelihood function $L(\boldsymbol{\Theta}; \mathbf{x}) = \prod_{i=1}^{k} f(x_i; \boldsymbol{\Theta})$, i.e., use the hypothesized population distribution *directly* to specify the likelihood function, and estimate $\boldsymbol{\Theta}$ accordingly. One then computes $\hat{p}_i = \int_{z \in \Delta_i} f\left(z; \hat{\boldsymbol{\theta}}\right) dz$ to calculate the $w$-statistic above. In this case, it turns out that $W$ does *not* have a limiting $\chi^2$ distribution at all. However, it can be shown that the distribution of $W$ in this case is bounded *between* a $\chi^2_{m-1}$ and a $\chi^2_{m-1-k}$ distribution (H. Chernoff and E.L. Lehmann, (1954), "The Use of Maximum Likelihood Estimates in $\chi^2$ Tests for Goodness of Fit," *Ann. Math. Statist.*, p. 579). Unfortunately, the limiting distribution is difficult to use, and in practice one can use $\chi^2_{m-1;\alpha}$ to define a size $\leq \alpha$ test, or else use $\chi^2_{m-1-k;\alpha}$ to define a test that is more likely to reject a true $H_0$ than the size $\alpha$ would indicate.

It can be shown that all of the preceding tests are *consistent* for any alternatives that imply multinomial probabilities different from those implied by $H_0$. On the other hand, the tests are generally *biased* to some degree.

One operational problem remains. How does one choose the sets $\Delta_i$, $i = 1,\ldots,m$? Rules of thumb developed from both empirical and theoretical considerations suggest that intervals be defined that have equal probability, i.e., $1/m$, based on $f(x;\boldsymbol{\Theta}_0)$ for simple $H_0$'s or on $f(x; \hat{\boldsymbol{\theta}})$ for composite $H_0$'s. Furthermore, in order that the asymptotics be a reasonable approximation, the intervals chosen should be such that $np_i(\boldsymbol{\Theta}_0)$ or $np_i(\hat{\boldsymbol{\theta}}) \geq 5 \ \forall i$.

**Example 10.28 *Testing Whether a Population Distribution Is an Exponential PDF***

Forty observations on the waiting times, in minutes, between customer arrivals at a service station/convenience store in a mid-size city were as follows:

| | | | | | | | |
|---|---|---|---|---|---|---|---|
| 1.37 | 1.96 | 0.74 | 0.42 | 2.23 | 1.11 | 1.73 | 0.26 |
| 0.12 | 0.61 | 1.98 | 1.76 | 1.77 | 1.35 | 2.91 | 0.93 |
| 1.73 | 3.32 | 1.44 | 2.46 | 1.50 | 2.72 | 1.73 | 0.59 |
| 0.34 | 2.31 | 2.14 | 2.11 | 0.36 | 0.24 | 2.68 | 0.30 |
| 2.84 | 2.47 | 1.25 | 0.66 | 0.10 | 2.75 | 1.68 | 0.88 |

We will define a (approximate) size .05 $\chi^2$ goodness of fit test of the null hypothesis that the waiting times are exponentially distributed versus some other family of distributions, i.e., $H_0$: $z \sim \theta^{-1}\exp(-z/\theta)I_{(0,\infty)}(x)$, $\theta > 0$ versus $H_a$: not $H_0$.

The ML estimate of $\theta$ based on the exponential population distribution assumption is $\hat{\theta} = \bar{x} = 1.494$. We let $m = 8$, and choose $\Delta_i$, $i = 1,\ldots,8$ so that $P(z \in \Delta_i; \hat{\theta}) = 1/8\ \forall i$. The boundaries of the intervals are found as follows:

$$
\begin{aligned}
.125 &= \int_0^{c_1} (1.494)^{-1}\exp(-z/1.494)dz \Rightarrow & c_1 &= .1995,\\
.125 &= \int_{.1995}^{c_2} (1.494)^{-1}\exp(-z/1.494)dz \Rightarrow & c_2 &= .4298,\\
. & & c_3 &= .7022,\\
. & & c_4 &= 1.0356,\\
. & & c_5 &= 1.4654,\\
. & & c_6 &= 2.0711,\\
.125 &= \int_{2.0711}^{c_7} (1.494)^{-1}\exp(-z/1.494)dz \Rightarrow & c_7 &= 3.1067.
\end{aligned}
$$

The intervals and the number of observations that actually occurred in each are given by:

| $i$ | $\Delta_i$ | $n_i$ |
|---|---|---|
| 1 | (0, .1995] | 2 |
| 2 | (.1995, .4298] | 6 |
| 3 | (.4298, .7022] | 3 |
| 4 | (.7022, 1.0356] | 3 |
| 5 | (1.0356, 1.4654] | 5 |
| 6 | (1.4654, 2.0711] | 9 |
| 7 | (2.0711, 3.1067] | 11 |
| 8 | (3.1067, ∞] | 1 |

Since $n\hat{p}_i = 40(.125) = 5\forall i$, the value of the $\chi^2$ statistic is given by

$$w = \sum_{i=1}^{8} \frac{(n_i - 5)^2}{5} = 17.2.$$

Adopting a conservative (toward $H_0$) stance and using the critical value $\chi^2_{m-1-k;.05} = \chi^2_{6;.05} = 12.6$, $w > 12.6$ and the exponential family of distributions is *rejected*. Note if the more liberal critical value of $\chi^2_{m-1;.05} = \chi^2_{7;.05} = 14.1$ is used, $w > 14.1$ and $H_0$ is still rejected, and we know that the significance level of this test is $\leq .05$. □

*10.8.1.2 Kolmogorov-Smirnov and Lilliefors Tests*

The Kolmogorov-Smirnov (K-S) test is an alternative procedure for testing whether a random sample outcome was drawn from a specified population distribution. The test procedure is based on the empirical distribution function (EDF) and the Glivenko-Cantelli theorem (recall Chapter 6) which implies that the EDF will converge functionally across all points of comparison to the true population CDF associated with a random sample. The basic idea of the K-S test is to reject $H_0$ when there is significant discrepancies between an EDF and a hypothesized CDF as revealed through the probability integral transform of the data (recall Section 6.8).

Before examining the test procedure in more detail, we note some advantages and limitations of the K-S test relative to the aforementioned $\chi^2$ test. On the positive side, the K-S test is fully applicable in the case of small samples whereas the $\chi^2$-test is only an asymptotically valid test. The K-S test deals with sample observations directly whereas the data must be summarized into categories for analysis via the $\chi^2$ approach, so that information is potentially lost in the categorization process. The K-S test assumes the population distribution is continuous and thus provides a more refined analysis specific to this case. Limitations include the fact that the K-S approach cannot be easily adjusted to allow for estimation of unknown parameters as in the $\chi^2$ case, so that in practice the K-S test is generally restricted to testing simple null hypotheses. Also, the applicability of K-S is limited to cases involving continuous distributions. We note that the K-S test has been modified by H. Lilliefors, (1967), ("On the Kolmogorov-Smirnov Test for Normality with Mean and Variance Unknown," *JASA*, pp. 399–402) to accommodate the estimation of $\mu$ and $\sigma^2$ when the null hypothesis is that of *normality*. We will examine this case below. Work has also been done on extending the K-S approach to exponential and Weibull cases where parameters are unknown.[9]

In order to identify the test procedure, we begin with the null hypothesis $H_0$: $Z \sim F_0(z)$ versus $H_a$: $Z \nsim F_0(z)$, where $F_0(z)$ is a *completely* specified CDF. Let $X_1, \ldots, X_n$ be a random sample, and let $\hat{F}_n(z)$ be the EDF based on the random sample. The **Kolmogorov-Smirnov test statistic** is given by[10]

$$d_n = \sup_z |\hat{F}_n(z) - F_0(z)|.$$

The value of $d_n$ is seen to be the largest distance between the hypothesized CDF, $F_0(z)$, and the EDF estimate of the true CDF, $\hat{F}_n(z)$. The larger is $d_n$, the greater is the largest numerical discrepancy between the estimated and hypothesized CDFs. Note also that $y_0 = F_0(z)$ is effectively the probability integral transform of the random sample as stated in $H_0$, while the EDF can be viewed as *estimating* outcomes from the true probability integral transform $y = F(z)$. Thus, an alternative interpretation of $d_n$ is the largest discrepancy between hypothesized and estimated probability integral transforms of the data.

It is useful to note for computational purposes that $d_n = \max(d_n^+, d_n^-)$, where

$$d_n^+ = \max_{i \in (1,\ldots,n)} \left( \frac{i}{n} - F_0\left(x_{(i)}\right) \right), \; d_n^- = \max_{i \in (1,\ldots,n)} \left( F_0\left(x_{(i)}\right) - \frac{(i-1)}{n} \right),$$

[9] See M.A. Stephens, (1974), *JASA*, p. 730; and Chandra, et. al, (1981), *JASA*, p. 729.

[10] As always, sup can be replaced by max when the maximum exists.

and $x_{(1)}, x_{(2)}, \ldots, x_{(n)}$ are the random sample outcomes ordered from lowest to highest (i.e., they are the *order statistics*). This computational approach makes the calculation of $D_n$ a simple matter on a computer through sorting and differencing operations.

In order to decide whether a discrepancy is significant, the sampling distribution of $d_n$ is required. The exact distribution of $d_n$ for $n \leq 40$ has been tabled by J.D. Gibbons, op. cit., p. 400. It was shown by A.N. Kolmogornov, (1933), (*Giorn. Inst. Ital. Attuari*, 4, pp. 83–91) that the limiting CDF of $n^{1/2}\, d_n$ can be represented as

$$\lim_{n\to\infty} P\left(n^{1/2} d_n \leq t\right) = 1 - 2\sum_{i=1}^{\infty} (-1)^{i-1} \exp(-2i^2 t^2),$$

which results in the approximate critical points

| $\alpha$ | .10 | .05 | .01 |
|---|---|---|---|
| $d_{n;\alpha}$ | $1.224n^{-1/2}$ | $1.358n^{-1/2}$ | $1.628n^{-1/2}$ |

Stephens (see footnote 9) has analyzed adjustments to these approximate critical values that make them more accurate for small $n$, the adjustments being to divide each of them by the factor $(1 + .12n^{-1/2} + .11\, n^{-1})$.

A size $\alpha$ K-S test of $H_0$: $Z \sim F_0(z)$ versus $H_a$: $Z \nsim F_0(z)$ is given as follows:

$$d_n \begin{Bmatrix} \geq \\ < \end{Bmatrix} d_{n;\alpha} \Rightarrow \begin{Bmatrix} \text{reject } H_0 \\ \text{do not reject } H_0 \end{Bmatrix}.$$

From the Glivenko-Cantelli theorem and the fact that $d_{n;\alpha} \to 0 \; \forall \; \alpha \in (0,1)$, it follows that the test is consistent.

**Example 10.29** ***Testing Whether a Population Distribution Is a Uniform PDF***

Revisit Example 10.28, and examine the null hypothesis that the distribution of waiting times is uniform over the interval (0, 3.5), i.e., $H_0$: $Z \sim F_0(z)$ versus $H_a$: $Z \nsim F_0(z)$ with $F_0(z) = z/3.5\, I_{(0,\,3.5)}(z)$. The value of $d_n = \sup_{z\in(0,3.5)} |\hat{F}_n(z) - (z/3.5)| = .1443$, the maximum difference occurring for $z = 2.47$ which has an EDF value of .85 and a (hypothesized) CDF value of $2.47/3.5 = .7057$. Stephens' adjusted critical value for $n = 40$ and a size $\alpha = .05$ test is given by $d_{40;.05} = [1.358/(40)^{1/2}]/(1 + .12/(40)^{1/2} + .11/40) = .2102$. Because $d_n = .1443 < d_{40;.05}, = .2102$, we do *not* reject the hypothesis that waiting times are uniformly distributed over the interval (0, 3.5). □

Lilliefors provided an adjustment to the K-S test that justifies its use for testing the null hypothesis $H_0$: $Z \sim N(\mu, \sigma^2)$ versus $H_a$: $Z \nsim N(\mu, \sigma^2)$ when $\mu$ and $\sigma^2$ are *unknown* and must be estimated from the data. In particular, he tabled the distribution of $D_n$ under the null hypothesis for $n = 1, \ldots, 30$ via Monte Carlo methods (see the previous Lilliefors reference), and presented approximate critical values that are accurate for $n > 30$ as follows:

| $\alpha$ | .10 | .05 | .01 |
|---|---|---|---|
| $d^*_{n;\alpha}$ | $.805n^{-1/2}$ | $.886n^{-1/2}$ | $1.031n^{-1/2}$ |

The test proceeds by first unbiasedly estimating $\mu$ and $\sigma^2$ using the estimators $\bar{x}$ and $\hat{\sigma}^2 = ns^2/(n-1)$ (using the sample variance itself would lead to the same test size asymptotically). Then $N(\bar{x}, \hat{\sigma}^2)$ is used for the distribution under the null hypothesis, and $d_n = \sup_z |\hat{F}_n(z) - \int_{-\infty}^{z} N(z; \bar{x}, \hat{\sigma}^2)dz|$ is calculated. Finally, the K-S test is performed as before but using Lilliefors' critical values, $d^*_{n;\alpha}$, in place of $d_{n;\alpha}$.

**Example 10.30** ***Testing Whether a Population Distribution Is a Normal PDF via Lilliefor's Test***

We revisit Example 10.21, and use the K-S test, as adjusted by Lilliefors, to test the hypothesis $H_0$: $Z \sim N(\mu, \sigma^2)$ versus $H_a$: $Z \nsim N(\mu, \sigma^2)$ for some $\mu$ and $\sigma^2$. We use $\alpha = .05$.

The estimates of $\mu$ and $\sigma^2$ are respectively $\hat{\mu} = \bar{x} = 1.494$ and $\hat{\sigma}^2 = .8338$. Then using $N(1.494, .8338)$ for the distribution of Z under the null hypothesis, the value of the K-S statistic can be calculated as

$$d_n = \sup_z |\hat{F}_n(z) - \int_{-\infty}^{(z-1.494)/.9131} N(z; 0, 1)dz| = .0955,$$

which occurs for $z = .74$, in which case the EDF value is .30 and $\int_{-\infty}^{.74} N(z; 1.494, .8338)dz = .2045$. Using Lilliefors' critical value of $.886/(40)^{1/2} = .1401$, the hypothesis of normality *cannot* be rejected with this sample outcome. □

While the Lilliefors K-S test is usually very effective in detecting a false $H_0$, in this case it has failed. It turns out that the waiting times data was generated by a uniform distribution (based on uniform random numbers generated by a computer), so that the Lilliefors test failed to reject a false $H_0$. Note that for large enough $n$, we would have rejected the hypothesis, since this K-S test is consistent. We note that Lilliefors' own Monte Carlo calculations indicated that among all of the alternative distributions he examined his test had the *lowest* power against the uniform distribution, which may be why the test had difficulty in the preceding example.

#### *10.8.1.3 Shapiro-Wilks Test*

The Shapiro-Wilks (SW) test procedure was designed *specifically* to test the hypothesis that a random sample is from a $N(\mu, \sigma^2)$ population distribution, the null and alternative hypotheses being $H_0$: $Z \sim N(\mu, \sigma^2)$ and $H_a$: $Z \nsim N(\mu, \sigma^2)$, respectively, for some (unspecified) $\mu$ and $\sigma^2$. The test procedure has been shown to perform especially well in comparison to other tests of normality when applied to the calculated residuals $\hat{\mathbf{e}} = \mathbf{y} - \mathbf{xb}$ of a least squares estimate of the GLM $\mathbf{Y} = \mathbf{x}\boldsymbol{\beta} + \boldsymbol{\varepsilon}$. The null hypothesis in this case is that $\varepsilon_i \sim N(0, \sigma^2)\ \forall i$ (see C. Huang and B. Bolch, (1974) op. cit.) The procedure fairs well in more general settings as well. The SW test was originally devised by S. Shapiro and M. Wilks, (1965), in "An Analysis of Variance Test for Normality," *Biometrika*, pp. 591–611.

The SW test statistic is calculated as

$$w = \left[\sum_{i=1}^{m} a_{n-i+1}\left(x_{(n-i+1)} - x_{(i)}\right)\right]^2 / ns^2, \quad m\begin{cases} n/2 & \text{if } n \text{ is even,} \\ (n-1)/2 & \text{if } n \text{ is odd,} \end{cases}$$

where $(x_{(1)}, x_{(2)}, \ldots, x_{(n)})$ are the sample observations ordered from smallest to largest (i.e., the *order statistics*), $s^2$ is the sample variance, and the $a_i$'s are coefficients that are tabulated in the article by SW, pp. 603–604. The distribution of the test statistic is quite involved, but it has been tabulated by SW for sample sizes $\leq 50$ (p. 605 of SW article) and for various levels of tail probabilities. The test rule is given by

$$w\begin{Bmatrix} \leq \\ > \end{Bmatrix} w_{n;1-\alpha} \Rightarrow \begin{Bmatrix} \text{reject } H_0 \\ \text{do not reject } H_0 \end{Bmatrix}.$$

The test procedure can be applied for small samples and is also a consistent test of $H_0$ versus $H_a$.

The heuristic motivation for the test lies in the fact that under the null hypothesis of normality, both the numerator and denominator of the SW statistic can be shown to be estimating the same quantity, $\sigma^2$, apart from constants. Under the alternative hypothesis, the denominator will still (apart from $n^{-1}$) be estimating $\sigma^2$, but such will not be the case for the numerator (in general). The difference in behavior of the numerator and denominator under $H_a$ is exploited by SW in the design of the test procedure, and further motivation can be found in their article.

**Example 10.31** ***Testing Residuals of a GLM for Normality Using Shapiro Wilks Test***

The relationship between varying levels of advertising expenditures and the quantity demanded of a product being advertised was studied over a 20-week period, and 20 weekly observations were used to estimate the GLM $\mathbf{Y} = \mathbf{x}\boldsymbol{\beta} + \boldsymbol{\varepsilon}$, where $\mathbf{Y}$ represents the vector of 20 observations on quantities sold, and $\mathbf{x}$ represents a matrix of variables explaining the demand for the product, including the level of advertising expenditures. The calculated residuals, $\hat{\mathbf{e}}$, from the least squares fit of the model yielded the following values, $\hat{\mathbf{e}}_*$, ordered from lowest to highest row-wise:

| | | | |
|---|---|---|---|
| −7.0890 | −5.6489 | −5.1510 | −4.9672 |
| −4.8390 | −3.9812 | −3.9161 | −3.6296 |
| −3.5914 | −3.2087 | −2.9827 | −2.7810 |
| −1.2857 | −1.1390 | 0.8321 | 5.3253 |
| 6.2581 | 11.3091 | 14.4254 | 16.0611 |

We use the SW test to assess the hypothesis that $H_0$: $\boldsymbol{\varepsilon} \sim N(\mathbf{0}, \sigma^2 \mathbf{I})$ versus $H_a$: not $H_0$ at the level $\alpha = .05$.

In this case, $ns^2 = \sum_{i=1}^{n} \left(\hat{e}_i - \bar{\hat{e}}\right)^2 = \sum_{i=1}^{n} \hat{e}_i^2 = 906.0822$ (since $\bar{\hat{e}} = 0$). Because $n$ is even, $m = n/2 = 10$. To calculate the numerator of the SW statistic we need the appropriate values of $a_{n-i+1} = a_{21-i}$, $i = 1,\ldots,10$, which are given in SW's table as

$a_{20} = .4734$ $a_{15} = .1334$
$a_{19} = .3211$ $a_{14} = .1013$
$a_{18} = .2565$ $a_{13} = .0711$
$a_{17} = .2085$ $a_{12} = .0422$
$a_{16} = .1686$ $a_{11} = .0140$

Then

$$\begin{aligned}\sum_{i=1}^{10} a_{21-i}\left(x_{(21-i)} - x_{(i)}\right) &= .4734(16.0611+7.0890) + .3211(14.4254+5.6489)\\ &+ .2565(11.3091+5.1510) + .2085(6.2581+4.9672)\\ &\vdots \qquad\qquad\qquad \vdots\\ &+ .0422(-2.7810+3.5914) + .0140(-2.9827+3.2087)\\ &= 26.8090.\end{aligned}$$

The critical value of $w$ is found from SW's table to be $w_{20;.05} = .905$. Then, since $w = (26.8090)^2/906.0822 = .7932 \leq .905$, $H_0$: $\boldsymbol{\varepsilon} \sim N(\mathbf{0}, \sigma^2\mathbf{I})$ is *rejected* at the .05 level. □

*10.8.1.4 Jarque-Bera Test*

The Jarque-Bera (JB) test[11] assesses the null hypothesis that the observations are *iid* from a normal population distribution $N(\mu, \sigma^2)$, versus the alternative hypothesis that the observations follow some other non-normal distribution. The test can also be used to test for normality of the residuals of a GLM that has an intercept term as part of the specification. The test can be altered to accommodate GLM model specifications that do not have intercept terms.

The rationale for the test is to assess whether the skewness and the kurtosis of the sample of observations are consistent with data that would be generated by a normal population distribution. The test statistic is defined by

$$JB = \frac{n}{6}\left(\left(\frac{\hat{\mu}_3}{\hat{\sigma}^3}\right)^2 + \frac{1}{4}\left(\frac{\hat{\mu}_4}{\hat{\sigma}^4} - 3\right)^2\right)$$

where

$$\frac{\hat{\mu}_3}{\hat{\sigma}^3} = \frac{n^{-1}\sum_{i=1}^{n}(x_i - \bar{x})^3}{\left(n^{-1}\sum_{i=1}^{n}(x_i - \bar{x})^2\right)^{3/2}} \text{ and } \frac{\hat{\mu}_4}{\hat{\sigma}^4} = \frac{n^{-1}\sum_{i=1}^{n}(x_i - \bar{x})^4}{\left(n^{-1}\sum_{i=1}^{n}(x_i - \bar{x})^2\right)^2}.$$

[11] C. Jarque and A. Bera, "A Test for Normality of Observations and Regression Residuals", *International Statistical Review*, pp.163–172, 1987.

Jarque and Bera demonstrated that the JB statistics has an asymptotic Chisquare distribution with 2 degrees of freedom. Based on its asymptotic distribution, the test would be conducted as

$$JB\left\{\begin{array}{c} > \\ \leq \end{array}\right\}\chi^2_{2;\alpha} \Rightarrow \left\{\begin{array}{c} \text{reject } \mathrm{H}_0 \\ \text{do not reject } H_0 \end{array}\right\}.$$

However, there is some Monte Carlo evidence suggesting that the convergence of the statistic to its asymptotic distribution is rather slow, so that relatively large sample sizes are needed for critical values set by the Chisquare distribution to be accurate. For smaller sample sizes, some analysts resort to the Lilliefors test mentioned above, while others (e.g., see the MATLAB implementation,[12] when $n < 2{,}000$) have used Monte Carlo simulations to approximate the appropriate critical values to use with the test. Jarque and Bera, in their article (see footnote 3) describe how computer simulation can be used to estimate the small sample distribution of the JB statistic, from which critical values for testing can be derived when $n$ is relatively small.

### 10.8.2 Testing the iid Assumption

In this section we examine a test of the hypothesis that $(X_1,\ldots,X_n)$ is an *iid* random sample from some population distribution, (i.e., a simple random sample), the alternative hypothesis being that $(X_1,\ldots,X_n)$ is a sample from some general (non-*iid*) experiment. That is, $H_0$: $X_i$'s are *iid* versus $H_a$: $X_i$'s are *not iid*. The test we will present, the Wald-Wolfowitz runs test, depends on the concept of a *run* in the outcome of a random sample.

#### *10.8.2.1 Runs*

To be able to define a run, we need to categorize the data according to some dichotomous criteria, which results in the sample outcomes being transformed into a collection of *iid* Bernoulli outcomes. A **run** is defined to be a succession of one or more identical values (1's or 0's), preceded and followed by a different value, or else no value at all if the run occurs at the beginning or end of the sample sequence. For example, in the sequence of 0's and 1's given by

0 0 1 1 1 0 1 0 1 1 0 0 1 1

there are eight runs which we differentiate by vertical lines as

00|111|0|1|0|11|00|11.

The basic idea of using runs to test the *iid* assumption is that if the random variables are truly *iid*, then there should be neither too few nor too many runs

[12] Analysis of the JB-Test in MATLAB. MathWorks. http://www.mathworks.com/access/helpdesk/help/toolbox/stats/jbtest.html. Retrieved May 2, 2012.

observed in any given outcome of the sample. Too few runs could be indicative of grouping, clustering, or trending. Too many runs could indicate a systematic alternating pattern. The Wald-Wolfowitz Runs test exploits this idea rigorously.

*10.8.2.2 Wald-Wolfowitz Runs Test of the* iid *Assumption*

The Wald-Wolfowitz[13] (WW) rums test is concerned with testing whether or not $(X_1,\ldots,X_n)$ can be considered an *iid* random sample from some population distribution. According to some dichotomous characteristic, we transform the outcome of $X_i$ into a 1 or 0, and we apply the same dichotomous characterization to all of the $x_i$'s. Note that the characteristic can be something inherent to the experiment, such as the male/female characteristic in a case where $X_i$'s refer to some measurement of consumer's response in a (hypothesized) random sample of consumers. Alternatively, a dichotomy could be imposed on random variables whose outcomes are in the form of some numerical response by assigning a 1 to values exceeding a specified value (e.g., the median value) and a 0 to values below the specified value.

However the dichotomy is defined, note that if there are $n_1$ responses equaling 1 and $n_0$ responses equaling 0, then there are $\binom{n_1+n_0}{n_1}$ different ways to rearrange the $n_1$ 1's and $n_0$ 0's. If the $x_i$'s are truly *iid* (i.e., $H_0$ is true), then each of these rearrangements is equally likely, and the probability of each rearrangement is $\binom{n_1+n_0}{n_1}^{-1}$ by classical probability.

Now let $w_1$ and $w_0$ represent the number of runs involving 1's and involving 0's, respectively. Note, by the definition of a run, it must be the case that $|w_1 - w_0| \leq 1$. Examine the case where the total number of runs, $w = w_1 + w_0$, is *even* so that $w_1 = w_0 = w/2$, and consider the definition of sequences consisting of $n_1$ 1's and $n_0$ 0's that represent $w_1$ runs of 1's and $w_0$ runs of 0's. How many different ways can we define the sequence of $w_1$ sets of 1's in the collection of $w$ runs? This number is equivalent to the number of different ways $w_1 - 1 = w/2 - 1$ vertical lines can be inserted in the $n_1 - 1$ spaces between the $n_1$ 1's (recall our previous use of vertical lines to delineate runs), and equals $\binom{n_1-1}{(w/2)-1}$. Similarly, the number of different ways we can define the sequence of $w_0$ sets of 0's in the collection of $w$ runs is $\binom{n_0-1}{(w/2)-1}$. Finally, since the sequence of $w$ runs can begin with either a group of 0's or a group of 1's, there are $2\binom{n_1-1}{(w/2)-1}\binom{n_0-1}{(w/2)-1}$ different ways of obtaining $w$ runs when $w$ is even, and thus

[13] Wald, A. and Wolfowitz, J., (1940), *On a test whether two samples are from the same population.* Ann. Math. Statist. 11, pp. 147–162.

$$P(w;n_1,n_0) = \frac{2\binom{n_1-1}{(w/2)-1}\binom{n_0-1}{(w/2)-1}}{\binom{n_1+n_0}{n_1}}, \text{ for even w.}$$

Now examine the case were $w = w_1 + w_0$ is *odd*, so that either $w_1 = (w + 1)/2$ and $w_0 = (w-1)/2$ if the sequence begins and ends with a group of 1's, or else $w_1 = (w-1)/2$ and $w_0 = (w+1)/2$ if the sequence begins and ends with a group of 0's. Using logic analogous to the case where $w$ was even, we have that the number of different ways of obtaining $w$ runs when $w$ is odd is given by the numerator of

$$P(w;n_1,n_0) = \frac{\binom{n_1-1}{(w-1)/2}\binom{n_0-1}{(w-3)/2} + \binom{n_1-1}{(w-3)/2}\binom{n_0-1}{(w-1)/2}}{\binom{n_1+n_0}{n_1}}, \text{ for odd w.}$$

We define $P(w; n_1, n_0) = 0$ in all other cases.

The discrete density function $P(w; n_1, n_0)$ can be used to define upper and lower critical values, and thus upper and lower rejection regions for the test statistic outcome $w$. As usual, the size of the test will be determined by the choice of critical values and the associated upper and lower tail probabilities as indicated by $P(w; n_1, n_0)$.

It can be shown that the mean and variance of $P(w; n_1, n_0)$ are given by

$$E(W) = \frac{2n_1n_0}{n_1+n_0} + 1 \text{ and } \mathrm{var}(W) = \frac{2n_1n_0(2n_1n_0 - n_1 - n_0)}{(n_1+n_0)^2(n_1+n_0-1)}.$$

Furthermore, as $n_1 \to \infty$ and $n_0 \to \infty$, it follows that $[W - E(W)]/[\mathrm{var}(W)]^{1/2} \xrightarrow{d} N(0,1)$, allowing asymptotically valid size $\alpha$ test of the *iid* assumption to be defined in terms of standard normal critical values. The rate of convergence to the limiting distribution is very rapid, so that for $n_1 > 10$ and $n_0 > 10$, the normal distribution approximation is very good. A table of critical values based on $P(w;n_1,n_0)$ for $n \le 20$ has been published by Swed and Eisenhart, (1943), *Annals of Mathematical Statistics*, (14), pp. 66–87.

A size $\alpha = \alpha_\ell + \alpha_h$ runs test can be defined as follows:

$$w\left\{\begin{matrix}\in\\ \notin\end{matrix}\right\}[0, w_{n_1,n_0;1-\alpha_\ell}] \cup [w_{n_1,n_0;\alpha_h}, n] \Rightarrow \left\{\begin{matrix}\text{reject } H_0\\ \text{do not reject } H_0\end{matrix}\right\}.$$

**Example 10.32**
***Testing the iid Assumption via the Wald-Wolfowitz Runs Test***

We will test whether the observations on the waiting times between customers given in Example 10.28 can be viewed as *iid* observations from some population distributionusing a test of size .05.

The observations occurred sequentially row-wise in the data matrix of Example 10.28, and a median of the observation is $(1.50 + 1.68)/2 = 1.59$.

Classifying the sample outcomes as 1's or 0's according to whether the outcome is > or < 1.59 yields the following results, with runs delineated by vertical lines

0|1|0000|1111|0|1|0|11111|00|1|0|1|0|1|0|1|00|11|000|1|00|11|0.

In this case, there are $w = 23$ runs. Since $n_1$ and $n_0$ are each $> 10$, we use the normal approximation to the distribution of $W$ to conduct the test. Note $n_1 = n_0 = 20$ in this case, so that

$$\mathrm{E}(W) = \frac{2(20)(20)}{40} + 1 = 21 \quad \text{and}$$

$$\mathrm{var}(W) = \frac{2(20)(20)(2(20)(20) - 20 - 20)}{(20+20)^2(20+20-1)} = 9.7436.$$

The size $\alpha$ runs test based on the asymptotic normal distribution for $W$ is

$$z = \frac{w - \mathrm{E}(W)}{(\mathrm{var}(W))^{1/2}} \left\{ \begin{matrix} \in \\ \notin \end{matrix} \right\} (-\infty, -z_{\alpha/2}] \cup [z_{\alpha/2}, \infty) \Rightarrow \left\{ \begin{matrix} \text{reject } H_0 \\ \text{do not reject } H_0 \end{matrix} \right\}.$$

Since in the case at hand $z = (23 - 21)/(9.7436)^{1/2} = .6407$ and $z_{.025} = 1.96$, then $z \notin (-\infty, -1.96] \cup [1.96, \infty)$, and the *iid* hypothesis is *not* rejected. □

One can consider using the runs test in the GLM context to check the validity of the *iid* assumption for the disturbances. Unfortunately, even if the $\boldsymbol{\varepsilon}$ vector of $\mathbf{Y} = \mathbf{x}\boldsymbol{\beta} + \boldsymbol{\varepsilon}$ consists of *iid* random variables, $\mathbf{e} = \mathbf{Y} - \mathbf{x}\hat{\boldsymbol{\beta}}$ does not, since $\mathbf{Cov}(\mathbf{e}) = \sigma^2\left(\mathbf{I} - \mathbf{x}(\mathbf{x}'\mathbf{x})^{-1}\mathbf{x}'\right)$. However, as $n$ increases, $\mathbf{e} = \left(\mathbf{I} - \mathbf{x}(\mathbf{x}'\mathbf{x})^{-1}\mathbf{x}'\right)\boldsymbol{\varepsilon} \approx \boldsymbol{\varepsilon}$ if $(\mathbf{x}'\mathbf{x})^{-1} \to \mathbf{0}$, which occurs very generally (see Section 10.8.2) and so for large samples, the runs test may serve as an approximate test.

There are a number of alternative nonparametric tests of randomness that have been proposed in the literature, including the signs test, the Mann-Kendall test, and Bartel's rank test, among others. The interested reader can continue her reading in this area by consulting books devoted to nonparametric statistical inference, such as the book by J. D. Gibbons and S. Chakraborti, (1992) *Nonparametric Statistical Inference*. New York: Marcel Dekker.

## 10.9 Noncentral $\chi^2$- and *F*-Distributions

### 10.9.1 Family Name: Noncentral $\chi^2$-Distribution

*Parameterization*: $v$ (degrees of freedom) and $\lambda$ (noncentrality parameter).
*Definition*: $f(x; v, \lambda) = \sum_{j=0} (\lambda^j/j!)\, e^{-\lambda} h(x; v+2j) I_{(0,\infty)}(x)$ where $h(x; v+2j)$ is the (central) $\chi^2$-density with $v+2j$ degrees of freedom.
*Moments*: $\mu = v + 2\lambda$, $\sigma^2 = 2(v + 4\lambda)$, $\mu_3 = 8(v + 6\lambda)$
*MGF*: $M_X(t) = (1-2t)^{-v/2} \exp\left(\frac{2\lambda t}{1-2t}\right)$ for $t < 1/2$

The noncentral $\chi^2$ distribution is the distribution of $Y = \sum_{i=1}^{v} Z_i^2$, where the $Z_i$'s are independent normally distributed random variables with unit variances and means $\phi_i$, $i = 1,\ldots,v$, i.e., $Z_i \sim N(\phi_i, 1)$, $i = 1,\ldots,v$ with $Z_1,\ldots,Z_n$ being independent random variables. The density definition can be found via the change of variables approach of Section 6.6. The noncentrality parameter $\lambda$ is related to the means of the $Z_i$'s as $\lambda = \phi'\phi/2$.

Note that the density definition can be recognized as a Poisson-weighted sum of (central) $\chi^2_{v+2j}$ density functions, $j = 0,1,2,\ldots$, and in effect can be thought of as the expected value of a $\chi^2$-density with $v + 2J$ degrees of freedom, the "random variable" $J$ having a Poisson density. This type of density definition, where the parameters of one density function family are effectively being treated as random variables and an expectation is taken with respect to another family of density functions, produces what is known as a **mixture distribution**. That is, members of one family are effectively being "mixed" using weights, applied to parameters, that are provided by another density family.

When $\lambda = 0$, the *central* $\chi^2$-distribution is defined, which we discussed in Section 4.2. When $\lambda > 0$, the mean of the distribution moves to the right, and the variance and skewness increase so that the density has less height and has a fatter, more pronounced right tail compared to the central $\chi^2$ density. In particular, for any $c < \infty$,

$$\lim_{\lambda\to\infty} (P(x \geq c; \lambda)) = \lim_{\lambda\to\infty} \left( \int_c^\infty f(x; v, \lambda) dx \right) = 1$$

and $P(x \geq c; \lambda)$ is monotonically increasing as a function of $\lambda$.

A principal use of the noncentral $\chi^2$-distribution is in analyzing the power function of statistical tests based on $\chi^2$-statistics whose distribution under $H_a$ is noncentral $\chi^2$. Power in these cases is generally represented as a function of the noncentrality parameter as

$$\pi_{C_r}(\lambda) = \int_{x\in C_r} f(x; v, \lambda) dx.$$

Examples are given by the GLR, LM, and Wald statistics, all of which have asymptotically valid noncentral $\chi^2$-distributions under local alternative hypotheses in $H_a$.

Integration of the noncentral $\chi^2$-distribution is straightforwardly accomplished on personal computers using software such as GAUSS, SAS, or MATLAB. The reader must be warned that the parameterization of noncentrality is *not* standard in the literature. Other parameterizations include noncentrality equal to $2\lambda$ and $\sqrt{2\lambda}$. The parameterization we use is more prevalent in the econometrics literature.

For further information, see N.L. Johnson and S. Kotz (1970), *Continuous Univariate Distributions*, II, New York: Wiley, pp. 130–148.

### 10.9.2 Family Name: Noncentral F-Distribution

*Parameterization*: $\nu_1$ (numerator degrees of freedom), $\nu_2$ (denominator degrees of freedom), $\lambda$ (noncentrality parameter)
*Definition*:

$$f(x;\nu_1,\nu_2,\lambda) = \sum_{j=0}^{\infty}\left(\frac{\lambda^j}{j!}\right)e^{-\lambda}\left[\frac{(\nu_1/\nu_2)\left(\frac{\nu_1}{\nu_2}x\right)^{(\nu_1/2)+j-1}}{B((\nu_1/2)+j,\ \nu_2/2)\left(1+\frac{\nu_1}{\nu_2}x\right)^{.5(\nu_1+\nu_2)+j}}\right] I_{(0,\infty)}(x)$$

where $B(a,b)$ is the Beta function (recall the Beta distribution discussion in Section 4.2.3).
*Moments*:

$$\mu = \frac{\nu_2(\nu_1+2\lambda)}{\nu_1(\nu_2-2)} \text{ for } \nu_2 > 2,$$

$$\sigma^2 = 2\left(\frac{\nu_2}{\nu_1}\right)^2 \frac{(\nu_1+2\lambda)^2+(\nu_1+4\lambda)(\nu_2-2)}{(\nu_2-2)^2(\nu_2-4)} \text{ for } \nu_2 > 4$$

$\mu_3 = g(\nu_1,\nu_2,\lambda)$ increasing in $\lambda$ and $\nu_1$, decreasing in $\nu_2$ for $\nu_2 > 6$.

*MGF*: Does not exist.

The noncentral *F*-distribution is the distribution of $Y = (Z_{\nu_1}/\nu_1)/(Z_{\nu_2}/\nu_2)$, where $Z_{\nu_1} \sim \chi^2_{\nu_1}(\lambda)$ and $Z_{\nu_2} \sim \chi^2_{\nu_2}$ are independent noncentral and central $\chi^2$ random variables with $\nu_1$ and $\nu_2$ degrees of freedom, respectively. The density of $Y$ can be established via transformation from the joint density of $(Z_{\nu_1}, Z_{\nu_2})$, which is the product of a noncentral and a central $\chi^2$ density.

When $\lambda = 0$, $Y$ is the ratio of two independent central $\chi^2$-random variables, each divided by their degrees of freedom, so that $Y$ has the (central) *F*-distribution as we derived in Section 6.7. When $\lambda > 0$, the mean of the distribution moves to the right, and the variance and skewness also increase so that the density has less height and has a fatter, more pronounced right tail in comparison to the central *F*-distribution. In particular, for $c < \infty$,

$$\lim_{\lambda\to\infty}(P(x \geq c;\lambda)) = \lim_{\lambda\to\infty}\left(\int_c^{\infty} f(x;\nu_1,\nu_2,\lambda)dx\right) = 1,$$

and $P(x \geq c;\lambda)$ is monotonically increasing as a function of $\lambda$.

As in the case of the noncentral $\chi^2$-distribution, a principal use of the noncentral *F*-distribution is in analyzing the power function of statistical tests based on test statistics that have a noncentral *F*-distribution when $H_a$ is true. Power in these cases is generally represented as a function of the noncentrality parameter as

$$\pi_{C_r}(\lambda) = \int_{x\in C_r} f(x;\nu_1,\nu_2,\lambda)dx.$$

An example of the use of the noncentral $F$-distribution was given in the discussion of testing $H_0$: $\mathbf{R\beta}=\mathbf{r}$ in the GLM context under normality, in which case the power function was representable in terms of a noncentral $F$-distribution.

Integration of the noncentral $F$-distribution is straightforwardly accomplished on personal computers using software such as GAUSS, SAS, and MATLAB. The warning regarding the parameterization of noncentrality given in our discussion of the noncentral $\chi^2$-distribution applies equally well here. Further information can be found in N.L. Johnson and S. Kotz, *Continuous Distributions*, pp. 189–200.

## 10.10 Appendix: Proofs and Proof References for Theorems

**Theorem 10.5: Proof** (This proof is somewhat involved. For a related proof, see T. Amemiya, (1985), *Advanced Econometrics*, Harvard University Press, Cambridge, pp. 142–144. The original work of A. Wald, (1943), "Tests of Statistical Hypotheses Concerning Several Parameters when the Number of Observations is Large," *Transactions of the American Mathematical Society*, pp. 426–482 provides an alternative, albeit more difficult and restrictive proof of the asymptotic distribution result).

Expand $\ln(L(\mathbf{\Theta}_0;\mathbf{X}))$ in a second order Taylor series around the ML estimator $\hat{\mathbf{\Theta}}$ as[14]

$$\ln\left(L\left(\hat{\mathbf{\Theta}};\mathbf{X}\right)\right)-\ln(L(\mathbf{\Theta}_0;\mathbf{X}))=-\frac{1}{2}\left(\hat{\mathbf{\Theta}}-\mathbf{\Theta}_0\right)'\frac{\partial^2\ln(L(\mathbf{\Theta}_*;\mathbf{X}))}{\partial\mathbf{\Theta}\partial\mathbf{\Theta}'}\left(\hat{\mathbf{\Theta}}-\mathbf{\Theta}_0\right), \tag{10.1}$$

where $\mathbf{\Theta}_* = \tau\hat{\mathbf{\Theta}} + (1-\tau)\mathbf{\Theta}_0$ *for* $\tau\in[0,1]$ and the first-order term is dropped since $\partial L\left(\hat{\mathbf{\Theta}},\mathbf{X}\right)/\partial\mathbf{\Theta}=\mathbf{0}$ by the first order conditions defining the MLE. Let $\mathbf{R}(\mathbf{\Theta})=\mathbf{r}$ be used to define $q$ of the entries in $\mathbf{\Theta}$ as functions of the remaining entries, and without loss of generality, assume the entries in $\mathbf{\Theta}$ have been ordered so that the first $q$ $\Theta_i$'s are functions of the remaining $(k-q)$ $\Theta_i$'s, say as $\mathbf{\Theta}^a=\mathbf{g}(\mathbf{\Theta}^b)$, where $\mathbf{\Theta}^a=(\Theta_1,\ldots,\Theta_q)$ and $\mathbf{\Theta}^b=(\Theta_{q+1},\ldots,\Theta_k)$. Then under the constraint $\mathbf{R}(\mathbf{\Theta})=\mathbf{r}$, the feasible $\mathbf{\Theta}$-vectors can be characterized as

$$\mathbf{\Theta}=\begin{bmatrix}\mathbf{g}\left(\mathbf{\Theta}^b\right)\\ \mathbf{\Theta}^b\end{bmatrix}=\mathbf{h}\left(\mathbf{\Theta}^b\right).$$

Making this substitution in the likelihood function defines the restricted likelihood function $L_r(\mathbf{\Theta}^b;\mathbf{X})\equiv L(\mathbf{h}(\mathbf{\Theta}^b);\mathbf{X})$. To simplify notation, *define* $\boldsymbol{\eta}\equiv\mathbf{\Theta}^b$. Assuming $H_0$: $\mathbf{R}(\mathbf{\Theta})=\boldsymbol{r}$ to be true, then $\mathbf{\Theta}_0=\mathbf{h}(\boldsymbol{\eta}_0)$, and we can expand $\ln(L_r(\boldsymbol{\eta}_0;\mathbf{x}))$ in a second order Taylor series around the ML estimate $\hat{\boldsymbol{\eta}}$, analogous to the preceding expansion, as

$$\ln(L_r(\hat{\boldsymbol{\eta}};\mathbf{X}))-\ln(L_r(\boldsymbol{\eta}_0;\mathbf{X}))=-\frac{1}{2}(\hat{\boldsymbol{\eta}}-\boldsymbol{\eta}_0)'\frac{\partial^2\ln(L_r(\boldsymbol{\eta}_*;\mathbf{X}))}{\partial\boldsymbol{\eta}\partial\boldsymbol{\eta}'}(\hat{\boldsymbol{\eta}}-\boldsymbol{\eta}_0). \tag{10.2}$$

[14] R.G. Bartle, op. cit., p. 371.

Now note that since $\hat{\Theta} \xrightarrow{p} \Theta_0$ and $\hat{\eta} \to \eta_0$, then $\Theta_* \xrightarrow{p} \Theta_0$ and $\eta_* \to \eta_0$. Then rewriting the right hand side of (1) as

$$-\frac{1}{2}\left(\hat{\Theta}-\Theta_0\right)' n^{1/2}\left[n^{-1}\frac{\partial^2 \ln(L(\Theta_*;\mathbf{X}))}{\partial\Theta\partial\Theta'}\right] n^{1/2}\left(\hat{\Theta}-\Theta_0\right),$$

note that the square bracketed expression converges in probability to $\mathbf{M}(\Theta_0)$, *a* positive definite symmetric matrix. Then $\ln(L(\hat{\Theta};\mathbf{X})) - \ln(L(\Theta_0;\mathbf{X}))$ and $-(1/2)(\hat{\Theta}-\Theta_0)'\, n^{1/2}\, \mathbf{M}(\Theta_0)\, n^{1/2}(\hat{\Theta}-\Theta_0)$ will share the same limiting distribution by Theorem 5.7. An analogous argument applied to (2) indicates that $\ln(L_r(\hat{\eta};\mathbf{X})) - \ln(L_r(\eta_0;\mathbf{x}))$ and

$$-\frac{1}{2}(\hat{\eta}-\eta_0)' n^{1/2} M_r(\eta_0) n^{1/2}(\hat{\eta}-\eta_0)$$

share the same limiting distribution. Letting $\stackrel{d}{=}$ indicate equivalence in terms of limiting distributions, it follows from $L(\Theta_0;\mathbf{x}) = L_r(\eta_0;\mathbf{x})$ that

$$-2\ln(\lambda(\mathbf{X})) \stackrel{d}{=} \left(\hat{\Theta}-\Theta_0\right)' n^{1/2}\mathbf{M}(\Theta_0) n^{1/2}\left(\hat{\Theta}-\Theta_0\right) - (\hat{\eta}-\eta_0)' n^{1/2}\mathbf{M}_r(\eta_0) n^{1/2}(\hat{\eta}-\eta_0) \quad (10.3)$$

We now relate the limiting distributions of $n^{1/2}\left(\hat{\Theta}-\Theta_0\right)$ and $n^{1/2}(\hat{\eta}-\eta_0)$. It follows from the proof of Theorem 8.17 that

$$n^{1/2}\left(\hat{\Theta}-\Theta_0\right) \stackrel{d}{=} \mathbf{M}(\Theta_0)^{-1} n^{-1/2}\frac{\partial \ln(L(\Theta_0;\mathbf{X}))}{\partial\Theta} \xrightarrow{d} N\left(\mathbf{0}, \mathbf{M}(\Theta_0)^{-1}\right)$$

and

$$n^{1/2}(\hat{\eta}-\eta_0) \stackrel{d}{=} \mathbf{M}_r(\eta_0)^{-1} n^{-1/2}\frac{\partial \ln(L_r(\eta_0;\mathbf{X}))}{\partial\eta} \xrightarrow{d} N\left(0, \mathbf{M}_r(\eta_0)^{-1}\right).$$

Because

$$\frac{\partial \ln(L_r(\eta_0;\mathbf{X}))}{\partial\eta} = \frac{\partial h(\eta_0)}{\partial\eta}\frac{\partial \ln(L(\Theta_0;\mathbf{X}))}{\partial\Theta},$$

it follows that

$$n^{1/2}(\hat{\eta}-\eta_0) \stackrel{d}{=} \mathbf{M}_r(\eta_0)^{-1}\, n^{-1/2}\frac{\partial h(\eta_0)}{\partial\eta} n^{1/2}\left[n^{-1/2}\frac{\partial \ln(L(\Theta_0;\mathbf{X}))}{\partial\Theta}\right].$$

Substituting for $n^{1/2}\left(\hat{\Theta}-\Theta_0\right)$ and $n^{1/2}(\hat{\eta}-\eta_0)$ on the right hand side of (10.3), and letting
$\mathbf{Z}_n = n^{-1/2}(\partial \ln(L(\Theta_0;\mathbf{X})/\partial\Theta))$ to simplify notation,

$$-2\ln(\lambda(\mathbf{X})) \stackrel{d}{=} \mathbf{Z}'_n\left[\mathbf{M}(\Theta_0)^{-1} - \mathbf{G}'\mathbf{M}_r(\eta_0)^{-1}\mathbf{G}\right]\mathbf{Z}_n, \quad (10.4)$$

where $\mathbf{G} = \partial h(\eta_0)/\partial\eta$.

Define $\mathbf{V}_n = \mathbf{M}(\Theta_0)^{-1/2}\,\mathbf{Z}_n$, so that $\mathbf{V}_n \xrightarrow{d} \mathbf{V} \sim N(\mathbf{0},\mathbf{I})$, and rewrite (10.4) *as*

$$-2\ln(\lambda(\mathbf{X})) \stackrel{d}{=} \mathbf{V}'_n\left[I - \mathbf{M}(\Theta_0)^{1/2}\mathbf{G}'\mathbf{M}_r(\eta_0)^{-1}\mathbf{G}\mathbf{M}(\Theta_0)^{1/2}\right]\mathbf{V}_n. \quad (10.5)$$

The bracketed matrix is idempotent, which can be demonstrated by multiplying the bracketed matrix by itself and using the condition $\mathbf{G}\ \mathbf{M}(\boldsymbol{\Theta}_0)\mathbf{G}' = \mathbf{M}_r(\boldsymbol{\eta}_0)$. Furthermore, the trace of the idempotent matrix is seen to be $k - (k - q) = q$. Then representing the quadratic form (10.5) in terms of the characteristic roots and vectors of the idempotent matrix as $\mathbf{V}'_n\ \mathbf{P}'\ \boldsymbol{\Lambda}\ \mathbf{P}\ \mathbf{V}_n$, analogous to the proof of Theorem 6.12b, we finally obtain

$$-2\ln(\lambda(\mathbf{X})) \xrightarrow{d} \mathbf{V}'_n\mathbf{P}\boldsymbol{\Lambda}\mathbf{P}'\mathbf{V} = \sum_{i=1}^{q} W_i^2 \sim \chi^2_{q'}$$

where $\mathbf{W}=\mathbf{P}'\mathbf{V} \sim N(\mathbf{0}, \mathbf{I})$ and $\boldsymbol{\Lambda}$ is a diagonal matrix *having* $q$-1's and $(k - q)$-0's along the diagonal. ■

**Theorem 10.6: Proof** Let $\hat{\boldsymbol{\Theta}}_r$ and $\hat{\boldsymbol{\Theta}}$ represent the MLEs for $\boldsymbol{\Theta}\in H_0$ and $\boldsymbol{\Theta}\in H_0 \cup H_a$, respectively, and expand $\ln(L(\hat{\boldsymbol{\Theta}}_r;\mathbf{X}))$ in a second order Taylor series around the point $\hat{\boldsymbol{\Theta}}$ to obtain

$$-2\left[\ln\left(L\left(\hat{\boldsymbol{\Theta}}_r;\mathbf{X}\right)\right) - \ln\left(L\left(\hat{\boldsymbol{\Theta}};\mathbf{X}\right)\right)\right] = -2\ln(\lambda(\mathbf{X}))$$

$$= \left(\hat{\boldsymbol{\Theta}}_r - \hat{\boldsymbol{\Theta}}\right)'\left[\frac{\partial^2 \ln(L(\boldsymbol{\Theta}_+;\mathbf{X}))}{\partial\boldsymbol{\Theta}\partial\boldsymbol{\Theta}'}\right]\left(\hat{\boldsymbol{\Theta}}_r - \hat{\boldsymbol{\Theta}}\right),$$

where we have used the fact that $\partial\ln\left(L(\hat{\boldsymbol{\Theta}};\mathbf{X})\right)/\partial\boldsymbol{\Theta} = \mathbf{0}$ in the first order term of the Taylor series and $\boldsymbol{\Theta}_+ = \tau\hat{\boldsymbol{\Theta}} + (1-\tau)\hat{\boldsymbol{\Theta}}_r$ for some $\tau\in[0,1]$. We know from the proofs of Theorems 8.18 and 8.19 that

$$n^{1/2}\left(\hat{\boldsymbol{\Theta}} - \boldsymbol{\Theta}_0\right) \stackrel{d}{=} \mathbf{M}(\boldsymbol{\Theta}_0)^{-1}\left[n^{-1/2}\frac{\partial\ln(L(\boldsymbol{\Theta}_0;\mathbf{X}))}{\partial\boldsymbol{\Theta}}\right]$$

(recall $\stackrel{d}{=}$ means equivalence in terms of limiting distributions).

Under the sequence of local alternatives $H_{an}$: $\mathbf{R}(\boldsymbol{\Theta}) = \mathbf{r} + n^{-1/2}\ \boldsymbol{\phi}$, $\hat{\boldsymbol{\Theta}}_r \xrightarrow{p} \boldsymbol{\Theta}_0$, and it can be shown that (see the subsequent proofs of Theorems 10.7 and 10.8)

$$n^{1/2}\left(\hat{\boldsymbol{\Theta}}_r - \boldsymbol{\Theta}_0\right) \stackrel{d}{=} \mathbf{M}(\boldsymbol{\Theta}_0)^{-1}\left[n^{-1/2}\frac{\partial\ln(L(\boldsymbol{\Theta}_0;\mathbf{X}))}{\partial\boldsymbol{\Theta}}\right] + \mathbf{M}(\boldsymbol{\Theta}_0)^{-1}\mathbf{G}'\left[-\mathbf{G}\mathbf{M}(\boldsymbol{\Theta}_0)^{-1}\mathbf{G}'\right]^{-1} \times$$
$$\left[\mathbf{G}\mathbf{M}(\boldsymbol{\Theta}_0)^{-1}\left(n^{-1/2}\frac{\partial\ln(L(\boldsymbol{\Theta}_0;\mathbf{X}))}{\partial\boldsymbol{\Theta}}\right) - \boldsymbol{\phi}\right]$$

where $\mathbf{G} = \partial\mathbf{R}(\boldsymbol{\Theta}_0)/\partial\boldsymbol{\Theta}'$. Then

$$n^{1/2}\left(\hat{\boldsymbol{\Theta}}_r - \hat{\boldsymbol{\Theta}}\right) = n^{1/2}\left(\hat{\boldsymbol{\Theta}}_r - \boldsymbol{\Theta}_0\right) - n^{1/2}\left(\hat{\boldsymbol{\Theta}} - \boldsymbol{\Theta}_0\right)$$
$$\stackrel{d}{=} \mathbf{M}(\boldsymbol{\Theta}_0)^{-1}\mathbf{G}'\left[-\mathbf{G}\mathbf{M}(\boldsymbol{\Theta}_0)^{-1}\mathbf{G}'\right]^{-1}\left[\mathbf{G}\mathbf{M}(\boldsymbol{\Theta}_0)^{-1}\left(n^{-1/2}\frac{\partial\ln(L(\boldsymbol{\Theta}_0;\mathbf{X}))}{\partial\boldsymbol{\Theta}}\right) - \boldsymbol{\phi}\right],$$
$$\stackrel{d}{=} \mathbf{M}(\boldsymbol{\Theta}_0)^{-1}\mathbf{G}'\mathbf{Z}_n,$$

where $\mathbf{Z}_n \xrightarrow{d} N([\mathbf{G}\mathbf{M}(\boldsymbol{\Theta}_0)^{-1}\ \mathbf{G}']^{-1}\ \boldsymbol{\phi}, [\mathbf{G}\mathbf{M}(\boldsymbol{\Theta}_0)^{-1}\ \mathbf{G}']^{-1})$.

Noting that $-2\ln(\lambda(\mathbf{X})) \stackrel{d}{=} n^{1/2}(\hat{\Theta}_r - \hat{\Theta})' \mathbf{M}(\Theta_0)(\hat{\Theta}_r - \Theta_0)n^{1/2}$ because

$$-n^{-1}\frac{\partial^2 \mathbf{L}(\Theta_+;\mathbf{X})}{\partial\Theta\partial\Theta'} \xrightarrow{p} \mathbf{M}(\Theta_0),$$

it follows that

$$-2ln(\lambda(\mathbf{X})) \stackrel{d}{=} \mathbf{Z}'_n\left[\mathbf{G}\mathbf{M}(\Theta_0)^{-1}\mathbf{G}'\right]\mathbf{Z}_n \xrightarrow{d} \chi^2_q(\lambda)$$

with $\lambda = \frac{1}{2}\boldsymbol{\phi}'\left[\mathbf{G}\mathbf{M}(\Theta_0)^{-1}\mathbf{G}'\right]^{-1}\boldsymbol{\phi}$ because

$$\left[\mathbf{G}\mathbf{M}(\Theta_0)^{-1}\mathbf{G}'\right]^{1/2}\mathbf{Z}_n \xrightarrow{d} N\left(\left[\mathbf{G}\mathbf{M}(\Theta_0)^{-1}\mathbf{G}'\right]^{-1/2}\boldsymbol{\phi},\ \mathbf{I}\right)$$

(see Section 10.9 on properties of the noncentral $\chi^2$ distribution). ■

**Theorem 10.7: Proof** Expanding both $\partial\ln\left(L\left(\hat{\Theta}_r;\mathbf{X}\right)\right)/\partial\Theta$ and $\mathbf{R}\left(\hat{\Theta}_r\right) - \mathbf{r}$ in a first order Taylor series around the true $\Theta_0$ allows the first order conditions of the ML problem to be written as

$$\frac{\partial\ln L(\Theta_0;\mathbf{X})}{\partial\Theta} + \frac{\partial^2\ln L(\Theta_*;\mathbf{X})}{\partial\Theta\partial\Theta'}\left(\hat{\Theta}_r - \Theta_0\right) - \frac{\partial\mathbf{R}\left(\hat{\Theta}_r\right)}{\partial\Theta}\boldsymbol{\Lambda}_r = \mathbf{0},$$
$$\frac{\partial\mathbf{R}\left(\hat{\Theta}_+\right)'}{\partial\Theta}\left(\hat{\Theta}_r - \Theta_0\right) = \mathbf{0}$$

where $\Theta_*$ and $\Theta_+$ each lie between $\Theta_0$ and $\hat{\Theta}_r$.[15] Note the second equation incorporates the fact that the first term in the Taylor series, $\mathbf{R}(\Theta_0) - \mathbf{r}$, is zero under $H_0$. Premultiplying the first of the preceding equations *by* $n^{-1/2}$ and the second by $n^{1/2}$, leads to the partitioned matrix equation

$$\left[\begin{array}{c|c} -n^{-1}\dfrac{\partial^2\ln L(\Theta_*;\mathbf{X})}{\partial\Theta\partial\Theta'} & \dfrac{\partial\mathbf{R}\left(\hat{\Theta}_r\right)}{\partial\Theta} \\ \hline \dfrac{\partial\mathbf{R}(\Theta_+)'}{\partial\Theta} & \mathbf{0}\end{array}\right]\begin{bmatrix} n^{1/2}\left(\hat{\Theta}_r - \Theta_0\right) \\ n^{-1/2}\boldsymbol{\Lambda}_r\end{bmatrix} = \begin{bmatrix} n^{-1/2}\dfrac{\partial\ln L(\Theta_0;\mathbf{X})}{\partial\Theta} \\ \mathbf{0}\end{bmatrix} \quad (*)$$

[15]We must alert the reader to a technical point that we suppress notationally regarding the use of Taylor series representations of *vector* functions. Specifically, such a representation is actually a collection of Taylor series representations, one for each entry in the vector function, say $\mathbf{f}(\mathbf{z})$. As such, the point of evaluation of the final derivative terms in each Taylor series can differ for each coordinate function. For example, if $\mathbf{f}(\mathbf{z})$ is $(j\times 1)$, then in the Taylor series representation

$$\mathbf{f}(\mathbf{z}) = \mathbf{f}(\mathbf{z}_0) + \frac{\partial\mathbf{f}(\mathbf{z}_*)'}{\partial\mathbf{z}}(\mathbf{z} - \mathbf{z}_0)$$

it can be that each *row* of $\frac{\partial\mathbf{f}(\mathbf{z}_*)'}{\partial\mathbf{z}}$ must be evaluated at a *different* $\mathbf{z}_* = \tau\mathbf{z} + (1-\tau)\mathbf{z}_0$, $\tau\in[0,1]$. Having alerted the reader to this situation, we will tacitly assume henceforth that this is understood. What is most important for our purposes is that $\mathbf{z}_* \to \mathbf{z}_0$ as $\mathbf{z} \to \mathbf{z}_0$, so in this case, in the limit, all rows of $\frac{\partial\mathbf{f}(\mathbf{z}_*)'}{\partial\mathbf{z}}$ will be evaluated at the same point.

Observe that the right hand side of the matrix equation (*) converges in distribution to the vector $\begin{bmatrix} \mathbf{Z} \\ \hline \mathbf{0} \end{bmatrix}$, where $\mathbf{Z} \sim N(\mathbf{0}, \mathbf{M}(\boldsymbol{\Theta}_0))$.

In examining the asymptotic behavior of $n^{1/2}(\hat{\boldsymbol{\Theta}}_r - \boldsymbol{\Theta}_0)$ and $n^{-1/2}\boldsymbol{\Lambda}_r$ it will be permissible by Theorem 5.9 to write the first matrix of (*) as

$$\left[\begin{array}{c|c} \mathbf{M}(\boldsymbol{\Theta}_0) & \dfrac{\partial \mathbf{R}(\boldsymbol{\Theta}_0)}{\partial \boldsymbol{\Theta}} \\ \hline \dfrac{\partial \mathbf{R}(\boldsymbol{\Theta}_0)'}{\partial \boldsymbol{\Theta}} & \mathbf{0} \end{array}\right]$$

because $\hat{\boldsymbol{\Theta}}_r$, and thus $\boldsymbol{\Theta}_*$ and $\boldsymbol{\Theta}_{+}$, $\xrightarrow{p} \boldsymbol{\Theta}_0$, $\partial \mathbf{R}(\boldsymbol{\Theta})/\partial \boldsymbol{\Theta}$ is continuous, and

$$n^{-1}\frac{\partial^2 \ln L(\boldsymbol{\Theta}_*; \mathbf{X})}{\partial \boldsymbol{\Theta} \partial \boldsymbol{\Theta}'} \xrightarrow{p} -M(\boldsymbol{\Theta}_0).$$

Using partitioned inversion,[16] $n^{-1/2}\boldsymbol{\Lambda}_r$ can be solved for, yielding

$$n^{-1/2}\boldsymbol{\Lambda}_r = \left[\frac{\partial \mathbf{R}(\boldsymbol{\Theta}_0)'}{\partial \boldsymbol{\Theta}}\mathbf{M}(\boldsymbol{\Theta}_0)^{-1}\frac{\partial \mathbf{R}(\boldsymbol{\Theta}_0)}{\partial \boldsymbol{\Theta}}\right]^{-1}\frac{\partial \mathbf{R}(\boldsymbol{\Theta}_0)'}{\partial \boldsymbol{\Theta}}\mathbf{M}(\boldsymbol{\Theta}_0)^{-1}\left[n^{-1/2}\frac{\partial \ln(L(\boldsymbol{\Theta}_0; \mathbf{X}))}{\partial \boldsymbol{\Theta}}\right]$$

If $\mathbf{G} = \partial \mathbf{R}(\boldsymbol{\Theta}_0)/\partial \boldsymbol{\Theta}'$ has full row rank, it follows by Slutsky's theorems that

$$n^{-1/2}\boldsymbol{\Lambda}_r \xrightarrow{d} N\left(\mathbf{0}, \left[\mathbf{G}\mathbf{M}(\boldsymbol{\Theta}_0)^{-1}\mathbf{G}'\right]^{-1}\right),$$

and

$$\left[\mathbf{G}\mathbf{M}(\boldsymbol{\Theta}_0)^{-1}\mathbf{G}'\right]^{1/2} n^{-1/2}\boldsymbol{\Lambda}_r \xrightarrow{d} N(\mathbf{0}, \mathbf{I}).$$

Because the limiting distribution is unaffected by replacing $\mathbf{G}$ with $\partial \mathbf{R}\left(\hat{\boldsymbol{\Theta}}_r\right)/\partial \boldsymbol{\Theta}'$ and $\mathbf{M}(\boldsymbol{\Theta}_0)$ by $\left[-n^{-1}\left(\partial^2 \ln\left(L\left(\hat{\boldsymbol{\Theta}}_r; X\right)\right)/\partial \boldsymbol{\Theta} \partial \boldsymbol{\Theta}'\right)\right]$ we finally have (recall $\overset{d}{=}$ means equivalent in limiting distribution)

$$W = n^{-1}\boldsymbol{\Lambda}'_r\left[\mathbf{G}\mathbf{M}(\boldsymbol{\Theta}_0)^{-1}\mathbf{G}'\right]\boldsymbol{\Lambda}_r \overset{d}{=} \boldsymbol{\Lambda}'_r \frac{\partial \mathbf{R}\left(\hat{\boldsymbol{\Theta}}_r\right)'}{\partial \boldsymbol{\Theta}}\left[-\left(\frac{\partial^2 \ln L\left(\hat{\boldsymbol{\Theta}}_r; \mathbf{X}\right)}{\partial \boldsymbol{\Theta} \partial \boldsymbol{\Theta}'}\right)\right]^{-1}\frac{\partial \mathbf{R}\left(\hat{\boldsymbol{\Theta}}_r\right)}{\partial \boldsymbol{\Theta}}\boldsymbol{\Lambda}_r,$$

$$\xrightarrow{d} \chi^2_q \text{ under } H_0.$$

The size $\alpha$ test indicated in part (2) of the theorem follows immediately from the limiting distribution of $W$. Consistency can be demonstrated by retracing the previous argument beginning with the initial Taylor series expansion of the first order conditions, but with $\mathbf{R}(\boldsymbol{\Theta}_0) - \mathbf{r} = \boldsymbol{\phi} \neq \mathbf{0}$ so that $\mathbf{0}$ is replaced by $-n^{1/2}\boldsymbol{\phi}$

[16] One can use the following result for symmetric matrices (Theil, H., (1971), *Principles of Econometrics*, John Wiley, NY, p. 18:

$$\left[\begin{array}{c|c} \mathbf{A} & \mathbf{C} \\ \hline \mathbf{C}' & \mathbf{B} \end{array}\right]^{-1} = \left[\begin{array}{c|c} \mathbf{A}^{-1} + \mathbf{A}^{-1}\mathbf{C}(\mathbf{B} - \mathbf{C}'\mathbf{A}^{-1}\mathbf{C})^{-1}\mathbf{C}'\mathbf{A}^{-1} & -\mathbf{A}^{-1}\mathbf{C}(\mathbf{B} - \mathbf{C}'\mathbf{A}^{-1}\mathbf{C})^{-1} \\ \hline -(\mathbf{B} - \mathbf{C}'\mathbf{A}^{-1}\mathbf{C})^{-1}\mathbf{C}'\mathbf{A}^{-1} & (\mathbf{B} - \mathbf{C}'\mathbf{A}^{-1}\mathbf{C})^{-1} \end{array}\right]$$

in the vector on the right hand side of the equality in (*). The net result is that $P(w > \tau) \to 1$ *as* $n \to \infty \; \forall \tau > 0$, and so $H_0$ is rejected for any rejection region of size $\in(0,1)$ of the form $\left[\chi^2_{q;\alpha}, \infty\right)$. See Silvey, S.D. op. cit. pp. 387–407, for further discussion.

The alternative representation of $W$ in part (3) of the theorem follows immediately from the first order conditions $\partial \ln\left(L\left(\hat{\boldsymbol{\Theta}}_r; \mathbf{X}\right)\right)/\partial\boldsymbol{\Theta} = \left(\partial R\left(\hat{\boldsymbol{\Theta}}_r\right)/\partial\boldsymbol{\Theta}\right)\boldsymbol{\Lambda}_r$. ■

**Theorem 10.8: Proof** Assuming $\mathbf{R}(\boldsymbol{\Theta}) - \mathbf{r} = n^{-1/2}\,\boldsymbol{\phi} \to \mathbf{0}$, as implied by the sequence of local alternatives, it follows that the sequence of restricted MLEs of $\boldsymbol{\Theta}_0$ is such that $\hat{\boldsymbol{\Theta}}_r \xrightarrow{p} \boldsymbol{\Theta}_0$. Matrix equation (*) in the proof of Theorem 10.7 applies except that $\mathbf{0}$ in the vector on the right hand side of the equality changes to $-\boldsymbol{\phi}$ (since the Taylor series expansion of $\mathbf{R}\left(\hat{\boldsymbol{\Theta}}_r\right) - \mathbf{r} = \mathbf{0}$ is now

$$n^{-1/2}\boldsymbol{\phi} + \frac{\partial \mathbf{R}(\boldsymbol{\Theta}_+)}{\partial\boldsymbol{\Theta}'}\left(\hat{\boldsymbol{\Theta}}_r - \boldsymbol{\Theta}_0\right) = \mathbf{0}.$$

Following the proof of Theorem 10.7 in applying partitioned inversion and the convergence result of Theorem 5.9, the revised matrix equation (*) yields

$$n^{-1/2}\boldsymbol{\Lambda}'_r \stackrel{d}{=} \left[\frac{\partial \mathbf{R}(\boldsymbol{\Theta}_0)'}{\partial\boldsymbol{\Theta}}\mathbf{M}(\boldsymbol{\Theta}_0)^{-1}\frac{\partial \mathbf{R}(\boldsymbol{\Theta}_0)}{\partial\boldsymbol{\Theta}}\right]^{-1}\left[\frac{\partial \mathbf{R}(\boldsymbol{\Theta}_0)'}{\partial\boldsymbol{\Theta}}\mathbf{M}(\boldsymbol{\Theta}_0)^{-1}\left(n^{-1/2}\frac{\partial \ln(L(\boldsymbol{\Theta}_0;\mathbf{X}))}{\partial\boldsymbol{\Theta}}\right) - \boldsymbol{\phi}\right].$$

By Slutsky's theorem, the right-most bracketed term converges in distribution to

$$N\left(-\boldsymbol{\phi}, \frac{\partial \mathbf{R}(\boldsymbol{\Theta}_0)'}{\partial\boldsymbol{\Theta}}\mathbf{M}(\boldsymbol{\Theta}_0)^{-1}\frac{\partial \mathbf{R}(\boldsymbol{\Theta}_0)}{\partial\boldsymbol{\Theta}}\right),$$

and letting $\mathbf{G} = \partial\mathbf{R}(\boldsymbol{\Theta}_0)/\partial\boldsymbol{\Theta}'$, it follows that

$$\left[\mathbf{G}\mathbf{M}(\boldsymbol{\Theta}_0)^{-1}\mathbf{G}'\right]^{1/2} n^{-1/2}\boldsymbol{\Lambda}_r \xrightarrow{d} N\left(-\left[\mathbf{G}\mathbf{M}(\boldsymbol{\Theta}_0)^{-1}\mathbf{G}'\right]^{-1/2}\boldsymbol{\phi}, \mathbf{I}\right)$$

Finally, since

$$\partial\mathbf{R}\left(\hat{\boldsymbol{\Theta}}_r\right)/\partial\boldsymbol{\Theta} \xrightarrow{p} \mathbf{G} \text{ and } n^{-1}\left(\partial^2 \ln\left(L\left(\hat{\boldsymbol{\Theta}}_r;\mathbf{X}\right)\right)/\partial\boldsymbol{\Theta}\partial\boldsymbol{\Theta}'\right) \xrightarrow{p} -\mathbf{M}(\boldsymbol{\Theta}_0)$$

then

$$\left[\frac{\partial\mathbf{R}\left(\hat{\boldsymbol{\Theta}}_r\right)'}{\partial\boldsymbol{\Theta}}\left[-n^{-1}\frac{\partial^2 \ln L\left(\hat{\boldsymbol{\Theta}}_r;\mathbf{X}\right)}{\partial\boldsymbol{\Theta}\partial\boldsymbol{\Theta}'}\right]^{-1}\frac{\partial\mathbf{R}\left(\hat{\boldsymbol{\Theta}}_r\right)}{\partial\boldsymbol{\Theta}}\right]^{1/2} \xrightarrow{p} \left[\mathbf{G}\mathbf{M}(\boldsymbol{\Theta}_0)^{-1}\mathbf{G}'\right]^{1/2}$$

so that by Slutsky's theorem,

$$W = \boldsymbol{\Lambda}'_r\frac{\partial\mathbf{R}\left(\hat{\boldsymbol{\Theta}}_r\right)'}{\partial\boldsymbol{\Theta}}\left[-\frac{\partial^2 \ln L\left(\hat{\boldsymbol{\Theta}}_r;\mathbf{X}\right)}{\partial\boldsymbol{\Theta}\partial\boldsymbol{\Theta}'}\right]^{-1}\frac{\partial\mathbf{R}\left(\hat{\boldsymbol{\Theta}}_r\right)}{\partial\boldsymbol{\Theta}}\boldsymbol{\Lambda}_r \xrightarrow{d} \chi^2_q(\lambda),$$

with $\lambda = (1/2)\boldsymbol{\phi}'\left[\mathbf{G}\mathbf{M}(\boldsymbol{\Theta}_0)^{-1}\mathbf{G}'\right]^{-1}\boldsymbol{\phi}$ (see Section 10.9 for properties of the noncentral $\chi^2$ distribution. ■

## Keywords, Phrases, and Symbols

Asymptotic confidence intervals and regions
Asymptotically equivalent tests
Asymptotic pivotal quantity
Asymptotic power
Asymptotically unbiased tests
Confidence coefficient
Confidence interval, confidence region
Confidence level
Consistent confidence region
Duality between confidence and critical regions
Generalized likelihood ratio (GLR) test
Heuristic principle of test construction
Jarque-Bera Test
Kolmogorov-Smirnov test statistic
Lagrange multiplier (LM) test
Lilliefors Test
Local alternatives
Mixture distribution
Noncentral $F$ distribution
Noncentral $t$-distribution
Noncentral $\chi^2$ distribution
Pittman drift
Pivotal quantity
Pivotal quantity method
Restricted likelihood function
Runs
Scoring test
Shapiro-Wilks Test
Unbiased confidence region
Uniformly most accurate (UMA)
Uniformly most accurate unbiased (UMAU)
Upper and lower confidence intervals
Wald Test
Wald-Wolfowitz (WW) Runs Test
$\chi^2$ Goodness of fit test

## Problems

**1.** A manufacturer of breakfast cereals has been accused of systematically underfilling their cereal packages. The manufacturer claims that the complaint must stem from settling of the cereal that occurs in shipment, and that the air space at the top of the package is normal settling of the contents. They claim that the filling process is normally distributed, and that the system is *under control* at a mean fill rate of $\mu = \mathrm{E}(Z) = 16.03$ ounces and a standard deviation of $\sigma = .01$, so that it is highly improbable that a package is filled below its stated contents of 16 ounces. They want you to test the hypothesis that the filling system is under control at a level of significance equal to .10. They provide you with the following random sample outcome of 40 observations on fill weights:

| | | | |
|---|---|---|---|
| 15.97 | 15.75 | 15.90 | 15.87 |
| 15.96 | 15.90 | 16.05 | 16.04 |
| 16.13 | 15.92 | 15.70 | 15.89 |
| 15.79 | 15.74 | 15.88 | 15.86 |
| 16.01 | 15.89 | 15.83 | 15.97 |
| 15.92 | 15.88 | 15.84 | 15.95 |
| 15.82 | 16.07 | 16.01 | 16.04 |
| 15.92 | 15.81 | 15.71 | 15.95 |
| 15.88 | 15.81 | 15.85 | 15.84 |
| 15.79 | 16.03 | 15.80 | 15.80 |

(a) Define a size .10 GLR test of the hypothesis $H_0$: $\mu = \mu_0$, $\sigma = \sigma_0$ versus $H_a$: not $H_0$. Test the hypothesis that the filling process is under control. You may use the asymptotic distribution of the GLR if you wish.

(b) Define a size .05 LM test of the same hypothesis as in (a) above, and test the hypothesis at significance level .10. Are the two tests different? Are the two test decisions in agreement?

(c) Test the two hypotheses $H_0$: $\mu = 16.03$ and $H_0$: $\sigma = .01$ *individually* using size .05 tests. Use whatever test procedures you feel are appropriate. Interpret the outcomes of the tests individually. Interpret the tests jointly using a Bonferroni approach.

(d) Can you define a Wald test for the hypothesis in (a)? If so, perform a Wald test of the joint null hypothesis at significance level .10.

**2. Testing for Differences in Two Populations: Variances.** After the firm in question (1) was inspected by the Dept. of Weights and Measures, a new random sample of forty observations on fill weights was taken the next day, resulting in the following summary statistics:

$m'_1 = 16.02753$ and $s^2 = .9968 \times 10^{-2}$.

You may continue to assume normality for the population distribution.

(a) Define a GLR size .05 test of the $H_0$: $\sigma_1^2 \le \sigma_2^2$ versus $H_a$: not $H_0$, where $\sigma_1^2$ and $\sigma_2^2$ refer to the variances of the populations from which the first and second sample were taken. Test the hypothesis.

(b) Repeat (a) for the hypothesis $H_0$: $\sigma_1^2 \ge \sigma_2^2$ versus $H_a$: not $H_0$.

(c) Define a GLR size .05 test of the $H_0$: $\sigma_1^2 = \sigma_2^2$ versus $H_a$: not $H_0$. Either go to the computer and perform this test, or else define an approximation to this test based on "equal tails" and perform the test.

(d) Define the power function for each of the hypothesis testing procedures you defined above. With the aid of a computer, plot the power functions and interpret their meanings.

**3. Testing for Differences in Two Populations: Means (Equal Variances).**
Referring to the data obtained in both preceding questions, consider testing whether the two populations means are the same. *Assume that the population variances are identical in the two cases–did you find any evidence to contradict this in your answer to (2) above?*

(a) Define a GLR size .05 test of the $H_0$: $\mu_1 \leq \mu_2$ versus $H_a$: not $H_0$, where $\mu_1$ and $\mu_2$ refer to the means of the populations from which the first and second sample were taken. Test the hypothesis.

(b) Repeat (a) for the hypothesis $H_0$: $\mu_1 \geq \mu_2$ versus $H_a$: not $H_0$.

(c) Define a GLR size .05 test of the $H_0$: $\mu_1 = \mu_2$ versus $H_a$: not $H_0$ . Test the hypothesis at the .10 level of significance.

(d) Define the power function for each of the hypothesis testing procedures you defined above. With the aid of a computer, plot the power functions and interpret their meanings.

**4. Testing for Differences in Two Populations: Means (Unequal Variances) (The Behrens-Fisher Problem).**
Consider the hypotheses in Problem (3) in the case where it is *not* assumed that the variances of the two populations are equal.

(a) Attempt to find a GLR test of the hypothesis $H_0$: $\mu_1 = \mu_2$ versus $H_a$: not $H_0$ based on a t-statistic similar to what you found in (3) above. Is this approach valid? The problem of testing hypotheses concerning the means of two normal population distributions when it is not assumed that the variances are equal is known as the Behrens-Fisher problem, which as of yet has no universally accepted solution.

One *reasonable* solution to this problem is to use the statistic

$$t = \frac{\bar{x}_1 - \bar{x}_2}{\left(s_1^2/n_1 + s_2^2/n_2\right)^{1/2}}$$

where $n_1$ and $n_2$ refer to the respective sample sizes of samples from the two populations. It can be shown that a test of size $\leq \alpha$ of any of the hypotheses concerning the means of the two population distributions given in (3a)–(3c) can be performed by defining rejection regions as follows:

| $H_0$ | $H_a$ | $C_r$ |
|---|---|---|
| $\mu_1 = \mu_2$ | $\mu_1 \neq \mu_2$ | $(-\infty, -t_{\alpha/2}(m)] \cup [t_{\alpha/2}(m), \infty)$ |
| $\mu_1 \geq \mu_2$ | $\mu_1 < \mu_2$ | $(-\infty, -t_\alpha(m)]$ |
| $\mu_1 \leq \mu_2$ | $\mu_1 > \mu_2$ | $[t_\alpha(m), \infty)$ |

The value of $t_\alpha(m)$ is a typical critical value of the student t distribution found in *t*-tables, except the degrees of freedom parameter $m = \min(n_1, n_2)$. The procedure is sometimes referred to as the *Hsu procedure*. Further discussion of the Behrens Fisher problem can be found in H. Scheffe, (1970), "Practical Solutions of the Behrens-Fisher Problems," *JASA*, 65, pp. 1501–1508.

(b) Using the Hsu procedure discussed above, test the hypothesis of the equality of the means in the population distributions referred to in Problems (1) and (2) above.

**5.** Regarding the asymptotically valid Wald and *Z*-tests for the value of $\sigma^2$ in the case of nonnormally distributed populations and the GLM as discussed at the end of Section 10.6, justify their asymptotic size and consistency.

**6.** Recall the analysis of the demand for textiles in the Netherlands performed by renown econometrician Henri Theil, as discussed in Problem 8.9. With reference to the least squares-based estimate of the constant elasticity demand function generated from the data in the problem, respond to the following, using size .10 tests when testing is called for:

(a) Test the hypothesis that the matrix of explanatory variables is significant in explaining changes in the logarithm of textile demand.

(b) Test whether income is significant in explaining textile consumption. Plot the power function of the test, and interpret its meaning.

(c) Test whether price is significant in explaining textile consumption.

(d) Test whether the price elasticity is inelastic. Test whether the income elasticity is inelastic.

(e) Is the demand equation homogeneous degree zero in price and income? Plot the power function of the test, and interpret its meaning.

(f) Calculate a confidence interval that has .95 a confidence coefficient for the income elasticity. Interpret the meaning of this confidence interval.

(g) Calculate a confidence interval that has .95 a confidence coefficient for the price elasticity. Interpret the meaning of this confidence interval.

(h) Calculate a confidence region having confidence coefficient .90 for the income and price elasticities. With the aid of a computer, graph the confidence region. Superimpose the intervals you calculated in (f) and (g) on this graph. The regions are different–interpret the difference.

(i) Test the least squares residuals for normality, preferably using the Shapiro-Wilks test (you'll need tables for this), or else use the chisquare goodness of fit test.

**7.** Testing The Equality of Two Exponential Population Distributions. The Reliable Computer Co. is considering purchasing CPU chips from one of two different suppliers to use in the production of personal computers. It has two bids from the suppliers, with supplier number offering the lower bid. Before making a purchase decision, you want to test the durability of the chips, and you obtain a random sample of 50 chips from each supplier. It is known that both CPUs have operating lives that are exponentially distributed, and you want to test the equality of the expected operating lives of the chips.

(a) Define a size .10 test of the equality of the means of the two population distributions, i.e., a test, of $H_0$: $\theta_1 = \theta_2$ versus $H_a$: not $H_0$.

(b) The respective sample means of operating lives for the two sample were $\bar{x}_1 = 24.23$ and $\bar{x}_2 = 18.23$. Conduct the test of the null hypothesis. Does the test outcome help you decide which supplier to purchase chips from?

(c) How would your test rule change if you wanted to test a one-sided hypothesis concerning the means of the population distributions?

(d) Consider using the LM test procedure for this problem. What is the test rule? Can you perform the test with the information provided?

(e) Consider using the WALD test procedure for this problem. What is the test rule? Can you perform the test with the information provided?

**8.** The Huntington Chemical Co. has been accused of dumping improperly treated waste in a local trout lake, and litigation may be imminent. The central allegation is that the trout population has been severely reduced by the company's dumping practices. The company counters with the claim that there still remain at least 100,000 trout in the lake, and this being a substantial number for a lake of its size, they are innocent of all charges. In order to investigate the size of the trout population, 500 trout are captured, tagged, and then returned to the lake. After the fish have redistributed themselves in the lake, 500 fish are captured, and the number of tagged fish are observed. Letting $\bar{x}$ represent the proportion of tagged fish, can a size $\leq .10$ test of the company's claim be defined? Would an outcome of $\bar{x} = .14$ support or contradict the company's claim?

**9.** Use a Wald-Wolfowitz runs test to assess whether the data provided by the cereal manufacturer in Problem 1 is the outcome of a random sample from some population distribution. The observations occurred sequentially *row-wise*. Use a size .05 test.

**10.** Test whether the data of the cereal manufacturer provided in Problem 1 can be interpreted as an outcome of a random sample from the specific normal population distribution claimed, i.e. test $H_0$: $Z \sim N(16.03, .0001)$ versus $H_a$: not $H_0$. Perform another test that assesses the more general hypothesis that the data is the outcome of a random sample from *some* normal population distribution, i.e. test $H_0$: $Z \sim N(\mu, \sigma^2)$ versus $H_a$: not $H_0$. Use size .05 tests.

**11. Paired Comparisons of Means.** A family counseling service offers a 1-day program of training in meal planning which they claim is effective in reducing cholesterol levels of individuals completing the course. In order to assess the effectiveness of the program, a consumer advocacy group randomly samples the cholesterol levels of 50 program participants both one day before and one month after the course is taken. They then summarize the pairs of observations on the individuals by reporting the sample mean and sample standard deviation of the *differences* between the before and after observations of the 50 program participants. Their finding were $\bar{d} = -11.73$ and $s = 3.89$.

(a) Assuming that the pairs of observations are *iid* outcomes from some bivariate normal population distribution with *before* and *after* means $\mu_b$ and $\mu_a$, define the appropriate likelihood function for the mean $\mu = \mu_a - \mu_b$ and variance $\sigma^2$ of the population distribution of *differences* in pairs of cholesterol measurements. Use this likelihood function to define a GLR size .05 test of the hypothesis that the meal planning program has no effect, i.e. a test for $H_0$: $\mu = 0$ versus $H_a$: not $H_0$. Test the hypothesis.

(b) Can you define an LM test of the null hypothesis in (a)? Can you test the hypothesis with the information available? If so, perform the test–if not, what other information would you need?

(c) Describe how you might test the hypothesis that the observations on paired differences are from a normal population distribution. What information would you need to test the hypothesis?

(d) Suppose that normality of the observations is *not* assumed. Can you define another test of the effectiveness of the meal planning program? If so, use the test to assess the effectiveness of the meal planning program. Discuss any differences that are required in the interpretation of the outcome of this test compared to the test in (a) that you could perform if the observations were normally distributed.

**12.** The production function for a commodity can be approximated over a restricted range of relevant input levels by a linear relationship of the form $\mathbf{Y} = \mathbf{x}\boldsymbol{\beta} + \boldsymbol{\varepsilon}$. The columns of $\mathbf{x}$ include, in order, a column of 1's, and three column vectors representing observations on labor, energy, and capital input levels. Thirty observations on weekly production levels yielded the following summary statistics:

$$\mathbf{b} = (\mathbf{x}'\mathbf{x})^{-1}\mathbf{x}'\mathbf{y} = \begin{bmatrix} 13.792 \\ 3.005 \\ 1.327 \\ 6.385 \end{bmatrix}, \hat{s}^2 = 3.775,$$

$$(\mathbf{x}'\mathbf{x})^{-1} = \begin{bmatrix} 4.002 & -.136 & -.191 & -.452 \\ & .020 & .006 & -.007 \\ & & .041 & .001 \\ \text{(symmetric)} & & & .099 \end{bmatrix}$$

In answering the following questions, be sure to clearly define any assumptions that you are making.

(a) Test the joint significance of the input variables for explaining the expected level of production using a size .05 test.

(b) Test the significance of the input variables individually. Which input variables contribute significantly to the explanation of the expected level of output? Be sure to explain the basis for your conclusion.

(c) Define confidence interval outcomes for each of the marginal products of the inputs. Use .95 confidence coefficients. What do these confidence intervals mean?

(d) Test the hypothesis that expected output is $\leq 35$ when labor, energy and capital are applied at levels 8, 4, and 3, respectively. Use a size .05 test.

(e) Calculate a confidence interval outcome for the expected level of production at the input levels indicated in (d). Use a .95 confidence coefficient. What is the meaning of this confidence interval?

(f) Test the hypothesis that the variance of the production process is $\geq 6$ at a .10 level of significance. Assume normality in conducting this test. What information would you need to conduct an asymptotically valid test of the hypothesis in the absence of normality?

**13.** The production department of a paper products manufacturer is analyzing the frequency of breakdowns in a certain type of envelope machine as it contemplates future machinery repair and replacement policy. It has been suggested that the number of breakdowns per day is a Poisson distributed random variables, and the department intends to test this conjecture. Forty days worth of observations on the number daily machine breakdowns yielded the following observations which occurred in sequence row-wise:

| | | | |
|---|---|---|---|
| 4 | 8 | 5 | 4 |
| 3 | 5 | 8 | 2 |
| 7 | 4 | 6 | 8 |
| 7 | 9 | 8 | 8 |
| 6 | 2 | 5 | 0 |
| 8 | 1 | 7 | 3 |
| 6 | 2 | 0 | 2 |
| 0 | 4 | 3 | 5 |
| 1 | 3 | 9 | 5 |
| 3 | 4 | 3 | 0 |

(a) Test the hypothesis that the observations are a random sample outcome from some population distribution. Use a .10 level test.

(b) Test the hypothesis that the observations are from a Poisson population distribution. Use a size .10 test.

(c) Test the hypothesis that the observations are from a Uniform population distribution. Use a size .10 test.

(d) Test the hypothesis that the expected number of daily breakdowns for this equipment is $\geq 8$ using a significance level of .10. Use whatever test procedure you feel is appropriate. Based on this test procedure, calculate a .90 level confidence interval for the expected number of daily breakdowns.

**14.** The relationship between sales and level of advertising expenditure is hypothesized to be quadratic over the relevant range of expenditure levels being examined, i.e., $Y_t = \beta_1 + \beta_2 a_t + \beta_3 a_t^2 + \varepsilon_t$, where $y_t$ is the level of sales in period t, $a_t$ is the level of advertising expenditure in period t, and the $\varepsilon_t$'s are disturbance terms assumed to be *iid* normally distributed. The least squares-based estimate of the relationship using 30 periods worth of observations resulted in the following:
$\mathbf{b} = (103.27\ \ 2.71\ \ -.13)$, $\hat{s}^2 = 1.27$

$$(\mathbf{x}'\mathbf{x})^{-1} = \begin{bmatrix} 111.17 & 1.71 & .013 \\ & .64 & .0017 \\ \text{symmetric} & & .004 \end{bmatrix}$$

(a) Test the hypothesis that advertising expenditure has a significant impact on the expected sales level. Use any significance level you feel is appropriate.

(b) Test whether the relationship between advertising expenditures and sales is actually a linear as opposed to a quadratic relationship.

(c) Define a confidence interval with confidence coefficient .95 for the expected level of sales expressed as a function of advertising level $a_t$. Plot the confidence interval as function of advertising expenditures (use expected sales on the vertical axis and advertising expenditures on the horizontal axis). Interpret the plot.

(d) Test the hypothesis that the level of advertising expenditure that maximizes expected sales is $\geq 15$.

(e) Calculate a confidence interval with confidence coefficient .95 for the level of advertising expenditure that maximizes expected sales.

**15.** An analyst is investigating the effect of certain policy events on common stock prices in a given industry. In an attempt to isolate abnormal from normal returns of firms in the industry, the following returns-generating equation was estimated via least squares: $R_t = \beta_1 + \beta_2\, R_{mt} + \varepsilon_t$ where $R_t$ is actual return by a firm on day $t$ and $R_{mt}$ is the return on a large portfolio of stocks designed to be representative of market return on day t. Abnormal returns is defined to be $\text{AR}_t = \varepsilon_t$, and is estimated from an estimate of the returns-generating equation as

$$\widehat{\text{AR}}_t = \text{R}_t - \hat{\text{R}}_t, \quad \text{where } \hat{R}_t = b_1 + b_2 R_{mt}.$$

Thus, the estimated abnormal returns are effectively the estimated residuals from the least squares estimate of the returns generating equation. A summary of the estimation results from an analysis of 43 observations on a given firm is given below:

$$\mathbf{b} = [.03\ \ 1.07]', \quad (\mathbf{x}'\mathbf{x})^{-1} = \begin{bmatrix} .0001 & .2 \times 10^{-4} \\ .2 \times 10^{-4} & .81 \end{bmatrix},$$

$$\hat{s}^2 = 1.0752,$$

$\hat{\mathbf{e}} =$

| | | | |
|---|---|---|---|
| −0.44 | 1.49 | −0.46 | −0.90 |
| 0.73 | −0.45 | −1.04 | −0.24 |
| −0.68 | −0.40 | −0.070 | −2.44 |
| −1.21 | 1.53 | 0.21 | −0.30 |
| −0.23 | −0.85 | 0.55 | −0.58 |
| 0.030 | 0.16 | −0.74 | −0.30 |
| 1.52 | 2.39 | 0.28 | 0.66 |
| −0.10 | −0.75 | 1.59 | 0.97 |
| −2.5 | −0.62 | −1.24 | −0.24 |
| 0.14 | 0.76 | 0.84 | 0.16 |
| 0.18 | −0.99 | −1.74 | |

(a) Use a runs test to provide an indication of whether the residuals of the returns generating equation, and hence the abnormal returns, can be viewed as *iid* from some population distribution. Use a size .05 test. Comments on the appropriateness of the runs test in this application.

(b) Test to see whether the abnormal returns can be interpreted as a random sample from some normal population distribution of returns. Use a size .05 level of test.

(c) Assuming normality, test whether the expected daily return of the firm is proportional to market return. Use a size .05 test.

(d) Define and test a hypothesis that will assess whether the expected return of the firm is greater than market return for any $R_{mt} \geq 0$. Use a size .05 test.

**16. Testing for Independence in a Bivariate Normal Population Distribution.** Let $(X_i, Y_i)$, $i = 1,\ldots,n$ be a random sample from some bivariate normal population distribution $N(\boldsymbol{\mu}, \boldsymbol{\Sigma})$ where $\sigma_1^2 = \sigma_2^2$. In this case, we know that the bivariate random variables are independent *iff* the correlation between them is zero. Consider testing the hypothesis of independence $H_0$: $\rho = 0$ versus the alternative hypothesis of dependence $H_a$: $\rho \neq 0$.

(a) Define a size .05 GLR test of the independence of the two random variables. You may use the limiting distribution of the GLR statistic to define the test.

(b) In a sample of 50 observations, it was found that $\sigma_x^2 = 5.37$, $s_y^2 = 3.62$, and $s_{xy} = .98$. Is this sufficient to reject the hypothesis of independence based on the asymptotically valid test above?

(c) Show that you can transform the GLR test into a test involving a critical region for the test statistic $w = s_{xy}/\left[\left(s_x^2 + s_y^2\right)/2\right]$.

(d) Derive the sampling distribution of the test statistic W defined in c) under $H_0$. Can you define a size .05 critical region for the test statistic? If so , test the hypothesis using the exact (as opposed to asymptotic) size .05 GLR test.

**17. Test of the Equality of Proportions.** Let $X_1$ and $X_2$ be independent random variables with Binomial distributions Binomial$(n_i, p_i)$, $i = 1,2$, where the $n_i$'s are assumed known. Consider testing the null hypothesis that the proportions (or probabilities) $p_1$ and $p_2$ are equal, i.e., the hypothesis is $H_0$: $p_1 = p_2$ versus $H_a$: $p_1 \neq p_2$.

(a) Define an asymptotically valid size $\alpha$ Wald-type test of $H_0$ versus $H_a$.

(b) If $n_1 = 45$, $n_2 = 34$, $x_1 = 19$, and $x_2 = 24$, is the hypothesis of equality of proportions rejected if $\alpha = .10$?

(c) Define an asymptotically valid size $\alpha$ GLR-type test of $H_0$ versus $H_a$. Repeat part (b).

**18. Testing Whether Random Sample is From a Discrete Uniform Population Distribution.** Let $X_i$, $i = 1,\ldots,n$ be random sample from some discrete integer-valued population distribution. Let the range of the random variables be $1,2,\ldots,m$.

(a) Describe how you would use the $\chi^2$ goodness of fit size $\alpha$ test to assess the hypothesis that the population distribution was a discrete uniform distribution with support $1,2,\ldots,m$.

(b) The Nevada Gaming Commission has been called in to investigate whether the infamous Dewey, Cheatum, & Howe Casino is using fair dice in its crap games. One of the alleged crooked dice is tossed 240 times, and the following outcomes were recorded:

| *X* | Frequency |
|---|---|
| 1 | 33 |
| 2 | 44 |
| 3 | 47 |
| 4 | 39 |
| 5 | 31 |
| 6 | 46 |

Does this die appear to be fair? What size test will you use? Why?

**19.** Describe how you would test one-sided and two-sided hypotheses, using both finite sample and asymptotically valid testing procedures, in each of the following cases. Use whatever procedures you feel are appropriate.

(a) Testing hypotheses about the mean of a binomial distribution based on a random sample of size 1 from the binomial distribution.

(b) Testing hypotheses about the mean of a Poisson distribution based on a random sample of size $n$ from a Poisson population distribution.

**20.** The weekly proportion of storage tank capacity that is utilized at a regional hazardous liquid waste receiving site is an outcome of a Beta$(\theta,1)$ random variable. At the end of each week the storage tank is emptied. A random sample of a year's worth of observations of capacity utilization at the site produced the results displayed in the table below, reported in sequence row-wise.

(a) Define a GLR size $\alpha$ test of $H_0$: $\theta = \theta_0$ versus $H_0$: $\theta \neq \theta_0$.

(b) Test the hypothesis that the expected weekly storage tank capacity utilized is equal to .5 at a significance level of .95. If the exact distribution of the GLR test appears to be intractable, then use an asymptotic test.

(c) Test that the observations can be interpreted as a random sample from some population distribution. Use a size .05 test.

(d) Test that the observations can be interpreted as a random sample from a Beta($\theta$,1) population distribution. Use a size .05 test.

(e) Would the use of an LM test be tractable here? If so, perform an LM test of the hypothesis.

| | | | |
|---|---|---|---|
| 0.148 | 0.501 | 0.394 | 0.257 |
| 0.759 | 0.763 | 0.092 | 0.155 |
| 0.409 | 0.257 | 0.586 | 0.919 |
| 0.278 | 0.076 | 0.019 | 0.513 |
| 0.123 | 0.390 | 0.030 | 0.082 |
| 0.935 | 0.607 | 0.075 | 0.729 |
| 0.664 | 0.802 | 0.338 | 0.539 |
| 0.055 | 0.984 | 0.269 | 0.069 |
| 0.679 | 0.382 | 0.549 | 0.028 |
| 0.770 | 0.296 | 0.105 | 0.465 |
| 0.194 | 0.675 | 0.696 | 0.068 |
| 0.091 | 0.132 | 0.156 | 0.050 |
| 0.477 | 0.754 | 0.164 | 0.527 |

**21.** Follow the proof of Theorem 10.14 to demonstrate that when random sampling from the continuous PDF $f(z;\Theta)$ with scalar parameter $\Theta$,

$$-2\sum_{i=1}^{n}\ln[1-F(X_i;\Theta)]\sim\chi^2_{2n}$$

is a pivotal quantity for $\Theta$.

**22.** Regional daily demand for gasoline in the summer driving months is assumed to be the outcome of a $N(\mu,\sigma^2)$ random variable. Assume you have 40 *iid* daily observations on daily demand for gasoline with quantity demanded measured in millions of gallons, and $\bar{x}=43$ and $s^2=2$.

(a) Show that $N(\mu,\sigma^2)$ is a location-scale parameter family of PDFs (recall Theorem 10.13).

(b) Define a pivotal quantity for $\mu$, and use it to define a .95 level confidence interval for $\mu$. Also, define a .95 lower confidence bound for $\mu$.

(c) Define a pivotal quantity for $\sigma^2$, and use it to define a .95 level confidence interval for $\sigma^2$.

(d) A colleague claims that mean daily gasoline demand is only 37 million gallons. Is your answer to (b) consistent with this claim? Explain.

**23.** The population distribution of income in a populous developing country is assumed to be given (approximately) by the continuous PDF $f(x;\Theta)=\Theta\,(1+x)^{-(\Theta+1)}\,I_{(0,\infty)}(x)$ where $\Theta>0$. A summary measure of a random sample outcome of size 100 from the population distribution is given by $\sum_{i=1}^{100}\ln(1+x_i)=40.54$, where $x_i$ is measured in 1,000's of units of the developing countries currency.

(a) Define a pivotal quantity for the parameter $\Theta$ (Hint: Theorem 10.13 and Problem 10.21 might be useful here).

(b) Define a general expression for a level $\gamma$ confidence interval for $\Theta$ based on n *iid* observations from $f(x;\Theta)$.

(c) Calculate an outcome of a .90 level confidence interval for $\Theta$ based on the expression derived in (b). Note: For large degrees of freedom > 30,

$$\chi^2_{v;\alpha}\approx v\left[1-\frac{2}{9v}+z_\alpha\left(\frac{2}{9v}\right)^{1/2}\right]^3,$$

where $\int_{z_\alpha}^{\infty}N(x;0,1)\,dx=\alpha$, is a very good approximation to critical values of the $\chi^2$ distribution (M. Abramowitz and I.A. Stegun, *Handbook of Mathematical Functions*, Dover, New York, 1972, p. 941)).

# Math Review Appendix: Sets, Functions, Permutations, Combinations, Notation, and Real Analysis

## A1. Introduction

In this appendix we review basic results concerning set theory, relations and functions, combinations and permutations, summation and integration notation, and some fundamental concepts in real analysis. We also review the meaning of the terms definition, *axiom, theorem, corollary, and lemma*, which are labels that are affixed to a myriad of statements and results that constitute the theory of probability and mathematical statistics. The topics reviewed in this appendix constitute basic foundational material on which the study of mathematical statistics is based. Additional mathematical results, often of a more advanced nature, will be introduced throughout the text as the need arises.

## A2. Definitions, Axioms, Theorems, Corollaries, and Lemmas

The development of the theory of probability and mathematical statistics involves a considerable number of statements consisting of definitions, axioms, theorems, corollaries, and lemmas. These terms will be used for organizing the various statements and results we will examine into these categories:

1. Descriptions of meaning;
2. Statements that are acceptable as true without proof;
3. Formulas or statements that require proof of validity;
4. Formulas or statements whose validity follows immediately from other true formulas or statements; and
5. Results, generally from other branches of mathematics, whose primary purpose is to facilitate the proof of validity of formulas or statements in mathematical statistics.

More formally, we present the following meaning of the terms.

**Definition:** A statement of the meaning of a word, word group, sign, or symbol.

**Axiom (or postulate):** A statement that has found general acceptance, or is thought to be worthy thereof, on the basis of an appeal to intrinsic merit or self-evidence, and thus requires no proof of validity.

**Theorem (or proposition):** A formula or statement that is deduced from other proved or accepted formulas or statements, and whose validity is thereby proved.

**Corollary:** A formula or statement that is immediately deducible from a proven theorem, and that requires little or no additional proof of validity.

**Lemma:** A proven auxiliary proposition stated for the expressed purpose of facilitating the proof of another proposition of more fundamental interest.

Thus, in the development of the theory of probability and mathematical statistics, axioms are the fundamental truths that are to be accepted at face value and not proven. Theorems and their corollaries are statements deducible from the fundamental truths and other proven statements and thus are *derived truths*. Lemmas represent proven results, often from fields outside of statistics per se, that are used in the proofs of other results of more primary interest.

We elaborate on the concept of a lemma, since our discussions will implicitly rely on lemmas more than any other type of statement, but we will generally choose not to exhaustively catalogue lemmas in the discussions. What constitutes a lemma and what does not depends on the problem context or one's point of view. A fundamental integration result from calculus could technically be referred to as a lemma when used in a proof of a statement in mathematical statistics, while in the study of calculus, it might be referred to as a theorem to be proved in and of itself. Since our study will require numerous auxiliary results from algebra, calculus, and matrix theory, exhaustively cataloging these results as lemmas would be cumbersome, and more importantly, not necessary given the prerequisites assumed for this course of study, namely, a familiarity with the basic concepts of algebra, univariate and multivariate calculus, and an introduction to matrix theory. We will have occasion to state a number of lemmas, but we will generally reserve this label for more exotic mathematical results that fall outside the realm of mathematics encompassed by the prerequisites.

## A3. Elements of Set Theory

In the study of probability and mathematical statistics, sets are the fundamental objects to which probability will be assigned, and it is important that the concept of a set, and operations on sets, be well understood. In this section we review some basic properties of and operations on sets. This begs the following question: What is meant by the term *set*? In modern axiomatic developments of set theory, the concept of a set is taken to be primitive and incapable of being defined in terms of more basic ideas. For our purposes, a more intuitive notion of a set will suffice, and we avoid the complexity of an axiomatic development of the theory (see Marsden,[1] Appendix A, for a brief introduction to the axiomatic development). We base our definition of a set on the intuitive definition originally proposed by the founder of set theory, Georg Cantor (1845–1918).[2]

---

[1] J.E. Marsden, (1974), *Elementary Classical Analysis*, San Francisco: Freeman and Co.

[2] Cantor's original text reads: "Unter einer 'Menge' verstehen wir jede Zusammenfassung M von bestimmten wohlunterschiedenen Objekten m unserer Anschauung oder unseres Denkens (welche die 'Elemente' von M genannt werden) zu einem ganzen," (*Collected Papers*, p. 282). Our translation of Cantor's definition is, "By set we mean any collection, *M*, of clearly defined, distinguishable objects, *m*, (which will be called elements of *M*) which from our perspective or through our reasoning we understand to be a whole."

**Definition A.1**
***Set***

A **set** is a collection of objects with the following characteristics:

1. All objects in the collection are *clearly defined*, so that it is evident which objects are members of the collection and which are not;
2. All objects are *distinguishable*, so that objects in the collection do not appear more than once;
3. *Order is irrelevant* regarding the listing of objects in the collection, so two collections that contain the same objects but are listed in different order are nonetheless the *same set*; and
4. *Objects in the collection can be sets themselves*, so that a *set of sets* can be defined.

The objects in the collection of objects comprising a set are its **elements**. The term **members** is also used to refer to the objects in the collection. In order to signify that an object belongs to a given set, the symbol $\in$, will be used in an expression such as $x \in A$, which is to be read "$x$ is an element (or member) of the set $A$." If an object is *not* a member of a given set, then a slash will be used as $x \notin A$ to denote that "$x$ is not an element (or member) of the set $A$." Note that the slash symbol, /, is used to indicate negation of a relationship. The characteristics of sets presented in Definition A.1 will be clarified and elaborated upon in examples and discussions provided in subsequent subsections.

### Set Defining Methods

Three basic methods are used in defining the objects in the collection constituting a given set: (1) **exhaustive listing;** (2) **verbal rule;** and (3) **mathematical rule**. An *exhaustive listing* requires that each and every object in a collection be individually identified either numerically, if the set is a collection of numbers, or by an explicit verbal description, if the collection is not of numbers. The object descriptions are conventionally separated by commas, and the entire group of descriptions is enclosed in brackets. The following are examples of sets defined by an exhaustive listing of the objects that are elements of the set:

**Example A.1** $S_1 = \{\text{HEAD, TAIL}\}$ Here $S_1$ is the set of possible occurrences when tossing a coin into the air and observing its resting position. Note the set can be equivalently represented as $S_1 = \{\text{TAIL, HEAD}\}$. □

**Example A.2** $S_2 = \{1,2,3,4,5,6\}$ Here $S_2$ is the set of positive integers from 1 to 6. Note that the set $S_2$ can be equivalently represented by the listing of the positive integers 1 to 6 in any order. □

A *verbal rule* is a verbal statement of characteristics that only the objects that are elements of a given set possess and that can be used as a test to determine set membership. The general form of the verbal rule is {x: *verbal statement*}, which is to be read "the collection of all $x$ for which *verbal statement* is true." The following are examples of sets described by verbal rules:

**Example A.3** $S_3 = \{x: x \text{ is a college student}\}$ Here $S_3$ is the set of college students. An individual is an element of the set $S_3$ *iff* (if and only if) he or she is a college student. □

**Example A.4** $S_4 = \{x: x \text{ is a positive integer}\}$ Here $S_4$ is the set of positive integers 1,2,3, . . . . *A* number is an element of the set $S_4$ *iff* it is a positive integer. □

A *mathematical rule* is of the same general form as a verbal rule, except the verbal statement is replaced by a mathematical expression of some type. The general form of the mathematical rule is {x: *mathematical expression*}, which is to be read "the collection of all $x$ for which *mathematical expression* is true." The following are examples of sets described by mathematical rules:

**Example A.5** $S_5 = \{x: x = 2k + 1, k = 0,1,2,3,\ldots\}$ Here $S_5$ is the set of odd positive integers. *A* number is an element of the set $S_5$ *iff* the number is equal to $2k + 1$ for some choice of $k = 0,1,2,3,\ldots$. □

**Example A.6** $S_6 = \{x: 0 \leq x \leq 1\}$ Here $S_6$ is the set of numbers greater than or equal to 0 but less than or equal to 1. *A* number is an element of the set $S_6$ *iff* it is neither less than 0 nor greater than 1. □

The choice of method for describing the objects that constitute elements of a set depends on what is convenient and/or feasible for the case at hand. For example, exhaustive listing of the elements in set $S_6$ is impossible. On the other hand, there is some discretion that can be exercised, since, for example, a verbal rule could have adequately described the set $S_5$, say as $S_5 = \{x: x \text{ is an odd positive integer}\}$. A mixing of the basic methods might also be used, such as $S_5 = \{x: x = 2k + 1, k \text{ is zero or a positive integer}\}$. One can choose whatever method appears most useful in a given problem context.

Note that although our preceding examples of verbal and mathematical rules treat $x$ as inherently one-dimensional, a vector interpretation of $x$ is clearly permissible. For example, we can represent the set of points on the boundary or interior of a circle centered at (0,0) and having radius 1 as

$$S_7 = \{(x_1, x_2) : x_1^2 + x_2^2 \leq 1\}$$

or we can represent the set of input–output combinations associated with a two-input Cobb-Douglas production function as

$$S_8 = \left\{(y, x_1, x_2) : y = b_0\, x_1^{b_1}\, x_2^{b2}, x_1 \geq 0, x_2 \geq 0\right\}$$

for given numerical values of $b_0$, $b_1$, and $b_2$. Of course, the entries in the $\mathbf{x}$-vector need not be numbers, as in the set

$$S_9 = \{(x_1, x_2) : x_1 \text{ is an economist}, \ x_2 \text{ is an accountant}\}.$$

### Set Classifications

Sets are classified according to the number of elements they contain, and whether the elements are countable. We differentiate between sets that have a finite number of elements and sets whose elements are infinite in number, referring to a set of the former type as a **finite set** and a set of the latter type as an **infinite set**. In terms of countability, sets are classified as being either countable or uncountable. Note that when we count objects, we intuitively place the objects in a one-to-one correspondence with the positive integers, i.e., we identify objects one by one, and count "1,2,3,4,... ." Thus, a **countable set** is one whose elements can be placed in a one-to-one correspondence with some or all of the positive integers-any other set is referred to as an **uncountable set**.

A finite set is, of course, always countable, and thus it would be redundant to use the term "countable finite set." Sets are thus either **finite**, **countably infinite**, or **uncountably infinite**. Of the sets, $S_1$ through $S_9$ described earlier, $S_1$, $S_2$, $S_3$, and $S_9$ are finite, $S_4$ and $S_5$ are countably infinite, and $S_6$, $S_7$, and $S_8$ are uncountably infinite (why?).

### Special Sets, Set Operations, and Set Relationships

We now proceed to a number of definitions and illustrations establishing relationships between sets, mathematical operations on sets, and the notions of the universal and empty sets.

**Definition A.2** *Subset*

$A$ is a **subset** of $B$, denoted as $A \subset B$ and read *$A$ is contained in $B$*, *iff* every element of $A$ is also an element of $B$.

**Definition A.3** *Equality of Sets*

Set $A$ is equal to set $B$, denoted as $A = B$, *iff* every element of $A$ is also an element of $B$, and every element of $B$ is also an element of $A$, i.e., *iff* $A \subset B$ and $B \subset A$.

**Definition A.4** *Universal Set*

The set containing all objects under consideration in a given problem setting, and from which all subsets are extracted, is the **universal set.**

**Definition A.5** *Empty or Null Set*

The set containing no elements, denoted by $\emptyset$, is called the **empty**, or **null set.**

**Definition A.6** *Set Difference*

Given any two subsets, $A$ and $B$, of a universal set, the set of all elements in $A$ that are *not* in $B$ is called the **set difference** between $A$ and $B$, and is denoted by $A-B$. If $A \subset B$, then $A-B = \emptyset$.

**Definition A.7**
***Complement***

Let $A$ be a subset of a universal set, $\Omega$. The **complement** of the set $A$ is the set of all elements in $\Omega$ that are not in $A$, and is denoted by $\bar{A}$. Equivalently, $\bar{A} = \Omega - A$.

**Definition A.8**
***Union***

Let $A$ and $B$ be any two subsets of a universal set, $\Omega$. Then, the **union** of the sets $A$ and $B$ is the set of all elements in $\Omega$ that are in *at least one* of the sets $A$ or $B$, it is denoted by $A \cup B$.

**Definition A.9**
***Intersection***

Let $A$ and $B$ be any two subsets of a specified universal set, $\Omega$. Then the **intersection** of the sets $A$ and $B$ is the set of all elements in $\Omega$ that are *in both* sets $A$ and $B$, and is denoted by $A \cap B$.

**Definition A.10**
***Mutually Exclusive (or Disjoint) Sets***

Subsets $A$ and $B$ of a universal set, $\Omega$, are said to be **mutually exclusive** or **disjoint** sets *iff* they have no elements in common, i.e., *iff* $A \cap B = \emptyset$.

We continue to use the slash, /, to indicate negation of a relationship (recall that / was previously used to indicate the negation of $\in$). Thus, $A \not\subset B$ denotes that $A$ is not a subset of $B$, and $A \neq B$ denotes that $A$ is not equal to $B$. We note here (and we shall state later as a theorem) that it is a logical requirement that $\emptyset$ is a subset of any set $A$, since if $\emptyset$ does not contain any elements, it cannot be the case that $\emptyset \not\subset A$, since the negation of $\subset$ would require the existence of an element in $\emptyset$ that was not in $A$.

**Example A.7** Let the universal set be defined as $\Omega = \{x: 0 \leq x \leq 1\}$ and define three additional sets as

$A = \{x: \ 0 \leq x \leq .5\}, B = \{x: \ .25 \leq x \leq .75\}$ and $C = \{x: \ .75 < x \leq 1\}$.

Then we can establish the following set relationships:

$$\begin{aligned}
\bar{B} &= \{x : 0 \leq x < .25 \text{ or } .75 < x \leq 1\}, \\
A \cup C &= \{x : 0 \leq x \leq .5 \text{ or } .75 < x \leq 1\}, \\
\bar{B} &\subset A \cup C, \\
C \cap A &= C \cap B = \emptyset, \\
\bar{C} &= A \cup B = \{x : 0 \leq x \leq .75\}, \\
A - B &= \{x : 0 \leq x < .25\}, \\
A \cap B &= \{x : .25 \leq x \leq .5\}.
\end{aligned}$$

□

Note that although our definitions of subset, equality of sets, set difference, complement, union, and intersection explicitly involve only two sets $A$ and $B$, it is implicit that the concepts can be applied to more complicated expressions involving an arbitrary number of sets. For example, since $A \cap B$ is itself a set, we can form its intersection with a set $C$ as $(A \cap B) \cap C$, or form the set difference, $(A \cap B) - C$, or establish that $(A \cap B) =$ or $\neq C$, and so on. The point is that the concepts apply to sets, which themselves may have been constructed from other sets via various set operations.

**Example A.8** Let $\Omega$, $A$, $B$, and $C$ be defined as in Example A.7. Then the following set relationships can be established:

$$A \cup B \cup C = \Omega = \{x : 0 \le x \le 1\},$$
$$(A \cup C) \cap B = \{x : .25 \le x \le .5\},$$
$$(A \cap B) \cap C = \overline{(A \cup B)} \cap \bar{C} = \emptyset,$$
$$(B \cup C) - A = \{x : .5 < x \le 1\},$$
$$((A \cup B) - (A \cap B)) \subset \bar{C} \cap (\bar{A} \cup \bar{B}).$$

Can $\subset$ be replaced with $=$ in the last relationship? □

It is sometimes useful to conceptualize set relationships through illustrations called *Venn diagrams* (named after the 19th-century English logician, John Venn). In a Venn diagram, the universal set is generally denoted by a rectangle, with subsets of the universal set represented by various geometric shapes located within the bounds of the rectangle. Figure A.1 uses Venn diagrams to illustrate the set relationships defined previously.

### Rules Governing Set Operations

Operations on sets must satisfy a number of basic rules. We state these basic rules as theorems, although we will not take the time to prove them here. The reader may wish to verify the plausibility of some of the theorems through the use of Venn diagrams. One of DeMorgan's laws will be proved to illustrate the formal proof method for the interested reader.

**Theorem A.1** ***Idempotency Laws*** $A \cup A = A$ and $A \cap A = A$

**Theorem A.2** ***Commutative Laws*** $A \cup B = B \cup A$ and $A \cap B = B \cap A$

**Theorem A.3** ***Associative Laws*** $(A \cup B) \cup C = A \cup (B \cup C)$ and $(A \cap B) \cap C = A \cap (B \cap C)$

**Theorem A.4** ***Distributive Laws*** $A \cap (B \cup C) = (A \cap B) \cup (A \cap C)$ and $A \cup (B \cap C) = (A \cup B) \cap (A \cup C)$

**Theorem A.5** ***Identity Elements for $\cap$ and $\cup$*** $A \cap \Omega = A$ ($\Omega$ is the identity element for $\cap$) $A \cup \emptyset = A$ ($\emptyset$ is the identity element for $\cup$)

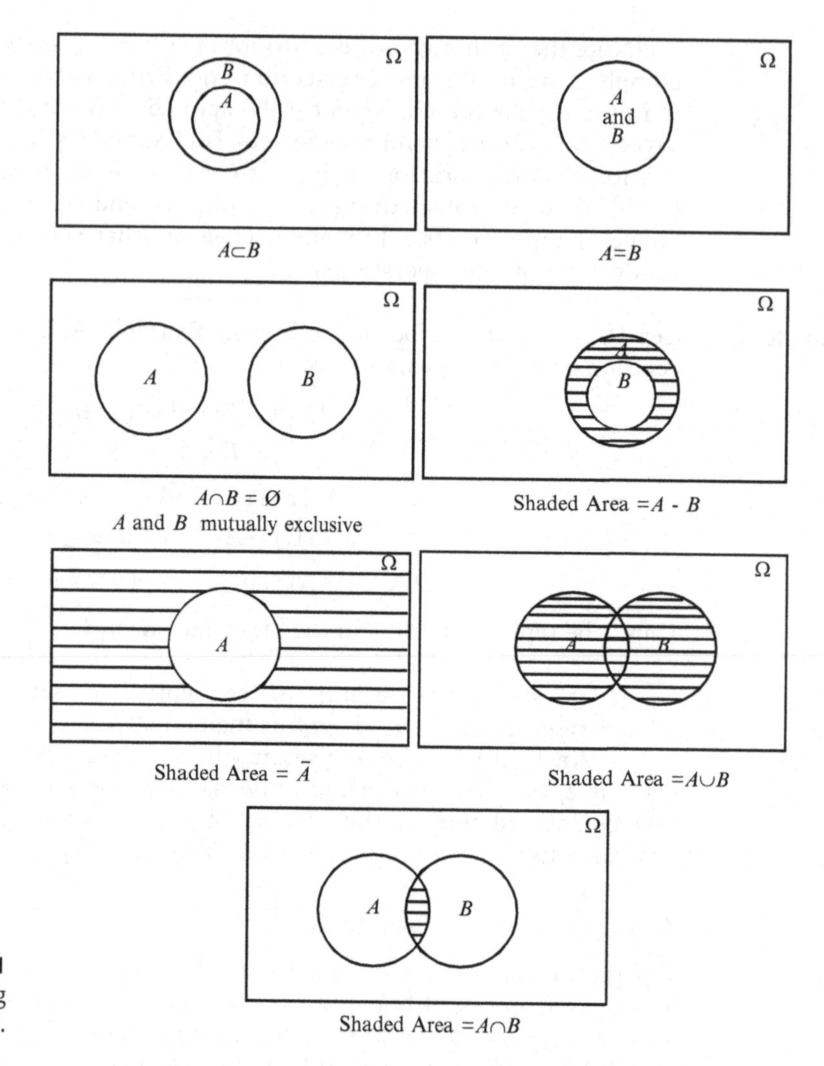


**Figure A.1** Venn diagrams illustrating set relationships.

**Theorem A.6** ***Intersection and Union of Complements*** $A \cup \bar{A} = \Omega$ and $A \cap \bar{A} = \emptyset$

**Theorem A.7** ***Complements of Complements*** $\left(\bar{\bar{A}}\right) = A$

**Theorem A.8** ***Intersection with the Null Set*** $A \cap \emptyset = \emptyset$

**Theorem A.9** ***Null Set as a Subset*** If $A$ is any set, then $\emptyset \subset A$

**Theorem A.10** ***DeMorgan's Laws*** $(\overline{A \cup B}) = \bar{A} \cap \bar{B}$ and $(\overline{A \cap B}) = \bar{A} \cup \bar{B}$

**Example A.9** Formal Proof of $(\overline{A \cap B}) = \bar{A} \cup \bar{B}$. By definition of the equality of sets, two sets are equal *iff* each is contained in the other. We first demonstrate that $\overline{A \cap B} \subset \bar{A} \cup \bar{B}$. By definition, $x \in \overline{A \cap B}$ implies that $x \notin A \cap B$. Suppose $x \notin \bar{A} \cup \bar{B}$. This implies

$x \notin \bar{A}$ and $x \notin \bar{B}$, which implies $x \in A$ and $x \in B$, i.e., $x \in A \cap B$, a contradiction. Therefore, if $x \in \overline{A \cap B}$, then $x \in \bar{A} \cup \bar{B}$, which implies $\overline{A \cap B} \subset \bar{A} \cup \bar{B}$. We next demonstrate that $\bar{A} \cup \bar{B} \subset \overline{A \cap B}$. Let $x \in \bar{A} \cup \bar{B}$. Then $x \notin A \cap B$, for if it were, then $x \in A$ and $x \in B$, contradicting that $x$ belongs to at least one of $\bar{A}$ and $\bar{B}$. However, $x \notin A \cap B$ implies $x \in \overline{A \cap B}$, and thus $\bar{A} \cup \bar{B} \subset \overline{A \cap B}$. □

We remind the reader that since the sets used in Theorems A.1–A.10 could themselves be the result of set operations applied to other sets, the theorems are extendable in a myriad of ways to involve an arbitrary number of sets. For example, in the first of DeMorgan's laws listed as Theorem A.10, if $A = C \cup D$ and $B = E \cup F$, then by substitution,

$$\left(\overline{C \cup D \cup E \cup F}\right) = \left(\overline{C \cup D}\right) \cap \left(\overline{E \cup F}\right).$$

Then by applying Theorem A.10 to both $(C \cup D)$ and $(E \cup F)$, we obtain a generalization of DeMorgan's law as

$$\left(\overline{C \cup D \cup E \cup F}\right) = \bar{C} \cap \bar{D} \cap \bar{E} \cap \bar{F}.$$

Given the wide range of extensions that are possible, Theorems A.1–A.10 provide a surprisingly broad conceptual foundation for applying the rules governing set operations.

### Some Useful Set Notation

Situations sometimes arise in which one is required to denote the union or intersection of a large number of sets. A convenient notation that represents such unions or intersections quite efficiently is available. Two types of notations are generally used, and they are differentiated on the basis of whether the union or intersection is of sets identified by a natural sequence of integer subscripts or whether the sets are identified by subscripts, say $i$'s, that are elements of some set of subscripts, $I$, called an **index set**.

**Definition A.11** ***Multiple Union Notation***

**a.** $\cup_{i=1}^{n} A_i = A_1 \cup A_2 \cup A_3 \ldots \cup A_n$.

**b.** $\cup_{i \in I} A_i =$ union of all sets $A_i$ for which $i \in I$.

**Definition A.12** ***Multiple Intersection Notation***

**a.** $\cap_{i=1}^{n} A_i = A_1 \cap A_2 \cap A_3 \ldots A_n$.

**b.** $\cap_{i \in I} A_i =$ intersection of all sets $A_i$ for which $i \in I$.

**Example A.10** Let the universal set be defined as $\Omega = \{x: 0 \le x \le 1\}$, and examine the following subsets of $\Omega$:

$$A_1 = \{x: 0 \le x \le .25\}, A_2 = \{x: 0 \le x \le .5\},$$

$A_3 = \{x: 0 \le x \le .75\}, A_4 = \{x: .75 \le x \le 1\}$.

Define the index sets $I_1$ and $I_2$ as

$I_1 = \{1, 3\}$, and $I_2 = \{1, 3, 4\}$.

Then,

$$\cup_{i=1}^4 A_i = \cup_{i=2}^4 A_i = \cup_{i=3}^4 A_i = \{x: 0 \le x \le 1\} = \Omega,$$
$$\cup_{i \in I_1} A_i = A_1 \cup A_3 = \{x: 0 \le x \le .75\},$$
$$\cup_{i \in I_2} A_i = A_1 \cup A_3 \cup A_4 = \{x: 0 \le x \le 1\} = \Omega,$$
$$\cap_{i=1}^4 A_i = \cap_{i=2}^4 A_i = \emptyset,$$
$$\cap_{i=3}^4 A_i = \{.75\},$$
$$\cap_{i \in I_1} A_i = A_1 \cap A_3 = \{x: 0 \le x \le .25\},$$
$$\cap_{i \in I_2} A_i = A_1 \cap A_3 \cap A_4 = \emptyset.$$

□

Whenever a set $A$ is an interval subset of the *real line* (where the **real line** refers to all of the numbers between $-\infty$ and $\infty$), the set can be indicated in abbreviated form by the standard notation for intervals, stated in the following definition.

**Definition A.13**
***Interval Set Notation***

Let $a$ and $b$ be two numbers on the real line for which $a < b$. Then the following four sets, called intervals with endpoints $a$ and $b$, can be defined as:

**(a) Closed interval:**

$[a, b] = \{x: a \le x \le b\}$,

**(b) Half-open (or half-closed) intervals:**

$(a, b] = \{x: a < x \le b\}$, and

$[a, b) = \{x: a \le x < b\}$,

**(c) Open interval:**

$(a, b) = \{x: a < x < b\}$.

Note that *weak* inequalities, $x \le$ or $\le x$, are signified by brackets ] or [, respectively. *Strong* inequalities, $x <$ or $< x$, are signified by parentheses, ) or (, respectively. Note further that whether the interval set contains its endpoints determines whether the set is closed.

As we have already done, $(x,y)$ will also be used to denote coordinates in the two-dimensional plane. The context of the discussion will make clear whether we are referring to an open interval $(a,b)$ or pair of coordinates $(x,y)$.

## A4. Relations, Point Functions, Set Functions

The concepts of point function and set function are central to a discussion of probability and statistics. We will see that probabilities can be represented by set functions and that in a large number of cases of practical interest, set functions can in turn be represented by a summation or integration operation applied to point functions. While readers may be somewhat familiar with point functions from introductory courses in algebra, the concept of a set function may not be familiar. We will review both function concepts within the broader context of the theory of relations. The relations context facilitates the presentation of a very general definition of "function" in which inputs into and outputs from the function may be objects of any kind including, but not limited to, numbers. The relations context also facilitates a demonstration of the significant similarities between the concept of a set function and the more familiar point function concept.

### Cartesian Product

The concept of a relation can be made clear once we define what is meant by the Cartesian product of two sets $A$ and $B$, named after the French mathematician Rene Descartes, (1596–1650).

**Definition A.14** *Cartesian Product of A and B*

Let $A$ and $B$ be two sets. Then the **Cartesian product** of $A$ and $B$, denoted as $A \times B$, is the set of ordered pairs $A \times B = \{(x,y): x \in A, y \in B\}$.

In words, $A \times B$ is the set of all possible pairs $(x,y)$ such that $x$ is an element of the set $A$ and $y$ is an element of the set $B$. Note carefully that the pairs are *ordered* in the sense that the first object in the pair must come from set $A$ and the second object from set $B$.

**Example A.11** Let $A = \{x: 1 \leq x \leq 2\}$ and $B = \{y: 2 \leq y \leq 4\}$. Then $A \times B = \{(x,y): 1 \leq x \leq 2$ and $2 \leq y \leq 4\}$. (see Figure A.2). □

**Example A.12** Let $A = \{x: x$ is a man$\}$ and $B = \{y: y$ is a woman$\}$. Then $A \times B = \{(x,y): x$ is a man and $y$ is a woman$\}$, which is the set of all possible man-woman pairings. □

In later chapters of the book we will have use for a more general notion of Cartesian product involving more than just two sets. The extension is given in the definition below.

**Definition A.15** *Cartesian Product (General)*

Let $A_1,\ldots,A_n$ be $n$ sets. Then the Cartesian product of $A_1,\ldots,A_n$ is the set of ordered $n$-tuples

$$\times_{i=1}^{n} A_i = A_1 \times A_2 \times \ldots \times A_n = \{(x_1,\ldots,x_n) : x_i \in A_i, i = 1,\ldots,n\}.$$

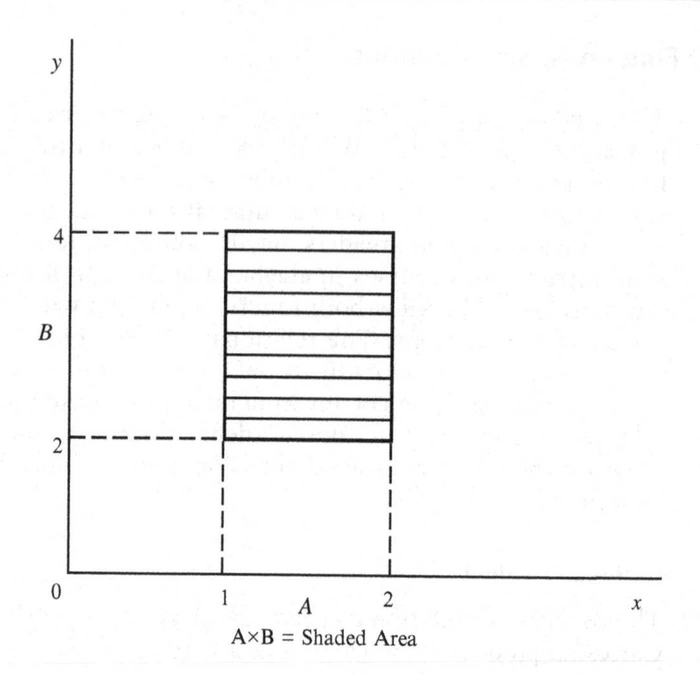


**Figure A.2**
$A \times B$ = shaded area.

In words, $\times_{i=1}^{n} A_i$ is the set of all possible $n$-tuples $(x_1,\ldots,x_n)$ such that $x_1$ is an element of set $A_1$, $x_2$ is an element of set $A_2$, and so on. Note that should the need arise, a general Cartesian product of sets could also be represented by the notation $\times_{i\in I} A_i$, where here the product is taken over all sets having subscript $i$ in the index set $I$ (recall Definitions A.11 and A.12, and the use of index set notation).

In certain cases, we may be interested in forming a Cartesian product of a set $A$ with itself. While we might represent such a Cartesian product by the notation $\times_{i=1}^{n} A = \{(x_1,\ldots,x_n): x_i \in A,\ i = 1,\ldots,n\}$, such a Cartesian product is generally denoted by the notation $A^n$, and so for example, $A^2 = A \times A$.

### Relation (Binary)

We now define what we mean by the term "binary relation."

**Definition A.16**
***Binary Relation***

Any subset of the Cartesian product $A \times B$ is a **binary relation from $A$ to $B$.**

Note that the adjective *binary* signifies that only two sets are involved in the relation. Relations that involve more than two sets can be defined, but for our purposes the concept of a binary relation will suffice.[3] Henceforth, we will use the word relation to mean binary relation. We should also mention that in the

[3] A higher order relation could be defined by taking a subset of the Cartesian product $\times_{i=1}^{n} A_i$, for example.

case where $B = A$, we will simply remain consistent with Definition A.16 and refer to a subset of $A \times A$ as a *relation from A to A*, although in this special case some authors prefer to call the subset of $A \times A$ a relation *on* $A$.

Now let $S \subset A \times B$. Thus by definition, $S$ is a relation from $A$ to $B$. We emphasize at this point that the choice of the letter $S$ is quite arbitrary, and we could just as well have chosen any other letter to represent a subset of $A \times B$ defining a relation from $A$ to $B$. If $(x,y) \in S$, we say that **x is in the relation** $S$ **to** $y$ or that **x is** $S$**-related to** $y$. An alternative notation for $(x,y) \in S$ is $xSy$. Also, we use $S: A \rightarrow B$ as an abbreviation for "the relation $S$ from $A$ to $B$."

As it now stands, the concept of a relation no doubt appears quite abstract. However, in practice, it is the context provided by the definition of the subset $S$ and the definitions of the sets $A$ and $B$ that provide intuitive meaning to $xSy$. That is, $x$ will be $S$-related to $y$ because of some property satisfied by the $(x,y)$ pair, the property being indicated in the set definition of $S$. The real-world objects being related will be clearly identified in the set definitions of $A$ and $B$. A few examples will clarify the intuitive side of the relation concept.

**Example A.13** Let $A = [0,\infty)$, and form the Cartesian product $A^2 = \{(x,y): x \in A \text{ and } y \in A\}$. The set $A^2$ can thus be interpreted as the nonnegative (or first) quadrant of the Euclidean plane. Then $S = \{(x,y): x \geq y, (x,y) \in A^2\}$ is a relation from $A$ to $A$ representing the set of points in the nonnegative quadrant for which the first coordinate has a value greater than or equal to the value of the second coordinate. The defining property of the relation $S$ is "$\geq$." This is displayed in Figure A.3. □

**Example A.14** Let $A = \{x: x \text{ is an employed U.S. citizen}\}$ and $B = \{y: y \text{ is a U.S. corporation}\}$. Then $A \times B = \{(x,y): x \text{ is an employed U.S. citizen and } y \text{ is a U.S. corporation}\}$ is the set of all possible pairings of employed U.S. citizens with U.S. corporations. The relation $S = \{(x,y): x \textit{ is employed by } y, (x,y) \in A \times B\}$ from $A$ to $B$ is the collection of U.S. citizens who are employed by U.S. corporations paired with their respective corporate affiliation. The defining property of the relation is the phrase "is employed by," and $xSy$ *iff* $x$ is a U.S. citizen employed by a U.S. corporation, and $y$ is his or her corporate affiliation. □

### Function

We are now in a position to define what is meant by the concept of a function. As indicated in the following definition, a function is simply a special type of relation. In the definition we introduce the symbol $\forall$, which stands **for every** or **for all**, the symbol $\exists$ which means **there exists.**

**Definition A.17** ***Function***

> A function from $A$ to $B$ is a relation $S: A \rightarrow B$ such that $\forall a \in A \; \exists$ one unique $b \in B$ such that $(a,b) \in S$.

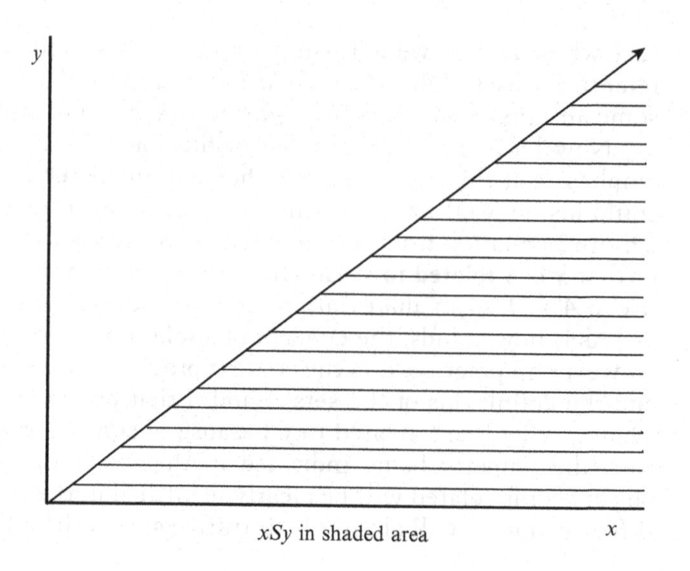


**Figure A.3**
*xSy* in shaded area.

A relation satisfying the above condition will often be given a special symbol to distinguish the relation as a function. A popular symbol used for this purpose is "$f$," where $f$: $A \rightarrow B$ is a common notation for designating the **function $f$ from $A$ to $B$**. As we had remarked when choosing a symbol to represent a relation, the choice of the letter $f$ is arbitrary, and when it is convenient or useful, any other letter or symbol could be used to depict a subset of $A \times B$ that represents a function from $A$ to $B$. In the text we will often have occasion to use a variety of letters to designate various functions of interest.

The unique element $b \in B$ that the function $f$: $A \rightarrow B$ associates with a given element $x \in A$ is called **the image of x under** $f$ and is represented symbolically by the notation $\boldsymbol{f(x)}$. If $f(x)$ is a real number, the image of $x$ under $f$ is alternatively referred to as **the value of the function $f$ at x**. In the following example, we use $\mathbb{R}$ to denote the set of real numbers $(-\infty, \infty)$, i.e., $\mathbb{R}$ stands for the **real line**. Furthermore, the nonnegative subset of the real line is represented by $\mathbb{R}_{\geq 0}$, $= [0,\infty)$.

**Example A.15** Let $f$: $\mathbb{R} \rightarrow \mathbb{R}_{\geq 0}$ be defined by $f = \{(x,y): y = x^2, x \in \mathbb{R}\}$. The *image of* $-2$ *under* $f$ is $f(-2) = 4$. The *value of the function f at* 3 is $f(3) = 9$. □

Associated with a given function, $f$, are two important sets called the **domain** and **range** of the function.

**Definition A.18** ***Domain and Range of a Function***

The **domain** of a function $f$: $A \rightarrow B$ is defined as $D(f) = A$. The **range** of $f$ is defined by $R(f) = \{y: y = f(x), x \in A\}$.

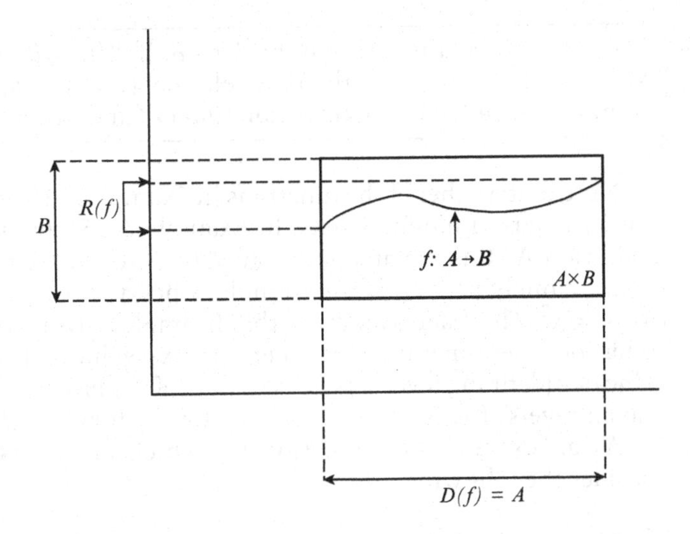


**Figure A.4**
Function with domain $D(f)$ and range $R(f)$.

Thus, the domain of a function $f$: $A \to B$ is simply the set $A$ of the Cartesian product $A \times B$ associated with the function. The range of $f$ is the collection of all elements in $B$ that are images of the elements in $A$ under the function $f$. It follows that $R(f) \subset B$. Figure A.4 provides a pictorial example of a function, including its *domain* and *range*.

Note that the concept of a function is completely general regarding the nature of the elements of the sets $D(f) = A$, $R(f)$, and $B$. The elements can be numbers, or other objects, or the elements can be sets themselves. For our work, it will suffice to deal only with *real-valued functions* meaning that $R(f)$ is a set of real numbers.

**Definition A.19**
***Real-Valued Function***

A function $f$: $A \to B$ such that $R(f) \subset \mathbb{R}$ is called a **real-valued function.**

The function defined in Example A.15 is a real-valued function. The following is another example.

**Example A.16** *Examine the Cobb-Douglas production function* $f$:$\mathbb{R}^2_{\geq 0} \to \mathbb{R}_{\geq 0}$ defined as $f = \{((x_1, x_2), y) : y = 10x_1^2x_2, (x_1, x_2) \in \mathbb{R}^2_{\geq 0}\}$. Interpreting $(x_1,x_2)$ as inputs into a production process, and $y$ as the output of the process, we see that the domain of the production function is $d(f) = \mathbb{R}^2_{\geq 0}$, i.e., any nonnegative level of the input pair $(x_1,x_2)$ is an admissible input level. The associated range of the production function is $\mathbb{R}(f) = \mathbb{R}_{\geq 0}$, i.e., any nonnegative level of output, $y$, is possible. Since $\mathbb{R}(f) = [0,\infty) \subset \mathbb{R}$, the production function is a real-valued function. □

In some cases the relation from $A$ to $B$ that defines the function $f$: $A \to B$ also defines a function from $B$ to $A$. This relates to the concept of an **inverse function.** In particular, if for each element $y \in B$ there exists precisely one element $x \in A$ whose image under $f$ is $y$, then such an *inverse function* exists.

**Definition A.20**
***Inverse Function of f***

Let $f$: $A \to B$ be a function from $A$ to $B$. If $\mathbb{R}(f) = B$ and $\forall y \in B\ \exists$ a unique $x \in A$ such that $y = f(x)$, then the relation $\{(y,x)\text{: } y = f(x),\ y \in B\}$ is a function from $B$ to $A$ called the **inverse function of** $f$ and denoted by $f^{-1}$: $B \to A$.

Note that neither of the functions in Example A.15 or Example A.16 are such that an inverse function exists. In Example A.15, the uniqueness condition of Definition A.20 is violated since $\forall y \neq 0$ there exist two values of $x$ for which $y = x^2$, namely $x = \pm\sqrt{y}$. For example, when $y = 4$, $x = 2$ and $-2$ are each such that $y = x^2$. The reader can verify that Example A.16 also violates the uniqueness condition where an infinite number of $(x_1, x_2)$ values satisfy $y = 10x_1^2x_2$ for a fixed value of $y$ (defining level sets or isoquants of the production function). Also, note that an inverse function does not exist for the function illustrated in Figure A.4.

As an example of a function for which an inverse function does exist, consider the following.

**Example A.17** Let $f$: $\mathbb{R} \to \mathbb{R}_+$ be defined as

$$f = \{(x, y) : y = e^x, x \in \mathbb{R}\}, \text{ where } \mathbb{R}_+ = (0, \infty).$$

Note that the *inverse function* can be represented as

$$f^{-1} = \{(y, x) : x = \ln(y), y \in \mathbb{R}_+\}$$

so that $f^{-1}$:$\mathbb{R}_+ \to \mathbb{R}$. It is clear that $\forall y \in \mathbb{R}_+$, $\exists$ one and only one $x \in \mathbb{R}$ such that $x = \ln(y)$. □

The final concept concerning functions that we will review here is the **inverse image** of $y \in R(f)$ or of $H \subset R(f)$. The inverse image of $y$ is the set of domain elements $x \in D(f)$ such that $y = f(x)$, i.e., the collection of all $x$ values in the domain of $f$ whose image under the function $f$ is $y$. The inverse image of $y$ can be represented as the set $\{x\text{: } f(x) = y\}$, and when the inverse function exists, the inverse image of $y$ can be represented as the value $f^{-1}(y)$. In Example A.17 above, the inverse image of 5 is $f^{-1}(5) = \ln(5) = 1.6094$, in Example A.15, the inverse image of 4 is $\{-2,2\}$, and in Example A.16, the inverse image of 3 is the isoquant $\{(x_1, x_2) : 3 = 10x_1^2x_2, (x_1, x_2) \in \mathbb{R}^2_{\geq 0}\}$. Similarly, the inverse image of $H \subset R(f)$ is the set of $x$ values in the domain of $f$ whose images under $f$ equal some $y \in H$, i.e., the inverse image of $H$ is $\{x\text{: } f(x) = y,\ y \in H\}$, and if the inverse function exists, $\{x\text{: } x = f^{-1}(y),\ y \in H\}$.

### Real-Valued Point Versus Set Functions

Two types of functions – point functions and set functions – are utilized extensively in modern discussions of probability and mathematical statistics. The reader should already have considerable experience with the application of real-valued point functions, since this type of function is the one that appears in elementary algebra and calculus courses and is central to discussions of utility, demand, production, and supply that the reader has encountered in his or her study of economic theory. Specifically, a **real-valued point function** is a

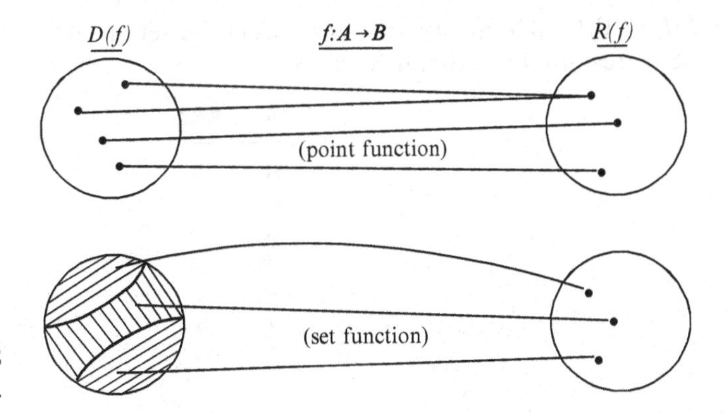


**Figure A.5**
Point versus set function.

real-valued function whose domain consists of a collection of points, where points are represented by coordinate vectors in $\mathbb{R}^n$. We have encountered examples of this type of function previously in Example A.15 through Example A.17. A typical ordered pair associated with a real-valued point function is of the form $(\mathbf{x},y)$, where $\mathbf{x}$ is a vector in $\mathbb{R}^n$ and $y$ is a real number in $\mathbb{R}$.

A set function is more general than a point function in that its domain consists of a collection of *sets* rather than a collection of points.[4] A typical ordered pair belonging to a **real-valued set function** would have the form $(A,y)$, where $A$ is a set of some type of objects and $y$ is a real number in $\mathbb{R}$. If the sets in the domain of the set function are contained in $\mathbb{R}^n$, i.e., they are sets of real numbers, then a real-valued set function is assigns a real number to each *set* of points in its domain in contrast to a real-valued point function which would assign a real number to each *point* in its domain. A pictorial illustration of a set function contrasted with a point function is given in Figure A.5.

Examples of set functions are presented below.

**Example A.18** Let $\Omega = \{1,2,3\}$, and let $A$ be the collection of all of the subsets of $\Omega$, i.e., $A = \{A_1, A_2, \ldots, A_8\}$, where

$A_1 = \{1\}, A_2 = \{2\}, A_3 = \{3\}, A_4 = \{1,2\}, A_5 = \{1,3\}, A_6 = \{2,3\},$
$A_7 = \{1,2,3\}$, and $A_8 = \emptyset$

The following is a real-valued set function $f$: $A \to \mathbb{R}$:

$$f = \left\{ (A_i, y) : y = \sum_{x \in A_i} x,\ A_i \subset A \right\},$$

where $\sum_{x \in A_i} x$ signifies the sum of the numerical values of all of the elements in the set $A_i$, and $\sum_{x \in \emptyset} x$ is defined to be zero. The range of the set function is

[4]Note that, in a sense, a point function can be viewed as a special case of a set function, since points can be interpreted as singleton (single element) sets. The set function concept is introduced to accommodate the case where one or more sets in its domain are *not* singleton.

$R(f) = \{0,1,2,3,4,5,6\}$ and the domain is the set of sets $D(f) = A$. The function can be represented in tabular form as

| $A_i$ | $f(A_i)$ |
|---|---|
| $A_1$ | 1 |
| $A_2$ | 2 |
| $A_3$ | 3 |
| $A_4$ | 3 |
| $A_5$ | 4 |
| $A_6$ | 5 |
| $A_7$ | 6 |
| $A_8$ | 0 |

□

**Example A.19** Let $A = \{A_r : A_r = \{(x,y): x^2 + y^2 \leq r^2\},\ r \in [0,1]\}$, so that $A$ is a set of sets, with typical element $A_r$ represents the set of points in $\mathbb{R}^2$ that are on the boundary and in the interior of a circle centered at (0,0) with radius $r$. The following is a real valued set function $f: A \rightarrow \mathbb{R}$:

$$f = \{(A_r, y) : y = \pi r^2,\ A_r \subset A\}.$$

Note the set function assigns a real number representing the area to each set, $A_r$. The assignment is made for circles having a radius anywhere from 0 to 1. The range of the set function is given by $\mathbb{R}(f) = [0,\pi]$, and the domain is the *set of sets* $D(f) = A$. □

A special type of set function called the **size-of-set function** will prove to be quite useful.

**Definition A.21** ***Size of Set Function***

Let $A$ be any set of objects. The **size-of-set function**, $N$, is the set function that assigns to the set $A$ the number of elements that are in set $A$, i.e., $N(A) = \sum_{x \in A} 1$.[5]

Applying the size-of-set function in Example A.18, note that $N(A) = 8$. In Example A.19 note that $N(A) = \infty$.

Another special (point) function that will be useful in our study is the **indicator function**, defined as follows:

**Definition A.22** ***Indicator Function***

Let $A$ be any subset of some universal set $\Omega$. The indicator function, denoted by $I_A$, is a real-valued function with domain $\Omega$ and range $\{0, 1\}$ such that

$$I_A(x) = \begin{Bmatrix} 1 & \text{if} & x \in A \\ 0 & \text{if} & x \notin A \end{Bmatrix}.$$

[5] Note that $\sum_{x \in A} 1$ signified that a collection of 1's are being summed together, the number in the collection being equal to the number of elements in the set $A$. If $A = \emptyset$, effectively no 1's are being added together, and thus $N(\emptyset) = 0$.

Note that the indicator function *indicates* the set $A$ by assigning the number 1 to any $x$ that is an element of $A$, while assigning zero to any $x$ that is not an element of $A$. The main use of the indicator function is notational efficiency in defining functions, as the following example illustrates:

**Example A.20** Let the function $f:\mathbb{R} \to \mathbb{R}$ be defined by

$$f(x) = \begin{bmatrix} 0 & \text{for} & x \in (-\infty, 0] \\ x & \text{for} & x \in (0, 2] \\ 3 - x & \text{for} & x \in (2, 3] \\ 0 & \text{for} & x \in (3, \infty) \end{bmatrix}$$

Utilizing the indicator function, we can alternatively represent $f(x)$ as

$$f(x) = xI_{(0,2]}(x) + (3 - x)I_{(2,3]}(x).$$

□

As a final note on the use of functions, we (as do the vast majority of other authors) will generally use a shorthand method for defining functions by simply specifying the relationship between elements in the domain of a function and their respective images in the range of the function. For example, we would define the function in Example A.19 by $f(A_r) = \pi r^2$ for $A_r \subset A$, or define the function in Example A.15 by $f(x) = x^2$ for $x \in \mathbb{R}$. In all cases, the reader should remember that a function is a set of ordered pairs $(x, f(x))$, or $(A, f(A))$. The reader will sometimes find in the literature phrases like **the function $f(x)$** or **the set function $f(A)$**. Literally speaking, such phrases are inconsistent, because $f(x)$ or $f(A)$ are not functions, but rather *images* of elements in the domain of the respective functions. In fact, the reader should *not* take such phrases literally, but rather interpret these phrases as shorthand for phrases such as *the function whose values are given by $f(x)$* or *the set function whose values are given by $f(A)$*.

## A5. Combinations and Permutations

In a number of situations involving probability assignments, it will be useful to have an efficient method for counting the number of *different* ways a group of $r$ objects can be selected from a group of $n$ distinct objects, $n \geq r$. Obviously, if two groups of $r$ objects do not contain the same $r$ objects, they must be considered *different*. But what if two groups of $r$ objects do contain the same objects, except the objects in the group are arranged in different orders? Are the two groups to be considered *different*? If difference in order constitutes difference in groups, then we are dealing with the notion of **permutations**. On the other hand, if the order of listing the objects in a group is not used as a basis for distinguishing between groups, then we are dealing with the notion of **combinations**.

In order to establish a formula for determining the number of permutations of $n$ distinct objects taken $r$ at a time, the following example is suggestive.

**Example A.21** Examine the number of different ways a group of three letters can be selected from the letters, *a,b,c,d*, where difference in order of listing is taken to mean difference in groups. Note that the first letter can be chosen in four different ways. After we have chosen one of the letters for the first selection, the second selection can be any of the remaining three letters. Finally, after we have chosen two letters in the first two selections, there are then two letters left to be potentially chosen for the third selection. Thus, there are $4{\cdot}3{\cdot}2 = 24$ different ways of selecting a group of three letters from the letters *a,b,c,d* if difference in the order of listing constitutes difference in groups (The reader should attempt to list the 24 different groups).

The logic of the preceding example can be applied to establish a general formula for determining the number of permutations of $n$ distinct objects taken $r$ at a time:

$$(n)_r = \frac{n!}{(n-r)!} = n(n-1)(n-2)\cdots(n-r+1),$$

where ! denotes the **factorial operation**, i.e.,[6]

$$n! = n(n-1)(n-2)(n-3)\ldots 1.$$

So, for example, $4! = 4{\cdot}3{\cdot}2{\cdot}1 = 24$. In Example A.21, $n = 4$ and $r = 3$, so that $(4)_3 = 4!/1! = 24$.

In order to establish a formula for determining the number of combinations of $n$ distinct objects taken $r$ at a time, examine the number of different ways a group of three letters can be selected from the letters *a,b,c,d*, where difference in order of listing does *not* imply the groups are different. Recall that we discovered that there were 24 permutations of the four letters *a,b,c,d* selected three at a time. Now note that any three letters, say *a,b,c*, can be arranged in $(3)_3 = 6$ different orders, which represents "overcounting" from the combinations point of view. Reducing the number of permutations by the degree of "overcounting" results in the number of combinations, i.e., there are $24/6 = 4$ combinations of the four letters taken three at a time, namely $(a,b,c)$, $(a,b,d)$, $(a,c,d)$, and $(b,c,d)$. □

In the preceding example, the number of permutations of $n(=4)$ objects taken $r(=3)$ at a time was reduced by a factor of $r!(=3!)$, where the latter value represents the number of possible permutations of $r$ objects. This suggests the general formula for the number of combinations of $n$ objects taken $r$ at a time:

$$\binom{n}{r} = \frac{(n)_r}{r!} = \frac{n!}{(n-r)!\,r!}$$

[6] By definition, we take $0! = 1$.

In Example A.21, we have

$$\binom{4}{3} = \frac{4!}{1!\,3!} = 4$$

as the appropriate number of combinations.

The concept of combinations is useful in determining the number of subsets that can be constructed from a finite set $A$. Note that in counting the number of subsets, changes in the order of listing set elements does not produce a different set, e.g., the sets $\{a,b,c\}$ and $\{c,a,b\}$ are the same set of letters (recall the definition of a set). Then the total number of subsets of a set $A$ containing $n$ elements is given by the number of different subsets defined by taking no elements (i.e., the null set) plus the number of different subsets defined by taking one element, plus the number of different subsets defined by taking two elements, ..., and finally, the number of different subsets defined by taking all $n$ elements (i.e., the set $A$ itself). Thus, the total number of different subsets of $A$ can be written as

$$\sum_{r=0}^{n} \binom{n}{r} = \sum_{r=0}^{n} \frac{n!}{(n-r)!\,r!}$$

This sum can be greatly simplified by recalling that

$$(x+y)^n = \sum_{r=0}^{n} \binom{n}{r} x^r y^{n-r}, n = 1,2,\ldots,$$

which is the **binomial theorem**. Then letting $x = y = 1$, we have that

$$2^n = \sum_{r=0}^{n} \binom{n}{r}$$

so that $2^n$ is the number of different subsets contained in a set $A$ that has $n$ elements. The set containing all $2^n$ subsets of a set, $A$, of $n$ elements is called the **power set** of $A$.

**Example A.22** ***Power Set*** In Example A.18, recall that we identified a total of eight subsets of the set $\Omega = \{1,2,3\}$. This is the number of subsets we would expect from our discussion above, i.e., since $n = 3$, there are $2^3 = 8$ subsets of $\Omega$. □

It should be noted that $\binom{n}{r}$ is *defined* to be 0 whenever $n < r$, or whenever $n$ and/or $r$ are $< 0$ or are not integer valued. The rationale for $\binom{n}{r} = 0$ in each of these cases is that there is no way to define subsets of size $r$ from a collection of $n$ objects for the designated values of $n$ and $r$.

When $n$ is large, the calculation of $n!$ needed in the previous formulas pertaining to numbers of permutations or combinations can be quite formidable. A result known as Stirling's formula can provide a useful approximation to $n!$ for large $n$.

**Table A.1** Summation and integration notation

| Notation | Definition |
|---|---|
| $\sum_{i=\ell}^{n} x_i$ | Sum the values of $x_\ell, x_{\ell+1},\ldots,x_n$, i.e., $x_\ell + x_{\ell+1} + \ldots + x_n$ |
| $\sum_{i\in I} x_i$ | Sum the values of the $x_i$'s, for $i \in I$ |
| $\sum_{x\in A} x$ | Sum the values of $x \in A$ |
| $\sum_{x=a}^{b} x$ | Sum the values of $x$ in the sequence of integers from $a$ to $b$, i.e., $a + (a+1) + (a+2) + \ldots + b$ |
| $\sum_{i=\ell}^{n}\sum_{j=k}^{m} x_{ij}$ | Sum the values of the $x_{ij}$'s for $i = \ell, \ell+1,\ldots,n$ and $j = k, k+1,\ldots,m$ |
| $\sum_{i\in I}\sum_{j\in J} x_{ij}$ or $\sum_{(i,j)\in A} x_{ij}$ | Sum the values of the $x_{ij}$'s for $i \in I$ and $j \in J$, or for $(i,j) \in A$ |
| $\sum_{x_1\in A_1}\cdots\sum_{x_n\in A_n} f(x_1,\ldots,x_n)$ | Sum the values of $f(x_1,\ldots,x_n)$ for $x_i \in A_i$, $i = 1,\ldots,n$ |
| $\sum_{(x_1,\ldots,x_n)\in A} f(x_1,\ldots,x_n)$ | Sum the values of $f(x_1,\ldots,x_n)$ for $(x_1,\ldots,x_n) \in A$ |
| $\sum_{x_1=a_1}^{b_1}\cdots\sum_{x_n=a_n}^{b_n} f(x_1,\ldots,x_n)$ | Sum the values of $f(x_1,\ldots,x_n)$ for $x_i$ in the sequence of integers $a_i$ to $b_i$, $i = 1,\ldots,n$ |
| $\int_a^b f(x)dx$ | Integral of the function $f(x)$ from $a$ to $b$ ($a$ and/or $b$ can be $-\infty$ and $\infty$) |
| $\int_{x\in A} f(x)dx$ | Integral of the function $f(x)$ over the set of points $A$ |
| $\int_{x_1\in A_1}\cdots\int_{x_n\in A_n} f(x_1,\ldots,x_n)dx_n\ldots dx_1$ | Iterated integral of the function $f(x_1,\ldots,x_n)$ over the points $x_i \in A_i$, $i = 1,\ldots,n$ |
| $\int_{(x_1,\ldots,x_n)\in A} f(x_1,\ldots,x_n)dx_1\ldots dx_n$ | Multiple integral of the function $f(x_1,\ldots,x_n)$ over the points $(x_1,\ldots,x_n) \in A$ |
| $\int_{a_1}^{b_1}\cdots\int_{a_n}^{b_n} f(x_1,\ldots,x_n)dx_n\ldots dx_1$ | Iterated integral of the function $f(x_1,\ldots,x_n)$ for $x_i$ in the (open, half open-half closed, or closed) interval $a_i$ to $b_i$, for $i = 1,\ldots,n$ |

**Definition A.23**
***Stirling's Formula***[7]

$n! \approx (2\pi)^{1/2}\, n^{n+.5}\, e^{-n}$ for large $n$.[8]

A logical question to ask regarding the use of Stirling's formula is how large is "large $n$"? Stirling's formula invariably underestimates $n!$ but the percentage error is $\leq 1$ percent for $n \geq 10$, and monotonically decreases as $n \to \infty$.

## A6. Summation and Integration Notation

We will use a number of variations on summation and integration notation in this text. The meaning of the various types of notation are presented in Table A.1.

We illustrate the use of some of the notation in the following examples.

**Example A.23**
***Summation Notation***

Let $A_1 = \{1,2,3\}$, $A_2 = \{2,4,6\}$, $A = A_1 \times A_2$, $B = \{(x_1,x_2)\!: x_1 \in A_1;\ x_2 \in \{x_1, x_1+1,\ldots,3x_1\}\}$, $y = (y_1,y_2,\ldots,y_n)$, and $f(x_1,x_2) = x_1 + 2x_2^2$. Then

[7] See Feller (1968), *An Introduction to the Theory of Probability and Its Applications*, 3rd ed., pp. 52–54.

[8] Note that $\approx$ means "approximately equal to."

$$\sum_{x\in A_1} x = 1+2+3 = 6, \sum_{x\in A_2} x^2 = 2^2+4^2+6^2 = 56,$$

$$\sum_{i\in A_2} y_i = y_2+y_4+y_6, \sum_{i\in A_1} y_i = \sum_{i=1}^{3} y_i = y_1+y_2+y_3$$

$$\sum_{x_1\in A_1}\sum_{x_2\in A_2} f(x_1,x_2) = \sum_{x_1\in A_1}\sum_{x_2\in A_2}(x_1+2x_2^2) = 354,$$

$$\sum_{(x_1,x_2)\in A} f(x_1,x_2) = \sum_{x_1\in A_1}\sum_{x_2\in A_2} f(x_1,x_2) = 354,$$

$$\sum_{(x_1,x_2)\in B} f(x_1,x_2) = \sum_{x_1=1}^{3}\sum_{x_2=x_1}^{3x_1}(x_1+2x_2^2) = 802.$$

□

**Example A.24** ***Integration Notation*** Let $A_1 = [0,3]$, $A_2 = [2,4]$, $A = A_1 \times A_2$, $B = \{(x_1,x_2): x_1 \in A_1, 0 < x_2 < x_1^2\}$, and $f(x_1, x_2) = x_1x_2^2$. Then

$$\int_{x\in A_1} 2x dx = \int_0^3 2x dx = \frac{2x^2}{2}\Big|_0^3 = 9,$$

$$\int_{x\in A_1\cap A_2} x^2 dx = \int_2^3 x^2 dx = \frac{x^3}{3}\Big|_2^3 = \frac{19}{3},$$

$$\int_{x_1\in A_1}\int_{x_2\in A_2} f(x_1,x_2)dx_2dx_1 = \int_0^3\int_2^4 f(x_1,x_2)dx_2dx_1$$

$$= \int_0^3 \frac{x_1x_2^3}{3}\Big|_2^4 dx_1 = \int_0^3 \frac{56}{3}x_1 dx_1$$

$$= \frac{56x_1^2}{6}\Big|_0^3 = 84,$$

$$\int_{(x_1,x_2)\in A} f(x_1,x_2)dx_1dx_2 = \int_{x_1\in A_1}\int_{x_2\in A_2} f(x_1,x_2)dx_2dx_1 = 84,$$

$$\int_{(x_1,x_2)\in B} f(x_1,x_2)dx_1dx_2 = \int_0^3\int_0^{x_1^2} f(x_1,x_2)dx_2dx_1$$

$$= \int_0^3 \frac{x_1x_2^3}{3}\Big|_0^{x_1^2} dx_1 = \int_0^3 \frac{x_1^7}{3}dx_1$$

$$= \frac{x_1^8}{24}\Big|_0^3 = 273.375.$$

□

Regarding matrix differentiation notation, we utilize the following conventions. Let $g(\mathbf{x})$ and $\mathbf{y}(\mathbf{x})$ be a scalar and $(n \times 1)$ vector function of the $(k \times 1)$ vector $\mathbf{x}$, respectively. Then

| Derivative | Matrix dimension | (*i,j*)th entry |
|---|---|---|
| $\frac{\partial g(\mathbf{x})}{\partial \mathbf{x}}$ | $(k \times 1)$ | $\frac{\partial g}{\partial x_i}$ |
| $\frac{\partial g(\mathbf{x})}{\partial \mathbf{x}'} = \frac{\partial g(\mathbf{x})'}{\partial \mathbf{x}}$ | $(1 \times k)$ | $\frac{\partial g}{\partial x_j}$ |
| $\frac{\partial g(\mathbf{x})}{\partial \mathbf{x} \partial \mathbf{x}'}$ | $(k \times k)$ | $\frac{\partial^2 g}{\partial x_i \partial x_j}$ |
| $\frac{\partial \mathbf{y}(\mathbf{x})}{\partial \mathbf{x}}$ | $(k \times n)$ | $\frac{\partial y_j}{\partial x_i}$ |
| $\frac{\partial \mathbf{y}(\mathbf{x})}{\partial \mathbf{x}'} = \frac{\partial \mathbf{y}(\mathbf{x})'}{\partial \mathbf{x}}$ | $(n \times k)$ | $\frac{\partial y_i}{\partial x_j}$ |

## A7. Elements of Real Analysis

In this section, we present a number of prerequisite results from real analysis that facilitate the development and understanding of various types of asymptotic probability behavior. In particular, the concepts of sequences, limits, continuity of a function, and orders of magnitude of a sequence will be examined.

### Sequences of Numbers and Random Variables

We begin with the notion of a **sequence**. In the definition, we refer to the set of **natural numbers**, which is simply the set of positive integers in their natural order, 1, 2, 3, ... .

**Definition A.24** ***Sequence***

Let $A$ be any set. $A$ sequence in $A$ is a function having the natural numbers, $N$, for its domain, and its range contained in $A$, i.e., $f: N \rightarrow A$, is a sequence in $A$.

When utilizing the concept of a sequence, we (and others) will often suppress the function aspect of its definition and concentrate on the ordered collection of image elements of the function. Thus, given a sequence defined by $\{(n,y): y = f(n), n \in N\}$, we will equivalently refer to the collection of image elements $\{y_1, y_2, y_3,\ldots\}$ as the sequence, where $y_n = f(n)$. The subscripts on the elements of the set $\{y_1, y_2, y_3,\ldots\}$ serve to define the order of the elements in the sequence. Furthermore, we will utilize the notation $\{y_n\}$ as an abbreviation for the sequence $\{y_1, y_2, y_3,\ldots\}$[9] In the following examples of sequences, we continue to use $N$ to denote the set of natural numbers.

[9]Another common abbreviated notation that is sometimes used to denote a sequence is given by $(y_n)$. The notation we have adopted is more prevalent in the statistics literature. While there will be no confusion in this text, in general, the reader will have to rely on the context of a discussion to determine whether $\{y_n\}$ refers to a sequence or to a set containing the single element $y_n$.

**Example A.25** ***Sequences in*** $\mathbb{R}$

**(a)** $\{2,4,8,\ldots\}$, which is defined by the function $y = 2^n$, $n \in N$.

**(b)** $\{1,1/3,1/9,1/27,\ldots\}$, which is defined by the function $y = (1/3)^{n-1}$, $n \in N$.
**(c)** $\{-3,-1,1,3,\ldots\}$, which is defined by the function $y = 2n - 5$, $n \in N$. □

**Example A.26** ***A Sequence of Matrices*** Let $\mathbf{x}_n$ be an $(n \times 2)$ matrix whose $i$th row is defined by the $(1 \times 2)$ vector $[1\ i]$, so that

$$\mathbf{x}_n = \begin{bmatrix} 1 & 1 \\ 1 & 2 \\ \vdots & \vdots \\ 1 & n \end{bmatrix}.$$

Then

$$\left\{ \begin{bmatrix} 1 & 1 \\ 1 & 1 \end{bmatrix}, \begin{bmatrix} 1 & \frac{3}{2} \\ \frac{3}{2} & \frac{5}{2} \end{bmatrix}, \begin{bmatrix} 1 & 2 \\ 2 & \frac{14}{3} \end{bmatrix}, \ldots \right\}$$

is a sequence of *matrices* $\{\mathbf{y}_1,\mathbf{y}_2,\mathbf{y}_3,\ldots\}$ defined by the function $\mathbf{y}_n = \frac{1}{n}\mathbf{x}'_n\mathbf{x}_n, n \in N$, where the $n$th element of the sequence is defined as

$$\mathbf{y}_n = \begin{bmatrix} 1 & \frac{\left(\sum_{i=1}^n i\right)}{n} \\ \frac{\left(\sum_{i=1}^n i\right)}{n} & \frac{\left(\sum_{i=1}^n i^2\right)}{n} \end{bmatrix} = \begin{bmatrix} 1 & \frac{(n+1)}{2} \\ \frac{(n+1)}{2} & \frac{(n+1)(2n+1)}{6} \end{bmatrix}.$$ □

In statistical practice one frequently encounters **sequences of random variables.** In this case, the set $A$ in Definition A.24 is a collection of random variables, and the function $f$: $N \to A$ defining the sequence places the random variables in $A$ in a specific order. That is, the sequence of random variables $\{Y_1, Y_2, Y_3,\ldots\}$ is simply an ordered collection of random variables. In our study of asymptotics, the elements in the sequence of random variables will often be defined as functions of other random variables, such as $Y_n = g_n(X_1,\ldots,X_n)$, and we will be interested in studying the characteristics of the sequence of probability distributions associated with the sequence of $Y_n$'s as $n \to \infty$.

The following are examples of sequences of random variables. We introduce the notation $Y \sim f(y)$ to indicate that $Y$ **has probability density** $f(y)$, or that $Y$ **is distributed as** $f(y)$. This notation can also be used as $Y \sim F(y)$ to denote that Y has the CDF $F(y)$. The acronym ***iid*** used ahead stands for **independent and identically distributed**, meaning that the random variables in a collection are independent and each of the random variables has the same PDF or probability distribution.

**Example A.27** Let $X_1,\ldots, X_n$ be *iid* random variables, each with PDF $N(\mu, \sigma^2)$, where $X_i$ represents the miles per gallon obtained from the $i$th automobile of a certain type tested for fuel efficiency. Examine the sequence of random variables

$\{Y_1, Y_2, Y_3, \ldots\}$, where $Y_n = n^{-1} \sum_{i=1}^{n} X_i, n \in N$.

Note that the $n$th element of the sequence represents the *average* miles per gallon obtained from $n$ of the automobiles tested, and

$$Y_n \sim N\left(\mu, \frac{\sigma^2}{n}\right),$$

so that we can define a **sequence of probability density functions** associated with the sequence of random variables as

$$\left\{N(\mu, \sigma^2), N\left(\mu, \tfrac{\sigma^2}{2}\right), N\left(\mu, \tfrac{\sigma^2}{3}\right), \ldots\right\}.$$

□

**Example A.28** Let $X_1, \ldots, X_n$ be *iid* Bernoulli-type random variables each with density function $p^z(1-p)^{1-z} I_{\{0,1\}}(z)$, where $X_i$ indicates whether the $i$th customer entering a store makes a purchase $(x_i = 1)$ or not $(x_i = 0)$. Examine the sequence of random variables $\{Y_1, Y_2, Y_3, \ldots\}$, where $Y_n = \sum_{i=1}^n X_i$, $n \in N$. Note that the $n$th element of the sequence represents how many of the first $n$ customers make a purchase, and $Y_n$ has a binomial distribution with parameters $n$ and $p$, as BIN($n$,$p$) or

$$Y_n \sim \binom{n}{y_n} p^{y_n} (1-p)^{n-y_n} I_{\{0,1,2\ldots n\}}(y_n).$$

The sequence of probability density functions associated with the sequence of random variables is given by {BIN(1,$p$), BIN(2,$p$), BIN(3,$p$),...}. □

#### Limit of a Real Number Sequence

We now examine the concept of the **limit of a real number sequence**. We begin with a sequence whose elements are scalars and then extend the result to a sequence whose elements are vectors of real numbers.

**Definition A.25** ***Limit of a Real Number Sequence***

Let $\{y_n\}$ be a sequence whose elements are scalar real numbers. Suppose there exists a real number, $y$, such that for every real $\varepsilon > 0$ there exists an integer $N(\varepsilon)$ for which $n \geq N(\varepsilon) \Rightarrow |y_n - y| < \varepsilon$. Then $y$ is the **limit of the sequence** $\{y_n\}$, and the sequence $\{y_n\}$ is said to **converge to** $y$ as $n \to \infty$. The existence of the limit is denoted by $y_n \to y$ or $\lim_{n\to\infty} y_n = y$. If the limit does not exist, the sequence is said to be **divergent**.

The definition of the limit implies that for a sufficiently large choice of $n$, $y_n$ (and $y_{n+1}, y_{n+2}, y_{n+3}, \ldots$) becomes arbitrarily close to the number $y$. This is so since, by definition, we can choose $\varepsilon > 0$ to be arbitrarily small and yet there exists an $n$ large enough (namely $n \geq N(\varepsilon)$) such that $y - \varepsilon < y_n < y + \varepsilon$. Figure A.5 provides a graphical illustration of the limit concept.

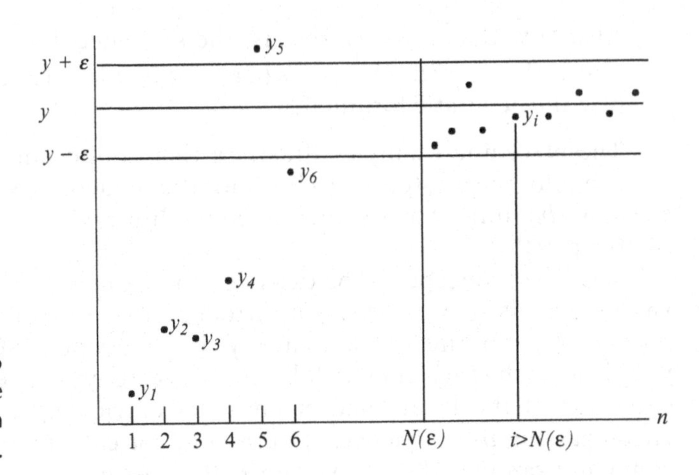


**Figure A.6** Illustration of the sequence $\{y_n\}$ for which $\lim_{n \to \infty} y_n = y$.

In the figure, it is seen that for all elements of $y_i \in \{y_n\}$ for which $i$ is large enough, i.e., for $i > N(\varepsilon)$, the value of $y_i$ is contained in the interval $(y - \varepsilon, y + \varepsilon)$. In other words, $y_i$ is within $\varepsilon$ – distance of $y$ for $i > N(\varepsilon)$. Furthermore, for every choice of $\varepsilon > 0$, no matter how small, there exists an $N(\varepsilon)$ for which a figure such as Figure A.6 applies.

It can be shown that for the limit of a sequence of real numbers to exist, it is necessary (but not sufficient) that the sequence is **bounded** (Bartle, *The Elements of Real Analysis*, 2nd Edition, John Wiley and Sons, New York, p. 93).

**Definition A.26**
***Bounded Sequence of Real Numbers***

The sequence of real numbers $\{y_n\}$ is **bounded** *iff* there exists a finite number $m > 0$ such that $|y_n| \leq m \; \forall \; n \in N$; otherwise the sequence is said to be **unbounded.**

Thus, for a sequence of real numbers to be bounded, there must exist a positive number that is larger than the absolute value of each and every number in the sequence. For a sequence that has no limit and is *also* unbounded, we write $y_n \to \infty$ (or $y_n \to -\infty$), denoting that the sequence **diverges to infinity** (or **diverges to negative infinity**), if $\forall m > 0$ there exists a positive integer $N(m)$ such that $y_n > m$ (or $y_n < -m$) $\forall n > N(m)$.

**Example A.29**
***Boundedness and existence of a Limit for Real Number Sequences***

**(a)** $y_n = 3 + n^{-2}, n \in N$. This sequence is bounded, since $|y_n| \leq 4 \; \forall \; n \in N$. Also, the sequence has a limit, where $y_n \to 3$. This follows since, $\forall \; \varepsilon > 0$, $|y_n - 3| < \varepsilon \; \forall \; n > \varepsilon^{-1/2}$, and there always exists an integer $N(\varepsilon) \geq \varepsilon^{-1/2}$ (e.g., trunc $(\varepsilon^{-1/2}) + 1$).

**(b)** $y_n = \sin(n)$, $n \in N$ (let $n$ be measured in degrees). The sequence is bounded, since $|\sin(x)| \leq 1 \; \forall \; x$. The sequence does *not* have a limit, since $\sin(x)$ cycles between the values of $-1$ and $1$.

**(c)** $y_n = n^2 - 3n + 1$, $n \in N$. The sequence is not bounded, since there does not exist a finite number $m > 0$ for which $n^2 - 3n + 1 \leq m \; \forall \; n \in N$. Since the sequence is unbounded, the sequence does not have a limit.

Also note that $y_n \to \infty$, that is, the sequence diverges to infinity, because $\forall\, m > 0,\ n^2 - 3n + 1 > m$ when $n > N(m) = \text{trunc}\left(\left[3 + \sqrt{5 + 4m}\,\right]/2\right) + 1$ (use the quadratic formula).□

The preceding examples illustrate that boundedness of a sequence is not sufficient for the existence of a limit for the sequence. We add that it is proper to speak of *the* limit of a sequence since the limit will be *unique* if it exists at all (Bartle, p. 93).

The limit concept can be extended to a sequence whose elements are real-valued vectors or matrices. We introduce the notation $y_n[i,j]$ to indicate the $(i,j)$th element of the $(q \times k)$ matrix $\mathbf{y}_n$ in a sequence of matrices, and similarly $y_n[i]$ denotes the $i$th element of the $(q \times 1)$ vector $\mathbf{y}_n$ in a sequence of vectors. The extension of the limit concept amounts to viewing the matrix sequence as encompassing $mk$ sequences, $\{y_n[i,j]\}$, one for each matrix element, with each element examined for convergence (by Definition A.25). Limits of vector sequences follow by letting $k = 1$.

**Definition A.27**
***Limit of a Real-Valued Matrix Sequence***

Let $\{\mathbf{y}_n\}$ be a sequence whose elements are $(q \times k)$ real-valued matrices. Suppose there exists a $(q \times k)$ matrix of real numbers $\mathbf{y}$ such that $y_n[i,j] \to y[i,j]$ for $i = 1,\ldots,q$ and $j = 1,\ldots,k$. Then the matrix $\mathbf{y}$ is the **limit of the matrix sequence** $\{\mathbf{y}_n\}$, and the sequence of matrices $\{\mathbf{y}_n\}$ is said to **converge to the matrix y** as $n \to \infty$. The existence of the limit is denoted by $\mathbf{y}_n \to \mathbf{y}$, or by $\lim_{n\to\infty}\mathbf{y}_n = \mathbf{y}$. If the limit does not exist, the sequence is said to be **divergent**.

The definition of the limit implies that for a sufficiently large choice of $n$, the matrix $\mathbf{y}_n$ (and $\mathbf{y}_{n+1}, \mathbf{y}_{n+2}, \mathbf{y}_{n+3},\ldots$) becomes arbitrarily close to the matrix $\mathbf{y}$, element by element. The definition also implies that for the real-valued matrix to have a limit, the sequence of matrices must be bounded elementwise, i.e., $|y_n[i,j]| \leq m\ \forall\, n \in N$ and $\forall i$, $j$ or else $y_n[i,j] \not\to y[i,j]$ for some $i$ and $j$, and then $\mathbf{y}_n \not\to \mathbf{y}$. Regarding divergence to infinity (or negative infinity), since there are essentially $qk$ convergence conditions involved when examining sequences of $(q \times k)$ matrices, patterns of divergence and convergence of the various elements of the matrices can be quite diverse.

**Example A.30**
***Boundedness and Limits of Matrices***

**(a)** Recall the sequence of matrices in Example A.26. In this case, only the sequence $\{y_n[1,1]\}$ is bounded. All other sequences of matrix elements are unbounded and, in fact, diverge to infinity, i.e., $y_n[i,j] \to \infty$ for $(i,j) \neq (1,1)$. Since all of the sequences of matrix elements must be bounded for the matrix sequence to converge, the matrix sequence does not have a limit.

**(b)** Let $\{\mathbf{y}_n\}$ be a sequence of matrices such that $\mathbf{y}_n = \begin{bmatrix} 3n^{-1} & n^{-1} \\ 3 & 1+n^{-1} \end{bmatrix},\ n \in N.$ All four sequences of matrix elements are bounded, since $|3n^{-1}| \leq 3$, $|n^{-1}| \leq 1$, $|3| \leq 3$, and $|1 + n^{-1}| \leq 2\ \forall\, n \in N$. Furthermore, limits exist for all four sequences of matrix elements, since $3n^{-1} \to 0$, $n^{-1} \to 0$, $3 \to 3$, and $1 + n^{-1} \to 1$. Thus, $\mathbf{y}_n \to \mathbf{y} = \begin{bmatrix} 0 & 0 \\ 3 & 1 \end{bmatrix}$. □

One might be interested in a sequence $\{y_n\}$ that is defined via a function of the elements of other sequences. For example, we may be interested in the sequence $\{y_n\}$ that is defined by adding corresponding elements in the sequences $\{x_n\}$ and $\{z_n\}$, as $y_n = x_n + z_n$. Of course, we can analyze the properties of the sequence $\{y_n\}$ directly to establish whether the sequence converges, but if the convergence properties of $\{x_n\}$ and $\{z_n\}$ are known, the following lemma can expedite the analysis if the functions are defined via addition, subtraction, or multiplication. We precede the statement of the lemma with definitions for adding, subtracting, and multiplying sequences. Note that $x[.,i]$ refers to the $i$th column of the matrix $\mathbf{x}$.

**Definition A.28**
***Adding, Subtracting, Multiplying Sequences***

Let $\{\mathbf{x}_n\}$ and $\{\mathbf{z}_n\}$ be sequences of conformable real-valued matrices.

**(a)** *Summation*: The summation of $\{\mathbf{x}_n\}$ and $\{\mathbf{z}_n\}$, $\{\mathbf{x}_n\} + \{\mathbf{z}_n\}$, is a sequence $\{\mathbf{y}_n\}$ defined by $\mathbf{y}_n = \mathbf{x}_n + \mathbf{z}_n$, $\forall n$.

**(b)** *Difference*: The difference between $\{\mathbf{x}_n\}$ and $\{\mathbf{z}_n\}$, $\{\mathbf{x}_n\} - \{\mathbf{z}_n\}$, is a sequence $\{\mathbf{y}_n\}$ defined by $\mathbf{y}_n = \mathbf{x}_n - \mathbf{z}_n$, $\forall n$.

**(c)** *Product*: The product of $\{\mathbf{x}_n\}$ and $\{\mathbf{z}_n\}$, $\{\mathbf{x}_n\}\,\{\mathbf{z}_n\}$, is a sequence $\{\mathbf{y}_n\}$ defined by $\mathbf{y}_n = \mathbf{x}_n\mathbf{z}_n$, $\forall n$.

**Lemma A.1**
***Combinations of Sequences***

Let $\{\mathbf{x}_n\}$ and $\{\mathbf{z}_n\}$ be convergent sequences of conformable real-valued matrices such that $\mathbf{x}_n \to \mathbf{x}$ and $\mathbf{z}_n \to \mathbf{z}$. Then

**(a)** $\mathbf{x}_n + \mathbf{z}_n \to \mathbf{x} + \mathbf{z}$,

**(b)** $\mathbf{x}_n - \mathbf{z}_n \to \mathbf{x} - \mathbf{z}$,

**(c)** $\mathbf{x}_n\mathbf{z}_n \to \mathbf{x}\mathbf{z}$,

**(d)** if $\{a_n\} \to a$, then $a_n\mathbf{x}_n \to a\mathbf{x}$,

**(e)** if $\{b_n\} \to b \neq 0$, then $b_n^{-1}\mathbf{x}_n \to b^{-1}\mathbf{x}$,

**(f)** $\sum_{i=1}^{k} \mathbf{x}_n[.,i] \to \sum_{i=1}^{k} \mathbf{x}[.,i]$

**(g)** if $\{\mathbf{z}_n\}$ is a sequence of nonsingular matrices that converges to the nonsingular matrix $\mathbf{z}$, then $\mathbf{z}_n^{-1} \to \mathbf{z}^{-1}$ and $\mathbf{z}_n^{-1}\mathbf{x}_n \to \mathbf{z}^{-1}\mathbf{x}$.

**Proof** Bartle, pp. 100–101, and Definitions 5.4 and 5.5.

Note that since the sequences $\{\mathbf{x}_n\}$ and $\{\mathbf{z}_n\}$ can themselves be defined in terms of combinations of other sequences, the lemma actually implies convergence results involving more than just two sequences. For example, letting $\mathbf{z}_n = \mathbf{a}_n + \mathbf{b}_n \to \mathbf{a} + \mathbf{b}$, then $\mathbf{a}_n + \mathbf{b}_n + \mathbf{z}_n \to \mathbf{a} + \mathbf{b} + \mathbf{z}$ and $(\mathbf{a}_n + \mathbf{b}_n)\mathbf{z}_n \to (\mathbf{a} + \mathbf{b})\mathbf{z}$.

**Example A.31**
***Convergence Properties of Sequences***

**(a)** Let $\{x_n\}$ and $\{z_n\}$ be defined as $x_n = 3 + n^{-1/2}$ and $z_n = 2\exp(-2/n)$ for $n \in N$, respectively. Note that $x_n \rightarrow 3$ and $z_n \rightarrow 2$. Then using Lemma A.1, $x_n + z_n \rightarrow 5$, $x_n - z_n \rightarrow 1$, and $x_n z_n \rightarrow 6$. Let $\{a_n\}$ be defined by $a_n = 5(n+1)/n$ for $n \in N$, and note that $a_n \rightarrow 5$. Also define the vector sequence $\{\mathbf{y}_n\}$ by $\underset{(2\times1)}{\mathbf{y}_n} = \begin{bmatrix} x_n \\ z_n \end{bmatrix}$ so that $\mathbf{y}_n \rightarrow \begin{bmatrix} 3 \\ 2 \end{bmatrix}$. Then from Lemma A.1, $a_n\mathbf{y}_n \rightarrow \begin{bmatrix} 15 \\ 10 \end{bmatrix}$ and $a_n^{-1}\mathbf{y}_n \rightarrow \begin{bmatrix} 3/5 \\ 2/5 \end{bmatrix}$.

**(b)** Let $\{\mathbf{w}_n\}$ be a matrix sequence defined by

$$\mathbf{w}_n = \begin{bmatrix} 2+n^{-1} & \dfrac{2}{n} \\ 3 & \dfrac{(n+1)}{n} \end{bmatrix} \text{ for } n \in N,$$

and let $\{\mathbf{x}_n\}$ be a vector sequence defined by

$$\mathbf{x}_n = \begin{pmatrix} 1+n^{-1} \\ 2\exp(n^{-1}) \end{pmatrix} \text{ for } n \in N.$$

Note that $\mathbf{w}_n \rightarrow \begin{bmatrix} 2 & 0 \\ 3 & 1 \end{bmatrix}$ and $x_n \rightarrow \begin{bmatrix} 1 \\ 2 \end{bmatrix}$. Using Lemma A.1, it follows that $\mathbf{w}_n^{-1} \rightarrow \begin{bmatrix} 2 & 0 \\ 3 & 1 \end{bmatrix}^{-1} = \begin{bmatrix} .5 & 0 \\ -1.5 & 1 \end{bmatrix}$, $\mathbf{w}_n\,\mathbf{x}_n \rightarrow \begin{bmatrix} 2 \\ 5 \end{bmatrix}$, and $\mathbf{w}_n^{-1}\mathbf{x}_n \rightarrow \begin{bmatrix} .5 \\ .5 \end{bmatrix}$.

Note further that $\mathbf{w}_n[.,1] + \mathbf{w}_n[.,2] + \mathbf{x}_n \rightarrow \begin{bmatrix} 3 \\ 6 \end{bmatrix}$. □

### Continuous Functions

Continuous functions play a prominent role in a number of important asymptotic results. We define the concept of a continuous function below using two alternative but completely equivalent characterizations. We remind the reader that $d(\mathbf{x},\mathbf{w}) = [(\mathbf{x}-\mathbf{w})'(\mathbf{x}-\mathbf{w})]^{1/2}$ is the distance between points $\mathbf{x}$ and $\mathbf{w}$.

**Definition A.29**
***Continuous Functions***

The function[10] $g\colon A \rightarrow \mathbb{R}$, for $A \subset \mathbb{R}^m$, is continuous at the point $\mathbf{x} \in A$ iff either:

**(a)** $\forall \varepsilon > 0$, $\exists \delta(\varepsilon) > 0$ such that $\mathbf{w} \in A$ and $d(\mathbf{x},\mathbf{w}) < \delta(\varepsilon)$ implies $|g(\mathbf{w}) - g(\mathbf{x})| < \varepsilon$

**(b)** $\forall$ sequence $\{\mathbf{x}_n\}$ in $A$ for which $\mathbf{x}_n \rightarrow \mathbf{x}$, it is also true that $g(\mathbf{x}_n) \rightarrow g(\mathbf{x})$.

[10] This definition can be altered to provide definitions for *continuity from the right* and *continuity from the left*. For continuity from the right, the condition $\mathbf{w} \geq \mathbf{x}$ is added in part (a). The condition $\mathbf{x}_n \geq \mathbf{x}\ \forall n$ is added to part (b). For continuity from the left, the conditions become $\mathbf{w} \leq \mathbf{x}$ and $\mathbf{x}_n \leq \mathbf{x}\ \forall n$.

The $k \times 1$ *vector* function $\mathbf{g}: A \to \mathbb{R}^k$ is continuous at the point $\mathbf{x} \in A$ *iff* each coordinate function $g_j(\mathbf{x})$ is continuous at the point $\mathbf{x}$, $j = 1,\ldots,k$. The function $\mathbf{g}$ is said to be continuous on the set $B \subset A$ if the function is continuous at every point in $B$.

Intuitively, a function is continuous at a point $\mathbf{x}$ if, when the function is evaluated at domain elements that are closer and closer to $\mathbf{x}$, the value of the function is closer and closer to the value of the function at $\mathbf{x}$ (in an elementwise vector comparison sense if $k \geq 2$). In the simple case where $m = k = 1$, the graph of a function that is continuous on an interval set $B = (a,b)$ can be drawn $\forall\, x \in B$ "without lifting the pencil from the paper," i.e., the graph is an unbroken curve.

**Example A.32** ***Continuity Properties of Functions***

**(a)** Let $f{:}(0,\infty) \to \mathbb{R}$ be defined by $y = x^{-1}$. Intuitively we expect $f$ to be continuous on $(0,\infty)$, since its graph is an unbroken curve. To demonstrate formally that $f$ is continuous on $(0,\infty)$, let $x_0 \in (0,\infty)$, and note that

$$|f(x) - f(x_0)| = |x^{-1} - x_0^{-1}| = \left|\frac{x_0 - x}{x x_0}\right| = \frac{|x_0 - x|}{x x_0}$$

where we have eliminated the denominator from the absolute value operator because in the domain of $f$, $x > 0$. It follows that if $|x - x_0| < \delta$, then $|f(x) - f(x_0)| < \delta / x x_0$.

Now choose any $\varepsilon > 0$. For $f$ to be continuous at $x_0$, it must be the case that $\delta$ can be chosen such that $|x - x_0| < \delta \Rightarrow |f(x) - f(x_0)| < \varepsilon$. Such choices of $\delta$ exist, one being equal to $\delta = x_0^2\, \varepsilon / (1 + x_0\, \varepsilon)$. Since the argument can be applied $\forall x_0 \in (0,\infty)$ and $\forall\, \varepsilon > 0$, $f$ is continuous on $(0,\infty)$.

**(b)** Let $f: \mathbb{R}^2 \to \mathbb{R}^2$ be defined by the coordinate functions

$$\begin{bmatrix} y[1] \\ y[2] \end{bmatrix} = \begin{bmatrix} f_1(\mathbf{x}) \\ f_2(\mathbf{x}) \end{bmatrix} = \begin{bmatrix} x[1]^2 + 2x[2] \\ x[1] \end{bmatrix}.$$

Let $\mathbf{x}_* \in \mathbb{R}^2$, and let $\{\mathbf{x}_n\}$ be any sequence in $\mathbb{R}^2$ for which $\mathbf{x}_n \to \mathbf{x}_*$. To demonstrate continuity of $f$ at $\mathbf{x}_*$, we will show that $\mathbf{x}_n \to \mathbf{x}_* \Rightarrow f(\mathbf{x}_n) \to f(\mathbf{x}_*)$. Examine $f_1(x)$ first. Note that $f_1(\mathbf{x}_n) = x_n[1]^2 + 2x_n[2] = (x_n[1]x_n[1]) + 2x_n[2]$ can be interpreted as the summation of the sequence $\{x_n[1]\}\{x_n[1]\}$ and the sequence $2\{x_n[2]\}$. Since $x_n[1] \to x_*[1]$ and $x_n[2] \to x_*[2]$ by assumption, it follows from Lemma A.1 that $\{x_n[1]\}\{x_n[1]\} \to x_*[1]^2$, $2\{x_n[2]\} \to 2\,x_*[2]$, and thus $f_1(\mathbf{x}_n) \to x_*[1]^2 + 2x_*[2] = f_1(\mathbf{x}_*)$. Verifying convergence of the second component function is straightforward, because $f_2(\mathbf{x}_n) = x_n[1] \to x_*[1] = f_2(\mathbf{x}_*)$. Thus, $f$ is continuous at $\mathbf{x}_*$, and since the above argument holds for any $\mathbf{x}_* \in \mathbb{R}^2$, $f$ is continuous on $\mathbb{R}^2$.

**(c)** Let $f: \mathbb{R} \to \mathbb{R}$ be defined by $f(x) = I_{(0,\infty)}(x)$. Intuitively, since the graph of the function has a break at $x = 0$, we expect that the function is not continuous on $\mathbb{R}$. To formally demonstrate that $f$ is discontinuous at $x = 0$, it will be shown that there does *not* exist a $\delta(\varepsilon) > 0$ for every $\varepsilon > 0$ such that $|x - 0| < \delta(\varepsilon) \Rightarrow |f(x) - f(0)| < \varepsilon$. Choose any $\varepsilon \in (0,1)$, and note that $f(0) = 0$. Given *any* choice of $\delta(\varepsilon) > 0$, let $x = \delta(\varepsilon)/2$, so that $|x - 0| < \delta(\varepsilon)$.

Then $|f(x) - f(0)| = 1 \nless \varepsilon$. Therefore, $f$ is not continuous on $\mathbb{R}$. (Note: $f$ *is* continuous on $(0,\infty)$ and on $(-\infty,0)$, as the reader might wish to verify). □

### Convergence of a Function Sequence

**Convergence of a sequence of functions** refers to a case where the function definitions themselves can change as $n$ changes, so that we can conceptualize a sequence of image values $\{f_n(x)\} = \{f_1(x), f_2(x), f_3(x), \ldots\}$ for *each* value of $x$ (which could be a vector) in the *common* domain of the sequence of functions. Interest centers on whether there exists a "limit" function definition, $f(x)$, such that $f_n(x) \to f(x)$ for all $x$ in some subset of the common domain of the sequence of functions. The subset of $x$-values on which $\{f_n(x)\}$ converges to $f(x)$ could be the entire domain, or some smaller subset of points, or the null set (i.e., $\{f_n(x)\}$ does not converge for any $x$).

We formalize the concept of convergence of a sequence of functions in the next definition. In the definition, the notation $\{f_n\}$ refers to the sequence of function definitions, as opposed to $\{f_n(x)\}$ which denotes the sequence of image values generated by the sequence of function definitions when evaluated at the point $x$.

**Definition A.30** ***Convergence of a Function Sequence***

Let $\{f_n\}$ be a sequence of functions $f_n : D \to \mathbb{R}^\ell$ for $n = 1, 2, \ldots$, having common domain $D \subset \mathbb{R}^m$. Let $f : D_0 \to \mathbb{R}^\ell$ be a function with domain $D_0 \subset D$. The function sequence $\{f_n\}$ is said to converge on $D_0$ to $f$ if $f_n(\mathbf{x}) \to f(\mathbf{x})$, $\forall \mathbf{x} \in D_0$. If $\{f_n\}$ converges to $f$ on $D_0$, $f$ is called the **limiting function** of $\{f_n\}$ on $D_0$, and $\{f_n\}$ is said to be **convergent** on $D_0$.

Intuitively, if $f$ is the limit function of $\{f_n\}$ on $D_0$, then for large enough $n$, $f_n(\mathbf{x}) \approx f(\mathbf{x})$ for $x \in D_0$ since $f_n(\mathbf{x})$ converges to $f(\mathbf{x})$ on $D_0$. Then $f(\mathbf{x})$ can be viewed as an approximation to $f_n(\mathbf{x})$ on $D_0$ when $n$ is large.

**Example A.33** ***Convergence of Function Sequences***

**(a)** Let the function sequence $\{f_n\}$ be defined by $f_n(x) = n^{-1} + 2x^2$ for $x \in \mathbb{R}$. Define the function $f$: $\mathbb{R} \to \mathbb{R}$ by $f(x) = 2x^2$. Then $f$ is the limiting function of $\{f_n\}$ on $\mathbb{R}$. To see this, note that $f_n(x) \to 2x^2 = f(x)\ \forall\ x \in \mathbb{R}$. For large $n$, $f_n(x) \approx f(x)\ \forall\ x \in \mathbb{R}$.

**(b)** Let the vector function sequence $\{f_n\}$ be defined by

$$f_n(\mathbf{x}) = \begin{bmatrix} f_{1n}(\mathbf{x}) \\ f_{2n}(\mathbf{x}) \end{bmatrix} = \begin{bmatrix} \dfrac{(2x_1^2 + 3nx_1x_2)}{n} \\ x_1^2 + x_2^2 \end{bmatrix} \quad \text{for } (x_1, x_2) \in \mathbb{R}^2.$$

Define the function $f$: $\mathbb{R}^2 \to \mathbb{R}^2$ by

$$f(\mathbf{x}) = \begin{bmatrix} f_1(\mathbf{x}) \\ f_2(\mathbf{x}) \end{bmatrix} = \begin{bmatrix} 3x_1 x_2 \\ x_1^2 + x_2^2 \end{bmatrix} \text{ for } (x_1, x_2) \in \mathbb{R}^2.$$

Then $f$ is the limiting function of $\{f_n\}$ on $\mathbb{R}^2$. To see this, note that $f_{1n}(\mathbf{x}) = 2x_1^2/n + 3x_1 x_2 \to 0 + 3x_1 x_2 = f_1(\mathbf{x})$, and $f_{2n}(\mathbf{x}) = x_1^2 + x_2^2 = f_2(\mathbf{x})$ $\forall \mathbf{x} \in \mathbb{R}^2$, so that $f_n(\mathbf{x}) \to f(\mathbf{x})\ \forall\ \mathbf{x} \in \mathbb{R}^2$.

(c) Let the function sequence $\{f_n\}$ be defined by $f_n(x) = x - 2\exp(-nx)$ for $x \in \mathbb{R}$. Let $f: \mathbb{R} \to \mathbb{R}$ be defined by $f(x) = x - 2I_{\{0\}}(x)$ for $x \in D_0 = [0,\infty]$. Then $f$ is the limiting function of $\{f_n\}$ on the set $[0,\infty)$. The sequence $\{f_n\}$ does not converge for $x < 0$. To justify these conclusions, note that, by Lemma A.1, $\lim_{n\to\infty} f_n(x) = \lim_{n\to\infty} x - \lim_{n\to\infty} 2\exp(-nx) = x - 2I_{(0)}(x) \forall x \geq 0$. When $x < 0$, $2\exp(-nx) \to \infty$, and thus $f_n(x)$ does not converge. □

### Order of Magnitude of a Sequence

In analyses involving sequences, it is sometimes useful to be able to characterize or compare sequences and/or terms in a sequence relative to their order of magnitude. In particular, when the definition of a sequence contains a number of terms, the orders of magnitude of the terms will distinguish which terms make dominant contributions to the magnitude of the sequence as $n$ increases. The concept is defined below.

**Definition A.31**
***Order of Magnitude of a Sequence***

Let $\{x_n\}$ be a real number sequence, and let $\{\mathbf{w}_n\}$ be a real-valued matrix sequence.

(a) The sequence $\{x_n\}$ is said to be **at most of order** $n^k$, denoted by $O(n^k)$, if there exists a finite real number $c$ such that $|n^{-k} x_n| \leq c \; \forall n \in N$.

(b) The sequence $\{x_n\}$ is said to be **of order smaller than** $n^k$, denoted by $o(n^k)$, if $n^{-k}x_n \to 0$.

(c) If $\{w_n[i,j]\}$ is $O(n^k)$ (or $o(n^k)$) $\forall i$ and $j$, then the matrix sequence $\{\mathbf{w}_n\}$ is said to be $O(n^k)$ (or $o(n^k)$).

Intuitively, a sequence $\{x_n\}$ is $O(n^k)$ if the corresponding sequence $\{n^{-k}x_n\}$ is such that all elements $n^{-k}x_n$ are bounded in absolute value by some positive number $c$. A sequence $\{x_n\}$ is $o(n^k)$ if the product sequence $\{n^{-k}x_n\}$ converges to zero. Note that if $\{x_n\}$ is $O(n^k)$, then $\{x_n\}$ is $o(n^{k+\epsilon}) \; \forall \epsilon > 0$, and if $\{x_n\}$ is $o(n^k)$, then it is also $O(n^k)$. Notationally, the case $O(n^o)$ or $o(n^o)$ is most often represented as $O(1)$ or $o(1)$.

**Example A.34**
***Order of Magnitude of a Sequence***

(a) Let $\{x_n\}$ be defined by $x_n = 3n^3 - n^2 + 2$, for $n \in N$. Then $\{x_n\}$ is $O(n^3)$, since $n^{-3}x_n = 3 - n^{-1} + 2n^{-3}$ is bounded. Also, $\{x_n\}$ is $o(n^{3+\varepsilon})$ for any $\varepsilon > 0$ since $n^{-3-\varepsilon}x_n = 3n^{-\varepsilon} - n^{-1-\varepsilon} + 2n^{-3-\varepsilon} \to 0$.
(b) Let $\{x_n\}$ be defined by $x_n = 3 + n^{-1}$, for $n \in N$. Then $\{x_n\}$ is $O(1)$, since $x_n = 3 + n^{-1}$ is bounded, and $\{x_n\}$ is $o(n^{\varepsilon}) \; \forall \varepsilon > 0$, since $n^{-\varepsilon}x_n = 3n^{-\varepsilon} + n^{-1-\varepsilon} \to 0$.
(c) Let the vector sequence $\{\mathbf{x}_n\}$ be defined by $\begin{bmatrix} x_n[1] \\ x_n[2] \end{bmatrix} = \begin{bmatrix} 3n^{-1} \\ n^{-1} \end{bmatrix}$. Then the vector sequence $\{\mathbf{x}_n\}$ is $o(1)$ and $O(1)$, since $x_n \to \begin{bmatrix} 0 \\ 0 \end{bmatrix}$. □

Some useful results regarding the order of magnitude of sums and products of sequences are given in the following lemma.

**Lemma A.2**

Let $\{x_n\}$ and $\{z_n\}$ be real number sequences. The following relationships between orders of magnitude hold:

| IF | | THEN | |
|---|---|---|---|
| $\{x_n\}$ | $\{z_n\}$ | $\{x_n + z_n\}$ | $\{x_n z_n\}$ |
| $O(n^k)$ | $O(n^m)$ | $O(n^{\max(k,m)})$ | $O(n^{k+m})$ |
| $o(n^k)$ | $o(n^m)$ | $o(n^{\max(k,m)})$ | $o(n^{k+m})$ |
| $O(n^k)$ | $o(n^m)$ | $O(n^{\max(k,m)})$ | $o(n^{k+m})$ |

**Proof** H. White (1984), *Asymptotic Theory for Econometricians.* Orlando, Academic Press, p. 15.

Given that the sequences referred to in Lemma A.2 can themselves be functions of other sequences, the results can be extended in a myriad of ways to an arbitrary finite number of sequences. In the following lemma we state some useful extensions.

**Lemma A.3**

Let $\{\mathbf{x}_n\}$ and $\{\mathbf{w}_n\}$ be $(m \times \ell)$ and $(r \times m)$ matrix sequences, respectively.

**(a)** If $\{\mathbf{x}_n\}$ is such that $\{\mathbf{x}_n[.,i]\}$ is $O(n^{k_i})$ (or $o(n^{k_i})$) for $i = 1,\ldots,\ell$, then $\{\sum_{i=1}^{\ell} \mathbf{x}_n[.,i]\}$ is $O(n^{k\max})$ (or $o(n^{k\max})$) where $k\max = \max(k_1,\ldots,k_\ell)$.

**(b)** (Special case of (a)): If $\{\mathbf{x}_n\}$ is $O(n^k)$ (or $o(n^k)$), then $\{\sum_{i=1}^{\ell} \mathbf{x}_n[.,i]\}$ is $O(n^k)$ (or $o(n^k)$).

**(c)** If $\{\mathbf{x}_n\}$ is $O(n^k)$ and $\{\mathbf{w}_n\}$ is $\left\{\begin{matrix} O(n^d) \\ o(n^d) \end{matrix}\right\}$, then $\{\mathbf{w}_n\mathbf{x}_n\}$ is $\left\{\begin{matrix} O(n^{k+d}) \\ o(n^{k+d}) \end{matrix}\right\}$ and $\{n^{-v}\mathbf{w}_n\mathbf{x}_n\}$ is $\left\{\begin{matrix} O(n^{k+d-v}) \\ o(n^{k+d-v}) \end{matrix}\right\}$.

**(d)** (Special case of (c)): If $\{\mathbf{x}_n\}$ is $O(n^k)$ (or $o(n^k)$), then $\{\mathbf{x}_n'\mathbf{x}_n\}$ is $O(n^{2k})$ (or $o(n^{2k})$) and $\{n^{-v}\mathbf{x}_n'\mathbf{x}_n\}$ is $O(n^{2k-v})$ (or $o(n^{2k-v})$).

**Proof** This follows from Lemma A.2 and mathematical induction.

We illustrate the application of some of the preceding results in the following example.

**Example A.35** ***Orders of Magnitude for Combinations of Sequences***

**(a)** Let $\{x_n\}$ be defined by $x_n = 2n^{-1} + 5n$ and let $\{z_n\}$ be defined by $z_n = n^2 + 2n$. Note that $\{x_n\}$ is $O(n^1)$ and $\{z_n\}$ is $O(n^2)$. It follows immediately, from Lemma A.2 that $\{x_n + z_n\} = \{n^2 + 7n + 2n^{-1}\}$ is $O(n^2)$ (and $o(n^{2+\varepsilon})$ for $\varepsilon > 0$) and $\{x_n z_n\} = \{5n^3 + 10n^2 + 2n + 4\}$ is $O(n^3)$ (and $o(n^{3+\varepsilon})$ for $\varepsilon > 0$).

**(b)** Let $\{\mathbf{x}_n\}$ and $\{\mathbf{w}_n\}$ be defined by

$$\mathbf{x}_n = \begin{bmatrix} 7+n^{-1} \\ n^{-1} \end{bmatrix} \text{ and } \mathbf{w}_n = \begin{bmatrix} n^2+2n+1 & 3n^2+7 \\ n^2 & n^2+n \end{bmatrix}$$

so that $\mathbf{x}_n$ is O(1) and $\mathbf{w}_n$ is O($n^2$). It follows immediately from Lemma A.3 that

$$\{\mathbf{w}_n \mathbf{x}_n\} = \left\{ \begin{bmatrix} 7n^2+18n+9+8n^{-1} \\ 7n^2+2n+1 \end{bmatrix} \right\}$$

is O($n^2$) and o($n^{2+\varepsilon}$) for $\varepsilon > 0$, and $\{\mathbf{x}'_n\mathbf{x}_n\} = \{49+14n^{-1}+2n^{-2}\}^{\frac{1}{2}}$ is O(1) and o($n^{\varepsilon}$) for $\varepsilon > 0$. It also follows, for example, that $\{n\mathbf{x}'_n\mathbf{x}_n\}$ and $\{n^{-1}\mathbf{w}_n\mathbf{x}_n\}$ are both O($n^1$) and o($n^{1+\varepsilon}$) for $\varepsilon > 0$. □

## Keywords, Phrases, and Symbols

$\in$
$\mathbb{R}_{\geq 0}$, $\mathbb{R}_+$
$A^2 = A \times A$
Associative laws
Axiom (or postulate)
Binary relation from $A$ to $B$, $S$: $A \to B$
Binomial theorem
Cartesian product
Closed, open, half-open intervals
Combinations, $\binom{n}{r}$
Commutative laws
Complement, $\bar{A}$
Complements of complements
Contained in, $\subset$
Corollary
Countable set
Definition
DeMorgan's laws
Distributive laws
Element
Empty or null set, $\emptyset$
Equality of sets, =
Exhaustive listing
Finite set
For every, $\forall$
Function from $A$ to $B$, $f$:$A \to B$
Idempotency laws
Identity elements
*iff* (if and only if)
Image of $x$ under $f$
Index set
Indicator function, $I_A(x)$
Infinite set
Integration notation
Intersection and union of complements
Intersection with null set
Intersection, $\cap$
Interval set notation
Inverse function, $f^{-1}$:$B \to A$
Inverse image
Lemma
Mathematical rule
Multiple intersection notation
Multiple union notation
Mutually exclusive (disjoint)
Negation, /
Null set as a subset
Permutations, $n_r$
Point function
QED
Real line
Real-valued function
Set
Set difference, −
Set function
Size of set function
Stirling's formula
Subset
Such that (such that)
Summation notation
The function $f(x)$
The set function $f(A)$
Theorem (or proposition)
There exists, $\exists$
Uncountable set
Union, $\cup$
Universal set
Venn diagram
Verbal rule
$xSy$

## Problems

1. Using either an exhaustive listing, verbal rule, or mathematical rule, define the following sets:

(a) The set of all senior citizens receiving social security payments in the United States.

(b) The set of all positive numbers that are positive integer powers of the number 10 (i.e., $10^1$, $10^2$, etc.).

(c) The set of all possible outcomes resulting from rolling a red and a green die and calculating the values of $y - x$, where $y$ = number of dots on the red die, $x$ = number of dots on the green die.

(d) the set of all two-tuples $(x_1, x_2)$ where $x_1$ is any real number and $x_2$ is related to $x_1$ by raising the number $e$ to the power $x_1$.

**2.** Label the sets you have identified in Problem (1) as being either finite, countably infinite, or uncountably infinite, and explain your choice.

**3.** For each set below, state whether the set is finite, countably infinite, or uncountably infinite.

(a) $S = \{x$: $x$ is a U.S. citizen who has purchased a Japanese car during the past year$\}$.

(b) $S = \{(x,y): y \leq x^2, x \text{ is a positive integer}, y \in \mathbb{R}_{\geq 0}\}$.

(c) $S = \{p$: $p$ is the price of a quart of milk sold at a retail store in the U.S. on Friday, September 13, 1991$\}$.

(d) $S = \{x: x = 2y, y \text{ is a positive integer}\}$.

**4.** Let the universal set be $\Omega = [0,10]$, and define the following subsets of $\Omega$.

$A = [0,2), B = [2,7], C = [5,6], D = \{2\}$,

$E = \{x : x = y^{-1}, \text{ y is an even positive integer} \geq 4\}$.

(a) Define the following sets:

$A \cup B, A \cap B, \overline{A \cup C}, (A \cup D) \cap B, B - C, A \cap E, \bar{D} \cap B$

(b) For each of the sets in (a), indicate whether the set is finite, countably infinite, or uncountably infinite.

**5.** Let the universal set be defined by $\Omega = [-5,5]$, and define the following subsets of $\Omega$:

$A_1 = [-2, 1)$
$A_2 = (1, 2)$
$A_3 = [2, 5]$
$A_4 = [-5, -2]$

Also, define an index set $I = \{1,3,4\}$.

(a) Define $\cup_{i \in I} A_i$.

(b) Define $\cup_{i=1}^{4} A_i$

(c) Define $A_1 \cap A_2$

(d) Define $A_4 - A_1$

(e) Define $\bar{A}_4$

**6.** Define the universal set, $\Omega$, as $\Omega = \{x : 0 \leq x \leq 5$ or $10 \leq x \leq 20\}$, and define the following subsets of $\Omega$ as

$A_1 = \{x : 0 \leq x < 2.5\}$,
$A_2 = \{x : 15 < x \leq 20\}$,
$A_3 = \{x : 2.5 \leq x \leq 5 \text{ or } 10 \leq x \leq 20\}$,
$A_4 = \{x : 0 \leq x \leq 5 \text{ or } 10 \leq x \leq 15\}$.

In addition, define the following two index sets as:

$I_1 = \{1,3\}, I_2 = \{1,4\}$.

Define the following sets:

(a) $\cup_{i \in I_1} A_i$

(b) $\cap_{i=1}^{4} A_i$

(c) $\cap_{i \in I_2} A_i$

(d) $\cap_{i=1}^{2} A_i$

(e) $A_1 - A_2$

(f) $A_4 - A_3$

(g) $\bar{A}_3$

(h) $A_2 - \cup_{i \in I_1} A_i$

**7.** In each situation below indicate where the relation is a function. If so, determine the domain and range of the function.

(a) $A = [0,10)$, $B = [0, \ln(11)]$, $S = \{(x,y): y = \ln(1 + x), (x,y) \in A \times B\}$

(b) Consider $S^{-1}$, the inverse of $S$ in (a).

(c) $A = \mathbb{R}^2_{\geq 0}$, $B = [0,\infty)$, $S = \{((x_1, x_2), y): y = 5\, x_1 x_2^2, ((x_1, x_2), y) \in A \times B\}$

(d) Consider $S^{-1}$, the inverse of $S$ in (c).

**8.** For each relation below, state whether the relation is a function, and state whether an inverse function exists. Explicitly define the inverse function if it exists.

(a) Let $P = \{\$.01, \$.02,\ldots,\$1.00, \$1.01,\ldots\}$ represent a set of possible prices for a given commodity, and let $Q = [0,\infty)$ represent possible levels of quantity demanded. Define $S: P \rightarrow Q$ as $S = \{(p,q): q = 20p^{-1.5}, (p,q) \in P \times Q\}$

(b) Let $A = \{D: D = \{(x_1, x_2): x_1 \in [a_1, b_1], x_2 \in [a_2, b_2]\}, a_1 < b_1, a_2 < b_2\}$ be a set of rectangular sets. Define $S: A \rightarrow \mathbb{R}$ as

$S = \{(D, y) : y = \text{area of } D,\ (D, y) \in A \times \mathbb{R}_{+}\}$

**9.** Let $A = (0,\infty)$, and examine the following relation on $A$:

$$S = \{(x,y) : y = 2 + 3x,\ (x,y) \in A_2)\}.$$

(a) Is 2S8?

(b) Is $S$ a function?

(c) Does an inverse function exist?

(d) If you can, define $f(2)$ and $f^{-1}(5)$.

(e) What is D($S$)? What is R($S$)?

**10.** Define a universal set as $\Omega = \{x{:}0 \le x \le 5\}$, and consider the set function

$$P(A) = .5 \int_{x \in A} x dx + 12.5$$

where the domain of the set function is all subsets $A \subset \Omega$ of the form $A = [a, b]$, for $0 \le a \le b \le 5$.

(a) What is the image of the set $A = [0, 2]$ under $P$?

(b) What is the image of $\Omega$ under $P$?

(c) What is the image of set $A = [3,3]$ under $P$?

(d) What is the range of the set function?

(e) What is the inverse image of 2?

**11.** Define a set function that will assign the appropriate area to all rectangles of the form $[x_1, x_2] \times [y_1, y_2]$, $x_2 \ge x_1$ and $y_2 \ge y_1$, contained in $\mathbb{R}^2$. Be sure to identify the domain and range of the set function.

**12.** A statistics class has 20 students in attendance, and in the room where the class meets, there are 25 desks available for the students. How many different ways can the students leave five desks unoccupied?

**13.** There are 15 students in an econometrics class that you are attending.

(a) How many different ways can a three-person committee be formed to give a class report?

(b) Of the number of possible three-person committees indicated in (a), how many involve you?

**14.** Competing for the title of Miss America are 50 contestants from each of the 50 states plus one contestant from the District of Columbia. How many different ways can the contestants be assigned the titles of Miss America, first runner up,..., fourth runner up?

**15.** Let $A_1 = \{x{:}\ x \text{ is a positive integer}\}$, $A_2 = \{1, 2, 3, 4, 5\}$, $B = \{(x_1, x_2){:}\ (x_1, x_2) \in A_1 \times A_2, x_1 \le x_2\}$, and $y_i = i^2$. Calculate the values of the following sums.

(a) $\sum_{x \in A_2} x$

(b) $\sum_{i \in A_2} y_i$

(c) $\sum_{x_1 \in A_1} \sum_{x_2 \in A_2} (1/2)^{x_1} x_2^2$

(d) $\sum_{x \in (A_1 - A_2)} (1/3)^x$

(e) $\sum_{(x_1,x_2) \in B} (x_1 + x_2)$

**16.** Let $A_1 = [0, \infty)$, $A_2 = [1, 10]$, and $B = \{(x_1, x_2){:}\ (x_1, x_2) \in A_1 \times A_2, x_2 > x_1\}$. Calculate the values of the following integrals.

(a) $\int_{x \in A_1} (1/2) e^{-x/2} dx$

(b) $\int_{x_1 \in A_1} \int_{x_2 \in A_2} x_2 e^{-x_1} dx_2 dx_1$

(c) $\int_{(x_1,x_2) \in B} (x_1 + x_2) dx_1 dx_2$

(d) $\int_0^2 \int_{x_2 \in A_1 \cap A_2} x_1 x_2^2 dx_2 dx_1$

# Useful Tables

**Table B.1** Cumulative normal distribution $F(x) = \int_{-\infty}^{x} \frac{1}{\sqrt{2\pi}} e^{-t^2/2} dt$

| $x$ | **0.00** | **0.01** | **0.02** | **0.03** | **0.04** | **0.05** | **0.06** | **0.07** | **0.08** | **0.09** |
|---|---|---|---|---|---|---|---|---|---|---|
| 0.0 | 0.5000 | 0.5040 | 0.5080 | 0.5120 | 0.5160 | 0.5199 | 0.5239 | 0.5279 | 0.5319 | 0.5359 |
| 0.1 | 0.5398 | 0.5438 | 0.5478 | 0.5517 | 0.5557 | 0.5596 | 0.5636 | 0.5675 | 0.5714 | 0.5753 |
| 0.2 | 0.5793 | 0.5832 | 0.5871 | 0.5910 | 0.5948 | 0.5987 | 0.6026 | 0.6064 | 0.6103 | 0.6141 |
| 0.3 | 0.6179 | 0.6217 | 0.6255 | 0.6293 | 0.6331 | 0.6368 | 0.6406 | 0.6443 | 0.6480 | 0.6517 |
| 0.4 | 0.6554 | 0.6591 | 0.6628 | 0.6664 | 0.6700 | 0.6736 | 0.6772 | 0.6808 | 0.6844 | 0.6879 |
| 0.5 | 0.6915 | 0.6950 | 0.6985 | 0.7019 | 0.7054 | 0.7088 | 0.7123 | 0.7157 | 0.7190 | 0.7224 |
| 0.6 | 0.7257 | 0.7291 | 0.7324 | 0.7357 | 0.7389 | 0.7422 | 0.7454 | 0.7486 | 0.7517 | 0.7549 |
| 0.7 | 0.7580 | 0.7611 | 0.7642 | 0.7673 | 0.7704 | 0.7734 | 0.7764 | 0.7794 | 0.7823 | 0.7852 |
| 0.8 | 0.7881 | 0.7910 | 0.7939 | 0.7967 | 0.7995 | 0.8023 | 0.8051 | 0.8078 | 0.8106 | 0.8133 |
| 0.9 | 0.8159 | 0.8186 | 0.8212 | 0.8238 | 0.8264 | 0.8289 | 0.8315 | 0.8340 | 0.8365 | 0.8389 |
| 1.0 | 0.8413 | 0.8438 | 0.8461 | 0.8485 | 0.8508 | 0.8531 | 0.8554 | 0.8577 | 0.8599 | 0.8621 |
| 1.1 | 0.8643 | 0.8665 | 0.8686 | 0.8708 | 0.8729 | 0.8749 | 0.8770 | 0.8790 | 0.8810 | 0.8830 |
| 1.2 | 0.8849 | 0.8869 | 0.8888 | 0.8907 | 0.8925 | 0.8944 | 0.8962 | 0.8980 | 0.8997 | 0.9015 |
| 1.3 | 0.9032 | 0.9049 | 0.9066 | 0.9082 | 0.9099 | 0.9115 | 0.9131 | 0.9147 | 0.9162 | 0.9177 |
| 1.4 | 0.9192 | 0.9207 | 0.9222 | 0.9236 | 0.9251 | 0.9265 | 0.9279 | 0.9292 | 0.9306 | 0.9319 |
| 1.5 | 0.9332 | 0.9345 | 0.9357 | 0.9370 | 0.9382 | 0.9394 | 0.9406 | 0.9418 | 0.9429 | 0.9441 |
| 1.6 | 0.9452 | 0.9463 | 0.9474 | 0.9484 | 0.9495 | 0.9505 | 0.9515 | 0.9525 | 0.9535 | 0.9545 |
| 1.7 | 0.9554 | 0.9564 | 0.9573 | 0.9582 | 0.9591 | 0.9599 | 0.9608 | 0.9616 | 0.9625 | 0.9633 |
| 1.8 | 0.9641 | 0.9649 | 0.9656 | 0.9664 | 0.9671 | 0.9678 | 0.9686 | 0.9693 | 0.9699 | 0.9706 |
| 1.9 | 0.9713 | 0.9719 | 0.9726 | 0.9732 | 0.9738 | 0.9744 | 0.9750 | 0.9756 | 0.9761 | 0.9767 |
| 2.0 | 0.9772 | 0.9778 | 0.9783 | 0.9788 | 0.9793 | 0.9798 | 0.9803 | 0.9808 | 0.9812 | 0.9817 |
| 2.1 | 0.9821 | 0.9826 | 0.9830 | 0.9830 | 0.9834 | 0.9838 | 0.9846 | 0.9850 | 0.9854 | 0.9857 |
| 2.2 | 0.9861 | 0.9864 | 0.9868 | 0.9871 | 0.9875 | 0.9878 | 0.9881 | 0.9884 | 0.9887 | 0.9890 |
| 2.3 | 0.9893 | 0.9896 | 0.9898 | 0.9901 | 0.9904 | 0.9906 | 0.9909 | 0.9911 | 0.9913 | 0.9916 |
| 2.4 | 0.9918 | 0.9920 | 0.9922 | 0.9925 | 0.9927 | 0.9929 | 0.9931 | 0.9932 | 0.9934 | 0.9936 |
| 2.5 | 0.9938 | 0.9940 | 0.9941 | 0.9943 | 0.9945 | 0.9946 | 0.9948 | 0.9949 | 0.9951 | 0.9952 |
| 2.6 | 0.9953 | 0.9955 | 0.9956 | 0.9957 | 0.9959 | 0.9960 | 0.9961 | 0.9962 | 0.9963 | 0.9964 |
| 2.7 | 0.9965 | 0.9966 | 0.9967 | 0.9968 | 0.9969 | 0.9970 | 0.9971 | 0.9972 | 0.9973 | 0.9974 |
| 2.8 | 0.9974 | 0.9975 | 0.9976 | 0.9977 | 0.9978 | 0.9979 | 0.9979 | 0.9979 | 0.9980 | 0.9981 |
| 2.9 | 0.9981 | 0.9982 | 0.9982 | 0.9983 | 0.9984 | 0.9984 | 0.9985 | 0.9985 | 0.9986 | 0.9986 |
| 3.0 | 0.9987 | 0.9987 | 0.9987 | 0.9988 | 0.9988 | 0.9989 | 0.9989 | 0.9989 | 0.9990 | 0.9990 |
| 3.1 | 0.9990 | 0.9991 | 0.9991 | 0.9991 | 0.9992 | 0.9992 | 0.9992 | 0.9992 | 0.9993 | 0.9993 |
| 3.2 | 0.9993 | 0.9993 | 0.9994 | 0.9994 | 0.9994 | 0.9994 | 0.9994 | 0.9995 | 0.9995 | 0.9995 |
| 3.3 | 0.9995 | 0.9995 | 0.9995 | 0.9996 | 0.9996 | 0.9996 | 0.9996 | 0.9996 | 0.9996 | 0.9997 |
| 3.4 | 0.9997 | 0.9997 | 0.9997 | 0.9997 | 0.9997 | 0.9997 | 0.9997 | 0.9997 | 0.9997 | 0.9998 |

*Source: Reprinted, by permission of the publisher, from A. M. Mood, F. A. Graybill, and D. C. Boes,* Introduction to the Theory of Statistics, *3d ed., New York: McGraw-Hill, 1974, p. 552*

**Table B.2** Student's *t* distribution. The first column lists the number of degrees of freedom ($v$). The headings of the other columns give probabilities ($P$) for $t$ to exceed the entry value. Use symmetry for negative $t$ values

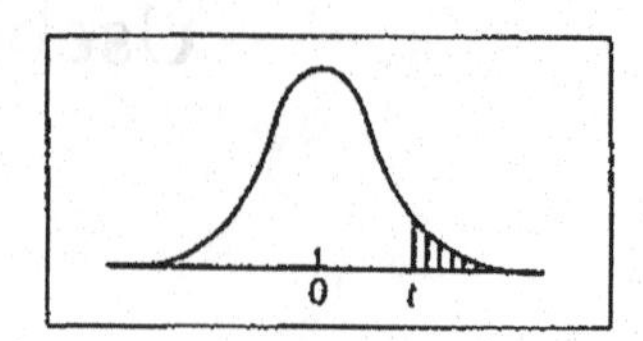


| $P$ / $v$ | 0.10 | 0.05 | 0.025 | 0.01 | 0.005 |
|---|---|---|---|---|---|
| 1 | 3.078 | 6.314 | 12.706 | 31.821 | 63.657 |
| 2 | 1.886 | 2.920 | 4.303 | 6.965 | 9.925 |
| 3 | 1.638 | 2.353 | 3.182 | 4.541 | 5.841 |
| 4 | 1.533 | 2.132 | 2.776 | 3.747 | 4.604 |
| 5 | 1.476 | 2.015 | 2.571 | 3.365 | 4.032 |
| 6 | 1.440 | 1.943 | 2.447 | 3.143 | 3.707 |
| 7 | 1.415 | 1.895 | 2.365 | 2.998 | 3.499 |
| 8 | 1.397 | 1.860 | 2.306 | 2.896 | 3.355 |
| 9 | 1.383 | 1.833 | 2.262 | 2.821 | 3.250 |
| 10 | 1.372 | 1.812 | 2.228 | 2.764 | 3.169 |
| 11 | 1.363 | 1.796 | 2.201 | 2.718 | 3.106 |
| 12 | 1.356 | 1.782 | 2.179 | 2.681 | 3.055 |
| 13 | 1.350 | 1.771 | 2.160 | 2.650 | 3.012 |
| 14 | 1.345 | 1.761 | 2.145 | 2.624 | 2.977 |
| 15 | 1.341 | 1.753 | 2.131 | 2.602 | 2.947 |
| 16 | 1.337 | 1.746 | 2.120 | 2.583 | 2.921 |
| 17 | 1.333 | 1.740 | 2.110 | 2.567 | 2.898 |
| 18 | 1.330 | 1.734 | 2.101 | 2.552 | 2.878 |
| 19 | 1.328 | 1.729 | 2.093 | 2.539 | 2.861 |
| 20 | 1.325 | 1.725 | 2.086 | 2.528 | 2.845 |
| 21 | 1.323 | 1.721 | 2.080 | 2.518 | 2.831 |
| 22 | 1.321 | 1.717 | 2.074 | 2.508 | 2.819 |
| 23 | 1.319 | 1.714 | 2.069 | 2.500 | 2.807 |
| 24 | 1.318 | 1.711 | 2.064 | 2.492 | 2.797 |
| 25 | 1.316 | 1.708 | 2.060 | 2.485 | 2.787 |
| 26 | 1.315 | 1.706 | 2.056 | 2.479 | 2.779 |
| 27 | 1.314 | 1.703 | 2.052 | 2.473 | 2.771 |
| 28 | 1.313 | 1.701 | 2.048 | 2.467 | 2.763 |
| 29 | 1.311 | 1.699 | 2.045 | 2.462 | 2.756 |
| 30 | 1.310 | 1.697 | 2.042 | 2.457 | 2.750 |
| 40 | 1.303 | 1.684 | 2.021 | 2.423 | 2.704 |
| 60 | 1.296 | 1.671 | 2.000 | 2.390 | 2.660 |
| 120 | 1.289 | 1.658 | 1.980 | 2.358 | 2.617 |
| $\infty$ | 1.282 | 1.645 | 1.960 | 2.326 | 2.576 |

*Source: Reprinted, by permission of the publisher, from P. G. Hoel,* Introduction to Mathematical Statistics, *4th ed., New York: John Wiley and Sons, Inc., 1971, p. 393*

**Table B.3** Chi-square distribution. The first column lists the number of degrees of freedom ($\nu$). The headings of the other columns give probabilities ($P$) for the $\chi_\nu^2$ random variable to exceed the entry value

| $P$ / $\nu$ | 0.995 | 0.990 | 0.975 | 0.950 | 0.900 | 0.750 |
|---|---|---|---|---|---|---|
| 1 | $392704 \times 10^{-10}$ | $157088 \times 10^{-9}$ | $982069 \times 10^{-9}$ | $393214 \times 10^{-8}$ | 0.0157908 | 0.1015308 |
| 2 | 0.0100251 | 0.0201007 | 0.0506356 | 0.102587 | 0.210720 | 0.575364 |
| 3 | 0.0717212 | 0.114832 | 0.215795 | 0.351846 | 0.584375 | 1.212534 |
| 4 | 0.206990 | 0.297110 | 0.484419 | 0.710721 | 1.063623 | 1.92255 |
| 5 | 0.411740 | 0.554300 | 0.831211 | 1.145476 | 1.61031 | 2.67460 |
| 6 | 0.675727 | 0.872085 | 1.237347 | 1.63539 | 2.20413 | 3.45460 |
| 7 | 0.989265 | 1.239043 | 1.68987 | 2.16735 | 2.83311 | 4.25485 |
| 8 | 1.344419 | 1.646482 | 2.17973 | 2.73264 | 3.48954 | 5.07064 |
| 9 | 1.734926 | 2.087912 | 2.70039 | 3.32511 | 1.16816 | 5.89883 |
| 10 | 2.15585 | 2.55821 | 3.24697 | 3.94030 | 4.86518 | 6.73720 |
| 11 | 2.60321 | 3.15347 | 3.81575 | 4.57481 | 5.57779 | 7.58412 |
| 12 | 3.07382 | 3.57056 | 4.40379 | 5.22603 | 6.30380 | 8.43842 |
| 13 | 3.56503 | 4.10691 | 5.00874 | 5.89186 | 7.04150 | 9.29906 |
| 14 | 4.07468 | 4.66043 | 5.62872 | 6.57063 | 7.78953 | 10.1653 |
| 15 | 4.60094 | 5.22935 | 6.26214 | 7.26094 | 8.54675 | 11.0365 |
| 16 | 5.14224 | 5.81221 | 6.90766 | 7.96164 | 9.31223 | 11.9122 |
| 17 | 5.69724 | 6.40776 | 7.56418 | 8.67176 | 10.0852 | 12.7919 |
| 18 | 6.26481 | 7.01491 | 8.23075 | 9.39046 | 10.8649 | 13.6753 |
| 19 | 6.84398 | 7.63273 | 8.90655 | 10.1170 | 11.6509 | 14.5620 |
| 20 | 7.43386 | 8.26040 | 9.59083 | 10.8508 | 12.4426 | 15.4518 |
| 21 | 8.03366 | 8.89720 | 10.28293 | 11.5613 | 13.2396 | 16.3444 |
| 22 | 8.64272 | 9.54279 | 10.3923 | 12.3380 | 14.0415 | 17.2396 |
| 23 | 9.26042 | 10.19567 | 11.6885 | 13.0905 | 14.8479 | 18.1373 |
| 24 | 9.88623 | 10.8564 | 12.4011 | 13.8484 | 15.6587 | 19.0372 |
| 25 | 10.5197 | 11.5240 | 13.1197 | 14.6114 | 16.4734 | 19.9393 |
| 26 | 11.1603 | 12.1981 | 13.8439 | 15.3791 | 17.2919 | 20.8434 |
| 27 | 11.8076 | 12.8786 | 14.5733 | 16.1513 | 18.1138 | 21.7494 |
| 28 | 12.4613 | 13.5648 | 15.3079 | 16.9279 | 18.9392 | 22.6572 |
| 29 | 13.1211 | 14.2565 | 16.0471 | 17.7083 | 19.7677 | 23.5666 |
| 30 | 13.7867 | 14.9535 | 16.7908 | 18.4926 | 20.5992 | 24.4776 |
| 40 | 20.7065 | 22.1643 | 24.4331 | 26.5093 | 29.0505 | 33.6603 |
| 50 | 27.9907 | 29.7067 | 32.3574 | 34.7642 | 37.6886 | 42.9421 |
| 60 | 35.5346 | 37.4848 | 40.4817 | 43.1879 | 46.4589 | 52.2938 |
| 70 | 43.2752 | 45.4418 | 48.7576 | 51.7393 | 55.3290 | 61.6983 |
| 80 | 51.1720 | 53.5400 | 57.1532 | 60.3915 | 64.2778 | 71.1445 |
| 90 | 59.1963 | 61.7541 | 65.6466 | 69.1260 | 73.2912 | 80.6247 |
| 100 | 67.3276 | 70.0648 | 74.2219 | 77.9295 | 82.3581 | 90.1332 |

**Table B.3** *(continued)*

| P / v | 0.500 | 0.250 | 0.100 | 0.050 | 0.025 | 0.010 | 0.005 |
|---|---|---|---|---|---|---|---|
| 1 | 0.454937 | 1.32330 | 2.70554 | 3.84146 | 5.02389 | 6.63490 | 7.87944 |
| 2 | 1.38629 | 2.77259 | 4.60517 | 5.99147 | 7.37776 | 9.21034 | 10.5966 |
| 3 | 2.36597 | 1.10835 | 6.25139 | 7.81473 | 9.34840 | 11.3449 | 12.8381 |
| 4 | 3.35670 | 5.38527 | 7.77944 | 9.48773 | 11.1433 | 13.2767 | 14.8602 |
| 5 | 4.35146 | 6.62568 | 9.23635 | 11.0705 | 12.8325 | 15.0863 | 16.7496 |
| 6 | 5.34812 | 7.84080 | 10.6446 | 12.5916 | 14.4494 | 16.8119 | 18.5476 |
| 7 | 6.34581 | 9.03715 | 12.0170 | 14.0671 | 16.0128 | 18.4753 | 20.2777 |
| 8 | 7.34412 | 10.2188 | 13.3616 | 15.5073 | 17.5346 | 20.0902 | 21.9550 |
| 9 | 8.34283 | 11.3887 | 14.6837 | 16.9190 | 19.0228 | 21.6660 | 23.5893 |
| 10 | 9.34182 | 12.5489 | 15.9871 | 18.3070 | 20.4831 | 23.2093 | 25.1882 |
| 11 | 10.3410 | 13.7007 | 17.2750 | 19.6751 | 21.9200 | 24.7250 | 26.7569 |
| 12 | 11.3403 | 14.8454 | 18.5494 | 21.0261 | 23.3367 | 26.2170 | 28.2995 |
| 13 | 12.3398 | 15.9839 | 19.8119 | 22.3621 | 24.7356 | 27.6883 | 29.8194 |
| 14 | 13.3393 | 17.1170 | 21.0642 | 23.6848 | 26.1190 | 29.1413 | 31.3193 |
| 15 | 14.3389 | 18.2451 | 22.3072 | 24.9958 | 27.4884 | 30.5779 | 32.8013 |
| 16 | 15.3385 | 19.3688 | 23.5418 | 26.2962 | 28.8454 | 31.9999 | 34.2672 |
| 17 | 16.3381 | 20.4887 | 24.4690 | 27.5871 | 30.1910 | 33.4087 | 35.7185 |
| 18 | 17.3379 | 21.6049 | 25.9894 | 28.8693 | 31.5264 | 34.8053 | 37.1564 |
| 19 | 18.3376 | 22.7178 | 27.2036 | 30.1435 | 32.8523 | 36.1908 | 38.5822 |
| 20 | 19.3374 | 23.8277 | 28.4120 | 31.4104 | 34.1696 | 37.5662 | 39.9968 |
| 21 | 20.3372 | 24.9348 | 29.6151 | 32.6705 | 35.4789 | 38.9321 | 41.4010 |
| 22 | 21.3370 | 26.0393 | 30.8133 | 33.9244 | 36.7807 | 40.2894 | 42.7956 |
| 23 | 22.3369 | 27.1413 | 32.0069 | 35.1725 | 38.0757 | 41.6384 | 44.1813 |
| 24 | 23.3367 | 28.2412 | 33.1963 | 36.4151 | 39.3641 | 42.9798 | 45.5585 |
| 25 | 24.3366 | 29.3389 | 34.3816 | 37.6525 | 40.6465 | 44.3141 | 46.9278 |
| 26 | 25.3364 | 30.4345 | 35.5631 | 38.8852 | 41.9232 | 45.6417 | 48.2899 |
| 27 | 26.3363 | 31.5284 | 36.7412 | 40.1133 | 43.1944 | 46.9630 | 49.6449 |
| 28 | 27.3363 | 32.6205 | 37.9159 | 41.3372 | 44.4607 | 48.2782 | 50.9933 |
| 29 | 28.3362 | 33.7109 | 39.0875 | 42.5569 | 45.7222 | 49.5879 | 52.3356 |
| 30 | 29.3360 | 34.7998 | 40.2560 | 43.7729 | 46.9792 | 50.8922 | 53.6720 |
| 40 | 39.3354 | 45.6160 | 51.8050 | 55.7585 | 59.3417 | 63.6907 | 66.7659 |
| 50 | 49.3349 | 56.3336 | 63.1671 | 67.5048 | 71.4202 | 76.1539 | 79.4900 |
| 60 | 59.3347 | 66.9814 | 74.3970 | 79.0819 | 83.2976 | 88.3794 | 91.9517 |
| 70 | 69.3344 | 77.5766 | 85.5271 | 90.5312 | 95.0231 | 100.425 | 104.215 |
| 80 | 79.3343 | 88.1303 | 96.5782 | 101.879 | 106.629 | 112.329 | 116.321 |
| 90 | 89.3342 | 98.6499 | 107.565 | 113.145 | 118.136 | 124.116 | 128.299 |
| 100 | 99.3341 | 109.141 | 118.498 | 124.342 | 129.561 | 135.807 | 140.169 |

*Source: Reprinted, by permission of the Biometrika Trustees from C. M. Thompson, "Tables of Percentage Points of the $\chi^2$ Distribution,"* Biometrika *32 (1941): 188–189*

**Table B.4** *F*-distribution: 5 % points. The first column lists the number of denominator degrees of freedom ($v_2$). The headings of the other columns list the numerator degrees of freedom ($v_1$). The table entry is the value of $c$ for which $P(F_{v_1,v_2} \geq c) = .05$

| $v_1$ / $v_2$ | 1 | 2 | 3 | 4 | 5 | 6 | 7 | 8 | 9 |
|---|---|---|---|---|---|---|---|---|---|
| 1 | 161.45 | 199.50 | 215.71 | 224.58 | 230.16 | 233.99 | 236.77 | 238.88 | 240.54 |
| 2 | 18.513 | 19.000 | 19.164 | 19.247 | 19.296 | 19.330 | 19.353 | 19.371 | 19.385 |
| 3 | 10.128 | 9.5521 | 9.2766 | 9.1172 | 9.0135 | 8.9406 | 8.8868 | 8.8452 | 8.8123 |
| 4 | 7.7086 | 6.9443 | 6.5914 | 6.3883 | 6.3560 | 6.1631 | 6.0942 | 6.0410 | 5.9988 |
| 5 | 6.6079 | 5.7861 | 5.4095 | 5.1922 | 5.0503 | 4.9503 | 4.8759 | 4.8183 | 4.7725 |
| 6 | 5.9874 | 5.1433 | 4.7571 | 4.5337 | 1.3874 | 4.2839 | 4.2066 | 4.1468 | 4.0990 |
| 7 | 5.5914 | 4.7374 | 4.3468 | 4.1203 | 3.9715 | 3.8660 | 3.7870 | 3.7257 | 3.6767 |
| 8 | 5.3177 | 4.4590 | 4.0662 | 3.8378 | 3.6875 | 3.5806 | 3.5005 | 3.4381 | 3.3881 |
| 9 | 5.1174 | 4.2565 | 3.8626 | 3.6331 | 3.4817 | 3.3738 | 3.2927 | 3.2296 | 3.1789 |
| 10 | 4.9646 | 4.1028 | 3.7083 | 3.4780 | 3.3258 | 3.2172 | 3.1355 | 3.0717 | 3.0204 |
| 11 | 4.8443 | 3.9823 | 3.5874 | 3.3567 | 3.2039 | 3.0946 | 3.0123 | 2.9480 | 2.8962 |
| 12 | 4.7472 | 3.8856 | 3.4903 | 3.2592 | 3.1059 | 2.9961 | 3.9134 | 2.8486 | 2.7964 |
| 13 | 4.6672 | 3.8056 | 3.4105 | 3.1791 | 3.0254 | 2.9153 | 2.8321 | 2.7669 | 2.7144 |
| 14 | 4.6001 | 3.7389 | 3.3439 | 3.1122 | 2.9582 | 2.8477 | 2.7642 | 2.6987 | 2.6458 |
| 15 | 4.5431 | 3.6823 | 3.2874 | 3.0556 | 2.9013 | 2.7905 | 2.7066 | 2.6408 | 2.5876 |
| 16 | 4.4940 | 3.6337 | 3.2389 | 3.0069 | 2.8524 | 2.7413 | 2.6572 | 2.5911 | 2.5377 |
| 17 | 4.4513 | 3.5915 | 3.1968 | 2.9647 | 2.8100 | 2.6987 | 2.6143 | 2.5480 | 2.4943 |
| 18 | 4.4139 | 3.5546 | 3.1599 | 2.9277 | 2.7729 | 2.6613 | 2.5767 | 2.5102 | 2.4563 |
| 19 | 4.3808 | 3.5219 | 3.1274 | 2.8951 | 2.7401 | 2.6283 | 2.5435 | 2.4768 | 2.4227 |
| 20 | 4.3513 | 3.4928 | 3.0984 | 2.8661 | 2.7109 | 2.5990 | 2.5140 | 2.4471 | 2.3928 |
| 21 | 4.3248 | 3.4668 | 3.0725 | 2.8401 | 2.6848 | 2.5727 | 2.4876 | 2.4205 | 2.3661 |
| 22 | 4.3009 | 3.4434 | 3.0491 | 2.8167 | 2.6613 | 2.5491 | 2.4638 | 2.3965 | 2.3419 |
| 23 | 4.2793 | 3.4221 | 3.0280 | 2.7955 | 2.6400 | 2.5277 | 2.4422 | 2.3748 | 2.3201 |
| 24 | 4.2597 | 3.4028 | 3.0088 | 3.7763 | 2.6207 | 2.5082 | 2.4226 | 2.3551 | 2.3002 |
| 25 | 4.2417 | 3.3852 | 2.9912 | 2.7587 | 2.6030 | 2.4904 | 2.4047 | 2.3371 | 2.2821 |
| 26 | 4.2252 | 3.3690 | 2.9751 | 2.7426 | 2.5868 | 2.4741 | 2.3883 | 2.3205 | 2.2655 |
| 27 | 4.2100 | 3.3541 | 2.9604 | 2.7278 | 2.5719 | 2.4591 | 2.3732 | 2.3053 | 2.2501 |
| 28 | 4.1960 | 3.3404 | 2.9467 | 2.7141 | 2.5581 | 2.4453 | 2.3593 | 2.2913 | 2.2360 |
| 29 | 4.1830 | 3.3277 | 2.9340 | 2.7014 | 2.5454 | 2.4324 | 2.3463 | 2.2782 | 2.2229 |
| 30 | 4.1709 | 3.3158 | 2.9223 | 2.6896 | 2.5336 | 2.4205 | 2.3343 | 2.2662 | 2.2107 |
| 40 | 4.0848 | 3.2317 | 2.8387 | 2.6060 | 2.4495 | 2.3359 | 2.2490 | 2.1802 | 2.1240 |
| 60 | 4.0012 | 3.1504 | 2.7581 | 2.5252 | 2.3683 | 2.2540 | 2.1665 | 2.0970 | 2.0401 |
| 120 | 3.9201 | 3.0718 | 2.6802 | 2.4472 | 2.2900 | 2.1750 | 2.0867 | 2.0164 | 1.9588 |
| $\infty$ | 3.8415 | 2.9957 | 2.6049 | 2.3719 | 2.2141 | 2.0986 | 2.0096 | 1.9354 | 1.8799 |

**Table B.4** *(continued)*

| $\nu_1$ / $\nu_2$ | 10 | 12 | 15 | 20 | 24 | 30 | 40 | 60 | 120 | ∞ |
|---|---|---|---|---|---|---|---|---|---|---|
| 1 | 241.88 | 243.91 | 245.95 | 248.01 | 249.05 | 250.09 | 251.14 | 252.20 | 253.25 | 254.32 |
| 2 | 19.396 | 19.413 | 19.429 | 19.446 | 19.454 | 19.462 | 19.471 | 19.479 | 19.487 | 19.496 |
| 3 | 8.7855 | 8.7446 | 8.7029 | 8.6602 | 8.6385 | 8.6166 | 8.5944 | 8.5720 | 8.5494 | 8.5265 |
| 4 | 5.9644 | 5.9117 | 5.8578 | 5.8025 | 5.7744 | 5.7459 | 5.7170 | 5.6878 | 5.6581 | 5.6281 |
| 5 | 4.7351 | 4.6777 | 4.6188 | 4.5581 | 4.5272 | 4.4957 | 4.4638 | 4.4314 | 4.3984 | 4.3650 |
| 6 | 4.0600 | 3.9999 | 3.9381 | 3.8742 | 3.8415 | 3.8082 | 3.7743 | 3.7398 | 3.7047 | 3.6688 |
| 7 | 3.6365 | 3.5747 | 3.5108 | 3.4445 | 3.4105 | 3.3758 | 3.3404 | 3.3043 | 3.2674 | 3.2298 |
| 8 | 3.3472 | 3.2840 | 3.2184 | 3.1503 | 3.1152 | 3.0794 | 3.0428 | 3.0053 | 2.9669 | 2.9276 |
| 9 | 3.1373 | 3.0729 | 3.0061 | 2.9365 | 2.9005 | 2.8637 | 2.8259 | 2.7872 | 2.7475 | 2.7067 |
| 10 | 2.9782 | 2.9130 | 2.8450 | 2.7740 | 2.7372 | 2.6996 | 2.6609 | 2.6211 | 2.5801 | 2.5379 |
| 11 | 2.8536 | 2.7876 | 2.7186 | 2.6464 | 2.6090 | 2.5705 | 2.5309 | 2.4901 | 2.4480 | 2.4045 |
| 12 | 2.7534 | 2.6866 | 2.6169 | 2.5436 | 2.5055 | 2.4663 | 2.4259 | 2.3842 | 2.3410 | 2.2962 |
| 13 | 2.6710 | 2.6037 | 2.5331 | 2.4589 | 2.4202 | 2.3803 | 2.3392 | 2.2966 | 2.2524 | 2.2064 |
| 14 | 2.6021 | 2.5342 | 2.4630 | 2.3879 | 2.3487 | 2.3082 | 2.2664 | 2.2230 | 2.1778 | 2.1307 |
| 15 | 2.5437 | 2.4753 | 2.4035 | 2.3275 | 2.2878 | 2.2468 | 2.2043 | 2.1601 | 2.1141 | 2.0658 |
| 16 | 2.4935 | 2.4247 | 2.3522 | 2.2756 | 2.2354 | 2.1938 | 2.1507 | 2.1058 | 2.0589 | 2.0096 |
| 17 | 2.4499 | 2.3807 | 2.3077 | 2.2304 | 2.1898 | 2.1477 | 2.1040 | 2.0584 | 2.0107 | 1.9604 |
| 18 | 2.4117 | 2.3421 | 2.2686 | 2.1906 | 2.1497 | 2.1071 | 2.0629 | 2.0166 | 1.9681 | 1.9168 |
| 19 | 2.3779 | 2.3080 | 2.2341 | 2.1555 | 2.1141 | 2.0712 | 2.0264 | 1.9796 | 1.9302 | 1.8780 |
| 20 | 2.3479 | 2.2776 | 2.2033 | 2.1242 | 2.0825 | 2.0391 | 1.9938 | 1.9464 | 1.8963 | 1.8432 |
| 21 | 2.3210 | 2.2504 | 2.1757 | 2.0960 | 2.0540 | 2.0102 | 1.9645 | 1.9165 | 1.8657 | 1.8117 |
| 22 | 2.2967 | 2.2258 | 2.1508 | 2.0707 | 2.0283 | 1.9842 | 1.9380 | 1.8895 | 1.8380 | 1.7831 |
| 23 | 2.2747 | 2.2036 | 2.1282 | 1.0476 | 2.0050 | 1.9605 | 1.9139 | 1.8649 | 1.8128 | 1.7570 |
| 24 | 2.2547 | 2.1834 | 2.1077 | 2.0267 | 1.9838 | 1.9390 | 1.8920 | 1.8424 | 1.7897 | 1.7331 |
| 25 | 2.2365 | 2.1649 | 2.0889 | 2.0075 | 1.9643 | 1.9192 | 1.8718 | 1.8217 | 1.7684 | 1.7110 |
| 26 | 2.2197 | 2.1479 | 2.0716 | 1.9898 | 1.9464 | 1.9010 | 1.8533 | 1.8027 | 1.7488 | 1.6906 |
| 27 | 2.2043 | 2.1323 | 2.0558 | 1.9736 | 1.9299 | 1.8842 | 1.8361 | 1.7851 | 1.7307 | 1.6717 |
| 28 | 2.1900 | 2.1179 | 2.0411 | 1.9586 | 1.9147 | 1.8687 | 1.8203 | 1.7689 | 1.7138 | 1.6541 |
| 29 | 2.1768 | 2.1045 | 2.0245 | 1.9446 | 1.9005 | 1.8543 | 1.8055 | 1.7537 | 1.6981 | 1.6377 |
| 30 | 2.1646 | 2.0921 | 2.0148 | 1.9317 | 1.8874 | 1.8409 | 1.7918 | 1.7396 | 1.6835 | 1.6223 |
| 40 | 2.0772 | 2.0035 | 1.9245 | 1.8389 | 1.7929 | 1.7444 | 1.6928 | 1.6373 | 1.5766 | 1.5089 |
| 60 | 1.9926 | 1.9174 | 1.8364 | 1.7480 | 1.7001 | 1.6491 | 1.5943 | 1.5343 | 1.4673 | 1.3893 |
| 120 | 1.9105 | 1.8337 | 1.7505 | 1.6587 | 1.6084 | 1.5543 | 1.4952 | 1.4290 | 1.3519 | 1.2539 |
| ∞ | 1.8307 | 1.7522 | 1.6664 | 1.5705 | 1.5173 | 1.4591 | 1.3940 | 1.3180 | 1.2214 | 1.0000 |

*Source: Reprinted, by permission of the Biometrika Trustees from M. Merrington and C. M. Thompson, "Tables of Percentage Points of the Inverted Beta(F) Distribution,"* Biometrika *33 (1943): 80–81*

**Table B.5** *F*-distribution: 1 % points. The first column lists the number of denominator degrees of freedom ($v_2$). The headings of the other columns list the numerator degrees of freedom ($v_1$). The table entry is the value of $c$ for which $P(F_{v_1,v_2} \geq c) = .01$

| $v_1$ / $v_2$ | 1 | 2 | 3 | 4 | 5 | 6 | 7 | 8 | 9 |
|---|---|---|---|---|---|---|---|---|---|
| 1 | 4,052.2 | 4,999.5 | 5,403.3 | 5,624.6 | 5,763.7 | 5,859.0 | 5,928.3 | 5,981.6 | 6,022.5 |
| 2 | 98.503 | 99.000 | 99.166 | 99.249 | 99.299 | 99.332 | 99.356 | 99.374 | 99.388 |
| 3 | 34.116 | 30.817 | 29.457 | 28.710 | 28.237 | 27.911 | 27.672 | 27.489 | 27.345 |
| 4 | 21.198 | 18.000 | 16.694 | 15.977 | 15.522 | 15.207 | 14.976 | 14.799 | 14.659 |
| 5 | 16.258 | 13.274 | 12.060 | 11.392 | 10.967 | 10.672 | 10.456 | 10.289 | 10.158 |
| 6 | 13.745 | 10.925 | 9.7795 | 9.1483 | 8.7459 | 8.4661 | 8.2600 | 8.1016 | 7.9761 |
| 7 | 12.246 | 9.5466 | 8.4513 | 7.8467 | 7.4604 | 7.1914 | 6.9928 | 6.8401 | 6.7188 |
| 8 | 11.259 | 8.6491 | 7.5910 | 7.0060 | 6.6318 | 6.3707 | 6.1776 | 6.0289 | 5.9106 |
| 9 | 10.561 | 8.0215 | 6.9919 | 6.4221 | 6.0569 | 5.8018 | 5.6129 | 5.4671 | 5.3511 |
| 10 | 10.044 | 7.5594 | 6.5523 | 5.9943 | 5.6363 | 5.3858 | 5.2001 | 5.0567 | 4.9424 |
| 11 | 9.6460 | 7.2057 | 6.2167 | 5.6683 | 5.3160 | 5.0692 | 4.8861 | 4.7445 | 4.6315 |
| 12 | 9.3302 | 6.9266 | 5.9526 | 5.4119 | 5.0643 | 4.8206 | 4.6395 | 4.4994 | 4.3875 |
| 13 | 9.0738 | 6.7010 | 5.7394 | 5.2053 | 4.8616 | 4.6204 | 4.4410 | 4.3021 | 4.1911 |
| 14 | 8.8616 | 6.5149 | 5.5639 | 5.0354 | 4.6950 | 4.4558 | 4.2779 | 4.1399 | 4.0297 |
| 15 | 8.6831 | 6.3589 | 5.4170 | 4.8932 | 4.5556 | 4.3183 | 4.1415 | 4.0045 | 3.8948 |
| 16 | 8.5310 | 6.2262 | 5.2922 | 4.7726 | 4.4374 | 4.2016 | 4.0259 | 3.8896 | 3.7804 |
| 17 | 8.3997 | 6.1121 | 5.1850 | 4.6690 | 4.3359 | 4.1015 | 3.9267 | 3.7910 | 3.6822 |
| 18 | 8.2854 | 6.0129 | 5.0919 | 4.5790 | 4.2479 | 4.0146 | 3.8406 | 3.7054 | 3.5971 |
| 19 | 8.1850 | 5.9259 | 5.0103 | 4.5003 | 4.1708 | 3.9386 | 3.7653 | 3.6305 | 3.5225 |
| 20 | 8.0906 | 5.8489 | 4.9382 | 4.4307 | 4.1027 | 3.8714 | 3.6987 | 3.5644 | 3.4567 |
| 21 | 8.0166 | 5.7804 | 4.8740 | 4.3688 | 4.0421 | 3.8117 | 3.6396 | 3.5056 | 3.3981 |
| 22 | 7.9454 | 5.7190 | 4.8166 | 4.3134 | 3.9880 | 3.7583 | 3.5867 | 3.4530 | 3.3458 |
| 23 | 7.8811 | 5.6637 | 4.7649 | 4.2635 | 3.9392 | 3.7102 | 3.5390 | 3.4057 | 3.2986 |
| 24 | 7.8229 | 5.6131 | 4.7181 | 4.2184 | 3.8951 | 3.6667 | 3.4959 | 3.3629 | 3.2560 |
| 25 | 7.7689 | 5.5680 | 4.6755 | 4.1774 | 3.8550 | 3.6272 | 3.4568 | 3.3239 | 3.2172 |
| 26 | 7.7213 | 5.5263 | 4.6366 | 4.1400 | 3.8183 | 3.5911 | 3.4210 | 3.2884 | 3.1818 |
| 27 | 7.6767 | 5.4881 | 4.6009 | 4.1056 | 3.7848 | 3.5580 | 3.3882 | 3.2558 | 3.1494 |
| 28 | 7.6356 | 5.4529 | 4.5681 | 4.0740 | 3.7539 | 3.5276 | 3.3581 | 3.2259 | 3.1195 |
| 29 | 7.5976 | 5.4205 | 4.5378 | 4.0449 | 3.7254 | 3.4995 | 3.3302 | 3.1982 | 3.0920 |
| 30 | 7.5625 | 5.3904 | 4.5097 | 4.0179 | 3.6990 | 3.4735 | 3.3045 | 3.1726 | 3.0665 |
| 40 | 7.3141 | 5.1785 | 4.3126 | 3.8283 | 3.5138 | 3.2910 | 3.1238 | 2.9930 | 2.8876 |
| 60 | 7.0771 | 4.9774 | 4.1259 | 3.6491 | 3.3389 | 3.1187 | 2.9530 | 2.8233 | 2.7185 |
| 120 | 6.8510 | 4.7865 | 3.9493 | 3.4796 | 3.1735 | 2.9559 | 2.7918 | 2.6629 | 2.5586 |
| $\infty$ | 6.6349 | 4.6052 | 3.7816 | 3.3192 | 3.0173 | 2.8020 | 2.6393 | 2.5113 | 2.4073 |

**Table B.5** *(continued)*

| $v_1$ / $v_2$ | 10 | 12 | 15 | 20 | 24 | 30 | 40 | 60 | 120 | ∞ |
|---|---|---|---|---|---|---|---|---|---|---|
| 1 | 6,055.8 | 6,106.3 | 6,157.3 | 6,208.7 | 6,234.6 | 6,260.7 | 6,286.8 | 6,313.0 | 6,339.4 | 6,366.0 |
| 2 | 99.399 | 99.416 | 99.449 | 99.458 | 99.458 | 99.466 | 99.474 | 99.483 | 99.491 | 99.501 |
| 3 | 27.229 | 27.052 | 26.872 | 26.690 | 26.598 | 26.505 | 26.411 | 26.316 | 26.221 | 26.125 |
| 4 | 14.546 | 14.374 | 14.198 | 14.020 | 13.929 | 13.838 | 13.745 | 13.652 | 13.558 | 13.463 |
| 5 | 10.051 | 9.8883 | 9.7222 | 9.5527 | 9.4665 | 9.3793 | 9.2912 | 9.2020 | 9.1118 | 9.0204 |
| 6 | 7.8741 | 7.7183 | 7.5590 | 7.3958 | 7.3127 | 7.2285 | 7.1432 | 7.0568 | 6.9690 | 6.8801 |
| 7 | 6.6201 | 6.4691 | 6.3143 | 6.1554 | 6.0743 | 5.9921 | 5.9084 | 5.8236 | 5.7372 | 5.6495 |
| 8 | 5.8143 | 5.6668 | 5.5151 | 5.3591 | 5.2793 | 5.1981 | 5.1156 | 5.0316 | 4.9460 | 4.8588 |
| 9 | 5.2565 | 5.1114 | 4.9621 | 4.8080 | 4.7290 | 4.6486 | 4.5667 | 4.4831 | 4.3978 | 4.3105 |
| 10 | 4.8492 | 4.7059 | 4.5582 | 4.4054 | 4.3269 | 4.2469 | 4.1653 | 4.0819 | 3.9965 | 3.9090 |
| 11 | 4.5393 | 4.3974 | 4.2509 | 4.0990 | 4.0209 | 3.9411 | 3.8596 | 3.7761 | 3.6904 | 3.6025 |
| 12 | 4.2961 | 4.1553 | 4.0096 | 3.8584 | 3.7805 | 3.7008 | 3.6192 | 3.5355 | 3.4494 | 3.3608 |
| 13 | 4.1003 | 3.9603 | 3.8154 | 3.6646 | 3.5868 | 3.5070 | 3.4253 | 3.3413 | 3.2548 | 3.1654 |
| 14 | 3.9394 | 3.8001 | 3.6557 | 3.5052 | 3.4274 | 3.3476 | 3.2656 | 3.1813 | 3.0942 | 3.0040 |
| 15 | 3.8049 | 3.6662 | 3.5255 | 3.3719 | 3.2940 | 3.2141 | 3.1319 | 3.0471 | 2.9595 | 2.8684 |
| 16 | 3.6909 | 3.5527 | 3.4089 | 3.2588 | 3.1808 | 3.1007 | 3.0182 | 2.9330 | 2.8447 | 2.7528 |
| 17 | 3.5931 | 3.4552 | 3.3117 | 3.1615 | 3.0835 | 3.0032 | 2.9205 | 2.8348 | 2.7459 | 2.6530 |
| 18 | 3.5082 | 3.3706 | 3.2273 | 3.0771 | 2.9990 | 2.9185 | 3.8354 | 2.7493 | 2.6597 | 2.5660 |
| 19 | 3.4338 | 3.2965 | 3.1533 | 3.0031 | 2.9249 | 2.8442 | 2.7608 | 2.6742 | 2.5839 | 2.4893 |
| 20 | 3.3682 | 3.2311 | 3.0880 | 2.9377 | 2.8594 | 2.7785 | 2.6947 | 2.6077 | 2.5168 | 2.4212 |
| 21 | 3.3098 | 3.1729 | 3.0299 | 2.8796 | 2.8011 | 2.7200 | 2.6359 | 2.5484 | 2.4568 | 2.3603 |
| 22 | 3.2576 | 3.1209 | 2.9780 | 2.8274 | 2.7488 | 2.6675 | 2.5831 | 2.4951 | 2.4029 | 2.3055 |
| 23 | 3.2106 | 3.0740 | 2.9311 | 2.7805 | 2.7017 | 2.6202 | 2.5355 | 2.4471 | 2.3542 | 2.2559 |
| 24 | 3.1681 | 3.0316 | 2.8887 | 2.7380 | 2.6591 | 2.5773 | 2.4923 | 2.4035 | 2.3099 | 2.2107 |
| 25 | 3.1294 | 2.9331 | 2.8502 | 2.6993 | 2.6203 | 2.5383 | 2.4530 | 2.3667 | 2.2695 | 2.1694 |
| 26 | 3.0941 | 2.9576 | 2.8150 | 2.6640 | 2.5848 | 2.5026 | 2.4170 | 2.3273 | 2.2325 | 2.1315 |
| 27 | 3.0618 | 2.2956 | 2.7827 | 2.6316 | 2.5522 | 2.4699 | 2.3840 | 2.2938 | 2.1984 | 2.0965 |
| 28 | 3.0320 | 2.8959 | 2.7530 | 2.6017 | 2.5223 | 2.4397 | 2.3535 | 2.2629 | 2.1670 | 2.0642 |
| 29 | 3.0045 | 2.8685 | 2.7256 | 2.5742 | 2.4946 | 2.4118 | 2.3253 | 2.2344 | 2.1378 | 2.0342 |
| 30 | 2.9791 | 2.8431 | 2.7002 | 2.5487 | 2.4689 | 2.3680 | 2.2992 | 2.2079 | 2.1107 | 2.0062 |
| 40 | 2.8005 | 2.6648 | 2.5216 | 2.3689 | 2.2880 | 2.2034 | 2.1142 | 2.0194 | 1.9172 | 1.8047 |
| 60 | 2.6318 | 2.4961 | 2.3523 | 2.1978 | 2.1154 | 2.0285 | 1.9360 | 1.8363 | 1.7263 | 1.6006 |
| 120 | 2.4721 | 2.3363 | 2.1915 | 2.0346 | 1.9500 | 1.8600 | 1.7628 | 1.6557 | 1.5330 | 1.3805 |
| ∞ | 2.3209 | 2.1848 | 2.0385 | 1.8783 | 1.7908 | 1.6964 | 1.5923 | 1.4730 | 1.3246 | 1.0000 |

*Source: Reprinted, by permission of the Biometrika Trustees from M. Merrington and C. M. Thompson, "Tables of Percentage Points of the Inverted Beta(F) Distribution,"* Biometrika *33 (1943): 84–85*

# Index

## A

N

## S

**Y**



**Z**